EUROPA REFERENCE BOOKS
for Automotive Technology

Modern
Automotive Technology

Fundamentals, service, diagnostics

D1719598

1st English edition

The German edition was written by technical instructors, engineers and technicians

Editorial office (German edition): R. Gscheidle, Studiendirektor, Winnenden – Stuttgart

VERLAG EUROPA-LEHRMITTEL · Nourney, Vollmer GmbH & Co. KG
Düsselberger Strasse 23 · 42781 Haan-Gruiten · Germany

Europa No.: 23018

Original title: Fachkunde Kraftfahrzeugtechnik, 28th edition 2004

Authors:

Fischer, Richard	Oberstudienrat	Polling – München
Gscheidle, Rolf	Studiendirektor	Winnenden – Stuttgart
Heider, Uwe	Kfz-Elektriker-Meister, Trainer Audi AG	Neckarsulm – Oedheim
Hohmann, Berthold	Oberstudienrat	Eversberg – Meschede
Keil, Wolfgang	Studiendirektor	München
Mann, Jochen	Dipl.-Gwl., Studienrat	Schorndorf – Stuttgart
Pichler, Wolfram	Ing. (grad.), Studiendirektor	Pullach – München
Schlögl, Bernd	Dipl.-Gwl., Studienrat	Rastatt – Gaggenau
Siegmayer, Paul	Dipl.-Ing., Studiendirektor	Langenalb – Pforzheim
Wimmer, Alois	Oberstudienrat	Stuttgart
Wormer, Günter	Dipl.-Ingenieur	Karlsruhe

Head of working group and editorial office:
Rolf Gscheidle, Studiendirektor, Winnenden – Stuttgart

Illustrations:
Drawing office of Verlag Europa-Lehrmittel, Leinfelden-Echterdingen

English edition: Modern Automotive Technology - Fundamentals, service, diagnostics

1st edition 2006
Impression 5 4 3 2 1
All impressions of the same edition can be used in parallel, as they do not differ from each other except with regard to the correction of printing errors.

ISBN 3-8085-2301-8

Translation: STAR Deutschland GmbH, Member of the STAR Group
Typesetting: STAR Deutschland GmbH, Member of the STAR Group
Print: Media Print Informationstechnologie, D-33100 Paderborn, Germany

Foreword

"Modern AutomotiveTechnology" is a standard work covering the subject of automotive technology.This first English edition is based on the 28th German edition of the title "Fachkunde Kraftfahrzeugtechnik". It has for many years proven to be a highly popular textbook used for training and further education. It provides apprentices, trainees, teachers and all those interested in this subject with the necessary theoretical knowledge in order to gain a firm grasp of the practical and technical skills involved. Fundamental, technical connections between individual systems are presented in a clear and compre-hensible way.

The book is intended to be used as a reference work by employees in the automotive industry and in motor-vehicle service outlets, by teachers, apprentices, trainees and automotive-technology students to help them look up information and supplement their technical knowledge. The work is intended to be used by all those interested in automotive technology as a means of extending their technical knowledge through private study.

The 22 chapters are logically arranged by subject and in their objectives are geared towards the changes in content that have occurred in the field of automotive technology.The book is particularly suitable for practically orientated training in all matters pertaining to motor vehicles.

This work covers the latest developments in automotive technology, such as, for example, service and maintenance of vehicle systems, management, communication, FSI engines, supercharging technology, common-rail systems, twin-clutch gearboxes, electronic transmission control, electronic brake systems, compressed-air monitoring systems, adaptive cornering lights, high-frequency technology, electromag-netic compatibility and comfort and convenience systems such as adaptive cruise control, parking assis-tance and navigation. A large chapter is devoted to the subject of electrical engineering. Here, the detailed coverage of the fundamentals of electrical engineering forms the basis for all the crucial issues and topics pertaining to automotive electrics, up to and including data transmission in motor vehicles. A separate chapter is devoted to the increasing importance in engineering of comfort and convenience technology.

Reference is made to German and European standards in the chapters on environmental protection and occupational safety, emissions-control engineering, braking technology and motorcycle engineering. However, the standards applicable in the respective individual countries are binding.

The work features numerous coloured pictures, drawings and system diagrams as well as particularly clearly and comprehensibly laid-out tables. These will help the reader to digest and comprehend the complex subject matter.

The work has been written and compiled – in close co-operation with the automotive trade and industry – by a team of educationally experienced vocational-school teachers, engineers and master tradesmen. The authors and the publishers will be grateful for any suggestions and constructive comments.

We would like to thank all the companies and organisations who have kindly contributed pictures and technical documents.

The Authors of the AutomotiveTechnologyTeam   Summer 2006

Abbreviations

A/C	Air conditioning	CPU	Central processing unit	EMS	Electronic engine management system
A/F	Air/fuel (mixture)	CR	Common rail	Eo	Exhaust valve opens
ABC	Active body control	CS	Crankshaft	EOBD	European on board diagnosis
ABS	Antilock braking system	CSR	Conti support ring		
ABV	Automatic braking-force distribution (German: Automatische Bremskraftverteilung)	CV	Commercial vehicle	EP	Exhaust passage
		CV	Check valve	EPHS	Electrically powered hydraulic steering
		CVlft	Check valve left		
		CVrt	Check valve right	EPS	Electro-pneumatic control system
AC	Alternating current	CVT	Continuous variable transmission		
ACC	Adaptive cruise control			ESP	Electronic stability program
ACEA	Association des Constructeurs Européens de l'automobile	DA	Drive axle		
		DC	Direct current	ETC	Electronic throttle control
ACS	Automatic clutch system	DI	Direct injection	ETN	European type number
AD	Analogue-digital (converter)	DME	Digital motor electronics	EV	Exhaust valve
		DOHC	Double overhead camshaft	FA	Front axle
ADSL	Asymmetrical digital subscriber line			FB	Function button
		DOT	Department of Transport	FDI	Fuel direct injection
AGM	Absorbing glas mat	DSC	Dynamic stability control	FF	Freeform (reflector)
ALDBFR	Automatic load-dependent brake-force regulator	DSG	Direct-shift gearbox	FH	Flat hump
		DSP	Dynamic shift-program selection	FL	Front left
				FOC	Fibre-optic cable
ALSD	Automatic limited-slip differential	DSST	Dunlop self-supporting technology	FOT	Fibre-optical transceiver
				FR	Front right
AM	Amplitude modulation			FSI	Fuel stratified injection
API	American Petroleum Institute	EBS	Electronic braking system	FWD	Four wheel drive
ASC	Anti-stability control	Ec	Exhaust valve closes		
ASTM	American Society for Testing and Materials	ECE	Economic Commission for Europe	GDI	Gasoline direct injection
				GFRP	Glass-fibre-reinforced plastic
ATF	Automatic transmission fluid	ECM	Electronic clutch management		
				GI	General inspection
ATS	Adaptive transmission control (system)	ECS	Electronic clutch system	GMR	Automatic regulation of yaw moment (German: Giermomentregelung)
		ECU	Electronic control unit		
		EDC	Electronic diesel control		
BAS	Brake assistant	EDP	Electronic data processing	GPS	Global positioning system
BDC	Bottom dead centre			GVWR	Gross vehicle weight rating
		EDTC	Engine-drag torque control		
CA	Crankshaft angle				
CS	Camshaft	EEPROM	Electrically erasable programmable read-only memory	HF	High frequency
CAN	Controller area network			HFM	Hot-film air-mass meter
CBS	Combined brake system			HGV	Heavy goods vehicle
CC	Cruise control	EGR	Exhaust gas recirculation	HNS	Homogeneous numerically calculated surface
CDI	Capacitive discharge ignition	EGS	Electronic gearbox control unit (German: Elektronisches Getriebesteuergerät)		
				HS	High-solid (paints)
CFPP	Cold filter plugging point			HTHS	High temperature, high shear
CFRP	Carbon-fibre-reinforced plastic	EH	Extended hump		
		EHB	Electro-hydraulic braking system	HV	Hybrid vehicle
CH	Combination hump				
CIH	Camshaft in head	EI	Emissions inspection		
CIP	Continuous improvement process	ELSD	Electronic limited-slip differential	IC	Integrated circuit
				Ic	Inlet valve closes
CN	Cetane number	EMC	Electro-magnetic compatibility	IC	Individual control
CNG	Compressed natural gas			IDI	Indirect injection

Abbreviations

IHPF	Internal high-pressure forming
Io	Inlet valve opens
IP	Inlet passage
IPO	Input/Processing/Output (principle)
IS	Input shaft
ISAD	Integrated starter alternator damper
IV	Inlet valve
IVlft	Inlet valve left
Ivrt	Inlet valve right
LA	Lifting axle
LD	Low density
LDR	Light depending resistor
LED	Light emitting diode
LEV	Low-emission vehicle
LF	Low frequency
LI	Load index
LIN	Local interconnect network
LNG	Liquefied natural gas
LS	Limited slip
LSG	Laminated safety glass
LU	Logical unit
LW	Long wave
MAF	Mass air flow
MAG	Metal-active-gas (welding)
MC	Microcomputer
MC	Main cylinder
ME	Motor electronics
MED	Motor electronics direct injection
MG	Motor generator
MIG	Metal-inert-gas (welding)
MIL	Malfunction indicator lamp
MON	Motor-octane number
MOST	Media-oriented system transport
MPI	Multi-point injection
MS	Medium-solid (paints)
MW	Medium wave
NF	Non-ferrous
NLGI	National Lubrication Grease Institute
NLS	Needle lift sensor
NTC	Negative temperature coefficient
OBD	On board diagnosis
OD	Outside diameter
OHC	Overhead camshaft
OHV	Overhead valves
ON	Octane number
OV	Outlet valve
OVlft	Outlet valve left
PBC	Parking-brake circuit
PC	Planet carriers
PCU	Pump control unit
PDA	Personal digital assistant
PEM	Proton exchange membran
PES	Poly ellipsoid system (reflector)
PIN	Personal identification number
PM	Particulate matter
POF	Plastic optical fibre
POT	Plastic optical transceiver
PR	Ply rating
PTC	Positive temperature coefficient
PWM	Pulse width modulation
QA	Quality assurance
QM	Quality management
RA	Rear axle
RDS	Radio data system
RHD	Right-hand driver
RL	Rear left
RLFS	Return-less-fuel system
RON	Research-octane number
ROP	Roll-over protection
ROV	Rotating high voltage distribution (German: Rotierende Hochspannungs-verteilung)
RR	Rear right
RRC	Radio remote control
RUV	Static high voltage distribution (German: Ruhende Hochspannungs-verteilung)
SAC	Self-adjusting clutch
SAE	Society of Automotive Engineers
SAM	Signal acquisition and actuation module
SBC	Sensotronic brake control
SC	Signal conditioning
SCR	Selective catalytic reduction
SCV	Solenoid control valve
SDC	Semi-drop centre
SE	Sensor
SI	Safety inspection
SLC	Select-low control
SoC	State of charge
SPI	Single-point injection
SRR	Short-range radar
SRS	Safety restraint systems
SSlft	Speed sensor left
SSR	Self-supporting run-flat tyres
SSrt	Speed sensor right
SV	Solenoid valve
SV	Side valve
SW	Short wave
SWR	Stationary wave ratio
Tc	Transfer passage closes
TCS	Traction control system
TDC	Top dead centre
TIG	Tungsten-inert gas
TL	Tubeless
To	Transfer passage opens
TP	Transfer passage
TPC	Tyre-pressure check
TSG	Toughened safety glass
TWI	Treadwear indicator
UIS	Unit injector system
UPS	Unit pump system
VDC	Vehicle dynamics controller
VDR	Voltage-dependent resistor
VF	Variable focus (reflector)
VHF	Very high frequency
VT	Viscosity temperature
VTec	Variable valve timing and lift electronic control
VTG	Variable turbine geometry
WIG	Wolfram-inert-gas (welding)

We wish to thank the companies listed below for providing technical advice, information, photographs and illustrations.

Alfa-Romeo-Automobile
Mailand/Italien

Aprilia Motorrad-Vertrieb
Düsseldorf

Aral AG, Bochum

Audatex Deutschland, Minden

Audi AG, Ingolstadt – Neckarsulm

Autokabel, Hausen

Autoliv, Oberschleißheim

G. Auwärter GmbH & Co
(Neoplan) Stuttgart

BBS Kraftfahrzeugtechnik, Schiltach

BEHR GmbH & Co, Stuttgart

Beissbarth GmbH Automobil Servicegeräte
München

BERU, Ludwigsburg

Aug. Bilstein GmbH & Co KG
Ennepetal

Boge GmbH, Eitdorf/Sieg

Robert Bosch GmbH, Stuttgart

Bostik GmbH, Oberursel/Taunus

BLACK HAWK, Kehl

BMW Bayerische Motoren-Werke AG
München/Berlin

CAR-OLINER, Kungsör, Schweden

CAR BENCH INTERNATIONAL.S.P.A.
Massa/Italien

Continental Teves AG & Co, OHG, Frankfurt

Celette GmbH, Kehl

Citroen Deutschland AG, Köln

DaimlerChrysler AG, Stuttgart

Dataliner Richtsysteme, Ahlerstedt

Deutsche BP AG, Hamburg

DUNLOP GmbH & Co KG, Hanau/Main

ESSO AG, Hamburg

FAG Kugelfischer Georg Schäfer KG aA
Ebern

J. Eberspächer, Esslingen

EMM Motoren Service, Lindau

Ford-Werke AG, Köln

Carl Freudenberg
Weinheim/Bergstraße

GKN Löbro, Offenbach / Main

Getrag Getriebe- und Zahnradfarbrik
Ludwigsburg

Girling-Bremsen GmbH, Koblenz

Glasurit GmbH, Münster/Westfalen

Globaljig, Deutschland GmbH
Cloppenburg

Glyco-Metall-Werke B.V. & Co KG
Wiesbaden/Schierstein

Goetze AG, Burscheid

Grau-Bremse, Heidelberg

Gutmann Messtechnik GmbH, Ihringen

Hazet-Werk, Hermann Zerver, Remscheid

HAMEG GmbH, Frankfurt/Main

Hella KG, Hueck & Co, Lippstadt

Hengst Filterwerke, Nienkamp

Fritz Hintermayr, Bing-Vergaser-Fabrik
Nürnberg

HITACHI Sales Europa GmbH
Düsseldorf

HONDA DEUTSCHLAND GMBH
Offenbach/Main

Hunger Maschinenfabrik GmbH
München und Kaufering

IBM Deutschland, Böblingen

IVECO-Magirus AG, Neu-Ulm

ITT Automotive (ATE, VDO,
MOTO-METER, SWF, KONI, Kienzle)
Frankfurt/Main

IXION Maschinenfabrik
Otto Häfner GmbH & Co
Hamburg-Wandsbeck

Jurid-Werke, Essen

Kawasaki-Motoren GmbH, Friedrichsdorf

Knecht Filterwerke GmbH, Stuttgart

Knorr-Bremse GmbH, München

Kolbenschmidt AG, Neckarsulm

KS Gleitlager GmbH, St. Leon-Rot

KTM Sportmotorcycles AG
Mattighofen/Österreich

Kühnle, Kopp und Kausch AG
Frankenthal/Pfalz

Lemmerz-Werke, Königswinter

LuK GmbH, Bühl/Baden

MAHLE GmbH, Stuttgart

Mannesmann Sachs AG, Schweinfurt

Mann und Hummel, Filterwerke
Ludwigsburg

MAN Maschinenfabrik
Augsburg-Nürnberg AG
München

Mazda Motors Deutschland GmbH
Leverkusen

MCC – Mikro Compact Car GmbH
Böblingen

Messer-Griesheim GmbH
Frankfurt/Main

Metzeler Reifen GmbH
München

Michelin Reifenwerke KGaA
Karlsruhe

Microsoft GmbH, Unterschleißheim

Mitsubishi Electric Europe B.V.
Ratingen

Mitsubishi MMC, Trebur

MOBIL OIL AG, Hamburg

NGK/NTK, Ratingen

Adam Opel AG, Rüsselsheim

OSRAM AG, München

OMV AG, Wien

Peugeot Deutschland GmbH
Saarbrücken

Pierburg GmbH, Neuss

Pirelli AG, Höchst im Odenwald

Dr. Ing. h.c. F. Porsche AG
Stuttgart-Zuffenhausen

Renault Nissan Deutschland AG
Brühl

Samsung Electronics GmbH, Köln

SATA Farbspritztechnik GmbH & Co
Kornwestheim

SCANIA Deutschland GmbH
Koblenz

SEKURIT SAINT-GOBAIN
Deutschland GmbH, Aachen

Siemens AG, München

SKF Kugellagerfabriken GmbH
Schweinfurt

SOLO Kleinmotoren GmbH
Maichingen

Stahlwille E. Wille
Wuppertal

Steyr-Daimler-Puch AG
Graz/Österreich

Subaru Deutschland GmbH
Friedberg

SUN Elektrik Deutschland
Mettmann

Suzuki GmbH
Oberschleißheim/Heppenheim

Technolit GmbH, Großlüder

Telma Retarder Deutschland GmbH
Ludwigsburg

Temic Elektronik, Nürnberg

TOYOTA Deutschland GmbH, Köln

VARTA Autobatterien GmbH
Hannover

Vereinigte Motor-Verlage GmbH & Co KG
Stuttgart

ViewSonic Central Europe, Willich

Voith GmbH & Co KG, Heidenheim

Volkswagen AG, Wolfsburg

Volvo Deutschland GmbH, Brühl

Wabco Westinghouse GmbH
Hannover

Webasto GmbH, Stockdorf

Yamaha Motor Deutschland GmbH
Neuss

ZF Getriebe GmbH, Saarbrücken

ZF Sachs AG, Schweinfurt

ZF Zahnradfabrik Friedrichshafen AG
Friedrichshafen/Schwäbisch Gmünd

Table of contents

19   Electrical engineering   476

20   Comfort and convenience technology 596

21   Motorcycle technology   618

22   Commercial vehicle technology   639

23   Keyword index   674

1 Motor vehicle

1

1.1 Evolution of the motor vehicle

1860 The Frenchman **Lenoir** constructs the first fully operational internal-combustion engine; this powerplant relies on city gas as its fuel source. Thermal efficiency is in the 3 % range.

1867 **Otto and Langen** display an improved internal-combustion engine at the Paris International Exhibition. Its thermal efficiency is approximately 9 %.

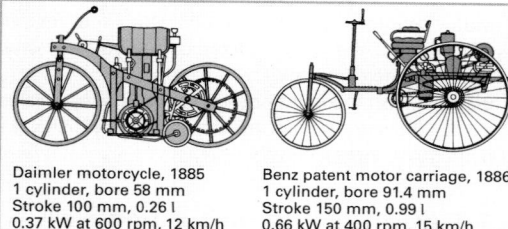

Daimler motorcycle, 1885
1 cylinder, bore 58 mm
Stroke 100 mm, 0.26 l
0.37 kW at 600 rpm, 12 km/h

Benz patent motor carriage, 1886
1 cylinder, bore 91.4 mm
Stroke 150 mm, 0.99 l
0.66 kW at 400 rpm, 15 km/h

Fig. 1: Daimler motorcycle and Benz motor carriage

1876 **Otto** builds the first gas-powered engine to utilise the **four-stroke compression cycle**. At virtually the same time **Clerk** constructs the first gas-powered **two-stroke engine** in England.

1883 **Daimler** and **Maybach** develop the first high-speed **four-cycle petrol engine** using a **hot-tube ignition system**.

1885 The first **self-propelled motorcycle** from **Daimler**. First **self-propelled three-wheeler** from **Benz** (patented in 1886) **(Fig. 1)**.

1886 First **four-wheeled motor carriage** with **petrol engine** from **Daimler (Fig. 2)**.

1887 **Bosch** invents the **magneto ignition**.

1889 **Dunlop** in England produces the first **pneumatic tyres**.

1893 **Maybach** invents the **spray-nozzle carburettor**.

1893 **Diesel** patents his design for a heavy oil-burning powerplant employing the self-ignition concept.

1897 **MAN** presents the first workable diesel engine.

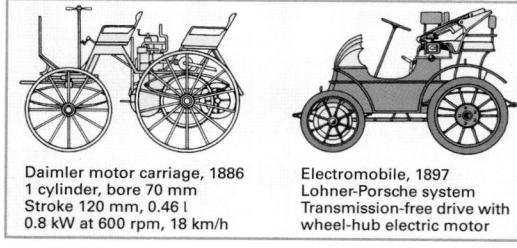

Daimler motor carriage, 1886
1 cylinder, bore 70 mm
Stroke 120 mm, 0.46 l
0.8 kW at 600 rpm, 18 km/h

Electromobile, 1897
Lohner-Porsche system
Transmission-free drive with
wheel-hub electric motor

Fig. 2: Daimler motor carriage and the first Electromobile

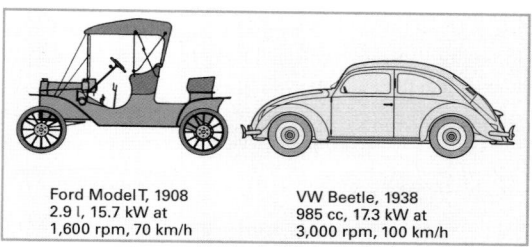

Ford Model T, 1908
2.9 l, 15.7 kW at
1,600 rpm, 70 km/h

VW Beetle, 1938
985 cc, 17.3 kW at
3,000 rpm, 100 km/h

Fig. 3: Ford Model T and VW Beetle

1897 First **Electromobile** from **Lohner-Porsche (Fig. 2)**.

1899 **Fiat Automobile Factory** founded in Turin.

1913 **Ford** introduces the **production line** to automotive manufacturing. Production of the **Tin Lizzy** (Model T, **Fig. 3**). By 1925, 9,109 were leaving the production line each day.

1916 The **Bavarian Motor Works** are founded.

1923 First **motor lorry** powered by a **diesel engine** produced by **Benz-MAN (Fig. 4)**.

1936 **Daimler-Benz** inaugurates series-production of passenger cars propelled by **diesel engines**.

1938 The **VW Works** are founded in Wolfsburg.

1949 First **low-profile tyre** and first **steel-belted radial tyre** produced by **Michelin**.

1950 First **gas-turbine** propulsion unit for automotive application makes its debut at **Rover** in England.

1954 **NSU-Wankel** constructs the **rotary engine** **(Fig. 4)**.

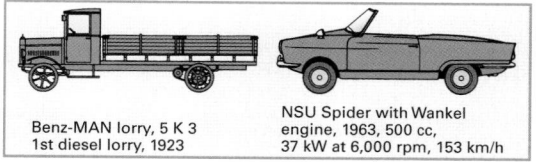

Benz-MAN lorry, 5 K 3
1st diesel lorry, 1923

NSU Spider with Wankel
engine, 1963, 500 cc,
37 kW at 6,000 rpm, 153 km/h

Fig. 4: Diesel-engined lorry
Passenger car with Wankel rotary engine

1966 **Electronic fuel injection (D-Jetronic)** for standard production vehicles produced by **Bosch**.

1970 **Seatbelts** for driver and front passengers.

1978 Initial application of the **ABS Antilock Braking System** in passenger cars.

1984 Debut of the **airbag** and **seatbelt tensioning system**.

1985 Advent of a **catalytic converter** designed for operation in conjunction with closed-loop mixture control, intended for use with unleaded fuel.

1997 Electronic **suspension control systems**.

1.2 Motor vehicle classifications

Roadgoing or highway vehicles is a category comprising all vehicles designed for road use, as opposed to operation on tracks or rails (**Fig. 1**).

The basic division is into two classes, motor vehicles and trailers. Motor vehicles always possess an integral mechanical propulsion system.

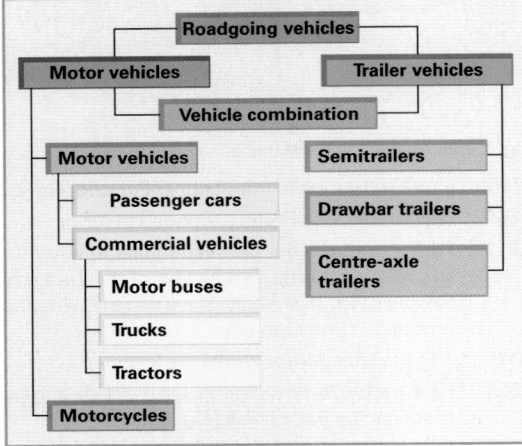

Fig. 1: Overview of roadgoing vehicles

Dual-track vehicles

Motor vehicles with more than two wheels can be found in dual-track and multiple-track versions. These include:

- **Passenger cars.** These are primarily intended for use in transporting people, as well as their luggage and other small cargo. They can also be used to pull trailers. The number of seats, including that of the driver, is restricted to nine.

- **Commercial vehicles.** These are designed to transport people and cargo and for pulling trailers. Passenger cars are not classified as commercial vehicles.

Single-track vehicles

Motorcycles are single-track vehicles with 2 wheels. A sidecar may be attached to the motorcycle, which remains classified as such provided that the tare weight of the combination does not exceed 400 kg. A motorcycle can also be employed to pull a trailer. Single-track vehicles include

- **Motorcycles.** These are equipped with permanent, fixed-location components (fuel tank, engine) located adjacent to the knees as well as footrests.

- **Motor scooters.** Because the operator's feet rest on a floor panel, there are no fixed components at knee level on these vehicles.

- **Bicycles with auxiliary power plants.** These vehicles exhibit the same salient features as bicycles, such as pedals (mopeds, motor bicycle, etc.).

1.3 Design of the motor vehicle

The motor vehicle consists of component assemblies and their individual components.

The layout of the individual assemblies and their relative positions is not governed by invariable standards. Thus, for example, the engine may be designed as an independent assembly, or it may be integrated as a subassembly within a larger powertrain unit.

One of the options described in this book is to divide the vehicle into 5 main assembly groups: engine, drivetrain, chassis, vehicle body and electrical system.

The relationships between the assemblies and their constituent components are illustrated in **Fig. 2**.

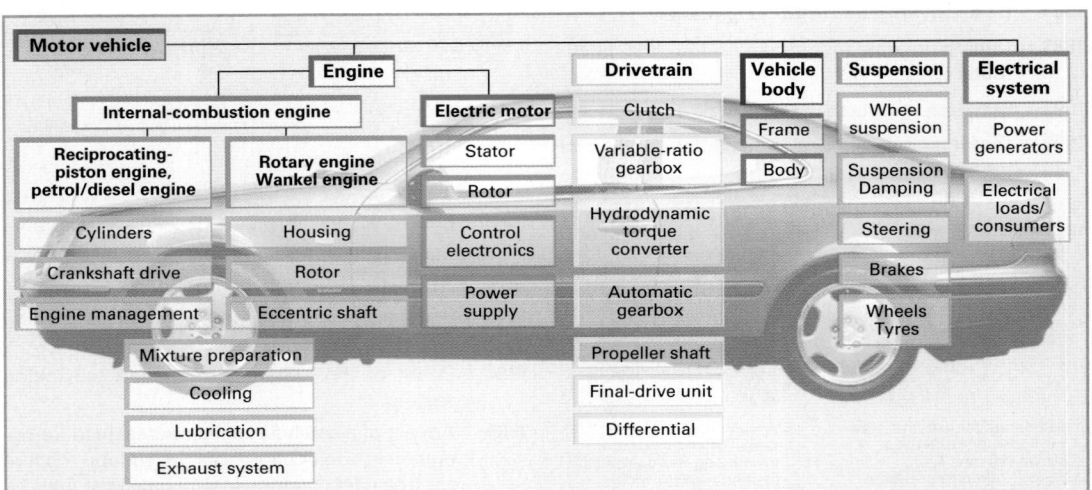

Fig. 2: Design of the motor vehicle

1.4 The motor vehicle as technical system

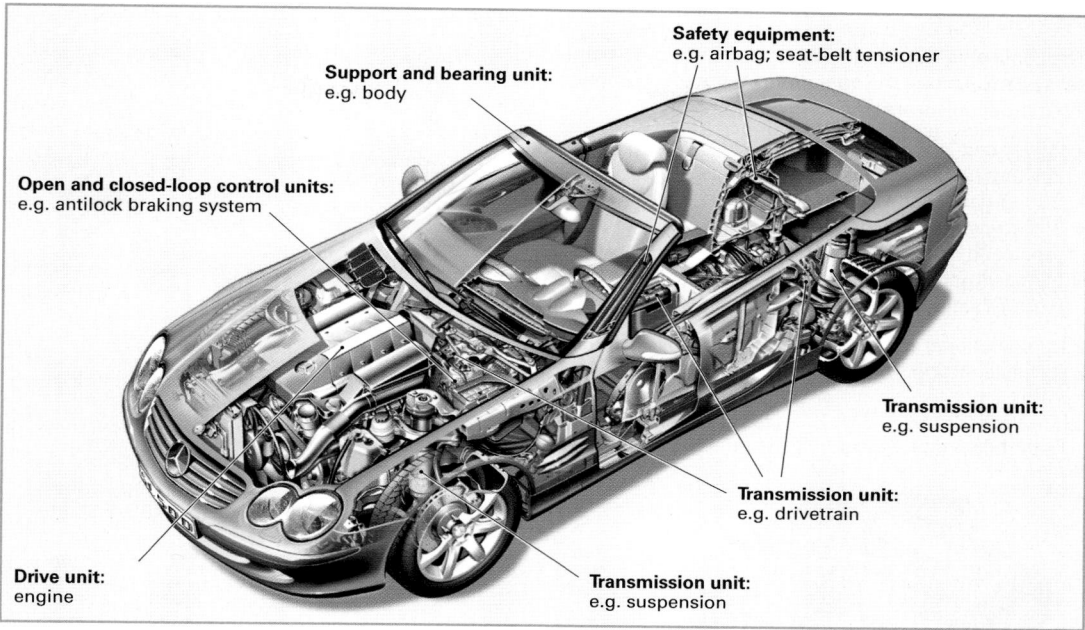

Safety equipment:
e.g. airbag; seat-belt tensioner

Support and bearing unit:
e.g. body

Open and closed-loop control units:
e.g. antilock braking system

Transmission unit:
e.g. suspension

Transmission unit:
e.g. drivetrain

Drive unit:
engine

Transmission unit:
e.g. suspension

Fig. 1: The motor vehicle as a system with operational units

1.4.1 Technical systems

Every machine forms a complete technical system.

> Characteristics of technical systems:
> - Defined system borders delineate their limits relative to the surrounding environment.
> - They possess input and output channels.
> - The salient factor defining system operation is the total function, and not the individual function, which is discharged internally, within the system.

A rectangle is employed in graphic portrayals of technical systems (**Fig. 2**).

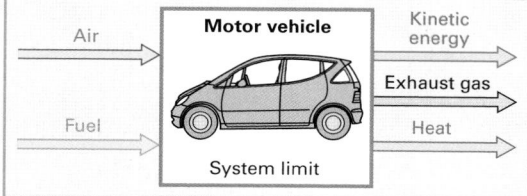

Fig. 2: Basic system portrait using a motor vehicle as an example

Input and output variables are represented by arrows. The number of arrows varies according to the number of input and output variables.

The rectangle symbolises the **system limit** (hypothetical boundary) that delineates the border separating each individual technical system from other systems and/or the surrounding environment.

> The distinctive, defining features of the individual system include:
> - **I**nput (input variables or parameters) entering from beyond the system limits
> - **P**rocessing within the system limits
> - **O**utput (output variables or parameters) issued and relayed to destinations lying outside the limits of the system (**IPO concept**)

1.4.2 Motor vehicle system

The motor vehicle is a complex technical system in which various subsystems operate in harmony to discharge a defined function.
The function of the passenger car is to transport people, while the function of the motor lorry, or truck, is to carry cargo.

Operational units within the motor vehicle

Systems designed to support operational processes are combined in operational units (**Fig. 1**). Familiarity with the processes performed in operational units such as the engine, drivetrain, etc. can enhance our understanding of the complete system represented by the motor vehicle in its implications for maintenance, diagnosis and repair.

The concept is suitable for application with any technical system. Among the **operational units** that comprise the motor vehicle are the:

- Power unit
- Power-transfer assembly
- Support and load-bearing structure
- Electro-hydraulic systems
  (open and closed-loop systems, etc.)
- Electrical and electronic systems
  (such as safety devices)

Each operational unit acts as a subsystem by assuming a specific function.

Operational unit: Power unit – engine

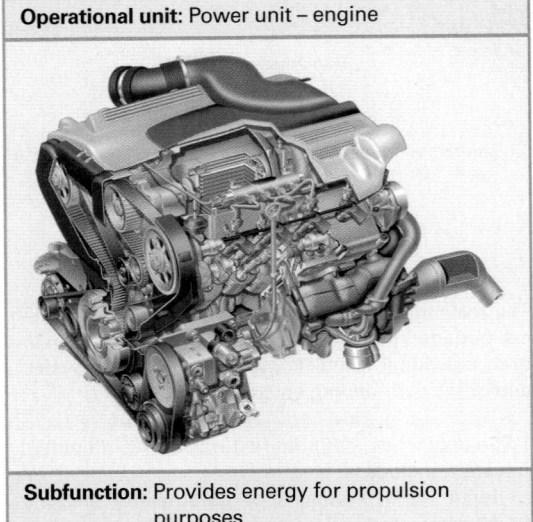

Subfunction: Provides energy for propulsion purposes

Operational unit: Power-transfer assembly, such as drivetrain

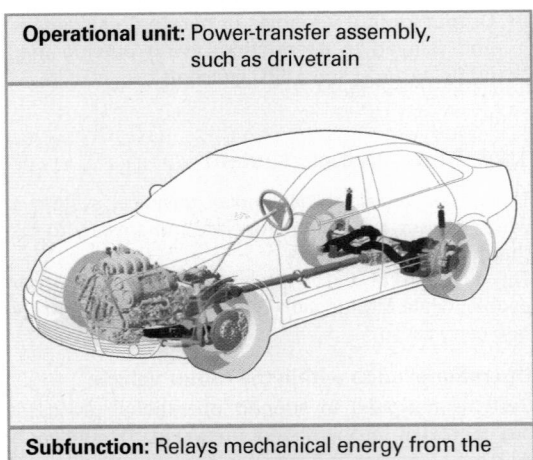

Subfunction: Relays mechanical energy from the power unit to the drive wheels

---

Operational unit: Vehicle structure as support structure, exemplified by body

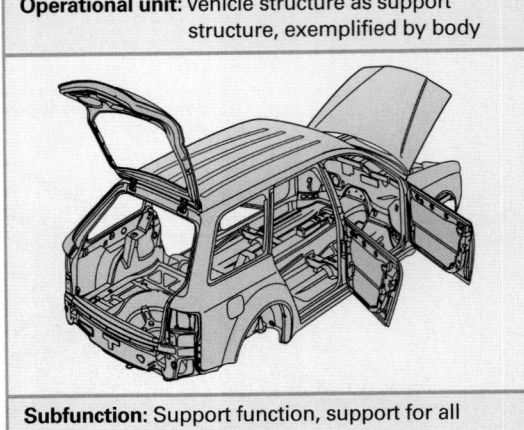

Subfunction: Support function, support for all subsystems

Operational unit: Electro-hydraulic systems (open and closed-loop control systems, such as ABS, ESP, etc.)

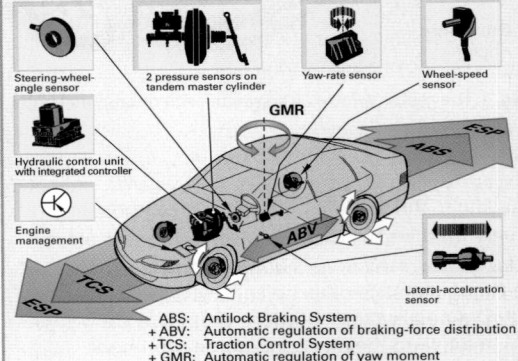

Steering-wheel-angle sensor

2 pressure sensors on tandem master cylinder

Yaw-rate sensor

Wheel-speed sensor

GMR

Hydraulic control unit with integrated controller

Engine management

ABV

TCS

ESP

ABS

ESP

Lateral-acceleration sensor

ABS:		Antilock Braking System
+ ABV:		Automatic regulation of braking-force distribution
+TCS:		Traction Control System
+ GMR:		Automatic regulation of yaw moment
= ESP:		Electronic Stability Program

Subfunction: Active occupant protection, improvements in dynamic response

---

Operational unit: Electr., electron. systems (safety and security devices, such as airbags, seatbelt tensioners)

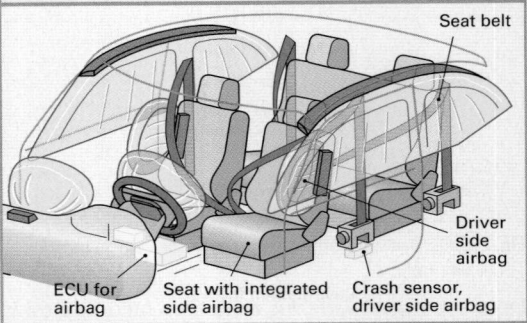

Seat belt

Driver side airbag

ECU for airbag

Seat with integrated side airbag

Crash sensor, driver side airbag

Subfunction: Passive protection for vehicle occupants

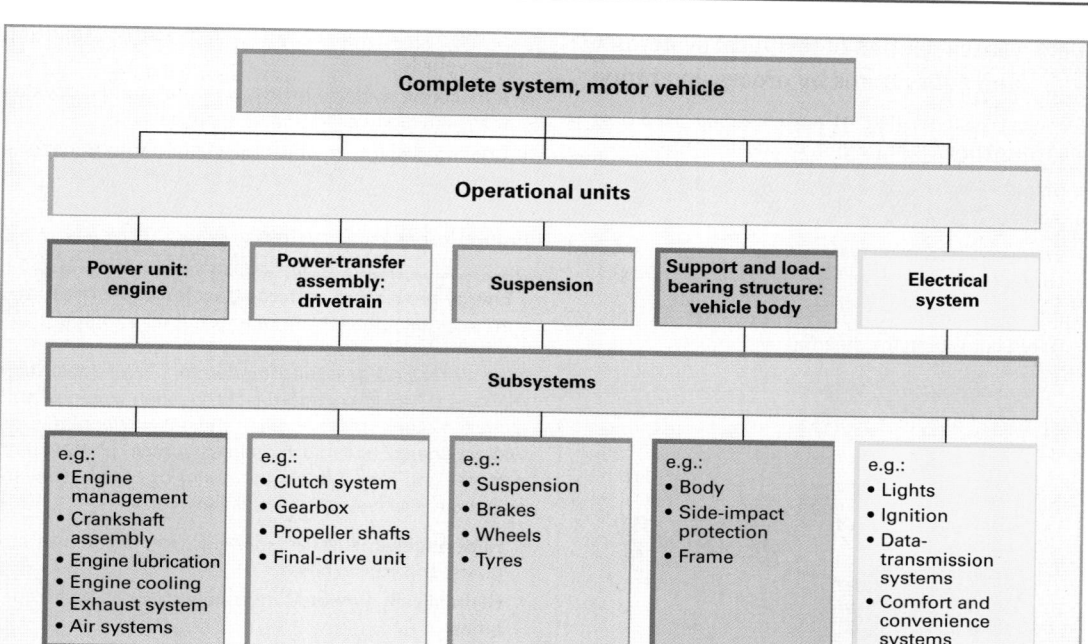

Fig. 1: The motor vehicle as composite system

Various subsystems must operate together for the motor vehicle to discharge its primary functions (**Fig. 1**). Reducing the scale of the system's limits shifts the focus to progressively smaller subsystems, ultimately leading to the level of the individual component.

The motor vehicle as a complete system

Defining the limits of the system to coincide with those of the overall vehicle produces boundaries in which the system's limits border on environmental entities such as air and the road surface. On the input side, air and fuel are the only factors entering from beyond the system's limits, while exhaust gas joins kinetic and thermal energy outside this boundary on the output side (**Fig. 2, Fig. 3**).

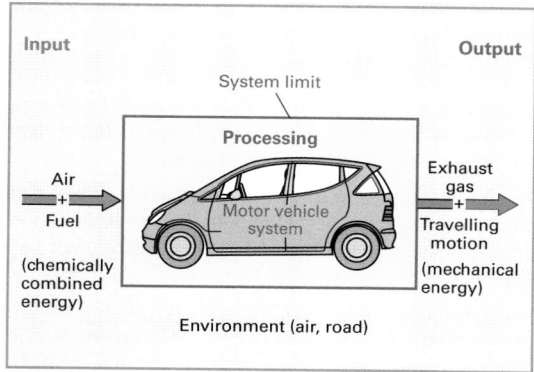

Fig. 2: System: Motor vehicle

1.4.3 Subsystems in the motor vehicle

Each subsystem is subject to the **IPO** concept (**Fig. 3**).

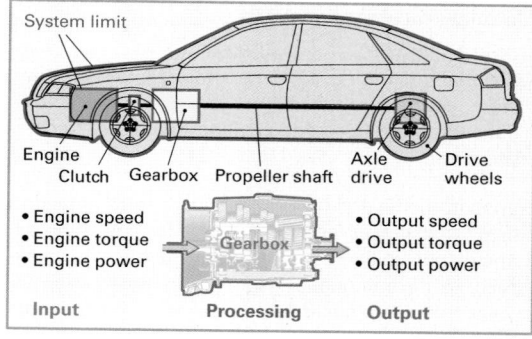

Fig. 3: Subsystem: Gearbox

Input. The factors operating on the input side of the gearbox are engine speed, engine torque and engine power.

Processing. The crankshaft's rotation speed and the torque it transfers undergo a transformation process within the gearbox.

Output. The elements exiting the subsystem on the output side include output-shaft speed, output torque and output power as well as heat.

Efficiency level. The efficiency of the drivetrain is reduced by energy losses sustained within the gearbox.

The "gearbox" subsystem is connected to the drive wheels via other subsystems, such as the propeller shaft, final-drive unit, and half shafts.

1.4.4 Classifications of technical systems and subsystems by processing mode

Technical systems (**Fig. 1**) are classified according to the type of processing within overall systems:

- Material-processing systems such as the fuel-supply system
- Energy-processing systems such as the internal-combustion engine
- Information-processing systems such as the on-board computer, the steering system, etc.

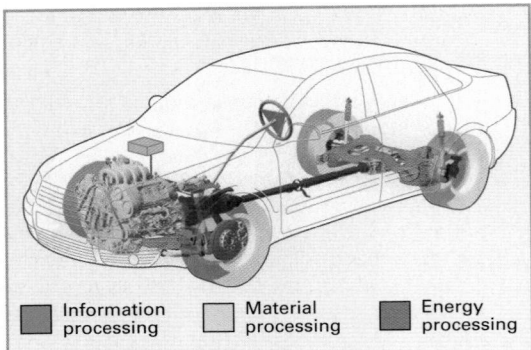

| Information processing | Material processing | Energy processing |

Fig. 1: Systems classified according to processing mode

Material-processing systems

> Material-processing systems modify materials in their geometrical configuration (reshaping) or transport them from one position to another (repositioning).

Transport media and basic machinery are employed to convey substances and materials. Machine tools assume responsibility for shaping materials. To cite an example: in the material-transport process, a pump induces motion in a static fluid (gasoline in the fuel tank) in order to transport it to the fuel-injection system. A precondition for this processing operation is provision of electrical energy to the operational machinery, such as a fuel pump, that is responsible for the process.

Overview of material-processing systems:

Machines for reshaping include machine tools such as drills, mills and lathes as well as the equipment found in foundries and stamping works such as metal presses.

Machines for repositioning include all conveyors, transporters and machines employed in the transport of solid materials (conveyor belts, fork lift trucks, trucks, passenger cars), liquids (pumps) and gases (fans, turbines).

Examples of material-processing systems within the motor vehicle:

- Lubrication system, in which the oil pump provides the motive power for material propulsion.
- Cooling system, in which the water pump transports a medium to support thermal transfer.

Energy-processing systems

> **Energy-processing systems** transform energy from an external source from one form into another.

This class embraces all manner of power-generation machines, including internal-combustion engines and electric motors, steam engines and gas power plants, as well as energy units such as heat pumps, photovoltaic systems and fuel cells. In the realm of energy conversion the operative distinction is between:

- **Heat engines,** such as spark-ignition and diesel engines, and gas turbines
- **Hydraulically powered machines,** such as water turbines
- **Wind-energy devices,** such as wind-powered generators
- **Solar-energy converters,** such as photovoltaic systems
- **Fuel cells**

Within the internal-combustion engine, the fuel's chemical energy is initially converted into thermal energy before undergoing a second transformation to emerge as kinetic energy (**Fig. 2**).

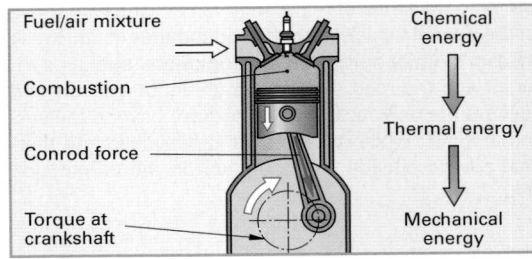

Fig. 2: Energy processing in the spark-ignition engine

This process can generate additional substances and information. Because these are of secondary significance in the operation of the energy-processing machine, they are not usually primary objects of attention.

The flow of substances and materials (entry of fuel and emission of exhaust gases) and the flow of information (fuel-air mixture, engine-speed control, steering, etc.) all assume the role of secondary functions.

Energy-processing system. The primary focus is on converting chemical energy contained in fuel into kinetic energy to propel the vehicle, with the **internal-combustion engine** serving as the energy-processing system.

Information-processing systems

> They monitor, process and relay information and data and support communications.

Information-processing and relay systems, such as electronic control units (ECU), CAN bus controllers and diagnostic equipment (testers) assume vital significance in the maintenance of modern vehicles.

Information. Knowledge concerning conditions and processes. Examples from within the vehicle include information on engine temperature, driving speed, load factor, etc. required to support vehicle operation. This information can be relayed from one electronic control unit to another. The data are registered in the form of signals.

Signals. Signals are data portrayed in physical form. Within the motor vehicle, sensors generate signals to represent parameters such as rotational speed, temperature and throttle-valve position.

Examples of information-processing systems in motor vehicles:

- **Engine control unit.** The engine-management ECU registers and processes an entire array of relevant data in order to adapt engine performance to provide ideal operation under any given conditions.
- **On-board computer.** Among its functions are to furnish the driver with information on average and current fuel consumption, estimated cruising range, average speed and outside temperature.

1.4.5 Using technical systems

Extensive familiarity with technical systems is essential for the operation and maintenance of motor vehicles. The manufacturer provides operating instructions (owner's manual) to help ensure that its vehicles operate with optimal safety, security and reliability, while also observing the interests of the natural environment.

Operating instructions contain, among other information:

- System descriptions
- Explanations of functions and operation
- System descriptions
- Operating diagrams
- Instructions on correct operation and use of the controls
- Maintenance and service inspection schedules
- Instructions for responding to malfunctions
- Information on approved fluids, lubricants and service materials, such as engine oils

- Technical data
- Emergency service addresses

Operation. Motor vehicles and machines should be operated by qualified and duly-authorised persons only.

Among the applicable stipulations ...

- ... the driver of a passenger car operating on public roads must be in possession of the required Class B driving licence.
- ... lift platforms and hydraulic hoists in automotive service facilities are to be operated exclusively by individuals over 18 years of age who have also received corresponding instruction in and authorisation for its use.
- ... the driver of a truck equipped with a crane must be in possession of a crane operator's licence.

This stipulation is intended to ensure that drivers of crane-equipped trucks have received the required training for operating lifts and hoisting equipment, and will provide the vehicle with the correct supplementary support **(Fig. 1)** whilst simultaneously observing all applicable accident-prevention regulations and operating the crane in a professional manner.

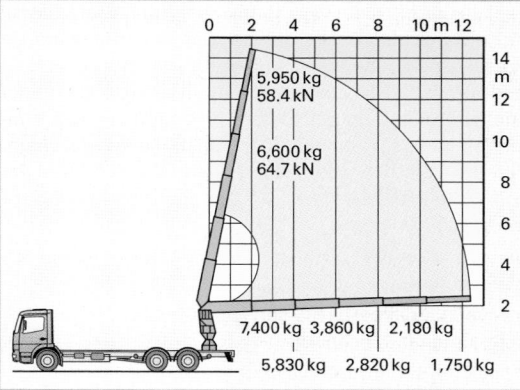

Fig. 1: Correct load distribution on a crane hoist

REVIEW QUESTIONS

1 What are the parameters that define a technical system?
2 What is the IPO concept?
3 What are the names of the operational units in the motor vehicle?
4 Name three subsystems in the motor vehicle, and describe the corresponding input and output variables.
5 What is the primary function of an energy-processing system?
6 What information is available in the operating instructions (vehicle owner's manual)?

1.5 Service and maintenance

> Professional-quality service and maintenance, performed in accordance with the manufacturer's instructions (by the factory service organisation, etc.) are vital elements in ensuring continued vehicle safety and in maintaining the validity of the manufacturer's warranty.

The manufacturer issues **service and maintenance schedules,** spare part catalogues and **repair instructions** to guide and support these activities. This documentation is available in many forms, including repair manuals, microfiche files and menu-guided computer programs designed to run on personal computers (PCs).

Service and maintenance. Service procedures include:
- Inspections, such as test procedures
- General maintenance, comprising oil changes, lubrication and cleaning
- Remedial action, such as repairs and component replacement

Aftersales service. Vehicle manufacturers and automotive repair operations offer professional service to their customers. Among the services offered by these facilities is to perform the prescribed preparations on new vehicles prior to delivery to the customer. Professional technicians also carry out service and maintenance processes that the vehicle operator may not be able to perform. In the official service and maintenance guidelines the manufacturer defines an action catalogue intended to ensure unrestricted functionality and maintain the vehicle's value. The individual procedures are contained in the service and maintenance schedules for the specific vehicles.

Service intervals can be defined according to the following criteria:
- Invariable, time-based service intervals (maintenance schedule)
- Flexible service intervals
- New service concepts

> Service, maintenance and inspection operations must be performed in accordance with defined schedules. Once operations have been carried out, they should be confirmed in a service record and signed by the responsible service technician.

Maintenance schedule

It furnishes information on the specified service and inspection intervals by specifying (for example) a major inspection for every 20,000 km or 12 months.

Service inspection schedule. This schedule defines the contents and lists the procedures included in the service inspection **(Fig. 1, Page 19).**

Flexible service intervals

Modern engine-management systems have allowed the advent of a new service concept characterised by adaptive scheduling. This concept reflects each individual vehicle's requirements based on its actual operating conditions. In addition to mileage, the system records and evaluates a variety of other factors (influencing variables) for inclusion in its calculations. A display then provides the driver with prompt notice as the inspection date approaches **(Fig. 1).** The process culminates with execution of the prescribed operations at the service facility in accordance with the service inspection schedule **(Fig. 1, Page 19).**

Oil change intervals. Two methods are available for defining oil change intervals:
- A virtual database, derived from such factors as mileage, overall fuel consumption and oil temperature curves, provides an index indicating how much the oil ages over a given period.
- The actual condition of the oil, meaning the quality and level of the oil as determined via the oil level sensor, in combination with the mileage and the registered engine load factors.

Brake pad wear. Brake pad wear is monitored electrically. When the brake pad reaches its wear limit a contact wire within the pad is perforated. The system then considers such factors as braking frequency, the duration of brake actuations and mileage in calculating the theoretically available mileage reserves, which are then reflected in the replacement intervals displayed to the driver.

Interior (passenger compartment) filter wear status. Data gleaned from the outside air temperature sensor, information on heater use, use of the recirculated-air mode, vehicle speed, fan blower speed, mileage and dates all flow into calculations to determine the period remaining until the dust and pollen filter will be due for replacement.

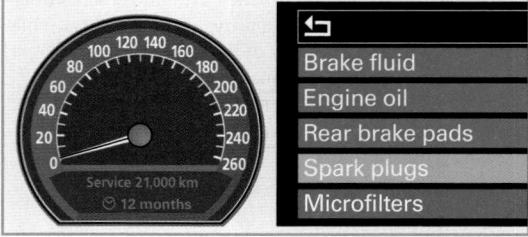

Fig. 1: Wear indicators

Sparking plug replacement intervals are still based on mileage, with new plugs specified after a specific distance, such as 100,000 km.

Replacement dates for **fluids and lubricants,** such as the coolant and brake fluid, are defined according to time, for instance, at intervals of 2 or 4 years.

New service concepts

The service date is calculated on the basis of data collected on the actual condition of wearing parts, fluids and lubricants, as well as information on the vehicle's operating conditions. When defined by this demand-based service concept, service and maintenance are carried out only when needed, for instance, when a component reaches its wear limit, or a fluid or lubricant has reached the end of its service life.

A new feature is provided by the on-board computer, which transmits coded data on the customer and the extent of the required service to the service facility.

This gives the service representative time to order any required replacement parts such as brake pads and to consult the customer in advance concerning a convenient service date.

Early recognition of potential problems is intended to help avoid repairs stemming from vehicle breakdowns. Additional advantages include:

- Precisely defined dates
- Minimal waiting times
- No information loss
- Flexible service

Service inspection schedule		
Job no.: 900109	**Vehicle model:** Passat	**Vehicle owner:** Smith
km reading/ mileage: 53,400	**Vehicle age:** 3	**Additional work, e.g. emissions inspection** ....

Servicing to be carried out	OK	not OK	Rectified
Electrical system			
Front lights. Check function: Parking lights, dipped beam, main beam, fog lamps, direction indicators and hazard-warning signals			
Rear lights. Check function: Brake lights, tail lights, reversing lights, fog warning lamp, number-plate lights, luggage-compartment light, parking lights, direction indicators and hazard-warning signals			
Interior and glove-compartment lights, cigarette lighter, signal horn and telltale lamps: Check function			
Self-diagnosis: Interrogate fault memories of all systems (insert printout at back of logbook wallet)			
Vehicle from the outside			
Door arresters and retaining bolts: lubricate			
Windscreen wash/wipe system and headlight washer system: Check function and spray-nozzle setting			
Windscreen wiper blades: Check for damage, check home position; in event of rubbing wiper blades: Check contact angle			
Tyres			
Tyres: Check condition, tyre tread pattern and inflation pressure, enter tread depth			
FL ____ mm FR ____ mm			
RL ____ mm RR ____ mm			
Vehicle from below			
Engine oil: Drain or draw off, replace oil filters			
Engine and components in engine compartment: Visually check for leaks and damage			
V-belts, ribbed V-belts: Check condition and tension			
Gearbox, final-drive unit and joint boots: Visually check for leaks and damage			
Manual gearbox / axle drive: Check oil level			

	OK	not OK	Rectified
Brake system: Visually check for leaks and damage			
Front and rear brake pads: Check thickness			
Undercoating: Visually check for damage			
Exhaust system: Visually check for leaks and damage			
Track-rod ends: Check play, mounting and sealing gaiters; axle joints: visually check sealing gaiters for leaks and damage			
Engine compartment			
Engine oil: Check oil level (during inspection service with filter change, change oil)			
Engine and components in engine compartment (from above): Visually check for leaks and damage			
Windscreen wash/wipe system: Top up fluid			
Cooling system: Check coolant level and antifreeze; setpoint value: −25 °C			
Actual value (measured value): _____ °C			
Dust and pollen filter: Replace filter element (every 12 months or every 15,000 km)			
Toothed belt for camshaft drive: Check condition and tension			
Air filter: Clean housing and replace filter element			
Fuel filter: Replace			
Power steering: Check fluid level			
Brake-fluid level (dependent on brake-pad wear): Check			
Battery: Check			
Idle speed: Check			
Headlight adjustment / documentation / final inspection			
Headlight adjustment: Check			
Service sticker: Enter date/mileage for next service (also brake-fluid renewal) on sticker and attach sticker to door pillar (B-pillar)			
Take vehicle for test drive			
Date / Signature (mechanic)			
Date / Signature (final inspection)			

Fig. 1: Service inspection schedule

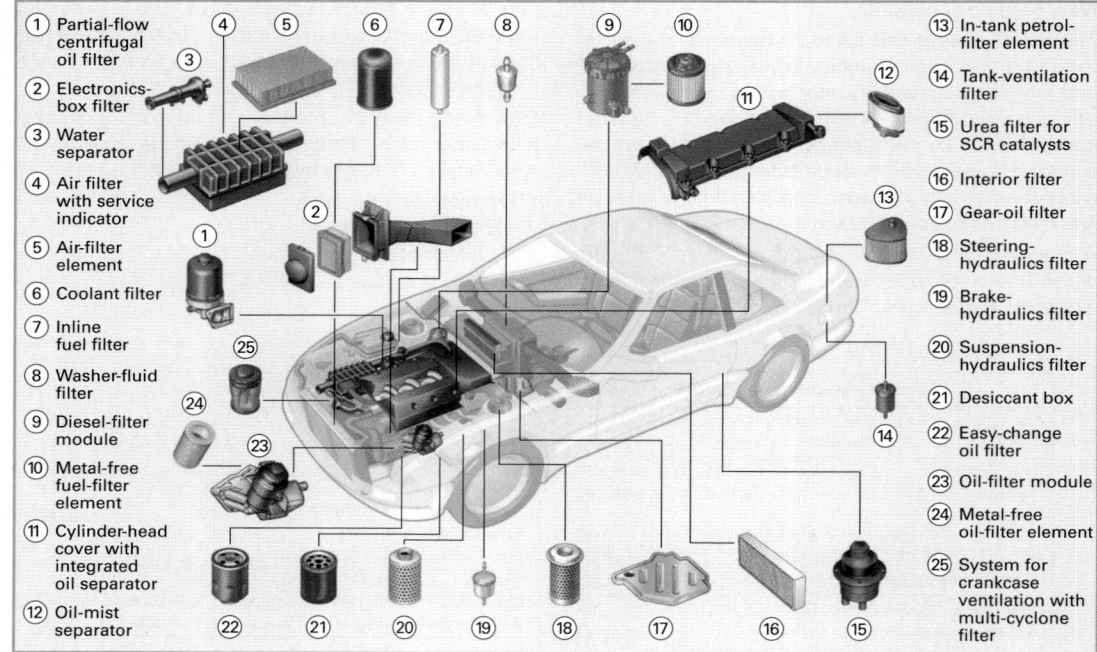

Fig. 1: Filters in modern motor vehicles

1. Partial-flow centrifugal oil filter
2. Electronics-box filter
3. Water separator
4. Air filter with service indicator
5. Air-filter element
6. Coolant filter
7. Inline fuel filter
8. Washer-fluid filter
9. Diesel-filter module
10. Metal-free fuel-filter element
11. Cylinder-head cover with integrated oil separator
12. Oil-mist separator
13. In-tank petrol-filter element
14. Tank-ventilation filter
15. Urea filter for SCR catalysts
16. Interior filter
17. Gear-oil filter
18. Steering-hydraulics filter
19. Brake-hydraulics filter
20. Suspension-hydraulics filter
21. Desiccant box
22. Easy-change oil filter
23. Oil-filter module
24. Metal-free oil-filter element
25. System for crankcase ventilation with multi-cyclone filter

1.6 Filter, body and maintenance

Filters installed in the motor vehicle guard against contaminants and impurities by providing protection for the engine, other vehicle components, and the vehicle's occupants.

Motor vehicle filters **(Fig. 1)** can be classified according to two criteria. These are the **filtration concept** and the **medium** being filtered.

Filtration concepts. Solid contaminants are filtered from flowing media such as air, oil, fuel and water by:
- Screen filtration, using sieve-type filter screens and fibre filters, etc.
- Adhesive filtration, including wet filters
- Magnetic filtration, as with magnetic separators
- Centrifugal filtration, with centrifugal filters, etc.

Strainers (filter screens). Filter mesh dimensions smaller than the contaminants facilitate filtration **(Fig. 2)**.

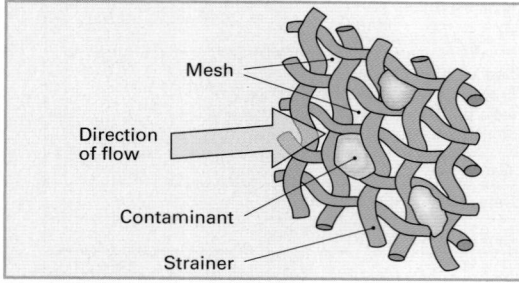

Fig. 2: Operational concept of the filter screen

Adhesive filters. These are usually wet air filters. Contaminants such as dust adhere to the filter surface on contact.

Magnetic filter. The filter (for instance, on the oil drain plug) attracts and retains ferromagnetic contaminants suspended in the flowing medium.

Centrifugal filter. The object medium (such as air) is placed in a state of rotation. Centrifugal force propels the contaminants onto the filter's walls, where they settle as deposits.

Filter types include
- Air and exhaust-gas filters
- Fuel filters
- Filters for lubricating oils
- Interior filters, such as pollen, smog and ozone filters
- Hydraulic filters, for ATF, etc.

1.6.1 Air filters

The purpose of the air filter is to cleanse induction air of impurities while simultaneously subduing induction roar.

Airborne dust particles are minute in size (0.005 mm to 0.05 mm). The air can also contain quartz. Dust concentrations vary considerably according to vehicle operating conditions (motorway, construction site). Should it enter the oil, this dust would form an abrasive film, leading to extreme wear, especially on the cylinder walls, the pistons and the valve guides.

Filter types

The following air filter types are in use:

- Dry air filter
- Wet air filter
- Oil-bath air filter
- Cyclone prefilter

Dry air filter. This type of filter generally retains dust by trapping it in replaceable elements consisting of folded paper. These filters are standard equipment on today's passenger cars and commercial vehicles. The filter element's service life is determined by the size of the paper surface and the dust content of the ingested air. Large surface areas are required to hold flow resistance to a minimum. The air filter also reduces induction roar.

Failure to replace or clean an air filter when service is due will lead to increased flow resistance, ultimately reducing effective cylinder charge volumes and engine performance. Minute particles that manage to pass through the filter contribute to sludge formation in the oil. Contaminated filters should always be replaced.

Wet air filters are sometimes still encountered on motorcycles. The filter element consists of a metal or plastic wire-mesh screen dampened with oil. As the air flows through the filter it impacts against the large, oil-covered surface. The filter retains the dust suspended in this incoming air. Service intervals are on the order of a mere 2,500 km. Once this period elapses the filter must be cleaned and recoated with oil.

Oil-bath air filter. A metal oil-bath screen is located within the filter housing, below the filter element **(Fig. 1)**. The incoming air impacts against the oil film to absorb droplets from the oil bath and then deposit them in the filter element. These droplets then drip downward, taking the accumulated dust with them into the oil bath. Owing to this self-cleaning effect, oil-bath air filters have longer service intervals than wet air filters.

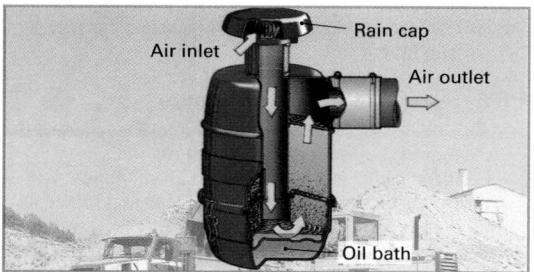

Fig. 1: Oil-bath air filter

Cyclone prefilters are essential components in engines forced to operate continuously in environments with dust-laden air. The assembly induces rapid rotation in the induction air **(Fig. 2)** to remove coarse dust particles with centrifugal force (coarse filter). More minute dust particles still remaining in the intake air are then removed by a dry air filter or similar device. This combination filter concept extends service intervals.

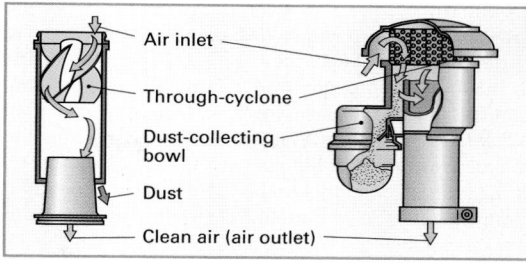

Fig. 2: Cyclone air filter

1.6.2 Fuel filters

The function of these filters is to guard the fuel system against contaminants and impurities, while also removing moisture in some cases.

The operative distinction is between:

- Coarse-mesh filters
- Inline filters
- Filter elements
- Replaceable filters

Coarse-mesh fuel filters. These filters are installed in the fuel tank, where they serve as intake prefilters. These are usually designed as strainers (filter screens) with close-mesh wire or polyamide fibre featuring a mesh size of roughly 0.06 mm.

Inline filters provide filtration for more minute particles. These are paper filters with a pore size of between 0.002 mm and 0.001 mm. Installed in the fuel line, these filters are replaced as complete units during maintenance.

Fuel filter elements. These replaceable elements are found in a dedicated housing installed on the engine. Felt and paper elements are used to obtain filtration of minute particles.

Replaceable fuel filters (box-type fuel filters) **(Fig. 3)**. These consist of a housing and a filter element and are replaced in their entirety as a single unit during vehicle maintenance.

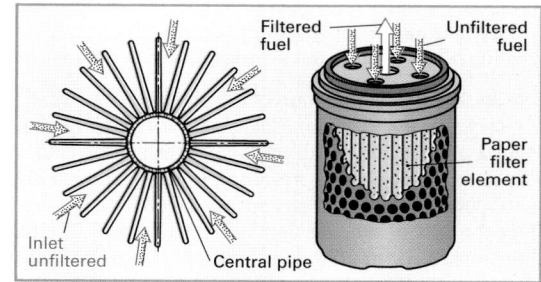

Fig. 3: Box filter with radial filter element

Here, as well, elements consisting of paper and felt are employed for filtration of minute particles. The radial filter employs a radially folded paper element mounted around the periphery of a central, perforated tube. The top and bottom of the folded paper element are enclosed by cover plates. The fuel flows from the outside to the inside of the filter (radial flow). The contaminant particles adhere to the filter's surface, and can then subside to the bottom. Water is unable to penetrate the microscopic pores of the filter. Its relatively high density in comparison to that of the fuel causes it to descend along the outside of the filter paper. It then collects in the filter housing's moisture trap. Meanwhile, the filtered fuel flows inward through the perforations in the central tube before proceeding upward.

Water separator (Fig. 1). These devices are employed in diesel-engined military, construction and off-road vehicles to filter out larger quantities of water. The amount of water trapped in the box-type fuel filter's moisture trap can be monitored through a transparent filter cover or with the aid of an integrated water level sensor (electronic conductivity sensor) designed to trigger a warning lamp in the instrument panel. The accumulated water can then be drained through a drain plug located on the filter housing.

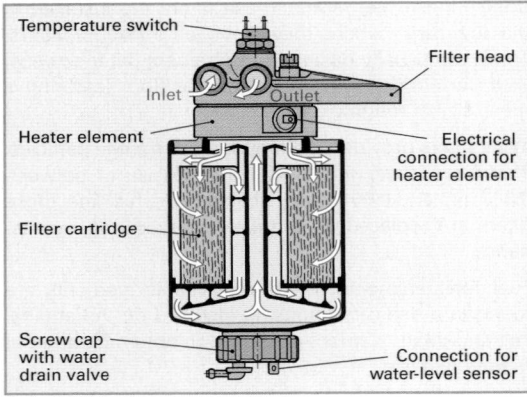

Fig. 1: Box filter with water separator

1.6.3 Oil filters

> They help prevent premature degradation of lubricating oil by filtering out suspended contaminants.

The structure and operation of the oil filter are basically the same as those of the replaceable fuel filter **(Fig. 3, Page 21)**. Filter elements remove particles down to a diameter of roughly 10 μm. Contaminants suspended within the oil, such as metal particles, soot and dust reduce the oil's quality, leading to increased wear. The oil filter makes it possible to extend oil-change intervals while simultaneously providing enhanced cooling for the circulating oil. However, oil filters are unable to remove contaminants that are in liquid form and/or have dissolved in the oil. They also have no effect on chemical and physical changes that may occur in the oil during engine operation, such as ageing.

1.6.4 Hydraulic filters

These sieves (filter screens) serve to remove contaminants from hydraulic fluids, including brake fluid as well as the automatic transmission fluid used in automatic gearboxes and power-steering systems.

Filter screens manufactured in synthetic materials are employed in applications such as the reservoirs for master cylinders. Automatic gearboxes are among the locations in which replaceable flat paper filters are used.

1.6.5 Interior filters

> They filter air for the occupants to protect them from dust, pollen and potentially hazardous gases such as smog and ozone.

Interior filters (Fig. 2) consist of three or four layers. The prefilter retains coarse-grained dirt particles. At the centrally-located microfibre layer an electrostatic charge retains even the most minute airborne particles. The third layer serves a support function for the others. A fourth layer may also be present. This layer employs activated charcoal to absorb incoming gaseous pollutants, such as ozone and exhaust gases. It is also here that the intense odours associated with some substances are neutralised.

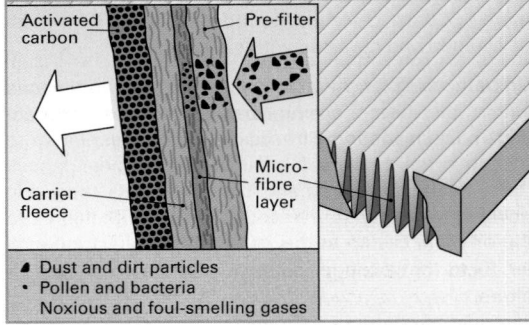

Fig. 2: Design structure of the interior (passenger compartment) filter

1.6.6 Service and maintenance

Service and maintenance instructions
- Perform filter replacements in accordance with the manufacturer's instructions (service intervals by elapsed time or mileage).
- Specified calendar intervals and/or mileages are contained in service and maintenance schedules, while filter changes are contained in service inspection schedules (see Chapter 1.5).
- Paper filters require replacement.
- Foam filters can be rinsed out; during cleaning with compressed air, the nozzle should be directed against the normal flow direction (that is, air should be directed from the inside out).
- The mixture of water and fuel contained in used fuel filters must be disposed of in accordance with environmental regulations.

1.7 Fluids and lubricants, auxiliary materials

> **Fluids, lubricants and standard maintenance items** are the fluids and substances essential to maintain operation of the motor vehicle. **Auxiliary materials** are used for repair, care and cleaning operations on the vehicle and its components.

Fluids and lubricants:

Liquid and gaseous fuels, such as petrol, diesel fuels, natural gas, hydrogen. During combustion within the engine these substances generate thermal energy, which is then converted into kinetic energy.

Oil and lubricants, such as motor oils, grease, graphite. These reduce friction and wear on moving parts.

Coolants and antifreeze, such as water, ethylene glycol, R 134a refrigerant, dry ice, liquid nitrogen. These protect engines against overheating and damage from freezing, or cool interiors and cargo compartments.

Brake fluids, for instance, glycol ethers. These fluids transfer substantial pressures within hydraulic brake systems and clutch-release systems. It is essential that they remain fluid and not evaporate into a gaseous state under the influence of high temperatures.

Fluids for force transfer, such as ATF, silicone fluids, hydraulic fluid. These are employed in hydrodynamic torque converters, power-steering systems, viscous-drive couplings and hydraulic jacking devices.

Auxiliary materials:

Cleaners for vehicle components, such as naphtha, grease remover, alcohol, plastic cleanser.

Products for cleaning and caring for vehicles, such as tar and insect remover, polishes for paint, chrome and aluminium components, waxes, windscreen washer fluid.

1.7.1 Fuels

Fuels consist of hydrocarbon compounds with various molecular structures. The structure and the size of the molecules join the relative proportions of hydrogen and carbon atoms as the essential determinants of the fuel's performance characteristics during combustion within the engine. Pure hydrogen is also suitable for use as a fuel.

Structure

Hydrocarbon molecules feature either cyclical or chain structures **(Fig. 1)**. Molecules with basic chain structures (paraffins and olefins) are extremely volatile and ignite easily. This leads to "knock" in spark-ignition engines.

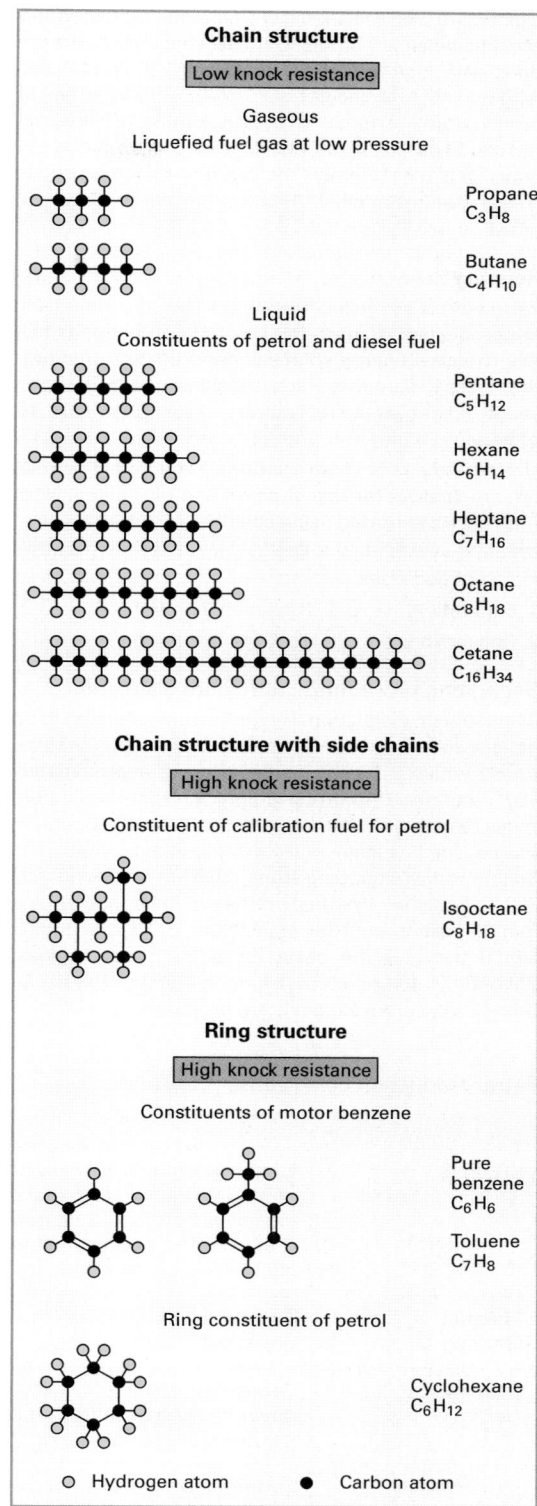

Chain structure

Low knock resistance

Gaseous
Liquefied fuel gas at low pressure

Propane
C_3H_8

Butane
C_4H_{10}

Liquid
Constituents of petrol and diesel fuel

Pentane
C_5H_{12}

Hexane
C_6H_{14}

Heptane
C_7H_{16}

Octane
C_8H_{18}

Cetane
$C_{16}H_{34}$

Chain structure with side chains

High knock resistance

Constituent of calibration fuel for petrol

Isooctane
C_8H_{18}

Ring structure

High knock resistance

Constituents of motor benzene

Pure benzene
C_6H_6

Toluene
C_7H_8

Ring constituent of petrol

Cyclohexane
C_6H_{12}

○ Hydrogen atom   ● Carbon atom

Fig. 1: Structure of the hydrocarbon molecule

Olefins are distinguished from paraffins by the double bond between 2 C atoms. On diesel engines, hydrocarbons with high ignitability produce ideal combustion with no knock. Molecules with side chains (isomers) or in ring form (aromatics, cycloparaffins) are less ignitable. They are knock-resistant in spark-ignition engines, but their ignition lag renders them subject to substantial knock when they are burned in diesel engines.

Refining

Petroleum is by far the most important raw material in the production of fuels. The current assumption is that this chemical energy source stems from sea creatures that died, decomposed and subsided over millions of years, serving as indirect storers of solar energy in the process.

Not all of the many hydrocarbons contained in petroleum are suitable for use in petrol and diesel fuel. Most must be transformed using chemical processes.

Within the refinery, the final product may be produced in one of two ways:

1. **Separation,**   e.g. distillation, filtration
2. **Conversion,**   e.g. cracking, reforming.

Separating procedures, petroleum distillation

Atmospheric distillation. The petroleum is heated in an airless, sealed environment. The components that vaporise within a boiling range of up to approximately 180 °C condense to produce light fuels, primarily benzenes, composed of a combination of normal paraffin (unbranched chains) and cycloparaffins (cyclic). The boiling range extending from 180 °C to approximately 280 °C furnishes the medium-heavy fuels (gas-turbine fuel, kerosine), and the range from 210 °C to roughly 360 °C provides the heavy fuels for diesel engines. Beyond this temperature begins the range at which lubricants and lubricating oils are produced.

Vacuum distillation. This process is employed to produce various lubricating oils. The procedure entails reheating residual materials remaining from atmospheric distillation before separating them according to individual boiling range in a distillation tower. The vacuum atmosphere lowers the boiling point. Bitumen is the final residual product from this process.

> This method of retrieving fuels according to their boiling ranges is also known as fractionating distillation **(Fig. 1)**.

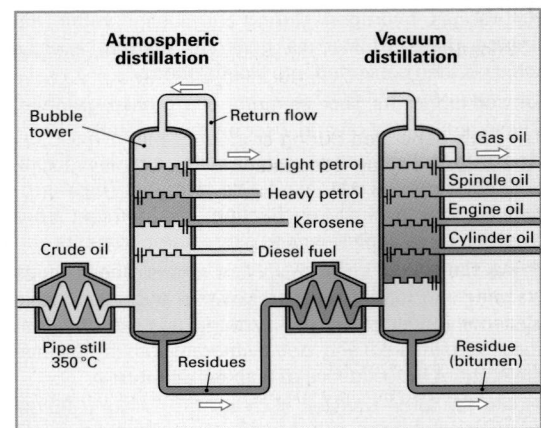

Fig. 1: Distillation of crude oil

The proportion of petrol fuel produced in standard distillation is much too modest for current requirements; with a RON of 62...64, it also displays inadequate knock-resistance. The response has been to develop processes that substantially raise the amount of fuel obtained for use in spark-ignition engines whilst simultaneously endowing petrol with greater resistance to knocking **(Table 1)**.

Petrol production using conversion processes

Table 1: Crack process		
Cracking	Large molecules from heavy oils with a higher boiling point are broken down into lighter and more knock-resistant isoparaffins and olefins (olefins are distinguished from paraffins by the double bond between 2 C atoms). The residual material consists of substances with extremely high boiling points, which are then available for subsequent processing.	**RON 88...92**
Thermal cracking	The conversion process proceeds at a temperature of 500 °C and under pressures of up to 20 bar.	
Catalytic cracking	Greater quantities of light fuels. An aluminium-silicate catalyst breaks down heavy fuels into valuable components for use in petrol at temperatures of 500 °C.	
Hydrocracking	Hydrogen is added to the dissolved molecules at pressures of 150 bar and temperatures of 400 °C. The result is high-quality, low-sulphur fuel.	

Table 1: Conversion processes		
Reformation	Catalysts (platinum: platforming process) transform paraffins emerging from distillation in the form of molecular chains into knock-resistant isoparaffins and aromatics.	**RON 93...98**
Polymerisation	Catalysts combine the gaseous hydrocarbons produced in the cracking and reforming processes to generate larger molecules, primarily isoparaffins. The process in which straight-chain paraffins are converted to isoparaffins is isomerisation.	**RON 95...100**
Hydration	Hydrogen atoms are added to unsaturated olefins to produce stable, knock-resistant isoparaffins.	**RON 92...94**
Alkylation	Mutual reactions induced in olefins and paraffins are employed to produce isoparaffins with higher levels of knock resistance.	**RON 92...94**

The fuels generated during cracking (**Table 1, Page 24**) do not yet display adequate knock resistance when they emerge from this process. Fuels for spark-ignition engines therefore include a mixture of components generated through cracking and various conversion processes (**Table 1**), combined to furnish the desired properties in the areas of knock resistance (RON) and ebullition (boiling characteristics).

Secondary processing

The knock-resistant petrol produced in the refinement process is then subjected to additional refining. Here the petrol's purity is enhanced (removal of gaseous residue, sulphur and dissolved resins). At this point additives enter the mixture. Their purpose is to counteract tendencies to form deposits, change colour and freeze, or to promote combustion knock or oxidation.

1.7.2 Fuels for spark-ignition engines

Boiling curves

Fuels must vaporise easily and completely within the spark-ignition engine. The tendency toward volumetric fuel evaporation is portrayed in the boiling curve (**Fig. 1**). While the volume of the fuel that evaporates between 40 °C and 50 °C must be large enough to ensure reliable cold starting in winter, it should not be so substantial as to promote formation of vapour bubbles in a hot engine. By the time the temperature reaches 180 °C approximately 90 % of the fuel's volume should be vaporised. This helps prevent unvaporised fuel from diluting the engine's lubricating oil, a particular hazard in cold engines. These factors make it necessary to produce fuels for summertime use that exhibit a higher boiling curve than those used in winter in order to accommodate climatic variations.

Cold-start properties

A fuel with a low boiling curve is required for cold starting and for smooth idling in the subsequent warm-up phase.

The 10 % vol. point indicates the temperature at which 10 percent (by volume) of the fuel evaporates.

At least 10 % of the fuel should evaporate at 40 °C ... 50 °C in the interests of reliable cold starting. However, an excessively low boiling range will promote formation of vapour bubbles.

Performance in the warm-up phase

In dealing with engines that have warmed to their normal operating temperature, as well as in summer, the priority shifts toward lower volatility (higher boiling curve) in the fuel. This requirement is satisfied through admixture of a certain proportion of fuels with higher boiling resistance; these contribute a second benefit in the form of a higher energy content. These added fuels are present in liquid form above the 90 % vol. point. At the corresponding temperature, once a fuel reaches the 90 % vol. point, 90 % of this fuel will have evaporated into a gaseous state. An excessively high proportion of boil-resistant fuel can promote formation of condensation on cylinder walls and cause consequent dilution of the oil.

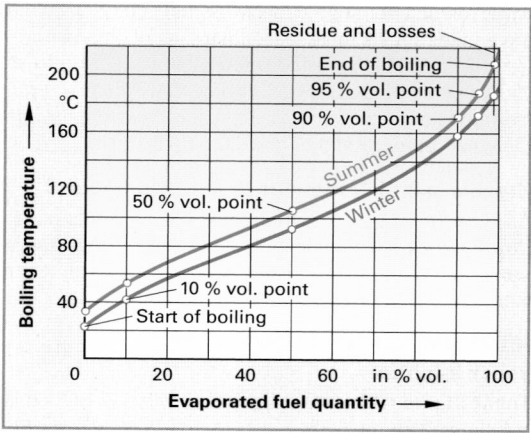

Fig. 1: Boiling curves of fuels for spark-ignition engines

Knock resistance (RON, MON)

The high self-ignition temperature of petrol corresponds to high levels of knock resistance. The indices of knock resistance are the Research Octane Number (RON) and the Motor Octane Number (MON). Both octane numbers are determined in a CFR engine (featuring a variable compression ratio) using an isooctane reference fuel (ON = 100) and normal heptane (ON = 0). A test fuel is defined as having an octane number of 95 when its knock resistance is identical to that of a mixture containing 95 % isooctane and 5 % normal heptane. The MON is lower than the RON owing to the fact that it is determined at higher rpm and with an induction mixture that has been preheated to approximately 150 °C.

Unleaded petrol

Vehicles equipped with a catalytic converter require unleaded petrol. If leaded petrol were used in such a vehicle, the lead compounds contained in the exhaust gases would gradually form a deposit layer on the catalytic material, rendering conversion of hazardous exhaust components into non-toxic elements impossible. In response, the lead content of "unleaded petrol" has been limited to 13 mg/litre of fuel.

Reduction in a fuel's lead content are accompanied by a substantial drop in its octane number. This makes it necessary to modify the fuel-production process, including more knock-resistant components for addition to the unleaded petrol by reforming, polymerisation and alkylation. Ultimately, however, **knock inhibitors** must be added to achieve the desired octane numbers.

Metallic knock inhibitors

Because these substances produce toxic components during combustion (lead, scavengers = bromine and chlorine compounds) they are no longer used in Germany.

Non-metallic knock inhibitors

Aromatics such as benzene, toluene and xylene lie within an octane range of between 108...112 (RON), and are added to fuels to increase their overall octane numbers. Benzene is limited to 5 % by volume owing to its carcinogenic properties. On the average, today's regular-grade petrol contains 2 % by volume benzene, while the volumetric content of premium fuels is 1 %.

Organic oxide compounds as knock inhibitors

Alcohols (methanol, ethanol), phenols and ether suffer from liabilities including poor solubility in fuel and a pungent odour, as well as economic disadvantages stemming from their low energy content.

MTB (Methyl-tertiary butyl ether) as knock inhibitor

Thanks to its high octane range of 110...115 RON this substance can exercise a substantial influence on the overall octane number. Its low boiling point of 55 °C furnishes improvements in knock resistance of particular significance at the lower end of the fuel's boiling range. Added to fuels at a ratio of roughly 10 %...15 %.

> Fuels for spark-ignition engines (petrol) have an ignition point of less than 21 °C, leading to their classification in Group A, Hazardous Substance Class I (highest hazard class).

1.7.3 Diesel fuels

In contrast with fuels for spark-ignition engines, diesel fuels should exhibit the highest possible level of ignitability in the interests of knock avoidance. The index of this ignition quality is the cetane number CN. The ignitability of a diesel fuel is proportional to its content in hydrocarbons of a molecular chain structure. The cetane number for diesel fuels should be in excess of 51, and current fuels reach cetane numbers as high as 62.

The cetane number is determined in a test engine by measuring the time that elapses between the start of injection and initial combustion (ignition lag). The cetane number of a given test fuel is defined by the mixture ratio of the reference fuel cetane (n-hexadecane, $C_{16}H_{34}$) **(Fig. 1)**, with its cetane number of 100, and methylnapthalene (CN 0). A higher cetane number translates into a more ignitable fuel.

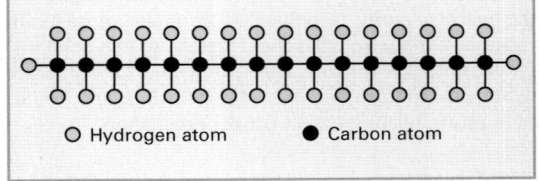

O Hydrogen atom   ● Carbon atom

Fig. 1: Cetane

In a drive to improve levels of sulphuric emissions emanating from combustion of diesel fuels, the sulphur content was limited to 300 mg per litre up to the year 2005. A reduction to 50 mg/l has been mandated for the subsequent period. The sulphur content must be below 50 ppm before a NO_x accumulator-type catalyst can be used.

Diesel fuels require additives to compensate for the resulting loss of lubricity. This not only lowers SO_2 emissions, but also reduces particulate generation.

Winter diesel fuels

Diesel fuels exhibit a general tendency to form paraffin crystals at low temperatures, and once these crystals reach a certain size the fuel loses its ability to flow through fuel filters. The filter is obstructed, and the engine stops running.

Winter diesel must remain suitable for passage through the filter down to a temperature of −15 °C. Suitability for filtration is defined by the Cold Filter Plugging Point (CFPP).

The Cold Filter Plugging Point indicates the temperature at which the paraffin crystals in the diesel fuel congeal to the point at which they prevent the fuel from flowing through a standardised filter screen within a defined period.

The only remedy is to dissolve the paraffin crystals with heat, for instance, by installing a filter heater.

While additives (flow improvers) do not prevent paraffin formation, they do slow and inhibit crystal growth, allowing fuel to flow through filters at temperatures below $-20\,°C$.

Addition of petrol to diesel is a practice to be avoided because …

- … lubrication for the high-pressure fuel-injection system is no longer ensured.

- … manufacturers do not usually issue official approval for the use of petrol, which means that engine damage may not be covered by the warranty.

- … the diesel fuel's reduced ignitability can affect the engine's oil (dilution).

- … fuel consumption increases.

- … the ignition point is changed, transforming the mixture into a fire hazard of Hazardous Substance Class I.

Operators adding kerosine to improve the flow properties of diesel fuels should always observe the manufacturer's recommendations.

Biodiesel fuel

This fuel is manufactured from rapeseed oil, a renewable raw material (rape methyl ester or RME). Before using biodiesel it is always important to verify that the engine's fuel system is designed to operate with this fuel. It acts as a potential source of swelling in rubber hoses and seals, which can cause leaks.

Biodiesel fuels are also hygroscopic, meaning they tend to absorb moisture. Because they act as a solvent on paint, these fuels should be wiped away at once when they come into contact with vehicle finishes.

During combustion these fuels exhibit several assets relative to conventional diesel fuels: they generate lower levels of unburned hydrocarbons, less carbon monoxide and fewer particles. The nitrous oxides are slightly higher. CO_2 emissions are cancelled by the CO_2 ingested by the rape plant as it grows.

Diesel fuels have an ignition point extending from above $55\,°C$ to $100\,°C$ and are included in Group A Hazardous Substance Class III.

REVIEW QUESTIONS

1 What are the raw materials used to produce engine fuels?

2 What are the two basic molecular structures found in the molecules of engine fuels?

3 What kinds of fuel molecule display a particular resistance to knock?

4 What is the meaning of RON?

5 Why is it necessary to convert fuel following distillation?

6 What is the significance of the 90 % point in the boiling-point diagram for petrol?

7 What means are available to endow fuels with the required octane number?

8 What are the distinctive properties of diesel fuels?

9 Name the aromatics employed as knock inhibitors.

10 How is the ignitability of diesel fuel indicated?

11 What is the maximum approved sulphur content for diesel fuels?

12 What are the distinguishing properties of biodiesel fuels?

1.7.4 Oils and lubricants

Production

Base oils for use in engines and gearboxes are produced by vacuum-distilling the residual materials remaining after atmospheric distillation of crude oil (**Fig. 1, Page 24**).

The long-chained hydrocarbon molecules contained in lubricating oils are sensitive to heat, and can degenerate into short-chained benzene molecules at temperatures as low as $330\,°C$. This is prevented by using vacuum conditions to lower the boiling point (vacuum distillation). As in atmospheric distillation, this produces distillates of varying viscosities (higher boiling temperatures correspond to shorter molecule chains).

Subsequent refinement processes are needed in order to transform these distillates into base oils suitable for use in manufacturing lubricants (**Table 1**).

Table 1: The functions of refinement
Removal of undesirable substances, such as sulphur
Enhance ageing stability
Formulate for a viscosity index of approximately 100
Paraffin extraction to lower the pour point to $-9\,°C … -15\,°C$

Selected additives are employed to endow the base oils with the desired properties in areas such as anti-oxidation protection and viscosity.

Hydrocarbon base oils

As is the case with the raffinates, **synthetic hydrocarbons** consist of carbon and hydrogen atoms, but have a molecular structure that differs from that of crude oil. Cracking is employed to convert the initial raw petrol into highly reactive gas molecules such as ethylene. These gas molecules are then synthesised into isoparaffins of the desired structure, polyalphaolefins (PAOs). This molecular structure endows synthetic hydrocarbons with a much higher viscosity index than the initial raffinates while simultaneously reducing evaporative losses and improving thermal response patterns.

Function and properties of lubricating oils

Lubricate	Clean
Cool	Protect against corrosion
Seal	Reduce noise

Viscosity. Viscosity serves as an index of an oil's flow properties by quantifying its internal friction. Highly-fluid, free-flowing oils have a low viscosity and correspondingly low resistance to displacement forces, while high-viscosity oils exhibit higher flow resistance.

This resistance that the fluid displays against the mutual displacement of two contiguous layers is also referred to as internal friction (shear stress). Viscosity varies according to oil grade, and falls as temperature rises **(Fig. 1)**.

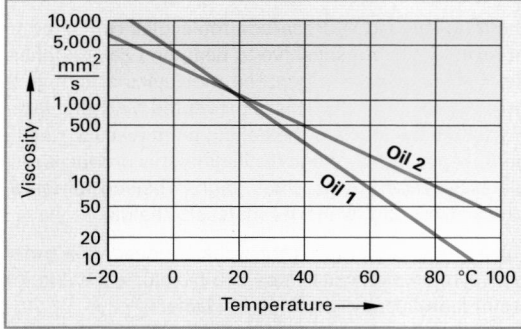

Fig. 1: Viscosity-Temperature diagram

Kinematic viscosity. This property is measured in the capillary viscometer **(Fig. 2)**. A defined quantity of the oil flows through a long, thin tube at a defined test temperature. The time it takes the oil to traverse the tube is used to define the viscosity in m^2/s or mm^2/s.

Dynamic viscosity. Low-temperature viscosity can be measured in pressurised capillary viscometers (especially for HTHS viscosity) or in a rotation viscometer **(Fig. 2)**. A rotor turns in a cylinder with test oil at a specified test temperature. The level of torque needed to induce rotation provides the essential index of viscosity in Pa·s (Pascal seconds) or, more usually, in mPa·s.

HTHS viscosity (High Temperature High Shear). The SAE, ACEA and various automotive manufacturers specify minimum viscosity curves at an oil temperature of 150 °C and a shear rate of 10^6 s^{-1}. The object is to ensure that the oil's ability to form a lubricating layer will remain undiminished at high engine speeds. A low HTHS can reduce fuel consumption.

Shear rate. This is the velocity of a moving object divided by the depth of the lubricant layer. Oil is subjected to various shear stresses at a lubrication orifice. While a piston's travel speed along a cylinder wall can reach as much as 36 m/s at high rpm, the layer of oil adhering to the wall's surface has a velocity of 0. At idle the shear rate is 10^5 s^{-1}, while it lies at 10^6 s^{-1} at maximum rpm. The lubricating film's layer depth under these conditions is between 3/100 mm and 4/100 mm.

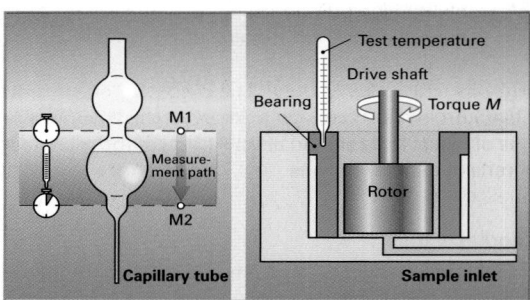

Fig. 2: Capillary viscometer, rotation viscometer

Viscosity index. The best oil for any engine is the one that displays the lowest variation in viscosity when it heats up (Oil 2 in **Fig. 1**), as this kind of oil will allow easy cold starting while simultaneously forming a reliable load-bearing lubricating layer after the oil becomes hot. The Viscosity Index VI provides information on the slope of the VT (viscosity-temperature) curves. A higher viscosity index corresponds to a lower response curve in the VT diagram. High-quality mineral oils have a viscosity index of 90 ... 100, and synthetic hydrocarbons can reach 120 ... 150, enhancing their ability to satisfy the demands of high-performance engines. The viscosity index number is defined based on the inclination angle of the curves in the VT diagram.

SAE viscosity grades. This classification system has been defined by the American **S**ociety of **A**utomotive **E**ngineers to assist in selecting engine and gear oils for use under various climatic conditions.

The basic distinction is between single-grade oils, such as SAE 10W, SAE 20W/20 (winter oils), SAE 30, SAE 50 (summer oils), and multigrade oils such as SAE 15W-50 for year-round use. The SAE viscosity classes start at 0 W and end at 50.

> Higher numbers correspond to thicker oils.

Multigrade oils are lubricating oils formulated to conform to the requirements of more than a single viscosity class. Thus SAE 15 W-50 meets the specifications for SAE 15W at −17.8 °C while also conforming to the requirements for SAE 50 at +98.9 °C, making it suitable for trouble-free cold starts while also endowing it with resistance to thermal stress.

Additives. Chemical additives respond to the fact that base oils cannot satisfy the multiplicity of demands placed on engine and gear lubricants. These additives improve an oil's characteristics and suppress undesirable properties **(Table 1)**. Additives are surface-active substances that can react with water and acids and are also soluble in oil. Potential proportions of additives range from less than 1 % to as much as 25 %.

HD oils (HD = Heavy Duty). These contain **dispersants** that surround contaminants and maintain them in suspension to prevent the sludge formation that results from congealed contaminants. Dispersants are present in all current oils.

The temperature ranges for which oils are suitable are presented in **Fig. 1**.

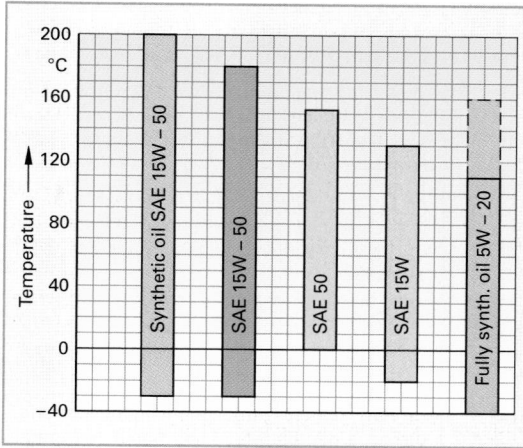

Fig. 1: **Motor oil temperature ranges**

API engine oil classifications

In an ongoing project, the **A**merican **P**etroleum **I**nstitute **(API)** co-operated with the SAE and **ASTM** (**A**merican **S**ociety for **T**esting and **M**aterials) in creating a classification system for engine oils that can be expanded to include new classes with more stringent requirements, without any need to revise the earlier system **(Table 1, Page 30)**.

Distinctions are defined according to S Classes (Service Classes) for use in service stations. They are suitable for use in spark-ignition engines. Oils for diesel engines are assigned to C Classes (Commercial Classes).

Table 1: Additives (selection)	
Anti-ageing additives **Anti-oxidation agents**	They prevent oil from responding to heat and oxygen by oxidising (ageing). Formation of a protective film also inhibits corrosion on metal surfaces.
Extreme pressure/ **Antiwear additives** **(EP/AW)**	High-pressure additives form thin, friction-inhibiting layers on sliding surfaces (bearings, pistons, cylinders, gears) to prevent direct contact between mutually opposed metallic surfaces.
VI improvers	These consist of long thin hydrocarbon molecules that intertwine with minimal dispersion in the cold oil. As the oil heats up these molecular structures separate and assume a larger volume, causing the oil to thicken. The viscosity index VI rises.
Pour point reducers	The pour point is the lowest temperature at which an oil still continues to flow while being cooled under defined conditions. Oil solidification is caused by precipitation and chain formation to produce paraffin crystals. Pour point reducers shift this process to lower temperatures.
Friction modifiers	A specific minimum coefficient of friction is required to ensure satisfactory operation of synchronised gearboxes, automatic gearboxes and limited-slip differentials. The required properties are obtained through admixture of selected friction modifiers.

Table 1: API Classes, Requirements

Spark-ignition engines

SG	More stringent demands in the areas of oxidation stability (ageing) and sludge formation
SH	The same requirements as SG, but higher levels of quality assurance provided by the specified test regimens (for instance, fuel economy test)
SJ	As SH, but reflecting recent insights regarding enhanced fuel economy. Phosphorus <0.1% for improved suitability for use with catalysts.

Diesel engines

CE	For older atmospheric-induction and turbocharged diesel engines
CF CF-4	CF for modern atmospheric-induction and turbocharged diesels in short-haul and long-distance operation. CF-4 with high limits for ash deposits to meet stringent emissions limits, with attendant sacrifices in oil-change intervals
CG-4	For ultra-low emissions diesel engines in commercial vehicles operated on long-distance routes using diesel fuels with a sulphur content <0.05%.
CH-4	For extremely extended oil-change intervals, wear protection to guard against the effects of soot.

It is always vital to observe the minimum requirements when selecting engine oils.

ACEA engine oil specifications

The new ACEA specifications have been in effect since January of 1996 (performance classes **Table 2**). The earlier CCMC specifications are no longer valid.

The **ACEA** (**A**ssociation des **C**onstructeurs **E**uropéens de l'**A**utomobile) defines minimum requirements for motor oils destined for use with spark-ignition and diesel engines in passenger cars as well as for diesel engines in commercial vehicles. These are determined using the world's most extensive, stringent and modern test procedures, and described in three performance groups. These include oils for ...

- ... spark-ignition engines in passenger cars, in ACEA A1, A2, A3 and A4 (composite ash deposits 1.5%).
- ... passenger-car diesel engines in ACEA B1, B2, B3, B4, B5 (composite ash deposits max. 1.8%).
- ... commercial vehicle diesel engines in ACEA E2, E3, E4 and E5 (composite ash deposits max. 2.0%).

Uniform requirements in the areas of foaming and compatibility with seal and gasket materials apply to all oils.

To limit carbon deposits on pistons and valves, evaporative losses in A1/2/3 and B1/2/3 oils may not exceed 13% ... 15%.

Evaporative losses are measured over a period of one hour at an oil temperature of 250°C.

Table 2: ACEA performance classes

Class	Requirements	Application
A1 B1	Less demanding specifications for shear stability, volatility and HTHS viscosity. Observe the vehicle manufacturer's approval guidelines.	Oils with high potential for enhancing fuel economy (free-flowing low-friction oils). Compliance with the supplementary requirements from the engine tests from A3 and B3/B4 makes these oils suitable for initial factory-fill use.
A2 B2	More demanding criteria for shear stability and HTHS viscosity than for A1 and B1 oils.	Especially well-suited for use in vehicle fleets, also in conjunction with E2 oils.
A3 B4	Highest-quality oil for passenger cars. Maximum shear stability, lowest vaporisation tendency, HTHS viscosity >3.5 mP·s to offer maximum reliability along with extended oil-change intervals.	The stringent limits applied in the engine test runs render these oils particularly suitable for use in continuous, extended operation at high rpm. B4 is formulated for application in diesel engines with common rail and unit injector systems.
E2 E3 E4 E5	All E oils comply with the same requirements regarding shearing stability, HTHS viscosity and evaporation resistance. They are specified in test runs with progressively higher limits. Tests focus on scoring, sludge, wear, piston contamination levels, oil consumption and viscosity shift.	E2 describes high-quality commercial vehicle oils for medium-term oil-change intervals intended for use in mixed-duty vehicle fleets (A2, B2 oils). E3 describes motor oils for extended oil-change intervals that, although less expensive, are still characterised by high quality. The focus is on high levels of internal engine cleanliness and a low tendency to congeal. E4/5 describes oils of the highest performance level, intended for oil-change intervals of up to 100,000 km.

The extent to which evaporative losses increase as a function of temperature depends upon the viscosity of the base oil, with lower viscosities promoting higher levels of evaporation. In addition to causing carbon deposits, these losses also lead to higher oil consumption.

Gear lubricants

The demands and specifications for gear oils vary from those applicable to engine oils in several respects:

- Wear protection on gear-tooth flanks and on bearing surfaces. Hypoid gears display an especially pronounced tendency to squeeze the protective film away from surfaces, leading to higher wear rates.
- Different friction properties. Within the synchronised gearbox, the synchronisation process relies on displacement of the oil film between synchroniser cone and ring.
- Protection against ageing extending throughout the service life.
- Suitability for use with gasket and seal materials, such as elastomers.

Both API classifications and SAE grades are used to describe gear lubricants (**Table 1**). The SAE classifications are not suitable for direct comparison with those used for motor oils. For instance, the viscosity of an SAE 80 gear lubricant corresponds to that of an SAE 20 motor oil.

Table 1: Gear oil performance classes		
Manual gearbox, final-drive unit without axle offset	API GL4	SAE 75, 80, 90
Manual gearbox (synchronisation not critical), Final-drive unit with large axle offset	API GL5	SAE 80, 90, 140 SAE 75W SAE 80W-90 SAE 85W-140

Low-friction gear lubricants. Highly viscous multigrade oils, such as SAE 75W - 90, with a high viscosity index. Friction modifiers are particularly effective at low temperatures, easing gear changes whilst also enhancing fuel economy.

Final-drive lubricants. To prevent displacement of the oil film between the gear teeth, oils with a high proportion of additives and extreme load capacity are especially important in hypoid-gear units. LS lubricants are used in limited-slip differentials, where they reduce adhesion loss to support the automatic locking function of the clutch cones' friction surfaces under pressure.

Automatic Transmission Fluid ATF (**A**utomatic **T**ransmission **F**luid). This fluid must meet supplementary demands extending beyond those applicable to basic gear lubricants:

- Lubrication for planetary gearsets and one-way clutches
- Actuation of brake bands and clutches
- Torque transmission from the pump to the converter turbine

- High viscosity index through an extended temperature range

ATF is a gearbox fluid featuring a low viscosity – comparable to that of an SAE 75W grade in standard gear lubricant – but also a high viscosity index as well as a pour point below −40 °C. The viscosity is contained in the specification, but not indicated on the container.

There is no universal mandatory standard for ATF. Instead, minimum requirements are defined in the corporate specifications issued by the automotive manufacturers, for instance, Dexron III for GM and Mercon for Ford.

> Always observe the approval and use guidelines issued by the automotive manufacturers.

Lubricating greases

Greases consist of an oil and a thickening agent. The thickening process creates absorbent structures in which oil is stored and then emitted again as needed.

> Lubricating greases are swelled thickeners in oil.

Composition of lubricating greases

Base oils. As with motor oils, the primary constituents are basic raffinates, hydrocrack oils and synthetic hydrocarbons (PAO). Synthetic esters and rapeseed oils are used when specifications call for biodegradable products.

Thickening agents. Soap thickeners, also known as metal soaps, are in general use as thickeners. These include lithium, calcium and sodium-based thickeners. Soap-free thickeners include gels and bentonite.

The ultimate consistency (rigidity) of the lubricating grease varies according to the type of thickener selected, the temperature, and the viscosity of the base oil.

Selection of lubricating greases

This depends upon the given operating temperatures as well as the loads to be encountered (as in bearings). Many greases respond to high temperatures by softening and dripping. The temperature at which a grease liquifies is its drop point. This factor depends on the base soap used (**Table 2**).

Table 2: Properties of lubricating greases			
Soap basis	Drop point °C	Water-resistant	Applications
Calcium (Calcium soap grease)	up to 200	yes	Automotive joint grease
Sodium (Sodium soap grease)	120...250	no	Roller bearing grease
Lithium (Lithium soap grease)	100...200	yes	Multipurpose grease

1

Sodium soap grease. The most common type of lubricating grease. Resists water and extreme temperatures, suitable for applications at –20 °C ... 130 °C.

Calcium soap grease. Water-resistant, low resistance to thermal stresses, temperature range –40 °C ... 60 °C.

Sodium soap grease. Not water-resistant, max. service temperature is 100 °C.

High-temperature grease. These greases are suitable for use at continuous temperatures in excess of 130 °C. We distinguish between:

- *Complex soap grease,* based on special Al, Ca, or Li metallic soaps (employed in the axle units of commercial vehicles)
- *Gel-type grease, bentonite grease,* based on soap-free thickening agents (application as hot-bearing and gear grease)

EP grease. This type of grease is formulated to withstand extreme pressures, and contains compounds of sulphur, phosphorus and/or lead.

EM grease. This type of grease contains molybdenum disulphide and is intended to provide residual lubrication under the extreme conditions encountered when the grease charge is lost.

Consistency. This is the resistance that a grease displays to deformation. The depth to which a standard sphere penetrates a grease serves as the basis for its NLGI classification (**N**ational **L**ubricating **G**rease Institute) 000, 00, 0, 1, 2, 3, 4, 5.

Consistency	Property, application
000 ... 1	Extremely soft, liquid grease, for central lubrication systems, etc.
2 ... 3	Soft, general-purpose grease for all other lubricating applications
4 ... 5	Hard, water pump grease

Designations for lubricating greases

Example: K PF 2 K – 30

K   Grease for roller bearings (G for gearbox)
PF   P for EP/AW additives, F for solid lubricants such as MoS$_2$
2   NLGI Class, 2 = automotive joint grease
K   Upper operating limit 120 °C
– 30 Lower operating limit in °C.

1.7.5 Antifreeze

Coolant is usually a mixture containing water (with mineral and calcium content as low as possible), antifreeze and additives including anti-oxidation agents as well as substances for lubrication (at the heater valve, etc.). Impurities within the coolant should be held to an absolute minimum, as calcium deposits, contaminants and grease not only reduce its thermal conductivity, but can also lead to obstructions in coolant lines and passages.

To prevent coolant from freezing and causing severe damage to the engine and radiator, its antifreeze concentration must be corrected to the prescribed ratio prior to the onset of cold weather. The factory-fill coolant usually contains between 40 % and 50 % antifreeze. The mixture ratio, and thus the freezing point, can be measured using a hydrometer or refractometer (**Fig. 1**).

This test provides results based on the coolant's specific gravity, which, in turn, is a function of its mixture ratio.

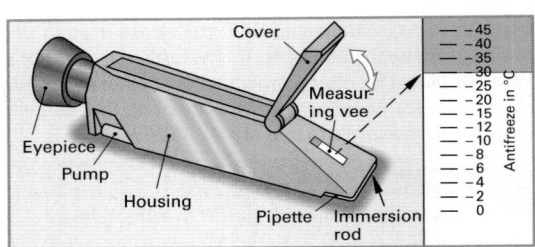

Fig. 1: Refractometer

The primary constituent of antifreeze is usually ethylene glycol, which lowers the freezing temperature. To prevent oxidation within the engine and radiator, antifreeze also contains various metallic components, known as corrosion inhibitors. These agents can display mutually antagonistic properties, promoting damage to metal components. This is why only factory-approved antifreeze should be used.

Coolant should be replaced at the intervals specified by the manufacturer. It should be collected, protected from contamination by other substances, and disposed of in accordance with environmental regulations.

1.7.6 Refrigerant

Automotive air conditioning systems rely on refrigerants to operate. Formerly, refrigerants based on chlorofluorocarbons (CFCs) such as freon (R12) were in common use. Because these substances are odourless, non-flammable and gaseous, they were ideal for air conditioners. In addition, they do not corrode metal. However, the chlorine contained in these products has been associated with depletion of the ozone layer. In response, substitutes such as R134a and R22, with physical properties similar to those of R12, have been approved for use.

As a partially halogenated CFC, R22 still poses a danger to the ozone layer, which is why manufacture of air conditioning systems that operated using refrigerants containing CFCs was discontinued on 1.1.2000. As a tetrafluoromethane, R134a is free of chlorine, and does not pose a hazard to the ozone layer.

Compressor oil. A special compressor oil is needed to lubricate the moving parts within the compressor. A portion of this oil (roughly 25 %) mixes with the refrigerant and recirculates through the circuit. The conventional oils used with R12 and R22 were mineral oils, but special synthetic oils (polyalkylglycol oils, PAG oils) have been developed for R134a.

> Conventional compressor oils do not dissolve in R134a and should not be used with this product. PAG oils are hygroscopic. Containers should always be sealed securely.

Working rules
- Avoid contact with refrigerants in their liquid state.
- Wear protective goggles.
- Never allow gaseous refrigerants to escape into the atmosphere.
- Asphyxiation hazard in work pits, because the gas is heavier than air.
- Do not expose refrigerant containers to temperatures in excess of 45 °C.

Purging (evacuating) the circuit
- Service equipment for R12 and R134a must be stored and used separately.
- Refrigerant is captured at extraction stations and recycling stations.
- Extraction station: Following extraction the refrigerant is pumped into disposal containers to be forwarded to the refrigerant supplier for subsequent handling.
- Recycling station: The refrigerant is cleaned (filtered of oil and moisture) and collected in a charge cylinder for subsequent use.

1.7.7 Brake fluid

Brake fluid must display the following properties:
- High boiling point (up to approximately 300 °C)
- Low pour point (roughly – 65 °C)
- Consistent viscosity
- Chemically neutral toward metal and elastomers
- Lubrication for moving parts in the master cylinder and wheel cylinders
- Suitable for mixing with comparable brake fluids

The boiling points defined in the **DOT standards** (U.S. **D**epartment **o**f **T**ransportation) ensure that fluids will absorb heat generated during braking without vaporising and forming bubbles.

Minimum boiling points for brake fluids:
DOT 3 205 °C, **DOT 4** 230 °C, **DOT 5.1** 260 °C.

Because brake fluid is composed of polyglycol compounds, it is hygroscopic, and tends to absorb water. The boiling point falls as the brake fluid's moisture water content rises. The "wet boiling point" **(Fig. 1)** indicates the boiling point with a water content of 3.5 %.

DOT 3 brake fluid reaches the hazardous wet boiling point at a relatively low 140 °C. Most moisture is absorbed through the brake hoses. Brake fluid reaches a moisture content of 3.5 % and the hazardous wet boiling point after a period of roughly 2 years. The heat generated during braking produces gas bubbles. Because these bubbles do not transfer the applied brake pressure, the result is brake failure. Brake fluid should be replaced every two years at the latest.

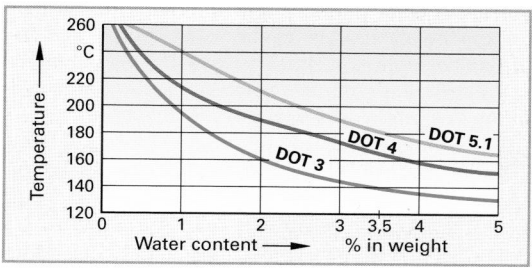

Fig. 1: Brake fluid boiling curves

Viscosity is also measured and defined at – 40 °C in order to ensure that brake fluid retains the properties needed for it to flow through the ABS solenoid valves at low temperatures. DOT 5.1 brake fluid offers the greatest safety reserves in the areas of wet boiling point and viscosity.

> Brake fluid is extremely toxic and also acts as a solvent on painted surfaces. Observe the manufacturer's recommendations when mixing and changing brake fluids.

2 Environmental protection, occupational safety*

2.1 Environmental protection in automotive service operations

2.1.1 Environmental pollution

Technical systems, including those associated with the industrial production of motor vehicles and with their subsequent operation, generate toxic gases, dust, chemical substances, waste water and noise, all of which combine to pose a growing threat to our natural environment. In concrete terms, environmental pollution can:

- Endanger the health and physical well-being of humans and animals, for instance, as with carcinogenic substances
- Damage plant life, exemplified by damage to forests from acid rain
- Destroy physical property, such as buildings constructed in sandstone
- Foster accumulations of dirt and contamination, for instance, from soot
- Damage the atmosphere and provoke associated climate change
- Consume irreplaceable natural resources

Air pollution is mainly caused by toxic and harmful substances stemming from combustion processes. Polluting substances posing a particular hazard to air quality include carbon monoxide (CO), unburned hydrocarbons (HC), nitrous oxides (NO_X), sulphur dioxide (SO_2), soot particles and minute particles containing heavy metals. Suitable measures and devices available for preserving air quality include unleaded fuels, catalytic converters, and particulate filters designed for use with diesel engines.

Water pollution is caused primarily by domestic sewage from homes and effluents emitted by industry. Domestic waste water consists primarily of human waste matter and of residue from soap and cleansers. Toxic materials and residual quantities of mineral oils are among the substances that pollute industrial waste water. Special-purpose filtration systems within individual commercial operations, such as oil separators and silt traps in automotive repair facilities, and water treatment systems in car washes, must remove these pollutants before industrial effluent can combine with domestic sewage at the water treatment plant. These steps must all be completed before the treated water can be returned to natural water courses.

Soil and ground water pollution is caused by petroleum products, e.g. oil and chemical cleansers, soaking into the soil. Other pollutants entering water through this path include heavy metals (such as lead) and toxic chemicals, which are joined by pollution stemming from excessive use of fertilisers and insecticides.

2.1.2 Disposal

Legal framework for environmental protection
To ensure that humanity will be able to continue living under acceptable conditions in the future we have no choice but to embrace action to reduce the amount of pollution entering the environment. Governments join in this effort by enacting laws, directives and guidelines aimed at reducing environmental damage. The following laws and regulations **(Table 1)** represent the foundation of environmental protection as it relates to automotive service operations.

Table 1: Environmental law *				
Law on waste disposal	**Water rights**	**Law on Chemicals**	**Highway and traffic regulations**	**Occupational health and safety regulations**
Closed Substance Cycle Waste Management Act	Water Resources Act	Chemicals Act	Hazardous Cargo Act	Equipment Safety Act
Ordinance on Waste Oils Ordinance of Codification of Waste Requiring Supervision Ordinance of Codification of Waste Requiring Special Supervision Ordinance on End-of-life Vehicles	Ordinance on Storage of Liquids Hazardous to Waters Ordinance on Equipment for Handling Materials Hazardous to Waters	Ordinance on Hazardous Materials and Substances	Ordinance on Hazardous Cargo Transportation	Ordinance on Flammable Liquids
Technical Instructions on Waste		Technical Guidelines on Hazardous Materials and Substances	Technical Directives	Technical Guidelines on Flammable Liquids

* The laws and ordinances mentioned in Chapter 2 relate to German and/or EU law.

Closed Substance Cycle Waste Management Act. This law regulates disposal of waste materials and used oil. Waste is defined as all portable, non-stationary materials of which their owners wish to dispose and/or materials and substances that must be disposed of through suitable channels in order to protect the environment. According to this law, waste disposal also includes recycling, storage, collection, transport, treatment and final storage of waste materials. The following principles have been recognised **(Fig. 1)**:

1. **Waste must be avoided** or at least reduced to the lowest possible level.
2. **Waste must be recycled** whenever possible.
3. **Waste must be separated and disposed of** when avoiding waste proves impossible and recycling is also not a viable option.

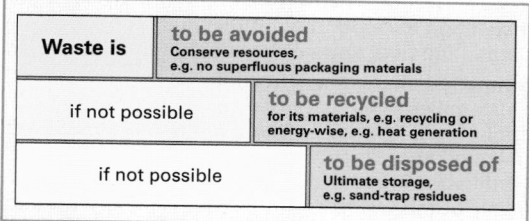

Fig. 1: **Basic principles of waste legislation**

The Closed Substance Cycle Waste Management Act distinguishes between two different types of waste **(Fig. 2)**.

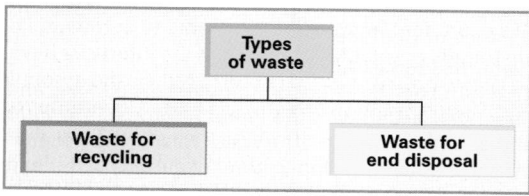

Fig. 2: **Classifications of types of waste according to the Waste Avoidance and Management Act**

Waste for recycling. This class consists of materials that are returned to the processing loop for recycling, and includes such substances as used oil (of known origin), brake fluid and non-ferrous scrap metal. In addition to producing recycled material for renewed industrial application, recycling can also culminate in energy generation. This process entails burning waste materials in a socially responsible manner and then either utilising the generated heat onsite or supplying it to third parties.

Waste for end disposal. This category embraces those materials that are not suitable for recycling. It includes sludge from filtration processes, used oil of unknown origin and mixed oil and water from parts cleaning. These must be disposed of in a socially responsible manner.

Ordinance on Waste Oils.
Used oil of known origin, such as engine oil and gearbox oil remaining from oil changes performed in the garage/repair shop and at filling stations.

These are collected and stored in special containers in accordance with the applicable regulations. This used oil is assigned to hazardous material class A III (applies to liquids with a flash point over 55 °C but not exceeding 100 °C).

Used oil of unknown origin consists of all used oil that persons performing home oil changes return to the garage/repair shop, filling stations and used-oil collection stations. It is impossible to guarantee that this oil will not have been contaminated with petrol, solvents, brake fluid, etc. In response, this oil is assigned to hazardous material class A I (applies to liquids with a flash point below 21 °C) **(Fig. 3)**.

> Used oil of known origin should never be mixed with used oil of unknown origin.

The Ordinance on Waste Oils also defines the oils classified as suitable for recycling. It prohibits recycling of used oil with a PCB (polychlorinated biphenyl) content of 20 mg or more per kilogram or a halogen content of 2 grammes or more per kg. These are disposed of through other channels, which may include socially responsible high-temperature combustion processes.

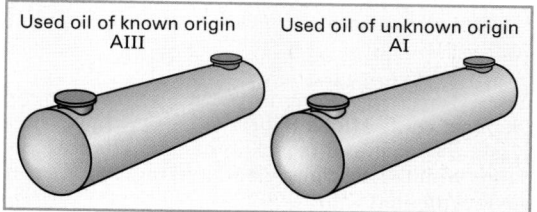

Fig. 3: **Used-oil classifications according to the Ordinance on Waste Oils**

Ordinance of Codification of Waste Requiring Supervision.
This law defines materials and substances capable of posing a serious environmental hazard if not transported, handled and stored according to regulations. This waste is recycled for subsequent use onsite or forwarded to companies that specialise in handling these materials. Documentation confirming compliance with regulations governing end disposal and recycling of waste materials must be available for presentation to the responsible authorities upon request **(Fig. 4)**.

Waste materials suitable for recycling include:

- **Without mandatory verification:** glass, paper, wood, cartons
- **With mandatory verification:** refrigerant, coolant, brake fluid, oil filters

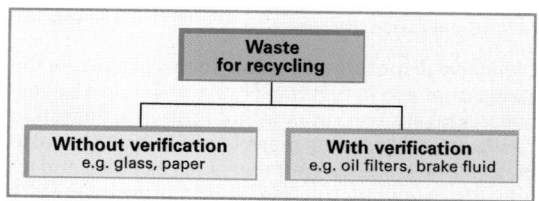

Fig. 4: **Classifications of waste for recycling**

2

Ordinance of Codification of Waste Requiring Special Supervision.

This category includes the types of waste for which disposal verification is absolutely mandatory. Disposal processes are facilitated by waste material classification numbers and the waste material designations contained in the General Administrative Provision to the EC List of Waste **(Table 1)**.

Table 1: Waste materials according to the Ordinance of Codification of Waste Requiring Special Supervision	
Waste material-classification number	Type of waste
16 01 07	Oil filters
13 02 04	Engine and gear oils (not suitable for recycling)
13 05 02	Petrol separator contents (sludge)
16 01 14	Coolant (ethylene glycol)
16 01 13	Brake fluid
14 06 03	Engine cleaner
16 06 01	Lead-acid batteries
15 01 10	Plastic containers with hazardous residue

Disposal verifications (Fig.1)

Waste disposal and recycling conducted in accordance with the officially mandated regulations can be documented with:

- Collective disposal verification
- Individual disposal verification
- Simplified disposal verification

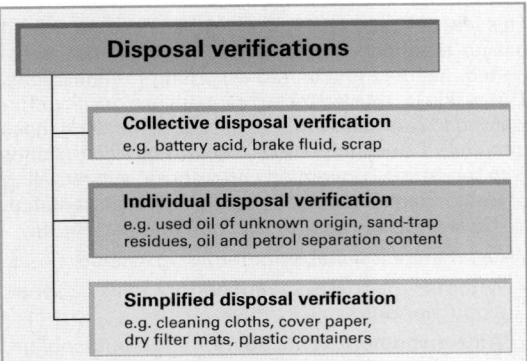

Fig. 1: Disposal verifications according to Ordinance of Codification of Waste

Collective disposal verification. In this procedure the collector of waste materials must maintain a collective disposal verification in the form of a collection certificate. When waste materials with a single waste material classification number, such as fluids and lubricants containing oil, are collected from an automotive service operation participating in a collective

disposal programme, the automotive service operations is not required to verify disposal.

Individual disposal verification

Strict regulations mandate accompanying documentation for the entire life cycle of the classified waste, extending from the producer to the agency responsible for final disposal. In this process the producer of the waste material, its transporter and the disposal agency are all required to identify the waste's class using individual, colour-coded certificates including the waste material's classification number and its designation. They must also provide a signature confirming that this information is correct. Some certificates are forwarded to the responsible government agency, while others are filed in documentation folders. These documentation folders are to be kept and remain available for inspection for a period of three years following the date of the final entry or last documentation paper.

Simplified disposal verification

This applies to all types of waste that are neither eligible for collection as domestic waste nor subject to mandatory documentation. Simplified disposal verification is based on certificates in which the producer of the waste assumes responsibility for and in which the disposal agency confirms acceptance of the material, with no direct participation by an official agency. This documentation must also be maintained in cases in which the producer of the waste material generates only limited quantities (< 2,000 kg) of the substance as defined in the Ordinance of Codification of Waste.

Water Resources Act

This furnishes the legal groundwork for all laws and directives issued with the intent of protecting water and aquatic resources. The Water Resources Act governs use (industrial application) of surface water and coastal waters as well as ground water. It stipulates that generation of waste substances containing hazardous pollutants must be followed by state-of-the-art cleaning processes to cleanse the water before it returns to public waterways or flows into a public drainage system.

Materials hazardous to water

This class contains all solid, liquid and gaseous materials with potentially negative effects on the physical, chemical or biological properties of water resources.

Water hazard classes (Table 2)

These classes indicate the degree to which various materials pose a hazard to water.

Table 2: Water hazard classes
Substances posing a severe hazard to water e.g. waste oil, lubricating oils, solvents, fuels for spark-ignition engines
Substances posing a hazard to water e.g. brake fluid, diesel fuels, heating oil
Substances posing a limited hazard to water e.g. as electrolyte, coolant, petroleum

Regulations for automotive service operations. Compliance with special storage regulations is mandatory. For instance, used batteries containing electrolyte must be stored in plastic tubs, used paint and lacquer in metal containers or barrels sealed with clamp rings, and used oil of known origin is kept in double-walled containers with a provision for sealing.

Waste water and liquid effluent are divided into two classes: that for which treatment is prescribed, and that for which no subsequent processing is required **(Fig. 1)**.

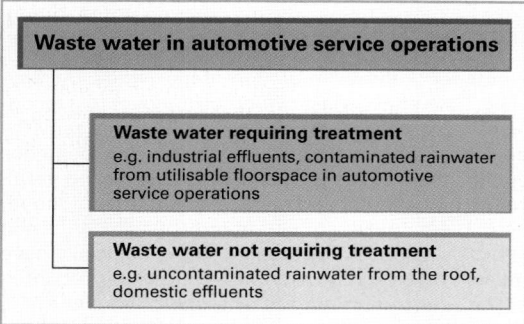

Fig. 1: Waste water and effluent classifications

Waste water requiring treatment may be contaminated with any of a variety of substances, including mineral oil, fuels and cleansers, as well as solids such as metal shavings, paint particles, contaminants in particulate form, etc. The liquid must be cleaned before being discharged into a public drainage system or into the surface water. A silt trap or petrol separator may be employed in this process **(Fig. 2)**.

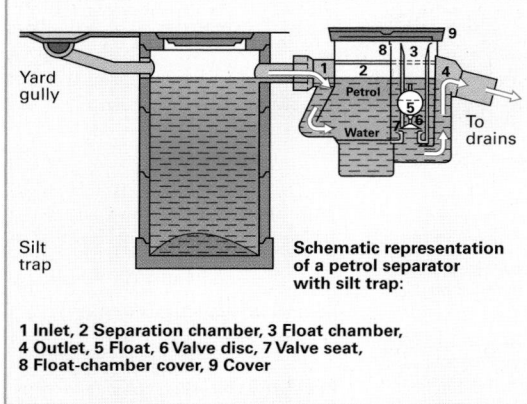

Fig. 2: Silt trap and petrol separator

The solids contained in the effluent subside and form deposits within the silt trap, while the petrol separator relies on the lower specific density of lighter fluids such as oils and petrol to skim them from the water.

Waste water not requiring treatment
This water may be allowed to flow into the public drainage system or into surface water without any preparatory processing **(Fig. 1)**.

Chemicals Act
This identifies the properties of hazardous substances and preparations that pose a potential hazard to human health or to the environment **(Table 1)**.

Table 1: Properties of hazardous substances and preparations as defined in the Chemicals Act	
● Explosive	● Irritant
● Oxidising	● Prolonged exposure
● Highly flammable	may increase sensitivity
● Easily flammable	● Carcinogenic
● Flammable	● Source of embryonic
● Extremely toxic	damage
● Toxic	● Source of genetic damage
● Mildly toxic	● Source of chronic illness
● Caustic	● Environmental hazard

Ordinance on Hazardous Materials and Substances
These regulations govern procedures for handling hazardous materials.

Among these hazardous substances are brush cleaners, solvents, fuels and acids.

When handling hazardous substances always observe the following:

● Hazardous materials must be identified as such.
● Hazardous materials must be used in accordance with regulations and guidelines.
● Hazardous materials must be stored in accordance with regulations.
● It is essential that containers employed for the storage of hazardous substances and materials be so identified that they will not be mistaken for containers for storing foodstuffs.
● **Never use beverage containers to store solvents.**
● Hazardous materials must be kept in locked storage, with controlled access limited to authorised persons only.

2.1.3 End-of-life vehicle disposal

Ordinance on End-of-life Vehicles
Owners and keepers of motor vehicles being permanently removed from use must furnish the registration agency with corresponding information regarding the vehicle. The regulations governing this action are set forth in the Ordinance on End-of-life Vehicles.

The following stipulations of the **Ordinance on End-of-life Vehicles** are of special significance when vehicles reach the end of their effective service lives:

- The last owner must provide the registration agency with a **recycling verification** (red certificate) when an end-of-life vehicle is permanently removed from use and disposed of or recycled **(Fig. 1)**.
- The last owner is obligated to turn the end-of-life vehicle over to a recognised acceptance agency or recycling company. A recycling verification will then be presented to the last owner.
- A recognised acceptance agency can also take possession of the end-of-life vehicle and issue the recycling verification on behalf of a recycling company.
- Recognition of automotive service operations as acceptance agencies for end-of-life vehicles is granted by the responsible automotive trade organisation.
- Should the last owner of an end-of-life vehicle remove it from service without turning it over for disposal and recycling, this individual will be required to present a **declaration of whereabouts** (brown certificate) concerning the vehicle. This procedure comes into play when (for instance) a passenger car remains in the owner's possession as a collector car or is taken off the road for extended repairs **(Fig. 1)**.

Voluntary undertaking

In Germany a voluntary, co-operative agreement exists between the government and several business organisations. Its intention is to ensure correct, environmentally aware disposal of end-of-life vehicles at the end of their service lives.

Among the points included in the voluntary undertaking are:

- Creation and expansion of a comprehensive infrastructure for recovering and recycling end-of-life vehicles as well as used components left from passenger car repairs.
- Environmentally aware processing of end-of-life vehicles, with fluid drainage and disposal followed by dismantling operations.
- Recycling and renewed application of all suitable materials to reduce the amount of residual waste stemming from end-of-life vehicles.
- Vehicle manufacturer's return acceptance obligation. Vehicles are accepted free-of-charge provided that the date of initial registration was prior to 1 April, 1998, and the vehicles are not more than 12 years old.

Permanent remove of an end-of-life vehicle by recycling. The vehicle's last owner turns it over to a recognised acceptance agency or recycling company for permanent removal from use. The vehicle acceptance agency then issues the last owner a recycling verification on behalf of and under the authority of the recycling company with ultimate responsibility for the vehicle's disposal. The owner can then officially have the vehicle's permanent removal from use recorded at the registration agency. The end-of-life vehicle is forwarded to the disposal company **(Fig. 1)**.

2.1.4 Recycling

Many of the substances and materials that flow into the manufacture of a motor vehicle possess considerable economic value and are suitable for repeated use in the production cycle. This reduces the expense associated with the production of new vehicles while simultaneously lowering waste-disposal costs **(Table 1, Page 39)**.

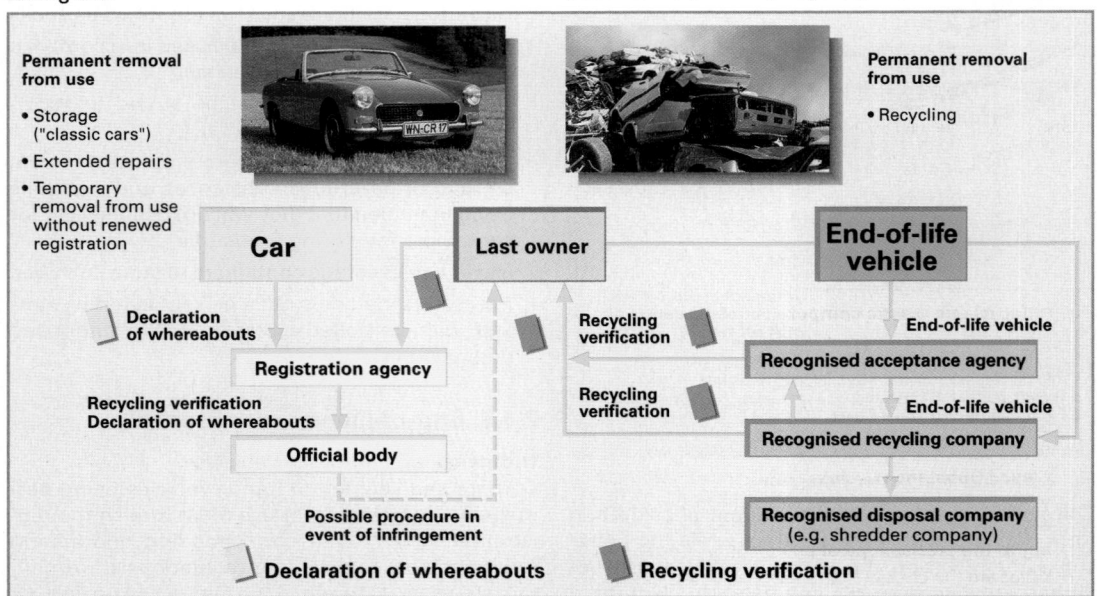

Fig. 1: Permanent removal from use of an end-of-life vehicle

Table 1: The end-of-life vehicle as source of raw materials in percent by weight	
Iron and steel	70 %
Rubber	9 %
Plastics and synthetic materials	8 %
Glass	3 %
Aluminium	3 %
Copper, zinc, lead	2 %
Other non-ferrous metals	1 %
Other	4 %

Used tyres furnish an example. Cold and hot curing processes can be employed to retread the tyres, or they can be dismantled and granulated, with the cord fabric being employed for acoustic-insulation barriers, while the rubber can be used in highway pavement. Coolant and brake fluid are cleaned and reprocessed. Battery casings can be processed into granulated plastic suitable for subsequent application in manufacturing extruded plastic components. The battery's electrolyte is purified and recycled, while the lead plates serve as a source of metal. In the realm of waste water disposal at automotive service operations, recycling systems can treat the water used in automatic car washes, after which most of this reprocessed water can be channelled back into the washing process for renewed use.

Catalytic converters are usually stripped of their casings for recycling. Refining or granulation of the ceramic bed can be employed to remove the noble metals (platinum, rhodium), which then emerge from processing in an extremely pure state. Ceramic slag is employed as an aggregating material in construction and steel and iron industry, and scrap metal is collected and remelted.

The manufacturers identify plastic components according to composition in order to facilitate subsequent separation and collection for recycling (**Fig. 1**).

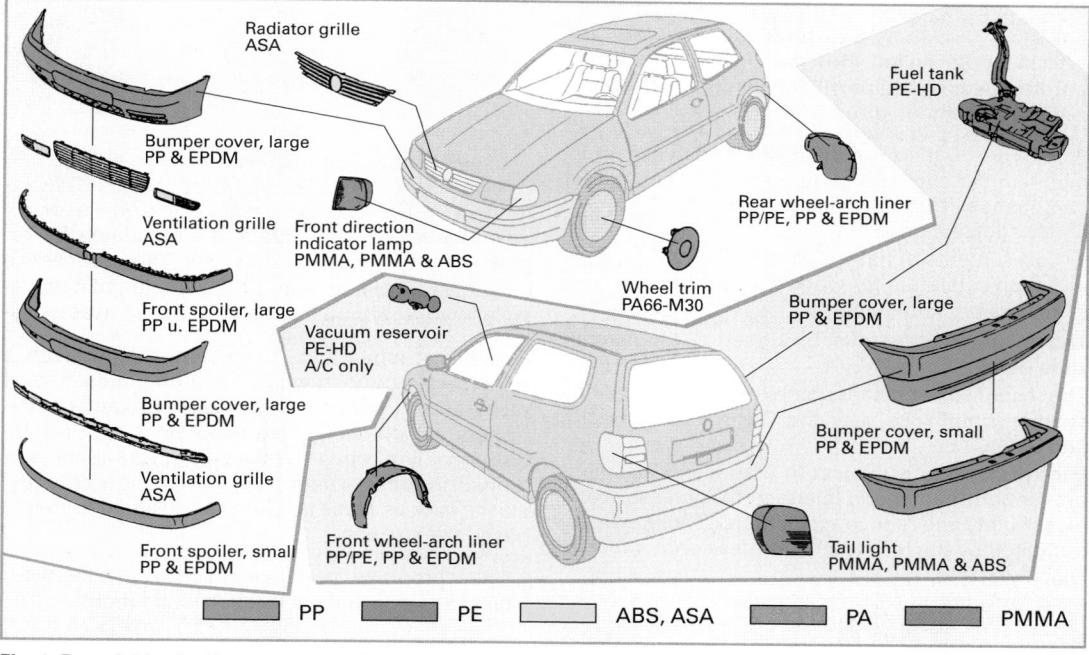

Fig. 1: Recyclable plastic components in motor vehicles

1 What specific acts and regulations form environmental law?

2 What three types of waste are defined in the Closed Substance Cycle Waste Management Act?

3 What method is employed to simplify separation of plastic motor vehicle parts for recycling according to the precise type of plastic they contain?

4 What are the characteristics of classified waste according to the Ordinance of Codification of Waste?

5 What are the water hazard classes into which pollutants are divided?

6 What are the contents of the Ordinance on End-of-life Vehicles?

7 What is the difference between a recycling verification and a declaration of whereabouts?

8 How is the catalytic converter processed for material recycling?

2.2 Occupational safety and accident prevention

Operation of machinery and technical systems joins processes involving use of technical materials and substances as a potential hazard source.

Accidents, job-related illnesses and disabilities can be caused by ...

- ... inadequate safety equipment and failure to observe safety precautions, as exemplified by failure to use prescribed equipment, such as safety glasses or goggles, during welding and grinding operations.
- ... the effects of potentially hazardous chemicals and physical factors, such as fuel vapours, acid fumes, dust particles (during grinding), and noise.
- ... excessive physical strain, when lifting objects, etc.

The regulations contained in the German Occupational Health and Safety Act and the Ordinance on Industrial Safety and Health obligate employers to identify potential sources of hazards to employees and to respond with appropriate action to enhance occupational safety.

> Accident-prevention measures in the workplace protect people, equipment, buildings and the natural environment against damage.

Employers with more than 10 employees must maintain the corresponding documentation, either personally or through a designated agent. This documentation includes the

- assessments of hazard potential
- action catalogue for worker protection
- results or examinations

The documentation must be signed by the responsible party.

The responsible business monitoring commissions ensure compliance with the Occupational Health and Safety Act.

Each profession is subject to mandatory accident-prevention regulations intended to increase workplace safety and reduce accident risk. These are promulgated by the responsible professional trade associations (**Fig. 1**).

Fig. 1: Signs representing the professional trade organisations

One of these is the metalworkers trade organisation, which is responsible for automotive repair facilities. The accident-prevention regulations must be clearly visible, placed or posted in a highly conspicuous location for convenient access.

> Every member of the company staff is obligated to strictly observe the accident-prevention regulations.

Failure to observe safety precautions can lead to serious or fatal injuries, cause illness, and result in serious damage to property. It can also pose an environmental hazard.

2.2.1. Safety signs

These are intended to enhance safety in the workplace. Standardised sign formats identify instructions, prohibitions, warnings and first-aid information.

Instruction signs (Fig. 2) are round, and coloured in blue and white. Symbols indicate that the required safety measures are strictly mandatory procedures. One example is the use of protective goggles and ear defenders required for employees performing cutting operations with grinders.

Fig. 2: Instruction signs

Prohibition signs (Fig. 3) are round signs with a white background in which the prohibited action is represented in black. A diagonal red bar and red periphery are the distinctive features of the prohibition sign. Fire, open flames and smoking are prohibited in areas where a high explosion potential exists, such as areas in which petrol and flammable gases are present.

> An appropriate prohibition sign must be displayed at a prominent location in the room.

Fig. 3: Prohibition signs

Warning signs (Fig. 1) are triangular notices featuring black symbols and a black periphery against a yellow background. Warning signs should always be placed in highly visible locations. They warn against specific hazards that may be encountered within an individual area. The appropriate warning signs are required in areas such as warehouses in which caustic sulphuric acid for use in starter batteries is being stored.

Industrial trucks	Substances harmful to health or irritants	Danger area
Flammable substances	Explosive substances	Dangerous voltage
Caustic substances	Toxic substances	Dangers posed by batteries

Fig. 1: Warning signs

First-aid and emergency exit signs (Fig. 2) are rectangular signs bearing white symbols on a green background. Arrows indicate the directions in which first-aid and lifesaving equipment such as stretchers are located. The also identify exits and directions for fast and secure escapes from hazardous areas.

> Escape paths must remain clear and unobstructed at all times, with no objects, locked doors or doors that open against the direction of flight blocking the exit route.

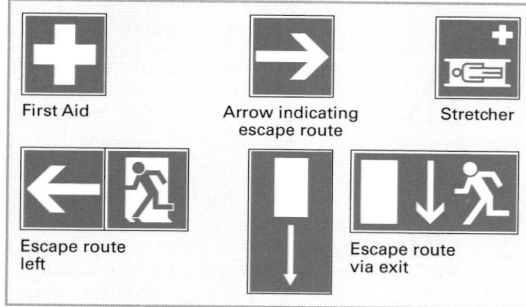

First Aid	Arrow indicating escape route	Stretcher
Escape route left		Escape route via exit

Fig. 2: First-aid and emergency exit signs

2.2.2 Accident causes

Even scrupulous care and comprehensive safety precautions will never be enough to prevent every possible accident. However, accident frequency can be reduced by the investigations and assessments of accident sources that serve as the basis for accident-prevention regulations. The operative distinction in assessing accident sources is between

- **Human error** stemming from failure to recognise a hazard, absent-minded conduct, negligence or failure to invest the required effort. Countermeasures to prevent these kinds of accidents include training and safety meetings, intended to foster safety-awareness, and technical safety equipment, such as guard screens, safety switches, etc.

- **Technical failure,** from material fatigue, unpredictable overloads, etc. Action against this type of accident is available in the form of technical safety measures, such as reinforcing a component whose breakage has caused an accident.

- **Acts of God** from unforeseeable outside sources, such as unusual meteorological phenomena.

2.2.3. Safety measures

Preventive safety measures can help avert many accidents while also reducing the severity of their consequences when they do occur. The following principles apply:

> Potential threats to safety must be avoided.

- Electrical equipment such as hand drills should never be used when the power cord is damaged.

- Safety glasses, goggles, and shields are prescribed to guard against hazards to the face and eyes, for instance, during welding, or from metal debris during grinding.

> Potentially dangerous areas must be shielded and clearly identified to prevent accidental entry.

- Shields, such as screens and covers, are prescribed for work pits, brake test stands, gearsets, spindles, shafts and moving components that mesh during operation.

- Containers containing hazardous substances, such as petrol, acids, flammable gas, etc., must bear the required identification as well as the appropriate warning signs and symbols **(Fig. 1)**.

> Potential hazards must be removed.

- Immediate steps must be taken to prevent machines and tools with safety-related defects from being used. Such equipment should immediately be scheduled for repair or removal from service.

- Sharp, pointed tools should never be carried in work clothes without being enclosed in a sheath or protective cover, etc.

- To prevent them from being caught by rotating machinery, rings, watches and jewellery should always be removed before work starts in areas where this danger is present.

- Travel and escape routes must always remain unobstructed.

2.2.4 Safe handling of hazardous materials

When not handled correctly, hazardous materials in the workplace can cause substantial damage to persons and property.

The Ordinance on Hazardous Materials and Substances obligates employers to draft special instructions for working in specific areas and with specified materials. Contents of the special instructions:

- Identification and labels for hazardous materials
- Hazards to human health and to the environment
- Protective action and rules of conduct
- Prescribed conduct in response to danger
- First aid
- Professional disposal practices

Once a year employees are to receive official safety instructions describing hazards and the corresponding safety measures in easily understood language, and to sign documentation confirming this instruction. The official safety instructions are to be maintained on display at a suitable location in the workplace.

Containers containing hazardous substances and materials are identified by hazard symbols, hazard notices (**R notices**) and safety recommendations (**S notices**) (**Table**).

Materials	Hazards (selection)	Symbol	Safety recommendations (selection)
Brake pads, clutch linings, grinding dust	Health hazard when dust particles are inhaled.	Irritant	Use vacuum extraction to remove dust, then bind it with a suitable substance and store it in securely sealed containers.
Solvents, cleaning agents for parts	Materials pose a potential health hazard when inhaled or swallowed.	Mildly toxic Easily flammable	Avoid direct contact with the skin. Use protective skin cream. Keep away from sparks and flame, do not smoke.
Fuels for spark-ignition engines	Explosive, highly flammable. Toxic when inhaled, ingested or in case of contact with the skin. Carcinogenic.	Highly flammable Toxic	Keep away from sparks and flame, do not smoke. Do not inhale fumes. Avoid contact with skin and eyes. Use protective skin cream. Never use as a cleanser.
Battery electrolyte	Caustic hazard to skin and eyes (danger of blindness). Irritates and damages mucous membranes in respiratory passages when inhaled.	Caustic	Avoid direct skin contact. Keep shipping containers closed. Store in original containers only. Wear gloves, as well as safety glasses or protective goggles and face protection where required. Ensure that the work area is well ventilated.
Engine oils of known and unknown origin, diesel fuel, gear oil	Avoid repeated skin contact extending over long periods. Engine oil of unknown origin is assigned to Hazardous Material Class AI.	Irritant Easily flammable	Use protective skin cream. In case of contact with skin or clothing, wash thoroughly using plenty of water. Keep well away from sparks and flame.
Box-section wax, undercoating, shipping wax, paint, paint residue, adhesives	Flammable. When inhaled, irritation and rashes affecting skin, mucous membranes, eyes and respiratory passages are possible; may also produce a numbing effect.	Easily flammable Mildly toxic	Keep well away from sparks and flame. Ensure that the work area is well ventilated. Stored in tightly sealed containers in a well-ventilated area. Use protective skin cream. Wear protective gloves, with safety glasses or protective goggles as indicated.

3 Business organisation, communication

3.1 Basics of business organisation

Each company relies on an internal organisational structure to process orders and to discharge its assigned responsibilities with prompt professionalism. This structure is based on a division of responsibilities along with precisely defined operational and commercial departments and sections.

> The business organisation assigns individuals and allots resources to ensure trouble-free processing of job orders and assignments. The objective is to achieve the best-possible results through optimal application of resources (staff, machinery, materials and time).

Priority is placed on the following objectives:

- Highest possible profit (economic objective)
- High levels of productivity (technical objective)
- Prestige and respect among customers and competitors alike (policy objective)
- Material and social responsibilities toward employees (social objective)

In addition to economic and social considerations, issues related to environmental protection must also be addressed.

The drive to achieve compliance with these objectives is usually supported by quality-assurance measures.

The business organisation must observe the following basic principles and priorities:

Goal-oriented processes. The objectives defined by management (such as the number of new vehicles sold within a specific period) provide the organisation with its primary points of orientation.

Clarity and logical structures. All organisational rules and directives must be clear and logically structured in both language and in the other forms in which they are conveyed, for instance, in organisational charts and job descriptions.

Unambiguous job assignments. Subsidiary assignments must be clearly defined according to areas of competence and responsibility, for instance, when carrying out legally mandated vehicle tests, such as emissions inspections.

Responsibility assignments. A clear and unambiguous area of responsibility must be assigned to each employee, i.e. as when designating an individual responsible for occupational safety.

Assignment co-ordination. Separate work procedures must be co-ordinated, as when defined procedures are employed to process repair orders.

Continuity and flexibility. Latitude must remain available to allow the defined organisational procedures and processes to adapt in response to specific contingencies, i.e. procedures for providing emergency services.

Monitoring. To minimise the error rate, arrangements must be in place to monitor working procedures, for instance, with checks carried out by the shop foreman.

Organisational sectors determine...	
WHO	does the processing,
WHEN	the processing is executed,
WHERE	the processing occurs,
WHICH	resources are employed for processing,
WHAT	is processed,
HOW	processing is carried out.

3.1.1 Organisation of a car dealership

The dealership consists of the management staff and of various separate business sectors or departments **(Fig. 1)**.

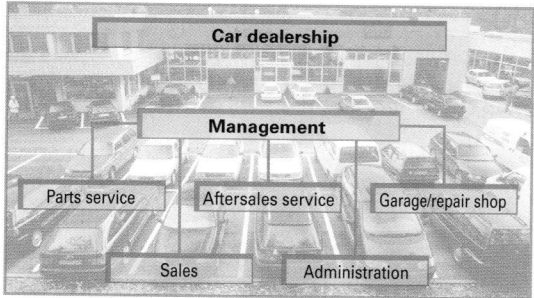

Fig. 1: Organisation of a car dealership

Management. Management monitors the activities of the various departments. It defines the dealership's objectives and determines business policy. Management assumes responsibility for supervising and organising the company as well as planning. Management also monitors all of the other departments and sectors of the company.

Various departments within the dealership are responsible for discharging specific duties and assignments.

Parts service (storeroom). The parts service manages the inventory of spare parts and accessories. This assignment embraces an array of individual activities including submission of parts orders and storage disposition as well as ongoing maintenance and monitoring of current part inventories. The parts service provides spare parts and accessories to the repair shop and also sells them directly to customers.

Aftersales service. The customer service section serves as the primary interface between the customer and the repair shop. It conducts technical consultations with the customer and accepts delivery of vehicles coming in for repairs. Following completion of the repairs it returns the vehicle to the customer. Customer service also assumes responsibility for processing warranty claims and for discretionary goodwill warranty work.

Garage/repair shop. The service section carries out repairs, maintenance and bodywork.

Sales. The sales department is responsible for selling new, employee-discount and used vehicles (with responsibilities also including leasing and financing arrangements), and also expedites pre-delivery handling of vehicles and their final delivery to the customer. This section is also responsible for assessing the value of and for purchasing used vehicles.

Administration. This section is responsible for commercial and business activities. Among these are accounting, processing transactions with suppliers and manufacturers, personnel disposition and payroll accounting and administration.

3.1.2 Aspects of the business organisation

Customer orientation is present when staff focus their thoughts and actions on the customer and the customer's requirements.

Various methods can be employed to obtain information concerning the customer's wishes **(Fig. 1).**

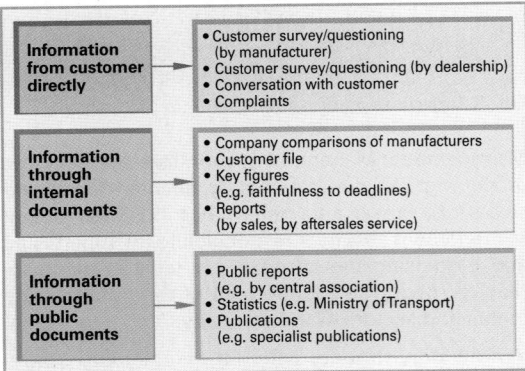

Fig. 1: Obtaining information on customer wishes

The interests and the requirements that the customer brings to the relationship with the organisation vary from one customer to the next.

Typical customer priorities may include:

- High-quality, professionally executed repairs and maintenance
- Immediate processing of complaints
- Compliance with stipulated deadlines
- Informed, professional advice from the staff
- Fast and smooth order processing
- Positive price-performance ratio
- Friendly service
- Equipment featuring state-of-the-art technology
- Priority service for urgent repairs
- Compliance with environmental-protection guidelines
- Dealership with a modern, professional image
- Pleasant atmosphere
- Clean showrooms and service facilities
- Vehicles that are clean when delivered to customers

Customer satisfaction is generated when the quality of the work performed is equal to the customer's expectations.

Customer satisfaction is the basis for loyalty toward the dealership. This effect is enhanced when the dealership not only meets the customer's expectations, but exceeds them. In addition, the satisfied customer will recommend the dealership to friends and acquaintances.

Dissatisfied customers can be easily recruited as new clients by the competition. And not all dissatisfied customers will bother to lodge complaints or claims for remedial action. Many customers seek to avoid the time, inconvenience and expense associated with submitting formal complaints. This is why customer satisfaction levels should be monitored on a regular, systematic basis, using such tools as customer surveys. The results of these surveys will furnish the starting points for efforts to improve the business organisation.

Customer satisfaction levels are affected by the following factors:

Technical product quality

- Build-quality and reliability of the vehicle
- Execution of maintenance and repair operations

Service quality

- Quality of the consultations and advice offered by vehicle salespersons, aftersales service advisors and other staff
- Procedures for dealing with complaints, for instance, with rapid response, or by granting approval for discretionary goodwill repairs
- Compliance with deadlines for repairs and deliveries

Quality of reputation

- Image of the vehicle manufacturer
- The dealership's good reputation
- The dealership's competence

Quality of personal relationship

- Relationships marked by trust between staff and customers
- Character of contact between staff and customers
- Congenial relations between staff and customers

Price perception

- Positive price-performance ratio, as reflected in repairs, new and used vehicles, spare parts and accessories, etc.
- Transparent invoices featuring logical structures and explanations, etc.
- Rebate offers
- Special no-cost extras

Customer loyalty. The objective behind all of these activities is to promote, maintain and enhance customer satisfaction, which will then serve as the basis for a long-term relationship between the customer and the dealership. This objective is served by first identifying and then focusing on those factors with the greatest potential for enhancing customer satisfaction. If customer satisfaction levels are low or non-existent, the bond between the customer and the dealership is in serious jeopardy **(Fig. 1)**.

Fig. 1: Effects of customer satisfaction on the company

Special service concepts represent an especially effective method for enhancing customer loyalty. Examples include:

- Complete range of services offered, e.g. sales, maintenance, repairs, leasing, insurance, etc.
- Programmes aimed at special target groups, such as private, business and high-volume customers
- Advertising, including special campaigns
- Warranties and discretionary goodwill warranties
- Mobility guarantee

- Vehicle retrieval and delivery service
- Customer club

Correct appraisal of the customer's importance to the dealership is vital for ensuring the dealership's commercial success. This appraisal assists in determining whether specifically defined offers for individual customers, such as special comprehensive service packages at discount prices, make sense from the dealership's standpoint.

Customer types

We distinguish between the following customer types:

- **Chance customer**
  This customer takes advantage of a passing opportunity, happens to be in the vicinity of the dealership by chance or is just passing through. The chance customer expects work to be completed quickly and reliably, but harbours no additional expectations. No bonds of loyalty connect this customer to the dealership. As a general rule this type of customer represents only limited potential as a source of turnover.

- **Standing customer**
  A bond of loyalty exists between this customer and the dealership. This customer takes advantage of the dealership's service offers when favourable circumstances present themselves, and responds to special offers and other incentives. The probable turnover from this customer is not especially high.

- **Regular customer**
  This customer has all service and repairs carried out in the repair shop, and places particular emphasis on personal contact with the staff. This customer expects personal service that reflects his or her personal value, and displays a high level of loyalty to the dealership. This customer can be expected to serve as a source of adequate profits and future growth. Yet another factor in this customer's favour is the high probability of recommendations for the dealership.

- **High-volume customer**
  This customer has an entire vehicle fleet serviced at the dealership. In most cases this customer will be a **company or government agency.** These customers generally receive a rebate. Here the decisive factor is fast, high-quality work, with the ultimate priority being firmly focused on minimising vehicle downtime. The dealership's relationship with this customer is characterised by a certain degree of dependence.

3.2 Communication

A vital factor for improving customer orientation and the business success that it produces is effective communication. While including communication between the dealership's management and its individual departments and sections, it also embraces interpersonal communication between employees as well as contact between staff and customers.

> Communication consists of an exchange of messages between a sender and a receiver.

3.2.1 Basics of communication

Interpersonal communication may be **verbal** (language-based) or **non-verbal** (without language).

Non-verbal means of communication are:
- **Gestures**
  (physical expressions, such as shaking one's head, nodding)
- **Facial expressions**
  (expressions such as a furrowed forehead, smiles, etc.)
- **Body language**
  (posture, such as upright, bowed, etc.)

Non-verbal communication can either reaffirm or refute the messages conveyed at the verbal level. Body language that reinforces the content of verbal messages fosters an impression of credibility. Body language that seems to contradict the content of verbal communication creates doubt regarding the sincerity of the spoken message.

> The first impression is of decisive importance in determining the subsequent course of communication.

The first impression, which takes shape within a matter of seconds, is defined by factors such as facial expression, gestures, body language, etc. Non-verbal conduct plays a highly significant role in shaping initial impressions at the outset of each interview. This is why it is important to recognise vital signals and to respond appropriately **(Table 1)**.

Table 1: Non-verbal conduct in dealing with customers	
Situation	**Possible reaction**
A customer enters the dealership while obviously searching for something.	• Eye contact • Nodding • Greeting the customer with a handshake • Turning toward the discussion partner
An enraged customer enters the dealership with a red face.	• Approach the customer in a friendly manner • Smile

Communication between senders and receivers proceed on different levels. As a result, a single statement can actually convey different messages. A listener can interpret a statement in different ways. This phenomenon is explained using the "4-ear model" **(Fig. 1)**.

> The truth is not what the sender says, but rather what the receiver (listener) hears.

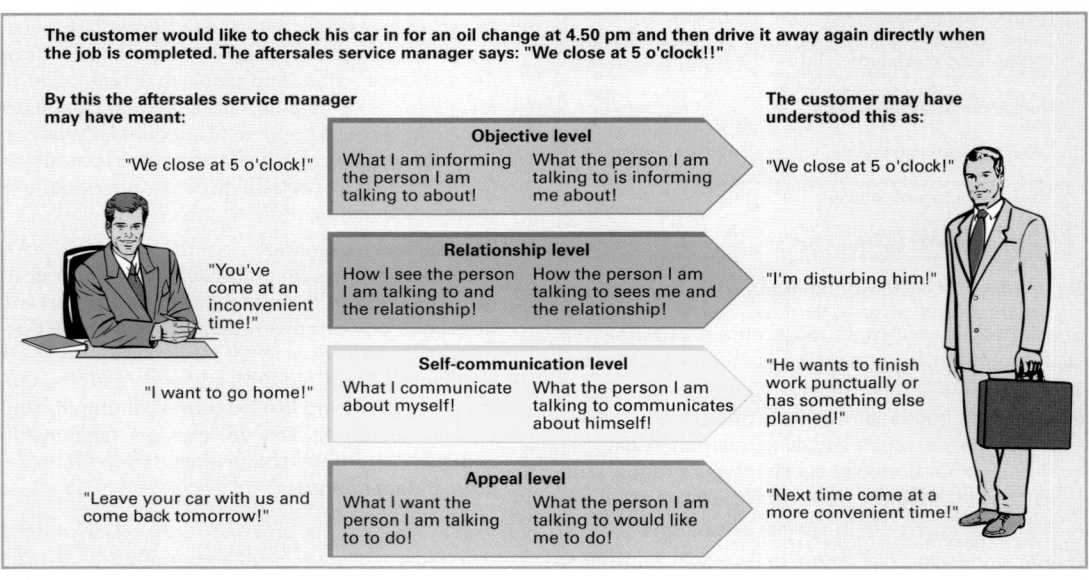

Fig. 1: Example of communication

The receiver (listener) who is capable of listening with "four ears" enjoys an advantage as communication proceeds

- by being more attentive ("active listening").
- by attaining a clearer understanding of his interlocutor's meaning, making it possible to respond with more empathy.
- by recognising potential conflicts promptly and responding accordingly.
- by communicating more effectively.
- by responding more effectively on the vital "relationship" level.

Transactional analysis. This concept is based on the assumption that the human personality is composed of three ego states (**Fig. 1**). These ego states describe the behavioural patterns in which people behave like children ("child ego"), like their own parents ("parents ego") or as rational adults ("adult ego"). The human personality is characterised by the presence of all three ego states.

This concept provides the basis for explaining communicative processes.

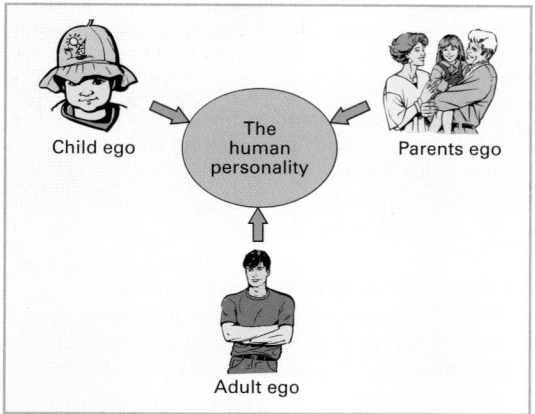

Fig. 1: Ego states of the human personality

Parents ego state

Individuals whose thought patterns, actions and feelings reflect those observed in their parents are in the parents ego state. The operative distinction is between the "controlling parents ego" (commands, prohibitions, criticism, prejudices) and the "nurturing parents ego" (motivation, helpfulness, care).

Examples:

The nurturing parents ego:
Noting that an apprentice is hesitating in the face of a new challenge, the shop foreman says, "Go ahead and give it a shot, you can do it!"

The controlling parents ego:
The foreman confronts an employee on the shop floor, "That oil change had better be done by 2 p.m. this afternoon!"

Adult ego state

Individuals who confront current reality, taking facts on board and dealing with them objectively, are in the adult ego state. It is in this state that all conscious personal experience is stored.

Example:
The head of the accounting department informs the shop foreman that "the most recent audit produced the following numbers."

Child ego state

Individuals who feel, think and act as they did in childhood are in the child ego state. The operative distinction is between the "adapted child ego", the "rebellious child ego" and the "free child ego".

Examples:

The adapted child ego:
The mechanic responds to a new assignment from a superior with, "Of course, I'll get it taken care of straight away!"

The rebellious child ego:
One apprentice remarks to another, "The foreman should do it himself!"

The free child ego:
A cheerfully whistling employee saunters through the shop and says, "I'm going out for a smoke."

3.2.2 Consultations

The first contact with the customer usually occurs with the initial consultation. The mutual impressions that both parties form during initial contact are of major importance. This is why the staff member should establish a good **personal rapport** with the customer as quickly as possible. This rapport should then be maintained throughout the subsequent course of the business process, starting with the advice and consultation phase, and then extending through actual execution of the work to the invoicing stage and, finally, the vehicle's return to the customer.

During consultations staff members must adjust themselves to the customer's personality.

3

The consultation can be divided into various phases:

1. Contact phase

- Active greeting:
  Give first and last name, offer a seat, smile, handshake
- Present the customer with a positive appearance (relationship level), positive first impression
- Conscious perception of verbal and non-verbal signals being sent by the discussion partner
- Small talk

Objective: Create a positive atmosphere! Don't argue! Don't inform!

2. Information phase

- Identify customer requirements/customer wishes
- Employ query techniques – ask targeted questions
- Listen actively and attentively

Objective: Use targeted questions to identify the wishes or problems that the customer has!

3. Negotiation phase

- Call attention to advantages and benefits for the customer
- Use positive formulations – present arguments leading to a logical conclusion
- Pay attention to the customer's objections and respond to them positively
- Result-oriented discussions
- Satisfy/arouse enthusiasm in the customer

Objective: Argumentation, effective response to any possible objections from the customer!

4. Termination phase

- Summarise positive aspects (create positive feelings)
- Clarify subsequent action
- Active farewell (offer as discussion partner)
- Provide business card
- Create a relationship for the future (small talk, good wishes)

Objective: Interview ends on a positive, future-oriented note!

Interviewing techniques

Interviewing techniques are of considerable significance during the information phase. Good interview techniques join attentive "active listening" as important elements in effective consultations.

In this process we distinguish between different kinds of questions:

The **closed question:**
- The response consists of "YES" or "NO".
- The question requests data or facts only.
- A series of consecutive closed questions tends to assume the character of an interrogation.

Example: "Are you satisfied with your winter tyres?"

The **open question:**
- The other party is encouraged to participate more actively in the conversation.
- The range of anticipated responses provides the interviewer with more information.
- The question starts with an interrogative adverb (which, how, what ...).

Example: "What brand of winter tyre do you prefer?"

The **leading question:**
- The anticipated reply is already contained within the question.
- A distinctive characteristic of this type of question is its use of words and expressions meant to foster agreement, such as "certainly" and "you doubtless agree that".

Example: "Of course you want to continue driving safely through the winter."

The **alternative question:**
- The interviewer formulates the question in such a way as to guide the respondent toward a specific decision.
- The interviewer places the desired response at the end of the question.

Example: "Do you prefer Brand X or Brand Y winter tyres?"

Active listening

Concentration and close attention to the customer's explanations are important in determining what the customer wants and in dealing with customer objections. Staff members should not interrupt customers while they are speaking. Because pauses furnish the customer with the opportunity to open up, speak freely and voice opinions, they represent a proven means of making discussions more interesting. Active listening encourages the customer to continue talking, and enhances the attention and the positive feelings directed toward the listener.

Characteristics of "active listening" are:

- Careful, deliberate silence
- Patient anticipation during silences (avoid speaking simply to fill the void of silence)
- The ability to withstand emotionally charged statements (avoiding the temptation to engage in intense verbal responses)
- Suppression of premature remarks
- Empathy, understanding the other person's point of view
- Accepting the discussion partner's situation
- Staying attentive and concentrated, following the other person's line of thought
- Recognising the "subliminal message"
- To make it clear to the other person that you are listening (by means of facial expression, nodding, etc.)

Targeted discussion

The negotiation phase marks an exchange process in which both parties make and defend assertions. The purpose of the discussion session is to bring the other person around to one's own point of view, and not to win a debate.

> Argumentation is an exchange of information in which the objective is to convince the other party that a particular opinion is correct.

Convincing arguments must meet two criteria:

- A basic willingness to modify initial conceptions held at the outset of the discussion must be present.
- The potential benefits, in the form of heightened personal satisfaction, must be clear.

A substantial number of the messages exchanged during discussion sessions proceed at the appeal level.

A three-step procedure works well in responding to objections (**Fig. 1**).

Reaction to customer objections. One way to respond to objections is with different query methods.

Response query. The objection is repeated in the form of a question as a means of obtaining additional information. This method can be used to buy time.

Examples:

- "Please tell me exactly what you mean."
- "Could you explain it to me in more detail?"

Yes, but. This method strives to replace the word "yes" with a substitute formulation whenever possible. Because the word "but" withdraws a previous statement or weakens its impact, it should be replaced by expressions such as "however".

Examples:

- "I understand your point, however..."
- "That's true, but when we consider every angle..."

Rhetorical questions. The objection is repeated in the form of a question. This simultaneously motivates your interlocutor.

Examples:

- "That's a good question, asking about the relationship between price and performance..."
- "That's a valid issue: the question of expense and return..."

Reversal. This process transforms an ostensible liability into an asset for your interlocutor. It is also called the boomerang method.

Examples:

- "It is precisely because..."
- "And that is just the reason..."

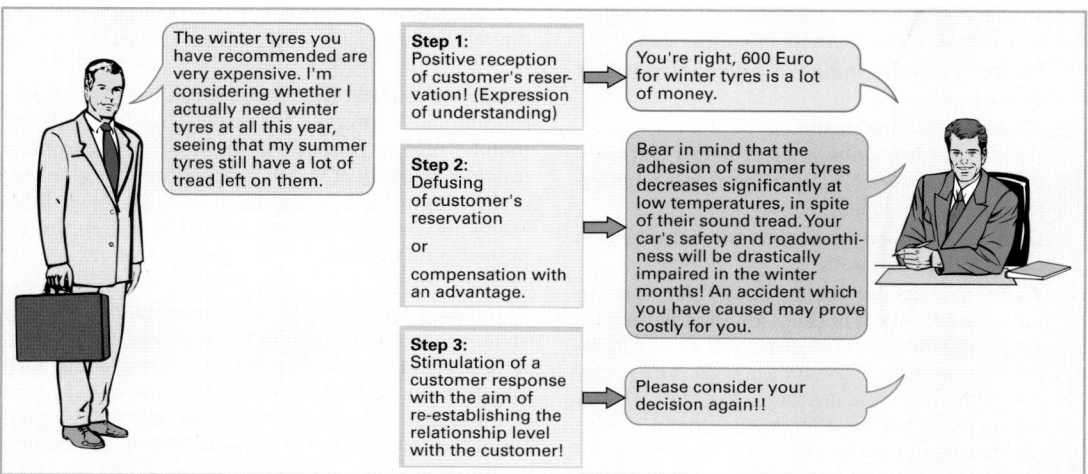

Fig. 1: Objection response strategies

3.2.3 Customer complaints and remedial action claims

Customer complaints and petitions for corrective action are not a pleasant business. At the same time, the ensuing discussion offers a golden opportunity for garnering information that could make an important contribution to improved service.

> A precondition for solving problems is understanding the other party's point of view.

Guidelines for conducting discussions stemming from complaints and remedial action claims:

- **Only two people should be present during the discussion.**
  Always try to avoid conducting this conversation in the presence of other customers. Instead, withdraw to (for example) a conference room. You can then ask a superior to come in and back you up as required.
- **Listen to the customer attentively.**
  ("Active listening").
- **Try to understand the customer.**
  Make absolutely certain that you have understood correctly. Repeat the gist of the customer's complaint in your own words: "Am I right in saying that you feel...?"
- **Frame targeted questions that deal precisely with the issue at hand.**
  Utilise the major query techniques (open questions, closed questions, alternative questions, leading questions).
- **Make sure that the customer can understand you.**
  Refrain from lecturing the customer, and avoid using technical terms and phrases. Explain the situation in the simplest terms possible, using language that the layman can understand.
- **Project a positive image of your company and of the product.**
- **Apologise for any mistakes.**
  But refrain from attempts to justify your actions, and never shift the blame for a problem to one of your colleagues.
- **Let the customer know that the complaint is going to be dealt with.**
- **Assume personal responsibility for dealing with the complaint or claim for remedial action.**
  Work together with the customer to find common ground for a solution that all parties can live with. Show the customer that you are acting on his or her behalf to ensure that the complaint is dealt with as effectively as possible.

Always honour your promises!

3.3 Personnel leadership

In itself, the organisational structure of a company says nothing at all about the type of leadership the employees are receiving. Leadership should always focus on achieving specific objectives, such as increasing production, reducing costs, etc.

> Management should provide leadership that inspires staff to adopt conduct that helps the company attain its defined objectives.

Leadership consists of exercising personal influence on other individuals to attain defined objectives. Leadership means convincing people of an idea's value and then giving them the motivation and the ability to transform their convictions into personal action.

We can distinguish between different leadership styles.

Authoritarian style. The characteristic attributes of this style are that the superior issues commands and directives, and the employee carries them out. In today's world the chances of attaining success with this style of management are very slim. Qualified staff will often be dissatisfied owing to limitations on their own latitude for making decisions and the restricted amount of responsibility that they are allowed to assume.

Co-operative style. Today's assignment-oriented leadership concept relies on flexibility, co-operation (joint action), an ability to delegate, and co-operation with and within groups. In the concept behind the co-operative leadership style the employees are partners, and action is directed toward achieving the company's goals. Responsibility is delegated, and employees engage in independent, critical thinking. Despite all this, authority is still present with the co-operative style.

Laissez-faire style. The term "laissez-faire" translates roughly as "let them do what they want", and ultimately means something approaching complete freedom of conduct for the employees. Decisions remain within the purview of individuals or groups. Management participation in decision-making processes is minimal.

Situational leadership style. Under this leadership style management adapts to each individual situation. Leadership relies on correct assessment of each situation and on appropriate responses. Based on this precept, the situational leadership style is authoritarian, co-operative or laissez-faire, depending on the individual contingency.

3.4 Staff conduct

A **positive attitude** to the work is an important precondition for professional success **(Fig. 1)**. Often enough, difficulties in relations with other people are not rooted in failure to communicate. Instead, negative attitudes, expectations and preconceptions condemn discussions and co-operative work to failure.

> Positive thinking leads to a positive attitude to one's work and to professional success.

Management and employees can make specific contributions to promoting a positive attitude in the workplace.

Positive attitude to the job. A personal interest in one's profession is the basic precondition for maintaining a positive attitude toward it. When superiors within the organisation make specific, defined efforts to foster employee development (for instance, with advanced training courses) this also has a positive effect on the employee's attitude toward the profession. In addition, it enhances job performance.

Positive attitude to the company. Here, employees who personally identify with the goals of their company are the most important factor. Management can also promote positive employee attitudes by adopting a co-operative leadership style, involving the employees in decision-making processes, etc.

Positive attitude to products. Every employee should be thoroughly familiar with the company's products, for instance, new vehicle models, as well as any new service offers at car dealerships. This enhances employee identification with the products, placing employees in a better position to represent the company's services and products to the customer effectively.

Positive attitude to customers. Employees should actively seek out contact with the customer. Attempts to avoid direct meetings with customers and to remain as inconspicuous as possible signal that confidence and competence are both lacking. Daily operations should be scheduled to include time for discussions with customers.

The manner in which staff meet and greet customers can also indicate a positive attitude, as exemplified, for instance, by

- upright posture
- appropriate facial expressions and body language
- well-groomed appearance, with clean clothes, etc.
- friendly appearance and manner, marked by politeness, etc.

Although dealing with pleasant customers poses no particular challenges, meetings with uncongenial individuals can be difficult and strenuous.

Unacceptable customer behaviour (for instance, insults) should be encountered with friendly yet unambiguous rejection. However, the same response cannot be employed when customers become belligerent or cause a scene. Conciliatory and even humorous reaction has a positive effect on the tenor of conversations with angry customers. Unpleasant situations can serve as a source of new experiences, furnishing insights regarding how one can improve one's own conduct.

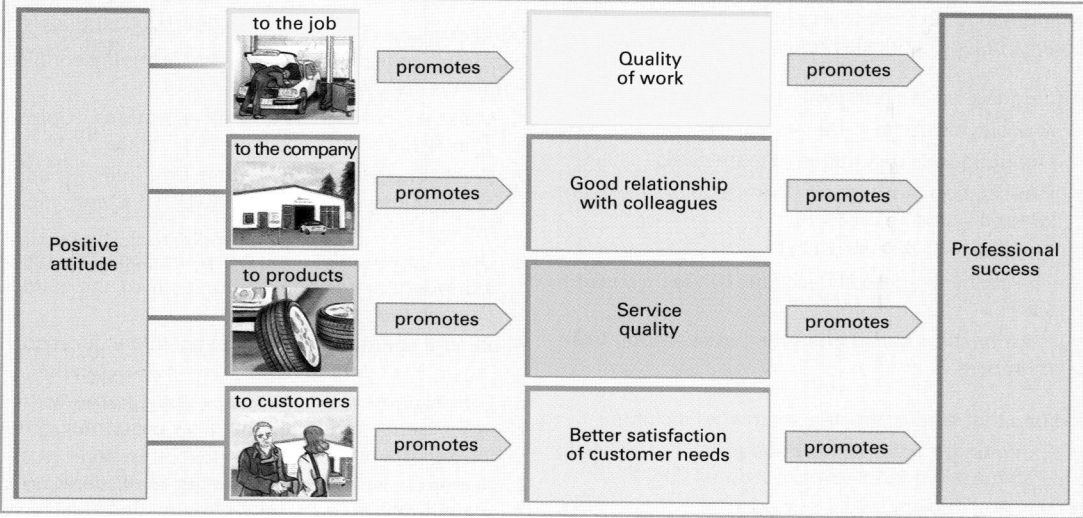

Fig. 1: Attitude to a staff member's assignments

3.5 Teamwork

The company's success does not depend exclusively upon how its staff members deal with customers. The way in which employees communicate and work together is also an important factor.

Within many firms problems are now being solved using teamwork, with several employees working in co-operation. It should be noted that the structures according to which teamwork is organised can vary from one company to the next. Employees at a small dealership or repair shop can view themselves as a team in the same way as a work group in a larger dealership that relies on team structures.

Ongoing technical advances are accompanied by increasing reliance on specialised staff. Owing to technical development, it can now be necessary for several employees to work together in order to carry out complex operations, such as diagnosis performed on electronic systems. Two individuals can have more than twice the potential problem-solving ability as one **(Fig. 1)**.

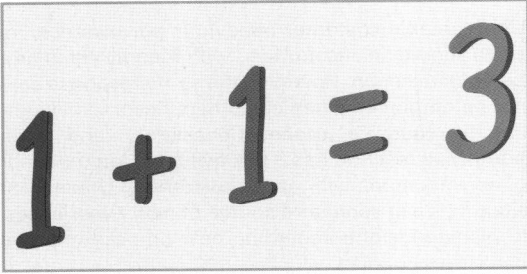

Fig. 1: Formula for group dynamics

In addition to professional competence, the ability to engage in teamwork also relies on capabilities.

Among the factors that foster teamwork are:

- A positive climate in the workplace
- Unambiguous definitions of team goals
- Participation among all members of the team when discussing problems
- The ability to be self-critical
- The ability to arrive at decisions based on consensus
- Relationships based on mutual trust among team members

Factors that hinder effective teamwork include:

- Exploiting a colleague's engagement in order to shift more work onto this individual **(Fig. 2)**
- Viewing colleagues as competitors
- Inhibitions about expressing one's own opinion

- Arrogance on the part of some team members
- Failure to communicate effectively among team members, as when conflicts remain undiscussed, etc.

Fig. 2: False conception of teamwork

Conflict resolution within the team. Over the long run it is inevitable that conflicts will arise within the team. In most cases emotions will be the dominant factor. This is why conflict resolution should focus exclusively on the objective aspects of any problem. Attempts to suppress or sublimate conflicts will not lead to success. Authoritarian conduct ("We'll see about that!") generally prevents effective, long-term conflict resolution.

Many different conflict sources can arise in the course of daily operations within the company. Frequently these are rooted in the personality structures of staff and/or customers, or in management's leadership methods. The company's organisational structure and the wider social environment can also play a role.

Conflict-resolution strategy. The following procedure has proven effective as a means of resolving conflicts within teams:

1. **Maintain control over your emotions.** Control your temper and keep a check on anger.
2. **Create an atmosphere of trust.** Promote formation of a genuine relationship between the team members.
3. **Open communication.** Work for collegial, problem-oriented relations among all of the participants in the process.
4. **Shared solutions.** Discuss suggestions for resolving an issue and then frame a final suggestion.
5. **Formal agreements.** When possible frame written agreements regarding how to encounter problems in the future.
6. **Personal resolution.** The conflict is officially laid to rest and the team's ability to act effectively is restored.

3.6 Order processing

Ideal arrangements for order processing depend upon mutual co-ordination of the following factors within the car dealership's organisational structure (**Fig. 1**):

- **Job content**
  - What needs to be processed?
- **Working time**
  - How long will it take to process the job order? When will it be finished?
- **Planning**
  - How should the job order be processed?
  - What spare parts and service equipment will be needed?
  - What individual procedures are included in the job order?
- **Work locations**
  - Which departments will carry out the individual procedures?

Guidelines for order processing in individual steps:

 Record customer and vehicle data

- The customer receives a friendly greeting. This includes learning the name of each new customer, and then addressing this customer by name.
- For regular customers: Access the data for the customer and the vehicle from the computer database and then go over the information together with the customer to verify that it is valid and up to date.
- For new customers: Take down data for the customer and the vehicle, for instance, based on the vehicle's registration papers, the service booklet, etc.
- Request or determine the vehicle's mileage.
- Ask or determine when the last service was carried out if this information is not available in the computer database.

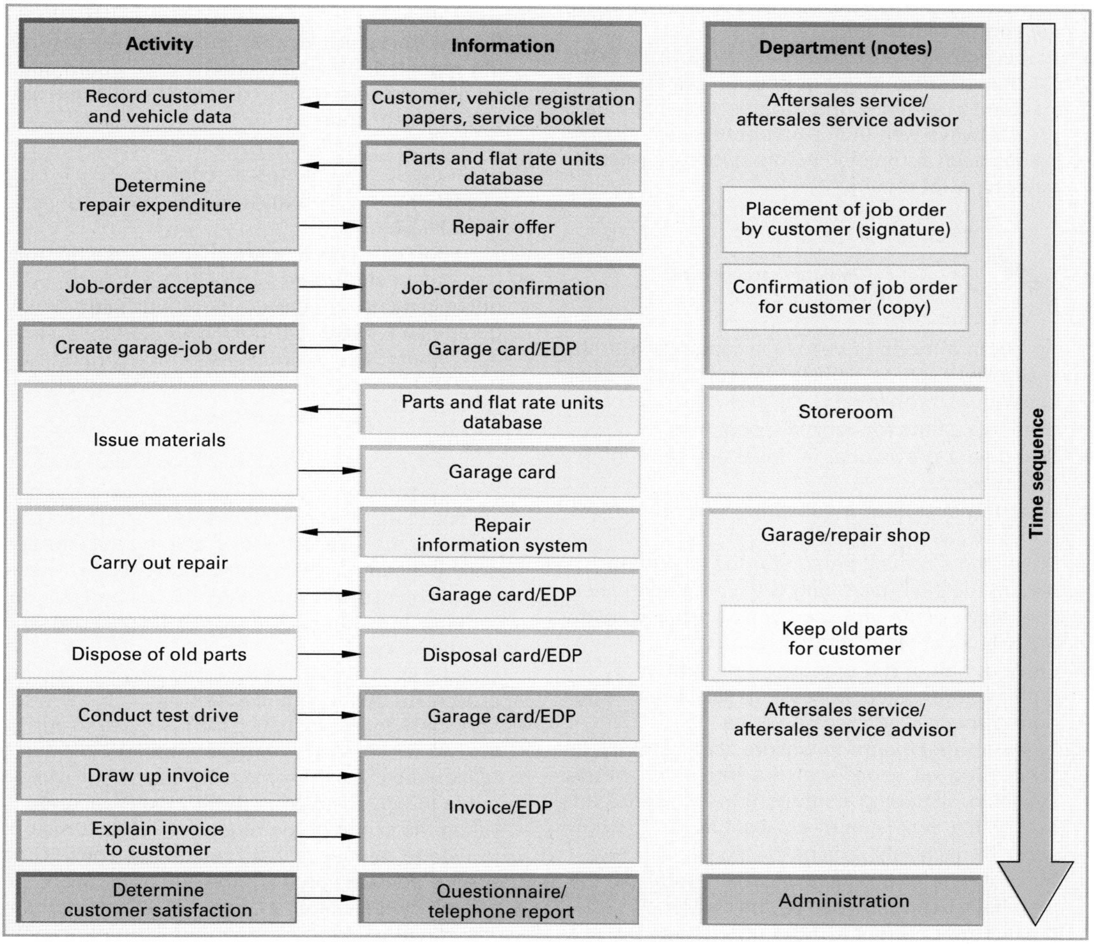

Activity	Information	Department (notes)	Time sequence
Record customer and vehicle data	Customer, vehicle registration papers, service booklet	Aftersales service/ aftersales service advisor	
Determine repair expenditure	Parts and flat rate units database		
	Repair offer	Placement of job order by customer (signature)	
Job-order acceptance	Job-order confirmation	Confirmation of job order for customer (copy)	
Create garage-job order	Garage card/EDP		
Issue materials	Parts and flat rate units database	Storeroom	
	Garage card		
Carry out repair	Repair information system	Garage/repair shop	
	Garage card/EDP		
Dispose of old parts	Disposal card/EDP	Keep old parts for customer	
Conduct test drive	Garage card/EDP	Aftersales service/ aftersales service advisor	
Draw up invoice	Invoice/EDP		
Explain invoice to customer			
Determine customer satisfaction	Questionnaire/ telephone report	Administration	

Fig. 1: Flow chart for a repair order

Determine repair expenditure

- Ask what the customer wants to have done (service inspection, repairs), and request information regarding problems with the vehicle or any sources of dissatisfaction.
- Use key questions to localise the problem. Examples might be: "When, and how often, does this problem crop up?", "When did this problem first appear?".
- Subject the vehicle to a thorough, systematic inspection, for instance, going over the vehicle together with the customer when it comes in.
- Document any damage to the vehicle, such as scratches.
- When necessary, conduct a test drive together with the customer, bringing along a technician or specialist as needed.
- When necessary, bring in other personnel to help localise the problem. These might be specialists, the shop foreman, etc.
- At this point the customer may be informed of the calculated price. Name a specific price only for standard operations and for jobs with a standard price. Always wait until a comprehensive diagnosis has been completed before quoting a price for other types of repairs.

Job-order acceptance

- Agree on a mode of vehicle acceptance with the customer, such as vehicle delivery during business hours, mobile retrieval, airport service, etc.
- Schedule a time for vehicle acceptance.
- Schedule a provisional retrieval time and inform the customer.
- Offer the customer a substitute means of transport, such as a rental car.
- Designate a contact person for the customer.
- Record the job order along with completion deadline, scope of the projected operations, etc., in the order book or the computer records.
- Carry out all of the operations prescribed by the manufacturer when processing recall vehicles and during special service campaigns.
- Make arrangements to ensure that capacity for processing all aspects of the job order will be available, including confirmation of spare parts availability, provision of a substitute vehicle, personnel, specialists, subcontracted services, etc.
- Make a note of products, such as motor oil, that may have been provided by the customer.
- Present any current special offers, such as seasonal campaigns, tyre changes, etc.

- Remind the customer to furnish the required documents, such as the service booklet, registration certificate, radio code, and keys for wheels and the vehicle itself.
- Agree on a mode of payment with the customer.
- Review all elements of the job order with the customer and run through all agreements one last time, schedule a deadline.
- Have the customer sign to confirm the job order!

Create garage-job order

- Have the shop foreman complete the existing or projected repair order, with information on the vehicle, applicable service instructions, etc.
- Provide information on the order to the responsible employees, such as mechanics and the representatives of the parts service, as well as any specialists whose services may be required.
- Plan any subcontracted activities, such as paintwork.
- Review the vehicle's repair record. If a repair is being repeated, obtain comprehensive information on the previous repair in order to identify the precise problem source.

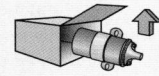
Issue materials

- Issue requested spare parts.
- Enter a record of the issued spare parts in the garage card or the computer database.

Carry out repair

- Proceed through all the items in the job order one step at a time, completing them fully and precisely.
- Observe the instructions in the manufacturer's documentation, which can include repair manuals, technical service bulletins, documents describing procedures for service inspections and maintenance.
- Use the special tools, testers and service supplies prescribed by the manufacturer.
- Check any defects identified but not repaired during the last service visit and repair these as indicated.
- Make a record of the repairs that have been carried out in the garage card or the computer records.
- When the scope of the required repairs turns out to extend beyond initial projections, consult the service advisor and have the additional work approved by the customer. Ensure that the mechanic's additional work is recorded with an entry in the garage card.

Dispose of old parts

- Dispose of old parts in accordance with official guidelines and regulations, then document this action on the disposal card. Have the parts service forward the disposal card or return the exchange part to the manufacturer.
- Retain any old parts needed for inspection.

Conduct test drive

- Conduct a test drive in accordance with service instructions.
- After carrying out service operations conduct a test drive in accordance with the manufacturer's instructions.
- Record the mileage at the end of the test drive.
- Should any defects be detected, initiate the required remedial action, while also informing the customer of any deviations from the agreed delivery deadline or the original cost estimate.

Draw up invoice

- Audit the job order: Check and compare labour and material costs, evaluate agreement with original cost estimate, check spare parts.
- When writing up the invoice ensure that it is clear and framed in language that the customer will understand.
- Make a note of any defects that have been detected but not repaired on the invoice, including a cost proposal as indicated.
- Organise the invoice according to labour charges and materials.
- Indicate the service advisor on the invoice, including the mechanic's name as indicated
- Record information on current special offers, special campaigns, product presentations and the date for the next service visit.
- Enclose documentation, such as the service inspection sheet, printouts of stored error codes from control unit readouts/electronics diagnosis, certificates for GI (general inspection), EI (emissions inspection) and/or SI (safety inspection).
- Put together a service pack, consisting of such items as the stamped service schedule, insert sheet, mobility guarantee, etc.; discretionary items include mirror tag with the mechanic's business card, the service advisor's business card, advertising materials, the dealership's own questionnaire.
- Prepare for the vehicle to be delivered to the customer: hold replaced parts in readiness for use in responding to any questions, check for service stickers and ensure that the vehicle is clean.

Explain invoice to customer

- When possible explain the completed operations directly at the vehicle.
- Explain the individual items on the invoice.
- Call the customer's attention to any problem areas that have been detected and will require repair in the near future, as related to brakes, tyres, etc., and suggest a specific date for scheduling the required repairs.
- Explain any supplementary operations extending beyond the scope of the original repair order.
- Show the customer the defective and replaced parts upon request.
- The service advisor calls on the telephone to present an offer the next day.
- Call the customer's attention to the mobility guarantee.
- Take delivery of the returned substitute transport vehicle.
- Accept the mode of payment agreed upon when the vehicle was initially brought in.
- Hand the vehicle papers, key and service pack over personally.

Determine customer satisfaction

- Make a report by phone (max. 1 week after service) or assess the data in the dealership's own questionnaire contained in the service pack.
- The service advisor should respond to any signs of customer dissatisfaction with a telephone call suggesting a solution; initiating steps for implementing the solution.
- Documentation of customer satisfaction.

REVIEW QUESTIONS

1 Explain the assignments and duties of the individual departments and sections in a car dealership.
2 Describe the factors that affect customer satisfaction.
3 What is the meaning of the expression "customer loyalty"?
4 Describe the different customer types and explain their significance for the company.
5 What are the phases that a consultation can be divided into?
6 What points need to be observed during customer discussions dealing with customer complaints?
7 Name the six phases of conflict resolution in teams.
8 Use key words to describe the sequence for processing repair orders.

3

3.7 Data processing in a car dealership

Electronic data processing provides support for the various departments and sections within the car dealership. The standard office programs with features for word processing, calculation tables, and Internet programs are usually supplemented by the manufacturer's own proprietary software applications. The EDP systems within the dealership's individual departments are generally networked **(Fig. 1)**.

The network links facilitate access to databases while also simplifying repeated activities.

EDP can furnish support for the following activities and functions within the individual departments:

Management
- Generation of statistics reflecting earnings for the day, month and year
- Generation of key statistics for the company
- Commercial monitoring functions

Parts service
- Managing parts and accessories
- Ordering and inventory
- Use-rate statistics
- Administration of data on automotive supply firms
- Administration of data on procurement and sales prices

Aftersales service
- Administration of customer and vehicle files
- Offer and invoice generation
- Calculations of repairs and bodywork
- Application of flat-rate-unit database
- Production of garage cards
- Invoice generation
- Scheduling
- Processing warranty and discretionary goodwill warranty claims
- Execution of advertising campaigns
- Administration of customer dates for GI, EI and SI

Garage/repair shop
- Administration of repair instructions, current flow diagrams, technical service bulletins

Sales
- Customer and vehicle files
- New-vehicle sales
- Used-car sales and appraisals
- Leasing
- Sales support

Administration/planning:
- Financial management and accounting
- Management of payment notices
- Personnel management
- Salary accounting

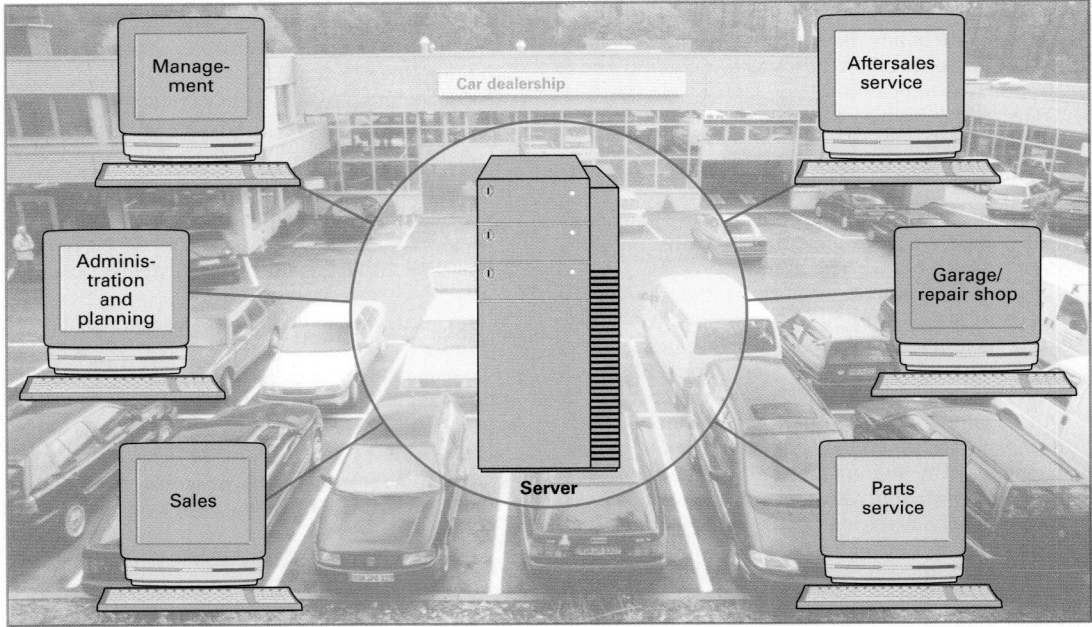

Fig. 1: The EDP network in a car dealership

Order processing with the aid of computer EDP

1. Records of customer and vehicle data:

The conversation with the customer combines with the vehicle registration certificate to furnish the required information regarding the customer and the vehicle. The data for regular customers will generally be stored for immediate file access **(Fig. 1)**.

```
ASS/SP orders        ASS/SP handling              Leatherhedge 17.01.06

                                    CSR TEXT NO.: 302
                                    YOU WERE ATTENDED TO BY MR WILLI HEIM..
COR. 1 ORDER NO.21171 ST  0    AA.N  CUS.NO.   3619
CHASSIS NO.    : WV2ZZZ25ZNG005304   REGISTRATION NO.: HSK-HO 900
NAME PART 1   : FALK, PAULA
NAME PART 2   :                      TEL. WORK  :
ROAD/STREET   : Stonybrook Road 9    TEL. HOME  :
POSTAL CODE   : 9999 9JH             TITLE-CODE :
TOWN/CITY     : Portville
PARTICULARS   :
FORM-OF-ADD. CODE: 1 MS    SP discount grades VW gar./rep. shop: 0
CUS. FEATURE  : 8020    LEASING VEHICLE: N       POR garage/repair shop: 0
PERM. REF. NO. :       REG. NO.     : HSK-HO 900 OPERATION      :     1
ACCEPTANCE DATE : 170103  TYPE/MODEL  : 255V92 11  SPF PERCENTAGE   : 1,000
KM READ./MILEAGE : 176122  DELIVERY    : 300792     SER.ORD./INV.NO.  : 85944
CUS. ADVISER  : H2      ENG. CODE LETT.: JX      LAST KM READ./MILEAGE: 176122
EI DUE        : 1003    GEARB. CODE LET: ASS     LAST VISIT    : 0300
GI DUE        : 1003    DEADLINE     : 1600     LAST SERVICE   : 0103
SERVICE Y/N   : N       KEY NO.      :          NO. OF VISITS  :    6
PAY EVAL. IN % :  0     SP EVAL. IN %  :  0
Message       : Vehicle may be subject to campaigns. Please check.

Connected to telnet                                            online
```

Fig. 1: Customer and vehicle data

The EDP system may also provide information on whether a vehicle is affected by a manufacturer's recall campaign or other special activity.

2. Job-order confirmation

The job-order confirmation contains vital customer and vehicle data as well as notices concerning repairs and maintenance **(Fig. 2)**. A copy of the job-order confirmation is presented to the customer. The customer should show this confirmation when the vehicle is returned.

Fig. 2: Job-order confirmation

3. Generation of the spare-parts list for the parts service

The program uses the job order as the basis for compiling a list of the required spare parts. The parts service then prepares the required spare parts using this list as a reference **(Fig. 3)**.

```
ASS/SP orders        ASS/SP handling              Leatherhedge 17.01.06

COR.  3  ORDER NO.21171  ST 13   AA.N   CUS.NO.          SHORT NAME
 3619   K FALK, PAULA           Stonybrook Road 9   9999 9JH Portville
CHASSIS NO.: WV2ZZZ25ZNG005304         REGISTRATION NO.: HSK-HO 900
                                   ENGINE CLET: JX   GEARBOX CLET.: ASS
SER.NO. PART NO.     PG FW TEXT/STOR.LOC.IQTY.UNIT PRICE SALE PRICE  K-M-E-EKZ
5010   026 198 025 C  2 VA SEALING KIT       1           15,200    0 1  1
5020   068 115 561 B  2 VA OIL FILTER        1           11,850    0 1  1
5030   060 129 620   2 VA FILTER ELEMENT 1               38,250    0 1  1
5040   1H0 127 401 C  2 VA FUEL FILTER       1           31,750    0 1  1
5050   ZUB 104       0 FO OIL 0 W-40    4,500            13,750    0 1  1
       ............

Connected to telnet                                            online
```

Fig. 3: Spare-parts list

The parts can be debited from inventory, and the system may also trigger a new order for parts from the manufacturer.

4. Using the electronic information system in the repair shop

The information system comprises such items as repair instructions, current flow diagrams and maintenance tables for service inspections, and also provides Internet access **(Fig. 4)**.

Fig. 4: Electronic information system

In most cases the electronic information system is linked to a system for vehicle diagnosis, testing and information. This system serves as a multilateral link connecting the vehicle's own self-diagnosis systems with the service test equipment and the technical documentation. There is also an option for guided problem diagnosis. The units are also equipped with multimeter and oscilloscope functions. They can discharge various functions related to vehicle service, such as reading out stored error codes from control units, programming software updates into control units, and resetting the service interval display.

The interfaces required for accessing service information on the local server as well as the manufacturer's Internet data are present. Current information available from the manufacturer via Internet supplements the standard information system with bulletins concerning recalls, etc.

The maintenance tables for service inspections contain the individual inspection operations that the mechanic performs **(Fig. 1)**.

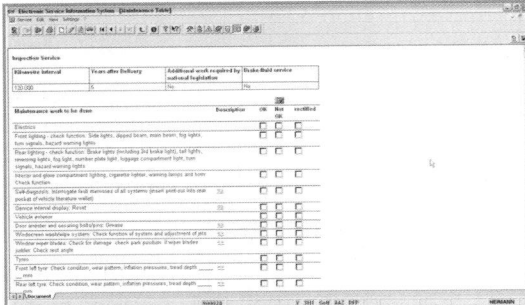

Fig. 1: Maintenance tables

The mechanic can access detailed instructions and information regarding individual inspection and test procedures, such as the battery check **(Fig. 2)** or the battery charge test **(Fig. 3)**.

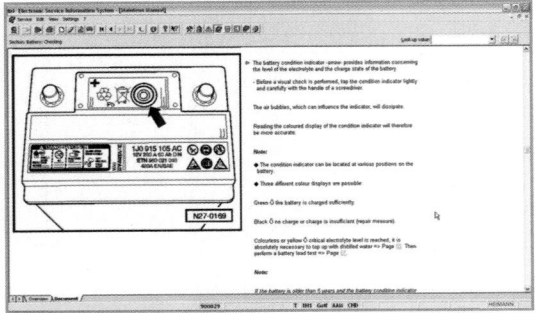

Fig. 2: Maintenance instruction

When a vehicle requires repairs, the mechanic can use the available repair instructions, which contain detailed descriptions of individual procedures.

In many cases the vehicle diagnosis system will include an option allowing tele-diagnosis (remote

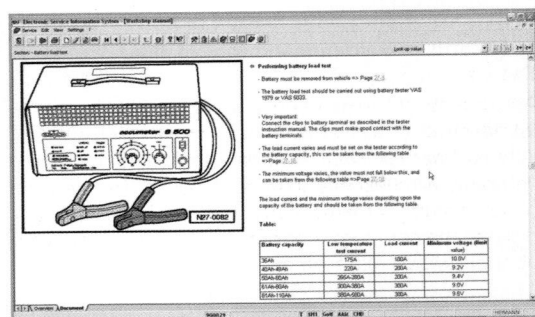

Fig. 3: Work-step instruction

diagnosis). When an especially difficult repair problem arises, this function lets the manufacturer's product support representative respond by monitoring the progress of the mechanic on the shop floor. The support representative can then furnish the mechanic with guidance, and even operate test equipment in some cases.

5. Invoice generation

Invoices are drawn up based on the data for the vehicle and the customer, data from the job order, the operations listed by the mechanic, and the spare parts, fluids and lubricants used **(Fig. 4)**.

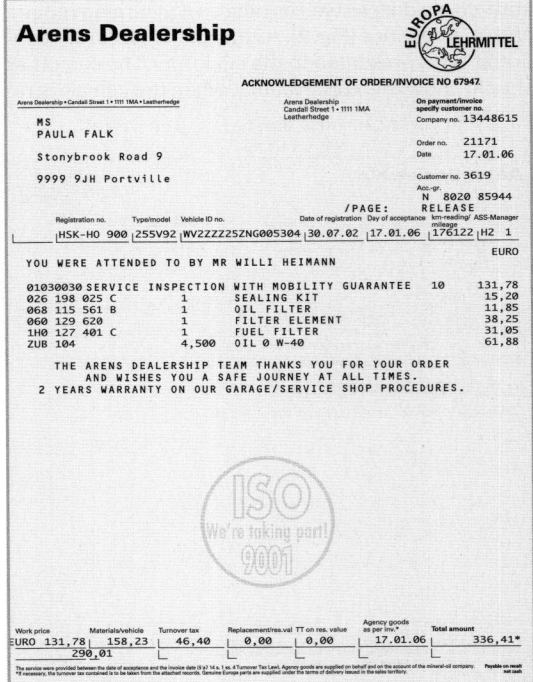

Fig. 4: Invoice

3.8 Quality management in automotive service operations

Basics

Appropriate management is essential for successful administration and leadership in the company. In this context "management" is understood as providing the company with leadership and guidance. The management system remains consistently focused on improving the company's performance. One of the components within this system is "quality management".

> **Quality management (QM).** QM comprises the entire range of co-ordinated management activities employed to supervise quality levels in the company.

Quality management defines:

- Quality policy
- Quality objectives
- Quality planning
- Quality assurance
- Quality enhancement

In the year 2000 DIN EN ISO 9000 ff. **(Fig. 1)** was issued to define and standardise the various terms employed in quality management.

DIN EN ISO 9000
General principles and terms
pertaining to quality management

DIN EN ISO 9001
Requirements of a quality-management system
with particular consideration of the effectiveness
of the QM system in satisfying customer
requirements and expectations

DIN EN ISO 9004
Guide with suggestions to the
QM system for companies wishing to go beyond
the requirements of DIN EN ISO 9001

Fig. 1: DIN-EN-ISO family

Meanings of the abbreviations in the standard's designations:

- **DIN:** German Standards Institute, Berlin
- **EN:** European Standards Organisation CEN (Comité Européen de Normalisation), Brussels
- **ISO:** International Organisation for Standardization, Geneva

DIN EN ISO standards are European standards that have been adopted in unrevised form for international application. The German-language versions of these standards assume the status of German standards.

Customer-oriented quality management. All of the company's activities remain consistently focused on the customer. The top priority of quality management is to improve quality in both products and services; the ultimate goal is to meet or exceed customer expectations.

> The benchmark in determining the degree to which objectives have been met is customer satisfaction in the areas of quality, deadline compliance and price.

Example:
When the customer receives the vehicle after a service inspection, in addition to being serviced it has also been cleaned.

Quality management system. This system embraces the organisational structure, responsibility assignments, procedures, processes and materials required to implement the company's quality policy. Quality policy is defined by company management. This policy expresses the company's intentions and objectives in the area of quality.

The basics and essential information concerning the quality management system are contained in the quality-management manual (QM manual). Procedural instructions, operational instructions and test instructions furnish more detailed information on individual activities **(Fig. 2)**. The manual also describes how quality management fits into the company's overall strategy, along with its implications for the organisation's structure.

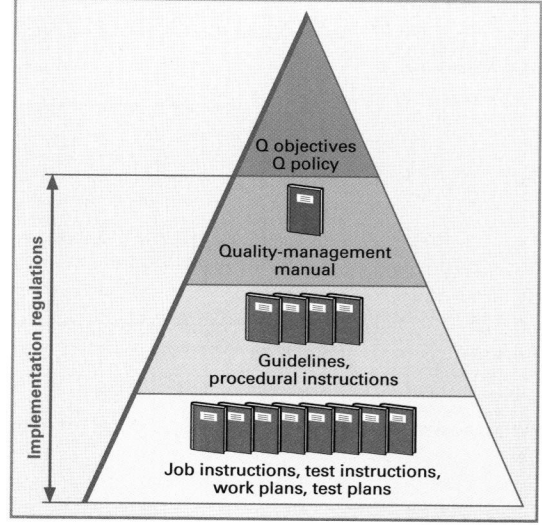

Fig. 2: Documentation pyramid of a quality-management system

> **Quality Assurance (QA).** As an element within the quality-management programme, it is directed toward generating trust in the ability to satisfy customer demand for quality.

Operational data and procedures, environmental-protection activities, etc., are documented in the quality-management manual as part of the quality-management process.

Objectives of the quality-management system. The company's management manual is the primary source of explanations regarding the ultimate objectives of the company's quality policy.

Among the goals of the quality-management system are:

- Optimal recognition and response to customer wishes
- Enhancing the quality of manufactured products and of services
- Improving the business organisation and the company's cost situation
- Improving employee qualification levels
- Improving the company's performance in the area of environmental protection

Process-oriented procedural concept. The company's **value-added processes** serve as the orientation point for its quality-management programme. **Added value** is the economic value created through activity (work). Typical processes found within an automotive operation are order processing for repairs and maintenance, and procurement activities for the parts inventory.

Process descriptions are generated as part of the quality-management programme. The purpose of these is to describe application ranges and procedures, such as those employed in order processing. All of the staff members who participate in the process, such as mechanics, employees in the parts service, etc., are obliged to orient their activities toward the individual points in the process descriptions. Job-order acceptance can serve as an example. The description explains how vehicle and customer data should be recorded as well as the points to be remembered when drawing up an order.

The quality-management programme also provides descriptions for "**support processes**". These processes might include:

- Service and maintenance on service equipment
- Selection, break-in procedures and qualification levels for personnel

The essential elements of quality management are interconnected in the process model **(Fig. 1)**. The example portrays the value-added processes for "repairing a customer's vehicle" and "procuring a spare part". Both processes are part of the quality-management control loop.

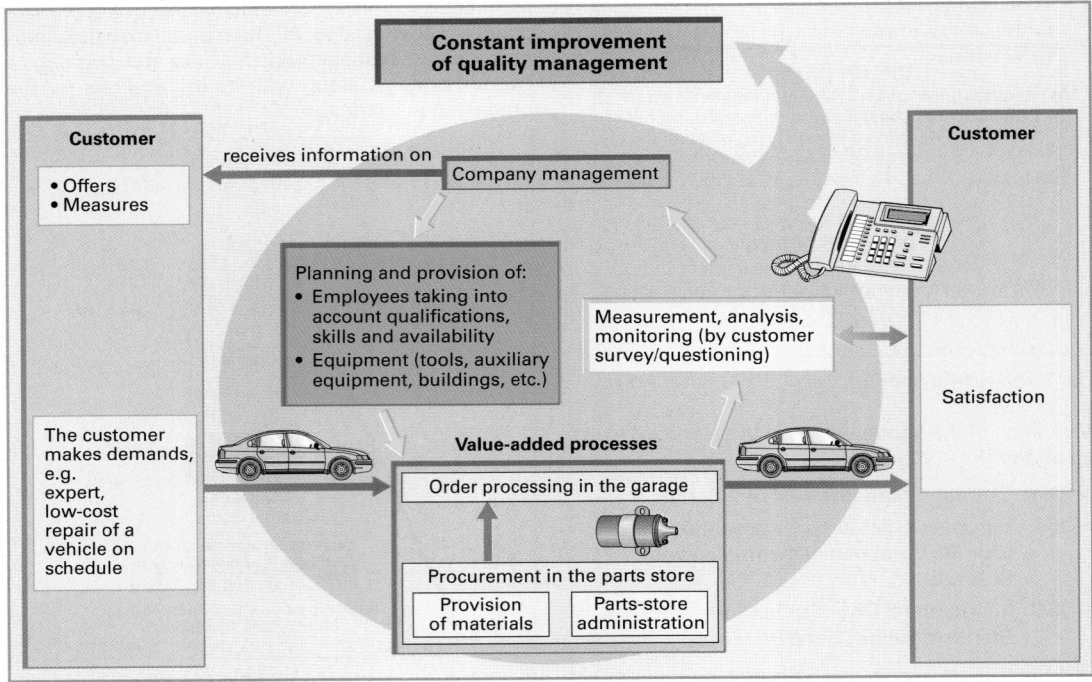

Fig. 1: Quality-management process model

Continuous Improvement Process (CIP). This embraces all action implemented within the company to increase benefit levels for the company and for the customer. An operational quality-management control loop can be expected to produce ongoing improvements in the company's performance.

> **Certification.** This process consists of an assessment of a firm's ability to deliver quality, as conducted by an independent professional.

As part of the auditing process, a checklist is used to rate the company on a point system. The company then receives certification once it earns a specific number of points. The certificate confirms achievement of the quality requirements. The certification process is usually repeated every three years. Annual internal monitoring audits assess quality in the interim periods between certification dates.

The assets that certification represents for a company include:

- Professionals conduct **checks on operational processes** affecting the company's quality potential
- **Enhanced quality awareness** on the part of the employees
- Potential evidence for use in **product liability** cases
- **A competitive advantage** relative to competitors lacking certification
- Optional **consultations and advice** in which the professional appraisal agent identifies liabilities within the company and proposes possible improvements

The auditing process. The automotive operation receives information on the test criteria in an initial preliminary notice. The next step usually consists of a trial run in the company seeking certification. This offers an opportunity to rectify any defects that may be discovered. This step is followed by an actual certification process in which the company is inspected over the course of one day.

The inspection, which consists of four sections, can include the following inspection criteria:

1. Records of operational data

- The size of the company
- Responsible individuals, such as management, the aftersales service supervisor, the parts manager, the sales manager
- Unique features of the particular company
- Number of annual new and used-car sales
- Number of daily service contacts

2. Records of a company's status

Overall impression of the company

- Does the external impression correspond to expectations, for instance, with regard to cleanliness in the repair shop, on the grounds and in the customer areas?
- Are clear and unambiguous signs present on company grounds to offer directions to customer parking, to order acceptance, and to part and accessory sales, etc.?
- Are the opening hours for the company and the repair shop clearly displayed?

Organisation of the company

- Is an organisational plan in place providing job descriptions and explaining the functions of management and of the personnel responsible for such activities as advertising, environmental protection, occupational safety and employee training?
- Is there an adequate number of service advisors?
- Are measures in place to monitor and ensure quality in the work being performed by subcontractors such as paintshops?
- Does the company offer an emergency service?
- Are regular, periodic customer surveys and telephone reports being conducted to determine customer satisfaction levels following repairs or aftersales service?
- Is negative feedback from customers being systematically recorded and evaluated, with results assessment and implementation of corrective action?
- Are employee training levels, and planning for ongoing and advanced training for the employees, being documented?
- What is the compliance level on safety guidelines, accident-prevention regulations and environmental protection directives? Are specific individuals responsible for these guidelines and regulations?

Customer contact area/job-order acceptance

- Are the prices for products and services (such as a fixed price for service inspections) on display?
- Are customers provided with an opportunity to scrutinise the conditions that apply to repairs, warranty claims and terms of payment?
- Does the company offer a retrieval and delivery service?
- Are substitute vehicles available in sufficient numbers? Are these vehicles clean and in flawless condition?
- Is a written customer order present for every repair and service order, and does the information on the invoice correspond to that on the order sheet?

- Does the vehicle-acceptance area or waiting area meet requirements in the areas of furnishings and equipment, cafeteria, materials with current product information, current magazines and newspapers, seating and a children's play corner?
- Is a final inspection/test drive carried out before vehicles are returned to the customer following maintenance and repairs? Are the results evaluated? Are the evaluations employed as the basis for defining, agreeing on and implementing corrective action?
- Are invoices framed in terms that customers can understand?
- Do personnel wear appropriate/recommended apparel? Do they wear name tags and carry business cards?
- Is a clear and logical system for scheduling deadlines and for daily resource allocation, based on forms, calendars, EDP, in place?
- Are systematic preparations for complying with schedules implemented once job orders are accepted?
- Are measures in place to ensure that the customer receives the vehicle personally, along with an explanation of the invoice, following repairs and maintenance?
- Does the company offer acceptance for repairs directly at the vehicle?
- Is advertising for current accessories present? Are accessories on display and available for sale in the customer-contact area?

Repair shop
- Are orders for service inspections and maintenance accompanied by the appropriate, valid forms, and do personnel comply with the prescribed procedures for working through the forms?
- Are customer vehicles treated carefully? Are covers for steering wheels and seats, etc., being used?
- How is the distribution of work among personnel and mechanics organised?
- Are the test records for equipment and tools being carefully maintained? Are the test deadlines for technical equipment and tools, such as platform hoists, compressors, torque wrenches, etc., being observed?
- How well is the service literature being maintained? Do the mechanics have access to documentation? Is an adequate number of work stations offering computer access available?
- Are the safety regulations for hazardous materials and fire prevention being complied with?
- Are measures for avoiding waste materials in place?
- How are environmentally hazardous materials stored and disposed of? Who is responsible?

Parts service
- Are factory replacement parts and the fluids, lubricants and service supplies prescribed by the manufacturer being used?
- Are incoming consignments and shipments being checked and are these checks being documented?
- Are used parts associated with warranty claims handled and stored separately?
- Are issued spare parts subjected to testing?
- Are all of the spare parts in the warehouse unambiguously identified?
- Are the storage arrangements for the parts selected to ensure that quality is not adversely affected by the storage process?

New and used-car sales
- Are the new and used cars in flawless condition?
- Are explanations that reflect the customer's requirements being provided when products and services are sold?
- Are regular, periodic advertising campaigns being carried out?

3. Assessment

Following evaluation of all these points the certification agent adds up the score. The results are then assessed in co-operation with the company's management. Firms achieving at least the minimum specified number of points while complying with the primary criteria receive the certificate. This documents that the company's ability to deliver quality has been assessed in accordance with DIN EN ISO 9001 and found to be in compliance with the standard.

4. Jointly approved action catalogue for enhancing quality

Additional action to improve quality is defined in co-operation with company management. When a company fails to achieve the minimum number of points and/or demonstrate compliance with the requirements in primary criteria, the outstanding issues must be resolved within a specified period of time. The certification agent then returns for a new inspection.

REVIEW QUESTIONS
1 **What is the meaning of quality management?**
2 **What are the objectives of a quality-management system?**
3 **What is certification?**
4 **How are audits carried out?**
5 **Name the primary criteria according to which automotive service operations are evaluated!**

4 Basics of information technology

Increasingly rapid processing is needed to accommodate continually expanding amounts of information and data. This trend places growing demands on the performance of the systems employed for electronic data processing in technology, science, business and administration. We distinguish between
- separate computers for individual work stations, e.g. personal computers
- networked systems, e.g. server-client networks

4.1 Hardware and software

Each EDP system consists of hardware and software.

Hardware is the generic term embracing all of the visible, concrete electronic and mechanical components physically present in the computer unit or combined in computer systems. This category includes all connection cables and plugs as well as data-storage media. Examples of hardware in a PC system (**Fig. 1**).
- System unit (computer) for processing data and
- Peripheral devices for
  - data entry, such as the keyboard and mouse
  - data displays and output devices, such as monitors and printers
  - external storage media, such as hard disks, floppy-disk drives, etc.

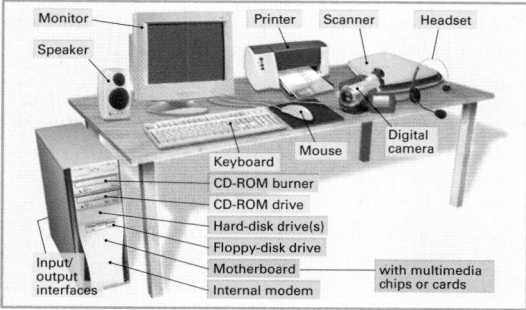

Fig. 1: PC system

Software is the generic designation for all programs and data.

Programs consist of command sequences. They translate the PC user's entries into a language that the machine can understand, control processing activities and generate data. We distinguish between system programs and application programs (**Table 1**).

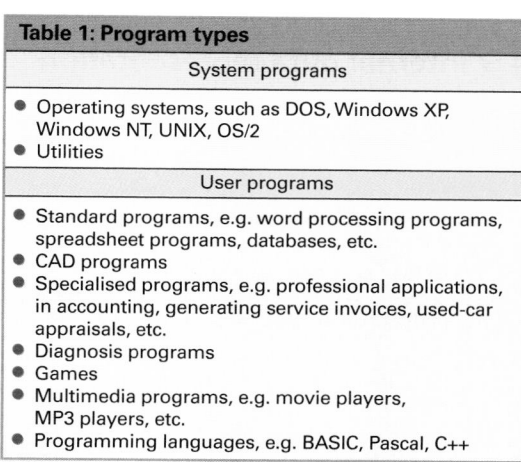

Table 1: Program types
System programs
• Operating systems, such as DOS, Windows XP, Windows NT, UNIX, OS/2 • Utilities
User programs
• Standard programs, e.g. word processing programs, spreadsheet programs, databases, etc. • CAD programs • Specialised programs, e.g. professional applications, in accounting, generating service invoices, used-car appraisals, etc. • Diagnosis programs • Games • Multimedia programs, e.g. movie players, MP3 players, etc. • Programming languages, e.g. BASIC, Pascal, C++

Data are units of information that the computer can register, process and issue as output. Here the distinction is between numerical, alphabetical and graphical data (**Fig. 2**). Combinations of various data types are called character strings.

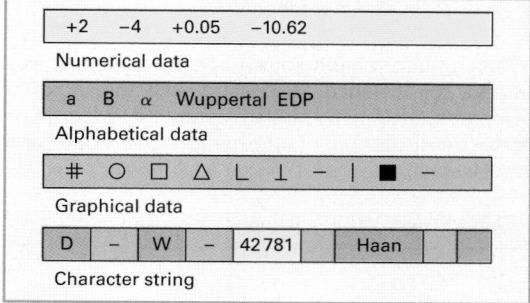

Fig. 2: Data types

4.2 IPO concept

As the data-processing sequence proceeds within the computer it always conforms to the concept embedded in the IPO structure (**Fig. 3**).

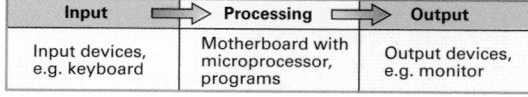

Input	Processing	Output
Input devices, e.g. keyboard	Motherboard with microprocessor, programs	Output devices, e.g. monitor

Fig.3: Computer operation

Input. Data are entered at the EDP work station, for instance, using a keyboard.

Processing. The computer's microprocessor processes the data with the aid of a program. The resulting data can be stored in the working memory.

Output. Program instructions can trigger display of these data in a recognisable form, for instance, on a monitor screen.

4.3 Internal data representation within the computer

Data portrayal within the computer is based on two electrical states. These electrical states are "on" and "off" **(Table 1)**. These conditions are employed to represent information using the binary system; the digit 0 represents "off" status and the number 1 corresponds to the "on" status.

Table 1: Electrical states in the binary system

Circuit		
Lamp	Off	On
Binary code	0	1

These two states represent the smallest units of information. The smallest individual unit is 1 bit (bit = binary digit).

> The information content of each bit is 1 or 0

When an information source consists of 2 bits, as it would, for instance, in dealing with the indicator lamps on a car, each bit can signify the information state 0 (indicator off) or information state 1 (indicator on) **(Table 2)**.

Table 2: Information in the binary system

Bit	Examples	Binary combinations	Number of information units
1	Light off　　Light on	0　1	$2^1 = 2$
2	Indicators off　Right indicator Left indicator　Hazard-warning flashers	00　01 10　11	$2^2 = 4$
3	Go　Stop　Stop　Get ready to pull away	000　001 010　011 100　101 110　111	$2^3 = 8$

Traffic signals employ three lamps to convey information to road users. This corresponds to an information content of three bits. This produces $2^3 = 8$ po-

tential switch states; although only 4 are actually used (combinations such as simultaneous red and green would be illogical in a traffic light, **Table 2**).

8 bits are required for processing the entire array of lower and upper case characters, numerals, special symbols and control characters.

> 8 bits are referred to as 1 byte

One byte can be used to portray $2^8 = 256$ states.

4.4 Numeric systems

The following numeric systems are employed in data processing:

- Decimal system
- Binary system
- Hexadecimal system

The decimal system consists of 10, the binary system of 2, and the hexadecimal system of 16 digits.

Information processing within the computer relies on the binary system. Owing to the length of the digital sequences generated by the binary system, information is often represented using the hexadecimal system.

Bus systems can use a combination of the binary and the hexadecimal systems.

As an example, **Table 1 on Page 65** shows how the number 123 is represented in various numbering systems.

The hexadecimal system's advantage over the decimal and binary system is its shorter length, e.g.

Decimal	Binary	Hexadecimal
13	1 1 0 1	D
	$1 \times 2^0 = 1$	
	$0 \times 2^1 = 0$	
	$1 \times 2^2 = 4$	
	$1 \times 2^3 = 8$	
	13	

Storage capacity. Indications of storage capacity available on internal and external memory are always furnished in bytes using the 8-bit code.

8 bits			=		1 byte
1 KB	=	2^{10} bytes	=		1,024 bytes
1 MB	=	2^{20} bytes	=		1,048,576 bytes
1 GB	=	2^{30} bytes	=		1,073,741,824 bytes

Table 1: Comparison of numbering systems

	Decimal	Binary	Hexadecimal
Basis	10	2	16
Numerals and characters	0, 1, 2, 3, 4, 5, 6, 7, 8, 9	0, 1	0, 1, 2, 3, 4, 5, 6, 7, 8, 9, A, B, C, D, E, F
Example	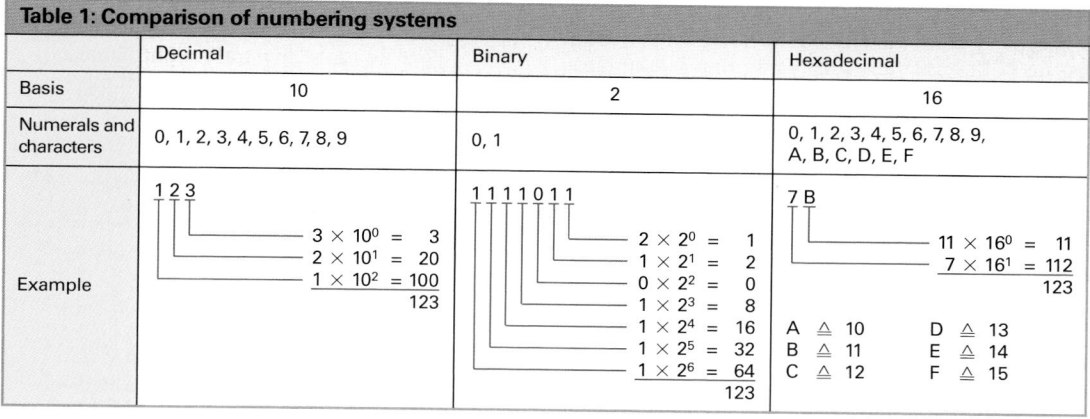		

For the Example row:

Decimal:
$$1\,2\,3$$
$$3 \times 10^0 = 3$$
$$2 \times 10^1 = 20$$
$$1 \times 10^2 = 100$$
$$123$$

Binary:
$$1\,1\,1\,1\,0\,1\,1$$
$$2 \times 2^0 = 1$$
$$1 \times 2^1 = 2$$
$$0 \times 2^2 = 0$$
$$1 \times 2^3 = 8$$
$$1 \times 2^4 = 16$$
$$1 \times 2^5 = 32$$
$$1 \times 2^6 = 64$$
$$123$$

Hexadecimal:
$$7\,B$$
$$11 \times 16^0 = 11$$
$$7 \times 16^1 = 112$$
$$123$$

A ≙ 10	D ≙ 13
B ≙ 11	E ≙ 14
C ≙ 12	F ≙ 15

4.5 Structure of the computer system

The computer system's hardware consists of the main unit with the motherboard and the connected devices (**Fig. 1**).

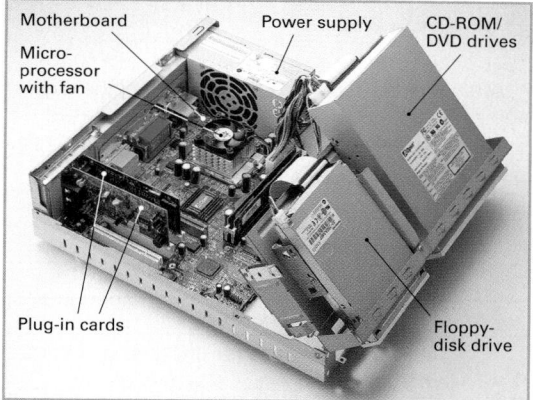

Fig. 1: Basic device

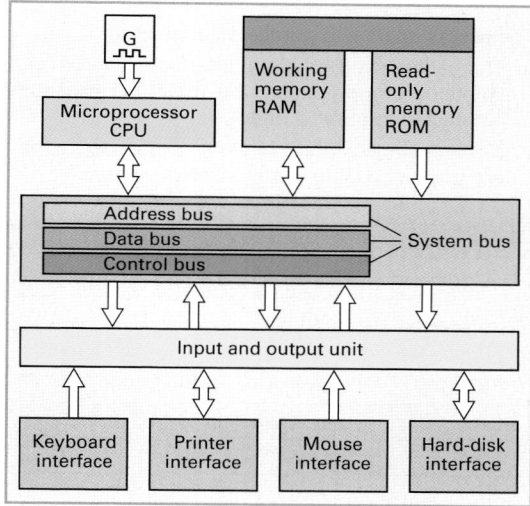

Fig. 2: Systematic diagram of a motherboard

Motherboard. The motherboard (**Fig. 2**) supports the following main component assemblies:

- Microprocessor
- Input and output unit
- Internal memory
- System bus

Microprocessor (CPU – Central Processing Unit). The CPU is responsible for sequential processing of program commands. The microprocessor's frequency defines the number of individual operations that it can execute in one second; a CPU running at 4 GHz thus executes 4,000,000,000 operations each second.

Internal memory. This consists of read-only memory and working memory.

Read-only memory (ROM – Read Only Memory). This type of storage capacity is where permanent, invariable programs and data reside, such as those employed for the computer's boot routine. When the computer is switched on this memory ensures that the operating system's programs are loaded to the working memory (RAM) from the external memory and then started. The data contained in the ROM remain intact when the computer is switched off.

Primary or working memory (RAM – Random Access Memory). This is where programs and data are stored for instantaneous access by the computer system. The data in this type of memory are lost when the computer is switched off, and can also be deleted using software commands.

System bus. It includes the control bus for the control signals, the address bus for accessing specific memory locations, and the data bus, to which data, commands and addresses are transmitted.

Input/output unit (IO interface). This administers multilateral data communications between the CPU, the working memory, and peripheral devices such as the mouse and keyboard.

Interfaces. These are the communications ports at which signals are transferred from one system (such as the PC) to another (such as a printer). Interfaces can assume the form of plug-in cards or plug-in connections **(Fig. 1)**. The operative distinction is between serial and parallel interfaces **(Table 1)**.

Parallel interfaces transfer 8 bits simultaneously (parallel) through separate, 8-strand cables; one example is the printer port. In this application the computer might be equipped with a 25-pin plug, while the printer could feature a 36-pin plug (Centronics plug).

Serial interfaces transfer individual bits sequentially (serial transmission) through a single wire; an example is the mouse connection. This external interface is also referred to as the V.24 or RS 232 interface. Plugs for these interfaces usually have 9 pins.

The **assets** relative to parallel interfaces:

- The cables used on serial interfaces can be substantially longer (up to 150 metres) than those connected to parallel interfaces.
- Because they exhibit a high input impedance they are not susceptible to electrical interference.

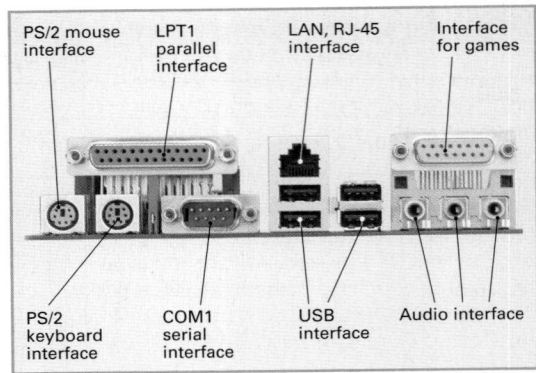

Fig. 1: Interfaces as plug-in port connections on the rear of a computer

- Fewer wire strands are required in the cable (only 2 on the RS232, for transmission and reception).

Interface data-transfer rates. This is usually indicated in Mbits/s or Mbyte/s.

1 Mbyte/s = 8 Mbit/s

Table 1: Computer interfaces			
Designation	Type	Data-transfer rate	Application
LPT 1	parallel	3 Mbyte/s	for printer and scanner connections
COM/RS232	serial	–	for modems and serial mice
LAN/RJ45	serial	10/100 Mbit/s	Ethernet, networks
IrDA	serial	up to 14.4 Kbyte/s	Infrared interface
PS/2	serial	–	for mouse, keyboard
USB	serial	1.5 Mbyte/s	Universal interface for printers, scanners, memory sticks
DIE, ATA	parallel	up to 150 Mbyte/s	for hard drives (on the motherboard)
SCSI	parallel	up to 160 Mbyte/s	for hard drives in professional use, servers
Bluetooth	serial	1 Mbit/s	Wireless communications interface
Firewire	serial	up to 400 Mbit/s	Universal interface for cameras, mobile phones
AGP	parallel	over 533 Mbyte/s	Interface for graphics cards
PCI	parallel	133 Mbyte/s	Interface for plug-in cards such as sound cards

4.6 Data communications

Data communications facilitate the exchange of information between computers, with the information assuming the form of data. Data networks provide links between individual computers.

Data networks. These are data lines through which data are transmitted, usually in the form of message frames. In addition to supporting data transfer, smaller data networks can also expedite control signals such as those used in motor vehicles. Data transfer is serial, i.e. the data bits are transmitted sequentially, and the information is available at all of the nodes on the network.

4.6.1. Data transfer

The systems for data exchange used within business organisations, in offices and in motor vehicles are called **LAN** (**L**ocal **A**rea **N**etworks). Data exchange between devices such as PCs, minicomputers, mainframe computers and all other peripheral devices are expedited at high data-transfer rates (roughly 10 Mbit/s ... 100 Mbit/s). This provides all stations on the network with virtually simultaneous access to all data for tasks such as editing and data revision. In addition to the data line, other elements that interlink computers include the plugs for the communications ports and the network cards that plug into the computers.

Network topologies. Depending upon the arrangement of the computers connected to a single data line, we distinguish between:

- Star topology • Ring topology • Bus topology

Star topology (Fig. 1). We distinguish between:
- Active star topologies
- Passive star topologies

Active star topology: This arrangement can be used to interconnect a large number of computers with only limited amounts of cable. In this structure each workstation is connected to a central node, usually a server or a hub. If one of the workstations on the network fails, the other devices remain unaffected. However, failure at a central node will cause the entire network to quit working.

Passive star topology: All workstations on the network have the same access privileges. The line node provides the link between the computers. This is why failure at one workstation or device will not cause collateral failures elsewhere in the network.

Ring topology (Fig. 2). A single cable provides a direct connection running from each workstation to the next. Data signals are relayed from each station to its neighbour in a single direction. Expansion is simply a matter of connecting additional computers. However, failure at a single computer within the ring network could cause the entire network to fail.

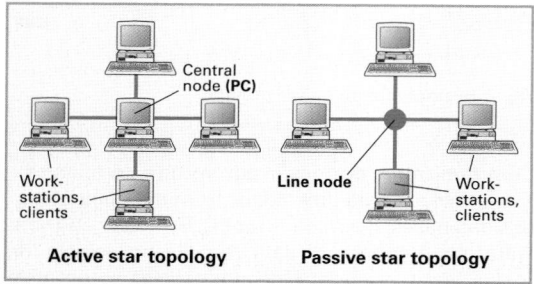

Active star topology   **Passive star topology**

Fig. 1: Star topologies in a network

Bus topology (Fig. 2). This structure can be used to transmit and receive data from individual stations using a single, shared, continuous cable. Because all stations can both transmit and receive, a suitable protocol is needed to arbitrate data communications. Any station may send a transmission whenever the bus is unoccupied. New workstations can be added for simple network expansion. In addition, failure at one station will not incapacitate the entire network. However, every transmission can be received, or accessed, by every station.

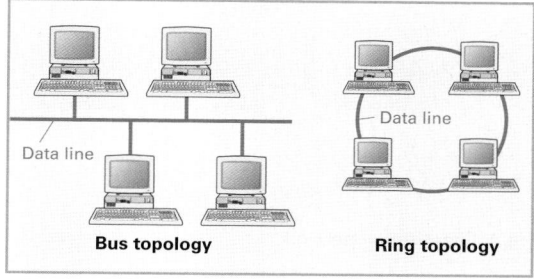

Bus topology   **Ring topology**

Fig. 2: Bus and ring topologies in a network

Peer to Peer network. This is a simple network featuring a bus topology on which a data line joins all computers as independent workstations with the same access priority. This allows such transactions as mutual data exchanges between two computers.

Other conditions for data communications:

Protocol. The protocol defines the conventions according to which data communications are expedited within the network.

Gateway. When two different networks are interlinked, the gateway translates the protocol from one network into that of the other.

Hub. The hub is the centralised distribution point in networks featuring star topology. All data are transferred to all of the computers connected to the network. A passive hub connects the network's bus trunks, while an active hub also amplifies signals.

Switch. This is a distribution point within networks with star topology. It routes data exclusively to those bus trunks on which the addressed computers are located.

Router. It interconnects two or several networks and furnishes "intelligent" data forwarding.

Server-client network

Server. The server furnishes the network with data, programs and computational reserves. The server can store substantial amounts of data, reducing the demands on the capacity of the client computers.

Client is the designation used to identify the other computers in the network. The clients take advantage of the server by using it to store data, to execute programs and to perform computational operations. These networks are used in large car dealerships **(Fig. 1)**.

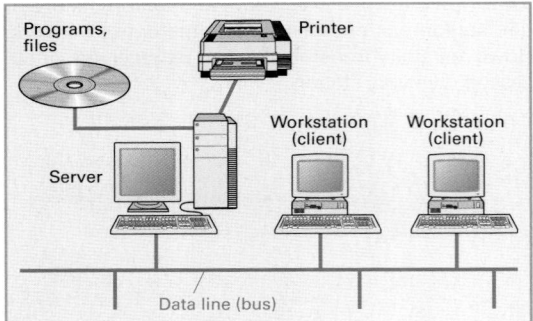

Fig. 1: Client-server network

4.6.2 Remote data transmission

It is used to exchange information, such as texts, images and data, between networks using cable connections or wireless links furnished by telecommunications services.

WAN (**W**ide **A**rea **N**etworks). These communications systems can extend over the entire globe. For instance, within Germany Telekom provides access for remote data transmissions through its telephone network. Links for telecommunications services rely on data transmission devices and end devices.

Data end device. These devices are needed to transmit and receive data. This category embraces equipment such as personal computers and terminals.

Data transmission device. This type of device is needed to adapt data signals between the data end device and the transmission path. This functionality is furnished by ISDN adapters and modems, e.g. ADSL modems. Suitable communication software supports data transfers between computers using serial or parallel interfaces.

Modem (**Mo**dulator/**Dem**odulator). This device converts (modulates) the digital signals from the computer into analogue signals for transmission. The signals can then be transmitted through the analogue telephone network. Following transmission

the modem on the receiving end converts the signals back into digital form suitable for subsequent processing in the computer (demodulation). The maximum transfer rate at which data can be transmitted through the telephone lines using modems is 56 Kbit/s. The devices usually incorporate auxiliary functionality in the form of fax units and answering machines.

ADSL (Asymmetric Digital Subscriber Line). This procedure relies on a modem and a suitable protocol to support high-speed transmission of data in digital form. The data-transfer rate is asymmetrical, with reception (downstream) at up to 768 Kbit/s, and transmission (upstream) at up to 128 Kbit/s.

ISDN (Integrated Services Digital Network). Remote data transmissions through ISDN adapters are expedited in digital form. Personal computers equipped with an appropriate interface in the shape of an ISDN plug-in card can be used directly as ISDN end devices. The data-transfer rate within digital networks is 64 Kbit/s per channel.

Online services
These service providers offer and sell information such as stock market quotations and news along with data and information on specific companies from their databases. They also offer customers a wide range of communication options for such activities as home banking, teleshopping, tele-learning, teleworking **(Fig. 2)**.

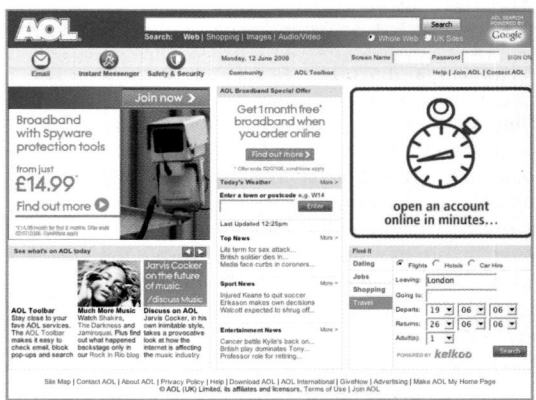

Fig. 2: Example for an online provider

Communication options in automotive companies
These include data downloads from the manufacturer's databases, downloads for ECUs, testers and software, maintenance of data records on vehicle life cycles, flash programming options for automotive service updates, etc.

4.7 Data integrity assurance and data protection

Increasingly widespread use of information technology and corresponding growth in the amount of data have been accompanied by a more urgent need to protect these vast data records and guard them against unauthorised access.

Growing numbers of local and public networks are especially vulnerable. They offer various users the potential to access available data records directly, increasing the risk that data could be subject to unauthorised revision. This is why data integrity assurance and data protection are gaining increasing significance.

4.7.1 Data integrity assurance

> Data integrity assurance comprises the entire array of activities, methods and equipment devoted to preventing data loss as well as unauthorised use or adulteration.

Intermediate and end storage. When a computer is in use, various factors such as user error, power failures, defective computer systems, system crashes, etc., can lead to loss of the data stored in the working memory. This is why intermediate and permanent data storage on external storage media such as diskettes and hard drives is required.

> The danger that data will be lost decreases in proportion to the frequency with which these data are stored.

Backup copies. These are copies of important files on storage media such as CD ROMs; the files are then available for renewed use should the originals be lost.

File overwrite protection. This feature prevents unintentional overwriting of files stored on external storage media. During the save operation the program issues a warning such as **"Delete earlier file?"** on the screen prior to overwriting the earlier file version during the storage process.

Password protection. Every user of an information-processing system has a password. The user must enter this password at the start of each work session. If the computer recognises the password it responds by affording the user access to defined sectors of the external data storage medium.

Antivirus programs. Antivirus programs are used to detect and delete virus programs as well as to prevent them from entering the system. Because new viruses can appear at any time, it is important to ensure that the latest updates of the antivirus programs are being used.

4.7.2 Data protection*

In Germany, the Federal Data Protection Act serves as the foundational law governing data privacy.

> The goal of data protection legislation is to guard against abuse of data in storage, transfer, revision and deletion.

For the purposes of privacy-protection legislation, **protected data** include such categories as **personal data related to natural individuals** unless such data are already available for universal access from a source such as telephone book entries.

Personal data are information regarding:

- **Personal status,** such as date of birth, age, nationality, religion, profession, illnesses, police records and convictions, political beliefs, personal records, consumer spending habits.

- **Material status,** including factors such as monthly income, personal wealth, debt, real estate holdings.

At the same time, storage of personal data has proven absolutely essential as a tool for rapid and rationalised processing in activities such as official administrative functions. This is why individuals must enjoy the following rights when their personal data are recorded:

- **Right to be informed** when data are recorded and stored.

- **General right of access** to stored data related to one's own person.

- **Right to seek revisions** when incorrect data are stored.

- **Right to have data deleted** should it be determined that the original storage was not legal, and also in cases in which the original grounds for storage have lapsed, for instance, when a loan is paid off, or when one ceases to transact business with a particular bank.

In many European countries the privacy protection guaranteed by the data protection laws is monitored by data protection officers.

* The regulations mentioned in chapter 4.7.2 relate to German and/ or European laws.

5 Open- and closed-loop control technology

5.1 Basics

Open- and closed-loop control systems ensure effective, co-ordinated operation of subsystems within an overall system. They also ensure that systems discharge their defined functions relative to their operating environments. A multiplicity of these open- and closed-loop control processes are active whenever a motor vehicle is operated.

Examples of open-loop control processes:
- Cams open and close valves to control gas flow.
- Wheel angles vary to steer the vehicle.

Examples of closed-loop control processes:
- The air-fuel ratio is regulated to a specified value, such as $\lambda \approx 1$.
- Cruise control.
- Braking force control from Antilock Braking System (ABS).
- Coolant temperature control provided by the thermostat.

5.1.1 Open-loop control

With open-loop control one or several input variables influence the output variables, with cause and effect being defined by the system's mode of operation. Open-loop control systems do not monitor whether the actual value of the output variable matches the specified value of the input variable. **The defining feature of the open-loop control is its open-process sequence.**

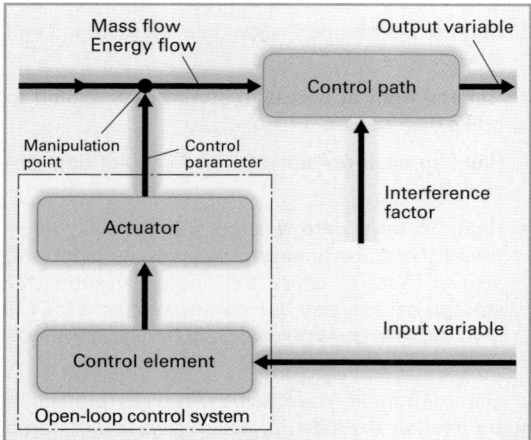

Fig. 1: Open-loop control sequence

[1] This analysis of control systems does not include the human element, which also represents a potential source of control inputs.

Open-loop control sequence (Fig. 1). The process flows sequentially through the open-loop control system's individual elements in a series where each element responds to a stimulus supplied by its predecessor. The open-loop control sequence is divided into the **open-loop control system** and **the control path.**

Example: Vehicle-speed control [1] (Fig. 2). The object is to maintain a constant speed of 80 km/h in a vehicle equipped with a spark-ignition engine.

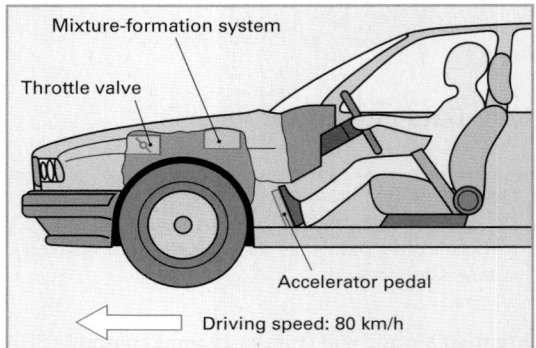

Fig. 2: Vehicle-speed control

Control parameters (Fig. 3)

The speed of 80 km/h represents the **object variable (Fig. 3).** Specified quantities of mixture must flow into the engine in order to maintain this speed under varying operating conditions. The driver controls the flow of mixture by depressing the accelerator pedal to the desired position. The pedal travel is, therefore, the **reference parameter (w)** (input variable). The motion of the accelerator pedal rotates the throttle valve

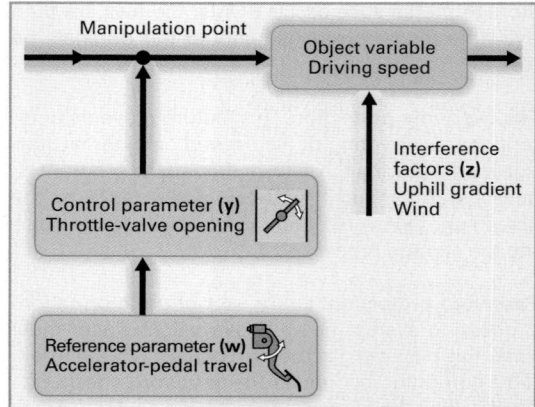

Fig. 3: Physical parameters of vehicle-speed control

within the intake manifold to a specific angle. The throttle-valve opening is the **control parameter (y)** that regulates the volumetric flow of mixture to the desired level.

Open-loop control system (Fig. 1). This system includes the **control element** and the **actuator.** These are the essential elements for manipulating the control path to achieve the desired effect.

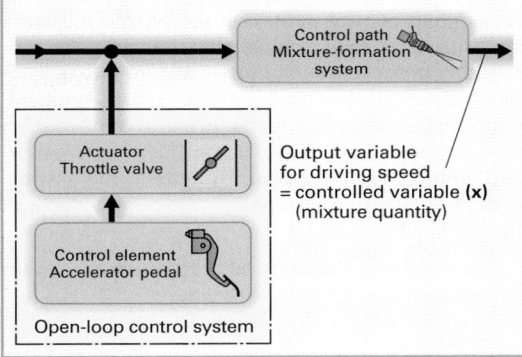

Fig. 1: Vehicle-speed control sequence

The control element is the accelerator pedal. The actuator is the throttle valve. The **reference parameter w** (pedal travel) is the input variable within the open-loop control system. The **control parameter y** (throttle-valve opening) simultaneously serves as the control system's output variable and the control path's input variable.

Control path. This consists of that portion of the system that must be subjected to active control intervention to achieve the object variable; in this case, the object variable is a specific driving speed. Here the control path is the mixture-formation system responsible for furnishing the supply of mixture needed to maintain the desired speed. The control path's output variable is defined as the **controlled variable x.**

The vehicle speed of 80 km/h will only be maintained for as long as no interference factors affect the system's operation. The speed will thus drop in response to factors such as uphill gradients. In our analysis of the control system's operation this gradient represents an **interference factor z.** Because variations in the object variable (vehicle speed) do not automatically generate feedback to initiate a response from the reference parameter (pedal travel), the open-loop control system has no way to react to the interference factor. The open-loop control system is thus characterised by an open process sequence. The vehicle returns to the desired speed once the interference factor disappears.

To compensate for the effects of the interference factor (uphill gradient) the driver must revise the reference parameter (pedal travel) being entered into the open-loop control system. As a result, the control element (accelerator pedal) and the actuator (throttle valve) generate a different output variable (mixture) in the

control path (mixture-formation system) in order to restore the object variable (speed = 80 km/h).

5.1.2 Closed-loop control

With the controlled system in active operation, the closed-loop control system continuously processes locally monitored feedback data representing the controlled object. The closed-loop control system then compares the monitored data with the specified closed-loop control parameter and responds to any deviations between specification and actual status by automatically adjusting the system to compensate. **The defining feature of the closed-loop control is its closed-process sequence (closed loop).**

Closed-loop control circuit (Fig. 2). This loop consists of the elements that maintain a closed control circuit to regulate system operation. The closed-loop control circuit consists of the **closed-loop control system** and the **closed-loop control path.**

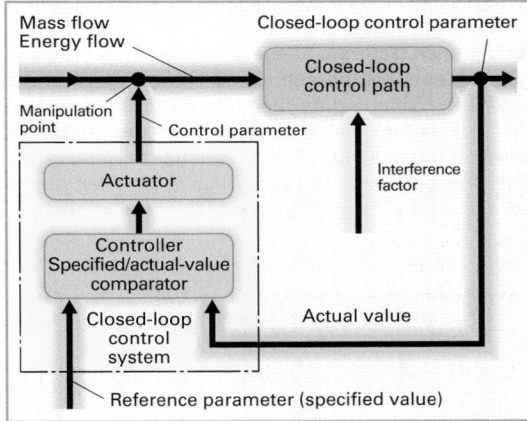

Fig. 2: Closed-loop control circuit

The terminology and designations have already been explained in the example based on cruise control.

The objective is to maintain a constant speed of 80 km/h **(Fig. 3)** in a vehicle equipped with a spark-ignition engine. The cruise control with which this vehicle is equipped will assist in attaining this objective.

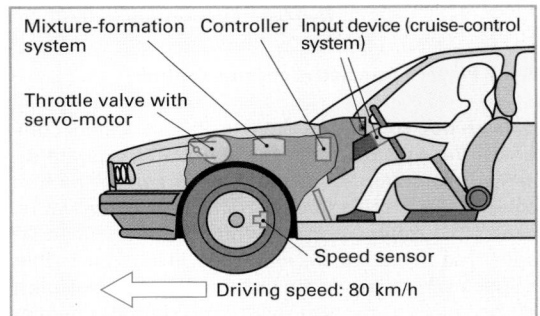

Fig. 3: Cruise control

5

Physical parameters (Fig. 1)

The vehicle's speed is the **closed-loop control parameter x**. Various means are available for activating the cruise-control system.

Usually the driver will use the accelerator pedal to bring the vehicle to a speed of 80 km/h and then use the column-mounted control stalk to activate the cruise control; the system then adopts the vehicle's current speed as its specified value. The specified value of 80 km/h is the **reference parameter w**.

The engine requires a specific quantity of mixture in order to maintain this speed. The system automatically adjusts the throttle valve to an aperture providing the mixture at the volumetric flow rate needed to maintain the desired speed. The throttle-valve opening is the **control parameter y** that provides the correct volumetric flow rate within the mixture-formation system. This flow rate causes the engine to generate a specific amount of power corresponding to the desired vehicle speed **(closed-loop control parameter x)**.

Interference factors z. These can assume the form of hills (ascents and descents) and wind, and affect the vehicle by changing its speed. In this case, however, the momentary speed is relayed to the controller, which compares this reading with the specified value. The closed-loop control system can then intervene to compensate for any interference factors that arise.

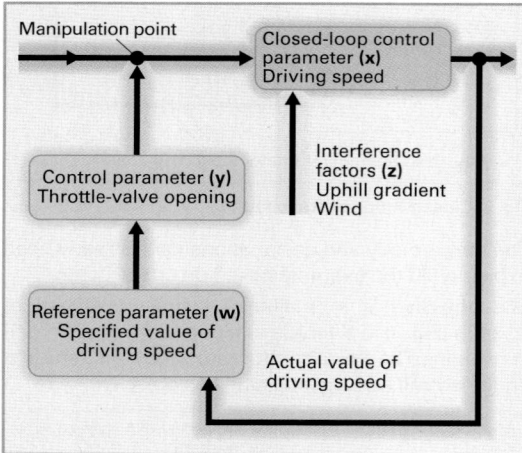

Fig. 1: Physical parameters of cruise control

Closed-loop control system (Fig. 2). This system comprises the **controller** and the **actuator**. These are essential elements for governing the control path to achieve the specified performance. The controller receives the reference parameter (specified value for speed) as an entry on its input side. In addition to this, a speed sensor sends the instantaneous actual value of the speed to the controller. The controller defines actuator signals based on its comparison of the actu-al and specified vehicle speed. The actuator consists of the servo-motor and the throttle valve. The actuator generates a **control parameter y** (for greater or reduced throttle-valve opening) that reflects the monitored level of control deviation.

The control parameter y simultaneously represents the closed-loop control system's output variable and the control path's input variable.

Closed-loop control path. This sector of the system comprises the mechanisms that are actively controlled in order to maintain the mixture at the volumetric flow rate corresponding to a consistent driving speed of 80 km/h. The mixture-formation system thus represents the control path. The mixture generated within this system corresponds to the specific level of engine output that will allow the vehicle to reach the defined speed.

The speed is the output variable for the control path (closed-loop control parameter x).

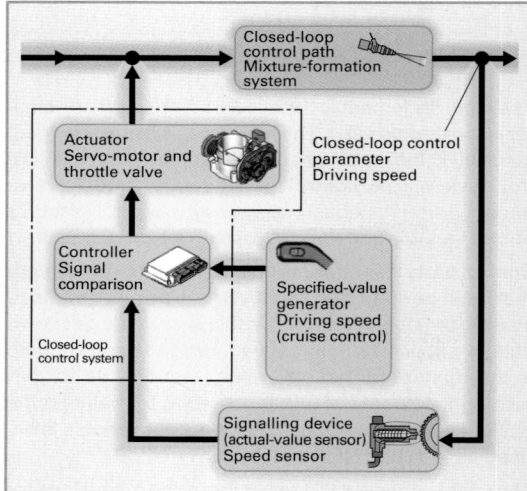

Fig. 2: Closed-loop control circuit for cruise control

Closed-loop control process. The closed-loop control system contrasts with the open-loop system by checking whether the control parameter's actual value (i.e. actual driving speed) corresponds to the specified value (= specified driving speed). The controller responds to deviations between the actual value of the speed and the specified value (= **control deviation**) by initiating a closed-loop control process using a different control parameter. This is why the system is referred to as a closed-loop control circuit.

Control limits. This term refers to the fact that closed-loop control is only possible within certain boundaries. As an example, cruise control is only available for operation within a specific range defined by the manufacturer, for instance, between 30 km/h and 210 km/h.

5.2 Structure and components of the open-loop control system

In terms of equipment, the open-loop control system can be divided into the following elements:

Signalling device, control element and actuator.

The elements in the open-loop control system form a circuit reflecting the system's operational sequence as they relay signals and commands toward the final control element according to the concept

| Input | → | Processing | → | Output. |

This signal flow sequence can be abbreviated as the **"IPO concept"**. **Fig. 1** illustrates this concept using the seatbelt tensioner as an example.

Pyrotechnical belt-tensioning systems are among the devices that manufacturers employ to protect vehicle occupants from injuries stemming from impact against the steering wheel, instrument panel and windscreen.

Signal input. An acceleration sensor supplies impact recognition. It registers variations in velocity instantaneously, and then relays a corresponding electrical signal to the triggering control unit.

Signal processing. The electronic circuitry within the triggering control unit determines whether the deceleration rate has exceeded the critical threshold defined for system activation. If it has, the triggering control unit responds by transmitting a pulse to activate the seatbelt tensioner's ignition system.

Signal output. The ignition capsule responds by igniting a solid propellant. A gas generator then tightens the safety belts using a piston and control cable.

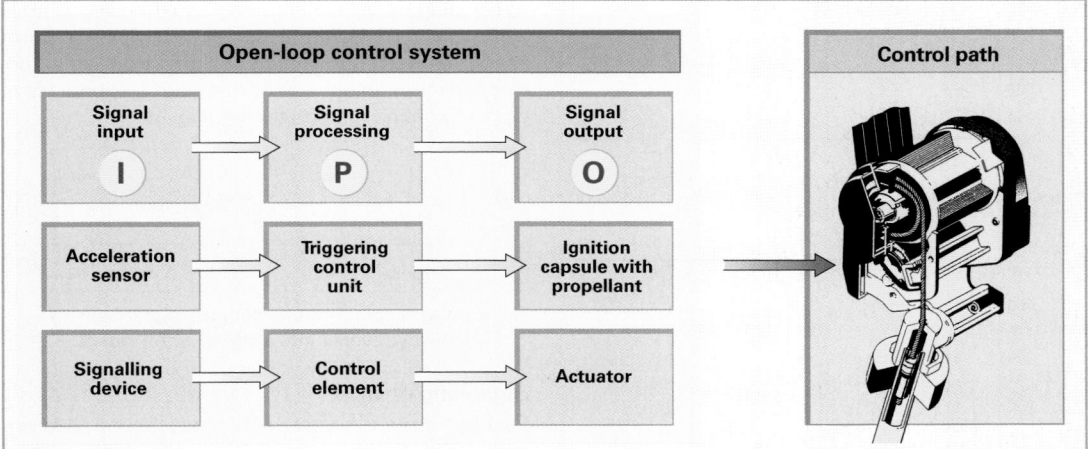

Fig. 1: Diagram showing control technology of a belt-tensioning system

5.2.1 Signalling devices, signal types and signal conversion

Signalling devices are also referred to as **sensors**. They monitor various kinds of physical parameters **(Fig. 2)** as the basis for generating input signals (in the form of voltages, etc.) for transmission to the control elements.

Signals can assume various forms. The operative distinction is between analogue, binary, digital and pulse-width-modulated signals.

Signal types

Analogue signals (Fig. 1, Page 74). These signals are registered and relayed in continuous, infinitely variable wave patterns.

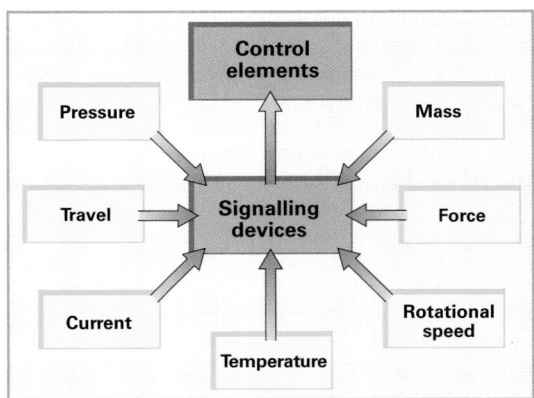

Fig. 2: Input signals derived from physical parameters

Example: The control trigger allows infinitely variable control of a drill's rotational speed. The drill's rotational speed displays a direct correspondence to the position of the control trigger (i.e. it is analogous). An infinite range of intermediate positions is available between the positions for "off" and for maximum speed.

Binary signals (Fig. 1). Binary signals rely on transmission and reception of only two signal states, such as **on** and **off**, or 0 and 1.

Example: A variable rotational speed is indicated by only two status markers, such as < 400 rpm (Status 0) or > 400 rpm (Status 1).

Digital signals (Fig. 1). These represent a special type of binary signal. Here, various intermediate values from analogue signals are recorded and relayed.

Example: A varying rotational speed is displayed in defined segments, such as 100 rpm.

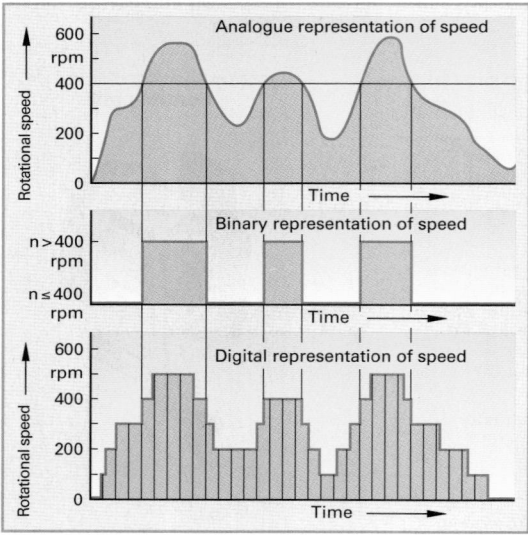

Fig.1: Signal types – analogue, binary, digital

Pulse-width-modulated signals (Fig. 2). These are voltage signals with a constant frequency and uniform activation voltage. The activation duration of the pulses is variable. They are used for such tasks as infinitely variable adjustment of solenoid valves featuring a variable effective port diameter.

The illustration shows a measurement to earth to monitor activation duration.

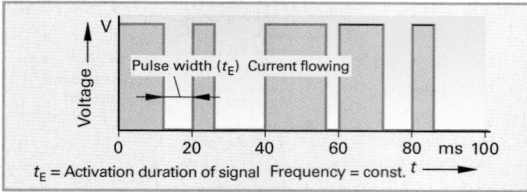

Fig. 2: Pulse-width-modulated signal

Signal conversion

It is frequently necessary to convert monitored data from sensors into specific signal forms for processing in control units and for other activities.

Analogue-digital converter (A/D). These devices convert analogue signals into digital signals. Example: The A/D converter transforms the continuously monitored temperature from an NTC temperature sensor into a digital signal **(Fig. 3)**.

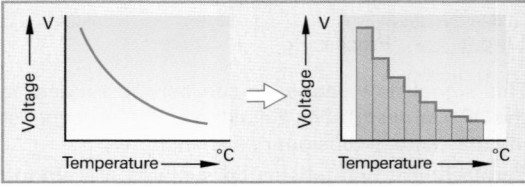

Fig. 3: Analogue-digital conversion of the signal from an NTC temperature sensor

Pulse shapers. These generate square-wave signals from any type of input signal. Example: A pulse shaper converts the analogue signal from an inductive sensor in an ignition system into a square-wave signal to define ignition timing **(Fig. 4)**.

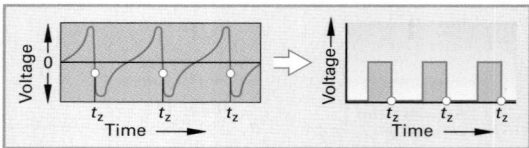

Fig. 4: Pulse shaping – Converting the signal from an inductive sensor into a square-wave signal

Signalling devices

Pushbuttons and **switches** are contact-controlled signalling devices.

Pushbuttons transmit signals only for as long as they remain depressed. A spring then usually returns them to their initial positions.

Switches are held in position by detents, or remain in the positions in which they have been placed. When pressed again they return to their initial positions, or they remain in the new position to which they are set. Electrical switches that close a circuit when activated are called normally open, or **NO contacts**; switches that open a circuit are referred to as normally closed, or **NC contacts (Fig. 5)**.

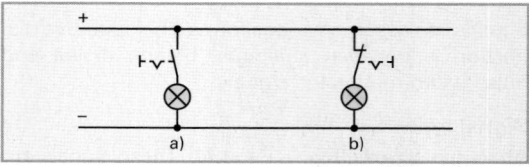

Fig. 5: a) Normally-open   b) Normally-closed
        (NO) contact   (NC) contact

Directional-control valves (Fig. 1) are used more frequently than pushbuttons and switches in pneumatic and hydraulic control systems.

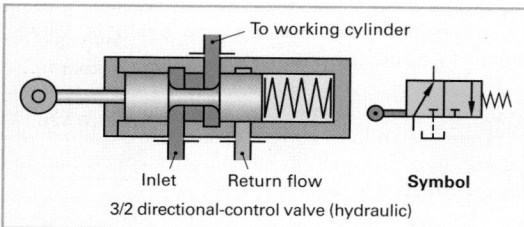

Fig. 1: **Limit switch**

Proximity sensors

These do not rely on external switch activation. They respond automatically.

Photoresistor (Fig. 2). This device reacts to variations in incident light with variations in its own internal resistance, making it suitable for controlling lights.

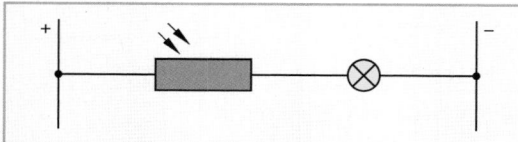

Fig. 2: **Photoresistor for controlling lights**

Inductive speed sensor (Fig. 3). These sensors are employed to measure engine speed, etc. The unit consists of a permanent magnet and an induction coil with a soft-iron core.

A ring gear is mounted on the monitored flywheel, where it serves as the pulse generator. Only a small gap separates the teeth on the ring gear from the speed sensor. As the ring gear turns each of its teeth produces flux in a magnetic field to generate inductive voltage in the coil.

The number of pulses registered within a specific period serves as an index of the flywheel's rotational speed.

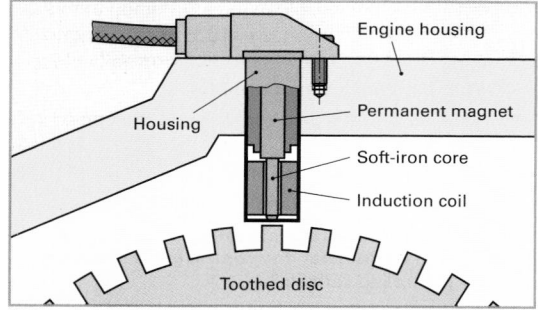

Fig. 3: **Speed sensor**

Temperature sensor (Fig. 4). These devices are used to monitored coolant temperature, etc. For example, in applications using an NTC resistor, the NTC device's resistance falls as temperature rises. The resultant decrease in voltage serves as an index of temperature variation.

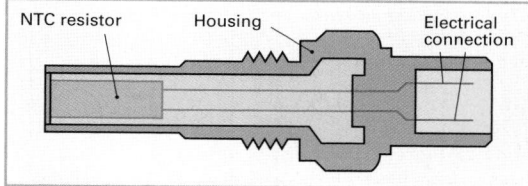

Fig. 4: **Temperature sensor**

5.2.2 Control elements

Control elements (control units) (Fig. 5) receive signals from the signalling devices and process them, then use them as the basis for generating control commands for transmission to actuators.

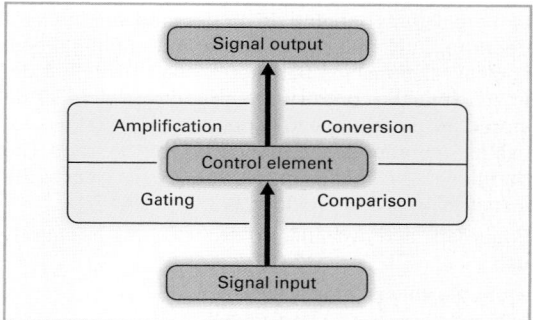

Fig. 5: **Signal processing in control elements**

Signal conversion. This process becomes necessary when a control element (control unit) is unable to process signals in the physical form in which they are originally generated at the signalling device, as well as when actuators rely on specially shaped signals transmitted at defined frequencies in order to operate. Thus, for instance, it may be necessary to convert a signal into **pulse-width-modulated** form within a control unit to render it suitable for use in providing infinitely variable control of a step motor.

Amplification. Input signals are often too weak or otherwise unsuitable for direct use as control signals in governing actuator operation. The signals must then be amplified within the system. Amplification may rely on assistance from a relay. Within the relay, a low control current flows through the coil in an electromagnet to close the contact switches governing the flow of the high main current. The high main current can then be used to power a light, etc. **(Fig. 1, Page 76).**

It is also possible to use a transistor instead of a relay. In these devices a minute control current (I_B) can be used to govern high main currents (I_C).
In technical applications these controlling elements are often referred to as amplifiers.

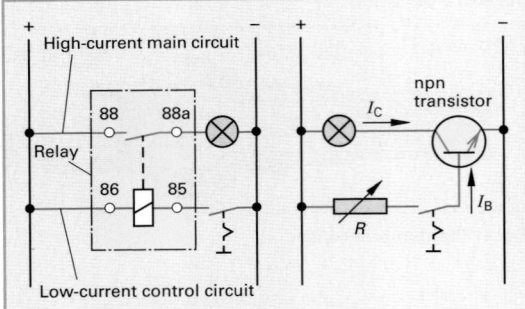

Fig. 1: **Relay and transistor as signal amplifiers**

Defining logical relationships between signals is required when a single output signal is to be generated from several input signals.

Example: Among the data received by the control unit in an electronically controlled fuel-injection system are signals defining driver demand, engine temperature, intake-air temperature and engine speed. Logical relationships between these signals are defined within the control unit. The control unit relies on stored program maps to generate voltage pulses for use as operational commands for the injectors. The duration of the voltage pulses varies according to the instantaneous demand for fuel.

Pneumatic and hydraulic control systems frequently use

● shuttle valves (logical OR) or
● bi-pressure valves (logical AND)

to define logical relationships between input signals **(Fig. 2).**

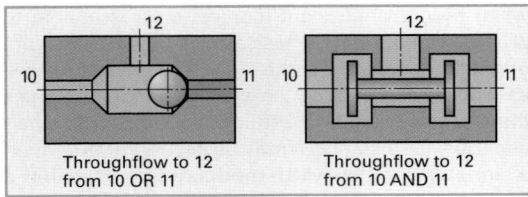

Fig. 2: **Shuttle valve and bi-pressure (AND) valve**

5.2.3 Actuators and drive elements

The actuator is the final element within the control system. It registers control commands issued by the control element (control unit) and responds with local intervention in the control path's mass flow or energy flow.

The actuator initiates a specified change in the object variable at the end of the control path.

Actuators include such devices as valves, throttle valves, cylinders, relays, motors, solenoid valves, transistors and thyristors.

In order to regulate the flow of substantial amounts of power, open-loop control systems are often divided into two components, the **control circuit** and the **power circuit (Fig. 3).**

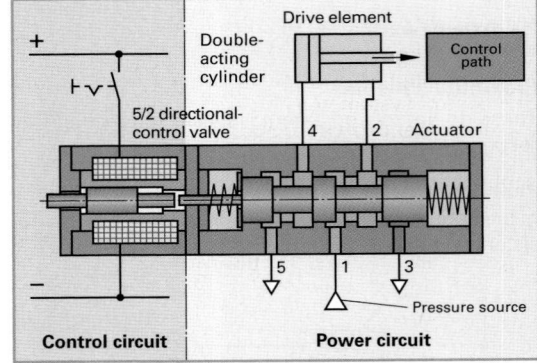

Fig. 3: **Actuator with drive element**

Only relatively small amounts of energy are required to govern operation of the actuators within the control circuit. The primary energy enters the system further downstream, at the drive element, where it is needed for intervention in the control path. Electromagnetic control systems often rely on solenoid-operated directional-control valves as actuators. They control operation of such devices as working cylinders, which serve as the **drive elements** that intervene directly in the control path.

Frequently employed drive elements include electric motors, pneumatic and hydraulic motors, and cylinders.

REVIEW QUESTIONS

1 **What is the difference between open-loop and closed-loop systems?**
2 **What are the main assemblies in the open-loop control sequence?**
3 **What is the function of control elements?**
4 **What is the function of actuators?**
5 **What components are contained in the open-loop control system?**
6 **What are the functions of signalling devices?**
7 **What do you understand by the IPO concept?**
8 **In the open-loop control system, what are the reference parameter, the control parameter and the controlled variable?**
9 **What different signal types exist?**

5.3 Control types

5.3.1 Mechanical control systems

> These control systems relay energy and control signals through mechanical transfer devices.

Some of the control mechanisms used in motor vehicles are purely mechanical. This class embraces the steering (on vehicles without power steering), the variable-ratio gearbox and the valve-control system (without hydraulic lifters).

Valve-control system (Fig. 1, Fig. 3). The valvetrain controls induction and discharge of the air-fuel mixture. The **output** or **controlled variable x** is gas flow. The control system consists of the camshaft, cams, rocker arms and valve springs. Because the port aperture defines (controls) the gas flow, the valve is the control path. The **reference parameter w** is defined by the rotation of the camshaft, maintained by power from the crankshaft. The **control parameter y** is the valve lift generated through the rocker arm, as well as the valve closure initiated by the valve springs. In this control system the **interference factors z** are represented by thermal expansion and mechanical clearances in components, etc.

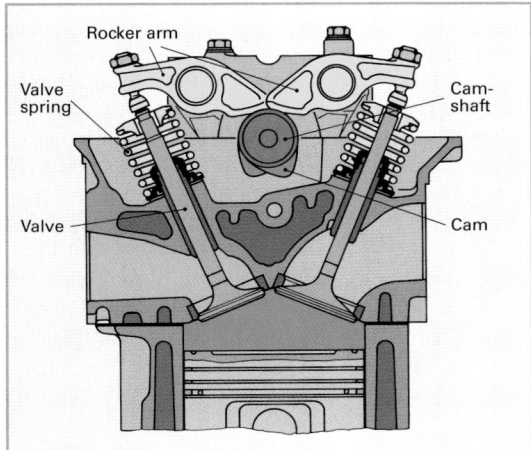

Fig. 1: Valve-control system

Vehicle-steering system (Fig. 2, Fig. 3). The steering system allows the driver to change the vehicle's direction of travel as desired.

Output or controlled variable x. This factor is controlled by the pivoting motion of the wheels. The control system consists of the steering wheel, steering shaft, steering gear, track rod and steering knuckles. Because their swivelling motion defines the vehicle's directional changes, the wheels serve as the control path.

The **reference parameter w** is defined by the driver's rotational input at the steering wheel. The steering shaft serves as the mechanical link that relays this ro-

tational input to a control element, such as the rack-and-pinion steering gear **(Fig. 2)** . The toothed rack converts the rotational motion from the shaft into linear travel. The toothed rack then transmits the steering motion to the wheels through the track rod and the steering knuckles (= control path). **Control parameter y** is the wheels' steering angle as transmitted through the track rod and steering knuckles. **Interference factors z** affecting the steering include external forces, etc.

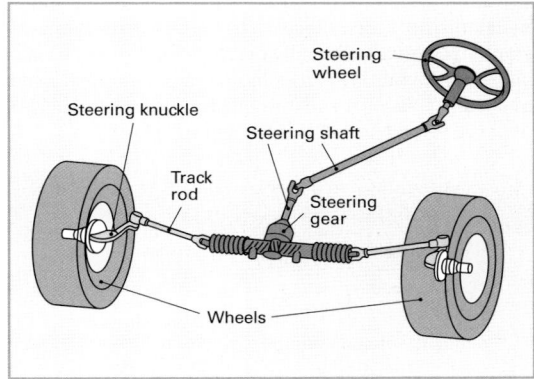

Fig. 2: Vehicle-steering system

Fig. 3 provides a schematic representation of the signal flows governing valve-control system and vehicle-steering system.

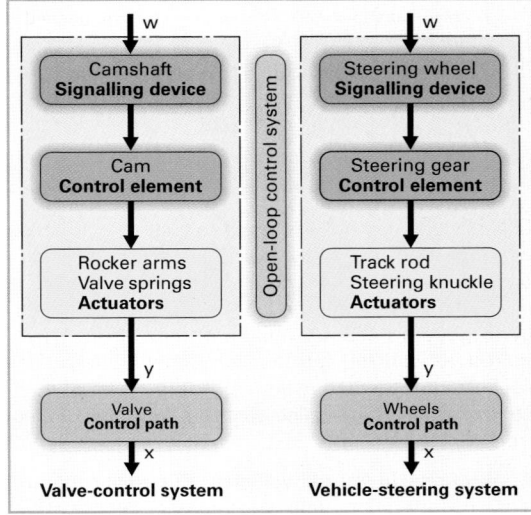

Fig. 3: Signal flows

Disadvantages of mechanical control systems. It is difficult to maintain control through extended transmission paths. These systems are also more susceptible to wear than others types of control system. This is why they are frequently replaced by pneumatic, hydraulic, electric or electro-pneumatic control systems.

5.3.2 Pneumatic and hydraulic control systems

> **Pneumatic control systems.** In these systems the energy conveyor is a gas, usually compressed air, with vacuum used less frequently.
> **Hydraulic control systems.** Hydraulic fluid is the energy-transfer medium.

An asset inherent in both of these energy-transfer mechanisms is their ability to transmit forces over substantial distances with only derisory friction losses. It is also easy to amplify forces (conversion) in these types of systems. **Table 1** lists the advantages and liabilities of compressed air relative to hydraulic fluid.

Table 1: Advantages and disadvantages of compressed air and hydraulic fluid as energy-transfer media

	Compressed air	Hydraulic fluid
Advantages	Compressible, easy to store	Cannot be compressed
	Equipment and systems with simple design structure	High pressures and forces available within a minimal space
	High cylinder and motor speeds possible	Possible to pull away from initial position under full load
	No return lines required	Consistent cylinder motion
Disadvantages	Only relatively small pressures are available	Return lines are needed
	Speed varies with load	Substantial complexity in design structure Emerging fluid poses an environmental hazard
	Noise generated by emerging air (discharge air)	

Often, electric control is used to trigger a function which is then physically executed by a hydraulic or pneumatic system. This concept makes it possible to employ relatively minute amounts of electrical energy to govern the substantial amounts of energy required to control specific component assemblies such as clutches and cylinders. These types of dual-control systems are called electro-pneumatic or electro-hydraulic systems.

Applications in motor vehicles. Pneumatic control systems operating with compressed air are often used in commercial vehicles. Sample applications include the air brakes and pneumatic suspension as well as the pneumatic control circuits employed to open and close doors. Pneumatic control systems that rely on vacuum for operation include the brake boosters used in automotive applications. While vacuum from the inlet manifold is available on vehicles with a spark-ignition engine, diesels require a sepa-

rate pump to generate the required vacuum pressure. In motor vehicles, hydraulic control systems are used in the brake system, dampers, power steering, differential locks, valve control systems featuring hydraulic valve lifters and the automatic gearboxes.

Energy generation

Pressurisation in pneumatic systems. A compressor generates the system pressure, which is then maintained in a compressed-air tank equipped with a bypass valve. A shop compressor can serve as an example for describing the design structure of a complete system for supplying compressed air **(Fig. 1)**.

The system consists of the filter, compressed-air tank, and service unit. The compressor's piston draws in air through a filter before compressing it for routing into the compressed-air tank. Once the tank is pressurised to the system's maximum operating pressure (such as 10 bar) the compressor motor is deactivated. When the amount of air discharged from the system causes the internal pressure to fall to the activation pressure the motor starts again. A pressure gauge on the compressed-air tank shows its internal air pressure. A bypass valve prevents excessive pressurisation. A drain valve is available for discharging water that accumulates from condensation within the tank.

The compressed-air tank is connected to a service unit. This consists of the filter and pressure-control valve along with the pressure gauge. Behind these is the discharge fitting, which supplies oil-free air for inflating tyres and painting vehicles, etc. The air supply used to power impact tools, etc., runs through an oiler.

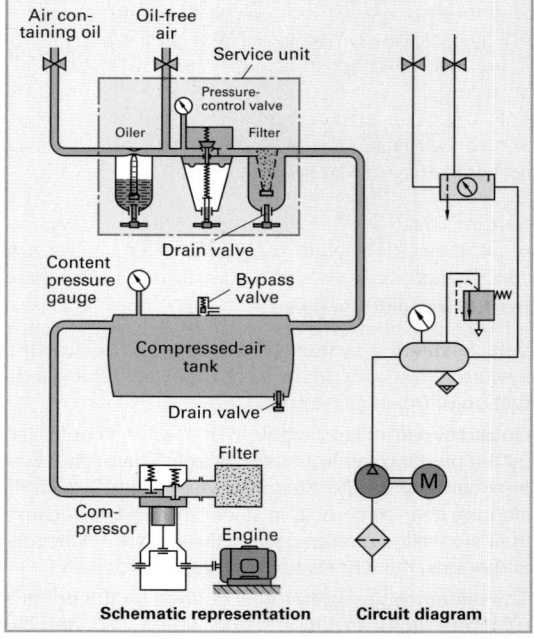

Fig. 1: Compressed-air supply system (shop compressor)

Pressurisation in hydraulic control systems

(Fig. 1). This process relies on a hydraulic pump, such as a gear pump, which is powered by a motor. It ingests hydraulic fluid from a fluid reservoir, then it transfers the pressurised fluid into the pressure line. To ensure safety, a bypass valve is installed downstream from the hydraulic pump. Should the system pressure exceed a defined maximum, the valve opens, allowing hydraulic fluid to flow back to the fluid reservoir through a return line. Return lines are also present on the hydraulic machinery to allow fluid to return to the reservoir.

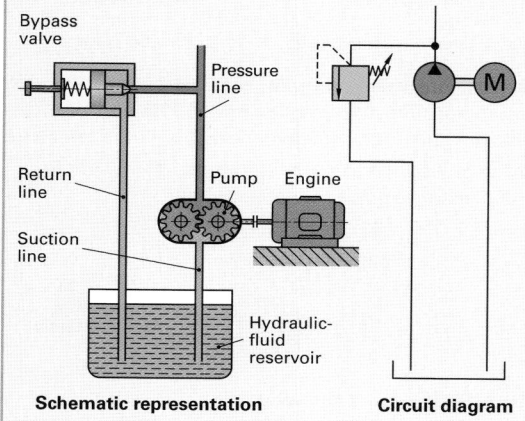

Fig. 1: Hydraulic system

Final-control elements

Design and operating principle of the final-control elements employed in hydraulic and in pneumatic control systems are broadly similar. However, the components in hydraulic machinery must be substantially stronger to withstand the higher operating pressures. The circuit diagrams for both types of system rely on a single set of standard, internationally recognised symbols.

Motors. These convert the energy stored in the compressed air or hydraulic fluid back into rotational motion. Pneumatic and hydraulic motors are manufactured in various designs, including rotary-piston motor and vane compressors as well as piston and gear motors.

Cylinders (Fig. 2). Cylinders serve to convert pneumatic and hydraulic energy into mechanical energy. They operate by moving through reciprocating linear travel paths. On single-acting cylinders with one connection, the piston responds to pressure from compressed air or hydraulic fluid by moving in just one direction. The integral return spring then returns it to its initial position. The double-acting cylinder features two connection fittings. This allows the pressurised medium to propel the piston in both directions.

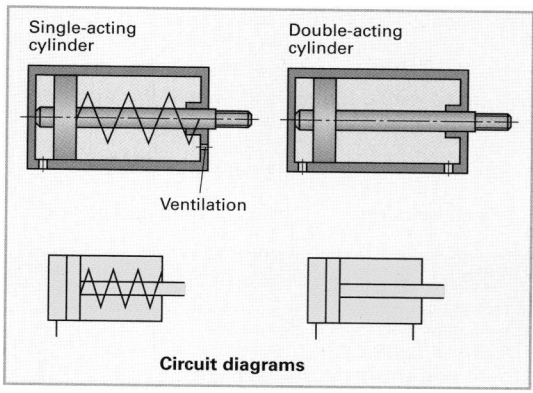

Fig. 2: Pneumatic and hydraulic cylinders

Actuators, signalling devices and control elements
Directional-control valves (Fig. 3).

> Directional-control valves regulate the flow of hydraulic and pneumatic energy by opening and closing flow paths.

In the circuit diagram each switch position is represented by a quadrant. The connection fittings for the lines are shown at the quadrants representing the initial positions. The flow paths and their directions are indicated by lines and arrows. Blocking mechanisms are indicated by a "T". Lower case characters, such as a, b and 0, show the switch positions.

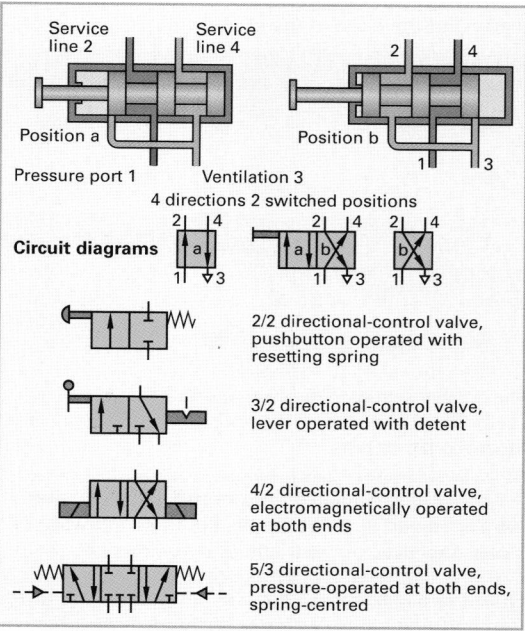

Fig. 3: Representation of directional-control valves

Directional-control valve designations (Fig. 1). The designations consist of two numbers joined by a hyphen. The first digit indicates the number of connection ports, while the second shows the number of control positions. Thus a

    4 / **3** **directional-control valve** has

    4 connections   and   **3 switched positions.**

Terminal designations	New	Old
Pressure port	1	P
Connection port for first service line	2	A
Discharge, return flow	3	R
Connection port for second service line	4	B

Control positions. In control position **a** , **1 with 2** and **4 with 3** are connected. Control position **b** links **1 with 4** and **2 with 3** . The control status is indicated by arrows. The lines are connected to the directional control valve in its initial position.

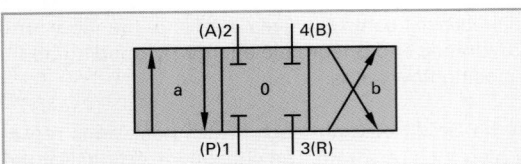

Fig. 1: 4/3 directional-control valve

Valve-actuation mechanisms (Fig. 2). Control mechanisms for operating valves can be by hand, by foot, mechanical, pneumatic, hydraulic, or electric. The type of control is indicated by a horizontal symbol leading to the edge of the quadrant.

By hand, by foot		Mechanically	
	Generally		By pushbutton
	By knob	WW	By spring
	By pedal	**By pressure**	
Electrically		-▷-	Directly
	By solenoid	-▶-	Indirectly via pilot valve

Fig. 2: Valve-actuation mechanisms (selection)

Non-return valves

1-way check valves, or non-return valves, prevent compressed air or hydraulic fluid from flowing in one direction, or from one of two directions.

One-way check valves (Fig. 3) feature one flow direction and one blocked direction; the hydraulic fluid or compressed air can travel in one direction only.

Shuttle valves (Fig. 4) have two inlet ports (10, 11) and one discharge port (12). Discharge port 12 is pressurised when pressure is applied to **10 or 11** .

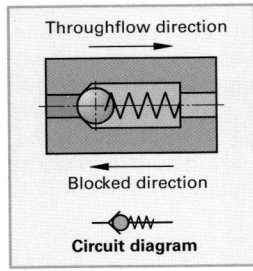

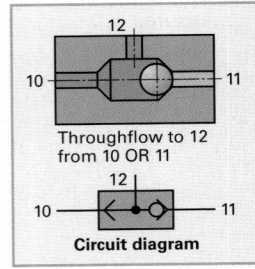

Fig. 3: One-way check valve   **Fig. 4: Shuttle valve**

Bi-pressure valves (Fig. 5), or AND valves, are also equipped with two inlet ports (10, 11) and one discharge port (12). However, on this valve pressure must be applied to both inlet ports **10 and 11** for the discharge port to be pressurised.

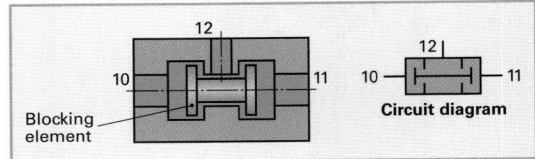

Fig. 5: Bi-pressure (AND) valve

Flow-control valves

Flow-control valves can be used to reduce or shut off the flow of a pressure-transfer medium in a line.

Throttle valves (Fig. 6). Throttle valves with both invariable and with variable restriction orifices are in use. Throttle valves can be used for such operations as reducing and regulating the travel speed of the piston in a single-acting cylinder.

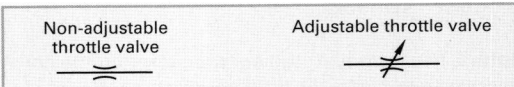

Fig. 6: Circuit diagrams of throttle valves

Non-return throttle valves (Fig. 7) feature free throughflow in one direction, while throughflow in the opposite direction is throttled. The throttle mechanism governing the level of restriction is adjustable.

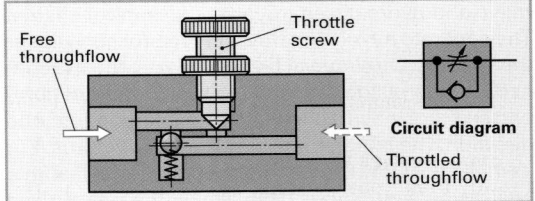

Fig. 7: Restricted non-return valve – adjustable

Delivery valves

Delivery valves can limit pressure, activate and deactivate actuators, and maintain operating pressures at a constant level.

Bypass valve (Fig. 1). This type of valve protects pressure lines, components and pressurised tanks (such as those used for pneumatic brakes) against excessively high pressures. If the internal pressure **P** within a line rises beyond a specified level, it exerts an opening force against the valve cone that exceeds the force of the spring tension holding it closed. The valve opens, allowing the pressure to escape and return to the hydraulic system's fluid reservoir. In pneumatic systems the valve vents the excess pressure into the atmosphere, whereby the air can be routed through a muffling device, etc. The broken line on the symbol indicates that a connection is established between P and T once a certain pressure threshold is reached.

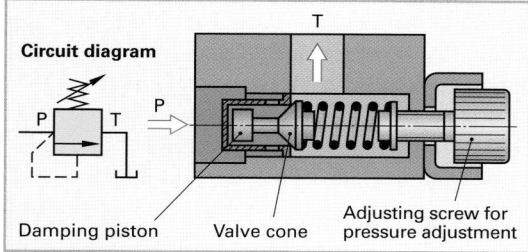

Damping piston   Valve cone   Adjusting screw for pressure adjustment

Fig. 1: Bypass valve

Pilot-controlled bypass valve (Fig. 2). Pilot-controlled bypass valves are used in systems featuring high volumetric flow rates. Maximum pressure is set at the pilot valve's conical-seat valve. The conical-seat valve responds to excessive pressure by opening. The restrictor, or throttle in the main valve produces a pressure differential between the top and bottom of the piston. The pressure counteracts the spring tension to slide the piston upward. The volumetric flow is channelled from **P** to **T**.

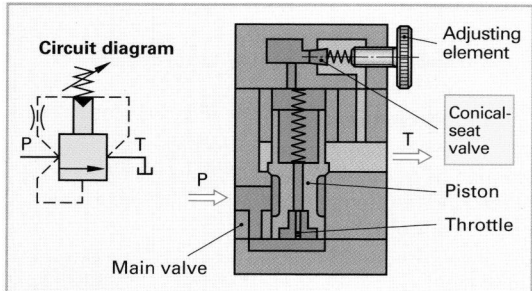

Adjusting element
Conical-seat valve
Piston
Throttle
Main valve

Fig. 2: Pilot-controlled bypass valve

Bypass valves are closed in their base positions; they then open in response to a specific, preset pressure.

Pressure-control valves (pressure-reduction valves). In their base position these valves are open. They maintain output pressure at an essentially constant level, even when the higher input pressure fluctuates.

These valves are required in such applications as ensuring smooth gear changes in automatic gearboxes, and for reducing operating pressures in pneumatic brake systems for trailers to specified levels.

Pressure-control valve with pressure-relief port (Fig. 3). This type of valve maintains a constant pressure in the line **A**. The hydraulic fluid flowing in from **P** is throttled as it flows through an annular orifice. This reduces the pressure in line **A** to a level below that of line **P**. Pressure reductions in line **A** are accompanied by a decrease in the force that the control line conveys to the bottom of the hydraulic piston. The spring's tension pushes the piston further to the left and the annular orifice's gap increases. Increasing pressure in line **A** reduces the effective gap of the annular orifice to compensate for the pressure rise. Should the pressure in **A** rise high enough to close the annular orifice, the pressure-relief port **T** will open. This limits the maximum pressure increase at the working port **A**.

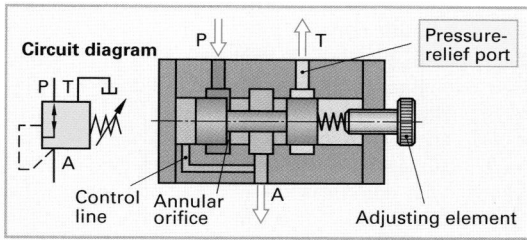

Control line   Annular orifice   Adjusting element

Fig. 3: Pressure-control valve with pressure-relief port

Pressure regulator with electric control (solenoid-control valve, Fig. 4). On this valve the pressure at the working port is proportional to the winding's current. Depending on the amperage of the current flowing through the coil, the annular orifice can open to serve as a pressure-relief port **T**. When no current is flowing, the annular orifice is closed, and the pressure at **A** is at its maximum level. As the current flowing through the coil rises, the force pulling the piston against the tension of the spring increases. The annular orifice to **T** increases in size, reducing the pressure at working port **A**.

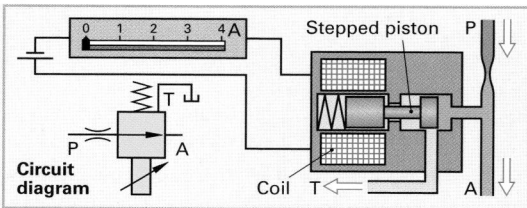

Stepped piston
Coil

Fig. 4: Pressure regulator with electric control

Basic circuits

Clear illustrations showing the structure and operating principle of hydraulic and pneumatic circuits are available in the form of circuit diagrams. The following conventions apply when drafting circuit diagrams:

- Elements are arranged from bottom to top, reflecting the direction in which the signals flow.
- Symbols are horizontal.
- Valves are shown in their initial positions.
- Primary service lines are drawn as solid lines, while control lines appear as broken lines.
- The energy source is portrayed as a triangle for enhanced clarity.
- Elements are identified with circuit number, component code letters and component list number, e.g. 1 V 2.
- Component code letters.

P Pumps and compressors	
S Signal receiver **V** Valves **A** Drive elements	
M Drive motors **Z** All other components	

- The designations should appear within a box, for instance 1V 2 .
- Supply elements should start with the number 0 when possible.

Direct cylinder control

Single-acting cylinder (Fig. 1a). This type of device is used for engine brakes on trucks, etc. To extend the cylinder, a pushbutton is used to activate the 3/2 directional-control valve, which shifts to position **a**. Ports 1 and 2 on the directional-control valve supply the single-action cylinder with compressed air. The spring returns the valve to its initial position **b** when the activation force is removed from the directional-control valve. This blocks the supply of compressed air to the cylinder. The resetting spring within the cylinder presses the piston back to its initial position. The air discharged during this process flows into the atmosphere through the discharge port 3 on the directional-control valve. The 3/2 directional-control valve serves as both signalling device and actuator within this control system.

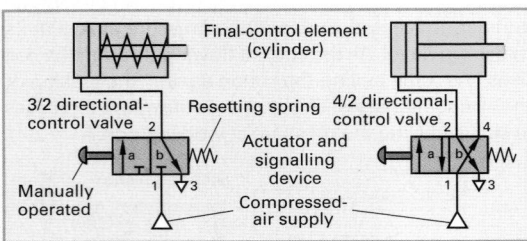

Fig. 1: a) Single-acting cylinder   b) Double-acting cylinder

Double-acting cylinder (Fig. 1b). Two lines connect this cylinder to the 4/2 directional-control valve, which serves as both the actuator and the signalling

device in this application. When the valve is in position **b** compressed air flows into the cylinder through ports 1 and 4. The piston remains in its retracted position. On the other side of the piston air can escape to the atmosphere through ports 2 and 3. When a pushbutton or other means is used to switch the directional-control valve to position **a**, compressed air flows through ports 1 and 2 on the left side of the piston, causing it to extend. Meanwhile, the air on the connecting rod side of the piston escapes into the atmosphere through ports 4 and 3.

Indirect cylinder control (Fig. 2)

Here the actuator for the cylinder 1A is a pneumatically actuated 4/2 directional-control valve 1V . Two 3/2 directional-control valves 1S1 and 1S2 , operated by pushbutton, serve as signalling devices to control operation of the actuator 1V . Brief activation of the signal valve 1S1 generates a pressure pulse (signal) to shift the directional-control valve 1V to position **a**. The piston is extended. It remains in this position even after activation of the signal valve 1S1 ceases. A signal from the directional-control valve 1S2 is required to return the actuator 1V to its initial position and retract the piston. The 4/2 directional-control valve 1V serves as a signal storage device.

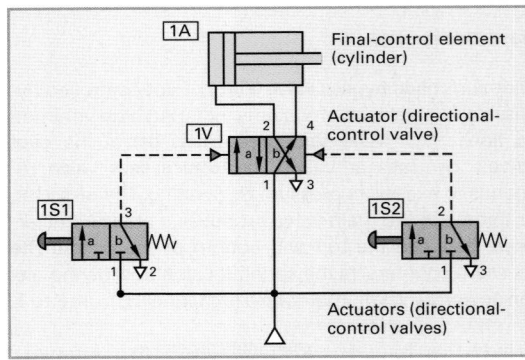

Fig. 2: Indirect control of a double-acting cylinder

Gate-logic circuits (Fig. 3)

Various valve assemblies can be linked in logical relationships to obtain logical AND and OR gates, etc., of the kind frequently required in pneumatic and hydraulic control technology.

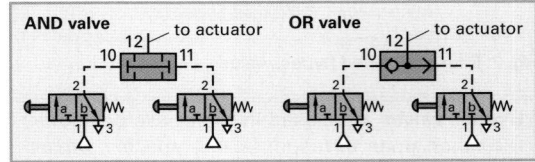

Fig. 3: AND gate   OR gate

5.3.3 Electric control systems

> These control systems use voltages and currents to relay energy.

Electric control systems can be designed for both security and simplicity, especially when only low voltages are being used. Simple systems can operate over considerable distances. A disadvantage is an intrinsic limitation on the levels of force that these systems can generate. A frequent response to this problem is to employ electric control in tandem with hydraulic and pneumatic final actuators.

Electrical equipment

Electrical equipment includes such components as switches, pushbuttons, relays and breakers. These are represented by standardised circuit symbols and designated with standardised code letters.

Switches and pushbuttons (Fig. 1). Depending upon the positions of their contacts during activation, switches can be NO contacts, NC contacts or 2-way switches. They remain in the positions in which the user places them, while pushbuttons return to their initial positions when released.

⊥	NO contact
⊥	NC contact
⊥	2-way switch
⊤	Button, pressure-operated
⊥	Button, manually operated

Fig. 1: Circuit symbols for switch, pushbutton

Relays (Fig. 2). Relays are switches that rely on electromagnetic activation to operate. Within the relay an electromagnet in the low-current control circuit activates a switch in the high-current main circuit.

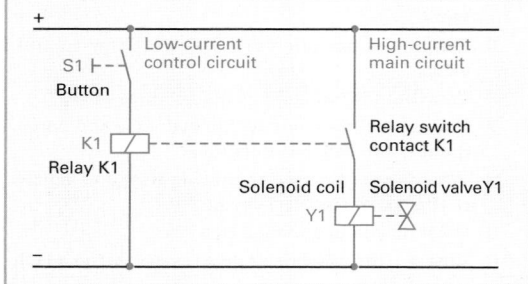

Fig. 2: Relay

Protective relays are used to handle power of more than 1 kW, etc.

Representation of electric control systems

Circuit diagrams provide clear illustrations of electric control systems. A vertical current flow is shown between the positive lead (+) and the negative lead (–) on each controlled device. The devices are identified with code letters, such as **K** for relay, **S** for switch, **Y** for solenoid valves. When several of the same devices are present, their code letters are supplemented by sequential numbers, e.g. **S 1, S 2**.

Basic electrical circuits

Direct control (Fig. 3), for instance, as used for signal lamps. This is provided by a NO contact. Signal lamp H is wired in series with the NO contact S.

Indirect control (Fig. 4). The NO contact is replaced by a **relay K,** for instance, with an NC contact (rheostat function). Here the signal lamp H is in the high-current main circuit. Activating S interrupts the high-current main circuit.

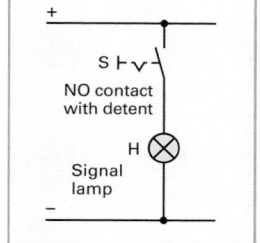
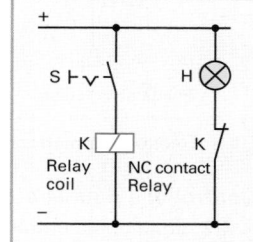

Fig. 3: Direct lamp control   **Fig. 4: Indirect lamp control**

Self-holding circuit (Fig. 5). This is used for such tasks as storing brief signals from pushbuttons. The pushbutton S1 closes the low-current control circuit to the relay K1. Even when the pushbutton S1 opens again, the relay keeps the low-current control circuit closed using the NO contact K1 wired parallel to the pushbutton S1. This stores the signal. The pushbutton S2, located upstream from the pushbutton S1 and the NO contact K1, remains available to cancel the self-locking effect.

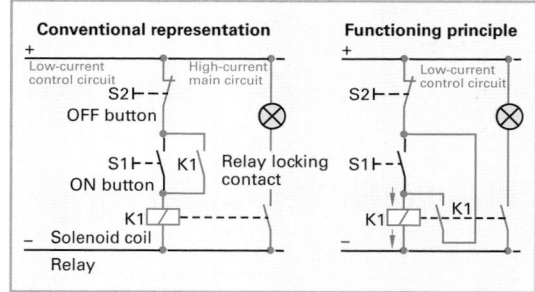

Fig. 5: Self-holding circuit

Electro-pneumatic control systems (Fig. 1)

Electro-pneumatic control systems consist of an electric control circuit and a pneumatic actuator circuit. The control signals are entered into the electric control circuit for processing. The electric control signals govern operation of the directional-control valve used as the actuator for the pneumatic final-control element. This actuator then controls the working cylinder.

The circuit diagram for the electrical subsection and the diagram for the pneumatic portion of the electro-pneumatic control system are shown in separate diagrams. This provides a clearer, more logical illustration.

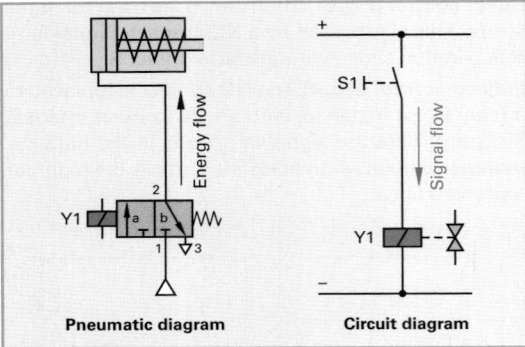

Pneumatic diagram   **Circuit diagram**

Fig. 1: Electro-pneumatic circuit

Control of a double-acting cylinder

A 4/2 directional-control valve is one of the means available for controlling operation of a double-action cylinder. The 4/2 directional-control valve must be equipped with two solenoids Y1 and Y2 or with one solenoid Y1 and one resetting spring.

4/2 directional-control valve with two solenoids

The 4/2 directional-control valve with two solenoids **(Fig. 2)** is a pulse valve, meaning that it responds to brief signals from the pushbutton by reciprocating between one position and the other. Brief activation of the pushbutton S1 triggers the solenoid Y1 to shift

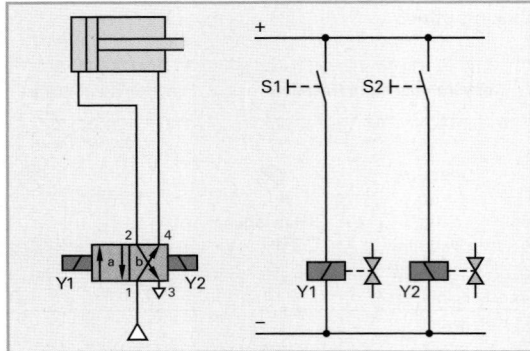

Fig. 2: 4/2 directional-control valve with solenoids

the valve to position **a**, where it remains. This extends the piston in the cylinder.

A pulse from the pushbutton S2 triggers the solenoid Y2 to return the directional-control valve to position **b**. This retracts the cylinder's piston. This 4/2 directional-control valve executes the signal storage function automatically. No electrical self-holding circuit is required.

4/2 directional-control valve with return spring

Because the spring prevents the 4/2 directional-control valve from storing the "on" signal, a latching relay is required **(Fig. 3)**. The self-holding circuit maintains the current flow through the closed switch K1 in the solenoid Y1. The solenoid valve remains at position **a**, and the signal remains stored until the pushbutton S2 is activated again. By interrupting relay K1's low-current control circuit, this also opens the high-current main circuit for the solenoid Y1. The spring now returns the directional-control valve to its original position **b**.

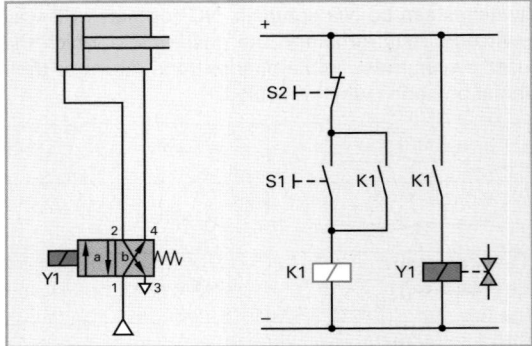

Fig. 3: 4/2 directional-control valve with return spring

REVIEW QUESTIONS

1. What are the assets and liabilities offered by compressed air as an energy-transfer medium?
2. What are the differences in the media used in hydraulic and pneumatic control systems?
3. Explain the design of a compressed-air supply system (shop compressor).
4. Name the components in pneumatic control systems.
5. Explain the operation of a 4/2 directional-control valve.
6. What are the functions of flow-control valves?
7. How are components within hydraulic and pneumatic control systems identified?
8. What means are used to provide clear and logical portrayals of electrical circuits?
9. What do you understand by a relay?
10. What are the different electrical circuits in the relay?
11. What is electro-hydraulic control?

5.3.4 Gate-logic control systems

> Gate-logic control systems rely on logical rela-
> tionships between two or more input signals to
> generate the required output signals.

The three basic functions of logical relationships are:
- **AND function**
- **OR function**
- **NOT function**

These basic functions can be portrayed using standardised circuit symbols. The truth table (**Table 1 in
Fig. 1**) facilitates identification and verification of a
gate-logic control system.

A sample truth table for two binary input signals
might consist of two switch states consisting of
$2^2 = 4$ lines, and $2^3 = 8$ lines for three binary signals.
The switch states are indicated by the binary bits 0
and 1.

0	No signal present

1	Signal present

AND function

With the AND function an output signal will be present when signals are present at all inputs. A sample electrical circuit would be that employed to control the fog lamp in a motor vehicle (**Fig. 1**).

The light switch S1 (NO contact) is connected in series with the fog-light switch S2 (NO contact) and
the fog lamps E1 and E2. Both switches, S1 and S2,
must be closed for the fog lamps E1 and E2 to light
up.

OR function

In OR function an output signal is present if at least
one, or several, or all input signals are present (**Table
2 in Fig. 2**). A sample of this function is provided by
the door contact switches employed to control interi-
or illumination in motor vehicles (**Fig. 2**). The interior
lamp E is activated by the two door contact switches
S1 and S2, which are wired in parallel. Making either
one of the switches or both simultaneously will cause
the lamp E to light up.

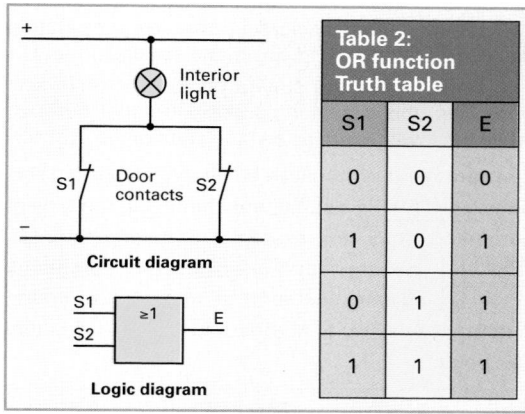

Table 2: OR function Truth table

S1	S2	E
0	0	0
1	0	1
0	1	1
1	1	1

Fig. 2: OR function, circuit diagram and logic circuit symbols, e.g. motor-vehicle door-contact control circuit

NOT function

The NOT function generates an output signal when
no input signal is present (**Table 3 in Fig. 3**). The NOT
gate is negation, meaning inversion of the signal
state. Within an electrical circuit a NO contact can
control switch operation, etc. Lamp E remains on
constantly for as long as the switch S is open (**Fig. 3**).

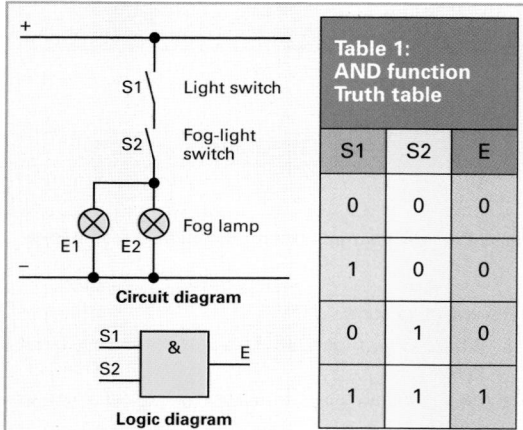

Table 1: AND function Truth table

S1	S2	E
0	0	0
1	0	0
0	1	0
1	1	1

Fig. 1: AND function, circuit diagram and logic circuit
symbols, e.g. motor-vehicle fog-lamp control circuit

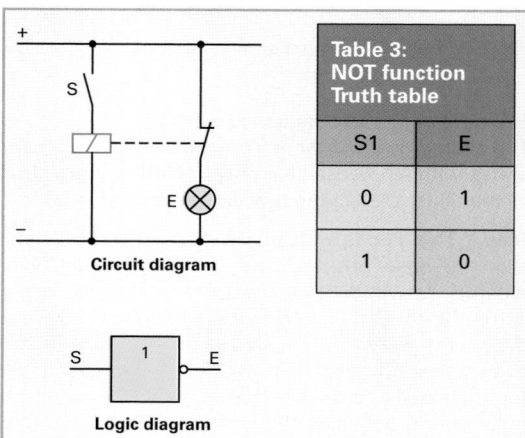

Table 3: NOT function Truth table

S1	E
0	1
1	0

Fig. 3: NOT function, circuit diagram and logic circuit
symbols

5.3.5 Process-sequence control systems

Process-sequence control systems employ a series of control phases which proceed, one step after the other, through a defined sequence. Each subsequent step in the series is initiated only after the conditions for proceeding have been met. Successive steps can be initiated based on time or on process execution.

Representation of process-sequence control systems
The control sequence can be portrayed in operating diagrams with step and command symbols **(Fig. 1)**.

The **step symbol** is a square divided into two sections. The upper section contains the step number, while the lower contains an indication of the process.

The command symbol can be divided into three sections. Field A is a statement identifying the type of command, thus, for example, **S** means "stored". Field B indicates the effect of the command. Field C indicates the abort point of the command output. For example, numbers in section C identify this point.

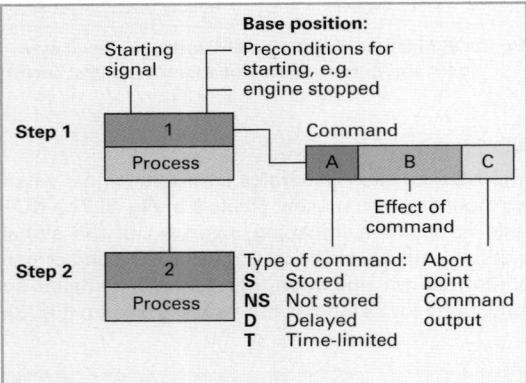

Fig. 1: Process diagram (schematic) with step and command symbols

Example of process-sequence control
The sliding-gear starter in a motor vehicle **(Fig. 2)** operates with a process-sequence control system such as that illustrated in the process diagram **(Fig. 3)**.

Step 1: This step is initiated when the control system is in its base position and the command ON is entered. Activation of the starting switch S1 relays current to the pull-in and holding winding in the engagement relay K1. The starter receives a reduced current through the pull-in winding, and responds by turning slowly. At the same time, the relay armature pulls in and activates the shift lever to engage the starter pinion with the flywheel's ring gear. This command remains stored until cancelled by a new command. Switch K1 does not close the relay until the re-

lay armature is all the way in. This process generates the signal for Step 2.

Step 2: Complete retraction of the relay armature initiates this step. The starter's excitation winding is now linked directly to B+. The starter motor can draw maximum current to turn the engine's flywheel until the engine starts. An overrunning clutch prevents the starter from turning at an excessively high rate once the engine starts.

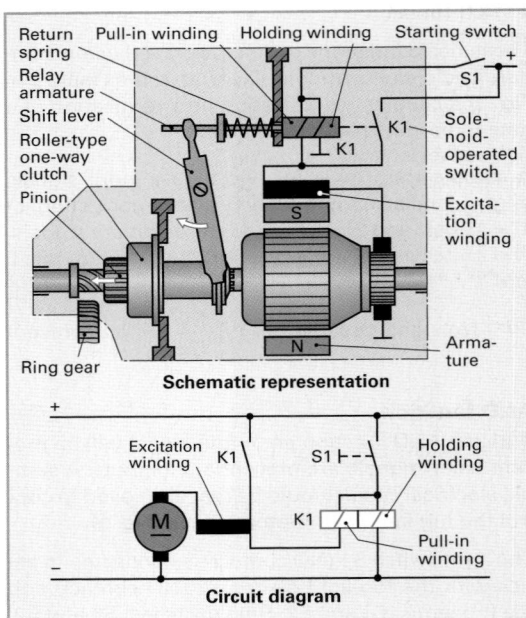

Fig. 2: Sliding-gear starter, circuit

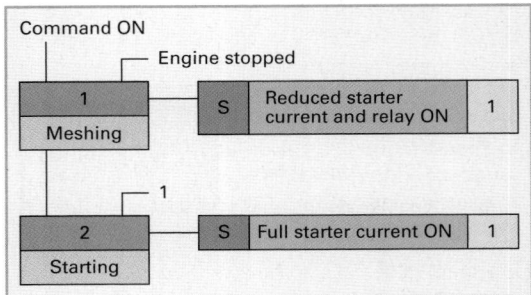

Fig. 3: Process diagram (excerpt) for sliding-gear starter

6 Test technology

In technological applications, testing represents the essential condition for manufacturing workpieces to precision tolerances, and it is also vital in detecting malfunctions when monitoring and maintaining tools and machinery.

In automotive technology, examination and inspection processes are often referred to as testing, such as ignition tests, emissions tests, brake tests.

> Testing determines whether a test object conforms to the specified status.

Subjective testing. This category embraces inspection procedures that rely on sensory perception, such as visual inspections, operational assessments and tactile evaluation.

Objective testing. These tests rely on measuring and test equipment, e.g. measuring (axle alignment checks) or gauging (feeler gauge).

6.1 Basics of linear test technology

6.1.1 Test types (Fig. 1)

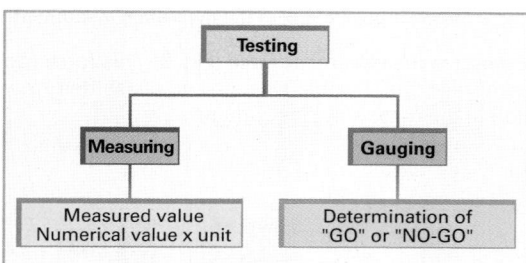

Fig. 1: Testing with linear test technology

> In measurement a measured quantity (such as length or angle) is compared with references on a measuring instrument. This process produces a measured value.

Measured value. This is the actual value of the measured quantity. It is indicated as the product of a numeric value and a unit, such as 15.00 mm **(Fig. 2)**.

> Gauge measurements employ a gauge to garner relative information on the test item. No numeric data are generated in this process.

Gauges are employed to determine whether the test item's dimensions and shape lie within a specified tolerance zone. The result assumes the form of a simple determination with no numeric indication, e.g. "GO" (good side) or "NO-GO" (reject side).

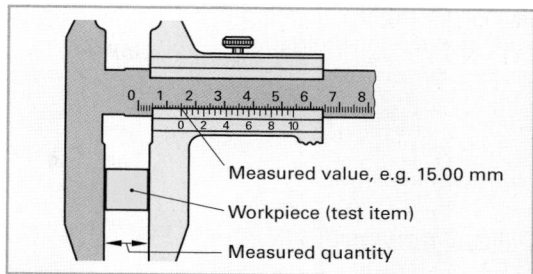

Fig. 2: Measurement designations

6.1.2 Measuring and test equipment

The applied measuring and test equipment **(Fig. 3)** include measuring instruments, gauges, and auxiliary equipment.

Measurement instruments can be material measures or measuring instruments with displays.

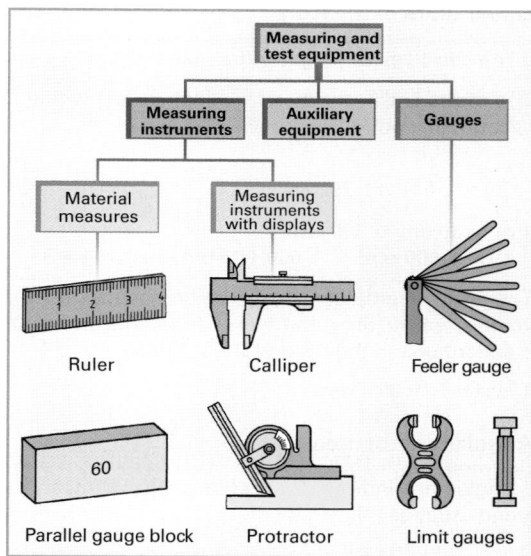

Fig. 3: Measuring and test equipment

Material measures. These feature the measured quantity in the form of defined graduations between lines (graduated linear increments), surfaces (gauge blocks) or angles between surfaces (angular increments).

Measuring instruments with displays can include callipers, micrometers, dial gauges and protractors, etc. They rely on display scales and a moving indicator or needle to show the measured value. Digital numeric displays provide the measured value in the form of numbers in a display panel.

Auxiliary equipment (Fig. 1) supports the measuring and test equipment or the workpiece during measurement. This category includes holders, measuring stands and prisms, etc., as well as devices for relaying dimensions during indirect measurements, such as extension probes.

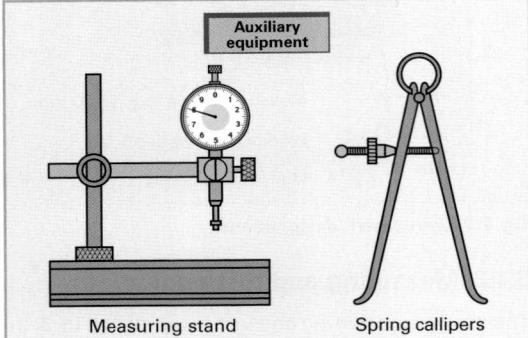

Fig. 1: Auxiliary equipment for measuring

6.1.3 Units of measured values

Linear units of measure

> The basic unit of length is the metre (m).

The **metre** is the distance that light travels within a vacuum in a period of

$$\frac{1}{299{,}792{,}458} \text{ seconds.}$$

1 m = 10 dm = 100 cm
= 1,000 mm = 1,000,000 µm (micrometer)

The **inch** is employed to specify dimensions for wheels, wheel rims and tyres. It is currently still a standard unit in the US and Great Britain.

1 inch = 1" = 25.4 mm

Angular unit of measure

> Angles are indicated in degrees (°), minutes (')
> and seconds (").

A complete circle comprises 360°. The subdivisions of the degree are minutes and seconds.
1° (degree) = 60' (minutes) = 3,600" (seconds)

6.1.4 Measurement errors

The measured value derived from a measurement procedure is usually imprecise, and characterised by a degree of deviation from the precise measured quantity.

Sources of measurement errors
- Imperfections in the measured object
- User error in handling measuring instruments

- Imperfections in the measuring instruments
- Environmental influences

Imperfections in the measured object
Traces remaining from manufacturing processes, such as burrs and scoring.

User error in handling measuring instruments (Fig. 2)
Measuring instrument positioned at an angle, as when it is at an angle while measuring diameter, or when a depth probe is not parallel to the measuring surface or a workpiece is tilted.

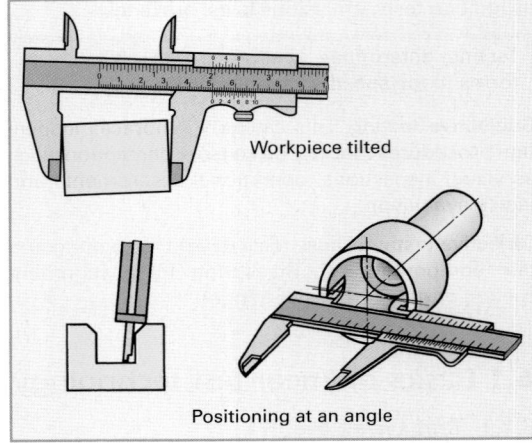

Fig. 2: Incorrect positioning of the measuring instrument

Excessive measurement force (Fig. 3). This force can compress thin, flexible sleeves and cause deformation in the measuring instrument.

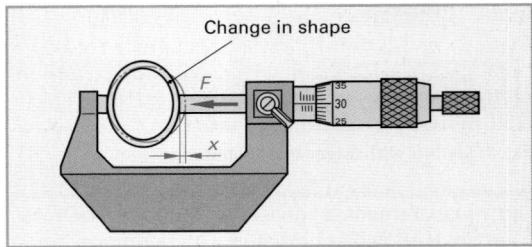

Fig. 3: Excessive measurement force

Parallax (Fig. 4). Another source of measurement error occurs when scale and indicator are not aligned in a single vertical plane perfectly perpendicular to the user's viewing direction.

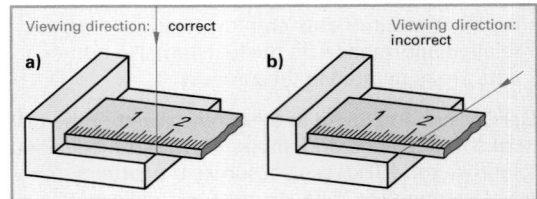

Fig. 4: Parallax error in direct measurement

Contaminants on measuring surfaces. Measuring surfaces on the measuring instrument and on the workpiece may be contaminated with dirt, shavings or grease.

Thermal influences (Fig. 1). Heat from hands, solar light and heat from work processes can also lead to measurement errors.

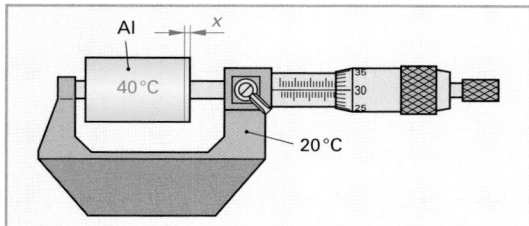

Fig. 1: Thermally induced measurement errors

> The reference temperature for defining the specifications for measuring tools and for workpieces is 20 °C.

Environmental influences. Barometric pressure and temperature can lead to inaccurate measurements during operations such as checking tyre inflation pressures, and provide indications of variations from specified value.

Baseline adjustment (calibration of measuring instruments). Prior to use, measuring instruments such as digital OD micrometers and pressure gauges must be calibrated with a gauge block in order to prevent systematic measurement errors from occurring.

Types of measurement error
This can assume the form of systematic measurement error, as is the case with detectable instrument deviations, or it may arise as random error, with possible variations in the degree of error, which can also occur without a temporal pattern.

Systematic measurement errors
Examples of this type of error include
● inconsistent graduations on scales
● pitch variations on threaded spindles
● uneven measuring surfaces
● changes in shape from consistent application of extreme force
● consistent deviations from measurement temperature

Because they consistently appear in uniform increments and with a single mathematical prefix + or –, it is possible to compensate for systematic measurement errors.

To obtain the precise measured quantity, the reading is corrected by an increment equal to the amount of the systematic error.

Random measurement errors
These errors produce **measurement uncertainty.** On measuring instruments with a display scale the measurement uncertainty of the measured value can equal up to one incremental graduation on the scale.

Among the potential sources of random measurement errors are:
● Reading errors arising from the parallax effect
● Inconsistent measuring instrument contact owing to dirt or burrs
● Thermal fluctuations not subject to monitoring
● Variations in friction and clearance resulting in inconsistent force application during measurement
● Angular deviations from excessive clearance and user error
● Measuring instrument applied at an angle owing to user error

6.1.5 Measurement procedures

Direct and indirect measurement (Fig. 2a). If, for example, a workpiece's length is to be measured against the scale of a linear instrument, the measured value can be read directly from the scale.

Indirect measurement (Fig. 2). When the measurement point has no direct access to the measuring instrument, auxiliary measuring instruments such as outside and inside probes must be used. These register the measured quantity, which can then be measured by the measuring instrument.

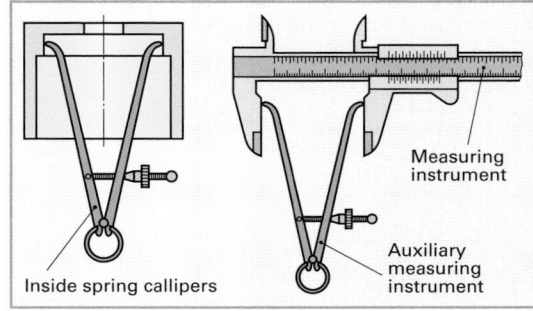

Inside spring callipers

Measuring instrument

Auxiliary measuring instrument

Fig. 2: Indirect measurement

6.2 Measuring instruments

Measuring instruments include material measures as well as display instruments, such as callipers, micrometers, dial gauges and protractors.

Measurement range. This usually coincides with the measuring instrument's display range. The range is the difference between the minimum and maximum readings. This range does not necessarily need to start with zero; for instance, to avoid excessively long measuring rods, the measurement range on micrometers is only 25 mm. OD micrometers with ranges extending from 0 mm to 25 mm, 25 mm to 50 mm, 50 to 75 mm, etc., are available for measuring greater lengths.

6.2.1 Material measures

Linear graduations (Fig. 1) are the most common material measures. Their material measure consists of a uniform line scale to allow direct readings of measured values. Steel rulers, working rulers, tape measures and folding rulers are all in use.

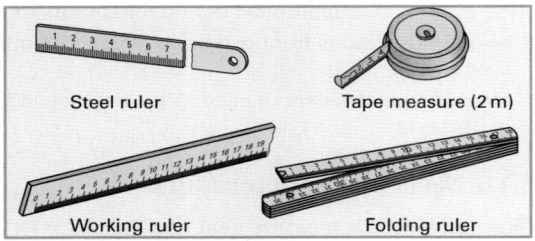

Fig. 1: Linear graduations

Parallel gauge blocks (Fig. 2) provide extremely precise readings of lengths between two parallel surfaces. They generally incorporate a rectangular cross-section, or, less frequently, a round cross-section. Their contact surfaces consist of steel or hard metal and are characterised by high quality. Gauge blocks are employed to check and adjust gauges as well as measuring instruments featuring displays, and to measure workpieces and adjust machinery. It is pos-

sible to press together a series of gauge blocks to obtain a desired dimension.

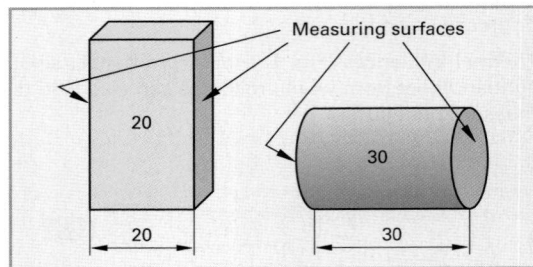

Fig. 2: Parallel gauge blocks

6.2.2 Calliper

Owing to this tool's wide range of potential uses, such as inside (bore), outside and depth measurements, callipers are one of the most frequently employed among those measuring instruments featuring displays and moving graduations.

The **calliper (Fig. 3)** consists of a rail connected to a sliding measuring jaw and to a cross tip. The main rule is located on the rail. This rail bears a scale in millimetres, with a second scale showing inches located above it. A second measuring jaw, connected to the tip, slides on the rail. This jaw is therefore referred to as a slide. On each side of the rail is a line scale, or nonius/vernier scale. A measurement rod mounted on the slide rail is used to measure depths. A clamping lever can be engaged to maintain the slider in position for accurate readings.

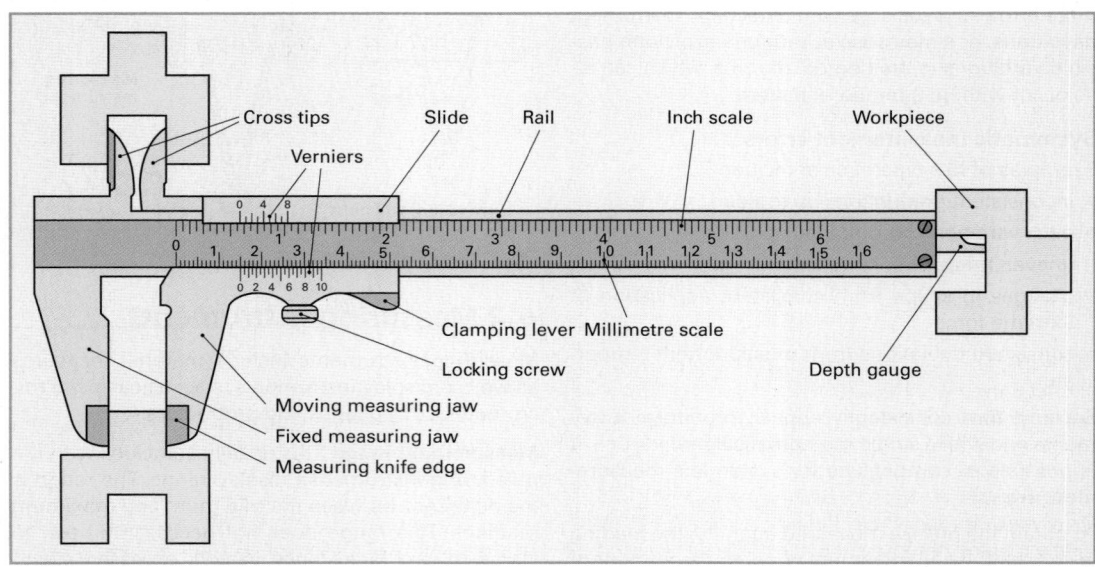

Fig. 3: Calliper

Vernier scale on callipers. This scale defines the accuracy of the readings. The measured value can be read to 1/10 mm = 0.1 mm, 1/20 mm = 0.05 mm and 1/50 mm = 0.02 mm.

With a 1/10 vernier scale **(Fig. 1)** 9 mm are divided into 10 sections. The vernier scale unit is thus 9 mm : 10 = 0.9 mm in length, while the division on the scale is 1 mm. Readings are thus possible to an accuracy of 1/10 mm.

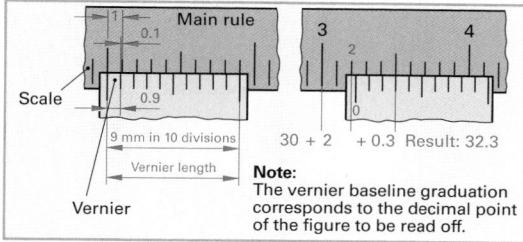

Fig. 1: 1/10 vernier scale

Reading the calliper
The baseline graduation on the vernier is the decimal point that separates the whole numbers from decimals.

The user starts by reading the whole millimetres on the left of the vernier scale's baseline graduation before proceeding to find the fractional graduation on the vernier that covers a graduation on the scale. This vernier graduation indicates the tenths. The graduation for zero should not be counted in adding the total.

On the standardised **1/10 nonius (Fig. 2)** the vernier scale is expanded, e.g. with 19 mm divided into 10 sections. Here a vernier graduation is 19 mm : 10 = 1.9 mm long and the vernier reading is 2 mm – 1.9 mm = 0.1 mm = 1/10 mm.

Vernier	Setting	Reading
1/10	4 5 6 ... 0 10	= 42.7
1/10	2 3 4 ... 0 10	= 23.5
1/20	6 7 8 ... 0 1 2 3 4 5 6	= 63.25

Fig. 2: Sample readings

With the **1/20 vernier scale** 39 mm are divided into 20 sections. In this case the vernier graduation is 39 mm : 20 = 1.95 mm long and the vernier reading is 2 mm – 1.95 mm = 0.05 mm = 1/20 mm.

Callipers with electronic numeric displays (Fig. 3) feature measurement graduations of 1/100 mm. The numeric display of the measured value allows users to read dimensions quickly, easily and without error.

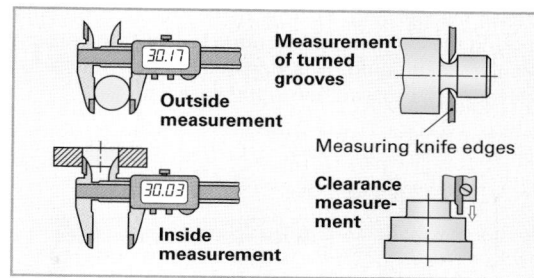

Fig. 3: Calliper measurement

A reset button allows the user to return the display to zero with the jaws at any position. This feature facilitates differential measurements.

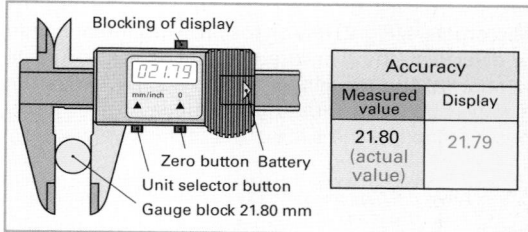

Accuracy	
Measured value	Display
21.80 (actual value)	21.79

Fig. 4: Calliper with numeric display

Measurement accuracy and display accuracy (Fig. 4) exhibit a mutual variation owing to manufacturing tolerances as well as rounding errors in the electronics (tolerance 2/100 – 3/100 mm).

Working rules
- Never attempt to measure **machine-turned parts** while the lathe is turning: this would pose an accident hazard as well as a potential source of damage to the callipers.
- **Outside measurements (Fig. 3).** Guide the measuring jaws over the workpiece, ensuring wide side clearance. Ensure that the measuring surfaces are clean and take care to apply the correct pressure while holding the callipers in the prescribed position. Use measuring knife edges to measure turned grooves, narrow grooves and root diameters.
- **Inside measurements** rely on cross tips. Start by pressing the solid tip against the bore wall, then slide out the opposed tip.
- Do not use the cross tips for scribing.
- **Depth and clearance measurements** are carried out with the depth probe. Avoid positioning at an angle. The recessed side of the depth probe should be against the workpiece to prevent dirt and machining transitions from obstructing direct surface contact.

Depth calliper (Fig. 1). This device has a slide with locking mechanism and a bridge to facilitate perpendicular positioning. It is particularly well-suited for measuring recessed bores and sockets.

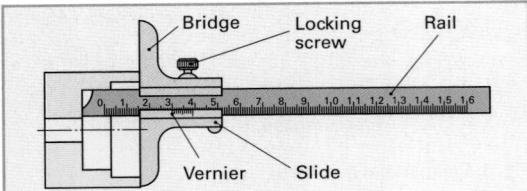

Fig. 1: Depth calliper

6.2.3 Micrometers

Micrometers **(Fig. 2)** rely on the pitch of a threaded rod to determine the measured quantity. Each complete rotation of the machined measuring rod varies the distance between measuring surfaces by an increment defined by the pitch of the threads on the rod.

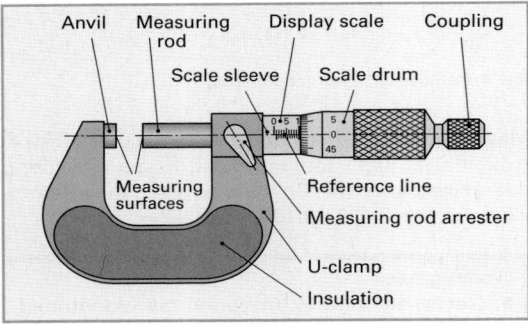

Fig. 2: Outside diameter micrometer

The pitch on the measuring rod is usually 0.5 mm. 50 linear graduations are indicated on the scale drum (thimble sleeve).

$$\text{Scale graduation value} = \frac{\text{Measuring rod thread pitch}}{\text{Linear graduations on scale drum}} =$$
$$= \frac{0.5 \text{ mm}}{50} = \frac{1}{100 \text{ mm}} = 0.01 \text{ mm}$$

The scale drum is permanently attached to the measuring rod. The measuring rod is screwed into the inside threads on the scale sleeve. To obtain a consistent application pressure that does not become excessive in response to the mechanical conversion provided by the threads, the measuring rod is equipped with a coupling to limit application force during measurement.

To keep measuring rod length down, the measurement range is usually only 25 mm.

Taking micrometer readings (Fig. 3)

Full and half millimetres are indicated on the scale sleeve, while hundredths can be read on the scale drum. If the scale drum exposes half a millimetre on the scale sleeve, this must be added to the hundredths.

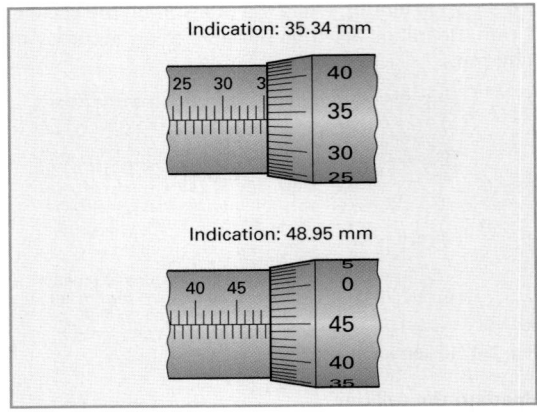

Fig. 3: Sample readings

Outside diameter (OD) micrometer (Fig. 2). This tool is used for external measurement. The components are: U-clamp with fixed jaw (anvil), insulation plate for thermal protection and scale sleeve; measuring rod with scale drum, coupling and locking mechanism. To minimise wear, the measuring rod and the anvil feature hardened contact surfaces or hard-metal surface coatings.

Electronic OD micrometer (Fig. 4). This provides a numeric display in addition to the standard round scale with scale graduations of 1/100 mm. The numeric display is graduated in increments of 1/1,000 mm.

The measuring instrument's electronics allow zero resets to facilitate differential measurements while also incorporating features for storing measured values and relaying data to a computer.

When measuring with electronic measuring instruments it is important to remember that the error range of the display does not directly correspond to that of the instrument itself (see Section Callipers with electronic numeric display).

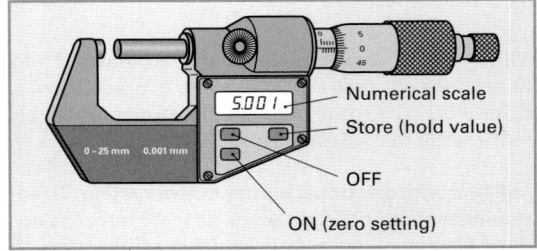

Fig. 4: Electronic OD micrometer

Bore (ID) micrometer (Fig. 1). This tool is employed for measuring internal diameters in cylinders and bores with diameters ranging from 3.5 mm to 300 mm. Bore micrometers should be self-centring. This is why they usually include three moving probe rods or bore rings to ensure concentricity within bores (3-point measurement). Extension probes are available for measuring deep bores.

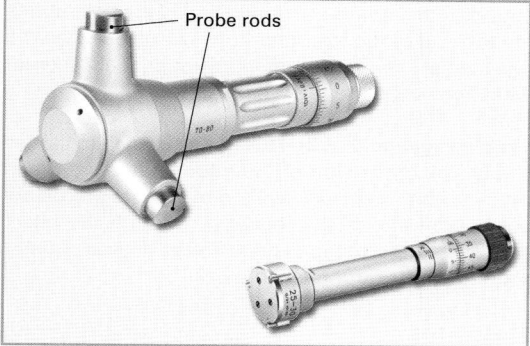

Fig. 1: Bore (ID) micrometer

6.2.4 Dial gauge

The dial gauge **(Fig. 2)** is used to check workpieces for runout (wheel bearing end float, shaft runout) and to determine the consistency of planar surfaces (surface smoothness on brake discs). It can be used together with an internal measuring mechanism to determine cylinder wear in engines.

The dial gauge is thus primarily employed for measuring differences. Instead of measuring an actual dimension, the user measures the variation from a preset actual value.

The probe rod's travel is relayed through a transfer rack and conversion gears to a needle on a graduated gauge for amplified display. The scale face can be rotated to allow zero resets regardless of the indicator needle's position. Two adjustable tolerance marks are used to mark the measurement sector.

> The dial gauge face is divided into 100 sections. Each rotation of the indicator needle represents a measuring probe rod rotation of 1 mm. Each scale graduation is thus 0.01 mm = 1/100 mm.

When the dial gauge's indicator needle travels through more than a single rotation, the number of rotations appears on a small millimetre scale, as each rotation of the needle corresponds to a measurement travel of 1 mm. Dial gauges with scale graduations of 0.001 mm are also available.

Friction and clearances in the transfer rack and gears can cause displays to vary according to the direction in which the probe rod is travelling. This display variation is the **hysteresis error**, which can amount to as much as 0.005 mm.

6.2.5 Protractors

Basic protractor (Fig. 3). This tool measures angles in degrees. The measurement range is 180°. The display reading does not always correspond to the measured value. Often the result must be calculated.

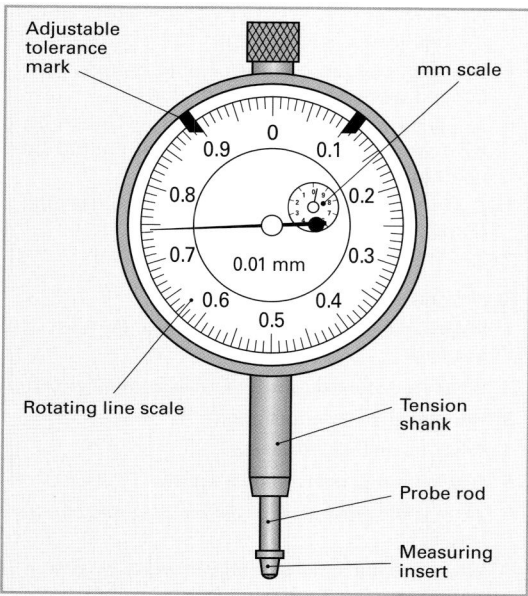

Fig. 2: Dial gauge

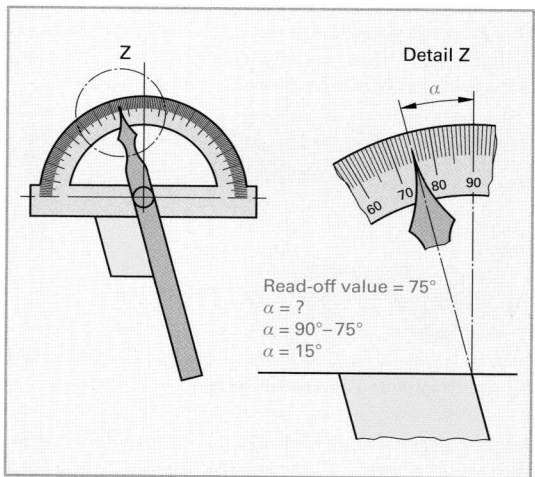

Read-off value = 75°
α = ?
α = 90° − 75°
α = 15°

Fig. 3: Basic protractor

6.3 Gauges

Gauges are measuring and test equipment that reflect the dimensions or shape, or dimensions and shape of the workpiece. No parts move on the gauge during the measurement process.

6.3.1 Dimensional gauges

Dimensional gauges include wire and sheet-metal gauges, bore gauges, feeler gauges (**Fig. 1**) and parallel gauge blocks.

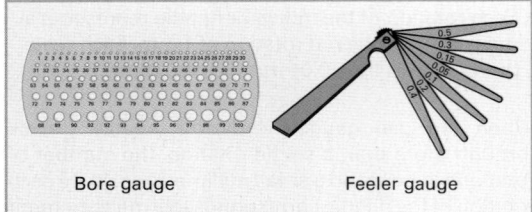

Bore gauge   Feeler gauge

Fig. 1: Dimensional gauges

Feeler gauge (Fig. 1). This tool consists of a number of steel blades in various thicknesses ranging from (for example) 0.05 mm to 1 mm. The nominal dimension of each blade is indicated on its surface. Feeler gauges are employed for such operations as determining clearances on bearings, pistons, valves and slide rails.

6.3.2 Form gauges

Shaped measuring tools such as the radius gauge **(Fig. 2),** thread pitch gauge, thread pitch gauge, angle gauge, straightedge **(Fig. 2)** are used to check the shapes of radii, profiles, angles and surface consistency based on gap. The gap should be as small as possible.

Thread pitch gauges are used to test the threads on workpieces. These usually comprise sets of steel discs with various thread profiles and pitches.

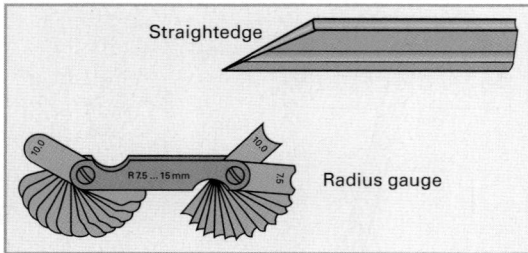

Straightedge

R 7.5 ... 15 mm   Radius gauge

Fig. 2: Straightedge and radius gauge

Angle gauges. These embody the shape of a specific angle. Flat squares, back squares and hair angles are in use.

6.3.3 Limit gauges

The operative distinction is between limit plug gauges and limit gap gauges. These have a "GO" and a "NO-GO" side; the reject side features a red paint marking to facilitate recognition.

Limit plug gauge (Fig. 3). This tool is employed to check bores. The dimension of the good side corresponds to the minimum dimension of the bore, while the top end of the tolerance zone is exceeded on the reject side. The measurement cylinder is also somewhat shorter on the reject side.

Limit gap gauge (Fig. 3). This gauge is used to assess shafts. The dimension of the good side is the maximum shaft dimension, while dimensions are below the bottom end of the tolerance zone on the reject side. The gauge is sloped on the reject side to prevent scoring marks during testing.

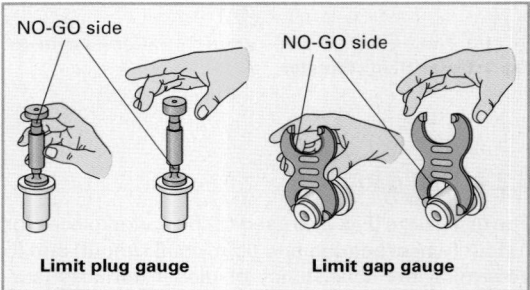

NO-GO side   NO-GO side

Limit plug gauge   Limit gap gauge

Fig. 3: Correct use of limit gauges

Working rules
- Never insert a plug gauge into a cylinder while the workpiece is still warm.
- Never press gap gauges or plug gauges over or onto workpieces with force.
- Apply limit gap gauges carefully. The tool's own weight is the only force that should be used to slide it over the workpiece.
- During measurements keep the straightedge **(Fig. 2)** perpendicular to the surface.

REVIEW QUESTIONS

1 **What is the difference between measuring and gauging?**

2 **What are potential causes of measurement error?**

3 **How can parallax errors occur?**

4 **What are material measures?**

5 **How does one read a vernier scale?**

6 **What is the procedure for determining micrometer readings?**

7 **Why is the micrometer equipped with a coupling?**

8 **What are dial gauges used for?**

6.4 Tolerances and fits

6.4.1 Purpose of standardisation

Technical and economic factors combine to render absolute conformity with specified dimensions impossible during manufacture of workpieces, standard parts and semifinished products. Applicable for all machining processes used in manufacture:

> The demand for precision in the workpiece should be just high enough to ensure that it fulfils its defined function.

In order to ensure satisfactory functionality in workpieces, they must be manufactured to actual dimensions lying between the maximum and minimum dimensional specifications. This range between these two dimensional tolerance limits is the tolerance zone **(Fig. 1)**.

If the workpiece is within the tolerance zone, it is a good ("GO") part. When a production part is outside the tolerance zone it must first be examined to determine whether it can be remachined, etc. If not, the part is a reject.

6.4.2 Terminology

The abbreviations in ( ) refer to designations that have not been standardised.

Nominal dimension (*N*). This is the dimension indicated in the blueprint that serves as the reference for the tolerance limits **(Fig. 1)**.

Actual dimension (*I*). This is the actual dimension of the finished workpiece as determined by measurement **(Fig. 1)**.

Dimensional tolerance limits. The **maximum dimension (*H*)** and the **minimum dimension (*M*)** on the workpiece are the dimensional tolerance limits. The **actual dimension (*I*)** must lie between the two dimensional tolerance limits **(Fig. 1)**.

Baseline (*NL*). This is the reference line for the specified nominal dimension **(Fig. 1)**.

Dimensional deviations. This is the difference between the tolerance limits and the nominal dimension or between the actual dimension and the nominal dimension. Tolerance limits are the upper dimensional deviation and the lower dimensional deviation. The approved deviations for shafts are indicated with the lower case letters *es* and *ei*, while the approved deviations for bores are designated with the upper case letters *ES* and *EI* **(Fig. 1)**.

Upper dimensional deviation *ES* or *es*.* This is the difference between the maximum dimension and the nominal dimension.

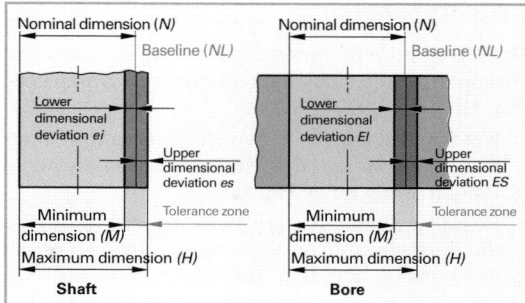

Fig. 1: Terminology

Lower dimensional deviation *EI* or *ei*.** This is the difference between the minimum dimension and the nominal dimension.

Dimensional tolerance, tolerance (*T*). This is the approved deviation from the specified nominal dimension. It is defined as the difference between the upper dimensional deviation *ES*, *es* and the lower dimensional deviation *EI*, *ei*.

Tolerance zone. This is the graphic illustration of tolerance. Its primary characteristics are the size of the tolerance (maximum and minimum dimensions) and their positions relative to the baseline. There are four possible positions for the tolerances relative to the baseline **(Fig. 2a to 2d)**:

- Both tolerance limits are positive (plus-side) **(2a)**.
- Both tolerance limits are negative (minus-side) **(2b)**.
- The tolerance limits have different mathematical signs **(2c)**.
- One tolerance limit is zero, while the other is positive or negative **(2d)**.

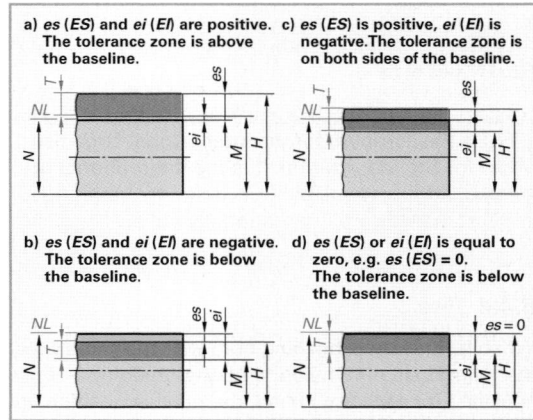

Fig. 2: Position of the tolerance zone relative to the baseline

---

*   *ES* or *es*   Ecart **S**upérieur (French) Upper dimensional deviation
** *EI* or *ei*   Ecart **I**nférieur (French) Lower dimensional deviation

6.4.3 Applications

The ISO system of dimensional tolerance limits and fits applies to the following types of fit on workpieces (**Fig. 1**):

- **Flat fits.** This applies to two workpieces with interlocking planar surfaces, such as those between a tongue and groove (**Fig. 1a**).
- **Regular-cylinder fit.** Here the workpieces are cylindrical with a circular diameter, e.g. between shaft and bushing bore (**Fig. 1b**).

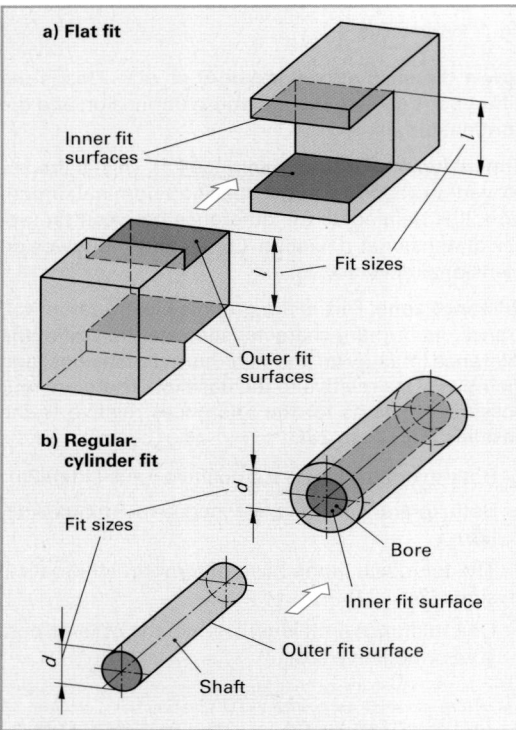

Fig. 1: Types of fit

ISO is the acronym for the Internatonal Organisation for Standardization. It is the international organisation responsible for defining standards. ISO standards are internationally valid.

6.4.4 Fits

When a shaft and a bushing bore with the same specified nominal dimension and the tolerance zones specified for each are joined, the relative positions of the respective tolerance zones can produce the following fits:

- Clearance fit
- Interference fit
- Transition fit

Two of the parts forming a joint share a single dimensional specification.

The fit is the dimensional difference between the dimension of the inner fit surface and the dimension of the outer fit surface.

Clearance fit. Constant clearance remains between the shaft and the bushing bore after assembly. The minimum dimension of the bushing bore is larger than or, under limit conditions, equal to the maximum dimension of the shaft (**Fig. 2**).

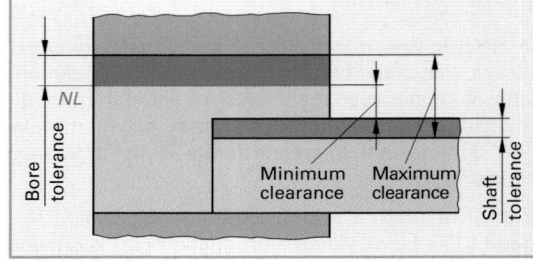

Fig. 2: Clearance fits

Minimum clearance. This is derived from the difference between the minimum dimension for the bore and the maximum dimension of the shaft.

Maximum clearance. This is derived from the difference between the maximum dimension for the bore and the minimum dimension of the shaft.

A clearance fit is characterised by a positive dimensional differential.

Interference fit. Clearance is negative, with excess shaft diameter, when the shaft is inserted into the bore. The maximum dimension of the bore is smaller then, or, under limit conditions, equal to the minimum dimension for the shaft (**Fig. 3**).

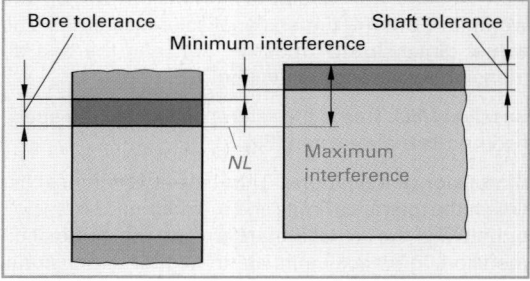

Fig. 3: Interference fits

Minimum interference. This is the difference between the maximum dimension for the bore and the minimum dimension for the shaft prior to joining.

Maximum interference. This is the difference between the minimum dimension for the bore and the maximum dimension for the shaft prior to joining.

An interference fit is present when the dimensional differential is negative.

Transition fit. Following joining, within their tolerance zones the bore and shaft exhibit either clearance or interference **(Fig. 1)**.

A transition fit is present when either a clearance fit or an interference fit may occur.

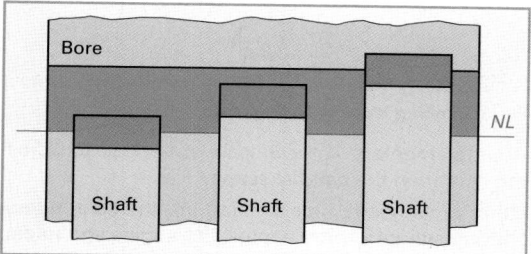

Fig. 1: Transition fits

6.4.5 Tolerance specifications

Tolerance specifications as tolerance limits

Tolerances can be defined as desired. The deviation limits are provided behind the specifications for nominal dimensions in the blueprints.
In most cases the upper dimensional deviation, regardless of mathematical sign, is defined as greater than the nominal dimension, while the lower dimensional deviation is less, for instance, $80^{+0.6}_{-0.2}$.

Tolerance specifications with ISO tolerances

ISO tolerances are indicated with letters and numbers.

Tolerance zone position. The position of the tolerance zone relative to the baseline is indicated for bores with upper case letters (A to ZC) and with lower case letters on shafts (a to zc).

The distance separating the tolerance zone from the baseline is proportional to the distance separating the applied letter from H or h.

Basic tolerance scale factors. These are designated with the numbers 01, 0, 1...18. This provides 20 basic tolerance scale factors. As the number of parts rises, so does the tolerance zone, i.e. the requirement for manufacturing precision decreases **(Fig. 2)**.

Smaller numbers for the basic tolerance scale correspond to higher levels of manufacturing precision.

The definition of tolerance consists of the nominal dimension specification and an ISO code comprising numbers and letters, thus bore 25 H7; shaft 25 n6; fit 25 H7/n6 or 25 $\frac{H7}{n6}$.

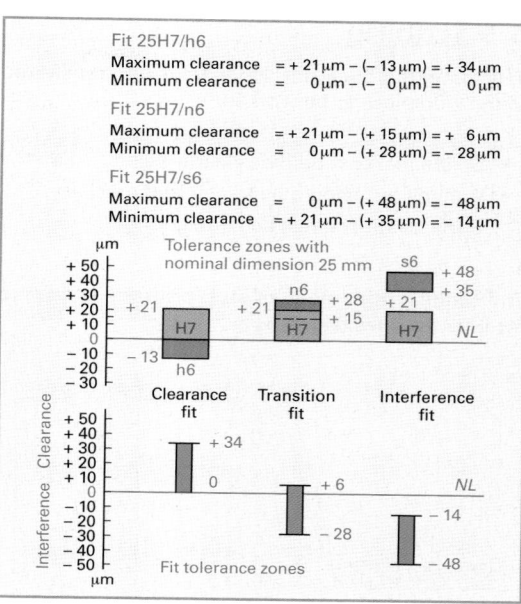

Fig. 2: Tolerance zones

6.4.6 Fit systems

The systems are the **basic hole** and the **basic shaft**.
Basic hole. The minimum dimension of the bore is equal to the nominal dimension. In other words, the lower dimensional deviation for the bore is zero. The shafts are larger or smaller by the clearances or interferences required for the extended fit **(Fig. 3)**.
Basic shaft. The maximum dimension for the shaft is equal to the specified nominal dimension; the shaft's upper dimensional deviation is zero. The bores are larger or smaller by the clearances or interferences required for the extended fit **(Fig. 3)**.

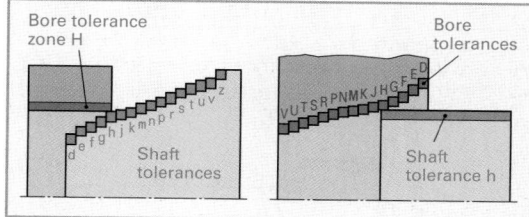

Fig. 3: Fit systems

REVIEW QUESTIONS
1 **What are the meanings of the terms nominal dimension, baseline and actual dimension?**
2 **Explain the terms tolerance limit and tolerance.**
3 **Explain the structure of the ISO tolerance code.**
4 **What parameters affect the tolerance?**
5 **What are the meanings of the terms clearance and interference?**

6.5 Scribing

In scribing the blueprint dimensions are transferred to the workpiece to be worked.

- Scribe marks should be clearly visible.
- Dimensions should be transferred precisely.
- Surfaces must never be damaged.

The **scribing iron**, made of steel or brass, is used to inscribe the lines **(Fig. 1)**.

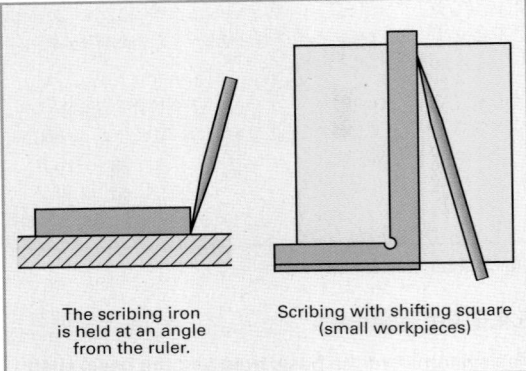

| The scribing iron is held at an angle from the ruler. | Scribing with shifting square (small workpieces) |

Fig. 1: Scribing with scribing iron

Brass scribing irons are used to inscribe scaled metal, extremely hard materials, and surfaces on which it is important to avoid fissure damage. A pencil is applied to the edge of the bead on light-alloy sheets to avoid the notching that could produce rupture during subsequent bending operations.

Pointed compasses (Fig. 2) and **beam compasses (Fig. 2)** are employed to transfer dimensions, scribe circles and mark off equal fractions of a distance.

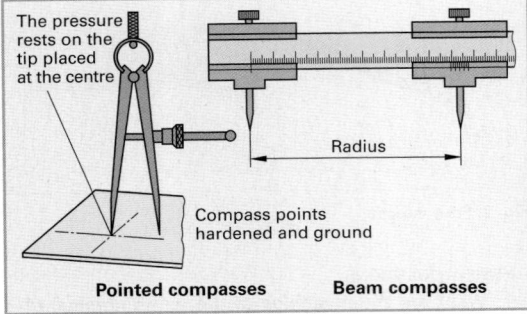

The pressure rests on the tip placed at the centre

Radius

Compass points hardened and ground

Pointed compasses   Beam compasses

Fig. 2: Pointed compasses and beam compasses

Scribing block. This tool can be used to mark lines parallel to the scribing plate at any desired height. A scale designed for fine adjustments is used for height adjustment **(Fig. 3)**.

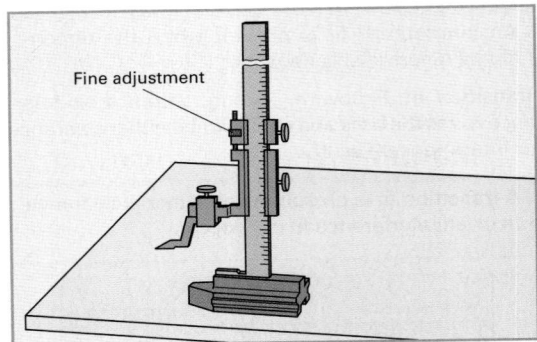

Fine adjustment

Fig. 3: Scribing block with line scale

A supplementary vertical scale is needed to adjust the height on the **parallel scripe**.

Other tools needed for scribing are the **steel rulers, shifting squares, mitre angles, flat angles** and **rulers**. Centre squares and centring bells are used to find the centre points of discs and shafts.

Scribing paint and copper sulphate can be applied to heighten the visibility of scribe marks on untreated metal surfaces. Dark materials can be marked with chalk.

The centre punch **(Fig. 4)** is used to mark centre points and scribe lines. Following machining half of the reference points should still be visible.

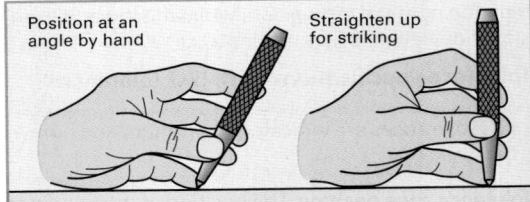

Position at an angle by hand   Straighten up for striking

Fig. 4: Using a centre punch

Working rules

- When clamping the scribing iron in the parallel scribe keep the protrusion at the tip as short as possible.
- The scribing plate should never be used to set up workpieces.
- Position the centre punch to the part surface at an angle to obtain a visible centre mark; apply it vertically to make centre marks.

REVIEW QUESTIONS
1 What is the function of scribing?
2 What are the most important scribing tools?
3 What are the advantages of brass scribing irons?
4 Why should the scribing plate not be used to set up workpieces?

7 Production engineering

7.1 Categorisation of manufacturing processes

Manufacturing and economic considerations determine the manufacturing processes, sequences and operations associated with the manufacturing and machining of workpieces.

> Production, or manufacturing, comprises all the operations which transform a workpiece from its raw state to a predetermined finished state. Prior to each operation, the workpiece is in its initial state and then in its final state.

With different manufacturing processes, the material cohesion required for the manufacture of a workpiece can first be created or, if it already exists, can be reduced or increased. The shape or form of a workpiece or the properties of materials can be changed.

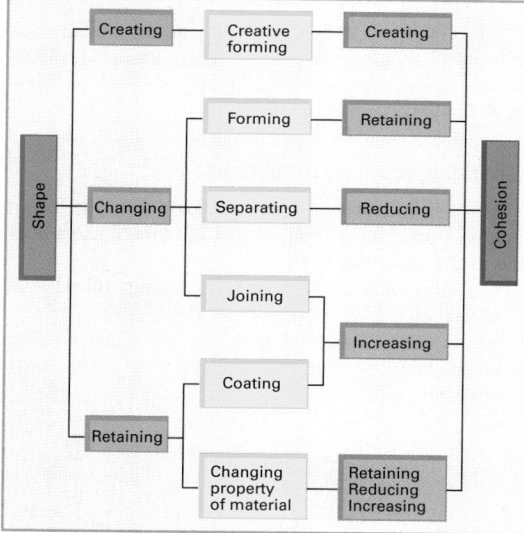

Fig. 1: Main categories of production engineering

7.1.1 Main categories of manufacturing processes

Manufacturing processes are divided into 6 main categories **(Fig. 1)**.
In the individual manufacturing processes the cohesion (property) and shape of the material can be created, changed, retained, increased or reduced.

- **Creating cohesion** involves formless substances or materials, e.g. powders, liquids, being creatively formed into geometrically defined solid bodies, e.g. by pressing, sintering, casting.
- **Retaining cohesion** involves an already formed part or workpiece being reformed during manufacture, e.g. by bending.
- **Reducing cohesion** involves the material or the workpiece parts being separated in its/their geometrical shape, e.g. by sawing.
- **Increasing cohesion** involves workpieces or materials being added, e.g. by screwing, bolting, deposit welding.

7.1.2 Subdivision of main categories

Creative forming

In the case of creative forming, a solid body with a defined shape is manufactured from a formless substance or material **(Fig. 2)**.

The forming process can be a completed operation in itself or a preliminary stage for further manufacturing processes.

Cohesion is created from the:

- Liquid state, e.g. by casting
- Paste-like or plastic state, e.g. by extruding (displacing), injection moulding
- Powdery or granular state, e.g. by pressing and heating during sintering
- Ionised state by electrolytic depositing, e.g. by electroforming

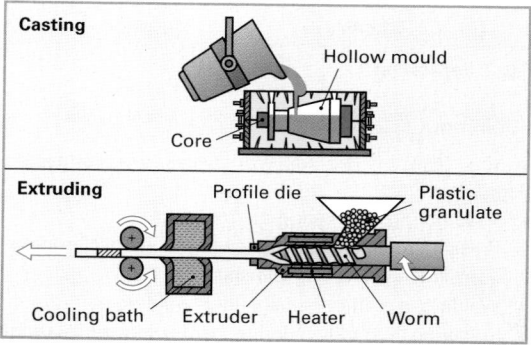

Fig. 2: Creative forming

Forming

In the case of forming, the mass and cohesion of a material are retained. The shape of a solid body (blank) is changed by plastic forming **(Fig. 1)** such as

- forming under compressive conditions, e.g. rolling, die forming
- forming under combination of tensile and compressive conditions, e.g. deep-drawing, compression
- forming under tensile conditions, e.g. widening, extending, stretch forming
- forming under bending conditions, e.g. edge folding, bending
- forming under shearing conditions, e.g. twisting, displacing.

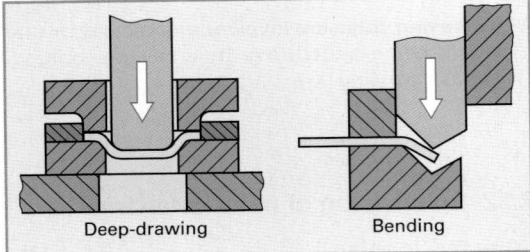

Deep-drawing   Bending

Fig. 1: Forming

Separating

In the case of separating, the shape of a solid body is changed **(Fig. 2)**. Here, the cohesion of a material is locally eliminated by:

- Severing, e.g. cropping, parting, breaking
- Machining-cutting, e.g. turning, grinding, drilling
- Removing, e.g. flame-cutting, eroding
- Disassembling, e.g. unscrewing, pressing out
- Cleaning, e.g. brushing, washing, degreasing

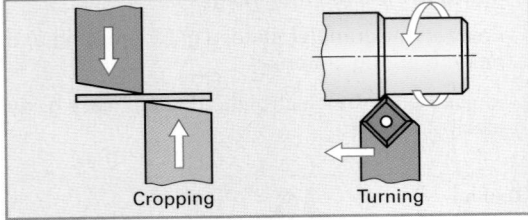

Cropping   Turning

Fig. 2: Separating

Joining

In the case of joining, workpieces are connected **(Fig. 3)** by:

- Assembling, e.g. laying on, hanging
- Pressing on and in, e.g. screwing, bolting, clamping
- Forming, e.g. riveting, clinching
- Welding, e.g. inert-gas-shielded welding
- Soldering, e.g. soft-soldering, hard-soldering (brazing)
- Bonding, e.g. cold-bonding, two-component bonding

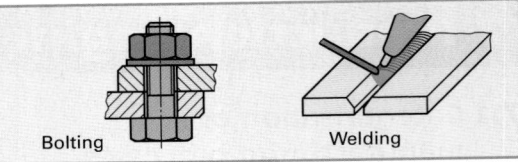

Bolting   Welding

Fig. 3: Joining

Coating

In the case of coating, a formless substance or material is applied as an adherent layer to a workpiece **(Fig. 4)**. Coating ...

- ... from the liquid or paste-like state, e.g. spray-painting, deposit welding.
- ... from the solid or powdery state, e.g. thermal spraying.
- ... by electrolytic or chemical depositing, e.g. electroplating.
- ... from the solid or granular state, e.g. whirl sintering.

Painting   Deposit welding

Fig. 4: Coating

Changing property of material

In the case of ageing, segregation or introduction of material particles, the material properties of solid bodies are changed **(Fig. 5)** by:

- Rearrangement of material particles, e.g. hardening, annealing
- Segregation of material particles, e.g. decarburising
- Introduction of material particles, e.g. carburising, nitriding

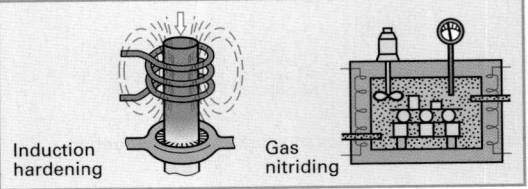

Induction   Gas
hardening   nitriding

Fig. 5: Changing property of material

REVIEW QUESTIONS
1 What are the main categories of production engineering?
2 Which manufacturing processes change the shape of a workpiece?
3 Which manufacturing processes create material cohesion?
4 Name processes which change the properties of a material.

7.2 Creative forming

> Creative forming is the manufacture of solid bodies from formless material. Cohesion is created.

Material cohesion can be created from a liquid material, e.g. during casting, or from a solid material, e.g. during sintering.

7.2.1 Casting

In the case of casting, a molten metal is cast into a mould. The melting mass fills the cavities in the mould. The original shape of the workpiece is created once the melting mass has solidified.

Table 1: Overview of forming and casting processes			
Casting in broken moulds with gravity	Casting in permanent moulds		
	with gravity	with pressure	with centrifugal force
Permanent patterns, e.g. wood, steel Broken patterns e.g. wax	Gravity diecasting, continuous casting	Pressure diecasting	Centrifugal casting

Casting with permanent pattern in broken sand mould

Patterns **(Fig. 1)** are used to produce broken moulds. The pattern serves to mould the outside contour of the workpiece. A broken mould is a casting mould which is rendered unusable after casting. Because casting metals contract as they cool, the pattern must be made approx. 0.5 % to 2 % bigger than the casting. Application: e.g. engine block made of cast iron EN-GJL-200, crankshaft made of nodular cast iron EN-GJS-700-2.

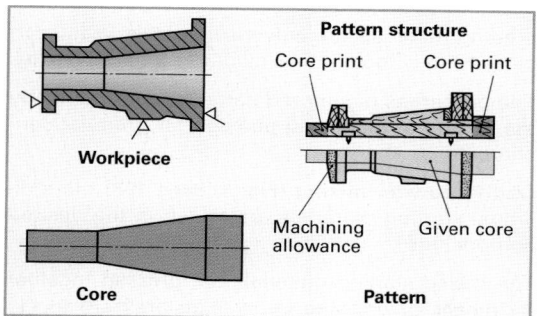

Fig. 1: Pattern and core

A core must be inserted into the mould in order to obtain a cavity. In order to cast **(Fig. 2)** the two-part pattern, the lower pattern half and the bottom box

are filled with moulding sand. The sand is compressed by tamping. The top box is placed on top of the bottom box after the latter has been turned over. Then patterns are inserted for the funnel and feeders and the moulding sand is added and tamped. When the top box is lifted off, both pattern halves and pattern parts are removed. Then chamfers and a runner are cut into the moulding sand and the core is inserted. Joining the top and bottom boxes together creates a cavity which corresponds to the shape of the workpiece.

The mould is filled through the funnel. Feeders allow air to escape during the filling process. The large feeder cross-sections enable liquid metal to continue flowing into the solidifying workpiece during cooling and this prevents blowholes. After cooling, the mould is destroyed and the casting removed.

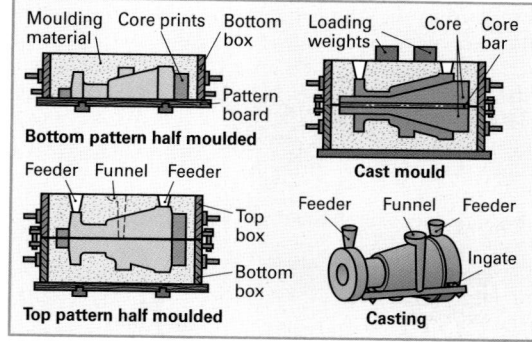

Fig. 2: Moulding

Pressure diecasting

> Pressure diecasting involves forcing metal swiftly in a liquid or paste-like state at high pressure into a permanent mould (steel mould).

Non-ferrous heavy-metal alloys (e.g. throttle-valve housings made of zinc casting alloys) and light-metal alloys (e.g. forged pistons and rims as well as housings made of aluminium or magnesium casting alloys) are often manufactured by means of pressure diecasting **(Fig. 3)**.

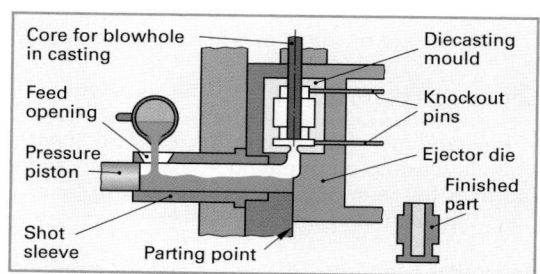

Fig. 3: Pressure diecasting

Advantages of pressure diecasting:

- Castings with maximum dimensional accuracy.
- Manufacture of finished parts possible because bores and threads can be cast. Only burrs and the funnel need to be removed.

Investment casting (pattern melt-out process)

> Investment casting is a casting process with broken (melted-out) patterns in a broken, one-part mould.

Patterns are manufactured from wax or plastic in accordance with a sample and combined into a pattern nest. The pattern nest is immersed repeatedly into a ceramic paste, covered with ceramic powder and then dried. The casting mould is burned in order to increase strength. The patterns melt out in the process and form blowholes **(Fig. 1)** for pouring in. Application e.g. piston rings, turbine wheels.

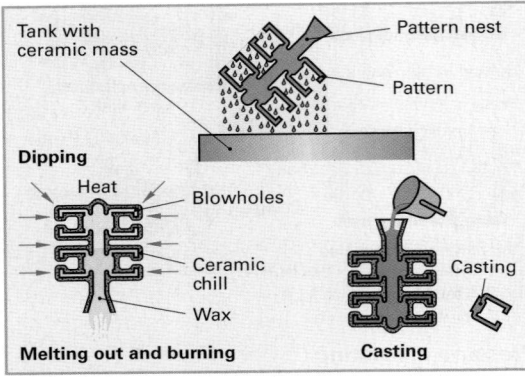

Fig. 1: Investment casting

Advantages of investment casting:

- Virtually all materials can be cast, even metals which are difficult to machine.
- For the smallest parts and low wall thicknesses.
- Very high dimensional accuracy.
- High surface quality, without burrs.
- Casting of finished parts possible, subsequent machining on locating surfaces only.
- Manufacture of complicated parts possible.

Centrifugal casting

> The melting mass is cast into a permanent mould (casting die) rotating at high speed and thrown by centrifugal force against the inner walls of the mould, where it solidifies.

Horizontal centrifugal casting. Used for example to manufacture liners and piston rings.

Vertical centrifugal casting (Fig. 2). Used to manufacture flat components such as e.g. gear wheels and pulleys.

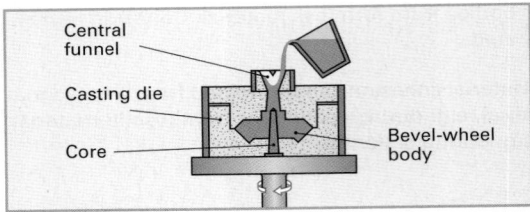

Fig. 2: Centrifugal casting

Advantages of centrifugal casting:

- Centrifugal force brings about a compressed structure of increased strength compared with casting with gravitational force.
- Structure is free of pinholes, blowholes and contaminants which have a lower density than the melting mass.

Gravity diecasting

> The melting mass is cast by gravity into permanent metal moulds (casting dies).

Compared with sand casting, the casting die produces increased surface quality. Heat dissipation can be specifically influenced. Thus, for example, when a camshaft is being cast, inserted casting dies at the subsequent bearings can effect such a rapid removal of heat that the surface is hardened. **Application** e.g. rims, gear wheels.

7.2.2 Sintering

> The manufacture of sintered parts involves the creative forming of workpieces from the solid state by compressing powders and subsequent sintering.

Starting materials are pulverised metals, metal carbides (with carbon), metal oxides or artificial resins.

The properties of sintered parts are determined by the mixing of individual powdery or granular constituents.

During **powder mixing (Fig. 1, Page 103)** the individual starting materials are mixed in the desired composition.

The mixed starting materials are pressed together by means of **pressing** under high pressure (up to 6,000 bar). In this process, the contact surfaces of the powder particles increase in size while the pore volumes decrease in size. Strain-hardening occurs at the contact surfaces. The die-formed part acquires its cohesion through adhesion of the powder particles.

Sintering is an annealing of pressed metal-powder particles, in the course of which a coalescent crystal structure is created by diffusion and recrystallisation **(Fig. 1)**.

During sintering the die-formed part acquires its final strength through the annealing of the metal powder particles under the influence of protective gas or in a vacuum. Diffusion occurs in the process, i.e. the atoms stray into neighbouring powder particles. Recrystallisation occurs at the strain-hardened points of contact. The sinter temperature is slightly below the melting temperature of the main constituent of the powder mixture. Sintering can even take place as early as during the pressing process (hot pressing). Workpieces which should be particularly high in density and strength are repressed and resintered again after sintering (double pressing, resintering).

Calibrating (Fig. 1) (subsequent treatment). If more stringent demands are placed on dimensional accuracy and surface quality, the sintered parts are repressed (calibrated) after sintering at a pressure of approx. 1,000 bar.

Types of sintered materials
Porous sintered materials are used for filters and self-lubricating friction bearings (Sint A). Starting materials are high-purity irons or iron-copper-tin alloys. Sintered friction bearings are impregnated with oil or have a lubricant reserve in the form of grease **(Fig. 3)**. They are maintenance-free and have excellent running and emergency-running properties.
Application e.g. bearing bushings of starters and water pumps.

Highly porous sintered materials (Sint AF) are, on account of their low volume ratio and their large pore volume of up to 27 %, used as filters.
Application e.g. fuel filters, gas filters.

Sintered frictional materials contain, among other things, CuSn, Pb and graphite constituents and mineral additives. They are wear-resistant and demonstrate excellent thermal conductivity and thermal resistance.
Application e.g. for linings for clutches and brakes (J 730).

Sintered materials for machined parts have as their starting materials iron, cast-iron or steel powder (sintered iron, sintered steel), to which alloyed metal powder can still be added. Heat treatment is frequently carried out in the case of sintered iron and sintered steel. Sintered steels containing carbon can be hardened and if necessary tempered.
Application e.g. for toothed-belt gears **(Fig. 2)**, shock-absorber pistons.

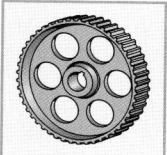 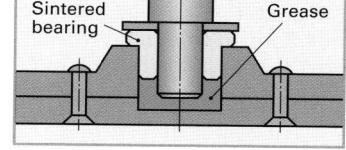

Fig. 2: Toothed-belt gear   **Fig. 3: Maintenance-free sintered bearing**

Sintered hard carbides are composite materials. Tungsten, titanium and tantalum carbides are mainly used as the starting materials; cobalt, because of its low melting point, is used as the binder. Hard carbides are for the most part brittle, susceptible to shocks and wear-resistant in their edges up to 900 °C (cutting edge remains hard).
Application e.g. for cutting materials, wearing parts.

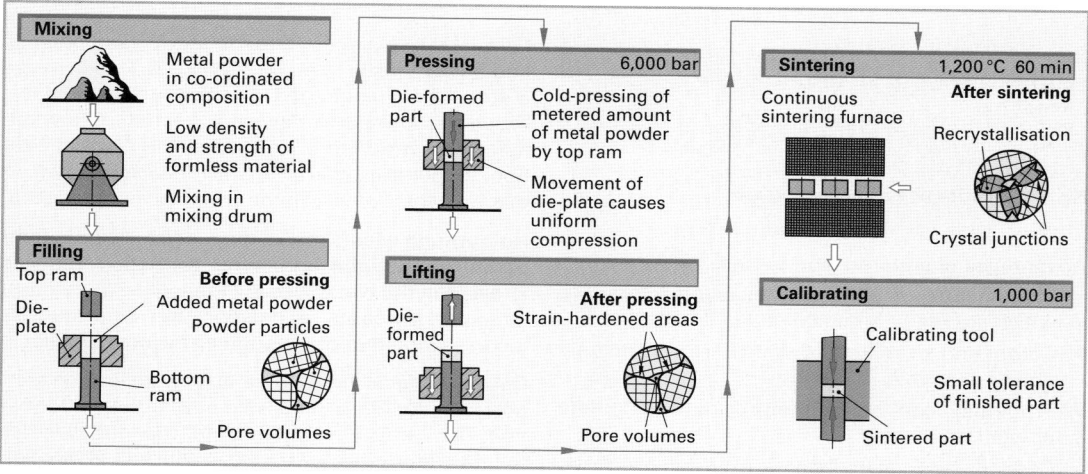

Fig. 1: Procedures for single pressing, sintering and calibrating

Oxide-ceramic composite materials. They consist primarily of special fused alumina (Al_2O_3), metal oxides (MgO, ZrO_2), carbides (TiC) and ceramic binders.

Application e.g. indexable inserts (**Fig. 1**). Because of their great hardness and resistance to wear, they enable very hard materials to be machined.

Fig. 1: Oxide-ceramic indexable inserts

Permanent magnetic materials are produced by sintering from iron, aluminium, nickel and cobalt (ALNICO), e.g. for starters.

Sintered forgings (Fig. 2). Sintered blanks are transformed from alloyed steel powders into their final shapes in the forging tool.

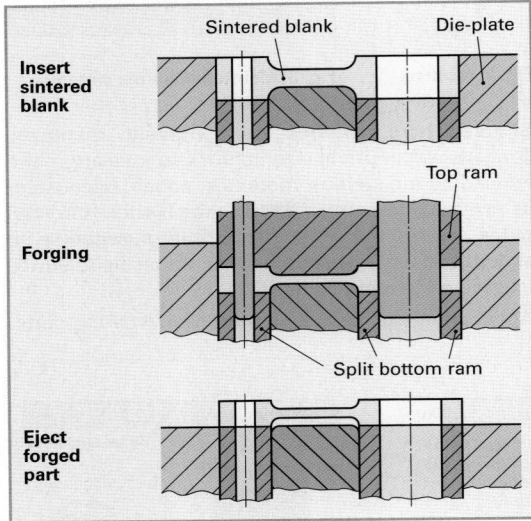

Fig. 2: Sinter forging a connecting rod

Advantages of sintered workpieces:

- Manufacture of parts with exacting tolerances which are ready for installation.
- Combining of materials which are not or only poorly alloyable.
- Properties can be influenced by appropriate powder mixture.
- Cost-effective manufacture of mass-produced parts since tools are not subject to wear.

7.3 Forming

Forming is a manufacturing process in which a plastically deformable, solid body acquires a new shape through the influence of external forces.

The prerequisite for each instance of forming is the plastic deformability of the material in question. As a result of the influence of external forces, the workpiece is subject to elastic deformation and, with higher forces, to plastic deformation. The mass and cohesion of the material are retained in this process, but the shape is changed, i.e. reformed.

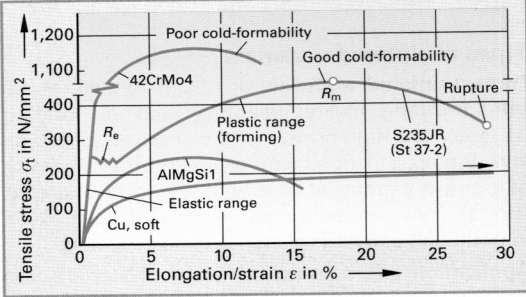

Fig. 3: Forming ranges in stress-strain diagram

Forming takes place in a range above yield point R_e and below tensile strength R_m (**Fig. 3**). This involves a structural change, and deformation takes place. When the formed workpiece is heated, a new, undeformed structure is created at the recrystallisation temperature characteristic for each material. In this way, the stresses in the workpiece are reduced.

Hot forming occurs above the recrystallisation temperature. The new structural formation reduces stresses in the workpiece. As the temperature rises, strength decreases, elongation and deformability increase, and the forming forces become smaller.

Cold forming occurs below the recrystallisation temperature, therefore there is no new structural formation. The partially great structural changes give rise to increased strength and reduced elongation (strain-hardening). The risk of cracking increases.

7

Advantages of a formed part:

- Fibre orientation **(Fig. 1)** is retained and therefore the notch effect is reduced.
- Strength can be increased considerably during cold forming.
- Low-loss material processing since the unmachined parts are frequently closer to the finished parts, little waste.
- Shorter production times compared with metal-cutting forming.
- Possibility of manufacturing parts ready for installation with high surface quality and small dimensional tolerances.

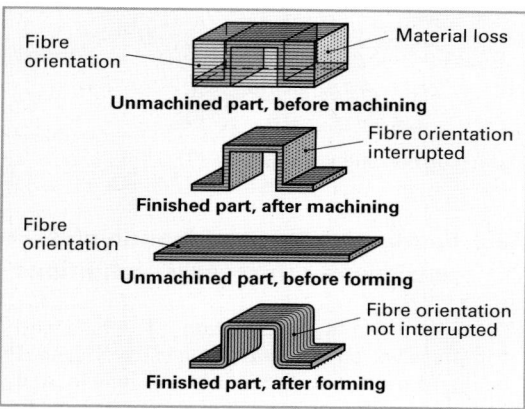

Fig. 1: Fibre orientation

Subdivision of forming processes

The processes can be subdivided according to the
- temperature (cold, hot forming)
- workpiece shape (massive forming, pressing)
- type of stress **(Fig. 2)** during forming

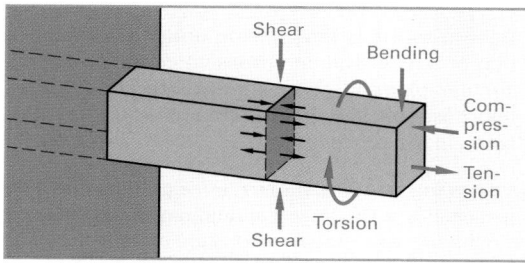

Fig. 2: Stresses in workpiece cross-section

According to the stressing of the workpiece cross-section, there are the following different processes:

- **Forming under bending conditions,** e.g. edge folding, bordering, beading, profiling
- **Forming under combination of tensile and compressive conditions,** e.g. deep-drawing

- **Forming under tensile conditions,** e.g. patent flattening **(Fig. 3)**
- **Forming under shearing conditions,** e.g. twisting **(Fig. 4)**
- **Forming under compressive conditions,** e.g. forging, rolling

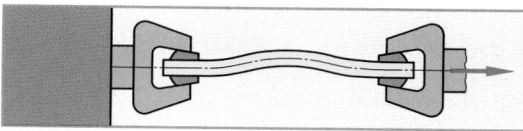

Fig. 3: Forming under tensile conditions, patent flattening

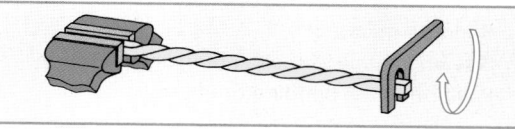

Fig. 4: Forming under shearing conditions, twisting

7.3.1 Forming under bending conditions

Forming under bending conditions (bending) is the forming of a solid body where the plastic state is primarily brought about by bending stress.

Prerequisites for bending:
- Material must be sufficiently extensible.
- Limit of elasticity of material must be exceeded.
- Breaking limit of material must not be reached.

During bending, a change to the workpiece cross-section occurs at the bending point. Specific bending radii must therefore not be undershot. The cohesion of the material is retained; only a part of the workpiece, the bending zone, is deformed.

Bending operation

When a workpiece is bent, the outer fibres are stretched (tensile stress) while the inner fibres are upset (compressive stress). In between the outer and inner fibres is a stress-free, neutral fibre, the length of which remains unchanged **(Fig. 5)**.

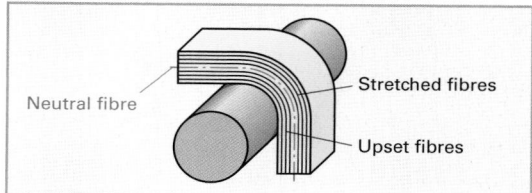

Fig. 5: Fibre orientation during bending

The deformation is elastic in the vicinity of the neutral fibre. In this way, the workpiece springs back slightly. This resilience must be taken into consideration during bending. When bending sheet metal, it is important to pay attention to the grain on account of the danger of cracking **(Fig. 1)**.

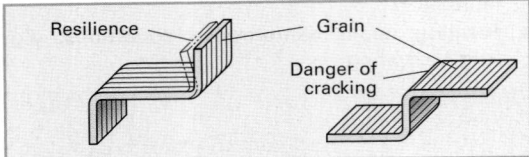

Fig. 1: Grain during bending

The bending force is dependent on:
- Extensibility and temperature of the material
- Bending radius
- Size and shape of the bending cross-section
- Position of the bending or neutral axis

Bending processes
- **Swage bending.** The workpiece is pressed with a punch into the bending swage.
- **Bending (Fig. 2)** with bending press.

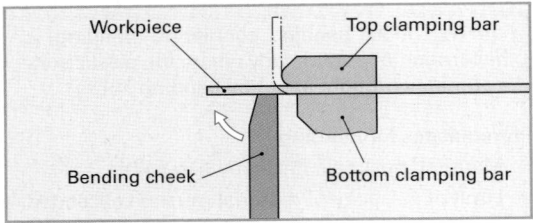

Fig. 2: Bending

- **Edge folding (Fig. 3)** produces a small bending radius.
- **Rounding** with bending rollers produces a large bending radius.
- **Bordering** serves to bend an edge on sheet metal.
- **Beading** stiffens a metal plate.
- **Profiling.** Metal strips are given their profiles during rolling.

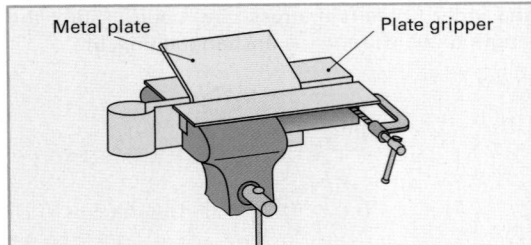

Fig. 3: Edge folding

Bending pipes
When pipes and other hollow sections are bent, in most thin-walled profiles the stretched zone (ten-

sile stress) and the upset zone (compressive stress) are very close to each other. When the workpiece cross-section is reduced, the workpiece can easily buckle. Pipes can be bent cold when they are filled with dry sand or a coil spring. The ends are sealed. The bending radius should not be less than three times the pipe diameter. In the case of lengthwise welded pipes, the seam must always be located in the neutral zone otherwise it will split open. Supporting the pipes laterally in the bending apparatus **(Fig. 4)** renders unnecessary the time-consuming need to fill the pipes prior to the bending process.

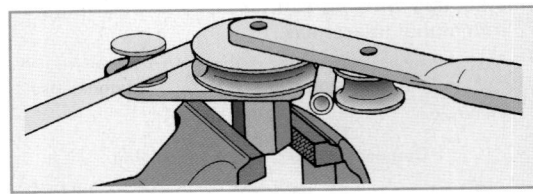

Fig. 4: Pipe-bending apparatus

7.3.2 Forming under combination of tensile and compressive conditions

Forming under a combination of tensile and compressive conditions involves workpieces being formed simultaneously by tensile and compressive forces.

There are the following different processes:
- Deep-drawing, e.g. of body sheets
- Compressing, e.g. compressing by means of a rotating shape
- Stripping, e.g. pipes, profiles, wire drawing

Deep-drawing

Deep-drawing is the forming under a combination of tensile and compressive conditions of a circular blank in one or more operations. The thickness of the metal remains roughly the same in the process.

Free-flowing materials such as e.g. deep-drawing sheet steel (e.g. DC 03) and aluminium sheets are used for deep-drawing. Body sections are manufactured.

Deep-drawing process
The flat circular blank **(Fig. 1, Page 107)** is placed on the drawing die and clamped in place by the blank holder. The drawing punch is now pressed downwards and draws the circular blank over the drawing edge into the opening of the drawing die. The material is subjected by the forming force to tensile and compressive stress. The blank holder presses the metal plate onto the drawing die and thereby

prevents the formation of wrinkles. The width of the drawing gap is slightly larger than the thickness of the metal plate. Lubricants reduce friction between the metal plate and the drawing die. When the deep-drawing process is finished, the upward-moving drawing punch releases the workpiece.

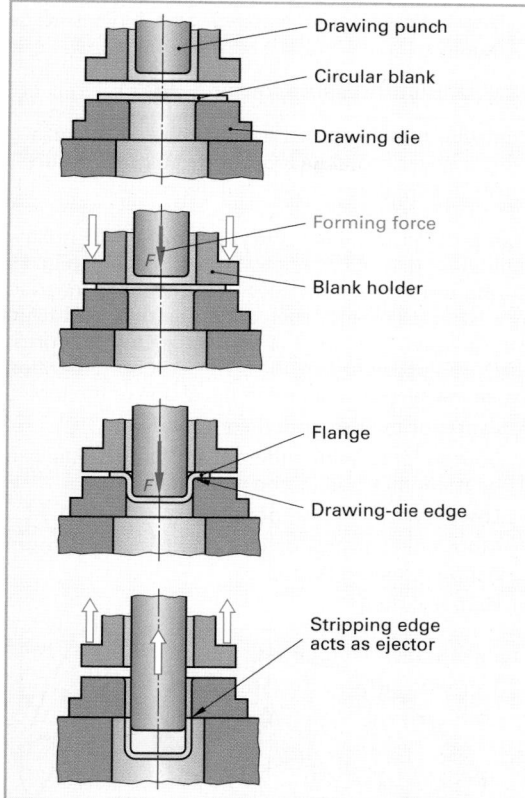

Fig. 1: Deep-drawing process

7.3.3 Forming under compressive conditions

> Forming under compressive conditions is the forming of a workpiece by compressive forces.

There are the following different processes:
- **Rolling (Fig. 2).** Sections, sheets, pipes and wires are produced between rotating rollers.
- **Extrusion.** In the case of extrusion moulding **(Fig. 3)**, heated materials are pressed through a die-plate and shaped into sections. In the case of **impact extrusion (Fig. 4)**, materials are pressed with a punch and die-plate into solid or hollow bodies.
- **Free forming** (forging).
- **Die forming** (drop forging).

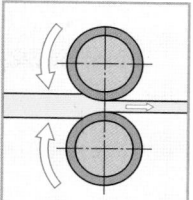

Fig. 2: Rolling

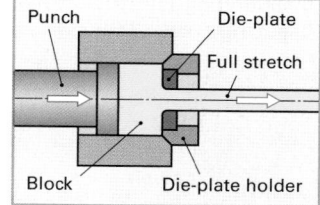

Fig. 3: Extrusion moulding

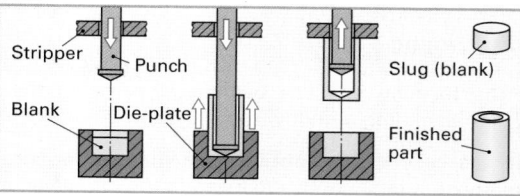

Fig. 4: Impact extrusion

Forging (free forming and die forming)

> Forging is the forming under compressive conditions of hot metals in a plastic state.

Forging involves forming workpieces without cutting for the most part while glowing by means of beating or compression. Upsetting and stretching the material changes the structure, in the course of which the fibre orientation is not interrupted **(Fig. 5)**. A tight structure and favourable fibre orientation increases the strength and stability under load of the forgings.

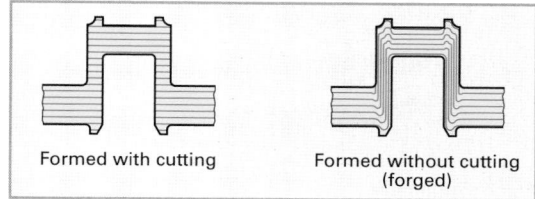

Fig. 5: Fibre orientation

Free forming

> In the case of free forming, the material can be freely displaced between an anvil and a hammer or press.

There are the following different processes:
- **Upsetting (Fig. 6).** Height gets smaller, cross-section gets bigger
- **Stepping (Fig. 7).** Forging a sharp-edged step

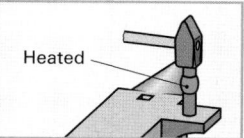

Fig. 6: Upsetting

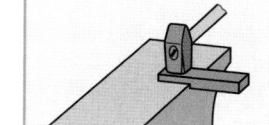

Fig. 7: Stepping

- **Extending** (stretching, **Fig. 1**). Extending by reducing cross-section

Fig. 1: Extending (stretching)

Die forming

Die forming involves beating or pressing the hot blank into a hollow mould (die).

In the case of die forming, the material is completely or significantly enclosed in the die while, in the case of free forming, it can be freely displaced in all directions.

Drop forging. Dies (**Fig. 2**) are usually made up of two halves, the upper die and the lower die. The cavities correspond to the shape of the finished drop forging. The workpiece is formed for the most part by repeated beating from the blank hot from forging. The volume of the blank here is somewhat greater than the volume of the finished part. This ensures that the hollow mould is filled completely. The burr created acts as a buffer and prevents hard beating of the die parts. The deformation rate is high.

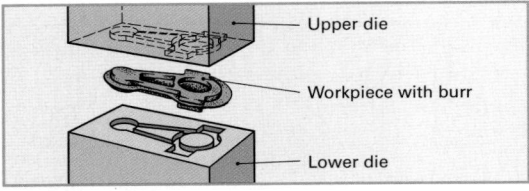

Fig. 2: Forging die

Pressure forging (Fig. 3). The forming force is applied by a forging press. The accuracy of manufacture is high thanks to the exact guidance of the upper and lower parts. The deformation rate is dependent on the material. It is, for example, lower for titanium than for steel.

Application: Titanium connecting rods for use in motor racing.

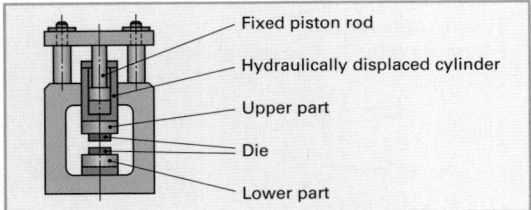

Fig. 3: Forging press

Advantages of (re)forming over metal-cutting shaping:

- Increased strength of workpieces
- Workpiece savings due to little waste
- Efficient production with high production numbers
- Good dimensional stability and high surface quality

Internal high-pressure forming (IHPF)

Internal high-pressure forming involves pressing a sheet-metal section under high pressure from the inside against a hollow mould.

This process facilitates the manufacture of hollow sections, e.g. roof frame of a light-alloy body **(Fig. 4)**. During production, a sheet-metal section is inserted in a two-part mould, sealed by cylinders and filled with a liquid. The liquid is then subjected to a pressure of approx. 1,700 bar. The forces generated cause the section to adapt to the shape of the tool.

Advantages over extruded components:

- Components with different cross-sections can be manufactured in one work step.
- Lower weight of components.
- More efficient production of expensive components in one work step.
- High accuracy.

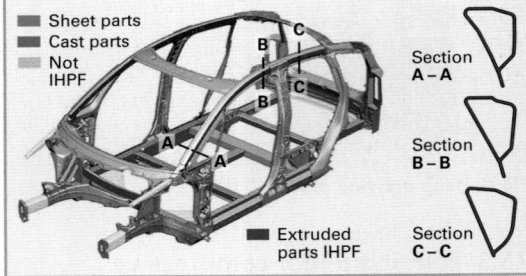

Fig. 4: Use of internal-high-pressure formed sections

Working rules for hammer forging
- Read and comply with the heat-treatment instructions of the material suppliers.
- Hold the aboutsledge close to the body. Beating out prohibited.

REVIEW QUESTIONS

1. **What forming processes are there?**
2. **What are the advantages of forming?**
3. **What prerequisites must be in place for bending?**
4. **What takes place in a deep-drawing process?**
5. **What do you understand by forging?**
6. **What are the advantages of die forming over cutting-shaping?**

7.3.4 Straightening

> **Straightening** is the elimination of unwanted deformations to re-establish the nominal shape of semi-finished products (e.g. sectional bars) and finished parts (e.g. vehicle frames, fenders/wings).

The prerequisite for straightening is the plastic deformability of the material in question. Straightening can be carried out while the material is cold or hot. Straightening is performed by **forming** (bending, stretching, upsetting, twisting) or by **thermal action** (flame-straightening).

Straightening by forming

Bent semi-finished products and workpieces with smaller cross-sections can be straightened on a straightening plate. They are placed with their hollow sides down **(Fig. 1)** on the straightening plate and straightened in small increments with a hammer.

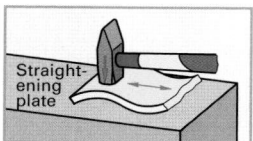

Fig. 1: Straightening flat rolled steel

Fig. 2: Straightening by stretching

The bulge is struck with the hammer. A wood, rubber or plastic hammer is used for thin panels and softer materials. Workpieces made of harder materials are straightened by stretching **(Fig. 2)**. Here the workpiece is stretched by closely concentrated strikes with the peen on its short (hollow) side. Parts which are bent edgewise are likewise straightened by stretching strikes in order to relengthen the side shortened during bending **(Fig. 3)**.

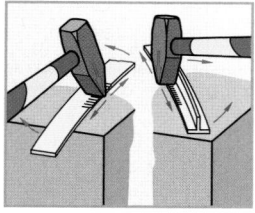

Fig. 3: Straightening profile sections

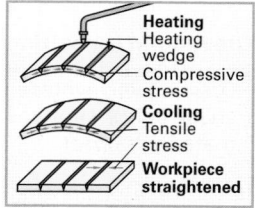

Fig. 4: Flame-straightening

Straightening by thermal action

In the case of flame-straightening **(Fig. 4),** the overlong side of the workpiece is lengthened further by means of heating wedges. High compressive stresses are created by the expansion of the heated metal. The metal in the heating wedge, which has turned plastic (pasty), contracts as it cools. The tensile stresses that arise upset the workpiece over its longer side to such an extent it straightens up.

Heat treatment can cause loss of strength on account of the structural transformation in the sheet metal.

7.3.5 Sheet-metal working processes

Metal sheets are manufactured by hot or cold rolling. The semi-finished products are delivered as strips (coils) or as plates.

Metal sheets are subsequently worked, for example, by:

- Forming under combination of tensile and compressive conditions (deep-drawing, compression)
- Forming under tensile conditions (stretch-forming)
- Forming under bending conditions (bending, swage bending)

Bending metal sheets

Grain. Metal sheets during manufacture are stretched by rolling predominantly along the grain and therefore acquire a fibre-like material structure. Sheets which are bent parallel to the grain are exposed to the risk of cracking. For this reason, sheets must wherever possible be bent vertically or diagonally to the grain **(Fig. 5)**.

Resilience. When a metal sheet has been bent, it is subject to a certain resilience, which is dependent on the material, grain, sheet thickness, bending angle and bending radius. The sheet must first be "overbent" so that the desired bending angle is obtained. The resilience angle **(Fig. 5)** is roughly 1 % ... 3 % of the bending angle.

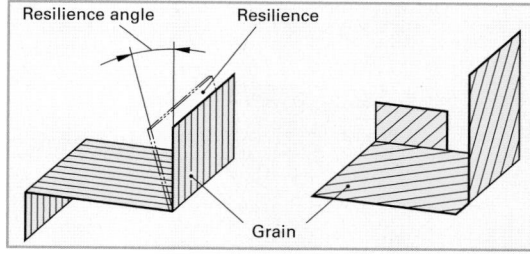

Fig. 5: Grain, resilience

Bending radius. When bending metal sheets, it is essential in order to avoid cracking not to drop below the minimum bending radii. Small bending radii give rise to large deformations and thereby high stresses in the sheet. The minimum bending radius is dependent on the material, grain and sheet thickness. The minimum bending radii of the most important metal sheets are standardised.

Blank lengths for bent sheet parts are calculated using the neutral fibre.

Edge folding

> Edge folding is the sharp-edged bending of a metal sheet along a straight edge with a very small bending radius.

Edge folding is mainly used to produce profile sections. It can be performed in a vice or on an anvil using a clamping rail **(Fig. 1)** or with a folding press.

In the case of forming and bending in a vice **(Fig. 1)**, the metal sheets are bent using bending shims. Protective hammers are used to protect the sheets against damage.

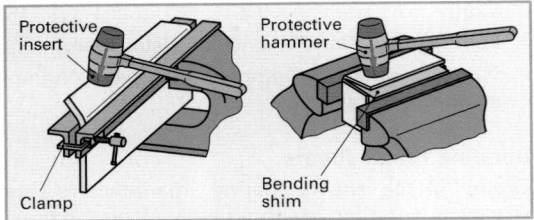

Fig. 1: **Edge folding with clamping rail, forming and bending**

In the case of bending with a bending press **(Fig. 2)**, the sheet is clamped between top and bottom clamping bars and bent with the swivelling bending cheek. The bending radius and shape of the bent metal sheet is produced by the shape and dimensions of the rail inserted in the top clamping bar.

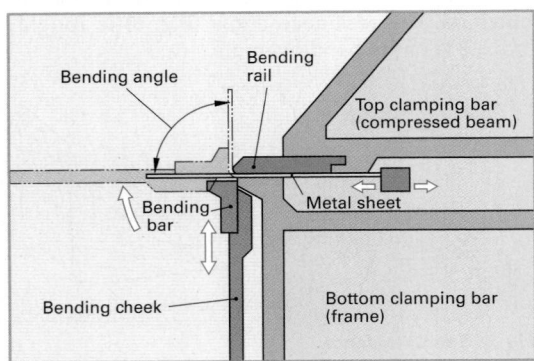

Fig. 2: **Bending**

Rounding

> Rounding is the bending of a metal sheet along a straight edge with a large bending radius or with large bends/curves.

Rounding by hand can be performed in a vice, on a stake **(Fig. 3)**, on the beak of an anvil or over a pipe.

Fig. 3: **Rounding by hand**

Curving (stretching)

> Curving involves specific parts of a metal sheet being stretched (extended) by hammering.

In the case of curving, the metal sheet is hammered and stretched on its edge. In the course of this process, the sheet thickness is reduced and the sheet bends **(Fig. 4)**.

The curving hammer or the peen is only permitted to come into contact with the edge of the sheet and should always point to the centre of the bend or curve.

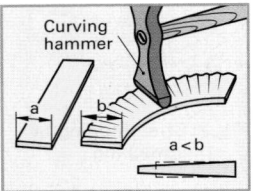 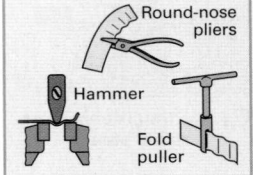

Fig. 4: **Curving**   Fig. 5: **Producing corrugations, drawing in**

Curving applications:
- Creating outer edges on round parts
- Rounding, bending or straightening profile sections
- Expanding pipes to facilitate interconnection

Drawing in (upsetting)

> Drawing in involves specific parts of a metal sheet being shortened (drawn in), where the material is upset.

In the case of drawing in, metal sheets are thickened at their edges by upsetting the material and thereby shortened. The sheet section to be shortened is turned into corrugations, which are then upset by specific hammering. In the case of drawing in by hand, the corrugations are produced with round-nose pliers, with a fold puller or using a vice **(Fig. 5)**.

Drawing-in applications:
- Creating inner edges on round parts
- Rounding or straightening profile sections (possibly in conjunction with curving)
- Drawing in pipes to facilitate interconnection

Smoothing

In the case of smoothing, the surfaces of deformed metal sheets are smoothed, small irregularities are eliminated, and the shape and appearance of sheet parts are improved.

Smoothing tools must have smooth, non-scored and wherever possible polished surfaces. Smoothing out **(Fig. 1)** involves delivering hammer blows to the inside of an arched part, while smoothing off **(Fig. 2)** involves delivering hammer blows to the outside of an arched part.

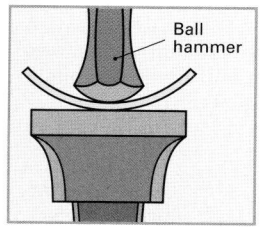 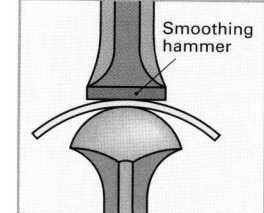

Fig. 1: **Smoothing out**   Fig. 2: **Smoothing off**

Chasing

Chasing involves forming metal sheets using the different metalworking processes by hand or with machines.

Chasing is primarily performed by stretching and upsetting the metal sheet and by smoothing and hammering. If the strain-hardening of the sheet becomes too great as a result of chasing, it must be eliminated by process-annealing (recrystallisation). Chasing **(Fig. 3)** is used to produce individual parts for effecting repairs to bodywork.

Raising. The metal sheet is shaped on a hard base in order to create flat arches **(Fig. 4)**.

 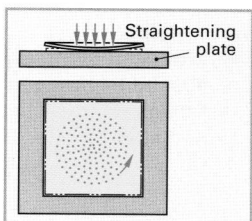

Fig. 3: **Chasing**   Fig. 4: **Raising**

Stepping

Stepping involves the edge strip of a metal sheet being stepped (bent in) by approximately one sheet thickness.

Stepping **(Fig. 5)** is primarily a preliminary task in order to obtain a flush outer surface for overlapping metal sheets, e.g. in the event of sectional repairs to bodywork. Stepping results in a stiffening of the metal sheet. Stepping a metal sheet by hand is performed using stepping pliers **(Fig. 6)**. The pliers illustrated can also be used to punch holes for welding spots.

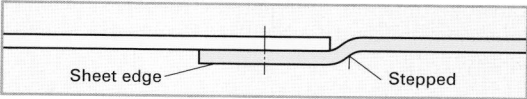

Fig. 5: **Stepping**

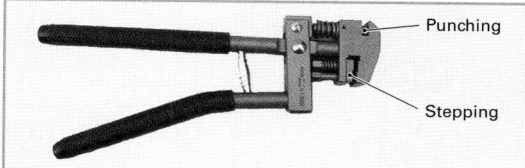

Fig. 6: **Combined stepping/hole-punching pliers**

Clinching

Clinching involves fashioning out within a metal sheet a partial surface which usually runs parallel to the sheet plane.

Clinching **(Fig. 7)** produces local stiffening of metal sheets, e.g. engine bonnet **(Fig. 8)**, rear lid, car doors.

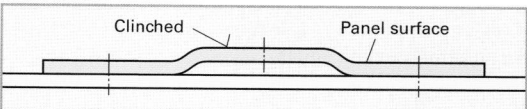

Fig. 7: **Clinching**

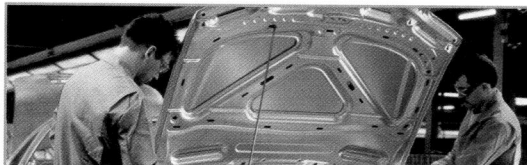

Fig. 8: **Bonnet**

Bordering (edge-forming)

Bordering involves rebending metal sheets cut in curves at their edges for the most part at right angles into narrow edges. The edges are called borders.

Edge folding is used to bend edges on straight-cut metal sheets. Bordering involves forming the edges of metal sheets and serves to stiffen the sheets or as a preliminary task for creating a seam by means of folding, riveting, soldering or welding.

7

Bordering in. In the case of bordering in **(Fig. 1)**, the metal sheet is bent inwards to an edge. Because there is too much material present, the sheet must be upset (drawn in) to create the inside edge.

Bordering out. In the case of bordering out **(Fig. 2)**, the metal sheet is bent outwards to an edge. Because there is too little material present, the sheet must be stretched (curved) to create the outside edge.

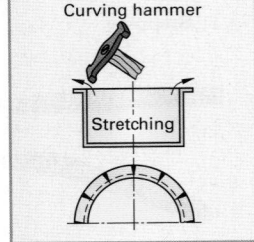

Fig. 1: Bordering in   **Fig. 2: Bordering out**

Strengthened edges. Strengthened edges **(Fig. 3)** are used to increase the strength of metal sheets and sheet edges. The risk of injuring oneself on sharp sheet edges is reduced.

Welts. The sheet edge is bent once or several times.

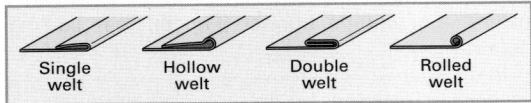

Fig. 3: Strengthened edges

Beading (strengthened metal sheets)

Beading involves forming primarily linear, groove-shaped parts of a metal sheet from its flat sheet surface.

Beads **(Fig. 4)** serve to strengthen metal-sheet surfaces, e.g. floor panels or roof surfaces.

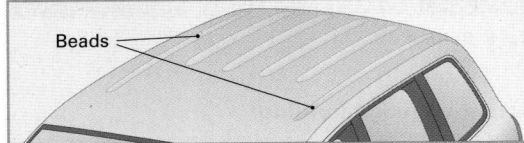

Fig. 4: Beads in roof surface

Folding

Folding is the joining together of metal sheets by hooking and then pressing the sheet edges together.

Folds are produced by bending by hand (edge folding, bordering) or with a machine **(Fig. 5)**.

Because of the high levels of strain to which the material is subjected, only easily workable and tough metal sheets may be used for folding.

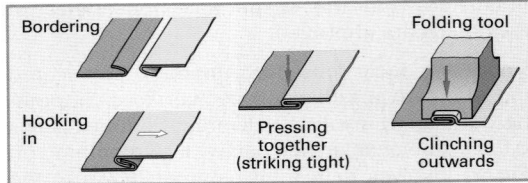

Fig. 5: Making a folded seam connection

Lock-forming is the joining together of metal sheets by interconnecting and then alternately bending or twisting the metal tabs.

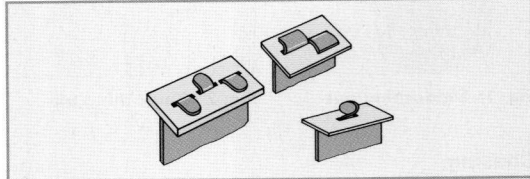

Fig. 6: Lock-form connection

Slits and tabs **(Fig. 6)** are manufactured using special machines. No additional material is required for joining. The joint is established only from an equivalent material.

Working rules
- When carrying out sheet-metal work, make sure that personal protective equipment, e.g. gloves, goggles, ear defenders, are provided and worn.
- Wear protective gloves when unloading and transporting metal plates.
- Deburr edges of metal sheets before metalworking.
- When drilling metal sheets on machines, it is absolutely essential to grip the metal sheets securely and secure them against being carried along by the drill bit.
- Do not wear gloves when drilling.

REVIEW QUESTIONS

1   **What must be borne in mind when metal sheets are being bent?**

2   **On what is the resilience of a metal sheet dependent during bending?**

3   **Why is it imperative not to drop below the minimum bending radius?**

4   **What is the advantage of stepping in the case of overlapping metal sheets?**

5   **What do you understand by clinching?**

7.4 Separating by cutting

Cutting is the mechanical separation of material particles. Here, the cohesion of a material is locally eliminated.

The processes for cutting-shaping are distinguished according to cutting motion and cutting-edge geometry. The cutting motion can be executed by the tool or the workpiece.

Table 1: Cutting-shaping (selection)

Cutting shape	Cutting motion	
	Linear	Circular
• Geometrically defined	Chiselling, filing, scraping, sawing (hacksaw)	Drilling, countersinking, reaming, sawing (circular saw), turning
• Geometrically undefined	Belt grinding, surface lapping	Parting by grinding, cylinder honing

The tools used for cutting-shaping remove chips from the material with their cutting edges **(Fig. 1)**.

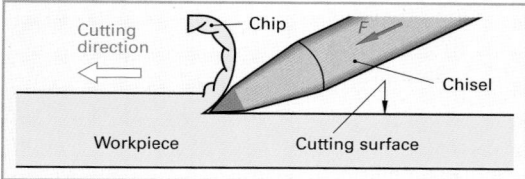

Fig. 1: Chip formation

The following four basic requirements must be satisfied during the cutting operation:
- The cut surface should become as flat as necessary.
- The cutting time should be as short as possible.
- The force exerted on the tool should be as small as possible.
- The life of the tool should be as long as possible.

7.4.1 Basics of cutting-shaping

The basic shape of the tool cutting edge for cutting-shaping is the wedge (cutting wedge).

Surfaces and angles on cutting wedge (Fig. 2)
Face is the surface of the wedge at which the chip is removed.
Flank is the surface of the wedge opposite the workpiece surface created (cutting surface).
Clearance angle α is the angle between the wedge and the workpiece surface. If the clearance angle is too small, the back of the wedge will rub on the workpiece surface.

Wedge angle β is the angle of the cutting wedge. It is formed by the flank and face on the cutting wedge.

Soft materials facilitate small wedge angles. Hard materials require large wedge angles.

Cutting angle γ is the angle between the face and an imaginary line perpendicular to the cutting direction. The chip moves along at this angle. The cutting angle can be positive or negative.

Cutting angle positive: cutting effect
Cutting angle negative: scraping effect

In the event of a negative cutting angle, the amount of material removed is minimal on account of the scraping effect of the tool **(Fig. 3)**. Clearance, wedge and cutting angles are always governed by:

$$\alpha + \beta + \gamma = 90°$$

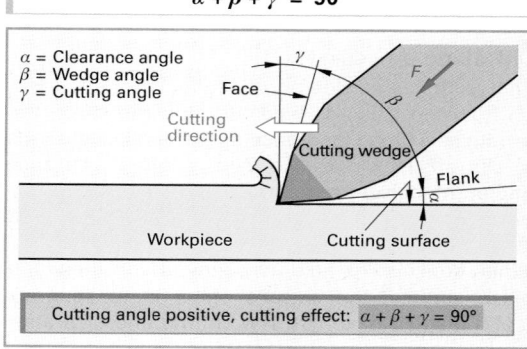

α = Clearance angle
β = Wedge angle
γ = Cutting angle

Cutting angle positive, cutting effect: $\alpha + \beta + \gamma = 90°$

Fig. 2: Surfaces and angles on cutting wedge

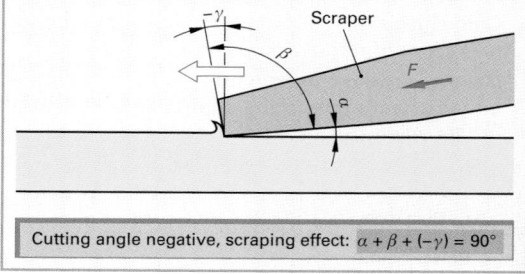

Cutting angle negative, scraping effect: $\alpha + \beta + (-\gamma) = 90°$

Fig. 3: Removal of material by scraping

7.4.2 Cutting-shaping by hand

7.4.2.1 Chiselling

A chisel serves to remove chips and to separate.

The **wedge angle** β of the cutting edge is between 40° and 70°. The optimal figure to choose for cutting medium-hard steel is roughly 60°.

A chisel has three distinct components: cutting edge, head and shank **(Fig. 1)**. The shank is rounded on its narrow sides or features, for instance, a square cross-section so that it rests easily in the hand. The head is tapered and crowned.

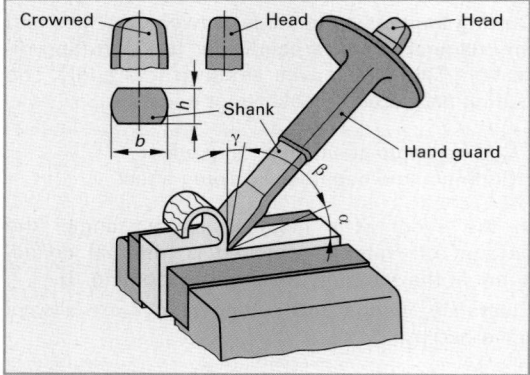

Fig. 1: Held chisel position

Chiselling process (Fig. 1)

The cutting and clearance angles are dependent on how the chisel is held. Holding the chisel flat produces a small clearance angle. The chisel tends to slip off the workpiece. With a larger clearance angle, the chisel penetrates deeper into the workpiece. The chip becomes thicker and the required cutting force greater.

When the chisel is held perpendicular to the workpiece surface, the chisel's effect is to separate (split) instead of to cut. The cutting and clearance angles when separating are 0°.

Chisel types (selection, Fig. 2)

- **Cold chisels** have a broad, straight cutting edge and serve to remove chips and to separate.
- **Groove-cutting chisels** with their narrow cutting edges at an angle to the shanks serve to chisel out narrow grooves.

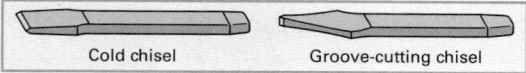

Cold chisel   Groove-cutting chisel

Fig. 2: Chisel types

Chisel attachments (Fig. 3)

Chisel attachments can be used in pneumatic chisel hammers to separate bodywork parts, release spot welds, cut and separate exhaust silencers and connections, and shear off rivets.

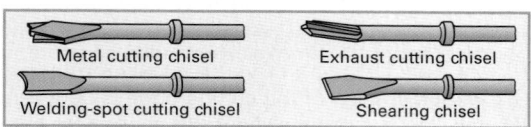

Metal cutting chisel   Exhaust cutting chisel

Welding-spot cutting chisel   Shearing chisel

Fig. 3: Chisel attachments

Working rules

- Use only chisels with a flawless cutting edge and a perfect hammer.
- The chisel head must not show any traces of burrs.
- Wear protective goggles and gloves; use face shields/protective screens to provide protection against flying chips and splinters.
- If possible, use a chisel equipped with a hand guard.
- Always keep your eye on the chisel cutting edge when chiselling.

1 How does the position in which the chisel is held affect the clearance and cutting angles?
2 On what is the wedge angle of the chisel cutting edge dependent?
3 What are the differences between cutting and separating?
4 What working and safety rules must be observed when chiselling?

7.4.2.2 Scraping

Scraping is cutting with a scraping tool to change the workpiece surface by removing the smallest material particles.

Scraping is used on metal workpieces to obtain a smooth, non-scored and uniformly bearing surface, as is usually required for sealing, sliding and guide surfaces.

Blunt and drawing scrapers are used for flat surfaces, while three-square and hollow-ground scrapers are used for curved surfaces **(Fig. 4)**.

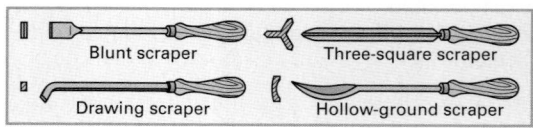

Blunt scraper   Three-square scraper

Drawing scraper   Hollow-ground scraper

Fig. 4: Scraper types

By holding the scraper correctly, the user obtains a negative cutting angle and thereby the proper scraping effect **(Fig. 5)**.

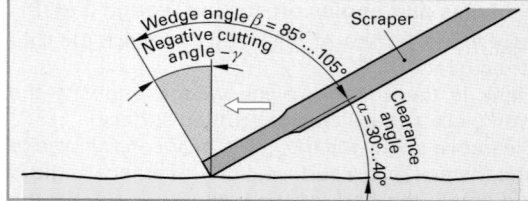

Fig. 5: Angles on blunt scraper – held scraper position

7.4.2.3 Sawing

> Sawing is cutting with a multitooth tool of small cutting width and with geometrically defined cutting wedges (saw teeth).

Sawing is used to
- separate materials or workpieces
- cut grooves or slits

How a saw works (Fig. 1)

The saw blade consists of numerous consecutive chisel-like cutting edges which engage the workpiece in succession and cut small chips. The cutting spaces (tooth gaps) collect the chips and remove them from the cut.

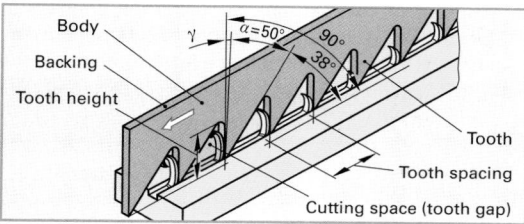

Fig. 1: How a saw works

In order to prevent the saw blade from jamming and to enable it to cut cleanly, the saw teeth are corrugated, crossed or upset **(Fig. 2)**.

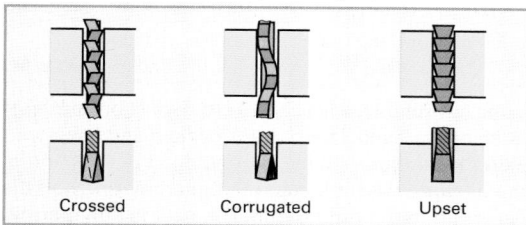

Fig. 2: Clean cutting of saw blade

Tooth spacing

The tooth spacing is the distance from tooth tip to tooth tip on the saw blade.

$$\text{Tooth spacing} = \frac{\text{Number of teeth}}{\text{Reference length}} = \frac{\text{Number of teeth}}{1 \text{ inch}}$$

1 inch = 25.4 mm

Table 1: Tooth spacing

Tooth spacing	Teeth Inch	Application
Coarse	... 16	Aluminium, copper, structural steel
Medium	... 22	Structural steel, cast iron, Cu-Zn alloys (brass)
Fine	... 32	Thin-walled pipes, sheet metal, steel, white cast iron

In the case of soft materials, e.g. aluminium, or long cuts, a large amount of chip material is removed. In this event, a saw blade with a coarse tooth spacing is required **(Tables 1 and 2)**. Otherwise there would not be enough room in the cutting spaces for the amount of chip material removed.

Table 2: Saw-blade selection

Tooth spacing	Material	Cut
Coarse	Soft	Long
Fine	Hard	Short

Tooth shape

Hand-saw blades usually have herringbone teeth **(Fig. 1)**. The cutting wedges of saw blades are determined by small cutting angles and large clearance angles. For sawing steel, the wedge angle of a tooth is approx. 50°, the clearance angle approx. 38° and the cutting angle approx. 2°.

Types of hand saws

Hacksaw (Fig. 3). Comprises frame and saw blade. The tooth tips on the saw blade must point in the direction of pushing.

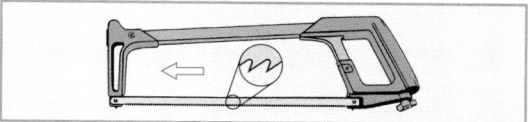

Fig. 3: Hacksaw

Short-stroke saw (electric hand pad saw, Fig. 4). Suitable for straight and slightly curving, non-warping cuts, e.g. for bodywork jobs for cutting out metal sheets in poorly accessible or confined spaces. Saw blades with fine tooth spacings, e.g. 32 teeth/inch, are used to saw metal sheets. The number and size of strokes can be adjusted to provide a burr-free cut.

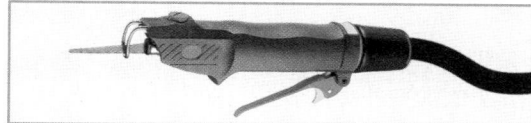

Fig. 4: Short-stroke saw

> **Working rules**
> - Attach the saw blade straight and tight, teeth pointing in the direction of pushing.
> - Select the tooth spacing to suit the material and the shape of the workpiece.

REVIEW QUESTIONS

1 **What are saws used for?**
2 **On what is the tooth spacing of a saw blade dependent?**
3 **How is the tooth spacing of a saw blade determined?**

7.4.2.4 Filing

> Filing is cutting by the repeated linear cutting motion of a multitooth tool with geometrically defined cutting wedges (file teeth).

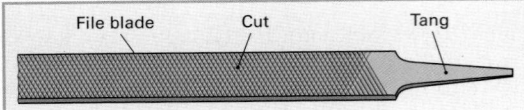

Fig. 1: Flat file – design

Design (Fig. 1). File body (file blade) with hewn cuts or milled teeth. Tang for attaching the file handle.

Distinguishing files

- **Size:** Hand files, key files, needlepoint files

- **Shape of cross-section and code letters (Fig. 2).**

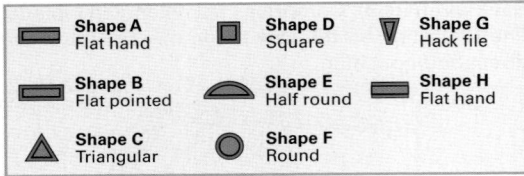

Fig. 2: Code letters for file cross-sections

- **Tooth shapes (Fig. 3) and type of manufacture**
  Hewn files have a scraping effect and are used for high-strength materials, e.g. steel, grey cast iron.
  Milled files have a cutting effect and are used for low-strength materials, e.g. aluminium, copper.

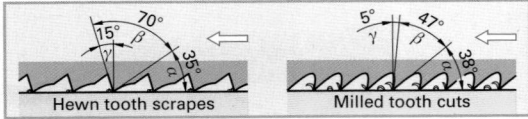

Fig. 3: Tooth shapes

- **Types of cut (Fig. 4)**
  Single-cut files are particularly suitable for working soft metals and for sharpening saws and other tools.
  Double- or cross-cut files are used for harder metals. The angle and spacing of the first and second cuts are different. In this way, the file teeth are offset as they come into contact with the workpiece and thereby prevent excessive striation.
  Rasp-cut files (pitted cut) are suitable for working e.g. wood, plastics, leather, cork and rubber.

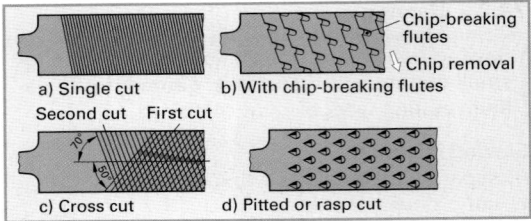

a) Single cut
b) With chip-breaking flutes — Chip-breaking flutes, Chip removal
c) Cross cut — Second cut, First cut
d) Pitted or rasp cut

Fig. 4: Types of cut

- **Cut number** is, for hewn files, the number of notches over a file length of 1 cm and, for cross cut, referred to the second cut. For rasp cut, the cut number corresponds to the number of pointed notches on $1 \, cm^2$ of hewn rasp surface.

- **Number of cuts per cm.** Indicates the fineness of the cut. The greater the number of cuts per cm, the finer the cut spacing (distance from tooth to tooth).

No. of cuts	Cut per cm	File name	Application
1	6...17	Bastard file	Preliminary filing
2	9...23	Middle-cut file	Preliminary smoothing
3	13...28	Bastard file	Smoothing
4	16...34	Second-cut file	Adjustment to fit
5...8	Not standardised	Ultrasmooth file	Ultrafine adjustment to fit

Table 1: Number of cuts per cm and cut numbers for hewn files (selection)

In the case of milled files, there are distinctions between toothings 1, 2 and 3 for coarse, medium and fine.

Filing technique. The file is moved in the direction of the file axis, where it is to be moved by one half the file width to the right or left. The user is only permitted to press on the file during the forward push.

> **Working rules**
> - Clean files with a file card.
> - Always ensure that the file handle is securely seated.

Milling pins (rotating or turbo-files, **Fig. 5**) are gripped in clamping chucks and electrically or pneumatically operated. They are used among other things for deburring and cleaning work, e.g. on weld seams.

Fig. 5: Milling pins

REVIEW QUESTIONS
1 What are the differences between milled and hewn files?
2 What standardised file cross-sections are there?
3 What does the number of cuts per cm indicate for files?

7.4.2.5 Reaming

> Reaming is the widening with low cutting thickness of a hole by a tool with geometrically defined cutting edges.

Reaming serves to finish- and fine-machine predrilled bores, e.g. bearing bushings, in order to obtain the required dimensional tolerance, shape tolerance and surface quality (roughness).

Reaming process. Chips are removed by the rotary cutting motion and axial feed motion of the reamer. Because the cutting angle is 0° or negative **(Fig. 1)**, chips are removed by scraping. The use of cutting oils improves the surface quality, reduces tool wear and thereby increases the life of the reaming tools. Cast iron is reamed dry.

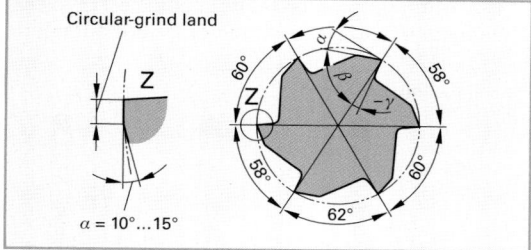

Fig. 1: **Cutting-edge angles and tooth spacing of reamers**

Unequal tooth spacings (Fig. 1) prevent chatter marks, since the subsequent cutting edges in the reaming process do not engage the workpiece at the same point.

Numbers of teeth. In the case of adjustable hand reamers, they are usually even-numbered to enable the reamer diameter to be measured easily with a micrometer gauge.

Hand reamer (Fig. 2). The **chamfer** (tapered) effects the removal of the chips. The **guide section** guides the reamer and smoothes the bore by means of circular-grind lands **(Fig. 1)**. The shank with square drive allows the reamer to be mounted, for example, in a tap wrench.

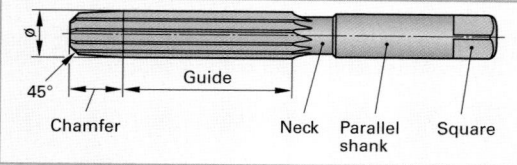

Fig. 2: **Hand reamer**

Machine reamer (Fig. 3). This has a short chamfer and a parallel or tapered shank. Thanks to their short chamfers, machine reamers can be used to ream blind holes almost down to the bottom.

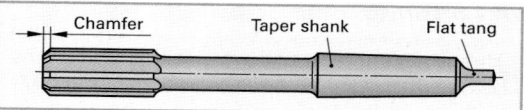

Fig. 3: **Machine reamer**

Teeth. Can be arranged straight or with a left-hand twist **(Fig. 4)**. The left-hand twist prevents the reamer from being pulled into the bore or being hooked into a longitudinal bore groove. At the same time the chips are removed by the left-hand twist in the feed direction.

Fig. 4: **Hand reamer with helical teeth**

Adjustable hand reamers (Fig. 5)

Slotted reamers can be spread by a tapered pin and thereby adjusted within narrow limits ($\approx$ 1/100 of the reamer diameter).

Reamers with inserted blades have a larger range of adjustment (1/10 ... 1/5 of the reamer diameter). When adjusted, the blades move inwards or outwards by means of two threaded rings on angled surfaces.

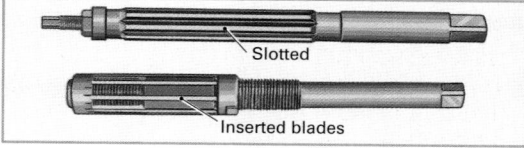

Fig. 5: **Adjustable hand reamers**

Working rules

● Insert the reamer and cut at right angles to the workpiece. Use cutting oil depending on the material.

● Screw in the reamer with uniform pressure in a clockwise direction and screw out in a clockwise direction.

● Never turn reamers in an anticlockwise direction as the trapped chips may break the cutting edge.

REVIEW QUESTIONS

1 What are the advantages of adjustable reamers?

2 Why must reamers not be turned anticlockwise?

3 Why do reamers with helical teeth have a left-hand twist?

7.4.2.6 Thread cutting/tapping by hand

> Thread tapping/cutting involves thread turns being cut-shaped on bolts or in bores with multiple-edged tools.

Threads **(Fig. 1)** can be produced as
- internal threads (female threads) with screw-taps
- external threads (male threads), e.g. with cutting dies

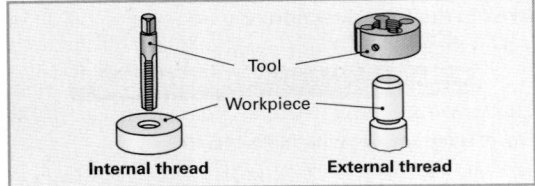

Fig. 1: Thread types

Internal threads (Fig. 2)

The core hole diameter for an internal thread must be drilled somewhat larger than corresponds to the core diameter of the internal thread. The screw-tap cuts the majority of the threads. However, part of the material is displaced inwards without cutting by the screw-tap. Here the thread is pressed open slightly and with it the core hole bore reduced. This process is known as cutting open.

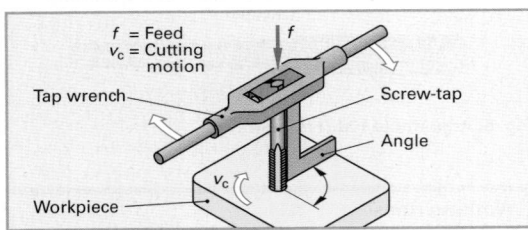

Fig. 2: Internal thread

Countersinking. Closed bores are countersunk on one side and through-holes on both sides to nominal thread diameter. This ensures that the screw-tap cuts better and the outer thread turns are not pressed out of the core hole bore.

Thread-tapping process. The screw-tap must be placed in position in the direction of the bore-hole axis and checked with an angle. The cutting motion is made up of the primary motion (turning of the screw-tap) and the feed motion (axial motion). Tap wrenches are used to turn screw-taps. While the thread is being tapped, the screw-tap should wherever possible not be turned back because frequent breaking off of the chips and recutting causes the cutting edges of the screw-tap to blunt premature-

ly. Exception: In the case of long-chipping materials and larger-sized threads, it is necessary to break off the chips repeatedly. This is done by turning the screw-tap back by roughly one quarter of a turn. At the same time, this allows fresh lubricant to reach the cutting edges.

Broken-off screw-taps are removed from the tapped hole with screw-tap removal device. They can also be loosened by gentle blows on a drift punch and twisted out with a pair of pliers.

Screw-taps. The following are used to tap internal threads:
- **3-part serial screw-taps (Fig. 3)** consisting of taper, second and bottoming taps, which are marked with 1, 2 and 3 or with rings. The chip-removal volume is spread over the three screw-taps (approx. 55 % : 25 % : 20 %). In this way, the thread-tapping tools are not subjected to excessive strain and clean thread turns are obtained.

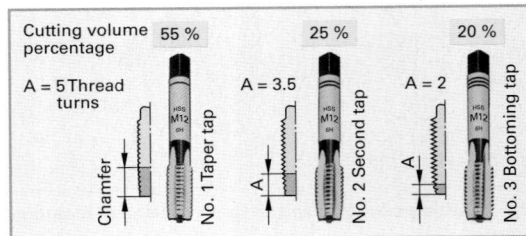

Fig. 3: 3-part serial screw-taps

- **2-part serial screw-taps** (no. 1 taper tap, no. 2 bottoming tap). These are used to tap fine-pitch threads.
- **Single-cut screw-taps (Fig. 4)** for tapping threads in metal sheets or workpieces the thickness of which is below 1.5 × nominal thread diameter. A smaller chamfer length is required due to the spiral face inclination.

Fig. 4: Single-cut screw-tap

- **Screw-taps for light alloys (Fig. 5)**. These have wider flutes and larger cutting angles.

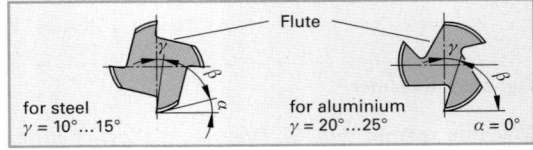

Fig. 5: Angles on screw-tap

The core holes must be drilled to dimensional accuracy for the purpose of tapping internal threads. For metric ISO threads (coarse- and fine-pitch threads) the core hole diameter d_c equals the thread diameter d minus the thread pitch P (**Table 1**).

$$d_c = d - P$$

Table 1: Dimensions in mm for metric ISO threads

Thread d	M3	M4	M5	M6	M8	M10	M12
Core hole d_c	2.5	3.3	4.2	5.0	6.8	8.5	10.2
Pitch P	0.5	0.7	0.8	1.0	1.25	1.5	1.75
Bolt dia. external thread	2.9	3.9	4.9	5.9	7.9	9.85	11.85

External threads

Cutting open also takes place when external threads (**Fig. 1**) are cut. The bolt diameter must therefore be slightly smaller than the thread diameter (**Table 1**).

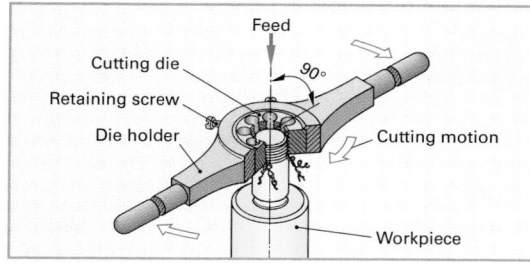

Fig. 1: External-thread cutting

In order to position the cutting die straight, the bolt must be chamfered at least to the core diameter. The chamfer simultaneously serves to protect the start of the thread.

The following are used to cut external threads:

- **Closed cutting dies (Fig. 2).** These finish-cut threads to dimensional accuracy in a single operation. Spiral face inclinations in the cutting dies make cutting easier and remove the chips cleanly in the cutting direction.

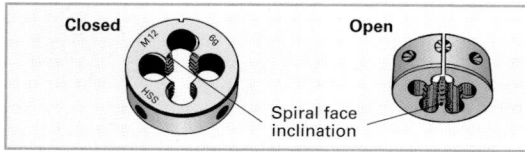

Fig. 2: Cutting dies – round

- **Open, slotted cutting dies (Fig. 2).** By adjusting with an adjusting screw or a pressure screw, it is possible to alter the thread diameter within exacting tolerances.
- **Hexagonal cutting dies (Fig. 3).** These are used to recut damaged threads or to cut threads in poorly accessible areas since they can be moved with wrenches or ratchets.

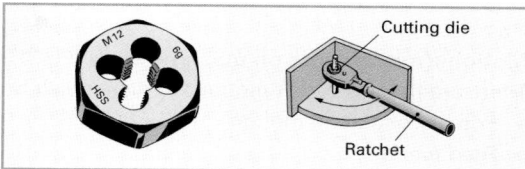

Fig. 3: Cutting dies – hexagonal

Thread-tapping/cutting lubricants
(**Table 2**)

It is important to use suitable lubricants to maintain the surface quality of the threads.

Table 2: Cooling lubricants for thread tapping/cutting

Cooling lubricants	Material (selection)
Cutting oil	Steel, Ti; Ti alloys
Water-mixable cooling lubricants	Grey cast iron, Cu; Cu alloys, Al, Al alloys, zinc

Working rules
- Countersink the thread core hole on both sides to the outside thread diameter. Chamfer the threaded bolt to core diameter.
- Position thread-cutting tools vertically when cutting. Continue to check that the tools are fixed at right angles.
- Wherever possible, do not turn thread-cutting tools back because frequent breaking off of the chips and recutting causes premature blunting. Risk of tool breakage!
- Use suitable cooling lubricant in sufficient quantities.

REVIEW QUESTIONS

1 **What tools are needed for thread tapping/cutting?**

2 **What types of screw-tap are used?**

3 **What do you understand by cutting open when making threads?**

4 **How is the core hole diameter determined for tapping internal threads?**

5 **How in external threads must the bolt diameter be compared with the thread diameter?**

7.4.3 Basics of cutting-shaping with machine tools

> Machine tools for cutting-shaping can machine flat, cylindrical, tapered and curved surfaces.
>
> To obtain a specific surface, it is necessary to move the workpiece and the tool towards each other accordingly.

Motions/movements on machine tools (Fig. 1)
There are three different motions to distinguish between:
- Primary or cutting motion
- Feed motion
- Positioning motion

Primary or cutting motion v_c. This is executed either by the tool or by the workpiece.

> The cutting speed v_c is the speed at which the chip is removed.

The cutting speed is generally given in m/min, but in the case of grinding is given in m/s.

Feed motion v_f. This can be executed manually (by hand) or automatically by the machine.

> The rate of feed v_f (mm/min) is the speed at which the workpiece and tool move towards each other during cutting.

Feed f is the travel of the tool, e.g. when drilling, grinding or milling, or the travel of the tool during one rotation of the tool, e.g. when turning.

Positioning motion a_p, a_e. This is the motion between the workpiece and the tool which determines the thickness of the chip to be removed.

The cutting speed, rate of feed and positioning motion are dependent on ...
- ... the operating process and the design of the machine.
- ... the material to be cut.
- ... the cutting material of the tool.
- ... the required surface quality.
- ... the cooling and lubrication of the tool's cutting edge.
- ... the required life of the tool.

Chip formation (Fig. 2)

> In each cutting-shaping operation, the material is upset, separated and removed in the form of a chip via the face by the penetrating cutting wedge of the tool cutting edge.

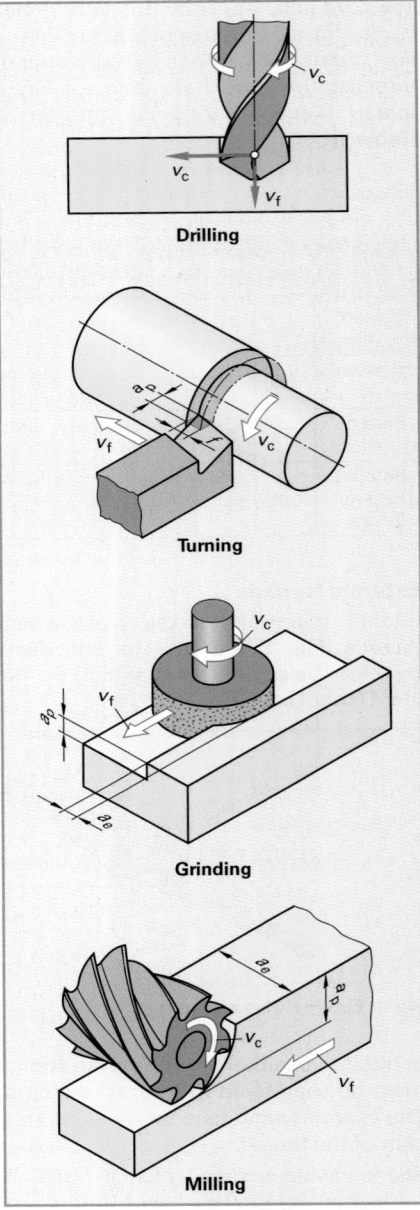

Fig. 1: Motions/movements on machine tools

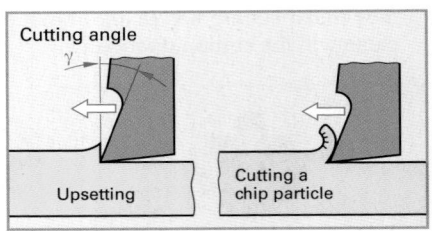

Fig. 2: Chip formation

Types of chip (Fig. 1)

Tearing chips are produced from brittle materials, e.g. grey cast iron, at small cutting angles, low cutting speeds and a large cutting depth. In the process, the surface becomes rough and loses its dimensional and geometrical accuracy.

Shearing chips are produced from tough materials at medium cutting angles and low cutting speeds. These chips flake off, bond with each other partially and for the most part form short helical chips. They are not an impediment to the work sequence.

Flowing chips are produced from tough materials at large cutting angles, high cutting speeds and low to medium cutting depth. Smooth workpiece surfaces of high surface quality are obtained. These types of chips are therefore the desired end result. Long, continuous flowing chips can hinder the work sequence, for instance, where automatic lathes are used.

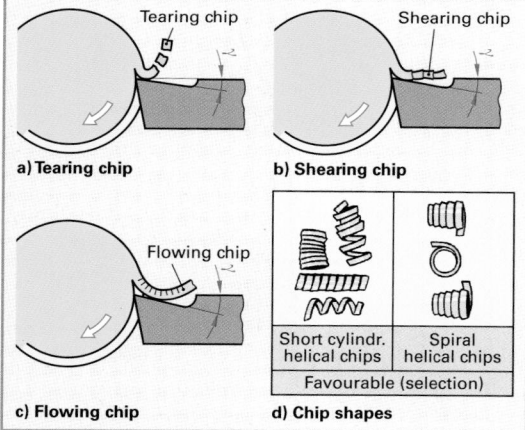

a) Tearing chip   b) Shearing chip

c) Flowing chip   d) Chip shapes

Short cylindr. helical chips	Spiral helical chips
Favourable (selection)	

Fig. 1: Types of chip, chip shapes

Built-up edge (Fig. 2). This is formed during the cutting process on the face of the tool. A built-up edge may be formed at excessively slow cutting speeds or in the event of insufficient cooling lubrication or an excessively rough tool face. The deposit of material particles alters the angles at the cutting wedge unfavourably and causes a rough workpiece surface. A built-up edge is not formed on oxide-ceramic or diamond tools.

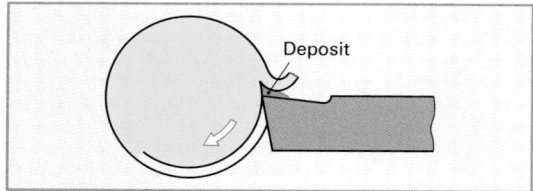

Deposit

Fig. 2: Built-up edge

Cooling and lubrication during cutting

During cutting-shaping, heat is generated as a result of friction at the tool's cutting edge and in the edge zone of the workpiece. If there is inadequate cooling lubrication during cutting, temperatures in excess of 1,000 °C may arise at the tool and in the edge zone of the workpiece.

Possible consequences of inadequate cooling lubrication:

- Premature tool wear
- Dimensional deviations
- Reduced surface quality
- Cracking in the workpiece edge zone
- Reduced strength.

During metal-cutting manufacturing processes, the following are used, depending on the cutting speed: non-water-mixable cooling lubricants, e.g. cutting oils with additives, and water-mixable cooling lubricants, e.g. drilling oil with water (cooling lubricant emulsion).

Low cutting speed → small cooling effect required → non-water-mixable cooling lubricants, e.g. thread-cutting.

High cutting speed → large cooling effect required → water-mixable cooling lubricants, e.g. drilling, turning, milling.

Disposal. Used cooling lubricants must be handled as hazardous waste.

REVIEW QUESTIONS
1 **What are the different movements/motions involved in cutting with machine tools?**
2 **What do you understand by cutting speed and rate of feed?**
3 **What are the potential consequences of inadequate cooling lubrication during cutting by machine?**

7.4.3.1 Milling

Milling is a machining manufacturing process with geometrically defined cutting edges. Here, flat and curved surfaces are produced with multiple-edged, rotating tools.

Applications. E.g. in the manufacture of: gear wheels, end faces **(Fig. 3),** solid faces (freeform milling) and helicoidal faces.

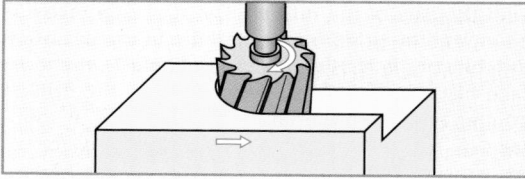

Fig. 3: Front milling with shell end mill

7.4.3.2 Drilling

> Drilling in metal technology is the cutting by machine with geometrically defined cutting edges and predominantly multiple-edged tools for producing cylindrical holes (bores).

Twist drill bit (Fig. 1). This is the most commonly used drilling tool. Its advantages are:

- Favourable angles at the cutting edges
- Good chucking capability
- Consistent diameter when regrinding
- Automatic chip removal
- Good supply of cooling lubricant

Drilling process. The primary or cutting motion is a rotary motion which is mostly executed by the drilling tool. At the same time, the tool is moved in the axial direction against the workpiece (feed). This results in continuous chip formation. The cutting speed is primarily dependent on the workpiece material and the drill-bit cutting material. The feed is dependent on the drill-bit diameter, on the material to be machined and on the drilling process.

Cutting-edge geometry of twist drill bit (Fig. 2)
Major cutting edges. Two helicoidal flutes form the major cutting edges on the tip of the drill bit and perform the actual cutting.

Minor cutting edges. These are formed by the flutes on the cutting wedge and smooth the bore.

Chisel edge. This hinders the cutting process as it scrapes as opposed to cuts.

Lands. These ensure that the drill bit is securely guided in the bore. They also reduce friction and thus the risk of the drill bit jamming in the bore.

Point angle. This is formed by the major cutting edges and is matched to the material to be machined. It is 118° for the machining of steel, cast steel, cast iron and malleable cast iron.

Clearance angle. This is produced by relief-grinding of the major cutting edges. The clearance angle enables the drill bit to penetrate into the material. A point angle of 118° produces a chisel-edge angle of 55° with correct relief-grinding. This corresponds to the correct clearance angle for the machining of steel.

Tool side rake γ. This is the angle of the faces (flutes) with the axis of the drill bit. The tool side rake cannot be altered by grinding the drill bit. In the case of twist drill bits of the **N, H and W types (Table 1),** the tool side rake has a specific size, depending on the drill-bit diameter and the material to be machined. A tool side rake of 19°...40° is suitable for machining, for example, steel or cast iron.

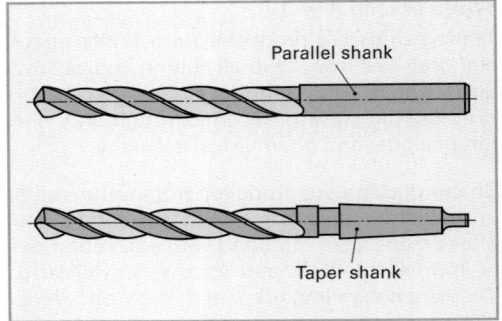

Fig. 1: Twist drill bits with parallel and taper shanks

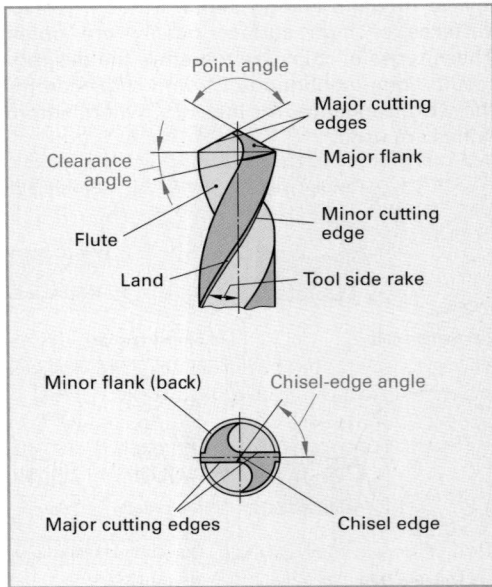

Fig. 2: Cutting-edge geometry of twist drill bit

Table 1: Tool side rake		
Drill-bit type		
N	H	W
118°	118°	130°
$\gamma = 19°...40°$	$\gamma = 10°...19°$	$\gamma = 27°...45°$
Normal-hard	Hard and tough-hard	Soft and tough
Metallic materials		

Grinding twist drill bit

Drill-bit grinders are used to precision-grind the drill-bit cutting edges.

Errors when grinding by hand (Table 1)
- Cutting edges unequal in length
- Point angles unequal
- Clearance angle too high, too low

Consequences of these errors
- Bore diameter too large
- Reduced drill-bit life

In order to avoid these errors, it is essential to check the grinding of the drill bit with a grinding gauge. If the clearance angle is too large, the cutting edges of the drill bit will break off because the tool wedge is weakened. If the clearance angle is too small, the friction between drill-bit flank and workpiece will be too great and the drill bit will burn out. Bores over 15 mm in diameter can be predrilled if necessary because an excessive feed force is required on account of the chisel edge. For this reason, drill bits are often pointed out, i.e. the chisel-edge length is shortened to roughly 1/10 of the drill-bit diameter.

Drills

Hand drills are suitable for bores for which low levels of precision are required. These are usually equipped with three-jaw drill chucks.

> Electric hand drills may only be used in perfect condition. Damaged cables, plugs or housings are potentially fatal dangers.

Bench and upright drills **(Fig. 1)** are suitable for drilling which requires high levels of precision and cutting performance.

Chucking drill bits (Fig. 2)

Small drill bits up to approx. 12 mm in diameter usually have parallel shanks and are chucked in three-jaw drill chucks, collet chucks or clamping sleeves. Larger drill bits generally have tapered shanks. They are non-positively connected to the inner taper of the drilling spindle by means of axial insertion. An ejector is required to release the drill bit from the drilling spindle.

Clamping workpieces

Caution must be exercised. It is essential to ensure that the workpieces, e.g. metal sheets, cannot be carried along by the drilling tool. The drill bit can easily catch as it emerges from the bore and this can cause accidents. Smaller workpieces must be clamped in a machine vice **(Fig. 1)**.

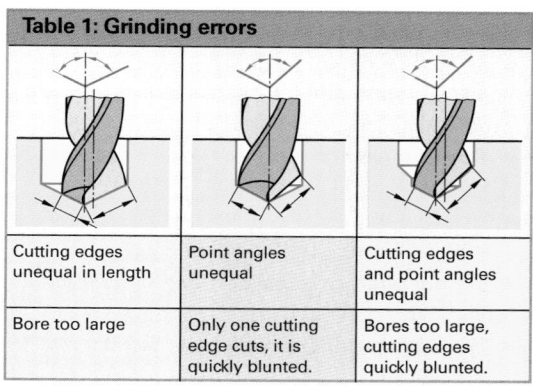

Table 1: Grinding errors		
Cutting edges unequal in length	Point angles unequal	Cutting edges and point angles unequal
Bore too large	Only one cutting edge cuts, it is quickly blunted.	Bores too large, cutting edges quickly blunted.

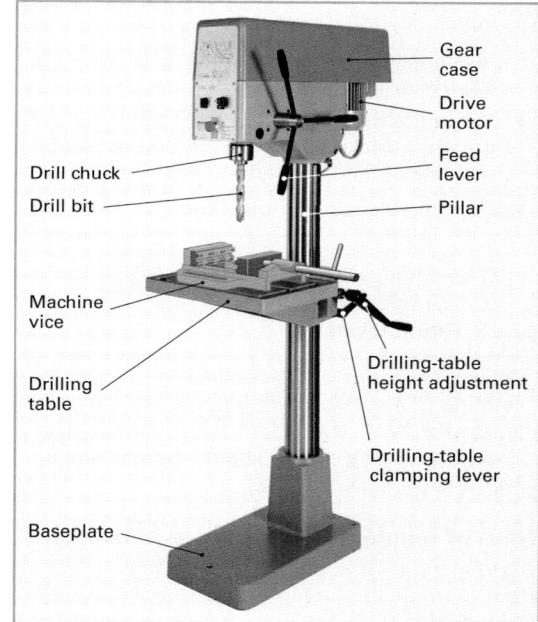

Fig. 1: Upright drill

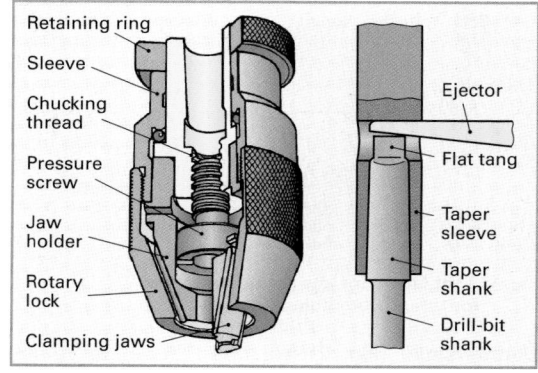

Fig. 2: Quick-action drill chuck

Drill bit with taper shank

Working rules

- Make sure that locating tapers, reducing sleeves and drill-bit shanks are absolutely clean.
- Do not force drill bits into the drill chuck with excessive force.
- Make sure that the drill bit is securely seated and runs true.
- Do not align chucked drill bits by knocking/ striking.
- Do not grip drill bits with taper shanks in the drill chuck.
- Mark the centre of the bore with a centre punch, predrill if necessary.
- Make sure that the workpiece is securely gripped when drilling.
- Select the right drill-bit type with the correct grinding. Make sure that the cutting speed and rate of feed are correct.
- Use suitable cooling lubricant.

Accident prevention

- Wear tight-fitting protective clothing with narrow sleeves.
- Wear suitable head covering if you have long hair, e.g. a hairnet.
- Remove the drill-chuck key or ejector from the drill spindle immediately after use.
- Wear protective goggles when drilling brittle materials.
- Secure workpieces against being carried along.
- All safety devices and equipment must be in place prior to the commencement of work.
- Remove drilling chips with a brush or by extraction.
- Belts may only be fitted when the machine is stopped.
- Have damage to the electrical system repaired immediately by a qualified electrician (do not carry out repairs yourself).

7.4.3.3 Countersinking

Countersinking is a special drilling process for creating in already existing bores end faces or tapered surfaces vertical to the rotational axis. Here, the cutting speed should be lower than for drilling.

Types of countersink and their applications

Spot facers have a fixed or exchangeable pilot in order to guide the tool in the bore. Spot facers for **end facing (Fig. 1)** are used, for example, to produce cylindrical countersinks for hexagon, TORX or Phillips screw heads.

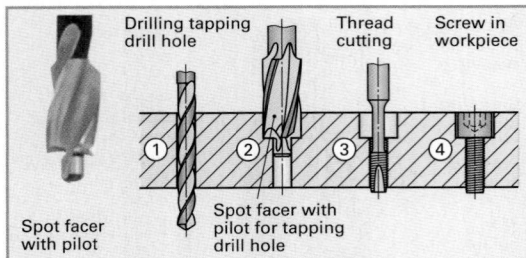

Fig. 1: Spot facers for end facing with pilot for thread core hole

Countersinks (Fig. 2) are one-, three- or multi-edged and come with and without pilots. These are used to deburr bores and to create tapered profile

countersink holes for rivet and screw heads. Their point angles are standardised, e.g.

- 60° for deburring
- 90° for countersunk-head screws and for internal threads
- 75° for rivet heads
- 120° for plate rivet

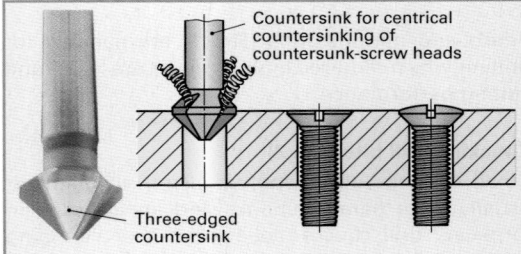

Fig. 2: Countersink for profile countersinking

1. How big are the point angle, chisel-edge angle and tool side rake for drilling steel?
2. What are excessively large bore diameters caused by?
3. Give possible causes for the burning out of drill bits.
4. What must be borne in mind when gripping/ chucking drilling tools and workpieces?
5. Why must protective goggles be worn when brittle materials are being drilled?
6. What angles do countersinks have?

7.4.3.4 Turning

> Turning is a machining manufacturing process with a geometrically defined cutting edge. Here, round or flat surfaces are produced with a one-edged tool.

Subdivision of turning processes
The processes are categorised in accordance with the ...

- ... position of the machining surfaces into outside and inside turning **(Fig. 1)**.
- ... feed direction into longitudinal turning (round turning) and lateral turning (facing) **(Fig. 2)**.
- ... created surface into round turning, facing, profiling, forming by turning and threading.

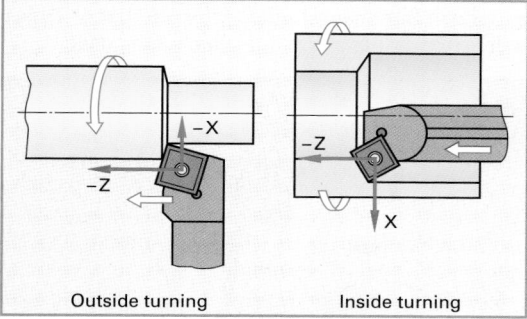

Fig. 1: Position of machining surfaces

Motion processes when turning (Fig. 2)
Cutting motion. This is performed by the workpiece gripped in the lathe, which executes the rotary motion. The cutting speed is derived from the diameter and the rotational speed. It is important when selecting the cutting speed to take into account:

- Material
- Cooling lubrication
- Cutting material
- Surface quality

Feed motion. This is performed with longitudinal turning in the Z-axis (workpiece axis) and with lateral turning in the X-axis (transversally to the workpiece axis). The **feed f** is given in mm per revolution.

Positioning motion. This is performed with longitudinal turning in the X-axis and with lateral turning in the Z-axis. The **cutting depth a** corresponds to the positioning of the lathe tool.

> The combination of cutting motion and feed motion with prespecified positioning results in a chip with the cutting cross-section A.

Cutting cross-section A (Fig. 2). This is the product of the feed f and the cutting depth a. If large amounts of material are removed, it is essential to turn in several stages, e.g. first roughturn, then smooth.

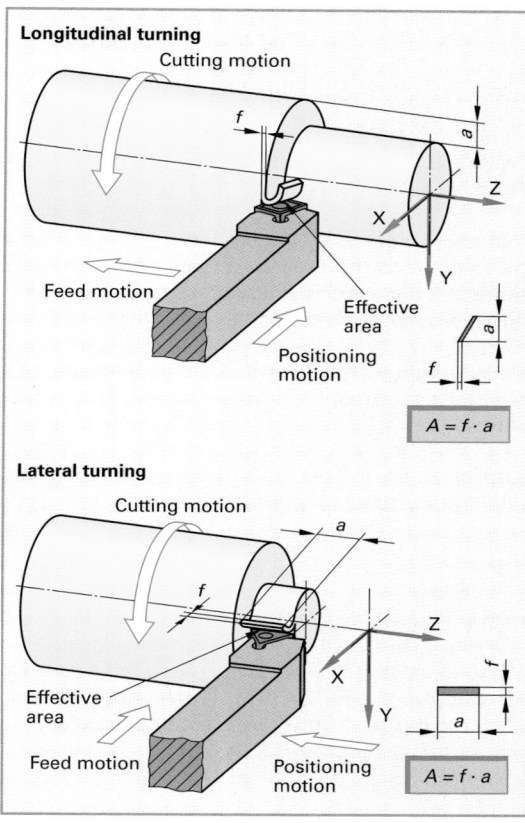

Fig. 2: Motion processes when turning

Shapes of lathe tools (Fig. 3). Lathe tools are distinguished according to

- cutting direction (**R** right-hand cutting, **L** left-hand cutting, **N** neutral)
- position of contact – outside and inside lathe tools **(Fig. 1)**

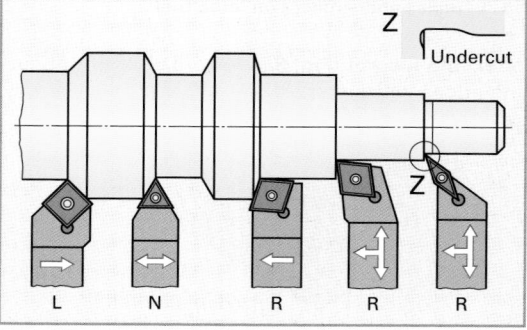

Fig. 3: Outside lathe tool

Cutting-edge geometry on lathe tool

Angles and surfaces (Fig. 1). In its basic shape, a lathe tool is a wedge with a clearance angle α, wedge angle β and cutting angle γ. The chip is removed at the tool's cutting surface by the lathe tool.

Major and minor cutting edges. The major cutting edge points to the feed direction and performs the actual cutting. The major and minor cutting edges together form a rounded nose, which affects the depth of the furrows created.

Clearance angle α. This is limited by the flank and the tangent to the cutting surface. Its size determines the friction or surface pressure between the workpiece and the lathe tool.

Wedge angle β. This is formed by the flank and the face. Its size depends on the material to be machined.

Cutting angle γ. This is formed by a horizontal plane through the rotational axis and the face.

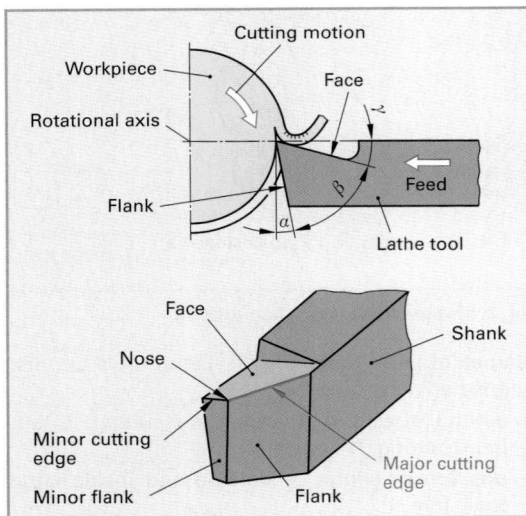

Fig. 1: Angles and surfaces on lathe tool

Indexable inserts (Fig. 2). These have several cutting edges which can be brought into place simply by turning.

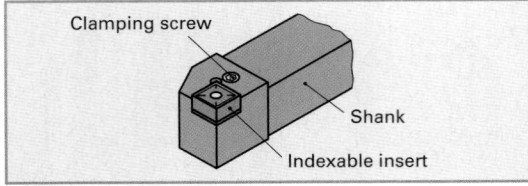

Fig. 2: Tool holder with indexable insert

Gripping lathe tools

The lathe tool must be gripped as briefly and securely as possible. The nose is normally set to the centre of the workpiece (height of the rotational axis). The clearance and cutting angles are correctly sized in this setting.

Gripping workpieces

Parts to be turned must be gripped on the lathe on the basis of how they are shaped.

The following means of gripping are used:

- Clamping chuck
- Collet chuck
- Lathe centres
- Surface plate.

Clamping chuck (Fig. 3). There are three- and four-jaw chucks. Cylindrical workpieces can be gripped in both types of chuck. Multiple-edge workpieces whose number of edges is divisible by **3** are gripped in three-jaw chucks. If the number of edges is divisible by **4,** then a four-jaw chuck is used.

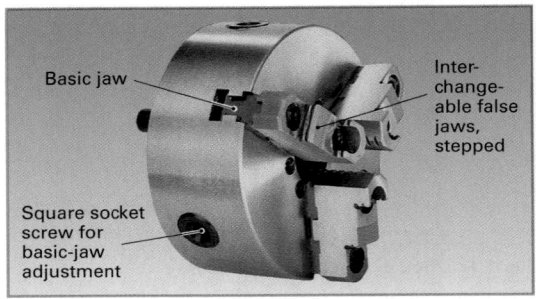

Fig. 3: Three-jaw chuck

Working rules

- Clamping jaws must not project far out of the chuck.
- The chucking force must be adapted to the workpiece and to the extent of the shear force.
- Always remove the chuck key immediately.
- Always grip the lathe tool firmly, securely and as briefly as possible at the centre of the workpiece.
- Do not grip or release the lathe tool while the machine is running.

REVIEW QUESTIONS

1 What is the feed direction for longitudinal and for lateral turning?

2 Why should the lathe tool be set to the centre of the workpiece?

3 What working rules must be followed when turning?

7.4.3.5 Grinding

> Grinding is a cutting manufacturing process with geometrically undefined cutting edges (Fig. 1).

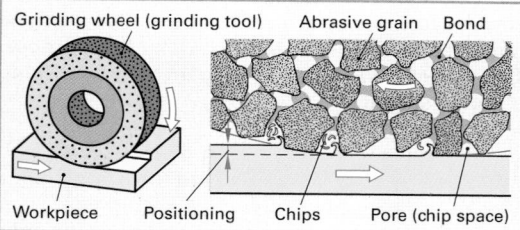

Fig. 1: Grinding process

High levels of dimensional and geometrical accuracy together with superior surface quality can be achieved in production engineering by grinding. Chief grinding operations in motor-vehicle applications:

- Parting by grinding, e.g. during repairs
- Surface grinding, e.g. superfinishing
- Face grinding, e.g. cylinder-head face
- Contour grinding, e.g. cam contour grinding
- Sharp grinding of tools, e.g. twist drill bits

Abrasive wheels. Comprise abrasives and binders.

Abrasives. Denoted by capital letters. Conventional abrasives are:
standard alumina, fused alumina, special fused alumina **(A)**, silicon carbide **(C)**, boron nitride **(B)** and diamond **(D)**.

Binders. There are **inorganic** bonds, e.g. ceramic bonds **(V)**, and **organic** bonds, e.g. artificial resin **(B)**, rubber **(R)**, fibre-reinforced rubber bond **(RF)**. The bond of the abrasive grains established by the binder is known as the degree of hardness. The degrees of hardness are denoted by letters (A, B, C, D ... X, Y, Z ≙ extremely soft ... extremely hard).

> Abrasive wheels with soft bonds must be selected for hard materials and abrasive wheels with hard bonds for soft materials.

Grain number. Corresponds in accordance with the US standard to the mesh size of a sieve 1 inch in length through the meshes of which the grains still just fall. In accordance with the European standard **(FEPA** standard = **F**ederation **E**uropéene de **P**roduits **A**brasifs), the grain is indicated by **P** and a **code number. Example:** Grain for fine grinding: P 500 (FEPA standard) corresponds to a grain number of approx. 320...360 according to the US standard.

Structure code number (Fig. 2). This denotes the ratio of abrasive grains, bond and pore volume in the abrasive wheel. The higher the structure code number, the more open the structure.

> The greater the removal of chips, the more open the structure must be.

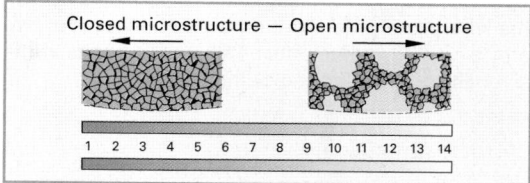

Fig. 2: Structure code number

7.4.3.6 Fine machining

The following properties can be achieved by the fine machining of motor-vehicle components:

- Low roughness
- High percentage contact area
- Good sliding behaviour
- High dimensional, geometrical and positional accuracy

Fine machining is performed with very small feeds and low cutting depth.

Lapping

> Lapping is fine machining by cutting with unbound (loose) abrasive wheels which have geometrically undefined cutting edges.

Lapping involves introducing a lapping mixture of lapping fluid and lapping powder (alumina, silicon or boron carbides) between the workpiece and the tool. Workpiece and tool move back and forth against each other under pressure and constant changes of direction **(Fig. 2)**.

Bore lapping (Fig. 3). Here, two components are matched to each other with the smallest gap dimension, e.g. pump elements of injection pumps, nozzle elements, switching valves in automatic gearboxes. The components can only be used in pairs.

Surface lapping (Fig. 3). Here, surfaces of components are fine-machined to such an extent that they are sufficiently sealed without the need for additional seals, e.g. gear oil pump surfaces.

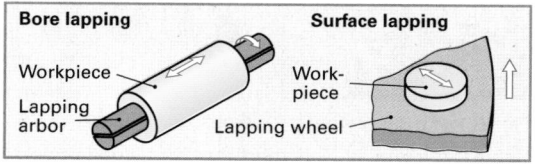

Fig. 3: Lapping process

Honing

> Honing is fine machining by cutting with bound abrasive wheels which have geometrically undefined cutting edges.

The following components, for example, are honed: master cylinders, wheel-brake cylinders, engine cylinders **(Fig. 1)**. When an engine cylinder is honed, the rotary and lifting motion of the honing tool creates a cross-grinding effect. This improves the adhesiveness of the oil to the cylinder wall.

Honing tool

Cylinder block

Honing stone

Honed cylinder surface

Fig. 1: Cylinder honing

7.4.3.7 Special processes for motor-vehicle repairs

> Special machining processes with special machines are used for the expert repairing of motor-vehicle components.

Valve-seat lathe (Fig. 2). This machine can be used to precision-turn valve seats. For this purpose, the pilot of the valve lathe tool must be inserted in the valve guide accurately and without play. With the lathe tool, which has three carbide cutting edges (valve-seat angle 45°, correction angles 15°, 75°), it is possible to perform three machining operations with a single chucking.

Lathe tool

Valves

Fig. 2: Valve-seat lathe   **Fig. 3: Valve lathe**

Valve lathe (Fig. 3). This machine can be used to precision-turn valve plates. The valve is pressed onto

the sprung centring taper in the headstock and chucked. The valve is guided in a steady rest. Positioning is performed by means of the tool holder.

Brake-lining lathe (Fig. 4). This machine can be used to round-turn overdimensioned brake linings while installed to the desired new dimension according to the brake-drum diameter. The machine must be mounted on the axle hub for this purpose. The following defects can be corrected by turning off on this lathe:

- Warped brake shoes
- Warping in the brake anchor plates
- Deviations in the roundness of the brake linings

> The dust must be drawn off when turning off on the lathe.

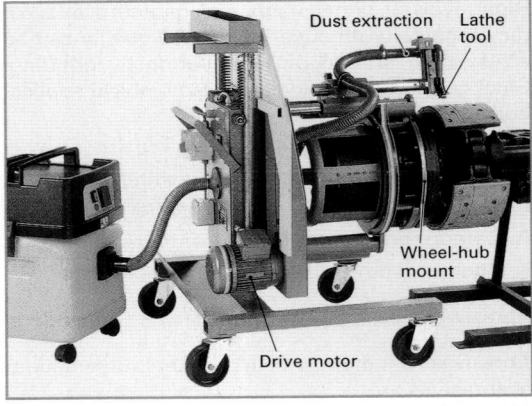

Dust extraction   Lathe tool

Wheel-hub mount

Drive motor

Fig. 4: Brake-lining lathe

Cylinder-block and cylinder-head surface grinder/ miller (Fig. 5). These are special machines which are suitable for machining sealing surfaces (facing) on cast-iron or aluminium-alloy cylinder heads and engine blocks. A rotating segment wheel gripped in the tool head executes the cutting and feed motions. The segment wheel is replaced by a milling head for Al alloys.

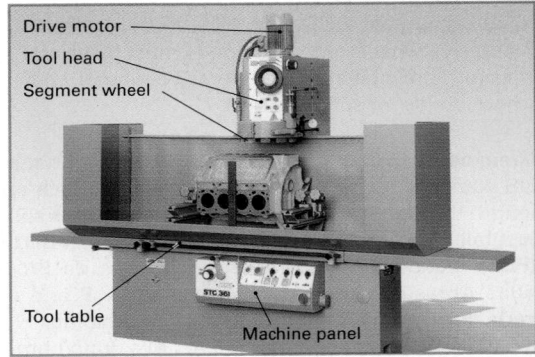

Drive motor

Tool head

Segment wheel

Tool table

Machine panel

Fig. 5: Cylinder-block and cylinder-head surface grinder/miller

7.5 Separating by dividing

Dividing is separating by mechanical means. This process leaves a remaining workpiece of predetermined shape without creating chips. There are two distinct types of separating by dividing: cropping and wedge-action cutting.

7.5.1 Cropping

Cropping or shearing involves dividing (without creating chips) between two cutting edges which move past each other. The material is shorn off in the process.

Cropping is divided into closed and open variations.

Closed cropping produces a cutting line which is closed in itself, e.g. when cutting out (punching) bores in mass-produced parts.

Open cropping produces an open cutting line. Here, the material is divided with shears, e.g. when cutting off a strip with a pair of sheet-metal shears.

Cutting with shears

When cutting with hand shears or shearing machines, the shearing blades cut through approximately 7/10 of the workpiece thickness while the remainder of the cross-section breaks in two completely.

The cutting edges of the shearing blades have a cutting angle of roughly 5°. This makes it easier for them to penetrate into the material. The clearance angle of 1.5° to 3° reduces the friction as the material is being cut through **(Fig. 1)**.

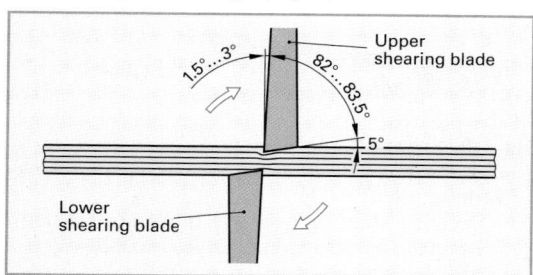

Fig. 1: Angles on shearing blades

Hand sheet-metal shears are used to cut sheet metal up to a maximum thickness of 1.8 mm. There are different types for different applications.

Straight sheet-metal shears (Fig. 2) are used for producing long, straight cuts. The pivot of the cutting edges is located above the metal sheet. The sheet passes underneath the user's hand, thereby

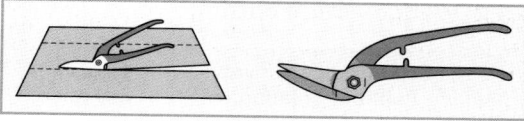

Fig. 2: Straight sheet-metal shears

eliminating any risk of injury. The cut piece and the remaining piece are not deformed.

Ideal sheet-metal shears (Fig. 3) for producing curved and straight cuts.

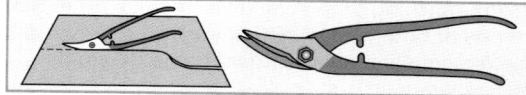

Fig. 3: Ideal sheet-metal shears

Hole-cutting shears are used to cut openings. The cutting operation usually starts off from a bore.

Nibblers (Fig. 4) cut thin sheets – including bent and corrugated sheets – without deforming the sheet surface. As it cuts, a nibbler creates a narrow metal strip which curls in a spiral shape. The edges of the cut are clean and without burrs. This enables the user to fashion figure cuts with narrow radii and rectangular cut-outs without chip obstruction.

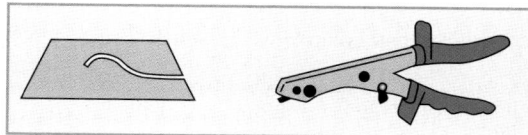

Fig. 4: Nibbler

Compressed-air nibblers (Fig. 5). These are used in bodywork repairs to produce straight and curved cuts.

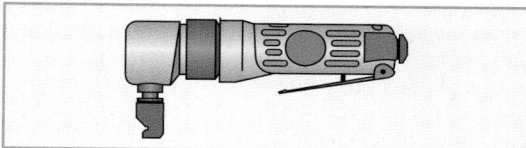

Fig. 5: Compressed-air nibbler

Sheet-metal slitting shears (Fig. 6) cut sheets up to a thickness of approx. 8 mm.

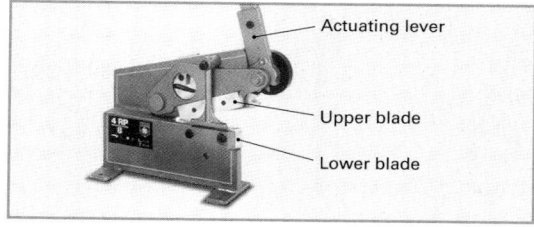

Fig. 6: Sheet-metal slitting shears

7.5.2 Wedge-action cutting

> Wedge-action cutting involves the mechanical dividing (without creating chips) of workpieces with one or two wedge-shaped cutting edges.

Wedge-action cutting is further categorised into
- cutting with a single cutting edge, e.g. with a hollow punch
- cutting with two approaching cutting edges, e.g. with side-cutting pliers.

Wedge-action cutting process (**Fig. 1**):
- Pre-notching of the workpiece by penetration of one or both wedge cutting edges
- Notching and displacement of the material
- Forcing apart and finally tearing of the material

The dividing process (**Fig. 1**) is determined by the wedge angle alone, because both faces of the cutting wedge are always engaged. There is therefore neither a clearance angle nor a cutting angle.

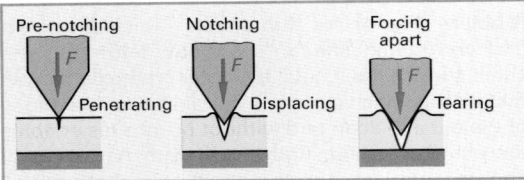

Fig. 1: Dividing process

Information pertaining to wedge-action cutting tools:

Wedge angle	Large	Small
Impact penetration force	Large	Small
Separating force	Small	Large
Tool life	Long	Short

Cutting with a single cutting edge

One-edged tools divide the workpiece with a wedge-shaped cutting edge from one side. When soft materials such as cork, for example, are cut with a single edge, the blade edge penetrates into the surface supporting the material. For this reason, the support surface must be made of a soft plate material itself so as to prevent damage to the blade edge. Typical one-edged dividing tools are, for example, vertical-action cold chisels and punches, hollow punches and pipe cutters (**Fig. 2**).

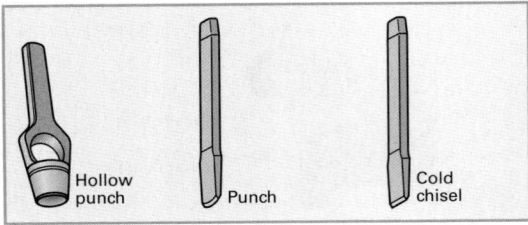

Fig. 2: One-edged tools

Pipe cutter (Fig. 3). This is used for cutting pipes, e.g. brake lines, square (i.e. at right angles). As the positioning screw is turned, the disc springs are tensioned and the cutting wheel is pressed into the pipe. As the pipe cutter turns, the cutting wheel cuts through the pipe by means of the pressing force of the disc spring. The spring tension is regulated by means of the positioning screw.

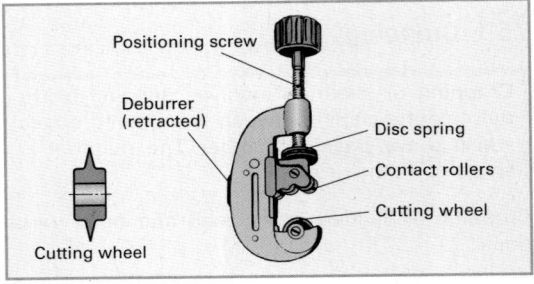

Fig. 3: Pipe cutter

Cutting with two approaching cutting edges

Two-edged tools (**Fig. 4**) use the approaching motion of the two wedge cutting edges to divide the workpiece. The workpiece is separated from both sides.

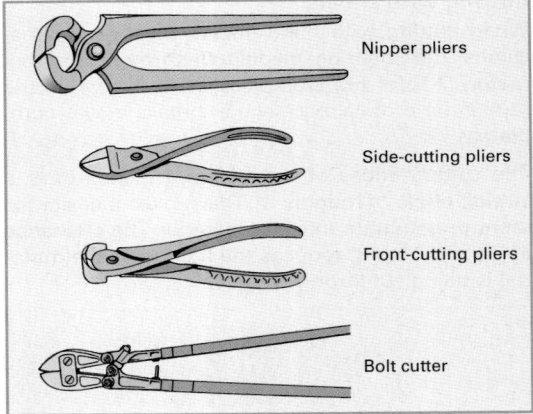

Fig. 4: Two-edged tools

> **Working rules**
> - Because sharp edges and burrs are created when sheet metal is cut, gloves must be worn.
> - Always deburr workpieces.

REVIEW QUESTIONS
1. What is the difference between cropping and wedge-action cutting?
2. For what tasks are straight sheet-metal shears, ideal sheet-metal shears and nibblers predominantly used?
3. What is the relationship between wedge angle and separating force in the case of wedge-action cutting?

7.6 Joining

Joining involves the connection of two or more workpieces with each other. In this way, cohesive strength at the joint is established and increased overall.

Depending on the properties that are required, connections such as, for example, weld joints, glued joints, riveted joints, screwed joints, splined connections and coatings are used on motor vehicles.

7.6.1 Categorisation of connections

Categorisation acc. to	Types, examples
Cohesive strength	**Friction**, e.g. screwed, clamped, press-fit joints **Keyed**, e.g. splined connection, fit bolts **Preloaded-keyed**, e.g. shaft-hub joint **Material-based**, e.g. weld, glued joint
Mobility	**Moving**, e.g. saddle guide **Fixed**, e.g. screwed, weld, riveted joints
Removability	**Removable**, e.g. screwed joint **Non-removable**, e.g. soldered, riveted, weld joints

Categorisation according to cohesive strength
Friction connections are, for example,
- screwed joints
- press-fit joints
- clamped joints
- friction clutches

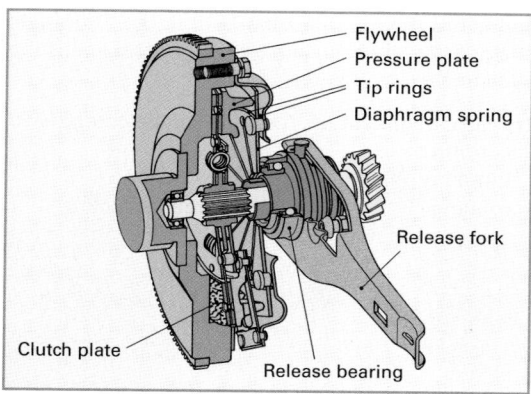

Fig. 1: Friction connection

The workpieces when joined are pressed onto each other in such a way that the friction at the contact surfaces is sufficient to transmit the forces arising during operation. The friction force between the contact surfaces must be greater than the greatest displacement force arising during operation.

Example: Single-plate clutch (Fig. 1). Here, the clutch plate is pressed by the clutch pressure plate against the flywheel in such a way that the peripheral force at the flywheel and the clutch pressure plate can be safely transmitted to the clutch plate.

Keyed connections are, for example,
- pin connections
- featherkey connections
- fit-bolt connections
- splined connections

The workpieces are connected by their interlocking geometrical shapes in such a way that the arising forces can be transmitted.

Example: Splined connection (Fig. 2). With this connection, the splines of the milled shaft section engage the grooves of the corresponding hub section and transmit the peripheral force from the shaft to the hub.

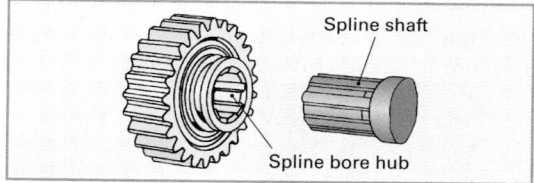

Fig. 2: Keyed connection

Preloaded-keyed connections are, for example, a shaft-hub connection with Woodruff keys **(Fig. 3).** Here the parts are connected by frictional and keyed means. The arising forces are initially transmitted by friction to the contact areas. If the static friction is no longer sufficient for this, the transmission of force is ensured by the additional positive fit.

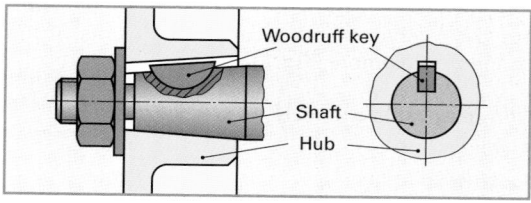

Fig. 3: Preloaded-keyed shaft-hub connection

Material-based connections are, for example,
- weld joints
- soldered joints
- glued joints

The components are connected in such a way that the forces arising during operation are transmitted by cohesion and adhesion.

Cohesion. In the case of weld joints, the contact surfaces of the joined components are fused together. In the case of soldered joints, the joining surfaces of the components are alloyed with the solder.

Adhesion. In the case of glued joints, the glue adheres to the joining surfaces of the components.

Categorisation according to mobility
Movable connections are, for example,
- saddle guides
- threaded spindles with nuts
- sliding pieces on articulated shafts
- link forks

Here the joined workpieces can change their positions in relation to each other within specific limits **(Fig. 1)**.

Fixed connections are, for example,
- screwed joints
- pin connections
- riveted joints
- press-fit connections

Here the joined workpieces cannot change their positions in relation to each other **(Fig. 2)**.

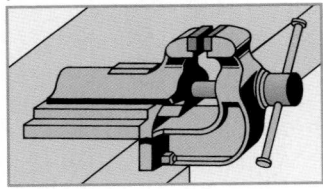 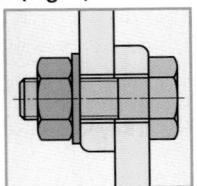

Fig. 1: Saddle guide, as a movable connection **Fig. 2: Bolt, as a fixed connection**

Categorisation according to removability
Removable connections are, for example,
- screwed joints
- featherkey connections
- clamped joints
- pin connections

They can be disassembled and reassembled without destroying the connected components of the connecting element **(Fig. 3)**.

Non-removable connections are, for example,
- weld joints
- soldered joints
- glued joints
- riveted joints

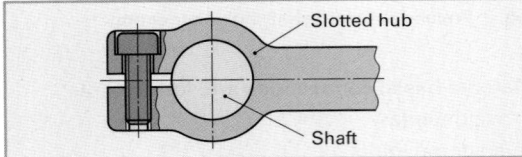

Fig. 3: Clamped joint, as a removable connection

They can be separated again only by destroying the joined components or disassembled again into their individual parts only by destroying the connecting element.

REVIEW QUESTIONS
1 **What are the different types of connections?**
2 **What do you understand by a friction connection?**
3 **What type of connection is established by a splined connection?**
4 **What type of connection is established by a weld joint?**

7.6.2 Threads

Components are often connected by screwing an external thread with an internal thread.

Helical line. This is created when an oblique plane is wound round a cylinder **(Fig. 4)**.

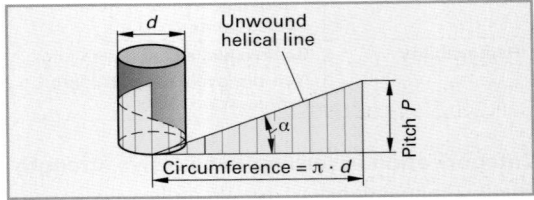

Fig. 4: Helical line

The base line of the oblique plane corresponds to the cylinder circumference and the height of the oblique plane corresponds to the pitch **P** of the helical line. The side opposite the right angle corresponds to the length of the helical line. The angle enclosed by the base line and the helical line is the pitch angle α.

Thread categorisation. A thread can be categorised according to
- thread profile
- running
- application
- thread structure

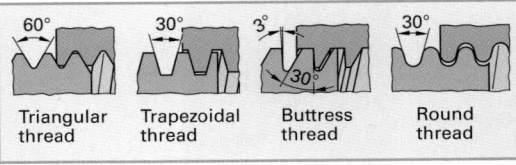

Fig. 5: Thread profiles

Thread profiles (Fig. 5). There are triangular, trapezoidal, buttress and round threads. Sheet-metal and wood screws have special profiles.

Fastening threads are predominantly triangular threads. The friction at the thread flanks gives rise to a self-locking action, i.e. the screws cannot release themselves; this is the case when the pitch angle is less than 15°.

Motion threads are predominantly trapezoidal, buttress and round threads. They can convert rotational into rectilinear motion (e.g. steering gear) or rectilinear into rotational motion.

Running (Fig. 1). There are two types of thread: single- and multiple-start. The running indicates how many thread turns (helical lines) pass round the cylinder. It is determined by the number of thread starts. Multiple-start threads have large pitches. A small turn produces a large axial motion together with high load capacity.

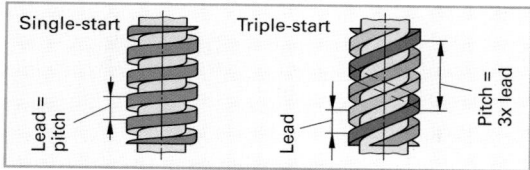

Fig. 1: Single- and multiple-start threads

Thread structure and standardisation. The standards for thread profiles record, among others, the following variables **(Fig. 2):**

- Outside diameter
- Core diameter
- Pitch diameter
- Thread angle
- Pitch
- Rounded areas, curved areas

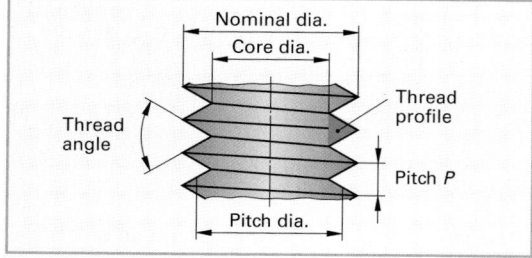

Fig. 2: Thread designations

Metric ISO thread, coarse-pitch thread **(Fig. 3).** The thread angle is 60°.

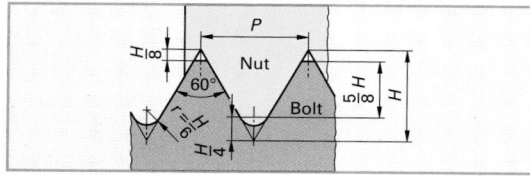

Fig. 3: Metric ISO thread

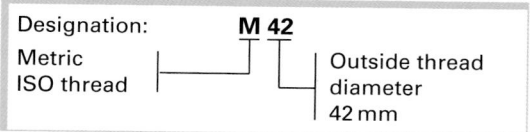

Designation:	**M 42**	
Metric		Outside thread
ISO thread		diameter
		42 mm

Metric ISO fine-pitch thread. While having the same thread angle as a coarse-pitch thread, this has a smaller thread depth. The smaller pitch results, at an identical tightening torque, in a greater initial tension than with coarse-pitch threads.

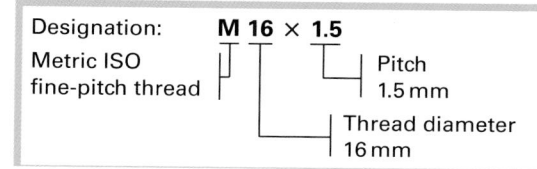

Designation:	**M 16 × 1.5**	
Metric ISO		Pitch
fine-pitch thread		1.5 mm
		Thread diameter
		16 mm

Advantages of fine-pitch threads over coarse-pitch threads:

- Greater sealing effect
- Greater initial force and self-locking with identical tightening torque

REVIEW QUESTIONS
1 How is a helical line created?
2 Why are multiple-start threads used?
3 Which are the most important thread types?
4 What do you understand by thread pitch?

7.6.3 Screwed joints

Screwed joints are usually removable friction connections, however some of them are keyed connections.

The bolts and screws used differ primarily in the shape of their heads and shanks.

The minimum screw-in depth must be taken into account for ensuring a sufficient transmission of force.

Bolts, screws and nuts

Hexagon bolts are used as through-bolts **(Fig. 4)** with nuts when the workpiece is provided with through-holes.

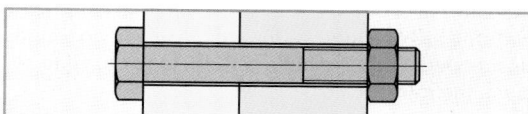

Fig. 4: Hexagon bolt as through-bolt

Hexagon bolts are used as pull-in bolts **(Fig. 1)** without nuts when the female thread is cut into the workpiece.

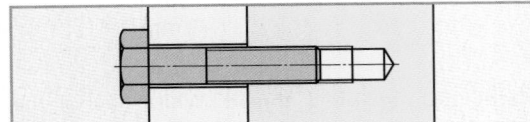

Fig. 1: Hexagon bolt as pull-in bolt

Hexagon-**socket-head cap screws (Fig. 2)** are space-saving alternatives thanks to their cylindrical screw heads. This enables the screwing distances to be kept small. The screw head is frequently sunk into the workpiece. Socket-head cap screws with internal serrations and spline sockets are also used as special forms.

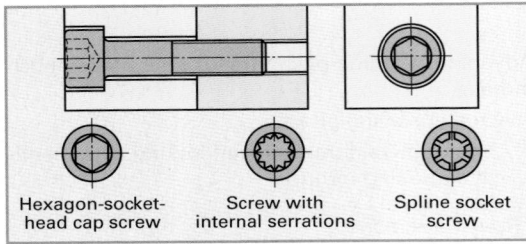

| Hexagon-socket-head cap screw | Screw with internal serrations | Spline socket screw |

Fig. 2: Socket-head cap screws

TORX screws (Fig. 3). These have a head with a six-tooth, star-shaped recess (inner TORX) or a corresponding crown (outer TORX) for engaging the screwing tools. The rounded transitions in the head profile and the flat contact of the screwing tools allow high tightening torques to be safely transmitted without overstressing the screw head and screwing tool.

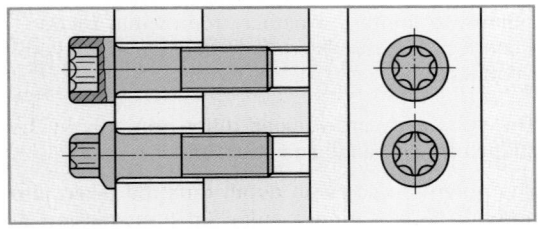

Fig. 3: TORX screws

Stud bolts (Fig. 4) are used when the screwed joint has to be frequently released and the internal thread in the workpiece would be worn by this frequent releasing, e.g. cylinder-head bolts in a light-metal block. The stud bolt is securely inserted with the short end of the thread into the internal thread with the aid of a stud setter and basically no longer unscrewed. Only the hexagon nut is released to release the joint.

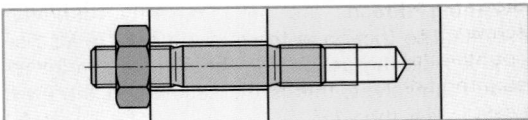

Fig. 4: Stud bolt

Fit bolts (Fig. 5) are, with their ground shanks which are larger in diameter than the thread diameter, fitted into the reamed bores of the workpieces. The bolt shank allows large transverse forces to be transmitted between the workpieces.

Furthermore, large friction forces can be transmitted at the joining surfaces of the workpieces with fit-bolt connections. A fit-bolt connection also guarantees the exact positioning of the workpieces in relation to each other.

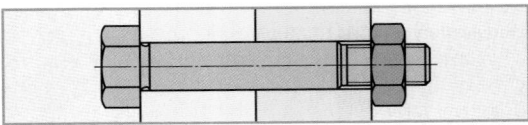

Fig. 5: Fit bolt

Anti-fatigue fit bolts (Fig. 6) are used when the screwed joint is exposed during operation to constant alternating loads, e.g. at the big end of a connecting rod.

Conventional bolts with shanks would work loose or break under constant alternating load after a certain amount of time as a result of fatigue, even if they were designed to be of sufficient strength.

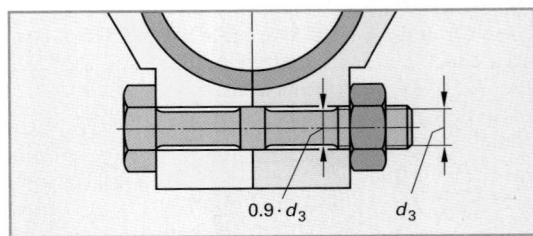

Fig. 6: Anti-fatigue fit bolt

The shank diameter of an anti-fatigue fit bolt is only approx. 90 % of the thread core diameter, except at those points against which the bolt is to rest in the bore. The thin shank renders the bolt elastic of shape.

When tightened with a torque wrench, the stress bolt is pretensioned with a tensile load that exceeds the yield point. In this way, the bolt is subjected to plastic deformation and cannot therefore be reused.

Stress bolts retain their initial stress themselves and therefore do not need to be locked.

Slotted screws and **cross-recessed (Phillips) screws (Fig. 1)** can take the form of cheese-head, countersunk-head, raised cheese-head or raised countersunk-head screws with slots or cross recesses. A screwdriver will centre more effectively in cross-recessed screw heads, enabling them to be tightened more firmly than regular slotted screw heads.

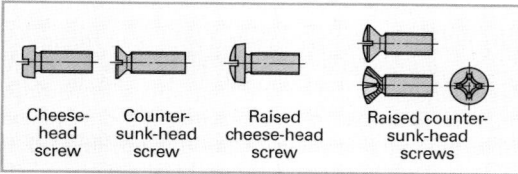

| Cheese-head screw | Counter-sunk-head screw | Raised cheese-head screw | Raised counter-sunk-head screws |

Fig. 1: Slotted screws

Headless setscrews (Fig. 2) are screws with threads over their full length. Depending on the application, they have differently shaped ends, e.g. cone point, full dog point or cup point. They are used to fasten or secure hubs or bearings.

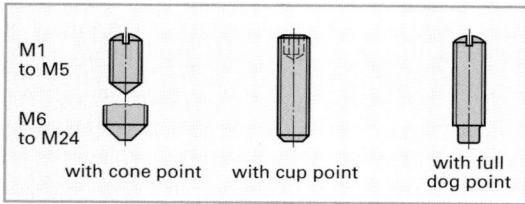

M1 to M5

M6 to M24

with cone point   with cup point   with full dog point

Fig. 2: Headless setscrews

Sheet-metal or self-tapping screws (Fig. 3) are used for connections with metal sheets. These can take the form of slotted, cross-recessed or hexagon screws and tap the female threads themselves as they are screwed in. The predrilled hole in the metal sheet should have roughly the same diameter as the screw core.

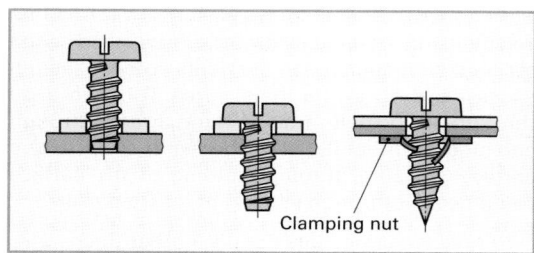

Clamping nut

Fig. 3: Sheet-metal screws

Screw-thread inserts (Fig. 4) are used when the female thread would otherwise have to be cut into a soft material and the screwed joint has to be repeatedly released or when the female thread in the workpiece is destroyed, e.g. spark-plug thread in a light-metal cylinder head.

Such screw-thread inserts are provided with internal and external threads and hardened. They have at their screw-in ends cutting slots or cutting bores, by means of which the thread is automatically cut in the workpiece as the insert is screwed in.

Another type of screw-thread insert is made of rhomboid-shaped steel wire. The steel wire is shaped into a resilient spiral which creates an internal and an external thread. A special screw-tap is used to cut a thread into the core hole of the workpiece and the screw-thread insert is screwed in under initial tension with an installation tool.

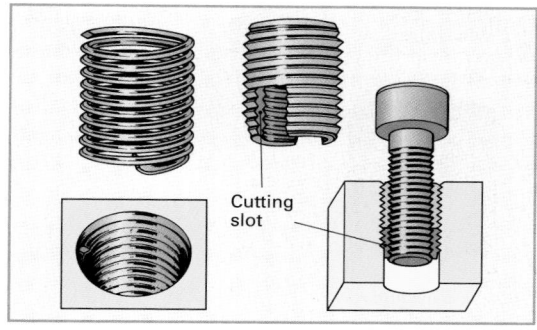

Cutting slot

Fig. 4: Screw-thread inserts

Nuts are made in different varieties to suit different applications. The following different types of nut are used:

Hexagon nuts (Fig. 5) have a height of roughly 0.8 $\times$ d or 0.5 $\times$ d (flat nuts).

Castle nuts with 6 or 10 slots are used when the connections are to be locked with split pins.

Cap nuts (Fig. 1, Page 136) cover the thread externally, protect it against damage, lend the screwed joint an attractive appearance and protect the user against injury.

Union nuts (Fig. 5) are used for pipe fittings.

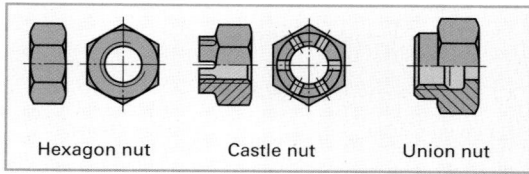

Hexagon nut   Castle nut   Union nut

Fig. 5: Nut varieties

Wing nuts (Fig. 1, Page 136) and knurled nuts can be tightened by hand without auxiliary tools.

Grooved nuts (Fig. 1, Page 136) are used, for example, to secure antifriction bearings on shafts.

Weld nuts and **captive nuts (Fig. 1)** are used in body manufacturing. Weld nuts are usually centred with their collar in the bore and secured by welding spots to the body. Captive nuts are either inserted loose in a metal retainer or suspended so that it can move in the metal retainer with the aid of plastic washers. The plastic washers prevent the nut from being electrostatically charged and thus the thread from being colour-coated during electrolytic dip priming. The metal retainer is welded to the body.

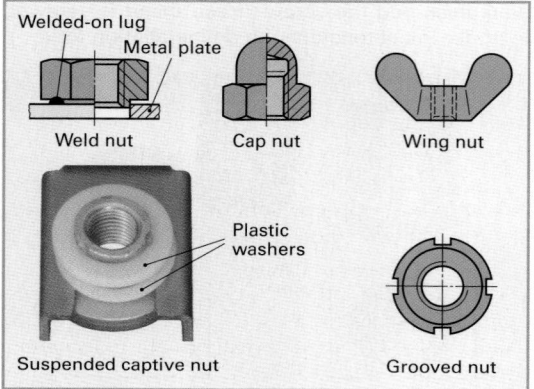

Welded-on lug
Metal plate
Weld nut   Cap nut   Wing nut
Plastic washers
Suspended captive nut   Grooved nut

Fig. 1: Nut varieties

Property (strength) classes of bolts, screws and nuts

Screws and bolts made of steel are marked with their manufacturer's symbol and property class. The property class is given in the form of two numbers separated by a decimal point, e.g. 10.9 **(Fig. 2)**.

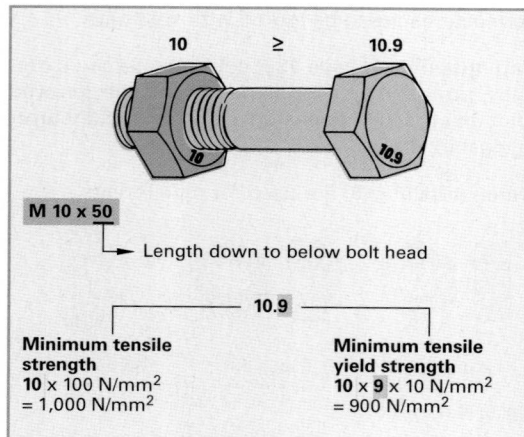

10   ≥   10.9

M 10 x 50
↳ Length down to below bolt head

10.9

Minimum tensile strength
10×100 N/mm^2 = 1,000 N/mm^2

Minimum tensile yield strength
$10 \times 9 \times 10$ N/mm^2 = 900 N/mm^2

Fig. 2: Denotation of property (strength) of bolts, screws and nuts

The first number denotes 1/100 of the minimum tensile strength of the screw/bolt material. The product of both numbers equates to 1/10 of the minimum tensile yield strength of the screw/bolt material. In the case of standard screws and bolts, the minimum tensile yield strength must not be exceeded. The manufacturers' permissible tightening torques must therefore be observed.

Screws and bolts for the most part with a strength rating of 8.8 to 12.9 are used in motor vehicles.

Nuts made of steel are marked with their manufacturer's symbol and property class. The property class denotes 1/100 of the test stress in N/mm^2. Thus the number 10 denotes that the nut can be subjected to a test stress of 1,000 N/mm^2 **(Fig. 2)**.

When pairing up a screw/bolt with a nut, it is essential to ensure that the property classes of nut and screw/bolt match up. E.g. screw/bolt 10.9, nut 10 **(Fig. 2)**.

Screw-locking arrangements

In the case of screwed joints which are exposed to static load, the self-locking effect of the thread is sufficient to secure the joint against loosening. However, alternating loads during operation, vibrations and shocks cause joints to be subjected to dynamic stress. Most of these screwed joints must be secured against unintentional loosening.

The following different locking arrangements are used:

* Friction screw locking
* Keyed screw locking
* Material-based screw locking

Friction screw-locking arrangements (Fig. 1, Page 137) achieve their effect by being installed as resilient elements under the screw head or the nut or by increasing the thread friction.

Spring lock washers, spring washers, toothed lock washers and serrated lock washers are used as the resilient elements. These elements compensate an excessive drop in the initial force. This can arise through plastic deformation in the thread, through creeping of the material in the event of excessive surface pressure, through settling of the surface roughness or through settling of inserted seals.

Two nuts tensioned against each other (lock nuts), nuts with inserted plastic rings, and pinched and slotted nuts are used to increase the thread friction. The high thread friction serves to secure the screwed joint against loosening.

Nut with plastic rings and pinched nuts must only be used once!

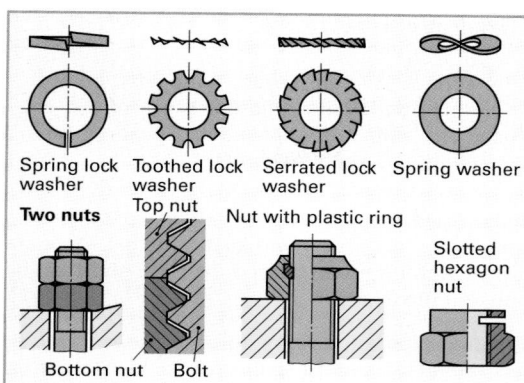

Fig. 1: Friction screw-locking arrangements

Keyed screw-locking arrangements (Fig. 2) achieve their effect through interlocking geometrical shapes, which prevent the joint from twisting loose. Castle nuts with split pins, tab washers, corrugated-head screws and nuts, and wire fasteners are used.

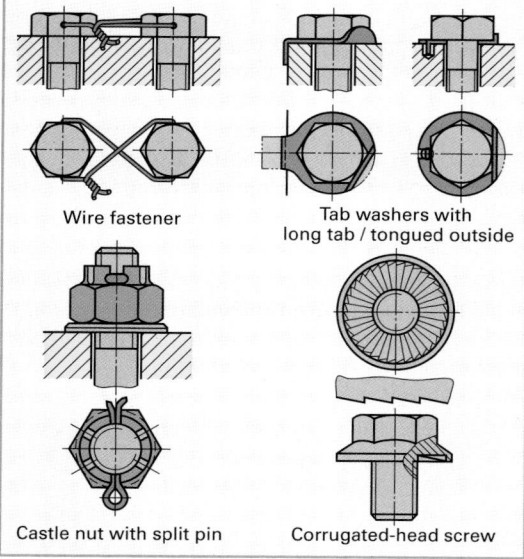

Fig. 2: Keyed screw-locking arrangements

Material-based screw-locking arrangements (Fig. 3) are created when the threads are glued to each other. This can be done with a single-component adhesive in two ways.

In the first scenario, a liquid adhesive is applied to the screw thread and then the screwed joint is established.

In the second scenario, the screws are already coated by the manufacturer with an adhesive base material containing so-called microcapsules. When the screws are inserted, the microcapsules burst open

and coat the base material. In both scenarios, the adhesive hardens on contact with metal under the exclusion of air and thereby secures the screwed joint.

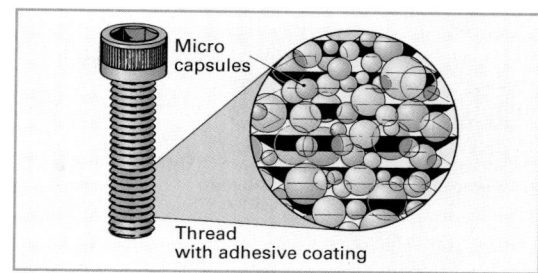

Fig. 3: Material-based screw-locking arrangement

Screwing tools (Fig. 4; Fig. 1, Page 138)

Wrenches must fit the heads of the screwed joints. The tightening torques set out in tables or manufacturer's instructions must be observed without fail!

This approach ensures that screws and nuts are tightened neither too tightly (overtightening could result in their being damaged) nor too loosely (undertightening could result in an inadequate connection).

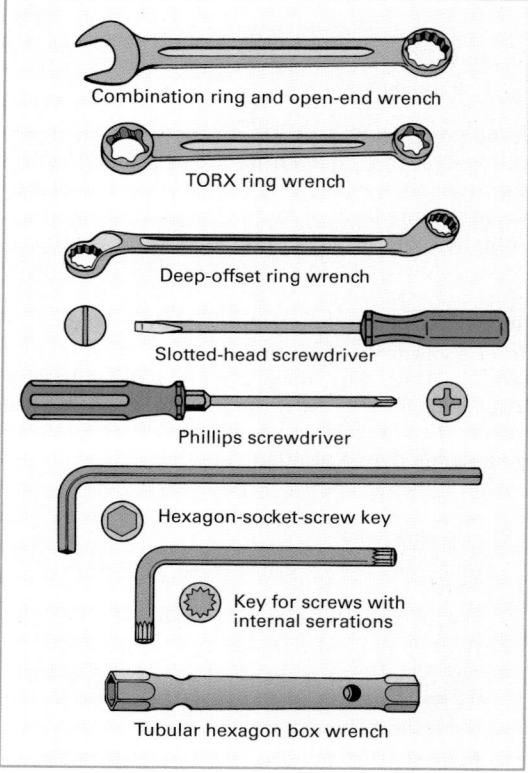

Fig. 4: Screwing tools

7

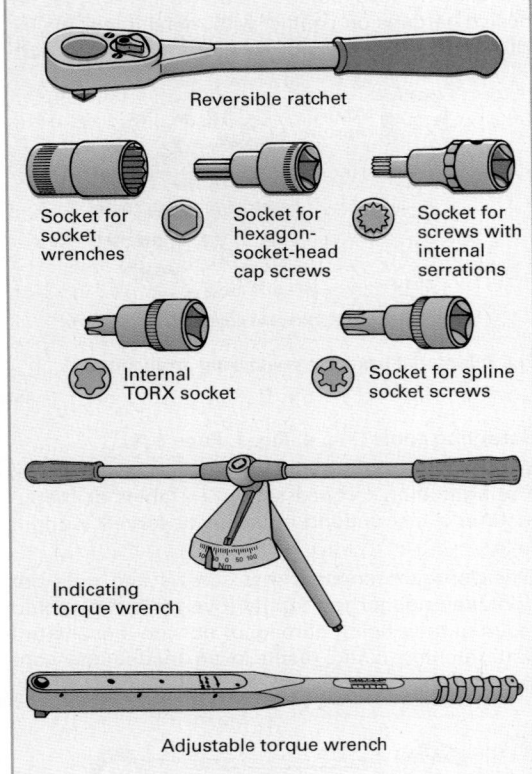

Fig. 1: Screwing tools

Torque wrench (Fig. 1). This is the only way of ensuring that the screwed joint is tightened to the correct initial tension. This type of wrench indicates on a scale the level of torque to which the screw is tightened, or torque wrenches are used which can be set to a specific torque. As soon as this specific torque is reached as the screw is being tightened, the wrench can be heard and felt to disengage.

Torque wrenches must be calibrated at regular intervals. After they have been used, adjustable torque wrenches must be set to the smallest torque. This ensures that the setting accuracy is maintained over a prolonged period of time.

7.6.4 Pin connections

Pin connections are keyed, removable connections. The following different types of pin are used to suit the application: locating pins, shearing pins and fixing pins.

Fixing pins connect two or more workpieces to each other by friction and keyed means. They are driven with interference into the bore and thereby transmit the forces.

Locating pins are designed not to transmit forces but rather to determine the precision position of two workpieces in relation to each other. They prevent any displacement of the workpieces, above all during installation, and thus simplify assembly.

Straight pins (Fig. 2) are manufactured with a chamfered end or with a rounded end. Straight pins with chamfered ends are manufactured from bright-drawn round steel (tolerance zone h8) and can usually be inserted with play into a reamed bore. Straight pins with rounded ends are ground (tolerance zone m6) and must normally be driven with interference into a reamed bore.

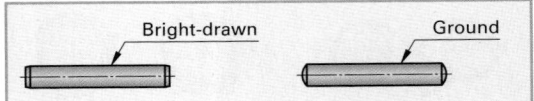

Bright-drawn   Ground

Fig. 2: Straight pins

Taper pins (Fig. 3) have a taper of 1:50 and can be used in reamed bores. This type of pin is driven in with a hammer until the end of the pin is flush with the workpiece.

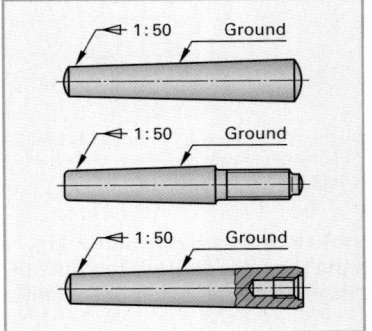
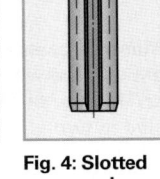

Fig. 3: Taper pins   Fig. 4: Slotted spring pin

When the connection is released, the taper pin must be driven out from through-holes from the opposite end. Taper pins for bottom holes have a threaded stem or an internal thread to enable them to be pulled out of the hole again.

Shearing pins protect sensitive components against overstraining. They transmit the entire driving force and are strained to shearing. The shearing pin is the predetermined rupture point of the connection. In the event of excessive strain, it is shorn off and with it the connection is broken.

Pins can be categorised by their shape into the following:

- Straight pins
- Taper pins
- Slotted spring pins
- Grooved pins

Slotted spring pins (Fig. 4, Page 138) are slotted hollow cylinders made of spring steel. Their diameter is 0.2 mm to 0.5 mm larger than the bore in the workpiece. As it is driven in, a slotted spring pin is subject to elastic deformation and generates the necessary contact pressure.

Grooved pins (Fig. 1) are cylindrical pins which are provided with three rolled-in grooves around their circumference. Different ways of rolling in the grooves produces different shapes of grooved pins. When such a pin is driven into the bore, the beads are partially pressed back into the grooves to produce a secure seat, even in non-reamed bores.

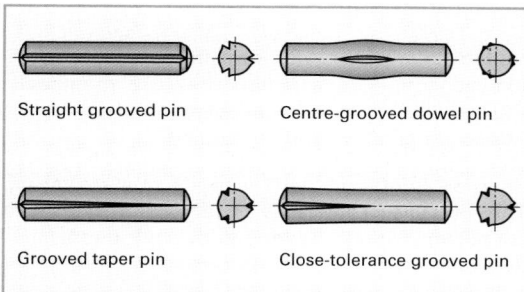

Straight grooved pin   Centre-grooved dowel pin

Grooved taper pin   Close-tolerance grooved pin

Fig. 1: Grooved pins

7.6.5 Riveted joints

> Riveted joints are non-removable connections. When they are manufactured, the protruding rivet shank is formed by upsetting or bordering into the closing head. Rivets can be categorised by the shape of their head and shank and by the riveting process.

Riveted joints are especially suitable for light-alloy construction because the strength of precipitation-hardenable aluminium alloys would be reduced if they were welded.

Approximately 1,800 punch-rivet joints are used, for example, in the light-alloy body of an Audi A2.

Riveting materials. The riveting material must be able to be easily formed and exhibit sufficient strength; it must not tear when forming the closing head. In the interests of avoiding corrosion, the rivets should as far as possible be made of the same material as the workpieces. Rivets composed of steel, copper, copper-zinc and aluminium alloys are generally used.

Riveting process (Fig. 2). The rivet used is placed with the swage-head on the fixed stay. The workpieces are pressed together with the rivet setter. The rivet shank is upset and bevelled on. The closing head is formed with the riveting die. In order to form the closing head, the protruding shank end must be of a certain length, e.g. 3 mm for hollow rivets with a shank diameter of 4 mm. The completed rivet comprises swage-head, shank and closing head.

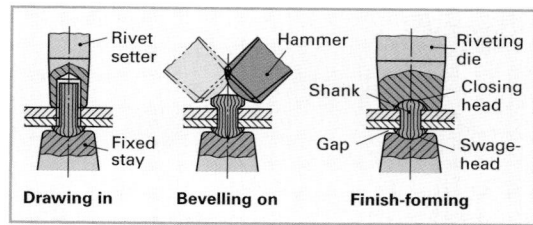

Drawing in   Bevelling on   Finish-forming

Fig. 2: Riveting operations

Rivet types

Rivets with full shanks (Fig. 3) are used when large forces are to be transmitted. They are used, for example, for connecting bearing components to frames of commercial vehicles.

Hollow rivets (Fig. 4) and rivets with spot-drilled shanks are used to rivet on clutch linings and brake linings. The closing head is formed by bordering the shank end.

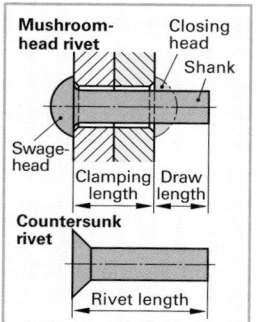
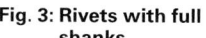

Fig. 3: Rivets with full shanks

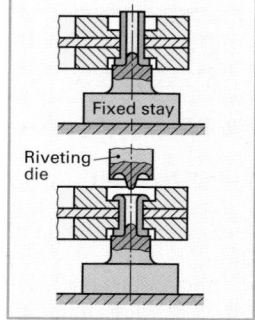

Fig. 4: Hollow rivet

In the case of rivets with full shanks, above all mushroom-head rivets and countersunk rivets are distinguished by the **shapes of their heads.** Hollow rivets usually have flat heads.

Blind rivets are, for example, mandrel rivets, blind-rivet nuts and body-bound rivets. These are used when the riveting point is only accessible from one side.

Mandrel rivets (Fig. 1) consist of a hollow rivet with mandrel. The protruding shank end is formed into the closing head by the mandrel using riveting tongs. The mandrel breaks at the predetermined rupture point.

Blind-rivet nuts (Fig. 2). A threaded mandrel, which draws in the protruding rivet shank during riveting with the riveting tongs and forms the closing head, is screwed into the internal thread of the rivet nut.

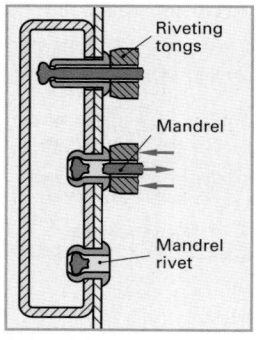
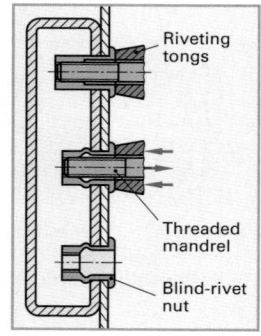

Fig. 1: Mandrel rivet   **Fig. 2: Blind-rivet nut**

Body-bound rivets (Fig. 3) are hollow rivets with slotted shank ends into which a grooved pin is inserted. The slotted shank end is expanded when the grooved pin is driven in.

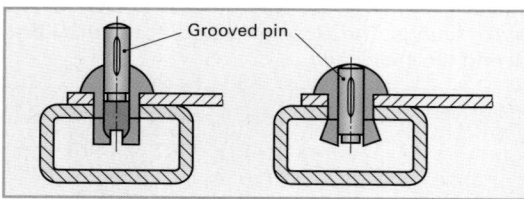

Fig. 3: Body-bound rivet

Punch riveting

Punch riveting **(Fig. 4)** involves a semi-hollow rivet, for example, being punched by the riveting-tool die through the metal-sheet layer on the die side. The second metal-sheet layer is plastically deformed by the rivet without being penetrated by it. The shape of the die-plate causes the foot of the semi-hollow rivet to expand and form a closing head.

Steel punch rivets are used to prevent contact corrosion.

Advantages of punch-rivet joints are:

- No prepunching of metal sheets required.

- No cutting through of the lower metal sheet and thus a tight joint.
- High strength.
- Requires less energy then spot welding.

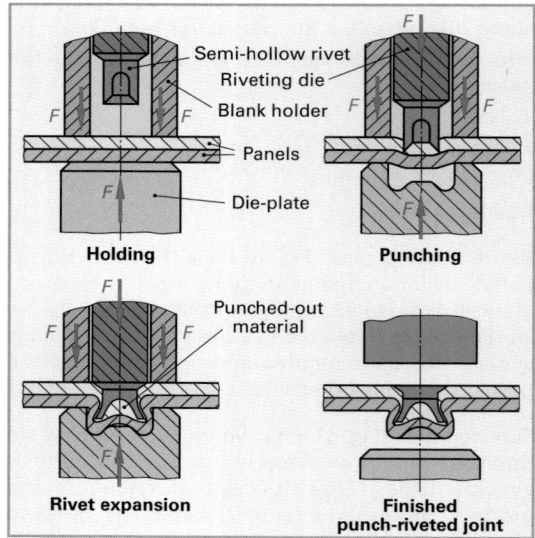

Fig. 4: Punch-riveting process

7.6.6 Clinching

Clinched joints are non-removable, friction and keyed connections. When they are manufactured, two or more metal sheets or profile sections are clinched and then upset.

Clinched joints are used in body manufacturing. This process produces a cheap, quick and clean connection for non-bearing body components.

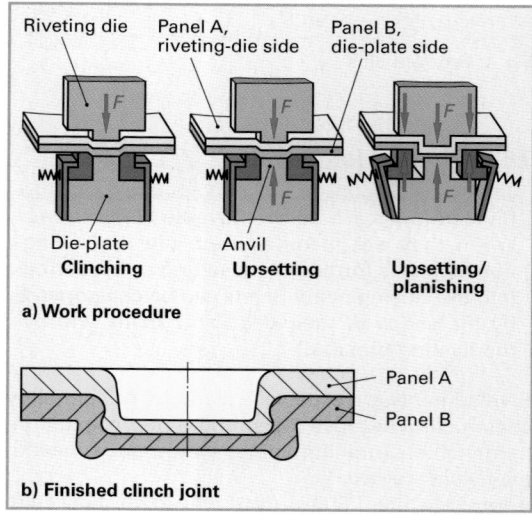

a) Work procedure

b) Finished clinch joint

Fig. 5: Clinching

Clinching process. The process is completed in three stages. In the **clinching** stage, the die **(Fig. 5, Page 140)** presses the two metal sheets into a die-plate until the latter contacts the anvil. In the **up-setting** stage, the anvil and the die press against each other and cause the material to strain-harden. The material diameter increases and the material thickness decreases in the upsetting area. The side sections of the die must therefore be movable.

In the **planishing** stage, the joint is further upset and at the same time levelled.

REVIEW QUESTIONS

1 **Which pins are distinguished by their shape?**
2 **What are the functions of locating pins?**
3 **In what situations are blind rivets used?**
4 **What are the advantages of punch-rivet joints?**
5 **For which body components is clinching suitable?**

7.6.7 Shaft-hub connections

Shaft-hub connections are for the most part keyed, non-removable connections. The keyed connection ensures safe transmission of the torque; removability simplifies installation and removal.

Shaft-hub connections of this type are

- featherkey connections
- serration profiles
- spline connections
- gear profiles
- woodruff key connections

Featherkey connections (Fig. 1). A longitudinal keyway is milled into the shaft. The hub has a longitudinal keyway of the same width in its bore. The featherkey inserted into the keyway transmits with its side faces the peripheral force from the shaft to the hub. A drive-type connection is created by the featherkey. The featherkey connection must be secured against axial displacement of the hub on the shaft.

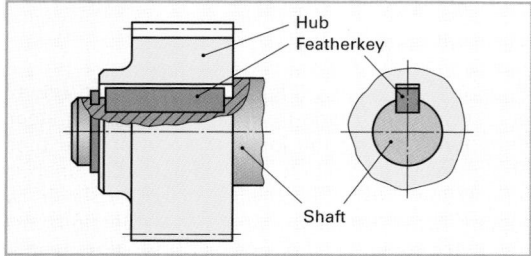

Fig. 1: Featherkey connection

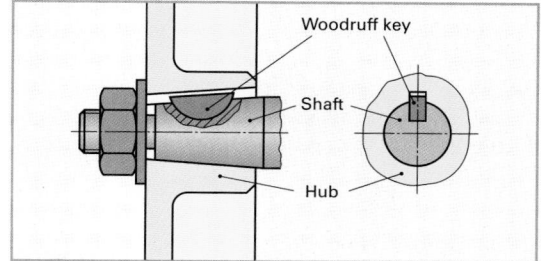

Fig. 2: Woodruff key connection

Woodruff key connections (Fig. 2). Here the shaft is greatly weakened by the keyway milled deep into its cross-section. Connections of this type can only transmit small torques. Woodruff key connections on tapered shaft ends have primarily a locking effect. The torque is transmitted by friction via the taper seat.

Spline connections (Fig. 3). Spline connections are used in the case of shocklike torque loads because the torque can be transmitted to numerous distributed gearings. Furthermore, spline connections are particularly suitable for moving shaft-hub connections, e.g. clutch plate on a gearbox input shaft or sliding section on a propeller shaft.

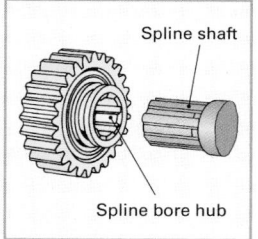
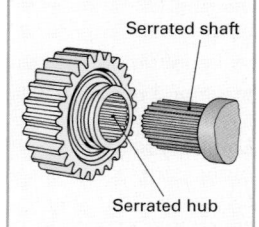

Fig. 3: Spline profile   **Fig. 4: Serration profile**

Serration profiles (grooved toothing, **Fig. 4)** with their finer profiles do not weaken shaft and hub as much by deep keyways as spline profiles and transmit the torque even more effectively to the circumference.

On account of the smaller pitch, the positions of shaft and hub can be optimally allocated to each other, e.g. the steering wheel on the steering spindle or the torsion bar in the spring arm.

Gear profiles (Fig. 1, Page 142) use, for example, involute gear teeth above all for the flexible connection of shaft and hub. This type of toothing is the same as the toothing normally used on gearwheels. Involute gear profiles are used, for example, in viscous couplings or multi-plate clutches to guide the plates and to transmit force between the plates and the clutch drum.

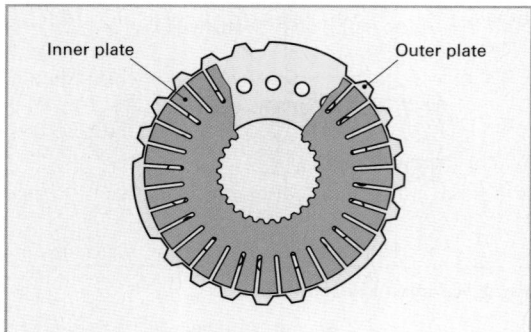

Fig. 1: Clutch plates with gear profiles

7

7.6.8 Press-fit joints

> Press-fit joints are friction connections. There is interference between the outer and inner parts prior to joining. The forces are transmitted by static friction between the joining surfaces.

Press-fit joints are used, for example, to join valve-seat inserts, ring gears, antifriction bearings and bearing bushings. The following different processes are used:

- Longitudinal pressing
- Lateral pressing

Longitudinal pressing (Fig. 2). Here, the inner and outer parts are pressed cold in an axial direction into each other, usually with the aid of a press. The roughness peaks of the joining surfaces are levelled slightly. This reduces the interference between the inner and outer parts.

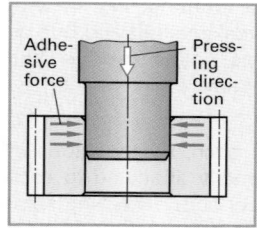
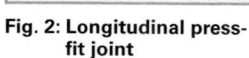
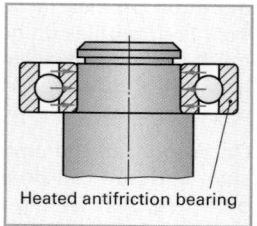

Fig. 2: Longitudinal press-fit joint

Fig. 3: Lateral press-fit joint, heating of outer part

Lateral pressing. Here, the interference is eliminated prior to joining by heating the outer part or supercooling the inner part. The outer and inner parts can then be joined with play and without pressure forces. In contrast to longitudinal pressing, there is no levelling of the roughness peaks.

Working rules

- Ensure that the components are heated uniformly. E.g. heat in an oil bath or on a steam plate.
- Always wear gloves when cooling, for example, with liquid nitrogen or dry ice (solid CO_2, $-78\,°C$).
- Make sure that adequate preparations have been made (e.g. provide suitable tools) because lateral pressing must be completed within the shortest possible time.

REVIEW QUESTIONS
1 What different types of shaft-hub connection are there?
2 What advantages does a serration profile have over a spline profile?
3 Where in a motor vehicle are press-fit joints used?
4 What are the differences between a lateral and a longitudinal press-fit joint?

7.6.9 Snap-in connections

> Snap-in connections are created by the elastic deformation of at least one of the joining parts. They come in both detachable and nondetachable forms.

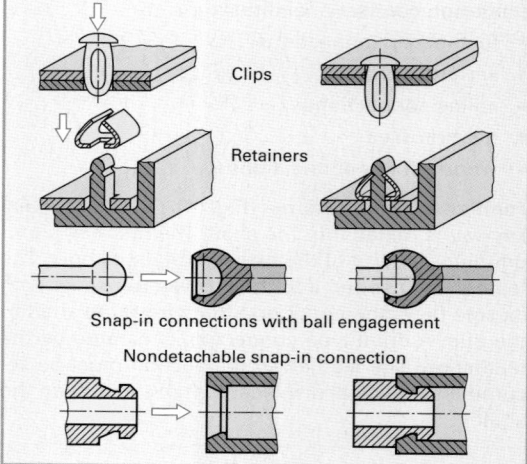

Fig. 4: Snap-in connections

Snap-in connections **(Fig. 4)**, e.g. clips and retainers, are usually made from plastic or spring steel. They bend during the joining operation and then straighten up again. If a nondetachable snap-in connection is to be established, this is achieved by means of an undercut. Snap-in connections are used, for example, for wheel caps, trim panels, valve-steam seals and moving linkages.

7.6.10 Soldering

Soldering is the material-based joining of metal workpieces by means of a molten filler metal (solder). The workpieces remain in a solid state. The melting temperature of the solder is lower than the melting temperature of the workpieces to be soldered.

Two different types of soldering process are used based on the melting temperature: soft soldering and hard soldering (brazing).

Soldering process	Melting temperature
Soft soldering	under 450 °C
Brazing	over 450 °C

Soldering processes

The soldering surface is heated to the soldering temperature. The added solder melts on the soldering surface. The solder should moisten the surface in the process over a spread-out area (**Fig. 1**). In this way, the solder can penetrate into the boundary layer and form an alloy with the material. It is essential to have clean, non-oxidised surfaces and a narrow solder gap (**Fig. 2**) for this to occur.

Fluxes are used to dissolve oxide layers on the soldering surfaces and prevent them from reforming.

Commonly used fluxes are, for example, FH 20 (multipurpose flux) for brazing and F-W 31 (non-corroding in paste form) for soft soldering.

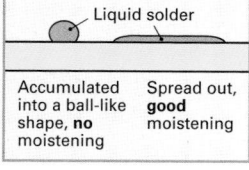

Fig. 1: Moistening of sol-
dering surface

Fig. 2: Capillary action
in solder gap

The soldering process is completed in three stages:

Moistening. The soldering is spread over the soldering surface.

Flowing. Liquid solder forces flux out of the solder gap.

Binding. Solder penetrates into the material and forms an alloy.

Soft soldering ($t < 450°$**).** Soldered joints are suitable for

- connections which must be tight, e.g. in radiator construction.
- connections where good electrical contact is required.

Primarily connections made of copper, copper alloy and zinc are soft-soldered.

Brazing ($t > 450°$**).** Brazed joints are suitable when …

- … exacting demands are made on strength, e.g. sheet-metal repairs.
- … the connection must be strong and tight, e.g. tanks/reservoirs.
- … the connection must be heat-resistant, e.g. lathe tools with carbide inserts.
- … welding processes cannot be used on account of the high temperatures or the materials used.

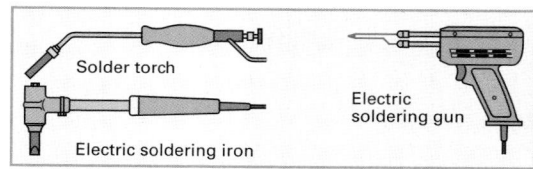

Fig. 3: Soldering iron and soldering gun

Depending on the application, solders with different properties must be chosen, for example, for soft soldering.

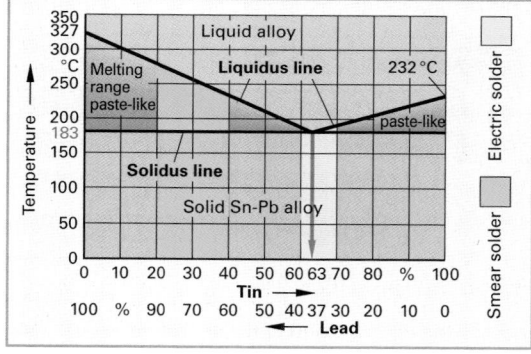

Fig. 4: Tin-lead constitutional diagram

Electric solder (Fig. 4). Soldering electrical components calls for as low a soldering temperature as possible in order to prevent the components in question, e.g. transistors, from being thermally overloaded.

Smear solder (Fig. 4). When smearing body sheet panels, it is important to choose a solder which can be worked (paste-like) over a wide temperature range.

| | Identifying abbreviation | | |
	New	Old	Application
Soft solder	S-Sn60Pb40	L-PbSn40	Electronics
	S-Pb98Sn2	L-PbSn2	Radiator construction
	S-Pb58Sn40Sb2	L-PbSn40Sb	Smear solder
Hard solder	AG 104	L-Ag45Sn	Steels, copper, Cu alloy
	CU 104	L-SFCu	Steels

Working rules
- Clean the soldering point before soldering.
- Before soldering, match up the workpieces with a thin solder gap.
- Heat the workpiece and solder quickly and uniformly.
- Select the correct working temperature.
- Solder must solidify in vibration-proof form.
- Remove remnants of corrosive flux after soldering.
- Avoid skin contact with the flux.
- Ventilate the work area adequately or provide a vapour-extraction facility.

7.6.11 Welding

Welding is a method of non-removable, material-based joining usually of materials of the same kind. The workpieces are connected at the joint through the application of heat in a liquid state or through the application of heat and pressure in a paste-like state.

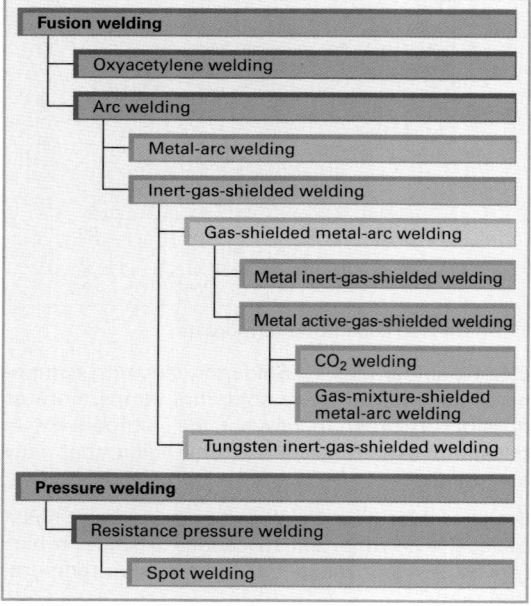

Fig. 1: Categorisation of welding processes

Weld joints are the most commonly used material-based connections. Two different processes may be used: **fusion welding** and **pressure welding** (**Fig. 1**).

It is essential when welding to take appropriate protective measures which suit the respective welding process. The welder must wear his/her protective clothing and equipment!

Gas-welding (oxyacetylene welding)

In the case of gas-welding (**Fig. 2**), also known as autogenous welding, the material is brought to melting by the heat of a fuel-gas/oxygen flame and joined with the aid of a filler metal. The fuel gas mainly used is acetylene (C_2H_2) because a high flame temperature of approx. 3,200 °C is achieved with acetylene.

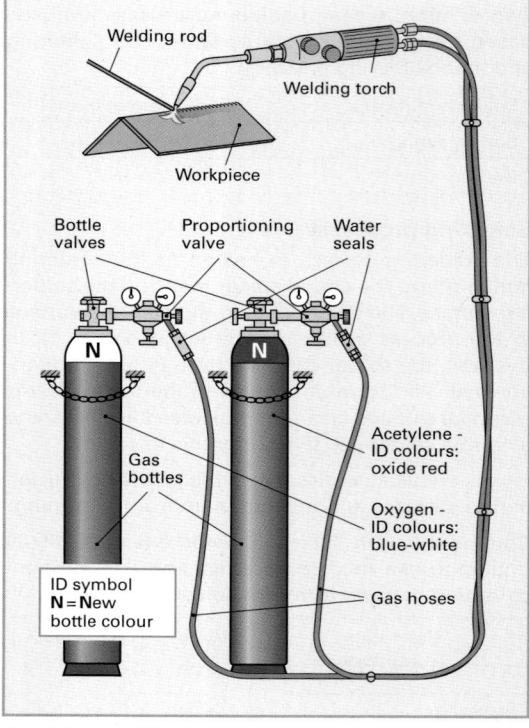

Fig. 2: Gas-welding apparatus

Oxygen bottle. Pure oxygen is stored under pressure in this bottle (cylinder). With a bottle volume of 40 litres and at a filling pressure of 150 bar, the capacity amounts to $150 \times 40\,l = 6,000\,l$ oxygen. It has an R $^3/_4$ inch connection and its identifying colour is blue, or from 1. 7. 2006 blue-white.

The bottle valve on the oxygen bottle must be kept free from grease and oil. Failure to meet this requirement may cause oxygen and hydrogen compounds of the grease to form detonating gas. This in turn gives rise to an explosion hazard.

Acetylene bottle. Acetylene cannot be stored under high pressure as this would cause it to decompose. Acetylene is dissolved in acetone when it is introduced into steel bottles.

An acetylene bottle incorporates a porous mass whose pores are filled with acetone; the acetylene is dissolved in this acetone. At a filling pressure of 18 bar, an acetylene bottle contains approx. 6,000 l acetylene. The acetylene is released again when the gas is extracted (pressure drop). An acetylene bottle has a cliplock and its identifying colour is yellow, or from 1. 7. 2006 oxide red.

> In order to prevent acetone from escaping, it is essential when welding not to lay the acetylene bottle on its side.
> Never extract more than 1,000 l per hour; if necessary, connect up several bottles.

Water seal (Fig. 2, Page 144). To protect against flash-back and gas flowback, the acetylene tube must in-corporate a water seal between the reducing regula-tor and the welding torch. This seal is connected as a non-return seal to the reducing regulator or as a sin-gle-bottle seal to the welding torch. A water seal is often connected to the oxygen hose as well.

> Prior to connecting the reducing regulator, open the bottle valve briefly in order to blow dirt particles out of the connection. Turn back the adjusting screw on the reducing regulator.

Reducing regulator (Fig. 1). The reducing regulator is connected to the bottle valve and serves to re-duce the high pressure in the bottle to the respec-tive working pressure. The volume pressure gauge registers the bottle pressure; the working pressure gauge regulates the working pressure, which can be adjusted at the adjusting screw. For welding, the working pressure is approx. 2.5 bar for oxygen and 0.25 bar to 0.5 bar for acetylene.

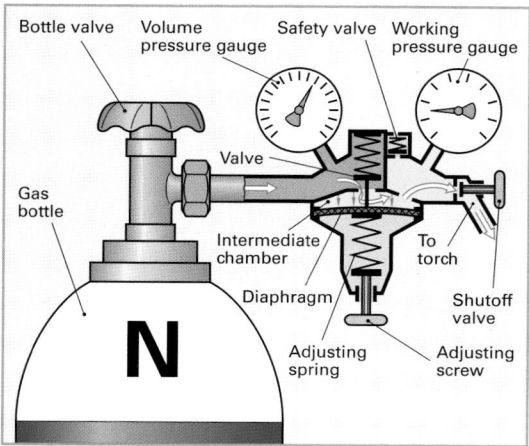

Fig. 1: Reducing regulator

Welding torch (Fig. 2). A low-pressure torch is pri-marily used for welding. The acetylene is drawn in-to the torch by the oxygen emerging under greater pressure.

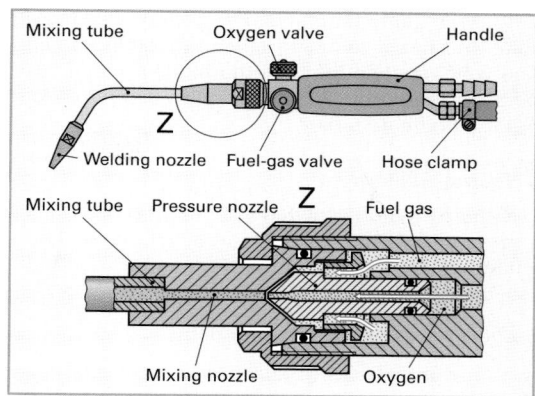

Fig. 2: Welding torch

The welding torch is made up of a handle and an interchangeable torch insert. The components of the torch insert are: pressure nozzle (injector noz-zle), mixing tube with mixing nozzle, welding noz-zle and union nut. Acetylene and oxygen are mixed in the mixing nozzle and mixing tube in or-der to burn ahead of the welding nozzle in the form of a jet flame.

The welding torch is connected by way of rubber hoses to the gas bottles. Red hoses for acetylene and blue hoses for oxygen are used. While having the same outside diameter, the oxygen hose has a smaller inside diameter than the acetylene hose.

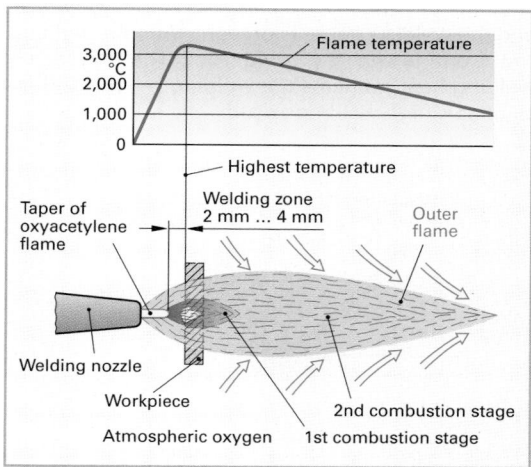

Fig. 3: Oxyacetylene flame

Oxyacetylene flame (Fig. 3). This is adjusted at the valves of the welding torch. Under normal adjust-ment conditions oxygen and acetylene are mixed in the ratio of 1:1. However, with this ratio, the oxygen is not sufficient to burn the acetylene com-pletely; this is only achieved with the oxygen from the ambient air. This creates an oxygen-free zone before the flame taper, which is called the welding zone and has a reducing effect (absorbs oxygen). The welding zone features the highest flame temperature of approx. 3,200 °C roughly 2 mm to 4 mm before the flame taper.

> **Neutral adjustment:**
> With an oxygen/acetylene mixture ratio of 1 : 1, the white flame taper is greatly limited; this flame adjustment is known as "neutral".

Steel is welded with a neutral flame.

In the event of excess acetylene, the flame taper flickers and is greenish in colour. In this case, the flame carries free carbon, which penetrates into the weld seam. This hardens due to enrichment with carbon.

In the event of excess oxygen, the flame taper shortens and is blueish in colour. The seam absorbs oxygen and becomes brittle; the flame has an oxidising effect.

Welding-torch and welding-rod guidance.
Both "left-hand" and "right-hand" welding can be performed.

Left-hand welding (Fig. 1) is used to weld thin metal sheets up to 3 mm thick. The welding flame points in the welding direction, the melting bath is situated outside the highest flame temperature, and the welding flame is prevented by the welding rod from fusing the weld root. The advancing welding heat preheats the welding groove and facilitates a high welding speed.

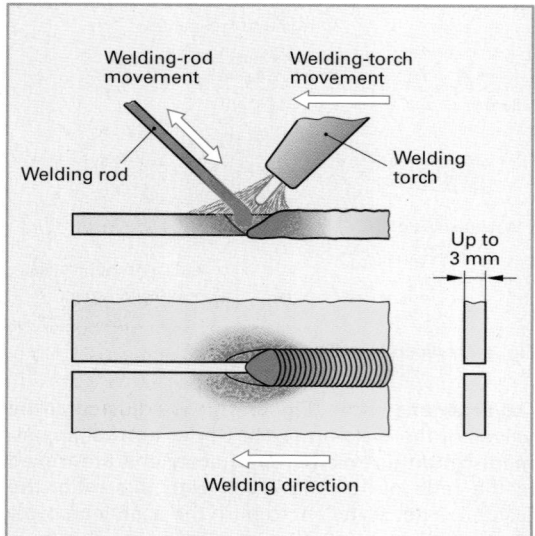

Fig. 1: Left-hand welding

Right-hand welding is used to weld thick metal sheets over 3 mm thick.

Gas-welding rods for joint welding are categorised according to their composition and suitability for the different steels into seven welding-rod classes, G I to G VII. Welding-rod classes G II to G IV are especially suitable for general structural steels.

The welding-rod class is engraved on each welding rod. Welding rods come in assorted diameters and their copper plating acts as corrosion protection.

Flame cutting of steel

Flame cutting utilises steel's property of burning in pure oxygen. A **cutting torch (Fig. 2)** is a type of welding torch, to which a further tube with a valve is attached in order to supply the cutting oxygen.

The cutting point is heated by the preheating flame of the cutting torch to ignition temperature (approx. 1,200 °C). When the cutting-oxygen valve is opened, the focussed oxygen jet of the cutting nozzle strikes the glowing point. Burning of the steel in pure oxygen proceeds very quickly. The slag is blown out of the cut by the pressure of the oxygen jet.

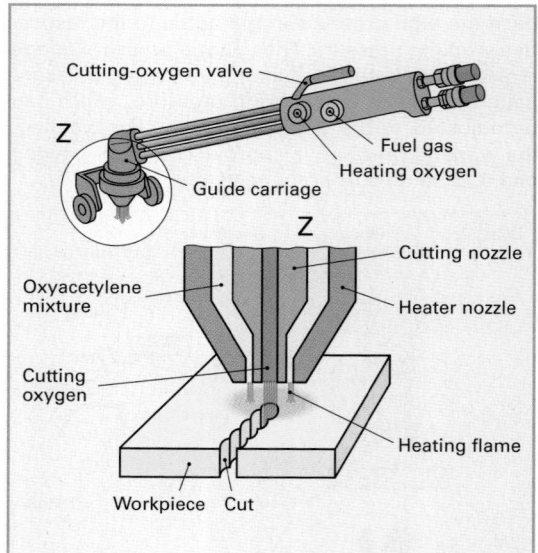

Fig. 2: Flame cutting of steel

Metal-arc welding

> In the case of metal-arc welding **(Fig. 1, Page 147),** the heat of the electric arc is utilised to melt the materials at the weld.

An arc is generated after a brief short circuit between electrode and workpiece and forms an electrically conductive gas trace of high temperature. The material melting off the electrode forms the welding bead with the melted-on workpiece material. The arc should be as short as possible (arc length ≈ electrode diameter) in order to keep the absorption of oxygen and nitrogen into the melting bath at a low level.

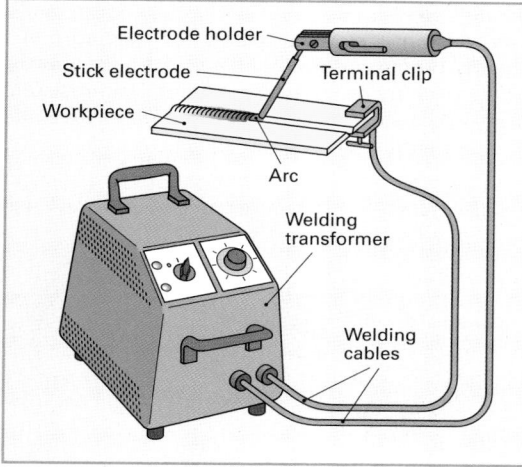

Fig. 1: Metal-arc welding

Welding-current sources. Welding transformers are used as the welding-current sources for welding with alternating current while welding rectifiers are used for welding with direct current. In garages/repair shops welding is often performed in confined spaces or between electrically conducting parts. In such situations the open-circuit voltage must not exceed 48 V when welding with alternating current or 113 V when welding with direct current. Such welding-current sources are specifically marked (**Table 1**).

Table 1: Marking of welding apparatus for welding under increased electrical hazard conditions		
Welding apparatus	Max. open-circuit voltage	Marking
Welding transformer	48 V	S
Welding rectifier	113 V	S

Stick electrodes (Fig. 2) consist of core wire and a coating.

The melting-off **core wire** forms the welding bead with the melted-on workpiece material. The **coating** melts off with the core wire and forms the slag on the weld seam. The slag slows down the rate at

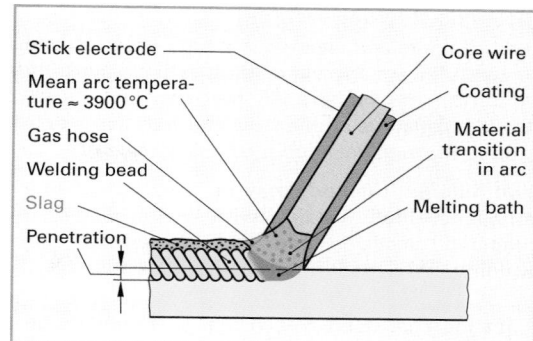

Fig. 2: Electric arc

which the seam cools. Contraction strains are reduced in this way. Part of the coating reduces to gas as it melts off and forms a volatile covering to shield the arc and the weld seam in the vicinity of the melting bath against ambient air and thereby reduces the melting loss of alloying constituents. The electrically conductive volatile covering also facilitates a uniform arc. When welding with alternating current, the arc would otherwise have to be constantly reignited since the current direction alternates constantly.

Welding tools (Fig. 3). In the electrode holder the handle and clamping fixture, apart from the contact surface for the electrode, are insulated to provide protection against voltage and burns. The pick hammer and wire brush are used to remove the slag. The welder's face shield is fitted with special shaded glass (protective filter), in front of which clear glass is usually fitted. Gauntlet gloves and aprons, usually made of leather, provide protection against jets, flying sparks and burns.

Always wear personal protective clothing and equipment when welding!

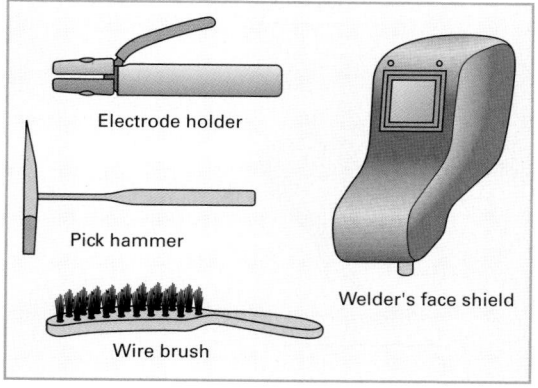

Fig. 3: Welding tools

Inert-gas-shielded welding

Inert-gas-shielded welding is a type of arc welding in which the arc and the melting bath are surrounded by an atmosphere of inert gas and thereby shield against the ambient air. The inert gas is delivered to the weld location by the welding torch.

Two different methods may be used: tungsten inert-gas-shielded welding with a non-melting-off tungsten electrode and gas-shielded metal-arc welding with a melting-off wire electrode.

> The inert gas used in each case is dependent on the welding process and the material to be welded.

The welding torch is guided by hand, by fully mechanised means or automatically. Welding torches for welding thinner metal sheets are air-cooled. The torches are water-cooled for thick sheets and high welding currents.

The advantages of inert-gas-shielded welding are:

- No ambient air in the melting bath
- No burning of alloying constituents
- No slag formation
- High welding speed
- Narrow heating zone
- Low distortion

Welding direction. Both "pricking" and "trailing" welding are used **(Fig. 1)**.

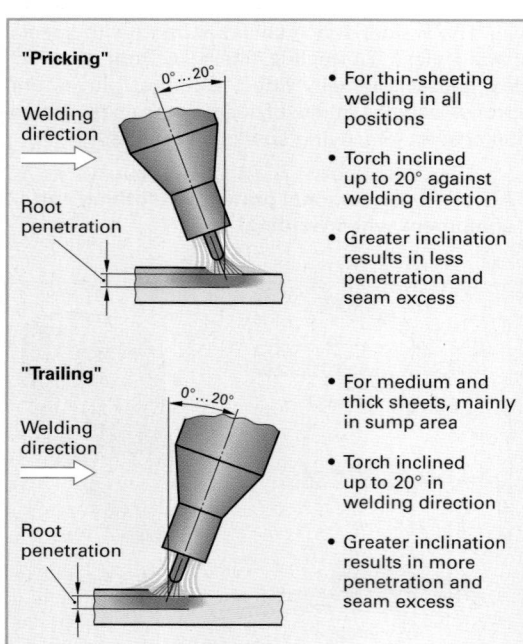

"Pricking"

0°...20°

Welding direction

Root penetration

- For thin-sheeting welding in all positions
- Torch inclined up to 20° against welding direction
- Greater inclination results in less penetration and seam excess

"Trailing"

0°...20°

Welding direction

Root penetration

- For medium and thick sheets, mainly in sump area
- Torch inclined up to 20° in welding direction
- Greater inclination results in more penetration and seam excess

Fig. 1: Pricking and trailing welding

Tungsten inert-gas-shielded welding (TIG welding, Fig. 2)

The arc burns between a tungsten electrode, which does not burn off, and the workpiece. The welding rod is guided by hand from the side into the melting bath. Depending on the workpiece material, welding with direct current or with alternating current is used. The slow-to-react noble gas argon or a mixture of argon and helium is used as the inert gas.

TIG welding is suitable mainly for welding metal sheets, profile sections and pipes up to approx. 5 mm thick made of heat-resistant, acid-resistant or stainless steels, of copper or copper alloys and of aluminium or aluminium alloys.

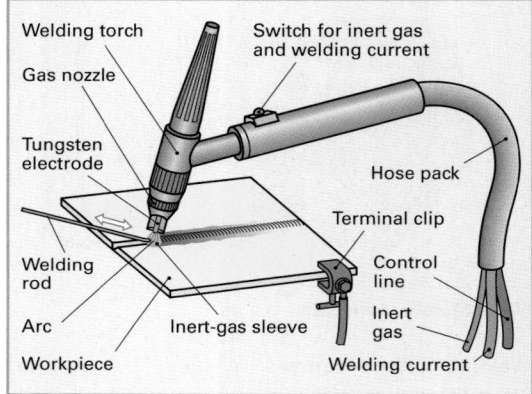

Welding torch

Gas nozzle

Tungsten electrode

Welding rod

Arc

Workpiece

Switch for inert gas and welding current

Hose pack

Terminal clip

Control line

Inert gas

Inert-gas sleeve

Welding current

Fig. 2: Tungsten inert-gas-shielded welding

Gas-shielded metal-arc welding (Fig. 3 and Fig. 1, Page 149)

The arc burns between the melting-off wire electrode and the workpiece. The wire electrode is

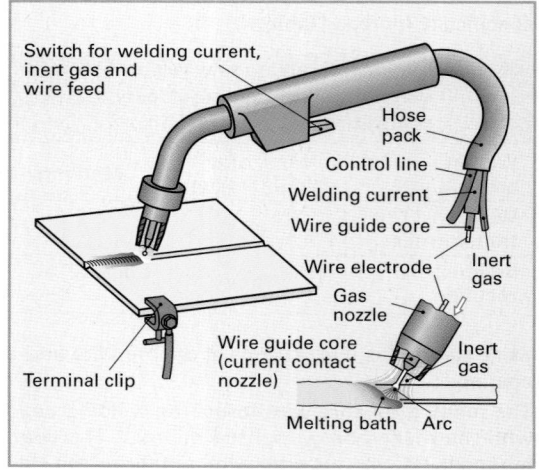

Switch for welding current, inert gas and wire feed

Hose pack

Control line

Welding current

Wire guide core

Wire electrode

Gas nozzle

Wire guide core (current contact nozzle)

Terminal clip

Inert gas

Inert gas

Melting bath   Arc

Fig. 3: Welding torch for MIG-MAG welding

wound on a wire reel and delivered to the welding torch with the wire feed motor through a flexible hose in the hose pack.

Gas metal-arc welding makes use of direct current which is fed to the wire electrode in the welding torch at the current contact nozzle shortly before the arc. The positive terminal is usually connected to the wire electrode. The small electrode cross-section provides for a high current density, high melt-off power, high welding speed and low penetration.

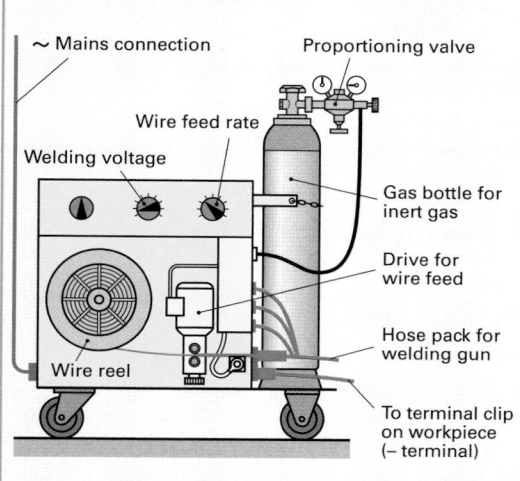

Fig. 1: MIG-MAG welding machine

Metal inert-gas-shielded welding (MIG welding, Fig. 3, Page 148 and Fig. 1)

In this case, inert welding gases (e.g. argon) which undergo no chemical reactions during the welding process are used.

MIG welding is suitable for welding thick sheets of high-alloy steels, of copper or Cu alloys and or aluminium or Al alloys. When light-alloy bodies are manufactured, thin sheets of Al alloys are MIG-welded with each other and with pressure diecastings and extruded sections made of Al alloys.

Metal active-gas-shielded welding (MAG welding, Fig. 3, Page 148 and Fig. 1)

Here gas mixtures of argon, carbon dioxide and oxygen or pure carbon dioxide are used as the inert gas.

MAG welding is a method of inert-gas-shielded welding for welding unalloyed and alloyed steels. With carbon dioxide and oxygen, the inert gases contain active constituents which react with the melting bath. The wire electrode therefore contains manganese and silicon as important alloying con-

stituents for deoxidising the melting bath. Both materials combine with oxygen, which either is released when the carbon dioxide decomposes or is present as a constituent of the mixture gas.

Inert-gas-shielded welding is carried out in garages/repair shops usually with one wire-electrode diameter only; mainly a 0.8 mm wire electrode is used, but 1 mm may also be used.

Laser welding

> In the case of this welding process, the heat required for melting on the material and the filler metal is introduced by a high-energy laser beam.

In a laser (**L**ight **A**mplification by **S**timulated **E**mission of **R**adiation), a medium, e.g. a helium/neon gas mixture, is brought to a higher energy level by "collision" with electrons. The energy is then given off again in the shape of an electromagnetic wave (e.g. as red light) in a highly concentrated form.

Welding process: The contact roller **(Fig. 2)** ensures that the parts to be joined are fixed in place and that the filler material is delivered to the weld. The thin laser beam enables a high-precision weld seam to be produced. The melting mass is protected by an inert gas to prevent it reacting with the ambient air.

Laser welding is used in industrial car manufacturing for bodyshells. Steels, light alloys and plastics can be welded.

Fig. 2: Laser welding

Features of laser welding:
- Clean seam
- Minimal material distortion thanks to low heat application during welding
- High productivity and rigidity
- Weight saving due to low or no overlapping
- High strength

Spot welding

> Spot welding is a form of resistance pressure welding. A non-removable, material-based connection is created when two metal sheets lying one on top of the other are joined in a paste-like state without filler metals at individual welding spots by means of heat and pressure.

The required pressure is exerted via the pin-shaped copper electrodes. A high current passes briefly through the copper electrodes and the metal sheets which are pressed together. The necessary welding heat is generated very quickly due to the high electrical resistance at the joint. Pressure, current intensity and welding time must be matched to each other.

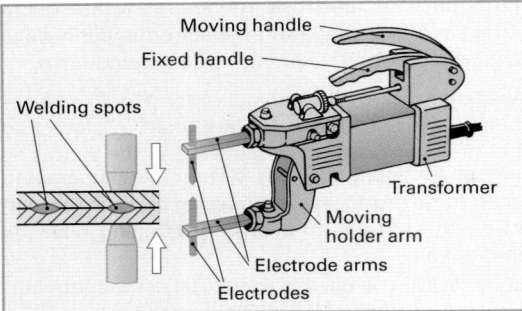

Fig. 1: Spot welding with welding tongs

Small, portable **spot-welding tongs (Fig. 1)** with an integrated transformer are used for motor-vehicle repairs. The electrodes are pressed together when the hand lever is actuated. Different types of electrode arms are available to enable welding spots to be applied even in otherwise inaccessible areas of the body. However, the weld must always be accessible to the welding tongs from both sides.

Joint spot-welder (Fig. 2). This can be used when the weld is only accessible from one side. For spot welding, the electrode of the welding gun is pressed against the weld in such a way that the two metal sheets touch each other and then the welding spot is applied.

The joint spot-welder has many uses:

- Spot welding from one side
- Removal of dents from metal sheets (in conjunction with a pulling hammer)
- Drawing in of metal sheets
- Welding on of stud bolts and pins

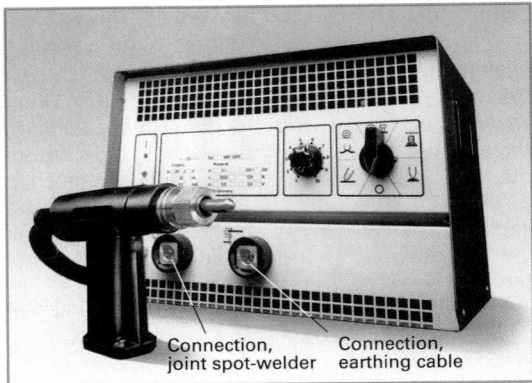

Connection, joint spot-welder   Connection, earthing cable

Fig. 2: Joint spot-welder

Working rules

- Always wear protective goggles for gas-welding and autogenous cutting.
- Open bottle valves slowly.
- Do not apply oil or grease to the lock/seal of an oxygen bottle (explosion hazard).
- Protect gas bottles against being hit and falling, and against heat and cold.
- Wear a face shield with side protection for electric-arc welding.
- When arc welding, screen the work area so that workers are not exposed to the damaging effects of radiation.
- Wear tight-fitting work clothes and gloves. They provide protection against splashes and the arc beam.
- Make sure that the work area is well ventilated.

REVIEW QUESTIONS

1. Why should oil and grease not be permitted anywhere near the locks/seals of oxygen bottles?
2. How can you recognise an acetylene bottle?
3. Which measured values are read off on the reducing regulator?
4. Why is the welding flame adjusted to neutral for the acetylene welding of steel?
5. What is flame cutting based on?
6. What different fusion-welding processes are there?
7. What is the function of the coating on an electrode?
8. Which welding directions are used in inert-gas-shielded welding?
9. What are the advantages of inert-gas-shielded welding over other welding processes?

7.6.12 Gluing

> Gluing is the non-removable, material-based joining of identical or different materials with the aid of a glue or adhesive (non-metallic material).

Stressing and shaping a glued joint

The strength of the joint is dependent on the **cohesion forces** in the glue layer and on the **adhesion forces** between the glue and the joining surfaces of the workpieces **(Fig. 1)**.

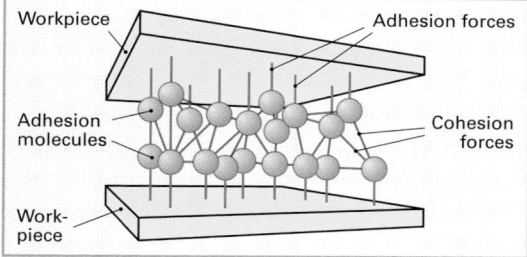

Fig. 1: Forces in glued joint

Whereas the cohesion forces in the glue layer are essentially dependent on the glue that is used, the adhesion forces are above all dependent on whether the joining surfaces have been carefully cleaned before the glue is applied for the purpose of exploiting the adhesive force of the glue to the best possible effect.

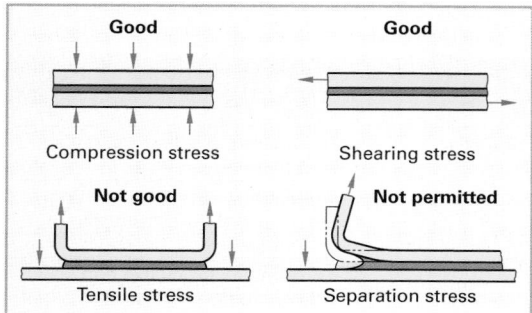

Fig. 2: Stressing of glued joint

Large joining surfaces are needed to transmit large forces. The glued joint should wherever possible only be subjected to compressive and shearing stresses **(Fig. 2)**. Tensile stress should be kept to a minimum and separation stress should be avoided completely as this could cause the glued joint to tear open.

Types of adhesive

Reaction adhesives (Table 1) harden due to the chemical reaction of their components. Adhesives are categorised by their composition into single-component adhesives and two-component adhesives, and by their application temperature into cold adhesives and hot adhesives.

Single-component adhesives contain all the components needed for gluing. They do not require the addition of a hardening agent.

Two-component adhesives consist of two separate components, the adhesive itself and a hardening agent. After mixing, the adhesive hardens due to chemical reaction as a function of the temperature. The adhesive must be applied within a certain period of time (potlife) otherwise it will have hardened too much and will no longer be able to coat the joining surfaces sufficiently.

Cold adhesives harden at room temperature. They are often used as sealing compounds.

Hot adhesives harden at temperatures of approx. 120 °C to 250 °C. Low temperatures result in a long hardening time.

The quality of a glued joint is significantly influenced by external factors, e.g. air humidity, temperature and dust. The application instructions must be observed without fail.

Table 1: Reaction adhesives for motor-vehicle repairs			
Adhesive		**Hardening by**	**Application**
Poly-urethane	Single-component adhesives	Air humidity	Body sealant
Anaerobic adhesives		Exclusion of air + metal contact	Screw locking, sealing
Cyano-acrylate		Air humidity	Superfast adhesive
Poly-urethane	Two-component adhesives	Hardening agent	Windscreen, body components
Epoxy resin		Hardening agent	Body components, e.g. sectional repair

REVIEW QUESTIONS

1 Which forces bring about the cohesive strength of a glued joint?

2 Why do glued joints usually have large joining surfaces?

3 What types of stress are favourable to a glued joint?

4 What do you understand by the potlife of an adhesive?

5 Name the factors which influence the quality of a glued joint.

6 What are the differences between single- and two-component adhesives?

7.7 Coating

> Coating is the application of an adherent layer of formless material on a workpiece.

Coating workpieces:

- Provides protection against corrosion, e.g. by spraying on zinc
- Improves their appearance, e.g. by coats of paint
- Improves the resistance to wear of base materials, e.g. by hard chrome plating, plating valves
- Provides insulation against electricity, heat and sound

Before being coated, the workpieces must be cleaned in order to obtain perfectly adherent coatings.

Cleaning processes

> Cleaning is the removal of contaminants or unwanted layers from the surfaces of workpieces.

Table 1: Cleaning processes

Mechanical dry	Brushing Grinding Hammering	Polishing Blasting
Mechanical wet	Washing Spray-cleaning Steam-jet cleaning	Ultrasonic cleaning
Chemical	**Degreasing,** e.g. with lyes (alkaline solutions), with organic solvents (fat solvents)	
	Corroding, e.g. with diluted sulphuric acid for removing layers of oxide or scale (then rinsing and drying)	

Mechanical dry cleaning. This can take the form of, for example, brushing, polishing, hammering, blasting (sandblasting) and grinding.

Mechanical wet cleaning. This can take the form of washing, spray-cleaning, steam-jet cleaning and ultrasonic cleaning. In the case of ultrasonic cleaning, the cleaning fluid is made to vibrate at high frequency, whereupon tiny vacuum bubbles are formed (diameter approx. 0.00001 mm). When the bubbles burst (implode), pressures of up to 1,000 bar are generated, resulting in even the smallest dirt particles being removed from pores and hairline cracks.

Chemical cleaning. This, for example, takes the form of degreasing or corroding. In the case of degreasing **(Fig. 1),** grease is broken down (dis-

persed) into the smallest grease particles by being immersed in or sprayed with organic solvents. These particles are distributed uniformly in the wash water (emulsified) and are then washed away with the wash water. Alkaline solvents, e.g. caustic soda, sodium carbonate, convert lubricants into soaps, which can then be washed away with water.

In the case of corroding, layers of oxide and scale on ferrous and non-ferrous metals are removed with diluted acids.

The disposal directives of the individual countries pertaining to chemical cleaning agents must be observed.

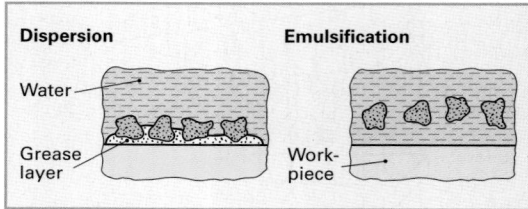

Fig. 1: Degreasing

Coating processes
Metal spraying (thermal spraying)

> In the case of metal spraying, the layer material is thrown in a dispersed, melted state onto the workpiece surface. The workpiece surface is partially melted on in the process.

Table 2: Coating processes

Coatings	Processes
Metallic	**Metal spraying,** e.g. flame spraying, plasma spraying
	Electroplating, e.g. chrome-plating, dipping, e.g. hot-dip galvanising of body floor panels
Non-metallic inorganic	**Application of a protective film,** e.g. phosphating, chromating, anodising
Non-metallic organic	**Paint coatings, lacquer coatings, plastic coatings,** e.g. whirl sintering, application of waxes

The coating sticks to the workpiece by adhesion and cohesion of the individual metal particles.

Different processes based on how the metal melts are used in metal spraying:

- **Melting-bath spraying.** Low-melting-point metals are melted in a melting pot and sprayed onto the workpiece.

- **Flame spraying (Fig. 1).** Metals with a melting point below 3,000 °C are melted in a fuel-gas/oxygen flame and sprayed by means of gas pressure onto the workpiece.
- **Arc spraying (Fig. 2).** The coating-metal wire is melted at over 5,000 °C and usually sprayed with compressed air onto the workpiece.
- **Plasma spraying (Fig. 3).** High-melting-point metals and also non-metals (e.g. aluminium oxide) are melted in powder form as the coating material at 10,000 °C ... 20,000 °C and thrown by means of a plasma beam onto the workpiece.

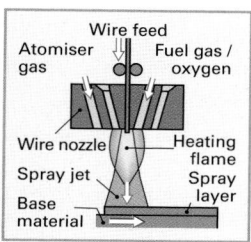

Fig. 1: Flame spraying   Fig. 2: Arc spraying

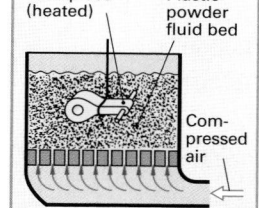

Fig. 3: Plasma spraying   Fig. 4: Whirl sintering

Electroplating

Electroplating is the electrochemical application of metals to metallic or electrically conducting layers using an electrolyte.

Motor-vehicle components can be coated as a means of corrosion protection by electroplating, e.g. electrolytic galvanising of body sheets, chrome-plating of trim parts.

Ceramic coatings

Ceramic materials are thrown onto the workpiece surface and only stick by means of toothing (positive locking). Ceramic coatings serve as thermal, corrosion and wear protection.

Oxidising

Oxidising metal surfaces involves bringing about artificial corrosion, e.g. by anodising. Oxidising causes the base material to convert on its surface; there is no actual layer build-up. Application: an-

odised aluminium parts, e.g. wheels, trim parts in the passenger compartment.

Plastic coatings

Plastic coatings are possible on practically all materials, e.g. by painting, flat coating, dipping, film coating, electrostatic coating, flame spraying and whirl sintering.
In the case of whirl sintering **(Fig. 4)**, the preheated workpiece (up to approx. 300 °C) is held in plastic powder shaken up by compressed air. The powder sticks to the material and forms the coating. Plastic coatings provide corrosion protection, acoustic, thermal and electrical insulation, and a decorative appearance.

Paint coatings

These coatings provide permanent corrosion protection and give the body an attractive appearance. The protective paint coating must be non-sensitive to scratching and resistant to atmospheric influences such as, for example, strong solar radiation (ultraviolet light) and acid rain. Aggressive liquids, e.g. bird droppings, should not cause damage. Paint applications consist of several layers **(Fig. 5)**. In the case of resprays, priming is followed by the reapplication of two (or, for metallic paints, three) outer layers.

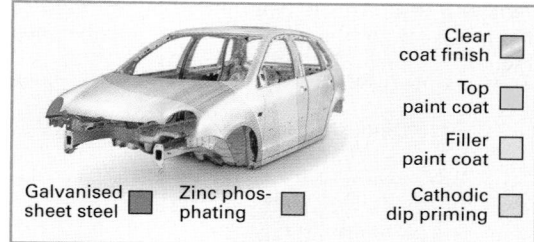

Fig. 5: Structure of vehicle paintwork

Wax coatings

Coatings containing wax are used for body-cavity preservation and as undercoatings. They provide protection against moisture and, in conjunction with additional corrosion inhibitors, long-lasting corrosion protection.

REVIEW QUESTIONS

1 **What do you understand by coating?**
2 **Why must the workpiece be cleaned before it is coated?**
3 **What are the advantages of plastic coatings?**
4 **What do you understand by metal spraying?**
5 **How is a respray structured for a metallic paint?**

8 Material science

8.1 Properties of materials

The choice of materials, to be used for example in a motor vehicle, is determined by the following conditions. Such materials should ...

- ... withstand the stresses and strains arising during regular operation.
- ... generate low material costs and manufacturing costs.
- ... be environmentally compatible and recyclable.

8.1.1 Physical properties

> Physical properties bring about a change in materials or substances; they denote the behaviour of materials.

Important physical properties are:
- Density
- Thermal expansion
- Thermal conductivity
- Melting temperature
- Hardness
- Electric conductivity
- Stress, strength
- Elasticity, plasticity
- Ductility
- Brittleness

Density ϱ (rho). This is determined by the ratio of the mass m to the volume V of a substance **(Fig. 1)**.

$$\text{Density } \varrho = \frac{\text{Mass}}{\text{Volume}} = \frac{m}{V}$$

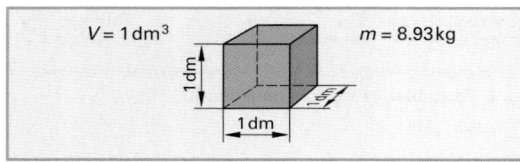

$V = 1\,dm^3$ $m = 8.93\,kg$

Fig. 1: dm³ cube of copper

Table 1: Density of substances			
Substance	Density kg/dm³	Substance	Density kg/dm³
Water	1.00	Titanium	4.54
Steel	7.85	Copper	8.93
Cast iron	7.25	Lead	11.30
Aluminium	2.70	PVC	1.40
Diesel fuel			0.82 ... 0.86
Premium-grade petrol			0.73 ... 0.78
Air at 0 °C and 1.013 bar			1.29 kg/m³

The density is given in kg/dm³, g/cm³ or t/m³ for solids and liquids, and in kg/m³ for gases **(Table 1)**.

Thermal expansion. Bodies generally expand in response to an increase in temperature. In the case of solid materials, only the linear expansion in one direction **(Fig. 2)** is measured for every increase in temperature of 1 Kelvin (K) and given as a mean coefficient of linear expansion α (alpha) in 1/K.
The linear expansion $\triangle l$ of a material when heated is dependent on ...

- ... the length prior to heating l_0 in m.
- ... the temperature difference $\triangle T$ in K.
- ... the coefficient of linear expansion α of the material in 1/K **(Table 2)**.

$$\triangle l = l_0 \cdot \alpha \cdot \triangle T$$

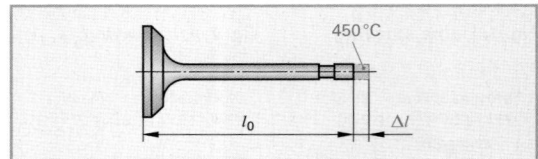

450 °C

l_0 $\triangle l$

Fig. 2: Thermal expansion

Table 2: Coefficient of linear expansion α	
Substance	α in 1/K
Steel, unalloyed	0.0000115
Copper	0.000017
Aluminium	0.0000238
Polyvinyl chloride (PVC)	0.00011

Thermal conductivity. This is the ability to conduct heat differently. It is denoted by the coefficient of thermal conductivity.

Good heat conductors are metals such as e.g. copper and aluminium.

Poor heat conductors are e.g. air, glass and plastics.

Ductility. This is the ability of a substance or material to be plastically deformed by external forces without breaking in the process. The material opposes deformation with great resistance. Ductile materials are e.g. structural steel, lead and copper.

Brittleness. Materials are classed as brittle if they break into pieces without a significant change in shape as the result of, for example, impact forces. Brittle materials are e.g. glass and flake-graphite cast iron.

Hardness. This is the resistance with which a material opposes penetration by a body, e.g. a steel ball **(Fig. 1)**. Hard materials are e.g. hardened steel, metal carbide and diamond.

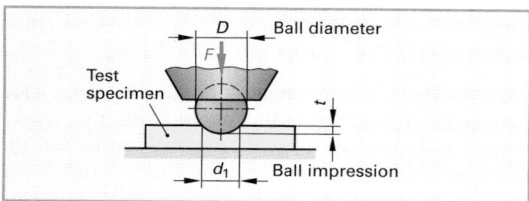

Fig. 1: Brinell hardness test

Stress. When external forces are exerted on a body, this gives rises to mechanical stress σ (sigma) in that body. This can be expressed as the ratio of the external force F to the cross-section S **(Fig. 2)**. The mechanical stress is usually given in N/mm².

Different stresses and strains, such as tensile, compressive, shearing, bending, buckling and torsional stress, may be created, depending on the direction of the external forces.

Tensile strength R_m. This is the greatest stress which occurs in the material of a workpiece. The tensile strength is ascertained by means of a tensile test carried out on a test bar **(Fig. 2)**.

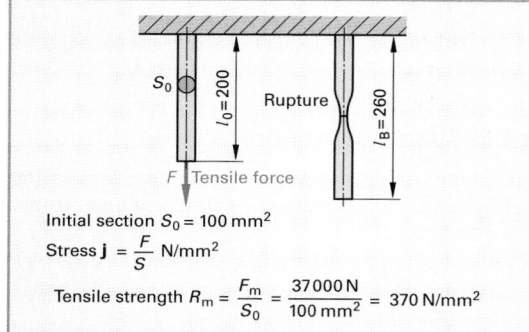

Initial section $S_0 = 100 \text{ mm}^2$

Stress $j = \dfrac{F}{S} \text{ N/mm}^2$

Tensile strength $R_m = \dfrac{F_m}{S_0} = \dfrac{37\,000 \text{ N}}{100 \text{ mm}^2} = 370 \text{ N/mm}^2$

Fig. 2: Stress and tensile strength

The rod is tensioned in the tensile-testing machine and subjected to load until it breaks. The tensile force and elongation are measured in the process **(Fig. 3)**.

At first, the elongation increases in the same proportion as the tensile force up to the yield strength R_e. In this range the material behaves elastically.

After the yield strength is exceeded and as elongation increases, the tensile force remains the same for the first time or even drops. As the tensile force continues to increase, the material is subject to plastic deformation and the permanent elongation increases rapidly. At point B the material's stability

under load, its breaking point, is reached. The tensile strength R_m is calculated from this value. It is referred to the initial section S_0 and given in N/mm².

> Tensile strength $R_m = \dfrac{\text{Greatest tensile force (breaking force)}}{\text{Initial section}}$
>
> $\qquad\qquad = \dfrac{F_m}{S_0}$

The bar contracts at the subsequent point of break and finally breaks off (Z).

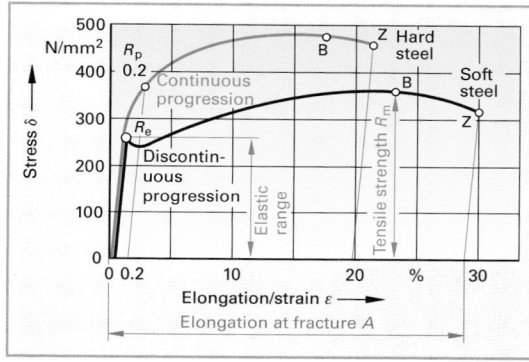

Fig. 3: Stress-strain diagram

Elasticity. A material is elastic if, after a load has been removed, it reassumes its original shape. A spring is compressed, for example, when subjected to load; when that load is removed, the spring reverts to its original shape. Steel **(Fig. 3)** behaves elastically when it is subjected to a force up to its yield strength (R_e).

Plasticity. If a material maintains its new shape after being deformed by a force, this property is referred to as plasticity.

Melting temperature. This is the temperature in °C at which a substance or material passes from the solid to the liquid state **(Table 1)**. Pure metals have a distinct melting point whereas metal alloys and compounds have a melting range.

Table 1: Melting temperatures	
Substance	Melting point in °C
Lead	327
Aluminium	660
Cast iron	1,200
Tungsten	3,410

Electric conductivity $\varkappa$ (kappa). This indicates how well or how badly a substance or material conducts electric current **(Table 1, Page 156)**. All metals conduct electricity. Non-metals, e.g. plastics, porcelain, are non-conductors and are therefore used as insulating materials.

Table 1: Electric conductivity	
Substance	$\varkappa$ in $\dfrac{m}{\Omega \cdot mm^2}$
Silver	60
Copper	56
Plastics	$10^{-15} \dots 10^{-20}$

8.1.2 Technological properties

The technological properties **(Fig. 1)** determine the suitability of a material for the different manufacturing processes.

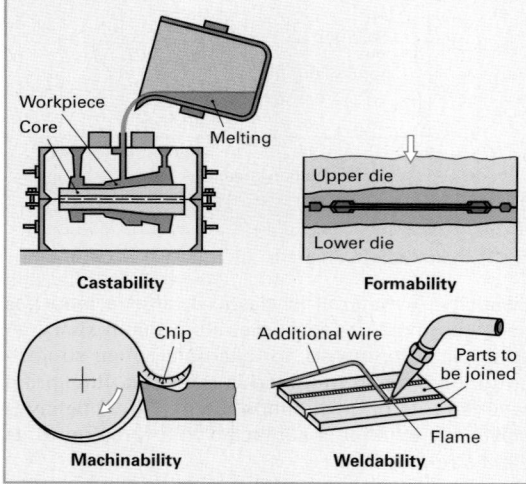

Castability   **Formability**

Machinability   **Weldability**

Fig. 1: Technological properties

Castability. A material has good castability properties if, during melting, it becomes liquid and absorbs hardly any gas, does not have an excessively high melting temperature and does not shrink excessively when it solidifies.

Cast iron and aluminium and copper-zinc casting alloys have good castability properties; unalloyed aluminium and copper have poor castability properties.

Formability. A material is formable if it can be plastically deformed into a workpiece under the influence of forces.

There are the following different types:

● Cold forming, such as e.g. cold rolling, bending, deep drawing

● Hot forming, such as e.g. hot rolling, forging

Low-carbon steel, lead, and copper, aluminium and wrought alloys have good formability properties; cast-iron materials and hard metals are not formable.

Machinability. This is the ability of materials to be machined by means of machining/cutting, such as e.g. turning, milling, drilling, grinding.

Materials of low ductility and average strength, such as e.g. unalloyed and low-alloy steels, cast iron, aluminium and its alloys, have good machinability properties.

Weldability. This is the ability of materials to be easily joined in a liquid or paste-like state into workpieces. Materials for motor-vehicle manufacturing, such as e.g. structural steels and aluminium wrought alloys, must have good weldability properties. Cast iron, for example, has poor weldability properties and can only be welded by means of special welding processes.

REVIEW QUESTIONS

1   According to which conditions is the material for a component chosen?

2   Name three physical properties.

3   On what is the linear expansion of a solid body dependent when heated?

4   In what way does the thermal expansion of steel and aluminium differ?

5   What does the density indicate?

6   What is the meaning of the yield strength R_e and the tensile strength R_m of a test bar?

7   In what case is a substance or material elastic?

8   What do you understand by hardness and brittleness?

9   Name three technological properties.

10   What types of formability are there?

8.1.3 Chemical properties

The chemical properties of materials relate to the behaviour of the materials or their changes under the influence of

● environmental factors (e.g. air humidity, water)

● aggressive substances (e.g. acids, alkalis, salts)

● heat (e.g. during annealing)

The following properties are derived from the behaviour of a material under the influence of the factors mentioned above:

● Corrosion resistance   ● Toxicity

● Resistance to heat   ● Combustibility

Corrosion resistance. This is the resistance to aggressive media (e.g. acids, alkalis) the influence of which must not result in any measurable changes to the workpiece surface.

Toxicity. Materials may have a toxic effect when they come into contact with food, e.g. fruit acids infected with zinc. Lead and cadmium have a toxic effect when they are absorbed through the mucous membranes.

Resistance to heat. Most steels oxidise when annealed at temperatures above 600 °C in a non-oxygen-free atmosphere.

Combustibility. This is low in the case of most metals. Exceptions are e.g. potassium, sodium and magnesium. Their ignition temperature is very low. Plastics also tend to burn on account of their low ignition temperature.

Corrosion. Corrosion refers to the reaction of a metallic material with the surrounding media, which result in a measurable change in the material and at the same time impaired functioning of the component in question.

> Environmental factors and aggressive substances can cause corrosion in metallic materials.

A distinction is made between electrochemical and chemical corrosion.

Electrochemical corrosion

This occurs when two different metals and an electrolyte (liquid containing acid, alkali or salt) come into contact. A galvanic element is formed. The extent of the voltage created is dependent on the position of the metals within the electrochemical series of metals **(Fig. 1)**.

> The voltage between two metals is higher the further apart these metals are in the electrochemical series of metals.

The corrosion is greater as the voltage increases. The baser metal is always destroyed or removed. The particles released from the baser material as a result of the electrochemical processes can form chemical compounds with the electrolyte. In addition, the electrolyte can react chemically with the material surface. In this case, chemical corrosion also takes place simultaneously.

Chemical corrosion

Most metals are chemically altered starting from the surface under the influence of acids, alkalis, saline solutions or gases (e.g. oxygen). A layer composed of the chemical compound of the metal and the acting substance is formed on the surface.

If the corrosion layer formed is nonporous, water-insoluble and impermeable to gas, it can inhibit the advance of the chemical corrosion and act as a protective layer, e.g. on aluminium. If the corrosion layer formed is porous, water-soluble or hygroscopic, the corrosion advances until the material is destroyed, e.g. rusting of steel.

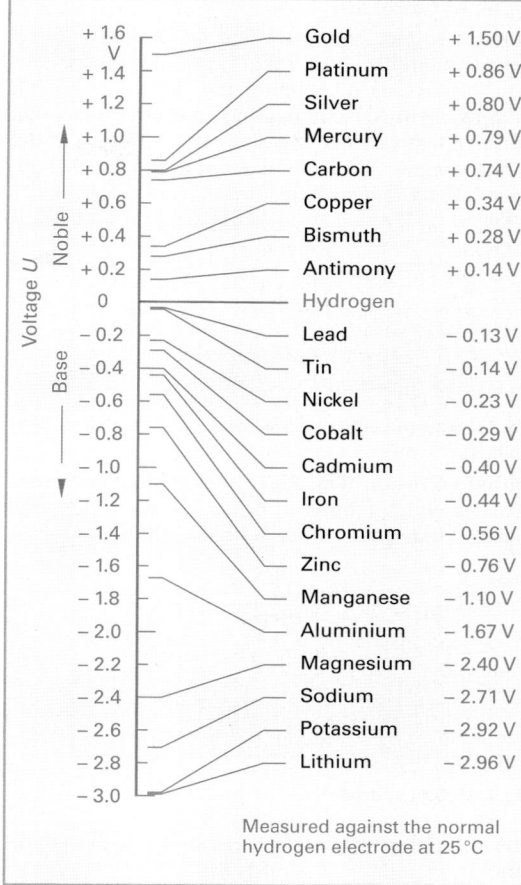

Fig. 1: Electrochemical series of metals

Influencing factors. The corrosion of the material can be influenced by ...

- ... the chemical composition, e.g. by alloying of stainless steels.
- ... the percentage purity, e.g. by unwanted alloying constituents in the processing of scrap metal.
- ... the surface quality, e.g. by polishing of the anodised aluminium surface.
- ... the composition of the corrosive medium, e.g. content of hydrochloric, oxo- and carbonic acids in the water, proportion of sulphur in liquids, proportion of dust and solid constituents in gases.
- ... the pressure and the temperature of the corrosive medium.

Types of corrosion

Uniform surface corrosion (Fig. 1). The metal is removed overall approximately parallel to the surface, regardless of whether the rate of corrosion changes. In the case of bearing steel structures, e.g. bridge constructions, the decline in strength of the component is taken into account in the form of appropriate dimensioning.

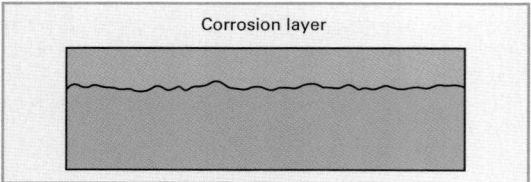

Fig. 1: Uniform surface corrosion

Pitting corrosion (Fig. 2). This is a local corrosion process which results in crater- or needle-shaped pits and in the final state in holing/pitting.

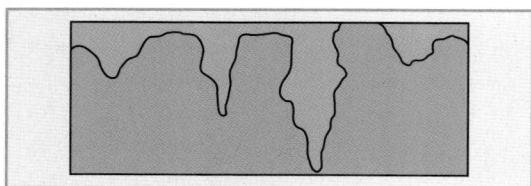

Fig. 2: Pitting corrosion

The pitted area is usually larger in depth than in diameter.

Contact corrosion (Fig. 3). This occurs when two metals which are far apart in the electrochemical series of metals come into contact and an electrolyte is added at the point of contact, e.g. at the contact gap of two components. The base metal is destroyed in the galvanic element formed. It is possible to prevent the element from forming by protecting the point of contact before the electrolyte.

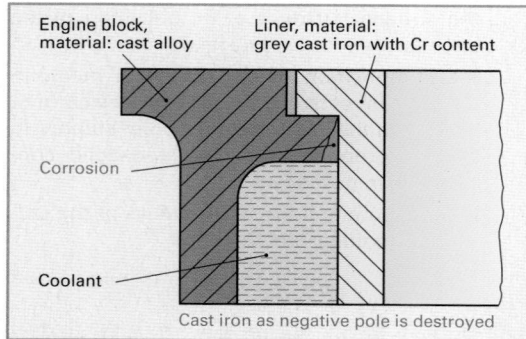

Engine block, material: cast alloy

Liner, material: grey cast iron with Cr content

Corrosion

Coolant

Cast iron as negative pole is destroyed

Fig. 3: Contact corrosion

Intercrystalline corrosion (Fig. 4). Electrochemical corrosion takes place on alloys between the different metal crystals along the grain boundaries, in the course of which fine, invisible hairline cracks arise.

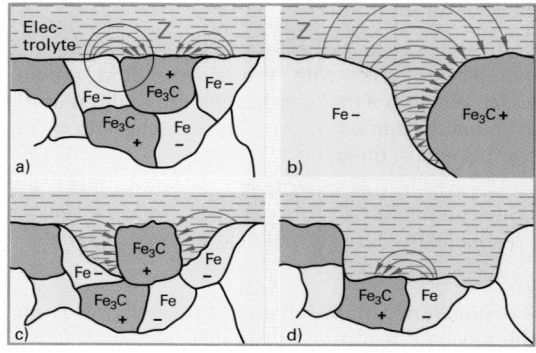

Fig. 4: Advancing intercrystalline corrosion

REVIEW QUESTIONS

1 **What do you understand by chemical and electrochemical corrosion?**

2 **How does intercrystalline corrosion come about?**

3 **By what can the corrosion of a material be influenced?**

8.2 Categorisation of materials

In order to obtain an overview of the large number of different materials, they are divided according to their composition or common properties into the three main categories of metals, non-metals and composite materials **(Fig. 1, Page 159)**. These main categories are then further divided into subcategories.

Cast-iron materials have good strength properties and can be easily cast into shapes. They are used for those workpieces which are best formed by casting, e.g. engine blocks.

Steels are high-strength ferrous products which are particularly suitable for forming by rolling, forging or machining. Profile sections, sheets, shafts and gear wheels, for example, are manufactured from steels.

Heavy metals (density greater than 5 kg/dm^3) are e.g. copper, zinc, tin, lead, chromium and nickel. Their use is primarily dependent on their particular properties. Copper, for example, is used for electric cables on account of its good electric conductivity.

Light metals (density below 5 kg/dm³) are aluminium, magnesium and titanium. Workpieces made of their alloys are low in weight with good strength properties. They are therefore used primarily in motor-vehicle and aircraft construction.

Natural materials are materials which occur in nature, such as e.g. leather, cork and fibrous substances. They are used in special instances, such as e.g. leather for upholstery.

Synthetic materials are materials which are manufactured synthetically using various processes or by transforming natural substances. These include, for example, plastics.

Composite materials. Here different materials are combined with each other in order to combine their properties. These include, for example, brake pads/linings, glass-fibre spatulas and printed circuit boards.

Operating fluids and process materials (Fig. 2)
Machines require operating fluids in order to work properly; motor vehicles, for example, need fuel, lubricants, coolant and brake fluid. They also require process materials for their manufacture and machining.

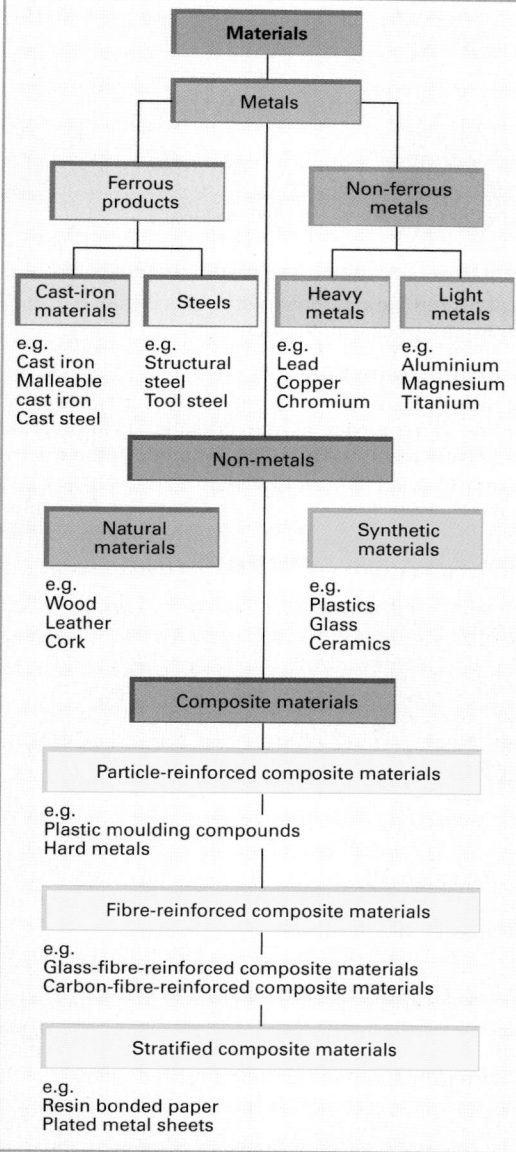

Fig. 1: Categorisation of materials – overview

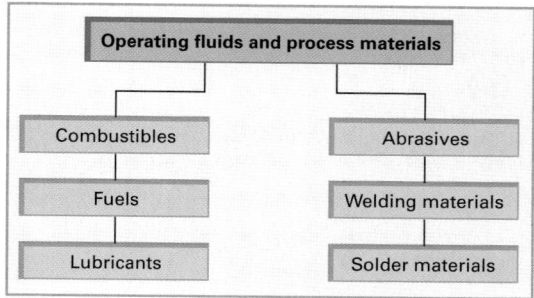

Fig. 2: Operating fluids and process materials

8.3 Structure of metallic materials

All metals form crystals when they solidify from the molten state. In this process the atoms assume their positions in the crystal in accordance with specific rules which are characteristic of the metal in question (Fig. 3).

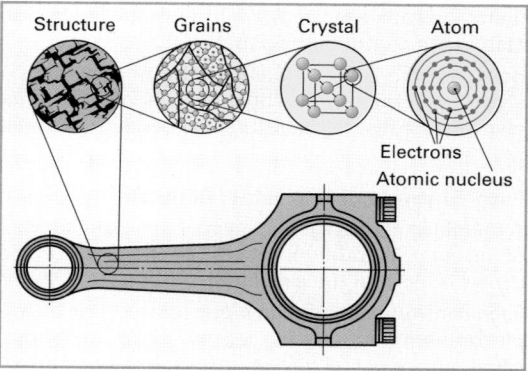

Fig. 3: Structure of metallic materials

On the other hand, the atoms of most non-metals arrange themselves without fixed rules (amorphously) next to each other when these materials solidify from the melting.

Metallic bond (Fig. 1). Metal atoms have permanently bound electrons at their nucleus and one or more "free" electrons on the outer electron shell.

The electric forces of attraction between the negatively charged electrons and the positively charged metal ions bring about the solid cohesion and thus the strength of the metallic material.

This type of bond between the metal ions and the free electrons is called a metallic bond because they are typical of all metals.

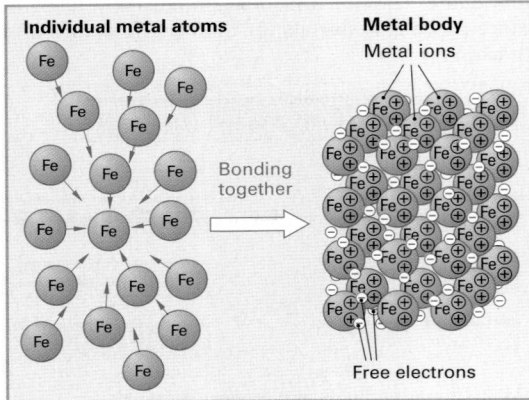

Fig. 1: Creation of metallic bond

8.3.1 Crystal lattice of pure metals

Metals crystallise into three shapes:
- Body-centred cubic crystal (b.c.c.)
- Face-centred cubic crystal (f.c.c.)
- Hexagonal crystal (hex)

Body-centred cubic crystal (Table 1)

The crystal is a cube, at the corners of which eight metal ions are arranged. A further metal ion is situated in the centre of the cube.

Body-centred cubic crystals are formed by, for example, chromium, molybdenum, vanadium, tungsten and steel at temperatures below approximately 723 °C.

Face-centred cubic crystal (Table 1)

A metal ion is arranged at each of the eight corners of the cube. In addition, a metal ion is situated in the centre of each of the six sides.

Face-centred cubic crystals are formed, for example by aluminium, lead, copper, nickel, platinum, silver and steel at temperatures above approximately 723°C ... 911 °C.

Hexagonal crystal (Table 1)

In the case of this crystal, the metal ions form a hexagonal prism each with a metal ion in the centre of the basic faces. The hexagonal crystal also

Table 1: Crystalline forms of metals

Spheroid model	Line model
Body-centred cubic crystal – b.c.c.	

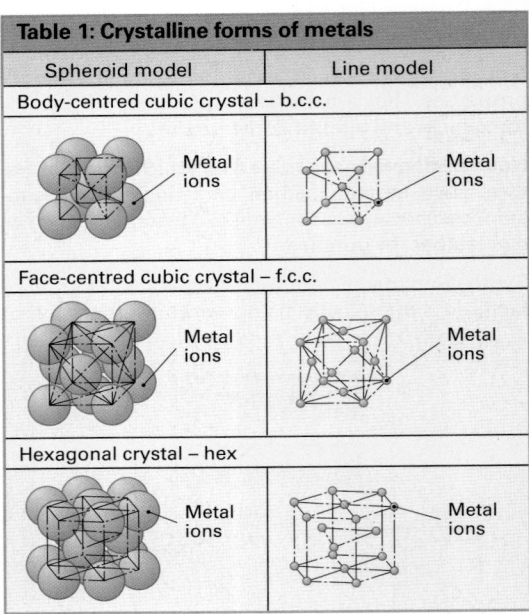

Face-centred cubic crystal – f.c.c.	

Hexagonal crystal – hex	

contains three metal ions within the prism.

Metals with hexagonal crystals are brittle, e.g. cadmium, magnesium, titanium, zinc.

Structure of metal

Metal consists of randomly limited crystallites or grains. The grains consist of crystals. The individual grains collide at the grain boundaries. In the micrograph of a polished and etched metal surface it is possible to make out under a microscope the grains and the grain boundaries, which appear as thin lines between the grains **(Fig. 2)**.

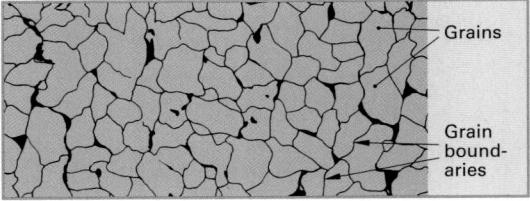

Fig. 2: Metal structure in micrograph

The configuration of the structure has a significant bearing on the properties of metals, e.g. on strength and hardness.

8.3.2 Crystal lattice of metal alloys

In engineering most metals are used not in their pure forms, but rather in the form of alloys.

The following different types of alloy are used: Mixed-crystal and crystal-mixture alloys.

Mixed-crystal alloys are created when, in the process of solidification, the atoms of the alloying element remain distributed uniformly in the crystal lattice of the base metal.

In the course of this process …

- … the atoms of the alloying element can settle in the position of a metal ion of the base metal **(exchange-mixture crystal, Fig. 1)**.

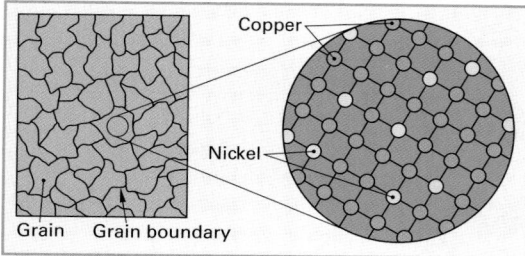

Fig. 1: Structure of exchange-mixture crystal

- … the atoms of the alloying element can arrange themselves between the metal ions **(intercalation-mixture crystal, Fig. 2)**.

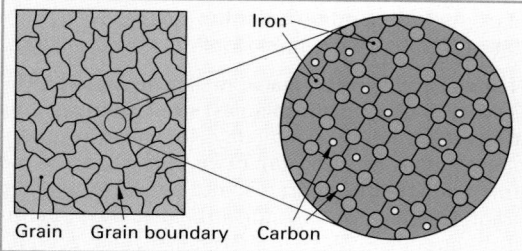

Fig. 2: Structure of intercalation-mixture crystal

Because of the distortion of the lattice by the intercalated alloying constituent, the hardness and strength of the mixed-crystal alloy are greater than those of the base metal. Mixed-crystal alloys are created, for example, when alloying iron and manganese or iron and nickel or copper and nickel.

Crystal-mixture alloys are created when, during solidification of the melting, the alloying constituents separate out. Each constituent forms its own crystals **(Fig. 3)**.

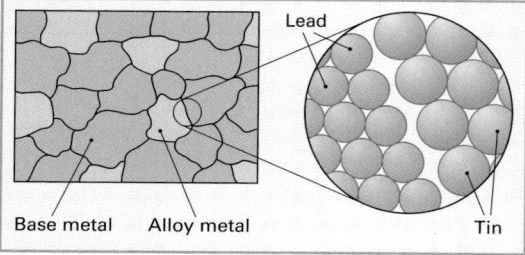

Fig. 3: Structure of intercalation-mixture crystal

Crystal-mixture alloys are created, for example, when alloying lead and tin (soldering tin, lubricating solder).

REVIEW QUESTIONS

1 What crystalline forms occur mainly in the case of metals?

2 What do you understand by metallic bond?

3 How are the grains and grain boundaries of a metal structure created?

4 In what form are metals usually used in engineering?

5 What are the differences between mixed crystals and crystal mixtures?

8.4 Ferrous products

These are metallic materials, the properties of which can be altered as required by different manufacturing processes, by alloying or by heat-treatment processes.

Ferrous products are, for example, high in strength and have good castability, formability, machinability and weldability properties.

There are the following different types:
- Steel
- Cast-iron materials

8.4.1 Steel

The carbon, silicon and manganese content in forge pig iron is reduced by oxidation. Phosphorus and sulphur are eliminated to a large extent, thereby making the ferrous product suitable for forging.

When not subsequently treated, steel is a forgeable ferrous product.

8.4.2 Cast-iron materials

Flake-graphite cast iron (grey cast iron)

Flake-graphite cast iron is cast iron in which the graphite is configured in lamellar form.

Flake-graphite cast iron	
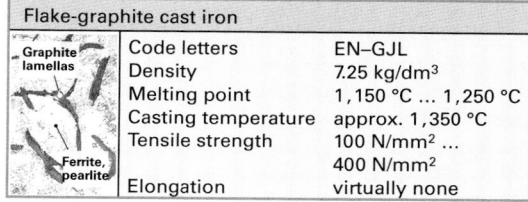 Code letters	EN–GJL
Density	7.25 kg/dm³
Melting point	1,150 °C … 1,250 °C
Casting temperature	approx. 1,350 °C
Tensile strength	100 N/mm² …
	400 N/mm²
Elongation	virtually none

Manufacture. Flake-graphite cast iron is melted from grey pig iron, cast-iron scrap and steel scrap. When the melting cools slowly after casting, the carbon separates out as graphite into lamellas and is intercalated in a fine distribution in the structure.

Properties. Graphite gives the surface of fracture a typical grey colour and provides good friction properties, easy machinability and good vibration damping. The high carbon content of 2.8 % to 3.6 % delivers a low melting point and good castability. The graphite lamellas in the cast iron create a notch effect, thereby greatly reducing tensile strength and elongation.

Application e.g. for cylinder blocks, piston rings, housings, exhaust manifolds, brake drums, brake discs, clutch covers, clutch pressure plates.

Nodular-graphite cast iron

Nodular-graphite cast iron is cast iron in which the graphite is configured in globular form.

Nodular-graphite cast iron		
Code letters	EN–GJS	
Density	7.1 kg/dm³ ... 7.3 kg/dm³	
Melting point	1,400 °C	
Tensile strength	400 N/mm² ... 800 N/mm²	
Elongation	15 % ... 2 %	

Manufacture. When magnesium is added to the molten cast iron in the crucible, the graphite separates out during cooling at the grain boundaries in globular form.

Properties. In contrast to lamellas, the graphite globules do not create a notch effect, and therefore elongation, flexural strength and tensile strength are higher. Nodular-graphite cast iron has high resistance to wear and can be heat-treated, e.g. hardened or quenched and drawn. It has good machinability properties and can be surface-hardened.

Application e.g. for crankshafts, camshafts, connecting rods, steering components, brake drums, brake discs, brake callipers.

Vermicular-graphite cast iron (EN–GJV)

Its structure has both fine, worm-shaped (vermicular) and globular graphite intercalations. Thus tensile strength and elongation at fracture are higher than for grey cast iron. Vermicular-graphite cast iron is used to manufacture thin-walled motor-vehicle components, e.g. exhaust manifolds, brake discs, gearbox casings, turbocharger housings, engine blocks.

Malleable cast iron

Malleable cast iron is cast iron which has been rendered ductile by annealing (fresh annealing or malleablising).

Manufacture. After casting, the castings are subjected to a lengthy annealing process. There are two different types, depending on the type of annealing used: decarburising and non-decarburising malleable cast iron.

Decarburising annealed malleable cast iron		
Code letters	EN–GJMW	
Density	7.4 kg/dm³	
Melting point	1,300 °C	
Tensile strength	350 N/mm² ... 650 N/mm²	
Elongation	1.5 % ... 3 %	

Non-decarburising annealed malleable cast iron		
Code letters	EN–GJMB	
Density	7.4 kg/dm³	
Melting point	1,300 °C	
Tensile strength	250 N/mm² ... 700 N/mm²	
Elongation	1 % ... 2 %	

Decarburising annealed malleable cast iron EN–GJMW (white-heart malleable cast iron). Carbon is extracted from the castings in an oxidising atmosphere by annealing over several days at approx. 1,000 °C. Because the decarburisation only takes place in the boundary layer (up to approx. 5 mm), this process is only suitable for thin-walled castings.

Non-decarburising annealed malleable cast iron EN–GJMB (all-black malleable cast iron). The castings are annealed over several days in a neutral atmosphere under the exclusion of air. In the process the iron carbide disintegrates into ferrite and flaky graphite, which settle in the structure. The structural change takes place in the entire workpiece and not just in the boundary layer.

Properties. Castings of malleable cast iron have properties similar to steel. They have good machinability, soft-solderability and brazability, heat treatability, surface hardenability and weldability properties.

Application in vehicle construction e.g. for connecting rods, steering columns, control forks and frame sleeves for motorcycles and bicycles.

Cast steel

Cast steel is steel cast in moulds for high-strength castings.

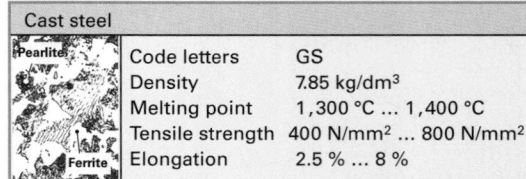

Cast steel		
Code letters	GS	
Density	7.85 kg/dm³	
Melting point	1,300 °C ... 1,400 °C	
Tensile strength	400 N/mm² ... 800 N/mm²	
Elongation	2.5 % ... 8 %	

Manufacture. The castings are annealed after casting in order to eliminate casting strains. The annealing temperature ranges between 800 °C and 900 °C, depending on the carbon content.

Properties. Cast steel has the properties of steel such as, for example, strength and ductility together with castability such that workpieces of complicated shapes can be manufactured.

Application e.g. for brake drums, brake discs, brake callipers, rear-axle housings, wheel hubs, trailer tow hitches for trucks, turbines, levers.

8.4.3 Influence of additives on ferrous products

Table 1: Influence of non-metals and metals on ferrous products

Element	Increased	Lowered	Example
Non-metals (companions to iron)			
Carbon C	Strength, hardness, hardenability, castability for cast iron	Melting point, ductility, elongation, weldability and malleability	C45
Silicon Si	Tensile strength, elasticity, corrosion resistance	Malleability, weldability, machinability	60SiCr7
Phosphorus P	Heat resistance, blue brittleness, fluidity for EN-GJL	Elongation, ductility, weldability	
Sulphur S	Cutting brittleness, red shortness during forging	Ductility, weldability, corrosion resistance	15S10
Metals (alloying elements)			
Chromium Cr	Tensile strength, heat resistance, corrosion resistance	Ductility, weldability, elongation	X5Cr17
Manganese Mn	Tensile strength, heat resistance, corrosion resistance	Resistance to wear, weldability, machinability	28Mn6
Molybdenum Mo	Tensile strength, heat resistance, edge-holding property, ductility	Weldability	20MoCr4
Nickel Ni	Tensile strength, heat resistance, corrosion resistance	Thermal elongation, weldability, machinability	36NiCr 6
Vanadium V	Fatigue limit, heat resistance, hardness	Weldability	115CrV3
Tungsten W	Tensile strength, hardness, heat resistance, edge-holding property	Resistance to wear, corrosion resistance	105WCr6

8.4.4 Designation of ferrous products

A ferrous product can be uniquely defined by a
* short name, e.g. S235JR
* material number, e.g. 1.0037

Short names. These provide in the form of letters and numbers details on the manufacture, properties and treatment of the material.

Material numbers. These make it easier for materials to be recorded with systems of electronic data processing (EDP), e.g. **S235JR** (short name) – **1.0037** (material number).

Designation of cast-iron materials
The short names for cast iron, malleable cast iron and cast steel can still be created following the previous designation system according to DIN.

Specifications for manufacture (Table 1). These short names are composed of the letter G and further letters indicating the casting type.

Table 2: Code letters for cast-iron materials

As per DIN	As per DIN-EN	Designation
G		Cast-iron material
GG	EN-GJL	Flake-graphite cast iron (grey cast iron)
GGG	EN-GJS	Nodular-graphite cast iron
GS	GS	Cast steel
GTS	EN-GJMB	Non-decarburising annealed malleable cast iron
GTW	EN-GJMW	Decarburising annealed malleable cast iron

Specification of strength. This is done using a code number which is multiplied by 10 (9.81 to be precise) to specify the minimum tensile strength in N/mm^2, e.g. GG-25 is flake-graphite cast iron with a minimum tensile strength of 250 N/mm^2. In the case of malleable cast iron, a number appended with a hyphen specifies the elongation at fracture in per cent, e.g. **GTS-65-02 (Fig. 1, Page 164).**

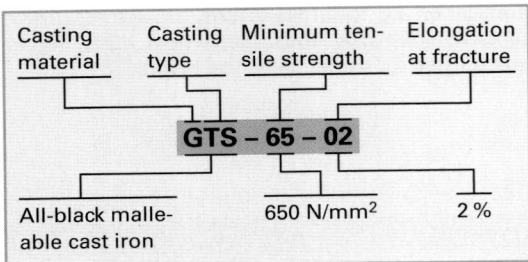

Casting material	Casting type	Minimum tensile strength	Elongation at fracture

GTS – 65 – 02

All-black malleable cast iron	650 N/mm²	2 %

Fig. 1: Designation of malleable cast iron

Designation system for steels according to EN

The short names for steels and cast steel are created in accordance with two main groups:

- Main group 1
- Main group 2

Main group 1: Short names with references to application and properties

The short names start with a code letter for the application or purpose of the steels, followed by details of the properties. In the main only engineering steels (code letter E) and steels for structural steel engineering (S) are taken into consideration for automotive engineering.

Main symbols

Code letters specify the application or purpose. In the case of cast steel, the letter G is used at the start.

The three-digit number following the code letter specifies the minimum yield strength R_e in N/mm², e.g. E350.

Table 1: Additional symbols for steel products (selection)

For special requirements	
+ C	Coarse-grained structural steel
+ F	Fine-grained steel
+ H	Steel of special hardness
For the type of coating	
+ AZ	Coated with Al-Zn alloy
+ S	Hot-dip tinned
+ Z	Hot-dip galvanised
+ ZE	Electrolytically galvanised
For the treatment status [1)]	
+ A	Soft-annealed
+ C	Strain-hardened
+ N	Normalised
+ Q	Quenched or hardened
+ QT	Quenched and drawn
+ U	Untreated

[1)] To avoid mix-ups with other symbols, it is possible to start with an S for the type of coating (e.g. +SA) or with a T for the treatment status (e.g. +TA).

Additional symbols

Additional symbols for steels are letters and letters with numbers. They are subdivided into group 1 and group 2.

Additional symbols for steel products **(Table 1)** can be appended to the short names with a plus sign (+).

Examples of steels of main group 1

1. Engineering steels

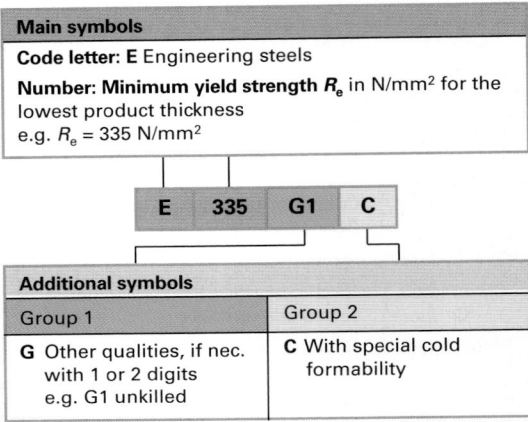

Main symbols

Code letter: E Engineering steels

Number: Minimum yield strength R_e in N/mm² for the lowest product thickness
e.g. R_e = 335 N/mm²

E	335	G1	C

Additional symbols

Group 1	Group 2
G Other qualities, if nec. with 1 or 2 digits e.g. G1 unkilled	**C** With special cold formability

2. Steels for structural steel engineering

Main symbols

Code letter: S Steels for structural steel engineering
Number: Minimum yield strength R_e in N/mm² for the lowest product thickness e.g. R_e = 235 N/mm²

S	235	JRG1	W

Additional symbols

Group 1				Group 2	
Notched-bar impact work in joules			Test temp.	**C** With special cold formability	
27J	40J	60J	°C	**E** For forging	
JR	KR	LR	+ 20	**N** Normalised	
J0	K0	L0	0	**Q** Quenched and drawn	
J2	L2	L2	– 20	**W** Weathering	
G Other qualities, if nec. with 1 or 2 digits e.g. G1 unkilled				**D** For hot-dip coatings	
Q Quenched and drawn					
N Normalised					

Main group 2: Short name with references to the chemical composition **(Table 1, Page 165).**

Main symbols

There are code letters and code numbers. The code letters are the symbols of the alloying elements; the code numbers provide information on the percentage contents.

Additional symbols for steels of main group 2 are subdivided into group 1 and group 2. Additional symbols of group 1 are only used for unalloyed steels with a Mn content of < 1 %. They are made up of letters or letters with numbers. Additional symbols for steel products **(Table 1, Page 164)** can be appended with a plus sign (+).

Table 1: Main group 2, multiplication factors	
Element	Factor
Cr, Co, Mn, Ni, Si, W	4
Al, Be, Cu, Mo, Nb, Pb, Ta, Ti, V, Zr	10
Ce, N, P, S	100
B	1,000

Examples of steels of main group 2

Unalloyed steels with an average Mn content < 1 %

C	35	E

Main symbols		Additional symbols
Letter	Code number	Group 1 (selection)
C for carbon	C content in $\frac{1}{100}$ % e.g. 0.35 % C	E Prescribed max. S content R Prescribed range of S content

Unalloyed steels with ≥ 1% Mn content, unalloyed free cutting steels, and alloyed steels when the content of the individual alloying elements is < 5%.

10	CrMo	9-10

Main symbols		Code numbers with hyphen
Code number	Letter	
C content in $\frac{1}{100}$ % e.g. 0.10 % C	Symbols for alloying elements e.g. Cr and Mo	Code number divided by factor **(Table 1)** produces the percentage content of the alloying element. E.g. Cr: 9/4 = 2.25 % Mo: 10/10 = 1 %

Alloyed steels when the content of an alloying element is ≥ 5% (without high-speed steels).

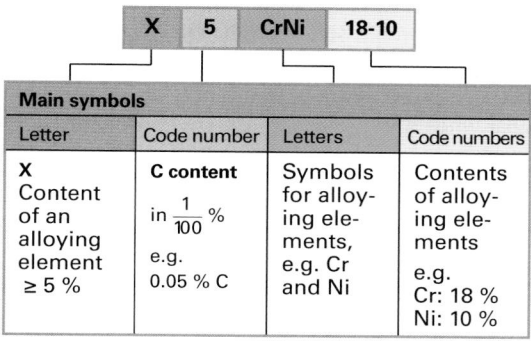

X	5	CrNi	18-10

Main symbols			
Letter	Code number	Letters	Code numbers
X Content of an alloying element ≥ 5 %	C content in $\frac{1}{100}$ % e.g. 0.05 % C	Symbols for alloying elements, e.g. Cr and Ni	Contents of alloying elements e.g. Cr: 18 % Ni: 10 %

High-speed steels

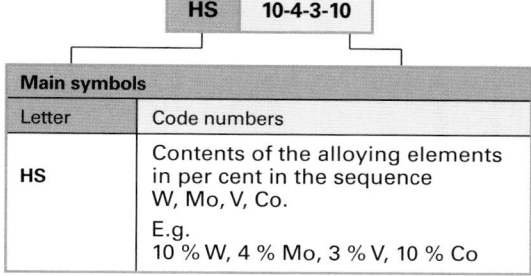

HS	10-4-3-10

Main symbols	
Letter	Code numbers
HS	Contents of the alloying elements in per cent in the sequence W, Mo, V, Co. E.g. 10 % W, 4 % Mo, 3 % V, 10 % Co

8.4.5 Categorisation and application of steels

Steels are categorised by their composition into unalloyed and alloyed steels and by their application or use into structural steels and tool steels **(Fig. 1, Page 166).** Alloyed steels can be subcategorised into low- and high-alloy steels. According to their purity and their application properties, they can also be referred to as ordinary low-carbon steels, high-grade steels or stainless steels.

Structural steels

Structural steels refer to steels which are used to build machines, vehicles and devices and in steel construction.

Unalloyed structural steels, e.g. S235JR (St 37-2), E335 (St 60)

The minimum yield strength and weldability are crucial to their application. Their carbon content ranges between 0.17 % and 0.5 %.

There are ordinary low-carbon steels and high-grade steels. They can be easily machined and are weldable. High-grade steels can be welded with all processes.

Application. E.g. for steel constructions, machine parts, sheets, screws, bolts, nuts, rivets.

Case-hardening steels, e.g. C15, 16MnCr5

These are used for components which are to have a hard and wear-resistant boundary layer and high strength by means of case-hardening while the core is to remain soft and ductile. These can be alloyed or unalloyed high-grade or stainless steels.

Application. E.g. for gudgeon pins, crankshafts, gear wheels.

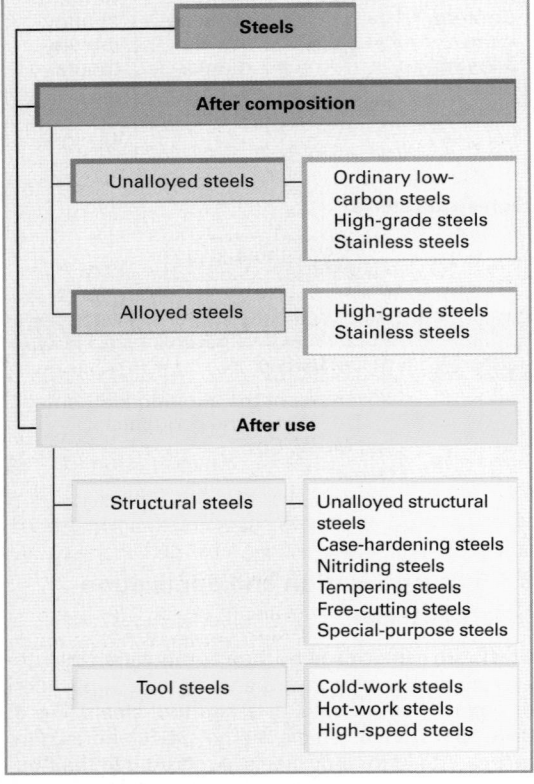

Fig. 1: Categorisation of steels

Nitriding steels, e.g. 31CrMoV9

These are alloyed with Cr, Al, Mo or Ni and hardened with nitrogen in the boundary layer.

Application. These are used for components which must not distort during hardening and which are not remachined, such as e.g. gudgeon pins, crankshafts, gear wheels, shafts, measuring instruments.

Tempering steels, e.g. C45E, 30CrNiMo8

These can be unalloyed, partly alloyed with Cr, Mn and Mo and occasionally even alloyed with V and Ni. When quenched and drawn, they acquire high strength while retaining good ductility down to the core. Their boundary layers are rendered wear-resistant by salt-bath nitriding.

Application. For components subjected to high strain, such as e.g. crankshafts, connecting rods, propeller shafts, steering knuckles, steering components.

Special-purpose steels

Special-purpose steels **(Table 1)** are steels with special properties for corresponding applications, such as, for example, stainless steels, heat-resistant steels, valve steels and spring steels. These are usually alloyed stainless steels.

Table 1: Special-purpose steels	
Short name	Application
Stainless steels	
X2CrTi12 X2CrNi18-9	Wheel trims Mufflers
Heat-resistant steels	
16CrMo4 X40CrNiCo 13-10	Exhaust pipes Turbine components
Valve steels	
X45CrSi9-3 X55CrNiMo20-8	Inlet valve Exhaust valve
Spring steels	
60SiCr7 X12CrNi17-7	Springs Valve springs

Stainless steels are resistant to acids. They are usually deep-drawable and weldable.

Heat-resistant steels on account of the addition of Cr, Mo, Ni, V or Si retain their strength even at high temperatures and are scale-resisting up to 1,100 °C.

Valve steels must be resistant to heat, melting loss, scaling, wear and corrosion while having good thermal conductivity. They are alloyed with Cr, Si, Ni and V.

Spring steels must have a high elasticity and fatigue limit. They usually have a higher Si content. Spring steels are quenched and drawn or hardened.

Tool steels

> Tool steels are steels of great hardness and strength, from which tools for separating (cutting) and forming are made.

These acquire their hardness for use by means of heat treatment. There are the following different types:

- Cold-work, hot-work and high-speed steels
- Unalloyed and alloyed tool steels

8.4.6 Commercial forms of steel

Steel-making plants usually put steels onto the market as semi-finished products in standardised shapes.

The commercial forms or semi-finished products are:

- Profile bars
- Tubes
- Steel wire
- Steel sheets
- Thin sheets
- Steel bars
- Small pieces

Profile bars (Fig. 1), such as angular, channel, T-section, double-T-section and Z-section steel, are usually made from S235J0. They are manufactured rolled and drawn in lengths of 3 m to 15 m.

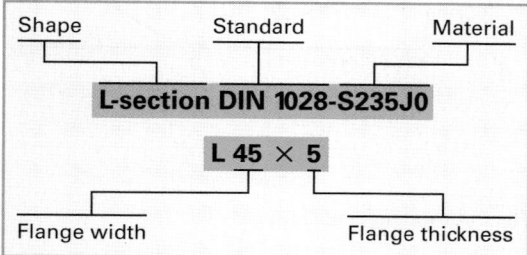

Shape	Standard	Material
L-section DIN 1028-S235J0		
L 45 × 5		
Flange width		Flange thickness

Fig. 1: Designation

Steel bars, such as round, flat, wide flat, square bar, hexagonal bar, half-round bar and strip steel, can be bright-drawn, ground and polished or hot-rolled.

All structural-steel grades are used as materials: above all 35S20+C for bright-drawn round and hexagonal bar steels and usually S235JRG1 for bright flat and square bar steels.

Steel wires are supplied rolled and drawn in rings or on drums.

Thin sheets are intended for forming work such as deep drawing and subsequent surface treatment such as painting or electroplating.

Tubes are manufactured butt-welded, lap-welded and solid-drawn.

Small pieces. These are screws, bolts, nuts, rivets, pins, springs, washers, screw-locking devices, split pins, nails.

Steel sheets are subdivided into extra-thin sheets and tinplate and into thin sheets, jobbing sheets and thick plates **(Table 1)**. The type of steel used for the sheets is dependent on the application and subsequent processing. Unalloyed structural steels, case-hardening and tempering steels or even stainless steels are used. Steel sheets are sold as black-plate, channelled plates, corrugated plates, scaled, pickled (descaled), perforated, galvanised, leaded and tinned plates (tinplate).

Table 1: Categorisation of steel sheets

Sheet type	Thickness in mm
Extra-thin sheet	under 0.5
Thin sheet	0.5 to 3
Jobbing sheet	3 to 4.75
Thick plate	over 4.75

REVIEW QUESTIONS

1 How can steels be categorised?
2 Explain the following workpiece designations: S275J2G1, C45E, 16MnCr5, X6CrMo17-7.
3 What are the special properties of tempering steels?
4 What do you understand by structural steel?
5 What are case-hardening steels used for?

8.4.7 Heat treatment of ferrous products

Heat treatment of ferrous products involves altering the material properties in order to improve hardness, strength and machinability.

These properties are dependent on the structure configuration, the carbon content and the alloying constituents.

The **iron-carbon diagram** shows the structure status of the steel as a function of the C content and the temperature **(Fig. 1, Page 168)**.

Steel with 0.8 % C (eutectoid steel). The standard structure is made up of ferrite grains run through by thin cementite strips. Because of its pearl-shell appearance, this structure is known as **pearlite**.

Steel with less than 0.8 % C (hypoeutectoid steel). The C content is not enough to create a pure pearlite structure. The structure is made up of ferrite and pearlite.

Steel with more than 0.8 % C (hypereutectoid steel). The C content is greater than is required for creating pearlite. Cementite is created as well as pearlite.

Structure of unalloyed steel

As well as pure iron (ferrite), unalloyed steel still contains up to 2.06 % carbon, which has bonded chemically with some of the iron to form iron carbide Fe_3C (cementite). Cementite is hard and brittle, pure iron is soft and ductile.

Heating of structure

When heated to a temperature in excess of 723 °C, the structure of the steel is transformed because the crystalline form of the iron changes and the cementite disintegrates. The body-centred cubic crystals **(Fig. 1)** of the ferrite shear into face-centred cubic crystals. A C atom of the disintegrating cementite can be intercalated in the empty cube area of the face-centred cubic crystal **(Fig. 1)**. Because the C atom in the crystal is intercalated in the solid state, the carbon in the iron is referred to as a solid solution. The mixed crystals created are called austenite.

The pearlite grains of the structure are transformed into austenite immediately at 723 °C. The ferrite and cementite still present depending on the C content is transformed into austenite on further heating up to the **G-S-E** line.

Cooling and quenching

In the case of **slow cooling**, the old structures continue to regress. However, the formation of pearlite is suppressed in the case of quenching from the temperature range above the G-S-K line. When the crystal lattice shears from the face-centred cubic into the body-centred cubic form, there is not enough time remaining for the C atoms to form cementite with the iron atoms. The C atoms are locked in the body-centred cubic crystals. This produces a distorted crystal lattice, which in turn produces very hard and brittle steel. The fine-needle structure created is called martensite.

Annealing

Annealing is a heat-treatment process in which the workpiece is heated to annealing temperature, maintained at this temperature for soaking purposes and then slowly cooled **(Fig. 2)**.

Soft annealing is conducted, depending on the carbon content, between 680 °C and 750 °C so that the workpiece can then be formed better with or without machining.

Normalising is conducted, depending on the carbon content, between 750 °C and 950 °C in order to obtain a uniform, fine-grained structure again after rolling or forging.

Stress-free annealing is conducted between 550 °C and 650 °C in order to eliminate internal stresses created by rolling, forging or welding.

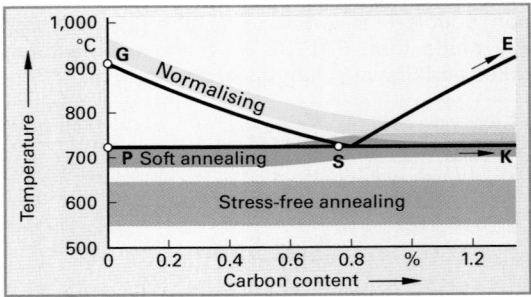

Fig. 2: Annealing temperatures

Hardening of tool steel

Hardening **(Fig. 1, Page 169)** is a heat-treatment process in which the workpiece is heated to hardening temperature, maintained at this temperature for the purpose of soaking the zone to be hardened and then quenched. The workpiece is tempered after it has been quenched.

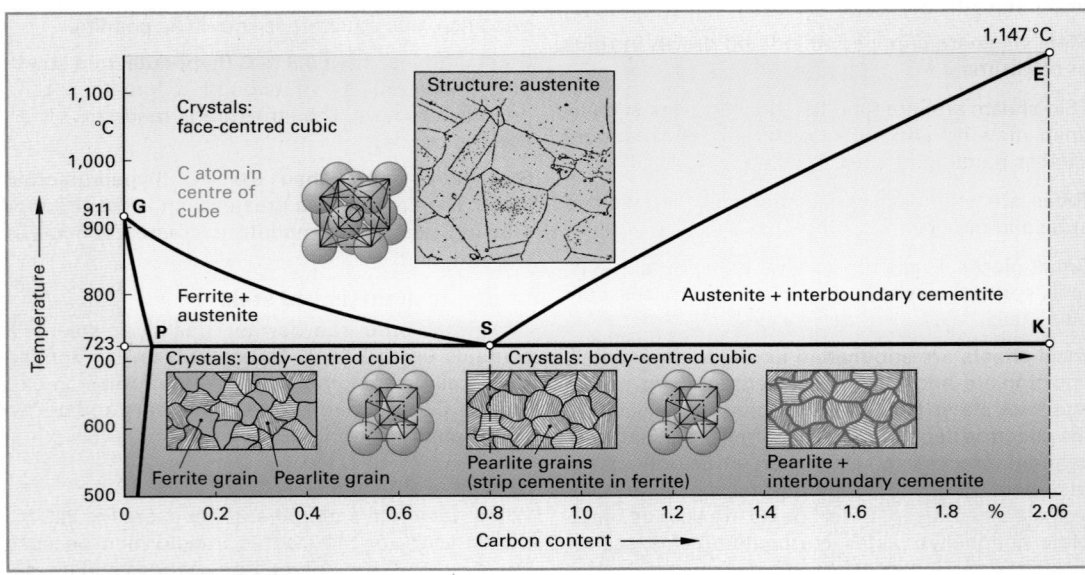

Fig. 1: Iron-carbon diagram (excerpt)

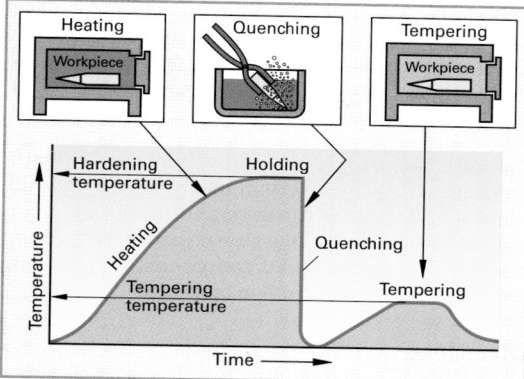

Fig. 1: Temperature curve during hardening

Hardening is intended to make the steel hard and wear-resistant. The hardening temperature of unalloyed tool steel ranges between 770 °C and 830 °C. Higher hardening temperatures are required for low- and high-alloy tool steels.

Different quenchants may be used:

- Water hardening, primarily for unalloyed tool steels.
- Oil hardening, primarily for low-alloy tool steels.
- Air hardening for high-alloy tool steels.

Tempering

> Tempering involves reheating and then cooling hardened workpieces.

The tools are rendered glasshard and brittle once they have been quenched. Tempering, for example at temperatures between 180 °C and 400 °C, improves ductility and reduces brittleness; hardness is also reduced slightly, and the steel acquires its hardness for use.

For the relevant steel grade, it is necessary to refer to the steel manufacturers' material specification sheets for details of hardening and tempering temperatures and quenchants.

Surface hardening

> Surface hardening is a heat-treatment process in which the boundary layer of the workpiece of heat-treatable steel is rapidly heated to hardening temperature and then immediately quenched.

There are the following different types:

- Surface hardening with carbon existing in sufficient quantities by structural transformation, e.g. flame hardening, induction hardening.
- Surface hardening with supplied carbon by structural transformation, e.g. case hardening.

- Surface hardening with supplied nitrogen by chemical bonding, e.g. nitriding, carbonitriding.

Surface hardening with existing carbon

The workpieces are normally made from tempering steel because this type of steel already contains the amount of carbon required for hardening (at least 0.45 %). The following different processes may be used here: flame hardening and induction hardening.

Flame hardening (Fig. 2). The boundary layer of the workpiece is briefly heated with a burner flame to hardening temperature. The workpiece is quenched with a water sprinkler just before the heat reaches the inside of the workpiece.

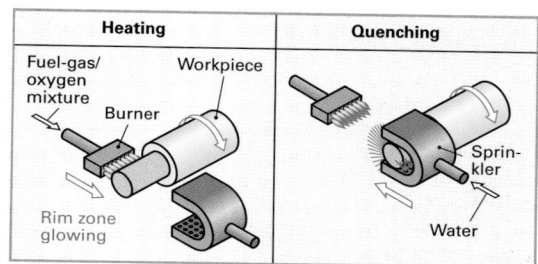

Fig. 2: Flame hardening

Induction hardening (Fig. 3). A high-frequency coil surrounds the workpiece at close distance. A high-frequency alternating current in the coil induces strong eddy currents, which rapidly heat the boundary layer of the workpiece to hardening temperature. The workpiece is quenched with a water sprinkler just before the heat reaches the inside of the workpiece.

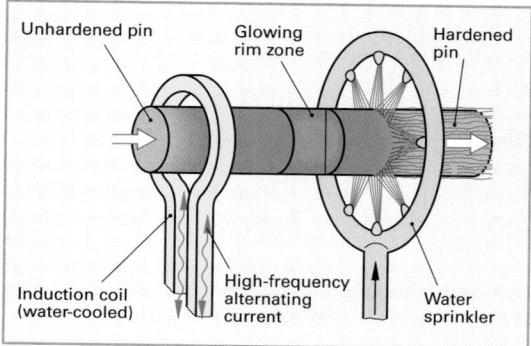

Fig. 3: Induction hardening

Surface hardening with supplied carbon

The workpieces are made from steel with a low carbon content < 0.2 %. Carbon is supplied to the surface to enable the boundary layer to harden.

Case hardening

> Case hardening is a heat-treatment process in which the rim zone of the workpiece of low-carbon steel is enriched with carbon and then hardened.

Case-hardening steels contain up to approx. 0.2 % C. In order to make the rim zone hardenable, the workpiece is annealed in agents that give off carbon.

This process is known as carburising **(Fig. 1)**, in which the form of the agent giving off carbon can be solid, liquid or gaseous.

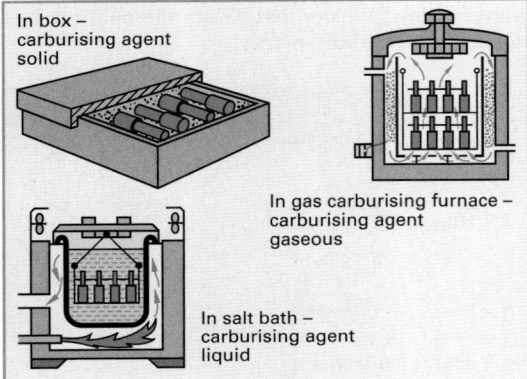

Fig. 1: Carburising during case hardening

The workpiece is hardened after it has been carburised. The carburised rim zone becomes hard and the core remains unhardened and ductile. The workpiece is then tempered.

Surface hardening with supplied nitrogen
The workpieces are made from steels which are alloyed with Al, Cr, Mo, Ti or V. Through the supply of nitrogen at higher temperatures, these alloying elements form very hard and stable nitrides, e.g. AlN, CrN.

Nitriding

> Nitriding is a heat-treatment process in which the rim zone of the workpiece is enriched with nitrogen.

Salt-bath nitriding. The rim zones of the workpiece are supplied with nitrogen in salt baths.

Gas nitriding. The rim zones of the workpieces are supplied with nitrogen in a gas furnace by way of ammonia gas.

The nitriding process gives nitriding steels a thin, very hard and wear-resistant rim zone. The hardness is created directly during nitriding by the cre-

ation of hard nitrides at annealing temperatures up to approx. 550 °C. The workpieces are not quenched or tempered; they are therefore not subject to scaling or distortion such that they can be finish-machined prior to nitriding.

Carbonitriding. This is a combination of case hardening and nitriding. Carbon and nitrogen are supplied simultaneously in a gas furnace. The workpieces are hardened by subsequent heating and quenching. Carbonitrided components are, for example, camshafts and valve levers.

Quenching and drawing

> Quenching and drawing is a heat-treatment process in which the workpiece after hardening and quenching is tempered to such high temperatures that it acquires great tensile strength together with good elongation and ductility instead of hardness **(Fig. 2)**.

During hardening, the workpiece acquires great strength and hardness but minimal elongation and ductility. When the workpiece is tempered to temperatures of 500 °C to 670 °C, hardness decreases dramatically, strength decreases slightly as well, and ductility and elongation increase.

During tempering at the lower temperature limit, strength is greater than at the upper temperature limit; ductility and elongation, on the other hand, are greater at the upper temperature limit.

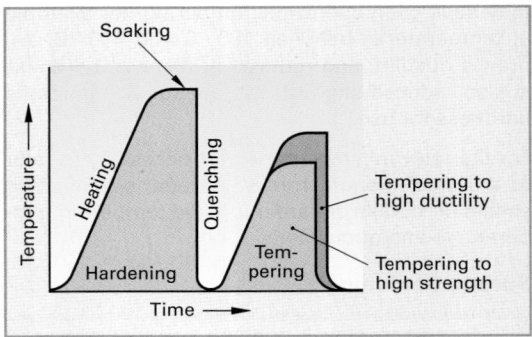

Fig. 2: Quenching and drawing

REVIEW QUESTIONS
1 What do you understand by hardening of steel?
2 Why are steels tempered after being quenched?
3 What do you understand by case hardening?
4 Why are workpieces which are to be surface-hardened usually made of tempering steel?
5 Why are workpieces tempered during quenching and drawing at high temperatures?

8.5 Non-ferrous metals

> Non-ferrous metals (NF metals) are all metals and their alloys with the exception of iron.

Non-ferrous metals are categorised according to their densities into heavy metals and light metals (**Fig. 1**).

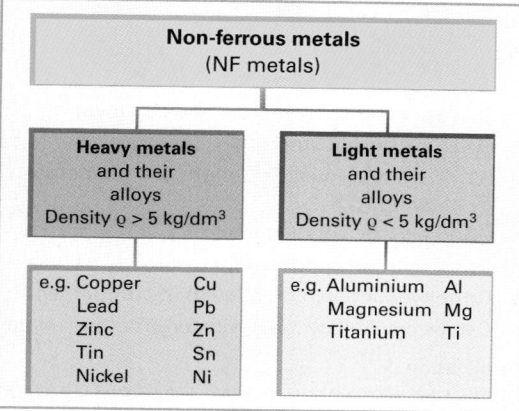

Fig. 1: Categorisation of NF metals

Most pure metals, such as e.g. copper, lead, aluminium, are very soft and have only minimal strength. Alloying is used to improve the properties of pure metals.

> Alloying means the mixing of two or more metals in a liquid state.

There are two different types of alloy: casting and wrought alloys.

Casting alloys. These have good casting properties and have greater strength and ductility than cast iron. Application: e.g. crankcases, cylinder heads.

Wrought alloys. The properties of these alloys can be altered within broad limits by means of appropriate alloying constituents and subsequent treatment, e.g. by forging. Wrought alloys are used, for example, as suspension arms for wheel-suspension systems and wheel rims.

8.5.1 Designation of NF metals

Designation of pure metals. After the chemical symbol for the metal comes the percentage purity, e.g. Al 99.99.

Designation of alloys. In the case of alloys, the chemical symbols for their constituents are given, in which the metal in the greatest proportion is given first without the percentage actually being stated; this is determined as the difference from the to-

tal of the other alloying additives. The base metal is followed by the symbols for the alloying additives with the percentage of each being stated. The order in which they appear is based on decreasing percentages. The percentage of an alloying additive is not stated if that additive is present in a small proportion.

Casting alloys are identified by code letters denoting manufacture and application (**Table 1**). These letters, separated by a hyphen, come before the alloy itself is specified. The alloy may be followed by identifying symbols denoting special properties (**Table 1**) or the treatment status, or the tensile strength in $1/10$ N/mm² is given as a numerical value with a preceding F.

Table 1: Designation of NF-metal alloys (selection)

Code letter for manufacture and application		
G- Casting (general)	Gl-	Anti-friction metal
GD- Diecasting	L-	Solder
GK- Gravity diecasting	Lg-	Bearing metal
GZ- Spun casting	V-	Master alloy
Identifying symbols for special properties, treatment status, tensile strength		
a Precipitation-hardened	g	Annealed
ka Precipitation-hardened cold	zh	Drawn
F + Code number for tensile strength in $1/10$ N/mm²		

Combined letters and numbers denoting diecasting alloys

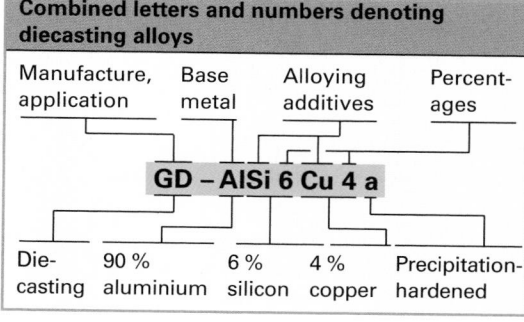

Wrought alloys have no code letters denoting manufacture and application, except in the case of Al wrought alloys.

Combined letters and numbers denoting wrought alloys

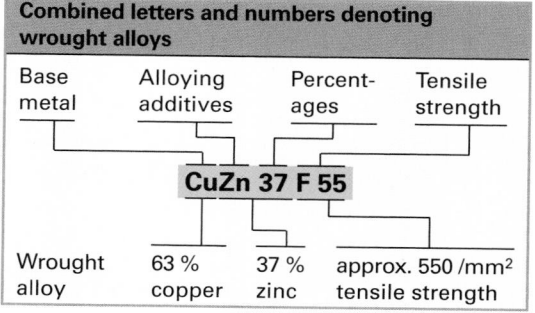

8.5.2 Heavy metals

The categorisation of heavy metals is shown in **Table 1**.

Table 1: Categorisation of NF heavy metals

Metal	Chem. symbol	Metal	Chem. symbol
Metals for use and their alloys			
Copper	Cu	Zinc	Zn
Lead	Pb	Tin	Sn
Nickel	Ni		
Alloy metals			
Molybdenum	Mo	Tantalum	Ta
Tungsten	W	Chromium	Cr
Cobalt	Co	Manganese	Mn
Vanadium	V	Bismuth	Bi
Antimony	Sb		
Precious metals			
Silver	Ag	Gold	Au
Platinum	Pt		

Copper (Cu)
Attributes
- Soft, ductile, flexible
- Reddish brown colour
- Good conductor of electricity and heat
- Corrosion- and fireproof
- Very good cold- and hot-forming properties
- Good soft-soldering and brazing properties
- Weldable
- Poor casting properties
- Can be cut well by tools with large cutting angles

Application
E.g. for electric cables, tubes/pipes for petrol, oil or water, radiators, seals and alloys **(Table 2)**.

Table 2: Copper alloys

Comb. letters & numbers	Application
Copper casting alloys	
GZ-CuSn 7 ZnPb	Small-end bushings
G-CuPb 10Sn	Multilayer bearings
G-CuAl 10Fe	Synchroniser rings, pinions
GD-CuZn 38Pb	Bright castings
G-CuZn 33Pb	Sand castings
Copper wrought alloys	
CuZn 39 Pb 3	Carburettor jets
CuZn 37	Radiator pipes
CuZn 31 Si	Valve guides
CuNi 44	Constantan

8.5.3 Light metals

Predominantly the following light metals are used in vehicle manufacturing: aluminium (Al), magnesium (Mg) and titanium (Ti).
Further light metals such as beryllium and the alkali metals lithium, sodium and calcium are used in some cases as alloying additives.

Aluminium (Al)
Properties
- Silvery white
- Corrosion-proof by means of oxide layer
- Soft, low tensile strength
- Hardness and tensile strength can be increased by alloying
- Good electric conductivity
- Good thermal conductivity
- Good workability and alloyability properties
- Can be cut well by tools with large cutting angles

Application
High-grade aluminium, e.g. for foil, tubes, cans, reflectors, trim mouldings, vehicle body components (as structural material and as alloying additive).

Aluminium alloys

Categorised into wrought and casting alloys. There are precipitation-hardenable and non-precipitation-hardenable alloys. They must not be heated.

Properties of alloys
Aluminium with copper has high strength but only low corrosion resistance. Aluminium with magnesium, silicon and manganese has good corrosion resistance as well as high strength. Aluminium with copper, magnesium and silicon is precipitation-hardenable and thereby acquires greater strength. Aluminium with a specific proportion of magnesium, silicon and manganese is non-precipitation-hardenable, high strength already being given.

Application
Wrought alloys (examples)

AlCuMg 2:	Wishbones/transverse control arms, wheel hubs, crankshaft and camshaft gears
AlMgSi 1:	Wheels, profile sections for vehicle body components
AlMg 2:	Sheet panels for body parts
AlSi 17 CuNi:	Pressed pistons

Casting alloys (examples)

G-AlSi 12:	Crankcases, oil pans, gearbox housings
G-AlSi 10 Mg:	Crankcases and water-cooled cylinder heads
GK-AlSi 12 CuNi:	Cast pistons

8.6 Plastics

> Synthetically produced materials are called plastics.

Plastics are manufactured from the raw materials crude oil, natural gas, coal, lime, air and water. They are organic materials because they are made up of carbon or silicon compounds. Further constituents may be, for example, oxygen (O_2), hydrogen (H_2), nitrogen (N_2), sulphur (S) and chlorine (Cl).

Typical properties of practically all plastics are:
- Low density
- Good workability and formability, can be dyed
- Resistant to corrosion, resistant to acids and alkalis
- Not electrically conducting, low heat conduction
- High thermal expansion, low heat resistance

There are four distinct groups of plastics: thermoplastics, duromers, elastomers and composite materials (**Fig. 3**).

8.6.1 Thermoplastics

> Thermoplastics are made up of long, intermatted filamentary molecules which are not cross-linked with each other (**Fig. 1**).

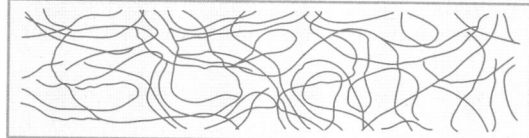

Fig. 1: Structure of thermoplastics

When heated, the molecules are induced to thermal vibrations, causing the structure to loosen and the material to melt and soften. Thermoplastics are hard and barely elastic at room temperature but become increasingly soft each time they are heated. When heated, they can be formed without cutting by casting, bending or welding. They are destroyed at very high temperatures. Through the addition of non-volatile solvents ("plasticisers"), they become ductile/tough, leathery or elastic (thermoelastics).

Examples of thermoplastics
Polyvinyl chloride PVC (Fig. 2)

Properties: Transparent, dyeable, glueable and weldable; resistant to oil, petrol and alkalis; sensitive to acetone.

Hard PVC: Hard, ductile, e.g. for trims/linings, entrance strips, number-plate reinforcements, strips on roof frames, tubes.

Soft PVC: Soft, flexible, rubbery or leathery through additives, e.g. for seals, foil, floor mats, cable insulation, leather, imitation leather.

Polycarbonate PC (Fig. 2)

Properties: Transparent, distortion-free, light-resisting, ductile, high strength, resistant to weak acids and alkalis, scratch-resistant.

Application: Glass for headlights and lamps, electrical switches.

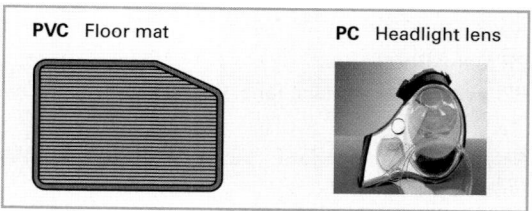

Fig. 2: Applications of PVC and PC

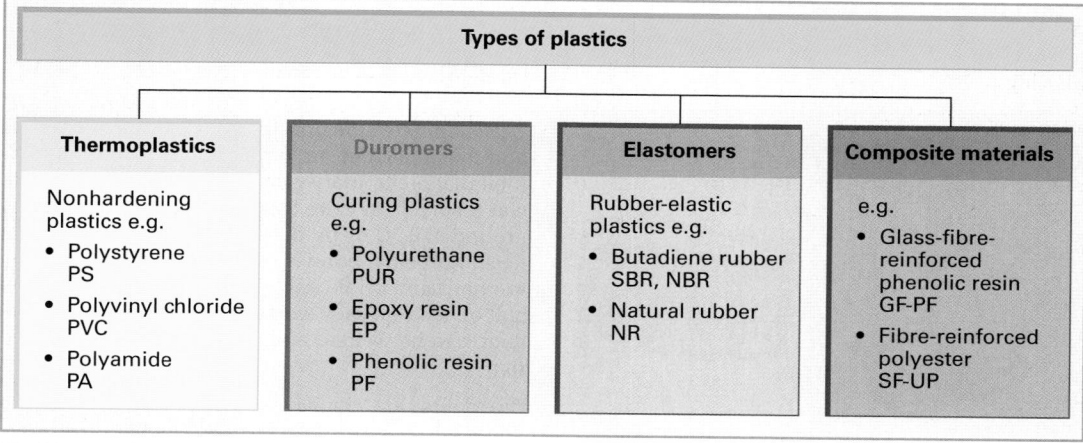

Fig. 3: Types of plastics

Styrene-butadiene SB (Fig. 1)

Properties: Opaque, impact-resistant, heat-resistant.

Application: Interior-trim parts, battery containers.

Acrylonitrile-butadiene-styrene ABS (Fig. 1)

Properties: Very impact-resistant, heat-resistant; resistant to acids and oil, but not to benzene.

Application: Radiator grilles, instrument frames, bumpers.

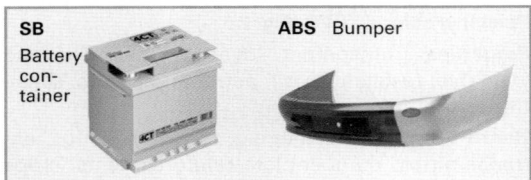

Fig. 1: Applications of SB and ABS

Polyolefins: polyethylene PE, polypropylene PP (Fig. 2)

Properties: Colourless to milky white, dyeable, unbreakable, waxy to touch, weldable, not glueable, resistant to acids, alkalis, petrol, benzene.

Application:

Soft PE: Axle sleeves, reservoirs, gaiter seals.

Hard PE: Fuel tanks, washer reservoirs, air-filter housings.

PE foam: Roofliners.

PP: Accelerator pedals, fans, wheel caps.

Polymethylmethacrylate (acrylic glass) **PMMA (Fig. 2)**

Properties: Transparent, polishable, hard, elastic, glueable and weldable; resistant to petrol, oil, acids.

Application: Direction-indicator and tail lamps, electrical parts, protective goggles, protecting glass, glazing.

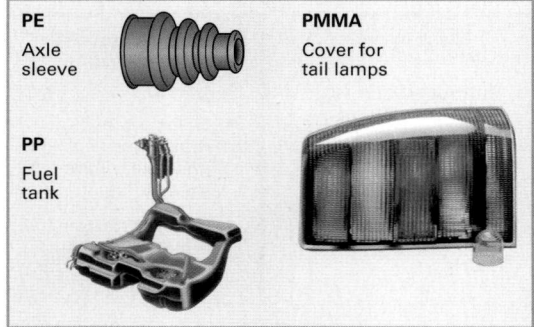

Fig. 2: Applications of PE, PP, PMMA

Polytetrafluorethylene PTFE (Fig. 3)

Properties: Milky white, waxy to touch, low friction, ductile, abrasion-proof, not glueable, resistant to chemicals, heat-resistant up to 280 °C.

Application: Coatings, seals, gaiter seals, heat-resistant cable insulation, maintenance-free bearings, diaphragms.

Polyamide PA (Fig. 3)

Properties: Milky white, hard, ductile, low friction, resistant to wear, resistant to oil, petrol, benzene, solvents, non-combustible as aramids (aromatic polyamides) and heat-resistant to over 260 °C.

Application: Bushings, friction bearings, fan impellers.

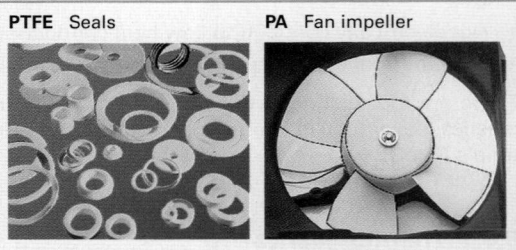

Fig. 3: Applications of PTFE and PA

8.6.2 Duromers

Duromers have filamentary macromolecules which are closely cross-linked with each other after curing (**Fig. 4**).

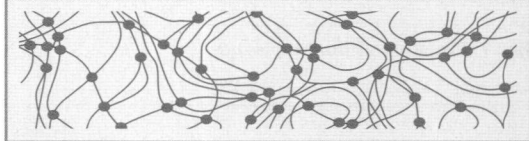

Fig. 4: Structure of duromers

This eliminates the possibility of thermal vibrations and these plastics no longer soften. Their liquid or meltable preliminary products are also called artificial resins. They cure by pressing at temperatures around 170 °C or by the addition of curing agents (casting resins, bonding resins). When cured, they are hard and brittle and no longer soften as the result of heating. They cannot be dissolved in any solvent, can be welded and are only formable by cutting.

Duromers are usually processed together with fillers into composite materials. The fillers improve the properties or serve as extenders.

Examples of duromers

Phenolic resin PF (Fig. 1)

Properties: Only dark colours, hard, brittle, glueable, not light-resistant, turns brown, smells of phenol.

Application (Fig. 1): With fillers as moulding compounds for dark moulded parts, moulded laminates, synthetic paints, casting resins.

Urea resin UR; melamine resin MF (Fig. 1)

Properties: Transparent, light-resisting, odourless, dyeable; hard, brittle, resistant to weak acids and solvents.

Application (Fig. 1): With fillers as moulding compounds for light moulded parts, moulded laminates, synthetic paints, hot-setting glues, cold-setting glues.

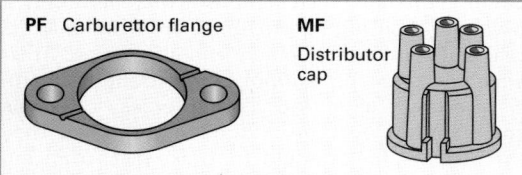

Fig. 1: Applications of PF and MF

Polyester resin UP (Fig. 2)

Properties: Transparent, can be hard, brittle, soft or elastic, good cast properties, good adhesiveness; resistant to oil, petrol, solvents, weak acids and alkalis.

Application (Fig. 2): Metal adhesives, knifing fillers, casting resins, glass-fibre-reinforced plastics.

Epoxy resin EP (Fig. 2)

Properties: Colourless to yellow, hard, impact-resistant, very good casting and good adhesive properties.

Application (Fig. 2): Glue, casting-in of electronic parts, glass-fibre-reinforced plastics.

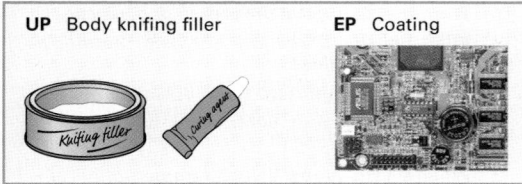

Fig. 2: Applications of UP and EP

Polyurethane resin PUR (Fig. 3)

Properties: Yellow, transparent, hard, ductile, soft or rubber-elastic, adhesive property, expandable.

Application:
Hard PUR: Bearings, gear wheels
Medium-hard to soft PUR: Bumpers, glues
PUR foam: Vehicle upholstery, integral foam for trim panels in motor vehicles.

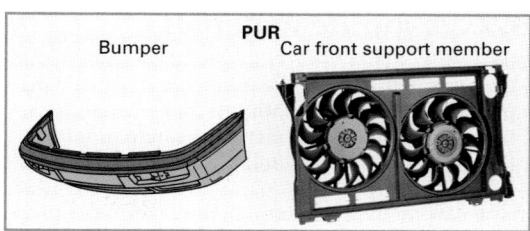

Fig. 3: Applications of PUR

8.6.3 Elastomers

Elastomers are made up of random filamentary molecules. The wide-meshed cross-linking occurs during vulcanisation (**Fig. 4**).

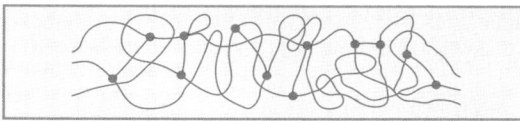

Fig. 4: Structure of elastomers

The preliminary stages of these plastics are synthetic rubbers or natural rubbers. They can be elongated by small forces and then spring back again. Under increasing temperatures, they do not melt; instead, they remain elastic up to the point of their destruction at higher temperatures. They have good strength together with high elongation and high elasticity. They are not meltable, not formable without cutting and not weldable; they can swell but they cannot dissolve.

Examples of elastomers (Fig. 5)

Rubber (natural rubber) NR

Properties: Elasticity decreases with increasing sulphur content, not resistant to oil, petrol, benzene or ageing.

Application (Fig. 5): Admixture in tyres, water hoses, seals, V-belts.

Styrene-butadiene rubber (synthetic rubber) SBR

Properties: Similar to rubber but more abrasion-proof, more non-ageing, less elastic; resistant to oil and petrol.

Application (Fig. 5): Admixture in tyres, sleeves, hoses.

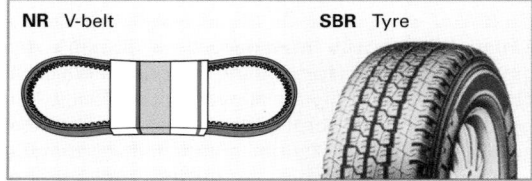

Fig. 5: Applications of NR and SBR

8.7 Composite materials

> Composite materials are materials in which two or more individual materials have been combined to form a new material. The individual materials are usually retained in recognisable form.

In the case of these materials, the favourable properties of the materials which are combined with each other, e.g. high strength, hardness, ductility or elasticity, are carried over to the new material.

In the case of glass-fibre-reinforced plastics, the low strength of the plastic, for instance, is increased by the strength of the glass fibres, while the ductility of the plastic compensates for the brittleness of the glass fibres **(Table 1)**.

Table 1: Composite-material properties

Glass fibre	+ artificial resin	→	glass-fibre-reinforced plastic
very strong	not very strong		very strong
but	but		and
brittle	ductile		ductile

Composite materials are distinguished by the form of their individual materials as follows:

- Particle-reinforced composite materials
- Fibre-reinforced composite materials

8.7.1 Particle-reinforced composite materials

Plastic-moulding compounds

> These are artificial resins mixed with fillers which a pressed into moulded parts.

Fillers can be rock dust, wood dust or soot, textile fibres, or paper shavings, wood chips or fabric cuttings. They influence the properties of the moulded parts such as strength, brittleness, heat conduction and insulating ability and also serve as expanders.

Application e.g. for steering wheels, gear-lever handles, brake and clutch linings, electric insulators and housings.

Sintered composite materials. It is possible by means of sintering to combine materials with each other which can only be alloyed with difficulty or not at all. Sintered composite materials are, for example, hard metals, oxide-ceramic cutting materials, permanent-magnet materials and carbon brushes, connecting rods.

8.7.2 Fibre-reinforced composite materials

> These are made up of fibres such as glass, carbon or aramid fibres embedded in plastic.

Their properties are influenced by the type and the arrangement of the fibres and by the type of artificial resin used for bonding. They have higher strength combined with lower weight than conventional metallic materials.

Glass-fibre-reinforced composite materials (GFRP) are manufactured by means of various processes, the most well-known being hand laminating, e.g. for body components or boats. Further applications are, for example, spring leaves, fan impellers, bucket seats and gear wheels.

Carbon-fibre-reinforced composite materials (CFRP). Owing to their high costs, these are used only occasionally in motor-vehicle manufacturing. Their advantages are, for example, high strength together with low weight, low inertial forces, low noise and vibration levels, good friction properties. They can be used, for example, for brake discs in racing cars, body components and moving engine components such as connecting rods **(Fig. 1)**.

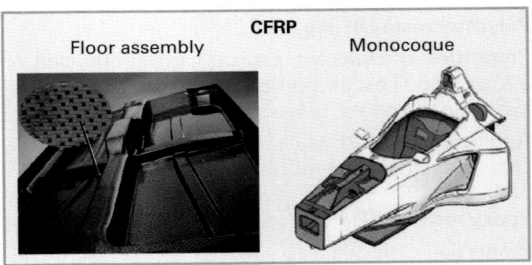

Fig. 1: Applications of CFRP

1 What are plastics?
2 What are the typical properties of plastics?
3 What are the differences between thermoplastics and duromers?
4 Name some important thermoplastics.
5 Where in a motor vehicle are thermoplastics used?
6 What are the properties of duromers?
7 Give examples of applications of duromers in a motor vehicle.
8 What are the special properties of elastomers?
9 Where in a motor vehicle are elastomers used?
10 What are composite materials?
11 Why do composite materials have more favourable properties than their starting materials?
12 What different types of composite materials are there?

9 Friction, lubrication, bearings, seals

9.1 Friction

When a body is moved on its base with force F, the friction force F_R opposes the direction of motion **(Fig. 1)**.

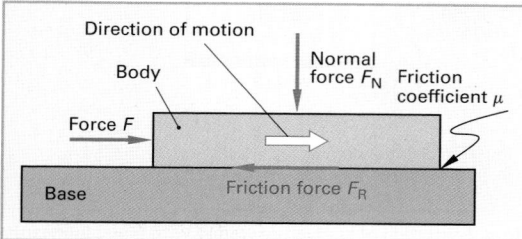

Fig. 1: Effective forces during friction

> Friction force F_R is the resistance against movement of one body on another body.

Its value is determined by:
- Normal force F_N
- Surface quality
- Pairing of materials
- Type of friction (static, sliding or rolling friction)
- State of lubrication
- Temperature

Normal force F_N. This always acts vertically to the friction surface.

The friction coefficient μ (mu) determined in tests records the other influencing variables. It is inserted as a constant value in the formula. The following formula applies:

> Friction force = normal force × friction coefficient
> $$F_R = F_N \times \mu$$

Thus, the friction force F_R is proportional to the normal force F_N. If the normal force F_N is increased, so the friction force F_R increases in the same proportion.

Types of friction

Static friction. This is the resistance which a body puts up to the movement on its base **(Fig. 1)**. Here, the force F is less than or equal to the friction force F_R. If a body is to be moved, the static friction must be overcome.

Sliding friction. This is the resistance which a body sliding on its base puts up to the sliding motion. The friction force F_R for sliding friction is less than it is for static friction and acts, for example, between a brake disc and the brake pads.

Rolling friction. This is the resistance which a body rolling on its base puts up to its motion. Rolling friction is significantly less than sliding friction. The value of the friction force F_R created by rolling friction is determined by the material of the bodies rolling on each other and the shape of their contact. For this reason, the rolling resistance of a ball bearing (point contact) is less than that of an antifriction bearing (line contact).

Force transmission by friction

In order for the wheels of a motor vehicle to be able to transmit peripheral forces F_U (motive or braking forces) and lateral forces F_S, static friction must exist between the tyres and the road surface. The transmissible force here is dependent on the normal force F_N (tyre load) and the friction coefficient ($\mu_{dry} \approx 0.9 ... 1$; $\mu_{ice} \approx 0.1$).

If, for example, the wheels lock during braking or spin during pulling away, there is sliding friction between the tyres and the road surface. In this case, no lateral forces can be transmitted to the road surface and the vehicle is no longer steerable.

The Kamm's friction circle depicted in **Fig. 2** illustrates this boundary condition. Here, the resulting force F_{Res} is the maximum force which the tyre can transmit in the event of static friction. It can be separated into two components:

- A peripheral force, e.g. when braking.
- A lateral force, e.g. when cornering.

When the peripheral force reaches a maximum limit, e.g. when the wheels spin, no lateral forces can be transmitted.

When the lateral force reaches its maximum limit, the tyres cannot transmit any peripheral forces.

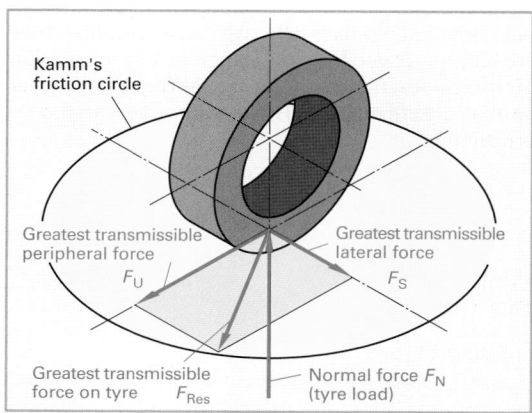

Fig. 2: Kamm's friction circle

9.2 Lubrication

Functions

Lubricants are used to keep the wear between moving parts as low as possible. They are intended to prevent the surfaces that slide on each other from coming into contact with each other.

They are also intended to

- reduce friction
- damp impact shocks
- protect against corrosion
- provide a fine seal
- dissipate heat
- remove wear particles
- damp noises

The following types of friction are distinguished, depending on the state of lubrication: dry friction, mixed friction and fluid friction.

Dry friction. The surfaces sliding on each other come into contact with each other directly without a separating lubricant film, e.g. in the case of piston seizures. This results in an increase in temperature and wear on the friction surfaces **(Fig. 1)**.

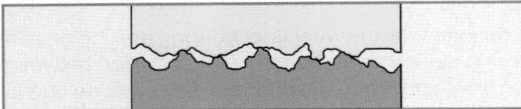

Fig. 1: Dry friction

Mixed friction. The surfaces sliding on each other come into partial contact with each other because they are only partially separated by a lubricant film, e.g. friction between piston and cylinder during cold starting, friction between the gearwheel flanks of a gearbox. The wear and the tendency to seize reduce **(Fig. 2)**.

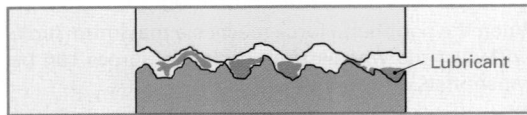

Fig. 2: Mixed friction

Fluid friction. The surfaces sliding on each other do not come into contact with each other because they are fully separated by a lubricant film, e.g. lubrication of the crankshaft during operation. There is no wear on the friction surfaces. The friction occurs in the lubricant **(Fig. 3)**.

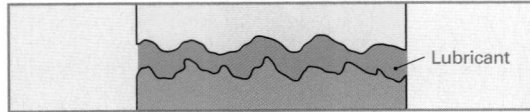

Fig. 3: Fluid friction

Lubricant types

Liquid, paste-like and solid substances can be used as lubricants in a motor vehicle. The right lubricant is used to suit the particular condition of use.

Liquid lubricants are mineral oils or synthetic oils alloyed with additives. Because the oil adheres to the surfaces sliding on each other, there is an oil film between these surfaces. A lubricating wedge, which lifts the sliding surfaces off each other, is created as a function of the sliding speed **(Fig. 4)**.

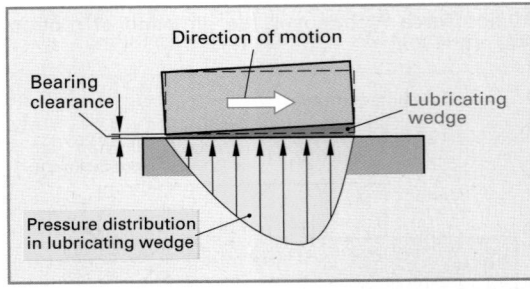

Fig. 4: Formation of lubricating wedge with liquid lubricants

Paste-like lubricants are greases which consist of a structure of lime, soda or lithium soaps into which mineral oils or synthetic oils are intercalated **(Fig. 5)**. When the soap structure is moved, oil droplets emerge to moisten the surface to be lubricated.

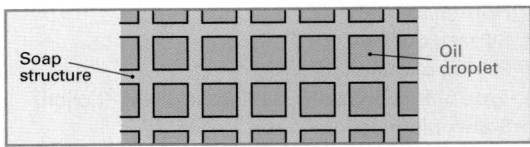

Fig. 5: Soap structure of a lubricating grease

The corresponding grease must be chosen to suit the condition of use. For instance, only antifriction-bearing greases may be used to lubricate antifriction bearings. Sealing is supported by the grease. Because of the flexing work that occurs and the associated heating, the housing cavities may only be filled with grease up to roughly half.

Solid lubricants. These are composed of fine-bladed, powdery graphite or molybdenum disulphide (MoS_2). The friction is reduced by the fact that the small sliding laminas in the lubricating gap slide on each other **(Fig. 6)**. When polytetrafluorethylene (PTFE) is used, tiny globular particles reduce the friction.

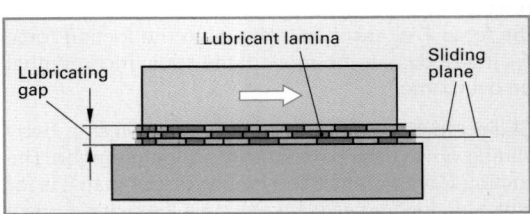

Fig. 6: Solid lubricant in lubricating gap

9.3 Bearings

> Bearings serve to guide and support shafts and axles, in the course of which as little friction and wear as possible are to occur in the bearings.

The following different bearings are used, depending on the magnitude and direction of the forces acting on the bearing:

- Radial bearings
- Thrust bearings

Radial bearings. These absorb the forces transversally to the bore axis. They are also known as supporting bearings.

Thrust bearings. These absorb the forces in the direction of the bore axis. They are also known as axial bearings. There are friction bearings and antifriction bearings for all load situations **(Fig. 1)**.

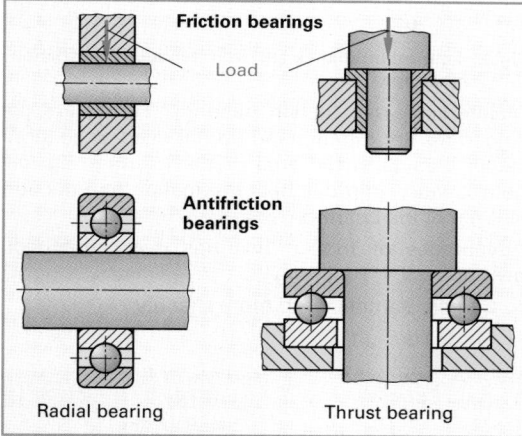

Fig. 1: Radial and thrust bearings

Friction bearings

In the case of a friction bearing, the shaft runs in bearing shells, bearing bushings or directly in the bearing body. A metallic contact between shaft and bearing is created under the effect of the bearing force. The heat generated in the process by friction would result in rapid seizing and thus in destruction of the bearing and the shaft.

Requirements

- Low friction
- High load capacity
- Good emergency-running properties
- High compressive strength
- Good sliding properties
- High resistance to wear
- Good thermal conductivity
- Good noise damping by lubricant layer

The requirements are satisfied by lubrication of the friction bearings and the friction-bearing material.

Lubrication of friction bearings. The function of the lubricating oil is to reduce the friction between shaft and bearing and to dissipate the heat generated so as to prevent the bearing from overheating and seizing. There must therefore be a film or layer of lubricating oil between shaft and bearing.

The lubricating-oil layer is created in hydrodynamically lubricated bearings by the rotary motion of the shaft journal. At the beginning of the rotary motion, the journal and bearing shell are not yet fully separated (mixed friction). As the rotational speed increases, a lubricating-oil wedge, which raises the shaft, is formed under the shaft journal. The shaft floats on the lubricating-oil layer (fluid friction) **(Fig. 2)**. The thickness of the lubricating-oil layer is dependent on the bearing clearance, the bearing load, the bearing pressure, the circumferential speed and the lubricant.

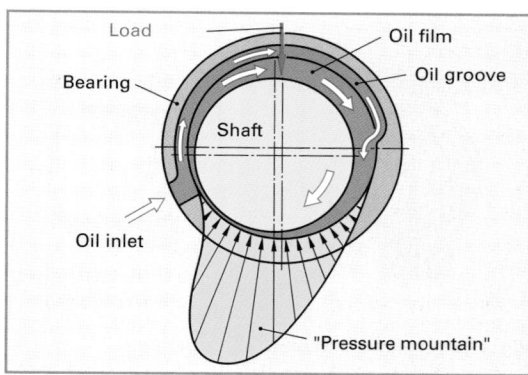

Fig. 2: Pressure distribution in a friction bearing

Friction-bearing materials. In the interests of keeping friction-bearing wear as low as possible during the starting process, the following properties are demanded of the bearing materials:

- Good emergency-running properties
- Good embeddability
- Good sliding properties
- Good penetration properties
- High resistance to wear
- Good thermal conductivity
- Good wettability/coatability

Suitable materials are

- NF-metal alloys, e.g. lead alloys, lead-tin alloys (babbitt metal), copper-tin and copper-tin-zinc alloys.
- sintered metals with pore percentages of 15 % to 25 % for absorbing lubricants.
- plastics, e.g. duromers (phenolplastics), thermoplastics (polytetrafluorethylene, polyamide) and composite materials.

Friction-bearing designs

Single-layer bearings. These are solid bearings which are made up of one material, e.g. CuZn alloys. They are used exclusively for bearing bushings, e.g. small-end bushings **(Fig. 1)**.

Multilayer bearings (Table 1). These consist of two or more layers, e.g. backing shell, carrier layer, liner **(Fig. 1)**. In the interests of improving the load capacity of the bearing, a carrier layer is applied to a steel backing shell by roll-bonding or sintering. The sliding properties of the bearing can be improved by a thin running layer (10 μm to 30 μm) of lead, tin, aluminium or copper alloys. These are applied by sintering, sputtering or galvanising. A nickel dam prevents tin atoms from diffusing from the liner into the carrier layer. This ensures that the properties of the liner are maintained over the entire service life of the bearing.

Sputtering. This process involves using cathodic evaporation to apply fine particles from a donor workpiece (e.g. AlSn20 Cu babbitt metal) to the liner.

Advantages of sputtering:

- Uniform application of the liner
- High bearing-load capacity

Low-maintenance friction bearings (Fig. 2) allow extended lubrication intervals in that they have excellent emergency-running properties.

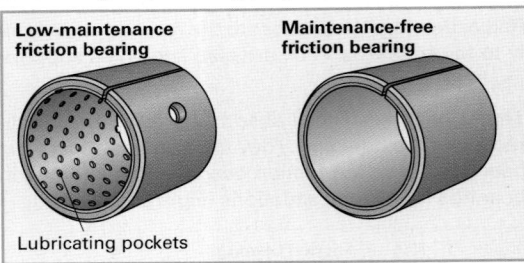

Fig. 2: **Low-maintenance and maintenance-free friction bearings**

Maintenance-free friction bearings (Fig. 2) are designed for dry running and do not need to be lubricated. Liquids in the area of the bearing, e.g. oil, petrol, water, improve heat dissipation and the service life of the bearing.

Advantages of friction bearings over antifriction bearings:

- Can be manufactured more cheaply
- Simple design

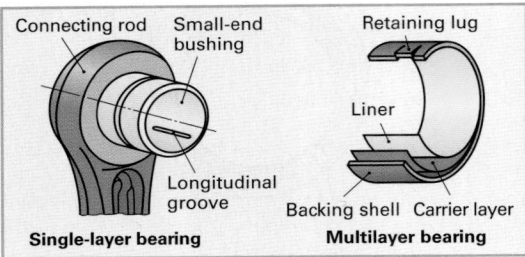

Fig. 1: **Single-layer and multilayer bearings**

Table 1: Multilayer-bearing types		
Multilayer bearings	**Design**	**Application**
Galvanised trimetal bearings	Backing shell: 1.5 mm of steel Carrier layer: 200-300 μm of lead bronze Separation layer: nickel dam 1 μm Liner: 12-20 μm of PbSnCu alloy	Crankshaft and conrod bearings for petrol engines and for turbocharged petrol and diesel engines
Plated aluminium bearings	Backing shell: 1.5 mm of steel Carrier layer: 20-40 μm of AlZn4.5 SiCuPb Liner: 200-400 μm of AlSn 20	
Bonderised bearings	Backing shell: 1.5 mm of steel Separation layer: high-grade aluminium Liner: 12-20 μm of AlZn4.5SiCu Penetration coating: zinc phosphate	
Sputtered bearings	Backing shell: 1.5 mm of steel Carrier layer: 200-300 μm of lead bronze Intermediate layer: 1-2 μm of NiSn alloy Liner: 12-20 μm of sputtered AlSn 20 alloy	Crankshaft bearings for turbocharged diesel engines (lower shell half)
Low-maintenance friction bearings (Fig. 2)	Backing shell: 0.5-2 mm of steel Carrier layer: 0.2-0.35 mm of bronze Liner: 0.05-0.1 mm of polyvinylidenefluoride (PVDF), polytetrafluoroethylene (PTFE) and lead (Pb)	Door hinges, pedal bearings, steering-knuckle bearings
Maintenance-free friction bearings (Fig. 2)	Backing shell: 0.5-2 mm of steel Carrier layer: 0.2-0.35 mm of tin-lead bronze Liner: 0.01-0.03 mm of polytetrafluoroethylene (PTFE) and lead (Pb)	Windscreen wipers, injection pumps, slideways in manual gearboxes

Antifriction bearings

In their simplest form, these consist of two bearing races (outer and inner race), the roll bodies and the roll-body cage (cage). The roll bodies roll on the tracks of the bearing races. In this way, the sliding friction is replaced by the much lower rolling friction. The cage maintains the roll bodies at a specific distance from each other. Antifriction bearings are practically maintenance-free.

There are two different types of antifriction bearing, based on the basic shape of the roll bodies: **ball bearings** and **roller bearings (Fig. 1)**.

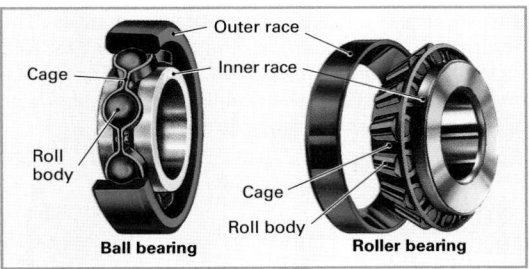

Fig. 1: Antifriction bearings

Ball bearings are suitable for high rotational speeds (deep-groove ball bearings up to 100,000 rpm). The load capacity of the bearing is low because the bearing pressure is transmitted at one point (point contact) **(Fig. 3)**.

Roller bearings are categorised by the roller shape into, for example, cylindrical roller bearings, needle bearings, tapered roller bearings and spherical roller bearings **(Fig. 2)**.

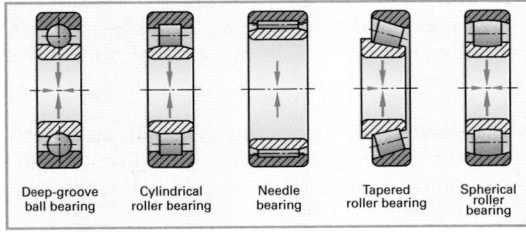

Fig. 2: Bearing types categorised by roll-body shape

They transmit the bearing pressure not at one point, but rather along a line **(Fig. 3)**. In this way, the load capacity of the bearing is significantly higher than for ball bearings, and bearing friction and heating increases during operation.

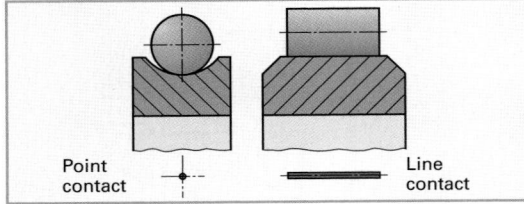

Fig. 3: Forms of contact in antifriction bearings

Bearing arrangement. In the case of a multiple-bearing shaft, the axial forces must only be absorbed by one bearing, the locating bearing. It must not be possible for the locating bearing to be axially moved. The manufacturing tolerances and the differing thermal expansion in the axial direction of shaft and housing must be compensated at the other bearings, the floating bearings, in order to prevent the roll bodies from being distorted in the bearing races. In the case of a locating bearing, both the outer race and the inner race must be securely seated in the housing or on the shaft. In the case of a floating bearing, only the inner or outer race may be securely seated, i.e. it must be possible for one race to move in the axial direction **(Fig. 4)**.

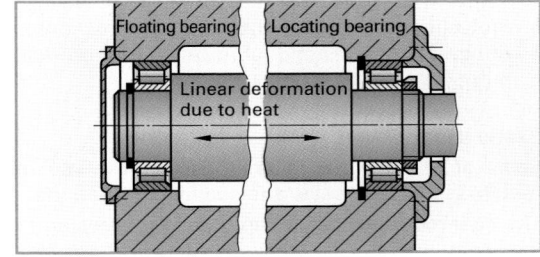

Fig. 4: Bearing arrangement

Spring-loaded bearings. The end float (axial play) can be adjusted according to the requirements – end float, zero end float or preloading – during installation. This is achieved by moving the inner race in an O arrangement and the outer race in an X arrangement up to the desired setting. If angular ball bearings or tapered roller bearings are used, the radial play is also reduced at the same time **(Fig. 5)**. Both bearings are installed as locating bearings in O or X bearing arrangements. They cannot therefore compensate axial linear deformations due to heat and are thus only suitable for short shafts.

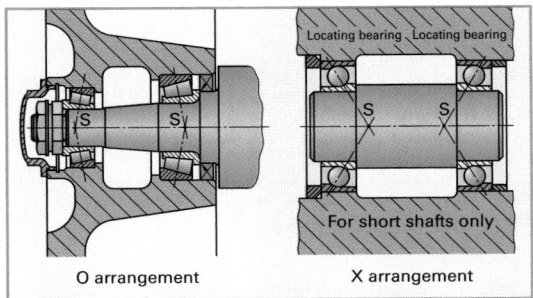

Fig. 5: Spring-loaded bearings

REVIEW QUESTIONS

1 How can antifriction bearings be categorised by the basic shape of the roll bodies?

2 What is the function of locating and floating bearings?

9.4 Seals

Seals are sealing parts on static or moving interfaces of machines, devices, pipes and tanks/reservoirs.

Functions

- Sealing spaces of different pressures against each other (e.g. combustion chamber and oilways).
- Separating spaces containing different operating fluids from each other (e.g. oilways and coolant ducts).
- Protecting spaces against penetration by foreign bodies (e.g. antifriction bearings against dust).
- Protecting machines and systems against losses of operating fluids and lubricants (fuel from fuel pump).

Static seals

Seals at static surfaces are possible by way of metallic seals, soft seals and sealing compounds.

The seal material must adapt itself by pressing to the irregularities of the sealing surfaces. It can also distribute uniformly the pressure forces issuing from the fastening fixtures (e.g. screws, snap fastenings).

Metallic seal. This is achieved by high fitting accuracy, high surface quality (low surface roughness) and high surface pressure at the sealing surfaces **(Fig. 1)**.

Soft seal. The seal material is deformed as a result of the surface pressure in such a way as to adapt to the surfaces to be sealed **(Fig. 1)**, e.g. in the case of valve-cover gaskets.

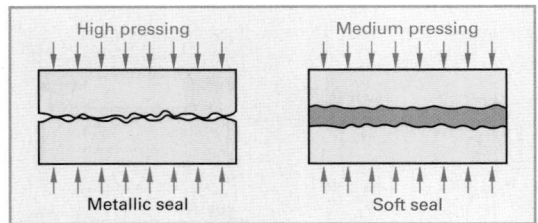

Fig. 1: Metallic and soft seals

Sealing compounds. These form, under the influence of the pressure forces, a self-shaping sealing element in accordance with a soft seal. The irregularities and surface roughness are filled by a liquid or paste-like substance **(Fig. 2)**.

Sealing compounds can also be used in combination with soft seals and metal seals **(Fig. 2)**. There are curing and permanently elastic sealing compounds.

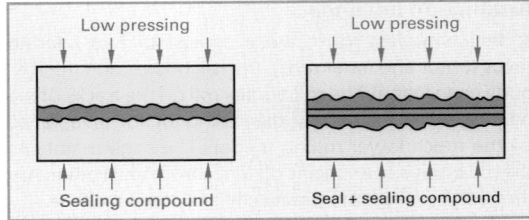

Fig. 2: Sealing compound and seal

Profile seals. The sealing effect is created by the deformation of the rubber-elastic sealing material. The deformed profile generates the necessary pressure force at the surface to be sealed, e.g. in the case of O-ring seals **(Fig. 3)** and plain sealing rings (rubber seal on the brake piston of a disc brake).

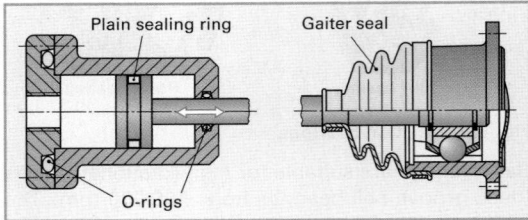

Fig. 3: Profile seals and gaiter seal

Gaiter seals. These are designed to protect bearings against the ingress of dirt and frequently contain the lubricating grease for lubricating the joints **(Fig. 3)**.

Dynamic seals

The seal material must keep as low as possible the inevitable leaking at sealing surfaces that move against each other.

Radial shaft seals. These are suitable for sealing rotating components. The shaft is sealed by pressing the sealing lip onto the shaft surface. This takes place by means of a spiral spring and full pressure (interference fit) at the outer sheath. As the shaft rotates, a sealing gap of approx. 1 μm is created at the sealing lip as a result of hydrodynamic lubrication. A small amount of oil passes through this gap to lubricate the sealing lip **(Fig. 4)**.

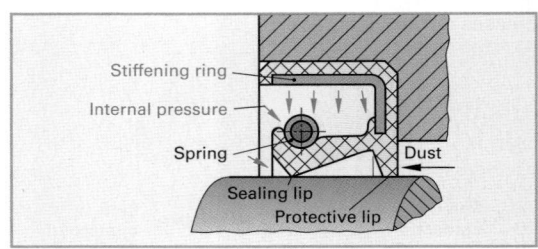

Fig. 4: Radial shaft seal

10 Design and operating principle of a four-stroke engine

Categorisation of combustion engines:

By mixture formation and ignition:

- **Spark-ignition engines.** These are run preferably on petrol and with external or even internal mixture formation. Combustion is initiated by externally supplied ignition (spark plug).
- **Diesel engines.** These have internal mixture formation and are run on diesel fuel. Combustion in the cylinder is initiated by auto-ignition.

By operating principle:

- **Four-stroke engines.** These have a closed (separate) gas exchange and require four piston strokes or two crankshaft revolutions for one power cycle.
- **Two-stroke engines.** These have an open gas exchange and require two piston strokes or one crankshaft revolution for one power cycle.

By cylinder arrangement **(Fig. 1)**:

- In-line engines
- Opposed-cylinder engines
- V-engines
- VR-engines

By piston stroke:

- Reciprocating engines
- Rotary engines

By cooling:

- Liquid-cooled engines
- Air-cooled engines

10.1 Spark-ignition engine

The spark-ignition engine is an internal-combustion engine which converts chemical energy into thermal energy by burning fuel and then converts the thermal energy into mechanical energy via a piston.

Design

The spark-ignition engine **(Fig. 2)** consists primarily of four assemblies and additional auxiliary installations:

- **Engine case** — Cylinder-head cover, cylinder head, cylinder, crankcase, oil pan
- **Crankshaft drive** — Piston, connecting rod, crankshaft
- **Engine timing** — Valves, valve springs, rocker arms, rocker-arm shaft, camshaft, timing gears, timing chain or toothed belt
- **Mixture-formation system** — Injection system or carburettor, intake pipe
- **Auxiliary installations** — Ignition system, engine lubrication, engine cooling, exhaust system, if necessary supercharging system

10

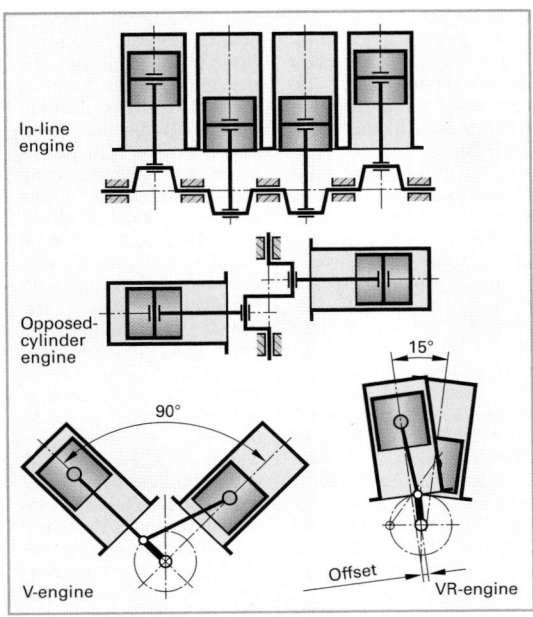

Fig. 1: Categorisation by type of cylinder arrangement

In-line engine

Opposed-cylinder engine

V-engine

90°

Offset

15°

VR-engine

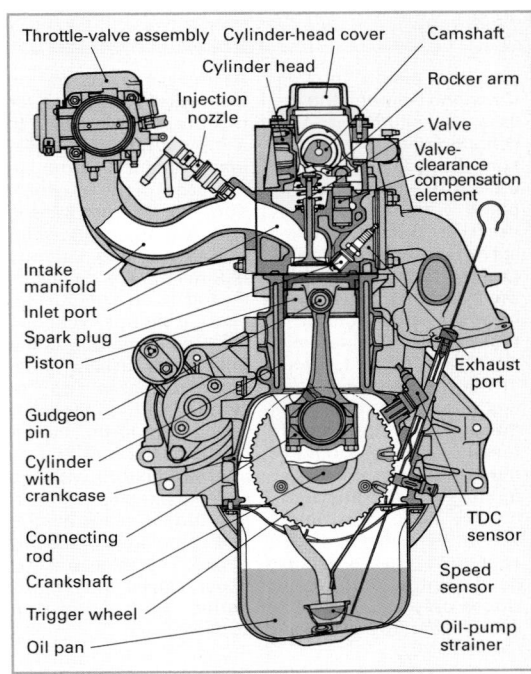

Fig. 2: Structure of a four-stroke spark-ignition engine

Throttle-valve assembly Cylinder-head cover Camshaft
Cylinder head
Injection nozzle
Rocker arm
Valve
Valve-clearance compensation element
Intake manifold
Inlet port
Spark plug
Piston
Exhaust port
Gudgeon pin
Cylinder with crankcase
Connecting rod
Crankshaft
TDC sensor
Speed sensor
Trigger wheel
Oil-pump strainer
Oil pan

Operating principle of spark-ignition engine

The four strokes of the power cycle are induction, compression, combustion and exhaust **(Fig. 1)**. One power cycle takes place in two crankshaft revolutions (720° crank angle).

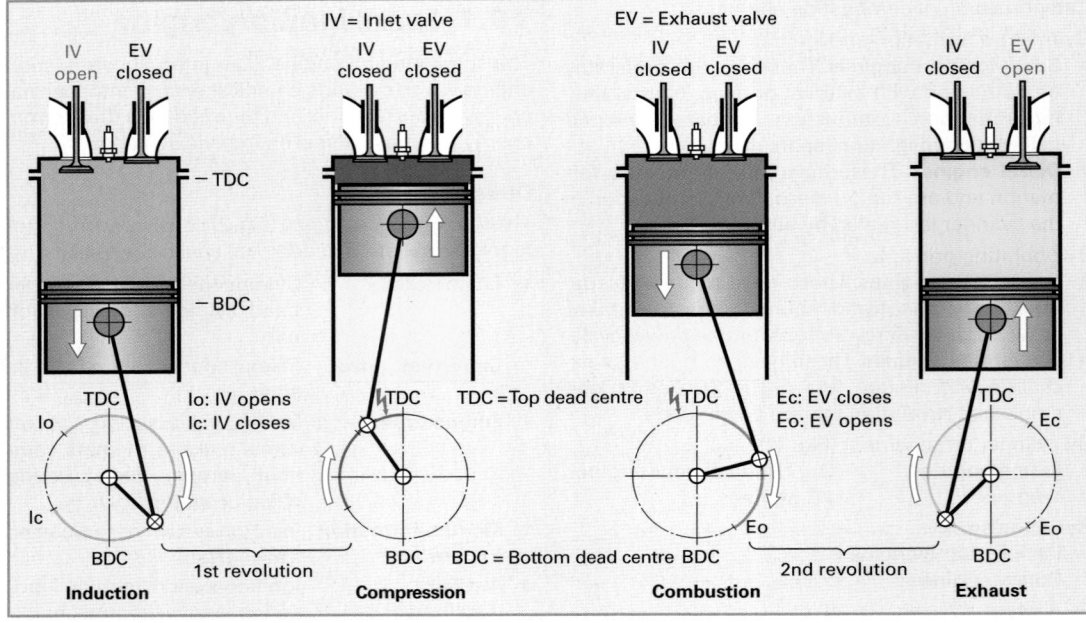

Fig. 1: The four strokes of a power cycle

1st stroke – Induction	2nd stroke – Compression	3rd stroke – Combustion	4th stroke – Exhaust
As the piston moves down the cylinder, the increased volume in the cylinder causes a pressure differential of – 0.1 bar to – 0.3 bar compared with the external pressure. Since the pressure outside the engine is greater than that inside the cylinder, air is forced into the induction system. The ignitable fuel-air mixture is formed either in the intake port or directly in the cylinder through the injection of fuel. In order to admit as much intake air or fuel-air mixture as possible into the cylinder, the inlet valve (IV) opens already at up to 45° CA before top dead centre (TDC) and closes only at 35° CA to 90° CA after bottom dead centre (BDC).	As the piston moves up the cylinder, the fuel-air mixture is compressed to a 7th to a 12th of the original cylinder volume. In the case of direct injection, air is compressed and the injection point can already begin shortly before TDC. The gas heats to 400 °C to 500 °C. Because it cannot expand at the high temperature, the final compression pressure increases up to 18 bar. The high pressure encourages further carburation of the fuel and its internal mixture with the air. This enables combustion to take place quickly and completely in the 3rd stroke. The inlet and exhaust valves are closed during the compression stroke.	Combustion is initiated by an ignition spark jumping across the electrodes of the spark plug. The length of time between the jumping of the spark and the complete development of the flame front is approx. 1/1,000 second at a combustion velocity of 20 m/s. For this reason, the ignition spark must jump across at 0° to approx. 40° before TDC, depending on the engine speed, so that the necessary maximum combustion pressure of 30 bar to 60 bar is available shortly after TDC (4° CA ... 10° CA). The expansion of the gases heated up to 2,500 °C forces the piston to bottom dead centre and thermal energy is converted into mechanical energy.	The exhaust valve opens at 40° to approx. 90° before BDC; this encourages the discharge of the exhaust gases and relieves the load on the crankshaft drive. The pressure of 3 bar to 5 bar still available at the end of the power stroke causes the exhaust gases heated up to 900 °C to be expelled from the cylinder at the speed of sound. As the piston moves up the cylinder, the remaining exhaust gas is discharged at a dynamic pressure of roughly 0.2 bar. To encourage the exhaust gases to be discharged, the exhaust valve closes only after TDC while the inlet valve is already open. This overlapping of the valve times encourages the draining and cooling of the combustion chamber and improves cylinder charge.

10.2 Diesel engine

Like the spark-ignition engine, the diesel engine is also an internal-combustion engine.

Design

The diesel engine **(Fig. 1)**, like the spark-ignition engine, consists primarily of four assemblies and additional auxiliary installations:

- Engine case
- Crankshaft drive
- Engine timing
- Fuel system with fuel-injection equipment, fuel-supply pump, fuel filter, high-pressure injection system, e.g.
  - common-rail system
  - unit-injector system
- Auxiliary installations
  Engine lubrication, engine cooling, exhaust system, if necessary supercharging system, e.g. with exhaust-gas turbocharger and intercooling, if necessary cold-starting system, e.g. preheating system

The diesel-vehicle engine is used as a fast-running engine with speeds up to approx. 5,500 rpm in passenger cars and light commercial vehicles. It is used as a slow-running engine (speeds up to approx. 2,200 rpm) in heavy commercial vehicles.

> Diesel engines consume up to 30 % less fuel than spark-ignition engines. Their efficiency can stretch up to 46 %.

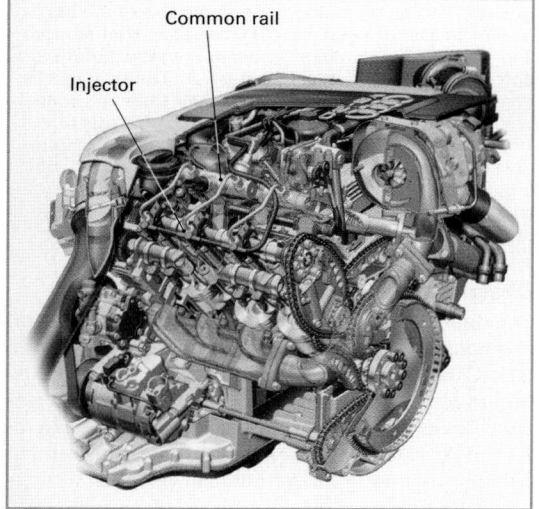

Common rail

Injector

Fig. 1: Diesel-vehicle engine for passenger cars

10.3 General physical and chemical principles

10.3.1 Features of a diesel engine

- **Running** on diesel or biodiesel fuel.

- **Internal mixture formation**
  Only air is admitted into the cylinder during the induction stroke. The fuel-air mixture is formed during the compression stroke by the injection of fuel under high pressure into the cylinder.

- **Auto-ignition**
  Immediately after being injected, the fuel is automatically ignited on the air, which has been rendered extremely hot by compression. The final compression temperature exceeds the ignition temperature.

- **Quality regulation**
  The naturally aspirated engine is unthrottled, i.e. there is no throttle valve before the intake ports. In this way, the engine is supplied over the entire speed range with an extensively constant air flow as the charge. Load control is effected by altering the quantity of fuel to be injected, which in turn alters the fuel-air mixture depending on the operating state.

Internal mixture formation
After the start of injection, the still liquid fuel must be converted into an ignitable mixture. **Table 1** sets out the time that elapses from the start of injection until auto-ignition. For internal mixture formation heat is removed from the hot air so that this air cools. But the air temperature must always be above the auto-ignition temperature of the fuel.

The time between the start of injection and the start of combustion is known as the ignition lag.

Table 1: Internal mixture formation and initiation of combustion		
Time requirement "ignition lag"	Removal of heat from the hot air	Fuel is injected in a fine-mist but still liquid state into the hot air.
		Fuel mist is heated to boiling temperature.
		Fuel evaporates at boiling temperature.
		Fuel vapours mix with the hot air.
		Fuel vapours heat up to ignition temperature.
		Fuel-air mixture ignites.
		Initiation of combustion.

10

Operating principle of diesel engine

> The four strokes of the power cycle are, as in a spark-ignition engine, induction, compression, combustion and exhaust **(Fig. 1)**. One power cycle takes place in two crankshaft revolutions (720° crank angle).

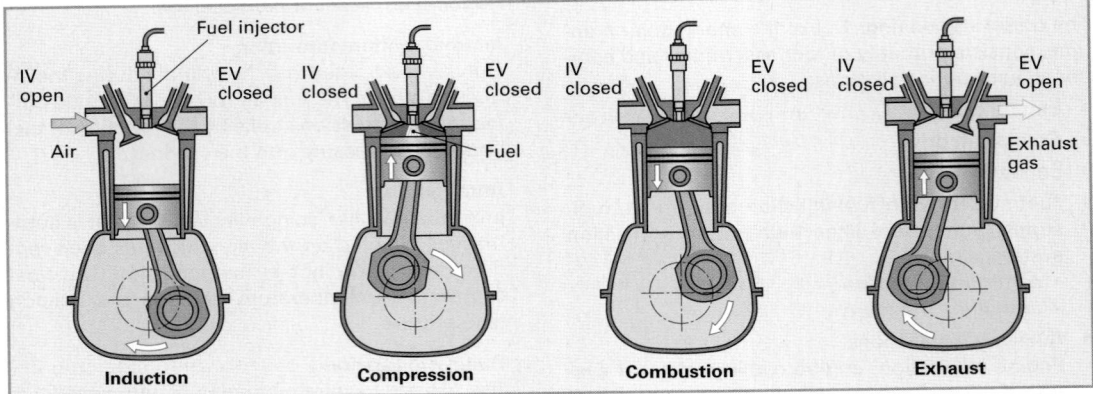

Fig. 1: The four strokes of a power cycle in a direct-injection engine

1st stroke – Induction	2nd stroke – Compression	3rd stroke – Combustion	4th stroke – Exhaust
As the piston moves down the cylinder, the increased volume in the cylinder causes a pressure differential p_a of – 0.1 bar to – 0.3 bar compared with the external pressure. Air is forced into the cylinder by the greater external pressure. The air is admitted unthrottled because there is no throttle valve. In order to admit as much intake air as possible into the cylinder, the inlet valve opens at up to 25° CA before TDC; it closes only at up to 28° CA after BDC in order to facilitate a subsequent flow of intake air. The air heats up to 70 °C to 100 °C in the cylinder.	As the piston moves up the cylinder, the air is compressed to a 14th to a 24th of the original cylinder volume. The air heats up to 600 °C to 900 °C in the process. Because the air cannot expand at the high temperature, the final compression pressure increases to 30 bar to 55 bar. Engines with secondary combustion chambers, such as a turbulence chamber for example, must be compressed to a greater extent because heat losses are generated by the larger combustion-chamber surface. The inlet and exhaust valves are closed during the compression stroke.	Towards the end of the compression stroke, at roughly 15° CA before TDC to 30° CA before TDC, finely atomised diesel fuel is injected under high pressure (up to 2,050 bar) into the combustion chamber. The fuel evaporates in the hot air and mixes with the air. Combustion is initiated due to the fact that the temperature of the compressed air is higher than the diesel fuel's auto-ignition temperature of 320 °C to 380 °C. The time between the start of injection and the start of combustion is known as the ignition lag. The high combustion pressure of up to 160 bar moves the piston towards BDC. Thermal energy is converted into mechanical work in the process.	The exhaust valve opens at 30° to approx. 60° before BDC; this encourages the discharge of the exhaust gases and relieves the load on the crankshaft drive. The pressure of 4 bar to 6 bar still available at the end of the power stroke causes the exhaust gases heated up to 550 °C to 750 °C to be expelled from the cylinder. As the piston moves up the cylinder, the remaining exhaust gas is discharged at a pressure of 0.2 bar to 0.4 bar. The exhaust valve closes slightly before or after TDC. The heat losses are lower than in a spark-ignition engine due to the lower exhaust-gas temperatures (greater efficiency).

Indirect-injection engines

The fuel is injected into secondary combustion chambers (turbulence, precombustion chambers). Because the split combustion chambers give rise to large surface areas, the correspondingly higher heat dissipation during the compression stroke must be compensated by greater compression in order for the ignition temperature of the diesel fuel to be safely exceeded. The compression ratio ε of indirect-injection engines is between 19 and 30.

Direct-injection engines (DI engines)

The fuel is injected directly into the combustion chamber. The air heated by compression to up to 900 °C dissipates little heat to the compact combustion-chamber surface, thus allowing lower compression. Direct-injection engines have a compression ratio ε of between 14 and 20 for passenger cars and of between 14 and 19 for commercial vehicles.

10.3.2 Features of a spark-ignition engine

- **Running** on petrol or gas.
- **Mixture formation**

  External mixture formation. The fuel-air mixture is formed in the carburettor or in the intake manifold outside the cylinder.

  Internal mixture formation. Initially only air is admitted into the cylinder during the induction stroke. The fuel-air mixture is formed during the induction or compression stroke by the injection of fuel into the cylinder.
- **Externally supplied ignition**
- **Constant-volume combustion**

  Combustion takes place in a virtually constant volume thanks to the sudden combustion of the fuel-air mixture.
- **Quantity regulation**

  The quantity of the fuel-air mixture is altered according to the position of the throttle valve (load state).

Charge

Charge refers to the mass of the gases (fuel-air mixture or air) flowing into the cylinder during the induction stroke.

Charge improvement. In order to improve charge and with it power, it is possible to extend the opening times of the inlet valves from 180° crank angle (corresponding to the piston stroke) to up to 315° crank angle (CA). During the exhaust stroke, the burned gases expelled at high speed generate a suction effect. If the inlet valve is opened before the piston has reached top dead centre, the mixture or the intake air can flow against the movement of the piston into the cylinder as a result of the vacuum pressure.

Valve overlap

> Both the inlet valve and the exhaust valve are opened in the transition phase from the exhaust stroke to the induction stroke.

If the inlet valve is left open until well into the compression stroke, the fuel/air mixture accelerated during induction to up to 100 m/s (360 km/h) can continue to flow into the cylinder on account of its mass inertia. This supercharging effect is terminated when the pressure generated by the upward-moving piston brakes the flowing-in mixture. The inlet valve must be closed again no later than at this point.

In spite of the induction time being extended, the cylinder charge reaches a maximum of 80 % in non-supercharged engines.

Volumetric efficiency

> The volumetric efficiency is the ratio of fuel-air mixture actually drawn in in kg to the theoretically possible (complete) cylinder charge with fuel-air mixture in kg.

In the case of internal mixture formation, the volumetric efficiency is the ratio of air mass drawn in to the theoretically possible air charge in kg.

$$\lambda_L = \frac{m_z}{m_{th}}$$

λ_L Volumetric efficiency
m_z Drawn-in mass of fresh-air or fuel-air mixture in kg
m_{th} Theoretically possible mass of fresh-air or fuel-air mixture in kg

In naturally aspirated engines, the volumetric efficiency ranges between 0.6 and 0.9 (charge 60 % to 90 %) while, in supercharged engines, a volumetric efficiency of 1.2 to 1.6 (charge 120 % to 160 %) is possible.

The charge can additionally be improved by a lower flow resistance of the fresh gases and by lower internal cylinder temperatures. This is achieved by:

- Optimally structured induction pipes
- Favourable combustion-chamber shapes
- Large inlet passages
- Several inlet valves per cylinder
- Good cooling

The charge deteriorates as a result of:

- The flow resistance of the throttle valve.
- Decreasing valve opening times at higher speeds.
- Lower air pressure, with an increase in altitude to 100 m engine power drops by roughly 1 %.

Compression ratio

Combustion chamber. This is the space enclosed by the cylinder, the cylinder head and the piston crown. Its size changes continually during a stroke. The combustion chamber is at its largest when the piston is at BDC and at its smallest when the piston is at TDC. The largest combustion chamber is composed of the swept volume and the compression chamber.

Compression space V_c. This is the smallest combustion chamber.

Swept volume V_h. This is the space between the two piston dead centres TDC and BDC.

Total swept volume V_H. This is derived from the sum total of the swept volumes of the individual cylinders of an engine.

Comparing the space above the piston before compression (swept volume V_h + compression space V_c) with the space above the piston after compression (compression space V_c) produces the compression ratio ε **(Fig. 1)**.

Compression ratio	=	Swept volume + Compression space / Compression space

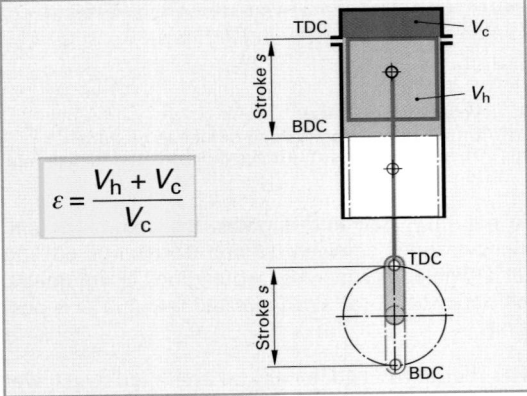

$$\varepsilon = \frac{V_h + V_c}{V_c}$$

Fig. 1: Compression ratio

Table 1: Comparison of compression ratios

Compression ratio	7	9
Final compression pressure	~10 bar	~16 bar
Maximum compression pressure	~30 bar	~42 bar
Pressure during opening of exhaust valve	~ 4 bar	~ 3 bar
Final compression temperature	400 °C	500 °C

The higher the compression ratio of a spark-ignition engine, the better the utilisation of fuel energy and thus the engine's efficiency.

In spite of the significantly increased compression work with $\varepsilon = 9$, the utilisation of the significantly increased pressure differential with the same fresh-gas charge results in a work gain or a power increase of more than 10 % and a reduction in fuel consumption of roughly 10 %.

Reasons for power increase:
- Better removal of burned gases from the smaller compression space.
- Higher temperature during compression, better and more complete carburation.
- On account of the high compression, the burned gases can expand to a larger volume, the exhaust-gas temperature decreases and less thermal energy is lost through the exhaust.

The final compression temperature rises as the compression ratio increases **(Table 1)**. The compression ratio is therefore limited by the auto-ignition temperature of the fuel.

In supercharged engines the compression is lower as the air is admitted in a highly compressed state into the cylinder.

Boyle-Mariotte's Law

The upward and downward movement of the piston in the cylinder causes the pressure and the temperature also to be altered with the volume.

Back in the 17th century the physicists Boyle and Mariotte had already discovered that volume and pressure in the cylinder change in inverse proportion with a **constant** temperature.

If, for instance, the volume is reduced to an 8th, so the pressure increases by a factor of 8 **(Fig. 2)**.

> The product of pressure and volume is constant.

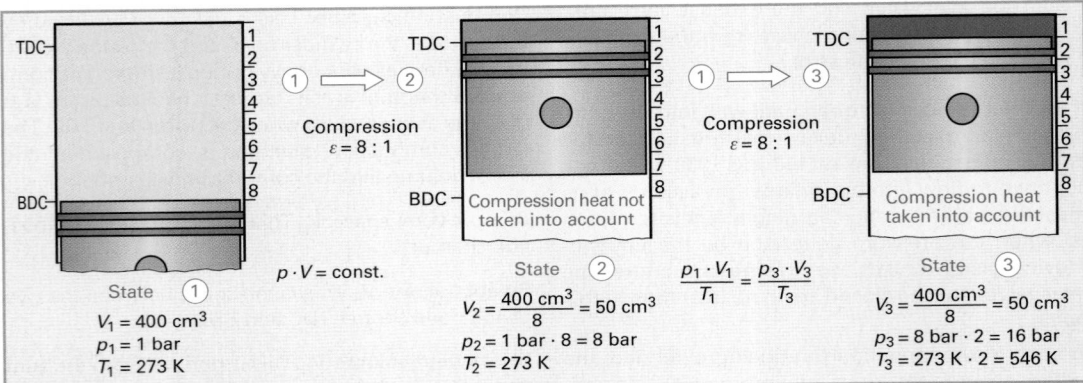

State ①
$V_1 = 400$ cm³
$p_1 = 1$ bar
$T_1 = 273$ K

$p \cdot V = $ const.

State ②
$V_2 = \dfrac{400 \text{ cm}^3}{8} = 50 \text{ cm}^3$
$p_2 = 1$ bar $\cdot 8 = 8$ bar
$T_2 = 273$ K

$\dfrac{p_1 \cdot V_1}{T_1} = \dfrac{p_3 \cdot V_3}{T_3}$

State ③
$V_3 = \dfrac{400 \text{ cm}^3}{8} = 50 \text{ cm}^3$
$p_3 = 8$ bar $\cdot 2 = 16$ bar
$T_3 = 273$ K $\cdot 2 = 546$ K

Fig. 2: Ratio of pressure, volume and temperature during compression

Gay-Lussac's Law

By including the temperature in the ratio of volume and pressure, the French physicist Gay-Lussac discovered the following regularity:

> If a gas is heated at constant pressure by 1 K (1 °C), it expands by a 273rd part of its volume.

When the gas is heated by 273 K, it expands to twice the volume.

When the gas is prevented from expanding, e.g. during compression **(Fig. 2, Page 188)**, the pressure is doubled. However, the final pressure is lower due to the dissipation of heat at the cylinder walls.

10.3.3 Combustion sequence of spark-ignition engine

Because only a very short time period is available for combustion of the fuel-air mixture (combustion is already completed shortly after TDC), fuel and oxygen molecules must be close to each other in the compressed mixture. The oxygen required for combustion is removed from the drawn-in air. Because the air contains only roughly 20 % oxygen, a proportionate amount of air must be admixed with the fuel. The minimum air quantity required for complete combustion, the theoretical air requirement, is roughly 14.8 kg air for 1 kg petrol (~ 12 m^3 at a density of ϱ = 1.29 kg/m^3).

The carbon contained in the fuel burns with the oxygen to form carbon dioxide (CO_2) and the hydrogen contained combines with the oxygen to form water vapour (H_2O). The nitrogen contained in the air does not assume a predominant role in the combustion. However, toxic nitrogen oxides (NO_x) are created at high pressures and combustion temperatures.

Complete combustion:

The chemical energy of the fuel is converted into thermal energy.

> $C + O_2 \rightarrow CO_2$ + thermal energy
> $2\,H_2 + O_2 \rightarrow 2\,H_2O$ + thermal energy

If, for example, only 13 kg air are available for 1 kg petrol, the fuel-air mixture will be too rich (1 : 13). Because there is not enough oxygen available, part of the carbon burns only incompletely to form carbon monoxide CO, which is toxic.

Incomplete combustion:

> $2\,C + O_2 \rightarrow 2\,CO$ + heat

If, for example, 16 kg air is available for 1 kg petrol, the fuel-air mixture will be too lean (1 : 16). Complete combustion can indeed occur, but because of the lower amount of available fuel which can evaporate, the interior cylinder chamber is cooled to a lesser extent with the result that the engine may overheat.

Knocking combustion

A spark-ignition engine will knock if the fuel-air mixture, instead of the combustion initiated by the ignition spark, ignites by itself **(Fig. 1)**.

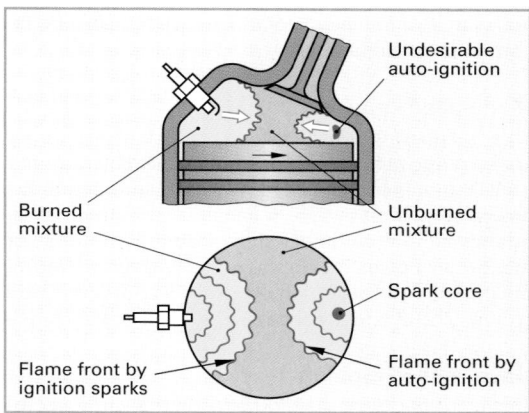

Fig. 1: Knocking combustion

This auto-ignition, which simultaneously initiates inflammation in several spark cores, results in a premature, sudden combustion, during which the globular flame fronts collide with each other. This gives rise to combustion velocities of 300 m/s to 500 m/s, which in turn result in excessively high pressures **(Fig. 2)**.

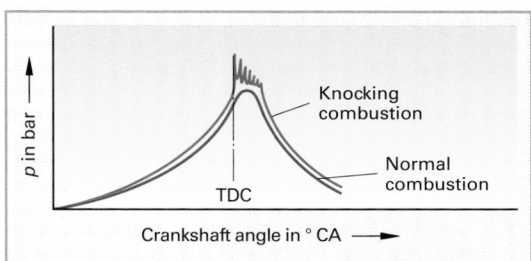

Fig. 2: Pressure characteristic during combustion

The knocking or often also pinging noise in the engine is caused by shock waves which are triggered by the different spark cores and result in individual engine components vibrating. Knocking results in increased mechanical and thermal load on the crankshaft drive and reduced power.

Causes of knocking

Apart from the use of unsuitable fuels, knocking can also be caused by:

- Excessively advanced ignition.
- Uneven mixture distribution in the cylinder.
- Poor heat dissipation due to carbon-residue deposits or faults in the cooling system.
- An excessively high compression ratio, e.g. when a thinner cylinder-head gasket is used.

Acceleration knocking

This occurs primarily when accelerating under full load from low engine speeds. It is usually caused by fuel with an insufficient octane number (RON) and incorrect spark adjustment.

High-speed knocking

This is knocking which usually occurs in the upper speed range at full load. It is often caused by fuel with a MON that is too low or fuel in which the difference between RON and MON (= sensitivity) is great. It frequently cannot be detected in good time because of the louder noises inside the vehicle. Overheating of the engine can cause damage such as burned piston crowns and cylinder heads as well as piston seizures.

Uncontrolled ignition

This is triggered by glowing parts in the engine combustion chamber already at a stage before the onset of normal ignition of the fuel-air mixture by the ignition spark (uncontrolled advanced ignition).

10.3.4 Combustion sequence of diesel engine

Ignition lag in a diesel engine

The period of time required for internal mixture formation up to the initiation of combustion is called the ignition lag.

> The ignition lag is the period of time between the start of injection and the initiation of combustion.

When the engine is a normal operating temperature, the ignition lag is normally approx. 0.001 s (1/1,000 s). It is substantially dependent on ...

- ... the structure of the fuel molecules (ignition quality, cetane number).
- ... the temperature of the compressed air before the start of injection.
- ... the degree of atomisation during injection (extent of the injection pressure, size of the fuel droplets).

The higher the pressure and the temperature, the shorter the ignition lag.

Injection takes place in the diesel engine in such a way that the main fuel quantity is only admitted into the combustion chamber when the initial parts of the fuel have ignited in the chamber with the result that further-injected fuel is continuously burned.

Diesel knocking

When the engine and intake-air temperatures are low, for example when the engine is started from cold, the time needed to form the internal mixture is prolonged. The ignition lag becomes too great (over 0.002 s) and the collected fuel burns suddenly with a loud noise, resulting in the diesel engine knocking. The sudden combustion is triggered by several spark cores which originate from accumulated fuel in the combustion chamber. The high pressure peaks thereby generated can result in damage to the crankshaft drive. The knocking can be reduced by preinjecting a small amount of fuel and by using diesel fuel with a higher cetane number in the winter months.

10.4 Pressure-volume diagram (p-V diagram)

Spark-ignition engine

The relationships between pressure, volume and temperature of gases can be carried over for the power cycle of a four-stroke spark-ignition engine into a pressure-volume diagram (p-V diagram). According to Boyle-Mariotte and Gay-Lussac, this produces an ideal diagram in which the volume does not change, i.e. remains constant, at the respective piston reversal points at BDC and TDC during the combustion process and the exhaust process.

> Constant-volume combustion: Sudden combustion takes place with a constant volume.

Ideal constant-volume combustion, as depicted in **Fig. 1, Page 191**, calls for the following preconditions:

- The cylinder contains only fresh gases and no residual exhaust gases.
- Complete combustion of the fuel-air mixture.
- Loss-free charge cycle.
- No heat transfer at the cylinder.
- Constant volume during combustion and cooling.
- The combustion chamber must be gastight (piston rings).

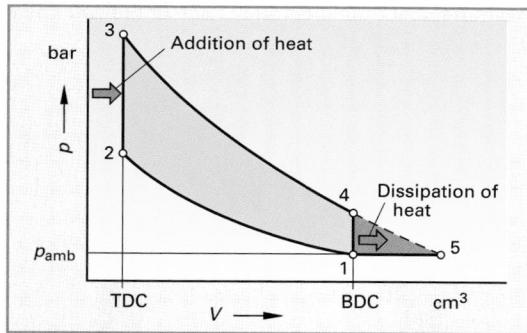

Fig. 1: Ideal constant-volume process (p-V diagram)

Process sequence

1 ➡ 2   **Compression** of the fuel-air mixture, pressure increase, no addition of heat.

2 ➡ 3   **Combustion** of the fuel-air mixture, pressure increase with constant volume, i.e. the piston remains for the brief period of combustion at TDC, addition of heat.

3 ➡ 4   **Expansion.** The gas under high pressure expands and moves the piston to BDC, the starting volume is reached. No dissipation of heat.

4 ➡ 1   **Cooling.** The process takes place with a constant volume. The pressure drops as a result of heat dissipation until the starting pressure at point 1 is reached again.

Energy gain, energy loss

The area created in the diagram (**Fig. 1**) with the corners 1-2-3-4 reflects the work gained during a power cycle (surface +).

The work gained could be greater if the exhaust valve were not to open already at point 4 but only after the gases have expanded down to the starting pressure at point 5.

However, this is not possible in practice because extending the expansion is associated with increasing the stroke (long-stroke engine). Thus the area 1-4-5 reflects the work lost.

The work gained can be increased by increasing the compression ratio.

Diesel engine

In contrast to a spark-ignition engine, the pressure theoretically does not change during the combustion process; this phenomenon is referred to therefore as constant-pressure combustion. In reality, neither the constant-volume nor the constant-pressure process takes place in ideal circumstances because the conditions cannot be maintained.

Actual p-V diagram

The pressure characteristic during the four strokes of a power cycle can be recorded with a piezoelectric indicator on the running engine and displayed as a curve on the screen. Here the differences from the ideal p-V diagram can be clearly seen. In practice, the curve shapes of a spark-ignition engine and a diesel engine still differ only in the extent of the pressures (**Fig. 2**).

Because of the significantly higher combustion pressure in a diesel engine and the subsequent expansion of the burned gases to 4 bar ... 6 bar, the exhaust gases cool more markedly than in a spark-ignition engine. This results in a reduction of the exhaust-gas losses, which in turn increases the work gained and with it the efficiency. The thermal load on the valves is lower. However, modern diesel engines are no longer able to deliver sufficient heat to heat the vehicles with the result that auxiliary heaters are necessary.

> The p-V diagram can be used to calculate the effective work W_{eff} of an engine by subtracting the lost work (area −) from the produced work (surface +) (**Fig. 1, Page 192**).

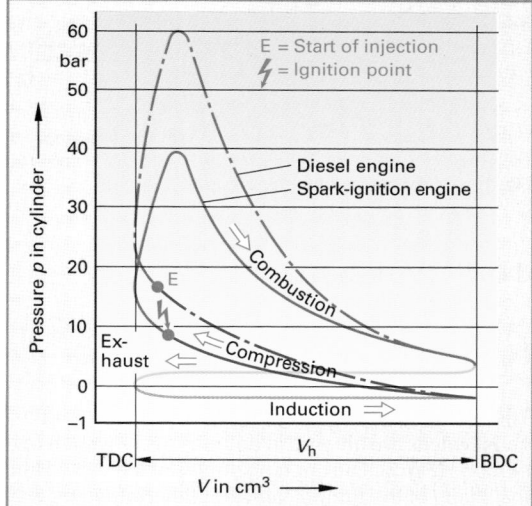

Fig. 2: Actual p-V diagram

Error detection in the p-V diagram

Larger deviations from the normal pressure characteristic enable errors in the engine settings (mixture formation, spark adjustment, compression) and above all errors arising from knocking phenomena to be detected (**Fig. 1, Page 192**).

Ignition point too advanced:

The highest possible pressure is already reached before the piston has arrived at top dead centre. The high pressures and temperatures created result in knocking combustion, poorer exhaust-gas values and loss of power, which can be identified from the smaller area in the diagram.

10

Ignition point too retarded:

Normal rise of the compression line up to top dead centre. After a short drop in the pressure after TDC, it rises again but can no longer reach the maximum combustion pressure because due to the retarded ignition point the piston has moved too far in the direction of BDC before the fuel-air mixture has burned fully.

The consequences are loss of power, higher fuel consumption and risk of overheating.

Leaking valves or piston rings:

Normal pressure build-up not possible, the rise of the compression line is flatter. The maximum combustion pressure cannot be reached even when the ignition point is correct. The consequences are loss of power and poorer exhaust-gas values.

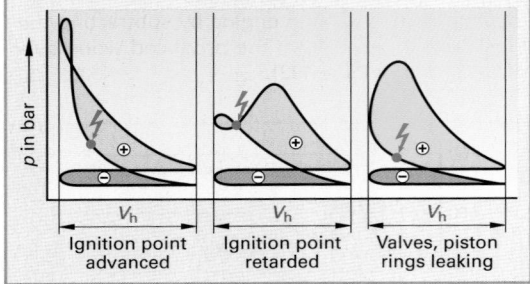

Fig. 1: *p-V* diagrams of faulty engines

10.5 Timing diagram

This provides an overview of the timing angles of the valves and the valve overlap.

The opening and closing angles of the inlet and exhaust valves are entered in degrees of crankshaft revolutions **(Fig. 2)**. The opening angles of the valves and the shape of the timing cams are determined by way of tests for each type in such a way that the engine delivers the best power possible. Because this is not possible over the entire speed range, engines are equipped with adjustable inlet camshafts. The opening and closing angles of the inlet valves can be changed by a specific adjustment angle (variable timing).

The timing angles of the individual engines deviate from each other to the extent that each engine has its own timing diagram, e.g. **Fig. 2.**

As a rule, the angles from the opening through to the closing of the valves are greater the higher the normal running speed of the engine is.

Symmetrical timing diagram. The angles lo before TDC and Ec after TDC are identical in size, as are the angles Eo before BDC and Ic after BDC.

Asymmetrical timing diagram. One of the two angle pairs is unequal.

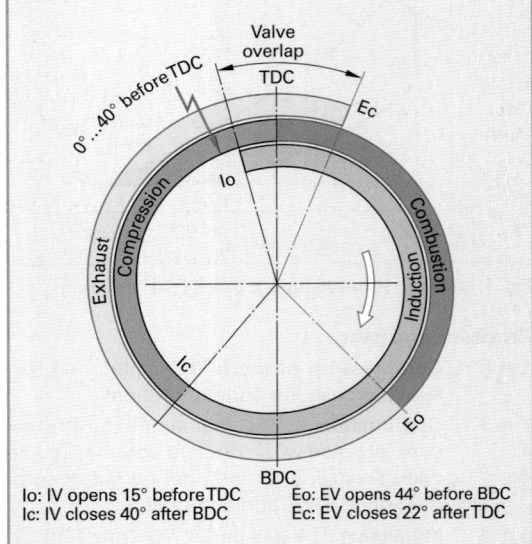

Io: IV opens 15° before TDC   Eo: EV opens 44° before BDC
Ic: IV closes 40° after BDC   Ec: EV closes 22° after TDC

Fig. 2: Timing diagram of a four-stroke spark-ignition engine

10.6 Cylinder numbering, firing orders

Cylinder numbering. The designation of the individual cylinders of an engine is standardised. The counting of the cylinders starts from the end opposite the output end. In the case of V-, VR- and opposed-cylinder engines, the counting starts on the left cylinder bank and then each bank is counted through **(Fig. 3)**.

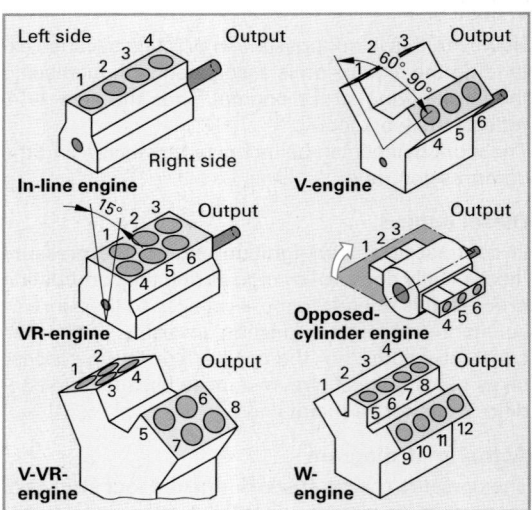

Fig. 3: Cylinder numbering

Firing order and ignition interval in multiple-cylinder engines (Fig. 1).

Firing order. This indicates the order in which the power strokes of the individual cylinders of an engine follow each other.

Ignition interval. This indicates the interval in crankangle degrees in which the power strokes or the firing operations of the individual cylinders follow each other. The greater the number of cylinders, the smaller the ignition interval. Engine operation becomes smoother and the torque output is more regular.

$$\text{Ignition interval} = \frac{720° \text{ CA}}{\text{Cylinder number}}$$

Example: In the case of a 5-cylinder engine, the ignition interval is calculated from 720° CA : 5 = 144° CA. A star diagram is drawn as a substitute for the crankshaft. Starting from the uppermost cylinder, which is designated as 1, the remaining cylinders are entered according to the firing order 1-2-4-5-3 against the direction of rotation at an interval of 144° CA. In this way, the firing order can be read off from each star diagram.

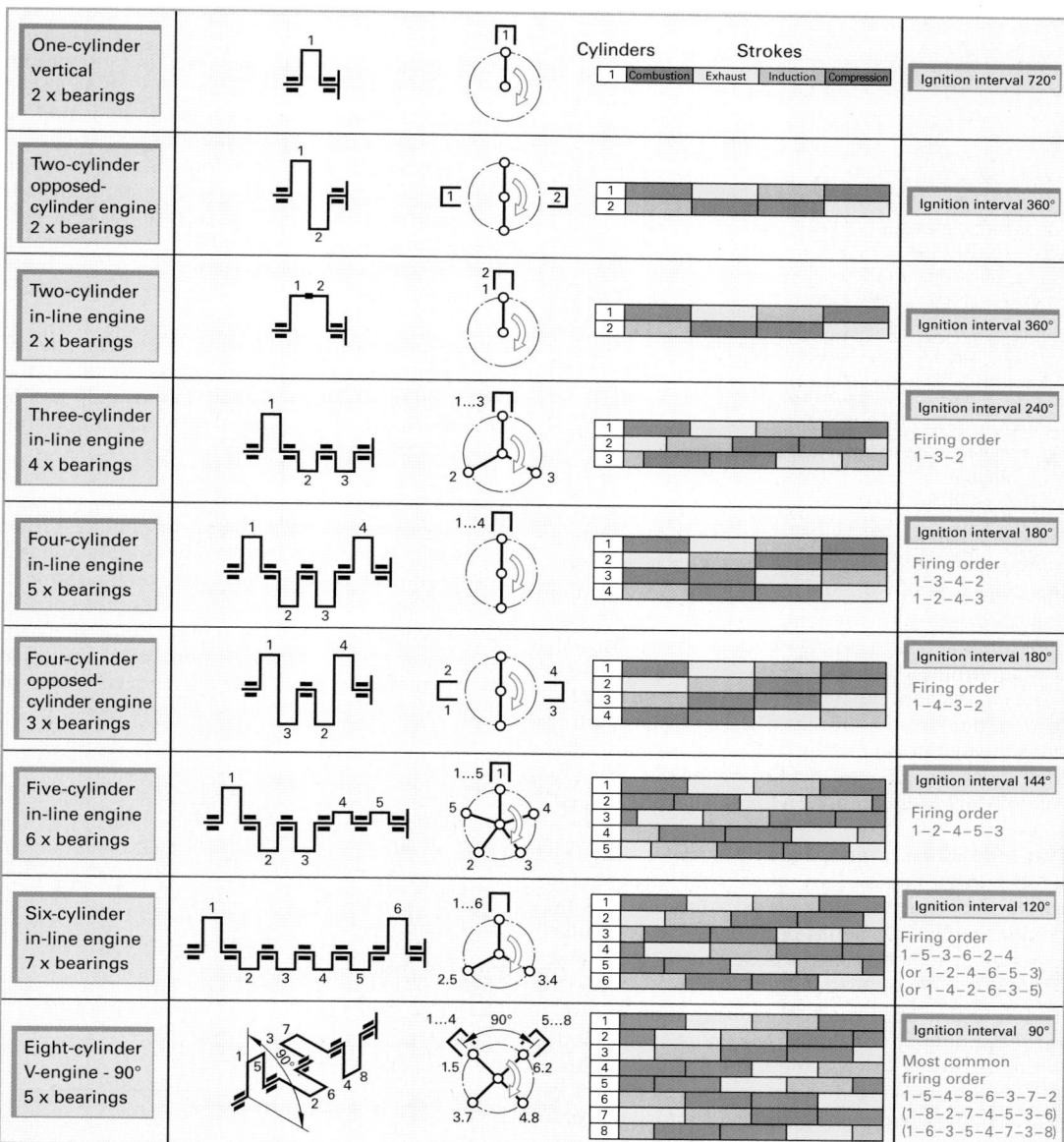

Fig. 1: Crankshaft designs, firing and power-stroke orders

10.7 Engine-performance curves

The characteristic of an engine is derived from the measured values for power, torque and specific fuel consumption determined on the test bench at different speeds.

When these measured values are plotted in a chart against the speeds, the curves located by the corresponding measuring points produce the engine's performance curves **(Fig. 1)**.

There are full-load and part-load curves.

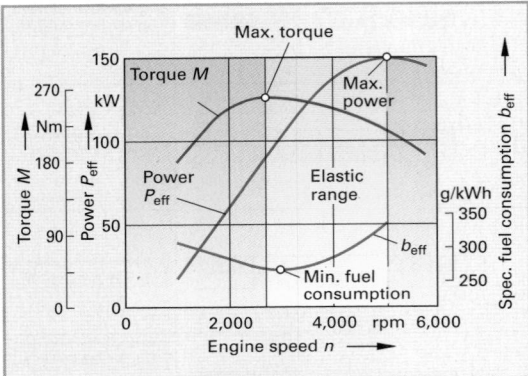

Fig. 1: Full-load curves of a four-stroke spark-ignition engine

Full-load curves. The engine at normal operating temperature is braked on a test bench with the throttle valve fully open.

Full load refers to the load which an engine can overcome at the respective engine speed. The values determined over the entire speed range under identical load are the basis for the curve shapes for torque, power and specific fuel consumption. It is possible to determine from these curve shapes the maximum torque, the maximum power and the minimum fuel consumption for an allocated speed.

Part-load curves. Measurements at part load are also important in view of the fact that an engine is rarely subjected to full load in everyday driving conditions. Various series of tests are carried out at different speeds for this purpose. The throttle valve is only partially opened in these tests.

Theoretically, both the fuel consumption and the torque should be constant in the entire speed range with the throttle valve in the same position since in actual fact always the same amount of energy of one cylinder charge should deliver the same rotatory force to the crankshaft. Accordingly the power should rise uniformly with the speed.

However, the power curve drops after the maximum power is reached because as the speed increases the loss of torque can no longer be compensated.

Causes of deviation from the ideal state:

- Fluctuating charge in the lower and upper speed ranges
- Air deficiency and poor swirling of the fuel-air mixture due to low flow velocity, thus slower and incomplete combustion
- Heat losses
- Friction losses

Elastic range. This lies between the maximum torque and the maximum power **(Fig. 1)**. As speed reduces, the decreasing power is compensated by an increasing torque. The maximum torque should where possible be situated before the middle speed range whereas the maximum power should be situated well into the upper speed range. This produces a broad elastic range, which has a favourable effect on gearbox tuning because the torque band increases in size.

Fuel-consumption map. Torque is plotted against speed at different specific fuel consumptions in the diagram **(Fig. 2)**. This results in curves with a constant specific fuel consumption which are sometimes closed in on themselves.

Because the curves resemble shells, they are also called **conchoids**. The diagram also contains further curves with constant effective power, from which it can be seen that the engine can deliver the same effective power with a completely different specific fuel consumption. Thus the engine in the diagram can deliver power of 60 kW both with a specific fuel consumption of 320 g/kWh and with one of 280 g/kWh together with increasing torque.

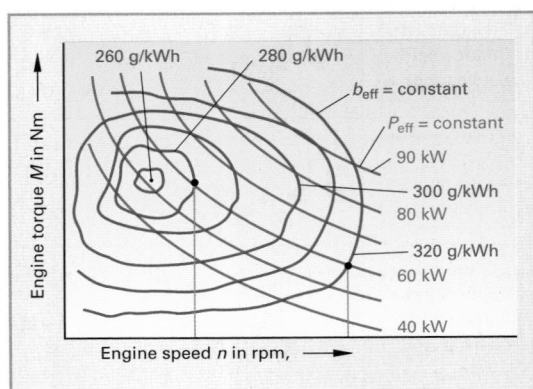

Fig. 2: Fuel-consumption map, conchoids

10.8 Stroke-to-bore ratio, power output per litre, weight-to-power ratio

Stroke-to-bore ratio

> This indicates the ratio of stroke to bore.

If the stroke is smaller than the bore, then the stroke-to-bore ratio is less than 1. It is more than 1 if the stroke is larger than the bore.

Short-stroke engines. In order for production engines to enjoy long service lives, it is important for them not to exceed a mean piston speed of 20 m/s. Nevertheless, short-stroke engines are built in order to achieve high engine speeds. These engines have a stroke-to-bore ratio of less than 1 (0.9 ... 0.7).

Long-stroke engines. The stroke-to-bore ratio is greater than 1 (1.1 ... 1.3). These engines are used mainly to power commercial vehicles and buses. High mileages and greater torques are achieved thanks to the lower engine speeds and the greater crank throw respectively.

Power output per litre

> The power output per litre indicates the greatest effective power of the engine per litre of swept volume.

Fast-running engines are more suitable for vehicle propulsion the greater their power is in proportion to the swept volume and the lower their structural weight is in proportion to the power.

The terms power output per litre and weight-to-power ratio have been introduced so that the individual engines can be compared with each other **(Table 1)**.

> The weight-to-power ratio of the engine indicates the structural weight of the engine per 1 kW of greatest effective power.

> The weight-to-power ratio of the vehicle indicates the weight of the vehicle per 1 kW of greatest effective power.

Table 1: Power output per litre, weight-to-power ratio

Engine type	Power output per litre	Weight-to-power ratio of engine	vehicle
	kW/l	kg/kW	kg/kW
Spark-ignition engines Motorcycles	30 ... 100	0.5 ... 3	2 ... 9
Passenger cars	35 ... 130	1.3 ... 5	4 ... 22
Racing cars	... 400	1 ... 0.2	1.5 ... 7
Diesel engines Passenger cars	20 ... 50	1.8 ... 5	12 ... 25
Diesel engines Commercial vehicles	10 ... 45	2.5 ... 8	60 ... 230
Supercharged engines Diesel passenger cars	30 ... 70	1 ... 4	9 ... 20
Supercharged engines Diesel commercial vehicles	18 ... 55	2 ... 7	50 ... 210

10

REVIEW QUESTIONS

1 Who built the first four-stroke engine and the first two-stroke engine respectively?

2 In what order do the strokes of a four-stroke engine take place?

3 What is the compression ratio of four-stroke spark-ignition engines?

4 Within which limits do the compression and combustion pressures of four-stroke spark-ignition engines lie?

5 What is the process of internal mixture formation in a diesel engine?

6 What is stated in Gay-Lussac's Law?

7 Which substances are formed during the combustion of the fuel-air mixture?

8 What are the causes of knocking in spark-ignition engines?

9 What do you understand by ignition lag in a diesel engine?

10 What do you understand by knocking in a diesel engine?

11 What special features apply to a diesel engine?

12 What are the four assemblies of a spark-ignition engine and of a diesel engine?

13 Which errors can be read off from a pressure-volume diagram?

14 What do you understand by a symmetrical timing diagram?

15 What are the differences between short-stroke and long-stroke engines?

16 What is the firing order of a 6-cylinder in-line engine?

17 How are the cylinders numbered in compliance with standards?

18 What does the power output per litre indicate?

19 What does the weight-to-power ratio of the engine indicate?

20 What do you understand by the greatest effective power of an engine?

21 What do you understand by full-load curves of an engine?

22 Why is the torque of a spark-ignition engine not of equal magnitude over the entire speed range?

23 What do you understand by the elastic range?

11 Mechanical engine components

11.1 Cylinder, cylinder head

11.1.1 Functions, stresses

Functions
- Form the combustion chamber together with the piston.
- Guidance of the piston by the cylinder.

Stresses
- High combustion pressures and temperatures.
- Large thermal stresses due to rapid temperature changes.
- Cylinder barrel subject to wear due to piston friction and combustion residues.
- Increased friction during cold starting, uncarburated fuel washes lubricant layer off the cylinder.

Properties of materials
- High strength and inherent stability
- Good heat conduction
- Low thermal expansion
- High resistance to wear
- Good sliding properties for the cylinder face

11.1.2 Cylinder types

Liquid-cooled cylinders

> The cylinders of liquid-cooled engines are usually combined to form a single block. Cooling ducts pass through the double-walled cylinder block; coolant is supplied by the water pump, cools the cylinder walls and flows through ducts into the cylinder head.

Engine block (Fig. 1). Cylinder block and crankcase upper half are cast in a single piece from
- **Flake-graphite cast iron (grey cast iron).** As well as good rigidity and strength and good sliding and wear performance, it has low thermal expansion and good noise damping.
- **Vermicular-graphite cast iron.** As the casting cools, the graphite precipitates not in the form of lamellas but rather in a vermicular form. The notch effect between the structure crystals is lower, thereby greatly increasing strength and rigidity. This facilitates higher internal cylinder pressures and thereby increased performance with thinner wall thicknesses (weight savings).
- **Al alloys.** Of particular benefit is the low density in comparison with grey cast iron and good thermal conductivity. The engine blocks are also finned in order to improve inherent stability. The wear properties of the cylinder barrels must be improved by means of special production processes or cylinder liners are used.

Closed-deck design (Fig. 1). The sealing surface of the engine block to the cylinder head is extensively closed around the cylinder bores; only the ducts e.g. for coolant are provided. This design is used almost exclusively for cast iron; for AlSi alloys (e.g. ALUSIL cylinders) these cylinder blocks are manufactured in gravity die-casting or in low-pressure casting.

Open-deck design (Fig. 2). The water jacket around the cylinder bores is open to the cylinder head. This makes it technically possible to produce cylinder blocks with cylinder faces in accordance with the LOKASIL concept in a die-casting process. The lower rigidity of open-deck cylinder blocks calls for metal instead of soft-material cylinder-head gaskets. Because of their low settling behaviour, they provide for a low initial force for the cylinder-head bolt connection. In turn, this reduces cylinder distortion and deck deformation.

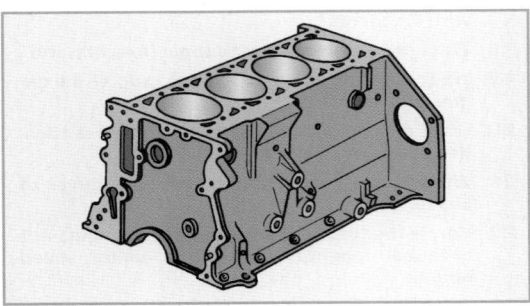

Fig. 1: Engine block

Fig. 2: Engine block, open-deck design

Cylinder liners

> Cylinder liners of superior-quality, close-grained cast iron (centrifugal casting) are drawn into cast-iron or aluminium-alloy cylinder blocks. They have a long service life in that they are more resistant to wear than the cylinder barrels of cast-iron cylinder blocks.

Two different types are used: wet and dry cylinder liners.

Wet cylinder liners (Fig. 1). Coolant flows around liners of this type directly, thereby providing for a good cooling effect. Wet liners can be replaced individually. However, the cylinder block is not as rigid and is distorted more easily. Liners of this type feature a collar at their top end. They must be sealed against the crankcase by sealing rings as otherwise coolant will get into the crankcase.

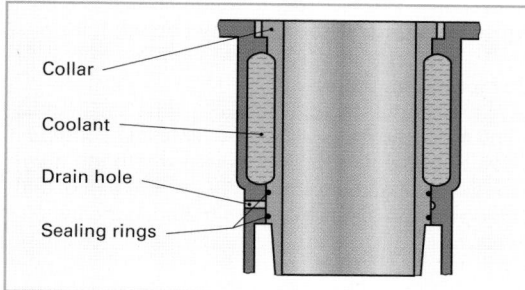

Fig. 1: **Wet cylinder liner**

Dry cylinder liners (Fig. 2). These are inserted with a sliding fit or with an interference fit as thin-walled liners into the cylinder block. Because they do not come into contact with the coolant, the transfer of heat to the coolant is not as good as in wet liners. Cylinder liners with a sliding fit are finish-machined prior to installation. Liners with an interference fit are pressed with a predrilled cylinder bore into the cylinder block. They are then fine-bored and honed.

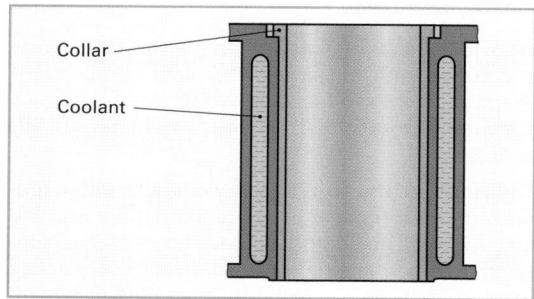

Fig. 2: **Dry cylinder liner**

Air-cooled cylinders

> Air-cooled cylinders are provided with cooling fins to increase the size of the surface area and thereby improve the cooling effect. They are bolted in the form of individually standing finned cylinders to the crankcase.

Air-cooled cylinders **(Fig. 3)** are cast predominantly from aluminium (Al) alloys. Because of the poor sliding and wear properties of the cylinder liners, they must, as in the case of liquid-cooled cylinders made from Al alloys, be improved by means of special production processes or liners are cast in, for example, in the Alfin process.

Iron-aluminium composite-casting process (Alfin process). The liners of flake-graphite cast iron are coated on the outside with a layer of iron aluminium ($FeAl_3$) and then cast in with the AlSi alloy for the finned cylinder. The Alfin intermediate layer establishes a connection with good heat conduction between the cast-iron liner and the AlSi alloy of the finned cylinder.

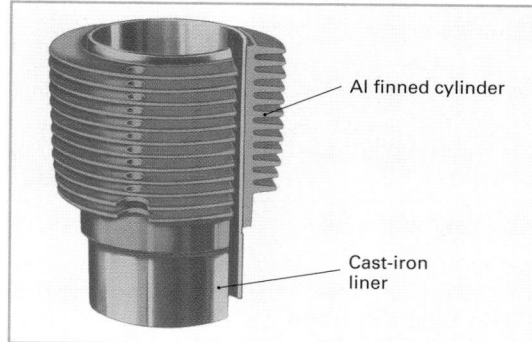

Fig. 3: **Air-cooled cylinder of Fe-Al composite casting**

Cylinder barrels of Al cylinders

ALUSIL process. The cylinder block is cast from an Al alloy with a high silicon content (up to 18 %) in gravity die-casting or low-pressure casting. In order for the silicon crystals to be formed above all on the cylinder faces, the cores which form the casting mould on the cylinders are cooled. After honing, the soft aluminium is removed around the silicon crystals by electrochemical etching. This creates a wear-resistant carrier barrel for piston and piston rings with spaces for the lubricating oil. Mostly ferrocoated pistons are used in order to reduce piston wear.

NIKASIL process. The cylinder barrel of AlSi alloy is galvanically coated with a wear-resistant layer of nickel with intercalated silicon-carbide crystals.

LOKASIL process. Silicon crystals are placed on the cylinder barrel with the aid of form bodies, also known as preforms. Preforms **(Fig. 1)** are highly porous and are made up of silicon crystals which are sintered with water glass (silicates dissolved in water). They are manufactured as hollow cylinders and inserted at that point in the casting mould where the face is located. In a special die-casting process, known as squeeze casting, the Al alloy is pressed slowly with low pressure from below into the vertical casting mould. The preforms must be preheated to 300 °C during insertion so that the Al alloy during penetration cannot cool down and in the process solidify before it has completely penetrated the preforms. Finally, the pressure is increased from 120 bar to 500 bar to close all the pores and eliminate air inclusions. Multistage honing is used to expose the silicon crystals in relief form, which in turn results in a wear-resistant cylinder barrel. Mostly ferrocoated pistons are used in LOKASIL cylinders.

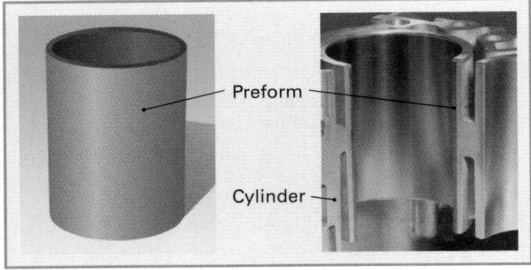

Fig. 1: LOKASIL process

11.1.3 Cylinder head

> The cylinder head seals off the combustion chamber at the top. It is secured by the cylinder-head bolts with the inserted cylinder-head gasket on the cylinder block.

Design. The cylinder head contains the fresh-gas and exhaust-gas ducts with their valve seats and usually also the compression space. It accommodates the spark plugs as well as the fuel injectors in the case of direct-injection engines together with the engine-timing components, e.g. the valves. The camshaft is often mounted on the cylinder head. The cylinder head is subjected to high loads as a result of combustion pressure and hot combustion gases and must therefore demonstrate high inherent stability, good heat conduction and low thermal expansion.

Liquid-cooled cylinder head (Fig. 2). This is predominantly cast from Al alloys for each cylinder in-

dividually or for the entire block in one piece. The coolant flows from the cylinder block into the cylinder head via flow ducts.

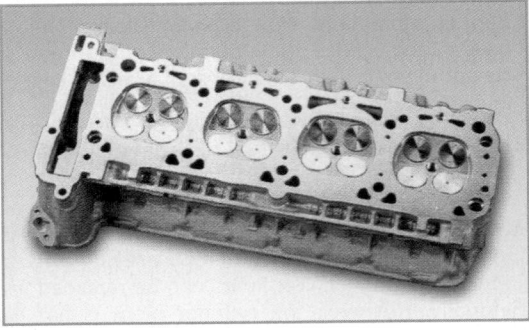

Fig. 2: Liquid-cooled cylinder head

Air-cooled cylinder head. This is manufactured entirely from Al alloys and is provided with cooling fins. Because the transfer of heat to air is less efficient than it is to coolant (cooling liquid), the cooling surface must be enlarged by cooling fins.

Compression space

> The compression space is the smallest combustion chamber. It is sealed towards the bottom by the position of the piston at TDC, where part of the compression space may be situated in the piston crown.

The geometrical shape of the compression space has a significant bearing on the engine's operating performance with regard to
- mixture swirl
- combustion process
- fuel consumption
- pollutant emissions
- knock resistance
- torque
- power output
- efficiency

The compression space is determined by
- compression ratio
- surface-to-volume ratio
- position of the spark plug or fuel injector
- valve arrangement

Wall heat losses increase when the surface is large and zones with an excessively cold combustion-chamber wall in the area of which the flame goes out may form easily. This results in increased HC emissions.

The length of the combustion distances differs depending on the position of the spark plug and the fuel injector. The engine would achieve its best efficiency levels when the combustion distances are as short as possible, i.e. with the spark plug or fuel injector in a central position.

The following properties are expected of an optimally shaped compression space:

- Facilitation of a high charge by means of appropriate opening cross-sections of the inlet valves.
- Swirling and mixing of inducted air and injected fuel with the aid of squish zones.
- Quick and as complete a discharge as possible of burned gases via the exhaust valves.
- Compact space, not fissured and without niches (small surface).
- Short combustion distances thanks to centrally positioned spark plug and fuel injector.

The ideal shape for a compression space would be hemispherical because then the combustion distances are the shortest and the surface is the smallest. It is necessary, however, to deviate from this ideal shape on account of the arrangement of the valves.

Roof-shaped compression space (Fig. 1). This is similar to the hemispherical shape. The inlet and exhaust valves are situated opposite each other at an angle in the cross-flow cylinder head. The inlet valve is usually larger in diameter than the exhaust valve and thereby facilitates an improved charge.

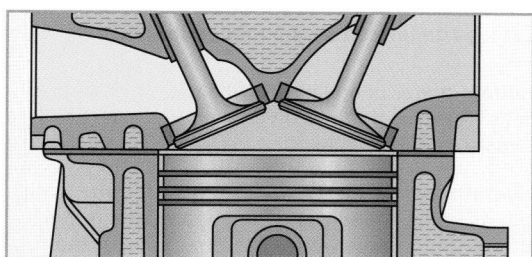

Fig. 1: Roof-shaped compression space

Compression spaces with squish zones (Fig. 2). These are usually particularly compact and can be roof-shaped, wedge-shaped or even bathtub-shaped. Shortly before TDC the air or the fuel-air mixture is pressed out of the squish zones. The swirling effect created ensures that the injected fuel is mixed with the air more intensively, which in turn results in rapid combustion. The necessary advanced ignition can be shortened, thus allowing for the possibility of a colder spark plug, a higher compression ratio or even the use of regular-grade petrol. Furthermore, the HC content in the exhaust gas is reduced due to complete combustion.

Multiple-valve cylinder head (Fig. 2). The compression space is roof-shaped or bathtub-shaped when 2 inlet valves and 2 exhaust valves are used. In addition, there are usually two opposing squish zones. The spark plug or injection nozzle can be po-

sitioned centrally, which results in short combustion distances and an increased rate of combustion. Further advantages are increased power, reduced fuel consumption and better exhaust-gas values. When 5 valves are used (3 inlet valves, 2 exhaust valves), the compression space is more spherical in shape because the diameter of the valve heads is smaller.

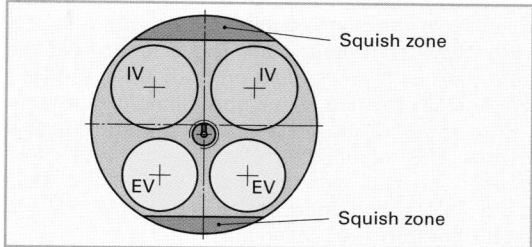

Fig. 2: Roof-shaped compression space with 4 valves and 2 squish zones

11.1.4 Cylinder-head gasket

The cylinder-head gasket **(Fig. 3)** is intended to seal the combustion chamber gas-tight and prevent coolant or engine oil from escaping from the flow ducts.

To obtain a good seal, it is necessary for the cylinder block and cylinder head to have a level surface.

Stresses. Fuel, exhaust gas, engine oil and coolant come into contact with the cylinder-head gasket in liquid and gaseous form, in cold and superheated states and partly mixed under high pressure and vacuum pressure with chemically active agents. The gasket must therefore satisfy the following requirements:

- Elastic adaptation of the sealing surfaces in all operating states.
- Low tendency to settle so as to facilitate cylinder-head tensioning without retightening.
- No sticking to sealing surfaces in order to simplify removal.

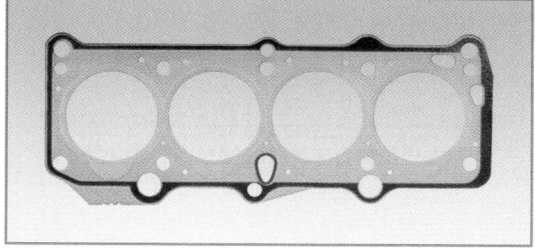

Fig. 3: Cylinder-head gasket

The diverse requirements are satisfied in both spark-ignition and diesel engines by ...

- ... metal/soft-material cylinder-head gaskets, which are most commonly used.
- ... metal cylinder-head gaskets for high-performance diesel engines, for CV diesel engines and increasingly for spark-ignition engines.

Metal/soft-material cylinder-head gasket (Fig. 1). A metallic carrier plate roughly 0.3 mm thick is provided with cramping teeth. These teeth hold the soft-material layer applied on both sides. The soft material is provided with a pore-filling plastic coating to improve its resistance to the surrounding media. The combustion-chamber passages are beaded in, e.g. with an aluminium-plated steel plate. The sealing effect at the liquid passages can be further improved by an elastomer coating.

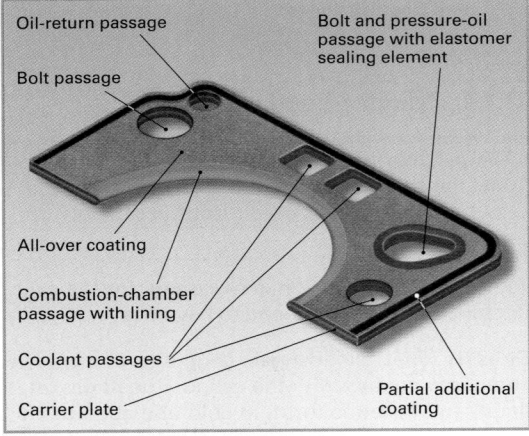

Fig. 1: Metal/soft-material cylinder-head gasket

Metal cylinder-head gasket (Fig. 2). This is usually manufactured as a multilayer cylinder-head gasket from steel plates. For the purpose of providing a secure gas seal, beads or plate linings are required to increase the local pressing. The sealing effect at the liquid passages is also increased by elastomer coatings.

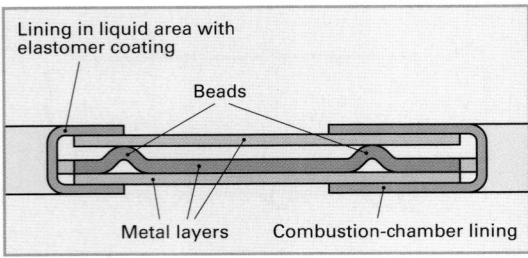

Fig. 2: Metal cylinder-head gasket

11.1.5 Crankcase

> The crankcase houses the crankshaft and sometimes also the camshaft. The cylinders are bolted to the crankcase.

Design. The crankcase is usually divided at the height of the crankshaft bearings. The crankcase upper half contains the bearing seats for the crankshaft. The bearing caps are secured from below with bolts. The advantage of this arrangement is that the crankshaft can be easily removed. The crankcase lower half is designed as the oil pan and is bolted oil-tight to the crankcase upper half.

The crankcase is manufactured from cast iron or from Al alloys and is used in air-cooled engines.

> **Engine block** = cylinder + crankcase

11.1.6 Engine suspension

> Engine mounts are intended to damp the transmission of vibrations between the engine and the body and take up the weight of the engine. They must also support the shear and tensile forces generated by the engine torque.

Rubber-metal connections (Silentbloc) or hydraulically damped mounts (hydro-mounts) are used as engine mounts.

Hydro-mount (Fig. 3). At idle the pressure which builds up as a result of engine vibrations in the hydraulic fluid in the upper chamber acts on the rubber diaphragm only. The diaphragm deforms and damps the vibrations. Air escapes from the air cushion through the opened solenoid valve. The solenoid valve is closed during vehicle operation. The pressure in the hydraulic fluid acts through the nozzle passage in the lower chamber on the rubber bellows, deforms them and thereby reduces the vibrations.

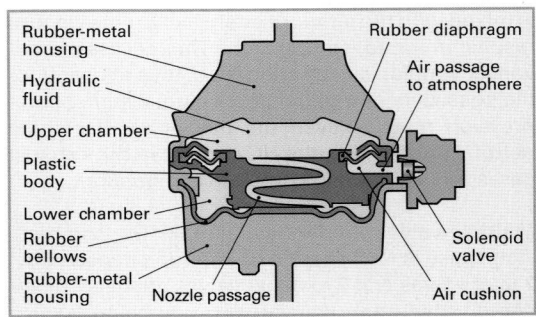

Fig. 3: Hydro-mount

Checking the compression pressure

The compression tester is used to carry out comparative measurements of the pressure conditions in the individual combustion chambers **(Fig. 1)**.

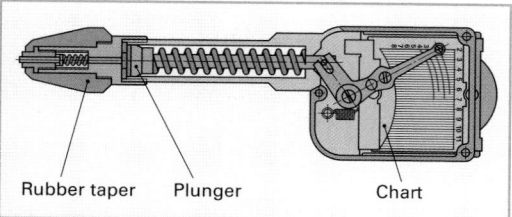

Rubber taper   Plunger   Chart

Fig. 1: Compression tester

Observe the following during the check:

- Carry out the check only when the engine is at normal operating temperature.
- Deactivate the electronic ignition system (observe the manufacturer's instructions).
- Insert the test chart in the compression tester so that the pointer is at cylinder 1.
- Unscrew all the spark plugs and briefly crank the engine with the starter to remove combustion residues.
- Press the rubber taper of the compression tester into the spark-plug hole and open the throttle valve fully.
- Crank the engine with the starter by approximately 10 revolutions.
- Vent the compression tester before the test chart is moved to the cylinder 2 position.

When the combustion chambers are in perfect condition, the compression pressures measured in the individual cylinders may only deviate from each other slightly (max. 2 bar).

Conclusions about fault causes:

- If a pressure of below 6 bar (12 bar in diesel engine) is indicated in all the cylinders, the engine is subject to regular wear.
- In the event of deviating compression pressures in the individual cylinders, it is possible to determine what has caused the pressure loss by spraying some engine oil into the combustion chamber. If the compression pressure increases during the subsequent measurement, the cylinder wall or piston rings is/are subject to wear. If there is no increase in the compression pressure, the valves, valve seats, valve guide, cylinder head or cylinder-head gasket may be defective.

- If an equal compression pressure is measured in two cylinders next to each other which is significantly lower than that in the other cylinders, there may be a crack in the cylinder head or a leaking cylinder-head gasket between the two cylinders.

Compression-loss test (Fig. 2)

This is carried out if there is a suspected leak in the cylinder chamber after the compression check.

When checking the relevant cylinder, adhere to the following procedure:

- The piston must be at TDC of the compression stroke.
- Connect the pressure-loss tester to the compressed-air system (5 bar … 10 bar) and calibrate using the knurled screw.
- Connect the tester via the spark-plug thread of the cylinder.
- The pressure loss created by the leak is indicated in per cent at the pressure gauge and must not exceed the values specified by the test-equipment manufacturer.
- In the event of larger leaks, the fault sources can be determined by identifying the air outlet.

Conclusions about fault causes:

- Blowing noises in the intake manifold or exhaust pipe: inlet or exhaust valve leaking.
- Blowing noises at the oil filler neck or at the dipstick opening: leaks due to wear of piston, piston rings and cylinder barrel, faulty cylinder-head gasket.
- Air bubbles in the radiator filler neck or blowing noises in the spark-plug hole of the adjacent cylinder: leaking cylinder-head gasket, cracks in the cylinder head.

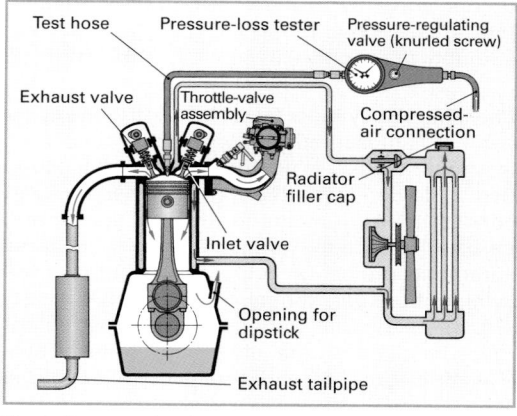

Fig. 2: Compression-loss test

CONTINUED FROM PAGE **201**

Wear to cylinder faces

In a new engine the cylinder faces are perfectly cylindrical. Increasing engine use is accompanied by a noticeable amount of wear to such an extent that the piston rings are no longer able to seal the combustion chamber fully. This results in engine oil and fuel getting into the combustion chamber (high oil consumption) and into the crankcase (oil dilution) respectively. In addition, compression decreases, engine power drops and fuel consumption increases. Engine operation also becomes noisier due to piston slap.

The cylinder barrel is not subject to uniform wear between the top and bottom dead centres because the lateral piston force decreases with the combustion pressure and because lubrication in the area of top dead centre is worse **(Fig. 1a)**.

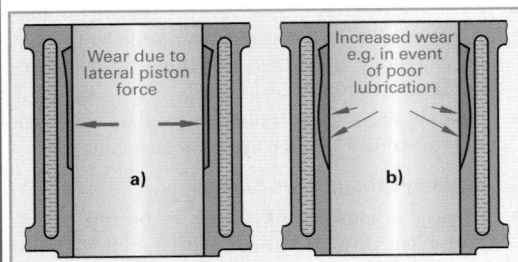

Fig. 1: Normal and abnormal wear to cylinder barrel

The wear is greatest in the area of top dead centre and the cylinder bore becomes conical.

In the event of abnormal wear, usually caused by failing lubrication, the bore becomes bulged **(Fig. 1b)**. The wear does not extend uniformly over the circumference of the cylinder, rather it occurs primarily in the direction of the lateral forces, intensified by piston slap. Normal wear amounts to roughly 0.01 mm over 10,000 km.

Checking the cylinder face. The wear to the cylinder face is measured with an internal measuring instrument which is equipped with a dial gauge **(Fig. 2)**. To determine the wear, it is necessary to conduct measurements in the direction of and perpendicular to the gudgeon-pin axis. The measurements are begun below the top edge of the cylinder bore and continued downwards in several operations. In the process, the instrument must be tilted in the direction of the arrows **(Fig. 2)** in order to avoid measurement errors. Measurement serves to determine wear, ovality and conical progression of the cylinder face.

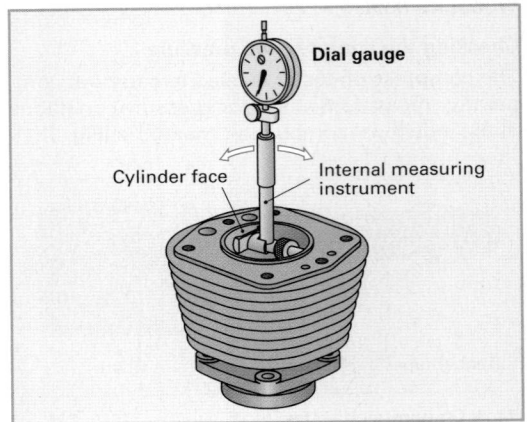

Fig. 2: Measuring a cylinder face

Machining the cylinder face. The cylindrical shape must be re-established by fine-boring or honing if the wear in the centre is approx.

- 0.5 mm in four-stroke engines,
- 0.2 mm in two-stroke engines,
- 0.8 mm in CV diesel engines.

Fine-boring is performed in accordance with the piston oversize in stages of 0.25 mm or 0.5 mm. Subsequent honing is performed on a honing machine.

Installing cylinder liners

Dry cylinder liners are usually finish-machined. The fit is configured as a sliding fit so that the liners can be pushed in with minimal pressure. Liners which are only predrilled are pressed with interference into the bore and finish-machined in the cylinder. The installed cylinder liner must not protrude. It should be level with the top surface or recede by up to 0.1 mm.

Wet cylinder liners are supplied ready for installation. They can usually be inserted with ease. The rubber rings must provide a good seal, but should not be too thick so that the liner does not deform under the pressure, which may cause piston seizure. The collar of the liner generally protrudes up to roughly 0.1 mm **(Fig. 1, Page 203)**.

The cylinder-head gasket must not feature an excessive flange; the flange must not press onto the inner liner edge as this will cause the collar to tear off when the cylinder-head bolts are tightened or the liner to distort.

The collar must not recede under any circumstances because then the liner would be able to move.

CONTINUED FROM PAGE 202

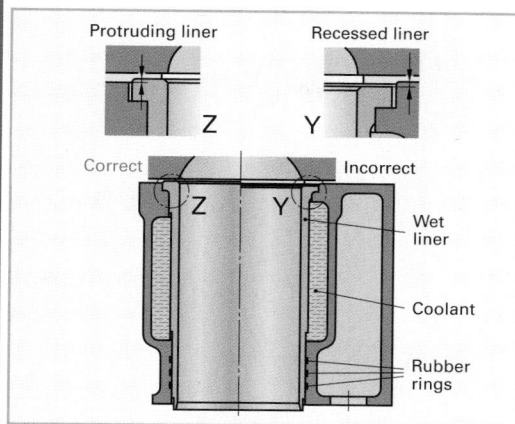

Fig. 1: Installation positions of cylinder liners

Replacing the cylinder-head gasket

An inadequate seal between the cylinder head and the cylinder block gives rise to the following detriments:

- Power loss, part of the gases is lost, the pressure in the cylinder drops
- Burning of the cylinder-head gasket
- Oil loss
- Engine damage caused by coolant entering the combustion chamber

The following points must be observed when the cylinder-head gasket is replaced:

- Before releasing the cylinder-head bolts, allow the engine to cool down in order to prevent the cylinder head from being distorted.
- Remove all gasket remnants which are stuck fast.
- The sealing surfaces of the cylinder head and the cylinder block must be level. Remachine irregularities on a surface grinding machine.

- The thickness of the cylinder-head gasket must conform to the manufacturer's specifications.
- The coolant and engine-oil passages must match up.
- Use a thicker gasket if the cylinder head and the cylinder block have been worn as otherwise the compression ratio would be changed.
- Combustion-chamber linings must not protrude into the combustion chamber as this could result in auto-ignition.

Tightening the cylinder-head bolts

These bolts are tightened in a particular sequence which is always specified in the repair handbooks. Deviations from this sequences will result in distortion of the cylinder head and in leakage.

Generally speaking, the cylinder-head bolts are tightened

- from the inside outwards in a spiral-shaped progression (Fig. 2) or
- from the inside outwards diagonally.

When the bolts are being tightened, the torque and if necessary the angle of rotation specified by the manufacturer must be observed. As a rule, a new set of bolts must be used.

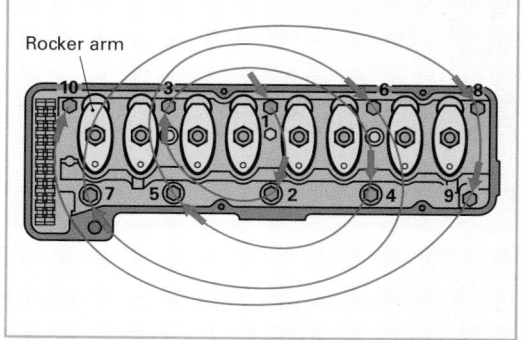

Fig. 2: Example of tightening cylinder-head bolts

REVIEW QUESTIONS

1 What are the functions of the cylinder and the cylinder head?

2 To what stresses are the cylinder and the cylinder head exposed?

3 Which properties is the cylinder or the cylinder liner meant to have?

4 What is the difference between wet and dry cylinder liners?

5 What are the advantages and disadvantages of cylinders made from light-metal alloys?

6 How can the running properties of light-metal cylinders be improved?

7 What are the advantages of compression spaces with squish zones?

8 What are the potential consequences of a defective cylinder-head gasket?

9 What must be borne in mind when installing wet cylinder liners?

10 How must cylinder-head bolts be tightened?

11.2 Engine-cooling systems

> The cooling system in internal-combustion engines has the following functions:
> To heat the engine quickly to optimal operating temperature and to dissipate excess heat during engine operation.

Roughly one third of the thermal energy generated during combustion is absorbed by the components, e.g. piston, valves, cylinder, cylinder head, exhaust-gas turbocharger and the engine oil, and must be dissipated owing to the limited heat resistance of the materials or the lubricating oil.

Therefore even economical direct-injection diesel and spark-ignition engines can only utilise max. 43 % of the energy stored in the fuel. The remainder of the energy is lost in the form of heat **(Fig. 1)**.

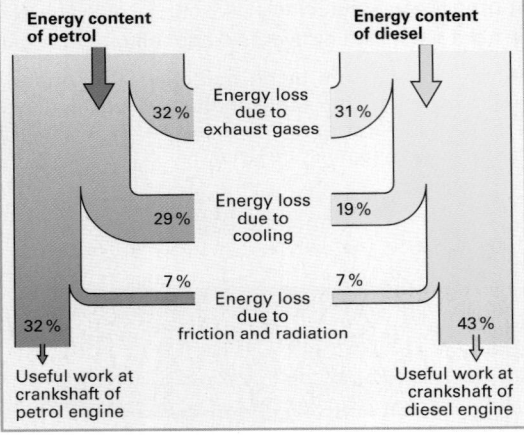

Fig. 1: Sankey diagrams

The thermal energy is generated in the process not just by the combustion of the fuel inside the cylinder. The friction of the moving components in the engine and the drivetrain also converts parts of the mechanical energy yielded back into thermal energy which can no longer be utilised as motive power **(Fig. 2)**.

Operating temperature

When a vehicle engine is started from cold, optimal mixture formation can only take place once the engine and with it the engine components have reached a specific temperature.

The upper temperature limit (maximum temperature) is dependent above all on the thermal load capability of the engine oil and of the engine materials. Vehicle engines usually run at a coolant temperature of 80 °C...90 °C **(Fig. 2)**.

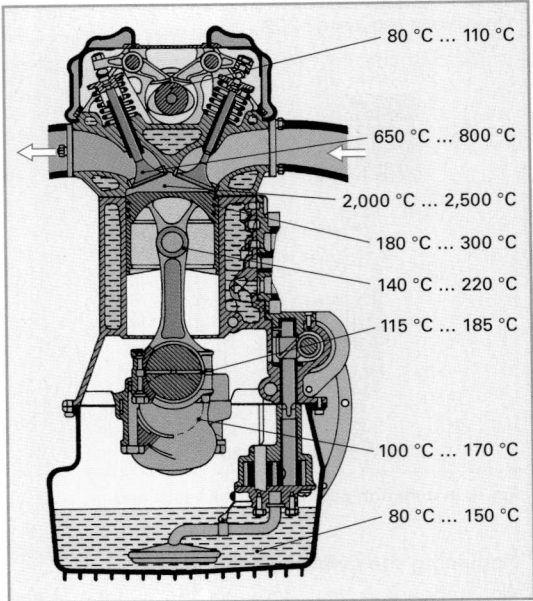

Fig. 2: Component temperatures at $T_{coolant}$ = 90 °C

New, map-controlled cooling systems allow operating temperatures of up to approx. 120 °C. This enables fuel consumption to be reduced, the reasons for this being the reduced friction and the improved mixture preparation at higher temperatures.

What is required of the cooling system:
- Adequate cooling effect
- Operating temperature quickly reached
- Low weight
- Uniform cooling of components in order to avoid thermal stresses
- Low energy demand for operation of the cooling system

Good cooling therefore provides for
- improved cylinder charge
- reduced tendency to knock in spark-ignition engines
- higher compression
- higher power with more economical fuel consumption
- more uniform operating temperatures

11.2.1 Types of cooling

The following different types of cooling are used:

Air cooling:
- Airstream cooling
- Forced air cooling

Liquid cooling:
- Thermocooling
- Pump cooling

Internal cooling of the combustion chamber by:
- Heat of fuel evaporation

Due to the increasing power, increased rotational speeds, the installation of the engines in a relatively sealed space and ever more compact dimensions, cooling systems are for the most part designed as forced-circulation cooling systems.

Today, air cooling is only used in isolated cases in motorcycle, aircraft and stationary engines.

11.2.2 Air cooling

> In the case of air cooling, the cooling heat to be removed is dissipated from the surfaces of the engine components directly to the ambient air flowing past.

Airstream cooling (Fig. 1). This is the simplest type of air cooling. It is frequently used on motorcycles since the airstream flows around their unfaired engines. In the interest of obtaining the greatest possible level of cooling, the transfer of heat to the ambient air is improved by increasing the effective cooling surface with cooling fins. For this reason, cylinders, cylinder head and often also the crankcase are fitted with cooling fins.

Fig. 1: Air-cooled engine with cooling fins

Forced air cooling. This provides for adequate cooling of engines around which the airstream does not flow. A fan is driven by the engine via a V-belt and cools the individual cylinders with the aid of baffles uniformly with the cooling air. Application: e.g. motor scooters.

Advantages of air cooling:
- Simpler design.
- Lower weight-to-power ratio.
- No coolant with anti-freeze required.
- Extensively maintenance-free.

Disadvantages of air cooling:
- Greater fluctuations in the operating temperature.

- Power requirement of the radiator fan is comparatively high (approx. 3%...4% of engine power).
- Louder noises emanating from the fan due to the lack of a coolant jacket.
- Greatly delayed and non-uniform passenger-compartment heating.
- Poorer heat transfer between cooling fins and air.
- Cannot be regulated.

11.2.3 Liquid cooling

> In the case of liquid cooling, the heat is removed by a cooling liquid (coolant). The coolant absorbs the component heat and dissipates it via the radiator to the ambient air.

With liquid cooling, the cylinder block and the cylinder head are twin-walled in design and are passed through by cooling ducts. A coolant circulates in these cavities and absorbs the heat to be removed from the walls. The coolant in the cooling circuit flows through hoses and pipes to a radiator around which air flows. It dissipates the absorbed heat in the engine via this surface to the ambient air. The coolant which has been recooled in this way then flows back to the engine, where it absorbs heat again.

If the flow operation is achieved exclusively by the change in coolant density, this is known as **natural-circulation cooling** or thermosiphon cooling. When a hot engine is switched off, this effect is utilised for further cooling. It is based on the fact that hot water has a lower density than cold water and therefore rises. Coolant circulation can take placed without a pump. This type of cooling is not suitable for today's engines. Today, cooling systems with pumps, so-called **pump cooling systems**, are used exclusively. In this case, the coolant is circulated by a pump.

Advantages of liquid cooling:
- Uniform cooling effect.
- Good damping of the combustion noises by the cooling jacket.
- Provides for good heating of the passenger compartment.

Disadvantages of liquid cooling:
- Relatively high weight.
- High maintenance requirements.
- Longer warm-up period until the operating temperature is reached.

11

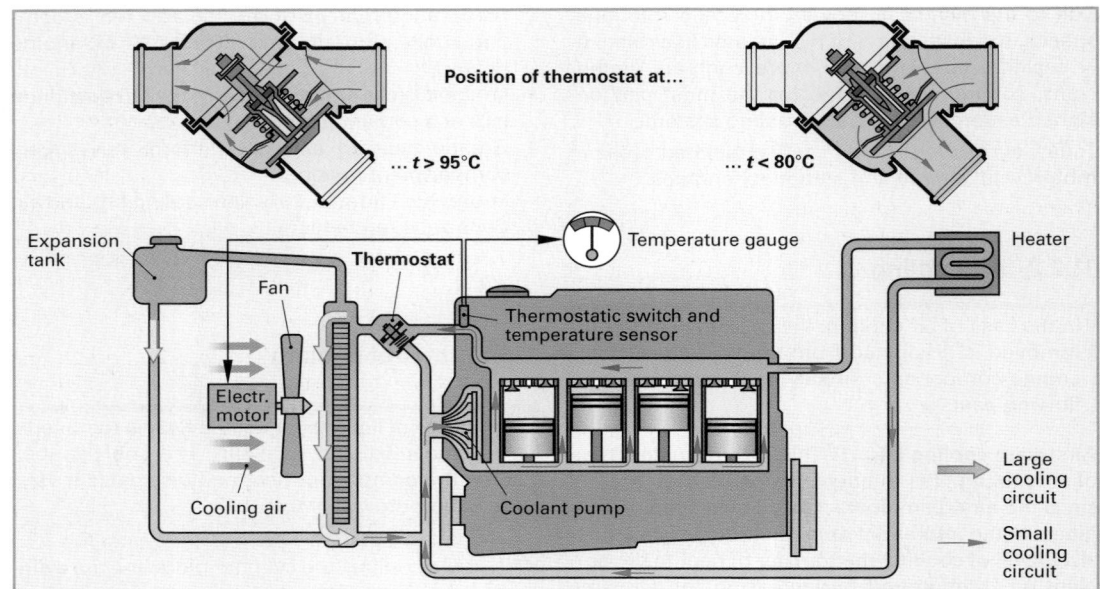

Fig. 1: Pump or forced-circulation cooling with expansion thermostat and double valve during cold starting

The coolant is directed with the aid of a thermostatic valve into a small or into a large cooling circuit.

Cold engine. When the engine is cold, the coolant pump pumps the coolant into the cooling jacket around the cylinders, where it flows round the cylinders before passing through through-holes to the cylinder head. From here it flows back to the pump via the thermostat, which is still closed. When the car heater is switched on, some of the coolant also flows, depending on the position of the heat control valve, via the heater's heat exchanger back to the pump **(small cooling circuit).**

Engine at operating temperature. When the engine operating temperature is reached, the radiator is cut into the circuit via the thermostat **(large cooling circuit).** The coolant heated by the engine now flows via the thermostat to the radiator and on to the coolant pump. The content of the expansion tank keeps the liquid level in the cooling system constant.

Coolant temperature. Depending on the vehicle operating state and the vehicle manufacturer, this is for
- passenger cars approx. 100 °C ... 120 °C
- commercial vehicles approx. 90 °C ... 95 °C

The maximum permissible overpressures in the cooling system are currently for
- passenger cars approx. 1.3 bar ... 2 bar
- commercial vehicles approx. 0.5 bar ... 1.1 bar

A higher pressure in the cooling system allows the coolant temperature to be increased without the coolant boiling. With the associated increase in engine temperature, it is possible to increase power and reduce consumption and emissions.

In spark-ignition engines, however, the rise in temperature is limited by the accompanying increase in the tendency to knock.

Coolant. This is usually a mixture of water with a low lime content and anti-freeze with additives for corrosion protection which is adapted to the relevant vehicle type **(see Page 32, Chapter 1.7.5).**

The amount of coolant introduced into the cooling circuit, which is roughly four to six times the engine displacement, is circulated approximately ten to fifteen times per minute. Thus, depending on engine power, between 4,000 l/h and 18,000 l/h and between 8,000 l/h and 32,000 l/h of coolant are circulated in passenger cars and commercial vehicles respectively. In this way, when the cooling heat is promptly removed, the temperature difference between coolant inlet to and outlet from the engine is roughly only 5 °C ... 7 °C. This in turn prevents thermal stresses in the engine.

11.2.4 Pump-cooling components

Coolant thermostat (temperature regulator). By means of stepless changeover between the small and large cooling circuits, this ensures that the engine quickly reaches its operating temperature and maintains it during operation with as minimal fluctuations as possible.

This regulation has an effect primarily on:

- Fuel consumption
- Exhaust-gas composition
- Wear

The thermostat can be installed both in an engine coolant fitting and in the supply or return line.

Thermostat with double valve (Fig. 1) (two-way thermostat). This is what is usually installed nowadays. This thermostat switches the coolant flow as temperature increases from the small to the large cooling circuit through the radiator. It closes the bypass line to the coolant pump in the process.

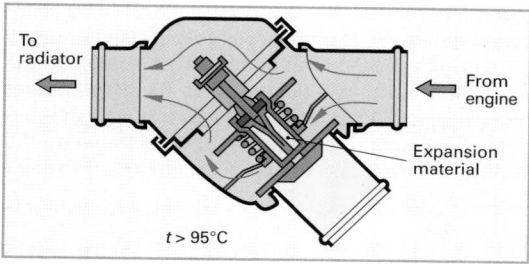

Fig. 1: Expansion thermostat with double valve, with cold engine

Expansion thermostat. In today's customary thermostat designs, an expansion element opens and closes the control valves. The expansion element **(Fig. 2)** is a pressure-proof metal container which is filled with a waxlike expansion material. When the engine is cold, the valve head mounted permanently on the working plunger closes off the throughflow to the radiator.

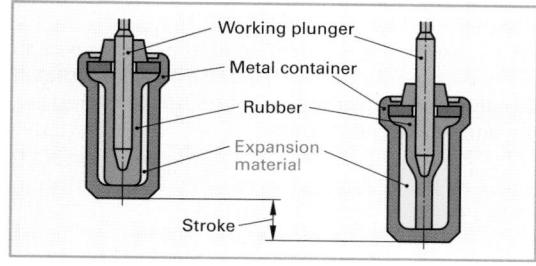

Fig. 2: Expansion element

A rise in coolant temperature to approx. 80 °C causes the expansion-material filling in the expansion element to melt. The volume of the expansion material increases significantly during the melting process. This causes the metal container on the fixed plunger to move; the valve controlling the throughflow to the radiator opens and at the same time the valve in the bypass line closes. The radiator is fully cut in at a coolant temperature of approx.

95 °C. When the coolant temperature drops again, a spring forces the metal container back over the plunger and thereby closes the valve controlling the throughflow to the radiator. At the same time the valve to the bypass line is opened again.

As a result of this permanent, alternating opening and closing, the coolant temperature fluctuates within a very narrow range only and in this way the engine temperature is kept largely constant. The expansion element operates to a large extent independently of the pressure in the cooling system. It achieves high actuating forces for valve actuation.

Coolant pump (Fig. 3). This is usually a centrifugal pump (radial pump, flow-type pump). The pump housing, which is filled with coolant, accommodates an impeller which operates at high speed. This impeller captures the liquid in the centre (suction side) and forces it outwards (pressure side). Cooled coolant flows constantly from the radiator or the thermostat to the centre of the impeller.

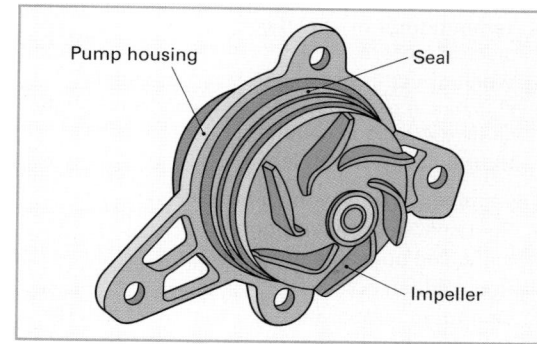

Fig. 3: Coolant pump

The coolant pump is normally driven by the crankshaft via a V-belt. It can also be driven by an electric motor **(see Chapter 11.2.5: Map-controlled cooling systems).** An electronically map-controlled pump drive dependent on the coolant temperature adapts the cooling-quantity requirement of the heat quantity to be dissipated by means of a variable pump delivery rate. The lower power input of the coolant pump results in reduced fuel consumption and emissions.

Fan. The function of the fan is to supply the radiator and engine compartment with an adequate amount of cooling air if the airstream proves insufficient, e.g. during slow driving or when the vehicle is stationary.

Rigid fan drive. The fan can be flanged to the water-pump shaft **(Fig. 1, Page 208)** or is driven by the crankshaft via a V-belt together with the water pump and possibly other accessories.

11

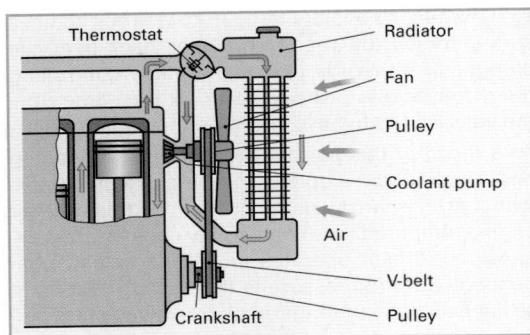

Fig. 1: Fan with rigid drive

Variable fan drive. This takes into account the fact that the required flow rate is very different, depending on the driving speed and the engine's operating state. For this reason, many engines use a fan that can be cut in or a variable-speed fan to save energy (in passenger cars approx. 2 kW...3 kW).

The following are used as variable drives:
- Temperature- or speed-controlled electric motors
- V-belt drives with temperature-dependent clutches, e.g. friction clutches, electromagnetic clutches, viscous couplings
- Hydraulically driven fans

Advantages of fans with variable drive:
- Reduced fuel consumption
- Increased utilisable drive power
- Reduced fan noise
- Operating temperature reached more quickly
- Constant operating temperature

Electrically driven fans (Fig. 2).
In this case, the fan is seated on the drive shaft of the electric motor.

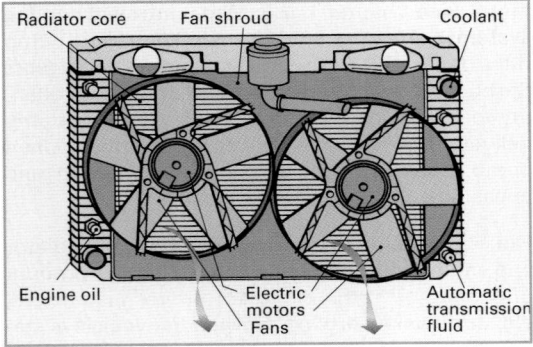

Fig. 2: Fan with electric drive

The electric motor can be switched on and off by a thermostatic switch around which coolant flows **(see circuit diagram, Page 514).** However, depend-

ing on the engine temperature, the speed of the fan motor can also be switched in multiple stages or even steplessly regulated.

Additional advantages of an electric-motor drive:
- The cooling-air flow of the fan can also be maintained after the engine is switched off to prevent overheating by post-heating.
- The radiator can be installed irrespectively of the position of the engine.

Viscous coupling (Fig. 3). This is another commonly used type of variable fan drive. In this case, the fan is bolted to the coupling body. Located in the fan hub of the viscous coupling are a working chamber and a supply chamber, separated by an intermediate disc. The drive disc, which is connected to the drive shaft driven by the V-belt, rotates in the working chamber. Silicone oil is used for power transmission. A bimetallically controlled valve enables the silicone oil to be exchanged between the supply chamber and the working chamber.

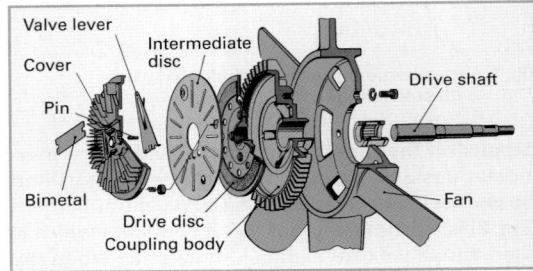

Fig. 3: Viscous fan coupling

The pump body mounted on the intermediate disc ensures that the oil is circulated from the working chamber to the supply chamber.

When the engine is cold **(Fig. 4),** the valve opening in the intermediate disc is closed off by the bimetallic valve and stops the circulation between the supply and working chambers.

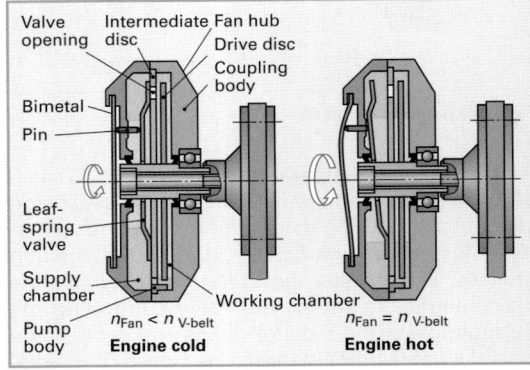

Fig. 4: Viscous coupling, operating states

The silicone oil is thus delivered by the pump body out of the working chamber and into the supply chamber. There the filling height increases in spite of the centrifugal forces. The drive disc is therefore no longer connected to the fan hub; the fan is disengaged. It continues to run only as a result of internal friction.

As the air (airstream) flowing through the radiator is increasingly heated, the bimetallic element at the front end of the viscous coupling heats up and opens the valve opening in the intermediate disc. The filling heights of the two chambers align themselves to each other. The power flow is steplessly established by the increasing friction of the silicone oil between the drive disc and the fan hub.

The pump body of the intermediate disc causes the silicone oil to be circulated on account of the constant speed differential between the drive disc (e.g. 2,000 rpm) and the coupling body (e.g. 1,900 rpm).

Hydraulic drive

The fan is driven by a hydraulic motor **(Fig. 1)**. A tandem hydraulic pump of the power-steering system usually serves to generate the oil pressure.

The first pump generates the pressure for the fan drive while the second pump generates the pressure for the power steering. A control valve clocked by the engine control unit can direct the quantity of oil, depending on engine temperature and vehicle speed, to the fan's hydraulic motor. In this way, the fan speed can be set with full variability by the control unit.

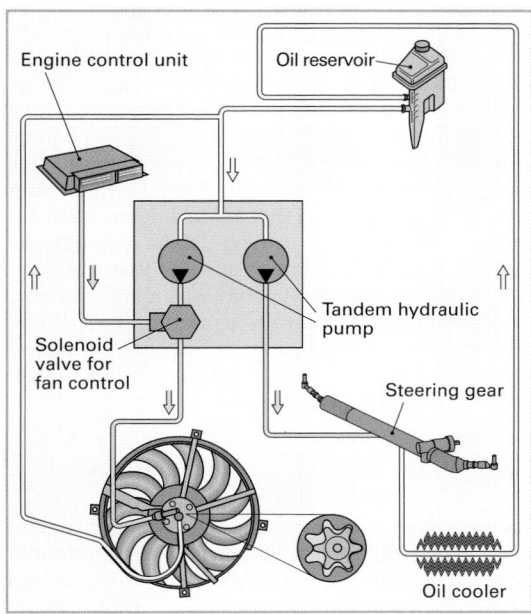

Fig. 1: Fan drive with hydraulic motor

Cooler assembly

> The cooler assembly is intended to dissipate the heat absorbed by the coolant to the air **(Fig. 2)**.

The following coolers are used:
- Radiator
- Intercooler
- ATF cooler
- Air-conditioning cooler
- Fuel cooler
- Gear-oil cooler
- EGR cooler
- Oil cooler

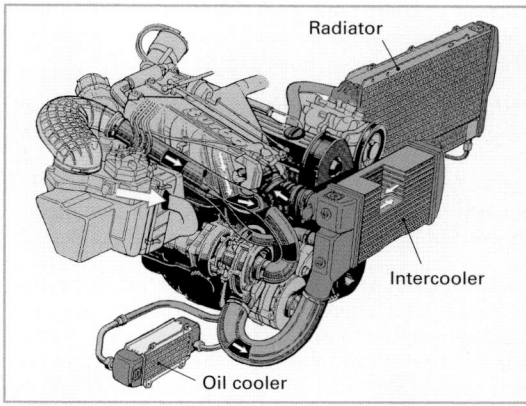

Fig. 2: Cooler arrangement

Coolant flows through the cooler assembly **(Fig. 3)** directly from top to bottom. This assembly consists of a bottom coolant tank and a top coolant tank, between which the radiator block is arranged with the radiator core.

The radiator block is formed from the:
- Coolant pipes
- Corrugated fins
- Pipe bases
- Side sections

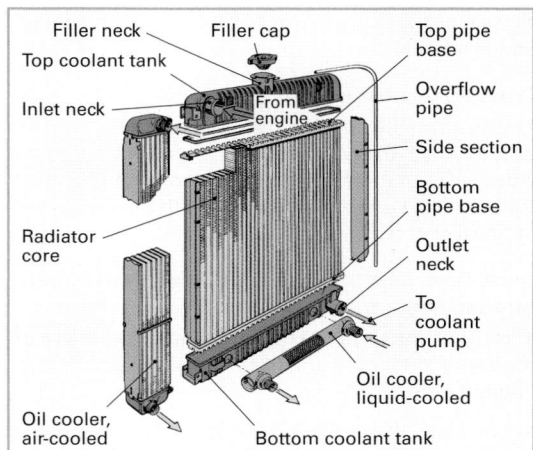

Fig. 3: Longitudinal-flow radiator

Top coolant tank. The inlet neck for the coolant flowing in from the engine is located here. This tank is often also fitted with a filler neck for replenishing the coolant. Attached to the tank is the overflow pipe, which diverts excess coolant and compensates any undesired overpressure or even vacuum pressure in the cooling system. The overflow pipe can be connected to an expansion tank so that the amount of coolant in the cooling system is permanently guaranteed. The filler neck is sealed by a filler cap **(Fig. 1, Page 211)**.

Bottom coolant tank. This tank features the outlet neck for the cooled coolant flowing to the engine. A drain plug or a drain cock can be fitted.

Today, radiator tanks are usually made from glass-fibre-reinforced plastic, but can also be made from light alloy or copper-zinc alloys. An elastomer gasket is inserted between the pipe base and the coolant tank for sealing purposes. The pipe base is connected by way of bead to the coolant tank.

Radiator core. A system of pipes and fins in the core of the radiator block creates as large a cooling surface as possible to enable the cooling air to extract a large amount of heat from the coolant. The bottom coolant tank can, usually in vehicles with automatic gearboxes, incorporate an oil cooler for the transmission fluid. Occasionally an engine-oil cooler may also be fitted, for example as a side section.

Radiator hoses. The radiator is resiliently connected to the engine normally by way of heat-resistant radiator hoses with fabric inserts. If greater gaps need to be bridged, e.g. rear engine and radiator at the front of the vehicle, metal or plastic tubes are also used.

Radiator mounting. The radiator must be installed in the vehicle so that it is protected against jolts and vibrations. It is therefore resiliently connected to the chassis or the body by way of rubber-metal connection elements.

Cross-flow radiator

This is a radiator design that is commonly used today, in which the coolant tanks are mounted on the sides of the radiator block. The coolant flows horizontally from one side to the other in the cross-flow radiator **(Fig. 1)**.

Cross-flow radiator with high- and low-temperature sections

If the inlet and outlet of the cross-flow cooler are on the same side, the coolant tank on this side is subdivided **(Fig. 2)**.

In this case, the coolant flows through the radiator in the top section for example to the right and in the bottom section in the opposite direction to the left.

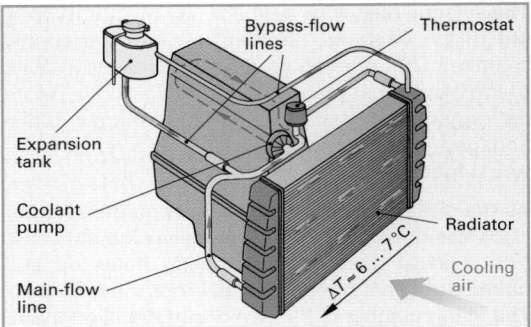

Fig. 1: Cooling system with cross-flow radiator

The coolant must flow twice through the width of the radiator. This improves the cooling effect. Furthermore, these radiators require a lower installation height.

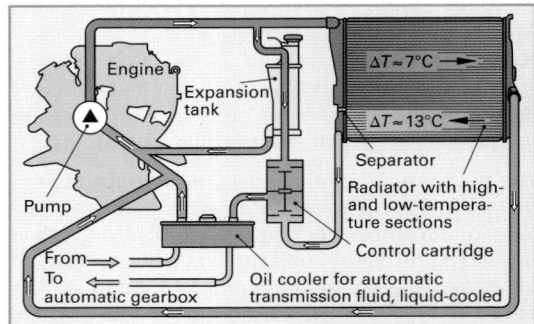

Fig. 2: Cooling system with cross-flow radiator with high- and low-temperature sections

Outlet connections arranged at different heights in conjunction with the installation of a separator allow for two different temperature zones. In the upper radiator area, there is a high-temperature zone with a temperature drop of roughly 7 °C for engine cooling. In the lower radiator area, it is possible to utilise a low-temperature zone with an overall temperature drop of roughly 20 °C via an additional heat exchanger for intensive cooling of the gearbox oil. Suitable thermostat control ensures that the gearbox oil is quickly heated by the coolant heat from the small cooling circuit via the expansion tank. When the operating temperature is reached, intensive cooling of the gearbox oil is guaranteed by coolant from the low-temperature section.

Filler cap

This is equipped with a pressure relief valve and a vacuum valve **(Fig. 1, Page 211)**.

Pressure relief valve. This serves to seal the cooling system gas-tight. Depending on how it is configured by the manufacturer, the pressure relief valve opens only at an overpressure in the cooling system of approx. 0.5 bar ... 2 bar.

The coolant temperature can increase on account of this overpressure to approx. 120 °C without the coolant boiling.

Vacuum valve. When the coolant cools, a vacuum pressure occurs due to the accompanying volume reduction in the cooling system. This can be compensated by the opening vacuum valve. This action prevents the radiator from buckling or the radiator hoses from contracting.

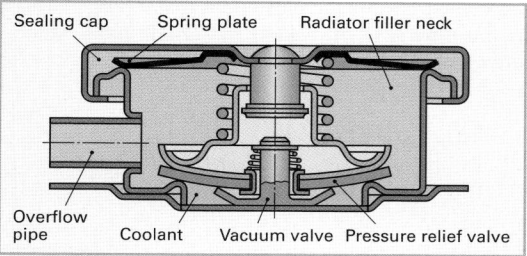

Fig. 1: Filler cap

11.2.5 Map-controlled cooling systems

Map-controlled temperature control reduces fuel consumption and emissions without compromising the engine's power output and service life. At the same time interior heating is improved and weight is reduced.

Features of this cooling system:

- The cold-running phase, which delivers unfavourable fuel-consumption and emission figures, is shortened by the fact that the engine and the catalytic converter are quickly heated up.
- The engine temperature is specifically increased to up to 120 °C in uncritical driving states, e.g. part load. Thanks to the improved viscosity of the engine oil, this results in reduced internal engine friction and significantly improved mixture formation.
- The engine temperature is reduced in critical driving states, e.g. full load, in order to avoid overheating of the engine components, retarded ignition due to knock control and charge losses.

11.2.6 Map-cooling components

Electronically controlled thermostat (Fig. 2).

An electric heating resistor alters the temperature in the expansion element. The heating resistor is activated by the ECU as a function of the input variables. It heats the expansion element in addition to the coolant heat. This heating gives rise to a larger stroke and thereby causes the thermostat to open more. This action reduces the coolant temperature.

The following input variables can be taken into consideration:

- Load
- Air temperature
- Coolant volume
- Air conditioning
- Driving speed
- Coolant temperature
- Interior heating

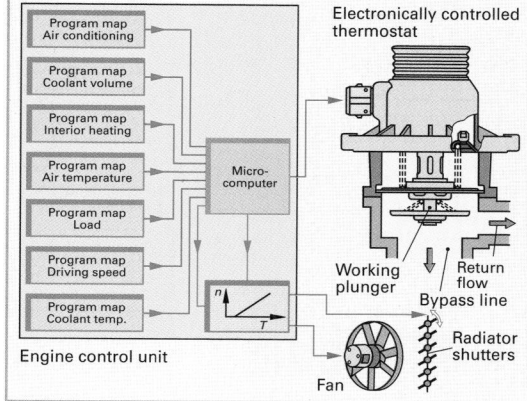

Fig. 2: Electronically controlled thermostat

Radiator shutters. Located in front of the fan is a set of shutters which is actuated electrically or via an expansion element. During cold starting the shutters remain closed in order virtually to excluded the possibility of cooling by the airstream. This enables the engine to reach operating temperature more quickly. In diesel vehicles the closed shutters also damp the loud combustion noises generated during cold starting. The shutters are opened as engine temperature increases.

Electrically driven coolant pump. Pumps of this type have the following advantages of V-belt-driven pumps:

- Coolant flow rate can be regulated independently of engine load and engine speed.
- Lower power consumption of on average approx. 200 W compared with 2 kW for a permanently driven pump.

CONTINUED FROM PAGE 211

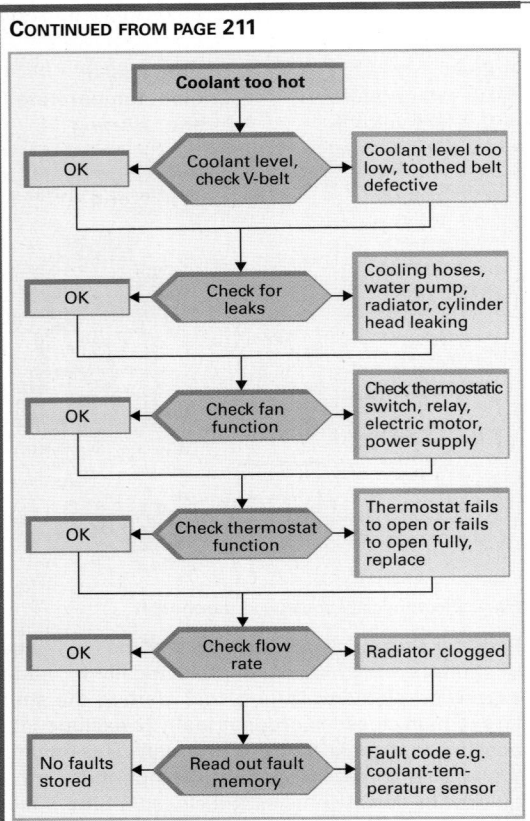

Fig. 1: Troubleshooting chart for cooling system

Leaking radiators must be repaired straight away because the damaged area will increase in size and this will be accompanied by the loss of coolant. Only older brass radiators can be repaired by soldering. In the case of aluminium radiators and very small leaks, sealant can be added to the coolant. The very fine sealant particles are carried by the flow of coolant to the area of the leak and settle there. After the repair has been completed, the coolant should first be brought up to operating temperature.

Replenishing coolant. Bear in mind that cold coolant may only be poured into a hot engine when the engine is running. The cold coolant must be poured in slowly so as to avoid dangerous stresses in the engine block and in the cylinder head.

Checking antifreezing property of coolant. The antifreezing limit can be determined with a measuring instrument (refractometer). It should be below – 25 °C. Replenish the system with manufacturer-recommended anti-freeze only.

Leak test. This test includes a visual inspection for leaks as well as an inspection of the hoses for porosity, brittleness and rodent bites. The engine must be at normal operating temperature when a leak test is conducted with a leak tester **(Fig. 2)**. This is the only way of picking out small leaks which are caused by thermal expansion.

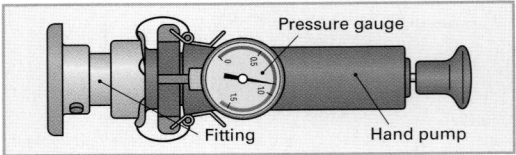

Fig. 2: Pressure tester

The tester is screwed on in place of the sealing cap. The hand pump is used to generate an overpressure which must be maintained for a period of two minutes. If the pressure drops, this is an indication of a leak, which must be located and repaired. This same tester can be used to check the opening pressure of the sealing cap.

CO tester. This is used to draw in air from the cooling system. If the blue test fluid turns yellow, the cylinder head or the cylinder-head gasket is leaking.

Thermostat. The function of the thermostat is tested when it is removed in a water bath. The water is slowly heated. The point at which the thermostat opens is checked with a thermometer.

Map-controlled thermostats. They are checked when they are removed by means of an actuator diagnosis with the system tester.

Clogged radiators must be cleaned. The use of mains water with a high lime content results in the build-up of scale, above all in the radiator. Scale obstructs the throughflow of coolant and impairs the dissipation of heat. For cleaning purposes, a chemical cleaning agent is usually added and the radiator is then thoroughly flushed through.

Fan with viscous coupling. This may only be stored upright after it has been removed.

REVIEW QUESTIONS

1　What are the advantages and disadvantages of air and liquid cooling respectively?

2　How can fans which can be cut in be driven?

3　What types of radiator are used?

4　What are the advantages of a map-controlled thermostat?

11.3 Crankshaft drive

11.3.1 Piston

Functions:

- Provides a moving seal for the combustion chamber against the crankcase.
- Absorbs the gas pressure generated during combustion and transmits it via the connecting rod as torsional force to the crankshaft.
- Transmits the heat given off by the combustion gases to the piston crown for the most part to the cylinder wall.
- Gas-exchange control in two-stroke engines.

Stresses

Piston force. The compression pressure acting in spark-ignition engines on the piston crown of up to 60 bar produces for a piston diameter of 80 mm a piston force of up to 30,000 N. The pin bosses are subjected by the gudgeon pin to a surface pressure of up to 60 N/mm^2.

Lateral force. The piston is alternately pressed against the cylinder wall. A lateral pressure of up to 0.8 N/mm^2 acts on the piston skirt in the process. This causes piston slap and therefore noises. Piston slap is reduced by means of a small piston clearance, a large skirt length and axial offset of the gudgeon pin.

Axial offset means that the gudgeon-pin axis is moved roughly 0.5 mm to 1.5 mm from the piston centre towards the pressure-loaded side **(Fig. 1)**. In the process, the piston changes its contact face already with the slowly building-up compression pressure before TDC and not just with the suddenly occurring combustion pressure shortly after TDC. Even the crankshaft can be moved from the centre.

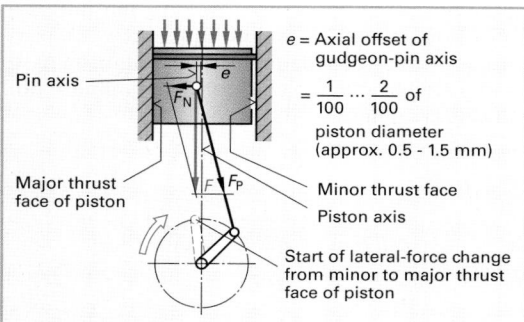

e = Axial offset of gudgeon-pin axis

$$= \frac{1}{100} \cdots \frac{2}{100} \text{ of}$$

piston diameter (approx. 0.5 - 1.5 mm)

Pin axis

Major thrust face of piston

Minor thrust face

Piston axis

Start of lateral-force change from minor to major thrust face of piston

Fig. 1: Forces on piston with axially offset pin

Friction force. The piston skirt, piston-ring grooves and pin bosses are subject to friction load. Friction and wear can be reduced by an appropriate choice of material, careful machining of the sliding surfaces and adequate lubrication.

Heat. Temperatures of up to 2,500 °C in the cylinder are produced by the combustion of the fuel-air mixture. A large proportion of the heat is given off to the cooled cylinder via the piston crown, the piston-ring belt and the piston rings. Even the lubricating oil removes heat. In spite of this, light-alloy pistons are still subject to an operating temperature of 250 °C to 350 °C at the piston crown and of up to 150 °C at the piston skirt **(Fig. 2)**.

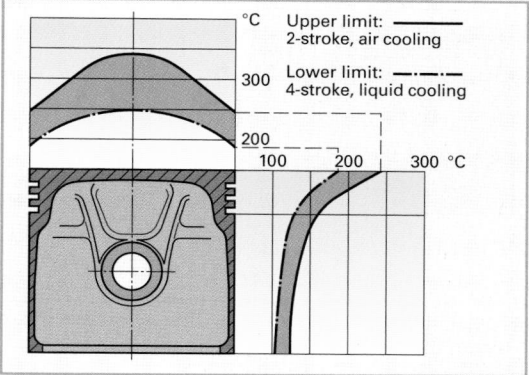

°C   Upper limit: ———— 2-stroke, air cooling

Lower limit: —·—·— 4-stroke, liquid cooling

Fig. 2: Operating temperatures at piston

Heating causes the material to expand, which in turn may cause piston seizure. Suitable shaping (e.g. tapered ring belt, oval piston cross-section) must be adopted to compensate the different thermal expansion at the various points of the piston.

Cold clearance. In cold conditions, therefore, as shown in **Fig. 3**, the clearance between piston and cylinder varies in extent. Thus, for example, at the skirt end the clearance is 0.088 mm in the pin direction while it is only 0.04 mm at 90° to the pin direction. This smallest clearance is the installation clearance because the diameter of the piston is at its largest at this point. The piston crown and ring

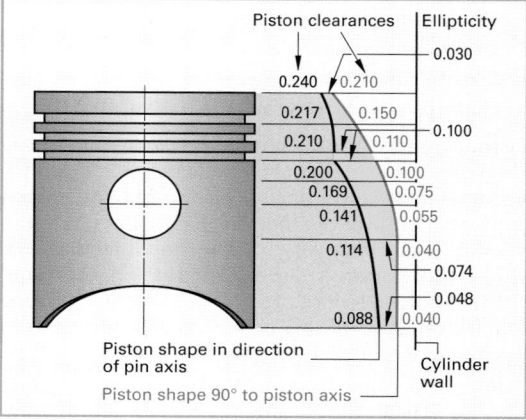

Piston clearances   Ellipticity

	Piston clearances	Ellipticity
		0.030
	0.240 0.210	
	0.217 0.150	
	0.210 0.110	0.100
	0.200 0.100	
	0.169 0.075	
	0.141 0.055	
	0.114 0.040	
		0.074
		0.048
	0.088 0.040	

Piston shape in direction of pin axis

Piston shape 90° to piston axis

Cylinder wall

Fig. 3: Piston shapes with installation clearances

belt are exposed to high temperatures and therefore expand more noticeably than the skirt such that the piston clearances must vary over the entire piston length. This is achieved by making the piston oval and not round as well as cambered or tapered but not cylindrical **(Fig. 1)**. The clearance differences, e.g. 0.088 mm − 0.04 mm = 0.048 mm in **Fig. 3, Page 213**, indicate the ellipticity (ovality) in the corresponding piston area.

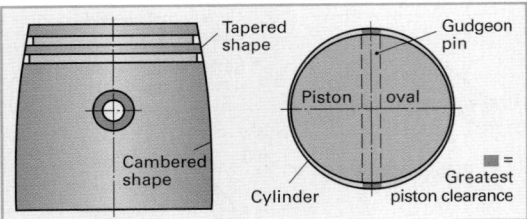

Fig. 1: Piston shape when cold

The installation clearance is the difference between the cylinder diameter and the greatest piston diameter.

Hot clearance. At normal operating temperature the piston assumes an approximately cylindrical shape and the piston clearance is reduced. The size of the hot clearance cannot be specified because the piston is deformed by the forces acting on it. Spare clearance must be available at any rate even when the permissible temperatures are exceeded for a brief period.

Piston materials.

The materials must demonstrate the following properties on account of the different types of stress and strain to which the piston is subjected:

- Low density (smaller acceleration forces)
- High strength (even at higher temperatures)
- Good thermal conductivity
- Low thermal expansion
- Low frictional resistance
- High resistance to wear

Pistons made from aluminium-silicon alloys are used on account of their low density ($\varrho \approx 2.7$ kg/dm³) and high thermal conductivity. The higher the silicon content, the lower the thermal expansion and wear, although machinability during manufacture becomes more difficult. Generally speaking, the material AlSi12CuNi suffices for these purposes; pistons made from AlSi18CuNi or AlSi25CuNi are used in the event of higher thermal loads.

Pistons are manufactured by gravity die-casting; pistons which must withstand particularly high pressures are pressed (forged).

Design and dimensions

A piston **(Figs. 2 and 3)** is made up of the following different sections:

- Piston crown
- Piston-ring belt with fire land
- Piston skirt and
- Pin bosses

Piston crown. This is either flat or slightly curved inwards (dished) or outwards (domed). The compression space can be partly transposed into the piston by the integration of a combustion recess in the piston crown. The shape of the piston crown is also influenced by the shape of the compression space and the arrangement of the valves.

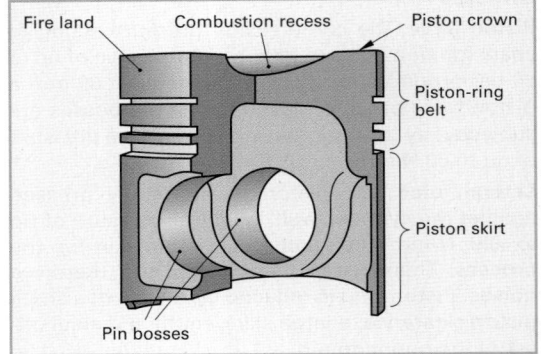

Fig. 2: Piston design

Fire land. This is intended to protect the uppermost piston ring against excessive heating. The markedly rounded transition into the piston-ring belt on the inside of the piston strengthens the piston crown and encourages heat dissipation. The piston skirt serves to guide the piston in the cylinder. It transmits the side forces to the cylinder wall. The pin bosses transmit the piston force to the gudgeon pin. The compression height influences the compression ratio; an adequate skirt length helps to keep piston slap in the cylinder to a minimum during side changing.

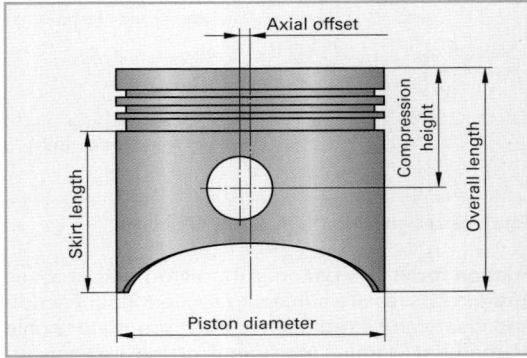

Fig. 3: Main piston dimensions

Piston types

Single-metal pistons. These are cast or pressed full-skirt pistons for spark-ignition and diesel engines and pistons for two-stroke engines **(Fig. 1)** which only consist of one material, e.g. AlSi alloy. In the case of higher combustion pressures, e.g. in a diesel engine, the piston crown, piston-ring belt and transition to the skirt are thicker in design.

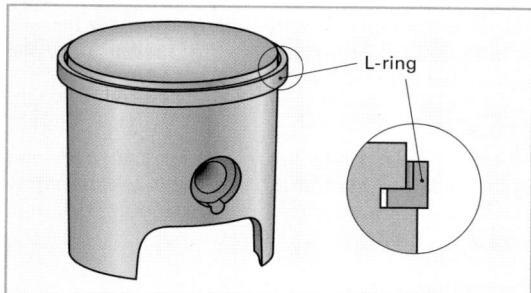

Fig. 1: Single-metal piston (two-stroke piston)

Steel-banded pistons. These have steel inserts, e.g. ring belts, steel struts **(Fig. 2)** or segment struts, which are cast into the light alloy. When the piston is heated, they can hinder the thermal expansion or direct it in a specific direction (bimetal effect). Good running performance and noise reduction together with a small installation clearance are achieved by better adaptation of the piston when cold and a normal operating temperature.

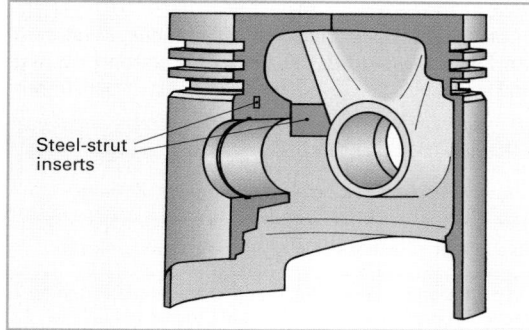

Fig. 2: Steel-banded (steel-strutted) piston

Bimetal effect (Fig. 3). The different expansion of steel and light alloy under the influence of heat causes the steel strut cast in the piston to bend, which in turn results in an increase in the piston diameter. Through the installation of steel struts in the area of the pin bosses, elongation is mainly directed towards the gudgeon-pin axis, in the course of which the piston diameter barely changes in the thrust direction (vertically to the pin axis). Thermal expansion can be compensated by oval machining of the piston in the area of the pin bosses.

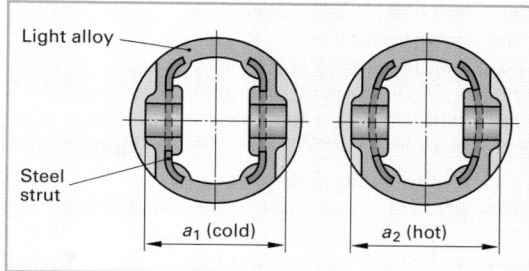

Fig. 3: Bimetal effect on steel-strutted piston

Ring-carrier pistons (Fig. 4). Because of the greater thermal and pressure loads, especially in the area of the fire land, a ring carrier is cast in for the first piston ring, thereby reducing wear by lining the groove. Additional anodising of the grooves also increases wear protection. Heat can also be removed via a cast-in cooling duct which is supplied with oil through an inlet hole. A brass bushing reinforces the pin boss.

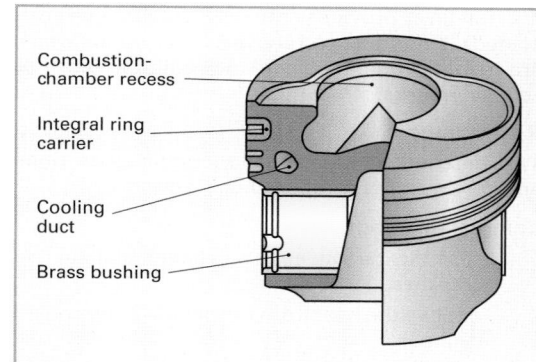

Fig. 4: Ring-carrier piston with cooling duct

Running-surface protection

The engine can be subjected to normal load already during the running-in period thanks to protective layers on the piston running surfaces because friction is reduced. In the event of temporarily impaired lubrication, piston seizure is avoided by limp-home properties. Most layers also offer anti-corrosion protection.

Tin layer. In a tin-salt bath tin is deposited on the light-alloy piston. Despite being low in thickness, the tin layer delivers good sliding and limp-home properties.

Lead layer. This has the advantage of a higher melting point (327 °C) compared with that of tin (232 °C) and is therefore the most commonly used layer.

Layer with main advantage of corrosion protection

Anodising layer. This provides high abrasion resistance but no limp-home properties. Anodised piston crowns have proven themselves under high thermal loads and as a means of corrosion protection.

Layers with main advantages of wear protection and corrosion protection.

Plastic layer. The skirt is coated with a mixture of resin, polyamide and wear-protection substances as well as graphite, Teflon, molybdenum sulphide or iron. This layer can be recognised from its gold-coloured appearance.

Graphite layer. Diesel pistons are provided with a graphite or molybdenum layer. The graphite layer is bound with phenolic resin, the molybdenum layer with epoxy resin. It offers good limp-home properties and outstanding protection against wear.

Iron layer. In the case of a "ferrocoated piston", the surface of the skirt is copper-plated and then coated with a 0.03 mm thick iron layer, which is roughly as hard as a chromium layer. A thin tin layer over the iron layer provides corrosion protection. Ferrocoated pistons are used above all in cylinders which are manufactured in accordance with the Alusil or Lokasil process.

Piston rings

Two different types are used: compression rings and oil-control or scraper rings **(Table 1)**.

Compression rings. These ...

- ... seal the piston in the cylinder against the crankcase.
- ... remove the heat from the piston to the cooled cylinder.

Oil-control rings. These ...

- ... scrape excess lubricating oil off the cylinder wall.
- ... return this lubricating oil to the oil pan.

Properties

Piston rings must be elastic and must not change their shape permanently when slipped over the piston and when compressed to their nominal dimension. The contact pressure against the cylinder wall is additionally intensified during operation by the gas forces behind the rings.

Materials

Normal piston rings are made of cast iron and quenched and tempered cast iron while highly stressed piston rings are made of nodular graphite iron or high-alloy steel.

Protective layers

Phosphate or **tin** improve the sliding properties and simplify running in.

Molybdenum is applied as an all-over coating to the piston-ring liner. Because of its good thermal conductivity, high melting point of 2,620 °C and good limp-home properties, the molybdenum layer prevents rings from seizing to a large extent.

Chromium-plated rings are resistant to corrosion and wear. These are mostly used as the uppermost piston ring because it is this ring which has the worst lubrication.

In order to reduce frictional resistances, small two-stroke engines often use only one piston ring, e.g. an L-ring.

Table 1: Piston rings				
Piston-ring shape		**Abbre-viation**	**Installation instruction**	**Purpose of shape**
Cross-section	**Designation**			
	Plain ring (compression ring)	P	Possible in both directions	Easy to manufacture
	Taper-face ring	TF	Ring flank designated with "Top" in direction of piston crown	Acceleration of running-in process (usually in uppermost groove)
	Keystone ring (half)	K	Tapered ring flank in direction of piston crown	Prevention of sticking in groove
	L-ring	LR	Large internal ring dia. in direction of piston crown or top ring edge = piston-crown edge	Intensified contact pressure due to combustion gases
	Stepped ring	S	Turned-out angle in direction of skirt end	Additional oil-scraping effect
	Slotted oil-control ring (normal)	SI	Possible in both directions	Scraping effect with oil passage to piston interior
	Slotted oil-control ring with spiral-type expander (bevel-edged ring)	SP	Possible in both directions	Higher contact pressure, better scraping effect

Gudgeon pin and gudgeon-pin circlip

Gudgeon pin. This connects the piston with the connecting rod. Its rapid movement back and forth together with the piston necessitates that it be low in mass.

The gudgeon pin is provided with a bore to reduce the acceleration forces. The abruptly alternating load necessitates that the pin material demonstrates toughness and fatigue strength under complete stress reversal. The small clearance in the pin bosses and in the connecting-rod eye calls for high surface quality and geometrical accuracy. The poor lubrication conditions call for high surface hardness in order to reduce wear.

Important shapes of gudgeon pins **(Fig. 1)**:

- Pin with cylindrical through-bore (normal shape)
- Pin with tapered widened bore ends (weight reduction)
- Pin with a closed bore in the middle or at one end to avoid scavenging losses in two-stroke engines

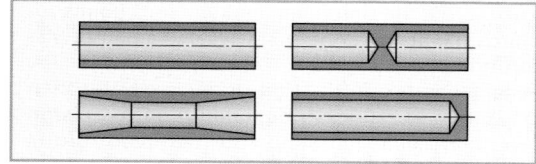

Fig. 1: Gudgeon-pin shapes

Materials for gudgeon pins are case-hardening and nitriding steels.

The case-hardening steel Ck 15 is used for normal stress applications while the case-hardening steels 17Cr3, 16MnCr5 and 31CrMo12 are used for higher-stressed spark-ignition and diesel engines.

Nitriding steels such as 31CrMo12 and 31CrMoV9 are used for maximum stress applications and greatest surface hardness.

Gudgeon-pin circlips (Fig. 2). If the gudgeon pin is mounted floating in the piston bosses, the circlips are intended to prevent the pin from moving and damaging the cylinder wall.

The circlips are radially sprung steel rings (Seeger circlip ring, wire circlip) which are inserted into the corresponding grooves in the pin bosses. To facilitate installation, Seeger circlip rings have eyelets with holes for the installation pliers while wire circlips have their ends bent into a hook shape.

Piston damage

Piston damage can be incurred if the pistons are incorrectly handled during assembly and installation in the engine, e.g. cylinder deformation due to unevenly tightened cylinder-head bolts, excessively tight fitting of the gudgeon pin in the connecting-rod eye, irregularities on the cylinder liners.
Piston damage may also be caused by the following:

- Auto-ignition: Use of fuel with too low an octane number, spark plugs with an incorrect thermal value.
- Knocking combustion: Unsuitable fuel, excessively high compression ratio due to the use of an incorrect cylinder-head gasket, excessively advanced ignition point, excessively lean mixture, engine overheating.
- Combustion malfunctions in diesel engines due to ignition lag, incomplete combustion and dribbling nozzles.
- Inadequate or no lubrication.
- Overheating due to inadequate engine cooling, retarded ignition and excessive fuel supply.

Selected damage patterns

1. Piston scoring (dry-running scoring, **Fig. 3**) concentrated on the piston skirt. The piston skirt is scored all round, the areas of scoring recognisable by the dark discolouration; the ring belt can be damaged by the highly lubricated piston material. Causes: The lubricant layer between piston and cylinder is removed by excessive engine overheating.

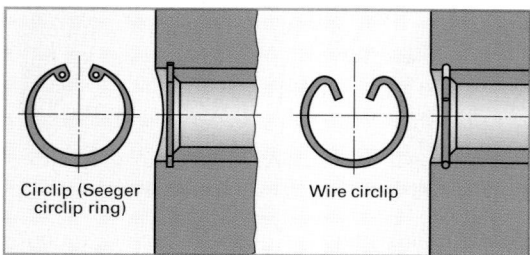

Circlip (Seeger circlip ring)   Wire circlip

Fig. 2: Gudgeon-pin circlips

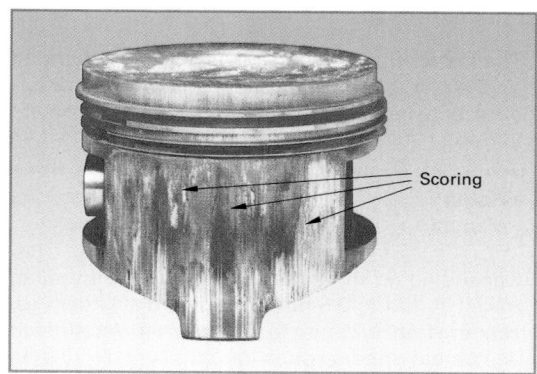

Scoring

Fig. 3: Dry-running scoring on piston skirt

This can be put down to faults in the cooling system (loss of coolant, deposits in the cooling ducts, defective thermostat), retarded ignition, incorrect oil grade or insufficient oil.

2. Hole in the piston crown (Fig. 1). A part of the material has melted out, a greater part has cut through downwards in a funnel shape (recognisable on the underside of the piston crown). The skirt and the ring belt are usually undamaged (no traces of scoring).

Causes: Within a few seconds the piston crown is locally heated by auto-ignition up to melting point. The combustion gases remove the softened mass, reducing the strength in this area in such a way that the combustion pressure forces through the remaining crown inwards in a funnel shape.

Fig. 1: Hole in piston crown

WORKSHOP NOTES

Checking piston diameter and installation clearance

Piston manufacturers supply ready-to-install pistons with the largest skirt diameter (piston diameter) marked in mm, e.g. 84.00, on the piston crown **(Fig. 2).** The diameter is measured at the end of the skirt, vertically to the pin axis.

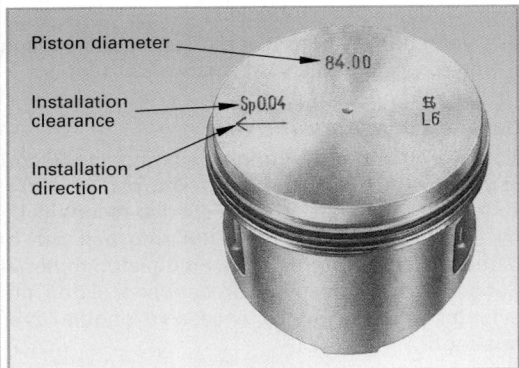

Fig. 2: Installation data on piston crown

The installation clearance, which is also stamped on the piston crown, e.g. 0.04, denotes the difference between the cylinder and piston diameters in mm at 20 °C.

Piston diameter	+ Installation clearance =	Cylinder diameter

Depending on the type, four oversizes ascending from 0.5 mm to 0.5 mm (for moped cylinders ascending from 0.25 mm to 0.25 mm) are determined as internal-grinding sizes for each cylinder diameter. Accordingly, there are four oversize pistons.

Installing the piston rings

Piston rings are supplied already installed in the piston. If individual piston rings are to be inserted, it is important to use the correct ring type and ensure that the ring flank marked "Top" points towards the piston crown. Always use a pair of piston-ring pliers **(Fig. 3).**

Axial clearance in the ring grooves greater than 0.025 mm to 0.04 mm depending on the piston type may cause jamming and "pumping" of the piston rings, i.e. if the ring groove is heavily worn the rings act like pumps supplying oil to the combustion chamber. Even when the piston is installed, piston rings should still have a gap clearance of 0.2 mm to 0.3 mm otherwise their spring effect could be impaired and the rings themselves could break. Gas losses occur if the clearance is too large. It is possible to check the clearance with a gauge by inserting the rings without the piston on a trial basis in a cylinder bore.

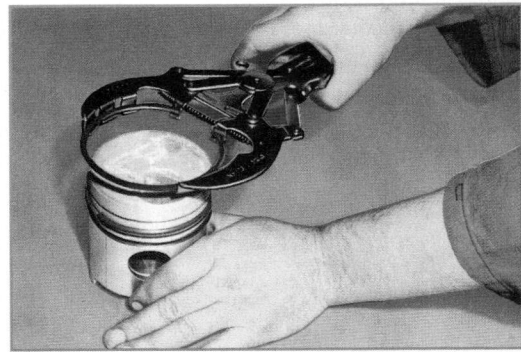

Fig. 3: Installing piston ring with special pliers

Details on installing the gudgeon pin are set out in the section entitled "Connecting rod" (11.3.2).

CONTINUED FROM PAGE 218

Fitting the gudgeon-pin circlips

Gudgeon-pin circlips may only be fitted with installation pliers so as not to damage the pin bore. Proceed with care when compressing the circlips so that the circlips are not permanently deformed and do not lose their initial tension and tight fit. The circlips are correctly installed when their openings are located precisely at the top or at the bottom in the gudgeon-pin boss. To check that the circlip is tightly seated in the groove, subject the circlip to a twisting test with a screwdriver - significant resistance must be felt during this test. Do not reuse old circlips once they have been removed.

Installing the piston

If necessary, clean the piston with installed rings thoroughly with clean petroleum spirit, dry with compressed air and then oil all the sliding surfaces well. Dirt can cause piston damage! Make sure there is sufficient clearance between the gudgeon-pin bosses and the connecting-rod eye so that the piston is not pressed on one side against the cylinder wall.

The connecting-rod eye and connecting-rod bearings must be parallel to the axis. Carefully clean the cylinder block and cylinders before installing the pistons. When inserting the thoroughly oiled piston into the cylinder, compress the piston rings with a ring sleeve **(Fig. 1)** to protect them from being damaged.

The ring joints must be arranged offset by 180° in each case on the piston circumference beforehand.

When the piston is correctly installed, there must be an equal space on both sides at the piston crown in the pin direction between the piston and the cylinder. This can be checked using a feeler gauge, which at the same time serves to check that the connecting rod is correctly squared.

Fig. 1: Installing the piston in the cylinder

Piston installation direction

Pistons with their pin axes offset from the middle must be installed in such a way that the gudgeon pins are offset on the major thrust face. In the case of these pistons and also pistons with special piston-crown shapes, an arrow symbol which usually points in the direction of travel is marked in the piston crown to indicate the installation direction **(Fig. 2, Page 218).** The arrow symbol can also be replaced by the designation "Front" or a crankshaft symbol (on transversely mounted engine or rear engines).

11

11.3.2 Connecting rod

Functions

- Connects the piston to the crankshaft
- Converts the linear motion of the piston into rotary motion of the crankshaft
- Transmits piston force to the crankshaft, where it generates torque

Stresses

- Pressure forces in the longitudinal direction as a result of the gas pressure on the piston crown
- Acceleration forces in the form of tensile and pressure forces in the longitudinal direction as a result of constantly changing piston speed
- Bending forces in the connecting-rod shank as a result of oscillating motion about the gudgeon-pin axis
- Buckling stress on account of the large pressure forces

In the interest of minimising the inertia forces, the mass of the connecting rod should be as small as possible.

Connecting-rod materials

Connecting rods (Fig. 1) are mainly manufactured from alloyed tempering steel which is drop-forged or from alloyed steel powder as a sinter forging **(Fig. 2)**. Sinter-forged connecting rods have better mechanical properties than their drop-forged counterparts. Their cross-sections can therefore be smaller and with it their weight lower. Weight tolerances practically never arise. The big end acquires its ultimate shape already during the manufacturing process. It is not sawn through, rather it is first slit with a laser and then cracked with a wedge. The connecting rod and the big-end bearing cap have the same granu-

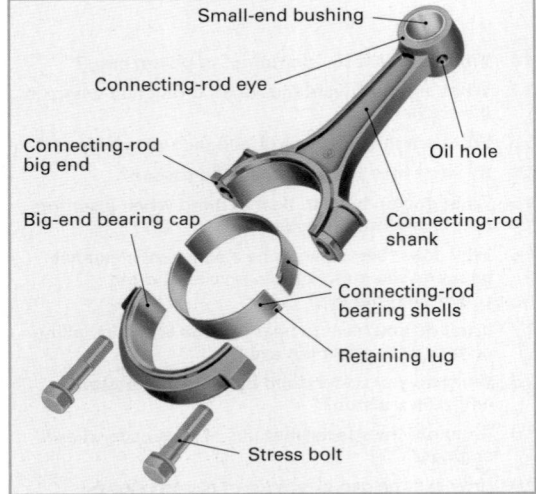

Fig. 1: Connecting rod with bearing shells

lar fracture surface and ensure that the big-end bearing cap is precisely seated on the big end during assembly. Fit bolts are no longer needed.

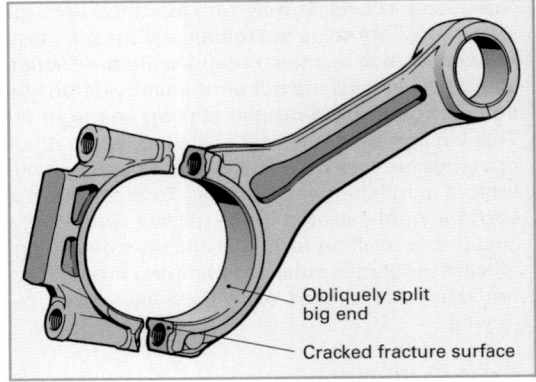

Fig. 2: Cracked connecting rod, obliquely split

Design

Connecting-rod eye. This accommodates the gudgeon pin. If the gudgeon pin is float-mounted, a bushing, usually made from a copper alloy (CuPb-Sn), is pressed into the connecting-rod eye. If the gudgeon pin is to be fixed in the connecting-rod eye with a shrink fit, it is shrunk directly into the eye.

Connecting-rod shank. This connects the connecting-rod eye to the big end. In order to increase its resistance to buckling, its cross-section is usually a double-T-profile section.

Connecting-rod big end. Together with the big-end bearing cap, this encloses the connecting-rod bearing, which is designed as a split plain bearing. The big-end bearing cap is usually secured to the big end with stress bolts.

Mounting of connecting rod on crankshaft. In the same way that the crankshaft is mounted in the crankcase, this mounting is achieved in multilayer bearing shells **(Fig. 1, Page 224)**. These shells are secured against sliding and turning by retaining lugs.

Bearing clearance. This is specified by the manufacturer and can be determined by measuring the connecting-rod bearing and the crankpin. When Plastigage is used **(Fig. 2, Page 224)**, each bearing must be measured individually.

Special connecting-rod designs

Trapezoidal connecting rods. The bottom half of the connecting-rod eye, which must absorb the combustion pressure, is wider while the top half remains narrower due to the lower levels of stress to which it is exposed. This arrangement creates the connecting-rod eye's trapezoidal shape.

Obliquely split connecting rods. Because of the higher pressures in diesel engines, the connecting-rod big end must be designed more sturdily with the result that its size outstrips the cylinder diameter. Withdrawal through the cylinder is only made possible by an obliquely split big end.

Single-piece connecting rods. In two-stroke single-cylinder engines the connecting-rod big end is often not split and the crankshaft must therefore be assembled from individual pieces. Antifriction bearings can be used instead of plain bearings.

Lubrication. The connecting-rod bearing is lubricated by engine oil which is supplied through a hole to the crankpin from the crankshaft journal. The small-end bushing with gudgeon pin is usually adequately lubricated by splash oil (oil hole in the connecting-rod eye, **Fig. 1, Page 220**).

WORKSHOP NOTES

Checking the weight tolerance. When connecting rods or pistons are replaced, it is important to ensure that the replacement parts are identical in weight so that unbalanced inertia forces do not hinder engine running. The permissible weight tolerance of the part sets (piston + connecting rod) is specified by the manufacturer. Minor excess weights are ground off from the connecting-rod big end.

Assembling piston and connecting rod

Floating mounting. If the gudgeon pin is mounted so that it floats in the small-end bushing and in the piston, it is important to ensure that the gudgeon-pin circlip is correctly fitted (see Page 218).

Close sliding or interference fit. If the gudgeon pin is to be subject to a close sliding or interference fit in the piston, the piston is heated prior to assembly, for example in an oil bath, to approximately 80 °C. For inserting the oiled, cold pin into the piston, the pin bosses and small-end bushing are centred by a guide pin so that the gudgeon pin is quickly pushed in and is not prematurely seated tightly in the piston.

Shrink fit. If the gudgeon pin is to be installed in the connecting rod with a shrink fit, the following procedure is adopted **(Fig. 1)**:

- Heat the connecting rod to approx. 280 °C to 320 °C (temperature check necessary).
- To facilitate installation, chill the gudgeon pin with dry ice or in a chest freezer to reduce its diameter.
- Carefully centre the piston on a form base with a stop mandrel.

- Place the heated connecting rod in a well centred position on the lower pin boss.
- Insert the cold gudgeon pin through the upper pin boss up against the connecting rod.
- Slide in the pin quickly in one operation up to the stop (limit position) on the stop mandrel.

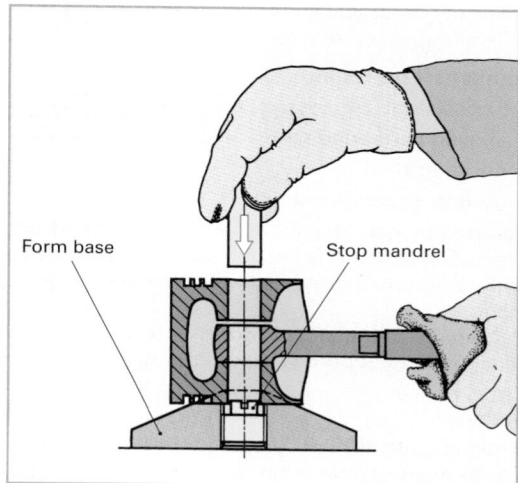

Form base   Stop mandrel

Fig. 1: Piston installation with shrink-fit gudgeon pin in the connecting rod

Assembling connecting rod and crankshaft. The connecting rod must have lateral clearance on the crankpin so that differences in the thermal expansion of the crankshaft and engine block can be compensated. The connecting-rod bolts, usually stress bolts, are tightened with a torque wrench to the manufacturer's specified torque.

REVIEW QUESTIONS

1 To which stresses is the connecting rod exposed?

2 What are the functions of connecting rods?

3 What are the advantages of the fracture technology in sinter-forged connecting rods?

4 How is the connecting rod mounted on the crankshaft and how is it lubricated?

5 What do you understand by a trapezoidal connecting rod?

6 How is the gudgeon pin mounted in the connecting-rod eye?

7 Why should the weight tolerance of the part sets (piston + connecting rod) not be exceeded?

8 Why should a guide pin be used when the piston and connecting rod are assembled?

11.3.3 Crankshaft

Functions

- Generates a rotary force and thus a torque from the connecting-rod force.
- Transmits one part of the torque via the flywheel to the clutch.
- Drives engine accessories with the other part of the torque.

Stresses

The connecting rods and pistons must be accelerated and decelerated again by the crankshaft during each stroke. This gives rise to high acceleration forces. High centrifugal forces also act on the crankshaft. The forces that are created cause the crankshaft to be subjected to torsional and bending stress and to torsional vibrations, and the bearings are also subjected to wear.

Crankshaft materials

The crankshaft is manufactured from

- alloyed tempering steel
- nitriding steel
- nodular graphite cast iron

Crankshafts manufactured from steel are drop-forged. The interconnected fibre orientation and the tight structure thereby achieved result in great strength.
Crankshafts manufactured from nodular cast iron have good vibration-damping properties.

Design

Crankpins, crankshaft journals. Each crankshaft **(Fig. 1)** features crankshaft journals lying along a single axis for mounting in the crankcase and crankpins for mounting the connecting-rod bearings. The boundary layers of these journals are hardened and ground.

Crank webs. The crankshaft journals and crankpins are connected to each other by crank webs. The crankpins and crank webs produce an unequal mass distribution. This is compensated by counterweights at the opposite ends to the crankpins. Oil holes (also known as oilways) lead from the crankshaft journals through the crank webs to the crankpins.

Crankshafts must be dynamically balanced. An accumulation of material at specific points can be eliminated by balance holes.

Thrust bearing. One of the crankshaft journals is provided with side thrust faces. A thrust bearing (locating bearing) is mounted on this crankshaft journal for the purpose of axially locating the crankshaft. This locating bearing prevents, for example, displacement of the crankshaft when the clutch is actuated.

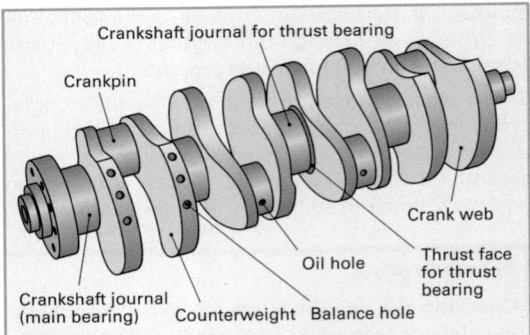

Fig. 1: Crankshaft designations

The flywheel, on which the clutch is usually mounted, is attached to the output end of the crankshaft. The opposite end of the crankshaft accommodates a gearwheel, a sprocket or a toothed-belt wheel (drive for camshaft, oil pump, etc.), a belt pulley and if necessary a vibration damper.

> The shape of the crankshaft is determined by the
> - number of cylinders
> - number of crankshaft bearings
> - size of the stroke
> - firing order
> - arrangement of the cylinders

Throw. This consists of a crankshaft journal and its two adjacent crank webs. Thus, for example, in straight-four engines, all the crankshaft throws lie in a single plane, while, in straight-six engines, the throws are offset by 120° to each other. Throws for parallel cylinders are always parallel. Those cylinders whose pistons are offset by a crankshaft angle of 360° to each other during a power cycle are called parallel cylinders.

Flywheel

A flywheel can store energy (available power) during the power stroke and release it again at a later stage. This flywheel energy is used to overcome the "idle strokes" and dead centres in the power cycle and compensate speed fluctuations. The ring gear with which the pinion for starting the engine meshes is usually shrunk onto or bolted to the circumference of the flywheel. The clutch transmits the engine torque from the flywheel to the gearbox.

The flywheel is manufactured from steel or special cast iron. The flywheel and crankshaft are usually dynamically balanced together so as to eliminate the risk of excessive imbalance at high rotational speeds. Excessive imbalance would cause the crankshaft to rotate irregularly and subject the bearings to excessive load.

Balancer shafts

> These are intended to compensate the inertia forces which are caused by the crankshaft-drive components and may cause the engine to vibrate.

Inertia forces as centrifugal forces. These are generated at the rotating components of the crankshaft drive and are compensated by uniform distribution of the crank throws, by counterweights and by careful balancing. If left uncompensated, they would subject the crankshaft bearings to additional load and cause the engine to vibrate.

Inertia forces from reciprocating motion of crank-shaft-drive components. Depending on the engine design, these can only be incompletely compensated. Whereas, for example, in straight-six engines the inertia forces from the reciprocating motion of crankshaft-drive components compensate each other inside the engine, this is not the case in straight-four engines. Inertia forces are generated in the direction of the cylinder axis which cannot be eliminated even by counterweights. Straight-four engines can be equipped with two balancer shafts to counter this drawback.

Balancer shafts. These are arranged on both sides of the crankshaft **(Fig. 1)**. They are provided with specific levels of imbalance whose imbalance forces act in the opposite direction to the uncompensated inertia forces of the crankshaft drive.
Because the vibrations to be damped have twice the frequency of the crankshaft speed, the balancer shafts are driven at twice the crankshaft speed. One balancer shaft rotates in the same direction as the crankshaft and the other rotates in the opposite direction.

These measures enable a straight-four engine to run as smoothly as a straight-six.

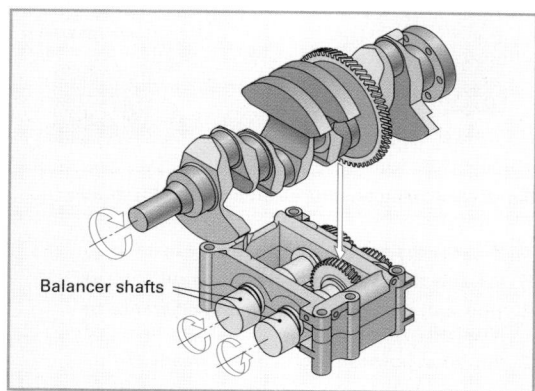

Fig. 1: Straight-four engine with balancer shafts

Vibration damper

> This is intended to counteract the torsional vibrations of the crankshaft which are caused by the combustion jolts in the individual cylinders.

If the torsional vibrations occur at specific speeds in the rhythm of the natural vibrations of the crankshaft, i.e. at the critical speeds, the vibrations can build up to such an extent that they may cause the crankshaft to break.

Design. The damping masses of the vibration damper **(Fig. 2)** are elastically connected by way of the rubber damper to the driving disc. The driving disc is mounted with torsional strength on the crankshaft. When the crankshaft starts to vibrate torsionally, the vibrations are damped by the inertia of the damping masses, in the course of which the rubber damper is subject to elastic deformation.

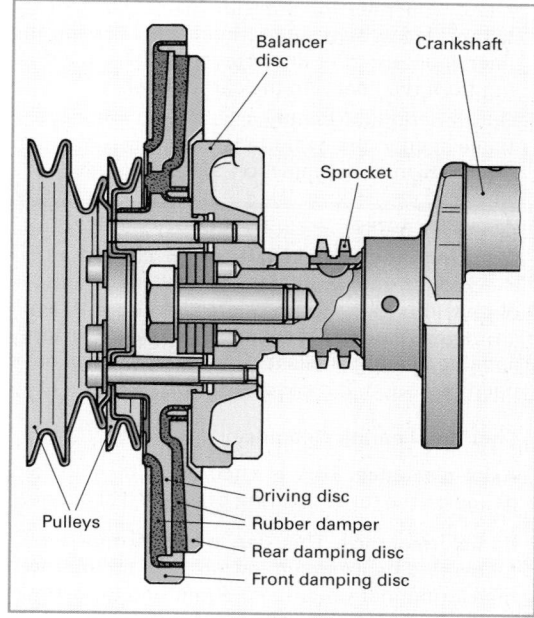

Fig. 2: Vibration damper

Crankshaft bearings

Main bearings. The crankshaft bearings are intended to support and guide the crankshaft in the crankcase. As little friction and wear as possible should occur in the bearings. The crankshaft bearings are usually designed as split plain bearings. The bearing seat is a part of the crankcase on which the bearing cap is bolted. The bearing seat and bearing cap together form the pocket hole into which the bearing shells are placed. All the pocket holes in the crankcase must be exactly flush.

The bearing shells are provided with retaining lugs to prevent them from sliding and turning.

Thrust bearing. One of the crankshaft bearings is designed for the purpose of axially locating the crankshaft as a thrust bearing (locating bearing) with a collar on both sides **(Fig. 1)** or with thrust washers. The pocket hole contains an appropriate recess on both sides for accommodating the thrust washers.

Trimetal bearings. These consist of a steel backing shell (1.5 mm thick), a thin carrier liner of the bearing material (usually PbSnCu alloy) 0.2 mm to 0.3 mm thick which is plated or sintered on, and the actual liner. Despite being only 0.012 mm to 0.020 mm thick, the liner should if at all possible be maintained over the entire life of the engine. The following different bearings may be used, depending on the liner:

- **Electroplate trimetal bearings** for medium to high loads. These have an electroplated liner, usually of a PbSnCu alloy. The liner has good embeddability properties for abrasion particles. A nickel barrier acting as a separating layer between the liner and the carrier liner prevents tin from diffusing from the liner into the carrier liner.

- **Sputter trimetal bearings** with high wear resistance even in the event of particularly high bearing loads in direct-injection diesel engines.

Cathode evaporation is used to sputter the liner in superfine distribution onto the carrier liner from a donor material (e.g. AlSn20Cu). A NiCr intermediate layer serves to provide binding between the liner and the carrier liner.

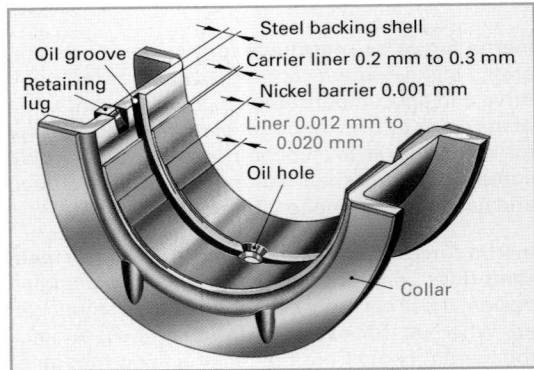

Fig. 1: Trimetal thrust bearing

Bearing lubrication. This is achieved by engine oil which is supplied by the oil pump via oilways through an oil hole to the crankshaft bearing. The bearing shells are usually still provided with a ring groove and a further oil hole **(Fig. 1)** through which the oil is directed to the connecting-rod bearings.

WORKSHOP NOTES

Checking the crankshaft. The crankshaft is checked for true running with a dial gauge and the dimensions of the journals are checked with a micrometer gauge. Manufacturers usually supply suitable bearing shells for crankshafts which have been reground to a specified lower size.

Checking bearing clearance

Axial clearance. This is checked with a feeler gauge on the thrust bearing or with a dial gauge.

Radial clearance. This can be determined by measuring bearing and journal with an internal measuring instrument and a micrometer gauge or also with Plastigage.

The Plastigage plastic thread is placed axially on the crankshaft journal **(Fig. 2)**. Once the bearing cap has been placed in position, the stress bolts are tightened in the elastic range to the specified low torque and then unscrewed again. The

pinched thread is compared with the scale printed on the bag; e.g. TYPE PG-1 shows a bearing clearance of 0.051 mm. Each bearing must be measured individually. During installation the stress bolts are finally tightened to the specified torque.

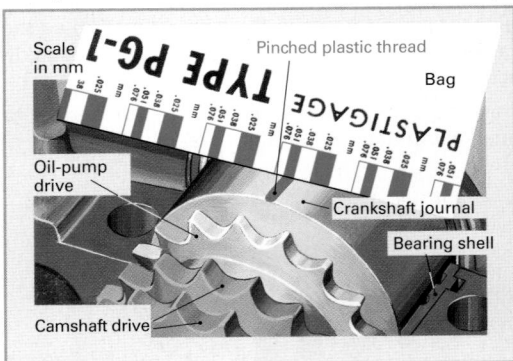

Fig. 2: Measuring bearing clearance with Plastigage

REVIEW QUESTIONS

1 What are the functions of the crankshaft?

2 To which stresses is the crankshaft exposed?

3 From which materials are crankshafts made?

4 What is the function of the thrust bearing?

5 What is the function of the vibration damper?

6 How can the crankshaft bearing clearance be checked?

7 For what purpose is a flywheel needed?

11.4 Dual-mass flywheel

This is intended to damp torsional vibrations which are generated by the periodic cycle of the four strokes and the firing order in the crankshaft and a conventional flywheel.

At certain rotational speeds torsional vibrations can result in gearbox noises (gearbox rattling) and body droning.

The conventional flywheel mass of an internal-combustion engine consists of the crankshaft-drive components, the flywheel and the clutch.

In the diagram **(Fig. 1)** the speed fluctuations of the engine and the gearbox at full load are plotted against time.

The vibrations of the engine output and gearbox input have virtually identical amplitudes and frequencies. In the event of superposition (resonant range), this results in gearbox noises and body droning.

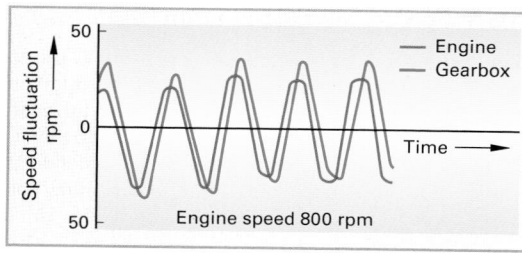

Fig. 1: Vibration curves for a conventional flywheel mass

Design (Fig. 2)

The conventional flywheel mass is divided into the **primary flywheel mass** (crankshaft drive, primary flywheel) and the **secondary flywheel mass** (secondary flywheel, clutch).

A torsional-vibration damper connects the two flywheel masses. The function of this damper is to isolate the flywheel-mass system of the engine from the gearbox and the drivetrain. A clutch disc without a torsion damper can therefore be used for the clutch.

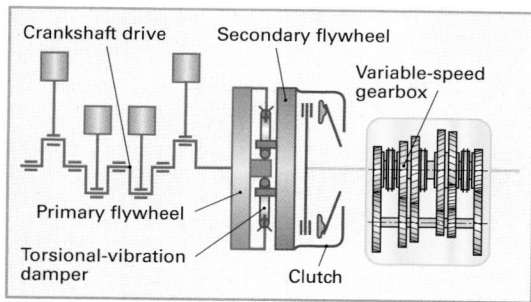

Fig. 2: Dual-mass vibration system

The dual-mass flywheel **(Fig. 3)** consists of
- Primary flywheel
- Secondary flywheel
- Inner damper
- Outer damper

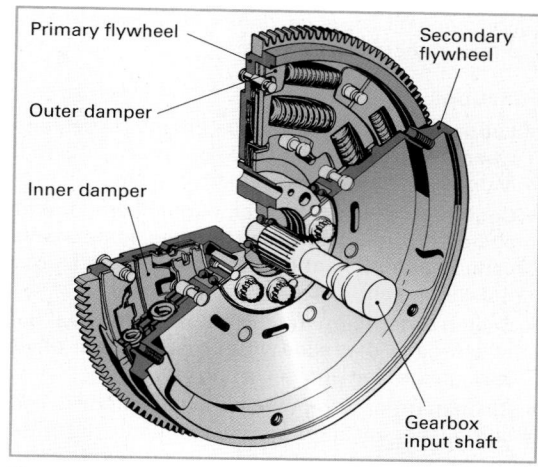

Fig. 3: Dual-mass flywheel

Operating principle

Dividing the flywheel mass into the primary mass on the engine side and the secondary mass on the gearbox side increases the mass moment of inertia of the rotating gearbox parts. In this way, the resonant range is below the engine's idle speed and thus not in the engine's operating range.

The diagram **(Fig. 4)** shows that the vibration curves of the engine output and gearbox input are clearly far removed from each other.

In this way, the torsional vibrations generated by the engine are isolated from the gearbox, and gearbox rattling and body droning no longer occur.

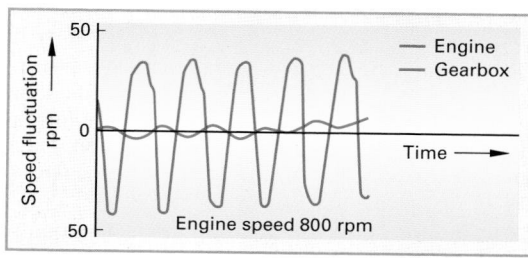

Fig. 4: Vibration curves of a dual-mass vibration system

Advantages
- Reduction of gearbox and body noises (rattling, chattering, droning).
- Protection of power-plant components.
- Lower synchromesh wear.
- Clutch disc does not require a torsion damper.

11.5 Engine lubricating systems

> The engine lubricating system must supply the engine components with an adequate amount of lubricating oil. The correct pressure must be guaranteed in the process.

Functions

- **Lubricate** in order to reduce energy losses and wear-inducing friction between those parts that slide on or against each other.
- **Cool** in order to protect the engine components against overheating because these components cannot give off heat directly to the coolant or to the cooling air.
- **Seal** in order to guarantee the precision seal between parts that slide on or against each other (e.g. piston ring against cylinder wall).
- **Clean** in order to remove abrasion, deposits and combustion residues or to bind them in the oil so as to render them harmless to the engine.
- **Protect against corrosion.**
- **Damp engine noises** because the lubricant layer has a noise- and vibration-damping effect.

Stresses of lubricating oil

> The lubricating oil is exposed to high thermal, chemical and mechanical stresses in the engine **(Fig. 2, Page 204).**

Mechanical contamination through dust, metal abrasion and combustion residues can be largely eliminated by suitable filters. Reasons for regular oil changes:

Oil ageing. Air and combustion gases (blow-by gases) force their way into the crankcase between the piston and the cylinder. This causes the oil to oxidise (age). Acids may be formed in the process.

Oil sludging. Separated oleoresins, road sweepings, metallic abrasion and released combustion residues cause oil sludge to form. Sludge formation is further encouraged by condensation water and if necessary coolant. Oil sludge can obstruct the oil circuit.

Oil dilution. The high-boiling fuel constituents, which get into the oil particularly when the engine is cold, result in oil dilution.

Oil thickening. Heavy oil oxidation, combined with the depositing of soot particles, occasionally cause the oil to thicken, mainly in diesel engines.

Oil consumption. Every engine has a certain level of normal oil consumption, which must be compensated. This comes about because the oil gets into the combustion chamber (e.g. oil film on the cylinder wall, valve guides), where it burns.

For this reason, it is necessary to check the oil level at regular intervals and if necessary to top up the oil. Oil changes must be carried out within the framework of vehicle servicing in accordance with the manufacturer's recommendations or in accordance with the flexible interval display.

Lubricating systems. Two different types of system may be used in four-stroke engines:
- Forced-feed lubrication
- Dry-sump lubrication

Likewise in two-stroke engines:
- Mixture lubrication
- Total-loss lubrication

Lubrication points. The following components in an engine must be supplied with oil by the lubricating system: e.g. crankshaft bearings, connecting-rod bearings, gudgeon-pin bearings, tappets, camshaft bearings, cam tracks, rocker arms, timing chain, chain tensioners, cylinder barrel and exhaust-gas turbocharger **(Fig. 1).**

11.5.1 Forced-feed lubrication

This is the most commonly used system. A pump draws in the supply of oil from the oil pan through an oil strainer and forces it through pipes and lubricating passages to the engine lubrication points. Several oil pressure and suction pumps can be installed in modern-day engines. Filters and occasionally oil coolers as well are connected in-between **(Fig. 1).**

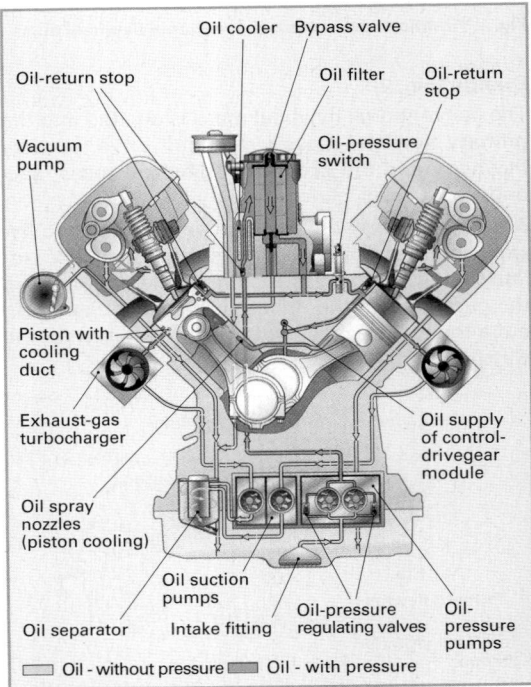

Fig. 1: Forced-feed lubrication

The oil trickles off from the lubrication points and flows back to the oil pan. The oil supply can be checked with a dipstick. Electric oil sensors, which enable the oil content and oil quality to be displayed in the instrument panel, are also being increasingly installed.

Dry-sump lubrication (Fig. 1). This is a special type of forced-feed lubrication system.

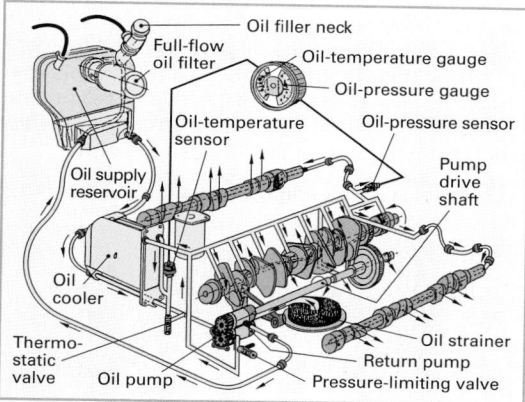

Fig. 1: Dry-sump lubrication

In this lubricating system, the oil returning to the oil pan is directed by a suction pump to a separate oil supply reservoir. A pressurised-oil delivery pump draws the oil off from this reservoir and forces it via a filter and if necessary an oil cooler to the lubrication points.

Advantages of dry-sump lubrication:
- The flat oil pan reduces the height of the engine and thus the vehicle's centre of gravity.
- Perfect lubrication is guaranteed in the event of large engine inclinations, e.g. offroad vehicles and motorcycles, or on bends/curves taken at high speed, e.g. in sports cars.
- Better oil cooling is achieved because the oil supply reservoir is isolated from the engine and thus from the heat of the engine.

Because dry-sump lubrication is more expensive than forced-feed lubrication, it is usually used only in low sports cars, offroad vehicles and motorcycles.

11.5.2 Engine-lubrication components

- Oil pan
- Pressure-limiting valve
- Oil-pressure gauge
- Overflow valve
- Oil pump
- Oil filter
- Ventilation
- Oil cooler

Oil pan. This holds the oil supply for the engine. The lowest point of the oil pan is frequently equipped with antirolling walls (baffle plates), which prevent the oil from flowing away from the suction point

during cornering, acceleration and braking. The surface of the oil pan also acts as a cooling surface for the oil supply. In small-sized engines, oil pans with cooling fins cast from light-metal alloy are increasingly being replaced by lighter and cheaper-to-manufacture plastic oil pans. The oil pan is sealed by flat gaskets or increasingly by liquid seals with silicone.

Two-part crankcase (Fig. 2). This enables the crank space to be sealed downwards against the oil pan. Thus, under the pistons there are changing, enclosed volumes which are connected via special oil return passages with the oil from the oil pan.

Advantages:
- Stiffer crankcase. The combination of the enclosed oil pan and the engine block reinforces the engine, which promotes a further weight reduction.
- Lower piston slap. In the enclosed crank space under the piston an overpressure is generated during the downward movement of the piston which expands again during the upward movement and thereby stabilises the piston.
- Reduced foaming of the engine oil caused by so-called blow-by gases.

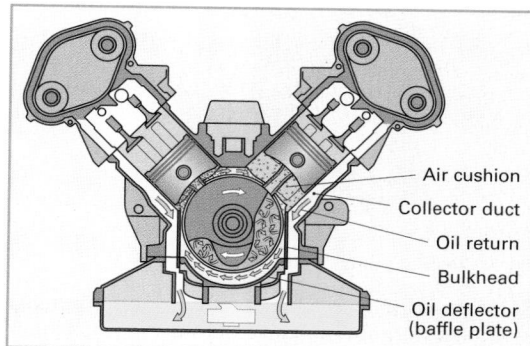

Fig. 2: Two-part crankcase

Oil pumps

These must ensure an adequate oil pressure in combination with a high delivery flow (approx. 250 l/h to 350 l/h). They deliver the oil, e.g. in tooth spaces, from the suction side to the pressure side.

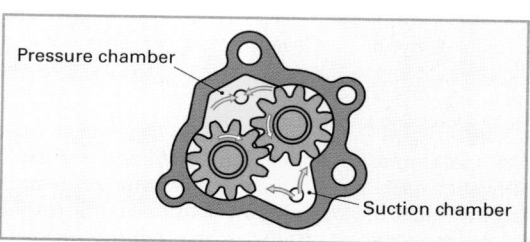

Fig. 3: Gear pump

Gear pump

In this type of pump, the oil is carried in the tooth spaces and delivered along the inner pump wall to the other side. The meshing of the teeth of the two gears prevents the oil from flowing back.

A vacuum pressure and an overpressure are generated on the suction side and the pressure side respectively.

Crescent pump (Fig. 1). This is a special type of gear pump. Its inner gear is usually seated directly on the engine crankshaft. An outer gear mounted in the pump housing is arranged eccentrically to the inner gear. This creates a suction chamber and a pressure chamber, which are separated from each other by a crescent-shaped spacer. The oil is delivered in the tooth spaces along both the upper and lower sides of the crescent-shaped spacer.

Advantage of a crescent pump over a conventional gear pump:

- Higher delivery rate, especially at low engine speeds

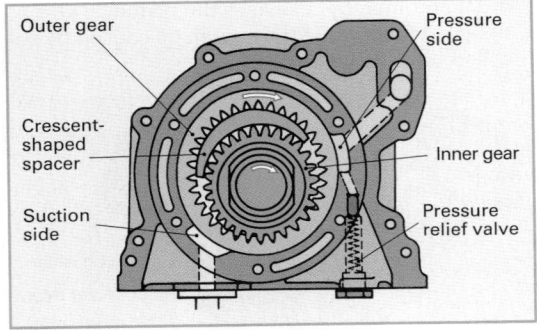

Fig. 1: Crescent pump

Rotor pump (Fig. 2). This consists of an internally toothed outer rotor and an externally toothed inner rotor. The inner rotor has one less tooth than the outer rotor and is connected to the drive shaft.

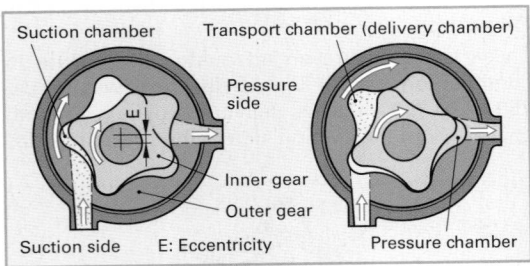

Fig. 2: Operating principle, rotor pump

The toothing of the inner rotor is shaped in such a way that each tooth touches the outer rotor and largely seals the chambers created. As the rotors rotate, the pump chambers on the suction side continually increase in size and the pump draws in oil.

The chambers on the pressure side decrease in size and the oil is forced into the pressure line.

The oil is simultaneously forced by several narrowing pump cells into the pressure line so that the rotor pump operates uniformly. It is able to generate high pressures with a high delivery flow.

Regulated rotor pump

Design. This has an additional control ring between the rotor ring and the pump housing which is turned depending on the oil pressure and the control spring **(Fig. 3).** This pump enables the oil pressure to be kept constant irrespective of the engine operating conditions. This results in constant lubricating conditions at all the lubrication points.

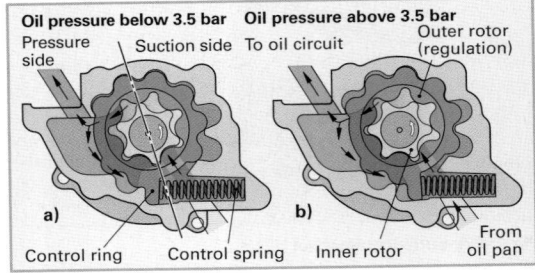

Fig. 3: Regulated rotor pump

Oil pressure too low. If the oil pressure drops below the limit pressure of for example 3.5 bar, the control spring can turn the control ring against the oil pressure **(Fig. 3a).** This results in an increase in chamber size between the inner and outer rotors. In this way, more oil is delivered from the suction side to the pressure side and the oil pressure rises.

Oil pressure too high. If the oil pressure rises above the limit value, the oil pressure presses against the control ring and the control spring is compressed **(Fig. 3b).** This results in a decrease in chamber size between the inner and outer rotors. In this way, less oil is delivered from the suction side to the pressure side and the oil pressure drops.

Oil-pressure gauge, oil-pressure telltale lamp

These components serve to monitor the oil pressure. Both are installed between the oil pump and the bearings.

Oil-pressure gauge. This allows the current oil pressure to be read off directly. A pressure sensor is required in the pressure line after the oil pump for pressure measurement.

Oil-pressure telltale lamp. This goes out during engine operation and indicates to the driver whether there is sufficient oil pressure available in the system. If oil from the pressure line presses on the switching-contact body, the earth/ground contact for the oil-pressure telltale lamp is interrupted and the lamp goes out **(Fig. 1, Page 229).**

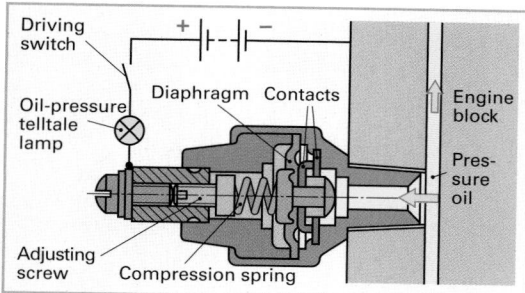

Fig. 1: Oil-pressure switch for oil telltale lamp

Pressure-limiting valve

This is connected downstream of the oil pump and prevents excessive oil pressure (> approx. 5 bar). A high oil pressure is not always proof of good lubrication. When the engine is cold, for example, lubrication is worse in spite of a high oil pressure than it is when the oil pressure is low in an engine which is at normal operating temperature. Oil pressure will also be high but lubrication poor when an oil line or a filter is clogged. An excessively high oil pressure poses a risk to seals, oil lines and oil tubes to the oil cooler and oil filter.

Oil filters

These are installed to prevent premature deterioration of the lubricating oil caused by solid impurities, e.g. metal abrasion, soot, dust particles (**see Chapter 1.6**). Oil filters, however, are unable to remove liquid contaminants or contaminants dissolved in the oil. Neither do they have any effect on chemical or physical changes to the oil during engine operation, e.g. caused by ageing. Two different types of oil filter are used, depending on where they are located in the oil flow: full-flow and partial-flow oil filters.

Full-flow oil filter (Fig. 2). Full-flow oil filtering ensures that no unfiltered oil can reach the lubrication points. In order to facilitate an adequate throughput of oil, it is important that the flow resistance of the filter (pore size) not be too high. This would limit the filter effect. Tiny contaminants would not be filtered out of the oil.

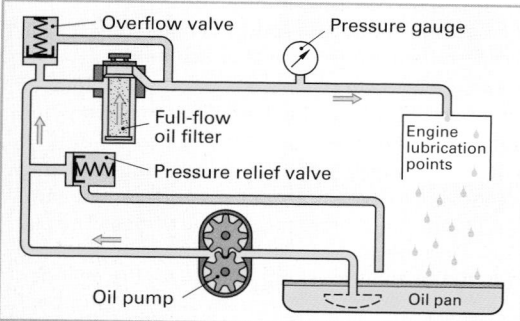

Fig. 2: Oil filter in full flow

Overflow valve (Fig. 3). If the filter is clogged, the oil can flow unfiltered through the overflow valve past the filter to the lubrication points.

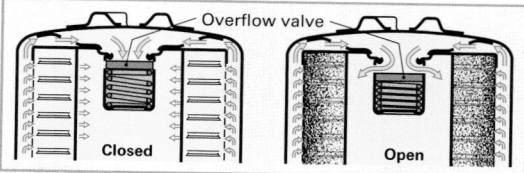

Fig. 3: Overflow valve

Return check valves (Fig. 4). These can additionally be installed in the filter inlet and outlet lines to prevent the full-flow oil filter from running dry when the engine is stopped.

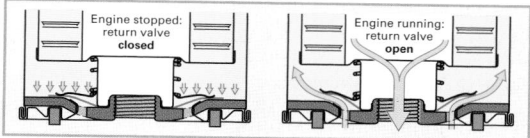

Fig. 4: Return check valve

Partial-flow oil filter (Fig. 5). This filter is arranged in a branch (partial flow) running parallel to the full flow. Thus only part of the oil delivery (5 % to 10 %) passes through this filter. In this way, only partially filtered oil reaches the lubrication points. The pore size of the filter can be reduced to such an extent that even tiny contaminants can be filtered out of the partial flow.

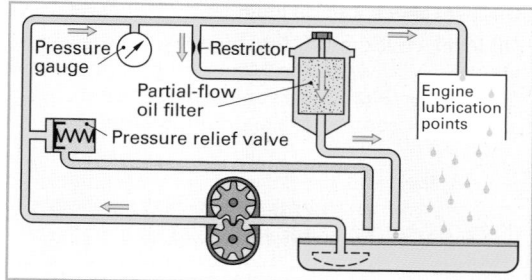

Fig. 5: Partial-flow oil filter

Combination of full-flow and partial-flow filters. This combination delivers the best filter effect. They are used, for example, in construction-site vehicles. For cost reasons, however, predominantly full-flow oil filters are used in passenger cars.

Crankcase ventilation. Spark-ignition engines and in particular turbocharged diesel engines are subject to blow-by gases which enter the crankcase from the combustion chamber. This gas, which is contaminated by superfine oil droplets, fuel residues, water vapour and soot, is returned to the engine intake air via an oil separator.

11

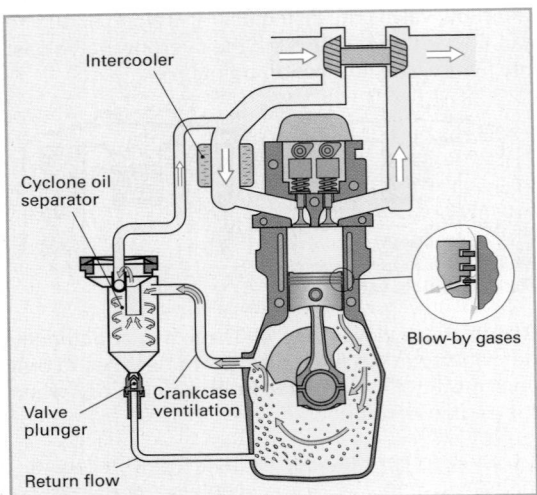

Fig. 1: Crankcase ventilation

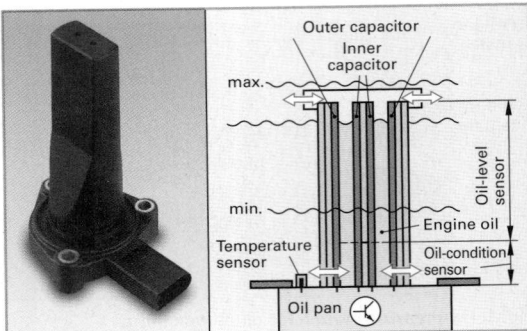

Fig. 2: Oil sensor

In order to lower oil consumption, protect sensitive engine components against contamination and thereby reduce emissions, it is essential for this gas to be filtered prior to recirculation. This operation is performed by an oil separator, e.g. labyrinth, cyclone or centrifugal oil separator **(Fig. 1)**.

Oil sensor. A sensor can be installed in the oil pan for accurate recording of the engine-oil level, oil temperature and engine-oil condition **(Fig. 2)**.

The sensor consists of two cylinder capacitors, one arranged on top of the other. The oil quality is evaluated in the lower section and the engine-oil level is determined in the upper section.

Measurement principle: If the condition of the oil changes as a result of wear and breakdown of additives, the capacitance of the oil-filled capacitor changes. The capacitance value is processed in the integrated electronic evaluation circuit into a digital signal and communicated to the engine control unit. The engine control unit processes this signal in order to calculate the next oil-change service. The driver is alerted to the oil level by means of a display.

Oil temperature. A temperature sensor (NTC) is integrated in the oil sensor to measure the oil temperature.

WORKSHOP NOTES

Oil-level check and oil change

Oil-level check with the dipstick. The fill level should be between the Max. and Min. marks on the dipstick **(Fig. 3)**. The vehicle must be parked on a level, horizontal surface during this check.

Fig. 3: Dipstick

Oil level too low: Oil must be topped up. Make sure in so doing that the Max. mark on the dipstick is not exceeded. The fill-quantity difference between the two marks is usually 1 litre in passenger cars.

Oil level too high: Vehicles which are frequently used for short journeys can be subject to oil dilution and as a result to a high oil level. In this case, the measurement must be repeated after extended vehicle operation with the engine at normal operating temperature. If an excessively high fill level is still recorded when the measurement is repeated, oil must be drained or drawn off.

Oil-level check via vehicle display: If the vehicle is equipped with an oil sensor, the current fill level must be determined in the display under the "Engine oil" menu item. This display also shows the remaining mileage until the next prescribed oil change.

Oil consumption

Normal oil consumption.
This is the case if no more than 0.1 ... 1.0 l oil is consumed over a distance of 1,000 km. New vehicles may be subject to higher oil consumption during running in.

Increased oil consumption.
If increased oil consumption is constantly registered, there must be a mechanical fault:

- **Valve-stem seals faulty.** Oil gets into the inlet or exhaust port here. Blue smoke is created when the oil is burned.

- **Axial clearance between piston ring and ring groove.** The clearance between these two components may increase as a result of wear. The piston ring moves back and forth in the ring groove as the piston moves up and down the cylinder. This reciprocal motion causes oil to be "pumped" into the combustion chamber **(Fig. 1, Page 231)**.

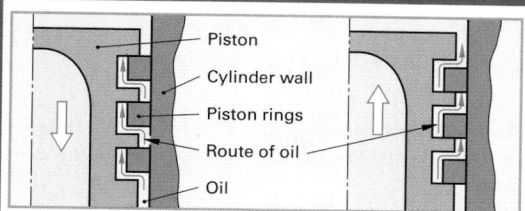

Fig. 1: "Pumping" piston rings

The oil passes into the combustion chamber, where it burns. The piston rings seize as a result of coked piston-ring grooves. This results in the creation of heavy blue smoke.

Damaged engine seals. These result in the loss of oil even when the engine is stopped. Patches of oil on the ground or oil fouling on the engine point to this fault.

Oil change

Vehicle manufacturers recommend an oil change at specific intervals. This change can be performed by draining or drawing off engine oil with the engine at normal operating temperature.

Draining engine oil. Here the oil flows almost completely out of the oil drain hole in the oil pan into the oil-collecting tray. The seal of the oil drain plug must be replaced. If the oil drain plug incorporates a magnetic separator, the metal abrasion sticking to the separator must be removed.

Drawing off engine oil. Here a suction probe is inserted into the oil pan and draws off the oil within 5 ... 10 mins. Roughly 0.5 litres of the old oil is left in the oil pan. This must be taken into account when the prescribed quantity of new oil is added, otherwise the oil level will be too high. If the probe cannot reach the lowest point of the oil pan, the contaminants that have collected there will not be drawn off.

Oil-filter replacement. As with changing the oil, oil-filter replacement at specific intervals is also recommended by the vehicle manufacturer.

Box-type oil filters. These filters must be released with the aid of special oil-filter wrenches (**Fig. 2**). Observe the following when installing box-type filters:

- Apply a light coating of oil to the O-ring seal of the new oil filter so that the O-ring does not stick to the sealing surface during removal.
- Tighten the box-type filter by hand.

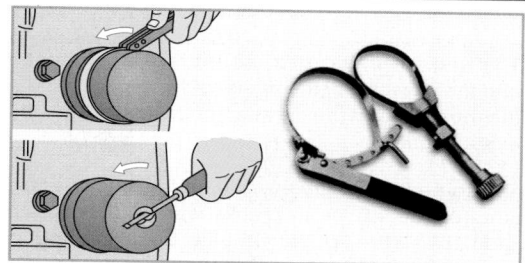

Fig. 2: Releasing a box-type filter; clamping bands

Throwaway oil filters. The cover of the filter module must be opened here. Make sure when inserting the new filter cartridge that it is correctly seated. The O-ring of the filter-module cover must be replaced.

Oil filling. Use only the manufacturer-approved oil grades. Carry out an oil-level check after adding the prescribed amount of oil. Check for leaks while the engine is running.

Used oil and used filters. These must be stored and disposed of in an environmentally compatible manner.

Oil-pressure check

Here the function of the oil-pressure switch and the oil pressure at a specific engine speed are checked. The oil-pressure switch is screwed into a test adapter. This adapter is screwed in in place of the oil-pressure switch. A test lamp is connected between the oil-pressure switch and the battery (**Fig. 3**).

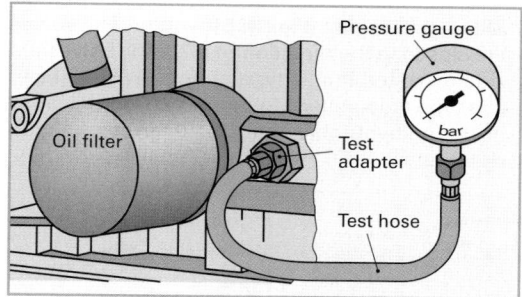

Fig. 3: Checking oil-pressure switch and oil pressure

The lamp must light up when the engine is stopped. It should go out if the pressure gauge indicates an oil pressure of 0.3 ... 0.6 bar when the engine is running. The oil-pressure switch must be replaced in the event of a negative result. The oil pressure should be at least 2 bar at an engine speed of 2,000 rpm and an oil temperature of 80 °C.

<small>REVIEW QUESTIONS</small>

1 **List the advantages of dry-sump lubrication.**

2 **What different types of oil pumps are there?**

3 **List two advantages of regulated oil pumps.**

4 **What different types of oil filter are there?**

5 **Which measured values can be recorded by an oil sensor? Explain their operating principle.**

11.6 Engine timing gear

> The engine timing gear controls the moment and the duration of the intake of fresh gases and the moment and the duration of the discharge of exhaust gases.

The moments are given as opening and closing points of the valves in crankshaft-angle degrees, e.g. Io 15° before TDC, Ic 42° after BDC (see timing diagram, **Page 192**).

11.6.1 Design of engine timing gear

The engine timing gear is driven from the crankshaft via toothed belt, roller chain or gears to the camshaft. The camshaft cams open the inlet and exhaust valves against the spring force of the valve springs via transfer elements, e.g. tappets. The valves are closed again by the spring force of the valve springs.

Because one power cycle stretches over four strokes, i.e. two crankshaft revolutions, and the valves are only actuated once in the process, the camshaft must rotate at half the speed of the crankshaft. The crankshaft gear therefore has half as many teeth as the camshaft gear.

> The speed ratio between crankshaft and camshaft is 2 : 1.

Arrangement of valves. The following different arrangements may be used:

- **Side-valve engine (Fig. 1), SV** engine. The valves are closed in the direction of BDC. The valves are side-mounted in this type of engine. Engines of this type are not used in motor vehicles because of the unfavourable shape of their combustion chambers.

- **Overhead-valve engine (Figs. 2 to 5).** The valves are closed in the direction of TDC. Engines of this type have overhead valves.

Arrangement of camshaft. The following different arrangements are possible in overhead-valve engines:

- **OHV** engine (**O**ver**H**ead **V**alves). Inverted valves in the cylinder head; the camshaft is located in the cylinder block or in the crankcase (**Fig. 2**).

- **OHC** Engine (**O**ver**H**ead **C**amshaft). The camshaft is located above the cylinder head (**Fig. 3**).

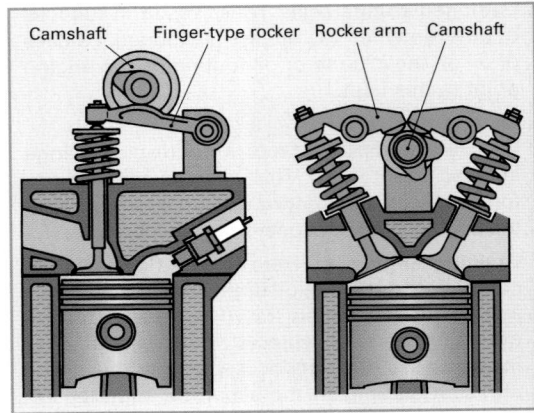

Fig. 3: OHC engine

- **DOHC** engine (**D**ouble **O**ver**H**ead **C**amshaft). Two camshafts are located above the cylinder head (**Fig. 4**).
- **CIH** engine (**C**amshaft **I**n **H**ead). The camshaft is located in the cylinder head (**Fig. 5**).

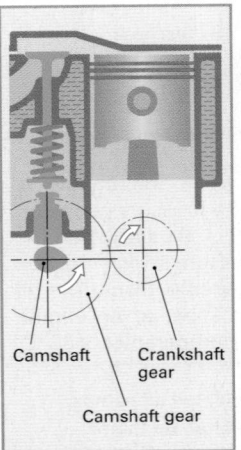

Fig. 1: Side-valve engine

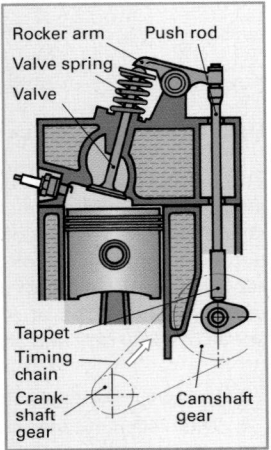

Fig. 2: Overhead-valve engine

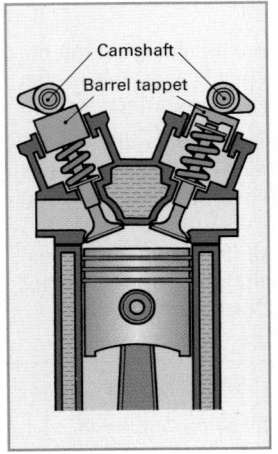

Fig. 4: DOHC engine

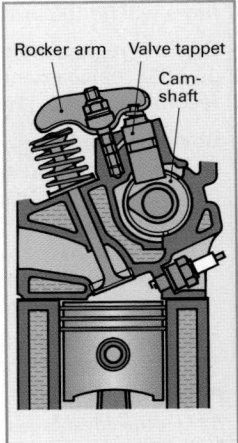

Fig. 5: CIH engine

11.6.2 Multiple-valve technology

In the interests of further improving the gas exchange in the cylinder, engines are also equipped with two or three inlet valves together with one or two exhaust valves.

Three-valve version (Fig. 1). Two inlet valves are arranged opposite one larger exhaust valve. If it is not possible to position the spark plug centrally, dual ignition with two off-centre spark plugs is used. This brings about improved burning of the mixture in the vicinity of the piston edge and at the fire land. A common camshaft controls the valves.

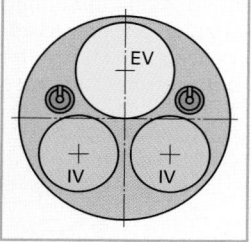 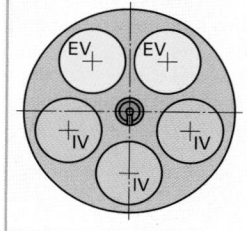

Fig. 1: Three-valve version   **Fig. 2: Five-valve version**

Four-valve version (Fig. 3). This is the most common multiple-valve-technology engine. Two frequently larger inlet valves are arranged opposite two exhaust valves. The spark plug can be positioned almost centrally. The inlet valves and the exhaust valves each require a separate camshaft.

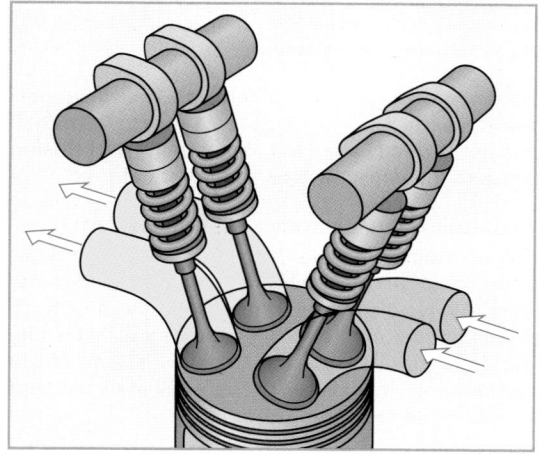

Fig. 3: Four-valve version

Five-valve version (Fig. 2). Three inlet valves and two exhaust valves offer the maximum through-flow cross-section. The spark plug can usually be positioned centrally. The inlet camshaft actuates the three inlet valves and the exhaust camshaft actuates the two exhaust valves.

11.6.3 Timing-gear components

Valves

Two different types are used: **inlet** and **exhaust valves**. The diameters of the valve heads and the valve lift must be sufficiently large as to facilitate as much as possible an unhindered gas exchange. The exhaust valve often has a smaller diameter than the inlet valve because the still high pressure of the exhaust gases when the exhaust valve opens ensures that the combustion chamber is quickly emptied.

Design (Fig. 4). A valve consists of a valve head and a valve stem. The valve head must in conjunction with the valve seat in the cylinder head seal the combustion chamber gas-tight. At the end of the valve stem is a recess or one or more grooves, with which the valve cones engage. The valve cones are pressed by the valve spring retainer into the recess or into the grooves of the valve stem.

Stresses. The valves are exposed to extremely high stress. They are raised roughly 4,000 times per minute and driven by the valve springs back onto the valve seats. The valve stem and the end of the stem are subject to friction wear.

Inlet valves (Fig. 4) may be permanently cooled by the fresh gases, but can still reach temperatures of up to approx. 500 °C.

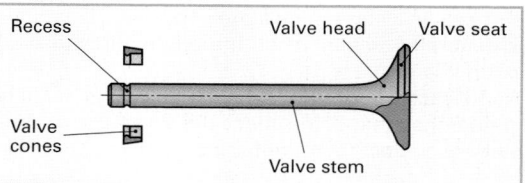

Fig. 4: Inlet valve

Inlet valves are usually single-metal valves. The valve seat, valve stem, recess for the valve cones and end face at the end of the stem can be hardened in order to reduce wear.

Exhaust valves (Fig. 1, Page 234). These are subjected to thermal load by the hot combustion gases (on the valve head up to approx. 900 °C) and to chemical corrosion. Exhaust valves are therefore usually manufactured as bimetal valves. Creep-, corrosion- and scale-resistant steel is used for the valve head and the lower part of the valve stem, which above all are exposed to the combustion gases. Such steels are not heat-treatable, have poor sliding properties, tend to seize in the valve guide and are poor conductors of heat.

The upper part of the stem is made from heat-treatable steel with good thermal conductivity. Both parts are connected to each other, for example by friction welding.

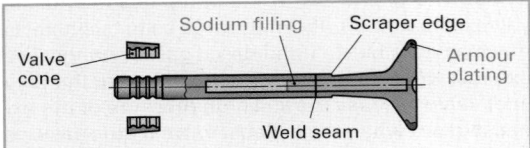

Fig. 1: Exhaust valve

Hollow-stem valves (Fig. 1). These are exhaust valves which in order to improve heat dissipation have a cavity which is filled by up to 60 % with sodium. Sodium melts at roughly 97 °C and has good thermal conductivity. The sloshing movement of the liquid sodium causes the heat to be dissipated more quickly from the valve head to the valve stem, thereby reducing the temperature of the valve head by roughly 100 °C. The valves are frequently armour-plated at the valve seat **(Fig. 1)**, e.g. with hard metal, in order to reduce wear and to prevent the seat from breaking on the valve head.

Valve clearance.
Inlet and exhaust valves expand during operation, depending on the temperature increase and material used. Linear deformations caused by wear also occur in the transfer elements of the timing gear. Clearance between the transfer elements is provided for so that the inlet and exhaust valves can close perfectly in all operating states. The valve clearance is generally greater when the engine is cold than when it is hot.
The clearance of the exhaust valves is usually greater than that of the inlet valves because the former are subject to hotter temperatures.

Valve clearance too small. The valve opens earlier and closes later.
The exhaust valve can become too hot because not enough heat can be dissipated from the valve head to the valve seat due to the shortened closing time. When the valve clearance is to small, there is also the danger that the exhaust valve or the inlet valve will no longer open when the engine is hot. In this situation, exhaust gas is drawn in through the gap in the exhaust valve and flames flash back through the gap in the inlet valve. This results in gas losses and power losses. The valves are overheated by the hot exhaust gases constantly flowing past, and this burns the valve head and valve seats.

Valve clearance too large. The valve opens too late and closes too early. This results in shorter opening times and smaller opening cross-sections, which in turn diminish charge and power. The mechanical stress on the valve and valve noises increase.

Adjustment of valve clearance. There are different ways of adjusting the valve clearance, depending on the engine type and make. It can be carried out when the engine is cold or even when it is hot, or when the engine is stopped or even when it is running at low speed.
In the case of an overhead camshaft and rocker arms, the valve clearance can be adjusted with an adjusting screw and a lock nut or, as shown in **Fig. 2**, by adjusting the ball-pressure pin in the self-locking thread on the bearing of the finger-type rocker. The valve clearance is checked at the gap between the cam base circle and the finger-type rocker.
In the case of an overhead camshaft and barrel tappets **(Fig. 2)**, hardened adjusting shims of varying thicknesses are placed in the tappet in order to adjust the correct valve clearance, which can be checked directly at the gap between the cam base circle and the adjusting shim.

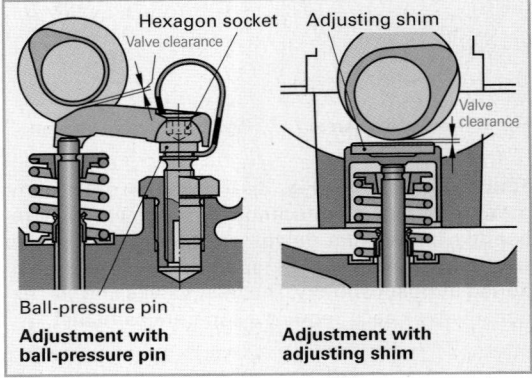

Adjustment with ball-pressure pin

Adjustment with adjusting shim

Fig. 2: Adjusting valve clearance

Depending on the manufacturer's specifications, the valve clearance is roughly 0.1 mm to 0.3 mm. If it is not correctly adjusted, the opening and closing times of the valves will be shifted or deferred.

Hydraulic valve-clearance compensation

It is no longer necessary to adjust the valve clearance on engines which are equipped with a hydraulic valve-clearance compensation facility. Such a facility uses hydraulically actuated transfer elements to compensate linear deformations of the components. In this way, the valve clearance is kept at zero while the engine is running.

Design. The clearance-compensation element is located in the barrel tappet. The valves are actuated directly by the overhead camshaft via barrel tappets **(Fig. 1, Page 235)**.
The hydraulic barrel tappet **(Fig. 1, Page 235)** is connected to the engine's oil circuit. Oil is supplied through a side bore in the tappet into the valve-tappet chamber and from there via the opening in the tappet base into the supply chamber via the pressure pin.

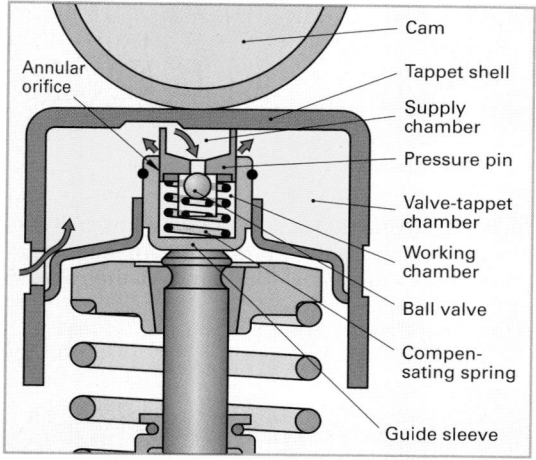

Fig. 1: Barrel tappet with hydraulic valve-clearance compensation

Operating principle

Trailing cam. The pressure pin is not subjected to load. The compensating spring forces the pressure pin upwards until the barrel tappet touches the cam or the cam base circle. As a result of the enlarged space below the pressure pin, oil flows from the supply chamber through the ball valve into the working chamber.

Leading cam. The pressure pin is subjected to load, the ball valve closes and the oil in the working chamber acts like a "rigid connection". The inlet or exhaust valve is opened via the guide sleeve. Excess oil can escape through the annular orifice between the pressure pin and the guide sleeve, e.g. in the event of thermal expansion of the timing-gear parts.

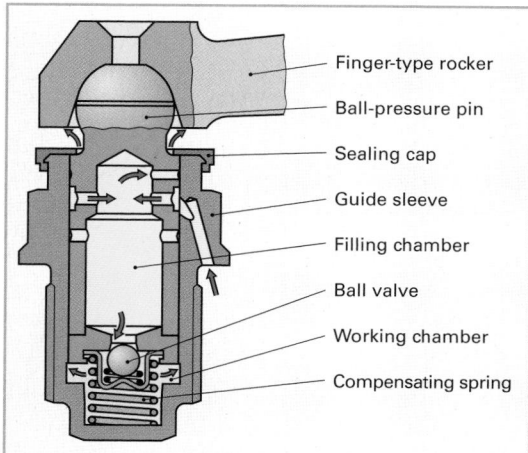

Fig. 2: Finger-type rocker bearing with hydraulic valve-clearance compensation

If, on the other hand, the valves are actuated by the camshaft via finger-type rockers, the compensating element is installed in the rocker bearing (**Fig. 2**). The operating principle is the same as in the barrel tappet.

Valve guide (Fig. 3)

Special valve guides with good sliding properties are pressed into cylinder heads made of Al alloys. These guides are usually made of cast bronze or special cast iron. The valve-stem seal at the top end of the valve guide must guarantee an adequate film of oil in the valve guide; however, it must prevent engine oil from getting through the valve guide into the intake or exhaust port. This would result in high oil consumption and oil-carbon deposits on the valve stem and even the effect of the catalytic converter could be impaired.

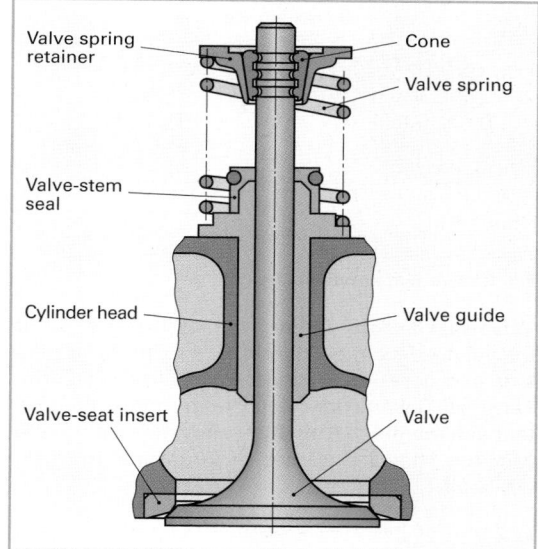

Fig. 3: Valve guide

Table 1: Possible faults in hydraulic valve-clearance compensation.
Chattering noises as a result of excessive valve clearance
• Compensating element runs dry because of excessive wear at the annular orifice.
• Oil restraint valve in the engine lubricating circuit faulty.
No valve-clearance compensation
• Faulty valve-clearance compensating element.
• Air in the valve-clearance compensating element caused by foaming oil due to an excessively high oil level.

Valve seat in cylinder head

The valve seats in the cylinder head (**Fig. 1**) usually have the same cone angle as the valve heads. The seat angle is 45°.

Correction angles. These are 15° and 75°. The correction angles improve the flow behaviour and serve to correct the valve-seat width.

Valve-seat width. This provides for a good combustion-chamber seal. It is approximately 1.5 mm for the inlet valve and approximately 2 mm for the exhaust valve in order to improve heat dissipation. Occasionally the seat angles on the valve head and in the cylinder head are slightly different, e.g. 44° on the valve head and 45° in the cylinder head. This creates a narrow sealing edge to the combustion chamber which increases in size during operation to the normal seat width.

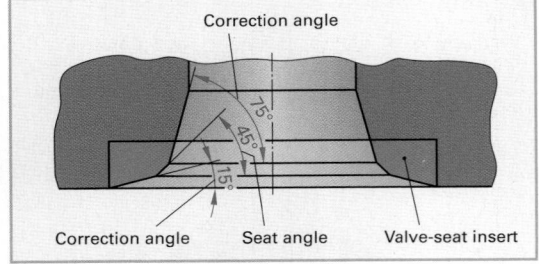

Fig. 1: Valve seat in cylinder head

Valve-seat inserts. These increase the strength of the valve seats in cylinder heads made from Al alloys and occasionally those made from cast iron. Valve-seat inserts are creep-, wear- and scale-resistant and are made from high-alloy steels or special cast iron. They are pressed or shrunk into the cylinder head.

Valve spring

This closes the valve at the end of the induction or exhaust stroke. Helical springs are used here. In the interests of avoiding a spring fracture caused by natural oscillation at high engine speeds, valve springs can be wound with a variable uphill gradient, in a tapered shape or with a decreasing wire diameter. Occasionally even two valve springs are arranged inside each other.

Camshaft

This must execute the lifting movement of the valves at the correct time and in the correct sequence and facilitate closing of the valves by the valve springs.

Cast camshafts (Fig. 2). These are manufactured from alloyed flake-graphite or nodular-graphite cast iron as clear chill castings.

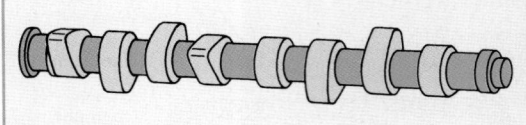

Fig. 2: Cast camshaft

Built-up camshafts (Fig. 3). Here the cams are manufactured individually from case-hardening, tempering or nitriding steel. The cams are then shrunk onto a steel tube.

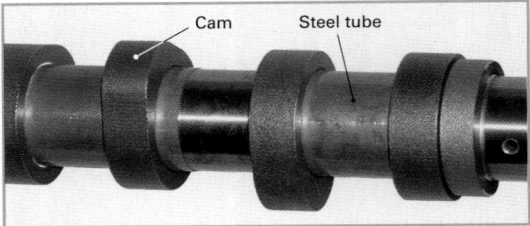

Fig. 3: Built-up camshaft

Cam shapes (Fig. 4). As the valves are opened and closed, these determine the

- Opening duration
- Lifting speed
- Height of the valve lift
- Sequence of movements

Pointed cam. The valve is slowly raised and closed and remains fully open for a short time only.

Asymmetrical cam. The flatter leading cam face effects a slower opening and the steeper trailing face a longer opening of the valve and a faster closing.

Steep (sharp) cam. The valve is quickly opened and closed and remains fully open for a longer time.

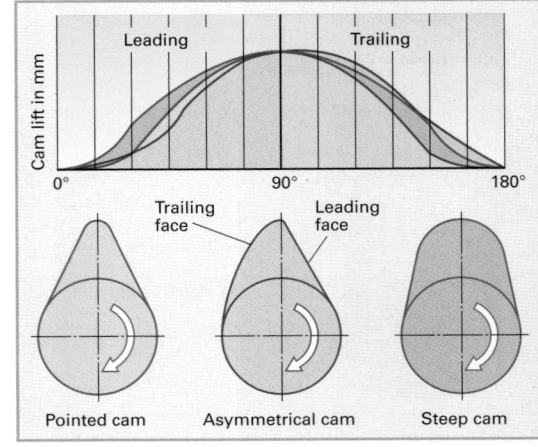

Fig. 4: Cam shapes and cam lift

Camshaft drives

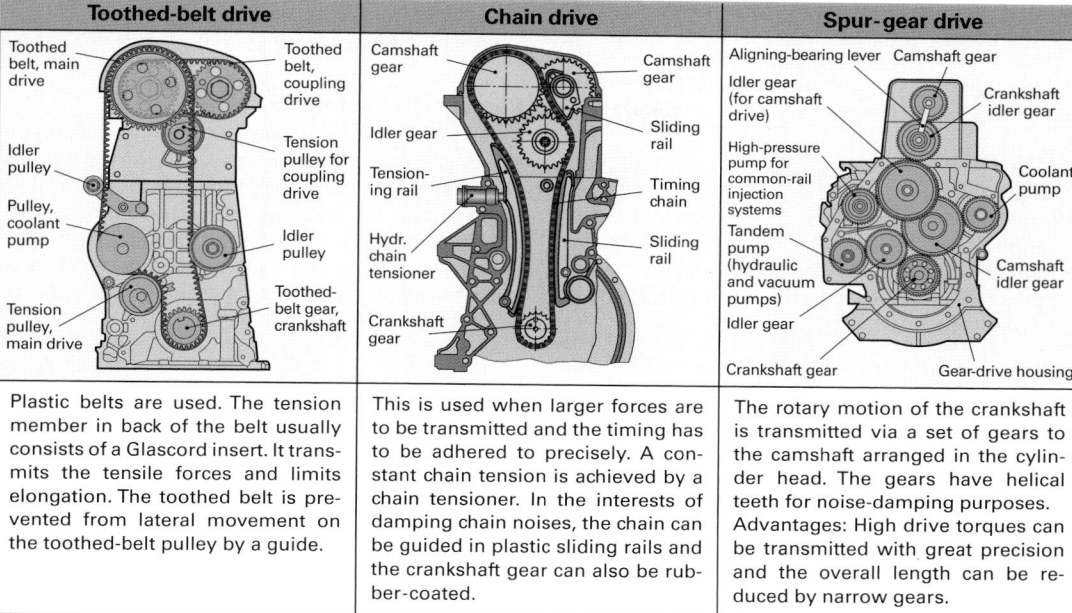

Toothed-belt drive	Chain drive	Spur-gear drive
Plastic belts are used. The tension member in back of the belt usually consists of a Glascord insert. It transmits the tensile forces and limits elongation. The toothed belt is prevented from lateral movement on the toothed-belt pulley by a guide.	This is used when larger forces are to be transmitted and the timing has to be adhered to precisely. A constant chain tension is achieved by a chain tensioner. In the interests of damping chain noises, the chain can be guided in plastic sliding rails and the crankshaft gear can also be rubber-coated.	The rotary motion of the crankshaft is transmitted via a set of gears to the camshaft arranged in the cylinder head. The gears have helical teeth for noise-damping purposes. Advantages: High drive torques can be transmitted with great precision and the overall length can be reduced by narrow gears.

Features of a toothed belt:

- Low mass.
- Silent operation.
- Low production costs.
- Requires only minimal initial tension.
- Requires no lubrication.
- Must be kept free from oil.
- Must not be kinked.
- Manufacturer's instructions must be observed when the belt is replaced.

Finger-type rockers, rocker arms

If the valves are not actuated directly by the camshaft via barrel tappets, they are opened by the camshaft via finger-type rockers or rocker arms.

Finger-type rockers are one-arm levers which rest at one end on a ball pin. At the other end they transmit the lifting movement of the cam to the valve. The friction between the cam and the rocker can be greatly reduced by using a roller cam follower **(Fig. 1)**.

Rocker arms are two-arm levers. The camshaft is situated below the rocker arm. The lifting movement of the camshaft is diverted by the rocker arm to the valve stem. The friction between the cam and the rocker arm can also be reduced here by using a roller cam follower.

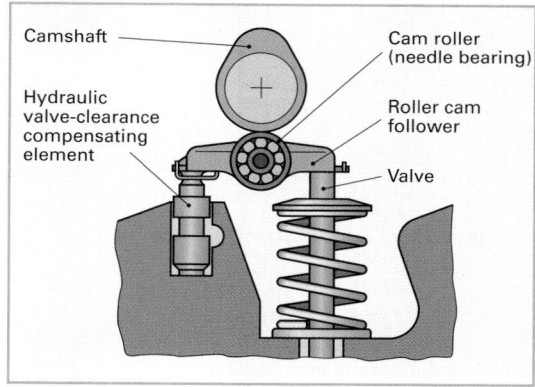

Fig. 1: Roller cam follower

REVIEW QUESTIONS

1 What do you understand by an overhead-valve engine?

2 What are the functions of camshafts?

3 Why is the speed of the camshaft only half as much as that of the crankshaft?

4 Why are the camshaft cams often asymmetrical?

5 What different types of camshaft drives are there?

6 Which faults may occur if the valve clearance is too small?

7 How is clearance-free and self-adjusting valve control achieved?

11.7 Charge optimisation

In the case of charge optimisation, the charge of the cylinders with fresh gases is improved by variable engine timing and/or supercharging within as wide a speed range as possible.

Advantages (Fig. 1)

- Greater power
- Improved torque curve over a specific speed range
- Reduced pollutants in the exhaust gas
- Lower fuel consumption thanks to better mixture formation

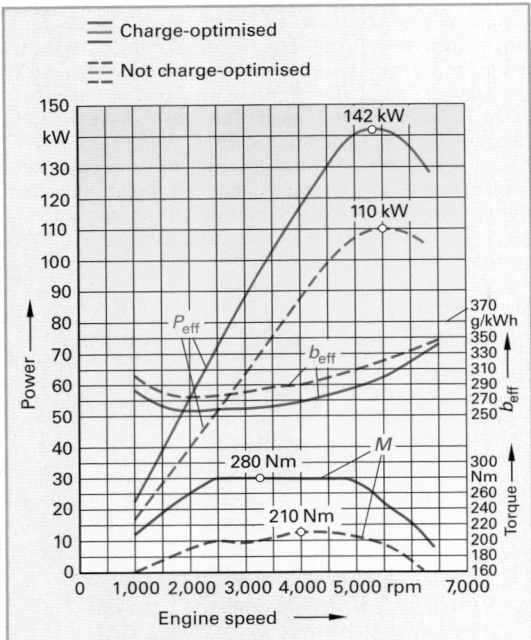

Fig. 1: Engine performance curves

11.7.1 Variable engine timing

In internal-combustion engines with conventional valve gears, the cylinder charge is only optimal within a specific speed range. The engine develops its greatest torque in this speed range. If the speed is further increased, the power does increase to a maximum value but the torque decreases due to the worsening cylinder charge.

Extending the opening time of the inlet valve improves the cylinder charge at high speeds. Torque and power increase. At low speeds the large valve overlap results in large scavenging losses and erratic engine operation. The pollutants in the exhaust gas also increase. These drawbacks can be prevented by variable engine timing.

The following different systems may be used:
- Camshaft adjustment (changing the timing)
- Variable valve gear (changing the timing, valve lift and valve-opening speed)

Camshaft adjustment

This system can be used to change the position of the inlet camshaft and if necessary of the exhaust camshaft in relation to the crankshaft. **Fig. 2** shows the opening and closing times of the valves of an engine with camshaft adjustment. A changed valve overlap is produced depending on how the opening and closing times are shifted in relation to the crankshaft.

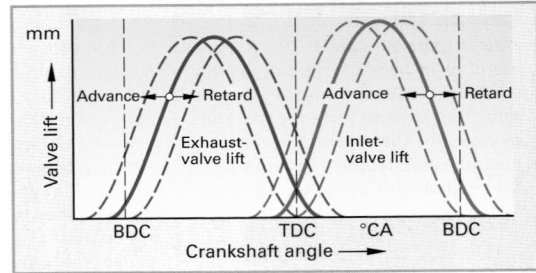

Fig. 2: Valve-lift curve

Map-controlled camshaft adjustment

The camshafts are adjusted as a function of load and speed using program maps which are stored in the engine control unit. The engine temperature, for example, can be used as the correction variable **(Fig. 3, Table 1, Page 239)**. Thus, timing advance or retardation is possible for example at a mean speed depending on the load.

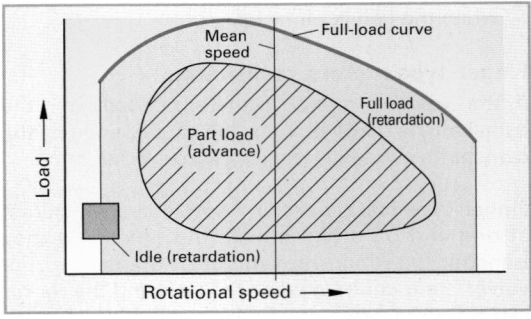

Fig. 3: Operating states during timing advance and retardation of inlet camshaft

The camshafts can be adjusted in different ways, for example by ...
- ... an adjustable chain tensioner, e.g. VarioCam.
- ... variable camshaft control, e.g. VANOS, vane-type adjuster (VaneCam).

Table 1: Adjustment of inlet camshaft as a function of operating state, Fig. 3, Page 238		
Operating range	Opening time IV	Effect
Idle	Retard	Valve overlap small, IV closes long after BDC ▶ no overflow of fresh gases into the exhaust port and exhaust gases into the inlet port, better combustion development ▶ higher torque at idle, idle speed can be reduced.
Part load	Advance	Valve overlap large, IV closes shortly after BDC ▶ fresh gases are not forced back into the inlet port, exhaust gases flow into the inlet port and are drawn in with the fresh gases, temperature of the combustion process decreases ▶ NO_x percentile decreases.
Full load	Retard	Valve overlap small, IV closes long after BDC, fresh gases continue to flow into the cylinder in spite of an upward-moving piston ▶ boost effect improves cylinder charge and torque.

Adjustable chain tensioner (VarioCam, Fig. 1).
The exhaust camshaft drives the inlet camshaft via a chain drive.

Chain tension.
This is generated by spring force.

Camshaft adjustment. In the normal setting the hydraulic cylinder of the chain tensioner is in the upper position and the inlet camshaft is in the retard setting. For the advance setting the hydraulic piston is displaced into the lower position; the lower chain section is lengthened and the upper chain section is shortened, which rotates the inlet camshaft into the advance setting.

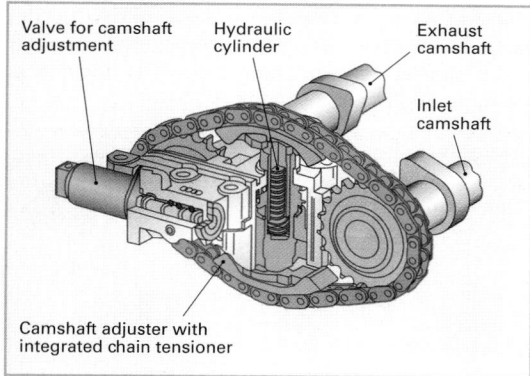

Fig. 1: Adjustable chain tensioner

Variable camshaft control (VANOS, Fig. 2).
The system consists of the following components:
- Hydraulic adjuster
- Mechanical adjuster
- Solenoid valve for hydraulic actuation

Operating principle. With this system the inlet camshaft is rotated in the opposite direction to the camshaft gear. Depending on the position of the solenoid valve, the hydraulic piston is displaced to the left or right. The axial movement of the hydraulic piston causes a camshaft adjustment in the "Advance" or "Retard" direction in the mechanical adjuster by means of the helical teeth. Adjustment can be performed steplessly.

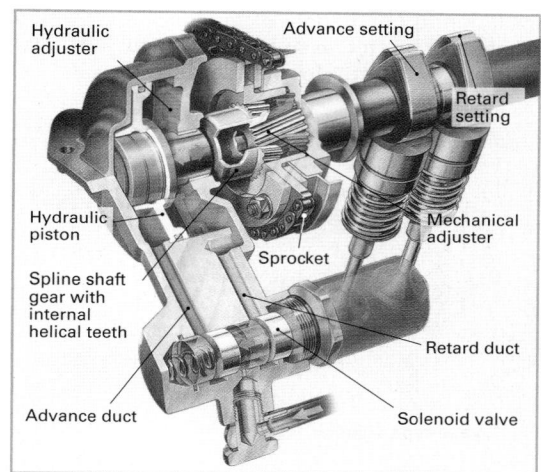

Fig. 2: Design of camshaft adjustment

Adjustment of inlet camshaft in "Advance" direction (Fig. 3). Here the engine oil pressure is directed through the advance duct. The hydraulic piston in the hydraulic adjuster is axially displaced to the right. The rotating spline shaft gear mounted in the hydraulic piston rotates the inlet camshaft in the "Advance" direction in the opposite direction to the sprocket.

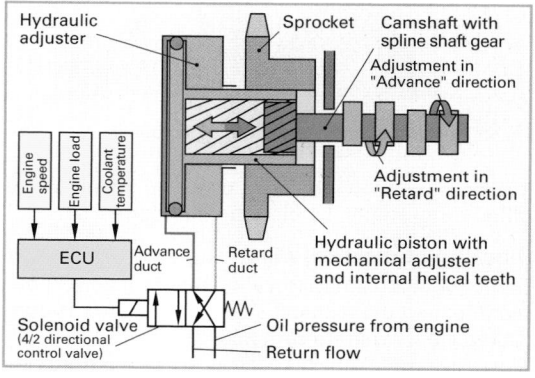

Fig. 3: Camshaft adjustment

11

Systems with adjustment of inlet and exhaust camshafts

Double VANOS. Additional adjustment of the exhaust camshaft brings about an increase in torque in the lower to middle speed range as well as in the middle to upper speed range.

Vane-type adjuster (VaneCam, Fig. 1). The outer rotor is permanently connected to the sprocket and the inner rotor to the camshaft respectively. The outer rotor and inner rotor can be rotated in opposing directions. The oil pressure in the respective oil chamber is variably controlled by the hydraulic valves. The camshafts are adjusted in this way. The maximum turning angle of the inner rotor to the outer rotor is, for example, 52 °CA for the inlet camshaft and 22 °CA for the exhaust camshaft.

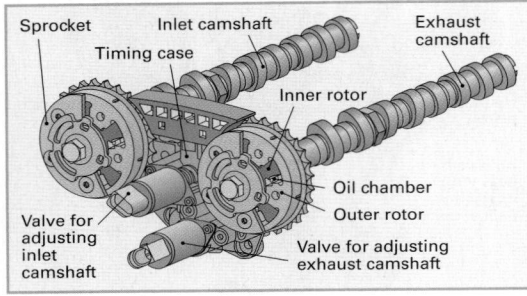

Fig. 1: Vane-type adjuster

Variable valve gear

Valve-opening times and valve-opening cross-section are adapted to the engine operating state.

The valve-opening time is changed by the shape of the cam, while the valve-opening cross-section is changed by the height of the cam (**Fig. 2**).

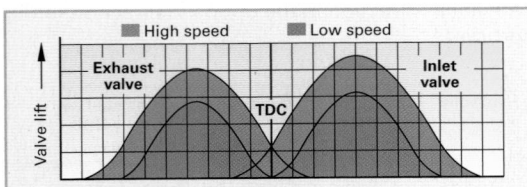

Fig. 2: Valve-lift curves at low and high speeds

It is possible to switch from one cam shape to another, for example, by blocking the rockers.

Blocking of rockers (VTec)

Three rockers (cam followers) are arranged on both the inlet and the exhaust sides. Each rocker is controlled via a separate cam (**Fig. 3**). The cam profile which actuates the two outer rockers differs from the cam profile which actuates the inner rocker. This enables the following variables to be varied at the valve gear:

- Valve overlap
- Opening speed
- Valve-opening time
- Valve lift

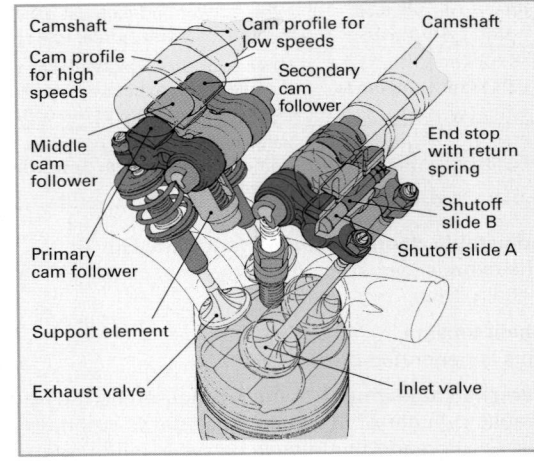

Fig. 3: Design of a variable valve gear

Valve actuation (Fig. 4)

Position 1. The rockers are released. The return spring holds the two shutoff slides A and B in the released position. The valves are actuated by the two outer rockers. This produces a small valve lift and a shorter valve-opening time. This position is ideal for small to low speeds.

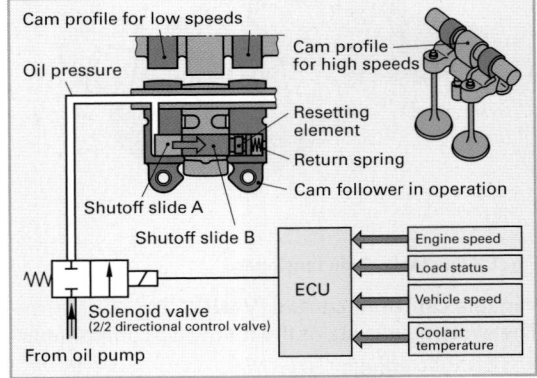

Fig. 4: Changeover from one cam shape to another

Position 2. At the changeover point the solenoid valve opens on a signal from the engine control unit. The engine-oil pressure acts on shutoff slide A. The two shutoff slides A and B are displaced to the right against the force of the return spring and lock the three rockers positively to each other. The valves are actuated in this position by the middle cam with the largest valve lift and the longest valve-opening time.

Fully variable electromechanical valve gear

> Valve lift and valve-opening angle are steplessly changed.

The camshaft acts on the intermediate lever. The angled underside of the intermediate lever actuates the rocker, which opens the valve. The intermediate lever reciprocates between the cam and the return spring as the camshaft rotates. The position of the pivot determines the extent of the reciprocating motion and thus the size of the valve lift **(Fig. 1)**.

Large reciprocating motion ⇒ **large valve lift**
Small reciprocating motion ⇒ **small valve lift**
Adjustment range: 0.3 mm to 9.85 mm

Advantage. Charge adjustment is performed via the valve-opening cross-section. There is no need for a throttle valve. There are therefore no throttling losses caused by a throttle valve. The valve-opening time is VANOS-controlled.

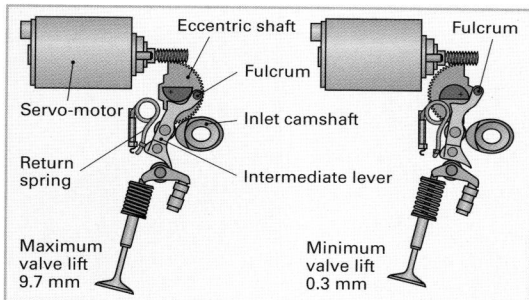

Fig. 1: Fully variable electromechanical valve gear

11.7.2 Supercharging (charge adjustment)

The extent of power output and the torque of an engine are significantly determined by the fresh-air percentage of a cylinder charge during induction. This is expressed by the volumetric efficiency.

> The volumetric efficiency specifies the ratio between the fresh-gas charge present in the cylinder and the theoretically possible fresh-gas charge of a cylinder for each power cycle.

Table 1: Volumetric efficiency of naturally aspirated engines and supercharged engines

Engine type	Volum. efficiency
Nat. aspirated engines, 4-stroke	0.7 to 0.9
Nat. aspirated engines, 2-stroke	0.5 to 0.7
Supercharged engines	1.2 to 1.6

The volumetric efficiency can be increased with the aid of supercharging systems. In this way, a greater mass of air is admitted into the combustion chamber so that more fuel can be burned.

Supercharging limits

Spark-ignition engines. Excessively high volumetric efficiency causes an excessively high final compression pressure and knocking combustion in supercharged spark-ignition engines. This in turn may cause damage, for example, to the pistons and bearings. For this reason, supercharged spark-ignition engines have lower geometric compression ratios than their naturally aspirated counterparts.

Diesel engines. Diesel engines can be subjected to such high mechanical loads by excessively high final combustion pressures on account of the high fresh-air percentage and the possible greater injected fuel quantity that the engines will be destroyed.

Internal-combustion engines differentiate between a geometric and an effective compression ratio. In this case, in order to prevent engine damage, it is essential not to exceed specific limit values. Boost-pressure limitations are therefore necessary in supercharged engines.

Geometric compression ratio ε_{geo}. This is the ratio of the largest combustion chamber to the smallest combustion chamber.

$$\varepsilon_{geo} = \frac{V_h + V_C}{V_C}$$

Effective compression ratio ε_{eff}. This can be calculated from the geometric compression ratio and the volumetric efficiency.

$$\varepsilon_{eff} \approx \varepsilon_{geo} \cdot \text{volumetric efficiency}$$

Supercharging systems

The following different systems may be used:
- Dynamic supercharging
- Independent supercharging

11.7.2.1 Dynamic supercharging

The fresh gases flowing in the intake manifold have kinetic energy. A returning pressure wave is triggered when the inlet valve is opened. The fresh gases flow back at the speed of sound and meet the static air at the open end of the intake manifold. There the pressure wave is reflected again and returns in the direction of the inlet valve. If the returning pressure wave reaches the inlet valve when the latter is just open, this brings about an improvement in the cylinder charge. A supercharging effect is created. The frequency of the vibration created is dependent on the intake-manifold length, intake-manifold cross-section and engine speed.

The following different systems are used, depending on the layout of the intake manifold and the associated supercharging:
- Ram-effect supercharging
- Tuned-intake pressure charging

Both systems can be combined with each other.

Ram-effect supercharging. Each cylinder has an intake manifold of a specific length. The gas-column vibration is excited by the induction work of the piston. The vibration is influenced by an appropriate choice of intake-manifold length in such a way that the pressure wave passes through the opened inlet valve and effects a better charge. Long intake manifolds with small cross-sections have proven efficacious in the lower speed range while short intake manifolds with large cross-sections are favoured in the upper speed range **(Fig. 1)**.

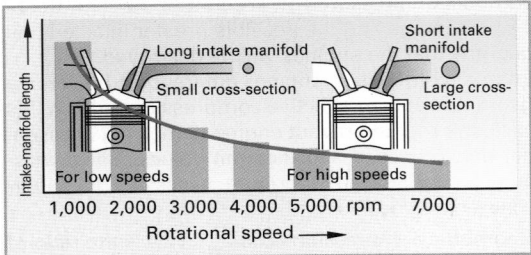

Fig. 1: Connection between ram-tube length and speed

Ram-tube systems

The following different types of intake-manifold system may be used:

- Switch-over intake manifolds
- Infinitely variable induction systems

Switch-over intake manifolds (Fig. 2). Short and long intake manifolds are combined here. In the lower speed range air flows through long ram tubes. The short induction passages are sealed by flaps or by rotating slides.

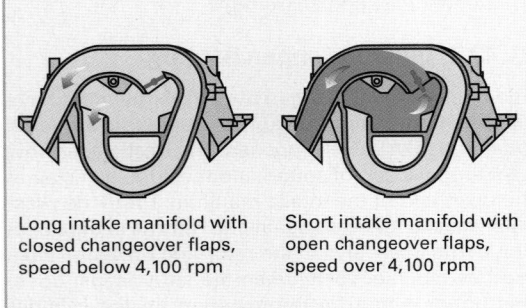

| Long intake manifold with closed changeover flaps, speed below 4,100 rpm | Short intake manifold with open changeover flaps, speed over 4,100 rpm |

Fig. 2: Ram-tube systems

At high speeds the flaps are electropneumatically or electrically opened; all the cylinders thus draw in air through short intake manifolds.

Fig. 3 shows that a higher and more uniform torque combined with greater power is achieved in the speed range up to 4,100 rpm with an engine with a switch-over intake manifold.

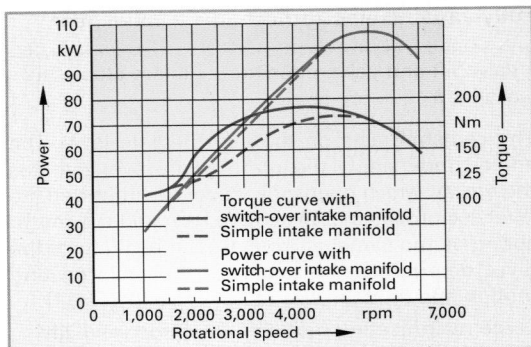

Fig. 3: Torque and engine power as a function of intake-manifold length

Infinitely variable induction system (Fig. 4). A rotor ring which changes the opening of a manifold volume is rotated as a function of speed and thereby the effective intake-manifold length is adapted to the speed. The ring is rotated by a stepping motor.

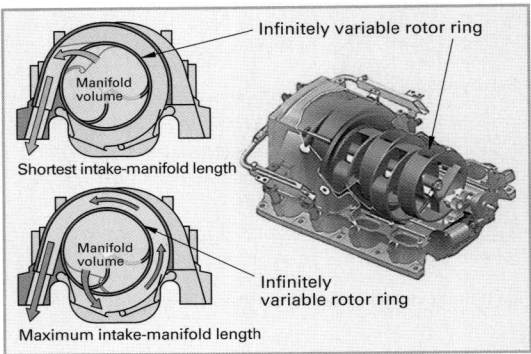

Fig. 4: Infinitely variable induction system

Tuned-intake pressure (resonance) charging. The frequency of the vibrating gas column is influenced by the valve-opening frequency **(Table 1)**.

Table 1: Valve-opening frequency and frequency of vibrating gas column		
Speed	Valve-opening frequency	Frequency of gas column
High	High	High
Low	Low	Low

Resonance occurs when the valve-opening frequency of the timing gear coincides with the frequency of the gas-column vibrations.

Resonance. This is the amplified natural oscillation of an oscillatory system.

The natural oscillation of a system is dependent on the magnitude of its vibrating mass. Large masses cause low-frequency, long vibrations while small masses cause high-frequency, short vibrations.

If a further mass is cut in to the vibrating gas column in an intake manifold by the opening of a resonance valve **(Fig. 1)**, its vibrating mass is increased and the frequency reduced. At low speeds this results in supercharging and thus a better charge as a result of the sympathetic (resonant) vibration.

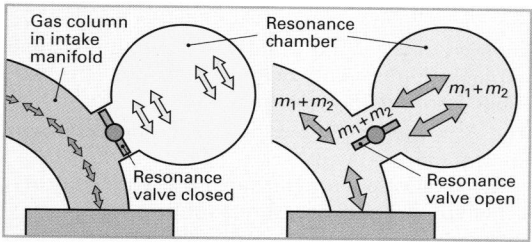

Fig. 1: Tuned-intake pressure charging

Tuned-intake pressure and ram-tube system (Fig. 2). Tuned-intake pressure and ram-tube systems are combined with each other so that the supercharging effects of the individual systems can be exploited. An improved charge is achieved, for example, in the lower to middle speed range by tuned-intake pressure charging and in the upper speed range by ram-effect supercharging **(Fig. 3)**. For this purpose, a flap is electrically or electropneumatically opened or closed in the intake-manifold system depending on the speed.

Example: Tuned-intake pressure charging in the lower to middle speed range, changeover flap closed.

When, for example, cylinder no. 2 draws in air, the chamber of cylinder bank 4, 5, 6 acts as an additional resonance chamber. In this way, the frequency of the vibrating mass is reduced and the valve-opening frequency is adapted.

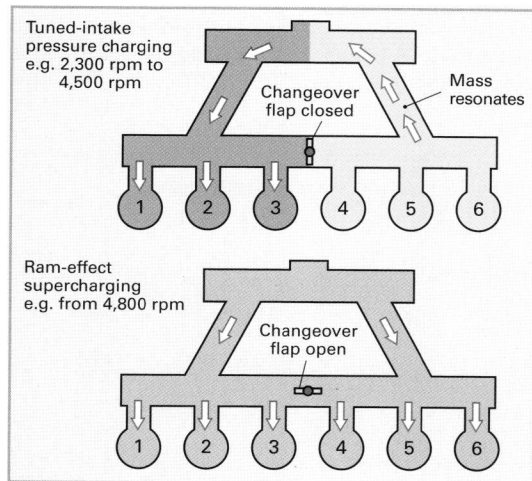

Fig. 2: Tuned-intake pressure and ram-tube system

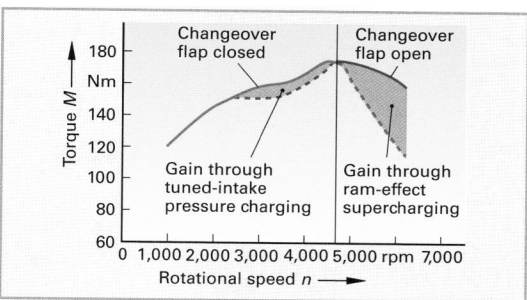

Fig. 3: Torque with combined tuned-intake pressure/ram-effect supercharging

11.7.2.2 Independent supercharging

As large a quantity as possible of fresh gas is delivered by a charger to the cylinder during the induction stroke. In addition, the fuel-air mixture or the air is precompressed completely or partially outside the cylinder. The following different types of charger are used:

- Charger without mechanical drive, e.g. exhaust-gas turbocharger **(Fig. 4)**
- Charger with mechanical drive, e.g. Roots blower, compressor, spiral-type charger, sliding-vane charger
- Pressure-wave charger, e.g. Comprex charger

Charger without mechanical drive

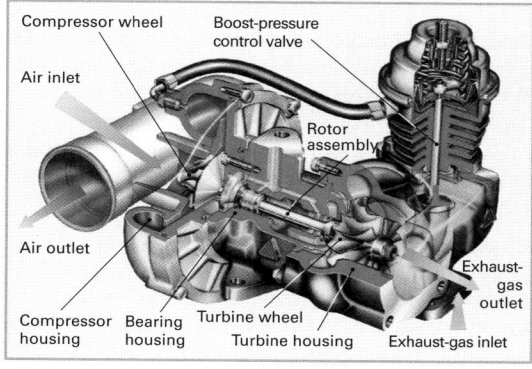

Fig. 4: Design of an exhaust-gas turbocharger

Exhaust-gas turbocharger.
Here the energy of the exhaust gas is used to deliver the fresh gases to the cylinders **(Fig. 4, Fig. 1, Page 244)**.

A significant boost effect is only achieved at middle to high speeds. Furthermore, these chargers respond with a slight delay to rapid changes in the accelerator-pedal position because the exhaust gases cannot follow the rapid load changes on account of mass inertia (turbo lag). These chargers operate virtually without any losses since they do not require any drive power from the crankshaft.

11

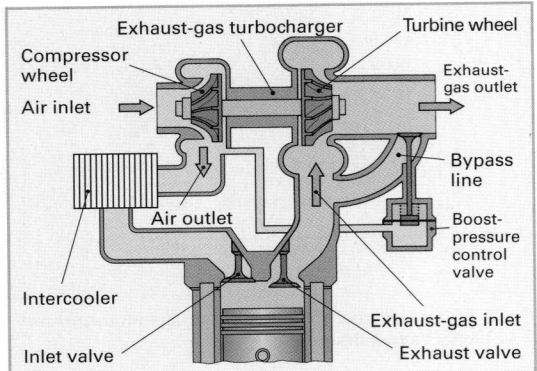

Fig. 1: Diagram of an engine with exhaust-gas turbocharger

Rotor assembly (Fig. 2). This consists of a turbine wheel with shaft and a compressor wheel.
Depending on the charger design, the rotor reaches continuous speeds of 50,000 rpm to 400,000 rpm.

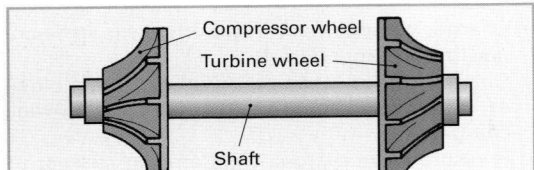

Fig. 2: Exhaust-gas-turbocharger rotor assembly

Operating principle. The exhaust gases from the engine drive the turbine wheel in the turbine, which drives the compressor wheel via the shaft.

The compressor draws in the fresh gases and supplies the engine with a precompressed fresh-gas charge. The charge air is heated by precompression by up to 180 °C.

Charge-air cooling and boost pressures. The heated air precompressed by the charger can be cooled by means of charge-air cooling before entering the cylinders. This increases the air density of the fresh-gas charge. The greater air mass allows a greater quantity of fuel to be injected. Engine power is increased. The boost pressures with and without charge-air cooling are set out in **Table 1**.

Table 1: Boost pressures as a function of charge-air cooling	
Supercharged engines	Overpressure in bar
Without charge-air cooling	0.2 to 1.8
With charge-air cooling	0.5 to 2.2

The boost pressures of an engine supercharged by an exhaust-gas turbocharger must not exceed the boost pressures stipulated by the manufacturer as this would destroy the engine.

Boost-pressure control. In addition to the danger of the engine being destroyed by excessively high boost pressures, the size of the turbocharger is established in such a way that a supercharging effect is also achieved at middle speeds and low exhaust-gas flows. The upshot of this is that at high engine speeds and with large exhaust-gas quantities either the turbocharger boost pressure becomes unacceptably high or the charger is operated in unacceptably high speed ranges. The boost pressure must therefore be regulated.

The following different types of boost-pressure control may be used:
- Mechanical-pneumatic boost-pressure control
- Electronic boost-pressure control
- Boost-pressure control with adjustable guide vanes

Mechanical-pneumatic boost-pressure control (Fig. 1).
Here a diaphragm pretensioned with a helical spring is pressurised with the boost pressure in a boost-pressure control valve **(Fig. 3)**. The valve opens as soon as the boost pressure overcomes the initial spring tension. The exhaust gases flow in the bypass line past the turbine into the exhaust pipe.

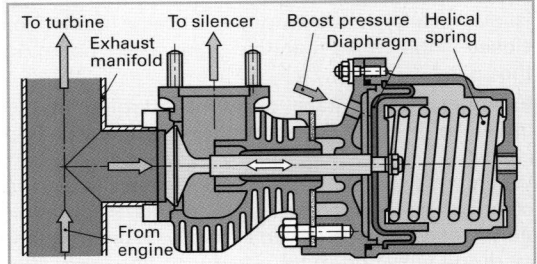

Fig. 3: Boost-pressure control valve

The boost-pressure control valve can be situated at any point in the exhaust system before the exhaust-gas turbine. A bypass flap can also be integrated in place of the boost-pressure control valve **(Fig. 4)**.

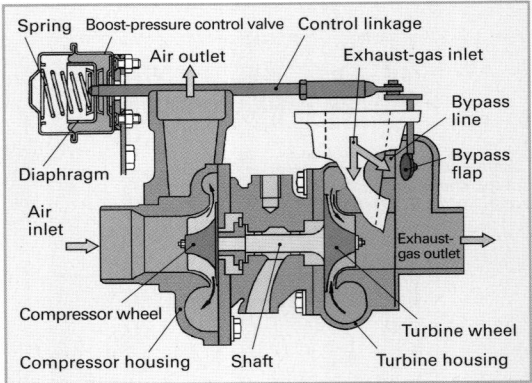

Fig. 4: Boost-pressure control with bypass flap

Here the flap, which opens and closes the bypass by way of a linkage, is connected to the control capsule, which is usually mounted on the compressor. Because the control capsule is located at a greater distance from the hot components, the thermal load of the plastic diaphragm is not as high, thereby minimising the risk of failure.

A high dynamic pressure is generated at the compressor in overrunning mode and when the throttle valve is closed. This pressure brakes the compressor wheel in such a way that deceleration occurs in the event of a sudden load change. In order to facilitate unrestricted continued operation of the compressor wheel in overrunning mode, it is possible for boost-pressure control systems to be equipped with an intake-manifold-pressure-controlled circulating-air valve (wastegate, **Fig. 1**). When the throttle valve is closed, the wastegate enables the precompressed air to be repumped from the compressor side to the intake side of the compressor.

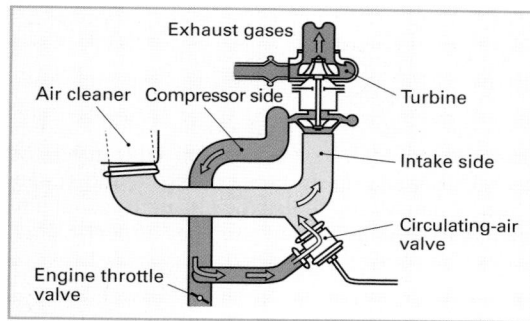

Fig. 1: Circulating-air valve

Electronic boost-pressure control (Fig. 2).
The optimal boost pressure is calculated by a boost-pressure ECU as a function of throttle-valve position and knock tendency. Intake-air temperature, engine temperature and rotational speed, for example, serve as the correction variables. Air-pressure fluctuations, e.g. when driving at high altitudes, are compensated because a height sensor in the engine control unit constantly measures the ambient-air pressure and takes it into account when calculating the boost pressure.

Operating principle
A pressure sensor records the boost pressure and the boost-pressure ECU activates a cycle valve **(Fig. 2)**. The duty factor controls the opening cross-section.

Boost pressure low. The cycle valve opens the connection between the pressure pipe and the suction side. A low boost pressure acts on the boost-pressure control valve, which remains closed. The turbine is driven by the entire exhaust-gas flow.

Boost pressure high. The boost-pressure sensor signals an excessively high boost pressure to the boost-pressure control ECU. The cycle valve closes the connection between the pressure pipe and the

intake manifold. The boost pressure in the control line rises. The boost-pressure control valve opens and the exhaust-gas flow to the turbine is reduced.

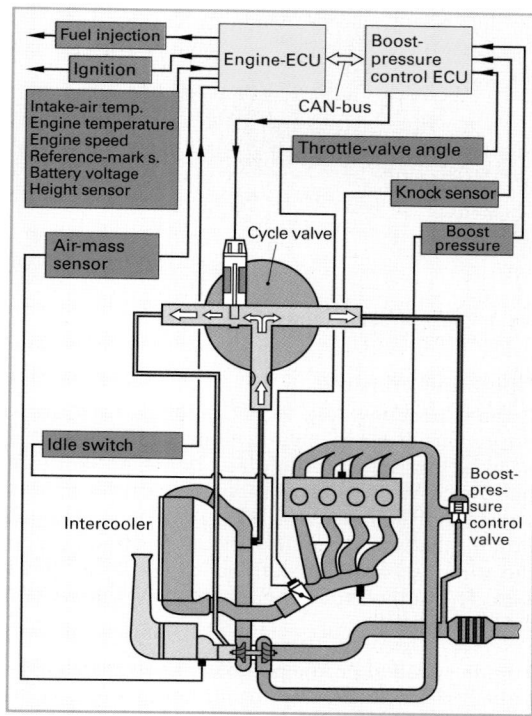

Fig. 2: Electronic boost-pressure control

"Overboost". This refers to a brief excessive increase in boost pressure for accelerating. When the accelerator pedal is quickly pressed to the floor (kickdown), the boost-pressure control valve is controlled by the cycle valve. The entire exhaust-gas flow is routed through the turbine and the boost pressure increases suddenly. The standard control operation is resumed after the desired driving speed has been reached.

Advantages of electronic boost-pressure control over mechanical-pneumatic boost-pressure control:

- Better response.
- Constant power output because it is air-pressure-independent (absolute-pressure control).
- Variable boost pressure which can be increased up to the knock limit.

Boost-pressure control with variable turbine geometry (VTG) (Fig. 1, Page 246)
In the case of this charger, the boost pressure is regulated by variable guide vanes. Regulation is performed irrespectively of the exhaust-gas flow determined by the engine speed.

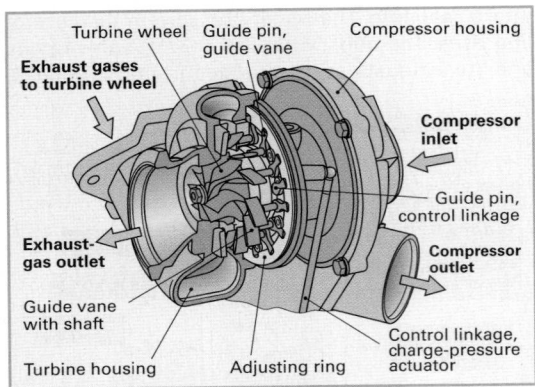

Fig. 1: Boost-pressure control with VTG

Operating principle

Engine speed low (Fig. 2). A high boost pressure is desired in order to have a high torque available even at low speeds. The guide vanes are set to a narrow inlet cross-section for this purpose. This constriction increases the speed of the exhaust-gas flow. At the same time the exhaust-gas flow acts on the outer area of the turbine vanes (large lever arm). The turbine speed and thereby the boost pressure increase.

Engine speed high. The guide vanes release a larger inlet cross-section in order to be able to accommodate the large exhaust-gas quantity at high speeds. In this way, the necessary boost pressure is reached but not exceeded. The exhaust-gas flow acts on the middle area of the turbine vanes.

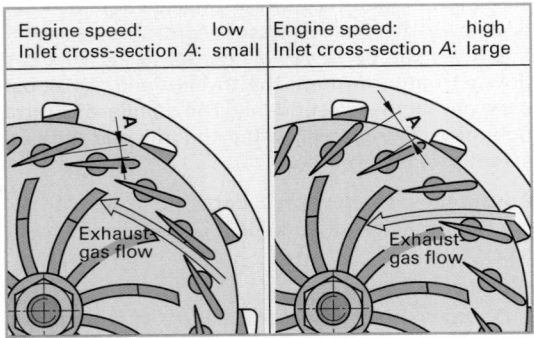

Fig. 2: Guide-vane setting

The change in the inlet cross-section can be used, for example, to achieve an additional, short-term increase in boost pressure at high speeds (overboost). There is no need for a bypass because the optimal boost pressure can be set for each operating state by adjusting the guide vanes. When the ECU signals engine limp-home operation, the guide vanes are controlled in such a way as to re-

lease the largest inlet cross-section, and boost pressure and engine power output are reduced.

Guide-vane adjustment (Fig. 1)

This is performed by means of a control linkage whose guide pin engages the adjusting ring. In this way, the adjusting ring can be rotated. This rotary motion is transmitted via the guide pin and the shaft to the guide vanes. All the guide vanes supported in the carrier ring are simultaneously and uniformly rotated into the desired setting. The guide vanes are adjusted electropneumatically.

Charger with mechanical drive

Screw-type supercharger (Roots blower, Fig. 3). The charger is driven by the engine via a magnetic clutch. The charger can be cut out for example at idle and cut in during acceleration and at full load via the magnetic clutch.

Advantages over an exhaust-gas turbocharger:
- No intervention in the engine's exhaust system.
- Rapid build-up of boost pressure.
- High torque, even at low speeds.

However, some of the additionally yielded engine power (up to 50 kW) must be utilised to drive the charger, depending on boost pressure and speed. Thus, engines with Roots blowers are less fuel-efficient than engines with exhaust-gas turbochargers.

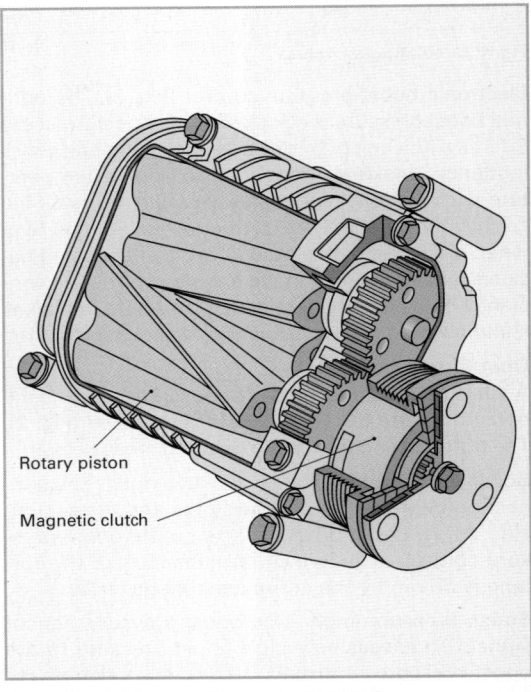

Fig. 3: Roots blower with magnetic clutch

12 Mixture formation

12.1 Fuel-supply systems in spark-ignition engines

12.1.1 Functions of systems

> The fuel-supply system **(Fig. 1)** is intended to supply the engine's mixture-formation system with sufficient fuel in all operating states.

For this purpose, the system must ...
- ... store fuel in the fuel tank.
- ... deliver bubble-free fuel.
- ... filter fuel.
- ... generate fuel pressure and keep it constant.
- ... return excess fuel.
- ... prevent fuel vapours from escaping.

12.1.2 Design of systems (Fig. 1)

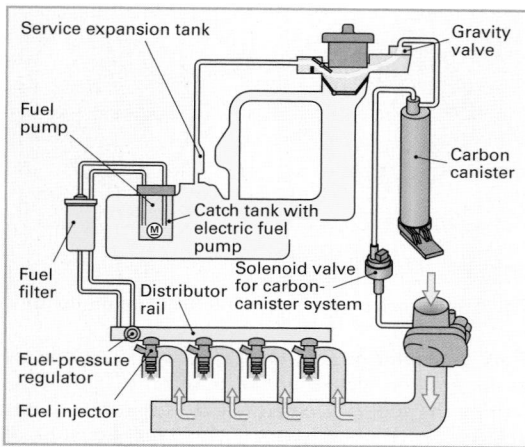

Fig. 1: Design of a fuel-supply system

The fuel is stored in the fuel tank. From there it is supplied under pressure by a fuel pump to the fuel injectors. A fuel filter is connected downstream of the fuel pump to retain any contaminants. The fuel pressure is kept constant or adapted to the intake-manifold pressure by a pressure regulator. In order to be able to provide sufficient fuel in all operating states, the system always supplies more fuel than is actually required at any time. The excess fuel returns from the pressure regulator to the fuel tank. An expensive ventilation system is required in view of the fact that neither fuel nor fuel vapours are permitted to escape into the environment and that pressure compensation must be created in the tank. The fuel vapours are temporarily stored in the carbon canister of the regenerating (purge) system

and specifically directed via the regenerating (purge) valve for combustion. The fuel tank must be monitored for leaks when On-Board Diagnosis II (OBD II) comes into force.

12.1.3 Components of systems

Fuel tank

Sheet-steel fuel tanks, on account of their simple structural shape and associated problem-free manufacture, are usually used in commercial vehicles. They are coated on both the inside and the outside with anticorrosion linings. In the case of large and partially filled fuel tanks, sudden, severe weight transfers may occur when the vehicle is cornering. This is prevented by the use of perforated partitions which divide the tank into several small compartments.

Steel fuel tanks are increasingly being manufactured for use in passenger cars as well. To obtain a categorisation as a low-emission vehicle (LEV) in the USA, it is necessary to limit greatly the emission of hydrocarbons, which also include vaporised fuels. (According to OBD II the evaporative losses in the fuel system must not exceed 2 g per day). These restrictive figures can be achieved more easily by using steel tanks rather than plastic tanks.

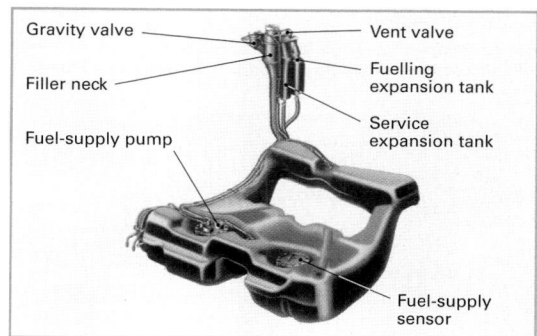

Fig. 2: PE fuel tank for a passenger car

For complicated fuel-tank shapes, as are commonly found in passenger cars, the tanks are manufactured predominantly from plastic, e.g. PE (polyethylene). These provide a high level of safety against bursting. (The tanks must be able to withstand a crash at 80 km/h.) However, there is the risk of plastic deformation at high fuel temperatures (more than 120 °C in diesel-injection systems) as well as the problem of heavy diffusion of the fuel vapours.

12

In extreme cornering situations or driving on steep terrain and with only a small amount of fuel left in the tank, the fuel is displaced to one side. Catch tanks **(Fig. 1, Page 247)** are used to ensure a sufficient supply of fuel to the fuel-supply pump and to be able to empty all the branched tank compartments. There are tanks inside the fuel tank which are filled by suction-jet pumps. The fuel-supply pumps are also located in the catch tanks (see also Fuel-delivery modules).

Fuel-supply pumps

Modern fuel-injection systems exclusively use electric fuel pumps for fuel supply and delivery. The delivery quantities of such pumps at nominal voltage range between 60 l/h and 200 l/h. In the process a pressure of 3 bar … 7 bar (as a presupply pump in direct-injection systems) must be achieved at 50 % … 60 % of the nominal battery voltage. Because this delivery at nominal voltage results in a multiple of the required fuel quantity being delivered at idle and part load, the electric fuel pumps go over to being activated by the ECU with pulse-width-modulated signals. In this way, the delivery quantity can be adapted to the operating conditions, which can save drive power, stop the fuel from being unnecessarily heated and extend the service life of the pumps.

A pump of this type consists of the
- fitting cover with electrical connections, non-return valve and pump outlet
- electric motor with armature and permanent magnet
- pump section

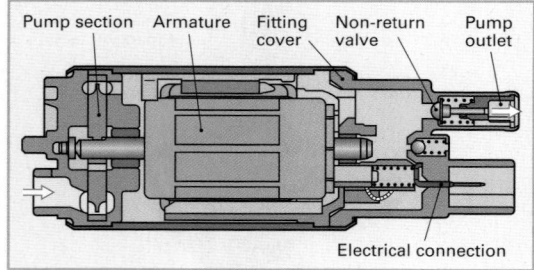

Fig. 1: Design of an electric fuel pump

Two different types are used, depending on their installation locations: inline pumps and in-tank pumps.

Inline pumps. These can be installed at any point in the fuel line. They are therefore easier to replace when faulty than in-tank pumps. However, they and in particular their electrical connections are exposed to increased corrosion when installed under the vehicle floor.

In-tank pumps. These usually form part of fuel-delivery modules which are installed in the fuel tanks. In-tank pumps are provided with extensive corrosion protection in the fuel tank. The noises generated by the pump in the tank are also damped.

The pumps are divided into positive-displacement and flow-type pumps, depending on the way in which they operate.

Positive-displacement pumps (Fig. 2). These are designed as either roller-cell pumps or internal-gear pumps. The fuel is drawn into the pump and delivered in a sealed chamber which decreases in size to the high-pressure side. Positive-displacement pumps facilitate system pressures of more than 4 bar and also have high delivery rates at low voltages. However, they do cause relatively strong pulsation noises. They are also subject to a marked drop in power if vapour bubbles are formed in the hot petrol. For this reason, these pumps usually have a flow-type pump as a preliminary stage for degassing.

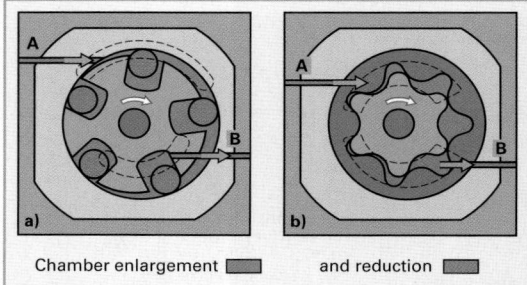

Fig. 2: Roller-cell pump (a) and internal-gear pump (b)

Flow-type pumps (Fig. 3). These are designed as peripheral or side-channel pumps. In flow-type pumps the fuel is accelerated by numerous vanes and pressure is built up by a constant exchange of pulses. Flow-type pumps operate with little noise because the pressure build-up is virtually pulsation-free and continuous. They are also insensitive to vapour-bubble formation as fuel in the vapour state can be separated via a degassing bore. However, these pumps only achieve system pressures of max. 4 bar.

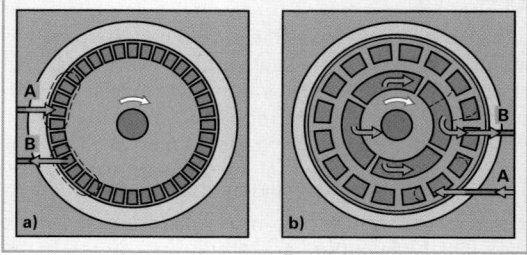

Fig. 3: Peripheral pump (a) and side-channel pump (b)

Two-stage electric fuel pumps (Fig. 1). Two-stage electric fuel pumps are used if high system pressures are required. A peripheral pump is connected upstream in order to prevent vapour bubbles from forming in the pump. It assumes the fuel presupply role and separates vapour bubbles. The downstream positive-displacement pump serves to build up the system pressure.

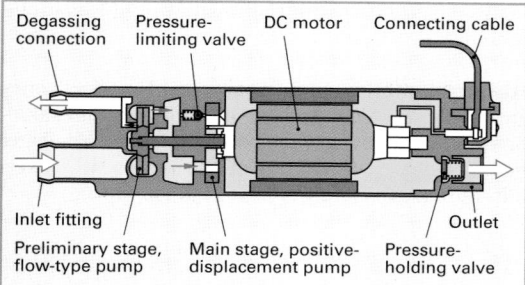

Fig. 1: Two-stage inline fuel pump

Suction-jet pumps (Fig. 2)

These are hydraulically driven pumps which serve to pump fuel inside the fuel tank. Thanks to the fuel flow of an electric fuel pump, fuel is drawn at the nozzle opening of a suction-jet pump for example out of the side chamber of a fuel tank. This fuel is then delivered to the catch tank.

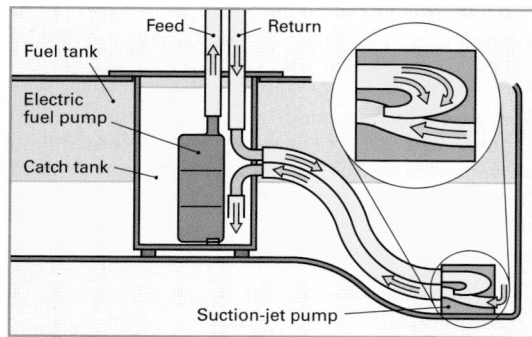

Fig. 2: Suction-jet pumps

Fuel lines

Steel pipes or hoses made from flame-retardant, fuel-resistant rubber or plastic are used as fuel lines. Because rubber and plastic hoses change chemically (age) when used for long periods, they become hard and porous. This may result in leaks.

It is important when laying fuel lines to ensure that ...

- ... they are able to withstand the torsion of the vehicle and the movements of the engine.

- ... they are protected against mechanical damage.
- ... the lines are not routed past hot parts - in order to avoid vapour-bubble formation.
- ... they are where possible laid in a steadily rising direction so that vapour bubbles can be quickly removed from the system.
- ... no fuel vapours can collect in the vehicle in the event of leaks.

Fuel filters

These are intended to protect the fuel system against contaminants because, for example, the fuel injectors of a petrol injection system can be destroyed even by tiny dirt particles.

Fuel-pressure regulator (two-line system)

The fuel-pressure regulator must keep the fuel differential pressure in the two-line system constant under all conditions.

The diaphragm-controlled fuel-pressure regulator **(Fig. 3)** with intake-manifold connection is located in two-line systems on the fuel rail. It consists of two chambers which are separated by a diaphragm: a spring chamber for housing the spring which acts on the diaphragm and a chamber for the fuel. When the preset fuel-system pressure is exceeded, a valve actuated by the diaphragm opens the opening for the return line, through which the excess fuel can flow back to the fuel tank.

Because the spring chamber is connected via a line to the intake manifold shortly after the throttle valve, the diaphragm is deformed by not only the

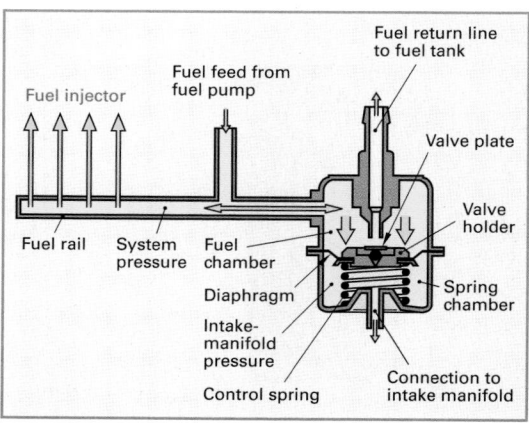

Fig. 3: Fuel-pressure regulator

fuel pressure but also by the vacuum pressure acting in the intake manifold. Thus the fuel-pressure regulator alters the system pressure in the fuel rail or at the fuel injectors in such a way that the differential pressure between the intake manifold and the fuel system remains constant.

Differential pressure = system pressure – intake-manifold pressure

If, for example, there is a vacuum pressure of – 0.6 bar in the intake manifold, the valve diaphragm is opened by fuel and intake-manifold pressure against spring force to such an extent that the system pressure drops for instance to 3.4 bar. The differential pressure Δp is therefore 3.4 bar – (– 0.6) bar = 4.0 bar.

Table 1: Examples of fuel pressures

	Differential pressure	System pressure	Intake-manifold pressure
Idle	4.0 bar	3.4 bar	– 0.6 bar
Part load	4.0 bar	3.7 bar	– 0.3 bar
Full load	4.0 bar	3.9 bar	– 0.1 bar

In **R**eturn-**L**ess **F**uel **S**ystems (**RLFS**) the almost identically designed pressure regulator is located in the fuel tank (**Fig. 1**). The fuel-system pressure is kept constant by the spring and the diaphragm. There is no intake-manifold connection. The excess fuel returns directly to the fuel tank, which is why there is no return line from the intake manifold.

Because the injected fuel quantities change as the intake-manifold pressures change, the ECU must adapt the injection time as a function of the intake-manifold pressure. It receives information on the intake-manifold pressure from an intake-manifold pressure sensor.

Fuel-delivery modules (Fig. 1)
The fuel-delivery components are combined in fuel-delivery modules, which are installed in the fuel tank.

Fuel gauge. Lever sensors or submerged-tube sensors are normally used to indicate the fuel level. These sensors pick off the conductor tracks of a potentiometer via a linkage. The voltage drop at the resistor is the measure of the amount of fuel in the fuel tank.

Fuel-consumption measurement. The fuel consumption is calculated by multiplying the valve-opening time by a valve constant. This specifies how much fuel flows out of the nozzle at a fixed differential pressure per unit of time.

12.1.4 Fuel-tank ventilation

It is necessary to ventilate the fuel tank to be able to create pressure compensation in the tank and to enable the vehicle to be refuelled without complications. Thus, it is essential under the influence of heat to ensure that expanding fuel and the increased gas pressure caused by this can be taken up in expansion tanks. On the other hand, the fuel tank must be ventilated when fuel is consumed during vehicle operation. Under no circumstances may fuel vapours be allowed to escape into the environment. The ventilation system comprises the following components (**Fig. 2**):

Service expansion tank. This takes up the fuel that expands as a result of heat. The volume, depending on the size of the fuel tank is 2 l … 5 l. The expansion tank is connected via a vent line to the carbon canister.

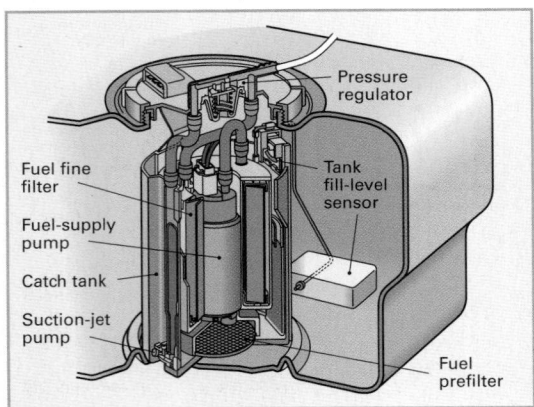

Fig. 1: Fuel-delivery module

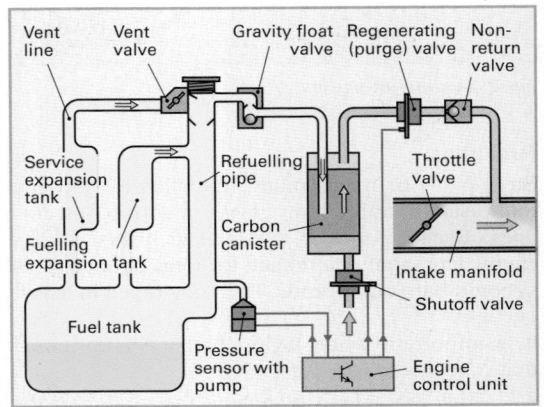

Fig. 2: Ventilation system

Fuelling expansion tank. The function of this tank is to take up briefly gases in the fuel tank which are displaced when the vehicle is being refuelled and to direct these gases via a vent line to the refuelling pipe. There these vapours are drawn off by the suction device of the fuel-pump nozzle.

Vent valve. This prevents fuel vapours from escaping from the service expansion tank into the environment or being drawn off. The valve is closed during refuelling.

Gravity float valve (Fig. 1) (rollover valve, safety valve). When the tank is absolutely full and the vehicle is in an inclined position or if the vehicle rolls over, fuel could escape to atmosphere via the carbon canister. To prevent this from happening, the line to the carbon canister is closed by the valve in such a situation.

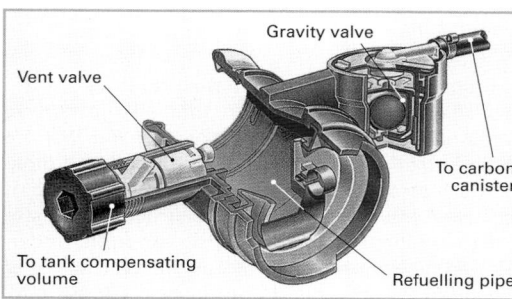

Fig. 1: Gravity and vent valve

Carbon canister. This filter stores gaseous hydrocarbons by adsorption on the activated carbon until they are drawn in by the vacuum pressure prevailing in the intake manifold when the regenerating valve is open and supplied for combustion in the cylinder.

Shutoff valve (from OBD II). When the engine is stopped, the line supplying incoming air to the carbon canister must be closed to prevent fuel vapours from escaping to the atmosphere. When the activated carbon is regenerated and the stored fuel vapours forwarded for combustion, the solenoid valve is clocked and opened by the engine ECU parallel to the regenerating valve.

Regenerating (purge) valve. This solenoid valve is clocked by the engine ECU depending on the operating status. When it opens, the fuel particles stored in the carbon canister are purged while the shutoff valve is also open by fresh air and drawn in by the intake-manifold vacuum pressure.

Diagnosis pump for fuel system, pressure sensor. The fuel system must be checked for leaks in accordance with OBD II. For this purpose, the fuel tank can be subjected to a pressure generated by a diagnosis pump. A pressure sensor transmits the pressure characteristic to the engine ECU. The ECU decides whether the system satisfies the requirements with regard to leaks.

WORKSHOP NOTES

Notes on servicing/maintenance:
- Regularly change the fuel filters (follow manufacturer's instructions).
- Visually inspect the system for leaks.
- Visually inspect the electrical connection (for corrosion, damage).

Faults and possible causes:
- Engine fails to fire:
  - No fuel in the tank
  - Pump not running
- Insufficient engine power:
  - Delivery rate too low
  - Delivery pressure too low due to kinked or crushed fuel line, lack of pump power supply, clogged filters, faulty pump (wear), vapour-bubble formation

Diagnostic options:
- Checking the delivery rate: measurement at the pressure regulator (in the return line).
- Checking the delivery pressure: measurement at the fuel rail (in the feed line).

- Pump power supply: faults in the electrical system are generally detected by self-diagnosis. Therefore the fault memory must be read out or an actuator diagnosis carried out. The positive and negative supply can be checked with a multimeter.

Fault causes: **(Fig. 1, Page 252)**
  - Faulty fuse in the fuel-supply pump (check continuity)
  - Faulty fuel-pump relay (to check, jumper term. 30 – term. 87)
  - Damaged leads and corroded contacts (measure voltage drop)

In addition to a faulty power supply, a fuel-supply pump which is not working may be caused by among others the following:

- Fuel-supply pump is faulty.
- Engine ECU receives no engine-speed signal after the ignition is switched on.
- Engine ECU receives no enable from the immobiliser.
- Engine ECU receives a crash-situation signal from a built-in crash-detection system (airbag).

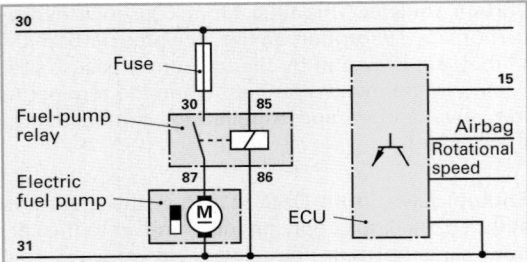

Fig. 1: Circuit diagram of fuel-supply pump

12.2 Mixture formation in spark-ignition engines

12.2.1 Basic principles

Spark-ignition engines can be run on petrol, methanol or liquefied petroleum gas. The compressed fuel-air mixture is ignited at the end of the compression stroke by a spark-ignition system.

Safety instructions

Petrol has a flash point of < 21 °C and therefore comes under danger class AI in accordance with the Directive on Combustible Liquids.

It is highly inflammable. It is therefore absolutely essential to follow the relevant safety precautions when welding, soldering or grinding.

Fuel vapours are heavier than air. They can therefore form hazardous mixtures in pits or drainage shafts.

Fuels contain benzene, methanol, toluene and xylene. These substances are toxic and must not be inhaled. Contact with the skin and mucous membranes must be avoided. Petrol should therefore never be used for cleaning purposes.

Function of mixture-formation systems

They should for each engine operating state form a fuel-air mixture which is sufficient in quantity and is combusted in the engine as fully as possible.

Complete combustion of a fuel-air mixture

This is understood to mean that all the carbon atoms and all the hydrogen atoms of the fuel are oxidised by the oxygen in the air into carbon dioxide (CO_2) or water (H_2O) under heat dissipation.

Because fuels, depending on the structure and size of their molecules, have differing amounts of carbon and hydrogen atoms, a quite specific air mass is needed for complete combustion of the fuel. Combustion deteriorates as the amount of deficient or excess air increases. The fuel will only combust incompletely. Combustion will not take place at all if specific limit values for the mixture ratio (ignition limits) are undershot or exceeded.

Mixture ratio

The mixture ratio describes the composition of the fuel-air mixture. There are two different types of mixture ratio: the theoretical and the practical.

Theoretical mixture ratio (stoichiometric ratio = theoretical air requirement). This specifies how many kg of air are required for the complete combustion of 1 kg of fuel. To burn 1 kg of petrol, this figure is roughly 14.8 kg or 10,300 l of air.

Practical mixture ratio. This deviates from the theoretical mixture ratio, depending on the engine operating state. A mixture with a larger proportion of fuel, e.g. 1 : 13, is known as a "rich" mixture (air deficiency). A mixture with a smaller proportion of fuel, e.g. 1 : 16, is known as a "lean" mixture (excess air).

REVIEW QUESTIONS

1   Which components make up the fuel-supply system?

2   Which types of fuel pump are used in a motor vehicle? In what way do they differ?

3   What is the function of the fuel-pressure regulator in a 2-line system?

4   Why do 1-line systems require an intake-manifold pressure sensor?

3   What are the functions of the intake-manifold pressure sensor?

6   What do you understand by a fuel-delivery module?

7   What important factors must be borne in mind when laying fuel lines?

8   Which components make up the ventilation system of a fuel-supply system?

9   What is the function of the service expansion tank?

10   For what purpose is a gravity float valve needed?

11   Why must fuels not be used for cleaning purposes?

Air ratio (λ = lambda)

The air ratio λ is the ratio of the actual air mass supplied for combustion to the air mass theoretically required for complete combustion.

$$\text{Air ratio } \lambda = \frac{\text{Supplied air mass in kg}}{\text{Theoretical air requirement in kg}}$$

With a theoretical mixture ratio of 1 : 14.8 the air ratio λ = 1 for petrol. Here the engine receives precisely the right amount of air that is needed for complete combustion of the fuel. If, on the other hand, 16 kg of air are supplied in the combustion of 1 kg of fuel, the air ratio is

$$\lambda = \frac{16.0 \text{ kg Luft}}{14.8 \text{ kg Luft}} = 1.08$$

i.e. a lean fuel-air mixture is formed containing more air than is needed for complete combustion. The excess air here is 8 %.

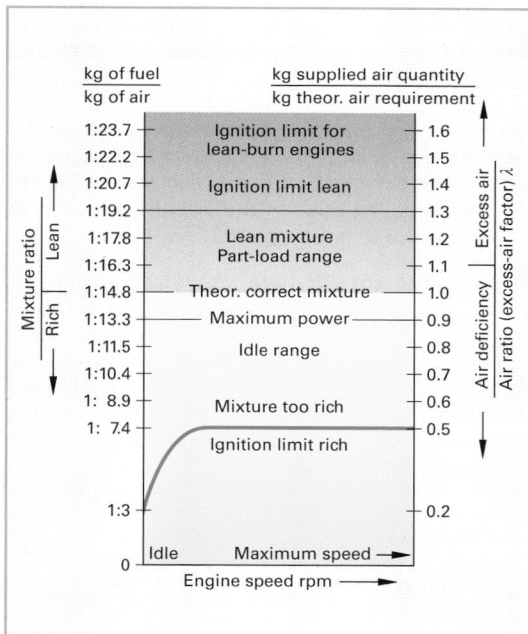

Fig. 1: Mixture ratios, air ratios for petrol

Consumption, power and exhaust-emission behaviour are dependent on the air ratio in the respective operating state of the spark-ignition engine.

The basic relationship between air ratio, torque and specific fuel consumption is shown in **Fig. 2**.

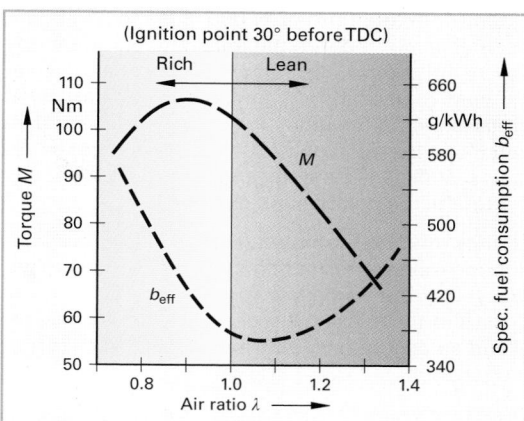

Fig. 2: Influence of air ratio

If engines with manifold injection are to be operated with an exhaust-gas catalyst, an air ratio of λ = 1 must be adhered to as closely as possible in order to obtain favourable exhaust-gas values.

Mixture composition

Homogeneous mixture. The mixture composition is the same in the entire combustion chamber. In order to achieve a homogeneous mixture composition, sufficient time must be made available for a uniform and thorough mixing of the fuel-air mixture. This is achieved by an advanced injection point during induction or by injection of fuel into the intake manifold.

Heterogeneous mixture. The combustion chamber has areas of differing mixture composition (stratified charge). A retarded injection of fuel into the cylinder during the compression stroke and precisely matched air turbulence facilitate a nonuniform mixture composition. There must be an air ratio of λ approximately equal to 1 to ensure safe ignition of the mixture in the spark-plug area in spark-ignition engines. The mixture is lean in the marginal areas of the combustion chamber.

Mixture formation

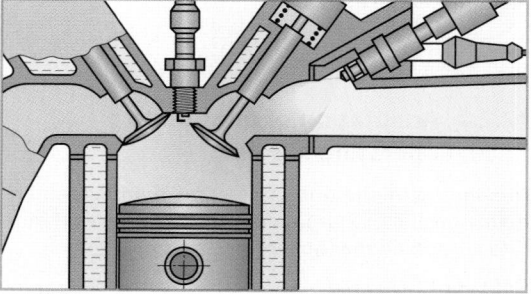

Fig. 3: Exterior mixture formation

Exterior mixture formation (Fig. 3, Page 253). Here the fuel is injected into the intake manifold shortly before the engine inlet valve, which is still closed at the start of injection. As a result of the admission process during the induction stroke and the subsequent compression of the fuel-air mixture, there remains sufficient time to create a homogeneous mixture in the combustion chamber.

Interior mixture formation (Fig. 1). In engines with interior mixture formation the fuel is injected directly into the combustion chamber. If this takes place shortly before the fuel-air mixture ignites, fuel and air cannot mix uniformly. The mixture is heterogeneous.

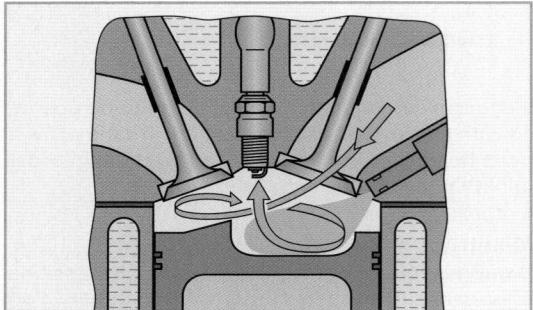

Fig. 1: Interior mixture formation

Power regulation

Quantity regulation. In engines with exterior mixture formation and a homogeneous mixture power regulation takes the form of the throttle valve being opened more or less depending on the load state. In this way, the amount of air inducted (quantity) is altered. The composition of the mixture must remain virtually the same here ($\lambda = 1$).

Quality regulation. In engines with interior mixture formation and a heterogeneous mixture power regulation takes the form of differing amounts of fuel being injected while the throttle valve is open. The amount of air inducted remains virtually the same here. In this way, the composition (quality) of the mixture in the combustion chamber changes depending on the load state.

12.2.2 Adaptation of mixture to operating states

Depending on the operating state, engines require quite specific mixture amounts (quantity) and mixture compositions (quality).

Cold start: In a cold engine only the low-boiling fuel constituents evaporate. The majority of the fuel condenses on the cold intake-manifold and cylinder walls. Thus these proportions of the fuel cannot be combusted or are only incompletely combusted. In order nonetheless to create an ignitable mixture in the combustion chamber, it is necessary to inject a very large amount of fuel (up to $\lambda = 0.3$). The amount of fuel injected is dependent here on the engine temperature.

More power must be generated in view of the fact that the frictional-resistance values are very high when the engine is cold, for instance due to the engine oil being cold. This is achieved by a greater amount of mixture.

Warming-up. This refers to the period from when the engine is started up to the point when normal operating temperature is reached. The fuel quantity is reduced as a function of temperature during the warming-up period. Enrichment of the mixture is reduced in stages as the condensation losses in the intake manifold and on the cylinder walls decreases as the engine warms up.

Transition, acceleration. The mixture is briefly leaned when the throttle valve is opened. More fuel must be injected briefly in order to prevent a momentary dip in power.

Full load. The operating condition in which the throttle valve is fully open is known as full load. In order to achieve maximum engine power in this operating state, the mixture is normally enriched to $\lambda = 0.85 \ldots 0.95$ **(Fig. 2, Page 253).**

Overrun fuel cut-off. Here the throttle valve is closed and the engine runs with increased revs. This occurs, for example, in downhill-driving situations or when the driver takes his/her foot off the accelerator pedal at high speed (overrun). In order to save fuel, no petrol is injected until the engine speed drops below a preprogrammed level or the throttle valve is reopened.

REVIEW QUESTIONS

1 **What do you understand by the theoretical mixture ratio?**

2 **Explain the air ratio λ.**

3 **What are the consequences in each case of a lean, a rich and a stoichiometric mixture?**

4 **Up to what mixture ratio or air ratio is a petrol-air mixture ignitable?**

5 **What characterises an interior mixture formation?**

6 **What do you understand by a homogeneous/ heterogeneous mixture?**

7 **What characterises an exterior mixture formation?**

8 **Why must the mixture be greatly enriched during cold-starting?**

12.3 Carburettor

> The carburettor is intended to atomise the fuel and mix it with air in the correct proportion. It must adapt the required amount of mixture to the respective engine operating state.

12.3.1 Basic operating principle

The air flow is drawn into the carburettor by the engine piston during the induction stroke. The velocity of the air flow is increased by narrowing the cross-section of the streamlined choke tube (venturi tube, **Fig. 1**). The highest flow velocity and the greatest vacuum (suction) occurs at the narrowest point, which is why the fuel outlet tube is located at this point. The fuel is carried by the air flow, atomised and mixed with the air flow in the mixing-chamber area. Fine atomisation is achieved by foaming the fuel with a supply of air through the air jet below the fuel level into a fuel-air mixture (preliminary mixture). The throttle valve serves to control the fuel-air mixture quantity (quantity control) and with it the engine power and speed.

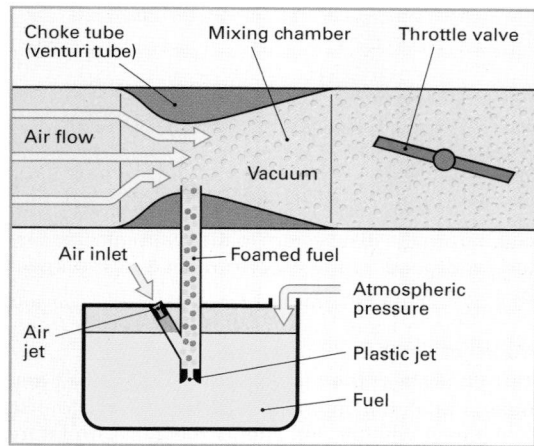

Fig. 1: Carburettor operating principle

12.3.2 Carburettor types

The following different types may be used, depending on the layout of the intake manifold on the engine and the direction of the suction flow in the carburettor itself: **down draught, side draught** and **semi-down draught carburettors**.

Down draught carburettors are normally used because here the fuel-air mixture descends into the cylinder in the direction of gravitational force. They are installed above the cylinder head.

Side draught and semi-down draught carburettors have very short intake paths. They are also used

where low installation heights are required and are installed below the cylinder head.

The following different types may be used based on the number and function of mixing-chamber bores:

- **Single-barrel carburettors (Fig. 2)** and **two-stage carburettors (Fig. 3)** (staged carburettors with stages opening in succession) for one intake manifold

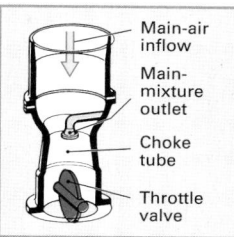

Fig. 2: Single-barrel carburettor

Fig. 3: Two-stage carburettor

- **Staged dual-barrel carburettors (Fig. 4)**
- **Dual-barrel carburettors (Fig. 5)**
- **Multiple carburettors** are used for separate intake manifolds

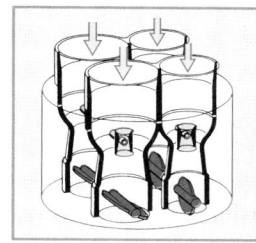
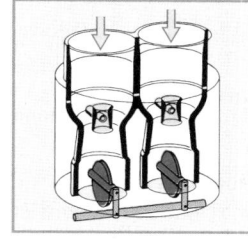

Fig. 4: Staged dual-barrel carburettor

Fig. 5: Dual-barrel carburettor

- **Constant-vacuum carburettors (Fig. 6)** operate with a variable choke-tube cross-section and virtually constant vacuum
- **Constant-vacuum slide carburettors (Fig. 7)** are used in motorcycles

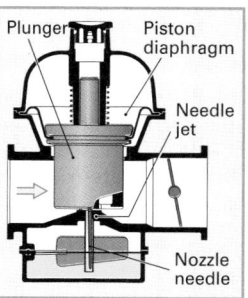
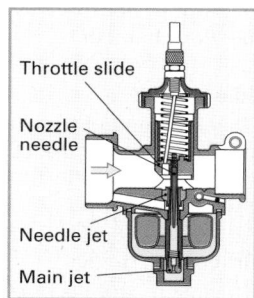

Fig. 6: Constant-vacuum carburettor

Fig. 7: Constant-vacuum slide carburettor

12.3.3 Design of a single-barrel carburettor

Carburettors usually consist of three main components: **throttle-valve assembly, carburettor housing and carburettor cover.** The following devices are housed in the main carburettor components (**Fig. 1**):

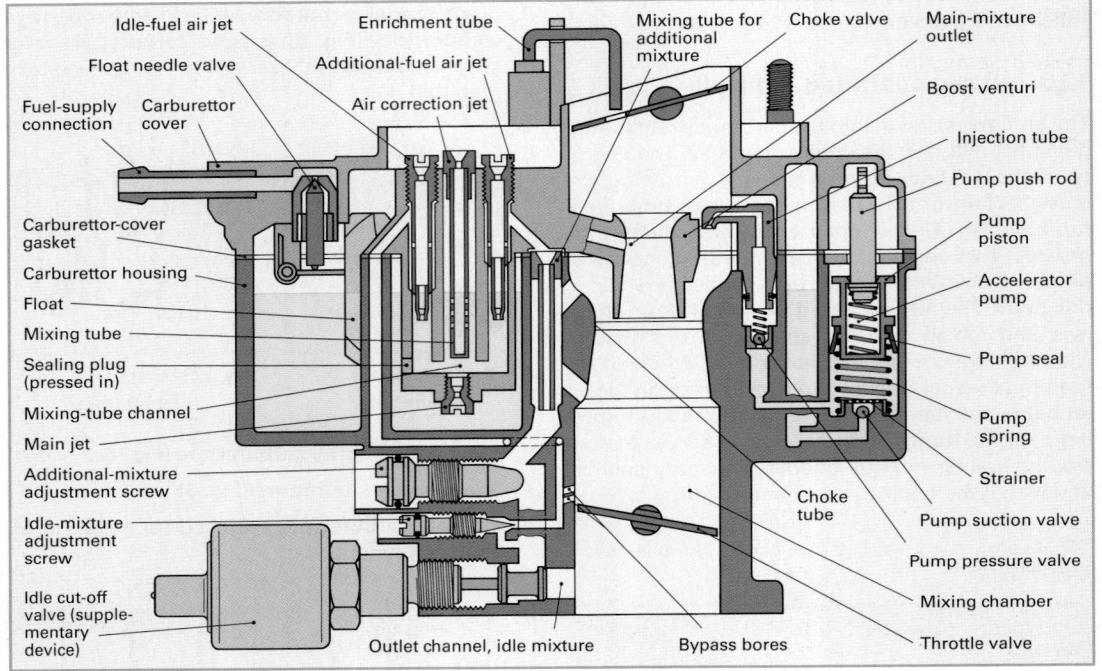

Fig. 1: Solex down draught carburettor 1B3, schematic section

Float device

The float device consists of the float housing, float and float needle valve. It is intended to regulate the fuel supply to the float chamber and keep the fuel level in the carburettor constant in all operating states.

Starting device

This is intended to induce a very rich fuel-air mixture of up to $\lambda \approx 0.2$ to be formed in the carburettor during cold-starting. This ensures that an ignitable mixture of approx. $\lambda = 0.9$ is available in the combustion chamber.

Idle device

This is intended to deliver the correct idle fuel-air mixture, ensure the correct idle speed and safeguard the transition from the idle system to the main-jet system.

Transition device

The idle mixture is briefly leaned when the throttle valve is opened. Additional fuel is therefore added to the air. This is intended to ensure a sound tran-

sition from the idle system to the main-injector system and good running performance in the lower part-load range.

Main-jet system

This consists of the main jet, air correction jet and mixing tube. It is intended to draw in and atomise fuel, mix it with air and supply the correct mixture ratio in the entire part-load range.

Supplementary devices

Supplementary devices can be used in mixture preparation in order to have a beneficial effect on drive comfort and fuel consumption.

Acceleration device

When the throttle valve is opened suddenly, leaning of the mixture must be prevented and additional fuel must be made available.

Enrichment device

This is intended to bring about the enrichment of the lean part-load mixture at full load and/or part load in order to obtain the greatest possible engine power.

12.4 Petrol injection

12.4.1 Basic principles of petrol injection

Functions of petrol-injection systems

> The function of petrol-injection systems is to spray fuel in finely atomised state into the inducted air. The required mixture quantity in each case and the mixture ratio must be adapted to the respective operating state in the process.

In fuel-injection systems the fuel is sprayed into the air in finely atomised state with the aid of nozzles and the pressure built up by the fuel pump. This increases the surface area of the injected fuel. This causes the fuel to carburate more quickly, which in turn leads to improved mixing with air, more complete combustion and better exhaust-gas values.

In the case of indirect injection (exterior mixture formation), the fuel injectors are arranged in such a way as to inject into the intake manifold or into the throttle-valve housing. In the case of direct injection (interior mixture formation), the fuel injectors are arranged in such a way as to inject into the combustion chamber.

Electronic control of the systems is intended optimally to adapt the fuel-air ratio (quality) and the amount of the mixture created (quantity) to the respective engine operating state.

The following objectives should be achieved:

- High engine torque
- High engine power
- Favourable engine performance curves
- Low fuel consumption
- Favourable exhaust-gas values

Types of injection

Indirect injection

With this type of injection, fuel and air already start to be mixed outside the combustion chamber. A uniformly distributed, homogeneous fuel-air mixture should be created in the whole combustion chamber during the induction and compression strokes.

The following different types of injection are used:

- **Single-point injection**
  (SPI) and
- **Multipoint injection**
  (MPI)

Single-point injection (Fig. 1). Here the fuel is injected centrally into the throttle-valve housing before the throttle valve. Atomisation in the throttle-valve gap and evaporation on hot intake-manifold walls or additional heater elements improve the preparation of the fuel-air mixture. Routes and manifolds of different lengths cause the fuel not to be distributed uniformly to all the cylinders. Peripheral turbulence and wall-applied film moistening especially in cold engines result in unequal mixture compositions. The have a negative effect on mixture formation. Single-point injection systems are much simpler in design than their multipoint counterparts.

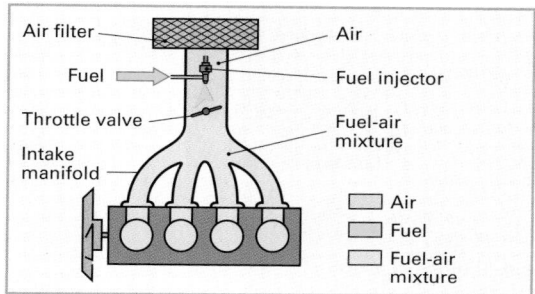

Fig. 1: Single-point injection

Multipoint injection (Fig. 2). Each cylinder is assigned a fuel injector. These injectors are situated in the intake manifold usually directly before the inlet valves. The mixture is therefore subject to intake paths of equal length and uniform distribution. An arrangement close to the inlet valves reduces the formation of wall-applied film when the engine is cold and reduces the build-up of noxious exhaust gases.

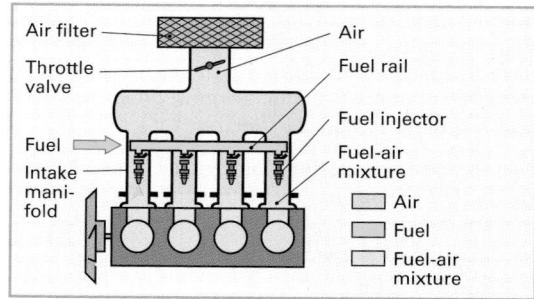

Fig. 2: Multipoint injection

Direct injection (Fig. 1, Page 258)

Systems with direct injection are always multipoint systems. The fuel is sprayed by the electrically actuated nozzles under high pressure (up to 120 bar) directly into the combustion chamber (interior mixture formation). There, depending on the engine layout and on the operating state, a homogeneous or heterogeneous mixture is formed with the inducted air.

12

Direct injection eliminates disruptive influences such as the formation of wall-applied film or unequal fuel distribution. This process, however, places very high demands on the electronic control of fuel-injection systems.

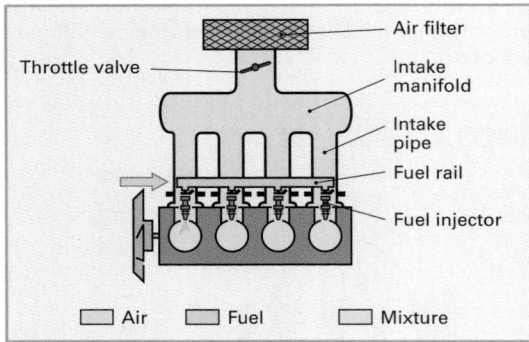

Fig. 1: Direct injection

Opening of fuel injectors

The fuel injectors are hydraulically opened by the fuel pressure or electromagnetically opened.

Continuous injection (see KE-Jetronic). The injectors are forced open by the fuel pressure and remain open for the entire time that the engine is in operation. They inject fuel continuously. The fuel is apportioned by a variable system pressure.

Intermittent injection. The injectors are electromagnetically opened for a brief period only and, once the calculated injection quantity has been injected, closed again. They are therefore only intermittently opened. The fuel is apportioned by a variable opening time of the fuel injectors.

Depending on how the fuel injectors are actuated by the ECU, there are four different types of intermittent injection:

- Simultaneous injection
- Group injection
- Sequential injection
- Cylinder-specific injection

Simultaneous injection (Fig. 2)

All the engine fuel injectors are actuated simultaneously. The time available for vaporising the fuel varies greatly for the individual cylinders. In order nonetheless to achieve as uniform a mixture composition as possible and good combustion, half of the fuel quantity required for combustion is injected in each case per crankshaft revolution.

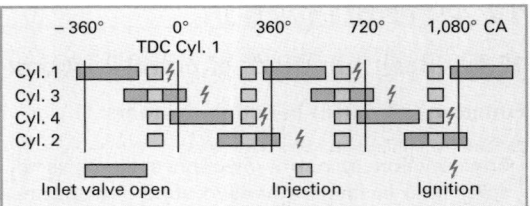

Fig. 2: Simultaneous injection

Group injection (Fig. 3)

The fuel injectors of cylinder 1 and cylinder 3, and of cylinder 2 and cylinder 4, are opened once per power cycle. In each case the entire fuel quantity is injected before the closed inlet valves. The times for vaporising the fuel vary in length.

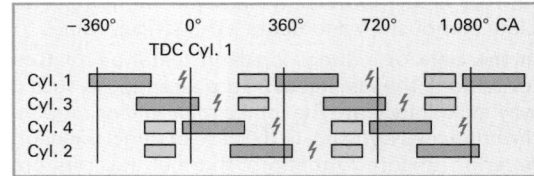

Fig. 3: Group injection

Sequential injection (Fig. 4)

The fuel injectors inject into the intake manifold the same entire fuel quantity in succession in firing sequence directly before the start of the induction stroke. This encourages optimal fuel-air mixture formation and improves internal cooling.

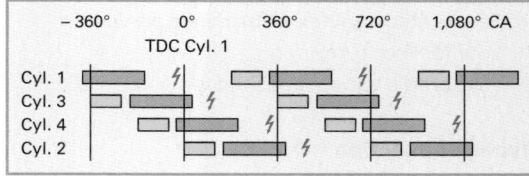

Fig. 4: Sequential injection

Cylinder-specific injection (Fig. 5)

This type of injection is a sequential-injection arrangement. Thanks to improved sensor technology and increased control sophistication, the ECU is able to apportion a specific fuel quantity to each individual cylinder.

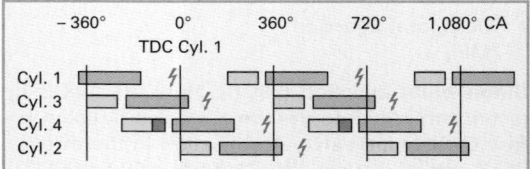

Fig. 5: Cylinder-specific injection

Table 1 shows the classification of electronic fuel-injection systems with intermittent injection

Table 1: Distinguishing features of electronic fuel-injection systems					
System	Central injection	L-Jetronic	LH-Jetronic	Intake-manifold pressure-controlled injection	Direct injection
External features	Central injection unit	Fuel rail with electrically actuated fuel injectors and			High-pressure fuel pump, pressure sensor and actuator
		Air-flow sensor	Air-mass meter	Intake-manifold pressure sensor	
Injection type		Indirect injection			Direct injection
Injection location	Before throttle valve	Before inlet valve			· Cylinder
Number of fuel injectors	One fuel injector single-point	According to number of cylinders, multipoint			
Intermittent injection	Clocked	Simultaneous or group injection	Sequential or cylinder-specific	Sequential	Cylinder-specific
Main controlled variables	• Throttle-valve angle • Speed	• Air flow • Speed	• Air mass • Speed	• Intake-manifold pressure • Speed	Requested torque (air mass, speed)

12.4.2 Design and function of electronic petrol injection

Electronic petrol-injection systems (**Fig. 1, Page 260**) consist of at least three subsystems:

- **Air-intake system**
  Air filter, intake manifold, throttle valve, individual intake pipes

- **Fuel system**
  Fuel tank, fuel pump, fuel filter, pressure regulator, fuel injector

- **Open- and closed-loop control system**
  – Sensors, e.g. temperature sensor
  – ECU and
  – Actuators, e.g. fuel-pump relay

The open- and closed-loop control system operates according to the **IPO** concept. This means:

Input: Sensors record and transmit information in the form of electrical voltage signals to the ECU.

Processing: The ECU processes the information contained in the voltage signals and compares the determined actual values with setpoint values usually stored in program maps. It calculates the actuation of the corresponding actuators.

Output: The relevant actuators, e.g. the fuel injectors, are supplied with power by the ECU. The desired system operating state is established.

Electronic petrol injection follows the sequence (function) below:

In the case of systems with homogeneous mixture formation, the engine draws in a quantity of air filtered in the air filter and regulated by the throttle valve. This air quantity is electronically recorded by a sensor.

The ECU uses stored program maps to calculate the **basic injection quantity** from the **engine speed** and the **air quantity (main controlled variables)**.

If mixture adaptation to special operating states, e.g. cold-starting, is required, the conditions (**correction quantities**) must be recorded by additional sensors and transmitted to the ECU again in the form of electrical signals. The ECU adapts the opening time of the fuel injectors to the changed operating conditions and supplies the injectors with power for the calculated time period.

The electromagnetic fuel injectors open and the fuel is injected at the pressure set by the pressure regulator. When the ECU terminates the supply of power, the injectors are closed by the closing spring. The injection process is completed.

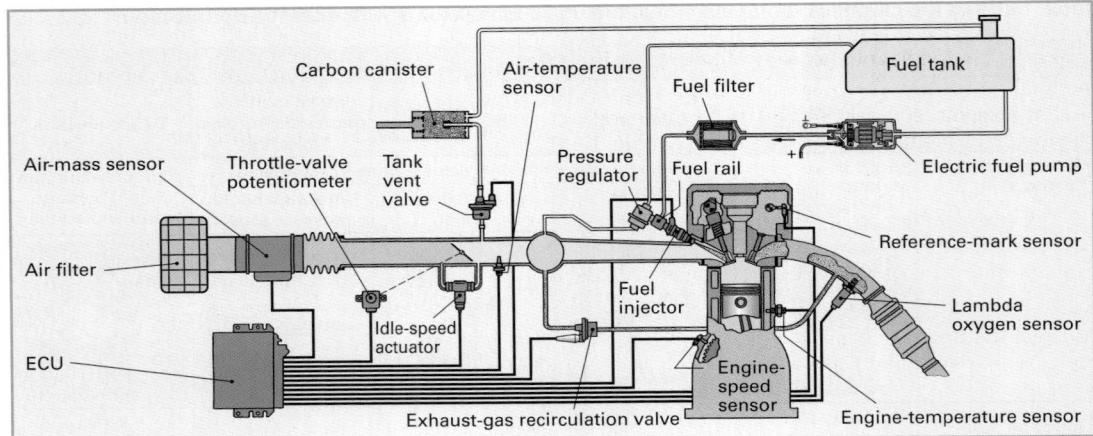

Fig. 1: Design of an electronic petrol-injection system

REVIEW QUESTIONS

1 What is the function of petrol injection?

2 What are the characteristic features of indirect injection and direct injection respectively?

3 Describe intermittent and continuous injection.

4 How does simultaneous injection differ from sequential injection?

5 A petrol-injection system consists of which subsystems? Name the main components.

6 What do you understand by the main controlled variables for electronic petrol injection?

7 What do you understand by the correction quantities of electronic petrol injection?

8 How can the ECU alter the injected fuel quantity for intermittent injection?

12.4.3 Operating-data acquisition

The ECU requires information from various sensors in order to correctly actuate the actuators contained in the fuel-injection system.

Load and engine speed are used to form the basic injection quantity. These quantities are known as **main controlled variables**. The signals from further sensors are required for the purpose of adapting the mixture to the respective operating states. These are known as correction quantities.

Main controlled variables

Load sensing. This can be performed by various sensors:

● Airflow sensor

● Hot-wire air-mass meter

● Hot-film air-mass meter

● Hot-film air-mass meter with return-flow detection

● Intake-manifold pressure sensor

● Throttle-valve potentiometer

Airflow sensor (Fig. 2). This incorporates the sensor flap, which is under coil-spring tension. The sensor flap is deflected against spring force by the air flow

during induction and moved into a specific angular position. This position is transmitted to a potentiometer. The ECU detects the sensor-flap position from the voltage drop at the resistor and calculates the inducted air quantity or flow with the aid of stored characteristic values. The compensation flap, which is permanently joined to the sensor flap, compensates mechanical vibrations acting from the outside in combination with the air cushion of the damping chamber.

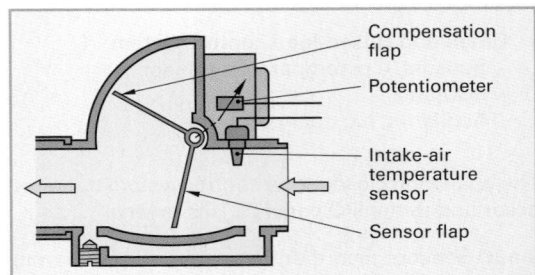

Fig. 2: Airflow sensor

Hot-wire air-mass meter (Fig. 1, Page 261). A hot wire tensioned in the air duct acts as the sensor. This wire is kept by electric current at a constant temperature of 100 °C above the intake-air temperature. More or less air mass is inducted in different driving states. This air mass cools the hot wire. The heat dissipated to the air must be compensated by the heating current. The magnitude of the required heating current is regulated by the corresponding voltage. Thus the heating current or the voltage required for the heating current is the measure of the air mass. The air mass is measured approximately 1,000 times per second. If the hot wire breaks, the ECU switches to emergency operation. The vehicle can continue to be driven under restricted conditions.

Because the hot wire is situated in the intake passage, deposits may form which can distort the measurement result. Every time the engine is switched off, therefore, the ECU sends a signal to heat the hot wire briefly to approx. 1,000 °C and thereby burn off any deposits.

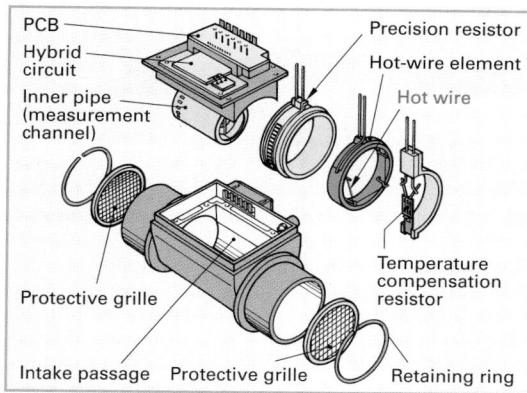

Fig. 1: **Hot-wire air-mass meter**

Hot-film air-mass meter (Fig. 2). A hot-film sensor is installed in the measurement channel situated in the intake passage.

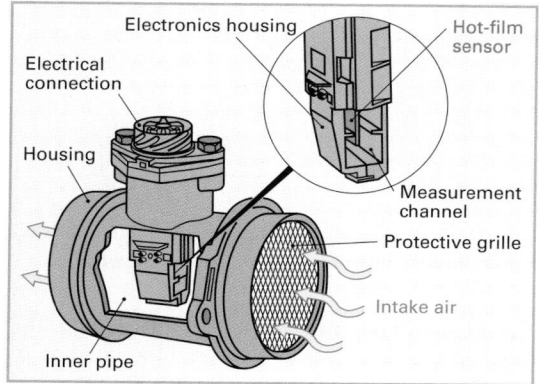

Fig. 2: **Hot-film air-mass meter**

This sensor is made up of three electrical resistors (NTC) **(Fig. 3)**.

- Heating resistor R_H (platinum-film resistor)
- Sensor resistor R_S
- Temperature resistor R_L (intake air)

The resistors, which are combined to form an electrical bridge circuit, are each attached as a thin film to a ceramic layer.

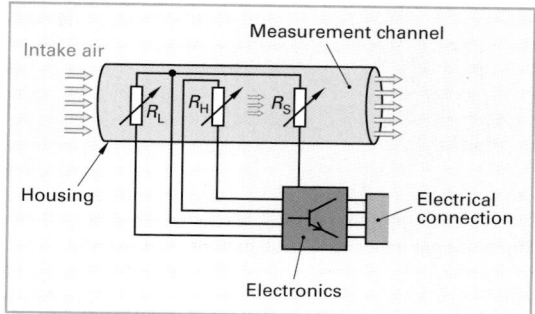

Fig. 3: **Bridge circuit of hot-film sensor**

The electronics in the hot-film air-mass meter regulates the temperature of the heating resistor R_H via a variable voltage in such a way that it is 160 °C above the intake air. The intake-air temperature is recorded by the temperature resistor R_L for this purpose. The temperature of the heating resistor R_H is determined by the sensor resistor R_S. The heating resistor is cooled to a greater or lesser extent in the event of an increased or reduced air-mass flow. The electronics regulates the voltage at the heating resistor by comparing the sensor resistor R_S and temperature resistor R_L in order to obtain the temperature difference of 160 °C again. From this control voltage the electronics generates a signal for the inducted air mass (air throughput).

Because this sensor is largely insensitive to contaminants, it is not necessary to burn off deposits as is the case with the hot-wire air-mass meter.

Hot-film air-mass meter with return-flow detection (Fig. 1, Page 262). Hot-film air-mass meters with return-flow detection are installed in order to minimise errors caused by the pulsating air column in the intake manifold. These sensors prevent the measurement result from being distorted by return flow. This enables the fuel to be apportioned more precisely (error max. +/– 0.5 %).

The sensors each contain a heating zone, which heats the inducted air flowing past. Thus, a higher temperature is measured at measuring cell *M2* than at measuring cell *M1*. When air flows back from the engine side, measuring cell *M2* is cooled and measuring cell *M1* is heated. Both flows, suction flow

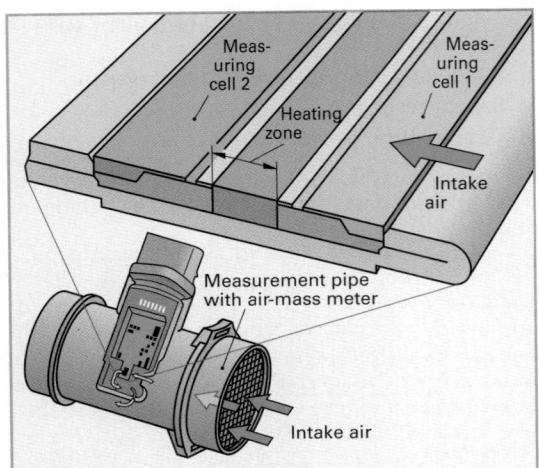

Fig. 1: Signal generation, hot-film air-mass meter

and return flow, thus have an effect on the measuring-cell temperatures. The temperature difference ΔT is converted in the evaluating circuit into a voltage, from which the ECU determines the inducted air mass.

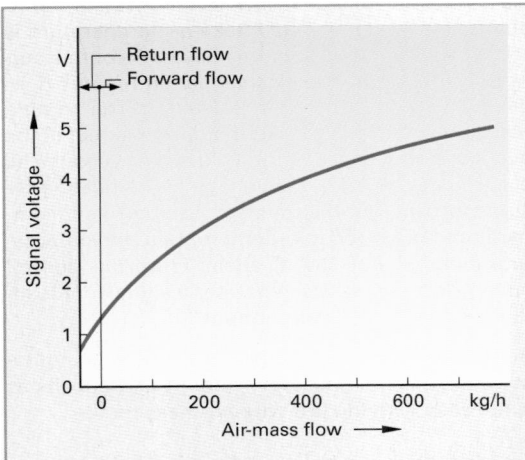

Fig. 2: Curve, hot-film air-mass meter with return-flow detection

The diagram (Fig. 2) shows that the signal voltage varies as a function of load between approx. 1 V (idle) and 5 V (full load).

Intake-manifold pressure sensor (Fig. 3). The function of this sensor is to record the pressure in the intake manifold. It can be mounted directly on the intake manifold or housed in the ECU. In the latter case, it is connected to the intake manifold by way of a hose. The sensor contains an evaluating circuit and a pressure cell with two sensor elements.

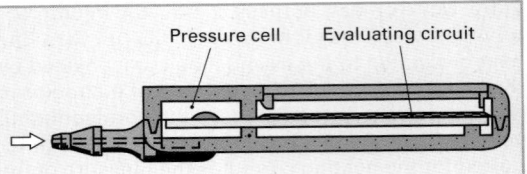

Fig. 3: Intake-manifold pressure sensor

Each sensor element consists of a diaphragm which contains a reference-pressure chamber with a specific internal pressure. On the diaphragm are resistors the conductivity of which varies as a function of pressure when they are exposed to mechanical stresses as the result of deformation of the diaphragm.

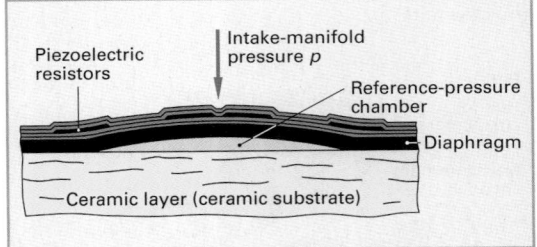

Fig. 4: Sensor cell of intake-manifold pressure sensor

The functions of the evaluating circuit are to:
- Amplify the voltage change generated by the resistance change
- Compensate temperature influences
- Generate as linear a curve as possible

The inducted air flow is determined from the voltage change (Fig. 5) by way of the intake-manifold pressure.

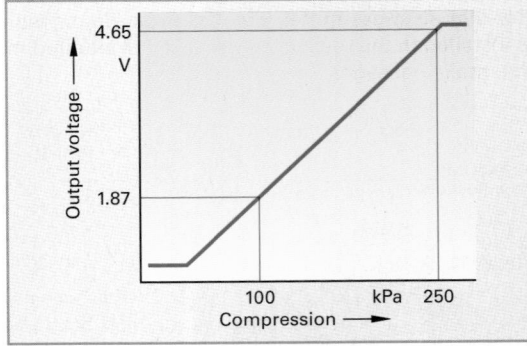

Fig. 5: Curve of intake-manifold pressure sensor

Throttle-valve potentiometer (Fig. 1, Page 263). The function of this sensor is to record the position of the throttle valve. When the throttle valve is opened, the throttle-valve shaft moves the wiper arms, which sweep the resistor paths. Due to the change in the voltage drop at the resistor paths, the ECU is able to determine the position of the throttle valve. Together with the speed and the intake-air temperature, the inducted air flow can be determined from the throttle-valve position.

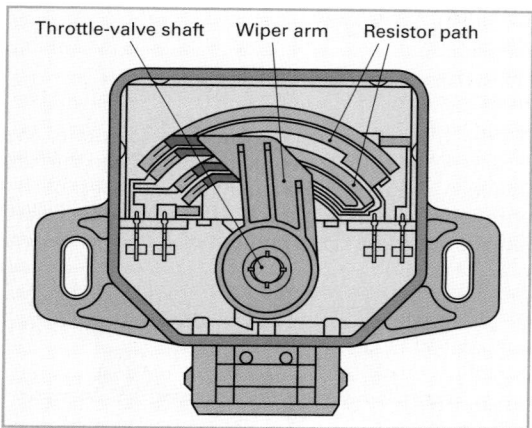

Fig. 1: Throttle-valve potentiometer

If the throttle-valve signal is to be used as the main load signal, potentiometers with twin resistor paths and two wiper arms are used. This increases the accuracy and reliability of the system. The voltages falling at the two potentiometers are then usually opposing **(Fig. 2)**.

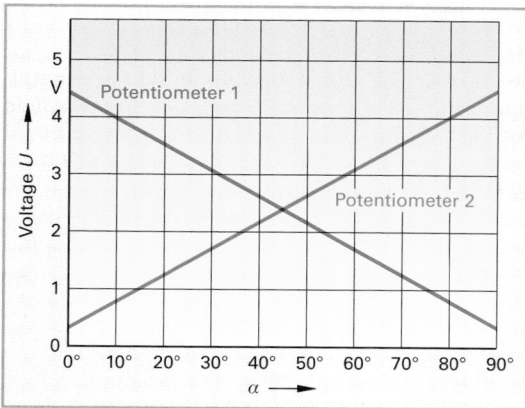

Fig. 2: Signal voltages with twin potentiometers

If the load is determined by sensors other than the throttle-valve potentiometer, it serves as the sensor for the dynamic function (opening speed of the throttle valve), for range detection (idle, part load, full load) and as a limp-home signal if the main load sensor fails.

The sensor housing frequently features an additional switch for detecting the idle position.

Speed recording: This can be performed by various sensors:

- Inductive speed sensor on the crankshaft
- Hall-effect sensor in the distributor (with diaphragm rotor)
- Hall-effect sensor on the camshaft (with magnet)
- Hall-effect sensor on the crankshaft (with pulse-generator wheel)

Inductive speed sensor (Fig. 3). A ferromagnetic pulse-generator wheel is mounted on the crankshaft. An inductive speed sensor, which consists of a soft-iron core with copper winding (sensor coil) and a permanent magnet, scans the tooth sequence. As the crankshaft rotates, the teeth of the pulse-generator wheel generate magnetic flux changes in the sensor coil, which induces an alternating voltage **(Fig. 4)**. The ECU can determine the engine speed from the frequency of the induced alternating voltage.

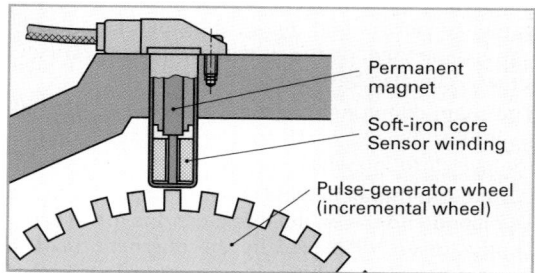

Fig. 3: Inductive speed sensor with pulse-generator wheel

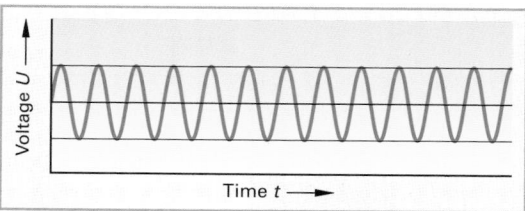

Fig. 4: Speed signal

If the crankshaft position is to be recorded by this sensor at the same time, a larger gap is incorporated on the pulse-generator wheel to act as a reference mark **(Fig. 5)**.

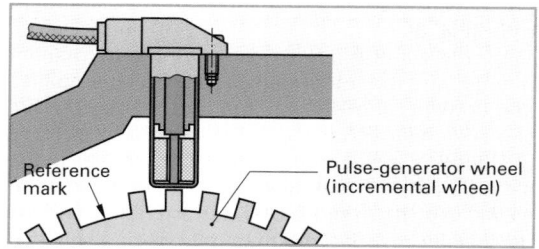

Fig. 5: Speed and reference-mark sensor

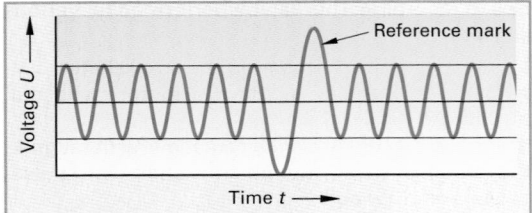

Fig. 1: Speed signal with reference mark

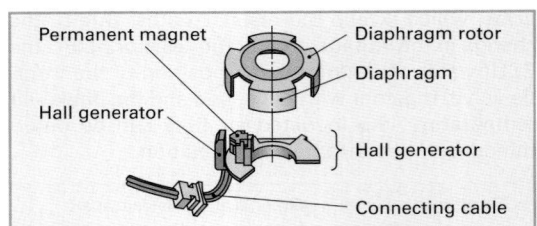

Fig. 3: Hall-effect sensor with diaphragm rotor

As the gap on the induction-type pulse generator rotates past, a higher voltage is induced on account of the greater magnetic flux change **(Fig. 1)**. Moreover, this voltage pulse has a lower frequency than the pulses generated for speed recording. It is the information for a specific crankshaft position. The reference mark indicates that the piston of cylinder no. 1 is, for example, at 108° CA before TDC.

The advantage of **Hall-effect sensors** over inductive speed sensors is that the level of their signal voltage is dependent on the speed. In this way, very low speeds can also be recorded.

The main component of such a sensor is the Hall generator **(Fig. 2)**, which consists of a semiconductor layer through which supply current I_V is passed. If there is a magnetic field (B) at right angles to the semiconductor layer, the free electrons in the semiconductor are displaced by the magnetic field to one side; the Hall voltage U_H is created. The level of the Hall voltage created is dependent on the strength of the magnetic field.

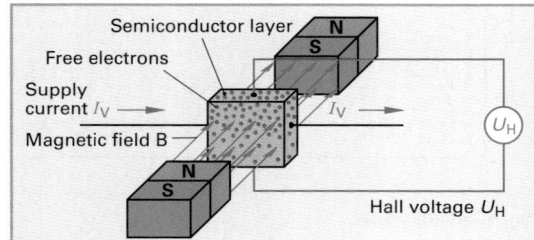

Fig. 2: Hall generator

The Hall principle is applied in different ways.

Hall-effect sensor in the distributor with diaphragm rotor (Fig. 3)
This consists of the Hall generator, the permanent magnet and the integrated circuit, which amplifies and converts the Hall voltage into a square-wave signal (sensor voltage U_G). The distributor rotor is designed as a diaphragm rotor which moves in the air gap between the Hall IC and the magnetic barrier. If a diaphragm slides between the Hall IC and the permanent magnet, the magnetic field is shielded and the Hall voltage U_H is zero **(Fig. 4)**.

As the diaphragm rotor continues to rotate, the magnetic field can penetrate the Hall IC and the Hall voltage is created. Because the number of diaphragms is stored in the ECU, the ECU can calculate the speed from the number of voltage changes.

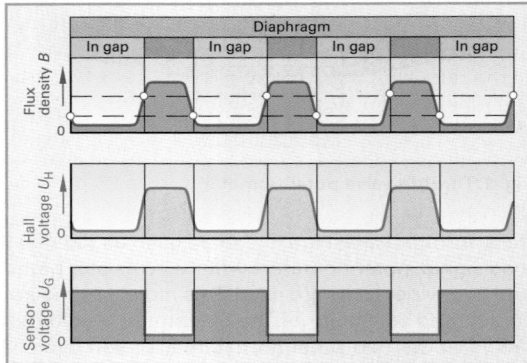

Fig. 4: Pulse shape of Hall-effect sensor with diaphragm rotor

Hall-effect sensor on the camshaft (Fig. 5). The sensor consists of the Hall generator and the integrated circuit for signal conditioning. The magnetic field for creating the Hall voltage U_H is generated by a magnetic plate mounted on the camshaft. When the magnetic plate moves past the sensor as the camshaft rotates, the Hall voltage U_H is created.

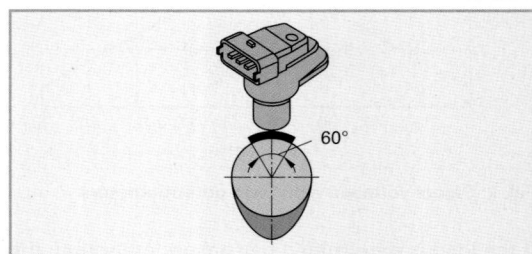

Fig. 5: Hall-effect sensor on the camshaft

The signal **(Fig. 1, Page 265)** of this sensor is only used to calculate speed in emergency operation if the engine-speed sensor malfunctions. However, when single-spark ignition coils are used or in the case of selective petrol injection, the engine ECU

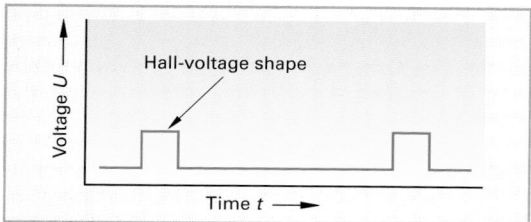

Fig. 1: Hall voltage by sensor on CA

requires the firing TDC of cylinder no. 1 to be clearly determined in order to actuate the correct ignition coil or the correct fuel injector. For this purpose, the signals of the engine-speed sensor on the crankshaft and the signal of the camshaft sensor are combined **(Fig. 2)**. If the reference marks of the TDC sensor and the speed sensor line up, the following TDC of cylinder no. 1 is the firing TDC. If only the reference mark of the speed sensor appears, the following TDC lies between the exhaust and induction strokes.

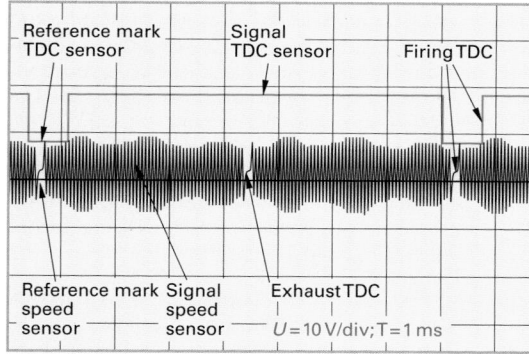

Fig. 2: Determining firing TDC

Hall-effect sensor on the crankshaft with pulse-generator wheel

This sensor consists of two Hall generators, a permanent magnet and the evaluation electronics, which evaluates the Hall voltages of the two Hall generators and amplifies these voltages for the sensor voltage. Like the inductive engine-speed sensor, it is mounted on the crankshaft and scans a pulse-generator wheel which is designed as an apertured diaphragm. When a diaphragm rotates past the sensor, the magnetic field is amplified differently, depending on the position of the diaphragm. In this way, the magnetic fields which act on the Hall generators vary in strength at times, which causes different Hall voltages **(Fig. 3)**. The evaluating circuit generates from the Hall voltages U_H created in each case the sensor voltage U_G **(Fig. 4)**. As with the inductive engine-speed sensor, a reference-mark signal can also be generated here increasing the opening in the apertured diaphragm.

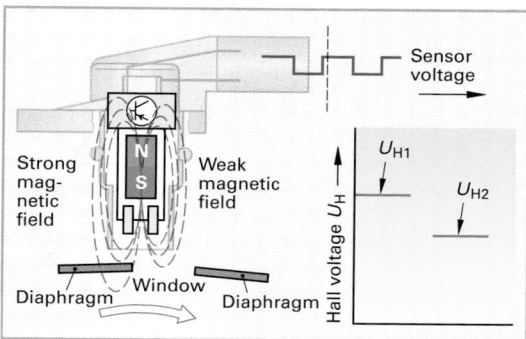

Fig. 3: Magnetic-field change by apertured diaphragms

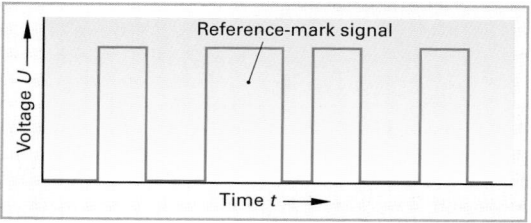

Fig. 4: Sensor voltage U_G

The signal of the Hall-effect sensor on the crankshaft can, like the signal of an inductive engine-speed sensor, be combined with the Hall-effect sensor on the camshaft to determine the firing TDC.

Correction quantities

The following are used to record the required correction quantities:

- Temperature sensors (NTC) for e.g. engine temperature, intake-air temperature
- Pressure sensors (piezo sensors) for e.g. ambient pressure, intake-manifold pressure
- Lambda sensors **(see Page 316)**

Lambda sensors **(see Page 316)**

12

REVIEW QUESTIONS

1 Which sensors are used to determine the load?
2 Which signals are generated by the respective sensors?
3 Which sensors for recording speed are used?
4 Which signals are generated by the respective sensors?
5 How in the case of selective injection does the ECU identify which fuel injector is to be actuated?
6 Which sensors are chiefly used to record the correction quantities?
7 Why are Hall-effect sensors increasingly being used instead of inductive sensors?

12.4.4 Single-point injection

> In the case of single-point injection all the cylinders of an engine are supplied with fuel by a centrally situated fuel injector.

Single-point injection systems are electronically controlled petrol-injection systems with a single electromagnetically actuated fuel injector (**SPI** = **S**ingle-**P**oint **I**njection). The injector is opened by the ECU for each power cycle in line with the number of cylinders in the engine. (See also Indirect injection, Single-point injection). The fuel is injected before the throttle valve.

12.4.4.1 Single-point injection subsystems

Air-intake system. The air inducted and filtered in the air filter flows through the central injection unit. There, the temperature of the air is recorded by the intake-air sensor, which transmits it in the form of an electrical voltage to the ECU. The throttle-valve actuator, also located in the central injection unit, regulates the required air flow rate at idle in such a way that a stored setpoint idle speed can be maintained. Fuel is injected before the throttle valve into the inducted air (exterior mixture formation). The mixture regulated in terms of quantity by the throttle valve flows through the intake manifold on account of the vacuum pressure acting in the cylinders. The intake-manifold walls or the inducted mixture are heated in order to counteract excessive condensate formation on the intake-manifold walls in cold engines. Finally, it passes through the opened inlet valves into the cylinder.

Fuel system. An electric fuel pump delivers the fuel from the fuel tank via a fuel filter to the central injection unit. The pressure regulator installed there in the return keeps the fuel pressure constant at approximately 1 bar (low-pressure system). When the electromagnetic fuel injector is supplied with power, fuel is injected before the throttle valve into the inducted air.

Regenerating system. The hydrocarbons temporarily stored in the carbon canister must be supplied for combustion in an appropriate operating state, e.g. part load. For this purpose, the regenerating valve is clocked by the engine ECU so that air and hydrocarbons can be drawn in by the vacuum pressure acting in the intake manifold.

Operating-data acquisition. The main information on the engine operating state is provided by throttle-valve angle α and engine speed n (main controlled variables, α-n-**system**). From them the basic injection quantity (quantity) and thus the basic injection time can be calculated in the ECU. In order to determine the precise fuel quantity (quality), the ECU must receive further information, e.g. air temperature, engine temperature and mixture composition, from the lambda sensor.

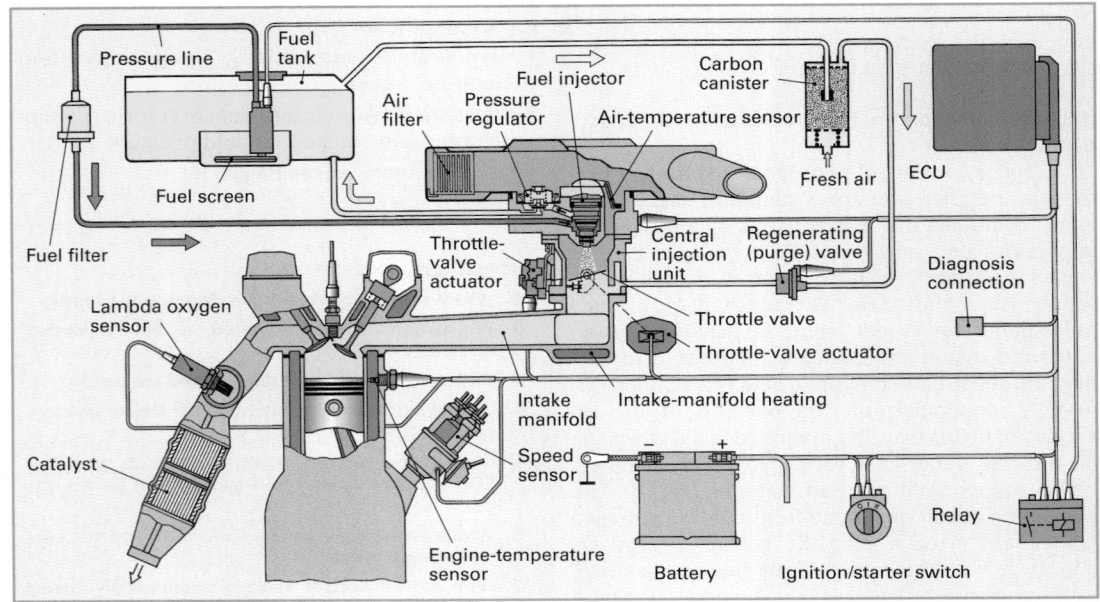

Fig. 1: Single-point injection

12.4.4.2 Single-point injection components

Central injection unit (Fig. 1). This comprises:
- Hydraulic section with fuel supply, return, fuel injector, pressure regulator, air-temperature sensor
- Throttle-valve section with throttle valve, throttle-valve potentiometer, throttle-valve actuator

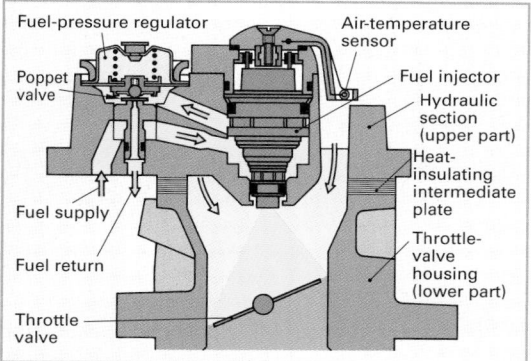

Fig. 1: Central injection unit

Fuel-pressure regulator (Fig. 1). This keeps the system pressure in the return constant at 1 bar. The injected fuel quantity is therefore dependent on the fuel-injector opening time. If the fuel-pump pressure exceeds the system pressure, the spring-loaded poppet valve opens and releases the fuel return. The fuel flowing to the pressure regulator flows through and around the fuel injector beforehand for cooling. This ensures a good hot-start response.

Throttle-valve actuator (Fig. 2). This is used for idle-speed control to a low speed level and stabilises the idle speed, for example, even when the air-conditioning system is switched on. The ECU supplies the actuating signal for positioning the throttle valve to the DC motor as a function of engine speed and engine temperature. The actuating push rod, which acts on the throttle valve, is extended and retracted by way of a screw thread.

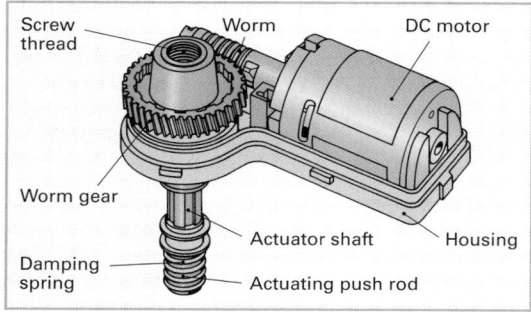

Fig. 2: Throttle-valve actuator

Central fuel injector (Fig. 3). This consists of the valve housing and the valve group. The valve housing accommodates the field winding with the electrical connection. The valve group consists of the valve body and the valve needle with solenoid armature guided in the body. The helical spring presses the valve needle into its seal seat with the assistance of the system pressure. When the field winding is excited, the pintle valve lifts off its seal seat by roughly 0.06 mm so that fuel can emerge from the annular orifice. The shape of the pintle nozzle provides for good atomisation together with a tapered injection jet. The fuel injector is triggered in time with the ignition pulses.

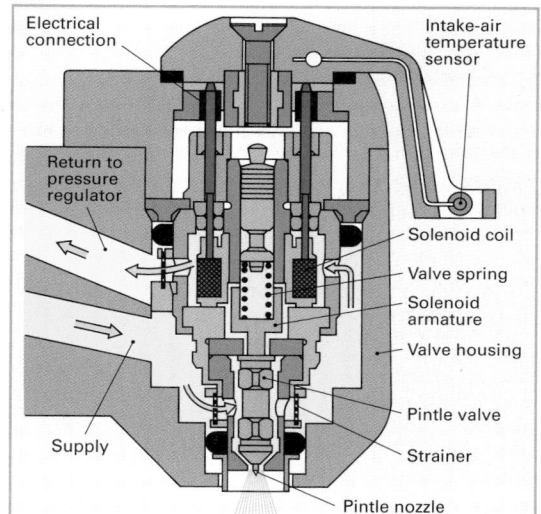

Fig. 3: Central fuel injector

12.4.4.3 Electronic control of single-point injection

The single-point injection system is electronically controlled in accordance with the IPO concept, i.e. the different operating states are recorded by sensors and transmitted to the ECU in the form of electrical signals. The ECU calculates the required starting values with the aid of various stored program maps and actuates the corresponding actuators by means of electrical signals (see block diagram, **Fig. 1, Page 268** and circuit diagram, **Fig. 1, Page 269**).

The single-point injection ECU thus demonstrates the following functions: starting, warm-up, acceleration and full-load enrichment, overrun fuel cutoff, lambda closed-loop control, hot-starting control, engine-speed limitation, adaptive idle-speed control, fuel-pump-relay activation, regenerating-valve activation, limp-home-mode function, self-diagnosis.

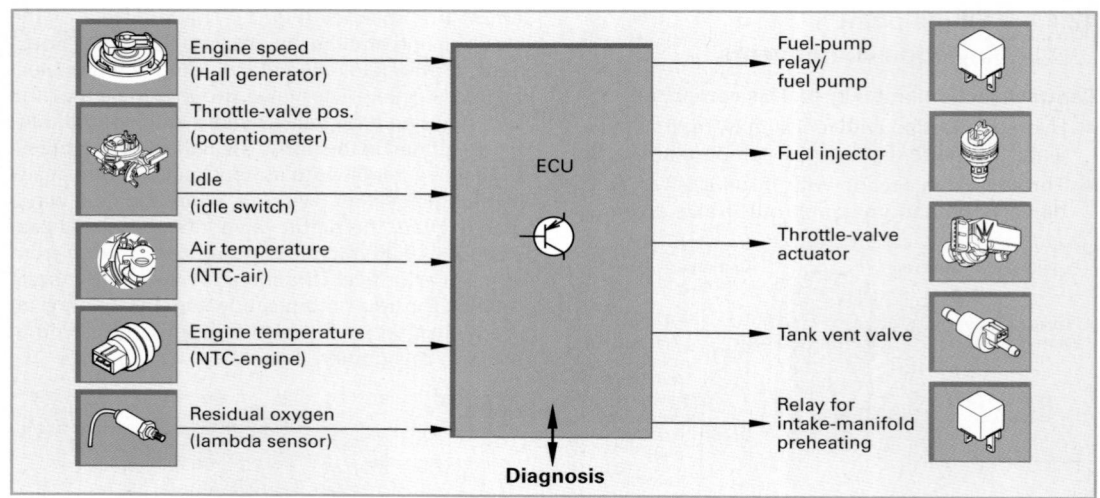

Fig. 1: Block diagram of single-point injection system

Engine speed. This is transmitted to the ECU by a Hall-effect sensor located in the distributor. The ECU uses the engine speed together with the throttle-valve position to calculate the length of time during which the fuel injector is supplied with power and thus the basic fuel quantity to be injected. If sensor **B5** fails, engine operation is no longer possible because the ECU is unable to calculate either the required injected fuel quantity or the number of injection operations. **B5** can be checked at pin 26 (terminal 7), pin 27 (terminal 8h) of the ECU and terminal 31 (terminal 31d).

Throttle-valve position. This is recorded by the throttle-valve potentiometer located in the central injection unit and transmitted to the ECU in the form of electrical voltages. From the level of these voltages the ECU is able to calculate with the aid of stored program maps the opening angle and together with the speed the inducted air quantity. If the voltages assume extreme values, the ECU detects full load from these values. In this case, lambda closed-loop control is shut down and the mixture to be formed is enriched. From the voltage change per unit of time the ECU is able to detect the driver's acceleration command. If a stored value is exceeded, it cuts out lambda closed-loop control and enriches the mixture. If sensor **B3** fails, limp-home operation can be maintained by the lambda sensor under certain circumstances. **B3** is checked at pin 7, pin 8, pin 18 of the ECU and terminal 31.

Idle position. The ECU receives this information from idle switch **Y2,** which is attached to the throttle-valve actuator. When the throttle valve is in the idle position, idle-speed control or overrun fuel cut-off is activated. If the signal fails, no idle-speed con-

trol or overrun fuel cut-off is possible because the idle position is no longer being detected. Component check **Y2**: pin 3, terminal 31M.

Intake-air temperature. This is recorded by an NTC resistor **B1** located in the central injection unit. The voltage drop at the resistor gets smaller as temperature increases. The signal is required so that more fuel (up to 20 %) can be injected at low temperatures. Increased contact resistances at e.g. corroded plug connections may result in an incorrect mixture formation. If the signal fails completely on account of an open circuit or a short circuit, the ECU can switch to a stored default value. The signal from **B1** can be checked at pin 14 and terminal 31.

Engine temperature. The signal from the engine-temperature sensor (NTC) is required so that the fuel quantity can be adapted as a function of the engine temperature when the engine is cold (correction quantity). Extending the injection time by up to 70 % prevents the mixture from leaning heavily as a result of condensation losses in the intake manifold and cylinder. As with the air-temperature sensor, an increased contact resistance at a plug connection can also result in an incorrect mixture formation here. The ECU can switch to a stored default value in the event of an open circuit or a short circuit. Component check **B2**: pin 2, terminal 31M.

Fuel-pump relay. The function of this relay is to supply the electric fuel pump with power. The relay control current flows if the ECU connects pin 17 to earth/ground. In this way, the operating current for the fuel-supply pump can flow from terminal 30 to the pump. If the ECU receives no signal from the

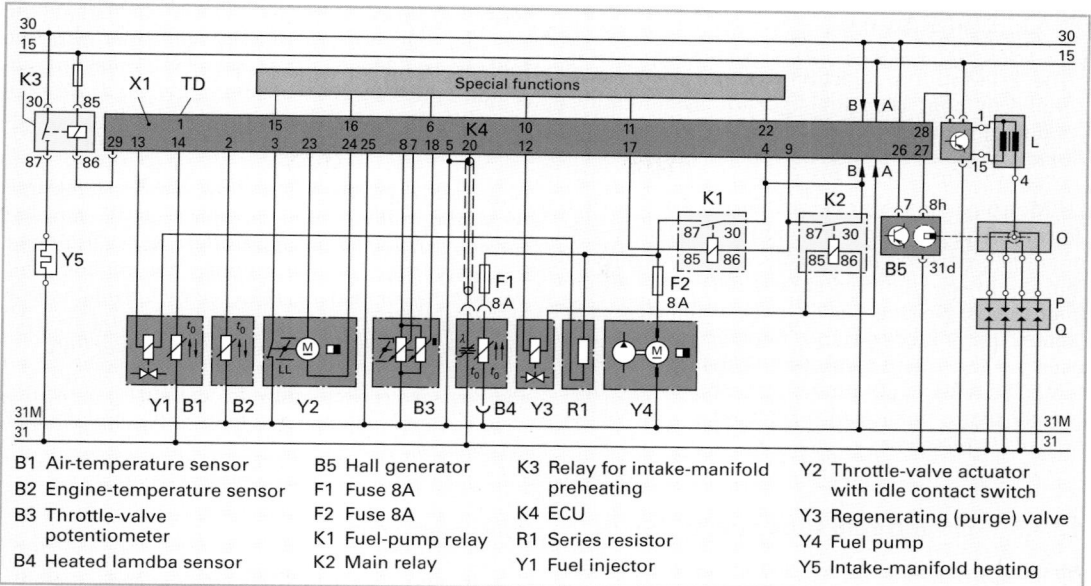

B1 Air-temperature sensor
B2 Engine-temperature sensor
B3 Throttle-valve potentiometer
B4 Heated lambda sensor
B5 Hall generator
F1 Fuse 8A
F2 Fuse 8A
K1 Fuel-pump relay
K2 Main relay
K3 Relay for intake-manifold preheating
K4 ECU
R1 Series resistor
Y1 Fuel injector
Y2 Throttle-valve actuator with idle contact switch
Y3 Regenerating (purge) valve
Y4 Fuel pump
Y5 Intake-manifold heating

Fig. 1: Circuit diagram of single-point injection system

engine-speed sensor for a period of three seconds, it interrupts the relay control current and the pump is deactivated. This feature is intended to prevent fuel from getting into the engine or escaping into the environment when the engine is stopped and the fuel injector is open (safety cut-out).

Fuel injector. The fuel injector sprays the fuel in finely atomised state before the throttle valve. The fuel is injected in each case on two crankshaft revolutions in accordance with the number of cylinders in the engine. The valve opens when ...

● ... fuel-pump relay K1 is closed and current flows from terminal 30 via the relay and the series resistor to the valve.

● ... the ECU connects pin 13 to earth/ground.

The length of time during which power is supplied determines the injected fuel quantity.

Throttle-valve actuator. The ECU uses this actuator to control the idle speed in such a way that a set-point value dependent on the engine temperature is maintained. When the ECU detects the idle position, it activates the throttle-valve actuator via pin 23 and pin 24 so that the throttle valve, depending on the actual value, is opened or closed further. For correct idle-speed control the ECU requires the signals from Hall-effect sensor **B5**, from engine-temperature sensor **B2** and from idle contact switch in throttle-valve actuator **Y2**.

Intake-manifold heating. This has the function of heating the intake-manifold walls when the engine

is cold. In this way, condensation of the fuel on the cold intake-manifold walls should be reduced or prevented. The intake-manifold heating relay closes if the ECU connects pin 29 to earth/ground. Thus current can flow from terminal 30 via relay **K3** to the intake-manifold heating (positive supply). The heating receives earth/ground from terminal 31.

12.4.4.4 Diagnosis

Whereas in older systems the stored faults were read out in the form of flashing codes, in newer systems stored faults can be read out by way of fault readout devices (engine testers). Actuator diagnosis is also possible for the throttle-valve actuator, for the intake-manifold preheating relay and for the regenerating valve.

REVIEW QUESTIONS

1 List the essential features of a single-point injection system.

2 Describe the fuel system of the single-point injection system.

3 From which subassemblies is the central injection unit made up? Explain their functions and effects.

4 Which sensors are required by the single-point injection system? Which variables/quantities do they record?

5 Which actuators are activated by the ECU?

6 Explain the function of the throttle-valve actuator.

12.4.5 LH-Motronic

> The LH-Motronic injection system is an electronically controlled fuel-injection system with multipoint injection, in which the air mass is used as one of the main controlled variables.

LH-Motronic is a further developed variant of L-Jetronic. The electromagnetic fuel injectors are sequentially actuated by the ECU. The fuel is injected into the intake manifold shortly before the engine inlet valves, which are still closed at the start of injection. Engine speed and inducted air mass are used as the main controlled variables (m/n-system). The latter is determined by a hot-wire or hot-film air-mass meter, which is also the external feature of LH-Motronic.

12.4.5.1 LH-Jetronic subsystems (Fig. 1)

Air-intake system. The air filtered by the air filter and inducted by the engine flows into the intake manifold. There the air mass is recorded by the air-mass meter and transmitted to the ECU in the form of a voltage signal. An NTC resistor, which can also be integrated in the air-mass meter, is used as the air-temperature sensor. The voltage drop at the thermistor is the measure of the intake-air temperature.

Fuel system. Two-line systems are usually used in LH-Jetronic. An electric fuel pump, which is located either in the fuel tank (in-tank pump) or on the vehicle underbody (inline pump), delivers the fuel from the fuel tank via a fuel filter to the fuel rail. All the fuel injectors are supplied with fuel from the fuel rail. At the end of the fuel rail is a pressure regulator, which keeps the differential pressure constant at approx. 3.5 bar. The excess fuel returns from the pressure regulator to the fuel tank.

Regenerating system. The hydrocarbons temporarily stored in the carbon canister must be supplied for combustion in an appropriate operating state, e.g. part load. For this purpose, the regenerating valve (tank vent valve) is clocked by the engine ECU so that air and hydrocarbons can be drawn in by the vacuum pressure acting in the intake manifold.

Exhaust-gas recirculation. An exhaust-gas recirculation system can be used to improve the exhaust-gas values.

Idle-speed control

> Its function is to keep the engine speed constant with the throttle valve closed at a setpoint value dependent on the engine temperature.

The internal-resistance levels of the engine when the engine is cold are greater than when it is hot on account of the viscous engine oil and increased friction. In order to overcome this resistance and facilitate stable idle speeds, the engine must generate more power. This is achieved by an increased amount of mixture. Furthermore, idle-speed fluctuations must be compensated by a loaded vehicle electrical system or by a cut-in A/C compressor.

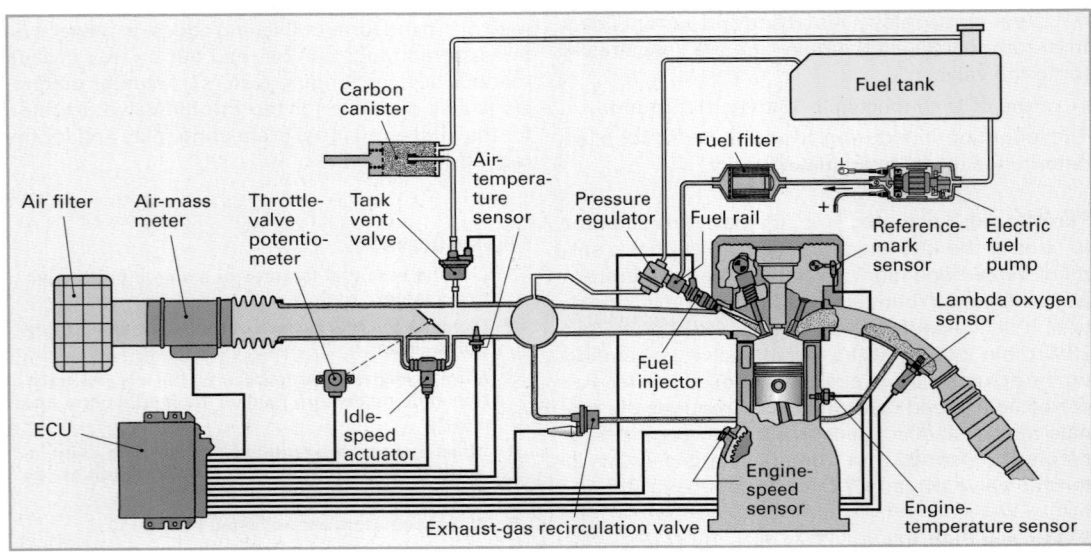

Fig. 1: LH-Jetronic

The ECU requires the signals from the following sensors for idle-speed control:

- Engine-speed sensor (actual speed)
- Engine-temperature sensor (determination of the setpoint speed)

One of the following actuators is used for speed control.

Idle-speed actuator (Fig. 1). This permits additional air depending on the requirement to flow in a by-pass around the closed throttle valve. For this purpose, it is actuated by the ECU by means of **pulse-width-modulated** signals, a process which opens the air duct to a lesser or greater extent.

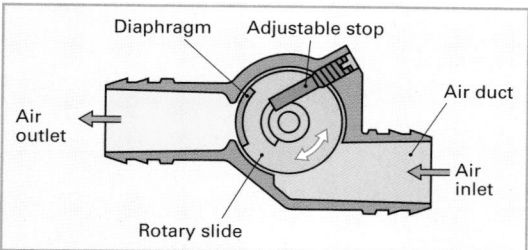

Fig. 1: Idle-speed actuator (rotary actuator)

Throttle-valve actuator (Fig. 2). This subassembly consists of an electric motor, a gearing and the throttle valve. At idle the electric motor is actuated by the engine ECU in such a way that it opens or closes the throttle valve depending on the actual speed so that a prespecified setpoint speed is maintained.

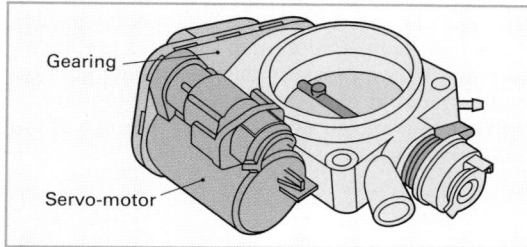

Fig. 2: Throttle-valve actuator

Overrun fuel cut-off

> No fuel is injected when overrun fuel cut-off is active.

When the engine is running with increased revs and with the throttle valve closed (overrunning, e.g. in downhill-driving situations), overrun fuel cut-off prevents fuel from being injected. Fuel injection resumes when the throttle valve opens or when the engine speed drops below a stored threshold, e.g. 1,200 rpm.

The ECU requires the following information for overrun fuel cut-off:

- Throttle-valve position from the throttle-valve switch or throttle-valve potentiometer
- Engine speed from the engine-speed sensor

Acceleration, full-load enrichment

> The mixture is enriched in order to facilitate maximum engine power output.

Engines with three-way catalysts are operated as far as possible in the $\lambda = 1$ range on account of exhaust-gas regulations. To be able to output the maximum engine power, the inducted mixture is enriched, depending on the engine, to lambda 0.85 to 0.95. Lambda closed-loop control must be cut out for this purpose. Enrichment begins when the throttle-valve potentiometer signals full load to the ECU or the voltage change per unit of time at the potentiometer exceeds a specific stored value. Extremely powerful engines do not necessarily require full-load enrichment.

Altitude adaptation

There is no need for special altitude adaptation in non-supercharged engines because the air-mass meter takes into account a reduced air density, for instance at greater altitudes.

Engine-speed limitation

> The function of this facility is to prevent the engine from overrevving.

Engine-speed limitation is activated when the ECU receives from the engine-speed sensor a signal from which it detects that the stored maximum speed has been reached. The moment of ignition is moved in the retard direction to limit the power and with it the maximum speed and also the top road-speed. Fuel injection is cut out in exceptional cases only.

LH-Jetronic as Motronic

All LH-Jetronic systems (fuel-injection systems) are essentially designed as Motronic systems, i.e. both mixture formation and ignition of the fuel-air mixture is controlled by a common engine ECU. Here, depending on the manufacturer's requirements and year of manufacture, different ignition systems can be combined with LH-Jetronic. By using Motronic systems, it is possible to reduce design complexity, increase operational reliability and improve the efficiency of the systems.

12.4.5.2 LH-Motronic fuel injectors

With LH-Jetronic each cylinder is assigned an electromagnetically actuated fuel injector **(Fig. 1)** which injects fuel sequentially into the intake manifold.

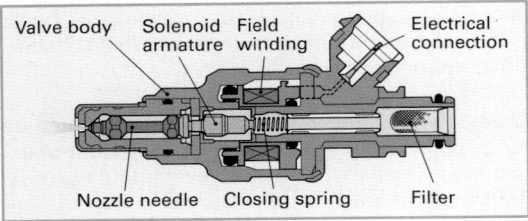

Fig. 1: Fuel injector

Function. When the valve field winding is supplied with power by the ECU, a magnetic field is generated in it which attracts the solenoid armature. This raises the nozzle needle off its seat and fuel is injected. The needle stroke, depending on the valve design, is 0.05 mm...0.1 mm. When the ECU stops supplying power as a function of the operating state (after 1.5 ms...18 ms), the magnetic field collapses and the closing spring forces the nozzle needle into its seat. Fuel injection is terminated. The mass of the injected fuel is dependent on ...

- ... the valve opening time.
- ... the injected fuel quantity per unit of time (valve constant).
- ... the fuel density.
- ... the fuel pressure.

Powering of valves. The fuel injectors are switched to negative by the ECU. In this way, the ECU can be protected against being destroyed by the short-circuit current in the event of a short circuit to earth/ground. The positive supply is provided via a relay switched by the ECU from terminal 15. The valve opening time can be determined by display-ing the injection operation on an oscilloscope **(Fig. 2).** The voltage peak during the closing operation is created by the switch-off induction of the field winding.

Types. With LH-Motronic different fuel injectors are used for engines with two- or multiple-valve technology. They differ in the shape of the fuel jet or spray and in the angle at which the fuel is injected by the nozzle **(Fig. 3).**

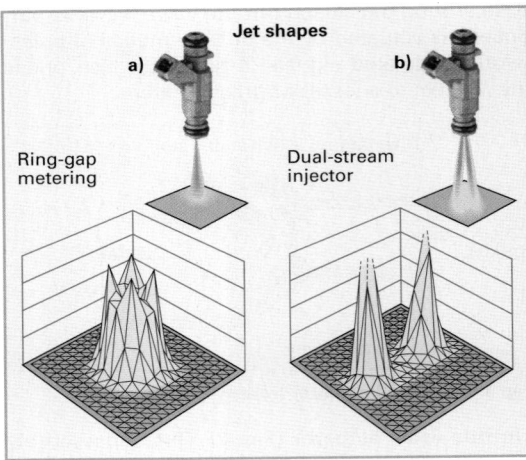

Fig. 3: Fuel injectors for two-valve technology (a) and multiple-valve technology (b)

Air-shrouded fuel injectors are used for finer atomisation of the fuel and for better mixing with air **(Fig. 4).** For this purpose, air is diverted before the throttle valve and routed via a line into the injector. In the narrow injector air gap the air is greatly accelerated by the pressure differential acting in the intake manifold at part load. The air emerging at high speed is mixed with the injected fuel and this process finely atomises the fuel.

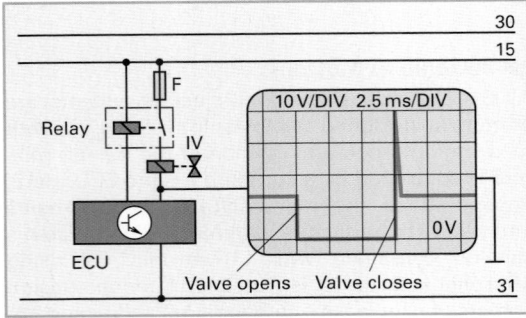

Fig. 2: Powering of fuel injector

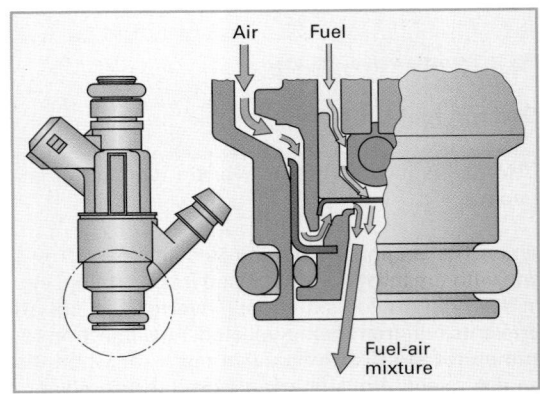

Fig. 4: Air-shrouded fuel injectors

When the fuel injectors are supplied with fuel from the fuel rail, this fuel is fed from the top (top-feed). The top end of the injector, sealed by an O-ring, is integrated in the fuel rail while the bottom end, also sealed by an O-ring, is integrated in the intake manifold. In the interests of saving space, the fuel injectors are often integrated in fuel-rail modules. In this case, so-called bottom-feed fuel injectors are used. The fuel is supplied from the side with these injectors. These injectors have good fuel-cooling properties and thus exhibit a good hot-start response.

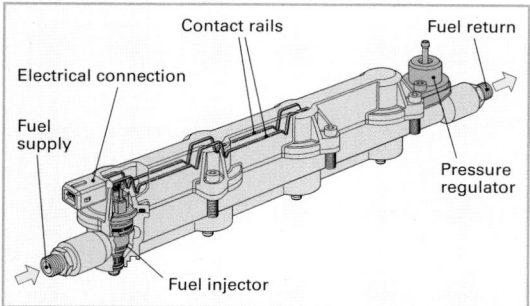

Fig. 1: Fuel-rail module with bottom-feed injectors

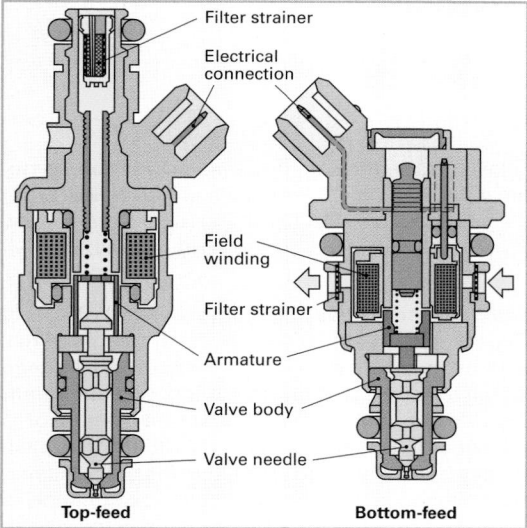

Fig. 2: Top-feed and bottom-feed fuel injectors

12.3.5.3 Electronic control of LH-Jetronic

The block diagram on **Page 274** and the circuit diagram on **Page 275** show in simplified form the design of electronic control of LH-Motronic. The following sensors and actuators are used here.

Hot-film air-mass meter B3. This determines the inducted air mass and transmits it in the form of a voltage signal to the ECU, which calculates the basic injection quantity (quantity) from it together with the engine speed. If the sensor fails, the system can generate a substitute signal from the throttle-valve position. The vehicle can continue to be driven under restricted conditions (limp-home operation). The air-mass meter is supplied with power from pin 10 and receives earth/ground from terminal 31. The voltage signal transmitted to the ECU can be picked off at pins 10 and 12.

Engine-speed sensor B1. The signal from this sensor serves first and foremost, together with the signal from the air-mass meter, to calculate the basic injection quantity. The system uses inductive speed sensors which are accommodated in the area of the crankshaft and scan a specific pulse-generator wheel. These sensors more often than not also supply the reference mark which is needed to determine the exact TDC of cylinder no. 1. The engine cannot be operated should this sensor fail. The signal is also needed for idle-speed control, overrun fuel cut-off and engine-speed limitation. Oscilloscope readings can also be taken at pin 6 and pin 7.

Throttle-valve potentiometer B4. This is located on the throttle valve and serves to record both the throttle-valve position and the opening speed. The integrated idle switch signals to the ECU when the throttle valve is closed. If the sensor fails, a default value stored in the ECU is taken as the basis for the minimum speed. This is usually expressed in an increased idle speed. Idle-speed control, overrun fuel cut-off and full-load and acceleration enrichment are no longer possible. The potentiometer signal can be picked off at pins 13 and 14 or pin 12. The throttle-valve switch is checked at pin 15 to terminal 31.

Some systems, especially when a throttle-valve actuator is utilised, use a double potentiometer for safety and accuracy reasons.

Intake-air temperature sensor B7. This is an NTC resistor, the function of which is to record the temperature of the intake air. It is located in the intake manifold. The ECU needs this signal to adapt the fuel quantity. The injection time can be extended by up to 20% when the air is very cold. If the signal fails, it is possible to switch to a stored default value. The resistance of the sensor can be checked at pin 18 and pin 19.

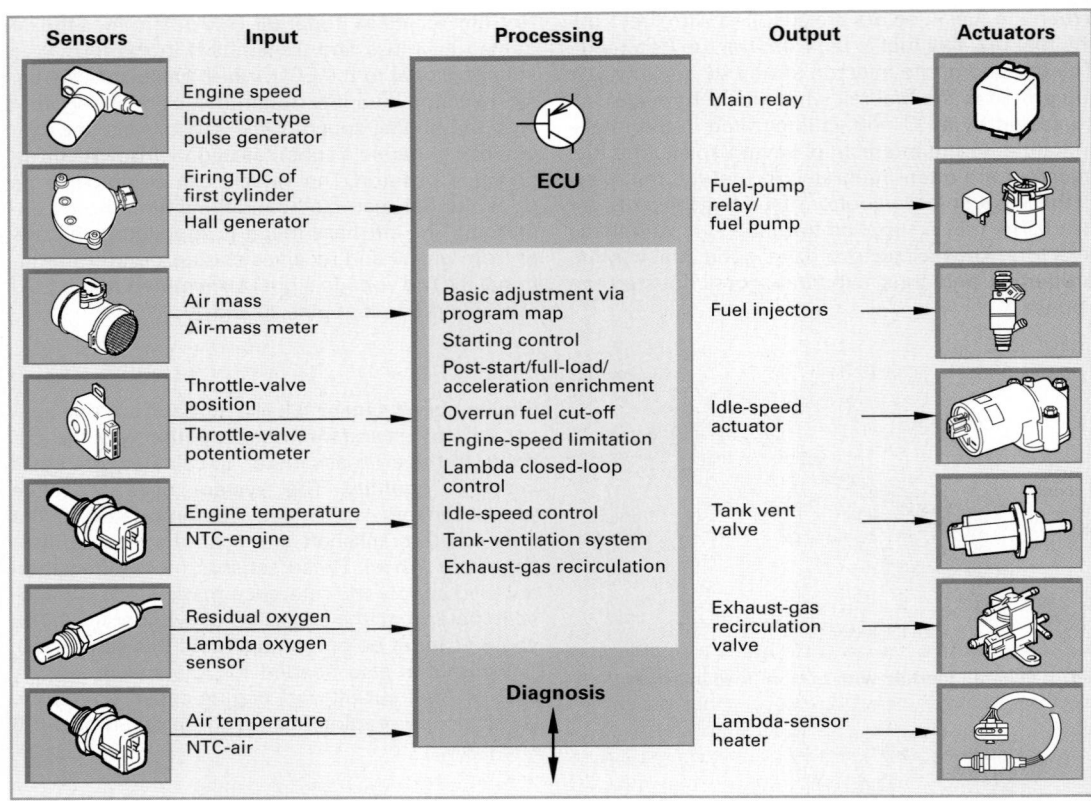

Fig. 1: Block diagram of LH-Jetronic sensors and actuators

Engine-temperature sensor B5. This NTC resistor records the engine temperature. Depending on the voltage drop at the resistor, the engine ECU adapts the injected fuel quantity to the operating state as a function of temperature. Thus the injection time is extended by up to 70% when the engine is cold. In addition, the moment of ignition, idle speed, exhaust-gas recirculation and knock control are modified when the engine is cold. The ECU can switch to a stored default value in the event of a signal interruption or a short circuit. An increased resistance, for example at a plug connection, is not detected however. This fault results in an enrichment of the mixture and thus among other things in increased CO emission. The resistance of the NTC resistor can be checked at the ECU plug at pin 12 and pin 16.

Reference-mark sensor B2. The signal from the inductive reference-mark sensor on the crankshaft and the signal from the Hall-effect sensor mounted on the camshaft are needed for the purpose of clearly identifying firing TDC. From both signals together with the engine speed the ECU calculates the correct moment for injection into the respective cylinder and the corresponding ignition angle. Os-

cilloscope readings of the sensor signal can be taken at pin 8 and pin 5. Terminals on the sensor: 7 (1) = signal positive; 31d (31d) = earth/ground supply. Power is supplied via pin 9 = terminal 8h (2).

Lambda sensor (voltage-jump sensor) B6. This registers the residual oxygen in the exhaust gas and, by means of feedback in the form of a voltage signal to the ECU, enables the injected fuel quantity to be regulated to $\lambda = 1$. Because the sensor is only operated at approx. 250 °C to 300 °C, it is electrically heated in order to achieve the quickest possible response. If the sensor fails, λ regulation is no longer possible. The failure is detected by the ECU. The mixture-formation system then operates as an open-loop control system. Oscilloscope readings of the sensor signal can be taken at pin 17 and terminal 31. The sensor heater receives positive from terminal 87 of K2 and negative from terminal 31.

Main relay K1. When the ignition is switched on, the main relay receives positive to terminal 85 from terminal 15 and negative to terminal 86 from ECU

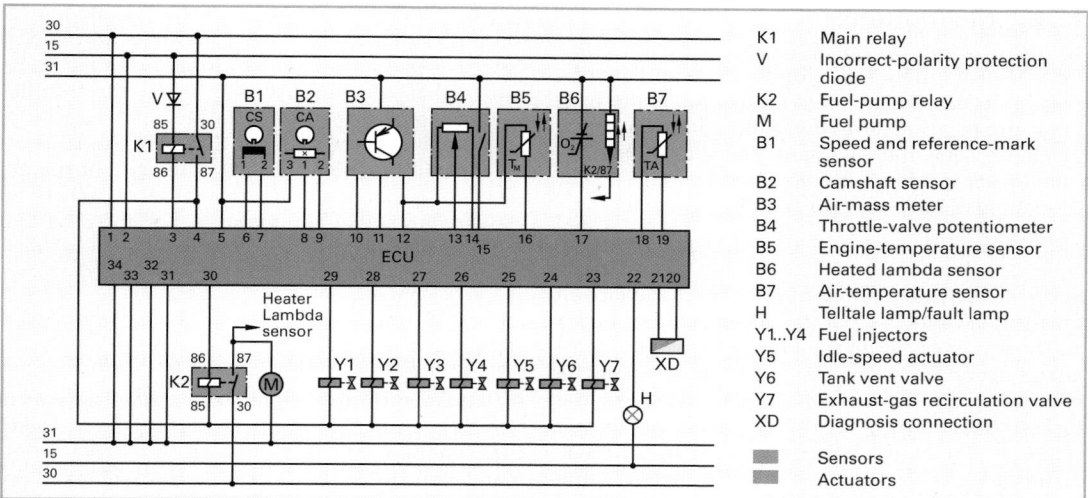

Fig. 1: LH-Jetronic circuit diagram

pin 3. In this way, the relay operating circuit closes and the ECU is supplied with power to pin 4. Likewise, solenoid valves Y1 to Y7 and the control circuit are supplied with power by K2 to terminal 85.

Fuel-pump relay K2. This relay closes when main relay K1 terminal 85 is supplied with positive and ECU terminal 86 is supplied with earth/ground. In order to establish the earth/ground connection, pin 30 must be connected to earth/ground. The operating circuit supplies fuel pump M and the lambda-sensor heater with power. The power supply is interrupted if the speed signal from the engine-speed sensor fails.

Fuel injectors Y1 to Y4. Like fuel-pump relay K2, these receive power from main relay K1. If the fuel injectors are top open, the ECU must connect in each case pins 26, 27, 28, 29 to earth/ground.

Idle-speed actuator Y5. The ECU uses this actuator to regulate the idle speed as a function of engine temperature. It is supplied with positive by K1 terminal 87. To facilitate stepless opening and closing of the bypass cross-section, the actuator is clocked by the ECU by means of **p**ulse-**w**idth-**m**odulated **s**ignals with negative.

Tank vent valve Y6. This solenoid valve opens and closes the connecting line between the intake manifold and the carbon canister. It is opened by **p**ulse-**w**idth-**m**odulated **s**ignals, during which the positive supply is provided by terminal 87 K1 and the negative supply by the ECU via pin 24. The valve remains closed if the signal fails.

Exhaust-gas recirculation valve Y7. This solenoid valve for exhaust-gas recirculation opens and closes the connecting line between the exhaust manifold and the intake manifold. It is opened by a **p**ulse-**w**idth-**m**odulated signal, during which it receives positive from terminal 87 K1 and negative from ECU pin 23. The valve closes if the signal fails.

REVIEW QUESTIONS

1   From which signals is the basic injection quantity calculated in LH-Motronic?

2   Which subsystems are featured in LH-Motronic?

3   Describe the various idle-speed control possibilities.

4   Which sensors does the ECU require for overrun fuel cut-off?

5   Explain the term "Motronic".

6   In terms of which features do LH-Jetronic fuel injectors differ?

7   What advantage do air-shrouded fuel injectors offer over conventional injectors?

8   Which sensors does LH-Motronic require and for what purpose are their signals used?

9   Which actuators are activated by LH-Motronic?

10   How is the fuel-pump relay actuated?

11   Explain the term "pulse-width-modulated signal".

12.4.6 ME-Motronic

ME-Motronic **(Fig. 1)** is a further development of LH-Motronic. A significant innovation is the replacement of mixture-formation control by so-called torque management. This has made it necessary to use electronic throttle control (ETC function). EOBD has also been integrated in the system.

In previous systems the driver opened and closed the throttle valve by operating the accelerator pedal. The inducted air mass and the fuel quantity injected accordingly determined together with the engine speed (main controlled variables) the torque requested by the driver. Additional torque requests, e.g. by the A/C compressor, occurred as disturbance values and had to be corrected by the system, e.g. by idle-speed control. Because of torque management, the accelerator-pedal position is now no longer the sole deciding factor for the torque to be generated. All the systems and components which influence the drive torque, e.g. automatic gearbox, A/C compressor, catalyst heaters, TCS/ASR, ESP, are used to calculate the engine torque to be generated. Motronic generates a substitute value, on which the requirements of the individual systems have an influence with different priorities. When, for instance, the A/C compressor is switched on, the drive torque is reduced. In order to avoid this, the ECU receives a signal before the A/C compressor is cut in. This causes the torque to be generated to be modified by the required amount by means of opening of the throttle valve, increased fuel injection and in other cases also a modified ignition angle. To facilitate this, it is necessary to isolate the throttle-valve position from the accelerator-pedal position. This is achieved by using an ETC (electronic throttle control) function. This also means that the accelerator-pedal position is from now on only to be viewed as a driver command, e.g. in the case of a TCS intervention.

12.4.6.1 ME-Motronic subsystems

Air-intake system. A significant, visible difference from LH-Jetronic is the introduction of the so-called **ETC function**. For this purpose, the driver command is recorded via an accelerator-pedal module. This is performed for safety reasons by two redundant potentiometers or Hall-effect sensors which are integrated in the module. The position and the rate of motion of the accelerator pedal are transmitted by the generated voltage signals to the engine ECU. The ECU uses stored program maps to calculate a necessary and useful torque and moves the throttle valve to a corresponding position by means of a servo-motor. This position is monitored by two potentiometers. Thus there is no longer any mechanical connection at all between the accelerator pedal and the throttle valve (drive by wire). In the event of faults within the system caused by unclear sensor signals, the throttle valve is moved into a limp-home position.

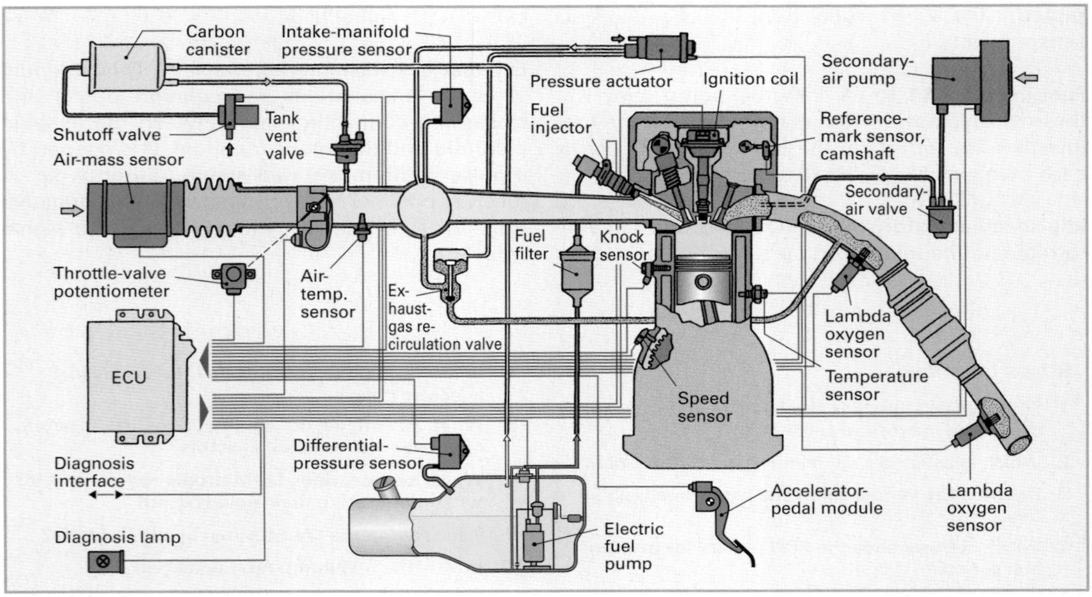

Fig. 1: ME-Motronic

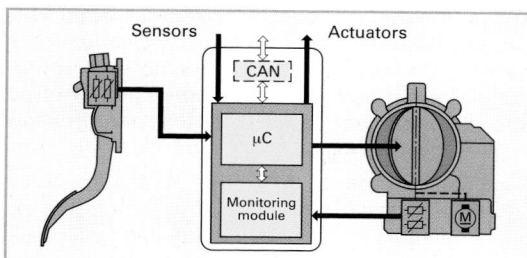

Fig. 1: ETC system

Fuel system. The fuel supply is increasingly supplied by one-line systems and delivery modules integrated in the tank. When one-line systems (**Re**turnless **F**uel **S**ystems) are used, the fuel supply pressure is usually kept constant at 3 bar in relation to the ambient pressure. As the intake-manifold pressure varies, so the differential pressure at the fuel injector changes, which results in different injected fuel quantities. This fault is corrected by a compensation function. For this purpose, the intake-manifold pressure is recorded by an intake-manifold pressure sensor and the injection time is extended or shortened by the ECU accordingly.

Pollutant-reducing systems

The increasingly stringent environmental-protection legislation passed over the years calls for pollutant-reducing subsystems to be elaborated and improved.

Mixture-formation system. More precise recording of the inducted air mass by hot-film air-mass meters with return-flow detection enable the engine to be operated in a narrower lambda window.

The recording of the lambda value by broadband lambda sensors enables the lambda value to be regulated more precisely than was previously possible with voltage-jump sensors.

The use of rapid-starting pulse-generator wheels on the camshaft enables the firing-TDC position to be detected earlier and therefore the engine to be started more quickly.

Tank-ventilation system. The fuel-supply system is outwardly sealed airtight. The carbon canister can be ventilated by a shutoff valve, which is connected in parallel to the regenerating valve.

Exhaust-gas recirculation. Cooling the exhaust gases recirculated in the combustion chamber improves NO_x reduction. An exhaust-gas recirculation cooler is installed for this purpose.

Secondary-air system. This consists of the secondary-air pump and valve. The system is used in the cold-starting phase to reduce CO and HC. It also heats up the catalyst very quickly to operating temperature (catalyst heating).

Introduction of OBD. Monitoring of all components which may cause changes in the exhaust-gas behaviour if they malfunction or are damaged must be guaranteed. Faults that have occurred must be stored and displayed.

12.4.6.2 Electronic control of ME-Motronic

In addition to the sensors and actuators used in LH-Motronic, the following components are used (see block diagram on **Page 278** and circuit diagram on **Page 279**).

Intake-manifold pressure sensor B9. The signal from this sensor is needed to record the intake-manifold pressure and to compensate the different differential pressure at the fuel injector by adapting the injection time. This signal is also used to calculate the purging flow of the carbon canister. If the air-mass meter fails, an approximately precise substitute signal for the inducted air mass can be generated via the intake-manifold pressure sensor. The sensor is connected via pins 49, 50 and 53 (earth/ground) to the ECU.

Differential-pressure sensor B10. Self-diagnosis is used to check the fuel tank for leaks by monitoring its internal pressure. The sensor is connected via pins 51, 52 and 53 (earth/ground) to the ECU.

Lambda sensor II B11. The post-catalyst sensor serves to monitor the catalyst function. It is also used to adapt the pre-catalyst sensor. If the sensor fails, the fault is detected and stored by OBD. Continued lambda closed-loop control by lambda sensor I is possible, however the catalyst function is no longer monitored. Oscilloscope readings of the sensor signal can be taken at pin 10 and pin 11. The sensor is heated via pin 9 (earth/ground) and K1 (positive).

Sensor for accelerator-pedal position B12. In the case of electronic throttle control, the driver command is determined from the position and rate of motion of the accelerator pedal. The necessary signal for the ECU is generated by two redundant potentiometers in the accelerator-pedal module. Potentiometer 1 is connected via pins 37, 38 and 39 and potentiometer 2 via pins 40, 41 and 42 to the ECU.

12

Sensor for throttle-valve position B4. The throttle-valve position must be recorded exactly by the ECU for the setpoint/actual-value comparison. As with the sensor for the accelerator-pedal position, two redundant potentiometers are also used here for safety and accuracy reasons. If the plausibility check of the four potentiometers by the ETC monitoring system reveals a deviation from the setpoint status, the system initially falls back on substitute signals. In emergency situations, e.g. two potentiometers on the throttle valve deliver different signals, the throttle valve is closed to such an extent as to permit only a low engine speed. The sensors can be checked at pins 31, 32, 33 (potentiometer 1) and at pins 31, 33, 34 (potentiometer 2) of the ECU plug.

ETC servo-motor B4. The throttle-valve servo-motor is actuated by the ECU via pin 35 and pin 36. The relevant throttle-valve position is calculated as a result

of the ECU determining the setpoint torque. This can be generated by a specific charge, for which in turn a quite specific throttle-valve position is necessary. If the motor fails, the throttle valve is moved into a limp-home position which only permits a low engine speed.

Secondary-air pump M1. This pumps fresh air shortly after the engine exhaust valve into the exhaust manifold as a function of the engine temperature under restricted time conditions. It is supplied with power via relay K3, where the positive supply is provided by K1 and the negative supply via terminal 31. The function of the pump is monitored by self-diagnosis.

Secondary-air valve Y9. This valve protects the secondary-air pump and prevents hot exhaust gases from flowing into the pump when the pump is stopped. It is opened by positive from K1 and negative from ECU pin 19.

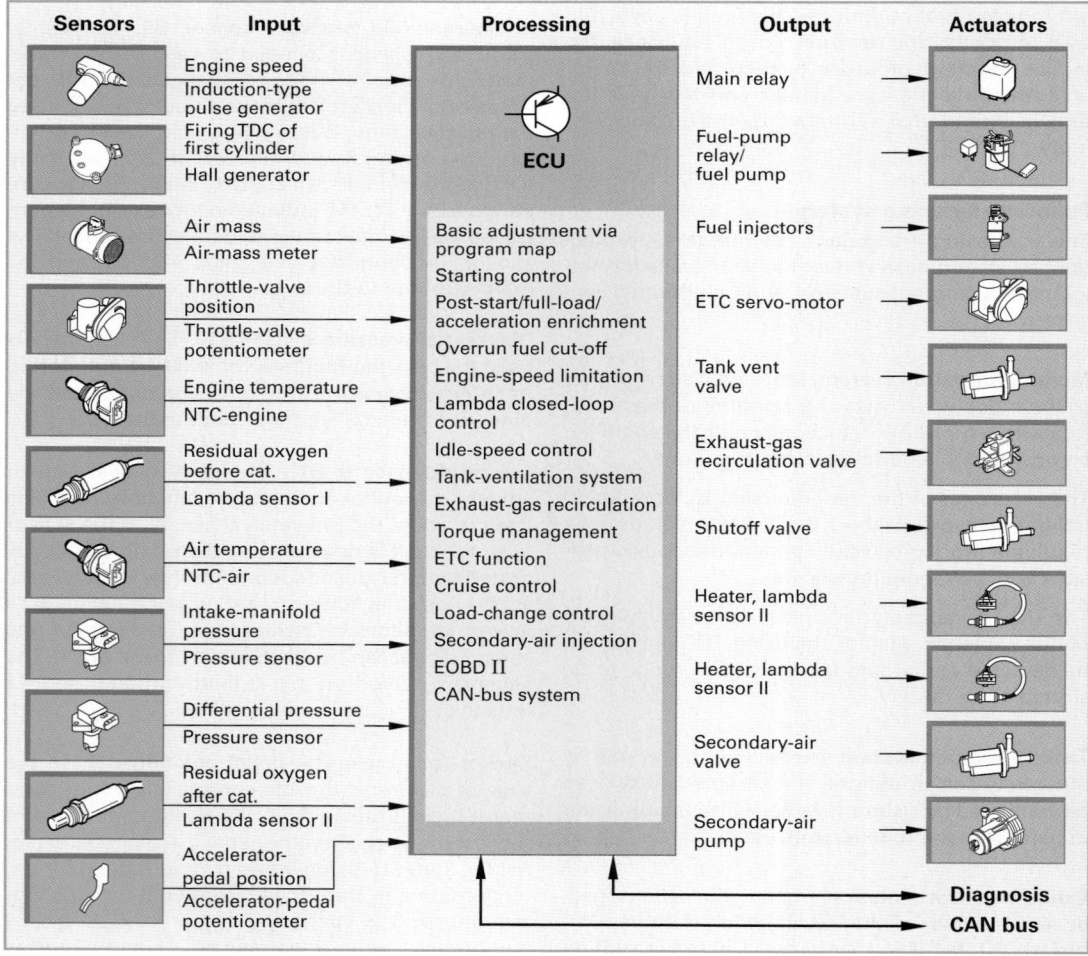

Fig. 1: ME-Motronic block diagram

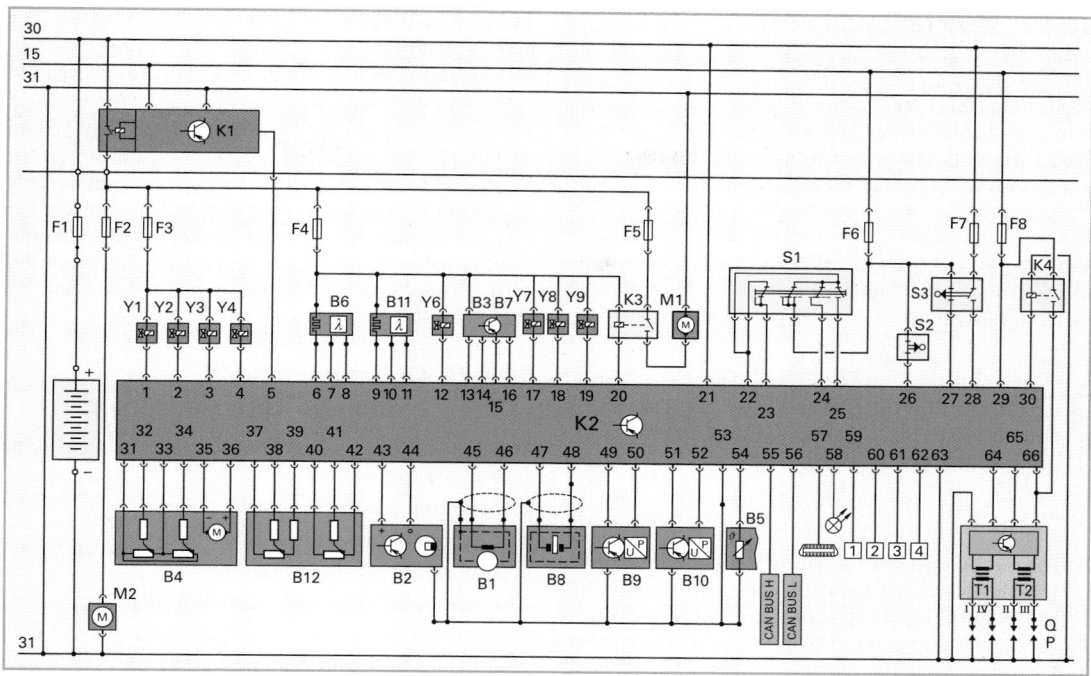

Fig. 1: ME-Motronic circuit diagram

Key to circuit diagram

B1	Crankshaft speed sensor	**B10**	Differential-pressure sensor	**S1** Switch for CC*
B2	Camshaft TDC sensor	**B11**	Heated lambda sensor II	**S2** Clutch-pedal switch
B3	Air-mass meter	**B12**	Sensor for accelerator-pedal position	**S3** Brake-pedal switch for CC*
B4	Sensor for throttle-valve position with ETC servo-motor	**F1...F8**	Fuses	**T1, T2** Twin-spark ignition coils
		K1	Fuel-pump relay	**Y1...Y4** Fuel injectors
B5	Engine-temperature sensor	**A**	ME-Motronic ECU	**Y6** Regenerating valve
B6	Heated lambda sensor I	**K3**	Relay, secondary-air pump	**Y7** Exhaust-gas recirculation valve
B7	Intake-air temperature sensor	**K4**	Relay, output stage, ignition system	**Y8** Shutoff valve
B8	Knock sensor			**Y9** Secondary-air valve
B9	Intake-manifold pressure sensor	**M1**	Secondary-air pump	**1...4** Inputs and outputs of other systems
		M2	Electric fuel pump	

* Cruise-control system

Shutoff valve Y8. The function of this valve is to shut off the air supply to the carbon canister when regeneration is deactivated. The shutoff valve is opened by positive from K1 and negative from ECU pin 18 parallel with the tank vent valve.

Connection of Motronic ECU with another system by CAN bus

All the data which are needed for a precise mixture formation in every operating state and in every operating situation must be made available to the engine ECU. For this purpose, all the ECUs which can influence the vehicle drive are interconnected by means of a high-speed bus system (CAN bus).

REVIEW QUESTIONS

1 **Which sensors and actuators are used in ME-Motronic?**

2 **At which pin can sensors B4, B9, B10, B11 and B12 be checked?**

3 **Describe the design and operating principle of an ETC system.**

4 **Which systems and measures for protecting the environment are used in ME-Motronic?**

5 **At which pin can actuators M1, Y9 and M2 be checked?**

6 **Which ignition system is used in the system shown in the system diagram?**

7 **What happens when K1 closes?**

12.4.7 MED-Motronic

> MED-Motronic (D = direct injection) is a Motronic system which is derived from ME-Motronic and adapted to the particular conditions and requirements of direct injection **(Fig. 1)**.

Virtually all manufacturers have taken to developing petrol direct injection because it offers the following advantages over indirect injection:

- Because the liquid fuel is injected directly into the combustion chamber, the liquid fuel vaporises in the combustion chamber only, which results in good internal cooling and increased engine power.

- Stratified-charge operation enables the engine to permit higher exhaust-gas recirculation rates.

- At partial load very much higher effective internal pressures are achieved on account of quality regulation. Throttling losses are reduced in stratified-charge operation by the fully opened throttle valve. This results in greater efficiency, increased power and reduced fuel consumption.

- During cold starting or acceleration the mixture has to be enriched less with direct injection than it does with indirect injection. This results in better exhaust-gas values and reduced consumption.

These advantages are offset by the following disadvantages:

- The significantly higher expenditure incurred in design and control.

- The increased emission of NO_x due to a lean mixture in stratified-charge operation, which cannot be reduced in the three-way catalyst. This makes it necessary to use a NO_x accumulator-type catalyst which must be regenerated at specific intervals. Furthermore, the sulphur in the fuel reduces the effectiveness of the catalyst.

12.4.7.1 Operating modes with petrol direct injection

The following operating modes can be used with petrol direct injection:

- Stratified-charge operation
- Homogeneous-charge operation
- Homogeneous lean operation
- Homogeneous stratified operation
- Homogeneous knock-protection mode
- Stratified-catalyst-heating operation

These operating modes are matched to each other so as to provide optimal mixture formation and combustion in every operating state. The control system must ensure that during vehicle operation neither power nor torque jumps are perceptible when switching from one operating state to another.

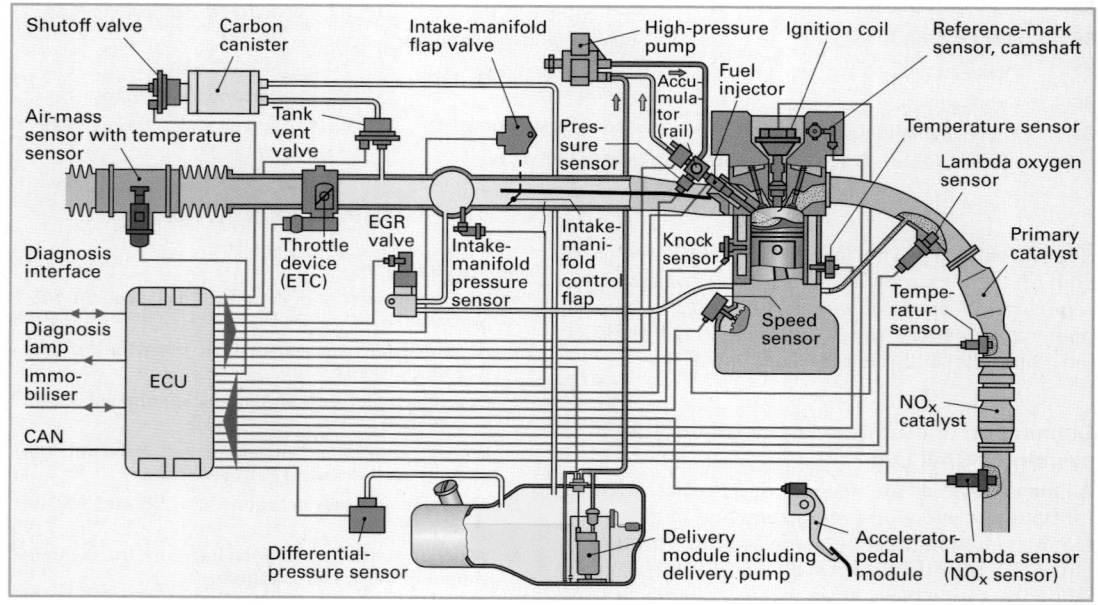

Fig. 1: MED-Motronic

Stratified-charge operation. Stratified-charge operation is possible in the lower torque and speed range up to approx. 3,000 rpm. Here the fuel is injected during the compression stroke into the combustion chamber shortly before the moment of ignition. Because of the short time period until it ignites, the fuel cannot mix uniformly with the air in the combustion chamber.

The fuel is transported in a cloud to the spark plug by an air swirl prevailing in the combustion chamber **(Fig. 1)**. Inside the mixture cloud the mixture ratio is approx. 0.95...1. Outside the cloud the mixture is very lean. A large amount of exhaust gas is recirculated into the combustion chamber in order to reduce the creation of NO_x by the altogether lean mixture.

Since the throttle valve is fully open during stratified-charge operation, the torque is generated during stratified-charge operation by quality regulation.

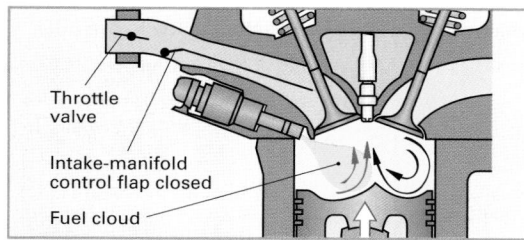

Throttle valve

Intake-manifold control flap closed

Fuel cloud

Fig. 1: Stratified-charge operation

If the torque demand were to become too high in this operating state, the increased injection would cause particles to form. Excessively high speeds give rise to turbulences in the combustion chamber, which no longer permit a steady mixture cloud of $\lambda = 1$. Consequences: faulty combustion, misfiring.

Homogeneous-charge operation. At high torque or high speed the engine is operated with a homogeneous mixture of $\lambda = 1$ or, to achieve maximum power, of $\lambda < 1$. For this purpose, the start of injection is moved to the induction stroke. The period of time thus available until the fuel-air mixture ignites

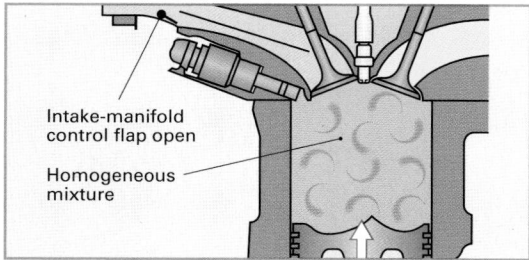

Intake-manifold control flap open

Homogeneous mixture

Fig. 2: Homogeneous-charge operation

allows the fuel to mix thoroughly with the inducted air and spread uniformly within the combustion chamber. In homogeneous-charge operation the torque is generated by quantity regulation. In other words, the quantity of inducted air is regulated by the throttle valve. Mixture formation and combustion take place accordingly as in manifold injection.

Homogeneous lean operation. The engine can be operated with a homogeneous lean mixture in a transition range between stratified- and homogeneous-charge operation. With operation at $1 < \lambda < 1.2$, fuel consumption is reduced in comparison with homogeneous-charge operation at $\lambda = 1$.

Homogeneous stratified operation. In this operating mode a homogeneous lean mixture is formed in the combustion chamber by advanced injection (approx. 75 % of the fuel) in the induction stroke. A second injection takes place in the compression stroke (double injection), which creates a zone of richer mixture in the area around the spark plug. This mixture is highly flammable and ensures complete combustion in the combustion chamber. This operating mode is chosen so that the torque can be better set during the changeover from homogeneous- to stratified-charge operation.

Homogeneous knock-protection mode. Thanks to double injection at full load it is possible to dispense with retarding the moment of ignition in order to avoid knocking. Charge stratification prevents the dangerous phenomenon of fuel auto-ignition.

Stratified-catalyst-heating operation. Another form of double injection allows the catalyst to be quickly heated. A lean mixture is formed here as in stratified-charge operation. After the fuel has ignited, fuel is injected once again during the power stroke. This share of fuel burns very late and heats up the exhaust-system branch intensively.

12.4.7.2 Combustion process of direct injection

With direct injection a distinction is drawn between how air and fuel are brought together in the combustion chamber. There are basically two different processes:

- Spray-directed combustion process
- Wall-directed process
  - Swirl flow
  - Tumble flow

Spray-directed combustion process (Fig. 1, Page 282). This process is characterised by the fact that the fuel is injected by the injection nozzle directly into the area of the spark plug and vaporises there.

Problem: When the spark plug is moistened by fuel particles, it is exposed to very high thermal load. Fuel particles which settle on the combustion-chamber surface do not combust or combust only incompletely.

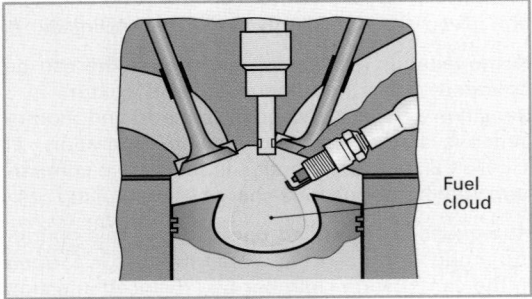

Fig. 1: Spray-directed combustion process

Wall-directed combustion process (Fig. 2). This process involves utilising a specific air flow to create a spatially delimited fuel cloud of $\lambda = 1$ which moves towards the spark plug.

The required swirl in the combustion chamber can be generated in two ways:

Swirl flow. The air flows through a spiral intake passage (swirl passage) into the cylinder and rotates in the combustion chamber about a vertical axis. The intake passage often has a double-flow design. The second passage (charge passage) is closed during stratified-charge operation by an intake-manifold control flap. For homogeneous-charge operation the intake-manifold control flap is opened in order to achieve a maximum combustion-chamber charge with air and thus greatest power.

Tumble flow. This process involves the creation of a cylindrical air flow. The air, which flows from above into the combustion chamber, is diverted by a pronounced recess in such a way as to move towards the spark plug again.

In practice, different measures for flow formation are often combined.

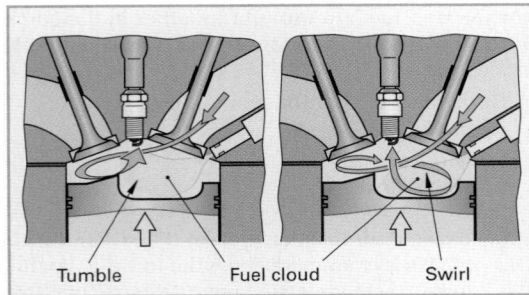

Fig. 2: Tumble and swirl flows

12.4.7.3 MED-Motronic fuel-supply system

The MED-Motronic fuel-supply system **(Fig. 3)** can be subdivided into a
- low-pressure circuit and a
- high-pressure circuit

The **low-pressure circuit** used in direct injection corresponds essentially to the fuel-supply system used in manifold injection. For the most part positive-displacement pumps are used as fuel-supply pumps because they can generate the required pre-delivery pressure of 3 bar...5 bar more easily. The integrated shutoff valve ensures that, for example during hot starting, the pressure can be briefly increased to 5 bar.

In the **high-pressure circuit** the fuel pressure is increased by a high-pressure pump to 50 bar...120 bar. The ECU regulates the pressure to the setpoint value by means of the pressure regulator. The closed-loop control circuit is closed by the pressure sensor, which communicates the actual value to the ECU.

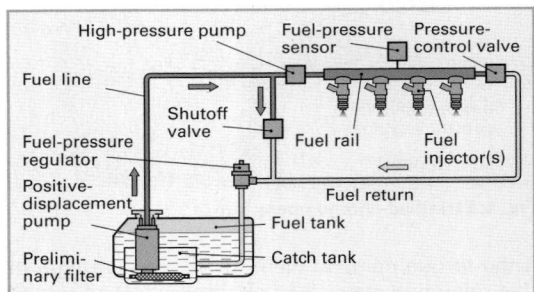

Fig. 3: MED-Motronic fuel-supply system

12.4.7.4 MED-Motronic components

High-pressure pump. The function of this pump is to pump the fuel delivered by the electric fuel pump at approx. 3 bar...5 bar into the rail at a pressure of

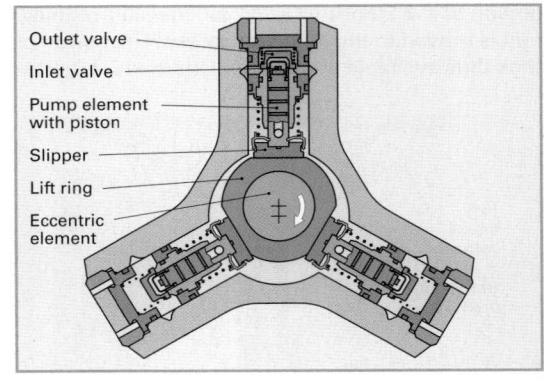

Fig. 4: Three-cylinder high-pressure pump

50 bar…120 bar. Three-cylinder pumps, which are similar in design to the pumps used in common-rail systems, are used here. Also pumps with only one cylinder are used for smaller delivery quantities.

Rail. The function of the rail is to take up the fuel delivered by the high-pressure pump and distribute it to the fuel injectors. The volume here must be large enough to compensate to a large extent the pressure pulsations caused by the pump.

The high-pressure sensor and the pressure-control valve are mounted on the rail.

Pressure-control valve (Fig. 1). This valve generates the desired pressure in the rail by the fact that the metering orifice to the low-pressure circuit is altered in accordance with demand. The pressure-control valve is closed at zero current. It is variably opened by the ECU by means of activation with **p**ulse-**w**idth-**m**odulated signals. A pressure-limiting function is incorporated to protect the system.

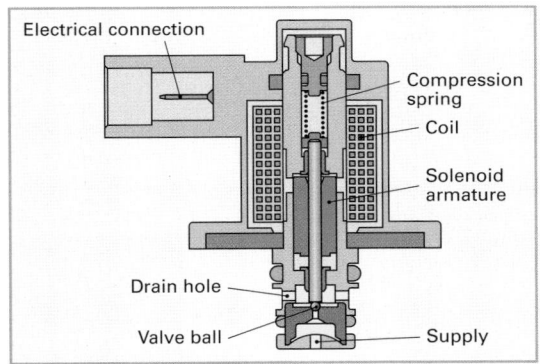

Fig. 1: Pressure-limiting valve

High-pressure fuel injector. The high-pressure fuel injector functions in the same way as that used in manifold injection. Increased demands are place on the injector by the completely different factors and conditions associated with direct injection. The injection pressure of up to 120 bar and the injection of fuel into the combustion chamber require the injector to be very strong and heat-resistant. Because

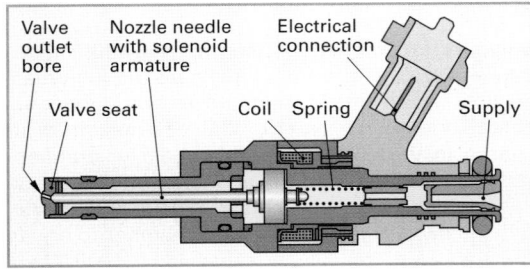

Fig. 2: High-pressure fuel injector

the time available for injection is considerably shortened, the injection process must be completed within 0.4 ms at idle up to 5 ms at full load. Faults caused by delayed injector opening therefore have much more serious ramifications than is the case with manifold injection. In order to open the injectors quickly, the field windings are actuated by high-power capacitors with up to 90 V.

12.4.7.5 Electronic control of MED-Motronic

In addition to the sensors used in ME-Motronic, the MED-Motronic control system contains the following sensors and actuators (see block diagram on **Page 284** and circuit diagram on **Page 285**).

NO_x sensor B14. The function of this sensor is to monitor the operation of the NO_x accumulator-type catalyst and to record both the NO_x and the oxygen content in the exhaust gas. The signal is evaluated by the ECU of the NO_x sensor (K6), which as required initiates the regeneration of the accumulator-type catalyst by switching to homogeneous rich operation.

Exhaust-gas temperature sensor B15. This sensor records the exhaust-gas temperature. The effective working range of the NO_x accumulator-type catalyst ranges between 250° C and 500° C. Therefore it is only permitted to switch to stratified-charge operation when the exhaust-gas temperature is between these limits. The sensor is connected to the ECU via pin 57 and pin 49.

Broadband lambda sensor B13. This is used to determine the oxygen content in the exhaust gas over a wide λ range. The injection time is corrected if the actual value generated by the sensor deviates from the setpoint value stored in a program map. The sensor is connected to the ECU via pins 24, 25, 26 and 27. The sensor heater receives positive from K5 and negative from pin 28.

Sensor for intake-manifold control-flap position B16. This uses a potentiometer to record the position of the intake-manifold control flap, which is closed during stratified-charge operation to generate the swirl and opened during homogeneous-charge operation. Because the position of the flap also influences ignition and exhaust-gas recirculation, it must also be monitored by On-Board Diagnosis. The potentiometer can be checked via pins 49, 52 and 54.

Fuel-pressure sensor B17. This records the fuel pressure in the rail. The information is transmitted in the form of a voltage signal to the ECU, which then sets the requested fuel pressure by means of the fuel-pressure control valve. The sensor receives positive

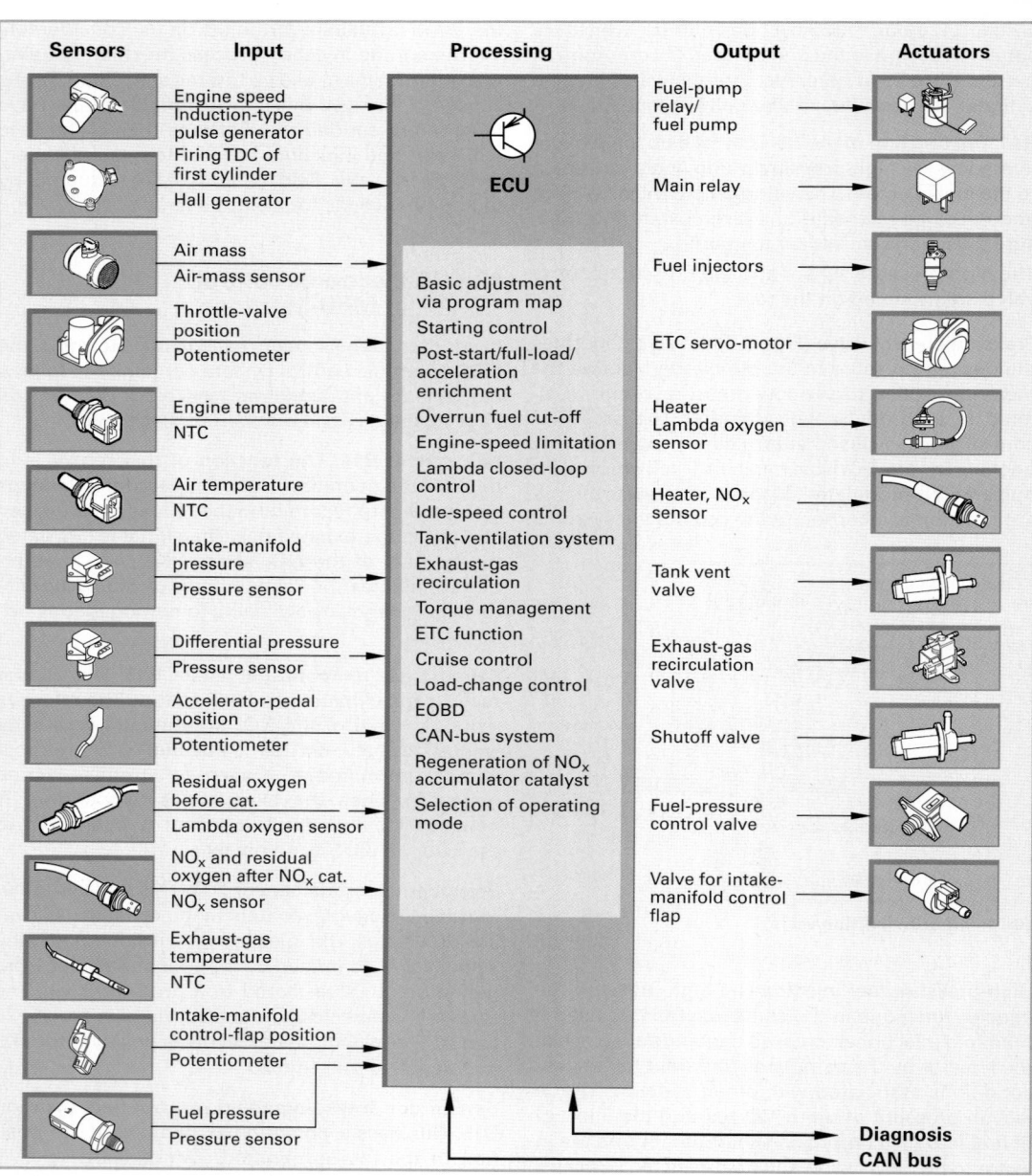

Fig. 1: MED-Motronic block diagram

from pin 12 and negative from pin 22. The signal is transmitted via pin 13.

On top of these, the MED-Motronic system requires the following additional actuators:

Fuel-pressure control valve Y11. This regulates the fuel pressure in the rail depending on the operating state to 50 bar to 120 bar. For this purpose, it is clocked by the ECU with earth/ground to pin 33. Positive is supplied by K5.

Valve for intake-manifold control flap Y10. The intake-manifold control flap opens up the full cross-section of the intake manifold in homogeneous-charge operation in order to achieve a maximum air charge in the combustion chamber. In stratified-charge operation it closes an intake-manifold passage, which increases the flow velocity and intensifies the swirl in the combustion chamber to create the fuel cloud. The valve is switched by the ECU via negative from pin 32. Positive is supplied by K5.

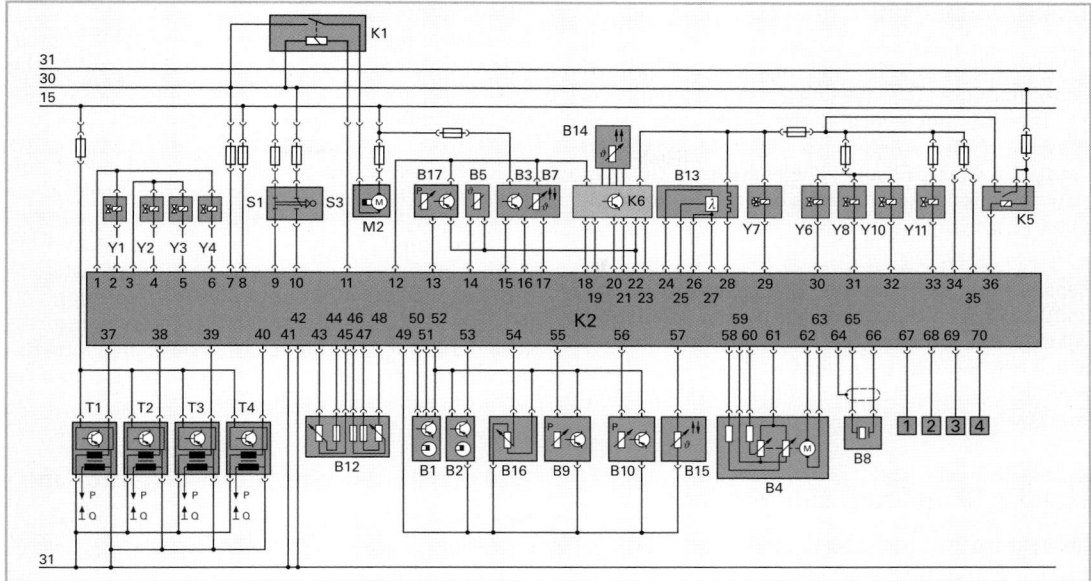

Fig. 1: MED-Motronic circuit diagram

Key to circuit diagram

B1	Crankshaft speed sensor	**B13**	Broadband λ sensor	**M2**	Electric fuel pump
B2	CamshaftTDC sensor	**B14**	NO_x sensor	**S1**	Switch for CC
B3	Air-mass meter	**B15**	Exhaust-gas temperature sensor	**S3**	Brake-pedal switch for CC
B4	Sensor for throttle-valve position with ETC servo-motor			**T1...T4**	Single-spark ignition coils
		B16	Intake-manifold control-flap potentiometer	**Y1...Y4**	Fuel injectors
B5	Engine-temperature sensor			**Y6**	Tank vent valve
B7	Intake-air temperature sensor	**B17**	Fuel-pressure sensor	**Y7**	Exhaust-gas recirculation valve
B8	Knock sensor	**F1...F12**	Fuses		
B9	Intake-manifold pressure sensor	**K1**	Fuel-pump relay	**Y8**	Shutoff valve
		K2	MED-Motronic ECU	**Y10**	Valve for intake-manifold control flap
B10	Differential-pressure sensor	**K5**	Motronic power-supply relay	**Y11**	Fuel-pressure control valve
B12	Sensor for accelerator-pedal position	**K6**	Control unit, NO_x sensor	**1...4**	Inputs and outputs of other systems

REVIEW QUESTIONS

1. What are the reasons in favour of introducing direct injection in petrol engines?

2. What are the disadvantages of petrol direct injection when compared with manifold injection?

3. Describe the individual operating modes of petrol direct injection.

4. Which combustion processes are used in petrol direct injection?

5. Describe how fuel is supplied and delivered with petrol direct injection.

6. Which components of petrol direct injection have been modified or added when compared with manifold injection? Explain their function and operation.

7. Which sensor types are used in petrol direct injection?

8. Which actuators are activated by the ECU in direct injection?

9. Why is a broadband λ sensor used instead of a voltage-jump sensor as the λ sensor?

10. Which components are supplied with power by the fuel-pump relay?

11. What is the function of relay K5? What conditions must be satisfied in order to enable the operating current to flow?

12. Why is the NO_x sensor needed?

13. How are the fuel injectors of the MED-Motronic system shown switched?

14. Which MED-Motronic components must be monitored by OBD?

15. What ignition system is used with the Motronic system shown?

12.4.8 KE-Jetronic

> With KE-Jetronic the fuel injectors are opened continuously. The basic injection quantity is formed by mechanical-hydraulic means. Adaptation of the mixture to different operating states is **electronically** controlled. KE = German abbreviation for continuous petrol injection with electronic control.

Because the fuel injectors in this system are continuously opened, it is not possible to regulate the injected fuel quantity, as for example is the case with LH-Motronic, by means of the opening time. With KE-Jetronic the injected fuel quantity is regulated by means of the injection pressure.

12.4.8.1 KE-Jetronic subsystems

Air system

The air filtered by the air filter, inducted by the engine and regulated by the throttle valve, flows into the intake manifold. Here the sensor plate, which is designed as a float, is raised to varying extents by the air flow. The deflection of the sensor plate is a measure of the inducted air flow. The deflection is transmitted as lift via a lever system to the control plunger in the fuel distributor.

Fuel system

An electric fuel pump generates the system pressure and delivers the fuel from the fuel tank to the fuel distributor. A fuel filter and a fuel accumulator are connected downstream of the pump.

The fuel accumulator is designed to reduce the pulsation noises of the roller-cell pump, which is for the most part used as a fuel pump. It is on the other hand intended, when the engine is switched off, to maintain a residual pressure in the fuel system in order to prevent vapour bubbles from forming and to improve hot-starting capability. The system-pressure regulator used limits the system pressure, depending on the design, to 4.8 bar to 5.6 bar and returns the excess fuel to the fuel tank.

12.4.8.2 KE-Jetronic mixture-control unit

This consists of (**Fig. 1**):
- Airflow sensor
- Fuel distributor
- Electrohydraulic pressure actuator

Airflow sensor. This sensor serves to determine the air flow inducted by the sensor. The flowing intake air raises the sensor plate. The control plunger in the fuel distributor is displaced by a lever mechanism in proportion to its deflection. A potentiome-

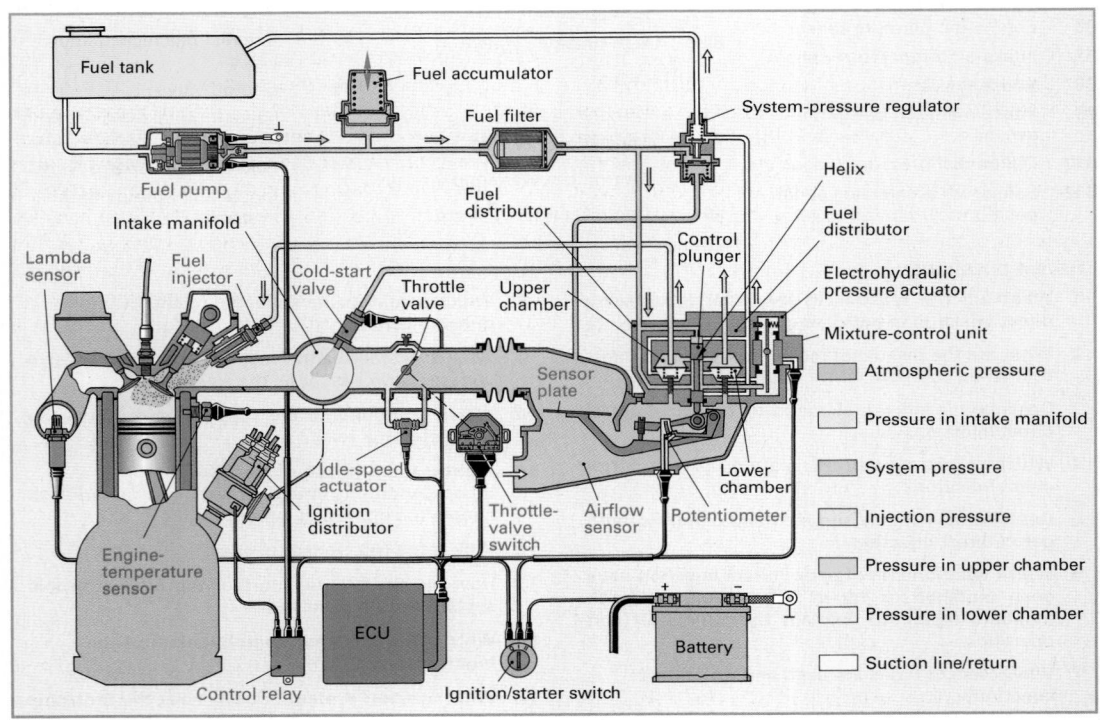

Fig. 1: KE-Jetronic system diagram

ter incorporated on the lever system records the sensor-plate lift and transmits it in the form of a voltage signal to the ECU.

Fuel distributor. This distributes the basic fuel quantity to the individual cylinders in accordance with the sensor-plate position. Depending on the position of the control plunger, which itself is dependent on the position of the sensor plate, the control-plunger helix opens up for example at idle a small discharge cross-section (orifice) in the slot barrel. In this way, only a small amount of fuel can flow into the upper chamber. At increased load the plunger is displaced upwards and the discharge cross-section is widened, thereby admitting an increased supply of fuel into the upper chamber **(Fig. 1)**.

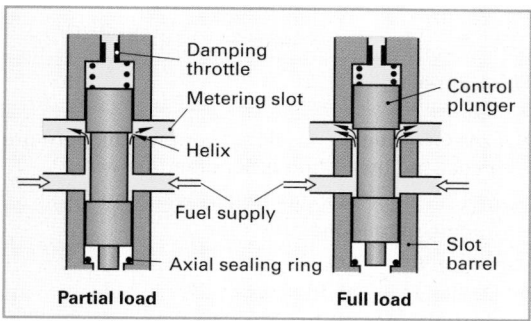

Fig. 1: Slot barrel with control plunger

Each cylinder in the engine has its own upper chamber and lower chamber. The upper chambers are separated from the lower chambers by a steel diaphragm. They are called differential-pressure valves because they are set in such a way that the pressure differential between upper and lower chamber is always 0.2 bar. When the pressure in the upper chamber increases on account of an increase in the fuel supply, the steel diaphragm deforms downwards against the spring pressure and the fuel pressure in the lower chamber. In this way, the

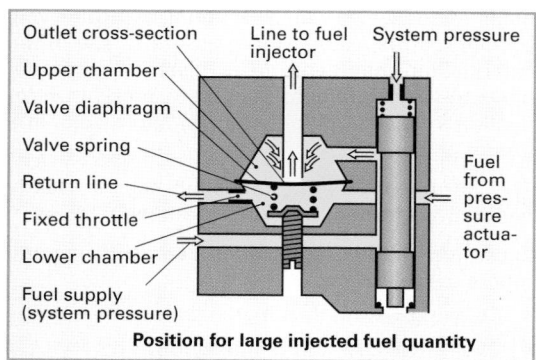

Position for large injected fuel quantity

Fig. 2: Differential-pressure valve

outlet cross-section to the fuel injector is widened, allowing more fuel to flow to the fuel injectors **(Fig. 2)**. The injectors are hydraulically opened as soon as the pressure at the valve exceeds 3.3 bar and close only when the engine is switched off or in the event of overrun fuel cut-off.

Electrohydraulic pressure actuator (Fig. 3). The ECU can adapt the mixture by means of this actuator to the respective operating states. Its pivoted baffle plate can be deflected by the ECU depending on the power supplied to the solenoid coil.

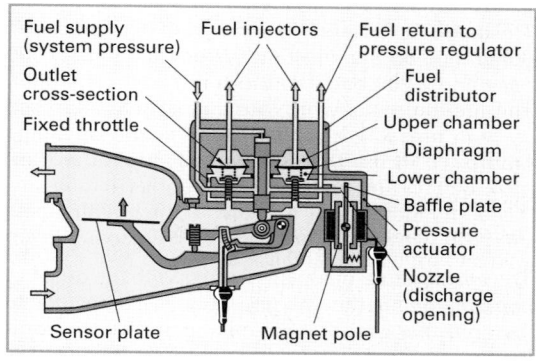

Fig. 3: Electrohydraulic pressure actuator

When the discharge opening of the nozzle is widened by the baffle plate moving away from the nozzle opening, increased fuel can flow into the pressure actuator. This causes the pressure to rise in the lower chambers of the differential-pressure valves, which are connected to the pressure actuator. This in turn causes the steel diaphragm to arch upwards slightly and the discharge cross-section in the upper chambers to narrow. In this way, less fuel flows to the injectors and the mixture leans. The mixture is enriched when the solenoid coil is energised in such a way that the baffle plate moves towards the nozzle opening.

The ECU requires the following signals for mixture adaptation:

- Throttle-valve switch → Idle, full load
- Speed sensor → Engine speed
- Ignition/starter switch → Start
- Engine-temperature sensor → Coolant temperature
- Lambda sensor → Mixture composition

In the event of electronic-control failure, the basic injection quantity can continue to be formed by the mechanical-hydraulic control system. The vehicle can therefore continue to be driven. However, the mixture can no longer be adapted.

12.5 Mixture formation in diesel engines

In contrast to spark-ignition engines with conventional manifold injection, diesel engines have interior mixture formation, i.e. the fuel is injected in liquid form at high pressure into the combustion chamber, where it combusts with the precompressed air.

12.5.1 Combustion sequence in a diesel engine

Complete combustion

The liquid fuel is injected under very high pressure, depending on the injection system and load at 180 bar up to 2,200 bar, into the combustion chamber. The fuel droplets, which are superfinely atomised by high pressure and very small nozzle openings (approx. 0.15 mm), are heated by the hot air (approx. 800 °C) and first begin to vaporise at their surfaces. The fuel vapour mixes with the hot air, heats up further and combusts with the oxygen in the air.

The temperature increases further with the onset of combustion in the mixture zone. This results in faster further evaporation, mixing and ideally combustion inside the fuel droplet.

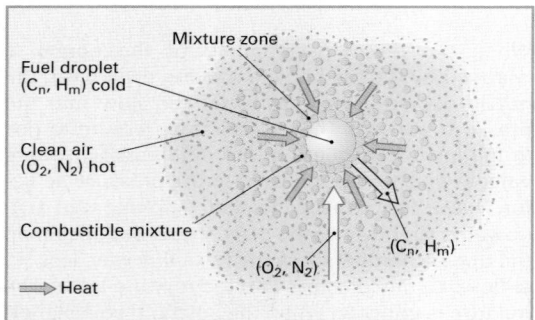

Fig. 1: Complete combustion at fuel droplet

12.5.2 Failures in combustion sequence

Incomplete combustion

Since the outer mixture zone on the fuel droplet reacts with the oxygen in the surrounding air while the core of the droplet has not yet reached the self-ignition temperature, an oxygen deficiency arises inside the evaporated fuel droplet. Combustion inside the droplet can only take place incompletely. The larger the fuel droplets, the greater the area in which there is an air deficiency and the more hydrocarbon molecules are only incompletely combusted.

Particle production. Incomplete combustion creates a core of soot on which further combustion

residues, e.g. sulphate particles and hydrocarbons, can accumulate **(Fig. 2)**. These are called soot particles. Thanks to modern fuel-injection systems, it has been possible to reduce the emission of such particles over the last few years by roughly 90 %.

However, the particles created now are so small that they can penetrate cell tissue if they are inhaled. They can cause cancer- and circulation-related illnesses.

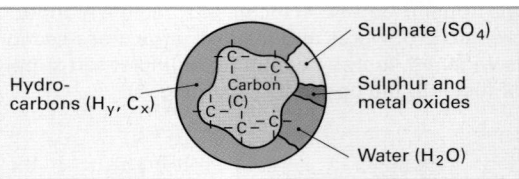

Fig. 2: Soot particle

Excessive particle emissions (soot production) are often caused by among others the following:

- Engine when cold starting or warming up
- Engine in full-load operation
- Clogged air filter
- Defective injection nozzle
- Faults in the combustion chamber or in the intake system

Ignition lag. The period of time between the discharge of the first fuel droplet from the nozzle opening and the start of combustion is known as the ignition lag.

If too long a time elapses between the discharge of the first fuel droplet and its subsequent ignition (more than 1 ms), this is known as increased ignition lag.

Excessive ignition lag causes a large amount of fuel to collect in the combustion chamber which heats up as a result of the retarded combustion and combusts suddenly. The large amount of explosively burning-off fuel causes the pressure to rise in the combustion chamber. This subjects the components to very high loads (similar to knocking combustion in spark-ignition engines) and results in a loss of power.

This abrupt onset of combustion can be clearly heard in the form of knocking combustion noise.

Excessive ignition lag is often caused by among others the following:

- Cold engine and thus high heat losses
- Over-advanced start of injection
- Poor fuel quality (cetane number too low)

- Deficient compression
- Dribbling injection nozzles

Measures for improving mixture formation

Mixture formation is essentially determined by the degree of air turbulence in the combustion chamber and by the magnitude of the injection pressure. The following measures are taken in particular:

- Operation of the diesel engine under excess-air conditions (up to $\lambda \approx 8$) and limitation of the maximum injected fuel quantity to $\lambda \approx 1.3$ to avoid air deficiency in the combustion chamber.

- Air turbulence by swirl passages **(see Inlet-passage control, Page 290)** and piston shape (direct injection) as well as swirl formation in the secondary combustion chambers (indirect injection) for better mixing of air and fuel.

- Optimisation of combustion-chamber geometry for improving the combustion process.

- Fuel preheating for finer atomisation and faster evaporation of the fuel.

- Glow control (pre- and post-glowing) for heating the combustion chamber and for reducing heat losses.

- Pre-injection of a small amount of fuel for heating the inducted air in order to reduce the ignition lag and achieve a smoother pressure increase.

- High injection pressures in order to create smaller and therefore faster- and more completely combusting fuel droplets.

- Post-injection for afterburning uncombusted particles.

12.5.3 Comparison of injection processes

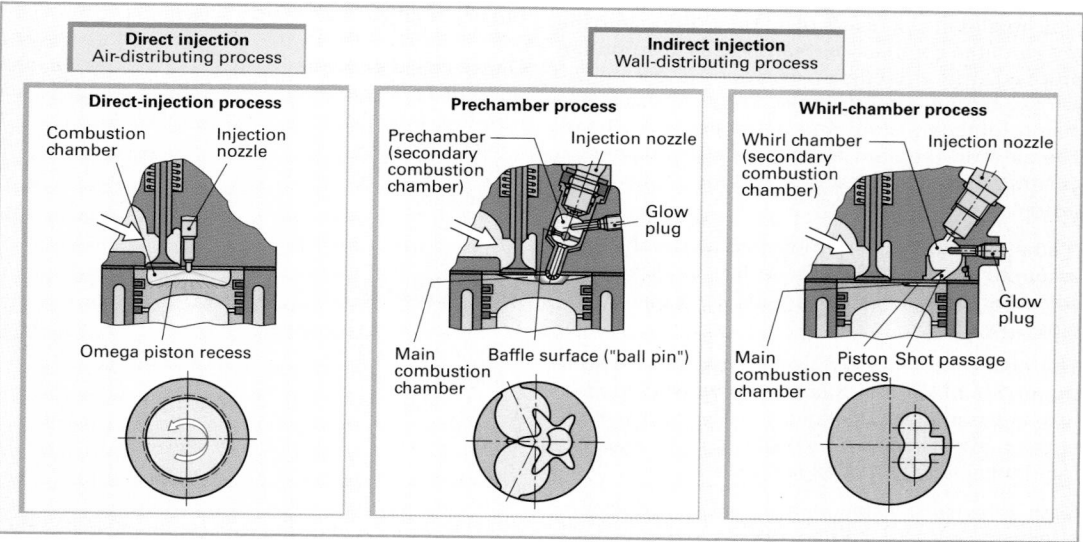

Fig. 1: Diesel injection processes

There are basically two different injection processes used in diesel engines:

- **Direct injection** into an undivided combustion chamber (**DI**)
- **Indirect injection** into the secondary chamber of a divided combustion chamber (**IDI**)

Indirect-injection diesel engines

Two different processes may be used, based on the shape of the secondary chamber: **prechamber** and **whirl-chamber process**.

The secondary chamber is located in the cylinder head and accommodates the nozzle holder with in-

jection nozzle and the glow plug. The secondary chambers are connected with the main combustion chambers by way of a shot passage (whirl chamber) or spray passages (prechamber) **(Fig. 1)**.

During compression air is forced through the passages into the prechambers, this operation setting the air in a state of swirl. Fuel is injected at a pressure of 180 bar to 450 bar by pintle nozzles into this air swirl. A significant amount of the fuel initially settles on the chamber surface **(wall-distributing process)**.

The fuel distributed in the air ignites. The heat generated by combustion allows the fuel deposited on the wall to vaporise off the combustion-chamber surface. The fuel heats up and also begins to burn.

The burning mixture is blasted by the combustion pressure in the secondary chamber into the main combustion chamber, where it combusts completely. This creates a delayed and thus "soft", two-stage combustion.

The large combustion-chamber surface in indirect-injection engines causes high heat losses. Indirect-injection engines therefore exhibit worse thermal efficiency than direct-injection engines.

During cold starting the heat losses result in insufficient mixture preparation and excessive ignition lag. This in turn results in starting difficulties at low outside temperatures, which is why a start-assist system must be installed.

Direct-injection diesel engines

In these engines the fuel is injected through hole-type nozzles at a pressure of up to 2,200 bar into the hot air of the combustion chamber (**air-distributing process**). The combustion chamber is essentially formed by the combustion-chamber recess in the piston (omega piston) and the flat cylinder head. The air swirl required for complete combustion is generated by swirl passages and the shape of the piston.

A small amount of fuel is injected before the main amount (pre-injection) in order to achieve a smoother pressure increase and thus quieter engine operation.

The heat losses are reduced by the combustion-chamber surface, which is smaller when compared with indirect-injection engines. This results in higher thermal efficiency. A cold-starting device is only required at very low outside temperatures.

Such a device is installed, however, to reduce pollutants during cold starting and warming up.

The principal advantage of direct-injection engines over their indirect-injection counterparts lies in their up to 20 % better fuel consumption. Originally rough engine operation (knocking combustion noise) can be eliminated to a large extent by pre-injection. Because of these advantages, direct-injection engines today enjoy a market share of almost 100 %.

12.5.4 Inlet-passage control (Fig. 1)

The charge inlet passages are closed and opened by flaps. They are controlled by an ECU as a function of a program map. A PWM signal is used to clock a servo-motor which moves all the flaps by means of a linkage.

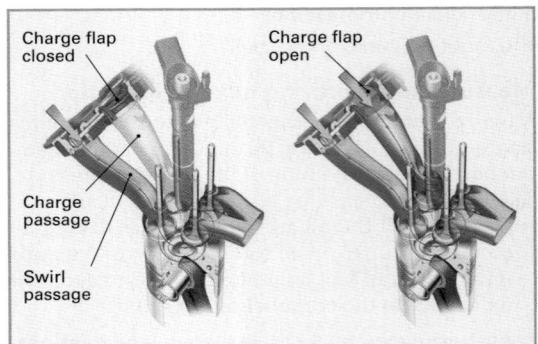

Fig. 1: Inlet-passage control

Lower speed and load range. All the charge flaps are closed here. The entire air mass flows in exclusively through the swirl passages. The resulting high air swirl ensures that the fuel is optimally mixed with the air and this results in better combustion. Particle production is reduced.

Upper speed and load range. Here the charge inlet passages are continuously opened so that the best possible ratio between air swirl and air mass is available for each engine operating point. In this way, exhaust-gas behaviour and engine power are optimised.

12.5.5 Start-assist systems

> The function of these systems is to help a diesel engine to start when cold, to provide smooth and stable idling and to reduce pollutant emissions.

The startability of diesel engines decreases as temperatures drop. Reduced compression pressures and heat losses caused by the cold combustion-chamber wall reduce the final compression temperature. Thus, starting without an additional start-assist device may no longer be possible under certain circumstances. This is accompanied by the risk of increased pollutant formation (white and black smoke).

Passenger cars utilise sheathed-element glow plugs as start-assist systems. In direct-injection engines glow filaments or heating flanges can additionally be installed in the intake manifold.

Sheathed-element glow plugs

Two different types of sheathed-element glow plug may be used:
- Self-regulating sheathed-element glow plugs
- Electronically regulated sheathed-element glow plugs

Self-regulating sheathed-element glow plug (Fig. 1)

Design. This type of glow plug consists of a control coil and a heating coil made from nickel wire and connected in series. Both coils have positive but differing temperature coefficients (PTC behaviour).

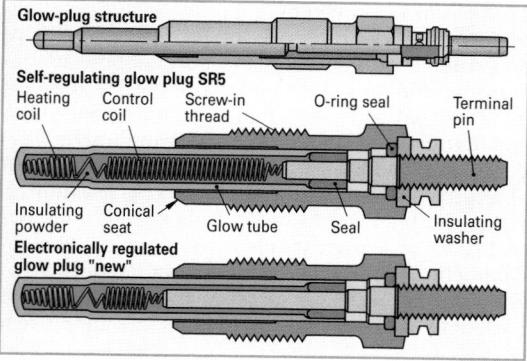

Fig. 1: Design of sheathed-element glow plugs

Self regulation. During pre-glowing a high current initially passes through the terminal pin and the control coil to the heating coil. The latter coil heats up quickly and makes the heating zone glow. The electrical resistance in the control coil increases as a result of the dissipating heat. The current is thereby reduced in such a way that the glow element does not overheat.

Self-regulating sheathed-element glow plugs usually operate with a nominal voltage of 11.5 V. They already reach the glow temperature required for triggering ignition of 850 °C after 2 to 7 seconds. They then continue to glow at a lower equilibrium temperature thanks to the PTC behaviour of the control coil. The power input is 100 W ... 120 W.

Glow control (Fig. 2)

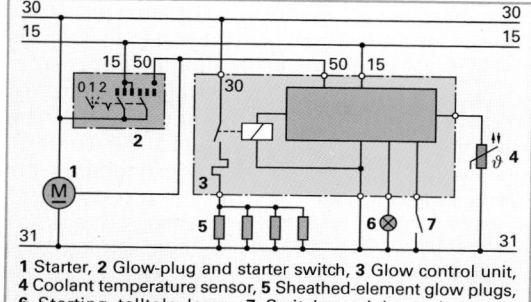

1 Starter, 2 Glow-plug and starter switch, 3 Glow control unit, 4 Coolant temperature sensor, 5 Sheathed-element glow plugs, 6 Starting telltale lamp, 7 Switch, pedal-travel sensor.

Fig. 2: Glow-plug starting system with glow control

This is used to switch the parallel-connected self-regulating sheathed-element glow plugs in such a way that mixture formation is optimally supported by the preheating system.

Design. The system essentially consists of an electronic module for controlling the glow curve, a device for indicating readiness for starting and a power relay for switching the glow-plug currents.

Operating principle. The glow process takes place in three phases **(Fig. 3)**.

- Pre-glowing
- Post-glowing
- Glow-plug start-assist

Pre-glowing. When the ignition lock is switched to position **1** (terminal 15), the glow control unit calculates the pre-glow time by way of the coolant-temperature sensor. No pre-glowing is performed at temperatures in excess of 60 °C.

Glow-plug start-assist. Pre-glowing is continued for a further 5 seconds after the pre-glow telltale lamp has gone out. The engine should be started within this period of time. The connection via terminal 50 ensures glowing for the entire duration of the starting process.

Post-glowing. Post-glowing begins after cold starting. Post-glowing is interrupted if the idle switch is opened and thus engine load is detected. Post-glowing is resumed on reversion to idle. Post-glowing is interrupted at temperatures in excess of 60 °C or in the event of a post-glow time of more than 180 seconds.

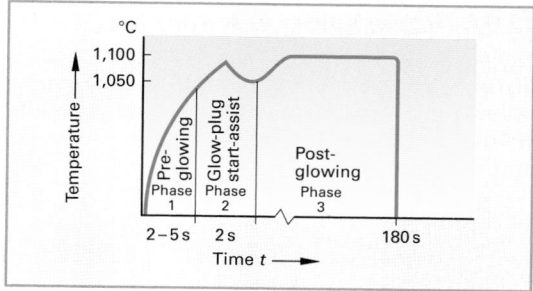

Fig. 3: Glow curve

Electronically regulated sheathed-element glow plugs

The control coil is shortened in length in order to obtain a minimal heating time. The glow elements designed for a nominal voltage of 5 V ... 8 V are briefly subjected under pulse-width-modulated conditions to an overvoltage of up to 11 V. This enables temperatures of 1,000 °C to be reached in 1 – 2 seconds. This facilitates a comfortable key-operated start (= starting without pre-glow delay) even at extremely low temperatures. Power semiconductors, which replace the conventional electromagnetic relay, are installed in the control unit to activate the glow plugs. Each glow plug can thus be individually activated, monitored and diagnosed **(Fig. 3)**.

Glow element and heating flange

These are installed in the intake manifold and consist of a heater element. The self-regulating PTC heater elements reach a temperature of 900 °C to 1,100 °C for intake-air preheating with an electrical power of 600 W.

WORKSHOP NOTES

Removing the glow plugs.
- Adhere to the specified releasing and tightening torques. Because deposits may form between the glow element and the cylinder head which oxidise at high combustion pressures, the glow plug can be screwed off during removal.
- If the glow plugs cannot be released with the specified torque, run the engine up to normal operating temperature. The glow plugs can then be removed.

Installing the glow plugs.
- Before installing new glow plugs, clean the glow passage with a special reamer.

Functional test.
- Glow plugs may only be operated at the specified voltage. However, when removed, they should only be operated for 1 – 3 seconds because they cannot dissipate any heat to the cylinder head and will therefore burn out after a short time.

12.5.6 Nozzle-holder assembly (Fig. 1)

The nozzle body of each injection nozzle is installed in a so-called nozzle-holder assembly in the engine cylinder head. There are two types: one- and two-spring nozzle-holder assemblies.

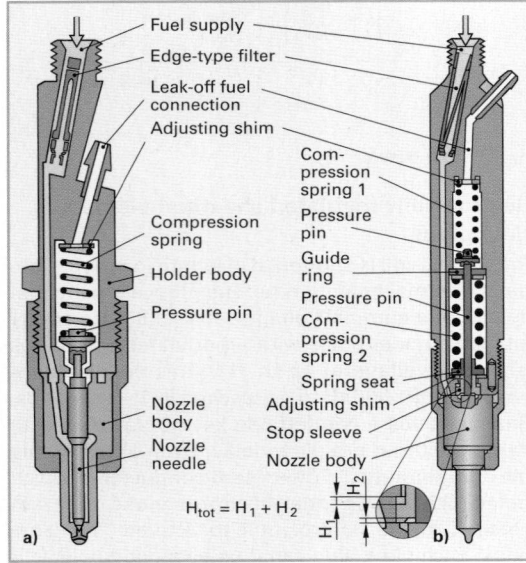

Fuel supply
Edge-type filter
Leak-off fuel connection
Adjusting shim
Compression spring
Holder body
Pressure pin
Compression spring 1
Pressure pin
Guide ring
Pressure pin
Compression spring 2
Spring seat
Nozzle body
Nozzle needle
Adjusting shim
Stop sleeve
Nozzle body

$$H_{tot} = H_1 + H_2$$

a)   b)

Fig. 1: One- and two-spring nozzle-holder assemblies

One-spring nozzle holder (Fig. 1a). The fuel delivered by the injection pump flows through the high-pressure line into the nozzle-holder supply inlet. The nozzle holder is equipped with a maintenance-free edge-type filter which is firmly pressed into the holder body. From there the fuel passes into the pressure chamber of the injection nozzle, where it presses on the pressure shoulder. If the resulting force of the fuel pressure acting on the pressure shoulder is greater than the spring force of the compression spring, the nozzle needle is lifted off its taper seat and fuel is injected. During each lift a small amount of fuel escapes past the nozzle needle to cool and lubricate the nozzle needle. This fuel returns through the return line to the fuel tank.

When the pressure in the injection line drops, the spring force increases again to overcome the resulting force acting on the pressure shoulder. In this way, the spring force presses the nozzle needle back onto its taper seat and the fuel injector closes.

Two-spring nozzle holder (Fig. 1b). This is equipped with two compression springs of differing hardness. These springs are matched to each other in such a way that initially with a low fuel pressure (approx. 180 bar) the nozzle needle is only raised against the force of the softer first spring until it reaches the stop of the harder second spring (lift H_2). The nozzle needle opens only slightly, thereby initially allowing a small amount of fuel to be injected into the combustion chamber **(pre-injection).**

As the fuel pressure increases (approx. 300 bar), the harder second spring is now compressed (main injection) and with it the nozzle needle lift is increased ($H_{tot} = H_2 + H_1$). Pre-injection provides for an altogether softer combustion from what would otherwise be a very quick, hard and abrupt combustion. This is accompanied by:
- Fewer combustion noises
- Stable idle speed
- Reduced pollutant formation

Thermal protection for injection nozzles. The temperature at the nozzle cones can rise to over 250 °C. This causes a deterioration in the material hardness of the nozzles, particularly in continuous operation, which in turn shortens the service life of the nozzles. Thermal-protection sleeves or plates **(Fig. 2)** can be used to dissipate and thereby reduce the temperature by approx. 50 °C. The sealing ring seals the combustion chamber.

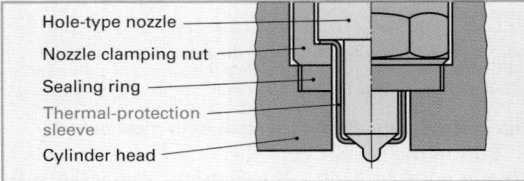

Hole-type nozzle
Nozzle clamping nut
Sealing ring
Thermal-protection sleeve
Cylinder head

Fig. 2: Thermal-protection sleeve

Injection nozzles

These are designed to inject fuel into the combustion chamber in such a way as to provide optimal mixture formation for the respective combustion-chamber geometry.

The injection pressure, the degree of atomisation and the shape of the fuel spray jets must be optimally matched to the combustion-chamber shape. The injection nozzles influence:
- Mixture formation and combustion sequence
- Engine power
- Exhaust-gas behaviour
- Combustion noises

Design. Injection nozzles each consist of a nozzle body and a nozzle needle **(Fig. 1)**. They are manufactured from high-quality steel and lapped. The tolerances are 0.002 to 0.003 mm. They must therefore only be replaced as a single unit.

Function. The nozzle needle is pressed into its seat by one or two compression springs in the nozzle-holder assembly **(see Nozzle-holder assembly, Page 292)**. The fuel pressure acting on the pressure shoulder of the nozzle needle creates a vertically acting lift force F_H. When it is greater than the spring force F_D, the nozzle needle is forced upwards. Fuel is injected. When the pressure drops, the spring force F_D increases again to overcome the lift force F_H. The nozzle closes.

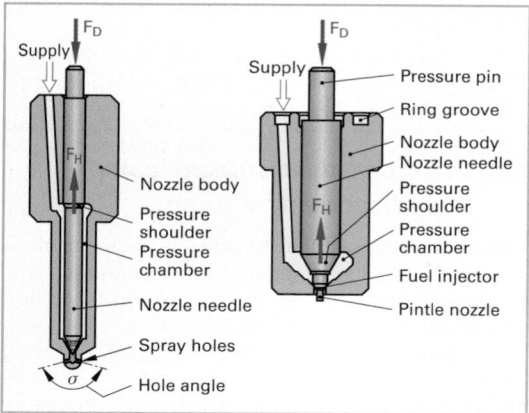

Fig. 1: Hole-type and pintle nozzles

Types. The following different types of nozzle may be used:
- Hole-type nozzle
- Pintle nozzle

Hole-type nozzle

Hole-type nozzles are used exclusively in direct-injection processes.

There are two types of hole-type nozzle **(Fig. 2)**:
- Blind-hole nozzle
- Sac-less (vco) nozzle

These have up to 8 symmetrically arranged spray holes.

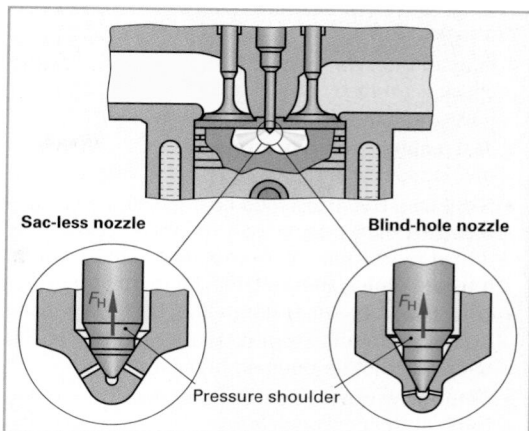

Fig. 2: Sac-less and blind-hole nozzles

Different sizes are used depending on the injected fuel quantity for each power stroke. The spray-hole diameter ranges between 0.15 mm (passenger cars) and 0.4 mm (commercial vehicles).

Features. To ensure a low emission of unburnt hydrocarbons, it is important for the residual volume below the nozzle seat to be as small as possible. This is best achieved with the sac-less (vco) nozzle.

Nozzle-opening pressures. These range, depending on the manufacturer, between 200 bar and 300 bar. However, this pressure is not to be confused with the maximum injection pressure. This is significantly higher as load and engine speed increase. In the course of the maximum pump-plunger stroke the injection pressure can rise to approx. 2,000 bar, depending on the injection system.

Pintle nozzle (Fig. 3).

This is used in engines with prechambers or whirl chambers.

The nozzle-opening pressure usually ranges between 80 bar and 125 bar. The nozzle needle has at its lower end a specially shaped pintle nozzle which projects into the spray hole of the nozzle body.

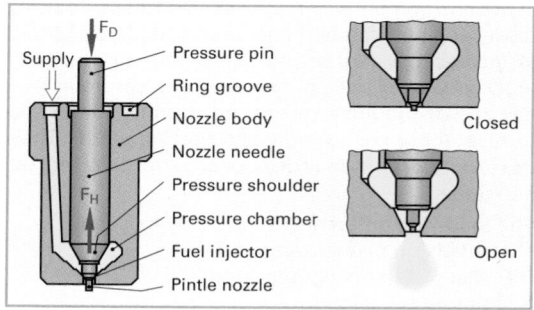

Fig. 3: Pintle nozzle

WORKSHOP NOTES

- **Injection-nozzle test rig (Fig. 1).** This is used to check removed nozzle-holder assemblies for leaks, opening pressure and spray shape.
- **Test requirements.** For an exact nozzle check, clean the nozzle first and then carry out a slide test.
- **Slide test.** The nozzle needle must slide naturally under its own weight into the nozzle body. Only the pressure pin of the nozzle needle may be touched during this test. Risk of corrosion!
- **Leak test.** A nozzle is tight when no fuel droplets drip out within 10 seconds at a pressure of 20 bar below the nozzle-opening pressure.
- **Setting the nozzle-opening pressure.** The closing force of the compression spring determines the opening pressure of the nozzle. Set the correct opening pressure for the nozzle-holder assembly by inserting adjusting shims under the compression spring.
- **Chatter test.** When setting the respective nozzle-opening pressure (varies from manufacturer to manufacture), check the spray pattern at the same time and listen out for a chattering noise when the nozzle needle is opened. While the injection nozzle is being opened, the pressure drops in such a way that the nozzle needle is pressed back onto its seat. Because fuel continues to be delivered, the pressure continues to rise and the nozzle needle lifts off its seat again. This opening and closing gives rise to a typical chattering noise. If this chattering cannot be clearly heard, then the nozzle guide is worn.

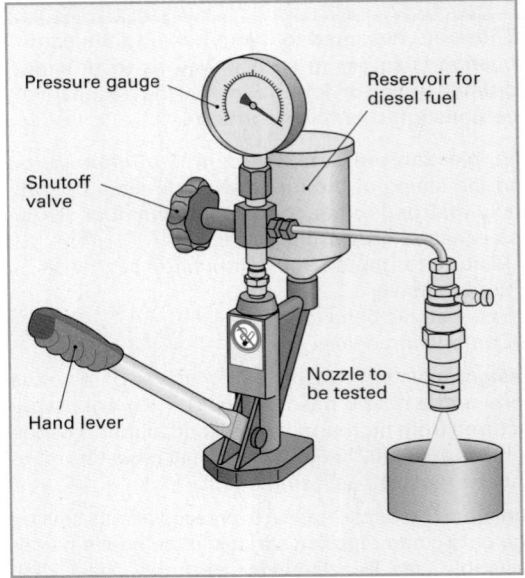

Fig. 1: Injection-nozzle test rig

Accident prevention.

- The high pressures involved mean that it is dangerous to touch the fuel spray because it can penetrate deep into the skin tissue. There is a danger of blood poisoning.
- Always wear protective goggles for all checks and tests. An extraction facility must be used because fuel vapours are hazardous to health.

12.5.7 Injection systems for passenger-car diesel engines

These systems must fulfil the following functions:
- Provide the necessary pressure.
- Inject the necessary amount of fuel (fuel-delivery control).
- Set the necessary start of injection (start-of-injection control).

In order to comply with the stricter emission limits for diesel engines, modern injection systems inject fuel at increasingly higher pressures and with increasingly greater precision. Mechanically controlled in-line and distributor-type injection pumps are unable to meet these increasingly exacting standards. They have therefore virtually disappeared from the market altogether.

The following different systems may be used:
- Axial-piston distributor pump
- Radial-piston distributor pump
- Unit-injector system
- Common rail

12.5.7.1 Axial-piston distributor pump with mechanical control

This is especially suitable for passenger-car diesel engines with 3, 4 and 6 cylinders. Particular features are:
- Low weight
- Compact construction
- Installation in any position possible
- Operates independently of the engine-lubricating circuit
- Only one high-pressure pump element needed
- Favourable engagement options for electronic control

Design. The distributor injection pump consists of the following subassemblies **(Fig. 1, Page 295):**
- Drive shaft
- Vane-type fuel pump
- Lifting-rotating facility for driving the pump plunger
- High-pressure pump element
- Control-lever system with control collar for fuel-quantity metering
- Mechanical governor
- Hydraulic timing device for start-of-delivery adjustment

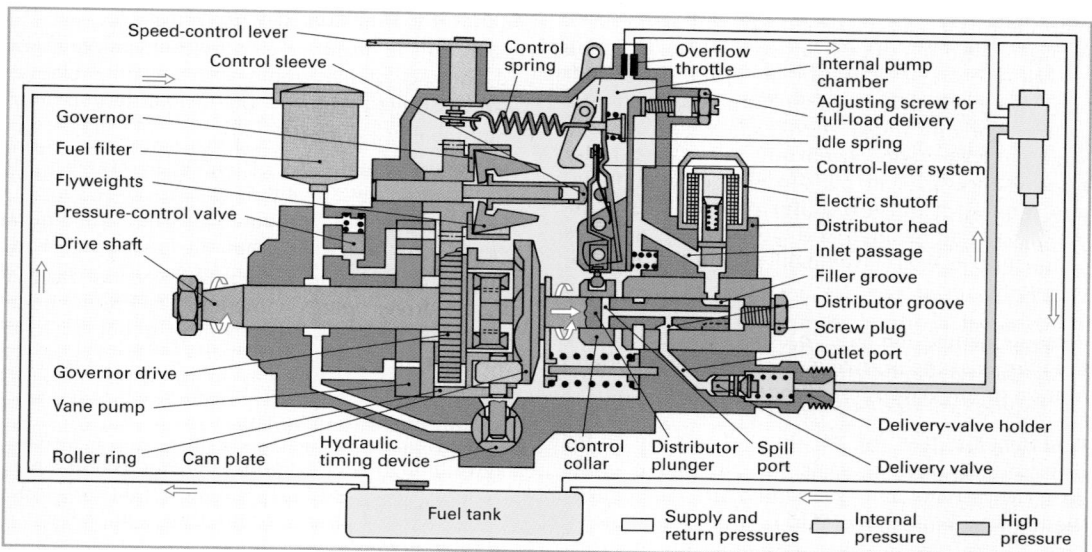

Fig. 1: Axial-piston distributor pump (VE)

The drive shaft of the distributor pump is mounted in the pump housing. The vane pump (fuel-supply pump), the governor drive (gear) and the cam plate, which rests on the roller ring, are mounted on the drive shaft.

The distributor head houses the high-pressure pump element with the control collar. The electric shutoff and the delivery-valve holder with the delivery valves are also screwed into the distributor head.

Vane pump (Fig. 2). With each revolution this pump delivers a constant fuel quantity from the fuel tank into the internal pump chamber. The delivery quantity (roughly 100 l/h ... 180 l/h) is sufficient to supply the high-pressure pump element with fuel for injection and to cool and lubricate the injection pump.

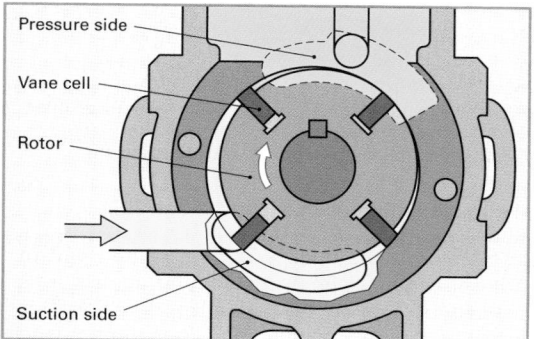

Fig. 2: Vane pump

Function. The vane-pump housing is eccentrically arranged around the impeller. This creates a suction chamber which increases in size in the direction of rotation and a pressure chamber which decreases in size. The fuel is delivered in this way into the inter-

nal pump chamber. As engine speed increases, the fuel pressure in the internal pump chamber rises.

Pressure-control valve (Fig. 3). This causes the internal-pump-chamber pressure to rise uniformly in proportion to the engine speed. The maximum pressure is limited to 12 bar. If the fuel pressure rises above this value, the valve plunger opens to allow fuel to return from the internal pump chamber to the suction side of the vane pump.

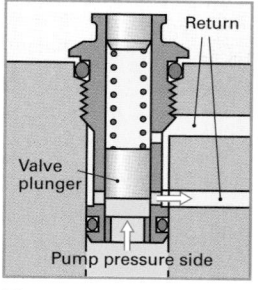
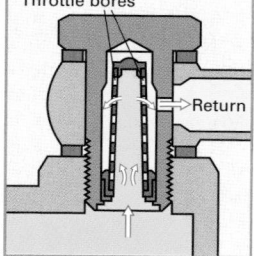

Fig. 3: Pressure-control valve   **Fig. 4: Overflow throttle**

Overflow throttle (Fig. 4). The overflow throttle (out screw) allows a variable amount of fuel to return through a small bore to the fuel tank.

The following internal pump pressures are set by the pressure-control valve and the overflow throttle:
- At idle speed approx. 3 bar
- At nominal speed up to approx. 8 bar

High-pressure pump element. From the internal pump chamber the fuel flows through the inlet passage and through the filler groove in the distributor plunger into the high-pressure chamber of the pump **(Fig. 1a)**. The distributor plunger is made to rotate by the drive shaft. It is also made to lift by the cam plate. The cam plate driven by the drive shaft has as many cam lobes as the number of cylinders in the engine. These cams move on radially arranged roller followers fixed to a rotating roller ring and allow the cam plate to move axially. The rotary motion of the distributor plunger causes its metering slots to open and close and establish a connection to the relevant spill port in the distributor head.

Pressure generation (Fig. 1b). This takes place after the inlet passage is closed as a result of the lifting motion of the distributor plunger.

Fuel delivery (Fig. 1b). This begins as soon as the distributor groove reaches the respective outlet port. The delivery valves are lifted off their seats by the high pressure generated and the fuel is delivered via the injection lines to the injection nozzles.

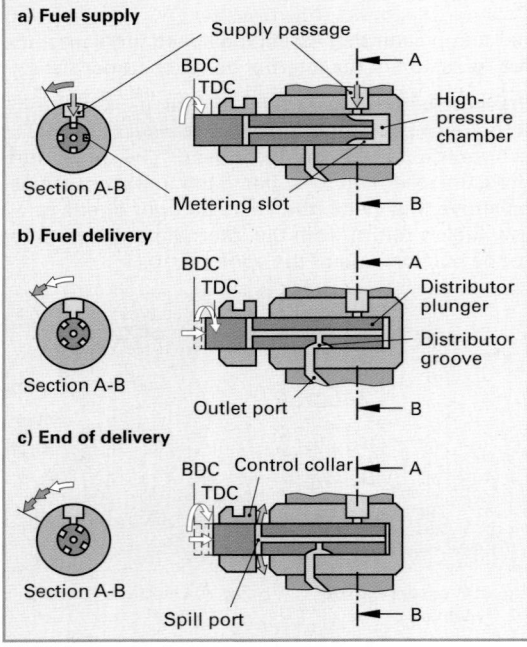

Fig. 1: Pump-element operating principle

End of delivery (Fig. 1c). This is reached when the control collar opens the transverse passage of the distributor plunger. The fuel flows back into the internal pump chamber during the remainder of the lift. After TDC is reached, the distributor plunger moves in the BDC direction again and closes the spill port by means of the control collar. The high-pressure chamber is again filled via the next metering slot in the direction of rotation.

Mechanical fuel-quantity governor

Design. This is designed as a mechanical governor and consists of flyweights and a control sleeve. These are driven via a gear pair with a step-up ratio.

Functions of the governor:

● **Idle-speed regulation.** Prevents the speed from dropping below the idle speed.

● **Maximum-speed regulation.** The maximum speed must be limited when the load is removed.

● **Intermediate-speed regulation.** Speeds desired by the driver between idle and maximum speed are kept constant even in the event of load changes.

Governor operation (Fig. 2).

Flyweights move the control collar via a linkage. In this way, depending on load and speed, the spill ports are opened and thus the delivery quantity limited.

Start/idle. Here the flyweights operate against the elastic force of the start/idle spring. The idle speed is thus kept constant.

Maximum speed. At higher speeds the flyweights operate against the control spring. They prevent the maximum permissible speed from being exceeded.

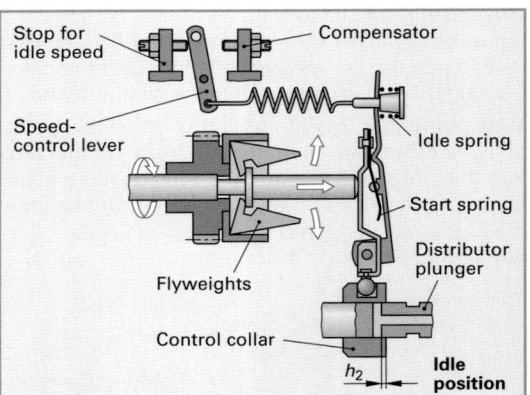

Fig. 2: Mechanical-governor operating principle

Hydraulic timing device (Fig. 1, Page 297).

This adapts the injection point to the relevant diesel-engine operating state and adjusts it as speed increases in the "Advance" direction. This delivers optimal power, good consumption and low amounts of pollutants in the exhaust gas.

Design. The hydraulic timing device is installed transversely on the underside of the pump and consists of a working cylinder with a spring-loaded hydraulic piston. This is connected by means of a bolt with the roller ring.

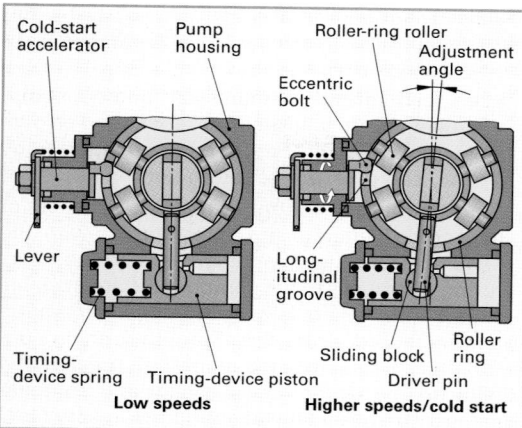

Cold-start accelerator · Pump housing · Roller-ring roller · Adjustment angle · Eccentric bolt · Lever · Longitudinal groove · Timing-device spring · Timing-device piston · Sliding block · Roller ring · Driver pin

Low speeds · **Higher speeds/cold start**

Fig. 1: Hydraulic timing device with cold-start accelerator

Operating principle. When the engine is stopped, the timing-device piston is held in the initial position by the preloaded timing-device spring. When the engine is running, the speed-dependent internal pressure overcomes the spring force and moves the timing-device piston. The axial piston movement is transmitted to the pivoted roller ring, which rotates for ignition-advance purposes against the pump piston's direction of rotation.

The cam plate is raised earlier by the roller-ring rollers; a more advanced injection point is achieved.

Supplementary devices

Manifold-pressure compensator. This is a diaphragm actuator for supercharged engines which alters the full-load fuel quantity as a function of boost pressure. Because the boost pressure rises as speed increases, more fuel can be injected and burnt without creating smoke in order to achieve greater engine power.

Cold-start accelerator (Fig. 1). The function of this device is to adjust the start of injection by a specific amount in the "Advance" direction when the engine is cold. In this way, more time is available for mixture formation when the engine is cold with increased ignition lag. Noise generation (diesel knock) and pollutant emissions are reduced. The adjustment is made either manually via a control cable or automatically by a temperature-sensitive control device.

Solenoid-operated shutoff valve. This consists of a solenoid valve which closes the inlet passage to the high-pressure chamber of the distributor plunger when the ignition switch is turned off in order to switch off the engine. It can also be used as an immobiliser.

WORKSHOP NOTES

Removing and installing the distributor injection pump

When removing the injection pump, first move cylinder no. 1 to the firing TDC position. The injection pump can now be removed. Reinstall the pump in reverse sequence, paying attention to the basic engine setting in relation to the pump with the aid of TDC markings. Check the start of delivery.

Setting the start of delivery

Adjust the start of delivery for the distributor injection pump with a dial gauge. Do not actuate the cold-start accelerator in the process. Rotate the crankshaft until the piston of cylinder no. 1 is at TDC. The TDC markings on the flywheel and on the clutch housing must be flush. Then remove the screw plug (**see Fig. 1, Page 295**) and install a dial gauge with the matching adapter (**Fig. 2**). Now rotate the crankshaft against the direction of engine rotation. If the needle on the dial gauge does not move, the pump plunger is at BDC. Now set the

dial gauge to "zero". Then crank the engine in the direction of rotation up to the reference mark. The dial gauge must now indicate a measurement in mm determined by the manufacturer for the pump-plunger lift. If this is not the case, release the pump flange and rotate the pump housing accordingly.

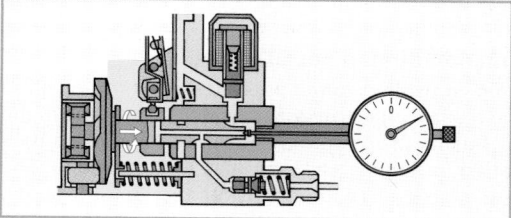

Fig. 2: Start-of-delivery setting

Idle and maximum-speed setting. These can be corrected at the corresponding adjusting screws with the engine at normal operating temperature. Do not actuate the cold-start accelerator in the process.

REVIEW QUESTIONS

1 Explain the start of injection and injection-quantity control of the distributor injection pump.

2 List the advantages of a two-spring nozzle holder over a one-spring nozzle holder.

3 Draw up a work schedule for setting the nozzle-opening pressure. Which accident-prevention regulations must be observed without fail in the process?

12.5.7.2 Electronic diesel control (EDC)

This is a map-controlled electronic fuel-injection system (**EDC** = **E**lectronic **D**iesel **C**ontrol). It is used to achieve exact control of the start of injection together with highly accurate fuel-quantity metering.

Advantages of a diesel injection system with EDC:
- Compliance with stricter emission limits
- Reduced fuel consumption
- Optimised torque and power
- Improved response
- Reduced engine noise
- Optimised quiet running
- Problem-free equipping of the vehicle with a vehicle-speed controller
- Simplified adaptation of an engine type to different vehicles

Design. EDC consists of:
- **Sensors.** These record the operating data, such as load, engine speed, engine temperature, boost pressure, and environmental conditions, e.g. intake-air temperature and air pressure.
- **Electronic control unit (ECU).** This is a microcomputer which, from the operating data and environmental information and while taking into account the setpoint values stored in program maps, determines the injected fuel quantity and the start of injection, and if necessary controls the exhaust-gas recirculation quantity and the boost pressure.
- **Actuators.** These allow interventions to be made in the fuel-injection system and if necessary in the exhaust-gas recirculation and supercharging systems.

Function of map-based control

Main controlled variables. A basic start of injection and a basic injection quantity are determined in the ECU using program maps from the two main controlled variables, i.e. load and engine speed. The load signal is recorded by an accelerator-pedal-travel sensor while the engine speed is recorded by the crankshaft speed sensor.

Correction controlled variables. These adapt the basic injection times optimally to the respective driving situation and ambient conditions. Further program maps are stored in the ECU for each correction controlled variable. Thus, for example, for:

- Engine temperature
- Boost pressure
- Fuel temperature
- Intake-air temperature

Setpoint/actual-value adjustment. Sensors inform the ECU whether the control intervention was sufficient. If necessary, the ECU effects a readjustment of the relevant actuators.

Particular features. In addition, the following further functions can be adapted (= applied) via EDC to the respective vehicle:

- **Idle-speed control.** For emission and consumption reasons the idle speed is kept as low as possible, regardless of torque requirements, e.g. by the A/C compressor and alternator.

- **Smooth-running control.** Not all the cylinders of an engine generate the same torque with the same injected fuel quantity. Wear and manufacturing tolerances of the components are possible causes. This results in irregular engine operation and increased emissions. Smooth-running control detects this from the crankshaft acceleration with the aid of the speed sensor and compensates it by specifically adapting the injected fuel quantity of the relevant cylinder.

- **Overrun fuel cut-off.** In overrunning situations, e.g. downhill driving, the injected fuel quantity is reduced to zero.

- **Active surge damping.** In the event of sudden load changes, the torque change of the engine induces bucking oscillations in the vehicle drivetrain. These oscillations/vibrations are detected from the engine-speed signal and damped by means of active control. The injected fuel quantity is reduced as speed increases and increased as speed decreases in order to counteract the speed oscillations/vibrations.

- **External torque intervention.** The injection time is influenced by other ECUs, e.g. transmission control, TCS/ASR, ESP. Independently of the driver command they inform the engine ECU via data bus whether and by how much the engine torque is to be altered.

- **Electronic immobiliser.** To prevent the vehicle from being driven without authorisation, it is only possible to start the engine when the engine ECU provides clearance.

- **Vehicle-speed controller.** This adjusts the vehicle speed to the desired value. The injected fuel quantity is increased or reduced until the measured actual speed corresponds to the set setpoint speed.

- **Diagnostic capability.** Input and output signals are checked. Detected faults are stored in the ECU and if necessary indicated to the driver.

- **Limp-home function.** The ECU activates an appropriate limp-home program to match the fault. The following different functions are available:
  ⇒ **30% power reduction,** e.g. on the lack of correction controlled variables
  ⇒ **Increased idle speed,** e.g. on the lack of main controlled variables
  ⇒ **Emergency OFF,** e.g. if there is a risk of engine damage

12.5.7.3 Electronically controlled axial-piston distributor injection pump (VE-EDC)

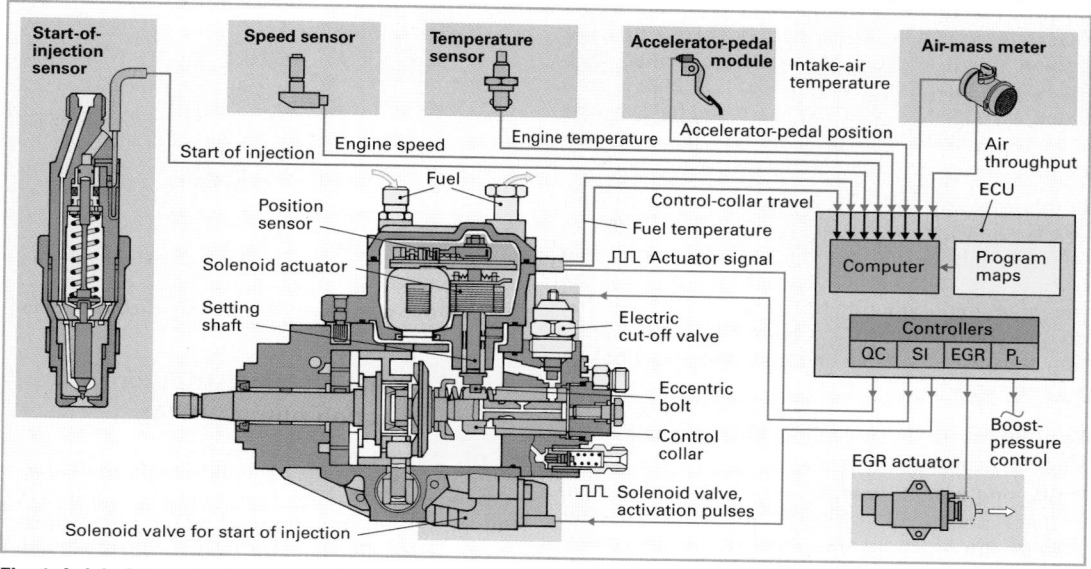

Fig. 1: Axial-piston distributor injection pump with EDC

This is a mechanical axial-piston distributor injection pump with a solenoid actuator for the control collar. The hydraulic timing device incorporates a solenoid valve for start-of-delivery correction (**Fig. 1**).

Control of start of injection (SI). The hydraulic timing device (**Fig. 2**) is influenced by a clocked solenoid valve (PWM signal).

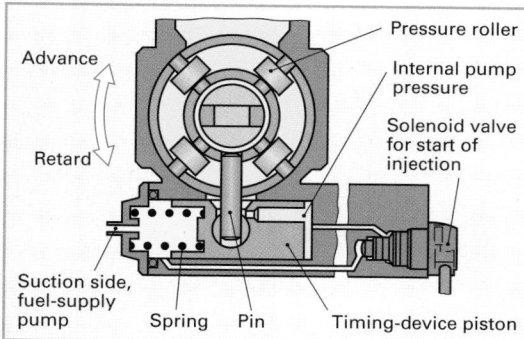

Fig. 2: Solenoid valve, timing-device piston

Ignition retard. When the valve is permanently open, the internal pump pressure applied at the timing-device piston is lowered and the roller ring rotates in the "Retard" direction.

Ignition advance. When the solenoid valve is closed, a higher internal pump pressure is applied and the roller ring rotates in the "Advance" direction.

Setpoint start of injection. This is dependent on the respective engine speed and different correction quantities.

Actual start of injection. This is recorded by the needle-lift sensor and transmitted to the ECU for setpoint/actual-value comparison. If the two values do not match up, the start of injection is corrected by continuous opening and closing of the solenoid valve.

Signal failure. If the timing-device solenoid valve is no longer actuated, the start of injection continues to be altered via the speed-dependent internal pump pressure. A fault is stored and the engine power is reduced by 30% (limp-home operation).

Needle-lift sensor (Fig. 1). This is seated in the start-of-injection sensor and is usually installed on one of the central injection nozzles. It records the opening and closing of the injection nozzle. Start of injection and end of injection can therefore be detected by the ECU.

Function. When the injection nozzle opens or closes, the moving nozzle needle creates a magnetic-field change in the coil and with it an induction voltage. When a threshold voltage is exceeded, this serves as a signal for the ECU for the start of injection (**Fig. 3**).

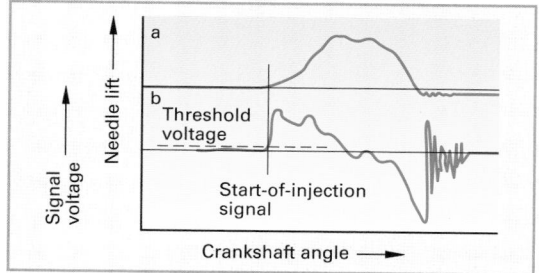

Fig. 3: Signal of needle-motion sensor

Control of injected fuel quantity (QC). The setpoint injected fuel quantity stored in the program map is determined from the accelerator-pedal position, the engine speed and the correction controlled variables. The solenoid actuator **(Fig. 1, S. 299)** is activated via a PWM signal. The magnetic field causes the shaft with the eccentric bolt to rotate. This bolt adjusts the control collar on the distributor plunger.

In this way, the spill port is opened earlier or later and the injected fuel quantity controlled. This can take place up to mid-range speeds so quickly that from one injection to another into the cylinder the fuel quantity is changed from stroke to stroke. The control-collar position is transmitted via the position sensor to the ECU, and adjusted and if necessary corrected.

Position sensor. The following different types of position sensor may be used:

- **Slip-ring potentiometer.** This consists of a wiper and a conductor track. The resistance changes when the wiper changes its position. It is subject to mechanical wear.
- **Sensor with semidifferential short-circuiting ring (Fig. 1).** This consists of a soft-iron core, a measuring coil, a reference coil and a rotating short-circuiting ring. When this ring rotates, it alters the magnetic flux and with it the measuring-coil voltage U_A. It is extremely accurate and is wear-resistant because it operates without contacts.

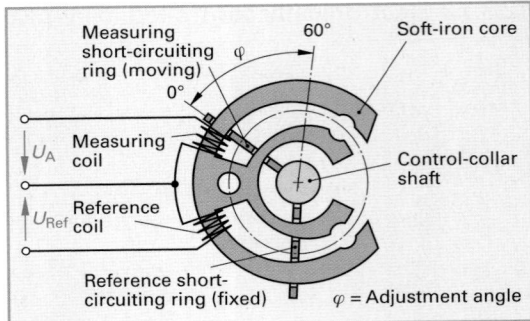

Fig. 1: Sensor with semidifferential short-circuiting ring

12.5.7.4 Radial-piston distributor injection pump (VP44)

This is an electronically controllable fuel-injection pump (EDC) with a pump ECU integrated in the pump housing. It generates injection pressures up to 1,900 bar and can be installed in any position **(Fig. 2)**.

Operating principle. The drive shaft drives at camshaft speed the distributor shaft with the high-pressure cylinders. The roller tappets of the high-pressure pistons roll during this rotary movement on the cam track of the cam ring and actuate the pistons in the process.

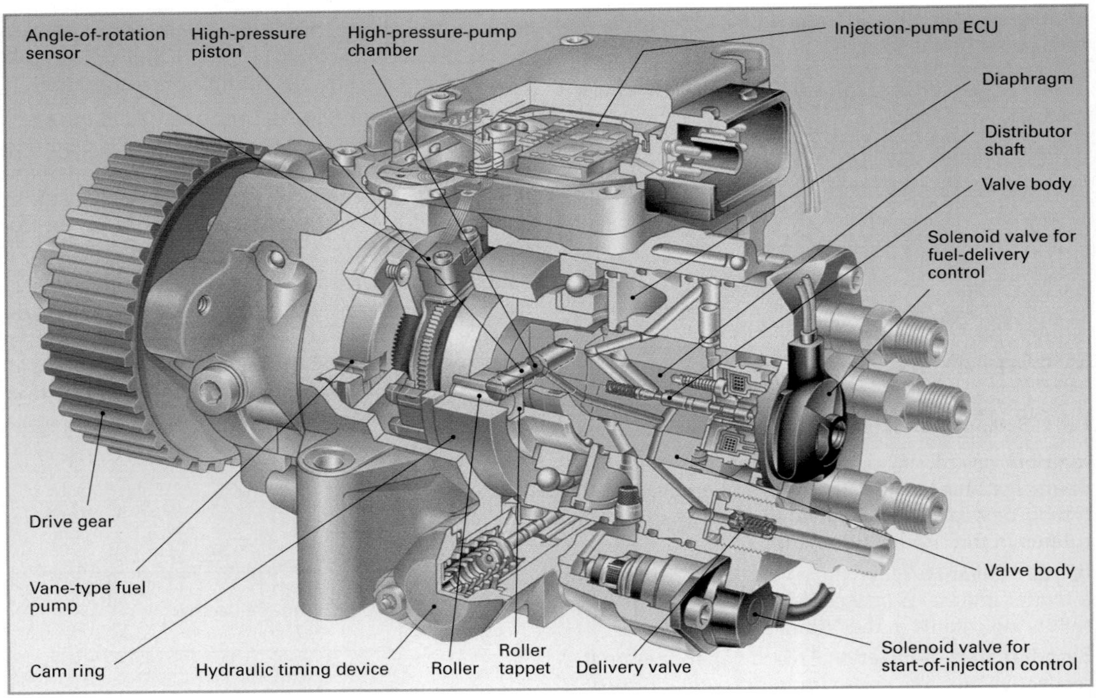

Fig. 2: Radial-piston distributor injection pump (VP44)

The high-pressure solenoid valve controls the actual start of injection and the injected fuel quantity.

The solenoid valve of the hydraulic timing device ensures that high pressure is available at the right time. This is done by rotating the cam ring. Then the solenoid valve can control the exact start of injection and the injected fuel quantity.

Functions. The VP44 assumes the following functions:

- Fuel delivery
- High-pressure generation and distribution
- Start-of-injection control
- Injected-fuel-quantity control

Fuel delivery. The fuel is delivered from the tank into the pump via a vane pump which is driven by the drive shaft **(Fig. 1)**. Unlike in the axial-piston distributor pump, the entire internal pump chamber is not filled; only the accumulator chamber behind a diaphragm is filled. This allows for higher pressures for filling the pump. The following internal pump pressures can be set, depending on pump speed:

- Idle approx. 3 to 4 bar
- Part load 4 to 15 bar
- Full-load speed approx. 15 to 20 bar

The pressure is limited via the pressure-control valve and the overflow throttle valve.

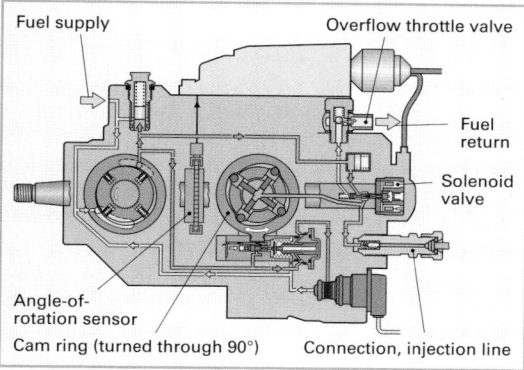

Fig. 1: Fuel flow in a VP44

Timing device (Fig. 2). When the cam ring rotates, the roller tappets move earlier or later on the cams and the high-pressure pistons move inwards. When the high-pressure solenoid valve is closed, high pressure is generated and injected. When it remains open, the fuel is forced back into the supply and return flows.

The position of the cam ring determines the time window in which fuel can be injected.

Function. The higher injection pressures of the VP44 require greater displacement forces at the cam ring

than in the axial-piston pump. Rapid and precise adjustments can be made thanks to the hydraulic transmission ratio in the timing-device piston.

Injection advance. In the rest position the timing-device piston is held by the resetting spring in the "Retard" position. When the solenoid valve is closed by a PWM signal, the internal pump pressure acts on the control plunger. This moves to the right with the control collar. The control collar opens an inlet passage to allow the fuel to flow under pressure into the working chamber of the timing-device piston and displace it in the "Advance" direction.

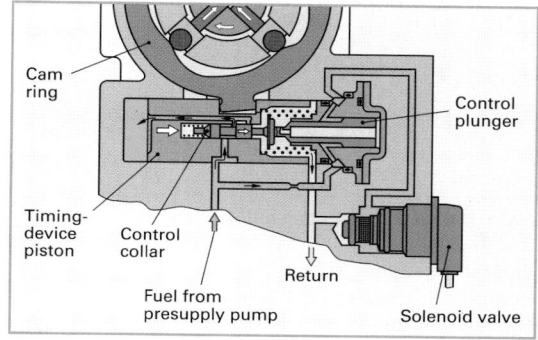

Fig. 2: Injection advance

Injection retard. The solenoid valve is clocked and thereby opened. The fuel pressure acting on the control plunger is reduced. The spring force moves the control plunger with the control collar to the left. The return passage is opened and the fuel pressure acting on the timing-device piston is reduced. The resetting spring displaces the timing-device piston in the "Retard" direction.

Delivery phases

Filling phase (Fig. 3). The roller tappets move off the cams. The pistons are forced outwards by the fuel pressure and the centrifugal forces in the process. The high-pressure solenoid valve is opened and therefore not supplied with power.

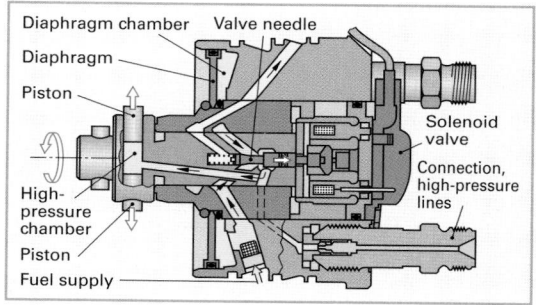

Fig. 3: Filling phase of high-pressure chamber

Delivery phase/start of injection (Fig. 1). The high-pressure solenoid valve shuts off the supply via a current pulse from the pump ECU (cut-in current = 20 A; holding current = 13 A). The high-pressure chamber is now sealed. Fuel is injected as soon as the fuel pressure has exceeded the nozzle-opening pressure at the beginning of the cam rise.

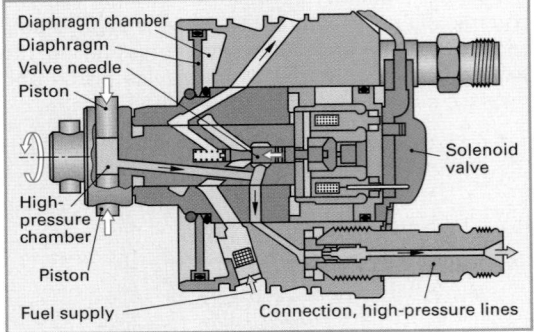

Fig. 1: Delivery phase/start of injection

End of injection. When the desired injected fuel quantity is reached, the pump ECU interrupts the current supply to the solenoid valve. The valve opens and the high pressure removed to the return flow and the diaphragm chamber. The high pressure peaks acting on the vane pump are damped by the diaphragm and the renewed filling process is accelerated by it.

Fig. 2 shows a diagram of the possible quantities and the actual quantity and start-of-injection control. If a sensor fails, a control process is derived from the closed-loop control circuit and the vehicle goes into a limp-home program.

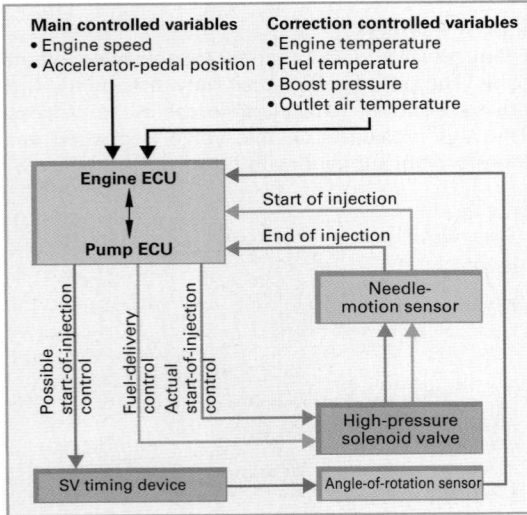

Fig. 2: Open- and closed-loop control circuits of a VP 44

12.5.7.5 Unit-injector system

This is an electronically controlled fuel-injection system in which each engine cylinder has a unit-injector element in its cylinder head **(Fig. 3)**. These injector elements allow top injection pressures up to 2,200 bar.

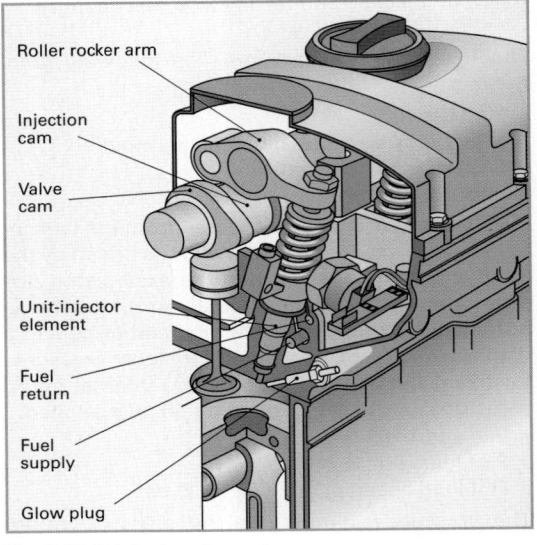

Fig. 3: Unit-injector element

Drive. The engine camshaft has an injection cam for each unit-injector element. The cam lift is transmitted via a roller rocker arm to the pump plunger. The injection cam has a steep leading face which forces the pump plunger downwards at high speed. This enables a high injection pressure to be achieved very quickly. The trailing face of the injection cam is flat. The pump plunger is slowly and uniformly forced upwards by the spring force.

Fuel supply. A fuel pump driven by the engine camshaft delivers the fuel to the unit-injector elements **(Fig. 4)**.

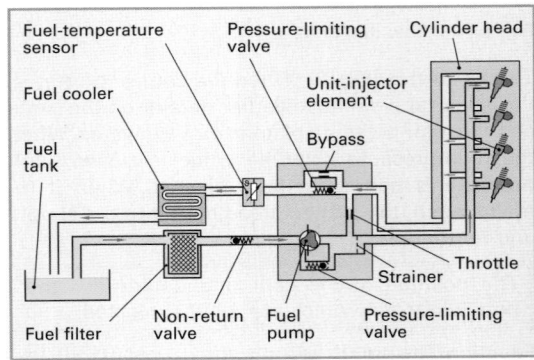

Fig. 4: Fuel supply

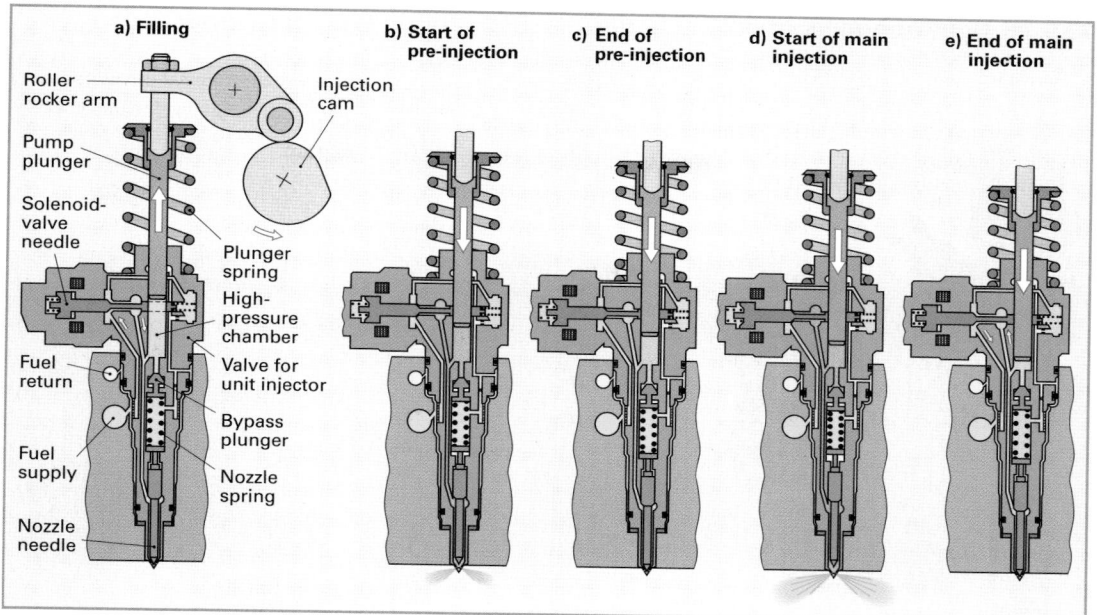

Fig. 1: Injection phases of a unit-injector element

The fuel not needed for injection cools the unit-injector elements. It flows through the return line from the cylinder head, past a temperature sensor, through a fuel cooler and back to the fuel tank.

Filling process. a) The pump plunger moves upwards in response to the force of the plunger spring and thereby increases the volume of the high-pressure chamber. The solenoid valve is not actuated. The solenoid-valve needle is in the rest position and opens the passage from the fuel supply to the high-pressure chamber. The fuel pressure in the supply forces the fuel into the high-pressure chamber.

Pre-injection starts. b) The pump plunger is forced downwards by the roller rocker arm and thereby forces the fuel out of the high-pressure chamber into the fuel supply. The injection process is initiated by the ECU. The ECU actuates the solenoid valve for this purpose. The solenoid-valve needle is pressed onto its seat and closes the passage from the high-pressure chamber to the fuel supply. This begins the build-up of pressure in the high-pressure chamber. At 180 bar the pressure is greater than the force of the nozzle spring. The nozzle needle is raised and pre-injection begins.

Pre-injection ends. c) Pre-injection ends immediately after the nozzle needle has opened. The bypass plunger moves downwards in response to the rising pressure and thereby increases the volume of the high-pressure chamber. The pressure drops for a brief moment and the nozzle needle closes. Pre-injection is completed. The nozzle spring is subject to greater load by the downward movement of the bypass plunger. A fuel pressure greater than that for pre-injection is therefore required to open the nozzle needle again for the subsequent main injection.

Main injection starts. d) The pressure in the high-pressure chamber rises shortly after the nozzle needle closes. The solenoid valve is furthermore closed and the pump plunger moves downwards. At approximately 300 bar the fuel pressure is greater than the force of the preloaded nozzle spring. The nozzle needle is raised again. The pressure rises to up to 2,200 bar because more fuel is displaced in the high-pressure chamber than can escape through the nozzle holes.

Main injection ends. e) End of injection is achieved when the engine ECU no longer actuates the solenoid valve.

The solenoid-valve needle opens the valve by means of the solenoid-valve spring. The fuel pressure can escape into the fuel supply and return. The pressure reduces. The nozzle needle closes and the bypass plunger is forced into its initial position by the nozzle spring. Main injection is now completed.

Electronic injected fuel quantity and start of injection control

The start of injection and the injected fuel quantity (injection time) are controlled by the ECU by means of solenoid-valve actuation.

12.5.7.6 Common-rail system

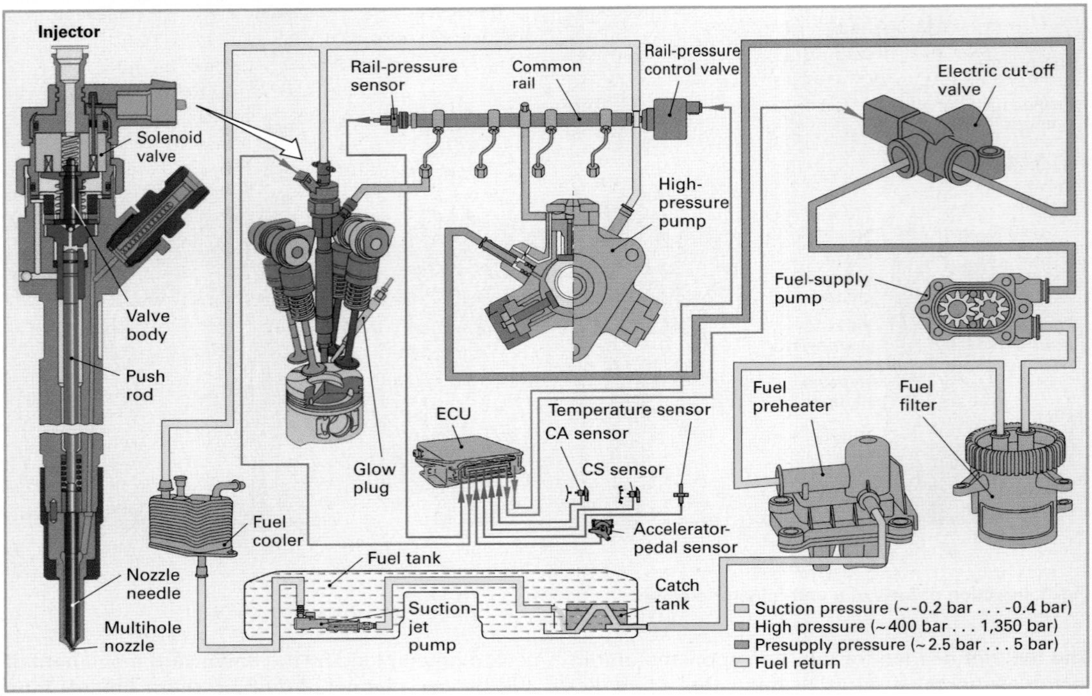

Injector

Rail-pressure sensor Common rail Rail-pressure control valve Electric cut-off valve

Solenoid valve

High-pressure pump

Valve body

Fuel-supply pump

Push rod

Fuel preheater Fuel filter

ECU

Temperature sensor
CA sensor

Glow plug

CS sensor

Fuel cooler

Accelerator-pedal sensor

Fuel tank

Nozzle needle

Catch tank

Multihole nozzle

Suction-jet pump

☐ Suction pressure (~-0.2 bar . . . -0.4 bar)
◼ High pressure (~400 bar . . . 1,350 bar)
◼ Presupply pressure (~2.5 bar . . . 5 bar)
☐ Fuel return

Fig. 1: Common-rail injection system

12

In a common-rail system the fuel is accumulated in a fuel rail under high pressure and injected into the combustion chambers under map-controlled conditions via solenoid valves on the injectors.

The injection pressure is generated by the high-pressure pump as a function of engine speed and accumulated in the rail. This is why it is also known as an accumulator injection system.
V-engines have a distributor rail in addition to the two rails for the two cylinder banks **(Fig. 2)**.

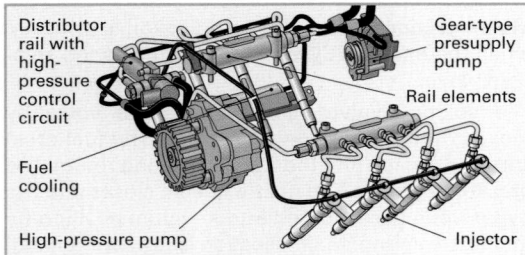

Distributor rail with high-pressure control circuit

Gear-type presupply pump

Rail elements

Fuel cooling

High-pressure pump

Injector

Fig. 2: Common-rail injection system (V8 engine)

The ECU (EDC) calculates, depending on the operating and ambient situation, optimal injection variables such as:
- Start of injection
- Injected fuel quantity
- Rate-of-discharge curve, e.g. pre-/post-injection

Design. The fuel-injection equipment **(Fig. 1)** comprises:

- **Low-pressure circuit.** This is divided into the suction-pressure area, the presupply-pressure area and the fuel return. It contains the fuel tank, fuel preheater, fuel filter, fuel-supply pump, electric cut-off valve and fuel cooler.

- **High-pressure area.** This consists of the high-pressure pump, the high-pressure lines, the rail and one injector for each cylinder.

- **Electronic control system.** This comprises the ECU, the sensor equipment, the rail-pressure control valve, the injector solenoid valves and the cut-off valve.

Components

Fuel-supply pump. This supplies the high-pressure piston pump with fuel. This pump is often a gear pump driven by the engine camshaft or an electrically driven roller-cell pump. It delivers between 40 l/h and 120 l/h of fuel as a function of engine speed. This is more than is needed for injection. The excess fuel flows through the fuel cooler in the return. Cooling is required because the fuel heats up intensely in the high-pressure area.

High-pressure pump (Fig. 1). This delivers the fuel to the rail at high pressure up to roughly 1,850 bar. It is usually a radial-piston pump, the drive shaft of which moves the three pump pistons of the pump elements up and down with the eccentric cams. The drive shaft can be driven indirectly via a toothed belt or gearwheel or directly by the engine camshaft.

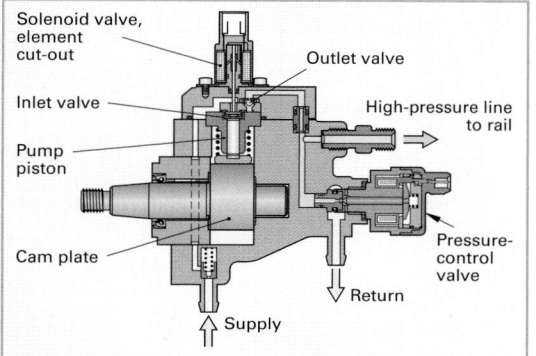

Fig. 1: 1st-generation high-pressure pump with element cut-out

Suction stroke. When the delivery pressure exceeds the inlet-valve opening pressure, fuel flows into the element chamber and the pump piston moves downwards.

Delivery stroke. When the bottom dead centre of the pump piston is exceeded, the inlet valve closes and the fuel can no longer escape. The outward-moving pump piston pressurises the fuel until the rail pressure is exceeded. Then the outlet valve opens and the fuel flows into the rail.

Element cut-out. A drive power of approx. 3.5 kW … 5 kW is required to drive the high-pressure pump at nominal speed. At idle and in the part-load range the pump delivers much more fuel at high pressure than is needed for injection. In order to save drive power and not to heat the fuel unnecessarily, the system limits the delivery rate by cutting out a pump element. The solenoid valve opens the inlet valve for this purpose. Thus the drawn-in fuel cannot be compressed.

Fuel rail. This is a thick-walled steel pipe with connections for the fuel lines, the pressure sensor and if necessary the pressure-control valve.

> The function of the rail is to accumulate the fuel under high pressure and to compensate pressure variations.

The rail pressure remains virtually constant due to the fact that the rail accumulator volume is signifi-

cantly greater than the injected fuel quantity or the pump delivery.

Rail-pressure sensor (Fig. 2). This is mounted on the rails and communicates the relevant rail pressure to the ECU. The pressure is recorded by the sensor element and transmitted via the integrated evaluation electronics in the form of a voltage signal to the ECU. The evaluation electronics is usually supplied with 5 V. As pressure rises the resistance of the sensor element drops. Thus the signal voltage is approx. 4.5 V in the case of high rail pressure and approx. 0.5 V in the case of low rail pressure. In the event of signal failure, the pressure-control valve is actuated with a fixed value and a limp-home function (increase idle speed) is activated.

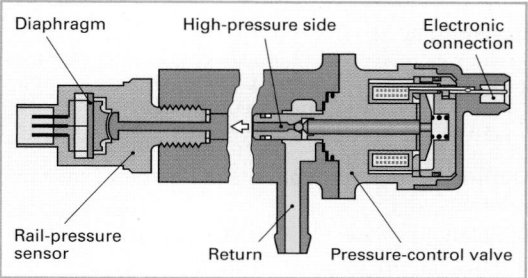

Fig. 2: Pressure sensor and pressure-control valve on the rail

Pressure-control valve (Fig. 2). This can be mounted on the rail or on the high-pressure pump.

> This adjusts the rail pressure as a function of the engine load state and operating conditions and maintains this pressure at a constant level.

Function. The solenoid valve is forced onto its seat by the valve spring and thereby seals the high-pressure side (rail) against the low-pressure side (return). This form of control is known as **return-quantity control**.

The spring is designed in such a way that a pressure of approx. 100 bar is set when the pressure-control valve is not actuated. In order to increase the pressure in the rail, it is necessary to build up a magnetic force in addition to the spring force. The rail pressure is set by way of a PWM signal to 400 bar at idle and to approx. 1,850 bar at full load.

2nd-generation high-pressure pump (Fig. 1, Page 306) Here the rail pressure is controlled when the engine is hot through metering of the fuel quantity drawn in by the high-pressure pump **(feed-quantity control)**. The metering valve allows only a small amount of fuel to flow to the high-pressure pump at idle, and more fuel accordingly at full load. This is performed by means of a proportioning valve in the pump feed line. Controlling the fuel quantity lowers the fuel temperature and the power input of the high-pressure pump. Furthermore, when the engine is cold, the return quantity is controlled with a pressure-control valve.

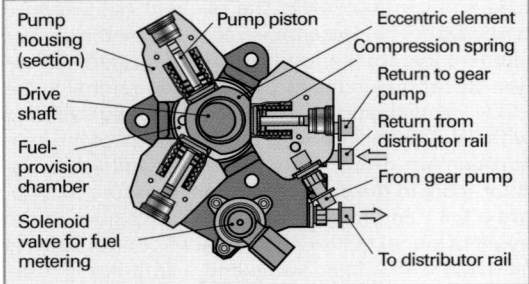

Fig. 1: 2nd-generation high-pressure pump

Injector (Fig. 2).

> The injector facilitates precise control of the start of injection and injected fuel quantity as well as pre- and post-injection.

Design. The injector consists of the following components:

- Solenoid valve
- Valve control chamber
- Valve control plunger
- Injection nozzle

Function

Injector closed (rest state). The fuel flows through the fuel-inlet passage into the valve control chamber and to the injection-nozzle pressure shoulder. The discharge throttle is closed by the valve ball. The pressure above the valve control plunger in the valve control chamber generates a force which is greater than the opposing force which acts on the nozzle-needle pressure shoulder. The nozzle needle thus remains closed, regardless of the rail pressure applied at the injector.

Injector open (start of injection). When the solenoid valve is actuated by the ECU, the armature is attracted, as a result of which the valve ball opens the discharge throttle. Now more fuel flows through the discharge throttle than through the inlet throttle. This causes a drop in pressure in the valve control chamber. Now the nozzle needle is raised by the force acting on its pressure shoulder. The nozzle needle opens and fuel is injected.

Injector closes (end of injection). When the solenoid valve is no longer supplied with power by the ECU, the valve spring forces the valve ball onto its seat. As a result, the return throttle is closed and the rail pressure builds up again abruptly in the valve control chamber. The force acting on the valve control plunger and the force of the nozzle spring acting from above exceed the force on the nozzle-needle pressure shoulder acting from below. The nozzle needle closes.

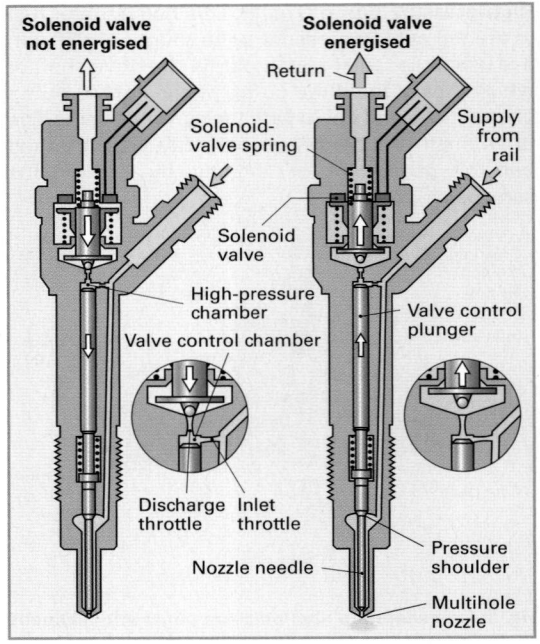

Fig. 2: Injector operating principle

Pre- and post-injection (Fig. 3). The solenoid valve is only supplied with power for a brief period in order to inject small quantities of fuel. The nozzle needle is raised slightly. The fuel injector opens briefly and in the process does not open the entire opening cross-section.

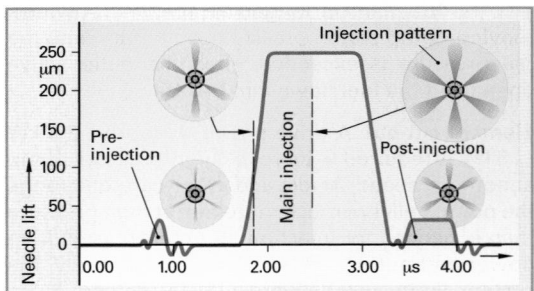

Fig. 3: Rate-of-discharge curve

Injector activation. Rapidly switching solenoid valves are required in order to be able to subdivide fuel injection into up to three sequences (pre-, main and post-injection). This can only be achieved with high currents of approx. 20 A and voltages of approx. 100 V. This so-called "booster voltage" is generated with the aid of induction when the solenoid valves are supplied with power and is stored in a capacitor in the ECU.

Piezo actuators. In the future solenoid valves will be replaced by faster piezo actuators. This will allow more injection operations per power stroke with simultaneously low energy consumption.

Sensors and actuators of EDC with common-rail system (Figs. 1 and 2)

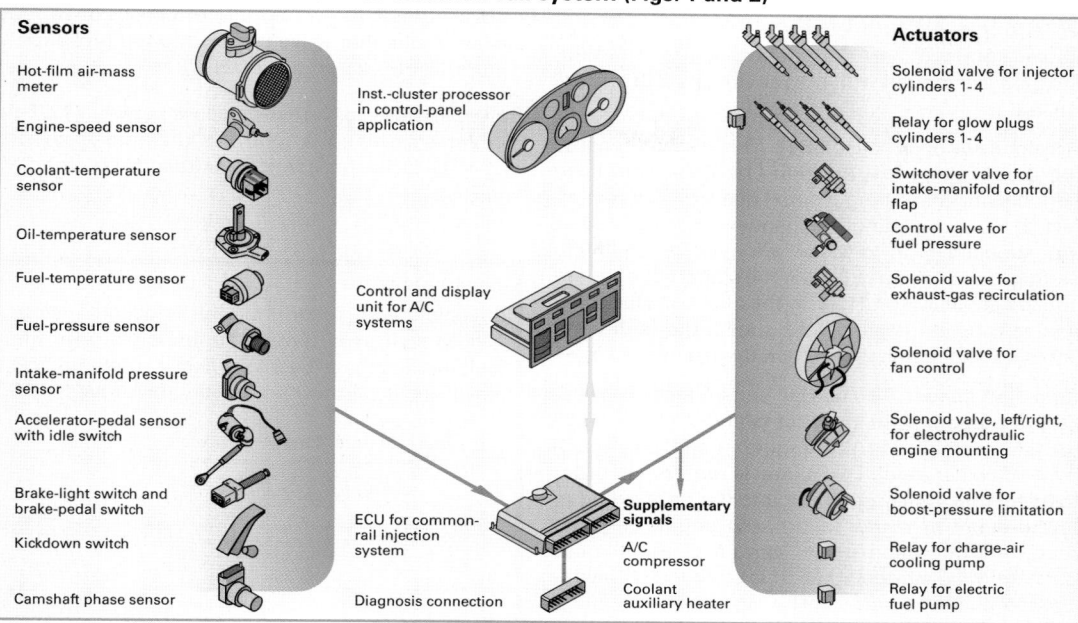

Sensors

Hot-film air-mass meter

Engine-speed sensor

Coolant-temperature sensor

Oil-temperature sensor

Fuel-temperature sensor

Fuel-pressure sensor

Intake-manifold pressure sensor

Accelerator-pedal sensor with idle switch

Brake-light switch and brake-pedal switch

Kickdown switch

Camshaft phase sensor

Inst.-cluster processor in control-panel application

Control and display unit for A/C systems

ECU for common-rail injection system

Diagnosis connection

Supplementary signals

A/C compressor

Coolant auxiliary heater

Actuators

Solenoid valve for injector cylinders 1-4

Relay for glow plugs cylinders 1-4

Switchover valve for intake-manifold control flap

Control valve for fuel pressure

Solenoid valve for exhaust-gas recirculation

Solenoid valve for fan control

Solenoid valve, left/right, for electrohydraulic engine mounting

Solenoid valve for boost-pressure limitation

Relay for charge-air cooling pump

Relay for electric fuel pump

Fig. 1: Block diagram of EDC with common-rail system

Fig. 2: Schematic diagram, common rail

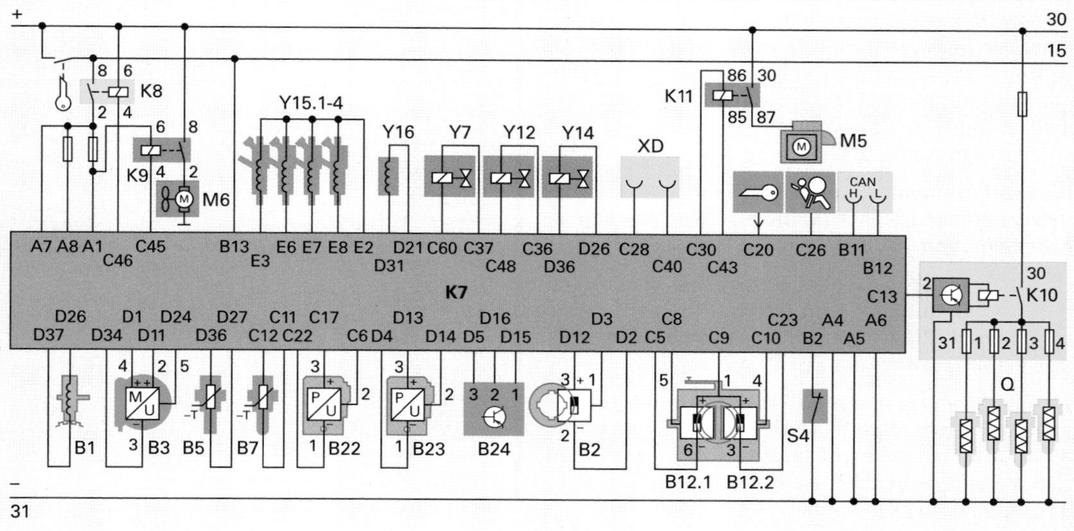

B1 Crankshaft speed sensor	**K7** ECU for common-rail injection	**S4** Clutch-pedal switch
B2 Camshaft sensor	**K8** Main relay	**Y7** Solenoid valve for exhaust-gas recirculation
B3 Air-mass meter	**K9** Relay for fan	
B5 Engine-temperature sensor	**K10** Glow control unit	**Y12** Solenoid valve for boost-pressure control
B7 Sensor for intake-manifold temperature	**K11** Relay for starter	**Y14** Electric cut-off valve
B12 Sensor for accelerator-pedal position	**M5** Starter	**Y15.1 ... Y15.4** Injectors
B22 Boost-pressure sensor	**M6** Fan motor	**Y16** Rail-pressure control valve
B23 Rail-pressure sensor	**Q** Glow plugs	**XD** Diagnosis connection
B24 Sensor for oil pressure, quantity and temperature		

12

Example of a circuit diagram for a common-rail system (Fig. 2, Page 307)

Power supply. When the ignition key is operated, the ECU is supplied with power via pins B13 and A7, A8, A1.

Accelerator-pedal sensor B12. This sensor records the accelerator-pedal position (*1st main controlled variable*) for calculating the fuel quantity. The Hall-effect sensors operating independently of each other are supplied with positive via pin C9 and with earth/ground via pins C5, C23. The signal voltages are applied at pins C8 and C10. In the event of a fault, the engine runs at increased idle speed. The fault lamp lights up and a fault is stored in the fault memory.

Camshaft sensor B2. The Hall sensor signals to the ECU the position of the first cylinder in the compression stroke and the engine speed (*2nd main controlled variable*). This signal is required in order to inject fuel at the right time into the right cylinder. In the event of signal failure, the engine remains ready for operation under certain circumstances but cannot be started again. It is supplied with voltage via pins D12 and D2. The signal voltage is applied at pin D3.

Rail-pressure sensor B23. This signals to the ECU the current fuel pressure in the rail. If an excessively low or high rail pressure is detected, the ECU identifies a fault in the system and shuts down the engine (emergency OFF). The signal voltage is applied at pin D13. Power supply of positive to D14 and earth/ground to D4.

Rail-pressure control valve Y16. This provides the map-controlled pressure in the rail. The PWM signal is applied by the ECU via pins D31 and D21.

Injectors Y15.1–Y15.4. These ensure that the correct injected fuel quantity is injected at the correct injection point. They are actuated by the ECU in each case via pin E2 and pins E3, E6, E7, and E8.

Hot-film air-mass meter B3. This determines the inducted air mass. With this value the EGR rate is determined as a matter of priority from a program map. Voltage is applied via pins D34 and D1, D11. The signal voltage is applied at pin D24.

Electric cut-off valve Y14. This is switched by the ECU via pins D26 and D36. In the event of serious system faults, the valve closes and the fuel supply is interrupted (emergency OFF).

Solenoid valve for boost-pressure control Y12. This facilitates continuous control of the boost pressure of a VTG turbocharger. It is actuated via pins C36 and C48.

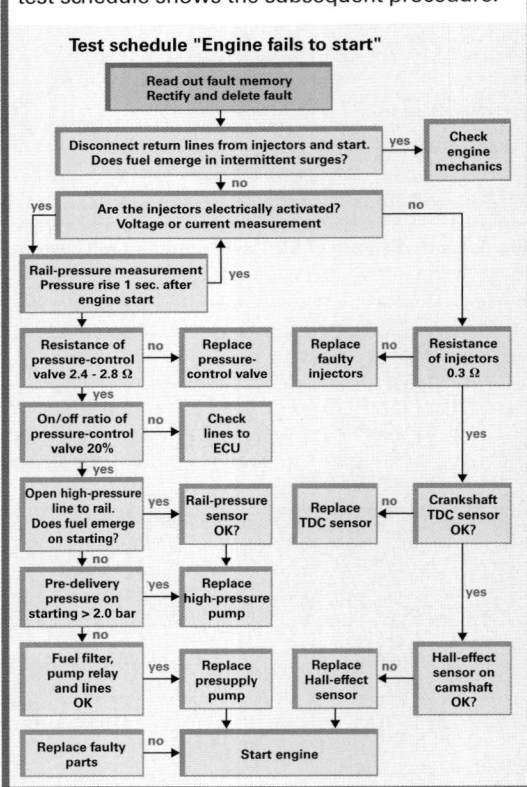

WORKSHOP NOTES

A vehicle with a common-rail injection system fails to start even though all the faults from the fault memory have been rectified and deleted. The test schedule shows the subsequent procedure:

Test schedule "Engine fails to start"

REVIEW QUESTIONS

1. Describe the electronic injection timing of the axial-piston distributor pump.

2. What is the function of the angle-of-rotation sensor of a radial-piston pump?

3. Explain the operating principle and function of an EDC needle-motion sensor.

4. What accident-prevention measures must be followed when working with an injection-nozzle test rig?

5. How is the UIS driven?

6. Name three features of the unit-injector system.

7. How is pre-injection effected in the UIS and on which quantity/variable does the duration of pre-injection depend?

8. Name and explain the function of 5 components of a common-rail system.

9. How is the injection pressure controlled in a common-rail system?

10. Compare the common-rail system with the UIS. What are the respective advantages? Give reasons.

13 Pollutant reduction

13.1 Exhaust system

Functions

- To muffle and relieve the exhaust gases emerging with strong pulses (bang) from the combustion chamber in such a way that a specific noise level is not exceeded.
- To divert exhaust gases safely in such a way as to prevent their entry into the vehicle interior.
- To reduce pollutants contained in the exhaust gas by means of a catalyst to the prescribed limit values.
- To influence the exhaust-gas flow during the muffling process in such a way that the engine power loss is as low as possible.
- To generate a sound which is appropriate for the vehicle in question (sound design).

Sound levels (Table 1). When the exhaust valve opens, the exhaust gases in the cylinder are still subject to an overpressure of 3 bar to 5 bar. Without a silencer, these exhaust gases would be discharged to atmosphere with a loud bang. A measure of the noise intensity is the sound level, which is measured in decibels (A) = dB (A), where (A) describes the method of measurement. A sound level of 0 dB (A) corresponds to the human threshold of audibility. An increase of 3 dB (A) equates to a doubling of the subjective loudness or volume. Noises in excess of 120 dB (A) are felt as pain. Continuous noise in excess of 130 dB (A) can have fatal consequences.

Table 1: Sound levels

Pneumatic hammer	130 dB (A)
Pain threshold	120 dB (A)
Discotheque	110 dB (A)
Engine without silencer	100 dB (A)
Machine room	90 dB (A)
Busy road/street	80 dB (A)
Perm. driving noise of a passenger car	74 dB (A)
Conversation	70 dB (A)
Living room	50 dB (A)
Bedroom	30 dB (A)
Very slight rustling of leaves	10 dB (A)
Threshold of audibility (0 decibels)	0 dB (A)

Driving/riding noise. The exhaust noise represents a significant part of the driving/riding noise of a motor vehicle. Other partial noises are, for example, engine-gearbox noises, rolling noises, and body and wind noises. The driving/riding noise must not exceed the permissible measurement. Limit values

Table 2: Driving/riding noises of motor vehicles (limit values)

Motor bicycle		70 dB (A)
Moped, moped w/kickstarter		72 dB (A)
Light motorcycle		75 dB (A)
Motorcycle	up to 80 cc	75 dB (A)
	up to 175 cc	77 dB (A)
	over 175 cc	80 dB (A)
Passenger car with spark-ignition or diesel engine		74 dB (A)
with direct-injection diesel engine		75 dB (A)
Bus/coach, HGV up to 3.5 t		76 dB (A)
with direct-injection diesel engine		77 dB (A)
HGV up to 75 kW		77 dB (A)
Bus/coach, HGV up to 150 kW		78 dB (A)
over 150 kW		80 dB (A)

for the generation of noise by the individual motor-vehicle categories are defined in German national Road Traffic Licensing Regulations and in EU Directives **(Table 2)**. These limits values have been repeatedly reduced in recent years.

Stresses

- High temperatures and marked changes of temperature above all in the front section of the exhaust system.
- External corrosion over the entire length of the exhaust system caused by exposure to the weather and road salt in winter.
- Internal corrosion caused by condensed combustion gases (water, sulphurous acid) particularly on the rear, colder components.
- Strong mechanical stresses to the exhaust system caused by stone impacts, body movements and engine vibrations.

The individual components are manufactured from a variety of materials in order to withstand these stresses. Because of the high operating temperatures involved, the front components of the exhaust system are primarily made from creep-resistant and nonscaling stainless steel which is also resistant to high-temperature corrosion. Silencers usually have a two-walled sandwich design. The inner metal jacket of a silencer is made of stainless steel on account of the aggressive condensates of the combustion gases while the outer metal jacket is made of unalloyed steel which is however coated with aluminium to provide protection against external corrosion. The exhaust pipes in the rear section of the exhaust system are also coated with aluminium.

13

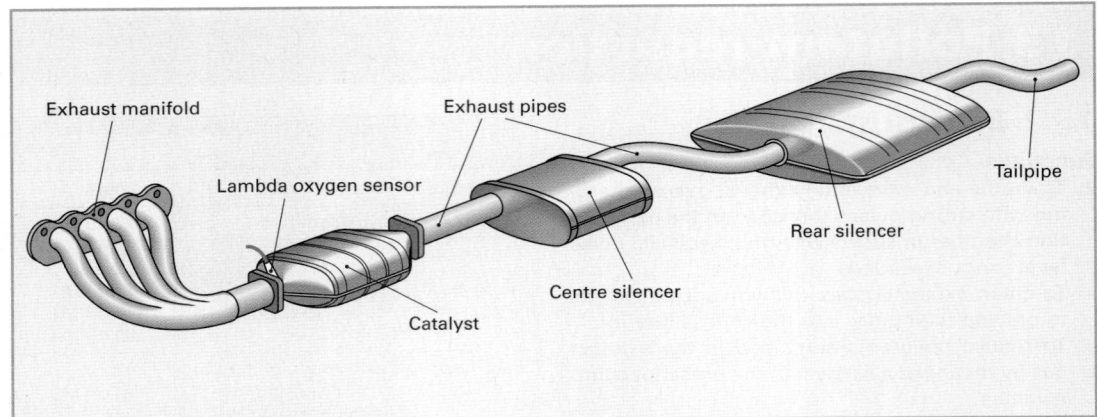

Fig. 1: Design of an exhaust system

Design of exhaust system

The exhaust system **(Fig. 1)** consists of the exhaust pipes, the catalytic converter (or catalyst) and one or more silencers, e.g. centre silencer and rear silencer. The front exhaust pipe is flanged to the exhaust manifold and leads into the catalyst. The catalyst is connected by way of connecting pipes to the silencers. From there the exhaust gas is discharged to atmosphere through the tailpipe.

The exhaust system must be gastight over its entire length so that combustion gases cannot get into the interior of the vehicle and so that the muffling process is not impaired.

The design and configuration of the silencers together with the length and cross-section of the connecting pipes are carefully matched to each other by the manufacturer. In this way the noise level of the combustion gases is reduced to the required level. On the other hand, the flow resistance of the exhaust system is kept low because engine power is impaired by the exhaust backpressure generated.

The exhaust noise is generated by the pulsating emission of gases from the cylinders. The sound energy of these gases can be damped by reflection and by absorption.

Reflection. Silencing by reflection involves placing obstacles (or baffles) in the way of the sound waves. In this way the sound waves are reflected and diverted. They are mutually extinguished partially like a dying echo. Sudden changes of cross-section in tubes and chambers also generate reflection.

Reflection silencer (Fig. 2). Chambers of different sizes are connected to each other by tubes which are open at both ends and offset in relation to each other. This forces the gas flow to be diverted in the silencer.

The tubes can also be perforated. The frequent changes in cross-section cause the sound waves to be reflected and thereby damped. Reflection silencers in the exhaust system are particularly suitable for damping medium and low frequencies.

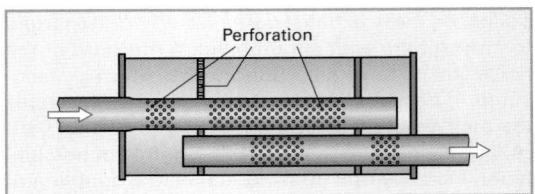

Fig. 2: Reflection silencer

Resonance effect. The sound waves travel back and forth repeatedly between the changes in cross-section and generate resonance under certain circumstances. This is referred to as a series resonator or a branch resonator, depending on whether the sympathetic or resonant vibrations are generated in the main flow or in a branch **(Fig. 1, Page 311).** Marked damping of specific frequencies can be achieved with resonators of this type.

Interference effect (Fig. 3). If the exhaust flow in the silencer is divided and then merges the sound waves again after they have travelled different lengths, the sound waves are partially extinguished mutually on merging.

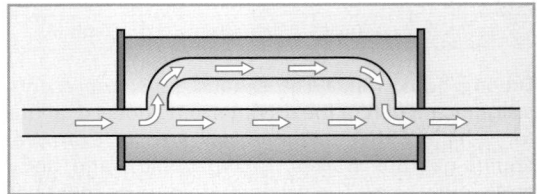

Fig. 3: Interference effect

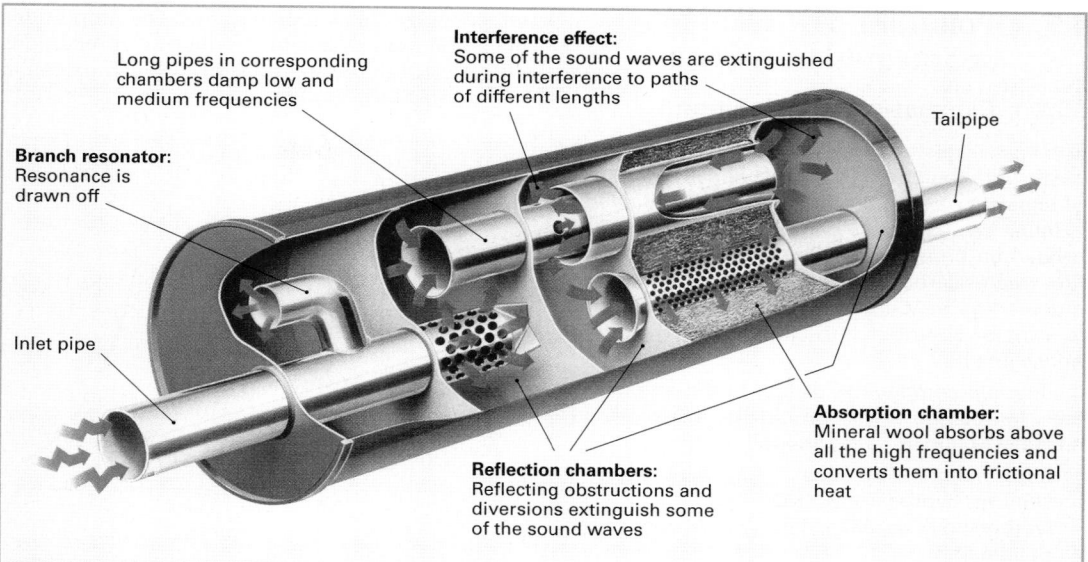

Interference effect:
Some of the sound waves are extinguished during interference to paths of different lengths

Long pipes in corresponding chambers damp low and medium frequencies

Branch resonator:
Resonance is drawn off

Tailpipe

Inlet pipe

Reflection chambers:
Reflecting obstructions and diversions extinguish some of the sound waves

Absorption chamber:
Mineral wool absorbs above all the high frequencies and converts them into frictional heat

Fig. 1: Composite-type reflection-absorption silencer

Absorption. Silencing by absorption involves directing the sound waves into a porous material. The sound energy is in practice "swallowed" because it is converted by friction into heat.

Absorption silencers (Fig. 2) consist of one or more chambers which are filled with rock wool or glass wool. The exhaust flow is directed through a perforated tube and passes through the silencer virtually unhindered. The sound waves however penetrate through the perforations into the mineral wool, which absorbs above all higher frequencies. Absorption silencers are mainly used as rear silencers.

Composite-type reflection-absorption silencer (Fig. 1). Reflection mufflers can be well matched to low frequencies. Absorption silencers are only effective in the upper frequency range. For this reason, both silencer types are used simultaneously, sometimes in separate silencers with both types of silencing.

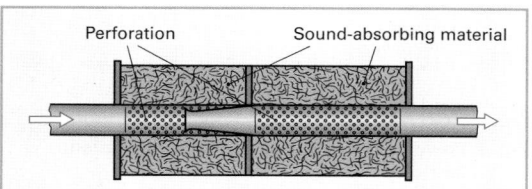

Perforation   Sound-absorbing material

Fig. 2: Absorption silencer

All exhaust-system components are matched to each other and must not be changed or modified. Installing non-approved parts will invalidate the type approval.

Working rules
- Check the exhaust system for leaks during each service-inspection. Traces of soot are signs of leaking or damaged areas.
- Check heat shields.
- Replace parts which have rusted through.
- Check all exhaust-system mountings.

REVIEW QUESTIONS

1 What are the functions of the exhaust system?

2 In which unit of measurement is the sound level given?

3 Why must the silencer system of a motor vehicle not be altered?

4 What is the maximum permissible driving noise of a passenger car?

5 What is the maximum permissible riding noise of a light motorcycle?

6 To which stresses is the exhaust system exposed?

7 An exhaust system is made up of which components?

8 Why must an exhaust system be gastight?

13.2 Pollutant reduction in a spark-ignition engine

13.2.1 Exhaust-gas composition

The high levels of air pollution caused by exhaust gases from road traffic have resulted in a great deal of legislation reducing the pollutants contained in exhaust gas. Fuels are composed primarily of hydrocarbon compounds. Complete combustion of fuel creates carbon dioxide, water vapour, nitrogen and inert gases. Carbon dioxide in large quantities is a major contributor to climatic change (greenhouse effect).

Incomplete combustion of fuel in the engine creates, in addition to carbon dioxide, water vapour, nitrogen and inert gases, the following:

- Carbon monoxide, CO
- Unburnt hydrocarbons, HC
- Nitrogen oxides, NO_x
- Solid matter

When a spark-ignition engine is at normal operating temperature, the proportion of CO, HC and NO_x before the catalyst at medium load and engine speed is approximately 1 % of the total exhaust-gas quantity **(Fig. 1)**.

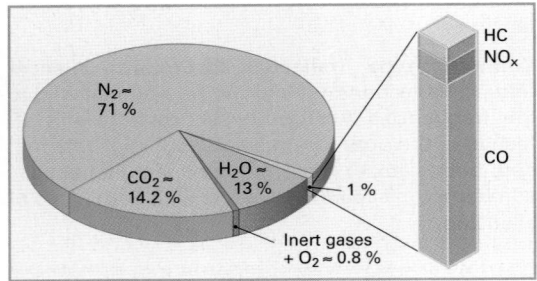

Fig. 1: Composition of exhaust gas before catalyst

Pollutants as a function of air ratio

> The proportion of individual pollutants in the exhaust gas is greatly influenced by the air ratio λ (lambda) **(Fig. 2)**.

Spark-ignition engines achieve their greatest power with an air deficiency of 5 % ... 10 % ($\lambda = 0.95 \ldots 0.90$; rich mixture) **(Fig. 3)**. The fuel is not fully utilised in the event of an air deficiency. The specific fuel consumption increases. The harmful exhaust-gas components – carbon monoxide and unburnt hydrocarbons – also increase.

Spark-ignition engines achieve their lowest fuel consumption with excess air of 5 % ... 10 % ($\lambda = 1.05 \ldots 1.1$; lean mixture). However, engine power is lower and the peak engine temperature is higher on account of, among other things, the lack of a cool-

ing effect by the evaporating fuel. The proportions of carbon monoxide and of unburnt hydrocarbons are low, but the proportions of nitrogen oxides in the exhaust gas very high.

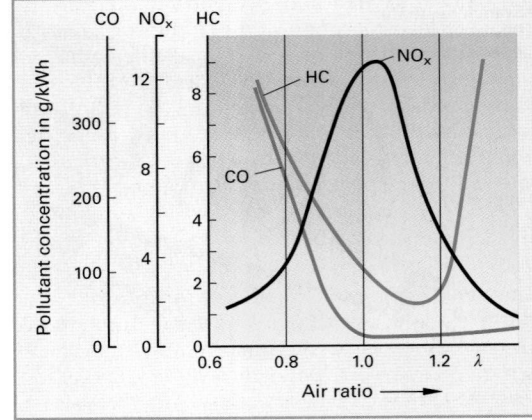

Fig. 2: Pollutants in exhaust gas with different air ratios

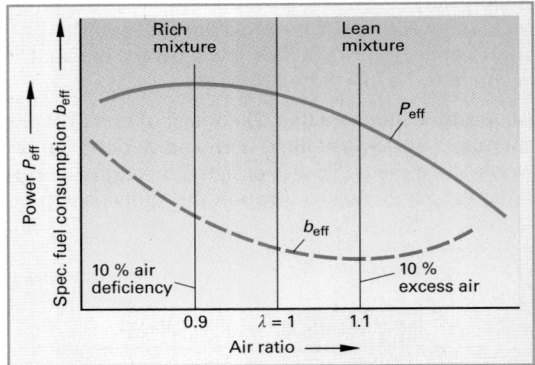

Fig. 3: Power and fuel consumption with different air ratios

Properties of harmful exhaust-gas components

Carbon monoxide CO. Carbon monoxide is a colourless and odourless gas. If inhaled, it blocks the flow of oxygen in the blood. Lower concentrations cause headaches, tiredness and impaired sensory perception. Extended exposure to concentrations greater than 0.3 % volume is fatal. Carbon monoxide is formed when fuel is incompletely burnt as a result of an air deficiency. The richer the fuel-air mixture, the greater the proportion of carbon monoxide in the exhaust gas. CO can also be formed in the event of excess air if the mixture of fuel and air is not homogeneous. There are therefore areas of rich and lean mixture in the combustion chamber.

Unburnt hydrocarbons, HC compounds. These comprise a large number of different compounds of carbon and hydrogen. Hydrocarbons irritate the mucous membranes, are bad-smelling and some are considered carcinogenic. In combination with nitrogen oxides, oxygen and insolation photochemical reactions lead to, among others, the formation of ozone (O_3) and thereby give rise to the damaging effects of so-called summer smog. Unburnt HC compounds are formed when the fuel-air mixture is incompletely burnt as a result of an air deficiency ($\lambda < 1$) or in the event of a very lean mixture ($\lambda > 1.2$). Unburnt HC compounds are also formed in those areas of the combustion chamber which are not fully touched by the flame, e.g. the gap on the fire land between the piston and the cylinder.

Nitrogen oxides, NO_x. The generic term nitrogen oxides is used for the different oxides of nitrogen (nitrogen monoxide NO, nitrogen dioxide NO_2, dinitrogen monoxide N_2O). Depending on the compound, nitrogen oxides can be colourless and odourless or reddish-brown with a pungent smell. Nitrogen oxides irritate the respiratory tracts, cause lung tissue to be destroyed in high concentrations, cause ozone to be created and result in damage to forests and woodland. Nitrogen oxides are formed above all at peak combustion-chamber temperatures.

Solid matter. This arises in the form of particulates when fuel is incompletely burnt (carbon nucleus/soot nucleus with attachments). The attachments of HC compounds to the carbon nucleus are categorised as carcinogenic. Unlike in a diesel engine, the particulates in a spark-ignition engine can be ignored (20 … 200 times fewer).

Legislation limiting harmful components in exhaust gas in spark-ignition engines

Legislative bodies have laid down maximum emission limits both during the type test **(Table 1)** for granting the general certification and during subsequent checks of pollutant emissions (exhaust-emissions inspection; OBD).

Table 1: Emission limits for passenger car/ estate car (M1) with spark-ignition engine in Europe

M1 ($\leq$ 2.5 t $\leq$ 6 seats)	CO (g/km)	HC (g/km)	NO_x (g/km)
Euro III Type approval 2000	2.30	0.20	0.15
Euro IV Type approval 2005	1.00	0.10	0.08

Determination of exhaust-emission values

The exhaust-emission values for the type test of motor vehicles are determined on chassis dynamometers. This involves imitating representative driving programs and analysing the pollutants ascertained in the process.

Europe test (European driving cycle, Fig. 1) for passenger cars up to a permitted vehicle weight of 2,500 kg and light HGV up to 3.5 t. The first part corresponds to a journey in urban traffic at a speed of 0 km/h to 50 km/h. The program is run four times without a break for 13 minutes, during which the engine is started as from EURO III at ambient-air temperature (20 °C). The subsequent second part corresponds to a 7-minute drive in extra-urban traffic, where the top speed is 120 km/h. During the entire test the exhaust gases are collected in accordance with predefined conditions and then the pollutants are analysed. The limit values in g/km must not be exceeded, regardless of the engine displacement. As from 2002 the exhaust-emissions test for vehicles with spark-ignition engines, type class M1, will be expanded to include a low-temperature test. The engine is started at – 7 °C; here the readings for HC and CO must be < 1.8 g/km and < 15 g/km respectively.

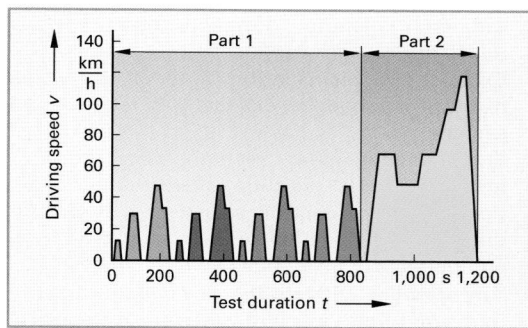

Fig. 1: Europe test (European driving cycle)

Emissions inspection (EI). This is required at specific intervals for motor vehicles already on the road. This involves checking the CO values when the vehicle is stationary with the engine running, depending on different operating parameters (operating temperature, engine speed). Further visual inspections and functional tests must also be carried out, e.g.

- checking of pollutant-relevant components
- checking of constricted tank filler neck
- control-loop check
- ignition-point check (if representable)

On-Board Diagnosis (OBD). With On-Board Diagnosis faults which occur in the engine-management system (fuel-injection system, ignition system), in the exhaust system (catalyst function, λ sensor) or in the fuel-supply system are stored in the ECU. An indicator lamp in the passenger compartment alerts the driver to malfunctions which result in significant increases in emissions. The malfunctions must be rectified immediately.

13

13.2.2 Procedures for reducing pollutants

> It is possible to reduce the pollutant constituents in the exhaust gases by using suitable fuel (low-sulphur, unleaded) in conjunction with on-engine measures or aftertreatment of the exhaust gases (secondary-air system, catalyst).

On-engine measures. A reduction in pollutants (untreated emissions) is achieved by burning the fuel-air mixture as completely as possible and by reducing the fuel consumption. The following on-engine measures can improve the exhaust-gas quality:

- **Suitable engine design:** Optimisation of combustion chamber and compression ratio; variable intake manifolds (in length and cross-section); variable valve control with regard to opening time and lift; dethrottling of the induction process.
- **Type and quality of mixture formation:** Exterior/interior mixture formation, homogeneous mixture; stratified charge.
- **Exhaust-gas recirculation:** Internal by valve overlap; external by exhaust-gas recirculation system.
- **Engine-management system:** Map-controlled ignition and fuel injection, overrun fuel cut-off; boost-pressure control; selective cylinder cut-out; checking function of exhaust-gas-relevant components, e.g. lambda sensors, catalyst.
- **Turbocharger with charge-air cooling:** Increase in power output per litre with simultaneous reduction of peak combustion-chamber temperature. This reduces the formation of NO_x.

External exhaust-gas recirculation (EGR, Fig. 1).

> External exhaust-gas recirculation involves a portion of the exhaust gas being removed shortly after the exhaust manifold and being remixed with the fuel-air mixture in the intake manifold.

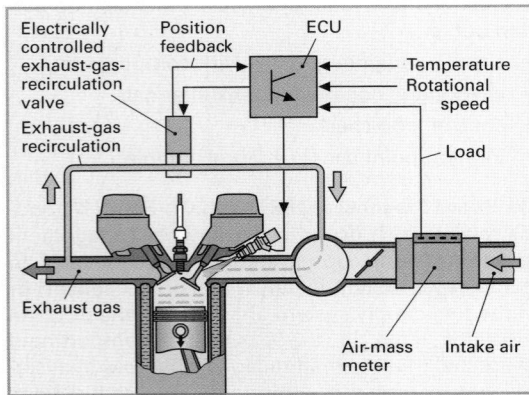

Fig. 1: Exhaust-gas recirculation

Exhaust-gas recirculation reduces the charge of fuel-air mixture supplied to the cylinders. The combustion temperature is reduced since the recirculated exhaust-gas components are no longer able to take part in combustion. In this way, significantly fewer nitrogen oxides are formed during combustion (up to 40%). As the exhaust-gas recirculation rate rises, both the content of unburnt HC compounds and the fuel consumption increase. These two factors determine the upper limit of the exhaust-gas recirculation rate (max. 20%). The smooth running of the engine is also compromised if the exhaust-gas recirculation rate is too high. Exhaust-gas recirculation is utilised with the engine at normal operating temperature in the part-load range and $\lambda \approx 1$. It is deactivated when rich fuel-air mixtures are being burnt during which few NO_x are formed, e.g. cold starting, warming up, accelerating, full load. Exhaust-gas recirculation is deactivated at idle on account of the smooth running of the engine. An exhaust-gas recirculation valve is installed in the exhaust-gas recirculation line between the exhaust manifold and the intake manifold for controlling exhaust-gas recirculation. The exhaust-gas recirculation rate is controlled as a function of engine temperature, load and engine speed.

Aftertreatment of exhaust gases in catalyst

> The exhaust gases are subjected to aftertreatment in order to convert the pollutants created during combustion after they have left the combustion chamber completely or partially into harmless substances.

At present the most effective method is exhaust-gas aftertreatment in a catalytic converter or catalyst. The catalyst chemically converts the pollutants into non-toxic substances without their being consumed in the process.

Design of catalyst. The main components of a catalyst **(Fig. 1, Page 315)** are:

- Ceramic substrate (aluminium-magnesium-silicate) or metal substrate
- Intermediate layer (wash-coat)
- Catalytically active layer

The substrate consists of several thousand fine channels, through which the exhaust gas flows. The channels of the ceramic or metal substrate are provided with a highly porous intermediate layer. This increases the effective surface of the catalyst by a factor of roughly 7,000. The catalytically active layer is attached to this intermediate layer. The catalytically active layer is dependent on the pollutant constituents which are formed depending on the engine concept and the associated mixture composition ($\lambda \approx 1$; $\lambda \gg 1$).

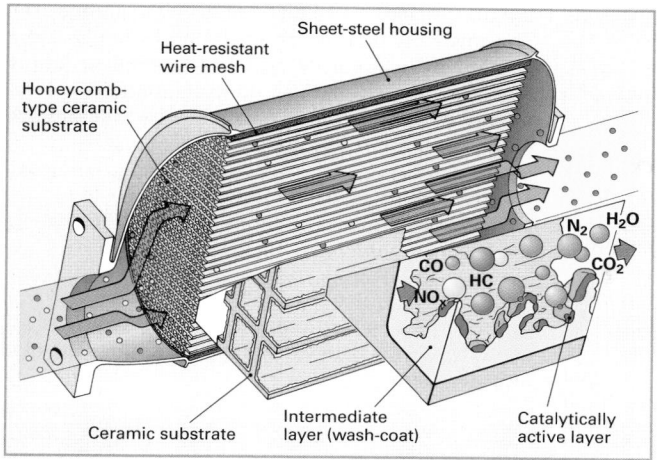

Fig. 1: Design and operating principle of a catalyst with ceramic substrate

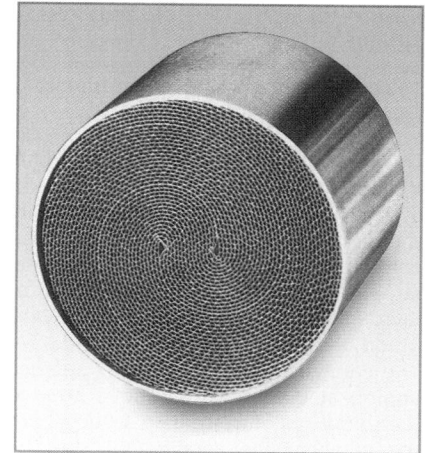

Fig. 2: Catalyst with metal substrate

Advantages and disadvantages of ceramic substrate compared with metal substrate (Fig. 1, Fig. 2)

Advantages: It has a more constant operating temperature. Also the noble-metal coating can be recovered more easily than from a metal substrate.

Disadvantages: The ceramic substrate is sensitive to jolts and vibrations. It must therefore be embedded in a heat-resistant wire mesh which is enclosed in a sheet-steel housing. Furthermore, the heating-up time is longer and the backpressure in the exhaust system is greater in the case of standard catalysts [cell density: 400 cpsi (cells per square inch)]. This results in reduced exhaust-gas conversion and reduced engine power shortly after engine starting. Ultrathin-wall substrates with a cell density of 600 cpsi ... 1,200 cpsi eliminate these disadvantages.

Operating principle of single-bed three-way catalyst (Fig. 1)

The single-bed three-way catalyst has a catalytically active layer of platinum (Pt), rhodium (Rd) and palladium (Pd). The term single-bed three-way catalyst indicates that three chemical conversions take place simultaneously in one housing. Depending on its operating temperature, the catalyst converts NO_x, CO and HC in the range $\lambda = 0.995 \dots 1.005$ (lambda window) up to 98 % into CO_2, H_2O and N_2.

- NO_x is reduced to nitrogen (oxygen O_2 is released).
- CO is oxidised into CO_2 (oxygen is used up).
- HC compounds are oxidised into CO_2 and H_2O (oxygen is used up).

Only an air ratio of $\lambda \approx 1$ delivers an exhaust-gas composition in which the oxygen released during the reduction of the nitrogen oxides is sufficient to oxidise the HC and CO content in the exhaust gas almost completely into CO_2 and H_2O. A richer mixture ($\lambda < 0.995$) results in an increase in the CO and HC content in the exhaust gas. A leaner mixture ($\lambda > 1.005$) results in an increase in nitrogen oxides (**Fig. 3**). Engines which operate in specific operating ranges at $\lambda \gg 1$, e.g. direct-injection engines, therefore require catalysts which are able to store NO_x intermediately.

13

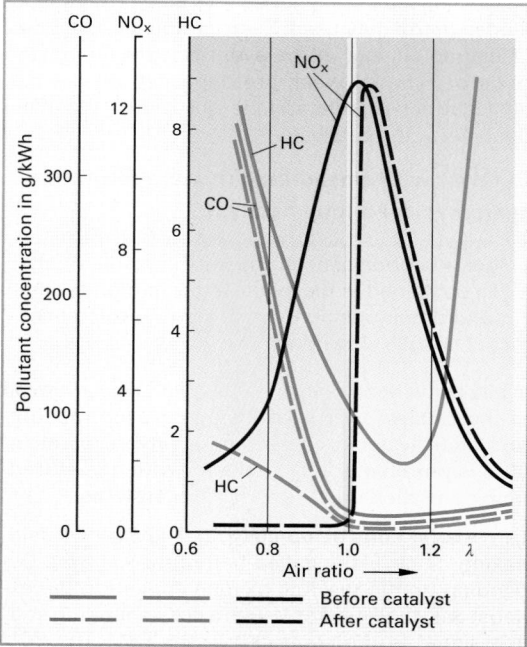

Fig. 3: Pollutants as a function of lambda

Operating conditions for single-bed three-way catalyst

> The optimal working or temperature range of the catalyst is 400 °C ... 800 °C.

Only from temperatures in excess of 300 °C does the catalyst have a conversion rate of more than 50 % ("light-off" temperature). The attainment of this temperature after cold starting can be significantly shortened by the following measures: upstream installation, catalyst heating, air-gap-insulated exhaust manifolds, marked reversal of the ignition point (up to 15°) and secondary-air injection. The catalytically active layer is subject to thermal ageing at temperatures above approx. 800 °C. If temperatures of more than 1,000 °C are reached in the catalyst, the catalyst will be thermally destroyed **(Fig. 1)**. This can occur for example as a consequence of misfiring. Here unburnt hydrocarbons enter the catalyst, where they burn with the accumulated residual oxygen.

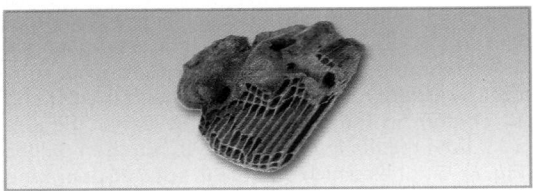

Fig. 1: Melted catalyst

In order that the catalytically active layer is not rendered ineffective ("poisoned") by deposits, only unleaded petrol may used. Even combustion residues of engine oil, e.g. in the event of defective piston rings or cylinder wear, are deposited on the catalytically active layer and thereby reduce the effectiveness of the catalyst.

Catalyst with closed-loop mixture-formation system (closed-loop catalyst)

> Here a lambda sensor utilises the residual oxygen contained in the exhaust gas to monitor the mixture composition within narrow tolerances $\lambda = 1 \pm 0.005$ (λ window).

While conversion rates of 94 % ... 98 % are achieved in the catalyst by closed-loop-controlled mixture-formation systems, on average a conversion rate of 60 % is possible in older vehicles with non-closed-loop-controlled mixture-formation systems.

Lambda (λ) control loop (Fig. 2). The λ sensor (signalling device) is installed before the catalyst. Depending on the residual-oxygen content in the exhaust gas, the λ sensor issues a corresponding voltage signal to the controller in the ECU. The ECU shortens the injection time if the amount of resid-

ual oxygen in the exhaust gas is low (rich mixture). The injection time is increased accordingly if the content of residual oxygen in the exhaust gas is high (lean mixture). The control operation is repeated at a prespecified frequency so that smooth engine operation is not affected.

> Condition for λ closed-loop control:
> - Sensor temperature higher than 300 °C
> - Engine in idle or part-load range
> - Engine temperature higher than 40 °C

A second λ sensor (monitor sensor), which is installed after the catalyst, serves to monitor the catalyst function.

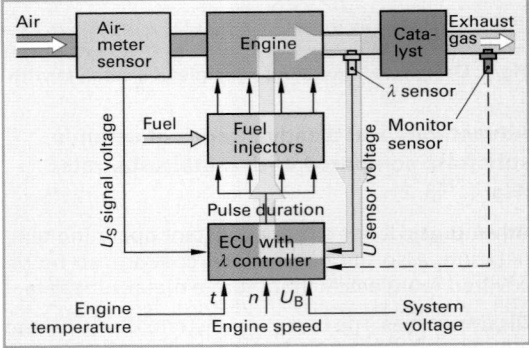

Fig. 2: Function diagram of an intermittently operating fuel-injection system with λ closed-loop control

Adaptive lambda closed-loop control. If, for example, the residual-oxygen value is constantly too low in a specific load range, i.e. the mixture is too rich, the basic injection quantity is reduced for this load range and stored as a pilot-control value in the ECU. In this way, disturbance values, such as e.g. incorrect fuel-system pressure, incorrect temperature values, engine ageing, unmetered air, can be corrected within a specific control range. At the same time the response time of λ closed-loop control is shortened and the exhaust-gas quality improved.

Types of λ sensors

Voltage-jump sensor (Fig. 1, Page 317)
Design. The λ sensor consists of a ceramic element, e.g. zirconium dioxide, which is coated on the inside and outside with a thin microporous platinum layer.

The sensor is heated to enable it to reach its response temperature as quickly as possible. The outer surface of the sensor is exposed to the exhaust-gas flow. It is connected by way of the platinum layer to the sensor housing and forms the negative pole (–). The inside surface of the sensor is in contact with the atmospheric air. It is connected by way of its platinum layer to the connection brought outwards and forms the positive pole (+).

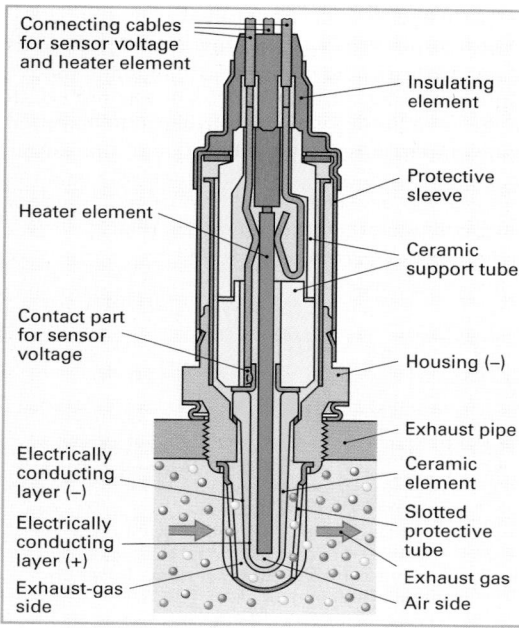

Fig. 1: Design of zirconium-dioxide sensor

Operating principle (Fig. 2). The ceramic material of the λ sensor becomes conductive for oxygen ions from roughly 300 °C. The oxygen content on the air side and the different contents of residual oxygen on the exhaust-gas side of the sensor give rise to a voltage jump between 100 mV (lean mixture) and 800 mV (rich mixture) at $\lambda \approx 1$. At $\lambda = 1$ the voltage is $\approx$ 450 mV ... 500 mV. The highest sensor temperature should not exceed 850 °C ... 900 °C.

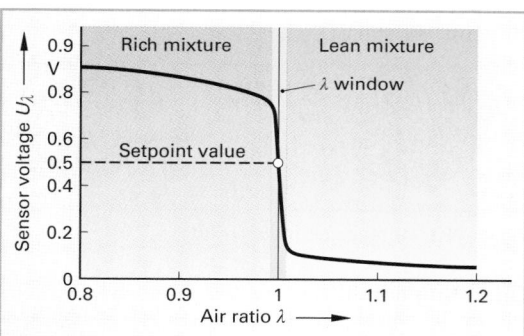

Fig. 2: Voltage curve of zirconium-dioxide sensor as a function of λ

Control frequency. This is usually greater than 1 Hz with an intact sensor at increased engine idle speed ($\approx$ 2,000 rpm), i.e. the voltage signal must fluctuate at least once per second between 0.1 V (lean mixture) and 0.9 V (rich mixture).

Resistance-jump sensor (Fig. 3).
The ceramic element of the sensor is made of tita-

nium dioxide which is coated with porous platinum electrodes.

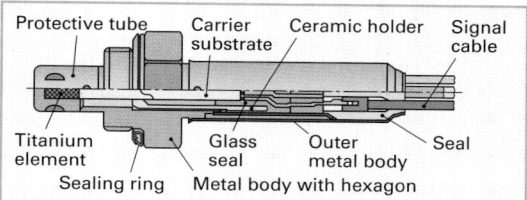

Fig. 3: Titanium-dioxide sensor

Operating principle (Fig. 4). Titanium dioxide alters its conductivity depending on the oxygen concentration in the exhaust gas and the ceramic-element temperature. It is less conductive with a lean mixture ($\lambda > 1$) than it is with a rich mixture ($\lambda < 1$).

> In the titanium-dioxide sensor the resistance value jumps at $\lambda \approx 1$ between 1 kΩ (rich mixture) and 1 MΩ (lean mixture).

A measuring shunt is connected in series to the sensor element in the ECU. The voltage drop at the measuring shunt varies between 0.4 V (lean mixture) and 3.9 V ... 5 V (rich mixture) as a result of the change in resistance of the titanium ceramic element as a function of the oxygen concentration in the exhaust gas. In contrast to the zirconium-dioxide sensor, reference air is not required. The control frequency is greater than 1 Hz with an intact sensor. The optimal working temperature is 600 °C ... 700 °C. Controllable sensor heating is required so that this temperature can be maintained. The sensor is ready for operation from 200 °C, at which point the control frequency is still too low for a precise correction of the mixture. The sensor may be destroyed if temperatures exceed 850 °C.

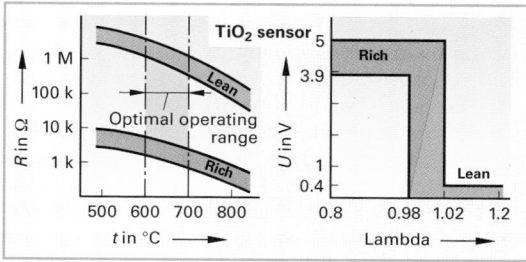

Fig. 4: Characteristic curves of titanium-dioxide sensor

Broadband lambda sensor

> λ values greater than > 0.7 can be infinitely measured with the broadband lambda sensor. This sensor is therefore suitable for continuous λ closed-loop control for lean-burn-concept spark-ignition engines, diesel engines and gas engines.

The optimal operating temperature of this sensor is 700 °C … 800 °C.

Design (Fig. 1). The sensor consists of two zirconium-dioxide voltage-jump sensors, where one sensor is the measuring cell (sensor cell) and the other is the pump cell. The two cells are arranged in such a way that there is a minimum diffusion gap (10 μm … 50 μm) between them. The diffusion gap serves as the measuring chamber and is connected with the exhaust gas via an inlet opening. The measuring cell incorporates a reference-air channel which is in contact with the outside air.

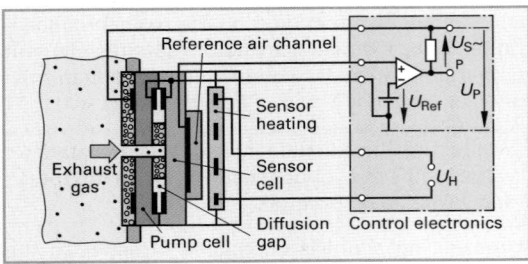

Fig. 1: Broadband lambda sensor

Operating principle of pump cell. It is possible by energising the solid electrolyte of the voltage-jump sensor to generate an oxygen-ion movement (= pump current) from a specific temperature. The direction of the oxygen-ion movement is dependent on the polarity (+/–) of the applied voltage.

Combined action of measuring cell – pump cell. The residual-oxygen content in the exhaust gas is determined by the measuring cell (sensor cell), which operates in accordance with the principle of the voltage-jump sensor. If the mixture is for example lean ($\lambda > 1 \rightarrow U_\lambda < 300$ mV), the control electronics applies a voltage to the pump cell ($U_{exhaust\text{-}gas\,side}$ $\boxed{+}$; $U_{measuring\,cell}$ $\boxed{-}$) in such a way that the oxygen ions move from the diffusion gap via the porous solid electrolyte towards the exhaust-gas side ("are pumped out"). This occurs until $\lambda = 1$ in the measuring cell. The required pump current here is proportional to the residual-oxygen concentration in the exhaust gas **(Fig. 2)**. The pump current thus serves as the measured quantity for the present lambda value. In this way, the engine control unit can permanently produce every desired mixture ratio depending on the program map stored.

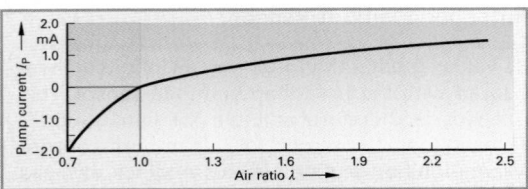

Fig. 2: Characteristic curve for broadband lambda sensor

NO$_x$ storage, reduction catalyst

In the case of direct-injection spark-ignition engines which are operated in specific operating ranges with charge stratification and $\lambda > 1$, it is not possible for a three-way catalyst to reduce the nitrogen oxides in these load states because of the excess air. In addition to the upstream three-way catalyst, a special NO$_x$ catalyst is deployed as an underfloor catalyst to treat the nitrogen oxides **(Fig. 3)**.

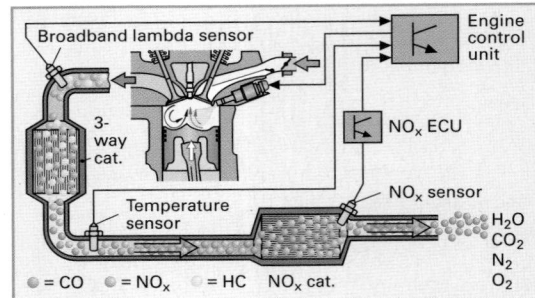

Fig. 3: Emission-control system for direct-injection spark-ignition engine at $\lambda > 1$

Design. An intermediate layer (wash-coat) is incorporated on a ceramic substrate. This layer is coated with barium oxide (BaO) or potassium oxide (KO) as the storage material.

Operating principle (Fig. 4). NO$_x$ storage. During lean mode the storage materials are able to bind (adsorb) the nitrogen oxides. The NO$_x$ sensor detects when the storage capacity is exhausted. **NO$_x$ reduction.** Thanks to periodic enrichment (1 … 5 seconds) the nitrogen oxides are released again and reduced by the noble metal rhodium to nitrogen with the aid of the unburnt exhaust-gas constituents HC and CO.

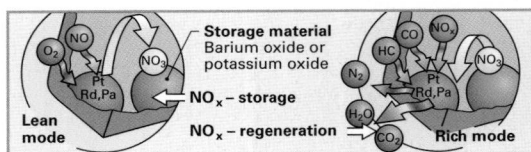

Fig. 4: NO$_x$ storage and regeneration

Operating conditions for NO$_x$ adsorber.

80 % … 90 % of the nitrogen oxides is reduced at working temperatures between 250 °C and 500 °C.

At temperatures in excess of 500 °C the catalyst is subject to high-temperature ageing. The exhaust gases must therefore be cooled if necessary, for example via bypass lines. The sulphur content in the fuel should be < 0.050 mg (< 0.050 ppm). Otherwise the storage capacity is significantly reduced ("sulphur poisoning").

Secondary-air system (Fig. 1).

Secondary-air injection serves to reduce HC and CO pollutant values in the engine cold and warming-up phases ($\lambda < 1$) by means of thermal afterburning.

The closed-loop three-way catalyst is not yet fully ready for operation in these engine operating states. Secondary-air injection involves supplying air to the exhaust manifold before the catalyst.

Advantages:

- The catalyst achieves operational readiness more quickly after cold starting.
- The catalyst can be installed at a greater distance from the exhaust port in order to increase its service life.

Operating principle (Fig. 1). In this system, a secondary-air fan and an electropneumatic changeover valve are activated by the ECU as a function of the engine temperature. Air is supplied via a shutoff valve and a non-return valve to the exhaust gas before the catalyst. The shutoff valve is activated by the elec-

tropneumatic changeover valve. The non-return valve ensures that the fan cannot be pressurised with exhaust-gas pressure and thereby damaged. The non-return valve also prevents exhaust-gas recirculation.

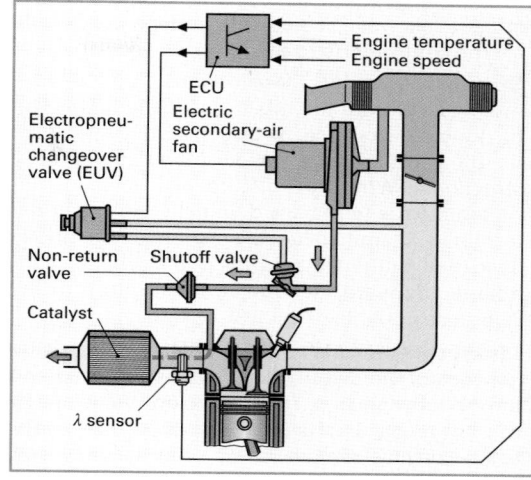

Fig. 1: Diagram, secondary-air system

13.2.3 Diagnosis and maintenance (EI)

Emissions inspection (EI). In the case of vehicles with spark-ignition engines (engines with externally supplied ignition) and at least 4 wheels, the **EI** must be carried out at specified intervals in accordance with statutory requirements. The engine type and exhaust-emission control are accordingly assigned to the relevant test procedure by means of code numbers and texts in **box 5** of the vehicle registration document **(Fig. 2)**. Details about the emission characteristics are also given in **box 1** of the registration document, e.g. xxxx**01** means LOW-EMISSION.

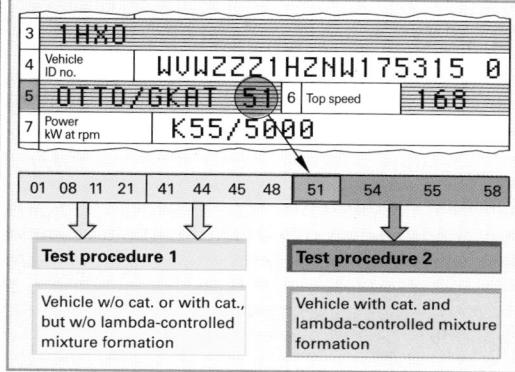

Fig. 2: Allocation of vehicle with spark-ignition engine to EI

The EI consists of a visual inspection of pollutant-relevant components and a functional test.

Visual inspection. The following components must be inspected for the purpose of ascertaining whether they are present, complete, leaking or damaged:

- Exhaust system
- Secondary-air system
- Exhaust-gas recirculation
- Crankcase ventilation
- Constricted tank filler neck
- Catalyst
- Lambda sensor
- Air filter
- Tank ventilation

Functional test. In the case of vehicles with closed-loop mixture-formation systems, this test covers:

- Pollutant-relevant setting data such as ignition point (where representable) and idle speed
- Emissions test of CO at idle speed and increased idle speed
- Lambda value at increased idle speed
- Control-loop check

Control-loop check. This is performed using disturbance-variable feedforward (basic procedure) or substitute/alternative procedures defined by the vehicle manufacturer. To be able to carry out this check, it is necessary for the engine to have reached the correct operating temperature. Using the example of the basic procedure, the control-loop check is described with two half-waves **(Fig. 1, Page 320)**. In λ-controlled engine mode the λ value must be between the statutory limits of $\lambda = 1.03$ and $\lambda = 0.97$. For the purpose of calculating the lambda value, the concentrations of the four exhaust-gas constituents CO, CO_2, HC and O_2 are continually checked by the emissions measuring

13

instrument at the end of the tailpipe. Once stable exhaust-gas values are achieved at the prescribed engine speed, e.g. idle speed, the lambda value measured by the test instrument, e.g. 0.997, is stored. This value serves as the reference variable for the control-loop check.

Disturbance-variable feedforward. E.g. by unmetered air. Here the mixture is leaned. This can be done, for example, by disconnecting a hose after the throttle valve (follow manufacturer's instructions). When the disturbance value of unmetered air is fed forward, initially the upper determined λ limit value of 1.03 must be briefly exceeded. Within 60 seconds of disturbance-variable feedforward the λ value must return to the reference variable of $\lambda = 0.997 \pm 0.01$.

Disturbance-variable removal. When the hose is reconnected, the disturbance value is removed and the mixture enriched. The lower λ limit value of 0.97 must be briefly undershot. Within 60 seconds of disturbance-value removal the λ value must be cor-

rected again to the reference variable of $\lambda = 0.997 \pm 0.01$. The control-loop function would thus be in order in accordance with the statutory requirements.

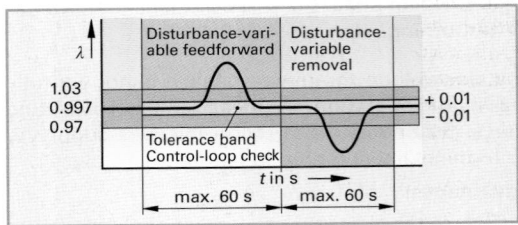

Fig. 1: Variation of λ value during control-loop check

Substitute procedure / alternative procedure.
These procedures are used if a control-loop check using disturbance-variable feedforward is not possible. Here the functional capability of the control loop is checked in accordance with the manufacturer's instructions, for example with a diagnosis tester.

13.2.4 European On-Board Diagnosis (EOBD)

This is a diagnosis integrated in the engine-management system. Here systems that influence exhaust gas are monitored during the entire vehicle operation. Any faults that occur are stored in the ECU and can be interrogated via a standardised interface. In addition, a fault signal is issued to the driver by way of an indicator lamp in the passenger compartment.

Fault indication (MIL = Malfunction Indicator Lamp)

 Faults which result in a significant increase in the exhaust-gas values **(Table 1)** or faults which occur during the self-test of the engine or gearbox control unit are indicated, for example, by a yellow, permanently lit warning lamp. Faults which may impair the function of or damage the catalyst, e.g. misfires, cause the MIL to flash.

Faults which cause the MIL to light up must be rectified immediately. The trip mileage after the warning lamp lights up is stored.

System monitoring

The following subsystems and sensors are, where installed, monitored permanently or once per driving cycle:

- Catalyst function, catalyst heating
- Function of lambda sensors
- Misfires
- Function of exhaust-gas recirculation
- Function of secondary-air system
- Fuel-tank ventilation system
- All electric circuits for exhaust-gas-relevant components
- If necessary, tank filler cap (if not permanently installed)

A driving cycle consists of engine start-up, driving at specific engine and road speeds, an overrun phase and engine stopping. Required coolant-temperature change: at least 22 °C to > 70 °C.

Catalyst monitoring

A second λ sensor **(Fig. 2)** after the catalyst serves to monitor the catalyst function. The engine control

Table 1: EOBD limits for Class M passenger cars				
$m_{perm} \leq 2{,}500$ kg; ≤ 6 seats	CO	HC	NO$_x$	PM[1]
Spark-ignition engine	3.2	0.4	0.6	–
Diesel engine	3.2	0.4	1.2	0.18
[1] Particulate matter				

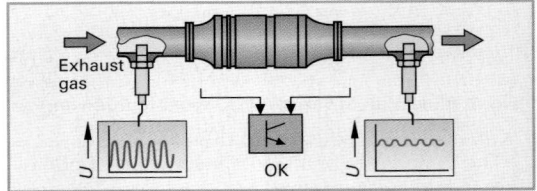

Fig. 2: Sensor signals for a high-efficiency catalyst

unit compares the signals of the two λ sensors. λ closed-loop control with the first sensor gives rise to rich-lean hunting. Due to the oxygen storage capacity of a high-efficiency catalyst, the voltage of the second sensor oscillates about the mean value. The catalyst loses its oxygen storage capacity as a result of ageing and can oxidise less CO and HC. In this way, the hunting of the λ sensor after the catalyst is similar to the hunting of the pre-catalyst sensor (**Fig. 1**). The ECU detects the inadequate effect of the catalyst, stores the fault and issues a fault signal to the OBD warning lamp.

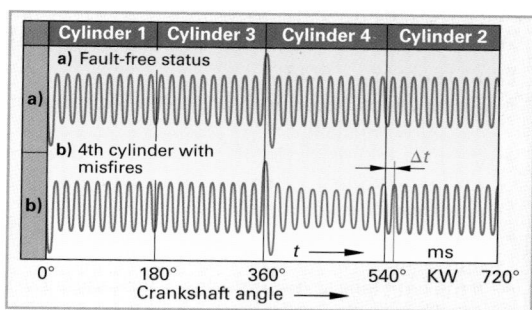

Fig. 2: Misfire detection in a four-cylinder engine

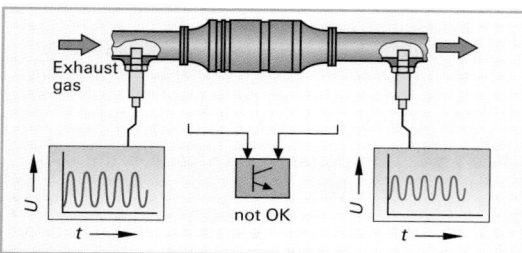

Fig. 1: Sensor signals for a low-efficiency catalyst

Sensor monitoring

The control frequency decreases in the event of an aged, defective pre-catalyst sensor. Furthermore, the amplitude level can decrease to such an extent that rich-lean detection is no longer possible. A fault signal is issued if limit values stored in the engine control unit are exceeded. It is also possible to detect from the voltage level whether there are short circuits to positive or earth/ground or open circuits. The function of the lambda-sensor heating can be checked by measuring the voltage drop at a series-connected resistor in the sensor-heating circuit.

Misfire monitoring

Each individual combustion generates a defined acceleration at the flywheel. An increment wheel (gearwheel with segments spread over its circumference) secured to the crankshaft and a sensor which forwards the signals to the engine control unit are used for monitoring purposes. If the combustion in all the cylinders is in order, the angular ignition spacing of all the segments is the same. Misfires give rise to torque fluctuations which in turn cause irregular engine running (**Fig. 2**). In this way, the flywheel and thus the increment wheel of the cylinder in question rotate at a slightly slower speed until the next ignition, i.e. the interval is extended by a fraction of a second. The fault lamp is activated if the misfires exceed a specific limit value. If there is a danger of the catalyst being damaged, the injection of fuel into the cylinder in question is cut out.

Function monitoring, secondary-air system

The secondary-air system can be monitored by means of the λ sensor voltage. The secondary-air pump is switched on as a function of load and engine speed when the engine is cold and during the engine warming-up phase. In the event of perfect air injection, the λ sensor voltage is in the lean range (≈ 300 mV…100 mV). This measurement is repeated in the cold-starting phase (approx. 90…150 seconds) at regular intervals. If in the process a sufficient number of low voltage values from the λ sensor is detected by the ECU, then the secondary-air system is classed as operational.

Function monitoring, exhaust-gas recirculation

This can be performed by activating (opening) the exhaust-gas recirculation valve during the overrun phase and measuring the intake-manifold pressure. If the exhaust-gas recirculation system is in order, the intake-manifold pressure must change in view of the fact that the connection between the exhaust manifold and the intake manifold is open. Otherwise a fault signal is issued.

Monitoring, tank-ventilation system

The tank-ventilation system can be checked by means of the λ sensor voltage. This usually occurs at idle. Initially the tank-venting valve is closed and the λ value determined. Then the tank-venting valve is opened. When the carbon canister is filled, the fuel-air mixture is enriched (U_λ = 800 mV … 900 mV). If there are no petrol hydrocarbons stored in the carbon canister, the fuel-air mixture is leaned (U_λ = 300 mV …100 mV). The ECU registers these values. This functional test is repeated several times. If a plausible value is obtained over a specific number of recurring functional tests, the tank-ventilation system is classed as operational.

13

Depending on the manufacturer, electrical checks are only carried out on the secondary-air system, the exhaust-gas recirculation valve and the tank-venting valve.

WORKSHOP NOTES– EOBD, Troubleshooting during lambda closed-loop control

Fault memory – EOBD

- Faults which occur once are deleted during the next driving cycle.
- If the fault is repeated, it is stored with its environmental data, e.g. engine speed, engine temperature ("freeze frame").
- If the fault has already been stored, the MIL lights up.
- If the fault fails to occur during 3 subsequent driving cycles, the indicator lamp goes out again.
- Faults which fail to occur after 40 driving cycles are deleted.

Reading out faults – OBD

> The interface, diagnostic record and diagnostic statements have been standardised for reading out exhaust-gas-relevant faults.

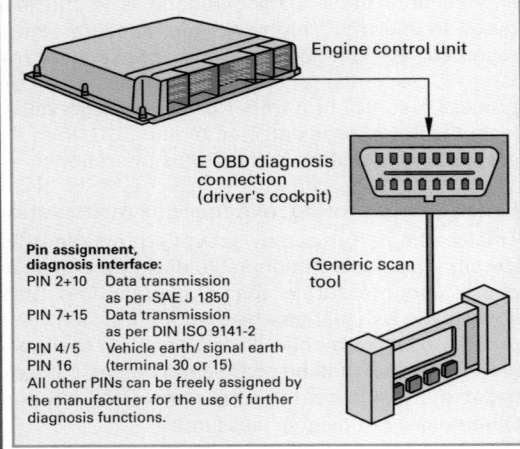

Engine control unit

E OBD diagnosis connection (driver's cockpit)

Generic scan tool

Pin assignment, diagnosis interface:
PIN 2+10   Data transmission as per SAE J 1850
PIN 7+15   Data transmission as per DIN ISO 9141-2
PIN 4/5   Vehicle earth/ signal earth
PIN 16   (terminal 30 or 15)
All other PINs can be freely assigned by the manufacturer for the use of further diagnosis functions.

Fig. 1: Reading out EOBD faults

Different test modes can be selected with the scan tool.

Test mode	Diagnostic function
1	Reading out of exhaust-gas-relevant actual values
2	Reading out of environmental fault data which were present when the exhaust-gas-relevant faults were present ("freeze frame")
3	Exhaust-gas-relevant faults which have occurred in two consecutive driving cycles (debounced faults)
4	Deletion of exhaust-gas-relevant diagnostic trouble codes
5	Measured values of lambda sensors
6	Measured values of non-permanently monitored systems, e.g. secondary-air injection
7	Reading out of sporadic exhaust-gas-relevant faults
8	Status indication whether system or component checks are completed (readiness code: yes = 0; no = 1)
9	Indication of vehicle and system information

Diagnostic trouble codes. These are 5-position alphanumeric codes, e.g. $\boxed{\text{P 0 150}}$.

$\boxed{\text{P}}$ 0 150   **P:** Powertrain (drive)

P $\boxed{0}$ 150   **0:** Codes are standardised irrespective of manufacturer
(**1:** Code is specified by the manufacturer.)

P 0 $\boxed{1}$ 50   **1:** Component group in which the fault occurs
  1/2 Fuel and air metering
  3 Ignition system or misfiring
  4 Additional systems for reducing emissions
  5 Cruise-control and idle-speed control systems
  6 Computers and output signals
  7 Gearbox

P 0 1 $\boxed{50}$ Component which causes the malfunction
  50 O_2 sensor before CATALYST

Troubleshooting during lambda closed-loop control

If it is ascertained during the control-loop check when the exhaust-gas constituents are being measured that λ closed-loop control is not in order, the fault can lie both in the electrical control loop and on the catalyst. For the purpose of locating a possible electrical fault in the control loop, there are test options which can be performed to a large extent irrespectively of different models and types.

Electrical check of λ closed-loop control in closed control loop for a voltage-jump sensor (Fig. 2)

With the ignition switched off a voltmeter or an oscilloscope is connected parallel to the signal voltage of the λ sensor. After the cold engine is started, a fixed voltage of 0.4 V ... 0.6 V is indicated (depending on the manufacturer). When the engine and λ sensor are at normal operating voltage, the voltage fluctuates between 0.1 V and 0.8 V.

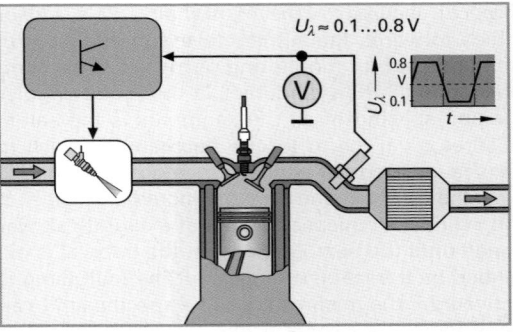

$U_\lambda \approx 0.1...0.8$ V

Fig. 2: Electrical check of λ closed-loop control in closed control loop

WORKSHOP NOTES CONTINUED FROM PAGE 322

If there is a fault such that the sensor voltage always shows a fixed value, e.g. ~0.1 V; ~ 0.5 V or ~0.8 V, first check whether, for example, the engine temperature, sensor temperature, temperature sensor and cable connections are in order. If this is the case, the fault can be further located by interrupting the control loop and electrical disturbance-variable feedforward.

Electrical check of λ closed-loop control in interrupted control loop (Fig. 1)

Prerequisite: Engine and λ sensor are at operating temperature, sensor heating working. Remove the coupler plug between lambda sensor and ECU and connect an adapter plug in between. Connect the measuring instrument.

Simulation: Rich mixture

The corresponding pin of the ECU is subjected to a voltage of approx. 0.8 V ... 0.9 V via a tester.

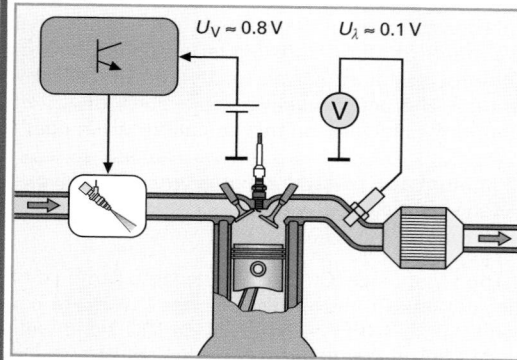

$U_V \approx 0.8\,V$ $U_\lambda \approx 0.1\,V$

Fig. 1: Electrical check of λ closed-loop control in interrupted control loop

The mixture leans if the ECU and the supply lead are in order. Smooth engine running deteriorates (engine bucks). The sensor voltage drops to approx. 0.1 V with an intact λ sensor. If this is not the case, the following faults may be present: engine-temperature sensor faulty, fault in the wiring harness or ECU. If leaning is performed by the ECU but the sensor voltage does not drop, the fault lies in the area of the λ sensor (earth/ground fault; sensor heating faulty; sensor aged/faulty).

Simulation: Lean mixture

Here a voltage of approx. 0.1 V is applied to the corresponding pin of the ECU. The engine speed should increase slightly for a brief period on account of the mixture enrichment brought about by the ECU. The λ sensor voltage should assume 0.8 V ... 0.9 V. If this is not the case, the same faults as described in the simulation of mixture enrichment may be present or unmetered air is inducted on account of leaks.

Important. In the case of titanium-dioxide sensors the voltage values oscillate between less than 0.3 V and greater than 3.9 V with an intact sensor.

Checking lambda-sensor heating. This can be checked by way of a resistance measurement. Customary values at room temperature: $2\,\Omega$ ... $14\,\Omega$, at values in excess of $30\,\Omega$ the sensor is faulty. Possible consequences: poor engine power, increased consumption or engine bucking.

Checking broadband lambda sensor. The function of these sensors can be checked using vehicle testers.

13

REVIEW QUESTIONS

1 What pollutants are created in the event of incomplete combustion of the fuel in the engine?
2 What are the effects of air deficiency and excess air on the exhaust-gas composition?
3 What are the basic procedures for reducing pollutants?
4 Which on-engine measures are used to reduce the formation of pollutants?
5 Explain the design and operating principle of an exhaust-gas recirculation system.
6 What do you understand by a single-bed three-way catalyst?
7 Explain the structure of a ceramic-substrate catalyst.
8 Which chemical conversions take place in a three-way catalyst?
9 Why should $\lambda = 1$ in exhaust systems with three-way catalysts?
10 Explain the λ control loop.

11 What does the λ sensor measure?
12 Between what values does the voltage jump of a titanium-dioxide sensor vary?
13 How is the λ value determined in a broadband lambda sensor?
14 What is the function of the NO_x catalyst?
15 Explain the engine concepts in which a NO_x catalyst may be required.
16 Explain the design and operating principle of the secondary-air system.
17 How is the control-loop check carried out in the EI?
18 How in OBD is the function of the catalyst monitored?
19 Under what conditions does the MIL light up?
20 How are misfires identified?
21 Under what conditions is the diagnostic trouble code automatically deleted in EOBD?
22 How can the lambda-sensor heating be checked?

13.3 Pollutant reduction in a diesel engine

13.3.1 Exhaust-gas composition

> In addition to nitrogen N_2 and oxygen O_2 as components of the residual air, the exhaust gas of a diesel engine contains assorted reaction products from carbon C, hydrogen H, oxygen O and nitrogen N.

Complete combustion. Under optimal conditions (not achievable with engine combustion) hydrocarbon (HC) compounds burn into carbon dioxide (CO_2) and water (H_2O).

Incomplete combustion. The diesel engine operates depending on engine load with differing degrees of excess air ($\lambda > 1$). At full load with a slight degree of excess air up to $\lambda \sim 1.3$). At part load and at idle with great degree of excess air up to $\lambda \sim 18$. In spite of excess air the fuel is only partially combusted.
The following pollutant compositions are created: carbon monoxide (CO), unburnt hydrocarbons (HC) and particulate matter (PM). The latter consist of a soot nucleus with attached contaminants such as e.g. metal oxides **(see Page 288)**.

On top of these come substances which result from contaminants or additives in the fuel or lubricant such as e.g. metal and sulphur compounds.

Nitrogen oxides NO_x (such as nitrogen monoxide NO and nitrogen dioxide NO_2) are formed at high peak combustion temperatures, high combustion pressures and high flame velocities. There are increased NO_x emissions on account of the excess air at idle and part load.

Emission limits for passenger-car diesel engines

In spite of the restrictions on emissions dating back to the mid-1970s, there has only been a slow decrease in vehicle emissions on account of increasing vehicle registrations. This has resulted in increasingly stringent emission limits.
Table 1 shows the European emission limits for new diesel passenger cars.

Table 1: Emission limits for passenger car/ estate car with diesel engine in Europe in (g/km)

	CO	CH + NO_x	NO_x	Particulates
Euro 2 since 1996	1.0	0.7	–	0.08
Euro 3 since 2000	0.64	0.56	0.5	0.05
Euro 4 from 2005	0.5	0.30	0.25	0.025
Euro 5 from 2008	0.5	?	0.08	0.0025

13.3.2 Procedures for reducing pollutants

In order to meet the emission limits stipulated by the Euro 4 and Euro 5 standards, inside- and outside-engine measures must be optimally coordinated.

Inside-engine measures. These comprise for example:
- Combustion-chamber optimisation
- Glow-time control
- Greater injection pressure
- Multiple-valve technology
- Intake-passage control
- Boost-pressure control
- Optimised start of injection and injected fuel quantity

Outside-engine measures. The following systems are currently used:
- Oxidation catalyst
- Particulate filter
- Exhaust-gas recirculation
- NO_x storage catalyst

Oxidation catalyst
This corresponds in design to a three-way catalyst. A coating of aluminium oxide is applied to a ceramic or metal substrate to enlarge the effective surface. The actual catalyst, consisting of $1 \dots 2$ g platinum, is located on this so-called "wash-coat".

> Platinum as a catalyst initiates two chemical oxidation processes without accompanying consumption.

Carbon monoxide (CO) is converted into carbon dioxide (CO_2) and unburnt hydrocarbons (HC) are converted at 90 % into carbon dioxide (CO_2) and water (H_2O). On account of the excess air, it is only possible to reduce nitrogen oxides (NO_x) to a slight extent.

Due to the high oxygen content in the exhaust gas, the effect of the oxidation catalyst already starts at 170 °C. The optimal operating temperature is 250 °C $\dots$ 350 °C.

Exhaust-gas recirculation (EGR, Fig. 1, Page 325)
Exhaust-gas recirculation serves to reduce NO_x emissions. The supplied oxygen content is reduced by adding exhaust gas to the intake air. The exhaust-gas components no longer take part in combustion and additionally absorb heat. This reduces the peak combustion temperature and with it nitrogen-oxide emissions (NO_x) down to approx. 60 %. The exhaust-gas recirculation rate can be up to 40 % vol.

If the exhaust-gas recirculation rate rises above this value, on the one hand the NO_x emissions continue to drop but on the other hand the diesel fuel no longer burns completely. In this way, the proportion of unburnt hydrocarbons (HC) and particulate matter (PM) continues to rise dramatically on account of the lack of oxygen.

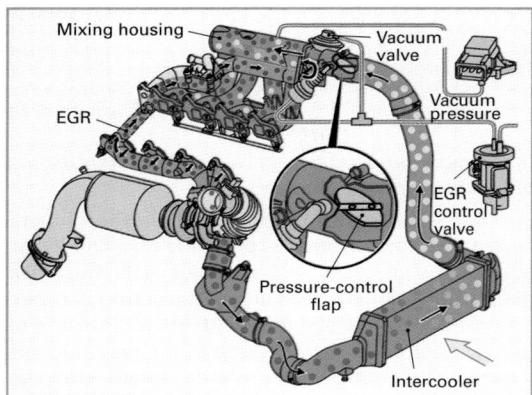

Fig. 1: Exhaust-gas recirculation with additional control flap

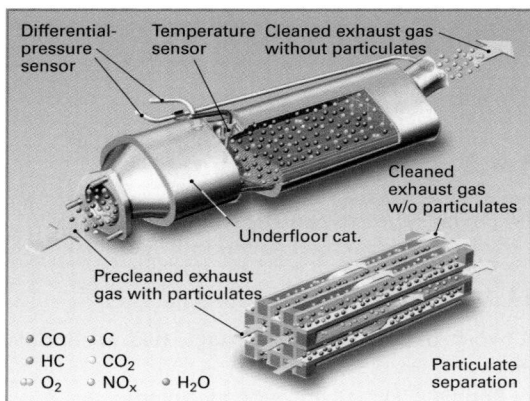

Fig. 2: Particulate filter

Regulation of EGR rate. This is controlled by the ECU via a vacuum valve or an electric servo-motor and is dependent on:

- Engine temperature
- Charge-air pressure
- Intake-air temperature
- Load/engine speed

> Exhaust-gas recirculation is activated when the diesel engine is hot in idle and part-load operation. Less air is inducted when exhaust-gas recirculation is active. The signal from the hot-film air-mass meter (HFM) serves to calculate the EGR quantity.

The EGR rate can be increased by cooling the recirculated exhaust gases. In addition, pressure-control flaps can be incorporated into the mixing housing. When the pressure-control flap is closed, a greater pressure differential between the intake manifold and the exhaust manifold is established and thus the EGR rate is increased.

In order to achieve full power and full torque, exhaust-gas recirculation is shut down at higher engine speeds and at full load. In addition, the air deficiency would result in excessive emission of particulate matter. The deterioration in smooth running at exhaust-gas recirculation rates at idle can be compensated by systems with idle-speed control.

Particulate filter (Fig. 2)

Design and operating principle. This consists of a ceramic honeycomb-type filter body. However, sintered-metal filter bodies can also be used. The channels of the particulate filter are alternately sealed. In this way, the exhaust gas has to pass through the porous filter walls.

The particulates are suspended in the process and slowly clog the pores. In this way, the exhaust back-pressure increases gradually. This results in increased fuel consumption and reduced power. The filter must be regenerated.

Regeneration. Here the stored particulates are converted into carbon dioxide (CO_2) and water vapour (H_2O). The particulate burn-off temperature is approx. 550 °C. Under normal conditions, however, only exhaust-gas temperatures of max. 400 °C are achieved. Two different systems are used to regenerate the filter:

- **Lowering the particulate burn-off temperature.** This is done with the aid of an additive which is mixed with the fuel in the tank via a metering unit. The particulate burn-off temperature can thus be lowered by approx. 100 °C.

- **Increasing the exhaust-gas temperature.** The exhaust-gas temperature is increased by specific secondary injection and by an increase in the torque request, e.g. by the A/C compressor and alternator.

Ash formation. A small amount of ash is created when the particulates are burnt. This ash collects with the particulates in the filter and clogs the filter with time. The filter must be removed and cleaned. Depending on the system and the driving style, this is conducted between 80,000 km and 240,000 km. An indicator lamp alerts the driver when this servicing work is due.

Regulation of regeneration process. The differential-pressure sensor **(Fig. 2)** records the pressure differential before and after the particulate filter. A high pressure differential is recorded when the particulate filter is full. Regeneration is initiated. The temperature during regeneration is recorded by the temperature sensor and must not exceed 700 °C.

13

Review questions

1 What is the purpose of exhaust-gas recirculation?
2 How does a particulate filter work?
3 What do you understand by regeneration?

14 Two-stroke spark-ignition engine, rotary engine

14.1 Two-stroke engine

> A power cycle in a two-stroke engine takes place during one crankshaft revolution (360°).

14.1.1 Design

A two-stroke spark-ignition engine (**Fig. 1**) consists of:

- **Engine case** — Cylinder head, cylinder, crankcase
- **Crankshaft drive** — Piston, connecting rod, crankshaft
- **Mixture formation** — Carburettor or fuel-injection system, intake manifold
- **Auxiliary devices** — Ignition system, engine cooling, exhaust system and lube-oil metering pump with total-loss lubrication (separate lubrication)

Because the gas exchange is usually controlled by the piston and via the ports in the cylinder wall, there is no need for any of the engine-management components which are used in a four-stroke engine.

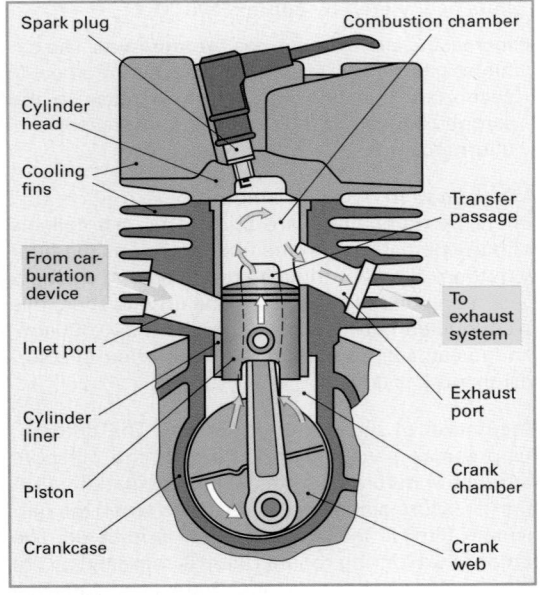

Fig. 1: Design of a two-stroke spark-ignition engine

Labels: Spark plug, Combustion chamber, Cylinder head, Cooling fins, From carburation device, Inlet port, Cylinder liner, Piston, Crankcase, Transfer passage, To exhaust system, Exhaust port, Crank chamber, Crank web

14.1.2 Operating principle

The power cycle of a two-stroke engine is made up, as in a four-stroke engine, of **induction, compression, combustion, exhaust**. However, the sequence

of the individual processes **(Table 1)** differs in terms of location and duration.

In order in a two-stroke engine to limit the power cycle to two piston strokes or to one crankshaft revolution, the processes of the power cycle must take place both in the cylinder and in the crank chamber. The crank chamber together with the bottom end of the cylinder and the underside of the piston forms a pump. The crank chamber must therefore also be gastight.

Table 1: Processes in a two-stroke engine

Processes in the cylinder (above the piston)	Transfer (scavenging) ← Compression Combustion Exhaust
Processes in the crank chamber (below the piston)	Preinduction Induction Precompression Transfer (scavenging) →

Three-port two-stroke engine (Fig. 1). This has three types of port for the purpose of gas control: one inlet port, one exhaust port and two opposing transfer passages.

Inlet port — This connects the carburettor with the crank chamber.

Transfer passage — This connects the crank chamber with the combustion chamber.

Exhaust port — This connects the combustion chamber with the exhaust system.

> The two-stroke engine has an open gas exchange.

This means that the exhaust port and the transfer passages are simultaneously open over a large part of the gas exchange. It is therefore inevitable that on the one hand mixing between fresh gases and exhaust gases takes place and on the other hand fresh-gas losses occur in a two-stroke engine.

Operating principle (three-port two-stroke engine)

> **1st stroke, crankshaft angle 0°... 180°**
> Piston travels from BDC to TDC
> **(Fig. 1, Page 327)**

Processes in the crank chamber

Preinduction. Once the piston has closed the transfer passage, the increase in volume creates a vacuum pressure of 0.2 bar ... 0.4 bar in the crank chamber. This process is known as preinduction.

Induction stroke. When the piston finally uncovers the inlet port, the actual process of inducting the fuel-air mixture into the crank chamber begins.

Processes in the combustion chamber

Compression stroke. Once the piston has closed the exhaust port, the process of compressing the fuel-air mixture in the cylinder begins. Ignition takes place shortly before TDC.

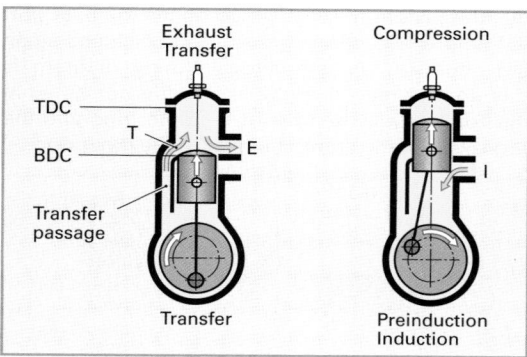

Fig. 1: 1st stroke

2nd stroke, crankshaft angle 180° ... 360°
Piston travels from TDC to BDC
(Fig. 2)

Processes in the combustion chamber

Power stroke. The pressure of the combustion gases moves the piston from TDC to BDC.

Processes in the crank chamber

Once the piston has closed the inlet port, precompression of the fuel-air mixture to approximately 0.3 bar... 0.8 bar gauge pressure begins.

Gas-exchange process

(processes under and above the piston)
The gas exchange takes place during the transition to the next power cycle.

Exhaust stroke. The top edge of the piston uncovers the slightly higher exhaust port and the exhaust gases rush out. It then uncovers the transfer passage and the precompressed fuel-air mixture as it passes from the crank chamber assumes the task of scavenging the cylinder and discharging the residual exhaust gases. The initial dynamic pressure in the exhaust pipe forces the residual exhaust gases initially back into the crank chamber when the transfer passage is uncovered. In this way, the precompression pressure of 0.3 bar increases to the scavenging pressure of approximately 0.8 bar. This in turn instigates the transfer of the fresh gases. The scavenging process is completed when the piston on its way to TDC has closed the transfer passage and then the exhaust port.

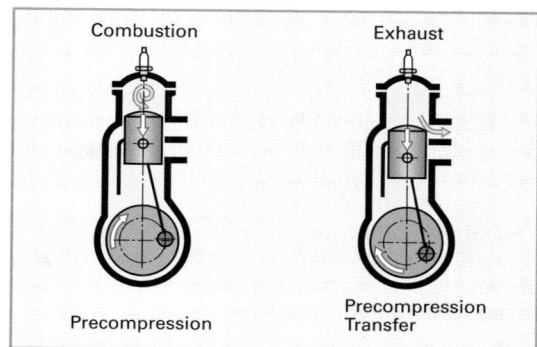

Fig. 2: 2nd stroke

Table 1: Gas pressures in bar			
Induction	Compression	Combustion	Exhaust
− 0.4 ... − 0.6	8 ... 12	25 ... 40	3 ... 0.1
Pre-induction	Pre-compression		Transfer
− 0.2 ... − 0.4	0.3 ... 0.8		1.3 ... 1.6

Scavenging process (reverse scavenging)

In the case of conventional Schnürle-type reverse scavenging (Adolf Schnürle, German engineer, 1896 – 1951), there is a transfer passage on either side of the exhaust port (**Fig. 3**). A piston window controls the inlet port. This form of scavenging is also known as "triple-flow scavenging".

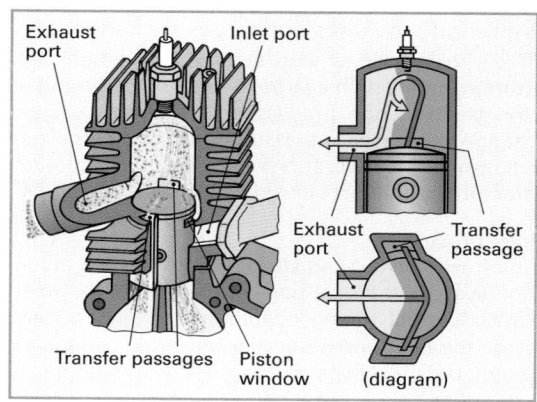

Fig. 3: Reverse scavenging

The scavenging flows are routed from the scavenging passages situated at an angle to the cylinder axis to the cylinder wall opposite the exhaust. There they mount up and push out the residual exhaust gases, following the cylinder wall, to the exhaust port. The scavenging flows therefore loop back in the cylinder. There can also be three or more transfer passages situated opposite the exhaust port or ports. In the case of four-passage reverse scavenging (**Fig. 1, Page 328**), the two main scavenging flows meet opposite the exhaust port and are diverted upwards. After being diverted, this movement being encouraged by the shape of the cylinder head, they force out the majority of the exhaust gas to the exhaust port. The two auxiliary scavenging ports are routed in such a way that they force and expel the exhaust-gas core still situated in the "dead area" of the cylinder to the exhaust port.

The looping of the main scavenging flows and the routing of the auxiliary scavenging flows reduce scavenging losses, expel the exhaust-gas core and improve volumetric efficiency.

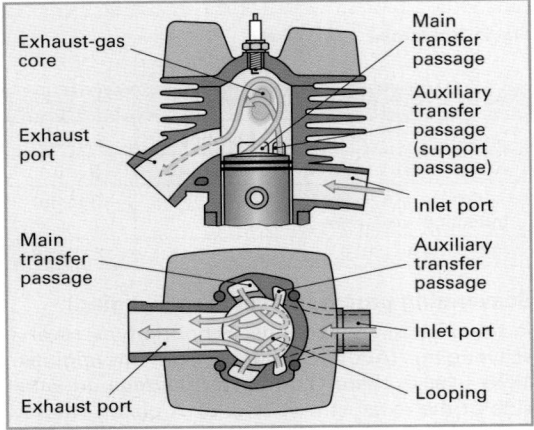

Fig. 1: Multiport reverse scavenging (loop scavenging)

Vibration processes during gas exchange

Two-stroke engines with symmetrical timing diagrams operate with a large overlap of the timing or the gas-exchange processes. The abrupt gas-exchange processes generate vibrations in the gas columns. These vibrations must be tuned to each other in the interests of reducing fresh-gas losses.

Inlet process

The fresh-gas column vibrates between the induction system, the inlet port and the crank chamber. Under correct tuning conditions the piston must close the inlet port when the fresh-gas column vibrates back to the crank chamber. The fresh gas is no longer able to flow back and the compression pressure increases.

Exhaust and scavenging process

The gas columns vibrate between the exhaust system, the cylinder and the crank chamber. The exhaust gas flowing out under pressure generates a pressure wave which is reflected by a baffle plate in the front silencer. This prevents fresh gas from flowing into the exhaust port. On account of these vibration processes the exhaust pipe with silencer and the intake line with air filter must be precisely matched to each other in order to avoid charge loss-

es. Inexpert or inappropriate modification or remachining will result in power losses and higher specific fuel consumption.

Symmetrical timing diagram (Fig. 2).

In a two-stroke engine with gas-exchange control by the piston the inlet and exhaust ports and the transfer passage are opened precisely as many degrees before TDC or BDC as they are closed. If, for example, the piston moving to TDC opens the inlet port 55° before TDC, the latter will also be closed again 55° after TDC. A symmetrical timing diagram is therefore produced. **Fig. 2** shows the processes in the combustion chamber in the outer ring and the processes in the crank chamber in the inner ring.

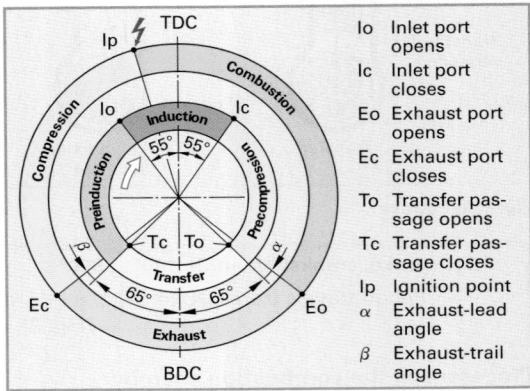

Fig. 2: Symmetrical timing diagram

Favourable exhaust lead. The piston moving to BDC opens first the exhaust port and then the transfer passage. The opening of the exhaust port is accompanied by a marked pressure drop such that the residual exhaust gases do not blow back so severely into the crank chamber and mix there with the precompressed fresh gas.

Harmful exhaust trail. The piston moving to TDC closes first the transfer passage and then the exhaust port. Fresh gases can be forced out to the exhaust port in the process.

Charge loss. A two-stroke engine has only roughly 130° crankshaft angle available for scavenging, corresponding to approximately one third of the gas-exchange time of a four-stroke engine.

Inlet control and/or exhaust control is used on account of these disadvantages. This results in asymmetrical timing diagrams.

Asymmetrical timing diagram

In the case of an asymmetrical timing diagram **(Fig. 1, Page 329)**, the opening and closing angles for the individual ports vary in size and are therefore no longer symmetrical to TDC or BDC.

Asymmetrical timing diagrams for inlet and exhaust control cannot be obtained with piston-dependent port control.

Useful supercharging/boosting. In two-stroke engines with asymmetrical timing diagrams the transfer passage can be closed after the exhaust port; the charge is improved by the mass inertia of the fresh gases.

"Useful supercharging/boosting" can only be achieved with highly elaborate design features, e.g. by inlet control using slides and by exhaust control using cam-controlled exhaust valves.

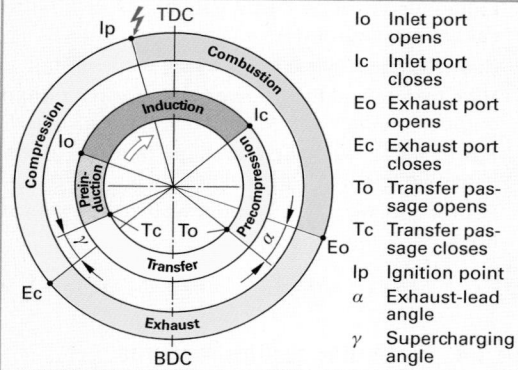

Io	Inlet port opens
Ic	Inlet port closes
Eo	Exhaust port opens
Ec	Exhaust port closes
To	Transfer passage opens
Tc	Transfer passage closes
Ip	Ignition point
α	Exhaust-lead angle
γ	Supercharging angle

Fig. 1: Asymmetrical timing diagram

The opening or/and closing angles can be "advanced" or "retarded" for the inlet process by means of diaphragm or rotating-slide control.

14.1.3 Types of control

Inlet control

Diaphragm control (Fig. 2). The supply of fresh gas is controlled by means of a diaphragm valve in the inlet port. When the piston moves to TDC (preinduction), a vacuum pressure is generated in the crank chamber. The diaphragm valve is opened by the differential pressure of the crank-chamber and atmospheric pressures. The fresh gas can flow during induction into the crank chamber until the precompression pressure generated by the pressure of the downward-moving piston and the pretensioned diaphragm close the inlet port. In this way, the diaphragm valve prevents the inducted fresh gas from flowing back into the induction system. An improved fresh-gas charge is achieved as a result.

Design of diaphragm valve (Fig. 2). The diaphragm strips are made of highly elastic, thin spring steel

and open already under the slightest differential pressure. The diaphragm stopper restricts the valve movement of the diaphragm strips and prevents them from swinging up.

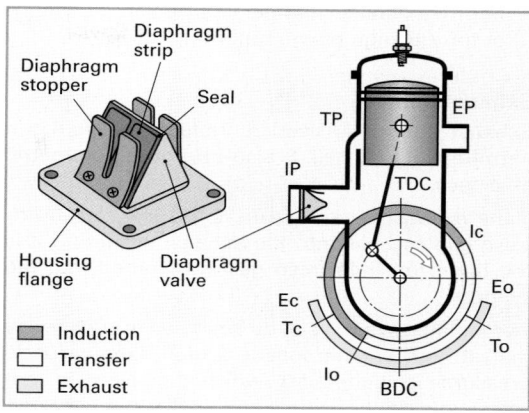

▨	Induction
☐	Transfer
☐	Exhaust

Fig. 2: Diaphragm control

Rotating-slide control (Fig. 3)

The inlet port is controlled by rotating slides. Unlike diaphragm control, the control angles cannot vary here. The opening of the inlet port into the crank chamber is opened and sealed by a rotating slide. The rotating slide rotates at crankshaft speed. It utilises the shape of its recess and the position in relation to the crankshaft to determine the inlet angle and thus the inlet time.

Crank webs with recesses can also be used as rotating slides.

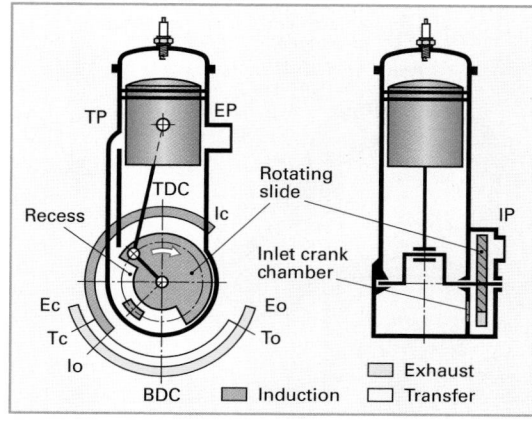

☐	Exhaust
▨	Induction
☐	Transfer

Fig. 3: Rotating-slide control

Features of inlet control

- Asymmetrical timing diagram.
- Control angles for "Inlet port opens" and "Inlet port closes" differ in size – control angles for transfer and exhaust are symmetrical to BDC.

- Variable inlet angle with diaphragm control is dependent on the crank-chamber vacuum pressure.
- Constant inlet angle with rotating-slide control.
- Improved crank-chamber charge and thus higher torque, high power output per litre.

Exhaust control

Exhaust control is used to reduce or avoid the harmful exhaust trail. It also improves volumetric efficiency.

If the exhaust backpressure is too low, too much fresh gas escapes into the exhaust system; if it is too high, too little fresh gas is admitted into the cylinder.

The exhaust system can be designed in such a way that at high engine speeds a high exhaust backpressure is generated which however is not achieved at low engine speeds. Within a very narrow speed range (resonance speed) the gas vibrations can be tuned in such a way as to reduce the scavenging losses and improve the volumetric efficiency.

Exhaust control with barrel valve (Fig. 1).

Exhaust control is performed with a barrel valve (power valve system). The barrel valve mounted transversally to the exhaust port has a segment-shaped cut-out with a sharp metering edge. The cross-section of the exhaust port is reduced as a function of engine speed by the rotation of the barrel valve.

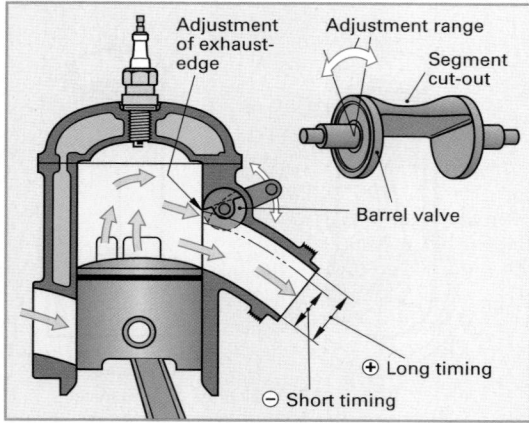

Fig. 1: Exhaust control with barrel valve

At low and medium engine speeds the top edge (metering edge) of the exhaust port is displaced downwards by the rotation of the barrel valve and the height of the exhaust-port cross-section is reduced. In this way, the exhaust control angle and

the exhaust timing are shortened to prevent fresh gas from flowing into the exhaust port. The effective stroke of the piston and the effective compression ratio are increased in the process. The barrel valve is rotated shortly before the top speed is reached so that the entire exhaust-port cross-section is opened. In this way, a larger exhaust control angle and longer exhaust timing are achieved.

The barrel valve can be positioned as a function of centrifugal force or by means of a servo-motor. The servo-motor records the number of ignition pulses as the reference variable.

Features of exhaust control

- Exhaust control by barrel valves has symmetrical timing diagrams.
- Reduced fresh-gas losses during scavenging.
- High torque and high power at low and medium engine speeds.
- Barrel valve subjected to high thermal loads, sensitive to carbon deposits.
- Poorer cooling of the cylinder wall in the exhaust area.

14.1.4 Particular design features

Crankcase

The crank chamber situated in the crankcase must be sealed to the outside so that it is pressure-proof and must be compact in design so that the necessary precompression pressure can be achieved.

Radial shaft seals as used to seal the crankshaft. In multiple-cylinder engines the crankshaft must also be sealed at the intermediate bearings. This prevents gas-exchange processes between the different cylinders and crank chambers.

Lubrication

Because the crank chamber serves to precompress the fuel-air mixture and therefore cannot be equipped with a forced-feed lubrication system, virtually all two-stroke engines have a mixture- or separate-lubrication system.

Mixture lubrication. Lube oil is mixed with the fuel (mixture ratio between 1 : 25 and 1 : 100 but usually 1 : 50). When the engine is at normal operating temperature the fuel evaporates and the lube oil precipitates on the bearings and the cylinder wall.

Total-loss lubrication (Fig. 1, Page 331). Fuel and oil are stored separately from each other in reservoirs (separate lubrication). From the oil reservoir the oil is supplied by a metering pump either to be mixed with the fuel or delivered to the intake port, where it is taken up by the fuel-air mixture. The crankshaft bearings can also be supplied directly with oil.

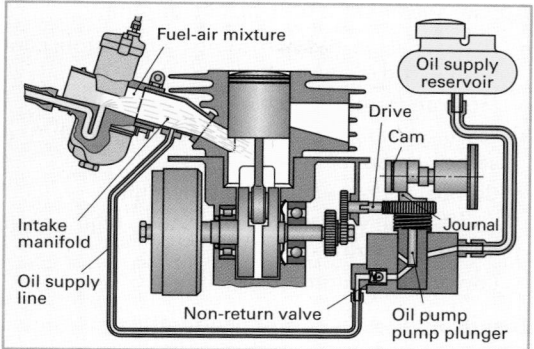

Fig. 1: Total-loss lubrication with metering pump

The pump element with pump plunger is rotated by the crankshaft and in this way the oil is delivered as a function of engine speed. The coil spring presses the plunger via the journal against the acting cam.

The throttle control alters the cam position and the oil is delivered as a function of load. Speed- and load-dependent metering help to achieve large oil savings (mixture ratio 1 : 100 and leaner).

Crankshaft and connecting rod

Antifriction bearings are used to support the crankshaft and the connecting rod.

If conventional non-split antifriction bearings (needle or roller bearings) are used, the crankshaft must be assembled from individual parts.

Piston with accessories

Because of double the number of working cycles and the control of the exhaust port, the piston in a two-stroke engine can become hotter than that in a four-stroke engine. The greater thermal expansion is compensated by larger installation clearances for the piston, the gudgeon pin and the piston rings. Two-stroke pistons incur greater wear due to the fact that they are exposed to higher levels of stress and strain.

Windows in the piston skirt (Fig. 2) partially assume the function of controlling the cylinder ports. They do however reduce the inherent stability of the piston.

Extremely light pistons are used in particularly fast-running two-stroke engines. In this way, the inertia forces that occur during operation can be kept low.

Closed gudgeon pins are used where hollow gudgeon pins would cause a bypass connection of the ports in the cylinder and thus scavenging losses.

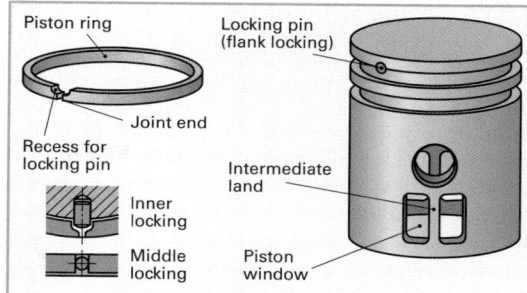

Fig. 2: Single-metal piston with windows

Wire snap rings without hook ends (detachable hooks) are used to lock the gudgeon pins axially. Hooks on account of their mass inertia could cause the snap rings to lift off in fast-running two-stroke engines (up to 16,000 rpm), thus jeopardising their secure seating in the groove.

Piston rings. Plain compression rings are normally used. Small two-stroke engines often have only one piston ring (an L-section ring) to reduce friction loss. This piston ring provides a particularly good seal due to the pressure of the combustion gases. Because of the low oil content in mixture-lubrication systems, two-stroke pistons do not have an oil control ring. Each piston-ring groove features a locking pin (Fig. 2) as an anti-rotation element.

If these locking pins were not fitted, the joint ends of the piston rings could rotate in such a way as to project and deflect into the cylinder ports and cause damage.

Cylinder

Gas passages, the openings of which are designed as "rectangular" ports, lead through the cylinder walls. The horizontal, curved port edges help the piston rings and piston to slide along without abrupt stresses.

Wide ports are interrupted by intermediate lands so that the closed piston-ring section cannot deflect too far. Carbon deposits can cause above all the exhaust ports to constrict. In this event, poor scavenging means that an ignitable fuel-air mixture is only created after every second scavenging process. The consequences of this are misfiring and therefore power losses.

Exhaust system

The scavenging process is a vibration process. The exhaust pipe with silencer and intake line with air filter must therefore be exactly matched to each other.

Alterations to the exhaust system contravene the statutory requirements and will result in the type approval being invalidated.

14.1.5 Application of two-stroke engines

The use of two-stroke engines is becoming increasingly more infrequent due to the constant tightening of exhaust-emission regulations. Even modern two-stroke engines such as the 50 cc engine with direct injection pictured below (**Fig. 1**) achieve at best Euro 2 ratings on account of their high HC emissions.

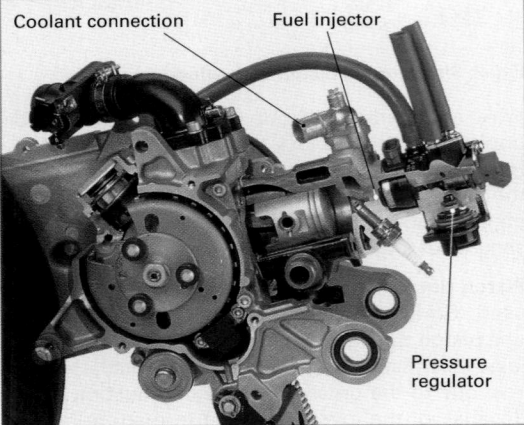

Fig. 1: Two-stroke direct-injection engine

The use of two-stroke engines will in future be confined to racing bikes in small-scale production quantities.

Advantages over a four-stroke engine:

- Simple design. Fewer moving parts (piston, connecting rod, crankshaft)
- More uniform torque, no idle strokes
- Lower-vibration, smoother running with the same number of cylinders
- Compact construction, lighter weight
- Low engine weight-to-power ratio, high power output per litre
- Low manufacturing costs

Disadvantages in comparison with a four-stroke engine:

- Poorer charge
- Exhaust-gas emissions, high HC values
- Higher thermal load, absence of idle strokes
- Lower mean piston pressures due to poorer cylinder charge
- Poor idle performance due to residual exhaust gas in the engine
- Higher specific fuel and oil consumption

WORKSHOP TASKS

- Use special two-stroke oils (self-mixing oil) only in accordance with the manufacturer's instructions and in the specified mixture ratio.
- Clean the air filter at regular intervals.
- Do not remove carbon deposits with sharp-edged objects, avoid making scratches.
- When cleaning, do not grind or sand the piston crown to a bright finish as this could cause overheating and increased carbon deposits. Do not damage the piston edges as this would alter the timing.

Faults

Drop in engine power due to e.g.

- Contaminated air filter; carbon deposits.
- Defective ventilation of the fuel tank.
- Insufficient fuel supply.

- Spark plug fouled by oil, coked or incorrect thermal value.
- Incorrect ignition-point setting.
- Poor compression.
- Leaking crank chamber.

Engine too hot due to e.g.

- Contaminated cooling fins.
- Fault in the cooling system.
- Use of excessively lean fuel-air mixture due to incorrect carburettor setting.
- Incorrect mixture ratio of fuel and two-stroke oil. Use of incorrect oil.
- Occurrence of auto-ignition due to incorrect spark plugs or carbon deposits.
- Excessive heat absorption due to a ground or sanded piston crown.

REVIEW QUESTIONS

1 What are the differences between a two-stroke and a four-stroke spark-ignition engine?

2 Why is reverse scavenging the most widely used scavenging process?

3 How does total-loss lubrication work?

4 What do you understand by an asymmetrical timing diagram?

5 What are the advantages of asymmetrical inlet control?

6 Why are there locking pins in the piston-ring grooves of two-stroke engines?

7 Why is it not permitted to change or modify anything on the induction and exhaust systems?

8 What are the advantages and disadvantages of the two-stroke spark-ignition engine in comparison with its four-stroke counterpart?

14.2 Rotary engine*

In a reciprocating engine the piston executes a reciprocating movement, which must be converted via the connecting rod in conjunction with the crankshaft into a rotary movement. In a rotary engine the rotor rotates and immediately generates rotary work during the expansion process. Here the centre of gravity of the uniformly rotating rotor describes a rotation. Greater power is achieved with the same engine weight due to the absence of acceleration and deceleration of the reciprocating masses.

The rotary engine operates in accordance with the

- **four-stroke principle,** because a closed gas exchange is in effect.
- **two-stroke principle,** because the rotor controls the gas exchange via ports in the jacket running surface and one eccentric-shaft revolution corresponds to one power cycle.

14.2.1 Design

The jacket **(Fig. 1)** is epitrochoidal in shape. A pinion permanently joined to a side section is located concentrically with the middle of the jacket.

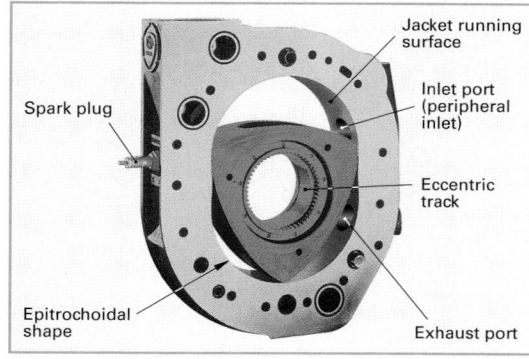

Fig. 1: Jacket with rotor

Passing through the middle of the jacket side sections is the eccentric shaft **(Fig. 2),** on the eccentric elements of which the rotors rotate. The rotor **(Fig. 3)** is sealed at all the contact surfaces.

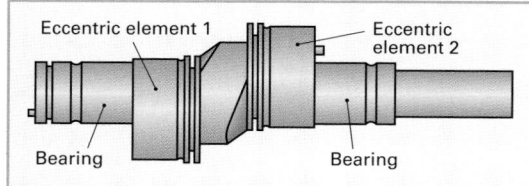

Fig. 2: Eccentric shaft of twin-rotor engine

---

* Invented by Felix Wankel, 1954

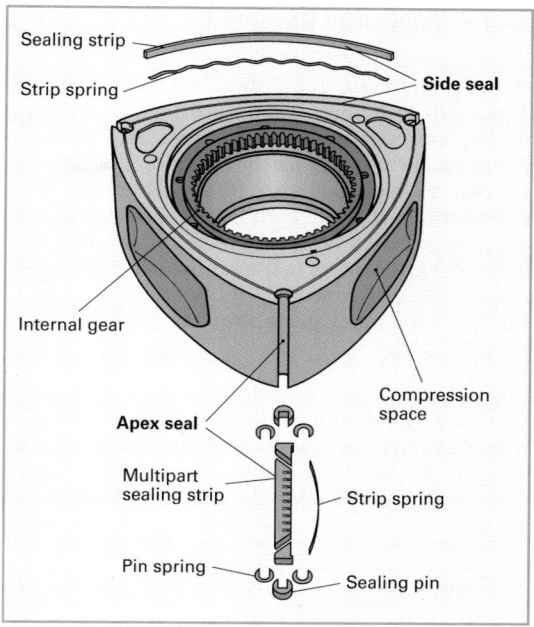

Fig. 3: Rotor with sealing elements

On one side of the rotor is an internal gear which is supported and rolls on the stationary pinion of the side section. This gear transmits no force; instead, it guides the rotor, which thus always rotates in the correct movement phase to the revolutions of the eccentric shaft and to the orbit pattern in the jacket.

The number of teeth of the stationary pinion and the internal gear of the rotor operate with a 2 : 3 conversion ratio. The rotor and the eccentric shaft rotate in the same direction, but the rotor rotates at a slower speed than the eccentric shaft.

Features of a rotary engine in comparison with a reciprocating engine:

- Excellent running smoothness since there are only rotating main parts (rotor and eccentric shaft), complete balancing of masses
- No engine-timing components
- Unconstricted gas-port cross-sections
- Fewer components, lower weight
- Low octane-number requirement
- Well suited for operation with hydrogen
- Less than optimal combustion-chamber shape, long flame paths
- High fuel and oil consumption
- High HC emission values
- Expensive seals on the rotor
- Higher manufacturing costs

14

14.2.2 Operating principle

The **rotary engine (Fig. 1)** is a three-chamber machine, the chambers of which are numbered 1, 2 and 3. The chambers increase or decrease in size during the movement of the rotor. A four-stroke power cycle takes place in all three chambers in succession in the course of three revolutions of the eccentric shaft: induction, compression, combustion and exhaust. When the rotor rotates anticlockwise, induction of the fuel-air mixture takes place in chamber 1 (a, b, c, d).

Compression takes place simultaneously in chamber 2 (a, b, c).

Ignition takes place at the end of compression (c). Then the expanding gases work in chamber 2 by rotating the eccentrically carried rotor anticlockwise (c, d). Here the rotor rests with its internal gear on the pinion permanently joined to the side section and exerts a rotary force on the eccentric shaft. The eccentric shaft in a rotary engine therefore assumes the function of the crankshaft in a reciprocating engine. Instead of acting via the connecting rod, the rotor force (rotor rotary force) acts directly on the eccentric shaft. Combustion takes place simultaneously in chamber 3 (a); then exhaust takes place (b, c, d). While the eccentric centre point ($\oplus$) rotates 270° anticlockwise (angle α), the rotor side A-B rotates only 90° in the direction of rotation (angle β). In three revolutions of the eccentric shaft, therefore, the rotor rotates only once with three power cycles.

This means that the rotor rotates forward in the jacket at only one third the speed of the eccentric shaft and accordingly lags behind the eccentric-shaft speed by two thirds. In this way, in spite of the high eccentric-shaft speeds, wear of the sealing elements, the jacket and the side sections is kept low. The speed of the eccentric shaft is specified as the engine speed.

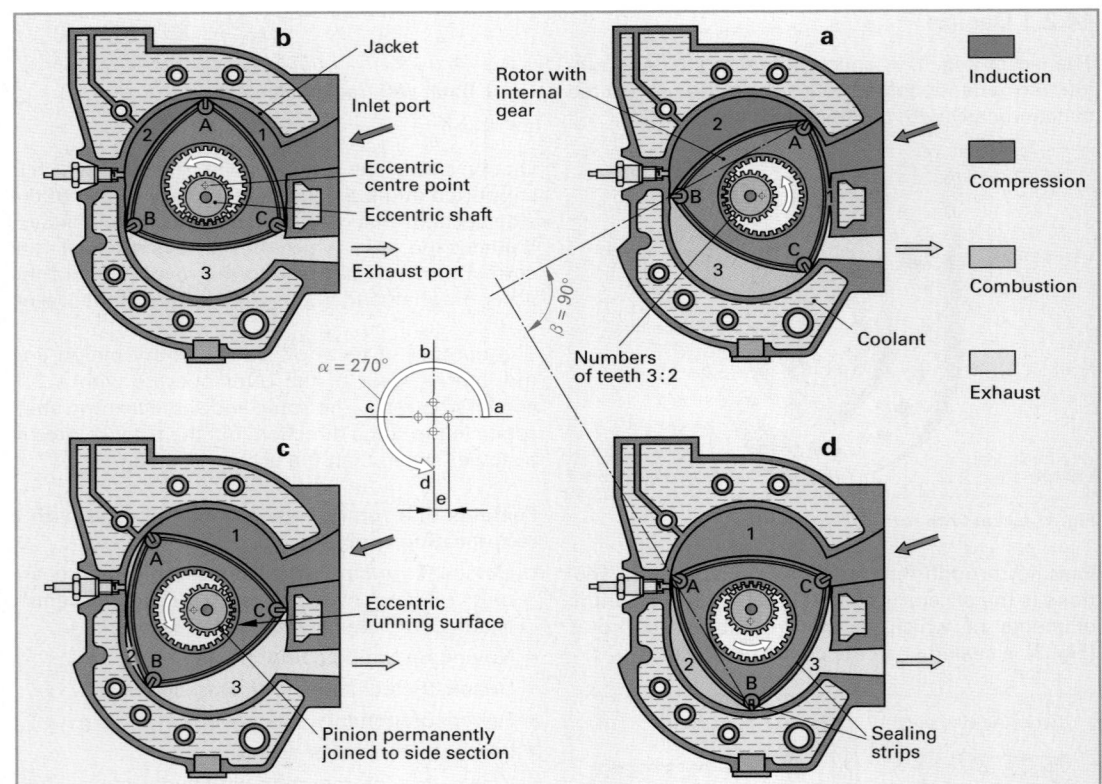

Fig. 1: Operating principle of rotary engine

REVIEW QUESTIONS

1 What are the advantages of the rotary engine?

2 How does the rotary engine differ from the reciprocating engine?

3 With which engine types is the rotary engine comparable with regard to a power cycle and the gas-exchange process?

15 Alternative drive concepts

Alternative drive concepts denote those drives which

- facilitate the use of alternative fuels, e.g. biodiesel (rapeseed oil methyl ester RME), natural gas or hydrogen, in conventional internal-combustion engines or
- provide for alternative types of drive, e.g. fuel-cell drives.

> The objective of alternative drive concepts is to reduce the consumption of fossil fuels and to minimise pollutant and noise emissions.

15.1 Alternative sources of energy

These comprise exhaustible and/or replaceable energies **(Fig. 1)**.

As well as the fuels produced from exhaustible energies (i.e. petrol and diesel), the following alternative fuels or energies can be used:

- Natural gas
- Methanol
- Electrical energy
- Hydrogen
- Fuels from biomass

Natural gas can also be used to create synthetic diesel fuels. These fuels have reduced amounts of sulphur and aromatics. The exhaust-gas quality of diesel engines is influenced to positive effect since both particulate matter and NO_x emissions are reduced.

15.2 Natural-gas drives

Natural gas is a fossil fuel consisting primarily of methane (CH_4). The percentage of actual methane contained in natural gas runs at 80 to 99 %, depending on the region where the gas is extracted. The rest is made up of carbon dioxide, nitrogen and low-grade hydrocarbons.

Natural gas can be stored in a vehicle either in liquid form at −162 °C as LNG (Liquefied Natural Gas) or in compressed form at pressures of up to 200 bar as CNG (Compressed Natural Gas). Natural gas is generally used in its compressed form because of the high cost of storing the liquefied version.

The high knock resistance of natural gas (approx. 140 RON) allows a compression of roughly 13:1. However, this advantage cannot be exploited in bivalent drives, i.e. in a combination of petrol and natural-gas drives, because the compression ratio must be tuned to petrol operation.

Advantages of natural-gas drive over spark-ignition and diesel engines:

- Very good combustion properties and low emissions of CO_2, NO_x, CO and virtually no particulate and sulphur emissions.
- Less carbon fouling of spark plugs and reduced contamination of engine oil.

Disadvantages of natural-gas drive in comparison with spark-ignition and diesel engines:

- Lower engine power due to the lower calorific value of natural gas.
- Expensive storage of natural gas necessary.
- Shorter cruising range with the same tank volume.
- Extensive safety regulations with regard to operating, servicing and repairing natural-gas vehicles.

15

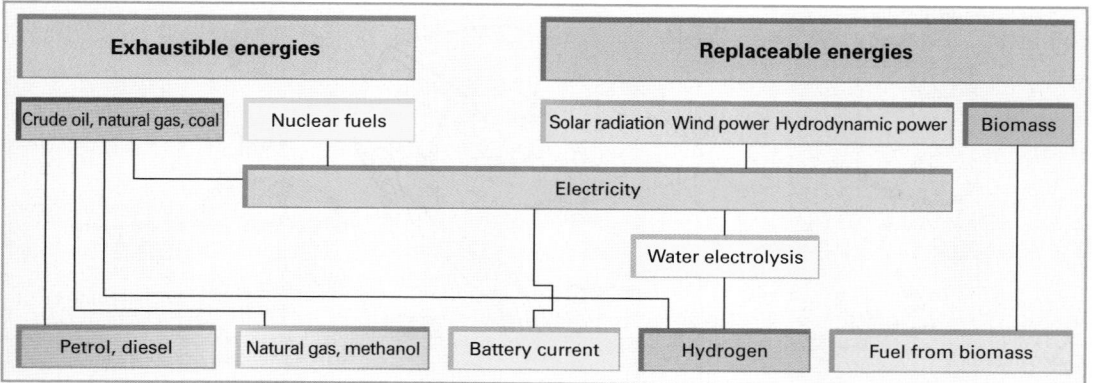

Fig. 1: Energies for driving vehicles

Design. Natural-gas drives are generally used in combination with petrol operation in spark-ignition engines (so-called bivalent drives). Various additional components must be installed in the vehicle for this purpose **(Fig. 1)**.

Operating principle. The natural gas stored at approx. 200 bar in the natural-gas tank flows to the gas-pressure regulator. This regulates the gas pressure in several reducing stages at approx. 9 bar. The gas injectors in the intake manifold are energised by the ECU as required and thereby opened. The gas mixes with the inducted air and then flows as a gas-air mixture into the combustion chamber.

Safety features.
Natural-gas drives pose certain risks to the environment, e.g. as a result of the unchecked discharge of gas or the danger of explosion caused by a rise in pressure. For this reason, these systems are equipped with various safety features.

- **Non-return valves.** These are located in the refuelling connection and on the tank shutoff valves and prevent the gas from flowing back via the refuelling valve.
- **Gastight sheathing.** This is wrapped round the lines and components routed inside the vehicle.
- **Screwed joints.** These are designed as double-clamping-ring screwed connections.
- **Natural-gas tanks.** These are made of steel or CFRP. Each tank is connected by two retainers to the vehicle. The burst pressures are approx. 400 bar for steel tanks and approx. 500 bar for CFRP tanks.
- **Solder fuse and thermal cut-out on the natural-gas tank.** These prevent an excessive pressure increase and thus the tank from exploding in the event of a fire.

- **Flow-rate limiter.** This prevents the natural-gas tank from draining suddenly in the event of a pipe breakage.
- **Electromagnetic shutoff valves.** This valve, which is mounted on the natural-gas tank, closes on changeover to petrol mode, in the event of a power failure, when the engine is stopped or in the event of an accident. A further shutoff valve is located on the pressure regulator.
- **Flexible gas lines.** These prevent breaks caused by fatigue failure on the low-pressure side, i.e. between the pressure regulator and the gas injectors.
- **Overpressure regulator.** This is mounted on the pressure regulator and protects the low-pressure side against excessive pressures.

WORKSHOP NOTES

If the high-pressure gas tank is replaced at regular intervals and the system is subjected to a service-inspection in accordance with the manufacturer's instructions, then there is no need to subject the gas system to the prescribed recurring checks.

The following components of the natural-gas system must be checked within the framework of a service-inspection:

- Natural-gas tanks and lines
- Electromagnetic shutoff valves
- Closing cap and natural-gas filler neck
- Vent lines on the natural-gas tanks

Leak tests must be carried out in accordance with the manufacturer's instructions, e.g. with a gas-leak detector.

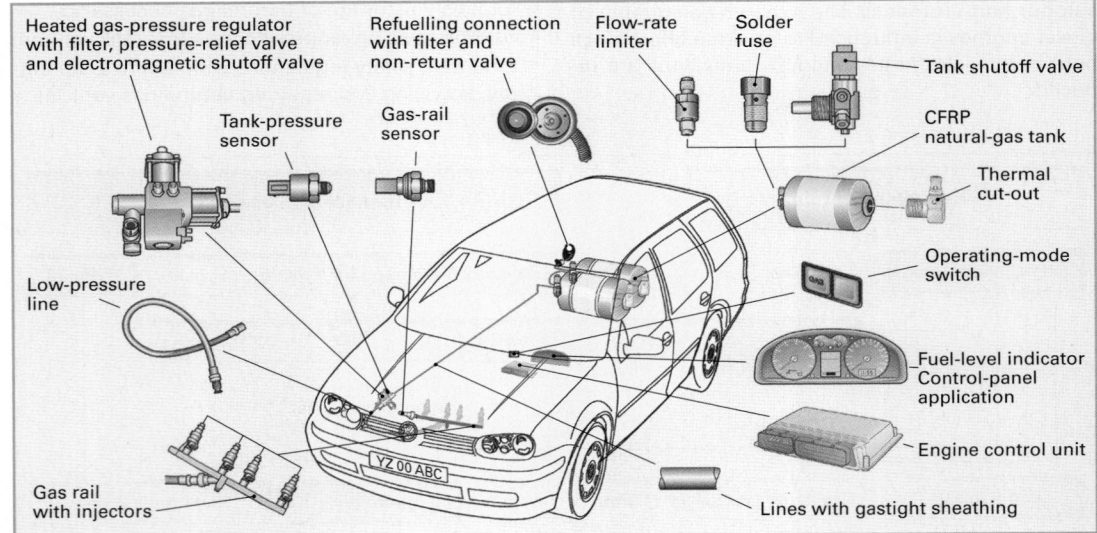

Fig. 1: Components for natural-gas operation

15.3 Hybrid drives

> Hybrid drives denote vehicle drives which have more than one drive source.

An internal-combustion engine and an electric motor are usually combined in hybrid drives.

Advantages of a hybrid system over spark-ignition and diesel engines:

- High torque at low speeds.
- Assistance to the low-torque internal-combustion engine during pulling away.
- Smaller internal-combustion engine with favourable consumption figures.
- Conversion of kinetic energy into electrical energy during braking.

In the case of hybrid drives a distinction is made between serial and parallel hybrid systems.

Serial hybrid system (Fig. 1). Here the drive is provided via an electric motor. An internal-combustion engine drives a generator for this purpose. The electrical energy generated is utilised by an electric motor as the drive. The excess electrical energy is stored in batteries and can be called on as required. An inverter converts the electrical energy generated in the generator as alternating current into direct current so that it can be stored in the battery. The inverter converts the direct current into alternating current to operate the electric motor.

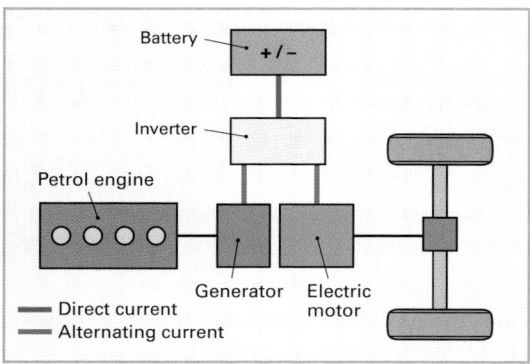

Fig. 1: Serial hybrid system (block diagram)

Parallel hybrid system (Fig. 2). Here the drive is provided either via an internal-combustion engine or an electric motor. The electric motor draws the electrical energy from a battery. The battery is charged by the electric motor in a process whereby the motor is switched over during driving with the internal-combustion engine or during braking and operates as a generator.

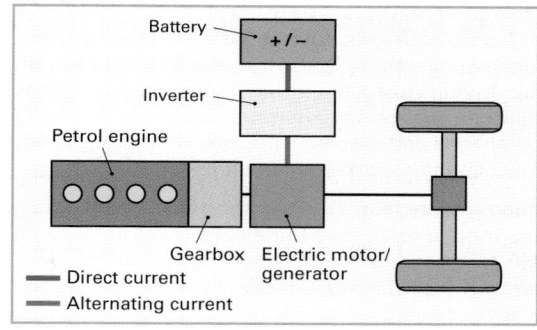

Fig. 2: Parallel hybrid system (block diagram)

Mixed hybrids

Design (Fig. 3). Conventional hybrid drives are a combination of serial and parallel systems. The hybrid drive consists for example of a spark-ignition engine and two three-phase synchronous motors with permanent magnets which can also be engaged as generators (motor generators **MG1** and **MG2**). The spark-ignition engine and the two electric motors are mechanically connected with each other via the planetary gear. MG2 and the differential for the drive wheels are combined by way of a drive chain and gearwheels.

Power is supplied by a sealed nickel-metal-hydride battery (Ni-MH) which consists of several cells combined into modules (each with a voltage of 1.2 V). Depending on the design, the HV battery (HV: Hybrid Vehicle) has a nominal voltage of 200 to 300 V. It is cooled by a fan so that the heat generated during charging and discharging can be dissipated.

The inverter converts the alternating current into direct current and vice versa. It has a separate liquid-cooling system.

The battery, the inverter and the two electric motors are connected by a heavy-current feeder cable.

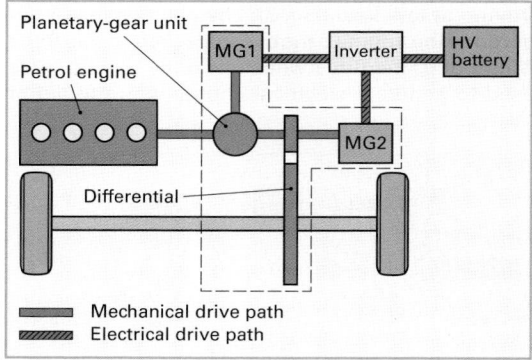

Fig. 3: Mixed hybrid (design)

15

Operating principle. The ECU for the hybrid system detects the driver's drive command via the accelerator-pedal sensor. It also receives information on the driving speed and the gearshift position. It uses this information to determine the vehicle driving conditions and regulates the motive forces of MG1, MG2 and the spark-ignition engine.

Pulling away (Fig. 1). When the vehicle pulls away, the motive force is made available exclusively by MG2. The engine remains switched off and MG1 rotates in the opposite direction without generating electrical energy.

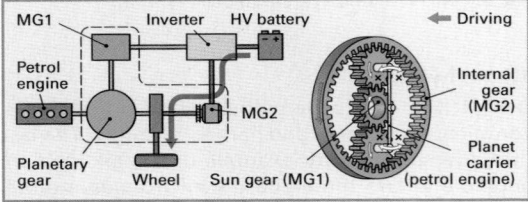

Fig. 1: Pulling away

Engine starting (Fig. 2). When the torque required for driving increases while the drive is only provided by MG2, MG1 is activated in order to start the engine. The engine is also started when for example the battery charge state or the battery temperature deviates from the prescribed level.

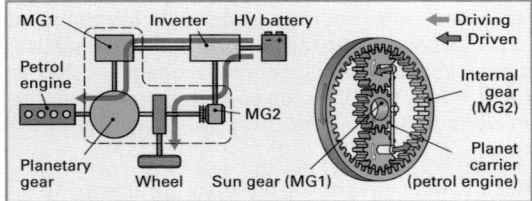

Fig. 2: Engine starting

MG1 operates as a generator once the engine has started. MG1 supplies the required electrical energy via the inverter to the HV battery.

Driving at low load (Fig. 3). The planetary gear apportions the engine's motive force. Part of this force is output to the drive wheels and the remainder is used to generate electrical energy with the aid of MG1.

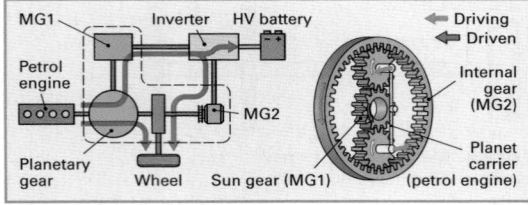

Fig. 3: Driving at low load

Acceleration at full throttle (Fig. 4). When the vehicle requires high drive power, the system increases the motive force from MG2 by drawing additional electrical energy from the HV battery.

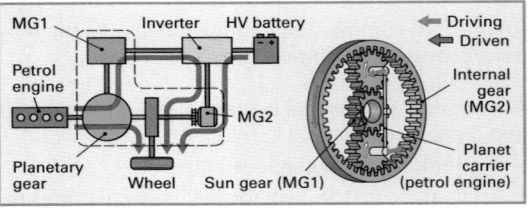

Fig. 4: Acceleration at full throttle

Deceleration (Fig. 5). The engine is switched off when the vehicle is braked. The drive wheels now drive MG2, which now operates as a generator and charges the HV battery.

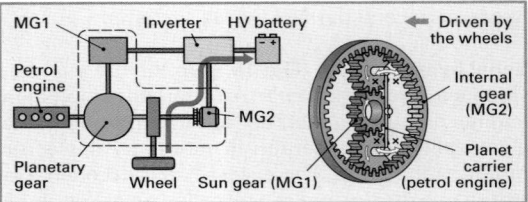

Fig. 5: Deceleration

When the vehicle is decelerated from a higher driving speed, the engine maintains a predetermined engine speed in order to protect the planetary gear.

Reversing. The drive is provided exclusively by MG2 when the vehicle is driven in reverse.

WORKSHOP NOTES

Servicing
The charge state and the charge differential of the individual modules of the HV battery must be checked in the course of a vehicle service-inspection. The charge state must not drop below a value specified by the manufacturer. An excessive charge differential of the battery modules can be compensated with the aid of a special battery servicing device.

The coolant for the inverter cooling system must be changed at regular intervals.

Safety measures
The hybrid system's high-voltage switching circuit can cause serious or even fatal injuries if it is handled improperly or negligently.

It is essential to follow the manufacturer's safety instructions when working on hybrid vehicles!

15.4 Drives with fuel cells

In fuel cells the chemically combined energy from hydrogen is converted in a combustion station directly into electrical energy (cold combustion $t \approx 80\ °C$).

Design. The core of the fuel cell consists of a proton-conducting plastic membrane (PEM: proton exchange membrane). This is coated on both sides with a platinum catalyst and graphite-paper electrodes (bipolar plates). Fine gas ports are milled into the bipolar plates; through these ports hydrogen is supplied on the one side and air or oxygen is supplied on the other.

Operating principle (Fig. 1). On one side of the fuel cell (cathode) hydrogen (H_2) is broken down by a catalyst into positive hydrogen ions (protons) and electrons. Only protons can pass through the plastic membrane (PEM: proton exchange membrane) to the other side of the cell (anode). Electrons cannot pass through the membrane. When the cathode and the anode are connected, the negatively charged electrons move to the positively charged side. This results in a flow of current which can drive a load/consumer, e.g. an electric motor. Oxygen, hydrogen ions and electrons combine on the anode.

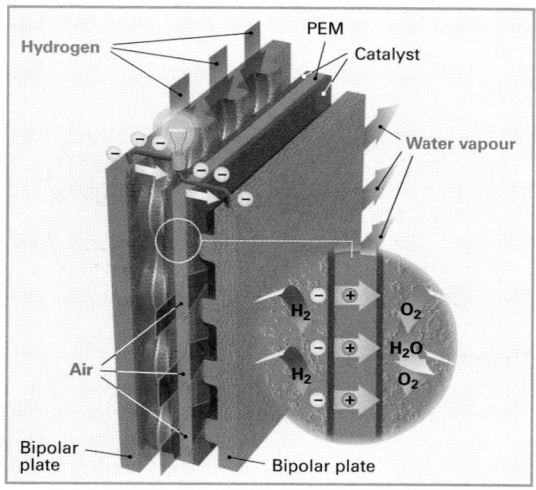

Fig. 1: **Design and operating principle of a fuel cell**

Hydrogen formation. Hydrogen can be produced by electrolysis outside the vehicle or by chemical processes on board the vehicle.

On-board hydrogen formation. The vehicle must be filled with liquid methanol (CH_3OH) for this purpose. The methanol is mixed with salt-free water, evaporated at 250 °C and converted in a reformer with a catalytic burner into hydrogen and CO_2. The purified hydrogen gas is then supplied to the fuel cell. Small quantities of CO_2 are created during the gas-purification process.

It is fundamentally possible to run spark-ignition engines on hydrogen.

15.5 Internal-combustion engines with hydrogen mode

In a hydrogen drive the chemically combined energy from hydrogen is ignited in combination with oxygen in the air by an ignition spark and converted into thermal energy (hot combustion).

Design. For this purpose spark-ignition engines use a special mixture-formation system which matches the hydrogen metering and the charge cycle to each other.

Hydrogen has a lower volume-specific calorific value than petrol. In order to be able to carry as much hydrogen as possible in the vehicle, hydrogen is liquefied and stored at approx. − 250 °C in an insulated fuel supply tank.

Operating principle. The hydrogen passes through filters, pressure reducers, shutoff valves and flow dividers to the injectors of the individual cylinders. Combustion always takes place with excess air. The additional air in the combustion chamber absorbs heat. The low combustion temperature prevents nitrogen oxides (NO_x) from forming to a large extent. The hydrogen engine operates without virtually any emissions.

15.6 Internal-combustion engines with vegetable-oil mode

Vegetable oil. It is fundamentally possible to use vegetable oil, e.g. rapeseed oil, in motor vehicles. However, the exclusive use of vegetable oil requires the engine to undergo expensive conversion work because this fuel has a lower cetane number and is more viscous than diesel fuel. In addition to installing a fuel preheater and an electrically heated auxiliary cleaner, it is necessary, depending on the engine type, also to modify certain components in the fuel-injection system.

Biodiesel (RME). Transesterification of rapeseed oil with the aid of methanol creates **R**apeseed oil **M**ethyl **E**ster (RME). This fuel, which is called biodiesel, has similar properties to conventional diesel fuel and is equally thin-bodied. Biodiesel behaves aggressively to plastics in seals, pipelines and injection pumps. For this reason, it must only be used in engines which have been approved by the manufacturer for biodiesel use.

15

16 Drivetrain

The drivetrain of a motor vehicle consists of the clutch, the gearbox, the propeller shaft, the final drive with differential and the drive shafts of the wheels **(Fig. 1)**. Together, these components can also be referred to as the powertrain or transmission line.

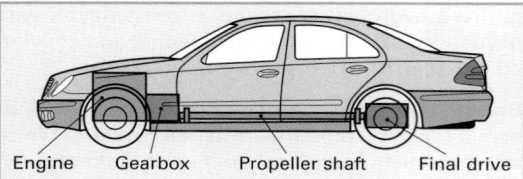

Engine   Gearbox   Propeller shaft   Final drive

Fig. 1: Drivetrain of a passenger car with front-engine drive

Functions

● Converts engine torque and engine speed
● Transmits torque to the drive wheels

Transmission losses are unavoidable in the above processes with the result that the power at the drive wheels is always less than the engine power (the overall efficiency of the drivetrain ranges between 92 % and 95 %).

16.1 Types of drive

The following different types of drive may be used in passenger cars and commercial vehicles: rear-wheel drive, front-wheel drive and all-wheel drive.

16.1.1 Rear-wheel drive

Front-engine drive

The engine is situated usually above or directly behind the front axle **(Fig. 1)** or, less commonly, in front of the front axle (overhang engine). The drive is effected via a propeller shaft to the rear wheels. The layout of the final drive on the rear axle results in a favourable distribution of weight between the front and rear axles. The cornering performance of a vehicle with such an engine is slightly understeering. A propeller-shaft tunnel must be incorporated in the vehicle interior to accommodate the propeller shaft between the variable-speed gearbox and the final drive.

Transaxle drive

This is a particular feature of the front-engine drive, in which the gearbox is situated on the rear axle. This layout results in a uniform distribution of weight between the two axles (50 % : 50 %) and a neutral cornering performance.

Rear-engine drive

Rear engines are situated above or behind the driven rear axle **(Fig. 2)**.

When an opposed-cylinder engine is used, the engine and gearbox take up a small amount of space inside the vehicle. Rear engines are rarely used because of the following factors: limited luggage-compartment size, problematic housing of the fuel tank, side-wind sensitivity and a tendency to oversteer when cornering.

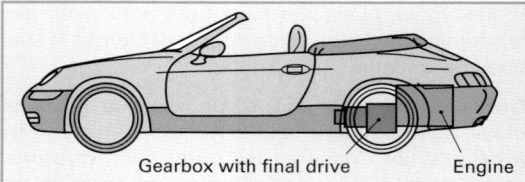

Gearbox with final drive   Engine

Fig. 2: Rear-engine drive

Mid-engine drive

This is used in sports and racing cars. The engine is situated in front of the rear axle **(Fig. 3)**. This layout results in an improved distribution of weight between the two axles and a favourable centre of gravity, which in turn contributes to a neutral cornering performance. On the minus side, the engine is difficult to access and the number of seats is limited.

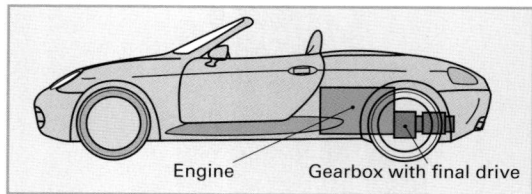

Engine   Gearbox with final drive

Fig. 3: Mid-engine drive

Underfloor-engine drive

This is particularly suitable for buses, coaches and heavy goods vehicles **(Fig. 4)**. The centrally situated, low-lying engine contributes to a favourable centre of gravity and a uniform axle-load distribution. Other advantages are the good utilisation of the vehicle interior in buses and coaches and the accessibility of the engine from below.

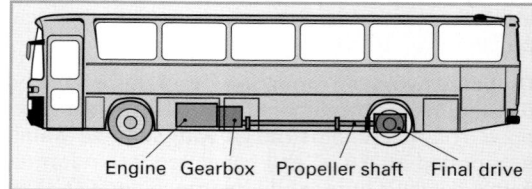

Engine   Gearbox   Propeller shaft   Final drive

Fig. 4: Underfloor-engine drive

16.1.2 Front-wheel drive

In the case of front-wheel drive, the engine is situated either in front of, above or behind the front axle **(Fig. 1)**.

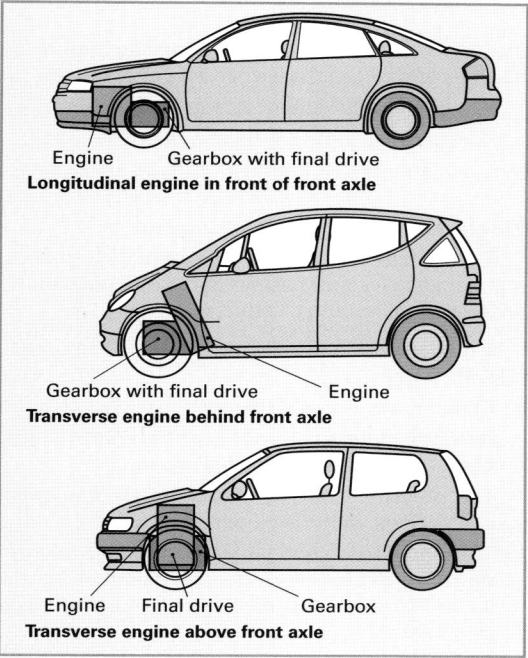

Engine   Gearbox with final drive
Longitudinal engine in front of front axle

Gearbox with final drive   Engine
Transverse engine behind front axle

Engine   Final drive   Gearbox
Transverse engine above front axle

Fig. 1: Front-wheel drive

Engine, clutch, variable-speed gearbox, final drive and differential are combined to form a single block (front drive assembly).

Advantages:
- Lower vehicle weight.
- Shortest path for the torque to cover from the engine to the drive wheels.
- No propeller-shaft tunnel.
- Large luggage compartment.
- When the engine is transversally mounted, simple final drive (spur gears), smaller front overhang and large useable footwell.
- Good straight-running stability since the vehicle is pulled and not pushed.

Disadvantages:
- Understeering when cornering at high speed.
- Unfavourable distribution of weight between front and rear axles.
- Greater tyre wear on the drive wheels.

> Vehicles with understeering behaviour drive through bends/curves with a greater radius than that corresponding to the steering angle.

16.1.3 All-wheel drive

Permanent all-wheel drive

Both axles are permanently driven. In a passenger car with front-wheel drive, the final drive of the rear axle is driven via a transfer box by a propeller shaft. A centre differential balances speed differences between the front and rear axles. This arrangement eliminates distortions in the drivetrain as well as wear in the drive and on the wheels.

Conventional switchable all-wheel drive

A propeller shaft leads to the final drives at both front and rear via a transfer box, which is flanged to the variable-speed gearbox **(Fig. 2)**. Normally, it is the rear axle which is driven while the front axle is engaged where necessary. The differentials can also be equipped with differential locks. If there is no centre differential, the all-wheel drive must not be activated on dry roads. Overrunning hubs on the front wheels prevent the drive and propeller shafts from also rotating when the axle is deactivated.

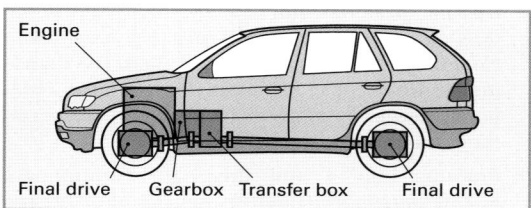

Engine

Final drive   Gearbox   Transfer box   Final drive

Fig. 2: Off-road vehicle with all-wheel drive

16.1.4 Hybrid drive

In the case of a hybrid drive*, two different power plants combined in one vehicle assume the drive function, e.g. a diesel engine for long-distance driving and an electric motor for emission-free urban driving **(Fig. 3)**. The electric motor is operated by batteries, which are charged by the 220 V system or partially during driving by the diesel engine. The electric motor acts as an alternator here. It is also possible to switch between the drive modes during driving without any problems.

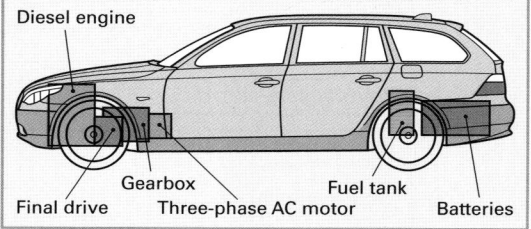

Diesel engine

Gearbox   Fuel tank
Final drive   Three-phase AC motor   Batteries

Fig. 3: Passenger car with hybrid drive

---

* hybrid (Greek/Latin) = product of mixing two or more different things

16

16.2 Clutch

The clutch is arranged in the drivetrain of a motor vehicle as an engaging and disengaging link between the engine and the variable-speed gearbox.

Functions

- **Transmits engine torque to the variable-speed gearbox.** The required torque must be directed to the variable-speed gearbox over the entire useful engine-speed range for all driving situations.
- **Facilitates smooth, jolt-free starting.** During starting, speed matching is performed between the rotating flywheel and the stationary gearbox input shaft by means of sliding friction (slip).
- **Facilitates swift, fault-free gear shifting.** In order to establish the synchronism of the gearbox components to be shifted, it is necessary for the frictional connection between the engine and the manual gearbox to be interrupted.
- **Damps torsional vibrations.** The rhythmic sequence of the engine's idle strokes gives rise to torsional vibrations at the crankshaft. The damping device in the clutch plate damps these vibrations. Gearbox noises, such as e.g. gearbox rattling, are thereby minimised.
- **Protects engine and drivetrain components against overload.** The transmission of excessive torques which occur, for example, when the engine is blocked is prevented by slip.

The following clutches are used as starting and interrupting clutches:

Starting and interrupting clutches
Friction clutch
Coil-spring clutch
Single-plate clutch
Multi-plate clutch
Diaphragm-spring clutch
Single-plate clutch
Double-plate clutch
Twin clutch
Multi-plate clutch
Centrifugal clutch
Magnetic-particle clutch

16.2.1 Friction clutch

Friction clutches transmit the engine torque by means of frictional forces from the engine to the input shaft of the variable-speed gearbox.

The transmissible clutch torque is dependent on the pressing forces. These can be generated by …

- … a central diaphragm spring (disc spring).
- … cylindrical coil springs (e.g. 6 or 9).
- … several flyweights.

Single-plate diaphragm-spring clutch

This is installed in both passenger cars and commercial vehicles and has almost completely superseded the coil-spring clutch.

It is made up of the following main components **(Fig. 1):**

- **Clutch cover** (housing) with

  pressure plate, diaphragm spring, spacer studs, fulcrum rings, tangential leaf springs.

  The pressure plate is connected via tangential leaf springs with the clutch cover.

  The diaphragm spring is supported on 2 fulcrum rings, which are secured by several spacer studs. It acts like a two-end lever with the fulcrum rings as a bearing.
- **Clutch plate** with clutch linings (2 friction-lining rings), lining carrier and hub.
- **Clutch operator** with release bearing.

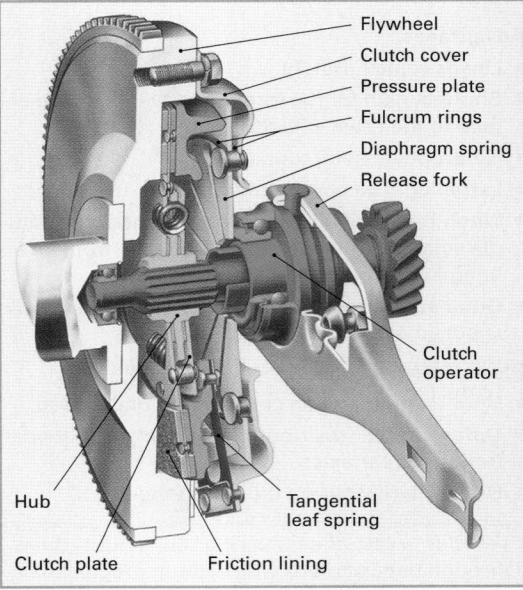

Flywheel
Clutch cover
Pressure plate
Fulcrum rings
Diaphragm spring
Release fork
Clutch operator
Hub
Tangential leaf spring
Clutch plate   Friction lining

Fig. 1: Single-plate diaphragm-spring clutch

16

Operating principle

Clutch engaged. The diaphragm spring is tensioned and presses the friction linings of the clutch plate onto the flywheel friction surface.

The generated frictional forces result in a torque at the middle rotary-force radius which is output via the clutch-plate hub to the gearbox input shaft.

The transmissible torque is dependent on the
- pressing force of the diaphragm spring
- coefficient of friction of the friction pairing
- effective rotary-force radius
- number of friction linings

Clutch disengaged (Fig. 1). When the clutch is disengaged, the clutch operator is pressed by the release lever against the inner edge of the diaphragm-spring tongues. In this way, the diaphragm spring is tipped between the fulcrum rings and the load is removed from the pressure plate. The tangential leaf springs, which connect the pressure plate and the clutch cover with each other, cause the pressure plate to lift off the lining. The power flow is interrupted, thereby creating a ventilation clearance between the linings and the contact surfaces.

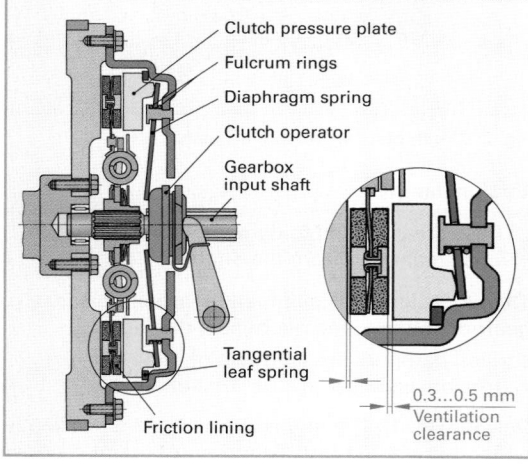

Fig. 1: Diaphragm-spring clutch with pressed release

Pressed release (Fig. 2). When the pedal is actuated, the diaphragm-spring tongues are **pressed** by the clutch operator in the direction of the flywheel. The two-end lever between the fulcrum rings moves the outer diameter area of the diaphragm spring in the opposite direction to the flywheel.

Pulled release (Fig. 2). The diaphragm-ring tongues engage in an all-round groove of the clutch operator and are **pulled** by the latter in the direction of the gearbox. The one-side lever produces a greater lever ratio a:b at the diaphragm spring and thus a smaller release force.

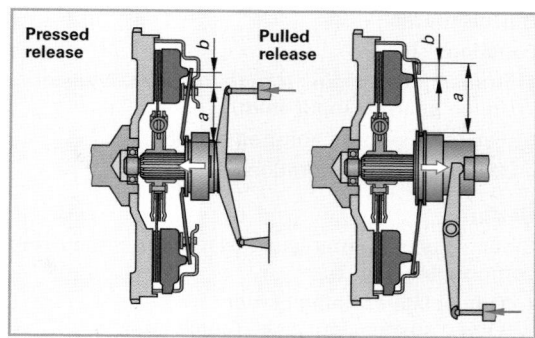

Fig. 2: Pressed and pulled release

Characteristic curves of coil-spring and diaphragm-spring clutches

In the diagram (**Fig. 3**) ...
- ... the pressing force of the pressure plate is plotted against the pressure-plate travel for the engaged state.
- ... the release force at the release bearing is plotted against the release travel.

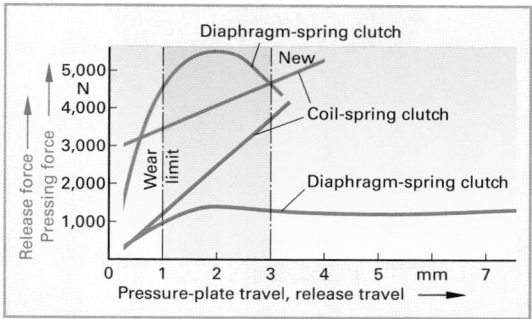

Fig. 3: Pressing force, release force

Coil-spring clutch

Pressing force. This decreases on a linear basis as lining wear increases.

Release force. This increases on a linear basis as the release travel increases.

Diaphragm-spring clutch

Pressing force. This increases initially with lining wear and drops off continuously. A marked drop in the pressing force can be observed from the wear limit.

Release force. The power required to release the clutch increases initially on a linear basis like the coil-spring clutch. The release force is reduced after the diaphragm spring is "tipped".

Features of the diaphragm-spring clutch
- Simple design
- Low actuating forces
- Pressing force is virtually independent of lining wear

Clutch plates

Functions

- Transmit the engine torque from the flywheel to the gearbox input shaft.
- Facilitate smooth, jolt-free starting.
- Damp torsional vibrations.

Design

Clutch plates primarily consist of the following components **(Fig. 1)**:

- Drive plate as lining carrier
- Lining suspension   • Clutch lining
- Hub with flange   • Torsion damper

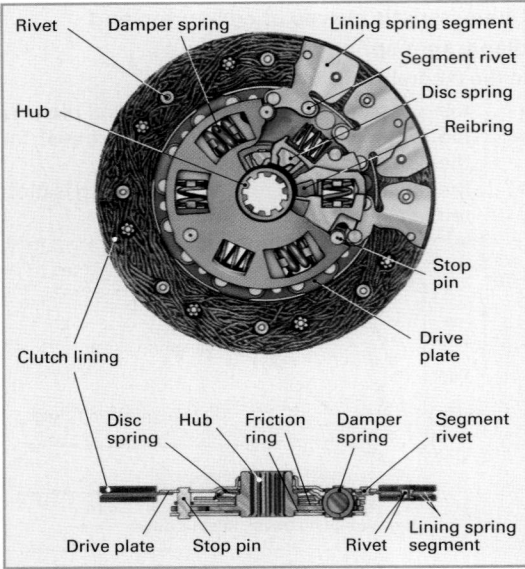

Fig. 1: Design of a clutch plate

Clutch linings

These are used as friction partners between the flywheel and pressure-plate friction surfaces.

They are required to demonstrate the following properties:

- Good heat resistance.
- High resistance to wear.
- High coefficient of friction which remains constant over as wide a temperature range as possible.

Types of clutch linings

Organic friction linings. These are made up of

- glass fibres, or aramide or carbon fibres
- inserts, e.g. copper or brass wires
- binders, e.g. phenolic resins
- fillers, e.g. soot, glass beads, barium sulphate

These organic linings are used in dry clutches for passenger cars and commercial vehicles.

Paper friction linings (wet-running paper). These are composed of wood or cotton fibres, carbon and glass fibres which are connected with each other by means of artificial resin (usually epoxy or phenolic). Application: wet-running multi-plate clutches in motorcycles.

Sintered-metal friction linings. These are composed of different metals (copper, iron) or metal alloys (bronze, brass). Hard friction components (e.g. metal oxides) and graphite are used as aggregates.
Properties: outstanding thermal resistance, good limp-home characteristics and good resistance to wear.
Application: wet clutches, e.g. multi-plate clutches in motorcycles and automatic gearboxes.

Sintered pads (Fig. 2) are friction linings of sintered metal with a high ceramic content, e.g. aluminium oxide.
Application: dry clutches in special vehicles, e.g. track-type vehicles and competition cars in motor sport.

Fig. 2: Clutch plate with sintered pads

Torsion dampers

> These are intended to damp the torsional vibrations emanating from the engine.

This is designed to eliminate gearbox noises, e.g. gearbox rattling, and gearwheel damage.

Torsion dampers are made up of
- torsion suspension   • friction device

Design (Fig. 1). The hub is pivoted and is resiliently supported via the hub flange and several damper springs against the drive plate and the counterplate. Under load, limited torsion is possible between the hub and the lining-carrying plate section.
The torque determined by the damper springs must be greater than the maximum engine torque in order to prevent the hub flange from striking against the stop pins.

Friction device. This is located in the hub section and consists of friction rings, disc spring, spring washer and support plate (see **Fig. 1**).
In the event of torsion, the friction device damps the torsional vibrations that occur by means of friction. The axial pressing force required for friction is achieved by the disc spring.

Lining suspension

> This facilitates smooth, jolt-free starting and is located between the friction linings.

The axial suspension is designed in such a way that the linings engage softly during starting and are in level contact when the clutch is fully engaged.

Clutch operator

> Its function is to interrupt the power flow between the engine and the gearbox via clutch pedal, cable or hydraulic control.

Two different types may be used: centrally guided mechanical and hydraulic clutch operators.

Centrally guided mechanical clutch operator. This is guided centrally on the gearbox input shaft by the guide sleeve mounted centrally on the gearbox **(Fig. 1)**.

The release bearing is an encapsulated ball bearing with a fixed outer race and a travelling inner race which rests on the diaphragm-spring tongues.

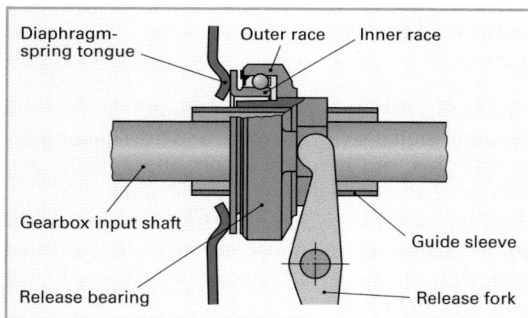

Fig. 1: Centrally guided mechanical clutch operator

Centrally guided hydraulic clutch operator

This combines with the slave cylinder of a hydraulic clutch-control mechanism to form a single structural unit and is mounted on the inside of the clutch housing **(Fig. 2)**.

Operating principle

The pressure piston is pressurised via the cup seal and the intermediate ring by the fluid pressure coming from the master cylinder and displaced on the guide sleeve.

In this way, the release bearing and the thrust ring are pressed against the diaphragm-spring tongues. The diaphragm spring tips and the pressure plate lifts off, thereby disengaging the clutch.

When the clutch is not actuated (no fluid pressure), the preload spring subjects the release bearing to an initial tension. In this way, the release bearing rests on the diaphragm-spring tongues and compensates clutch play.

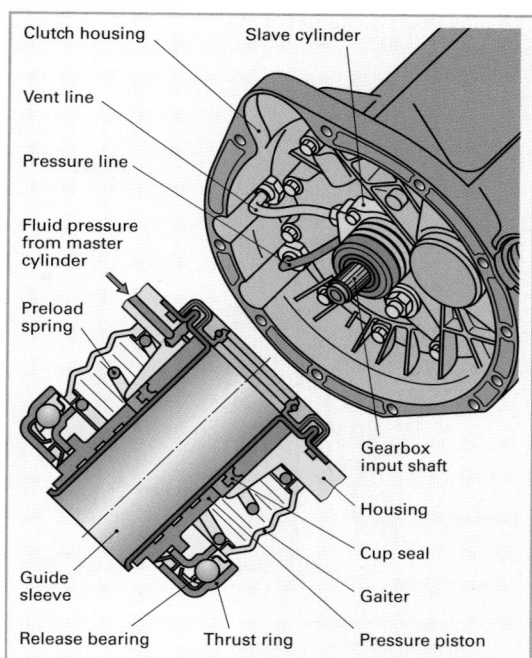

Fig. 2: Centrally guided hydraulic clutch operator

Clutch control

> The function of the clutch control is to boost the pedal force applied by the driver and transmit this force to the clutch operator.
>
> Clutches can be mechanically or hydraulically actuated.

Mechanical clutch control

Leg/foot force is transmitted via clutch pedal, cable or linkage and release lever to the clutch operator **(Fig. 1, Page 346)**.

The lever ratios in the clutch pedal and release lever are configured in such a way that the leg/foot force required for clutch disengagement is not too high and the pedal travel is not too long.

Mechanical clutch control without self-adjustment

In the case of this clutch control, there is play of 1 mm ... 3 mm between the clutch operator and the diaphragm-spring tongues and play of 10 mm ... 30 mm at the clutch pedal.

As the lining wears out, so the pressure plate moves towards the flywheel. The diaphragm-spring tongues of the diaphragm spring acting as a two-end lever stray outwards in the direction of the clutch operator. This reduces the play at both the clutch operator and the clutch pedal.

The play must be adjusted in good time as it disappears completely as lining wear increases and the diaphragm-spring tongues would in this event contact the clutch operator.

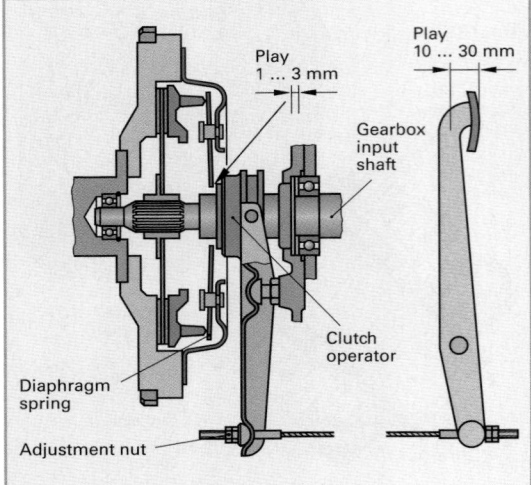

Fig. 1: Mechanical clutch control

Effects of excessively low clutch play:

- Slipping of the clutch due to the low pressing force of the diaphragm spring
- Overheating of the clutch linings
- Burning out of the diaphragm spring
- Scoring of the diaphragm-spring tips
- Overheating of the friction surface at certain points of the flywheel

The play is adjusted either at the release lever or at the clutch pedal by turning an adjustment nut.

Mechanical clutch control with automatic cable adjustment

In the event of lining wear, the clutch play between the clutch operator and the diaphragm-spring tongues is automatically set to zero by an adjusting device in the clutch cable.

The cable-adjusting device is located in the clutch cable between the pedal and the clutch operator.

Passenger-car diaphragm-spring clutch with automatic adjustment (SAC) (Fig. 2)

The **SAC** clutch (**S**elf-**A**djusting **C**lutch) adjusts itself automatically in the event of lining wear.

In contrast to a conventional diaphragm-spring clutch, the release forces, pedal forces and pressing forces remain constant over a greater wear travel

on the part of the clutch linings. This may increase the service life of the clutch.

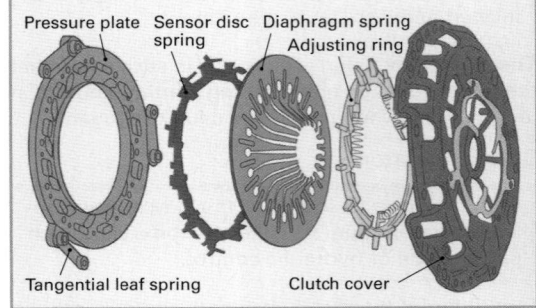

Fig. 2: Components of self-adjusting clutch (SAC)

Particular feature: The diaphragm-spring bearing is not permanently riveted to the clutch cover; instead, it is supported so that it pivots via the sensor disc spring and the adjusting ring.

Operating principle of SAC (Fig. 3)

As the lining wears, so the pressure plate moves towards the flywheel.

If the retaining force at the bearing point of the sensor disc spring is exceeded during clutch disengagement, the spring yields in the direction of the flywheel until the release force and the sensor-disc-spring force are identical again. The annular gap created is compensated by the adjusting ring.

If, during repairs to a damaged SAC, only the clutch plate is replaced, the adjusting ring must be turned back with the aid of a device so that the specified pressing force is achieved again.

The adjusting ring is already turned back on new SACs.

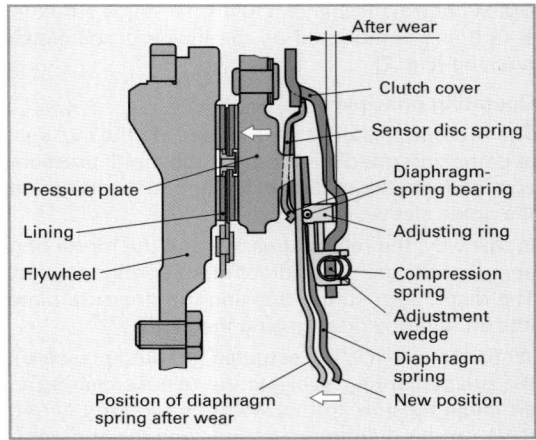

Fig. 3: Self-adjusting clutch (SAC)

Hydraulic clutch control

> This is intended to boost hydraulically the transmitted pedal force via a master cylinder and a slave cylinder and direct it to the clutch operator.

The hydraulic section consists of:
- Master cylinder
- Hose
- Hydraulic fluid
- Tubing
- Slave cylinder

Operating principle

Clutch disengagement. The leg/foot force is transmitted via the clutch pedal and linkage to the master-cylinder piston. In the process, the leg/foot force can be supported from a specific pedal travel by an over-centre helper spring.

The fluid pressure generated in the pressure chamber of the master cylinder propagates in the tubing and in the connecting hose. It exerts a force on the slave-cylinder piston, which actuates the clutch operator via a tappet and a lever and disengages the clutch.

Clutch engagement. The diaphragm spring forces the pistons of the slave and master cylinders back into their initial positions via the release bearing, release lever and tappet.

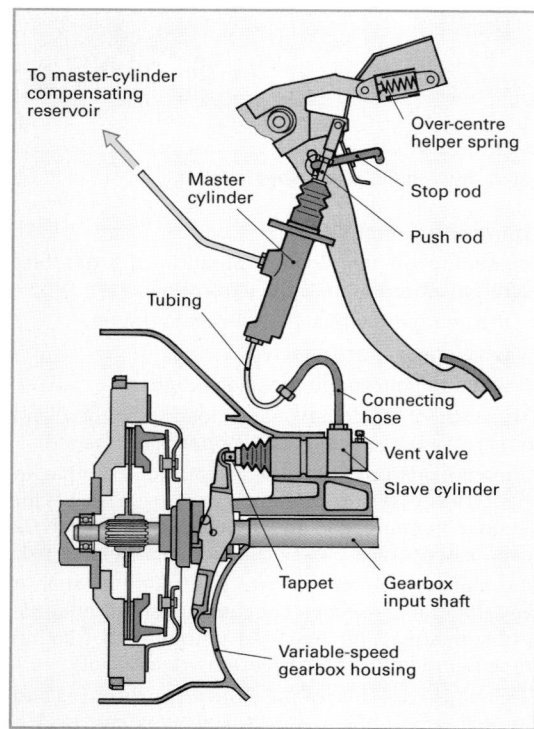

Fig. 1: Hydraulic clutch control

Advantages of hydraulic over mechanical clutch control
- Easier bridging of large distances between pedal and clutch
- Boosting of pedal force by hydraulic transmission possible
- Virtually loss-free transmission of force

Master cylinder (Fig. 2)

> Its function is to generate the fluid pressure for the hydraulic system.

This piston is designed as a double piston with primary and secondary cup seals.

The primary cup seal seals the pressure chamber while the secondary cup seal provides an outward seal. The chamber between the two cup seals is connected with the expansion tank by way of the balancing port.

When the piston is in its initial position, volume compensation is possible between the pressure chamber and the expansion tank via the balancing port.

The fluid pressure starts to build up in the pressure chamber as soon as the piston passes over the balancing port.

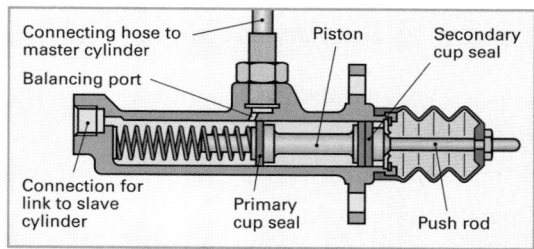

Fig. 2: Master cylinder

Slave cylinder (Fig. 3)

> Its function is to transmit the fluid pressure generated in the master cylinder as force to actuate the clutch operator.

This comprises
- housing
- piston with seal
- vent valve
- push rod

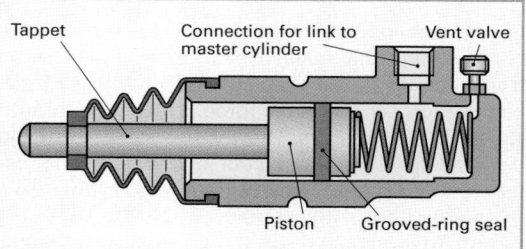

Fig. 3: Slave cylinder

16

16.2.2 Double-plate clutch

A double-plate clutch can transmit twice as much torque as a single-plate clutch with the same pressing force and the same lining diameters.

Design (Fig. 1)

The double-plate clutch comprises two clutch plates arranged in succession, a driving disc, a diaphragm spring and a release device.

Both clutch plates are connected with torsional resilience via their splined hubs with the gearbox input shaft.

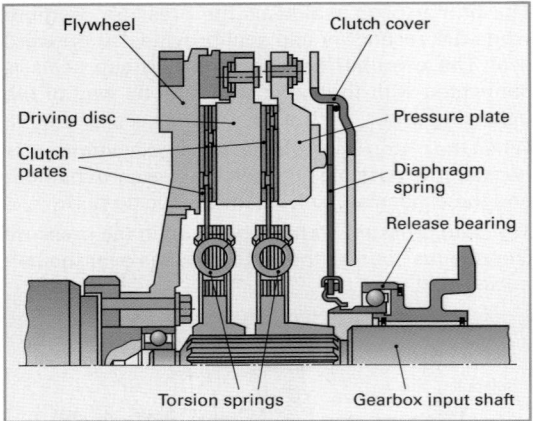

Fig. 1: Double-plate clutch

Operating principle

Clutch engaged (Fig. 1). The pressing force of the tensioned diaphragm spring presses the friction surfaces of the pressure plate, gearbox-side clutch plate, driving disc, engine-side clutch plate and flywheel against each other.

The total drive force effects transmission of the torque.

Power flow. The engine torque is directed via the flywheel, clutch cover, pressure plate and driving disc to the friction linings of the two clutch plates and via their hubs to the common gearbox input shaft.

Clutch disengaged. When the driver steps on the clutch pedal, the release bearing is pulled away from the hub of the clutch plate. The release force counteracts the tension force of the diaphragm spring in the process. The friction surfaces of the clutch plates lift off the friction surfaces of the pressure plate, driving disc and flywheel. In this way, the power flow is interrupted.

Double-plate clutches are used in heavy commercial vehicles.

16.2.3 Twin clutch

A twin clutch facilitates swift gear shifting without interrupting the traction force in automated manual gearboxes, for example, in a direct-shift gearbox (DSG).

Design (Fig. 2)

The twin clutch consists of two independently operating single clutches **C1** and **C2** which are accommodated in a housing. The hubs of the clutch plates of **C1** and **C2** are connected with two gearbox input shafts IS1 and IS2.

The fixed gear wheels for 1st, 3rd and 5th gears and those for 2nd, 4th and 6th gears are situated on IS1 and IS2 respectively.

The gearbox output shafts accommodate the idler gear wheels of the individual gears, which are shifted by selector sleeves.

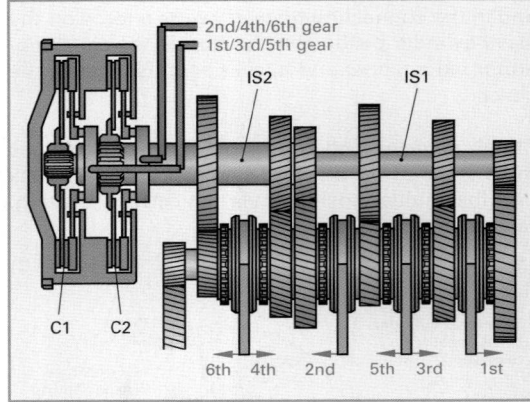

Fig. 2: Twin clutch, direct-shift gearbox

Operating principle

Depending on the driving situation, the gearbox software stored in the ECU determines

- the gear with which the vehicle is driven,
- which gear is preselected,
- when to shift to the preselected gear.

The clutch for the driven gear is closed (engaged) and that for the preselected gear is open (disengaged).

When the gearbox software decides, for example, to change from 1st to 2nd gear, clutch C1 responsible for 1st gear is disengaged and simultaneously clutch C2 responsible for the preselected 2nd gear is engaged.

The clutching processes are controlled in such a way that the respective clutches are simultaneously disengaged and engaged without the traction force being interrupted (overlapping control).

The release devices for the clutching processes and the actuations of the selector sleeves can be hydraulically or electrically controlled.

16

16.2.4 Multi-plate clutch

> Multi-plate clutches operate for example in an oil bath (wet clutch). They are used among others in motorcycles and automatic gearboxes.

Design (Fig. 1)
Several clutch plates are arranged in succession alternately as driving externally toothed plates (friction plates) and as driven internally toothed plates (steel plates). They usually operate in an oil bath.

The externally toothed plates engage the grooves of the cage while the internally toothed plates engage the external toothing of the hub. The pressure plate accommodates several compression springs and is connected via the hub with the gearbox input shaft.

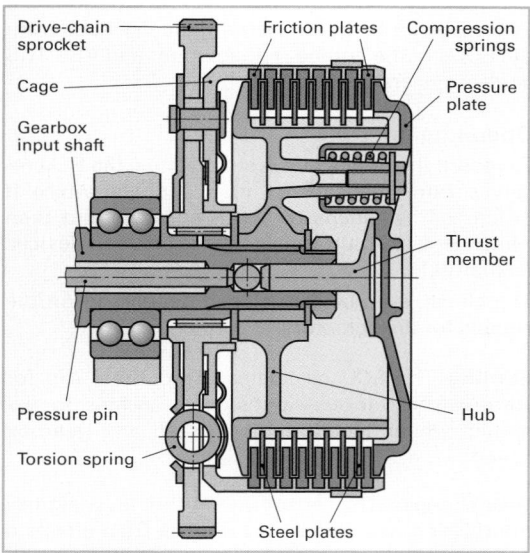

Fig. 1: Multi-plate clutch

16.2.5 Magnetic-particle clutch

> Magnetic-particle clutches are used in infinitely variable automatic CVT transmissions (e.g. automatic pushbelt) in low-power passenger cars as starting clutches.

Design (Fig. 2)
The clutch plate incorporates a solenoid coil which is connected to the main circuit.

Fine iron powder is situated in the annular orifice between the inner side of the drive plate (outer rotor) and the ring grooves on the outer circumference of the clutch plate (inner rotor).

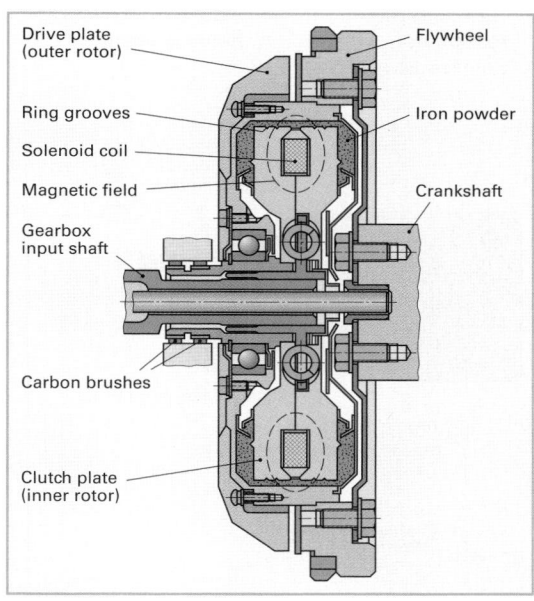

Fig. 2: Magnetic-particle clutch

Clutch engaged. The compression springs press the pressure plate and the externally and internally toothed plates together by adherence.
The externally toothed friction plates carry the internally toothed steel plates by friction. In this way, the clutch cage and clutch hub are connected.

Power flow. The engine torque is transmitted via the clutch cage, externally toothed friction plates, internally toothed steel plates, pressure plate and hub to the gearbox input shafts.

Clutch disengagement. The clutch operator presses via the pressure pin and the thrust member against the pressure plate. The pressure plate is lifted off the clutch plates against the force of the compression springs. The power flow is interrupted.

Operating principle
Current is supplied to the solenoid coil in order to establish a frictional connection between the inner and outer rotors. The amount of supplied current is electronically controlled by a control unit as a function of engine speed, driving speed and accelerator-pedal position.

The torque transmitted by the magnetic-particle clutch is dependent on the intensity of the electromagnetic field between the clutch plate (inner rotor) and the drive plate (outer rotor). The field intensity is determined by the supplied current.

Power flow. This passes from the flywheel via the drive plate, iron powder and clutch plate to the gearbox input shaft.

16.2.6 Automatic clutch system ACS

> This is an automatic clutch system in which opening of the clutch (disengagement) and closing of the clutch (engagement) are initiated by sensor signals.

The clutching operation by the driver is not applicable, thereby is no need for a clutch pedal.

Sensor signals which influence the control process are:

- Ignition switch
- Engine speed
- Gear recognition
- Gearshift-intention recognition
- Release travel
- Accelerator-pedal position
- Driving speed
- ABS/TCS signals
- Release speed

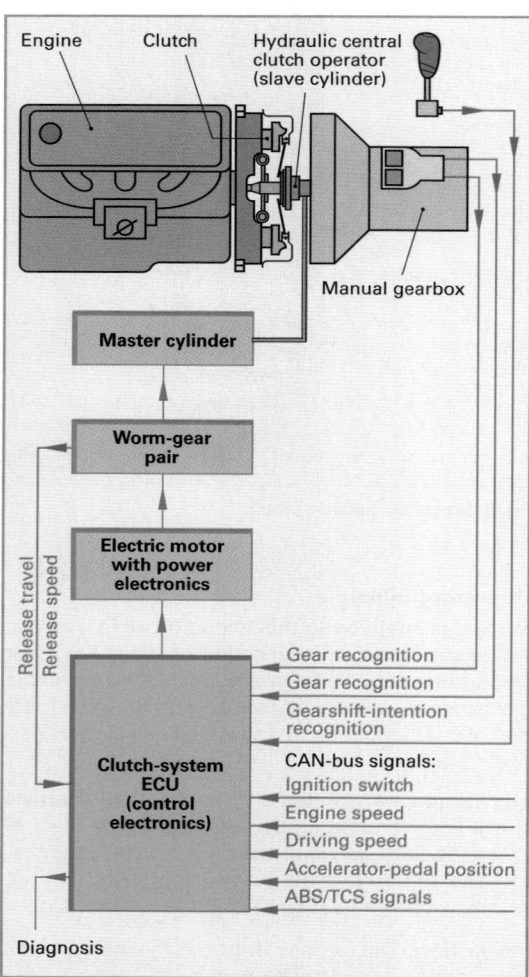

Fig. 1: Block diagram of automatic clutch system ACS

Design (Fig. 1)

Components of clutch system

- **Clutch:** Self-adjusting diaphragm-spring clutch (SAC) with hydraulic central clutch operator
- **Sensors** for gearshift-intention recognition, gear recognition, release travel and release speed
- **Clutch-system ECU**
- **Actuators**
  - Electric motor and worm-gear pair
  - Master cylinder, hydraulic central clutch operator (slave cylinder)

Gearshift-intention recognition. This is recorded by a sensor (rotary potentiometer) on the shift lever.
Gear recognition. This is recorded by two contactless angle-of-rotation sensors on the gearshift linkage in the gearbox. In addition to the sensor signals for gearshift intention and gear recognition, the ECU receives further signals via the CAN bus from the ECUs of the engine-management and ABS/TCS control systems.

Operating principle

To record the respective system status, the ECU receives from the sensors input signals, which it processes by means of clutch software and then transmits as output signals to the actuating devices (actuators).

The clutch is opened or closed depending on the signals for the actuators.

Starting. The ECU calculates the optimal slip for starting from various input signals, such as, for example, wheel speed, engine speed and gearbox speed.

Gear changes. The sensor on the shift lever signals the driver's gearshift intention. The ECU effects a build-up of pressure in the master cylinder via an electric motor with worm-gear pair. This pressure opens the clutch via the hydraulic central clutch operator (slave cylinder). After shifting, the gear-recognition sensors signal which gear has been selected.

Now the ECU sends a signal to the electric motor, which causes the clutch to close with a defined amount of slip.

The driver does not necessarily have to take his/her foot off the accelerator pedal during the gearshift. The injected fuel quantity is automatically reduced and then increased again.

Normal driving mode. In order to damp torsional vibrations, the ECU calculates the difference from the signals for engine speed and gearbox input speed so that a controlled amount of slip is set where necessary.

Load changes. When the accelerator pedal is actuated jerkily, the pitching of the vehicle (bonanza effect) is kept within limits since the clutch is opened briefly. Smooth acceleration is achieved in this way.

Downshifting on smooth road surfaces. The signal of the locking drive wheels is processed by the ECU in such a way that the clutch opens at the onset of wheel locking and releases the wheels.

Features

- No clutch pedal required.
- Favourable wear characteristics for clutch lining and release bearing.
- No engine stalling during starting and braking.
- Engine torsional vibrations are damped by slip in the clutch.
- No disruptive load-change reactions.

Examples of electronic clutch systems:

ECS **E**lectronic **C**lutch **S**ystem

ECM **E**lectronic **C**lutch **M**anagement

ACS **A**utomatic **C**lutch **S**ystem

16.2.7 Function checks on friction clutches

Several clutching operations should be performed during a test drive prior to the function checks so that the clutch is brought to normal operating temperature. Because of the danger of overheating, the clutch must not be heated by allowing it to slip when stationary.

Checking for slipping when starting

1. Engage 1st gear when the vehicle is stationary.

2. Increase engine revs to twice the idle speed.
3. Swiftly engage the clutch at the same time.

> The vehicle must accelerate smoothly and without jerks. If this is not the case, the clutch is slipping.

Checking for slipping with the handbrake applied

1. Press the clutch pedal, engage the highest forward gear.
2. Increase engine revs up to maximum torque.
3. Swiftly engage the clutch and at the same time press the accelerator pedal to the floor.

> The engine must be stalled in the process. If the clutch slips or if there is an increase in engine revs, the transmission capability of the clutch is not in order.

Checking the separation performance

1. Press the clutch pedal.
2. Wait approximately 3 to 4 seconds.
3. Engage gear and listen out for noises.

> No noises must be heard; if there are noises, the separation performance of the clutch is not in order.

Or
1. Raise the powered axle.
2. Engage the clutch and engage gear.

> The drive wheels must not turn.

REVIEW QUESTIONS

1 What 5 functions must be performed by the clutch in a motor vehicle?

2 Why does the power flow have to be interrupted for gearshifting in motor vehicles which have manual gearboxes?

3 What are the 3 main components of a single-plate friction clutch?

4 What are the functions of clutch operators?

5 Explain the design of a centrally guided hydraulic clutch operator.

6 On which four variables does the transmissible torque of a diaphragm-spring clutch depend?

7 Explain the operating principle of a diaphragm-spring clutch during disengagement.

8 What do you understand by the ventilation clearance of a friction clutch?

9 What are the advantages of a diaphragm-spring clutch over a coil-spring clutch?

10 What is the advantage in terms of torque transmission of a double-plate clutch over a single-plate clutch?

11 Explain the design of a clutch plate.

12 What are the properties that are required of clutch linings?

13 Clutch linings are made up of what?

14 What is the design and structure of hydraulic clutch control?

15 What are the advantages of hydraulic clutch control?

16 How is the torque transmitted in a magnetic-particle clutch?

17 What function checks are carried out on passenger-car friction clutches?

18 What are the features of an automatic clutch system?

16.3 Variable-speed gearbox

> The variable-speed gearbox is situated in the drivetrain of a motor vehicle between the clutch and the final drive and converts engine torques and engine speeds.

Functions

- Converts and transmits engine torque.
- Converts engine speed.
- Facilitates engine idling while the vehicle is stationary.
- Facilitates reversal of the direction of rotation for reversing.

Every internal-combustion engine operates between a minimum speed and a maximum speed and can only output a limited torque within this powerful speed range **(Fig. 1)**.

The speed range between maximum engine torque and maximum engine power is known as the **elastic range**.

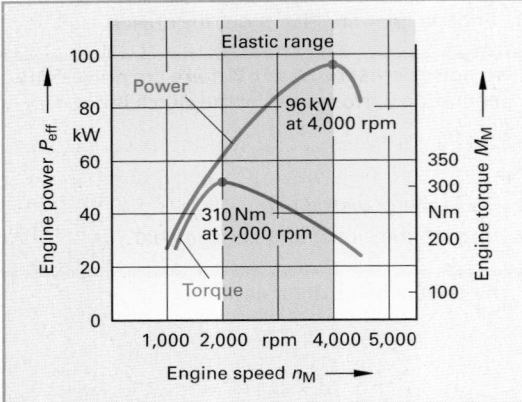

Fig. 1: Power and torque curve of an internal-combustion engine

The torques and speeds output by the engine within the powerful speed range are converted so that they can be utilised for driving a motor vehicle.

The desired drive speeds and torques are achieved at the drive wheels by the different transmission stages in the gearbox via an intermediate final drive.

Torque-speed conversion

This is effected in variable-speed gearboxes with the aid of gear wheels **(Fig. 2)**.

In the case of a gearwheel pair, the greater torque and the lesser speed always act on the larger gear wheel (larger lever arm, more teeth).

The leverage ratio r_2/r_1 corresponds to the ratio of the numbers of teeth z_2 of the driven gear wheel to the driving gear wheel z_1 or to the ratio of the driving speed n_1 to the driven speed n_2.

This ratio is known as the **gear ratio i** or **transmission ratio** and expresses the degree of torque and speed conversion.

> A **transmission ratio $i > 1$** results in a torque increase and a speed decrease, while a **transmission ratio $i < 1$** results in a torque decrease and a speed increase.

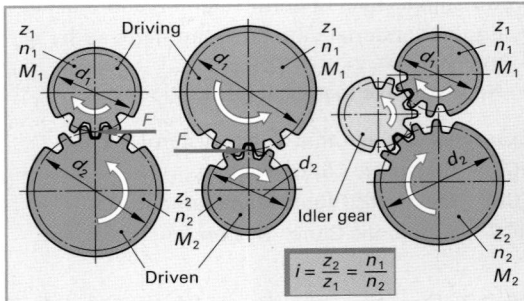

$$i = \frac{z_2}{z_1} = \frac{n_1}{n_2}$$

Fig. 2: Torque and speed conversion

Gearbox output curves (Fig. 3)

Different gear ratios achieve different output torques and output speeds. The gearbox output curves are plotted from the results.

The curve points are calculated from the values of the engine torque curve and the gearbox gear ratios.

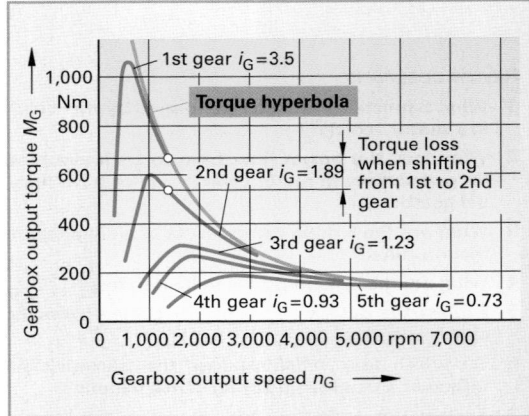

Fig. 3: Gearbox output curves

Torque hyperbola (Fig. 3). The gearbox output torques needed to propel a vehicle are plotted against the gearbox output speeds.

For example, a high drive torque is needed to accelerate a vehicle from a standstill. This is made possible, for example, by the gear ratio of 1st gear. The gearbox output curves should approach the torque hyperbola in order to achieve as low a torque loss as possible during the gear change. This is a feature of the favourable stepping of a gearbox.

Traction-force hyperbola. If instead of the output torques the traction forces are plotted against the driving speed, a traction-force hyperbola is obtained as the vehicle's driving chart.

Idle position

This effects the interruption of the power flow.

Reversal of direction of rotation

The German Road Traffic Licensing Regulations require motor vehicles with a permissible total weight of more than 400 kg to be equipped with a facility for reversing the direction of rotation in order to reverse/back up the vehicles.

This is achieved with an idler gear.

16.4 Manual variable-speed gearbox

Gearboxes of this type are differentiated according to ...

... the progression of the power flow in the gearbox **(Fig. 1)**:
- Identical-shaft manual gearbox
- Non-identical-shaft manual gearbox

... the installation position in the vehicle:
- Longitudinal gearbox (= identical-shaft)
- Transverse gearbox (= non-identical-shaft)

... the components which connect torque proof the shift gear wheels (idler gears) with their shafts:
- Selector-sleeve gearbox
- Shift-dog gearbox

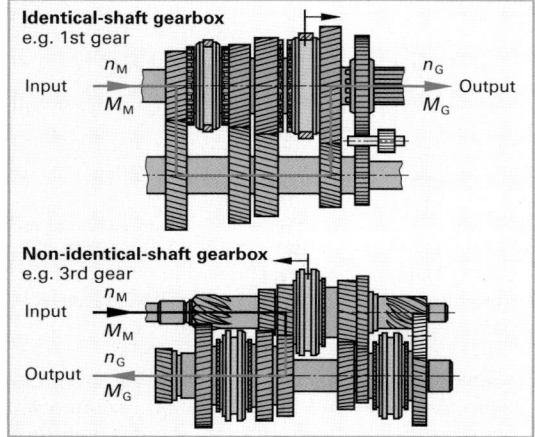

Fig. 1: Identical-shaft and non-identical-shaft gearboxes

16.4.1 Selector-sleeve gearbox

> The power flow between the shift gear wheel (idler gear wheel) and the gearbox shaft is established via a selector sleeve, which is connected torque proof for example via a synchromesh body with the gearbox shaft **(Fig. 2)**.

All the gearwheel pairs for the forward gears have helical teeth and constantly mesh with each other. This is only possible if for each non-shifted gearwheel pair one gear wheel is freely rotatable as an idler gear on the shaft.

Gearshift process. A selector sleeve is displaced in such a way that the corresponding shift gear wheel (idler gear) is connected torque proof with the shaft. Here the inner dogs of the selector sleeve are displaced via the shift teeth of the shift gear wheel.

Non-identical-shaft selector-sleeve gearbox

> In non-identical-shaft gearboxes, the gear ratios are achieved by means of one gearwheel pair in each case.

Gearboxes of this types are used, for example, in vehicles with engines mounted transversally to the direction of travel. The input and output shafts lie on different (unequal) planes **(Fig. 2)**. The output shaft is also known as the main shaft.

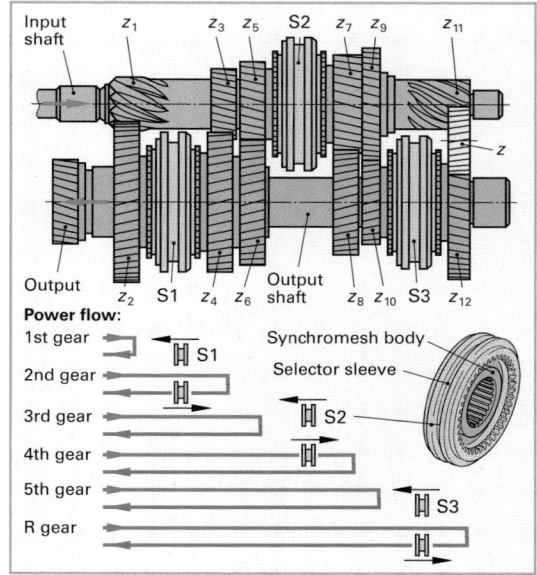

Fig. 2: Non-identical-shaft 5-speed gearbox

Example (gearshift, 3rd gear): Selector sleeve S2 is displaced to the left to the shift teeth of shift gear wheel z_5.

Power flow, 3rd gear: Input shaft $\rightarrow$ synchromesh body $\rightarrow$ selector sleeve S2 $\rightarrow$ shift teeth of z_5 $\rightarrow$ gear wheel z_5 $\rightarrow$ gear wheel z_6 $\rightarrow$ output shaft.

6-speed manual gearbox short-design

This is a non-identical-shaft gearbox and is used for transverse installation in front-wheel and all-wheel drives.

Design (Fig. 1):

- 1 input shaft **In** with 5 gear wheels (fixed gears).
- 2 output shafts **Out1** and **Out2** with shift gear wheels and 2 output gear wheels z_{out1}, z_{out2}, which act on a common final-drive spur gear.
- 4 selector sleeves S1, ... S4, 2 on each output shaft.

Gear wheels: z_1, z_3, z_5, z_7, z_9 are mounted as fixed gear wheels on the input shaft **In**.

Shift gear wheels: z_2, z_4, z_6, z_8 for 1st to 4th gears are mounted as idler gear wheels on output shaft **Out1**. The shift gear wheels z_{10}, z_{12}, z_{16} for 5th, 6th and R (reverse) gears are mounted on output shaft **Out2**.

Selector sleeves: S1, S2, S3, and S4 for the forward gears establish the torque proof connection of the shift gear wheels with output shafts **Out1** and **Out2**.

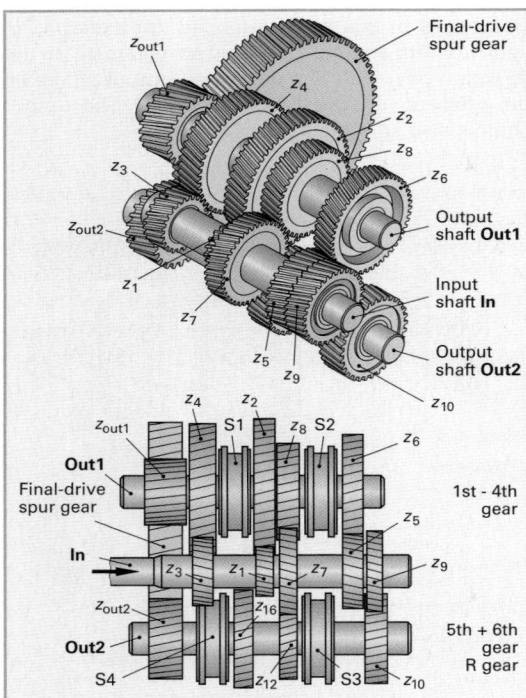

Fig. 1: 6-speed gearbox for transverse installation
Power flow:

1st gear: **In** $\rightarrow z_1$ $\rightarrow z_2$ $\rightarrow$ S1 $\rightarrow$ **Out1** $\rightarrow z_{out1}$
2nd gear: **In** $\rightarrow z_3$ $\rightarrow z_4$ $\rightarrow$ S1 $\rightarrow$ **Out1** $\rightarrow z_{out1}$
3rd gear: **In** $\rightarrow z_5$ $\rightarrow z_6$ $\rightarrow$ S2 $\rightarrow$ **Out1** $\rightarrow z_{out1}$
4th gear: **In** $\rightarrow z_7$ $\rightarrow z_8$ $\rightarrow$ S2 $\rightarrow$ **Out1** $\rightarrow z_{out1}$
5th gear: **In** $\rightarrow z_9$ $\rightarrow z_{10}$ $\rightarrow$ S3 $\rightarrow$ **Out2** $\rightarrow z_{out2}$
6th gear: **In** $\rightarrow z_7$ $\rightarrow z_{12}$ $\rightarrow$ S3 $\rightarrow$ **Out2** $\rightarrow z_{out2}$

Identical-shaft selector-sleeve gearbox

> In identical-shaft gearboxes, the input and output shafts are mounted in the same plane **(Fig. 2)**.

Gearboxes of this types are used in vehicles with front engines and rear-wheel drives installed in the direction of travel and are known as "3-shaft gearboxes" (input shaft, countershaft, output shaft).

Input shaft. This is connected with the clutch plate and drives the countershaft via z_1.

Countershaft. Together with gear wheels z_2, z_3, z_5, z_7, z_9 and z_{11}, this forms a cluster gear.

Output shaft (main shaft). Shift gear wheels (idler gears) z_4, z_6, z_8, z_{10}, z_{12} are mounted on this shaft. Selector sleeves S1, S2 and S3 are displaced to the left or right to shift the gears. In this way, one shift gear wheel (idler gear) in each case is connected torque proof via the synchromesh body with the output shaft.

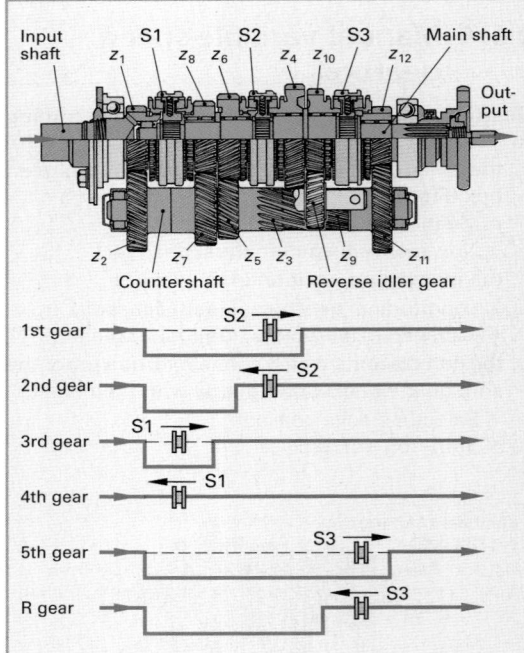

Fig. 2: Identical-shaft 5-speed gearbox

> The gear ratios i_G of the individual gears are, except in the direct gear, achieved in each case by means of two gearwheel pairs. The gearwheel pair z_1 / z_2 is always effective here.

Direct gear (4th gear): Selector sleeve S1 is displaced to the left and the input shaft is connected torque proof with the input shaft. There is no torque or speed change here.

16.4.2 Synchromesh devices of selector-sleeve gearboxes

> Synchromesh devices are designed to establish synchronism between the selector sleeve and the shift gear wheel (idler gear) and to facilitate noise-free and swift gear shifting.

Matching at different rotational speeds is effected by means of sliding friction between the conical friction surfaces of the synchroniser ring and the gear wheel. The matching process is known as synchronisation.

Types

- Single synchromesh devices (1 friction cone) with inner or outer synchronisation
- Multiple synchromesh devices (2 or 3 friction cones)

Single synchromesh device with inner synchronisation (Borg-Warner system)

Design

The synchromesh device (Fig. 1) consists of selector sleeve, synchromesh body, 3 thrust members, 2 holding springs, synchroniser ring and shift gear wheel.

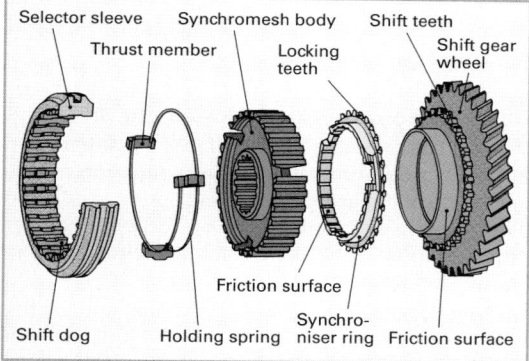

Fig. 1: Single synchromesh device with inner synchronisation (Borg-Warner system)

Selector sleeve. This features on its inner side shift dogs which mesh with the outer recesses on the synchromesh body. The 3 thrust members are guided by the synchromesh body and pressed by 2 holding springs against the shift dogs of the selector sleeve. In this way, the selector sleeve is held in a central position on the synchromesh body.

Synchromesh body. This is connected torque proof with the shaft of the shift gear wheel (gear wheel).

Synchroniser ring. This has a conical friction surface on the inside and locking teeth on the outside. 3 recesses in the synchroniser ring limit its twisting in relation to the thrust members.

Shift gear wheel. This has a conical friction surface on the outside on the side facing the synchroniser ring; after the friction surface come the shift teeth.

Operating principle

Neutral position (Fig. 2). When a gear is not selected, the selector sleeve is held by the thrust members on the synchromesh body. The shift gear wheel runs loose on the shaft.

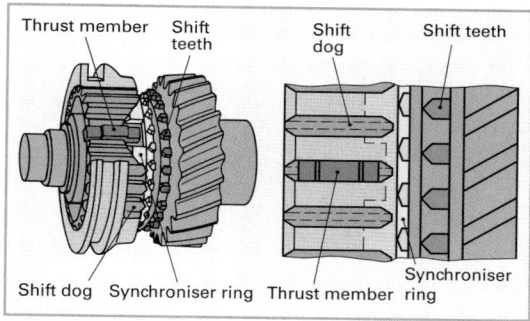

Fig. 2: Neutral position

Locking and synchronising position (Fig. 3). During shifting, the selector sleeve is displaced via the selector fork in the direction of the gear to be shifted. The 3 thrust members are pressed against the synchroniser ring. In this way, they displace the synchroniser ring axially and press it against the conical friction surface of the shift gear wheel.

As long as the selector sleeve and the shift gear wheel rotate at different speeds, a friction torque is generated which turns the synchroniser ring until its recesses contact the thrust members. In this way, the locking teeth are now in front of the shift dogs of the selector sleeve and prevent (lock) displacement of the selector sleeve.

The friction between the friction surfaces of the synchroniser ring and the shift gear wheel accelerates or brakes the shift gear wheel and thereby establishes synchronism between shift gear wheel, selector sleeve and shaft.

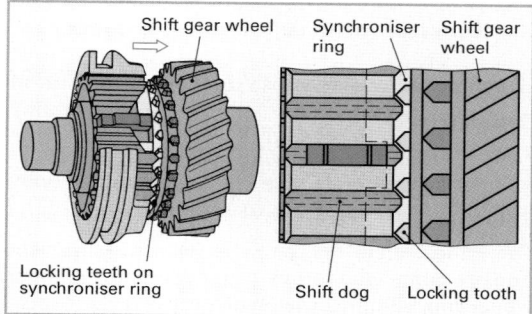

Fig. 3: Locking and synchronising position

Gear shifted (Fig. 1, Page 356). Once synchronism is established between the shift gear wheel and the selector sleeve, there is no longer any friction torque acting on the synchroniser ring. It can be ro-

16

tated back by the angled sections of the shift dogs. The selector sleeve is no longer locked and can now be displaced via the shift teeth of the shift gear wheel.

The connection between the gearbox shaft and the shift gear wheel is established.

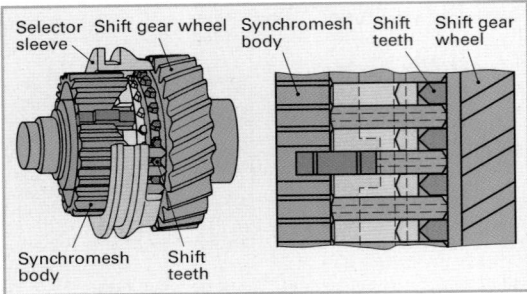

Fig. 1: Gear shifted

Synchromesh devices with multiple synchronisation

These are mainly used to shift low gears. Here the speed differences between the selector sleeve and the loose-running shift gear wheel are greater than in the high gears. Greater frictional forces are needed than in the higher gears to match the rotational speeds (acceleration or braking of the gear wheels).

In modern passenger-car manual gearboxes the following are used as synchromesh devices, for example, in 6-speed manual gearboxes:

three-cone synchronisation for 1st and 2nd gear,
dual synchronisation for 3rd and 4th gear,
single synchronisation for 5th, 6th and reverse gear.

Advantages of multiple synchronisation

- Greater frictional force with the same shifting force.
- Swifter and easier shifting.
- Less wear on the friction cones because the surface pressure on the friction surface is lower.

Synchromesh device with dual synchronisation

Design (Fig. 2)

- Inner synchroniser ring
- Outer synchroniser ring
- Intermediate ring
- Selector sleeve
- Synchromesh body
- Shift gear wheel

The intermediate ring and the inner synchroniser ring are connected torque proof with the shift gear wheel and with the outer synchroniser ring respectively.

2 friction pairings (inner synchroniser ring/intermediate ring and intermediate ring/outer synchroniser ring) are provided in the case of dual synchronisation. In this way, the total friction surface is almost twice as large as that with single synchronisation.

Operating principle. During synchronisation, the selector sleeve displaces the outer synchroniser ring on the intermediate ring and the intermediate ring on the inner synchroniser ring.

The outer and inner synchroniser rings are rotated by friction to such an extent that the locking teeth of the outer synchroniser ring prevents further displacement of the selector sleeve until synchronism is achieved.

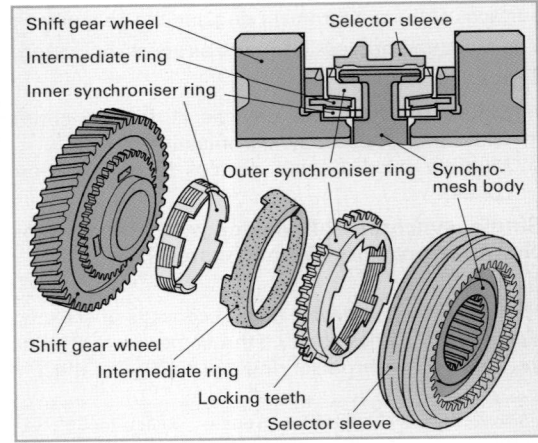

Fig. 2: Dual synchronisation

Outer-cone synchromesh device (Fig. 3)

Design. The conical friction surfaces are situated on the outer side of the synchroniser ring and on the inner side of the selector sleeve. Locking is effected by the 3 angled/inclined lugs on the synchroniser ring. The inner springs hold the synchroniser ring on the shift gear wheel.

Locking process: At different rotational speeds the synchroniser ring is rotated until the lugs prevent further shifting of the selector sleeve. When synchronism is achieved, there is no more friction torque and the lugs can be pressed into the grooves of the shift teeth. In this way, the selector sleeve can be displaced into the shift teeth. The greater friction radius facilitates easier and swifter shifting.

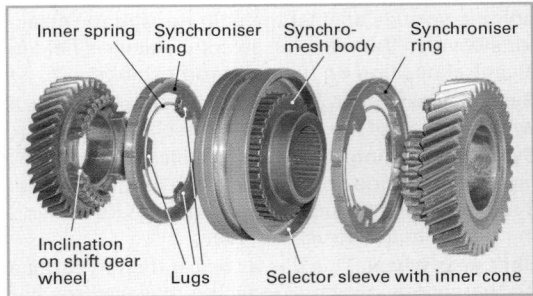

Fig. 3: Outer-cone synchromesh device

16.4.3 Servicing and troubleshooting on manual gearboxes

Essential servicing tasks

- Check the oil level, correct if necessary.
- Check the gearshift mechanism for ease of movement and correct operation.
- Perform an oil change, if specified, in accordance with the manufacturer's instructions.
- Check the gearbox housing for leaks.

Checks to locate faults and malfunctions

- Visual inspections, e.g. of gearbox suspension and gearshift linkage.
- Noise checks, e.g. gear and bearing noises at idle and during load changes.
- Function checks, e.g. synchronisation during gear changes.

Troubleshooting		
Fault/malfunction	Cause	Remedy
Gearshift sticks	Gearshift linkage bent Gearbox suspension defective Incorrect adjustment	Replace defective components Correct adjustment
Gear jumps out	Selector fork bent Shift teeth worn Gearshift locking defective Engine or gearbox suspension damaged	Replace selector fork Replace gear wheels Replace suspension components
Poor gearbox synchronisation	Synchroniser ring worn Incorrect gearbox oil used	Replace synchroniser rings Use recommended gearbox oil
Gearbox noises when driving under load	Gearbox mounts defective Teeth damaged	Replace gearbox mounts Replace gear wheels
Leaks from housing	Sealing rings, gaskets leaking	Replace defective components

REVIEW QUESTIONS

1 Why does a vehicle with an internal-combustion engine require a variable-speed gearbox?

2 What are the functions of the variable-speed gearbox in a motor vehicle?

3 What do you understand by the elastic range of an internal-combustion engine?

4 Explain the term gear ratio using the example of a simple gearwheel pair.

5 What are the effects of a gear ratio $i > 1$ on torque conversion and speed conversion in a gearbox?

6 What are the differences between identical-shaft and non-identical-shaft gearboxes?

7 A manual gearbox has a gear ratio of $i = 3.5$ in 1st gear and $i = 0.73$ in 5th gear. How are an engine torque of 100 Nm and an engine speed of 1,000 rpm converted by these gear ratios?

8 According to which factors are manual variable-speed gearboxes differentiated?

9 Under what condition is it possible for several gearwheel pairs to be constantly operating together in a gearbox?

10 How in a selector-sleeve gearbox is the power flow established between the shift gear wheel and the gearbox shaft?

11 Through how many gearwheel pairs does the power flow progress in an identical-shaft and in a non-identical-shaft variable-speed gearbox respectively?

12 What are the functions of synchromesh devices in selector-sleeve gearboxes?

13 Name the components of single cone synchronisation (Borg-Warner).

14 Explain the synchronising and locking processes in single cone synchronisation (Borg-Warner system).

15 What are the advantages of multiple synchronisation over single synchronisation?

16 Explain the design of a synchromesh device with dual synchronisation.

17 In which gears are single, dual and three-cone synchronisation used?

18 Which servicing tasks must be carried out on variable-speed gearboxes?

19 How can faults be located on variable-speed gearboxes?

20 Which faults may be present when a gear jumps out?

21 A gearbox is experiencing poor synchronisation. Give possible causes for this.

16

16.5 Automatic gearbox

Different types of gearbox are used: semiautomatic and fully automatic gearboxes.

Semiautomatic gearboxes

- The power flow is interrupted by automatic clutch disengagement and cut in by clutch engagement, e.g. automatic clutch system ACS.
- Shifting of gears to change the gear ratios and direction of rotation is performed manually by operating a shift lever.

Fully automatic gearboxes

- The power flow is automatically interrupted or cut in as required.
- Shifting of gears to change the gear ratios is performed automatically by electrohydraulic or electropneumatic control.

Fully automatic gearboxes		
Automated manual gearbox with: • Diaphragm-spring clutch • Spur gears The gear-ratio change is stepped.	**Converter transmission** with: • Hydrodynamic torque converter • Planetary-gear set The gear-ratio change is stepped.	**CVT automatic transmission** with: • Primary and secondary V-pulleys • Steel pushbelt or link chain The gear-ratio change is continuously variable.
E.g. Automated manual gearbox, Easytronic, direct-shift gearbox (DSG).	E.g. 5-, 6-, 7-speed converter transmission.	E.g. Ecotronic, Multitronic. (CVT = Continuously Variable Transmission).

16.5.1 Automated manual gearbox

Automated manual gearboxes are fully automatic gearboxes in which conventional 5- or 6-speed manual gearboxes with clutches are automatically shifted.

In the automated manual gearbox illustrated in **Fig. 1**, the clutch is actuated and the gears are shifted by electrohydraulic means.

It is also possible to shift the gears manually, for example by actuating shift paddles on the steering wheel or by means of the Tiptronic-function selector lever.

Automated manual gearbox

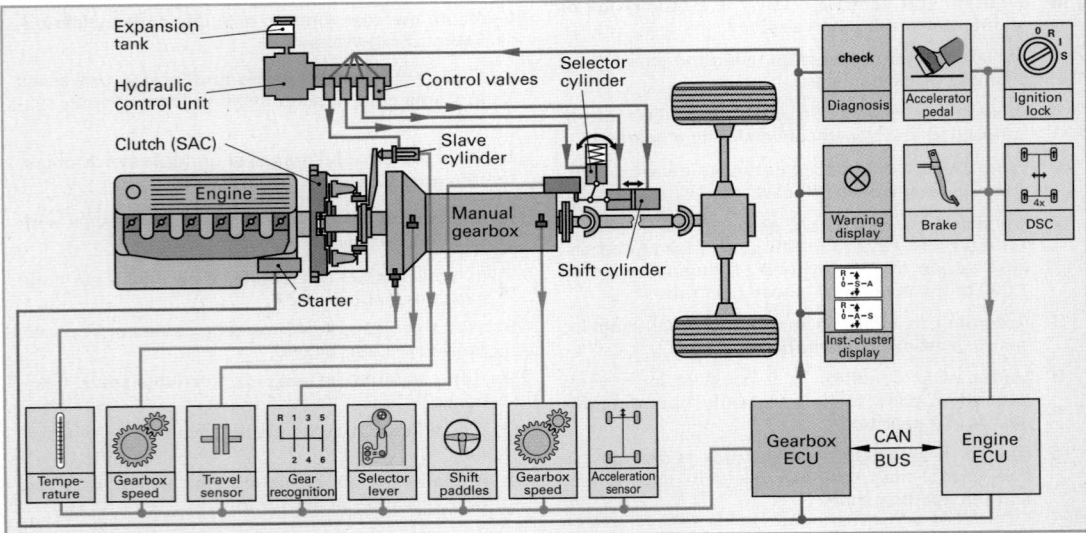

Fig. 1: Automated manual gearbox with hydraulic control unit

Main controlled variables for the automated gearshift sequence are: driving speed, selector-lever position, selected driving program and accelerator-pedal position. The gears can also be shifted manually by means of the Tiptronic lever or steering-wheel shift paddles.

Sensors are used to record engine speed, gearbox input speed and clutch travel for the purpose of implementing a slip-controlled, optimal engagement and disengagement process.

The longitudinal-acceleration sensor serves to detect uphill and downhill gradients and to record acceleration or deceleration.

The gears are engaged by selector and shift cylinders, the positions of which are recorded by sensors. The shifting points are influenced by the gearbox-oil temperature sensor.

System control. The automated manual gearbox ECU evaluates the input signals from the sensors with gearbox/clutch software. It uses stored program maps to determine the output signals for activating the clutch slave cylinder, selector cylinder and shift cylinder.

The gearshift sequence is divided into 3 phases:
- Disengagement
- Shifting
- Engagement.

The 3 phases are varied according to the respective driving situation in order to achieve the best possible shifting comfort and short shifting times.

Sensors, for example on the brake pedal and the door contacts, are fitted to act as safety devices.

The shifting processes take place sequentially, i.e. it is only possible to shift up or down one gear in each case. For this reason, these gearboxes are also known as sequential gearboxes.

Clutch and clutch actuator. The clutch used is a self-adjusting clutch (**SAC**). The clutch actuator consists of a slave cylinder with travel sensor.

Manual gearbox and gearbox actuators. A 6-speed manual gearbox, for example, is used as the gearbox, to which a selector cylinder and a shift cylinder are attached (add-on system). These cylinders are hydraulically actuated and execute the movements of the selector shaft and selector fork which are required to change gear.

Hydraulic control unit. The operating pressure is generated by an oil pump. Electrohydraulically actuated valves direct the pressure in accordance with the program maps stored in the ECU to the selector and shift cylinders and the clutch slave cylinder.

System networking. The automated shift gearbox control unit (ATS) is linked via the CAN with other systems installed in the vehicle, such as e.g. engine management and driving-dynamics control systems.

Direct-shift gearbox (DSG)

> The direct-shift gearbox (DSG) is an automated gearbox in which a 6-speed manual gearbox with twin clutch is automatically shifted by electric actuators.

The DSG can be operated both in automatic mode and manually in Tiptronic mode.

Design
- 6-speed manual gearbox with twin clutch
- Oil pump, oil cooler, oil filter
- Electrohydraulic transmission control
- Sensors for recording the input signals
- Electric actuators for actuating clutches C1 and C2 and for shifting the gears

Twin clutch. This consists of 2 wet clutches C1 and C2, the pressure of which is hydraulically controlled. Clutch C1 is used for starting and is connected with the gear wheels mounted on the hollow shaft for the odd gears (1st/3rd/5th gear) and the reverse gear. Clutch C2 is responsible for the even gears (2nd/4th/6th gear).

Gearshift process, e.g. from 1st to 2nd gear.
While the vehicle is being driven in 1st gear, clutch C1 is closed and 2nd gear is already preselected with opened clutch C2.

As soon as the transmission control detects from the input signals that the ideal shifting point is reached, clutch C1 responsible for 1st gear is opened and simultaneously clutch C2 responsible for 2nd gear is closed. This results in an overlap between the opening and closing of the two clutches. The entire gearshift process is completed within 3/100 ... 4/100 second. There is virtually no interruption of the power flow.

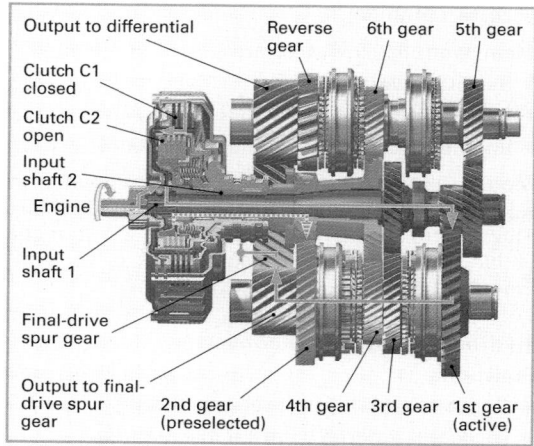

Fig. 1: **Direct-shift gearbox DSG (1st gear shifted)**

16.5.2 Stepped automatic gearbox with hydrodynamic converter

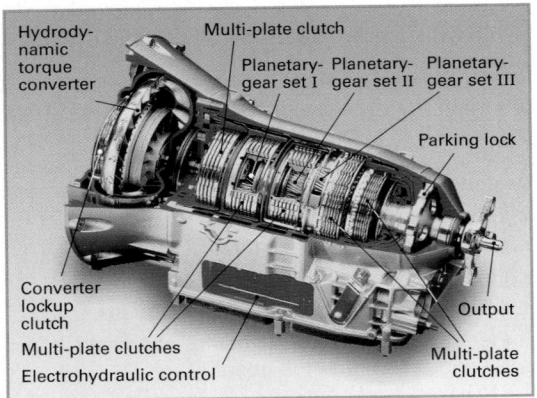

Fig. 1: 5-speed automatic gearbox with Wilson planetary-gear set

Gearbox components

- **Hydrodynamic torque converter.** This serves as a starting clutch and multiplies the torque in the conversion range.

- **Planetary gear.** Connected downstream of the hydrodynamic torque converter, this transmits torques and speeds and reverses the direction of rotation for the reverse gear.

The following are used as planetary gears:

- Ravigneaux set
- Simpson set
- Wilson set
- Lepelletier set

A simple planetary-gear set can also be connected upstream or downstream of a Ravigneaux set or Simpson set.

- **Electrohydraulic control.** The function of this control system is to effect the automatic upshifting and downshifting of the individual gears at the correct moment.

Main controlled variables are:

- **Selector-lever position**
- **Driving speed**
- **Engine load** (accelerator-pedal position)

Modern automatic gearboxes are equipped with adaptive self-learning transmission control (**ATS**). This control system ensures that the automatic shifting of the gearbox is adapted to the driver's driving style and to the driving conditions.

Hydrodynamic torque converter
Functions

- Converts and transmits engine torque
- Facilitates smooth, comfortable starting
- Damps engine torsional vibrations

Design

The hydrodynamic torque converter (**Fig. 2**) consists of:

- Impeller
- Stator with overrunning clutch
- Turbine wheel
- Lockup clutch

The impeller, turbine wheel and stator are designed as curved blade wheels and operate in an enclosed housing which is filled with hydraulic fluid.

The impeller is driven by the flywheel via the converter housing at engine speed.

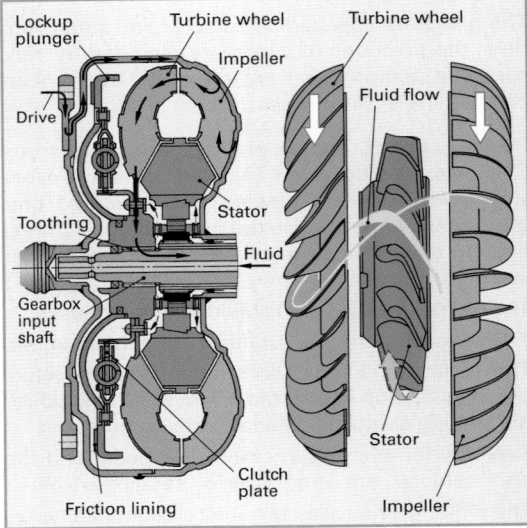

Fig. 2: Hydrodynamic torque converter

Fluid circuit (Fig. 3)

A fluid pump is driven by the torque-converter impeller. This pump ensures that a filling pressure of usually 3 bar to 4 bar is built up in the converter and the hydraulic fluid is circulated via a restrictor, a fluid cooler and a fluid reservoir.

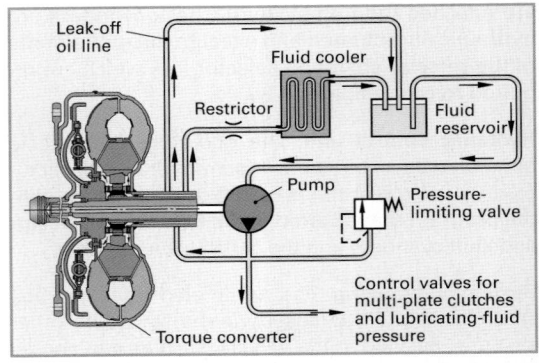

Fig. 3: Fluid circuit

The filling pressure in the hydrodynamic torque converter prevents cavitation (formation of bubbles), which impairs efficiency.

Operating principle

Conversion range (Fig. 1). During starting, the impeller rotates at engine speed, while the turbine wheel and the stator are stationary.

The fluid flows from the impeller to the turbine wheel, dissipates its energy to the latter and is deflected in the process **(Fig. 2, Page 360).**

The turbine wheel starts to rotate if the torque at the turbine wheel is greater than the section modulus (moment of resistance) at the gearbox input shaft. The fluid flow emerging from the turbine wheel strikes the stator blades and attempts to rotate the stator in the opposite direction to the impeller and turbine wheel. This direction of rotation is blocked by the overrunning clutch. The fluid is supported on the stator blades curved by approximately 90° and in the process generates a strong backpressure, which results in an increase in the rotary force at the turbine-wheel blades.

As a result of the increase in the rotary force, the torque at the turbine-wheel shaft (gearbox input shaft) is greater than the engine torque introduced into the torque converter.

The stator directs the fluid flow at a favourable angle to the impeller blades. This creates an enclosed fluid circuit.

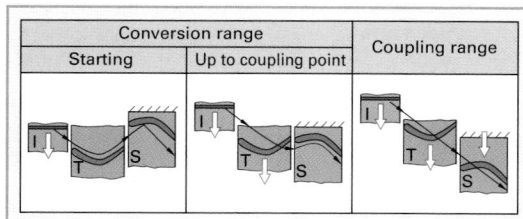

Fig. 1: Flow paths

As the speed of the turbine wheel increases, the speed difference between the impeller and the turbine wheel gets smaller. The fluid flow undergoes less deflection and strikes the stator blades at a smaller angle **(Fig. 1).** This results in a reduction of the supporting force and thus of the additional force on the turbine-wheel blades. Torque multiplication is reduced.

Coupling range. When the impeller and turbine wheel rotate at roughly the same speed (speed ratio $n_T/n_I \approx 0.85 ... 0.9$), the fluid flow strikes the stators from its reverse side, the overrunning clutch is released and the stator starts to rotate.

From this point no further backpressure force acts on the turbine wheel and thus there is no torque multiplication. This point is known as the **coupling point.**

Curves of hydrodynamic torque converter. Fig. 2 shows in a graph the converter curves for a drive torque M_I at the impeller of e.g. 200 Nm.

The shape of the curve of the torque M_T at the turbine wheel (= gearbox input torque) shows that the torque is at its largest at the starting point.

In the example chosen, the turbine-wheel torque is 500 Nm with a multiplication of $M_T/M_I = 2.5$.

The multiplication M_T/M_I decreases as the turbine-wheel speed increases.

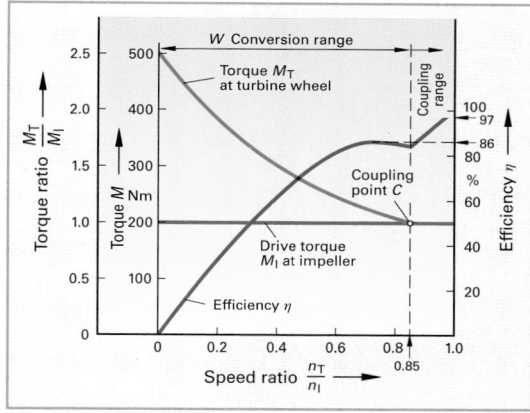

Fig. 2: Curves of a torque converter

Efficiency η. This is approximately 97 % with the hydrodynamic torque converter above the coupling point at high speeds. A slip of approximately 3 % is set in the process.

The speed difference between the impeller and the turbine wheel is referred to as slip.

An improvement in efficiency is achieved if the flow losses are cut out by a mechanical lockup clutch in the torque converter.

Features of hydrodynamic torque converter

- No mechanical wear.
- Smooth, comfortable starting process.
- Engine cannot be stalled during starting.
- Torque multiplication is adapted automatically and steplessly to the respective driving situation.
- Torque multiplication is at its greatest during the starting process.
- Engine torque jolts and torsional vibrations are damped by the hydraulic fluid.
- Low-noise operation.

Converter lockup clutch

> This is designed to eliminate flow losses by the hydrodynamic torque converter in the coupling range in order to save fuel.

The converter lockup clutch is cut in after the converter coupling point is exceeded.

Design (Fig. 1)
The outer plate carrier is connected with the impeller (converter housing), the inner plate carrier is connected with the turbine wheel. The fluid-pressure-controlled multi-plate clutch is engaged and disengaged by the lockup plunger.

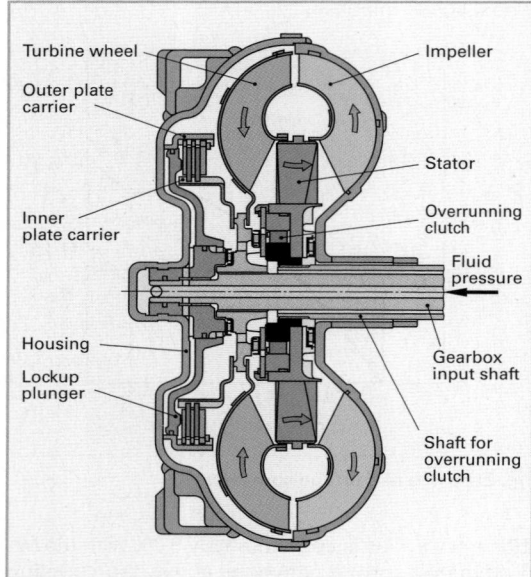

Fig. 1: **Converter lockup clutch with multi-plate clutch**

Operating principle

Lockup clutch open. The fluid flows through a bore in the input shaft to the right side of the lockup plunger and forces the plunger to the left. The multi-plate clutch is thus disengaged (opened).

Lockup clutch closed. The fluid flows to the left side of the lockup plunger. The plunger is displaced to the right and compresses the plate pack. The multi-plate clutch is engaged (closed).

The impeller and turbine wheel are now connected with each other by adherence and without slip.

Converter lockup clutches are, depending on engine load and driving speed, automatically cut in, for example in 3rd, 4th and 5th gears.
The operating temperature must be achieved prior to shifting. The converter lockup clutch is opened in trailing-throttle conditions and during braking.

Controllable converter lockup clutch
This lockup clutch provides for 3 operating states:
- Open
- Slipping
- Closed

Operating principle
In the case of slip-controlled converter lockup clutches, a limited slip is facilitated between the impeller and the turbine wheel by program maps stored in the ECU in specific engine operating states.

This slip prevents torsional vibrations from being transmitted from the engine to the gearbox. The clutch-plate torsion dampers are omitted from such systems.

The following influencing variables are taken into account, for example, in the program maps of the control software:
- Accelerator-pedal position gradient
- Gearbox shifting function
- Uphill/downhill
- Transmission-fluid temperature
- Engine temperature

16

Planetary gear

A simple planetary-gear set **(Fig. 1)** consists of

- sun gear
- internal gear
- planet gears
- planet carrier

The planet gears are supported with their axes in the planet carrier. They circulate on the inner teeth of the internal gear and on the outer teeth of the sun gear.

All the gear wheels constantly mesh with each other. The sun gear, internal gear or planet carrier can be both driven and braked. Output is effected via the internal gear or the planet carrier.

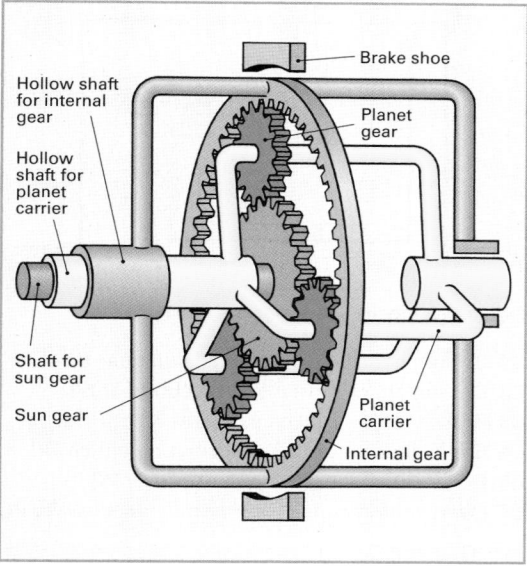

Fig. 1: Simple planetary-gear set

Operating principle

> The different gear ratios are achieved by either the sun gear, internal gear or planet carrier being driven. In the process, a non-driven part must be braked in each case. Output is effected via the component which is neither driven nor braked.

Gear-ratio stages. For the planetary gear shown, four gear-ratio stages in the same direction of rotation and one in the reverse direction of rotation are possible.

Drive. For this purpose, one component of the planetary-gear set is driven via a multi-plate clutch (driving clutch) and set in rotation.

Braking. The corresponding component is connected via a multi-plate clutch (brake clutch) or a brake band with the gearbox housing.

Example of a 3-gear planetary gear

1st gear (Fig. 2). The sun gear is the driving gear, the internal gear is braked. The planet gears circulate on the inner teeth of the internal gear. The planet carrier and the input shaft permanently connected with it rotate in the same direction as the driven sun gear. A large gearing down takes place.

2nd gear (Fig. 2). The internal gear is the driving gear, the sun gear is braked. The planet gears circulate on the outer teeth of the sun gear. The planet carrier and output shaft rotate in the same direction as the driven internal gear.

A smaller gearing down takes place.

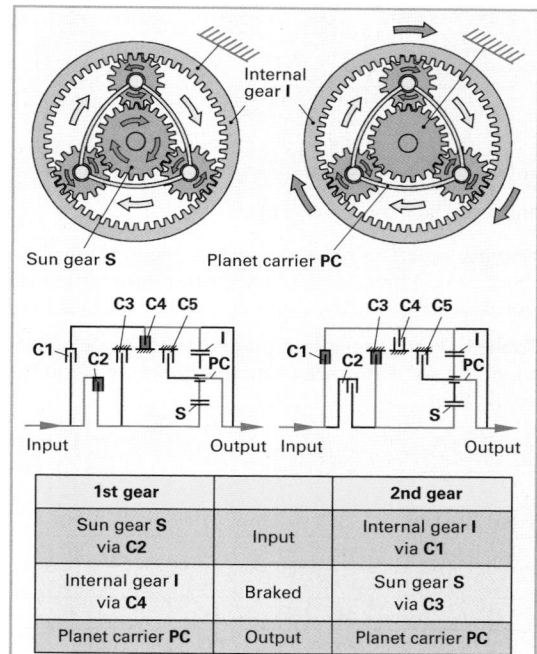

1st gear		2nd gear
Sun gear **S** via **C2**	Input	Internal gear **I** via **C1**
Internal gear **I** via **C4**	Braked	Sun gear **S** via **C3**
Planet carrier **PC**	Output	Planet carrier **PC**

Fig. 2: 1st gear and 2nd gear

3rd gear (Fig. 1, Page 364). The planetary-gear set is blocked by the drive of the sun gear and internal gear. The planet gears stop rotating and act as drivers. The output rotates in the same direction as the input and can in this case take place via the planet carrier.

Reverse gear (Fig. 1, Page 364). The sun gear is the driving gear, the planet carrier is braked. The planet gears reverse the direction of rotation of the internal gear in respect of the input.

A large gearing down is achieved.

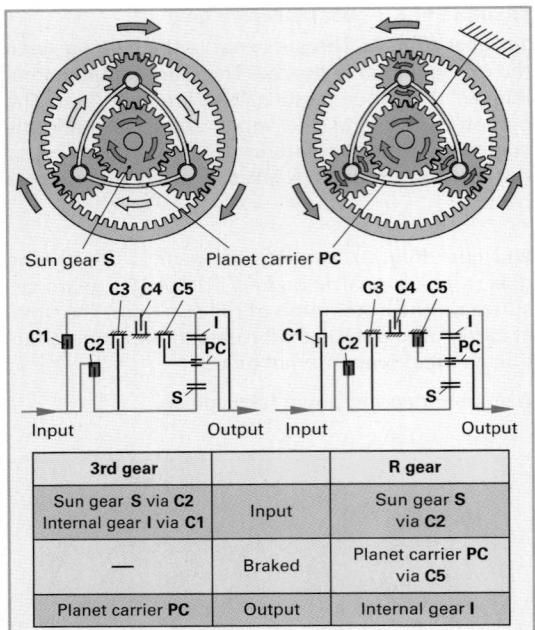

Fig. 1: 3rd gear and reverse gear

3rd gear		R gear
Sun gear **S** via **C2** Internal gear **I** via **C1**	Input	Sun gear **S** via **C2**
—	Braked	Planet carrier **PC** via **C5**
Planet carrier **PC**	Output	Internal gear **I**

Shifting logic. This shows the components of the planetary gear via which input and output take place and which clutches and overrunning clutches are effective.

Table 1 shows the shifting logic for a simple planetary-gear set with three forward gears and one reverse gear.

Table 1: Shifting logic, 3-gear planetary gear

Gear	Input	Braked	Output
1st gear	S	I	PC
2nd gear	I	S	PC
3rd gear	S + I	–	PC
R gear	S	PC	I

S Sun gear   **I** Internal gear   **PC** Planet carrier

One simple planetary-gear set cannot be used for automatic gearboxes because in practice it does not deliver sufficiently useable gear ratios and two output shafts are needed.
For this reason, two or three simple planetary-gear sets are shifted in succession.

Ravigneaux set (Fig. 2). This consists of
- a common internal gear,
- a common planet carrier,
- two different-sized sun gears and
- short and long planet gears.

The different gear-ratio stages are achieved as in the simple planetary-gear set by driving and braking specific parts or by blocking the entire planetary-gear set.

Output can be effected either via the internal gear or via the planet carrier.

The Ravigneaux set shown in **Fig. 2** facilitates 3 forward gears and 1 reverse gear.

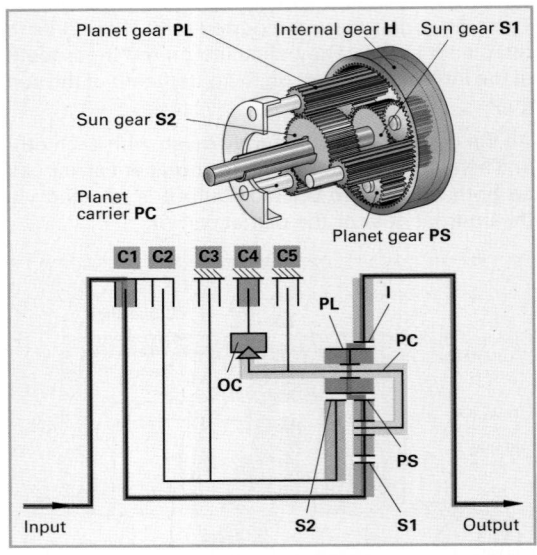

Fig. 2: Ravigneaux set, 1st gear shifted

C1 Driving clutch – drives small sun gear **S1**
C2 Driving clutch – drives large sun gear **S2**
C3 Brake clutch – brakes sun gear **S2**
C4 Brake clutch – brakes overrunning clutch **OC**
C5 Brake clutch – brakes planet carrier **PC**
OC Overrunning clutch – supports planet carrier **PC**

Table 2: Shifting logic, Ravigneaux set

Gear	C1	C2	C3	C4	C5	OC
1st gear	●			●		●
2nd gear	●		●			
3rd gear	●	●				
R gear		●			●	

Example of 1st gear:
C1 and C4 are shifted. C4 holds the planet carrier in one direction of rotation via overrunning clutch.
Power-flow progression:
Input ⇒ C1 ⇒ S1 ⇒ PS ⇒ PL ⇒ I ⇒ Output.

Simpson set (Fig. 1, Page 365). This consists of
- a common sun gear,
- two internal gears with identical diameters and
- two planet carriers with planet gears.

Output is effected via the outer internal gear (I1). The Simpson set is used for example in 4-speed automatic gearboxes in conjunction with a simple planetary-gear set.

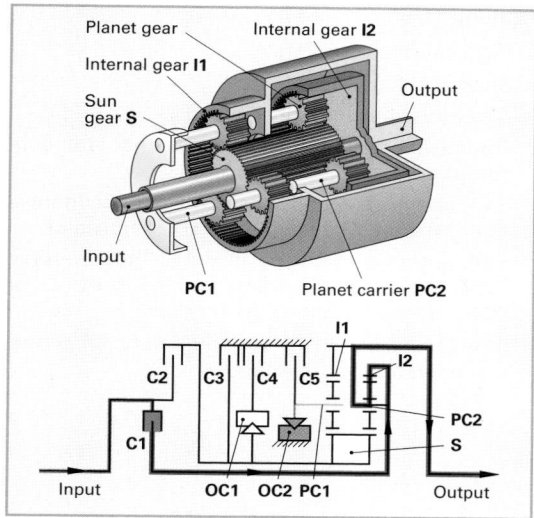

Fig. 1: Simpson set (1st gear shifted)

S	Common sun gear
I1, I2	2 internal gears (outer & inner internal gears)
PC1, PC2	2 planet carriers
P1, P2	Planet gears (same dimensions)
C1, C2	Driving clutches, **C1** drives I2, **C2** drives S
C3, C4, C5	Brake clutches, **C3** brakes S, **C4** brakes OC1, **C5** brakes PC1
OC1, OC2	Overrunning clutches, OC1 supports S when C4 is shifted. OC2 supports PC

Table 1: Shifting logic, Simpson set

Gear	C1	C2	C3	C4	C5	OC1	OC2
1st gear	●						●
2nd gear	●		●1)	●		●	
3rd gear	●	●		●			
R gear		●			●		

● By adherence for selector-lever position D

1) In selector-lever position 2 C3 is shifted, S is torsionally resilient

Combining planetary-gear sets

It is possible by combining planetary-gear sets, e.g. a Ravigneaux set with a downstream simple planetary-gear set or a Simpson set with a downstream simple planetary-gear set, to realise 4- and 5-speed automatic gearboxes.

1 A simple planetary-gear set is made up of which components?
2 How are the different gear-ratio stages achieved in a simple planetary-gear set?
3 How is 1st gear and reverse gear achieved with a simple planetary-gear set?

Wilson set (Fig. 2).

> The Wilson set consists of 3 simple planetary-gear sets which are arranged in succession. Output is effected in all gears via the planet carrier of the rear gear set.

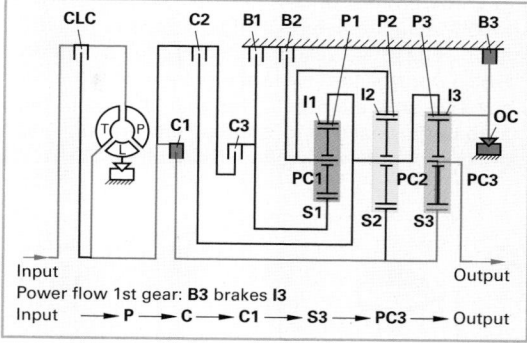

Power flow 1st gear: **B3** brakes **I3**

Input ──► P ──► C ──► C1 ──► S3 ──► PC3 ──► Output

Fig. 2: Wilson set, 1st gear shifted

S1, S2, S3	Sun gears **OC** Overrunning clutch
I1, I2, I3	Internal gears
PC1 ... PC3	Planet carriers
C1, C2, C3	Driving clutches
B1, B2, B3	Brake clutches
CLC	Converter lockup clutch

Gears	Shifting elements in action 1)								i_G	i_s 4)
	C1	C2	C3	B1	B2	B3	OC	CLC		
Neutral					●				–	
1st gear	●					●2)	●		3.57	
2nd gear	●				●			●3)	2.20	
3rd gear	●			●				●3)	1.51	4.46
4th gear	●	●						●3)	1.00	
5th gear	●		●	●				●3)	0.80	
R gear			●			●			– 4.10	

● By adherence   1) Selector-lever position D   4) Spreading
2) Must be shifted for trailing throttle
3) Shifting takes place with slip control

$$i_s = \frac{i_{G1}}{i_{G5}}$$

Lepelletier set

This consists of a simple upstream planetary-gear set and a downstream Ravigneaux set. This enables 6 useable gears to be shifted in the motor vehicle.

Features of planetary gears

- Gears can be shifted without interruption of the power flow.
- Lower tooth forces since the torque slides over several tooth engagements.
- Low-noise operation.

4 In which ways do a Ravigneaux set and a Simpson set differ in terms of design?
5 Explain the design of a Wilson set.
6 Which components in planetary gears drive the sun gear, internal gear and planet carrier and which components brake them?

16.5.3 Electrohydraulic transmission control

Electrohydraulic transmission control involves sensors recording specific operating states. These states are processed by the electronic gearbox control unit. Solenoid valves are electrically activated, depending on the driving situation. These valves actuate hydraulic valves, which control the hydraulic pressure to the respective shift elements. The gear change in the automatic gearbox is effected by driving and braking of different shift elements **(Fig. 1)**.

Features:

- High gearshift comfort.
- Short shifting times.
- Common utilisation of sensors.
- Optimisation of exhaust emissions and consumption.
- Shift-curve selection possible, e.g. Economic, Sport, Winter, Manual (Tiptronic, Steptronic).
- Shift-program matching to driver type possible (**ATS** – adaptive transmission control or **DSP** – dynamic shift-progam selection).
- Simple realisation of various safety functions, e.g. selector-lever interlock.

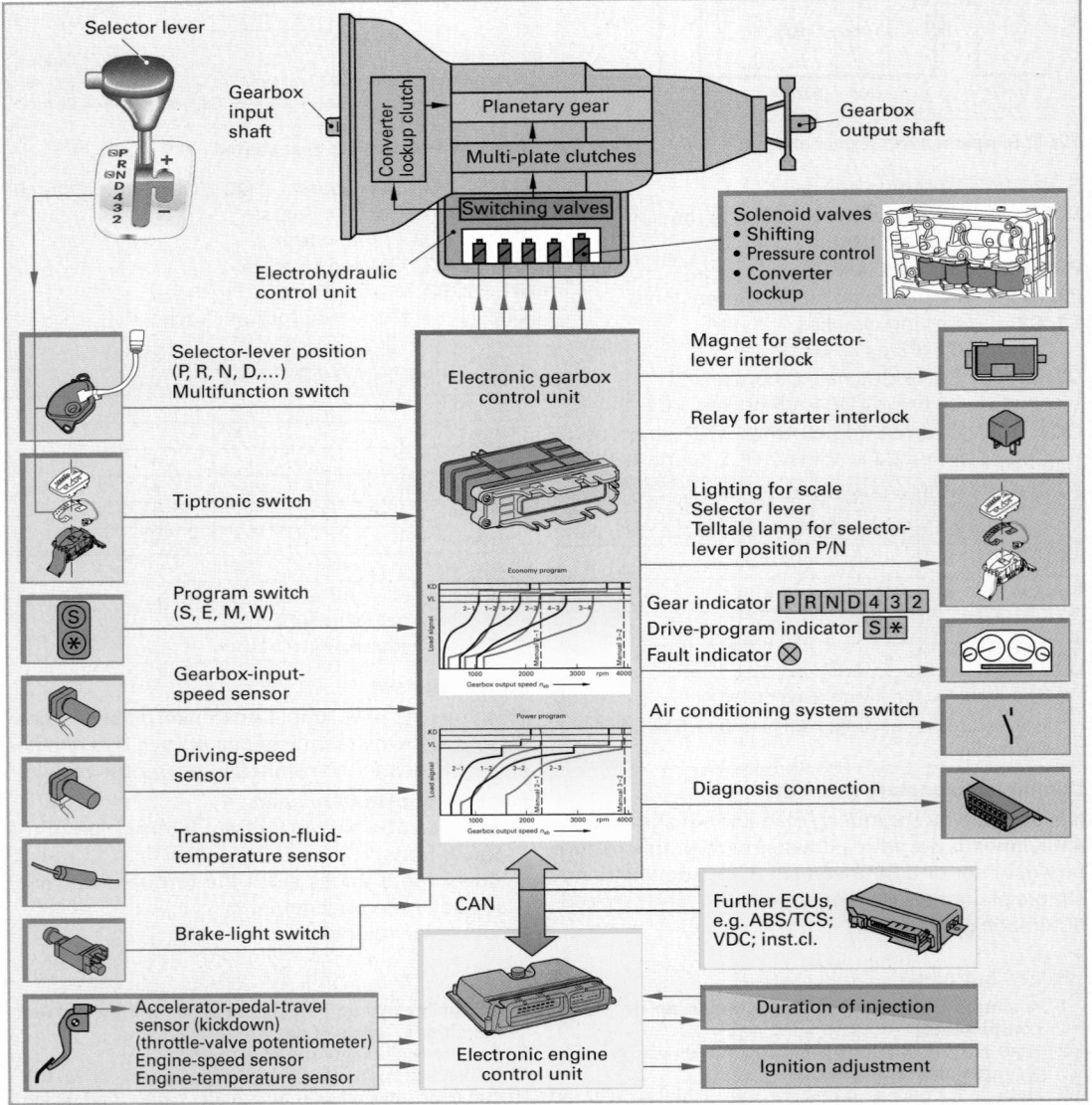

Fig. 1: System diagram – electronic/electrohydraulic transmission control

Design of control system (Fig. 1, Page 366)

The control system consists of

- Sensors, e.g. selector lever with multifunction switch, accelerator-pedal-travel sensor (load signal), speed sensor (these sensors form the main controlled variables)
- Electronic gearbox control unit, which among other things communicates via the CAN bus with other ECUs, e.g. engine control unit
- Electrohydraulic control unit with solenoid valves and hydraulic switching and control valves
- Shift elements, e.g. multi-plate clutches, band brakes, overrunning clutches

Basic operating principle

Electronic gearbox control unit (EGS). This processes the input signals of the different sensors and switches as well as signals from other ECUs via the CAN bus.

Vehicle-side signals:

- Selector lever: $\boxed{P}$ Park, $\boxed{R}$ Reverse, $\boxed{N}$ Neutral, $\boxed{D}$ Drive (all forward gears), $\boxed{4}$ 1st ... 4th gear, $\boxed{3}$ 1st ... 3rd gear, $\boxed{2}$ 1/2 gear
- Tiptronic function = manual shifting
- Program selector switch: $\boxed{S}$ Sport, $\boxed{E}$ Economic, $\boxed{W}$/* Winter – starting e.g. in 2nd gear
- Brake-light switch
- Signals from other vehicle systems, e.g. ABS/TCS, ESP, vehicle-speed controller

Gearbox-side signals:

- Gearbox input speed
- Gearbox output speed / driving speed
- Transmission-fluid temperature

Engine-side signals:

- Accelerator-pedal position with kickdown (throttle-valve position)
- Engine load (injection time)
- Engine speed
- Coolant temperature

The gearshift sequences are selected using stored program maps in the electronic gearbox control unit in accordance with the vehicle's current operating state. The relevant gearshift process and the control of the converter lockup clutch are effected by electric activation of solenoid valves in the **electrohydraulic control unit**.

Additional functions of electronic gearbox control unit EGS

Activation of display in instrument cluster. Gear, program and fault indicators.

Engine intervention. In order to improve shift quality and extend the service life of the shift elements (multi-plate clutches), the engine torque is reduced during the gearshift processes by brief ignition retardation in spark-ignition engines. In diesel engines, the injected fuel quantity is reduced briefly.

Downshift protection. Selector-lever downshifts are only executed if these do not result in excessively high engine speeds.

Selector-lever interlock – shift-lock. Only after the ignition is switched on can the selector lever be moved with the brake applied from the $\boxed{P}$ or $\boxed{N}$ position to a new position so that the vehicle cannot move off unintentionally. An actuator solenoid is activated by the electronic gearbox control unit for this purpose.

R/P interlock. At speeds generally in excess of 10 km/h, the selector lever cannot be moved from $\boxed{R}$ to $\boxed{P}$. This prevents mechanical gearbox damage.

Starter interlock. In order to start the engine, the selector lever must be in the $\boxed{P}$ or $\boxed{N}$ position and the brake pedal must be actuated. Otherwise the start-locking relay is not actuated by the electronic gearbox control unit.

Electrohydraulic system (Fig. 1).
The electrohydraulic system consists of:
- Fluid pump for pressure generation
- Pressure-control valve for operating-pressure control
- Control valves for shift-pressure control
- Hand selector slide for distributing fluid flow to the respective valves
- Switching valves for controlling multi-plate clutches, band brakes and converter lockup clutch

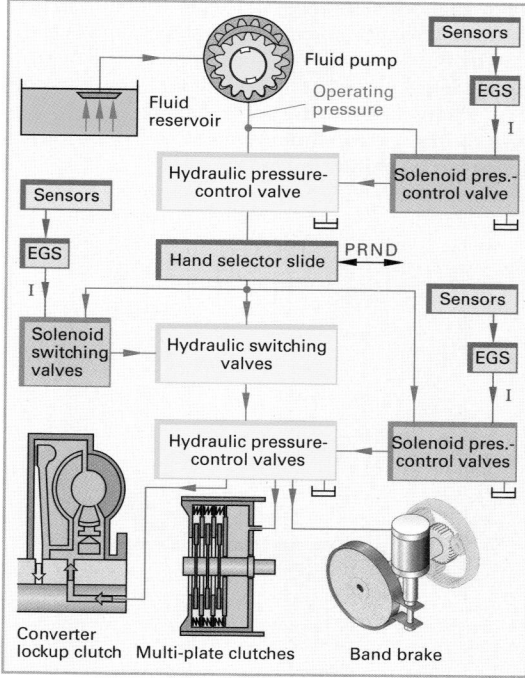

Fig. 1: Block diagram, electrohydraulic system

Shift-point control. Solenoid valves in the electro-hydraulic control unit control hydraulic directional-control valves for shifting the gears.

Shift-quality control. The shift pressure is controlled by means of pressure-control valves as a function of different operating parameters, e.g. load, engine speed, in order to guarantee comfortable shifting.

Control of converter lockup clutch. Pulse-width-modulated solenoid valves control, as a function of program maps, hydraulic control valves which activate and deactivate the converter lockup clutch. The converter lockup clutch can also be operated with slip control in order to damp torsional vibrations generated by the engine.

Pressure generation (Fig. 1, Page 367)

Fluid pump. This is driven by the torque-converter impeller and generates the operating pressure (main pressure).

> The operating pressure is the highest pressure in the hydraulic system (up to 25 bar). All further pressures, such as shift pressures, control pressures, converter filling pressure and lubricating-fluid pressure, are derived from this operating pressure.

Lubricating-fluid pressure. The fluid flows at this pressure through the torque converter and the fluid cooler and lubricates the bearings in the converter and the planetary gear.

Hand selector slide (Fig. 1). This is accommodated in the electrohydraulic control housing and is actuated by the driver via the selector lever. The operating pressure is applied to this slide. Depending on its position, it directs the controlled operating pressure to the corresponding valves. All the forward gears can be shifted in the **D** position.

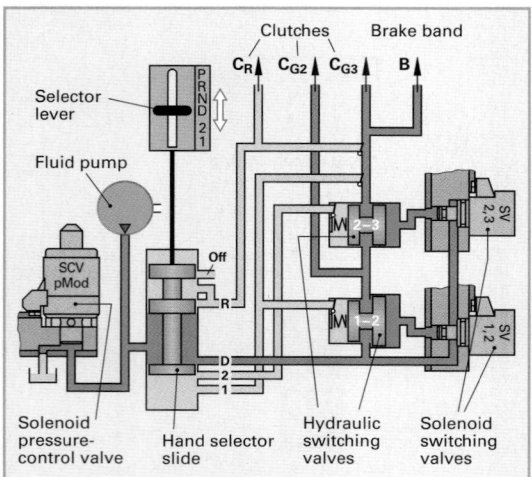

Fig. 1: Hand selector slide in electrohydraulic system

Parking lock. In the **P** position the output shaft of the automatic gearbox is mechanically arrested by a parking-lock pawl and prevents the vehicles from moving **(Fig. 2)**.

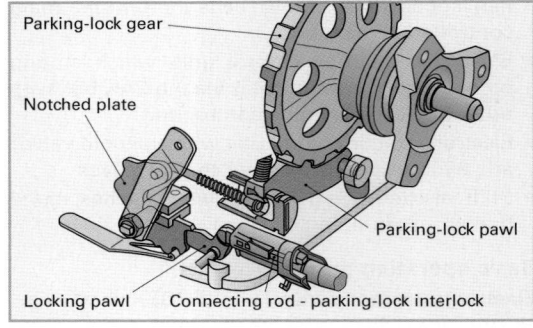

Fig. 2: Parking-lock mechanism

Switching valves. These are activated by solenoid switching valves and direct the controlled operating pressure to the shift elements (multi-plate clutches, band brakes).

Pressure-control valves. These control …

- … the operating pressure as a function of engine load.
- … a variable shift pressure (6 … 12 bar) for comfortable gear changing.
- … activation and deactivation of e.g. two multi-plate clutches by means of shift-pressure control in order to be able to implement overlap control.

Shift elements

> These connect or brake corresponding components of the planetary-gear set.

There are the following different types:
- Drive clutches (multi-plate clutches)
- Brake clutches or band brakes
- Overrunning clutches

Drive clutch (Fig. 1, Page 369)

Clutch closed. The operating pressure is directed by the switching valve and acts on the plunger. The plunger actuates the disc spring, which compresses the plate pack. The frictional/adherent connection is established.

Clutch released. There is no acting operating pressure; the plunger is pressed back by the disc spring. The power flow is interrupted.

By means of appropriate control of the shift pressure, the clutch can be both fully closed and operated with slip. This enables the shift quality to be improved.

The shift pressure is replaced by the operating pressure at the end of the shift process.

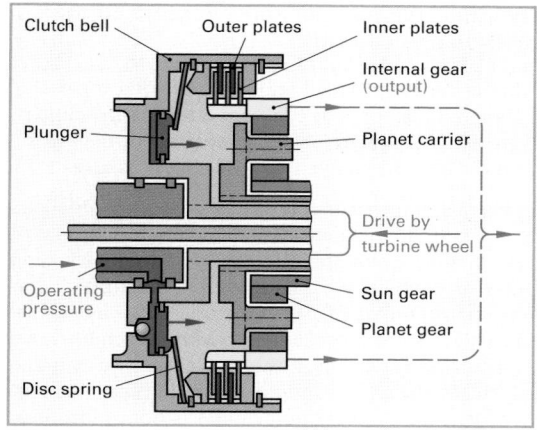

Fig. 1: Multi-plate clutch

Overrunning clutch. Its function is to connect with each other specific parts of the planetary-gear set. The clamping-body overrunning clutch shown in **Fig. 2** consists of an outer ring, an inner ring and the clamping bodies supported in a cage.
When the outer ring rotates in a clockwise direction while the inner ring is braked, the clamping bodies assume an upright position and establish the torque proof connection.
The connection is released when the outer ring rotates in an anticlockwise direction.

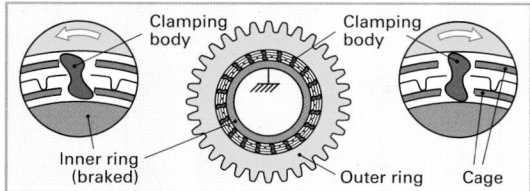

Fig. 2: Clamping-body overrunning clutch

Band brake (Fig. 3). This consists of a steel band, friction lining, plunger rod, plunger, housing, spring and adjuster.

Operating principle. When the operating pressure acts on the plunger surface from the right, the plunger rod tightens the brake band and brakes the brake drum.

The operating pressure acts on the plunger surface from the left to release the brake band.

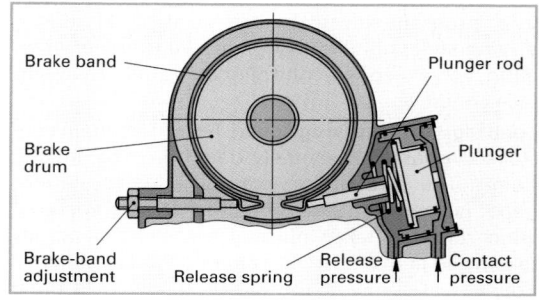

Fig. 3: Band brake

3-speed automatic gearbox with Ravigneaux set (Fig. 4)

The shifting logic **(Table 1)** shows which shift elements (clutches, band brake, overrunning clutch) are shifted in the different gears and which parts of the planetary-gear set are driven or braked by them.

Table 1: Shifting logic

Gear	Input	Fixed	Output	B	C_{G2}	C_{G3}	OC	C_R
1.	S1	S2	PC	●			●	
2.	I	S2	PC	●	●			
3.	S1 + I	–	PC			●	●	
R.	S1	I	PC			●		●

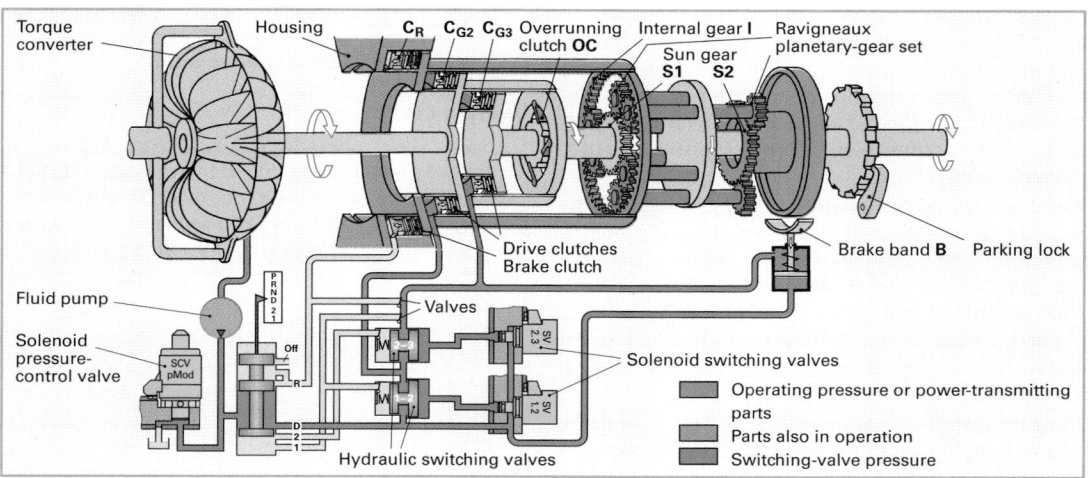

Fig. 4: System diagram, 3-speed automatic gearbox with electrohydraulic control – selector-lever position D, 3rd gear

Influence of different operating parameters on shift program and shift-point control

> The basic shift program with upshift and downshift points is dependent on the main controlled variables of selector-lever position, accelerator-pedal position and driving speed.

The shift program/shift-curve selection can be adapted to different operating parameters, e.g. transmission-fluid temperature, coolant temperature, program-selector-switch position, kickdown, driving style, uphill/downhill gradient, trailer operation, vehicle-speed-controller operation, road-surface condition.

Load signal and driving speed. These two main controlled variables essentially determine the shifting points. The further, for example, the accelerator pedal is pressed, the higher the driving speed at which shifting takes place. Downshifts generally take place at lower driving speeds than upshifts. In this way, a constant shifting back and forth (gear hunting) between two gears is avoided.

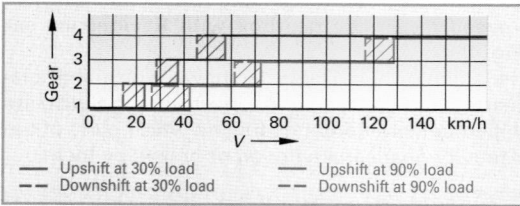

— Upshift at 30% load   — Upshift at 90% load
-- Downshift at 30% load   -- Downshift at 90% load

Fig. 1: Shift diagram – selector-lever position D

Program selector switch (Economy, Sport, Winter, Manual). In the Sport program as opposed to the Economy program upshifts are carried out only at higher driving speeds. In this way, the vehicle has a better acceleration response, but this is accompanied by an increase in fuel consumption. In the Winter program the vehicle is started in a higher gear, e.g. 2nd gear, in order to reduce the drive torque and thus prevent the wheels from spinning. In the Manual program the driver can shift up (**M+**) and down (**M-**) by means of a special selector-lever gutter by touching the selector lever. There is no automatic shifting.

Kickdown (forced downshift). When the accelerator pedal is fully depressed, either a kickdown switch is actuated or the signal is determined by the accelerator-pedal-travel sensor. Where possible, a downshift by one or two gears takes place. The shifted gears are then driven flat out up to the maximum engine revs in each case in order to improve the vehicle's acceleration response.

Transmission-fluid temperature. When a specific critical fluid temperature is reached, shifting is only performed at higher engine speeds. This increases the amount of repumped fluid.

Hydraulic diagram for shift-pressure control- shift-quality control

> In the interests of avoiding gearshift jolts, the multi-plate clutches and the converter lockup clutch are activated by means of solenoid control valves with a load-sensitive, metered shift pressure.

In **Fig. 2** shift-pressure control is shown schematically in a simplified electrohydraulic circuit diagram. At the shifting point the 3/2-way solenoid switching valve is electrically actuated by the electronic gearbox control unit (EGS). The hydraulic 3/2-way switching valve is then pressurised with switching-valve pressure and switches. It actuates a working cylinder, which actuates, for example, the multi-plate clutches. In order that the working cylinder is not pressurised immediately with full operating pressure, a solenoid control valve actuated by the EGS reduces the shift pressure during the shift phase. The extent of the operating pressure is then set as a function of load to the highest currently required pressure. A solenoid control valve with an upstream pressure-limiting valve is actuated by the EGS for this purpose.

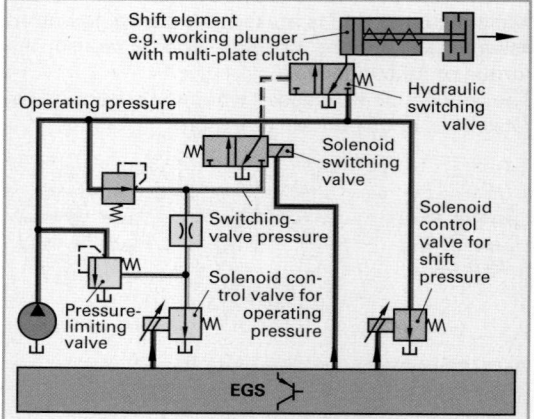

Fig. 2: Diagram for shift-pressure control

Overlap control (Fig. 3)

> Here the pressure is reduced in the shifted clutch **C1** and at the same time increased in the clutch to be shifted **C2**. Shifting can be performed with slip without interrupting the power flow.

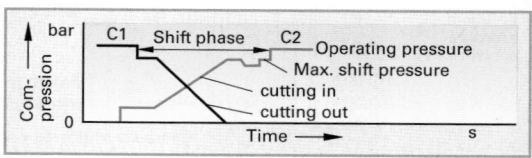

Fig. 3: Pressure control with overlap control

Operating parameters for controlling converter lockup clutch

> This is activated by a solenoid valve as a function of gearbox output speed (driving speed), engine speed, gearbox input speed, brake-light switch and engine temperature.

The converter lockup clutch is usually opened in order ...

- ... to achieve a high starting torque in low gears.
- ... to avoid vibrations in the drivetrain when the engine is cold and at low vehicle speeds.
- ... when operating the brake pedal to prevent the engine from stalling during braking.

It is possible to operate converter lockup clutches with slip control with the aid of electrohydraulic transmission control. In this way, vibrations in the drivetrain are avoided, while converter efficiency is improved.

Special functions

Interlock (keylock). Here the ignition key can only be removed from the ignition lock when the selector lever is in the **P** position. This is effected, for example, mechanically via a cable. This prevents the vehicle from moving/rolling after the ignition key has been removed.

Adaptive transmission control (ATS). This makes use of different criteria to select automatically a suitable shift program, e.g. optimised consumption or sporty, from a range of several programs.

Circuit-diagram example of electronic automatic-gearbox control (Fig. 1).

The circuit diagram shows a simplified example of electronic 4-speed automatic-gearbox control without a CAN bus with two solenoid switching valves, converter-lockup-clutch control and pressure control for the operating pressure.

Power supply. The ECU is supplied via pin 18 by terminal 30 with continuous positive and via pin 17 by terminal 15 (+) with power. Pins 22 and 35 are connected to terminal 31 (earth/ground).

Starting process. The vehicle can only be started with the selector lever in the **P** or **N** position. The start-locking relay is actuated via terminals J and K. At the same time the brake-light switch **S4** must be actuated by the foot brake. The ECU is actuated with positive via pin 11 in the process. This prevents the vehicle from being started unintentionally.

Selector-lever position (Table 1). The selector-lever position switch **S1** is connected via pins 9, 10, 27 and 28 to the ECU. Depending on its respective position, positive is connected via terminals A, B, C and E to the respective pin. The logic is specified in the circuit diagram.

Table 1: Pin activation with positive ⊕						
Pin	**P**	**R**	**N**	**D**	**3**	**2**	**1**
9	⊕	⊕			⊕	⊕	
10		⊕	⊕	⊕	⊕		
27	⊕			⊕		⊕	⊕
28				⊕	⊕	⊕	⊕

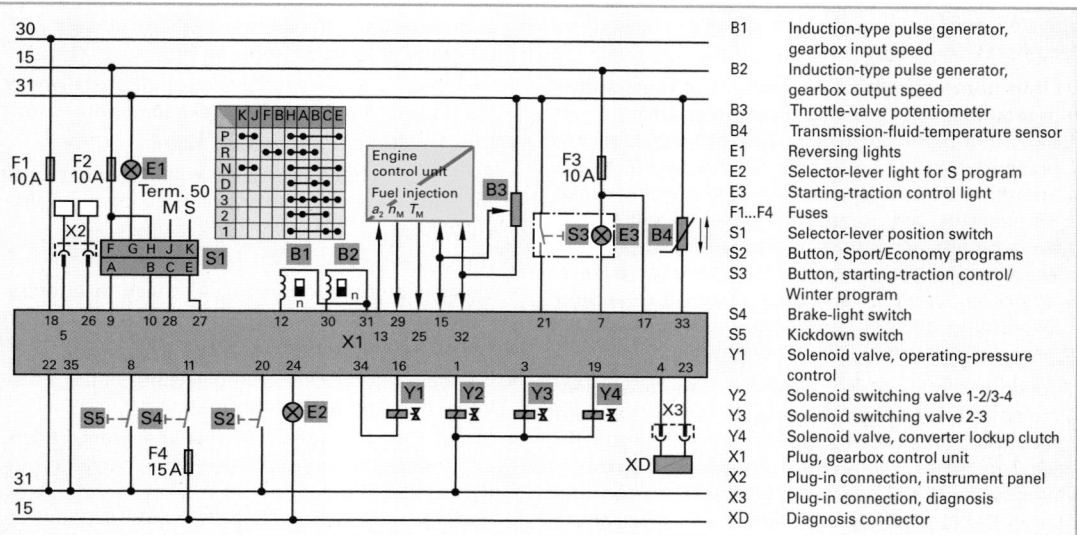

B1 Induction-type pulse generator, gearbox input speed
B2 Induction-type pulse generator, gearbox output speed
B3 Throttle-valve potentiometer
B4 Transmission-fluid-temperature sensor
E1 Reversing lights
E2 Selector-lever light for S program
E3 Starting-traction control light
F1...F4 Fuses
S1 Selector-lever position switch
S2 Button, Sport/Economy programs
S3 Button, starting-traction control/ Winter program
S4 Brake-light switch
S5 Kickdown switch
Y1 Solenoid valve, operating-pressure control
Y2 Solenoid switching valve 1-2/3-4
Y3 Solenoid switching valve 2-3
Y4 Solenoid valve, converter lockup clutch
X1 Plug, gearbox control unit
X2 Plug-in connection, instrument panel
X3 Plug-in connection, diagnosis
XD Diagnosis connector

Fig. 1: Circuit diagram, electronic transmission control (simplified)

Load signal from throttle-valve potentiometer B3. Pin 32 is activated with positive. This results in a constant voltage drop between term. 31 and pin 32. A voltage signal dependent on the throttle-valve position is issued via pin 15 to the ECUs.

Kickdown switch S5. When **S5** is actuated, a circuit is completed in the ECU on the earth/ground side via pin 8.

Sport/Economy button S2. When **S2** is actuated, a seal-in circuit is triggered for the Sport or Economy program via pin 20. When the Sport program is engaged, pin 24 is activated with negative; E2 lights up.

Starting-traction control/Winter-program button S3. When **S3** is actuated, starting-traction control is activated via pin 21. E3 lights up accordingly. The gearbox control unit then effects activation of solenoid switching valves Y2 and Y3 so that the vehicle can be started in a higher gear, e.g. 2nd gear.

Engine-speed signals (B1, B2, n_M). The ECU receives alternating-voltage signals at different frequencies from induction-type pulse generators via pins 12, 30, 31 and 29.

Transmission-fluid-temperature sensor B4 (NTC). As the transmission-fluid temperature increases, the voltage drop between term. 31 and pin 33 (positive) decreases on account of the decreasing resistance of B4.

Engine-temperature signal (t_M). The EGS is activated by the engine control unit via pin 25 for this purpose.

Output signals. Depending on the input signals, output signals are calculated by the gearbox control unit and the corresponding pin is clocked with positive or negative in each case via driver stages. For example

- solenoid switching valves **Y2** (pin 1) and **Y3** (pin 3)
- solenoid operating-pressure-control valve **Y1** (pins 16/34)
- solenoid valve, converter lockup clutch **Y4** (pin 19)

Testing. The following components, for example, can be tested at the connections of the control-unit plug with the aid of a pinbox and a multimeter:

Y1: pin 16 – pin 34	**B1:** pin 12 – pin 31
Y2: pin 1 – pins 22/35	**B2:** pin 30 – pin 31
Y3: pin 3 – pins 22/35	**S2:** pin 20 – pins 22/35
Y4: pin 19 – pins 22/35	**S3:** pin 21 – pin 35

Faults can be read out with the aid of a diagnostic tester via plug-in connection X3 and an actuator diagnosis carried out.

WORKSHOP NOTES

Limp-home operation on account of electrical faults. E.g. cable interruptions, solenoid switching valves defective, no sensor signals, gearbox electronics fails. The vehicle can continue to be moved in selector-lever position D only in one gear, e.g. 2nd gear, and in R. Safety functions such as e.g. selector-lever interlock are if necessary no longer active. On restart, it may not be possible for a selector-lever position to be selected. The faults are stored in self-diagnosis and must be deleted on completion of repairs.

Limp-home operation on account of mechanical-hydraulic faults (gearbox electronics OK). E.g. multi-plate clutch slips due to excessively low pressure build-up, multi-plate clutch worn. This is identified from engine-speed differences if these, for example, are greater than 3%. The last gear to be detected as good remains selected. The reverse gear can be engaged. The selector-lever interlock is active. The fault is reset on restarting. Depending on the manufacturer, these faults are not stored in the ECU's self-diagnosis.

Towing. It is absolutely essential when towing vehicles with automatic gearboxes to follow the manufacturer's instructions to the letter because the fluid pump is not driven. There is therefore insufficient lubrication in the gearbox. The selector lever must be in the **N** position. In vehicles with electromagnetically actuated parking locks, the parking lock must be mechanically unlocked. Towing speed as a rule ≤ 50 km/h; towing distance ≤ 50 km.

Fault diagnosis. To be able to carry out safe diagnosis of faults in the automatic gearbox, it is necessary to perform the following checks before removing the gearbox:

- Fluid level. An excessively high fluid level causes overhard gearshifts and possibly leaks. An excessively low fluid level results in an insufficient frictional connection and thus in slipping shifting points. **Make sure that the correct ATF oil grade is used for topping up!**
- Fluid quality. Burnt-smelling fluid indicates that the multi-plate clutches and/or brake bands are worn.
- Check self-diagnosis with a tester.
- Retrace the upshift and downshift points as a function of selector-lever position, load and driving speed.
- Check the selector-lever setting.
- On older automatic gearboxes – check the throttle-cable adjustment.
- Check the hydraulic pressures.
- If necessary, check the fluid strainer in the shift-valve housing for contamination.
- Check the stall speed. Follow the manufacturer's instructions to the letter when carrying out this check. The gearbox is subject to the risk of damage, e.g. leaks, clutch wear, on account of very intensive fluid heating.

16.5.4 Adaptive transmission control

> Adaptive transmission control (ATS) automatically selects, for example, sporty or optimised-consumption shift programs using differently weighted parameters **(Fig. 1)**. It can in principle be implemented in all fully automated, electronically controlled gearboxes.

Shift-program selection

This is essentially dependent on driver-type evaluation, environment recognition and driving-situation recognition. In addition, shift-program selection is influenced by manual driver-intervention recognition, e.g. by the program selector switch (if fitted) or by selecting the manual shift gutter.

Driver-type evaluation. Here the driver's driving behaviour is evaluated with a dynamic index and from this a suitable shift map is selected. The following features are used to determine the dynamic index:

- **Evaluation of starting process.** An appropriate shift program is selected the first time that the accelerator is pressed, e.g. moderate or forced.
- **Kick-fast evaluation.** The speed at which the accelerator pedal is moved is evaluated. When the accelerator pedal is quickly depressed, the system switches, for example, from an optimised-consumption to a sporty shift program.
- **Kickdown evaluation.** When the accelerator pedal is pressed to the floor, a sporty shift program is selected and, where possible, a downshift by one or two gear steps is performed.
- **Driving evaluation.** In steady-state driving situations, the system switches within a short period of time, for example, from a sporty to an optimised-consumption shift program and shifts to the highest possible gear.
- **Braking evaluation.** The speed reduction caused by braking is evaluated. The downshift points are influenced in this way.

Environment recognition. E.g. winter recognition. Here the system initially selects a performance-reducing shift program and also starts the vehicle in a higher gear. This is detected through comparison of the wheel speeds of the powered axle with the wheel speeds of the nonpowered axle.

Driving-situation recognition. E.g. uphill driving, trailer operation. The system selects optimised torque shift programs in order to avoid gear hunting between two gears.

Gear selection

In addition to shift-program selection, some parameters have a direct influence on gear selection. For example:

- **Cornering recognition.** When the vehicle is cornering at high speed, the system for example does not perform downshifts or upshifts in order to avoid load-change reactions.
- **Downhill-driving recognition.** Here upshifts are avoided so that the engine braking effect can be utilised to better effect.
- **Fast-OFF recognition** (throttling back by the driver). The gear is held when the driver takes his/her foot off the accelerator pedal quickly. In this way, the engine braking effect can be utilised to better effect.
- **Manual driver-intervention recognition** (Tiptronic/ Steptronic M+ / M-). There is no further automatic upshifting or downshifting.
- **ABS/TCS/ESP recognition.** In the event of control interventions by these systems, no gearshifts which have negative effects on traction control are performed.
- **Stop-and-go recognition.** The system does not shift down to 1st gear, thereby reducing fuel consumption.

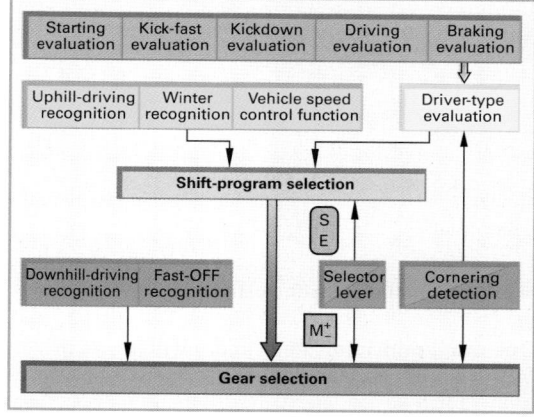

Fig. 1: Program structure of adaptive transmission control

REVIEW QUESTIONS

1 The control system of an automatic gearbox with electrohydraulic control is made up of which components?

2 Which signals are processed by the electronic gearbox control unit?

3 What are the functions of the fluid pump, hand selector slide, switching valves, pressure-control valves and shift elements in an automatic gearbox?

4 Which main controlled variables influence the shifting points in an automatic gearbox?

5 What do you understand by overlap control?

6 On which parameters does activation of the converter lockup clutch depend?

7 What do you understand by shift-lock, interlock?

8 On which parameters does shift-program selection depend for adaptive transmission control?

16.5.5 Continuously variable automatic transmission with pushbelt or link chain

> The gear ratios are varied continuously over the entire working range by the primary and secondary V-pulley pairs (variator).

Gearboxes of this types are also referred to as **CVT transmission** (**C**ontinously **V**ariable **T**ransmission).

Design (Fig. 1)
- Primary V-pulley
- Secondary V-pulley
- Pushbelt
- Multi-plate clutches
- Pressure cylinder
- Planetary-gear set

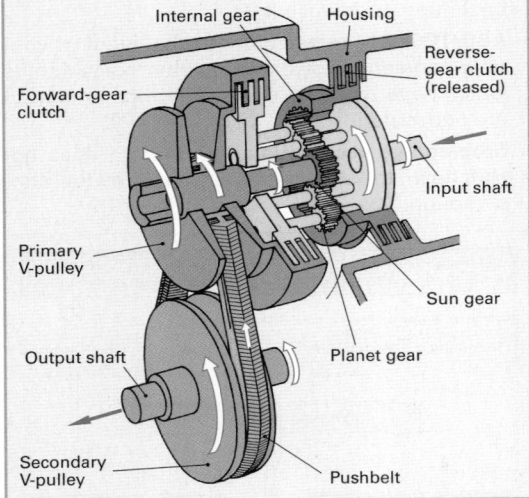

Fig. 1: Automatic CVT with pushbelt

Operating principle. The primary V-pulley is driven by the planetary gear when the forward- or reverse-gear clutch is engaged. It drives the secondary V-pulley via a pushbelt or a link chain **(Fig. 2)**.

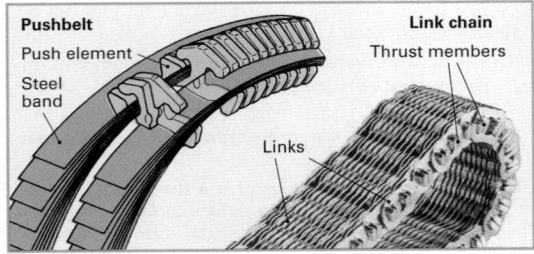

Fig. 2: Pushbelt, link chain

Variation of the gear ratio is achieved by axial displacement of a diagonally opposite pulley half in each case. In this way, the effective lever arms r_{w1} and r_{w2} are **continuously** varied in opposition, i.e. bigger or smaller **(Fig. 3)**. Displacement of the pulley halves is effected by the two pressure-controlled primary and secondary cylinders.

The greatest gear ratio is achieved when the pushbelt engages the smallest effective lever arm r_{w1} of the primary pulley and the biggest lever arm r_{w2} of the secondary pulley.

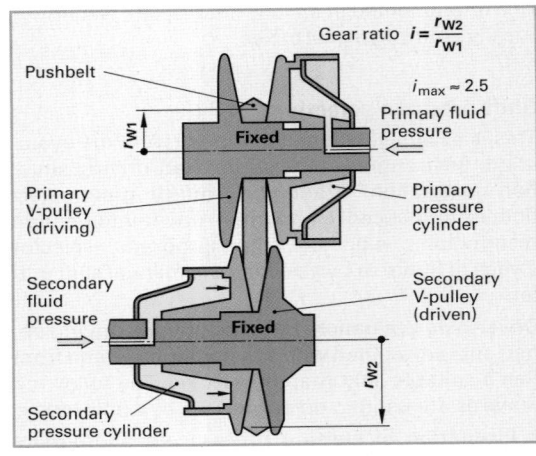

Fig. 3: Pulley position with greatest gear ratio

> The effective gear ratio i in each case is formed by the ratio of the lever arms of the secondary pulley r_{w2} to the primary pulley r_{w1}.

Selector-lever position N (Neutral) and P (Park). Both clutches are released. There is no power transmission. In the **P** position the secondary V-pulley is blocked by the parking lock.

Selector-lever position D (Drive) and L (Load). The forward-gear clutch is closed and the reverse-gear clutch released. The planet carrier, planet gears, internal gear and sun gear circulate as a single block and drive the secondary V-pulley via the primary V-pulley and pushbelt or link chain. The secondary pulley directs the torque to the output shaft.
The input shaft, primary and secondary V-pulleys and output shaft all rotate in the same direction.

Selector-lever position R (Reverse). The forward-gear clutch is released and the reverse-gear clutch closed. This brakes the internal gear on the transmission housing. The planet-gear pairs driven via the planet carrier reverse the direction of rotation and gear down.

Control. This is effected electrohydraulically as a function of selector-lever position, accelerator-pedal position and driving speed.
A magnetic-particle clutch, a hydrodynamic torque converter or a slip-controlled multi-plate clutch is used, for example, for starting.

Ecotronic-continuously variable automatic transmission

> This is an automatic transmission with 2 V-pulley pairs (variators) and a link chain, in which the gear ratios are continuously varied.

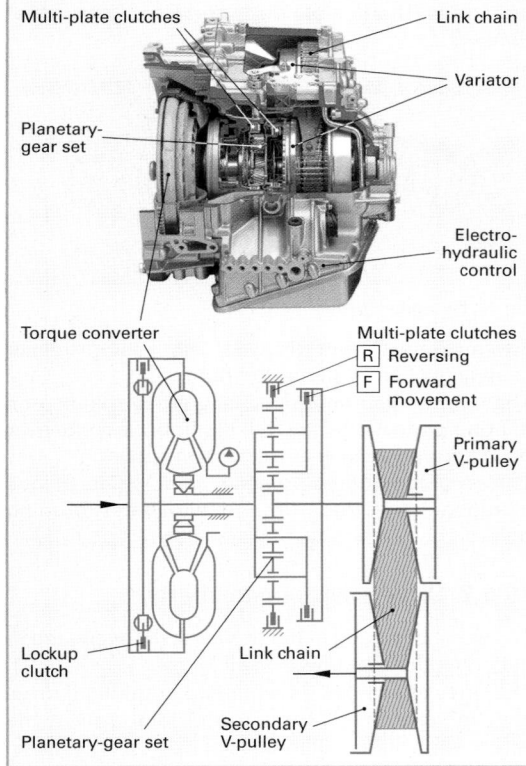

Fig. 1: Ecotronic CVT transmission

Operating principle

Power transmission. This passes from the primary V-pulley via a link chain to the secondary V-pulley. In contrast to the pushbelt, the link chain transmits the force not by thrust but rather by tension.

Gear-ratio variation. This is performed continuously by axial displacement of the opposite pulley halves of the variators. In the process, the effective lever arms become alternately bigger and smaller.

Transmission control. This is performed electrohydraulically and controls the hydraulic pressures, which effect axial displacement of the pulley halves. The main controlled variables are: selector-lever position, selected shift program, driving speed and accelerator-pedal position (load). Transmission control is adaptive. The correct gear ratio is selected on a self-learning basis.

Starting. A converter with lockup clutch is used for this purpose.

Multitronic-continuously variable automatic transmission

> In a Multitronic transmission a link chain with thrust members is used between the variators for power transmission.

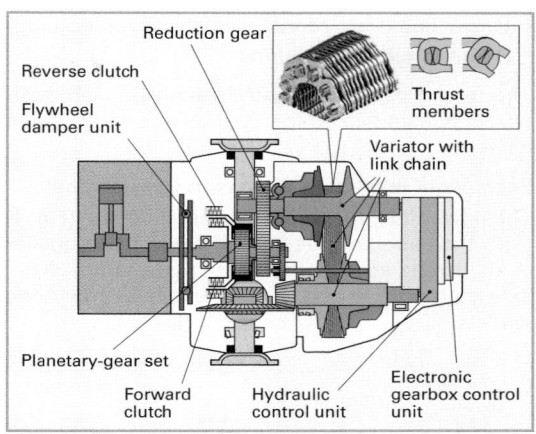

Fig. 2: Design of Multitronic, link chain

Power transmission. The engine torque is transmitted via the flywheel damper unit, multi-plate clutch, planetary-gear set and reduction gear to the primary V-pulley **(Fig. 2)**.
There is a virtually loss-free transmission of torque to the secondary V-pulley via a link chain with thrust members.

Gear-ratio variation. This is performed continuously by axial displacement of two diagonally opposed pulley halves of the variators.

Transmission control. This is performed electrohydraulically and selects adaptively (self-learning) the correct gear ratio from the program map.

Transmission chart (Fig. 3). The range for all possible gear ratios of the Multitronic CVT automatic transmission is situated in the control map between the most economical and the sportiest curves.
6 curves for the gears are available for the manually Tiptronic-shifted gears with stepped gear ratios.

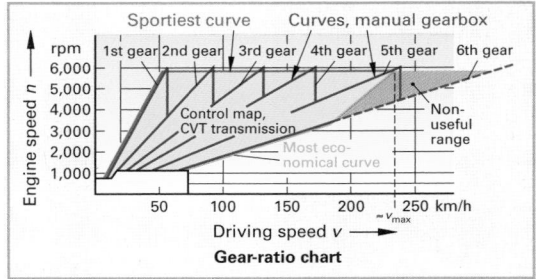

Fig. 3: Multitronic transmission chart

16.6 Propeller shafts, drive shafts, joints

Functions

- Transmit torques
- Facilitate angular variations
- Permit linear variations (axial displacement)
- Damp torsional vibrations

The torque converted by the variable-speed gearbox is transmitted to the final drive and the drive wheels.

Example of rear-wheel drive with front engine:

The power flow progresses in the drivetrain (**Fig. 1**) from the variable-speed gearbox via the propeller shaft (cardan shaft) to the final drive and on via the axle shafts and constant-velocity joints to the drive wheels.

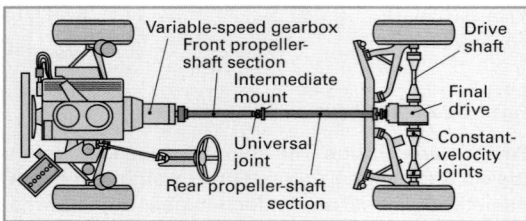

Fig. 1: Drivetrain, rear-wheel drive with front engine

Example of front-wheel drive with front engine and rear-wheel drive with rear engine:

The power flow progresses in the drivetrain from the variable-speed gearbox via the final drive, constant-velocity joints and drive shafts to the drive wheels.

No cardan shaft is required here.

The variable-speed gearbox and the final drive are accommodated in a single housing.

16.6.1 Propeller shafts

In vehicles with front engines and rear-wheel drives, these are situated between the variable-speed gearbox and the final drive in the vehicle longitudinal direction.

Propeller shafts consist of a shaft tube with slide and joints, e.g. two universal joints (**Fig. 2**).

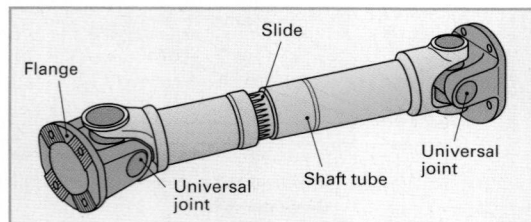

Fig. 2: Propeller shaft with two universal joints

If in vehicles with independent suspension a large distance has to be covered between the variable-speed gearbox and the final drive, a two-piece propeller shaft is used which is supported by an intermediate mount (**Fig. 3**).

Universal joints are deployed to facilitate an axis offset between the variable-speed gearbox and the final drive. The flexible discs serve to damp vibrations.

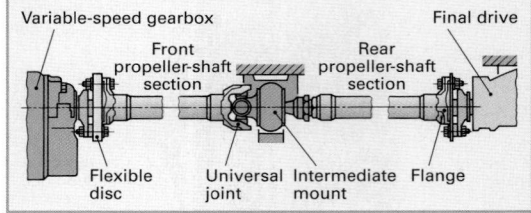

Fig. 3: Two-piece propeller shaft

Intermediate mount (Fig. 3). The split propeller shaft is resiliently supported here.

The intermediate mount is secured by means of a bearing pedestal to the vehicle floor. It contains a ball bearing which is embedded in rubber.

The separation of the propeller shaft results in low-vibration, quiet running and eliminates droning noises.

16.6.2 Drive shafts (axle shafts)

These are arranged in the drivetrain between the final drive and the drive wheels.

The drive shafts can be equipped at the final-drive end, for example, with a tripod joint and at the wheel end with a ball joint.

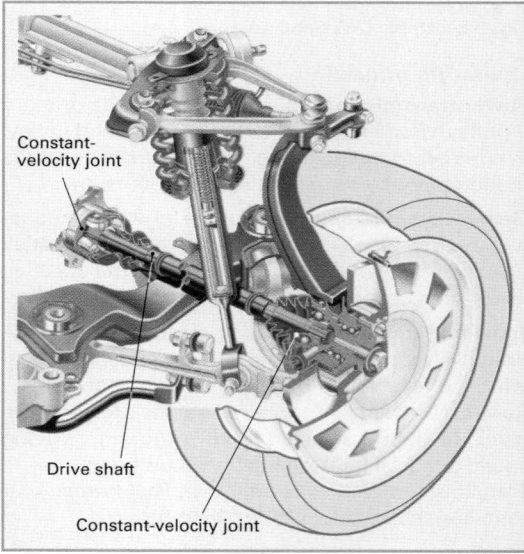

Fig. 4: Drive shaft in a front-wheel drive

16.6.3 Joins

The following are used:
- Flexible discs
- Tripod joints
- Universal joints
- Ball joints
- Double joints

Universal joints (Fig. 1). The link forks are flexibly connected with each other by the joint bolts arranged in the spider. The joint bolts are usually mounted in the link forks in fully encapsulated needle bearings (therefore requiring no maintenance).

In motor vehicles universal joints are used for diffraction angles up to 8°.

Special designs, e.g. for power takeoff units, permit greater diffraction angles.

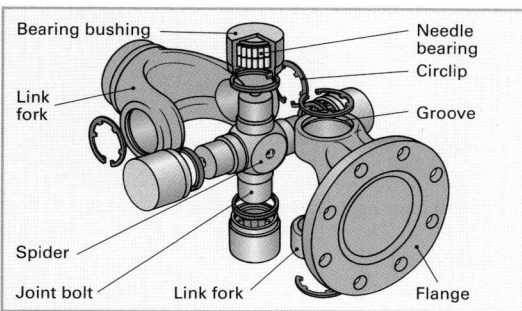

Fig. 1: Universal joint

When an angled universal joint is used, a non-uniform motion is generated at the output end.

If a diffraction angle β exists between the input and output of a universal joint **(Fig. 2)**, the output shaft with the input shaft at uniform rotational speed ω_1 executes a non-uniform motion with sinusoidally alternating rotational speed ω_2.

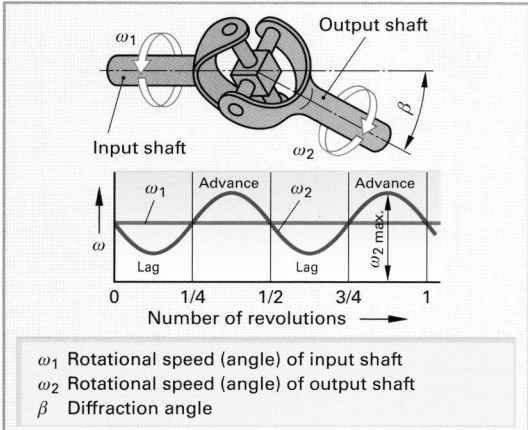

ω_1 Rotational speed (angle) of input shaft
ω_2 Rotational speed (angle) of output shaft
β Diffraction angle

Fig. 2: Universal joint with diffraction angle

Gimbal error. With 1 revolution of the input shaft, two advances and two lags occur at the output shaft (gimbal error) **(Fig. 2).**

A propeller or drive shaft with **one universal joint** may only be used when **small diffraction angles β** occur. When larger diffraction angles occur, for instance on vehicles with rigid axles, the propeller or drive shaft must be fitted with **two universal joints (Fig. 3).**

In this way, the so-called "gimbal error" of joint A is compensated by an equal but opposite "gimbal error" of joint B (ω compensation).

Conditions for compensating the "gimbal error":

- The diffraction angles β_1 of joint A and β_2 of joint B must be equal.
- The link forks of the connecting shaft must lie in the same plane. This must be observed particularly in the assembly of the intermediate shaft (slide).

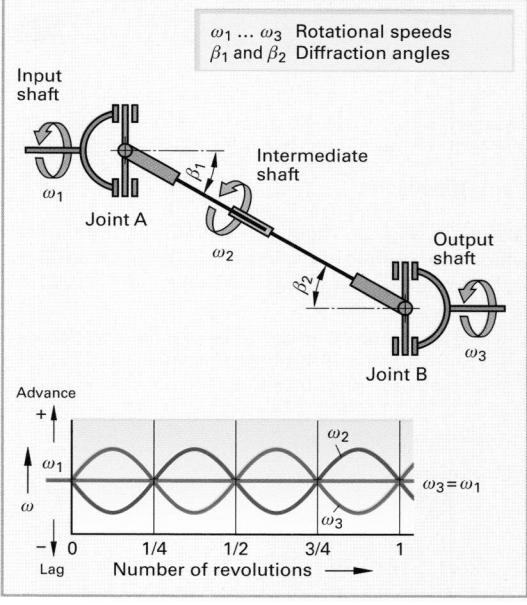

Fig. 3: Propeller shaft with two diffracted universal joints

The distance variations (linear variations) that occur during deflection between the universal joints are compensated by a **slide**.

Universal joints are used for example on propeller shafts between the variable-speed gearbox and the final drive; in commercial vehicles, they are also used on axle shafts.

Constant-velocity joints

> Constant-velocity joints (homokinetic joints) transmit the rotary motion uniformly even with larger diffraction angles.

Sliding constant-velocity joints

Tripod joints (Fig. 1). These can be used in the case of independent suspension both on powered front axles (front-wheel drive) and on powered rear axles (rear-wheel drive).

> Tripod joints permit diffraction angles up to 26° and axial displacement up to 55 mm.

The tripod star is always turned towards the final-drive end.

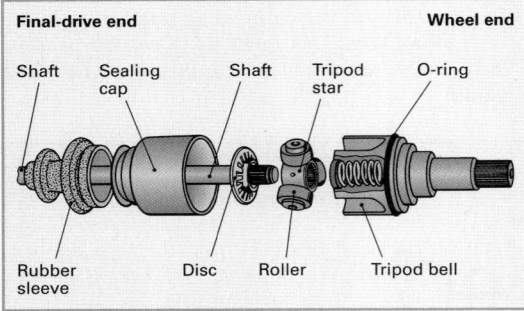

Fig. 1: Tripod joint

Pot joints (Fig. 2). These are ball joints, the balls of which are guided by a cage and run on **straight tracks** of the ball star and the ball shell.

> Pot joints permit diffraction angles up to 22° and axial displacement up to 45 mm.

Pot joints are situated at the final-drive end.

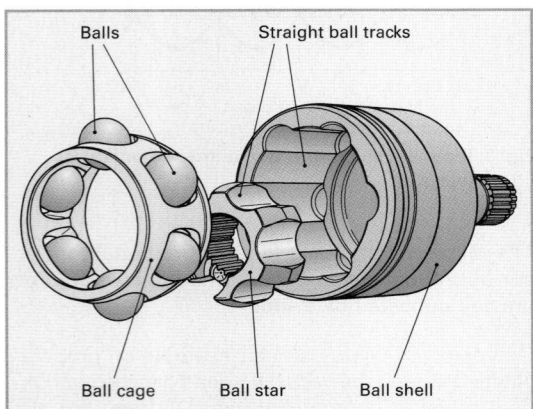

Fig. 2: Pot joint

Fixed constant-velocity joints
Ball joints

These consist of the ball star, ball shell, ball cage and balls **(Fig. 3)**.
The ball shell and ball star have curved tracks, on which the balls run.

> Ball joints permit diffraction angles up to 38° in their normal version and up to 47° in their special version. They do not permit any axial displacement.

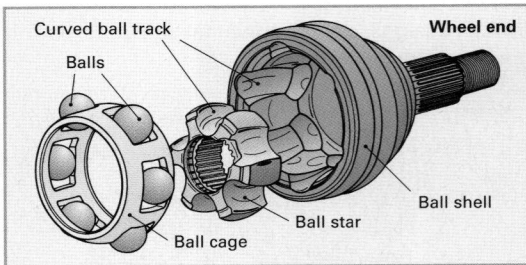

Fig. 3: Ball joint as fixed constant-velocity joint

Double joints

Two universal joints are combined to form a single joint **(Fig. 4)**. In order to ensure fault-free operation, the shaft ends to be connected are centred on the inside of the joint.
They are used in commercial vehicles.

> Double joints permit diffraction angles up to 50°. They do not permit any axial displacement.

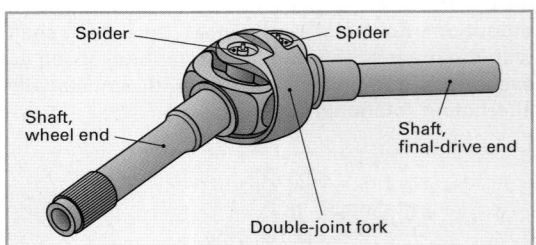

Fig. 4: Double joint

Flexible discs

Flexible discs are resilient, maintenance-free joints. They permit only small diffraction angles and linear variations. They are installed in the drivetrain primarily as flexible elements for damping vibrations and noises. Flexible discs are used in vehicles whose final drives are permanently connected with the body or frame.

There are the following different types:

- Hardy discs
- Silentbloc joints

Hardy discs (Fig. 1)

Several steel bushings (e.g. 6) are wrapped in textile cords so that one coil pack passes round two adjacent bushings. The textile cords and steel bushings are vulcanised in rubber.

Hardy discs are used in the drivetrain as flexible intermediate links, for example, on two-piece propeller shafts.

> Hardy discs permit diffraction angles up to 5° and axial displacement up to 1.5 mm.

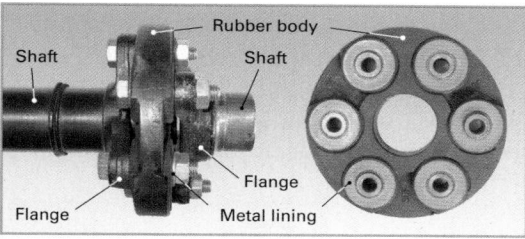

Fig. 1: Hardy disc

Silentbloc joints (Fig. 2). Several silentblocs (e.g. 6), consisting of rubber bodies with sleeve inserts, are combined in a metal jacket and bolted on both sides to three-arm flanges. Depending on the propeller-shaft connection, the centre section can be floating or centred.

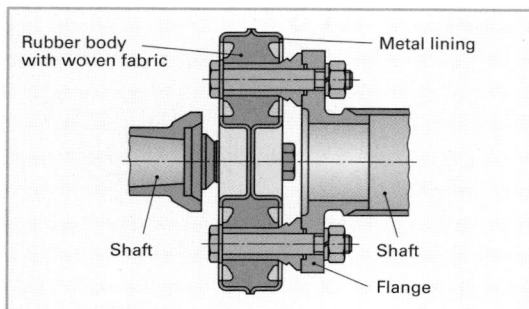

Fig. 2: Silentbloc joint

REVIEW QUESTIONS

1 **What are the functions of propeller shafts?**

2 **Into which categories are joints subdivided?**

3 **Which constant-velocity joints are used in vehicle manufacturing?**

4 **What is the function of the slide of a propeller shaft?**

5 **What are the functions of flexible discs?**

16.7 Final drive

Functions

- **Transmits and multiplies torque.** The torque converted by the variable-speed gearbox must be multiplied in the final drive so that sufficient torques are available at the drive wheels for all driving states.

- **Gears down engine revolutions.** The engine revolutions converted by the gearbox are geared down by the constant gear ratio of the final drive.

- **Diverts the power flow, if necessary.** If the engine is installed in the vehicle longitudinal axis, the power flow must be diverted through 90° by a bevel-gear final drive because the drive shafts are always arranged transversally to the vehicle longitudinal axis **(Fig. 3)**.

In vehicles with engines installed transversally to the vehicle longitudinal axis, the direction of the power flow does not have to be diverted. Spur-gear final drives are used here.

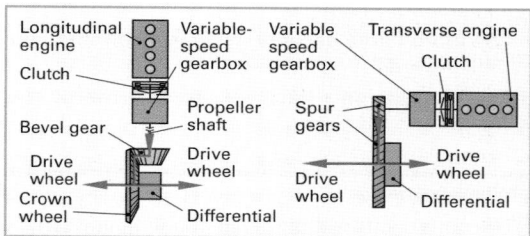

Fig. 3: Bevel-gear final drive, spur-gear final drive

16.7.1 Bevel-gear final drive

A bevel-gear final drive consists of a bevel pinion and a crown wheel.

A distinction **(Fig. 4)** is made between bevel-gear final drives

- with **non-offset axes** and
- with **offset axes** (hypoid drive).

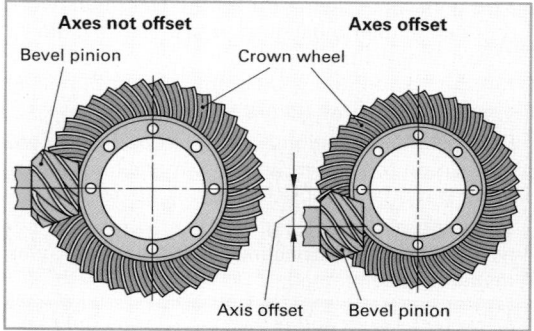

Fig. 4: Final drive with offset and non-offset axes (hypoid drive)

16

Advantages of hypoid drive

- **Increased smooth running** because a greater number of teeth meshes.
- **Higher load capability** because the diameter and the tooth widths of the bevel pinion are larger.
- **Takes up less space** because the crown wheel while being subject to the same load has a smaller diameter. Thus, the propeller shaft can be mounted in a lower position in vehicles with front engines and rear-wheel drive. The propeller-shaft tunnel and the centre of gravity are lower.

As a result of the axis offset, greater friction movements occur during rolling contact between the touching tooth flanks than is the case with non-offset axes. This makes it necessary to use particularly pressure-resistant hypoid oils.

Two different types of toothing may be used: Gleason toothing or Klingelnberg toothing.

Gleason toothing (Fig. 1)

- The tooth flanks of the crown wheel form parts of the arc of a circle.
- The tooth backs become narrower from the outside inwards.
- The tooth heights become smaller towards the inside.

Klingelnberg toothing (Fig. 1)

- The tooth flanks form parts of a spiral.
- The tooth backs maintain a constant width from the outside inwards.

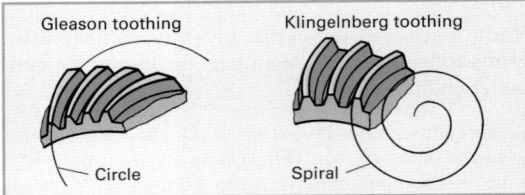

Fig. 1: Gleason and Klingelnberg toothing

16.7.2 Spur-gear final drive (Fig. 2)

This consists of a small input spur gear and a large output spur gear. Both gear wheels have helical teeth, which are cheaper to manufacture than spiral teeth.

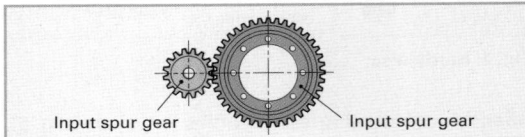

Fig. 2: Spur-gear final drive

WORKSHOP NOTES

The correct interaction of bevel pinion and crown wheel is essential to low-noise operation by and a long service life for the final drive. Because bevel pinion and crown wheel are matched in pairs to each other for fault-free operation, they are provided with markings by the manufacturers (**Fig. 3**). They are given a pair number p, which is marked on the face of the bevel gear and on the flange side at the top of the crown wheel.
R and T are design dimensions.

The dimensional deviations r and t from these design dimensions are determined by the manufacturer when the gears are run in. The gears work together at their most quiet with these deviations. These dimensional deviations r and t must therefore be taken into consideration when the bevel gear and crown wheel are adjusted.

Dimensional deviation t and tooth backlash z are marked on the crown wheel. Dimensional deviation r is marked on the face of the bevel pinion.

The teeth of the bevel gear and crown wheel between which the specified tooth backlash was measured are marked.

> The bevel gear and crown wheel may only be replaced in pairs.

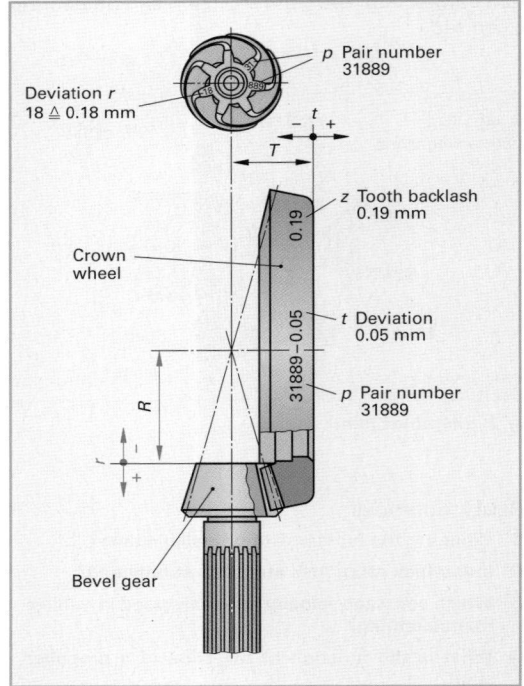

Fig. 3: Bevel pinion and crown wheel

WORKSHOP NOTES

Testing for true running and concentricity

Every change in the distance between the crown wheel and bevel gear results in a change in the backlash and the tip clearance (distance between tooth tip and tooth gullet).

This results in the bevel gears no longer working flawlessly together.

A lateral run-out of the bevel gears will also alter the backlash.

For this reason, it is essential to test above all the crown wheel (after it has been flanged to the differential housing) **laterally for true running** and **circumferentially for concentricity** with a dial gauge **(Fig. 1)**.

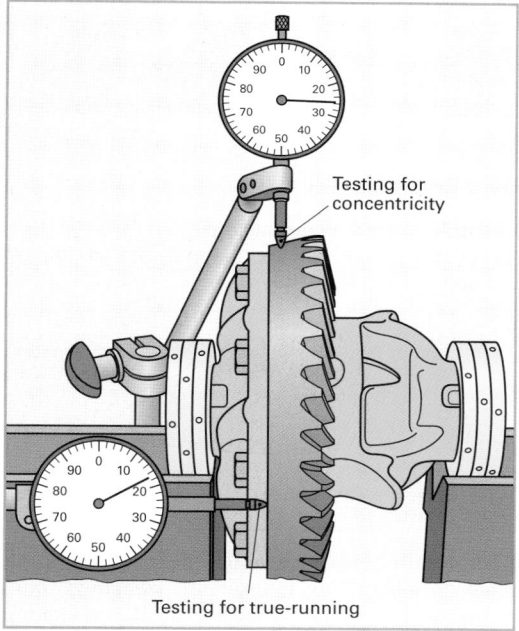

Fig. 1: **Testing the crown wheel for true-running and concentricity**

Adjustment work to bevel-gear final drives with offset axes

Prior to assembly, the correct position of the bevel gear and crown wheel must be ascertained.

Adjusting the bevel gear

Remove the crown wheel, fit the new bevel gear (bevel pinion) to be installed initially without compensating shims **(Fig. 2)**.

After inserting the plug gauge and attaching the measuring cylinder, measure the **difference in level** between plug gauge and measuring cylinder with a dial gauge mounted on the measurement rail.

This corresponds to the indicated measured value **M**.

The plug gauge and measuring cylinder can be obtained from the final-drive manufacturers.

Calculate the thickness **S** of the individual compensating shims from the measured value **M** and the deviation **r** entered on the face of the bevel gear ($\triangleq$ reference dimension K).

> Thickness **S** = measured value **M** – reference dimension **K**

From a set of compensating shims select a shim of the thickness calculated. Remove the bevel gear, insert the selected shim and reinstall the bevel gear. During the reference measurement the dial gauge must indicate the reference dimension **K** ($\triangleq$ deviation **r**).

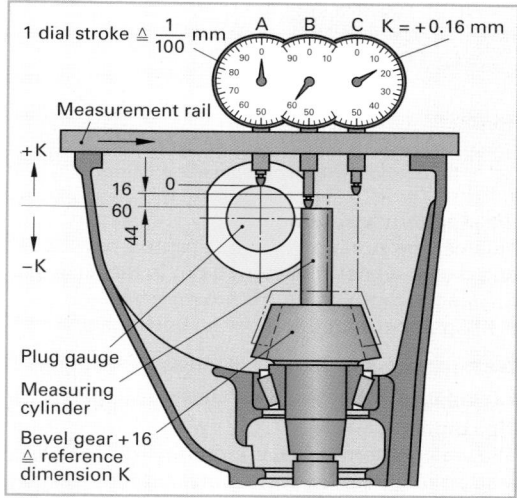

Fig. 2: **Measurement with installed bevel gear**

Example of measuring sequence (Fig. 2)

The crown wheel is removed

- Install the bevel gear without shims
- Insert the plug gauge and attach the measuring cylinder
- Install the dial gauge in the measurement rail and set to 0 at the highest point of the plug gauge (**A**)
- Read off the difference in level between plug gauge and measuring cylinder on the dial gauge (**B**), read off here: 60 units $\triangleq$ 0.6 mm
- Check reference dimension **K** (**C**)
- Calculate the thickness of the compensating shims:
  $S = M - K$ = 0.60 mm – 0.16 mm = 0.44 mm
  Thickness of shim to be inserted **S = 0.44 mm**
- Remove the bevel gear, insert a 0.44 mm thick shim, reinstall the bevel gear
- Carry out the reference measurement: the dial gauge must indicate the reference dimension **K** = 0.16 mm.

16

WORKSHOP NOTES

Adjusting the crown wheel

It is possible to correct the adjustment of the crown wheel by inserting compensating shims (**Fig. 1**) or by turning adjustment nuts.
Correct adjustment can be checked by measuring the tooth backlash.

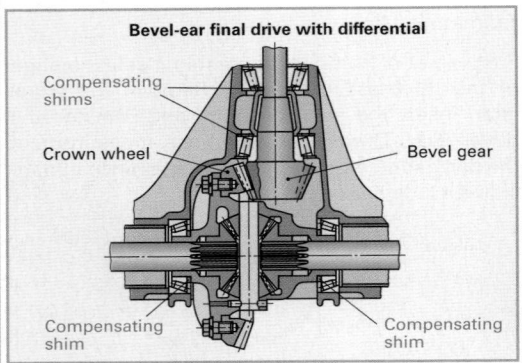

Fig. 1: Adjustment with compensating shims

Wear-pattern assessment

This can be done after adjusting the bevel gear and crown wheel. It is carried out in the case of
• Gleason toothing on the crown wheel
• Klingelnberg toothing on the bevel gear

Wear-pattern assessment of Gleason toothing

Examples of defective wear patterns (Fig. 2)

Tip contact

The wear pattern is in the tooth-tip area. To correct, position the bevel gear at a higher level by inserting a thicker compensating shim.

Root contact

The wear pattern is in the tooth-root area. To correct, position the bevel gear at a lower level by inserting a thinner compensating shim.

Heel contact

The wear pattern is in the tooth-heel area. To correct, move the crown wheel closer to the bevel gear.

Toe contact

The wear pattern is in the tooth-toe area. To correct, move the crown wheel away from the bevel gear.

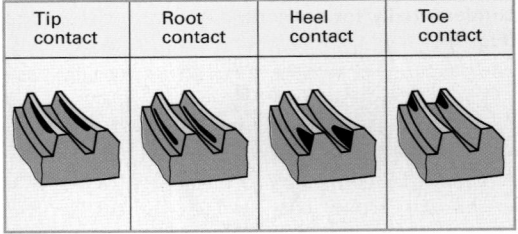

Tip contact	Root contact	Heel contact	Toe contact

Fig. 2: Defective wear patterns

Example: correct wear pattern

It has an oblong, convex shape in the middle of the active flank (**Fig. 3**).

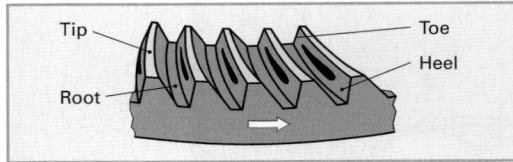

Fig. 3: Correct wear pattern

16.8 Differential

Functions

• Balances speed discrepancies between the drive wheels.
• Distributes torque in equal proportions to the drive wheels.

Balancing speed discrepancies between the drive wheels. When a vehicle is driving through a bend/curve, the outside wheels must cover a greater distance than the inside wheels. Different road surfaces also cause discrepancies in the distance covered. The wheels on one axle therefore rotate at different speeds.

> The differential balances the speed discrepancies between the drive wheels. The outside drive wheel during cornering rotates faster at the same rate that the inside drive wheel rotates more slowly.

Distributing torque in equal proportions to the drive wheels. The differential transmits equal torque to both drive wheels even if, for example during cornering, one drive wheel is rotating faster than the other.

> The amount of transmitted torque is determined by the drive wheel which has the poorer adhesion on the road surface.

Types

• **Bevel-gear differential**. This is accommodated together with the final drive in a single housing (differential).

• **Worm gear differential (Torsen differential)**. This is installed, for example, in all-wheel-drive vehicles as a transfer box between the powered axles and has an automatic locking effect.

Bevel-gear differential
Design (Fig. 1)

The bevel pinion is connected for example with the propeller shaft and drives the crown wheel, which is bolted to the differential housing. The differential bevel gears are mounted so that they can rotate in the differential housing. The differential bevel gears mesh with the axle-shaft gears, which are connected with the axle shafts.

Operating principle

Straight-ahead driving. Both drive wheels and axle-shaft gears rotate at the same speed. The differential bevel gears do not rotate. They act as drivers and transmit the drive torque in equal proportions to the left and right axle-shaft gears.

One wheel spins, the other is stationary. The axle-shaft gear of the spinning wheel causes the rotation of the differential bevel gears, which circulate on the stationary axle-shaft gear.

The speed discrepancy is balanced by the fact that the spinning wheel rotates twice as fast as the crown wheel.

Torque distribution is effected in equal proportions to the axle-shaft gears and is directed at the drive wheel with the poorer adhesion. Propulsion of the vehicle is not possible.

Cornering (example: left-hand bend, Fig. 1). The outside drive wheels cover a larger distance and rotate faster than the inside wheels.

This is made possible by the differential bevel gears, which balance speed discrepancies between the left and right axle-shaft gears.

Here the differential gears which are mounted in the differential housing rotate about their axes.

Fig. 1 shows the directions of rotation of the shafts and bevel gears on a left-hand bend.

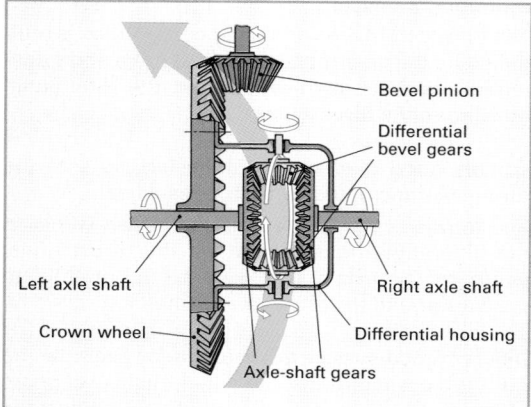

Fig. 1: Bevel-gear differential

The inside left drive wheel rotates more slowly at the same rate that the outside right drive wheel rotates faster. The speed discrepancy is balanced by the differential bevel gears.
Each drive wheel receives equal torque.

16.9 Differential locks

A distinction is made between
● interwheel locks and
● longitudinal locks

Interwheel lock. This locks speed balancing between the drive wheels of one axle.

> An interwheel lock allocates more torque to the drive wheel of an axle which has the better road-surface adhesion (traction).

Example: If for instance a drive wheel spins on an icy road surface or soft ground, this wheel transmits insufficient motive force to the road surface to propel the vehicle.

The differential has a disadvantageous effect here because the other drive wheel with good road-surface adhesion is allocated the same torque. More torque is allocated to the drive wheel with the better road-surface adhesion thanks to the differential lock. The allocated torque is dependent on the traction and the locking value of the installed differential lock.

Locking value

> The locking value **S** indicates how much torque discrepancy is possible between the left and right drive wheels on a powered axle or between two final drives of the front and rear axles of all-wheel-drive vehicles.

$$S = \frac{\text{Difference, torques}}{\text{Sum total, torques}} \cdot 100\ \%$$

The locking value **S** is given in %. It refers for example to the load torque applied to the crown wheel.

Example, locking value 40 %: Here the drive wheel with the better traction can transmit 40 % more torque than the other drive wheel.

Longitudinal lock. This locks speed balancing between the wheels of two powered axles.

> A longitudinal lock allocates more torque to the drive wheels of the axle which have the better road-surface adhesion (traction).

Example: If, for instance on an all-wheel-drive vehicle, the wheels on a powered axle spin, the wheels on the powered axle with the better traction are allocated more torque in accordance with the locking value of the installed differential lock.

16

16.9.1 Shiftable differential locks

The differential lock depicted in **Fig. 1** consists of a shift actuator and a dog coupling. The shifting can be performed for example mechanically by hand or pneumatically.

Operating principle. When engaged, the dog coupling connects the right axle shaft with torsional resilience with the differential housing and crown wheel.

A positive and torsionally resilient coupling is achieved between the right axle shaft and the differential housing via the internal teeth of the selector sleeve and the external teeth on the right side of the differential housing. In this way, the differential gears are unable to circulate on the axle-shaft gears. They act only as drivers. Balancing is 100 % locked. Shiftable differential locks must be released in the event of normal adhesion of the drive wheels with the road surface so that damage to the dog coupling and to the gears is avoided.

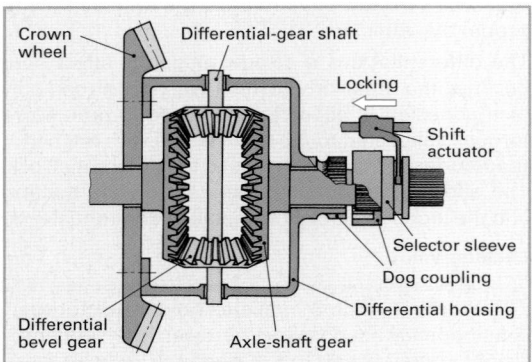

Fig. 1: Shiftable differential lock

16.9.2 Automatic differential locks

These cause speed balancing, for example between the drive wheels on one axle, to be automatically locked. The drive wheel with the better adhesion is allocated more torque.

Conventional locking values range between 25 % and 70 %.

Types
- Differential lock with multi-plate clutches
- Torsen differential • Viscous clutch
- Automatic limited-slip differential (ALSD)
- Electronic limited-slip differential (ELSD)
- Haldex clutch

Differential lock with multi-plate clutches
Design (Fig. 2)
The standard components of a differential are supplemented by two thrust rings and two multi-plate clutches.

The thrust rings feature on their lateral surfaces drivers which engage longitudinal grooves on the differential housing. They are connected with torsional resilience with the differential housing, but can be axially displaced in the housing.

The plates are arranged between the outer end faces of the thrust rings and the end faces of the differential housing.

The externally toothed plates engage the longitudinal grooves of the differential housing while the internally toothed plates engage the external toothing of the axle shafts.

The two thrust rings feature on their inside end faces four recesses in which the shafts of the differential bevel gears are mounted. Disc springs are installed to pretension the plates.

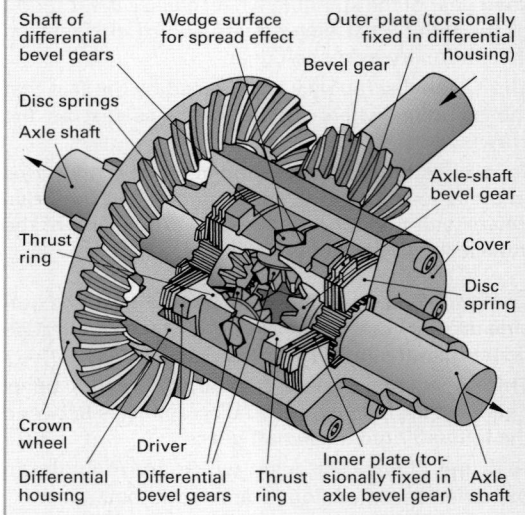

Fig. 2: Differential lock with multi-plate clutches

Operating principle
The torque from the variable-speed gearbox is multiplied by the gear ratio in the final drive and transmitted via the crown wheel and the differential housing to the thrust rings.

Equally good road-surface adhesion. 50 % of the torque is transmitted by each drive wheel.

The torque is transmitted via the crown wheel to the differential housing and via the thrust rings, which can be axially displaced, to the multi-plate clutches and to the toothed axle shafts.

Different road-surface adhesion. If for example the right drive wheel spins, the differential bevel gears rotate. Their shafts press the thrust rings against the two plate packs.

The pressing force generates a load-dependent friction torque between the faster-rotating internally toothed plates and the externally toothed plates of the right plate pack.

This friction torque is directed via the differential housing, the left plate pack and the toothing of the left axle shaft to the left drive wheel.

Here it acts in addition to the normal drive torque of the right drive side.

Torsen differential

> The basic principle of the Torsen differential is based on self-locking between the worm gear and the worm of a worm-gear pair.
>
> The extent of self-locking is dependent on the lead angle of the toothing of the worm and worm gear.
>
> Self-locking is eliminated when the worm drives the worm gear.

A Torsen differential (Torsen = torque sensing) distributes the torque from the variable-speed gearbox as a function of traction. It can be used as an interwheel lock in the axle differential and as a longitudinal lock in the centre differential of all-wheel-drive vehicles.

Design (Fig. 1)

A Torsen differential consists of two worm-gear drives. The spur gears positively connect the worm-gear drives with each other by way of the worm gears. They are mounted so that they can rotate in the differential housing.

Each worm is connected with torsional resilience with a drive shaft. The differential housing is bolted to the final-drive crown wheel.

Operating principle
Power flow in Torsen differential
The drive torque comes from the bevel gear via the crown wheel and is transmitted by the differential housing and the worm gears to the worms of the two drive shafts.

Equally good road-surface adhesion. When the drive wheels rotate at the same speed during straight-ahead driving, the worm gears do not rotate with the side spur gears, and act as drivers. Torque distribution is effected in equal proportions to the two drive shafts.

Different road-surface adhesion, cornering. Speed balancing is effected via the rotating spur gears and worm gears, in the course of which self-locking occurs.

If, for example, the left drive wheel rotates faster, the left worm drives its worm gears.

The spur gears on the left side transmit the rotary motion to the spur and worm gears on the right side. Self-locking corresponding to the locking value occurs between the worm gear and the worm.

The wheel with the better road-surface adhesion or the lower speed, here the right wheel, is allocated more torque.

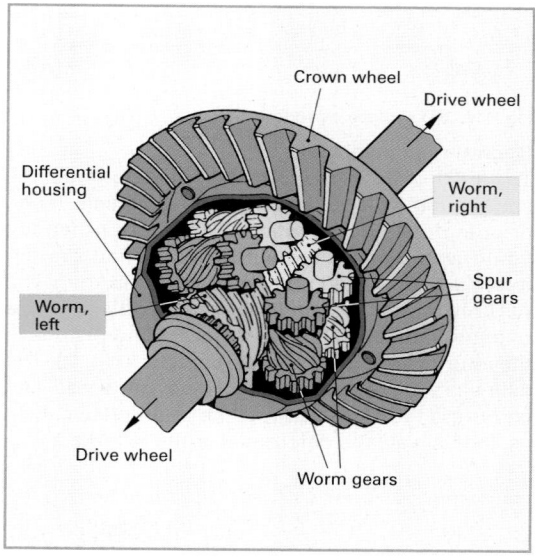

Fig. 1: Torsen differential as interwheel lock

16

1   What are the functions of final drives?

2   In what ways do bevel-gear final drives with offset axes differ from bevel-gear final drives with non-offset axes?

3   List the advantages of a hypoid drive.

4   In what ways do the shapes of the tooth flanks and tooth backs of Gleason toothing differ from those of Klingelnberg toothing?

5   List two possible ways of correcting the adjustment of the crown wheel.

6   What are the functions of differentials?

7   How is the thickness S of the compensating shims calculated during adjustment of the bevel gear?

8   According to which principle does a differential distribute torque to the drive wheels?

9   What are the differences between interwheel locks and longitudinal locks?

10 Name the different types of automatic differential lock.

Automatic limited-slip differential (ALSD)

The is an electrohydraulic system which is used as an aid to starting and locks the differential 100% when the wheels spin. Improved traction is achieved as a result.

Design (Fig. 1)
The system features the following assemblies:
- Differential with ring cylinder and multi-plate clutches
- Fluid pump, fluid reservoir, ALSD hydraulic control unit with pressure accumulator and solenoid valve
- Wheel sensors, ALSD ECU, function and fault indicators

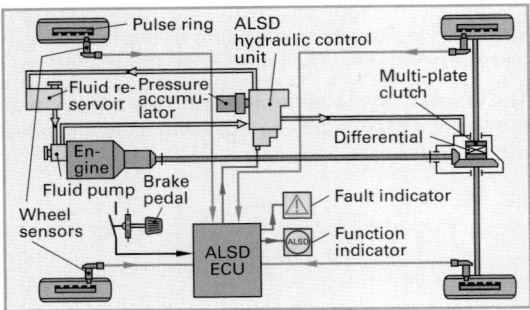

Fig. 1: ALSD system overview

Operating principle (Fig. 1).
The ECU uses wheel sensors to determine the speeds of the driven and non-driven wheels. If one or both drive wheels rotate 2 km/h faster than the non-driven wheels, the automatic limited-slip differential is activated up to a speed of 35 km/h. The ALSD hydraulic system is activated and a pressure of approximately 30 bar acts on the ring pistons in the ring cylinders of the axle shafts **(Fig. 2)**. The axle-shaft bevel gears are pulled outwards and compress the plate packs of the multi-plate clutches. In this way, the differential is 100% locked.

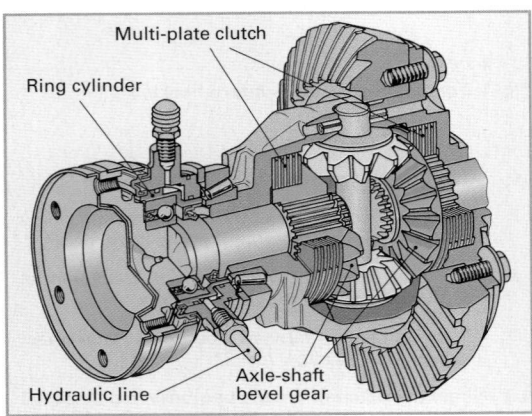

Fig. 2: Automatic limited-slip differential (ALSD)

At speeds in excess of 40 km/h, in overrunning mode or during braking, the lock is not activated or is released so that the vehicle is not prone to skidding.

In these driving states, the multi-plate clutches act like a self-locking differential with a specified locking value.

A diagnostic program integrated in the ECU monitors the electrical ALSD system and shuts it down in the event of malfunctions. The driver is alerted to this by the fault indicator.

Electronic limited-slip differential (ELSD)

This is an electrohydraulic system which is used as an aid to starting. The locking effect is created by brake intervention at the spinning drive wheel.

Design (Fig. 3)
The ELSD system is combined with an ABS system and comprises the following assemblies:

Hydraulic system. This consists of a hydraulic pump with suction and delivery valves, inlet and outlet valves, hydraulic changeover valve and check valve with pressure-limiting valve.

Electrical system. This consists of the ABS/ELSD ECU and speed sensors for the wheels.

Operating principle
If a drive wheel spins, this is detected by the ECU via the speed sensors for the drive wheels and the pressure build-up phase is initiated.

Pressure build-up. The hydraulic pump and check valve are activated. The check valve CV closes and the pressure generated by the hydraulic pump P brakes the spinning wheel.

Pressure holding. Here the pump is deactivated and the inlet valve IV closed.

Pressure reduction. If the wheel is no longer spinning, the inlet and check valves are opened and the pressure is reduced via the master cylinder to the expansion tank.

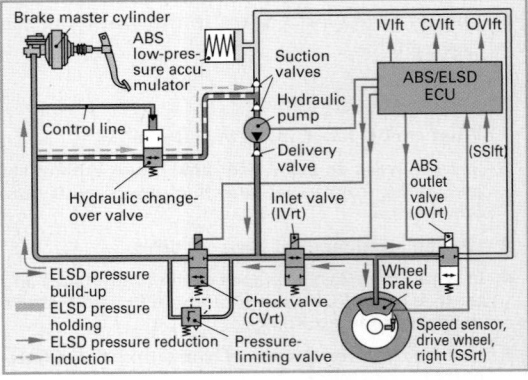

Fig. 3: ELSD brake circuit of a drive wheel

16.10 All-wheel drive

A drive wheel of a motor vehicle can only transmit as much motive force F_M as the friction force between the tyre and the road surface permits.

Starting out from a total weight of e.g. 2,800 kg for the vehicle shown in the picture, each wheel is subjected to a load of 700 kg with equal weight distribution. This equates to a wheel load of 7,000 N. On an icy road surface, for example, each wheel can transmit a maximum motive force of $F_M = F_N \times \mu_H = 7{,}000 \text{ N} \times 0.1 = 700 \text{ N}$.

This produces for

two-wheel drive $F_{M \text{ tot}} = 2 \times 700 \text{ N} = 1{,}400 \text{ N}$

four-wheel drive $F_{M \text{ tot}} = 4 \times 700 \text{ N} = 2{,}800 \text{ N}$

An all-wheel-drive vehicle with 4 driven wheels can, with the same drive power and weight distribution, transmit twice as much motive force as a vehicle with 2 drive wheels.

The following different all-wheel-drive systems may be used:

- **Conventional switchable all-wheel drive**
  Here the wheels on a powered axle are always driven, while the other wheels are engaged where necessary.

- **Permanent all-wheel drive**
  All the wheels are permanently driven.
  A centre differential for balancing different wheel speeds between the front and rear axles is required.

Design

The following assemblies are used in vehicles with permanent all-wheel drive **(Fig. 1)**:
- Transfer box with centre differential and longitudinal lock
- Front final drive with differential
- Rear final drive with differential and interwheel lock

Transfer box. This distributes the torque from the variable-speed gearbox in the proportion of e.g. 50 % to the front final drive and 50 % to the rear final drive.

Centre differential. This balances different speeds, for example when cornering. Distortions in the drivetrain are avoided.

If the drive wheels on an axle spin, the centre differential can be locked by means of a longitudinal lock. In this way, more torque is allocated to the wheels on the axle with the better road-surface adhesion.

Front and rear differentials

These balance different wheel speeds and distribute the torque in equal proportions to the drive wheels on one axle.

If a wheel spins, balancing can be locked by an **interwheel lock** and more torque is allocated to the wheel with the better road-surface adhesion.

1 longitudinal lock and 2 interwheel locks are required in order to be able to transmit a maximum torque under all driving conditions with an all-wheel-drive vehicle.

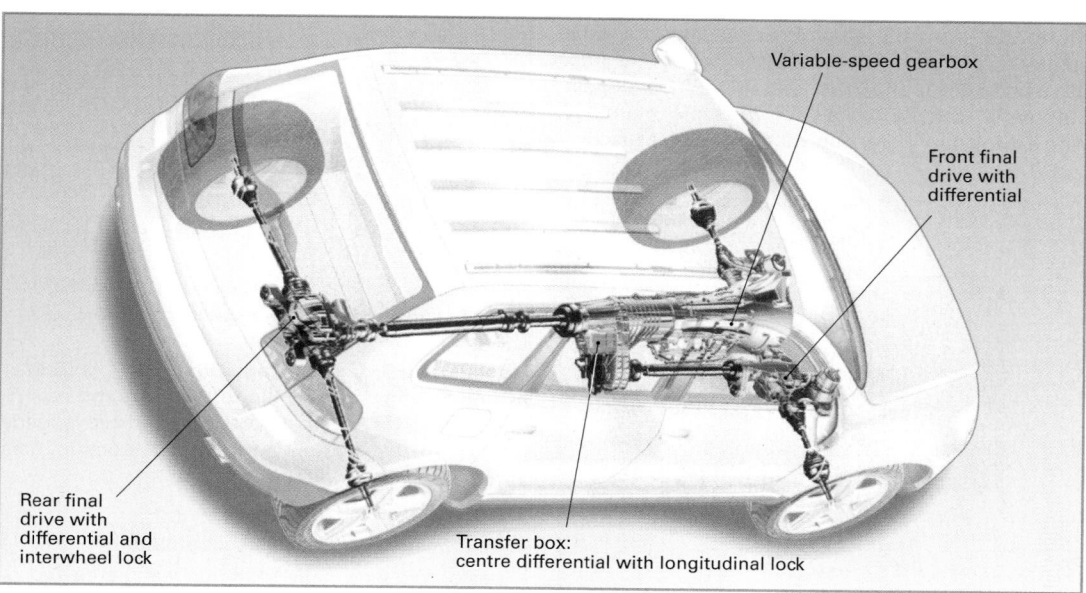

Variable-speed gearbox

Front final drive with differential

Rear final drive with differential and interwheel lock

Transfer box: centre differential with longitudinal lock

Fig. 1: All-wheel-drive vehicle with permanent all-wheel drive

Centre differential and transfer box

These are usually combined in a single structural unit. The following different types may be used:

- Bevel-gear differential
- Torsen differential
- Planetary gear
- Viscous clutch
- Haldex clutch

The possible torque distributions to the front and rear axles are set out in **Fig. 1**.

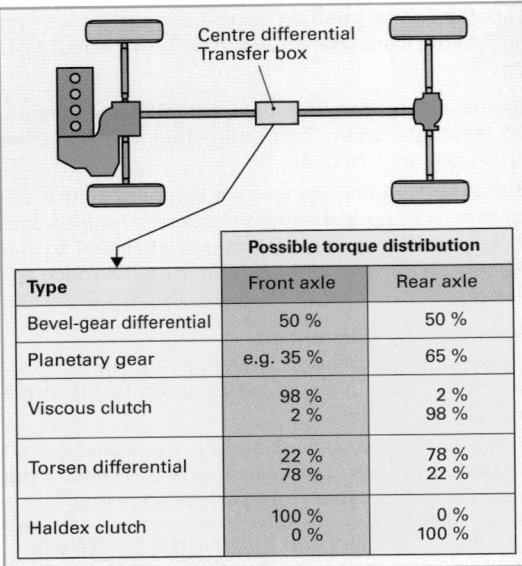

Type	Possible torque distribution	
	Front axle	Rear axle
Bevel-gear differential	50 %	50 %
Planetary gear	e.g. 35 %	65 %
Viscous clutch	98 % 2 %	2 % 98 %
Torsen differential	22 % 78 %	78 % 22 %
Haldex clutch	100 % 0 %	0 % 100 %

Fig. 1: Torque distribution of transfer boxes

Bevel-gear centre differential (Fig. 2). This balances different speeds of the drive wheels and distributes the torque from the variable-speed gearbox constantly in the proportions of 50 % to the rear final drive and 50 % to the front final drive. A bevel-gear differential can be locked for example manually with a dog coupling or automatically with a viscous clutch.

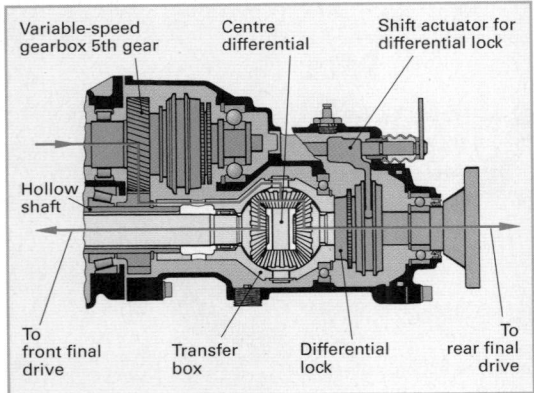

Fig. 2: Bevel-gear centre differential

Planetary gear as centre differential (Fig. 3). This balances different speeds of the front and rear wheels and distributes the torque with a constant ratio to the final drives on the front and rear axles.

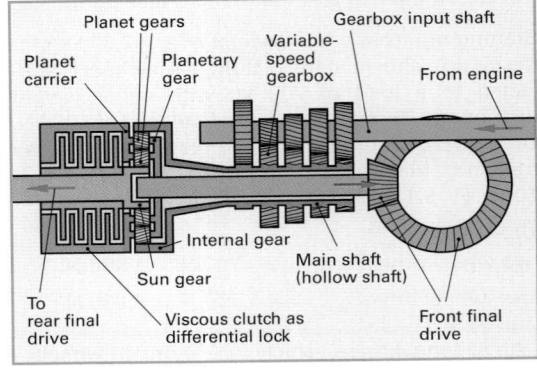

Fig. 3: Planetary gear as centre differential

The torque comes from the variable-speed gearbox via the hollow shaft (main shaft) to the internal gear of the planetary gear. From there it is distributed by the planet carrier to the front final drive and by the sun gear to the rear final drive. Torque distribution is effected on account of the different lever arms of the planet carrier and the sun gear **(Fig. 4)** in unequal proportions (asymmetrical), e.g. 65 % to the front and 35 % to the rear final drive. In the event of slip at the front or rear wheels, a viscous clutch locks as a function of traction and allocates more torque to the drive side with the better road-surface adhesion.

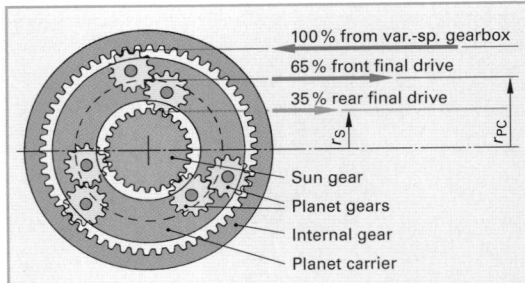

Fig. 4: Torque distribution by planetary gear

Viscous clutch (Fig. 1, Page 389). This distributes torque as a function of slip to the front and rear final drives, balances different final-drive speeds (wheel speeds) and, as a function of traction, acts automatically as a differential lock.

The viscous clutch comprises
- the housing filled with silicone fluid,
- perforated externally toothed steel plates and
- slotted internally toothed steel plates.

The outer plates engage the internal teeth of the housing while the inner plates engage the external teeth of the hub.

The locking effect in the event of speed differences between the drive wheels is achieved by the shear effect between the plates and the silicone fluid.

Under normal road-surface adhesion conditions, the majority (approx. 98 %) of the drive torque is allocated to the front final drive. As soon as slip occurs at the front drive wheels, the viscous clutch locks speed balancing and allocates more torque to the rear drive wheels.

The locking value of a viscous clutch is variable, ranging between 2 % and 98 %.

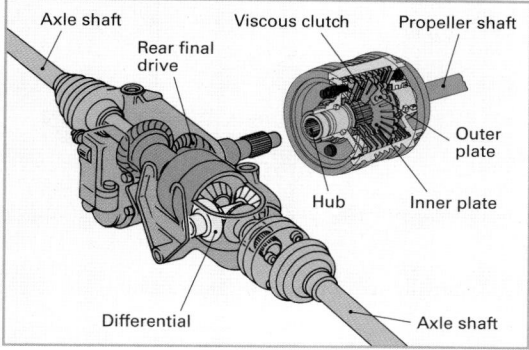

Fig. 1: Viscous clutch

Torsen differential. This is integrated for example in the variable-speed gearbox **(Fig. 2)** and distributes torque on a dry surface in equal proportions to the drive wheels on the front axle (50 %) and rear axle (50 %). As a centre differential, the Torsen differential balances different speeds of the front and rear final drives.

If slip occurs at the wheels on a powered axle, it acts as a function of traction as an automatic differential lock.

It allocates more torque to the wheels on the powered axle with the better road-surface adhesion. The locking value can stretch up to 56 %.

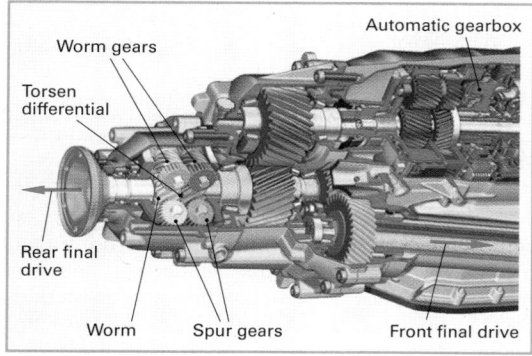

Fig. 2: Torsen differential

Haldex clutch (Fig. 3). This is flanged to the rear final drive and acts as a longitudinal lock.

In normal situations, e.g. in the event of sufficient traction, the front axle is allocated 100 % and the rear axle 0 % of the drive torque.

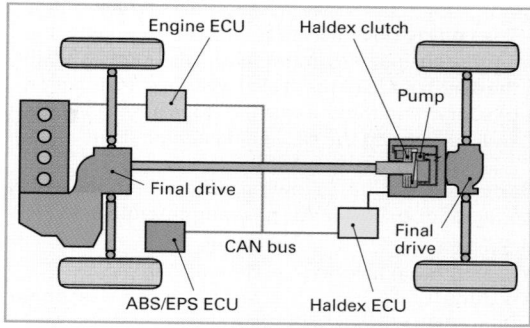

Fig. 3: All-wheel-drive system with Haldex clutch

Operating principle (Fig. 4). In the event of speed differences between the front and rear final drives, the rotary-piston pump is actuated by the cam plate and generates a fluid pressure. This pressure acts on the working plunger, which presses the externally and internally toothed plates of the plate pack against each other.

The ECU uses program maps to activate the control valve and thereby control the pressure for pressing the plates together.

The locking value of the Haldex clutch can range between 0 % and 100 %.

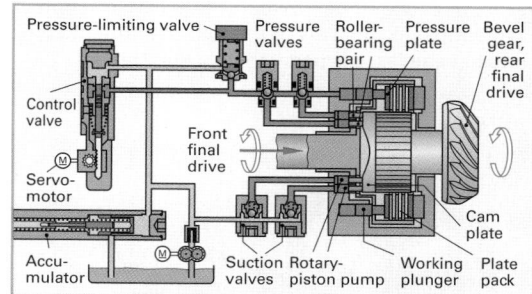

Fig. 4: Haldex clutch

REVIEW QUESTIONS

1 What are the advantages of an all-wheel drive with 4 drive wheels over a two-wheel drive?
2 Name the components of vehicles with permanent all-wheel drive.
3 What are the functions of the transfer box with integrated centre differential?
4 How does a viscous clutch react when slip occurs at the front drive wheels?
5 How does a Torsen differential distribute torque on a dry road surface?
6 How does a Haldex clutch work?

17 Vehicle body

17.1 Vehicle body/bodywork

> The vehicle body serves to protect occupants and goods against environmental influences and in the event of accidents. It also assumes the function of supporting the chassis and drive assemblies as well as carrying the occupants and payload.

Body shapes. The following different shapes may be used for example for passenger vehicles:

- Saloon
- Convertible saloon
- Coupé
- Limousine
- Special passenger vehicle, e.g. camper van
- Estate
- Convertible
- Multipurpose passenger vehicle

Body constructions. The following different constructions may be used:

- Separate construction
- Partially self-supporting construction
- Self-supporting construction

17.1.1 Separate construction

With this construction, the vehicle body is mounted on a frame **(Fig. 1)**. The further chassis groups, such as axles, steering etc., are also mounted on the frame. Because of its flexibility, this construction is used almost exclusively in the manufacture of commercial vehicles, offroad vehicles and trailers.

Fig. 1: Frame construction

The main body shape used here is the ladder-type frame. Here, two side members are riveted, bolted or welded to several cross-members. The steel members used with open sections (U-sections, L-[angle] sections) or closed sections (round or box sections) produce a frame with great flexural strength, great torsional elasticity and high carrying force.

17.1.2 Partially self-supporting construction

With this construction, a self-supporting structure assumes part of the overall supporting function in addition to a frame. When compared with the self-supporting construction, it is possible to realise different body-framing variants more easily.

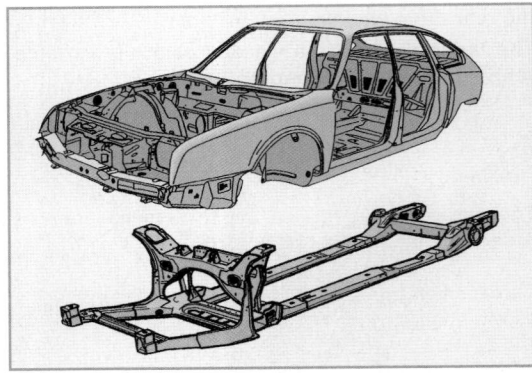

Fig. 2: Partially self-supporting construction

17.1.3 Self-supporting construction

The self-supporting construction is used in passenger cars and buses/coaches.
In passenger cars the frame is replaced by a floor assembly **(Fig. 3)**, which, in addition to the supporting components such as engine bearers, side members and cross-members also contains the luggage-compartment floor and wheel houses.

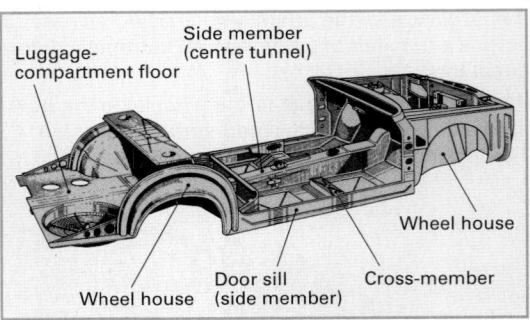

Fig. 3: Floor assembly

Further sheet-metal panels welded to the floor assembly, such as A-, B-, C- and D-pillars, roof frame, roof and wings, and bonded windscreen and rear window produce a self-supporting, monocoque-construction body **(Fig. 1, Page 391)**. In this case, the body is stabilised by beads, reliefs, closed sections and outer surfaces.

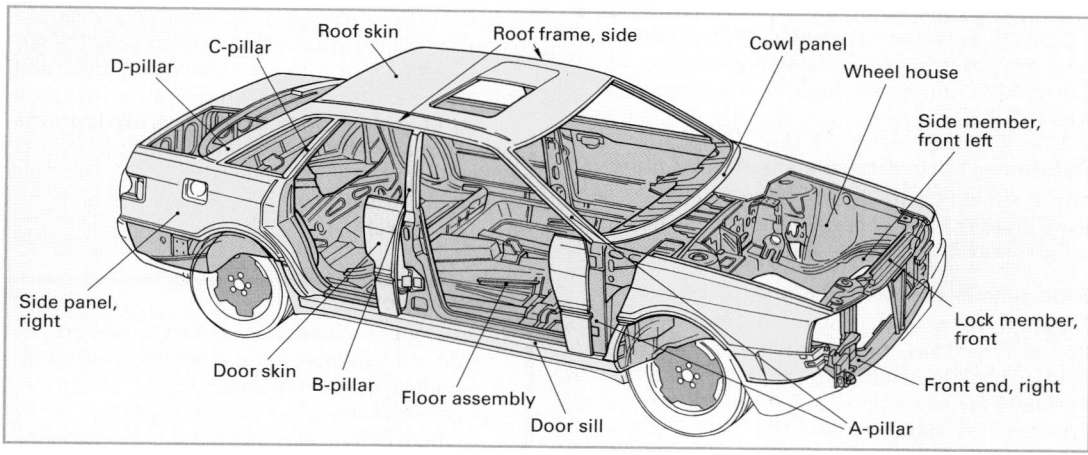

Fig. 1: Self-supporting monocoque-construction body

The carcass construction is used as well as the monocoque construction.

Carcass construction. This is often also referred to as the space-frame construction. A latticework-like rod system performs the main supporting function of the body. The outer surfaces can have a co-supporting function. This construction is used, for example, in passenger-car designs **(Fig. 2)** with aluminium bodies. Here, the frame structure is formed by differently shaped extruded and sheet-aluminium sections which are joined at highly stressed points by cast plates.

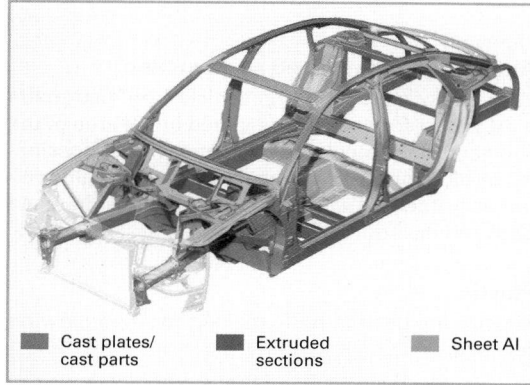

| Cast plates/ cast parts | Extruded sections | Sheet Al |

Fig. 2: Space frame of an aluminium passenger-car body

It is essential to follow the manufacturer's instructions to the letter when making repairs to self-supporting bodies. Using incorrect materials and incorrect repair methods and adding or omitting components alters the stability of the body and thereby reduces vehicle safety in the event of accidents.

17.1.4 Materials in body making

The materials predominantly used are sheet-steel plates, galvanised sheet-steel plates, sheet-aluminium plates, sections of these materials and plastics.

Sheet-steel plate

Self-supporting vehicle bodies are primarily manufactured from super-high-tensile and high-tensile steel-plate preforms **(Fig. 3)**. Super-high-tensile body panels have a yield point of approximately $400 \ N/mm^2$, whereas normal body panels have a yield point of $120 \ N/mm^2 \dots 180 \ N/mm^2$. Plate thicknesses vary from 0.5 mm to 2 mm.

Tailored blanks. These are sheet-metal blanks of different strengths and thicknesses. They are welded according to requirements to plates (= complete body component, e.g. side section).

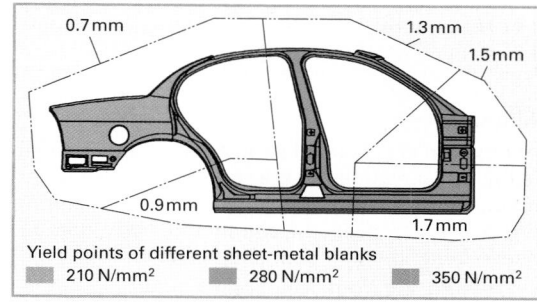

Yield points of different sheet-metal blanks

| $210 \ N/mm^2$ | $280 \ N/mm^2$ | $350 \ N/mm^2$ |

Fig. 3: Use of super-high-tensile sheet-steel plates in a body side section

Reforming super-high-tensile sheet-steel plates. These are reformed with greater difficulty and have a stronger resilience. In the transition from normal-tensile to super-high-tensile sheet-steel plates, normal-tensile sheet-steel plates require additional braces during reforming in order to avoid unwanted deformations.

> Super-high-tensile sheet-steel plates must not be straightened hot as they already lose more than 50 % of their strength from a temperature of 400 °C.

Reforming normal-tensile sheet-steel plates

These should normally be reformed cold. If, however, there is a risk of cracking, they may be heated up to a maximum temperature of 700 °C.

High-tensile sheet-steel plates

These have a yield point of > 400 N/mm² ... 950 N/mm². They should not be reformed either cold or hot. Depending on the manufacturer, they are used for example in the A- and B-pillar areas and help to significantly stiffen the body while keeping the weight low.

In the interests of keeping the introduction of heat as low as possible in the case of repairs, MIG soldering is often used as the joining method.

Galvanised sheet steel

Body panels can be galvanised for reasons of corrosion protection. Floor panels are hot-dip-galvanised. Electrogalvanising, on account of the surface quality, is used on panels for the body outer skin.

> **WORKSHOP NOTES**
> - Draw off toxic zinc oxide when welding.
> - Resistance spot welding processes are to be favoured over other processes because a protective zinc ring is formed again around the welding spot.
> - Before welding, apply paints containing zinc (zinc-dust paint) to overlapping areas.
> - Make sure that the zinc coating is not destroyed on new parts.

Aluminium

Aluminium is only used as an alloy in body making (alloying constituents are primarily silicon and magnesium). The following manufacturing processes are used on aluminium body components, depending on their shape and stress:
- Pressing, e.g. roof skin, bonnet, wing
- Extruding, e.g. space frame
- Pressure diecasting, e.g. spring-strut mount, cast plate

Whereas pressed parts and extruded sections can be partially repaired by reforming, this is not possible with pressure diecastings.

Properties. Aluminium alloys begin to lose strength significantly when subjected to temperatures starting from approx. 180 °C. If they come into contact

with other materials, e.g. steel, electrochemical corrosion will occur if an electrolyte is present. The surface of aluminium forms a tight oxide layer with high electrical resistance. Aluminium must therefore not be welded with standard workshop resistance spot welding equipment. Al alloys can be welded to good effect with TIG or MIG inert-gas welding processes (inert gas: 100 % argon or argon-helium mixture).

> **WORKSHOP NOTES**
> - On account of potential contact corrosion
>   - do not use machining tools for aluminium bodies on other metals,
>   - use only special-steel wire brushes,
>   - when using different joining techniques, e.g. bolting, riveting, use only the connecting elements approved by the manufacturer.
> - In order to eliminate strength losses, do not heat body components to temperatures in excess of 120 °C when straightening.
> - Welding and straightening tasks may only be carried out by specially trained personnel.
> - Because of the risk of cracking due to electrochemical reactions, do not tin sheet-aluminium plates.
> - Because of the risks to health and the danger of explosion, draw off or extract Al grinding dust immediately.

Aluminium-steel mixed construction

With this construction, the vehicle front end made of Al alloys, for example, is joined in the area of the A-pillars to the steel body. Because of the possibility of electrochemical corrosion between aluminium and steel, insulating adhesive fillings are used between the abutting parts.

Plastic

Plastics are used in body making for the following reasons:
- Low specific weight and thus significant weight savings.
- Corrosion resistance.
- Extensive freedom of shaping.
- Not susceptible to shocks.
- Manufacture of components without reworking.
- When damaged, they can be repaired at low cost with the appropriate knowledge.

Some areas in which plastics are used in body making are depicted in **Fig. 1** on Page 393.

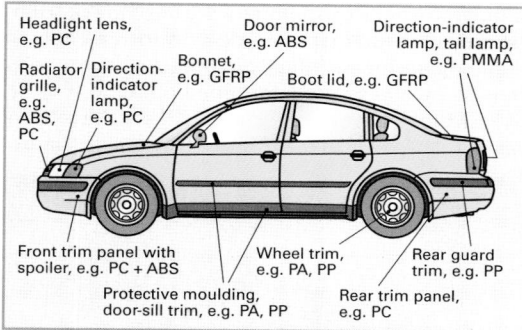

Fig. 1: Examples of plastic body components

Repairing plastic

Plastic parts can be repaired by welding, laminating or gluing with 2-component repair materials.

Welding. This process can only be used on thermoplastics, such as for example PA, PC, PE, PP, ABS, ABS/PC (for explanations of terms, see Chapter 8.6, Page 173).

Laminating (Fig. 2). Here, holes for example are repaired with hardener with the aid of glass-fibre mats (GFRP) and resin (polyester resin; epoxy resin). The area of damage must be bevelled in such a way that a connection can be established between each glass-fibre-mat layer and the original part. If necessary, the area of damage must be provided with a reinforcing layer prior to lamination.

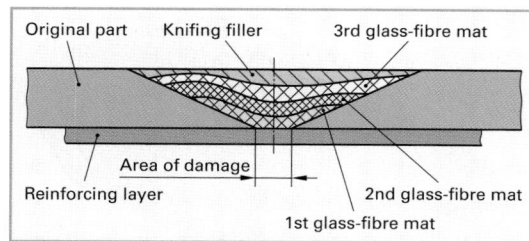

Fig. 2: Structure of a GFRP repaired area with reinforcing layer

Gluing with 2-component repair materials. Depending on the repair material used, holes, cracks and scratches can be repaired without identification of the plastic to be repaired. The basis is provided for example by a 2-component polyurethane glue in a twin cartridge which is mixed in the correct proportions via forced mixing tube. The glue is applied to the cleaned and prepared area of damage. The glued area can then be heated with a radiant heater. This makes it harden more quickly. The repaired area can then be further treated in the form of grinding and painting.

17.1.5 Safety in vehicle manufacturing

Structural measures in vehicle manufacturing are intended to minimise the risks of accidents to the greatest possible extent. A distinction is made in this respect between active and passive safety.

Active safety

> Active safety refers to the structural measures taken on a vehicle which help to prevent accidents.

Active safety can be divided into four areas.

Driving safety, e.g. by:
- Neutral driving behaviour on bends/curves.
- Straight-running stability of the vehicle.
- Light and precise steering.
- Greatest possible braking deceleration without wheel locking (ABS).
- Optimally tuned suspension and damping with the wheel suspension.
- Traction control system (TCS/ASR, VDC, ESP).

Perceptibility safety, e.g. by:
- Large windows, dipping rearview mirror.
- Headlights which ensure that the road surface is well illuminated.
- Acoustic warning devices.
- Heated windows and door mirrors.

Conditional safety, e.g. by:
- Ergonomic design of the driver's seat.
- Comfortable suspension.
- Good passenger-compartment ventilation, air conditioning.
- Soundproofing.

Operating safety, e.g. by:
- Clear, uncluttered layout of switches, warning lamps and instruments.
- Pedals designed to suit the driver.

Passive safety

> Passive safety refers to the structural measures taken on a vehicle which in the event of an accident minimise to the greatest possible extent the risks of injury and death (accident consequences) to road users.

A distinction is made here between interior and exterior safety.

Interior safety

This comprises essentially measures relating to
- the deformation behaviour of the body,
- the strength of the passenger cell,
- impact-protection measures (seat belt, airbag),
- fire protection and
- occupant extrication.

17

Accident analyses **(Fig. 1)** demonstrate that front-impact accidents at 60 % ... 65 % and side-impact accidents at 20 % ... 25 % are the most common causes of personal injuries.

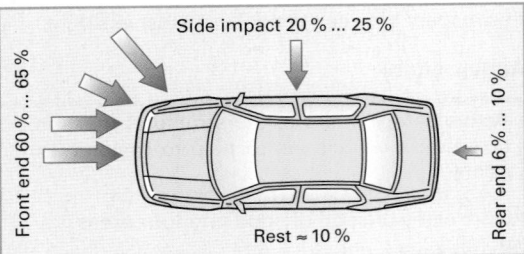

Fig. 1: Distribution of accidents to types of collision with personal injuries

On account of accident analyses computer calculations and defined crash tests **(Fig. 2)** are used to examine the behaviour of the body and the effects on the persons involved in accidents. The optimal vehicle body is determined in this way from the results. A standardised test involves for example engineering the frontal impact of a vehicle at roughly 50 km/h against a stationary obstacle. In order that the vehicle occupants are not exposed to any excessively critical deceleration values, the kinetic energy is converted via crumple zones into a specific deformation (deformation energy).

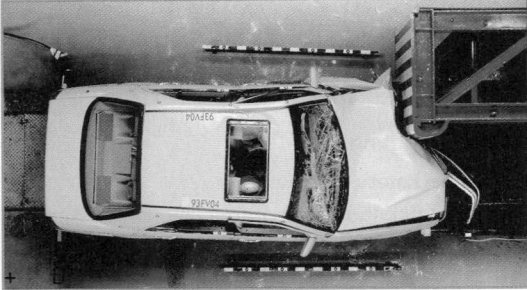

Fig. 2: Offset front-end collision
(≈ 50 % overlap, offset)

Safety body (Fig. 3). This consists of a stable passenger cell and crumple zones at the front and rear ends. Even in the event of serious accidents the passenger cell maintains its shape and thereby enables the occupants to survive.

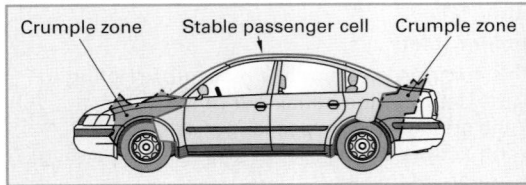

Fig. 3: Safety body

Longitudinal or side members are used in the **crumple zones**; these members deform for example in front-end accidents initially in the front lower body area by means of predetermined buckling **(Fig. 4)**. Only in the event of serious accident are the rear areas of the body also used for energy conversion.

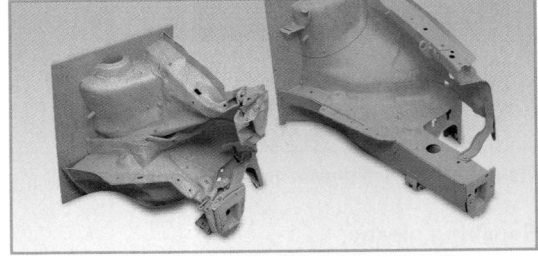

Fig. 4: Crumpling behaviour of a front side member

Belt line (Fig. 5). In vehicles in which the conventional areas of the crumple zone are not sufficient for energy conversion in the case of front-end accidents, parts situated in the area of the belt line are also utilised for defined deformation. This prevents the passenger cell from being too greatly deformed in the front area. The belt line runs from the front panel through the upper wing mounting, the A-pillar, the side-impact door strip, the B-pillar and, depending on the design, to the C-pillar. As a result of these structural measures, there is significantly greater deformation above the floor assembly after an accident.

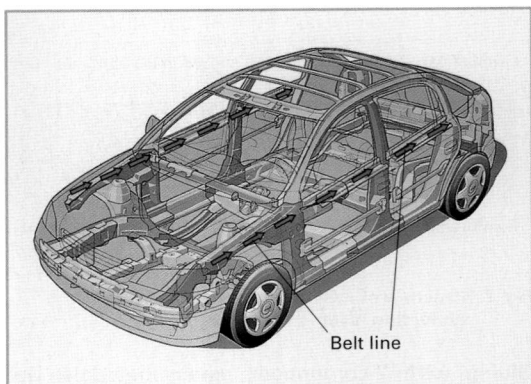

Fig. 5: Deformation travel in area of belt line

Side-impact protection (Fig. 1, Page 395). Reinforcements in the door area, cross-members between the two A-pillars at the height of the dashboard, reinforcements of the door sill, the B-pillar and the C-pillar and cross-members in the floor area influence the deformation behaviour of the body in side-impact accidents in such a way that the occupants are better protected against injury.

Fig. 1: Member structure with progression of force in event of side impact

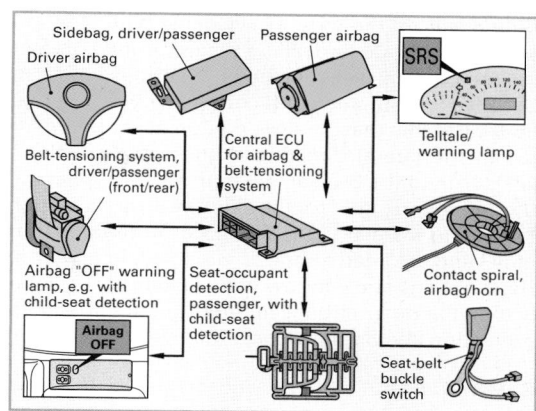

Fig. 3: Safety restraint systems (SRS) – networking of components

Doors and door locks. These should be able to be opened without tools from both the inside and the outside after the accident but should not open during the impact itself.

Fuel tank. This is usually mounted with protection against shocks above the rear axle. The filler neck and fuel lines are routed in such a way that the fuel cannot escape in the event of serious accidents and overturning of the vehicle.

Interior safety zone

This reduces the risk of injury in the interior of the passenger cell through the use of restraint systems and impact-protection measures.

In order to guarantee optimal protection of vehicle occupants, newer vehicles utilise so-called multiple restraint systems. Here, in the event of a crash, for example the driver airbag, front passenger airbag, side airbag, head airbag/window bag and belt-tensioning system are activated by a central ECU (**Fig. 2**).

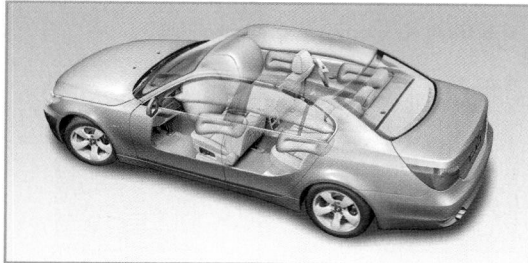

Fig. 2: Layout of airbags in vehicle

Activation for triggering the different restraint systems is performed independently of each other by a central ECU. The latter activates the individual firing circuits of the systems among other things depending on the type and severity of the accident (**Fig. 3**).

Triggering criteria. Vehicle acceleration in both the longitudinal and lateral directions is continuously recorded by acceleration sensors. The central ECU evaluates the voltage signals continuously. In crash and driving tests triggering thresholds for the respective restraint systems are determined depending on the vehicle type and stored in program maps in the central ECU.

Frontal impact. When the triggering threshold is reached or exceeded and this is confirmed by a safety sensor installed in the ECU (saving sensor), the relevant firing circuits of the belt-tensioning systems and if necessary of the driver and front passenger airbags are activated. In addition, if seat-occupancy mats are fitted (**Fig. 3**), it is possible to detect in the front passenger seat by way of pressure sensors whether the seat is occupied by an adult or by a child seat. If the passenger seat is not occupied or is occupied by a child seat, the front passenger airbag is not triggered.

Belt-buckle sensing. It is possible by means of microswitches or Hall sensors in the seat-belt buckle to ascertain whether the vehicle occupants have their seat belts fastened. If this is not the case, the triggering threshold for the Airbag is lowered.

Front airbags and belt-tensioning systems are triggered in the event of front-end accidents at up to ± 30° from the vehicle longitudinal axis (**Fig. 4**).

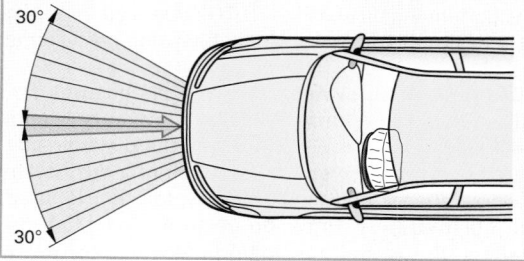

Fig. 4: Range of influence of front airbag and belt-tensioning system

Side crash. In the case of accidents with side impacts (approx. 20 % ... 25 %), a significantly greater number of vehicle occupants are seriously injured because of the reduced crumple zone (approx. 36 %). For this reason, acceleration sensors are fitted in the lateral direction in the cross-members which run to the B-pillar. When in a side impact the triggering threshold for the side airbag is exceeded, the airbag installed in the seat or in the door trim panel and the head airbag or window bag are fired. Head airbags and window bags deflate more slowly than the other airbags in order to offer adequate protection should the vehicle roll over.

Rear-end crash. In a rear-end crash the belt-tensioning system is activated when the triggering threshold is exceeded. If active head restraints are also installed in a vehicle, these are tilted forwards in the vehicle longitudinal direction. In this way, they counteract whiplash injuries.

Crash sensors. These are either situated in the ECU or installed separately and only function correctly when they are correctly aligned. In conjunction with a series-connected safety sensor, e.g. Reed contact, they prevent the systems from being inadvertently triggered.

ECU. The central ECU is equipped with a self-diagnosis function. If faults are identified in safety restraint systems **(SRS)**, the faults are stored and the SRS warning lamp is activated.

Seat belts and belt-tensioning systems. In order to minimise the risk of injury in the event of an accident, it is essential for all vehicle occupants to have their seat belts fastened. Thus, in the event of a frontal impact from 50 km/h and in spite of a crumple zone, the vehicle occupants are subject to a deceleration of 30 g ... 50 g (1 g = 9.81 m/s^2). For a person weighing 70 kg, this would require a supporting force of roughly 30 kN. If the vehicle occupants are not wearing their seat belts, they will collide with the braked passenger compartment (steering wheel, dashboard, windscreen). To ensure optimal belt efficiency, it is essential that the belt forces can be absorbed by the sternum and the pelvis. The belt must therefore be worn as tight-fitting as possible. This is achieved for example by three-point belts with belt-tensioning systems.

Belt-tensioning system (Fig. 1). This ensures that the belt fits to optimal effect and prevents what is known as belt slack. Belt slack is the distance covered by the belt strap until is fits tightly against the body. When a belt-tensioning system is triggered, the belt strap is tightened up to roughly 150 mm within 15 ms. Belt-tensioning systems operate by pyrotechnical means with explosive substances or in older systems by mechanical means.

Operating principle (Fig. 1). In pyrotechnically operating belt-tensioning systems, the deceleration values are recorded by acceleration sensors. If in the event of a frontal impact the deceleration is above the triggering threshold, e.g. deceleration greater than 2 g and vehicle speed greater than 15 km/h, the ECU detects critical deceleration values and effects via an electrical pulse the firing of a propellant charge (gas generator) by a firing pellet. Depending on the design of the belt-tensioning system, the gas generator acts on

- a plunger in a cylinder,
- steel balls or
- a Wankel rotor.

Belt-tensioning systems with Wankel rotors have three propellant charges, which are fired in succession. The belt-tensioning system is connected via a retractor to the seat belt. The belt is tensioned when the retractor is rotated.

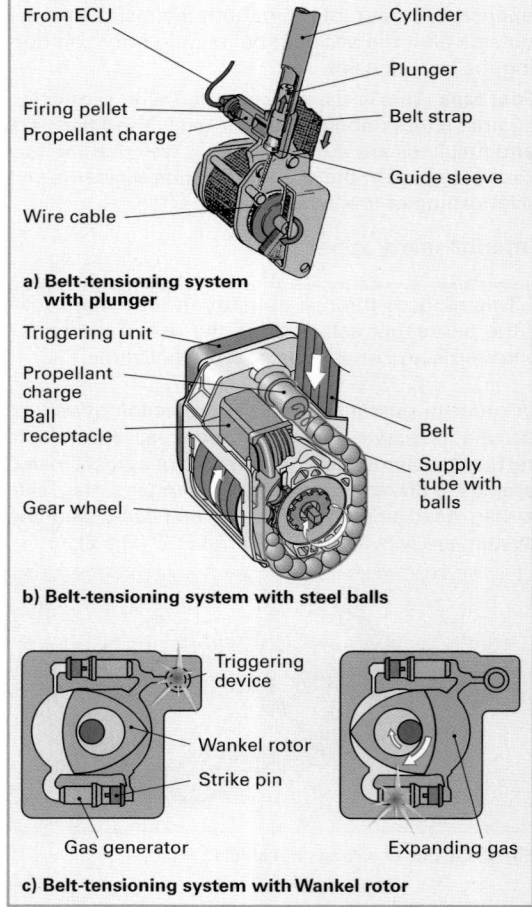

a) **Belt-tensioning system with plunger**

b) **Belt-tensioning system with steel balls**

c) **Belt-tensioning system with Wankel rotor**

Fig. 1: Pyrotechnically operating belt-tensioning systems

Belt-tensioning systems are inoperative after they have been triggered and must be replaced.

When vehicles are scrapped, pyrotechnically operating belt-tensioning systems must be triggered in accordance with the manufacturer's instructions by technically qualified personnel and thereby rendered inoperative.

Mechanically operating belt-tensioning system (Fig. 1). With this system, a preloaded spring is located in the seat-belt buckle. In the event of a front-end crash, the load can be removed from the spring by a release mechanism, thereby tensioning the belt via a control cable.

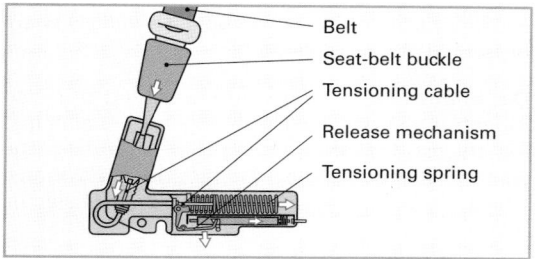

Belt
Seat-belt buckle
Tensioning cable
Release mechanism
Tensioning spring

Fig. 1: Mechanically operating belt-tensioning system

Belt-force limiter. The force exerted by the belt-tensioning system can cause injuries in the thorax and shoulder areas. Belt-force limiters among other things are therefore fitted. In this way, a torsion rod for example in the retractor is turned starting from a specific level of force acting on the belt strap and thereby limits the belt force to a maximum value.

Airbag (Fig. 2)

Design of front airbag. The airbag consists of a coated fabric. It is tightly folded behind a cover cap with a specified inflation control seam. When the airbag is triggered, the propellant charge in the gas generator (pot generator) is ignited via a firing pellet. The propellant charge is converted into gas (mainly nitrogen and carbon dioxide), which passes through a filter to fill the airbag.

The gas is then discharged to atmosphere via outlet openings facing away from the occupant's side. The electrical connection between the gas generator and the ECU is established in the steering-wheel unit by a volute spring in the contact unit.

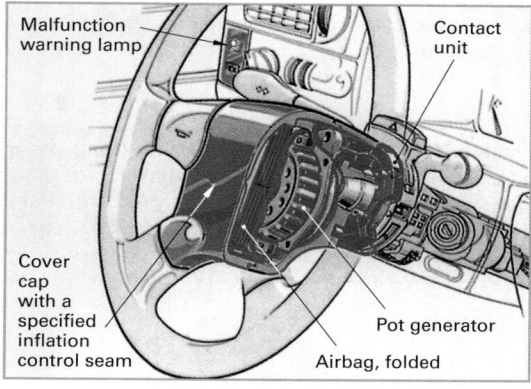

Malfunction warning lamp
Contact unit
Cover cap with a specified inflation control seam
Pot generator
Airbag, folded

Fig. 2: Cutaway view of driver-airbag system

Operating principle: Acceleration sensors transmit, depending on the accident, voltage signals to the triggering control unit. When specific deceleration values are reached, an electrical pulse triggered by the control unit ignites the propellant charges of the corresponding airbag.

Fig. 3 shows the time lapse of the function of a driver airbag in conjunction with a three-point belt in the event of a front-end crash. The entire process of start of the accident, airbag firing, airbag inflation and escape of the compressed gas mixture takes approximately 150 ms. If the airbag system is isolated from the vehicle electrical system in the event of an accident, a backup capacitor provides the firing pulse. The airbag inflates fully within 45 ms to 50 ms. In order to have a sufficient protective effect in front-end accidents, however, it is absolutely essential for the vehicle occupants to have their seat belts fastened.

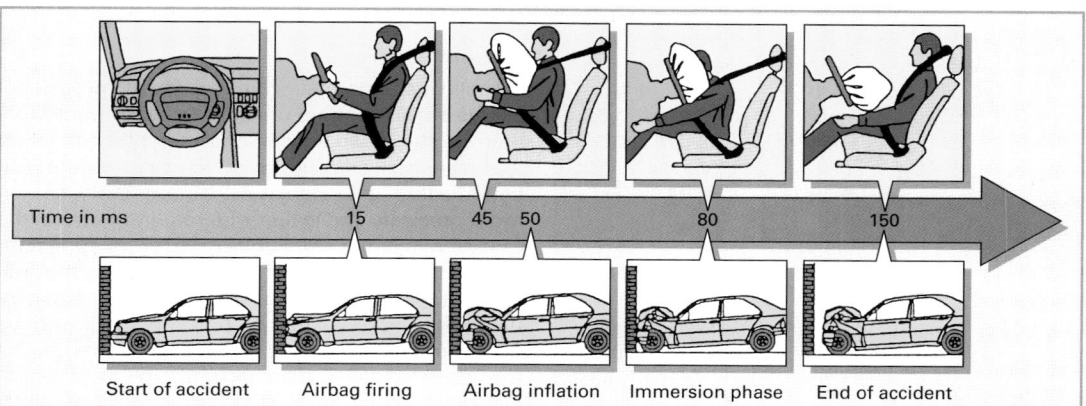

Time in ms   15   45 50   80   150

Start of accident   Airbag firing   Airbag inflation   Immersion phase   End of accident

Fig. 3: Time lapse of an airbag function in event of a front-end crash

Head airbag, window bag, side airbag. In the event of a side crash, the airbag must be fully operative within max. 20 ms. This is achieved by the fact that hybrid generators are used which can fill the airbag is a very short period of time. Furthermore, these airbags, when compared with front airbags, have a smaller volume of 10 l ... 15 l (front airbags have a filling volume of 30 l ... 75 l on the driver's side and 60 l ... 180 l on the passenger side).

Hybrid generator

Design (Fig. 1). This consists of a firing unit (firing pellet), a small amount of solid fuel and a pressure-gas bottle which is filled with inert gas. Argon and helium, precompressed to a pressure of 200 bar ... 500 bar, is generally used as the inert gas.

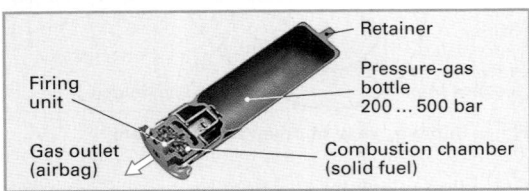

Firing unit — Retainer — Pressure-gas bottle 200 ... 500 bar — Gas outlet (airbag) — Combustion chamber (solid fuel)

Fig. 1: Hybrid generator

Operating principle. Output stages in the ECU use an electrical pulse to ignite a small amount of propellant (firing pellet). In this way, a larger amount of solid fuel is ignited. This causes the outlet opening of the pressure-gas bottle to open and a mixture of hot, pyrotechnically generated gas and cold inert gas to fill the airbag within a very short period of time.

Integrated restraint systems – Pre-Safe (Fig. 2)

These refer to measures which already activate safety systems prior to a crash, e.g. moving the seat into the correct position, tensioning the seat belt and if necessary closing the sliding sunroof.

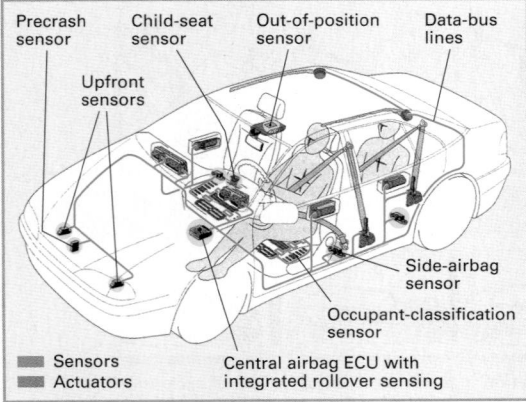

Precrash sensor — Child-seat sensor — Out-of-position sensor — Data-bus lines — Upfront sensors — Side-airbag sensor — Occupant-classification sensor — Sensors — Actuators — Central airbag ECU with integrated rollover sensing

Fig. 2: Integrated restraint systems

In order to achieve this, the Pre-Safe system is networked with the antilock braking system (ABS), the brake assistant (BAS) and the Electronic Stability Program (ESP). These vehicles also require pre-crash sensors and out-of-position sensors, which provide suitable data prior to a potential crash.

Precrash sensors use radar technology to measure the distance and the angle to a possible obstacle. These sensors form a virtual safety belt of ≤ 14m around the vehicle. The radar signals are also known as **S**hort **R**ange **R**adar (**SRR**). These signals can also be used to warn the driver against specific danger situations, for example if a vehicle is approaching at a blind angle.

Out-of-position sensors use ultrasound or video monitoring to detect whether a person is sitting at the correct distance and angle to the airbag. The seat can already be moved to the correct position if necessary prior to an accident by powerful electric motors.

Operating principle. Emergency driving manoeuvres, e.g. panic braking, skidding, are detected with the aid of sensors by the ABS, BAS and/or ESP ECU and signalled via the CAN BUS to the central airbag ECU. A very fast reduction of the distance to an obstacle, e.g. vehicle driving in front, is detected by the evaluation of the signals supplied by the pre-crash sensors by a computing program in the central airbag ECU. Already before a possible collision the central airbag ECU causes ...

- ... the driver and front passenger seat belts to be pretensioned by reversible belt-tensioning systems with the aid of an electric motor.
- ... the driver and front passenger seats to be moved to a better position in terms of their longitudinal positions, their cushions and their seat-back inclination.
- ... the sliding sunroof is automatically closed if there is a risk of skidding.

Upfront sensors

If a collision does ensue, the severity of the accident can be analysed more precisely by **upfront sensors** than by the crash sensor on the centre tunnel. In this way, it is possible to match the triggering of belt-tensioning system and airbag even earlier, more precisely and in line with requirements. In the case of two-stage airbag gas generators, for example in the event of a minor impact, only the first stage of the gas generator is fired and the airbag inflates with lower internal pressure. If a serious frontal impact is detected, the second chamber of the gas generator is activated with a delay of roughly 15 ms and the airbag inflates with higher pressure.

Safety instructions for airbag systems and pyrotechnically operating belt-tensioning systems

- Testing and installation tasks may only be carried out by qualified personnel.
- **Deactivation.** Before working on the airbag and belt-tensioning systems, turn off the ignition and disconnect and isolate the battery negative terminal. If necessary, to discharge a backup capacitor, wait 5 ... 20 minutes, depending on the manufacturer's instructions.
- Do not leave airbag and belt-tensioning units unsupervised during breaks in work.
- Always store airbag units so that the airbag outlet surface points upwards.
- Do not carry out any repairs to individual components.
- Airbag and belt-tensioning units must not be exposed to temperatures in excess of 100 °C and must, for example during body repairs, be protected against flying sparks.
- Do not install airbag and belt-tensioning units which have been dropped from greater heights (approx. 0.5 m) in a vehicle.
- Airbag and belt-tensioning units are inoperative after they have been triggered and must be replaced.
- Do not allow parts to come into contact with grease, oil or cleaning agents.
- The systems may only be electrically tested in installed condition without persons present in the passenger compartment. Resistance measurements and measurements with a test lamp are prohibited.

- When scrapping a vehicle, fire the airbag- and belt-tensioning-system gas generators with the vehicle doors closed from the outside with a firing device recommended by the manufacturer. Maintain a recommended safety distance (approx. 10 m).

Troubleshooting chart – airbag

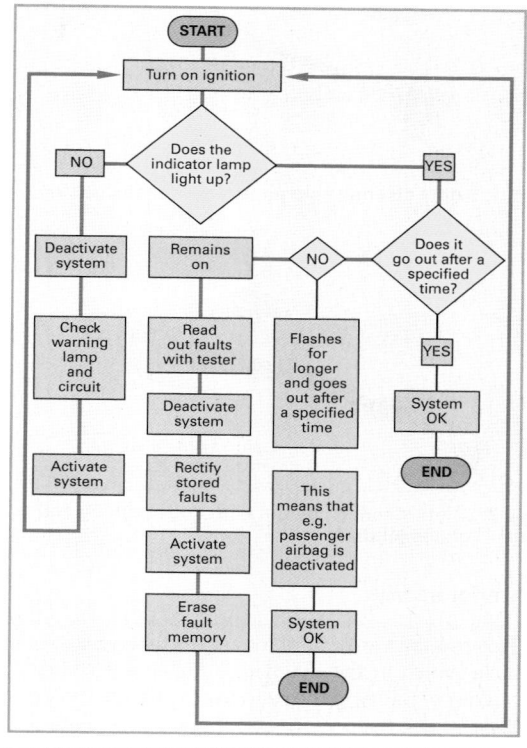

Safety glass. Two different types may be used: single-pane toughened safety glass **(TSG)** and laminated safety glass **(LSG)**.

Single-pane toughened safety glass. This glass is used for side windows and sometimes for rear windows. Owing to the initial tension of the glass, which is achieved through rapid cooling, small blunt-edged shards of glass are created when it breaks. The glass breaks over its entire surface, which means that it is unsuitable for use as a windscreen for the following reasons:

- When the glass breaks, the driver's field of vision is extremely impaired.
- In accidents the shards of glass entering the passenger compartment at high energy can injure the vehicle occupants.

Laminated safety glass (Fig. 1). This is used for windscreens and often also for rear windows. Here, two or more non-initially-tensioned glass panes are bonded with plastic intermediate layers in the middle. In the event of breakage, cracks shaped liked spider's webs are created, preserving however the majority of the driver's field of vision. Minor damage, caused for example by stone impacts, can be repaired.

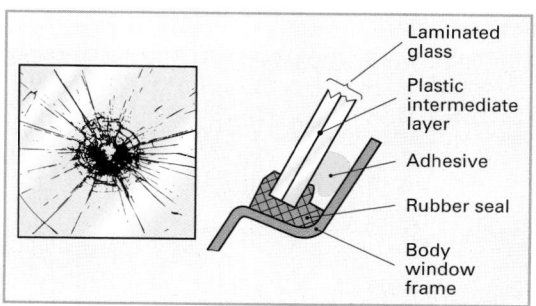

Fig. 1: Break structure and design of laminated safety glass

Safety steering column (Fig. 1, Fig. 2). This is intended to prevent the steering column from projecting into the passenger compartment in front-end accidents. Safety steering columns are designed in such a way as to deform, fold or retracting into themselves.

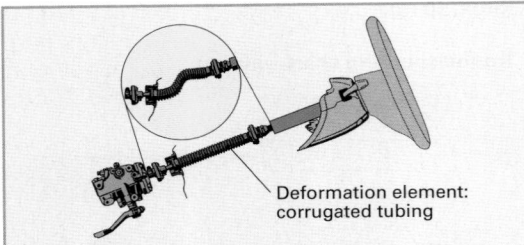

Deformation element: corrugated tubing

Fig. 1: Safety steering column with corrugated tubing

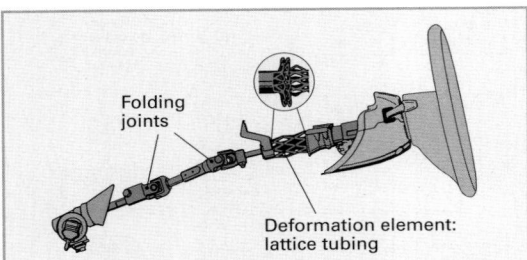

Folding joints

Deformation element: lattice tubing

Fig. 2: Safety steering column with lattice tubing and folding joints

Exterior safety

This refers to structural measures taken on a vehicle which in the event of an accident reduce the risk of injury to the road users outside the vehicle.

Examples:

- Protection of pedestrians and cyclists by means of deformable material at the front end (soft face), rounded outer edges, hideaway wipers, underride protectors in commercial vehicles.
- Protection of occupants of other, especially smaller-sized motor vehicles by means of matched deformation behaviour of the body **(Fig. 3)**.

Fig. 3: Effect of an energy-absorbing front underride protector integrated on a truck

1 **What different types of vehicle-body constructions are there?**
2 **What must be borne in mind in each case when machining super-high-tensile steels, galvanised sheet steel and aluminium?**
3 **What do you understand by active and passive safety?**
4 **Which measures in vehicle manufacturing improve active and passive safety?**
5 **How do belt-tensioning systems and airbags work?**
6 **Which measures on a vehicle constitute interior and exterior safety?**

17.1.6 Damage assessment and measurement

In the case of accident damage, body panels and frame components are subject to different stresses, e.g. due to upsetting deformation, extension, bending, twisting or buckling of the material. Depending on the type of collision involved, the following deformations of the frame, floor assembly or body are possible:

- **Sinking (Fig. 4),** e.g. in front- or rear-end impacts
- **Upward compression (Fig. 5),** e.g. in front-end collisions
- **Side distortion (banana damage) (Fig. 6),** e.g. in side accidents
- **Twisting (Fig. 7),** e.g. when a vehicle rolls over

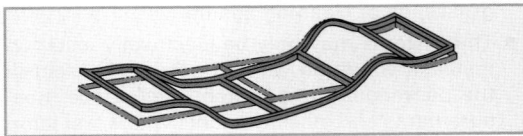

Fig. 4: Sinking

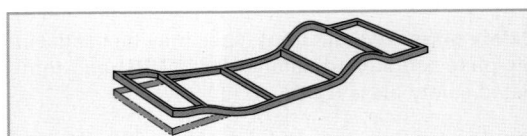

Fig. 5: Upward compression

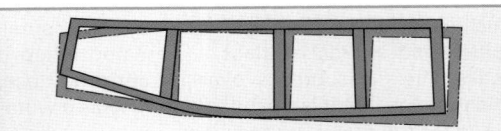

Fig. 6: Side distortion

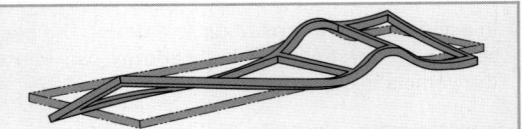

Fig. 7: Twisting

The materials can also demonstrate breaks or cracks.

To be able to assess accident damage exactly, it is necessary to conduct a visual inspection and, depending on the severity of the accident, a measurement of the body.

Damage assessment by visual inspection, determination of repair path and measurement of body

Visual inspection

This serves to establish what damage is present, whether a measurement of the vehicle is required and which repair jobs have to be carried out.

Depending on the severity of the accident, the vehicle must be examined in different areas for damage.

Exterior damage. The following must be checked within the framework of an all-round inspection of the vehicle:

- Deformation damage.
- Gap dimensions **(Fig. 1)**, e.g. on doors, bumpers, bonnet, boot. Altered gap dimensions indicate that the body has been distorted and renders a measurement necessary.
- Minor distortions, e.g. dents, buckling on larger surfaces; these can be identified by different light reflections.
- Glass damage, paintwork damage, cracks, widened seams.

Fig. 1: Visual inspection: gap dimensions

Floor-assembly damage. If upsetting deformation, buckling, twisting or symmetrical discrepancies can be made out, then the vehicle must be measured.

Interior damage. The following can be ascertained:

- Buckling, upsetting deformation (trim panels must often be removed for this purpose)
- Triggering of belt-tensioning systems
- Triggering of airbags
- Fire damage
- Dirt/contamination

Secondary damage. Here it is necessary to check whether further components have been damaged as a result of the accident, e.g. radiator, shafts, engine, gearbox, axles, axle suspension, steering, ECUs, electric cables.

Determination of repair path

The damage ascertained during the visual inspection is entered encrypted in the form of alphanumeric codes in technical specification sheets **(Fig. 2)**. Here the necessary repair operation, such as for instance replacement, sectional repair, part replacement, measurement, painting, is determined. The data are then processed with the aid of EDP calculation programs. This process is used to determine the proportion in which the repair costs relate to the present value of the vehicle.

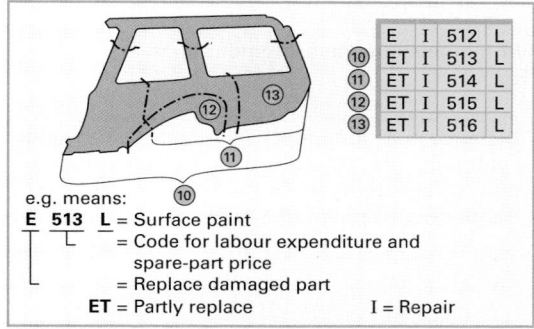

Fig. 2: **Excerpt from a technical specification sheet for costing in event of accident damage**

Measurement of body

It is necessary to measure a vehicle in order to establish whether the frame or floor assembly has been distorted. The following tools may be used as aids to this end: dividers, centring gauges, mechanical, optical or electronic measuring systems. The basis for such measurements is provided by the vehicle manufacturers' dimension tables or measurement sheets **(Fig. 3)**.

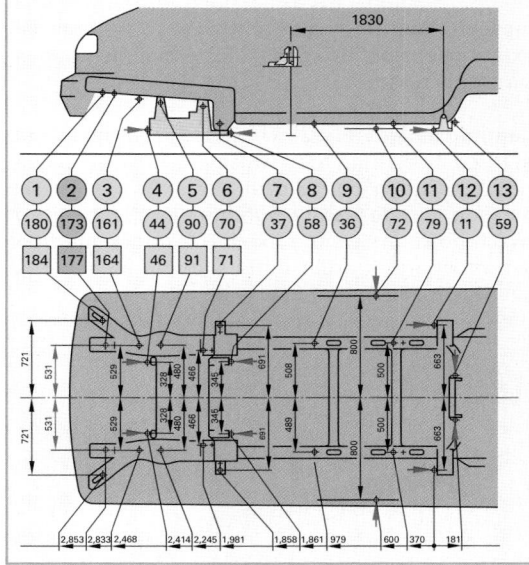

Fig. 3: **Excerpt from a measurement sheet for a floor-assembly measurement**

Explanation of Fig. 3, Page 401. Symmetry and height dimensions are given for the different measuring points. Two values are often given for the height dimensions:

- With installed assemblies
- Without assemblies

Thus, for example, measuring point **2** has a symmetry dimension of 531 mm and height dimensions of 173 mm (with installed assemblies) and 177 mm (without assemblies). These different height dimensions arise on account of body elasticity.

Two-dimensional measurement of body (Fig. 1)

> Distance measurements in length, width and symmetry are possible with two-dimensional body measurement. This is only suitable for rough measurement of a body.

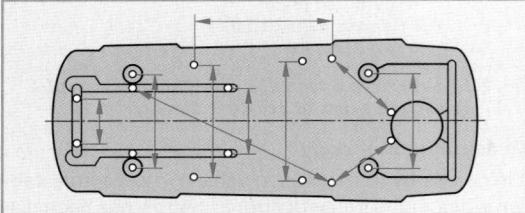

Fig. 1: Floor assembly with dimension reference points for two-dimensional measurement

Dividers. These are used to determine length, width and diagonal dimensions. If, for example, dimensional deviations are ascertained during diagonal measurement from the front right to the rear left axle suspension, this can indicate that the floor assembly is twisted.

Centring gauges (Fig. 2). These are usually made up of three measuring rods which are attached at specific measuring points of the floor assembly. The measuring rods feature sighting pins, by means of which positions can be fixed. The frame and floor assembly are in order if the sighting pins line up over the full length of the body.

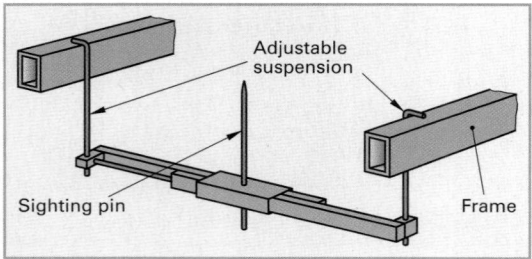

Adjustable suspension

Sighting pin   Frame

Fig. 2: Centring gauge

In **Fig. 3** the sighting pins do not line up, indicating that the vehicle has been displaced to the left in the middle (banana).

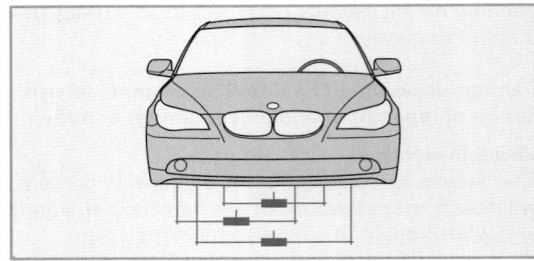

Fig. 3: Use of centring gauges

Three-dimensional measurement of body (Fig. 4)

> The body points can be determined in length, width and height with three-dimensional body measurement. This process is suitable for exact body measurement.

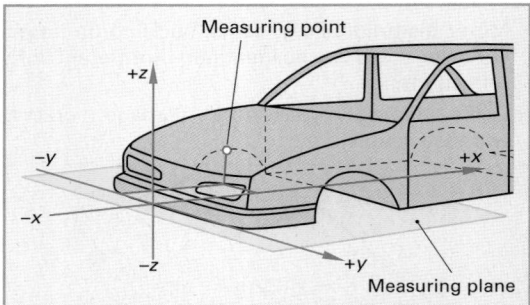

Measuring point

$+z$

$-y$

$-x$

$-z$ $+y$

$+x$

Measuring plane

Fig. 4: Three-dimensional measuring principle

Straightening bench with mechanical universal measuring system (Fig. 1, Page 403)

Here, the damaged vehicle is secured on the straightening bench with body clamps, which are attached to the door-sill join. Then the measuring bridge is slid under the vehicle and aligned. For this purpose, three undamaged body measuring points must be selected, two of which are parallel to the vehicle longitudinal axis. The third measuring point must be as far removed as possible.

Measuring cradles are mounted on the measuring bridge; these cradles can be precisely adjusted to the individual measuring points in order to determine the length and width dimensions. Each measuring cradle is fitted with telescopic measuring sleeves, onto which suitable measuring tips are attached.

When the measuring tips are extended, they are moved into the measuring points of the body, which also enables the height dimension to be precisely determined.

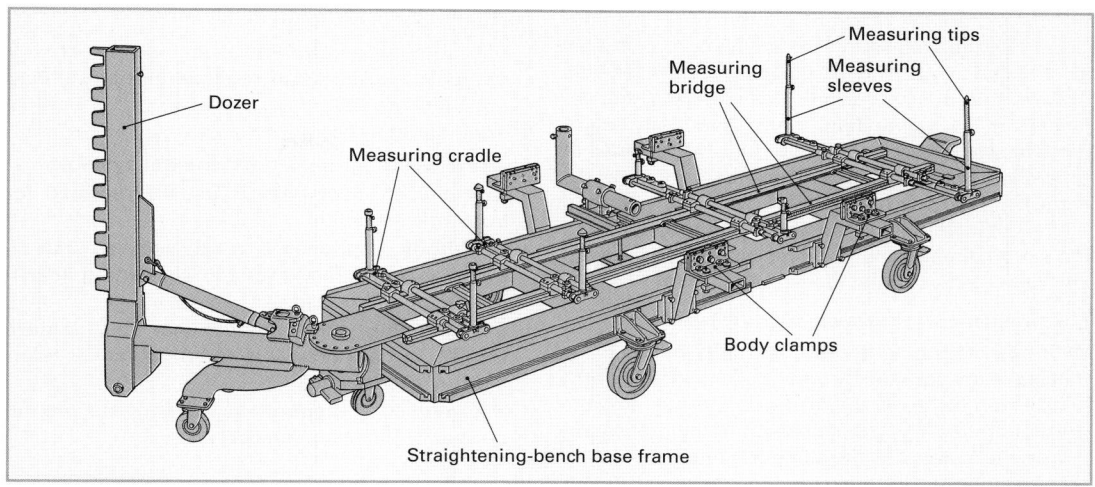

Fig. 1: Straightening bench with mechanical measuring system

Top-section measurement (Fig. 2). For this purpose, a portal frame with measuring devices is mounted for example on the measuring bridge, which is secured on a straightening-bench base frame. In this way, the top section can be measured in sections at defined points in accordance with the measurement sheets.

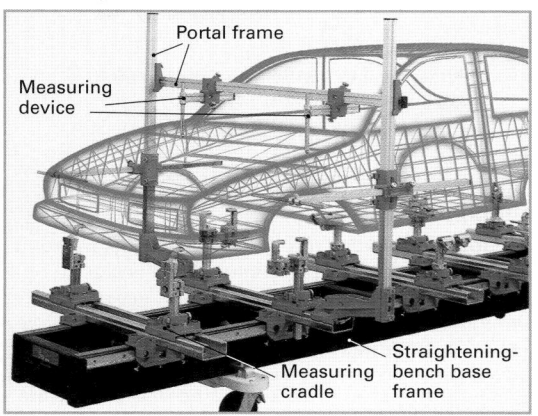

Fig. 2: Mechanical top-section measurement

Mechanical measuring systems are mounted on the straightening-bench base frame and can therefore make it difficult to attach straightening tools. They are also exposed to welding splashes, dust and mechanical influences and can be damaged in the process.

Optical measuring system (Fig. 3)

In the case of optical measurement of the body with the aid of light beams, the measuring system is mounted outside the straightening-bench base frame. Measurement is also possible without the straightening-bench base frame if the vehicle is placed on stands or raised on a lifting platform.

Two measurement rails which are erected at right angles to each other around the vehicle are used for measurement. These rails accommodate a laser unit, a beam splitter and several prism units. The laser unit generates small beams of light which are emitted in parallel. These light beams remain invisible until such time that they encounter resistance. The beam splitter directs the laser beam at right angles to the short measurement rail and simultaneously allows it to progress in a straight line. The prism units direct the light beam at right angles underneath the vehicle floor.

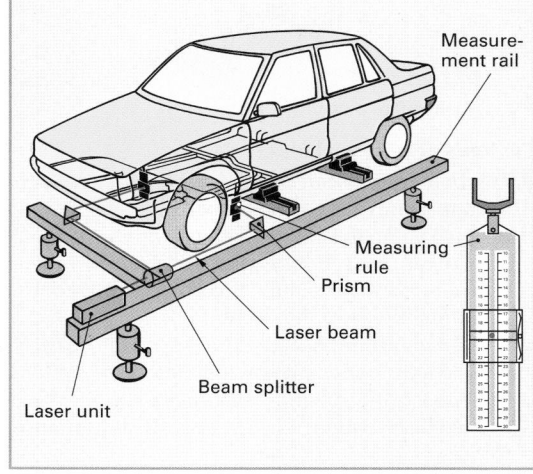

Fig. 3: Optical measuring system

Measuring rules made of transparent plastic with the associated connecting pieces are suspended and set in accordance with the measurement sheet at at least three undamaged body measuring points. Once the laser unit has been switched on, the measurement rails are altered until the light beam hits the area set on the measuring rules. This is identified by a red dot on the measuring rules. This ensures that the laser beam runs parallel to the vehicle floor. In order to determine further body height dimensions, further measuring rules are attached at different measuring points of the vehicle underbody. By moving the prism units, it is now possible to read off the height dimensions at the measuring rules and the length dimensions at the measurement rails and compare them with the measurement sheet.

Electronic measuring system (Fig. 1)

With this measuring system, the body points to be measured are selected by means of a measuring arm which runs on a measuring track and has suitable measuring tips. The exact position of the measuring points is calculated by the computer integrated in the measuring arm. The measured values are transmitted by radio signal to the measuring computer.

Measurement procedure. First three undamaged body points must be selected with the measuring tip in order to establish the basic position of the vehicle. Then the measuring tip must be positioned at the desired measuring points. The measured actual values are compared with the setpoint values stored in the measuring computer's program. In the event of dimensional deviations, an error message is output or automatic entry is made in the measurement record.

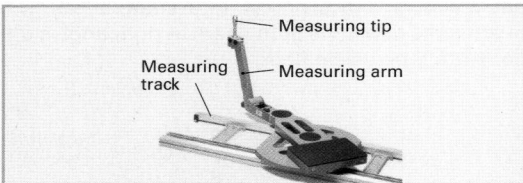

Fig. 1: Electronic measuring system

Features of universal measuring systems

- The body points can be measured with and without the removal of assemblies.
- Bonded vehicle window panes, even those which are broken, must not be removed prior to vehicle measurement since they absorb up to 30% of the torsional forces of the body.
- Special measurement sheets are available for each vehicle type, depending on the measuring system.
- The measuring systems are unable either to bear the vehicle weight or to absorb reformation forces.
- If new body components are to be welded in, special parts holders are required which are secured for example to the straightening-bench base frame.
- On measuring systems with laser beams, operators must not look directly into the laser beam.
- Universal measuring systems usually function as computer-aided systems.

Further body-measurement systems

- Straightening-attachment systems **(Fig. 2)** with one-part, two-part and multiple-part straightening attachments.
- Welding-jig systems.

Features of straightening-attachment systems

- Removal of assemblies is usually required for body measurement.
- New body components can be secured with dimensional accuracy to the straightening attachments for welding in.
- Straightening attachments bear the vehicle weight, but are only able to absorb small reformation forces.
- Different straightening-attachment sets are required for different vehicle types.

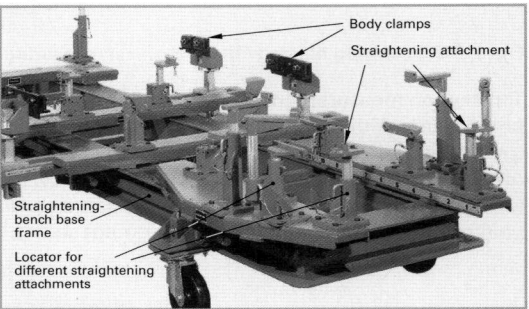

Fig. 2: Straightening bench with straightening attachments

17.1.7 Accident repairs to self-supporting bodies

Straightening

A body can convert large amounts of energy through the deformation of body panels in an accident. Suitably large tensile and pressure forces are required to straighten the body after an accident; these forces are exerted by hydraulic pulling and pressing tools.

> The reformation force should be in the opposite direction to the deformation force.

Hydraulic straightening tools (Fig. 3). These consist of a hydraulic press and a cylinder, which are connected by a high-pressure hose. In the pressing cylinder, the piston rod is extended by high pressure; in the puller cylinder, it is retracted. While the ends of the cylinder and piston rod can be well supported during pressing, puller clamps must be used when pulling or puller plates are welded onto the part to be pulled.

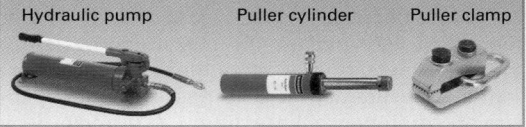

Fig. 3: Hydraulic straightening tools

Hydraulic dozer (Fig. 1). This consists of a horizontal beam and a pillar pivoted at its end which can be moved by a pressure cylinder. The straightening device can be used independently of straightening benches for minor to medium-level body damage where large tensile forces are not required.

For this purpose, however, the body must be secured on the horizontal beam at the points specified by the manufacturer with the aid of chassis clamps and support tubes.

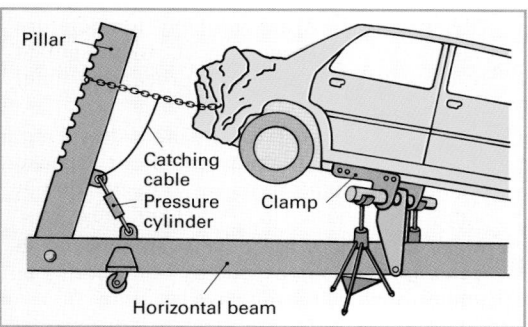

Fig. 1: Hydraulic dozer

Straightening bench with hydraulic straightening device (Fig. 2). The straightening bench consists of a sturdy frame, which absorbs the straightening forces. The vehicles are bolted on this frame at the bottom edge of the door-sill members with body clamps. The hydraulic straightening device can be quickly secured at any point of the straightening bench.

Even major body damage can be repaired with straightening benches. In this way, it is easier than it is with a dozer with a horizontal beam for reformation of the body to take place in the opposite direction to deformation. Hydraulic straightening tools which operate according to the vector principle can also be used. These are straightening tools which can pull or press deformed parts of the body in any spatial direction.

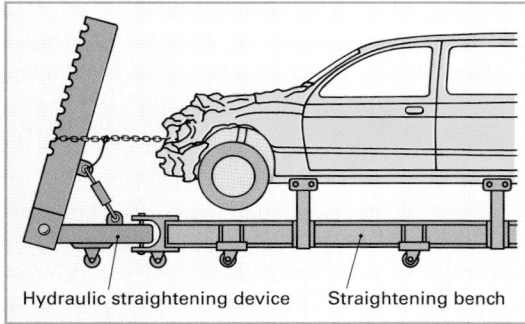

Fig. 2: Straightening bench with hydraulic straightening device

Deflection of reformation force (Fig. 3). If for example in an accident the body has in addition to horizontal deformation been displaced upwards, reformation by the straightening device must be performed by means of a deflection pulley. In this way, the tensile force acts in the opposite direction to the original deformation force. As shown in Fig. 3, for example, the body is pulled both downwards and forwards.

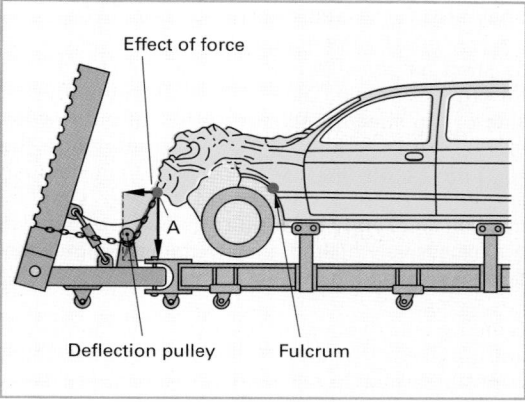

Fig. 3: Pulling arrangement for a body section displaced horizontally and upwards

---

WORKSHOP NOTES

- Carry out straightening work before cutting body components which can no longer be repaired.

- If straightening is possible, try to obtain the original state by cold reforming.

- If cold reforming is not possible without the risk of cracking, for normal-tensile body panels heat the deformed area over a large area for example with a gas welding torch. In so doing, owing to the possibility of structural transformation, do not exceed temperatures of 700 °C (dark-red colour). For high-tensile sheet metal and aluminium, follow the manufacturer's instructions to the letter.

- Check the position of the measuring points after each straightening operation.

- To enable frame components to achieve the exact body dimension free from stress, pull them a little beyond the setpoint dimension on account of the resilient behaviour of body panels.

- For safety reasons, always replace bearing components which are cracked or buckled.

- Secure pull chains with catching cables.

17

Part replacement and sectional repair (Fig. 1)

If in the event of heavily deformed sheet-metal parts a repair is not possible, is too expensive or is not permitted, a part replacement or a sectional repair can be performed in accordance with the manufacturer's instructions. In the case of part replacement, a body component, for example the rear left side section, is completely replaced.

In the case of sectional repair, only the damaged body area for example is cut out. The cutting lines are specified by the manufacturer. They are usually as short as possible and normally must not pass through reinforcing panels situated on the reverse side of the sheet-metal part to be cut out, for example on door hinges and seat-belt mountings. The repair panel must be accordingly cut to fit, stepped and inserted. MAG shielded arc welding is mainly used on steel body panels. In this case, the panel is not excessively heated, thereby minimising distortion and reworking.

Bonding and riveting are also used on aluminium and non-bearing parts. If aluminium needs to be welded, MIG or TIG shielded arc welding is used.

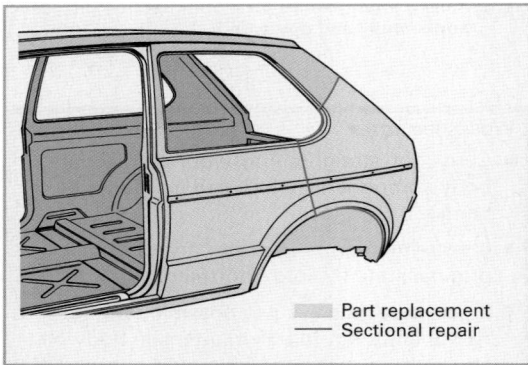

Legend:
- Part replacement
- Sectional repair

Fig. 1: Cutting lines for part replacement and sectional repair

Metal bonding in body making

The advantages of bonding over welding are:

- Only a minimum of reworking.
- Combustible materials do not have to be removed during the repair operation, e.g. fuel tank.
- No contact corrosion and good corrosion-protection effect in the repair area.
- Materials are not exposed to any thermal load during joining, e.g. Al parts.
- Joining of assorted materials possible.

WORKSHOP NOTES

- Grind bonding zones until metallically bright.
- Old and new parts must overlap at least 20 mm in order to create a sufficient bonding surface which will withstand later loads (**Fig. 2**).
- Bevel the outer panel (30 °, **Fig. 2**) so as to rule out any hairline cracks during subsequent painting.
- Coat both sides of the panel with adhesive. Observe the correct working temperature (usually room temperature, 20 °C), the potlife of the adhesive (maximum processing time) and the air-drying time.
- Fix the panel in position and press down, e.g. with prepared riveted joints or special clamps. If welding spots are to be set, do not apply any adhesive in the welding area.
- Make sure there are no movements in the bonding area, e.g. caused by other body operations, while the adhesive is curing.

Bonded joints may only be used in areas not subject to peeling stress.

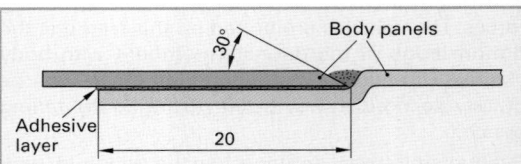

Fig. 2: Bonded joint of two panels

Replacing bonded vehicle windows. Bonded windows make a crucial contribution to the stability of the passenger cell. Great care must therefore be exercised when removing and installing window glass.

WORKSHOP NOTES – REMOVING A WINDOW

- Mechanically remove the adhesive layer between the body and the old glass for example with an oscillating blade. If the paint has to be protected against damage, tape off the edge of the glass.
- Lift out the glass with suction cups.
- If the old adhesive bead is in order, grind it down until smooth down to a remainder of approx. 1 mm to 2 mm. Otherwise, remove the bead completely.
- Clean the bonding edge of the vehicle frame with special cleaning agent.

When bonding windscreens or rear windows into position, open the side windows to rule out the possibility of an overpressure being created in the passenger compartment by the doors slamming shut or the glass being pressed out before the bonded joint has cured.

Surface working

Removing dents from sheet-metal parts

Depending on the size, accessibility and type of dent, different methods are used for planishing (removing dents), e.g.
- removing dents with a hammer and a holder-up
- pressing out dents according to the so-called MAGLOC procedure or with general-purpose spoons
- pulling out dents with the pulling-hammer procedure
- removing dents by means of heat technology

Removing dents with a hammer and a holder-up

(Fig. 1). The dent must be easily accessible from both sides for this purpose. Planishing dollies are used as the holders-up. Flat spoons are required in poorly accessible areas. Smaller dents can also be pressed out directly with flat spoons.

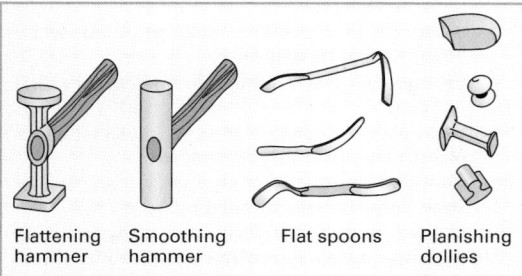

| Flattening hammer | Smoothing hammer | Flat spoons | Planishing dollies |

Fig. 1: Planishing tools

Procedure (Fig. 2). Medium- to larger-sized dents are hammered starting from the edge of the dent in a spiral shape to the middle of the dent. In the case of sheet steel, the holder-up must be constantly removed from the centre as the hammer strikes. The remaining smoothing of the surface can be performed by direct hammering (holder-up and hammer on a single axis) with a smoothing hammer.

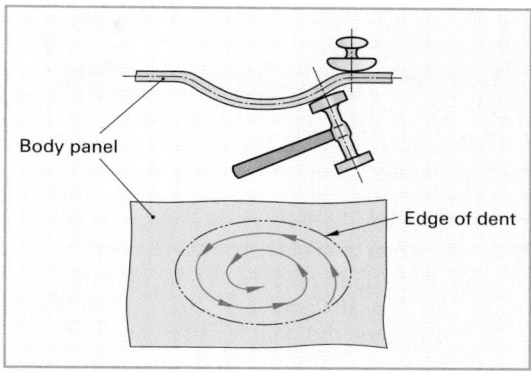

Body panel

Edge of dent

Fig. 2: Planishing procedure

MAGLOC procedure (Fig. 3). This procedure can be used to remove small dents (e.g. damage caused by hail) without damaging the paintwork. A pressing tool with a magnetic head is placed against the inside of the body panel. For the purpose of exactly locating the respective middle of the dent, a small steel ball is placed from the outside on the body panel. The steel ball is attracted by the magnetic head of the pressing tool. Once the ball has been centred in the middle of the dent, the dent itself can be pressed out.

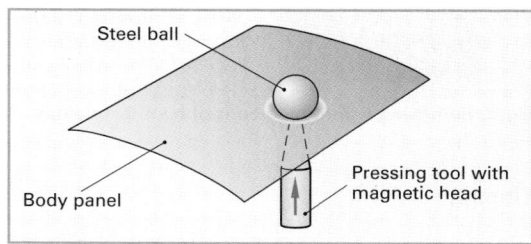

Steel ball

Pressing tool with magnetic head

Body panel

Fig. 3: MAGLOC procedure

Pressing out dents with general-purpose spoons (Fig. 4). This procedure can be used to press out smaller-sized dents in poorly accessible areas without damaging the paintwork.

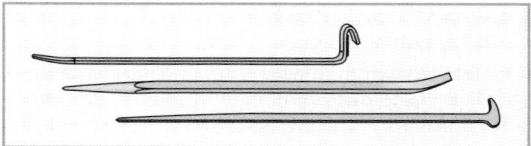

Fig. 4: General-purpose spoons

17

Removing dents with a pulling hammer (Fig. 1).
This procedure is used if dents are only accessible from one side, for example on double-walled panels. Here, perforated discs, for example, are welded onto the surface to be planished (Multispot). A rod with a handle and an impact weight is hooked into the perforated disc. The dent is pulled out by moving the impact weight towards the handle.

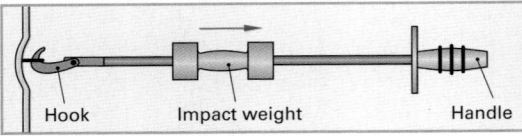

Fig. 1: Removing a dent with a pulling hammer

Removing dents by means of heat technology (Fig. 2).
Medium-sized dents are heated from the outside inwards with a weak flame in a spiral motion (**Fig. 2, a, b**). In this way, the dent is raised above its surrounding area and a small ring-shaped elevation is created. The heat can now be dissipated from the elevation with the aid of a cold panel file (**Fig. 2, c**). The panel contracts at these points and is smoothed to a large extent (**Fig. 2, d**). This procedure cannot be used on dents in the area of stiffening ribs, beads and welding spots.

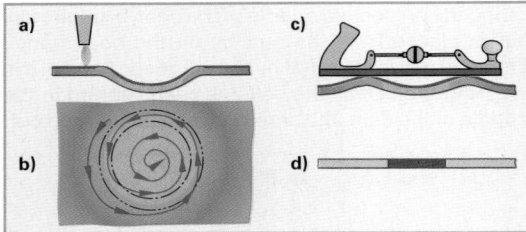

Fig. 2: Removing a dent by means of heat technology

Tinning

If larger irregularities remain after planishing, they can be corrected by tinning and smoothing by reworking.

WORKSHOP NOTES
- Grind the corresponding body area until metallically bright.
- Apply tinning paste with a brush to the bright body panel.
- Heat the tinning paste with a naked flame until it discolours (changes colour to brown). Wipe away the excess tinning paste with a clean cloth.

- Apply filling solder, e.g. L-PbSn25Sb, to the surface to be worked with the aid of a naked flame and smooth for example with a specially shaped piece of wood (soldering wood) dipped in beeswax.
- After the working surfaces have cooled, smooth this area with a panel file.

Due to the high lead content in filling solder, toxic vapours are formed during the tinning process and these must be drawn off.

Applying knifing filler

Smaller-sized irregularities can be corrected with knifing filler. Two-component knifing fillers of polyester or epoxy resin can be used for example. Larger-sized irregularities must be corrected by tinning because the application of an excessively thick layer of knifing filler may give rise to the following problems:

- Knifing filler flakes off the panel surface on account of the different degrees to which the materials expand.
- Cracking
  a) during curing, due to different heat zones within the knifing filler.
  b) because the knifing filler is not as elastic as the body panel.

Grinding dust must be drawn off because of the potential danger to skin and respiratory tracts.

REVIEW QUESTIONS

1 How is damage assessment by visual inspection carried out?
2 Which dimensions can be determined during two-dimensional measurement of the body?
3 How is three-dimensional measurement of a body with mechanical measuring systems carried out?
4 How is measurement of a body with an optical measuring system carried out?
5 How is electronic measurement of a body carried out?
6 What are the features of straightening-attachment systems?
7 What do you have to bear in mind when straightening by reforming?
8 What do you have to bear in mind when metal bonding?
9 What working rules must be observed when replacing a bonded vehicle window?
10 With which procedures can smaller- and larger-sized dents be straightened out?
11 How can larger- and smaller-sized irregularities, e.g. after planishing, be corrected?
12 Why must knifing filler not be applied too thickly?

17.2 Corrosion protection on motor vehicles

A distinction is made between active and passive corrosion protection.

17.2.1 Active corrosion protection

This can be performed:
- On the material, e.g. appropriate alloys on special steels.
- On the corrosive agent, e.g. by removal of moisture from the air.
- By external voltage, e.g. a steel tank is applied to the positive terminal of a battery (cathodic corrosion protection).

17.2.2 Passive corrosion protection

This can be performed by preservation or sealing, e.g. metallic and nonmetallic coatings.

Preservation procedures

These are used on a motor vehicle in undercoating and cavity sealing.

Undercoating. This has the following functions:
- To keep moisture away from the underbody
- To be insensitive to stone impacts
- To remain elastic
- To prevent natural vibration of the panels (anti-drumming effect)

Wax-, plastic- and bitumen-based preservative agents are used.

Cavity sealing. The preservative agent (sealant) is composed of film-forming oils, waxes, solvents and rust inhibitors. The cavity sealant is injected at a pressure of roughly 70 bar through openings at points precisely specified by the vehicle manufacturer or cavities are flooded and then sealed with plastic plugs. Rust inhibitors prevent rust from forming.

Metallic coatings

This will only provide lasting corrosion protection if they are nonporous, water-insoluble and impermeable to gas. If the protective layer consists of a baser material than the workpiece, e.g. zinc on steel **(Fig. 1)**, a local galvanic element is created if the protective layer is damaged which will cause the gradual destruction of the protective layer; the

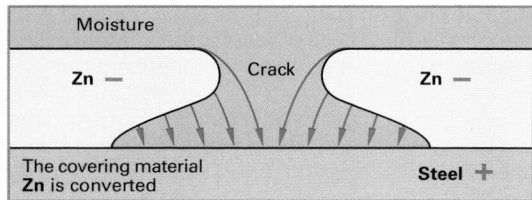

Fig. 1: Genuine protective material

workpiece does not rust for the time being. However, if the protective layer consists of a nobler material than the workpiece, e.g. nickel on steel, the workpiece is destroyed if the protective layer is damaged because it rusts. The protective layer remains **(Fig. 2)**.

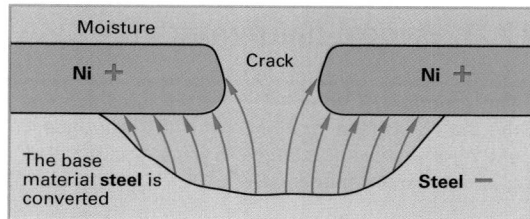

Fig. 2: Spurious protective material

Genuine protective material	▶ The coating metal is baser than the base metal.
Spurious protective material	▶ The coating metal is nobler than the base metal.

Coating in fused mass (hot-dip metalling). Hot-dip galvanising involves immersing the finished body component or the unfinished panel after appropriate pretreatment in a bath of liquid zinc.

Electroplating. Electrogalvanised sheets are used as deep-drawing body sheets in body making. The layer thickness is very uniform, amounting to roughly 7.5 μm. The surface is smooth. This facilitates a uniform top coat without expensive surface working.

Nonmetallic coatings

Phosphatising (bonderising, atramentising). The workpiece is immersed in an aqueous phosphate solution. A porous protective layer of iron phosphate is created on the surface. This forms the primer for paints.

Plastic coatings. Seals can be effected on joins and edges with the aid of permanently resilient plastics, e.g. PVC. Here, different metals can also be connected to each other without corrosion, e.g. aluminium panelling on steel frames **(Fig. 3)**.

17

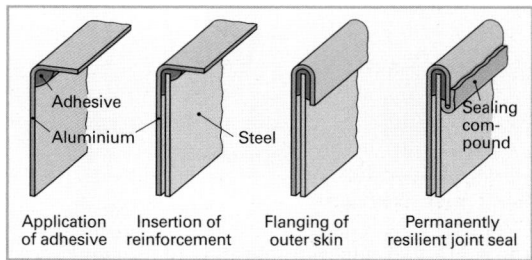

Fig. 3: Corrosion protection at join

Anodising. The surface of aluminium workpieces can be electrolytically oxidised. This produces layer thicknesses of 5 μm, where the shape and volume of the workpiece do not change. The corrosion-resistant surface can be coloured.

17.3 Vehicle paintwork

> The function of vehicle paintwork is to protect the body surface against external influences, e.g. aggressive substances in water and air, and against stone impacts.

Furthermore, the vehicle paintwork is intended to ...
- ... form a tight and cohesive protective film.
- ... be hard and at the same time elastic.
- ... be colourfast.
- ... generate a signal effect.
- ... be easily cleanable and maintainable.

Application processes
Paints can be applied by spraying, immersing or electric spraying processes.

Spraying. Spray guns **(Fig. 1)** operate for the most part with compressed air. Here, the paint is drawn in according to the injector principle by the air flowing past the injector and transported to the jet nozzle. As the paint emerges from the jet nozzle, a paint mist is created which settles on the surface.

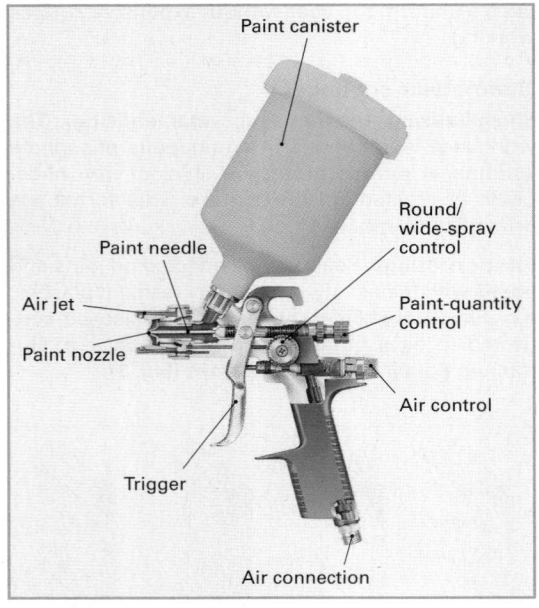

Fig. 1: Paint spray gun

There are two different methods of spraying: cold and hot spraying.

Cold spraying. The paint is thinned by a solvent (change in viscosity) until it can be easily sprayed. The solvent evaporates after painting. Excessively quick evaporation of the solvent may cause the painting surface to shrink.

Hot spraying. The paint is preheated by a heater in the paint canister to 50 °C to 120 °C. In this way, the viscosity of the paint is reduced to such an extent that it can be sprayed without a solvent.

Spraying with electrostatic charge. Electrostatic spraying is used in mass production. The positive pole and the negative pole of a direct-voltage source are applied to the body and to the paint-spraying nozzles respectively. The voltage can range up to 200,000 V. The negatively charged paint mist is attracted by the positively charged body. The loss of paint is thereby reduced.

In the electrostatic spraying process, the paint can be applied by high-rotation bells instead of by spray guns **(Fig. 2)**. Spraying robots are used to paint those areas of the body which cannot be reached by the paint mist from the high-rotation bells.

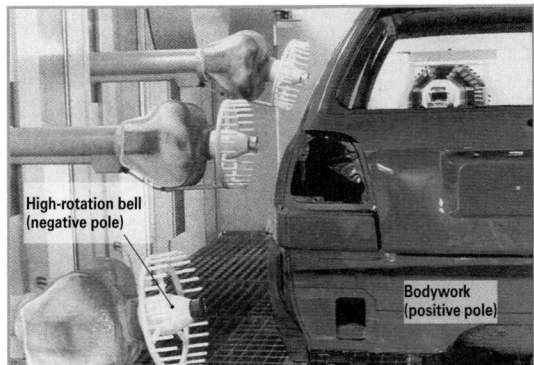

Fig. 2: Electrostatic spraying

Airless spraying (high-pressure spraying). The paint material is subjected to high hydrostatic pressure (100 bar to 200 bar). It atomises as it relaxes at the outlet of the spray nozzle. Airless spraying facilitates fine-mist atomisation even of viscous coating substances. To be able to work with a lower hydrostatic pressure (40 bar to 60 bar), it is possible to support the atomisation of the paint material with compressed air. These processes are primarily used to apply undercoating and corrosion inhibitors.

Immersing. In mass production the priming coat can be established by immersing the body in a tank filled with priming paint. Excess paint on the body can be removed by suspension and through drain holes.

Electrophoresis process (Fig. 1). The paint particles for example of a water/artificial-resin emulsion suspended in an electrolyte are electrically charged and transported to the opposite-charged body, on which they form a uniform coat of paint. The process continues until the last bare area is covered with paint, i.e. insulated. This process is only suitable for the first coat of paint, the priming coat. The positive hydrogen ions generated by electrolysis during the decomposition of water stray to the negatively charged body, where they prevent oxide formation on the panel/plate during the coating process.

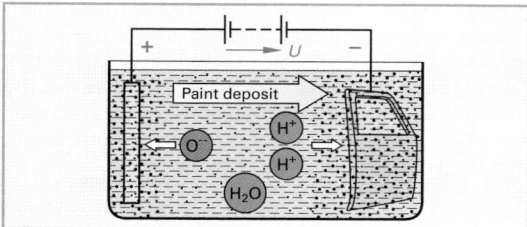

Fig. 1: Cataphoresis

Buildup of paint

The paintwork of a motor vehicle **(Fig. 2)** is made up of the following layers:

- Phosphate layer
- Electro-dipcoat-priming
- Intermediate anti-chip foundation
- Filler (spray foundation)
- Top coat (plain or metallic paint)

The body must be preheated before the necessary layers can be applied for the buildup of paint. For this purpose, it must be cleaned, degreased and then provided with a phosphate layer.

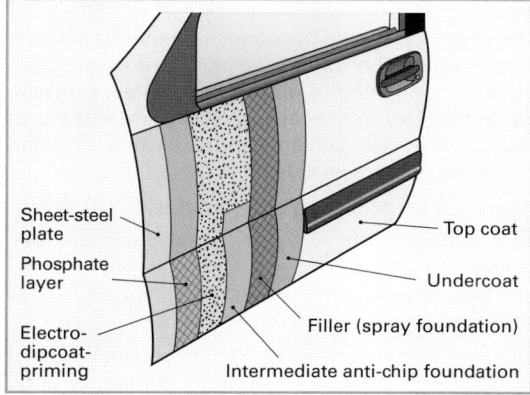

Fig. 2: Buildup of paint layers on a vehicle door by example

Phosphate layer. A porous iron-phosphate layer is created on the sheet-metal surface by phosphatising. This is essential to the sound adhesion of the subsequent layers and to very good corrosion protection.

Priming coat. This creates an adherent layer for the intermediate anti-chip foundation, the filler and the top coat. The priming coat is usually applied in immersing or electrophoresis processes.

Intermediate anti-chip foundation. This can be applied to outer-skin areas of the body particularly exposed to the risks of stone impacts, for example the side surfaces of the body up to the lower edges of the window apertures and the bonnet.

Filler (spray foundation). This serves to correct smaller irregularities, grinding marks and pores on the surface. The filler is usually applied by an electrostatic-spraying machine. It forms the background for the undercoat and the top coat. If the top coat is applied directly to the filler layer, the filler also assumes the function of the undercoat.

Top coat

A distinction is made between plain paintwork and metallic paintwork.

Plain paintwork

Four-layer paint structure. In addition to the priming and filler layers, two further layers are applied by electrostatic spraying.

- Spraying on of the undercoat and drying at approx. 140 °C.
- Spraying on of the top coat and drying at approx. 130 °C.

Three-layer paint structure. This consists of primer, filler and top coat. Here, the top coat is sprayed immediately onto the filler layer wet on wet and then dried. The advantage of the four-layer paint structure over its three-layer counterpart is that the overall thickness of the paint layer is very uniform over the entire body surface since the undercoat layer and the topcoat layer are of equal thickness.

Metallic paintwork

In contrast to plain paintwork, metallic paintwork involves the application of a metallic base paint as the colouring, effect-giving layer and a clear lacquer as the glossy, protective layer. The metallic base paint and the clear lacquer are applied by air atomisation and electrostatic spraying respectively. They are applied in a "wet-on-wet" process, i.e. the clear lacquer is sprayed onto the base paint without intermediate drying. The two paint layers are then dried at roughly 130 °C.

17

Paints

These consist of non-volatile and volatile components **(Table 1)**.

Table 1: Paint components

Non-volatile components	
Binders	Resins, film-forming media
Pigments	Insoluble paint particles
Additives	Catalysts, antioxidant agents, extenders, rust inhibitors
Volatile components	
Solvents	Thinners, reaction products

Binders. These form the paint film after the coating and drying processes. Here, the colouring pigments are combined with each other by the resins. Film-forming media speed up the layer-formation process and improve the processability of the paint.

Pigments. These lend the coating the desired colour appearance. Pigments are paint particles which are present in insolubly solid form in the paint.

Additives. Catalysts speed up the curing and drying processes. Antioxidant agents prevent the paint from forming a skin and thickening. The extenders improve the shine and film formation of the paint. The protective properties of the paint are improved by rust inhibitors.

Solvents. These dissolve the solid and viscous components of the paint and establish the viscosity required for processing. Solvents and reaction products evaporate during processing and during drying of the paint film. Reaction products are created during the stoving process and during the film-formation process, e.g. dehydration by polycondensation.

Types of paint

There are the following different types:
- Nitrocellulose paints
- Synthetic paints
- Effect paints
- Water paints
- High-solid paints
- Powder-based paints

Nitrocellulose paints. These are normally no longer used nowadays in painting vehicles. Nitrocellulose paints harden rapidly due to the evaporation of the solvent. They are highly inflammable, non-resistant to fuels and require regular upkeep in order to maintain the high-gloss surface. Nitrocellulose paints are mainly used today in painting classic cars.

Synthetic paints. These use, for example, duroplastics (alkyd resins, melamine resins) as binders. These paints cure under the influence of atmospheric oxygen. This is known as oxidative curing. Application: Top coat.

Acrylic-resin paints. These use acrylic resins (thermoplastics) as binders. They cure through physical drying, i.e. through evaporation of the solvents. These paints can be dissolved again with the aid of solvents; they are reversible. Silicone resins are used as binders for heat-resistant paints. Application: They are used both as priming coats and as top coats on vehicles. Acrylic-resin paints are subdivided into

- one-component paints (1C paints)
- two-component paints (2C paints)

One-component paints (1C paints). These usually cure under the influence of atmospheric oxygen through cross-linking of the molecules (polymerisation). Solvents and reaction products evaporate in the process. A high-gloss paint layer is created. The ultimate hardness of the paint layer is usually only established after a period of several weeks. The curing process can be sped up by stoving at temperatures between 100 °C and 140 °C.

Two-component paints (2C paints). These consist of extenders and hardeners. In mass-production painting these paints are usually mixed in the correct proportions in a spray gun. Between the two components a chemical reaction (polyaddition) takes place which gradually cures the applied paint film without reaction products even at room temperature.

The curing process can be sped up at temperatures up to roughly 130 °C. Acrylic-resin paints are resistant to chemicals, scratch-resistant and weatherproof.

Effect paints (metallic paints). In addition to the colour pigments, these contain mica or flakes of aluminium in the base paint. Because these additives reflect the incident light, a metallic effect is created on the surface. Once the base paint has been applied, a second layer of clear paint is applied wet on wet to protect the base paint.

Water paints. Plastic-based resins serve as binders. In the case of fillers and base paints the organic solvent components are almost completely replaced by water. Only in the case of clear paint is the proportion of organic solvents approximately 10 % and the water content as solvent is up to 80 %.

There are the following different types:

True water paints. The resin molecules are dissolved in water.

Water-thinnable paints (dispersion). The resin particles are finely distributed in the water.

After application, the water and solvent in the paint layer are evaporated in drying systems. A tight, water-proof and chemical-resistant paint layer is formed. However, the drying process takes longer due to the low solvent content. But damage to the environment by solvent emissions is lower.

High-solid paints (HS paints) and **medium-solid paints** (MS paints). These are paints which contain a high proportion of non-volatile components (solid content up to 70 %). The solvent content on the other hand is greatly reduced (20 % to 30 %) for environmental-protection reasons. These paints are primarily used in repair applications. They are characterised by very good coverage, rapid thorough drying and a high-gloss finish. Furthermore, the large layer thickness possible in each work operation greatly reduces the labour expenditure.

Powder-based paints. A plastic which is processed into powder with a grain size of 20 μm to 60 μm is used as the binder. Then the powder is sprayed with special spray guns onto the cold or warm workpiece to be coated. The paint powder sticks to cold materials electrostatically and to warm workpieces by melting on. Then the paint film must be established on the coated parts. By baking for example with infrared radiators at approx. 120 °C or in a stove furnace at temperatures above 130 °C, the powder melts and the macromolecules of the binder cross-link (polyaddition). During cooling, a tight, shockproof, chemical-resistant paint layer with a thickness of up to 120 μm is created. The advantage of this process lies in the fact that there is no emission of solvents. Furthermore, there are no spraying losses since the powder-based paint that does not stick (overspray) can be returned to the production process.

WORKSHOP NOTES

Note: For respraying operations it is important that the components of the respraying system match each other. For this reason, use a complete system from a single manufacturer instead of products from different manufacturers. To achieve an optimal result, follow the manufacturer's instructions to the letter.

Pretreatment. First clean the damaged area to remove dirt, rust, grease, old paint residues and silicone residues. Then remove the silicone residues with silicone remover. Grind the damaged area and grind old paintwork down to the healthy layers.

Applying knifing filler. It is possible to correct irregularities and damage in several work operations by applying knifing filler. Use 2C polyester knifing fillers for this purpose.

There are

- putties for the smallest irregularities and thus for a smooth, nonporous surface.
- filler and pulling pastes for larger irregularities.
- knifing fillers which can be used as both pulling pastes and putties. They are suitable for all backgrounds, e.g. also on galvanised sheet-steel plates. They possess good adhesive properties.

Priming coat. After grinding, paint over the entire repair area with a priming filler in several work operations. The crucial factor in choosing the filler is the way in which the rework and the top coat are carried out in the selected paint system.

The following different types of filler may be used:

- Colouring filler for good colour results
- Machine-grinding filler for subsequent regrinding with a grinder
- Hand-grinding filler for subsequent regrinding by hand
- Wet-on-wet filler for and immediate top coat

Top coat. After drying and finish-grinding the repair area, apply the top coat with the desired shade of colour. Plain paintwork in one or two layers, metallic paintwork in two layers. When applying top coats, always follow the manufacturer's instructions to the letter in order to avoid paintwork defects.

Accident-prevention regulations

Always wear protective goggles when carrying out grinding work. If possible, always work with a dust-extraction facility. Paints and additives contain hazardous components; for this reason, wear protective clothing and a breathing mask when carrying out painting work.

If paints and solvents whose flash points are below 21 °C are processed, the painting rooms are subject to explosion hazards. Such rooms must always have two clearly identified exits which must not be blocked or obstructed.

Ensure that there are no fireplaces or sparking machines or devices within 5 m of the painting area.

Always have on hand a sufficient number of hand-held fire extinguishers and fire blankets.

Always wear fresh-air breathing apparatus when working in well ventilated spray booths.

REVIEW QUESTIONS

1 What are the functions of paintwork?
2 What different types of paint are there?

3 What are powder-based paints?
4 What different ways of applying paint are used on a motor vehicle?

18 Chassis

The chassis of a motor vehicle includes:

- Wheel suspension
- Steering
- Suspension
- Brakes
- Wheels and tyres

They are responsible for dynamics and for the road safety of the vehicle.

18.1 Driving dynamics

Dynamics deal with the action of the forces affecting the vehicle while it is being driven and the resulting movements of the vehicle.

The movements occur about the longitudinal axis, transversal axis and vertical axis **(Fig. 1)**.

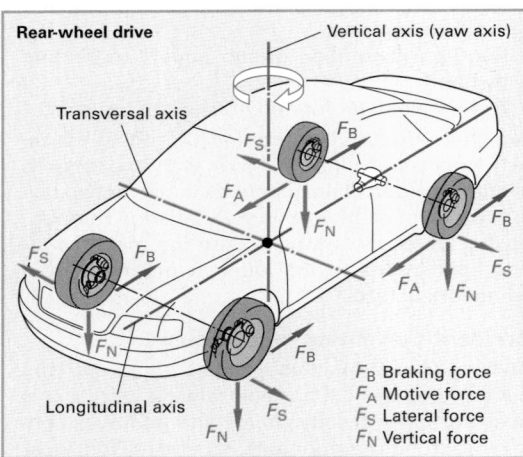

Fig. 1: Forces and axes on the vehicle

F_B Braking force
F_A Motive force
F_S Lateral force
F_N Vertical force

A distinction can be made between:

- Forces acting along the longitudinal axis: motive force, braking force, friction force
- Forces acting along the transversal axis: centrifugal force, wind force, lateral force
- Forces acting along the vertical axis: wheel load, forces created by jolts from a rough road surface

The movements resulting from all forces acting together express themselves in the drivability of the vehicle.

Factors influencing the drivability are:

- The location of the centre of gravity, roll centre, roll axis, driving axis
- The type of drive and the mounting location of the power plant
- The wheel suspension and the wheel positions
- The suspension and the oscillation damping
- The wheel control systems, such as ABS, TCS, ESP

Roll centre (instantaneous centre, **Fig. 2**). This is the point (**W**) on an imaginary perpendicular to the centre of the axle, about which the vehicle body rotates due to the action of lateral forces F_S.

The roll centre of a vehicle axle is located in the centre of the vehicle when viewed from the front. Its height depends on the wheel suspension.

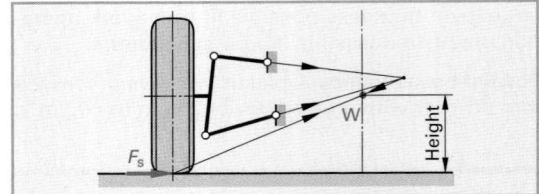

Fig. 2: Roll centre

Roll axis. This is formed by connecting the roll centres of front axle W_F and rear axle W_R **(Fig. 3)**. It usually slopes down towards the front of the vehicle, since the roll centre is lower at the front wheel suspensions than at the rear.

The closer the centre of gravity **S** lies to the roll axis, the less the vehicle tilts when cornering.

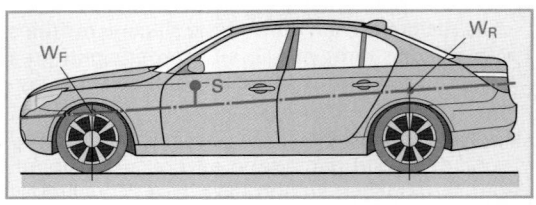

Fig. 3: Roll axis

Axis of symmetry. This runs in vehicle longitudinal direction through the centre of the front and rear axles **(Fig. 4)**.

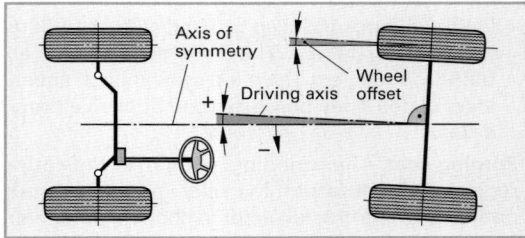

Fig. 4: Axis of symmetry, driving axis

Geometrical driving axis. This is formed by the position of the rear wheels and is the bisector of the toe-in angle of the rear wheels **(Fig. 4)**.

The **wheel offset** is the angle by which, for example, the two rear wheels are offset against each other towards the front (+) or towards the rear (–) **(Fig. 4)**.

Wheel-slip angle. If a vehicle is hit by a lateral interference factor while it is in motion (e.g. wind force, centrifugal force), lateral forces F_S act in the tyre contact patches of all four tyres. If the steering is not corrected, the direction of travel of the wheels changes, they run at an angle to the original direction of travel by an angle of α **(Fig. 1)**.

> Wheel-slip angle α is the angle between the wheel plane and the actual direction of wheel motion.

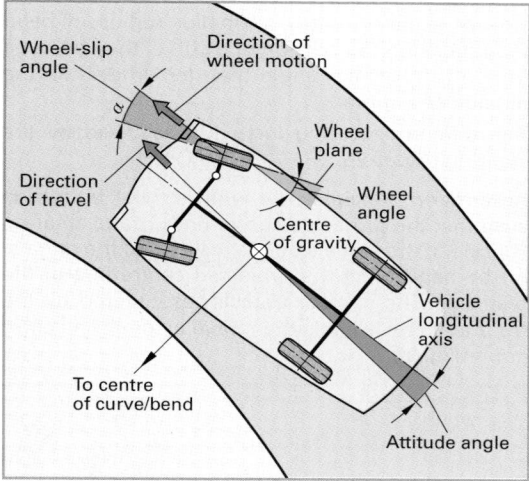

Fig. 1: Wheel-slip angle and attitude angle

Attitude angle. This relates to the whole vehicle **(Fig. 1)**.

> The attitude angle is the angle between the direction of travel (direction of motion of the vehicle) and the vehicle longitudinal axis.

Self-steering effect

To assess drivability, standard driving manoeuvres are performed, e.g. steady-state turn, and the self-steering effect of a motor vehicle is determined.

Up to the cornering limit speed, the adhesion between tyres and road surface is adequate for establishing the lateral forces required.

If the corner is taken at a higher speed, lateral slip occurs at the front or rear wheels or at all wheels.

A distinction is drawn between:

- **Understeer (Fig. 2).** Wheel-slip angles α_F of the front wheels are greater than those of the rear wheels α_R. The vehicle wants to steer a larger radius of bend than that corresponding to the lock on the front wheels and drifts outwards over the front wheels.

- **Oversteer (Fig. 3).** The wheel-slip angles of the rear wheels α_R are greater than those of the front wheels α_F. The vehicle wants to steer a smaller radius of bend than that corresponding to the lock on the front wheels and the vehicle starts to break away at the rear.

- **Neutral drivability.** The wheel-slip angle of the front and rear wheels is the same. The vehicle drifts evenly on all the wheels.

Vehicles with

- front wheel drive tend to understeer
- rear engines and rear-wheel drive tend to oversteer
- all-wheel drive tends towards neutral drivability

The aim is for neutral or slightly understeered drivability (with the exception of sports vehicles).

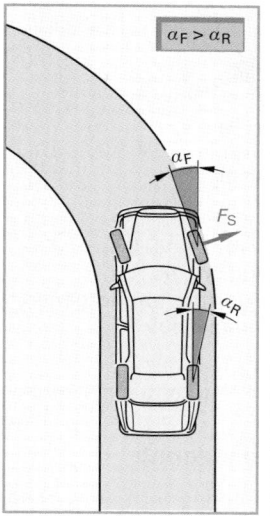
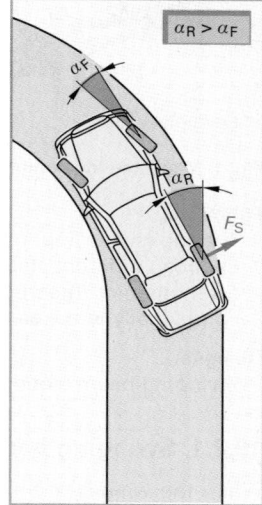

Fig. 2: Understeer   **Fig. 3: Oversteer**

Yawing is the rotational motion of the vehicle about its vertical axis (yaw axis) **(Fig. 1, Page 414)**. The yaw velocity is measured by yaw sensors on vehicles with ESP.

Rolling is the tipping movement about the roll axis **(Fig. 3, Page 414)**.

Pitching is the rotational motion of a vehicle about its transversal axis **(Fig. 1, Page 414)**.

REVIEW QUESTIONS

1 What are the 3 spatial axes of a vehicle and what are the movements about them called?
2 What is meant by roll centre (instantaneous centre)?
3 How is the roll axis of a vehicle formed?
4 What is the wheel-slip angle?
5 Explain the terms understeer, oversteer and neutral drivability.

18.2 Basic principles of steering

The main steering components in the motor vehicle are (Fig. 1):
- Steering wheel
- Steering spindle
- Steering gear
- Tie rod
- Tie-rod arm

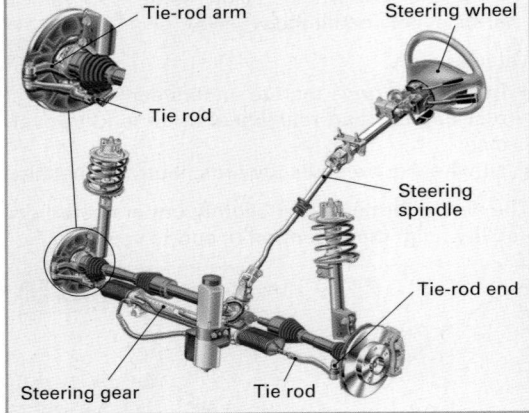

Fig. 1: Main steering components

Functions:
- Turning (swivelling) the front wheels.
- Enabling different steering angles.
- Strengthening (gearing up) the torque generated manually at the steering wheel.

Designs:
- Swinging beam steering
- Ackermann steering

18.2.1 Swinging beam steering

When the wheels of the steering axle are turned, they are pivoted about a common rotational axis (steering axis). The tendency to tilt increases due to the reduction in the size of the standing area. Swinging beam steering is used on twin-axle trailers. It offers good manoeuvrability.

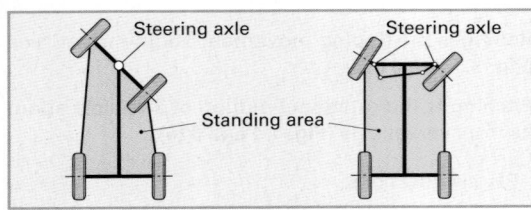

Fig. 2: Swinging beam steering, Ackermann steering

18.2.2 Ackermann steering

Each wheel is pivoted about its own axis, the steering axis. It is formed by the connection of the upper and lower mounting points of the wheel suspension (Fig. 2, Page 418) or by the longitudinal connection of the kingpins. Ackermann steering is used on all dual-track motor vehicles. When the wheels are turned about the steering axis, the standing area remains almost the same size.

Rolling motion of the wheels when cornering

In order for the wheels to be able to roll faultlessly when cornering, each steered wheel must be turned to an angle appropriate to the radius of bend. A greater wheel angle is required for a small radius bend than for a larger one.

Since on dual-track vehicles, the wheels on the inside of a curve follow a smaller radius of bend than those on the outside of a curve, they must be turned to a greater angle than the wheels on the outside of a curve.

The different steering locks are achieved by the steering trapezoid.

Ackermann principle. The wheels must be turned such that the projected centre lines of the steering knuckle of the wheels on the inside and the outside of the bend meet the projected centre line of the rear axle. The circular trajectories covered by the front and rear wheels then have a common centre point (Fig. 3).

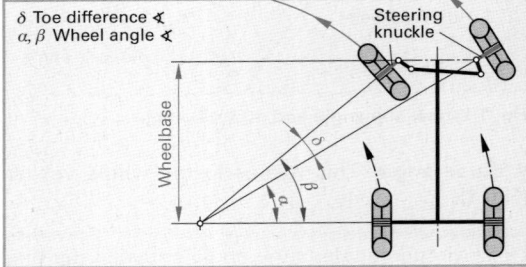

Fig. 3: Ackermann steering, toe-difference angle

Steering trapezoid

This is formed by the tie rod, the two tie-rod arms and the line through the two steering axles (Fig. 4) when the front wheels are set to the straight-ahead position.

> The steering trapezoid allows the front wheels to turn at different angles, the inside wheel being turned further than the outside wheel.

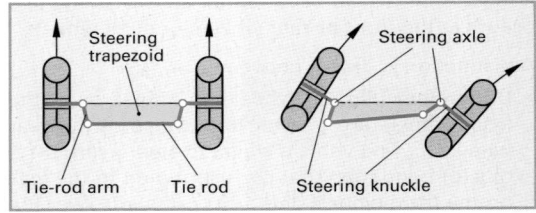

Fig. 4: Steering trapezoid

18.2.3 Steering linkage

Functions:
- Transfer of the steering movement produced by the steering gear to the front wheels.
- Guidance of the wheels in a particular toe-in angle.

Main components

Tie rod(s), tie-rod joints, tie-rod arms, possibly intermediate lever and steering rod.

Rigid front axle. Recirculating-ball steering gear is usually used as the steering gear on commercial vehicles. The movement is transferred by the steering-gear pitman arm via the steering rod to the intermediate lever and track arms (tie-rod arms). The latter is connected to the one-piece tie rod and the track arm of the other side of the axle by a tie-rod linkage **(Fig. 1)**.

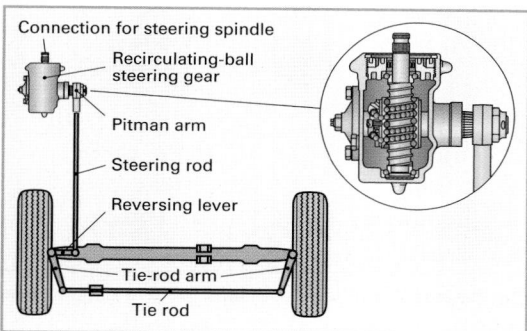

Fig. 1: Rigid axle with single-piece tie rod

18.3 Wheel adjustments

Wheelbase

> The wheelbase is the distance between the centre of the front wheels and the centre of the rear wheels **(Fig. 2)**.

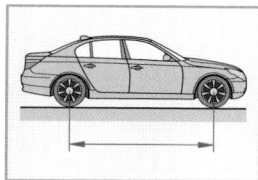

Fig. 2: Wheelbase　　　　**Fig. 3: Track width**

Track width

> The track width is the distance between the wheels on one axle, from the centre of one tyre to the centre of the other, measured on stand plane **(Fig. 3)**.

Wheelbase times track width gives the wheel contact area.

Toe-in

> The toe is the difference in length $l_2 - l_1$ between the front of the two wheels and the rear of the two wheels when set straight ahead.

The toe-in is measured at the hub height from rim flange to rim flange and may be given as the toe-in angle (for both wheels) either in millimetres or in degrees (°).

A distinction is drawn between:
- **Toe-in**
- **Zero toe**
- **Toe-out**

Toe-in $(l_2 - l_1) > 0$ (Fig. 4)
This is used with rear-wheel drive and positive kingpin offset. The wheels are pivoted outwards by the rolling resistance at the front.

Zero toe $(l_2 - l_1) = 0$

Toe-out $(l_2 - l_1) < 0$ (Fig. 5)
This is used with front-wheel drive and positive kingpin offset. The wheels are turned inwards by the motive force acting on the tyre contact patch.

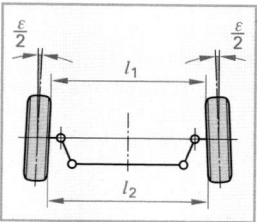
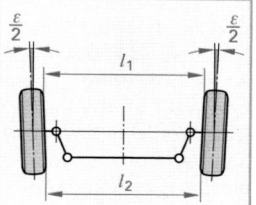

Fig. 4: Toe-in　　　　**Fig. 5: Toe-out**

Toe, camber, kingpin inclination, kingpin offset and castor are determined such that the following objectives are attained:
- Small and favourable self-steering effect
- Good straight-running stability
- Low tyre wear
- Compensation for play in the wheel location
- Little or no tendency of the wheels to wobble

Toe-difference angle

> The toe-difference angle δ is the angle by which the wheel on the inside of the bend is turned beyond the angle turned by the wheel on the outside of the bend **(Fig. 3, Page 416)**.

The toe-difference angle is determined at a steering angle of 20° on the wheel on the inside of a bend.
It is needed when checking the steering trapezoid for faults, e.g. if the track arms or tie rods are deformed.

To optimise a vehicle's handling characteristics in respect of the self-steering effect, straight-running stability, directional stability and of the tendency of the wheels to wobble, the various wheel settings, such as **camber, kingpin inclination, kingpin offset, castor and toe-in,** are coordinated. The aim of this is the least possible tyre wear.

Camber

> Camber is the angle of the wheel plane in relation to a vertical line at the wheel contact point at right angles to the vehicle longitudinal axis **(Fig. 1)**.

Camber angle γ is given in degrees and minutes. A distinction can be made between:

- Positive camber
- Negative camber

Positive camber. The wheel plane tilts outward at the top. Positive camber produces a cone effect. The wheel thereby tends to turn (pivot) outward. The greater the positive camber, the lower the lateral force when cornering.

Negative camber. The wheel plane is tilted inward at the top. The cone effect causes the wheel to tend to turn inwards.
Negative camber improves the lateral guidance when cornering, however it produces increased tyre wear on the inside of the tread.
Most vehicles have a camber of $-60'$ to $+30'$ at the steered front wheels when the wheels are in the straight-ahead position. Deviations of $\pm 30'$ are permitted.
Generally, a negative camber of $-30'$ to $-2°$ is used at the rear wheels.

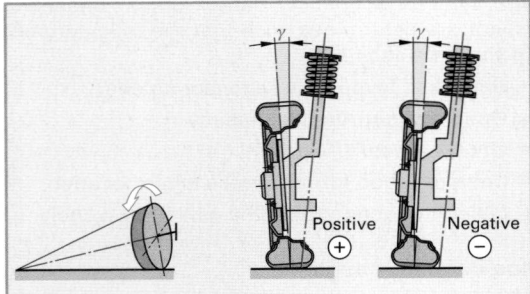

Fig. 1: Positive and negative camber

Kingpin inclination

> The kingpin inclination is the angle of the steering axis or kingpin at right angles to the vehicle longitudinal axis in relation to the vertical from the road surface **(Fig. 2)**.

The steering axis runs through the upper and lower wheel suspension points, for example.

The kingpin inclination δ is given in degrees and minutes. Kingpin inclinations of 5° to 10° are usual.
Kingpin inclination and camber together form an angle, the size of which remains the same during compression and rebound. If the kingpin inclination becomes smaller δ, the camber angle becomes larger and vice-versa.

With positive kingpin offset, the kingpin inclination causes the vehicle to be raised at the front when the wheels are turned.

The weight of the vehicle creates a torque, which causes the wheels to return automatically to the straight-ahead driving position.

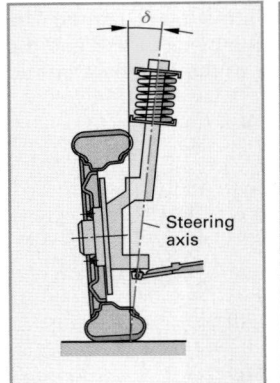 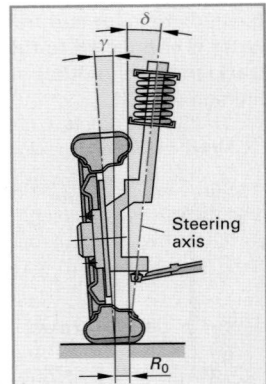

Fig. 2: Kingpin inclination   Fig. 3: Positive kingpin offset

Kingpin offset

> The kingpin offset R_0 is the lever arm on which the frictional forces which occur between the tyres and the road act **(Fig. 3)**. It is measured between the centre of the tyre contact patch and the intersection of the extended steering axis with the road surface.

The kingpin inclination and camber together influence the kingpin offset. A distinction can be made between:

- Positive kingpin offset
- Zero kingpin offset
- Negative kingpin offset

Positive kingpin offset (Fig. 3)

> The extended steering axis intersects with the road surface beyond the centre of the tyre contact patch towards the inside of the tyre.

If a braking force acts on the tyre, the wheel pivots outward. If the grip of the wheels is different, the wheel with better grip is pivoted further outward and the vehicle pulls to one side. The aim is a small kingpin offset, to keep the effect of outside forces on the steering to a minimum.

Negative kingpin offset

> The extended steering axis intersects with the road surface beyond the centre of the tyre contact patch towards the outside of the tyre (Fig. 1).

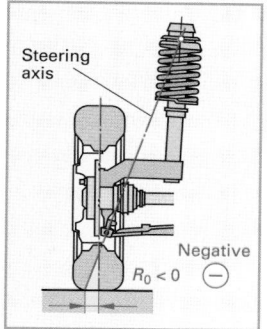 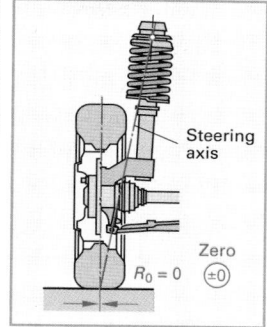

Fig. 1: Negative kingpin offset   **Fig. 2: Zero kingpin offset**

Negative kingpin offset is enabled by using dished wheels and floating-calliper disc brakes, for example.

The braking forces acting on a wheel produce a torque which pivots the wheel inwards at the front, since the pivot is located on the outer side of the wheel. If differing adhesion characteristics occur, e.g. during braking (one wheel on a dry road surface, the other on an icy one, or in the event of a puncture), the wheel with the greater grip is pivoted inward more. This creates an automatic countersteer, which counteracts the tendency of a vehicle to pull towards the side of the more heavily braked wheel (Fig. 3).

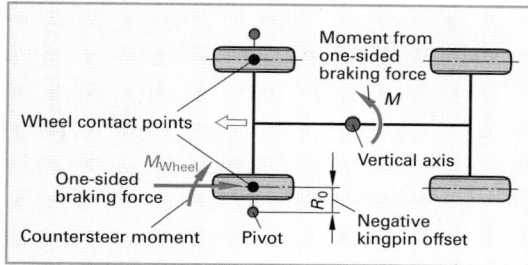

Fig. 3: Effect of negative kingpin offset

Zero kingpin offset

> The extended steering axis intersects with the road surface exactly in the centre of the tyre contact patch (Fig. 2).

Characteristics:

- Low action on the steering by interference factors while driving.
- The wheel moves when the steering lock is applied while the vehicle is stationary.

Castor

> Castor is the result of the steering axis or kingpin being angled along the vehicle longitudinal axis so that it is not perpendicular to the road surface (Fig. 4).

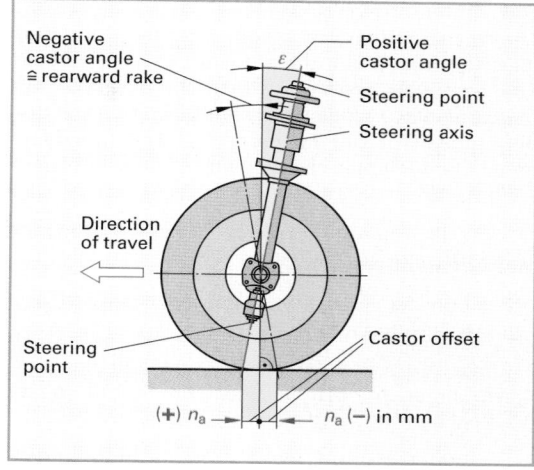

Fig. 4: Castor

Castor is usually expressed as an angle ε in degrees and minutes. Castor may also be given as a distance n_a in mm.

> **Positive castor.** The wheel contact point is behind the steering axis intersection with the road surface.

The wheels are pulled by positive castor. This is used with rear-wheel drive and helps to stabilise the steered wheels.

If the castor angle is positive, the wheel on the inside of a bend is lowered and the wheel on the outside of a bend is raised when the wheels are turned. This gives a steering aligning torque after cornering. A negative camber is also produced on the wheel on the outside of a bend.

> **Negative castor.** The wheel contact point is in front of the steering axis intersection with the road surface.

On vehicles with front-wheel drive, zero castor or small negative castor can be used. This causes a reduction in the return forces and prevents the wheels from being turned back to the straight-ahead position too quickly after cornering.

> Castor, kingpin inclination and kingpin offset jointly influence the return forces on the turned wheels. They have a stabilising effect on the steering.

18.4 Computerised axle alignment

For computer axle alignment, the wheel position dimensions of the motor vehicle are electronically detected and processed by a computer using measurement software.

The measurement is made, e.g. by 8 pickups (**Fig. 1**), which relay the signals to the computer. The computer processes the data received into digital display values which are output on the display screen or on the computer. The individual measured variables can be given to an accuracy of ± 5' to ± 10'.

Measurement process

- Position the vehicle on a horizontal surface, e.g. a measuring platform.
- Check the tyre wear profile, tyre and rim size, tyre pressure, condition of the tie-rod joints, wheel bearings and steering pins.
- Position the front wheels on rotating underplates, the rear wheels on sliding underplates.
- Compress the vehicle springs.
- Secure the angle sensor to the wheels using clamping fixtures.
- Establish communication between the angle sensor and the computer.
- Enter the vehicle data into the computer.
- Run rim run-out compensation, if necessary, by turning the angle sensor.
- Perform the measuring process for the individual wheel setting values and adjust if necessary.
- After adjustment work, perform a reference dimension check.
- Print out the result report.

The geometrical driving axis is automatically used by the system as the reference axis for computer axle alignment.

The geometrical driving axis is formed by the position of the rear wheels (**Fig. 1**).

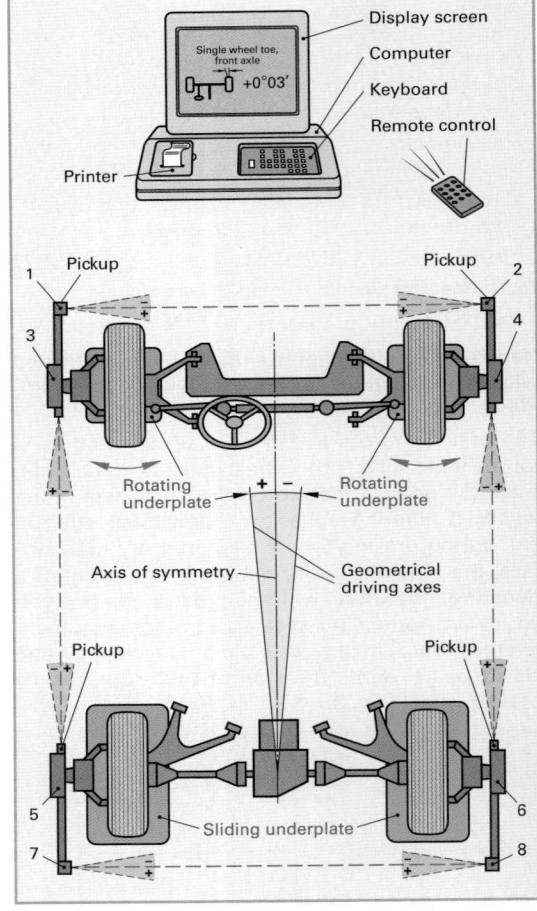

Fig. 1: Computer axle alignment

REVIEW QUESTIONS

1 What functions does the steering have?
2 Describe the trajectory of the front wheels on a vehicle with Ackermann steering when cornering.
3 What makes up the steering trapezoid?
4 What functions does the steering trapezoid have?
5 What functions does the steering linkage have?
6 Explain the terms toe-in and camber.
7 What does the term toe-difference angle mean?
8 Where is the steering axis of a wheel?
9 Which different wheel settings are there?
10 What do positive and negative camber mean?
11 Explain the term kingpin inclination.
12 What effect does kingpin inclination have on the vehicle when the front wheels are turned?
13 Explain the term kingpin offset.
14 What effect does negative kingpin offset have on the front wheels under braking with a one-sided braking force?
15 How is the toe-difference angle measured?
16 Describe an alignment process.

18.5 Steering gear

Functions:

- Conversion of the rotary motion of the steering wheel into displacement of the rack and/or moving the pitman arm.
- Amplification (gearing up) of the torque generated by hand at the steering wheel.

The transmission ratio in the steering gear must be designed such that the maximum force at the steering wheel, e.g. 200 N for vehicle category M3, is not exceeded.

The transmission ratio is up to $i = 19$ on passenger cars, up to around $i = 36$ on commercial vehicles.

Nowadays, rack-and-pinion steering gear (**Fig. 1**) is used on almost all passenger cars, whereas commercial vehicles generally use recirculating-ball steering gear.

Rack-and-pinion steering gear (mechanical)

Structure. A pinion fitted in the steering-gear housing sits on the steering spindle and engages with the rack by way of helical teeth. The rack is guided in bushes and continuously pressed against the pinion by a thrust member and disc springs to eliminate play (**Fig. 1**).

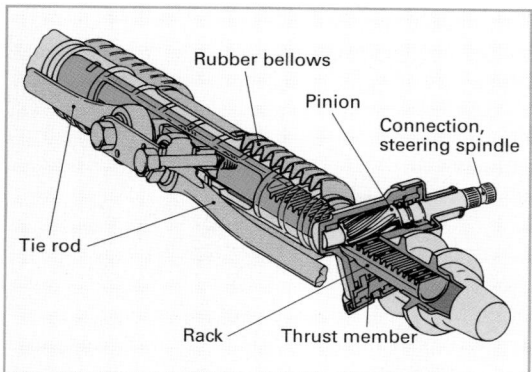

Fig. 1: Mechanical rack-and-pinion steering gear

Operating principle. When the steering wheel is turned, the rack is displaced axially by the rotary motion of the pinion and pivots the wheels via the tie rods, tie-rod arms and steering knuckles.

Rack-and-pinion steering gear features direct transmission ratio, easy return and flat design.

Constant transmission ratio. The tooth pitch is the same over the whole rack.

Variable transmission ratio. On mechanical steering gear without hydraulic assistance, the transmission ratio is designed such that the steering in the range of smaller deflections (central range) has a more direct effect than with larger deflections in the outer range (**Fig. 2**).

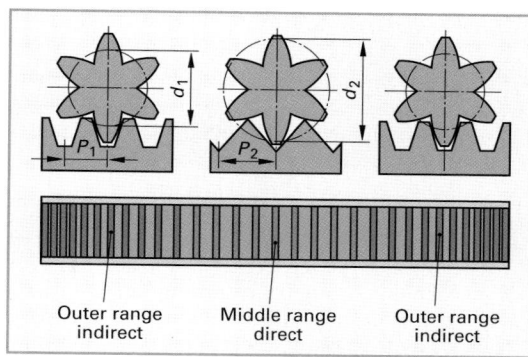

Fig. 2: Variable transmission ratio on mechanical rack-and-pinion steering gear

Advantages of the variable transmission ratio:

- More direct steering for fast straight-ahead driving.
- Low amount of effort required for large steering angles, e.g. when manoeuvring into a parking space.

18.6 Steering systems

A distinction is drawn between steering gear with

- hydraulic assistance, e.g. rack-and-pinion steering and recirculating-ball power steering,
- electro-hydraulic assistance (servo effect), e.g. Servotronic and Active steering, and
- electrical assistance, e.g. Servolectric and Active steering.

18.6.1 Hydraulic rack-and-pinion steering

Structure (Fig. 1, Page 422). This consists of:

- Mechanical rack-and-pinion steering gear
- Hydraulic working cylinders with working plungers
- Rotary slide as control valve
- Oil pump, pressure-limiting valve, oil reservoir

The rack is driven by the pinion, the drive applied to the tie rods is designed on both ends as a side output.

The housing for the rack constitutes the working cylinder, which is divided into two working chambers by a plunger.

Rotary slide valves (**Fig. 1, Page 422**) or rotating plunger valves are used as control valves.

The torsion bar is connected by 2 pins on one end with the control bushing and the drive pinion, at the other end it has a rigid connection to the steering spindle and the rotary slide valve.

The rotary slide valve is composed of the rotating slide and the control bushing. They have control grooves on their lateral surfaces. The grooves on the control bushing open into housing flutes which lead to the two ram chambers, to the vane pump and to the oil reservoir.

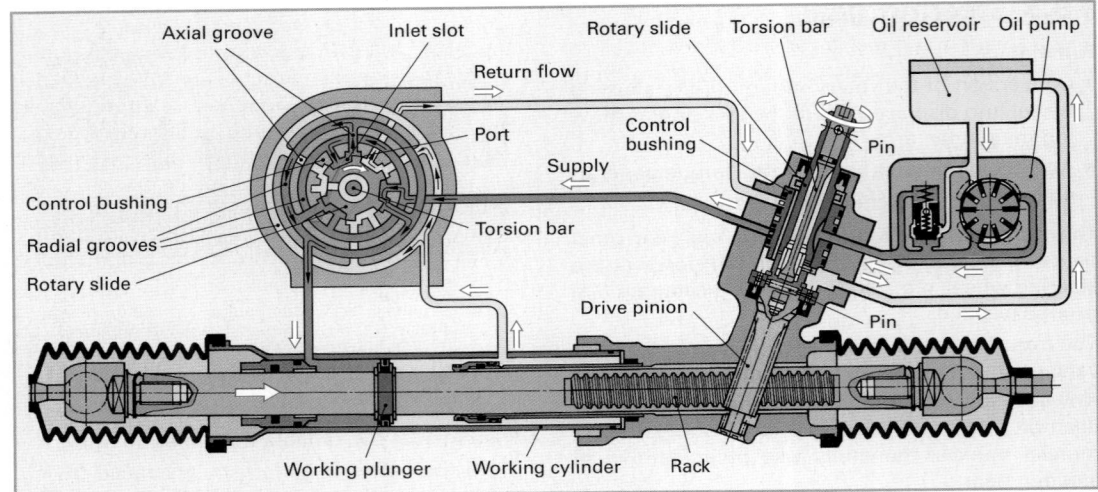

Fig. 1: Hydraulic rack-and-pinion steering with rotary slide valve steering right

Operating principle. When the steering wheel is turned, the steering force applied manually is transferred via the torsion bar to the drive pinion. At the same time, the torsion bar is stressed in proportion to the counterforce and twisted slightly. This causes the rotary slide to turn in relation to the control bushing surrounding it. This changes the positions of the control grooves in relation to one another. The inlet slots for the pressure oil supply are opened. The pressure oil coming from the oil pump flows through the inlet slots into the lower radial groove of the control bushing and is channelled into the relevant ram chamber.

The fluid pressure acts on either the right-hand or the left-hand side of the working plunger and generates the hydraulic assisting force here. It acts in addition to the steering force transferred mechanically from the pinion to the rack.

If the steering wheel is not turned any further, the torsion bar and rotary slide valve return to the neutral position. The ports to the ram chambers are closed, the ports for the return flow are opened.

The oil flows from the oil pump via the control valve back to the supply reservoir.

18.6.2 Electro-hydraulic power steering Servotronic

Servotronic is an electronically controlled rack-and-pinion steering system in which the hydraulic assisting forces are influenced by the driving speed.

At low driving speeds, the full assisting force of the hydraulic rack-and-pinion steering takes effect. The hydraulic assisting force is reduced as driving speed increases.

Structure (Fig. 2). Servotronic consists of:
- Electronic speedometer
- Hydraulic rack-and-pinion steering
- Electro-hydraulic converter
- ECU
- Oil reservoir
- Oil pump

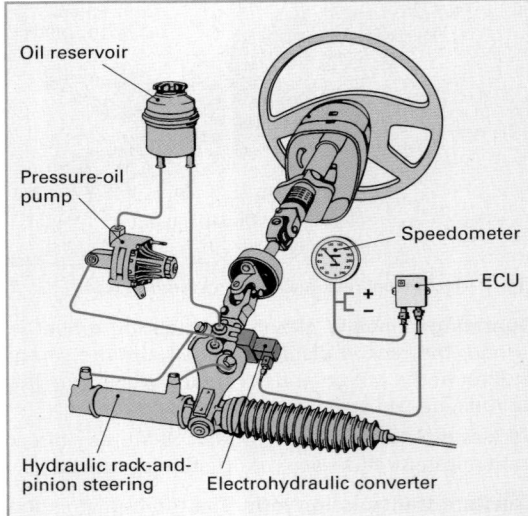

Fig. 2: Servotronic with hydraulic rack-and-pinion steering

Operating principle. At speeds below 20 km/h the solenoid valve controlled by the ECU remains closed.

As the speed increases, the solenoid valve is gradually opened.

Steering right at low speed.
If the steering spindle is turned clockwise, right valve plunger (6) is pushed down by the torsion bar and the lever fitted to it. The pressure oil flows into ram chamber (12), acts on the working plunger, thereby assisting the steering force.
At the same time, the oil flows through open non-return valve (8) into chambers (4) and (5).

Steering right at high speed. The solenoid valve is fully open. The pressure oil flows from ram chamber (12) via open non-return valve (8), throttle (10) and the open solenoid valve to the return flow.
As a result of the oil flowing in through non-return valve (8) and the throttle effect of throttle (10), the pressure in chamber (4) is greater than in chamber (5). This pushes the lever of plunger (6) upwards and produces a reaction torque on the torsion bar and steering spindle.
The steering power assistance thus decreases, the driver must apply more steering force to the steering wheel, the steering is more direct.

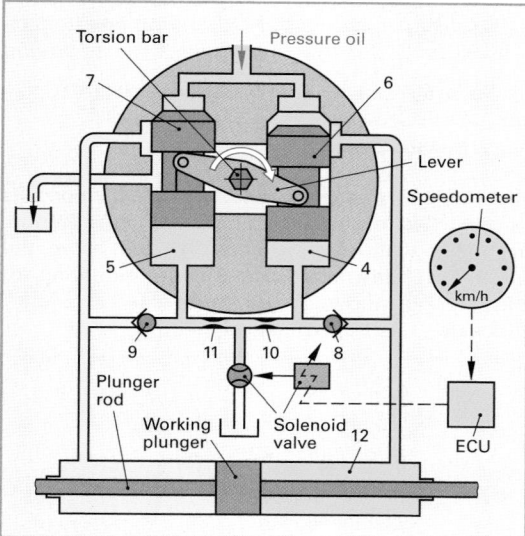

Fig. 1: Servotronic hydraulic system steering right and
v < 20 km/h

18.6.3 Electric power steering Servolectric

With Servolectric (**Fig. 2**), the assisting force is generated by an electronically controlled electric motor. The electric motor is only switched on when required.

Operating principle. The steering torque applied by the driver is measured via a torsion rod with a torque sensor and in addition the speed is mea-

sured via a speed sensor. The two signals are fed to the ECU. The ECU calculates the torque required and its force-transfer direction using stored program maps and sends the relevant output signals to the electric motor. The latter generates an assisting torque which is transformed by a worm-gear pair and transmitted via the steering spindle to the rack-and-pinion steering gear.

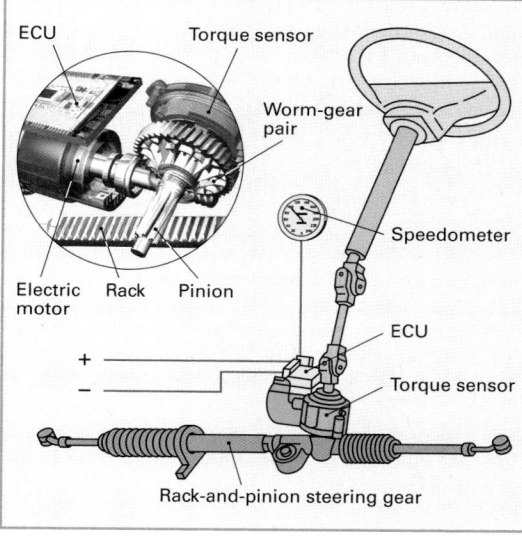

Fig. 2: Electric power steering Servolectric

18.6.4 Active steering

Active steering allows a steering movement to be made without any driver input.

This system primarily consists of:
- Hydraulic rack-and-pinion steering
- Electric motor
- ECU
- Planetary gearbox
- Sensors

REVIEW QUESTIONS

1 Name the functions of a steering gear.

2 What is a variable transmission ratio on a rack-and-pinion steering gear?

3 What are the different types of steering systems?

4 How is a hydraulic rack-and-pinion steering unit constructed?

5 Describe the operating principle of hydraulic rack-and-pinion steering.

6 How is the electric-hydraulic power steering Servotronic constructed?

7 How is the electric-hydraulic power steering Servotronic distinguished from Servolectric electric power steering?

18.7 Wheel suspension

> Wheel suspensions have the task of forming a connection between the vehicle body and the wheels. They must absorb high static forces (load) and dynamic forces (motive, braking and lateral forces).

The wheel geometry should change little or in the desired manner when the springs on the axles are compressed, to achieve a high degree of driving safety and comfort with low tyre wear. A distinction is drawn between

- rigid axles
- semi-rigid axles
- independent suspension

Rigid axles

The two wheels are connected to each other by a rigid axle and sprung against the body.

> On rigid axles, there is no change in the toe-in or camber during the compression and rebound of a wheel, which reduces tyre friction.

When the vehicle is driven over an obstacle on one side, the whole axle is tilted and the camber of the wheels is changed.

Rigid axle with integrated drive. The axle is usually designed as housing for the final-drive unit with differential and the axle shafts. Since the housing is generally made from cast steel, this results in relatively large unsprung masses, which reduce driving smoothness and driving safety. On commercial vehicles, it is easiest to secure to the frame or to the body using the leaf springs. In addition to the suspension, these can also take over wheel guidance in a longitudinal or lateral direction. When helical springs or air springs are used

- trailing arms transfer the wheel forces in a longitudinal direction.
- transverse struts (panhard rod) transfer the lateral forces **(Fig. 1)**.

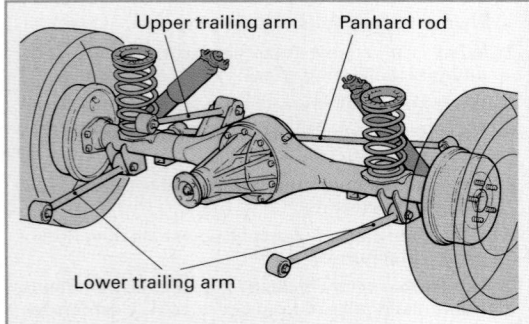

Fig. 1: Rigid axle with integrated drive

The use of several trailing arms can reduce diving under braking and rear squatting under acceleration.

Rigid axle with separate drive (De Dion axle). To reduce the large unsprung masses of the driven axle, the final-drive unit is separated from the axle and is fitted to the bodywork. Power is transmitted via propeller shafts, each having two homokinetic joints with additional length compensation. The rigid rear-axle tube twisted into a U-shape can be laterally guided by:

- Two transverse struts **(Fig. 2)**
- A Watt linkage
- A Panhard rod

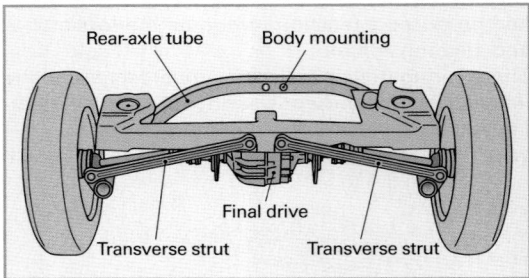

Fig. 2: De Dion axle

Rigid axle as a steering axle. This usually consists of quenched forging with a I-shaped cross section. To ensure that the engine has sufficient room, the axle is bent downward **(Fig. 3)**. As a mounting for the steering knuckle, a stub – **stub axle** – or a fork – **fork axle** – is forged on **(Fig. 4)**.

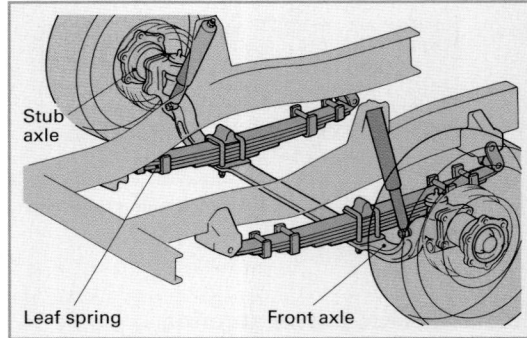

Fig. 3: Rigid axle as a steering axle

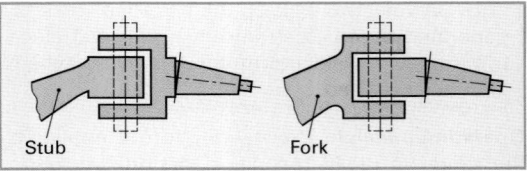

Fig. 4: Stub axle, fork axle

Semi-rigid axles

> On semi-rigid axles, the wheels are fixed rigidly to each other by axle supports. The wheels can move independently of each other to a certain extent due to the elasticity of the axle supports.

On vehicles with front-wheel drive, the use of semi-rigid axles is preferred, the rear axle can be of simple design so that the unsprung masses remain light.

> A semi-rigid axle acts like a rigid axle if both wheels are compressed at the same time and like independent wheel suspension if compressed at different times.

Torsion-beam axle. The rear wheels are suspended from the trailing arms, which are welded to a cross-member made of spring steel **(Fig. 1)**. The cross-member itself is screwed on to the body with rubber-metal bearings. If both wheels compress to the same extent, e.g. under load, the whole axle housing is pivoted evenly in the rubber-metal bearings. If only one wheel spring is compressed, the cross-member becomes twisted in itself and acts like an anti-roll bar. Only small toe and camber changes occur.

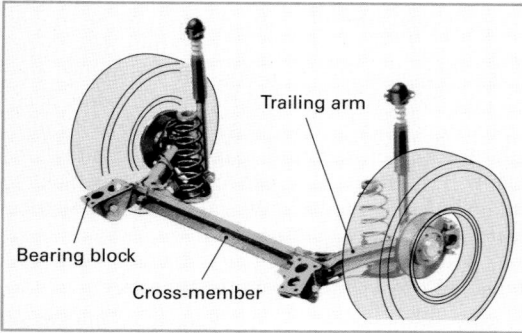

Trailing arm

Bearing block

Cross-member

Fig. 1: Torsion-beam axle

Independent suspension

> Independent suspension allows the mass of the unsprung parts to be kept small. The compression and rebound of a wheel has no influence on the other wheels.

The following are used for the front-wheel suspension:
- Double-wishbone axles
- Multiple suspension arms
- McPherson suspension strut with control arm

The rear wheels are predominantly suspended on:
- Trailing arms
- Semi-trailing arms
- Multiple suspension arms

Wheel suspension on double-wishbone axles. Two control arms, one on top of the other are each connected via a ball joint to the steering knuckle. Camber and toe changes can be controlled during operation by the length of the individual suspension arms.

Control arms are usually of the wishbone type, to increase rigidity in the direction of travel. They are secured to the chassis by two bearings.

Wheel suspension on unequal-length control arms (trapezium shape, Fig. 2). The upper control arm is always shorter than the lower one. This results in a negative camber and little toe change during compression and rebound, which improves stability when cornering.

Wheel suspension on equal-length control arms (parallelogram shape). The camber does not change during compression and rebound, however er there is a toe change.

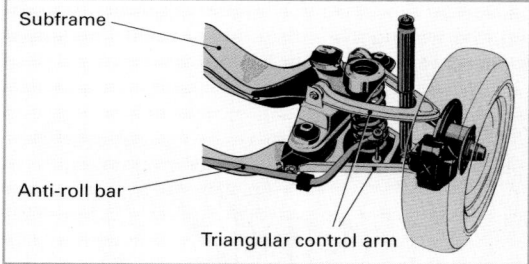

Subframe

Anti-roll bar

Triangular control arm

Fig. 2: Wheel suspension on double-wishbone axles

Wheel suspension with suspension strut and control arm (McPherson axle). The McPherson axle **(Fig. 3)** developed from the double wishbone axle. The upper control arm was replaced by a vibration-damper pipe, to which a steering knuckle is attached. The plunger rod of the damper is secured to the vehicle body in an elastic rubber bearing. There is a helical spring between this attachment point and the spring seat on the damper pipe. Due to the large braking, acceleration and lateral forces, the plunger rod and plunger rod guide are

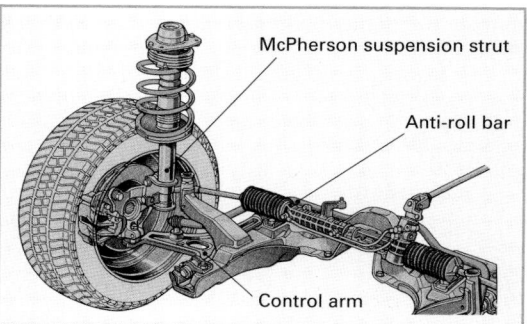

McPherson suspension strut

Anti-roll bar

Control arm

Fig. 3: McPherson axle

of a particularly sturdy design. The rubber bearing must absorb large axial forces and allow large angles of twist at the steering axles. The wheel housing is strengthened at the upper attachment point.

Wheel suspension on trailing arms. This is particularly suitable for vehicles with front-wheel drive, since the boot floor between the rear wheels can be lower. If the suspension rotational axis is lying horizontal, the track width, toe-in and camber do not change during compression and rebound.

Subframe (Fig. 1). To keep noises and vibrations away from the body more adequately, suspension arms are not attached directly to the body, but are attached to a subframe. This consists of 2 retainer arms which are connected to a horizontal tube. It is bolted to the body at 4 rubber bearings, with the front rubber bearings designed as hydro mounts. The two trailing arms are attached to the subframe by taper roller bearings. To minimise toe changes caused by the lateral forces created during cornering, the trailing arm has a tension bolt. These two together form a four-bar linkage.

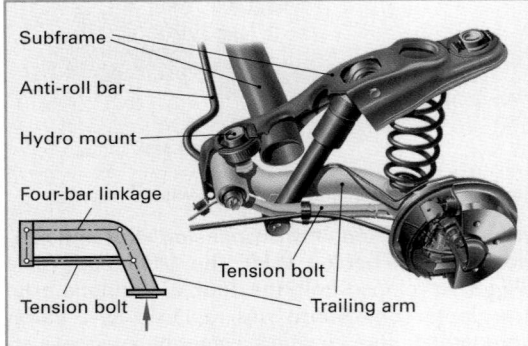

Fig. 1: Wheel suspension on trailing arm

Wheel suspension on semi-trailing arms. Semi-trailing link axles (**Fig. 2** and **Fig. 3**) consist of two wishbones, on which the rotational axis of the two mounting bearings runs diagonally to the transversal axis of the vehicle (α = 10° to 20°) and horizontally or slightly tilted towards the centre of the vehicle (β).

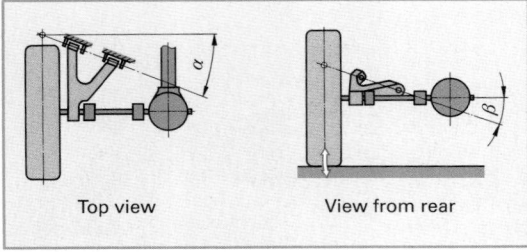

Fig. 2: Tilt angle on semi-trailing arms

The toe and camber changes during compression and rebound are dependent on the inclined position and slope of the semi-trailing arm. If angles α and β are increased, the wheels adopt greater negative camber during compression, which increases the lateral force when cornering.

With this type of wheel suspension, the drive shafts change length during compression and rebound, which necessitates 2 slip joints on each side each with length compensation.

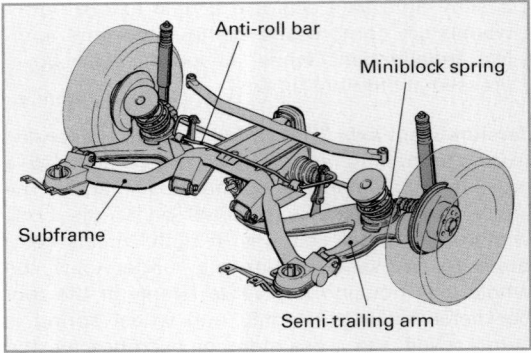

Fig. 3: Rear-wheel suspension on semi-trailing arms

Multiple suspension arm axles. All existing wheel suspensions permit undesirable steering movements while the vehicle is in motion due to the elastic suspension mounting on the body, subframe or wheel carrier. Steering movements occur when forces act on the wheel and move it out of the direction of travel by a steering angle towards toe-in or toe-out. This can cause the vehicle to change course significantly, e.g. if there is a crosswind.

Type and effect of forces on the wheels:

- **Motive forces** act in the centre of the wheel along the longitudinal axis of the vehicle and turn the wheel in the toe-in direction.
- **Braking forces** act in the centre of the tyre contact area along the longitudinal axis of the vehicle and turn the wheel in the toe-out direction.
- **Lateral forces** act just behind the centre of the tyre contact area at right angles to the vehicle longitudinal axis. When cornering, the wheel on the outside of a bend is steered into the toe-out direction, which reduces cornering safety. When cornering sharply, the tread of the tyre is deformed by the rolling movement of the body and by the lateral force which reduces the tyre's reserves of adhesion.
- **Vertical forces** act in the direction of the vehicle vertical axis. These occur if the road surface is uneven or if the vehicle is loaded and cause small toe and camber changes.

Elastic steering faults. Fig. 1 shows the steering angle created by the motive force. While the rear rod control arm is tensioned and elongates slightly due to the elastic suspension, the front rod control arm is placed under pressure, leading to a slight contraction. The wheel is turned out of the direction of travel.

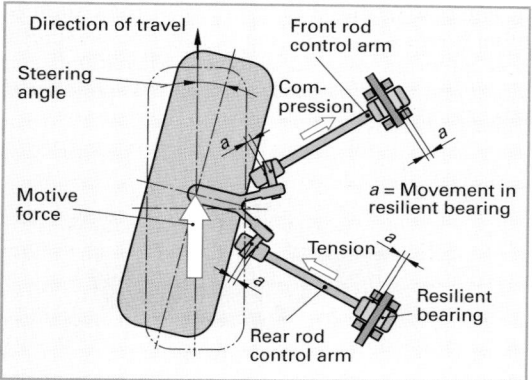

Fig.1: Generation of a steering angle

Multilink rear suspension. This compensates for elastic steering faults. It was developed from the twin-control-arm axle with anti-roll bar. The suspension arms which were originally rigidly coupled were broken down into 5 individual beam suspension arms, which lie in exactly fixed position in relation to each other in space and guide the wheel **(Fig. 2)**.

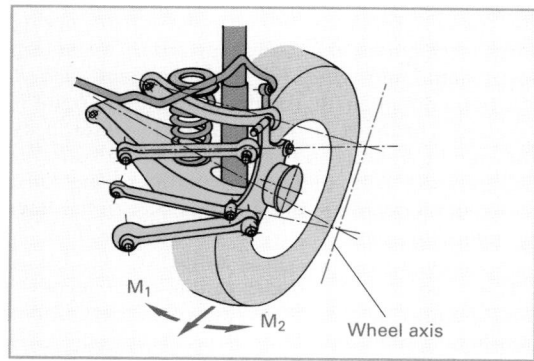

Fig. 2: Multilink rear suspension

The intersection point of the suspension-arm centre line lies outside the wheel midplane, so that the wheel, for example by the action of motive forces, steers exactly as far outwards (M_2) as is steered inwards by the elastic fault (M_1).

Kinematics of the multilink rear suspension. The critical factors for drivability are primarily the toe-in and camber changes, since the self-steering effect of the vehicle is determined by these. If there are changes in the toe angle, a lateral force is created which disrupts the straight-running stability. In **Fig. 3** it is possible to see that the toe-angle change during compression or rebound is almost zero. Camber changes in the middle zone of the corner (straight-ahead travel) should be as small as possible, in order not to create large lateral forces. A negative camber arises from compression during cornering, which improves lateral guidance.

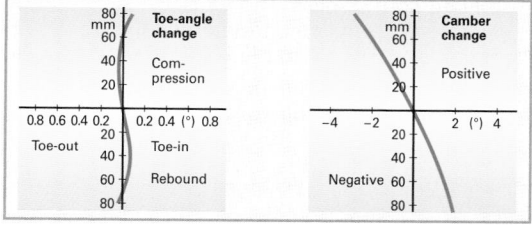

Fig. 3: Changes in toe angle and camber

Roll centre (instantaneous centre). This is the point about which the body, connected to the chassis by springs, tilts under the action of a lateral force. Instantaneous means that this point is only located in this position for a moment.

The higher the roll centre, the less distance to the centre of gravity of the vehicle, i.e. the lever arm on which the centrifugal force acts becomes smaller, the lateral tilt is reduced. However, the greater toe-width changes are a disadvantage as they cause uneven straight-running stability. The connecting line through the roll centres of the front and rear axles gives the roll axis. Its distance to the centre of gravity determines the lateral tilt of the body.

REVIEW QUESTIONS

1 What are the advantages and disadvantages of rigid axles?

2 What is a semi-rigid axle?

3 Name the most important types of independent suspension.

4 What advantage does the wheel suspension on a double-wishbone axle have?

5 What is an elastic steering fault?

6 What is a subframe?

7 How is a McPherson axle constructed?

8 Which forces act on the wheel when the vehicle is in motion and how does it react to them?

9 What are semi-trailing arms?

10 How is a multilink rear suspension constructed and what advantages does it have?

11 What is the roll centre?

12 How does a high roll centre affect the tilt of the body?

18.8 Suspension

18.8.1 Function of the suspension

Due to the unevenness of a road surface, the wheels of a vehicle must perform movements up and down in addition to their rotational motion. When the vehicle is being driven fast, these movements occur within a very short space of time, generating accelerations and decelerations perpendicular to the road surface which are a multiple of acceleration due to gravity. This causes significant, impulsive forces to act on the vehicle, which are greater if the inertia is greater.

> The suspension works together with the damping to absorb jolts from the road and to convert them into vibrations.

Suspension and damping are decisive for

- **driving smoothness.** The vibration of the body moderates the uncomfortable impacts which could cause injury to the occupants and fragile loads are protected.

- **driving safety.** If the road surface is very uneven, contact with the surface may be lost; if wheels are up in the air, they cannot transmit any forces, e.g. motive forces, braking forces.

- **cornering ability.** When the vehicle is cornering at high speeds, the low wheel grip on the wheels on the inside of a bend causes a reduction in the lateral force. To prevent the vehicle sliding out of a bend, the suspension must have shock absorbers and an anti-roll bar to ensure constant grip of the wheels.

The springs are fitted between the wheel suspension and the body. The action of the springs is supported by the tyres. An additional suspension, which is only of benefit to the occupants, however, is the seat suspension **(Fig. 1)**.

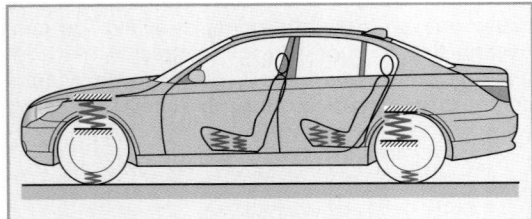

Fig. 1: Suspension in a passenger car

Lateral suspension. In addition to the vertical jolts from rough road surface, slight lateral jolts also occur. The suspension must therefore also be effective in this direction. In addition, lateral suspension can be provided in part by the tyres and by the rubber bearings, which serve to secure and guide the wheel suspension components.

18.8.2 Operating principle of the suspension

Due to the suspension, the motor vehicle becomes a vibratory structure with its own vibration frequency defined by the vehicle weight and by the spring (body-vibration frequency).

In addition to the jolts from rough road surface, other forces also act on the vehicle (motive forces, braking forces, centrifugal forces). Movements and vibrations can thus occur along the 3 spatial axes **(Fig. 2)**.

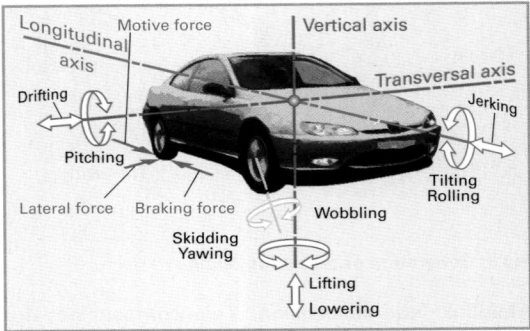

Fig. 2: Types of vibration acting on the motor vehicle

Vibrations

If the wheel of a motor vehicle travels over an obstacle, both the body and the wheel start to vibrate. The upwards movement of the wheel causes the helical spring to be compressed, the spring force accelerates the body upwards. The spring force generated when the spring expands slows the body down again, the upper reversing point is reached. The body is accelerated downwards by the weight, beyond the rest position. The spring is compressed (tightened), the resulting spring force slows the movement of the body down to the lower reversing point.

> The travel from the upper to the lower reversing point of a vibration is known as the amplitude of oscillation.

This motion sequence is repeated until the kinetic energy is converted into heat by spring and air friction **(Fig. 3)**.

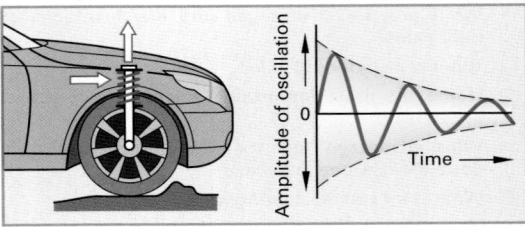

Fig. 3: Damped vibration

Resonance. The vibration is pitched if the body is jolted at the frequency of natural oscillation, e.g. when driving over rough roads, where the obstacles are equal distances apart one after the other **(Fig. 1)**.

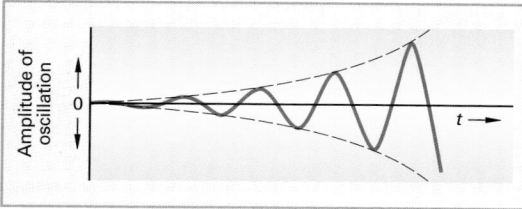

Fig. 1: Pitched vibration

Frequency. This is the number of vibrations per second. Since a body does not vibrate very quickly, the number of vibrations is also given per minute (vibration frequency, body-vibration frequency).

> A large mass and soft springs result in a low frequency (vibration frequency) and a large spring travel.

Spring rate. This indicates the properties of the spring (hard, soft). To check or compare springs, a load is applied to them and the resulting compression is measured. The ratio of force F to travel l is referred to as the spring rate c in N/m.

Spring characteristics. If the spring rate is the same over the whole range of spring (constant), as for a normal helical spring, for example, the spring has a linear characteristic **(Fig. 2)**.

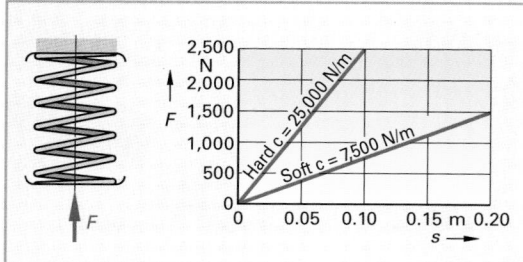

Fig. 2: Linear spring characteristics

If the spring rate increases as the range of spring increases, e.g. with multi-leaf springs or conical helical springs, the characteristic is plotted as a curve. The spring has a progressive characteristic **(Fig. 3)**.

Sprung masses, unsprung masses

On motor vehicles, a distinction is drawn between sprung masses (body with load) and unsprung masses (wheels with drum or disc brakes, parts of

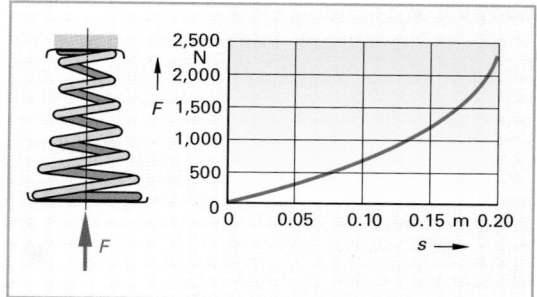

Fig. 3: Progressive spring characteristic

the wheel suspension). These different masses are connected (coupled) to one another by the springs. This causes feedback to one another, so that the two masses vibrate in different frequency ranges independently of each other **(Fig. 4)**. If a vibration damper (shock absorber) is fitted between the two masses, the amplitude of oscillation becomes smaller, the vibration dies out more quickly.

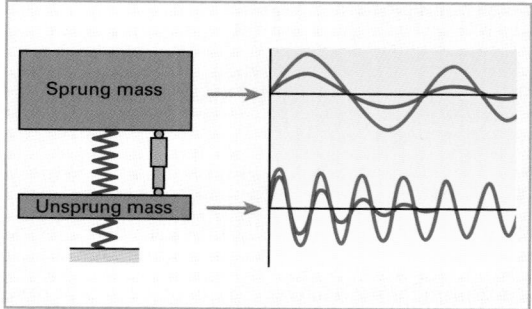

Fig. 4: Motion sequence when driving over an uneven road surface

If a vehicle is driven over a hump at high speed, the body initially remains balanced due to the large mass. The wheel, with its small mass in relation to the body, is accelerated upwards very rapidly, and in doing so it compresses the spring. Only the force corresponding to this spring travel is acting on the body.

On the other side of the hump, the wheel is accelerated downwards by the preloaded spring. Only the load relief of the spring corresponding to the bump acts on the body.

If the force from the wheel is greater than the initial tension of the spring, the wheel loses adhesion on the road surface for a short time, as the initial spring tension is insufficient to move the wheel downwards quickly enough.

> To achieve good driving safety and the best possible comfort, the unsprung mass should be as small as possible.

18

Body vibration frequencies

These can be determined by the vibration at the front or rear of the vehicle. A complete vibration consists of the spring compression and rebound process. The number of vibrations per minute then gives the body-vibration frequency. Vibration dampers do not control the vibration frequency, the amplitude of oscillation is downrated by the greater resistance. In contrast, the mass plays a large part. The heavier the vehicle or the larger the payload, the lower the vibration frequencies become.

Soft suspension: 60 vibrations per minute or lower can cause nausea. This can be rectified by stronger damping.

Hard suspension: 90 vibrations per minute or lower jar the spine. Hard springs are often required for high payloads on the rear axle, however, whereby a more moderate driving smoothness when unladen is achieved. This applies particularly to small vehicles, which must be equipped with sustainable, i.e. hard, springs due to the unfavourable ratio of net weight to maximum load.

18.8.3 Types of springs

18.8.3.1 Steel springs

Most motor vehicles are fitted with steel springs. These may be:

- Leaf springs
- Torsion-bar springs
- Helical springs
- Anti-roll bars

The spring effect is caused by the elastic deformation of spring steel (e.g. chrome-vanadium spring steel) up to the limit of elasticity. The spring characteristic is linear, but the design of the spring can cause it to be progressive.

Leaf springs

These have a minor role in passenger cars. However, in heavy vehicles, they are the most commonly used type of spring (see Chapter Commercial vehicle technology).

Helical springs

These are primarily used as compression springs in passenger cars.

Advantages: Low weight, low space requirements
Disadvantages: Almost no damping, no transmission of wheel forces (longitudinal and transverse forces).

Helical springs usually have a linear spring characteristic. Soft helical springs differ from hard helical springs in that they have a:

- Smaller wire diameter
- Larger spring internal diameter
- More loosely wound coil

Helical springs with a progressive characteristic must be fitted to allow a greater payload and adequate comfort when the vehicle is unladen. This can be achieved with the following:

- Different sizes of the internal diameter, e.g. taper shape, barrel shape, waist shape
- Different wire diameters **(Fig. 1)**

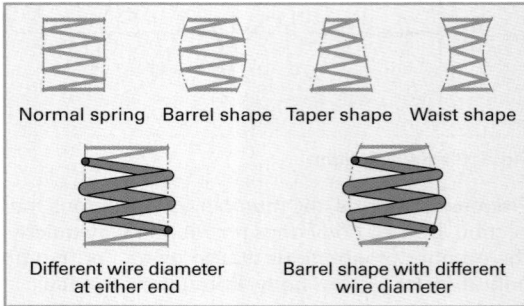

Normal spring   Barrel shape   Taper shape   Waist shape

Different wire diameter at either end   Barrel shape with different wire diameter

Fig. 1: Types of helical springs

The barrel-shaped miniblock spring has the advantage over the cylindrical helical spring that the spring coils cannot touch when the spring is compressed while the vehicle is in motion because each coil lies inside the larger ones forming a spiral **(Fig. 2)**. This means that the spring can be shorter without sacrificing a long spring range for a high load-carrying capacity. The miniblock spring incorporates all the options for a progressive spring.

Fig. 2: Miniblock spring

> Helical springs cannot transfer wheel guidance forces.

They are therefore only used in axle designs in which the motive, braking and lateral forces are transferred by other elements (control arm, trailing arm, McPherson-suspension strut). Vibration dampers are nowadays only rarely used inside the helical spring **(Fig. 2)** because fitting and removal are very time consuming.

Torsion-bar spring

A torsion-bar spring is a rod made of spring steel **(Fig. 1)** which is caused to twist by a lever on which the wheel is mounted.

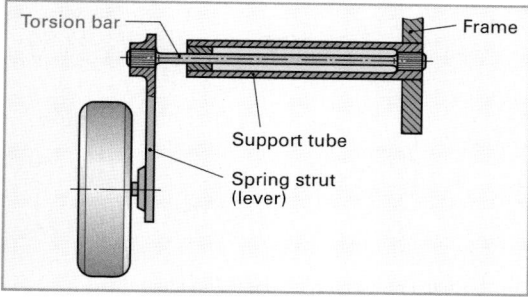

Fig. 1: Torsion-bar suspension

Torsion bars are mostly round rods, square bars and packages of flat bars. They may be arranged longitudinally or transversely. A longitudinal arrangement allows greater length and therefore a greater torsion angle. These springs are softer and have a longer travel.

Torsion bars cannot be subjected to bending. They are therefore often fitted in a tube which provides support against bending and which also provides protection.

The heads are usually interlocked. This toothed interlocking allows the initial tension to be changed and to be adjusted evenly on all wheels.

Anti-roll bar

This is a suspension element which helps to improve the roadholding. U-shape torsion bars are usually used **(Fig. 2)**.

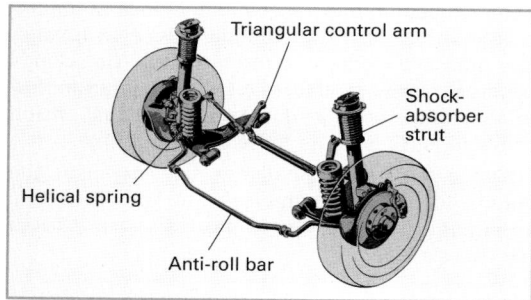

Fig. 2: Anti-roll bar

The centre section of the anti-roll bar is able to rotate in its mounting on the body and the two links are attached to the wheel suspension, e.g. control arms, via rubber elements.

When a wheel is lifted (compression), the twisting action of the anti-roll bar also raises the other wheel and lowers it when the wheel is lowered.

This counteracts the excessive rolling action (leaning to one side) of the body when the vehicle is cornering. The anti-roll bar has no effect if both wheels are compressed at the same time.

18.8.3.2 Rubber spring

Natural and synthetic rubber are very elastic and have high internal damping characteristics. Many different types of rubber springs are manufactured **(Fig. 3)** but are not actually used as vehicle springs. The high internal damping and elasticity of the rubber is used to intercept high-frequency vibrations and as noise insulation. To this end, the actual vehicle springs or mountings, e.g. the control arm, are mounted in rubber cushions. This also improves the transverse suspension.

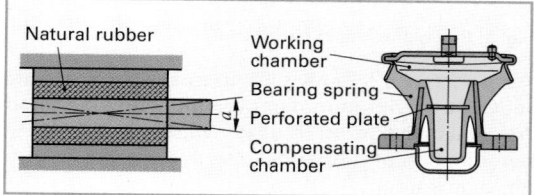

Fig. 3: Rubber spring   **Fig. 4: Hydro mount**

Hydraulically damped elastomer mountings **(hydro mount, Fig. 4)** are used instead of simple rubber springs to prevent vibrations of various frequencies from being transferred from the engine to the body. These consist of an elastic bearing spring made from natural rubber, which forms the mechanical connection between the engine and the body, and a hydraulic section, which consists of a working chamber and a compensating chamber and which is filled with hydraulic fluid. A perforated plate between the two chambers impedes the flow of fluid into the compensating chamber and damps any vibrations that have been transferred here (see also Chapter Mechanical Engine Components and Engine suspension).

18.8.3.3 Gas-filled spring

A gas-filled spring exploits the elastic properties of an enclosed volume of gas (air or nitrogen) for the purposes of suspension.

Air spring

These are the most commonly used, but they require a pressure generating system and are therefore primarily used in buses and commercial vehicles which already have one of these for the brakes (see Chapter Commercial vehicle technology).

The air spring has a progressive characteristic and has the advantage that the travel of the spring can be adjusted to the load by altering the air pressure. The height of the load area or entrance can also be set or maintained using level control.

On passenger cars, the body can be raised and lowered according to the speed of the vehicle. The angle of the body when cornering can be considerably reduced by control interventions.

To prevent pressure loss, the enclosed volume of air is sealed in fixed rubber bellows. This may be roll bellows or a gaiter seal (**Fig. 1**).

Fig. 1: Roll bellows and gaiter seal

Air only has a low level of internal damping. This means that vibration dampers must also be fitted or a suspension strut used which consists of a combination of rubber bellows and a gas-pressure shock absorber.

Air springs cannot transfer wheel forces and they are therefore fitted between suspension arms or axles, e.g. torsion-beam rear axles (**Fig. 2**), and the body.

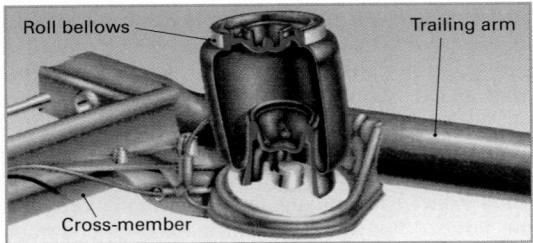

Fig. 2: Torsion-beam rear axle with roll bellows

Hydro-pneumatic spring

In principle, a hydro-pneumatic spring (**Fig. 3**) is a gas-filled spring combined with a working cylinder. It has the effect of both suspension and a shock absorber. A constant volume of gas (usually nitrogen) in a spring ball is compressed to a greater or lesser extent by pumping in or releasing hydraulic fluid. The gas and fluid are separated by a diaphragm. Gas and fluid are pressurised equally. The pressure is generated by a high-pressure pump and is approximately 180 bar.

Depending on the space available, the spring ball may be on the side next to the working cylinder or it may be completely separate from it.

The valves between the working cylinder and the spring ball throttle the flow of fluid in both directions and act like a vibration damper.

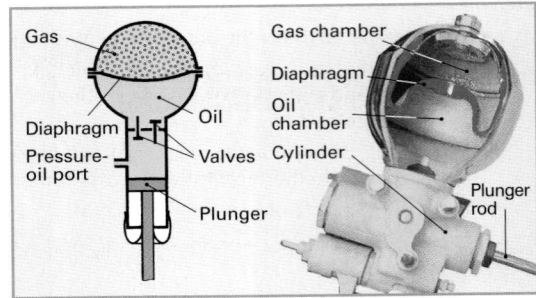

Fig. 3: Hydro-pneumatic suspension elements

All the suspension elements are interconnected by a network of lines. The hydraulic cylinder plunger rod is fixed to the trailing arm or the wheel-suspension control arm.

Level control. The ground clearance of the body can be adjusted, e.g. for travelling over rough ground or for changing a wheel, using a manually operated level-control valve. The level can be controlled automatically for all load conditions by a linkage which is fixed to the trailing arm and which acts on the ride-height controller plunger (**Fig. 4**). If the vehicle is more heavily laden, the rear sinks and the plunger rod in the cylinder moves in. At the same time, the plunger in the ride-height controller is moved by the trailing arm and linkage, thereby allowing the pressure oil to flow in. The plunger rod in the cylinder moves out until the old level is reached and the flow of fluid into the ride-height controller is shut off.

The increase in load causes an increase in the hydraulic pressure in the cylinder and a simultaneous pressure increase in the nitrogen. The springs become harder and, because the vibration frequency of the body also increases, the suspension characteristics become more uncomfortable.

Fitting a third spring ball per axle increases the volume of gas and therefore the volume of the spring; this improves the comfort characteristics of the chassis when driving in a straight line.

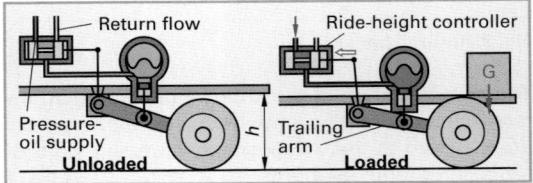

Fig. 4: Hydro-pneumatic suspension

Hydractive chassis

Structure. Additional components connect the hydro-pneumatic suspension system to a chassis which is able to ...

- ... reduce the lateral roll of the body when cornering.
- ... counteract diving of the front of the vehicle under braking and squatting of the rear of the vehicle under acceleration.
- ... change the ride comfort between soft and hard, regardless of whether comfort or sports tuning is selected.

The following additional components are required **(Fig. 1):**

- 2 anti-roll bars, each with 2 spring cylinders, for the front and rear axles
- A centre spring ball with hardness controller for both the front and rear axle
- Hydraulic block
- Height sensors for the front and rear axle
- Steering angle, accelerator-pedal and brake-pedal sensors

Problematic driving situations can occur when the vehicle corners or swerves suddenly. The angle of the body means that the load on the wheels on the inside of the bend is reduced so that smaller forces are transferred to the road surface. This can result in the vehicle breaking away at the rear or rolling. The cornering speed and the distance of the rolling axis from the vehicle's centre of gravity determine the lateral roll of the body.

The angle can be reduced by fitting anti-roll bars. If the wheels are compressed to a differing extent, the anti-roll bars are twisted, they act as additional torsion-bar springs and the suspension becomes generally harder and more uncomfortable.

Structure. On a hydractive chassis, the front spring cylinders are mounted vertically and attached to the anti-roll bar via coupling rods, the rear spring cylinders are mounted horizontally. Depending on the hydraulic pressure supplied in the spring cylinders, additional forces may act on the anti-roll bar and cause the spring action to be too hard.

There is a third spring ball and a height sensor between the spring cylinders on each axle. Spring movements cause the anti-roll bar to twist, which the height sensor reports to the ECU as a change of body attitude.

All the spring cylinders and spring balls are interconnected via the hydraulic block.

Hydraulic block. This consists of the hydraulic pump with electric motor, 4 solenoid valves and the ECU. The spring balls are supplied with hydraulic fluid by the hydraulic pump. The operating pressure of the system is between 80 bar and 140 bar. Two solenoid valves on both the front and rear axle control the supply and return flow of the hydraulic fluid. This means that the front end of the vehicle can be raised or lowered independently of the rear end and vice versa.

Operating principle of the hydractive chassis. By selecting the "Comfort" or "Sport" drive program, the driver can choose between a soft or hard suspension setting.

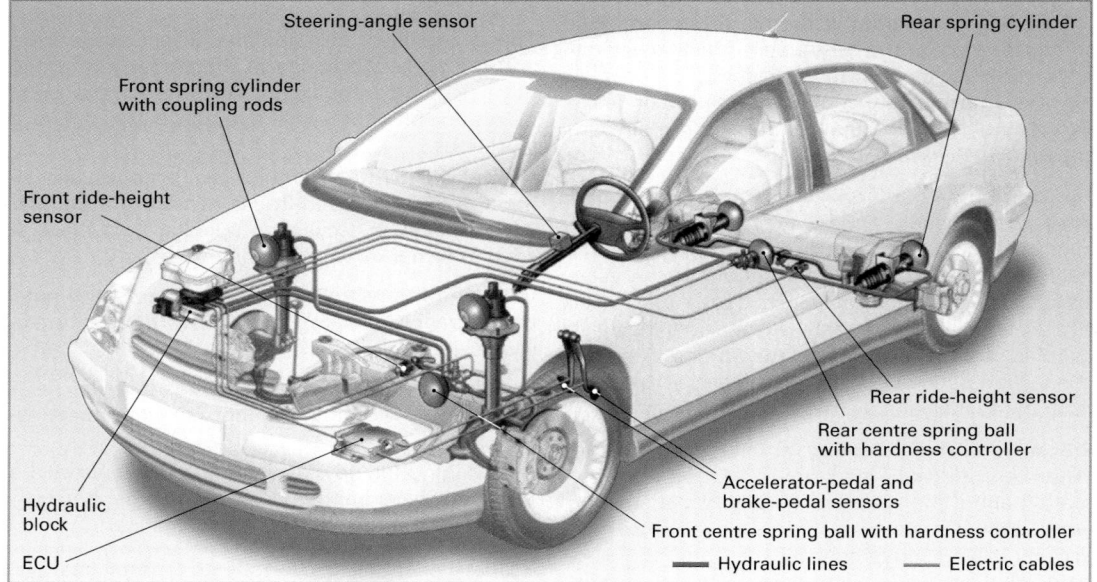

Fig. 1: **Hydractive chassis components and system**

Steering-angle sensor

Rear spring cylinder

Front spring cylinder with coupling rods

Front ride-height sensor

Rear ride-height sensor

Rear centre spring ball with hardness controller

Accelerator-pedal and brake-pedal sensors

Hydraulic block

Front centre spring ball with hardness controller

ECU

━━ Hydraulic lines   ━━ Electric cables

Depending on the driving conditions and driving style (e.g. rapid cornering), the ECU may also make the suspension hard in the "Comfort" drive program.

"Comfort" drive program. The three spring balls on each axle are interconnected. When the suspension is compressed, the spring-cylinder plunger rod moves in and pushes out the hydraulic oil, which can flow into the spring balls, press against the diaphragms and compress the nitrogen cushion. The 3rd spring ball provides an additional gas cushion which allows a softer spring action (**Fig. 1**).

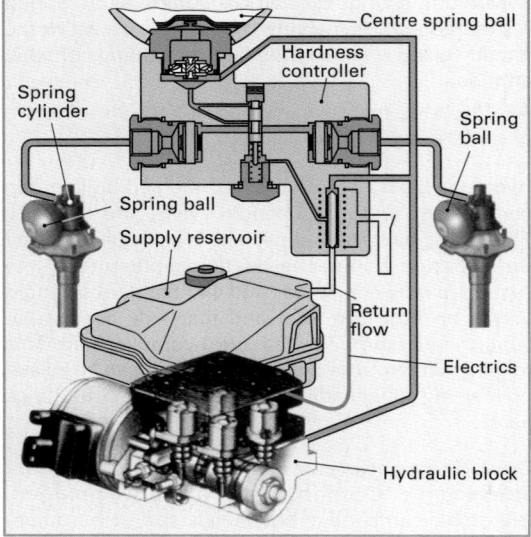

Fig. 1: Hydractive chassis in comfort setting

"Sport" drive program. If the hardness controller solenoid valve on the centre spring ball is activated by the ECU, the flow to these spring cylinders is blocked. This means that only the volume of gas in two spring cylinders is available and the suspension becomes harder.

Function of the hardness controller in the "Sport" drive program (Fig. 2). The solenoid valve is supplied with power, the return flow to the hydraulic fluid reservoir is opened, the bottom of the valve spool is at zero pressure. Because the top is still subject to the pressure of the suspension, the valve spool is pushed downwards, thereby breaking the connection between the suspension elements and from the suspension elements to the spring ball.

Processes during cornering, acceleration, braking. When the vehicle corners, the ECU receives information about the speed and angle of the steering wheel from the steering-angle sensor. To counteract any rolling movement by the body, the solenoid valve is supplied with power and the connection to the spring balls is established.

The suspension becomes harder and the tilt angle of the body is reduced. If this measure were not in place, the body would dive towards the wheels on the outside of the bend and the hydraulic fluid would flow into the suspension elements on the other side. To counteract the squatting of the rear of the vehicle under acceleration, the ECU uses information from the accelerator pedal sensor to disconnect the centre spring ball on the rear axle.

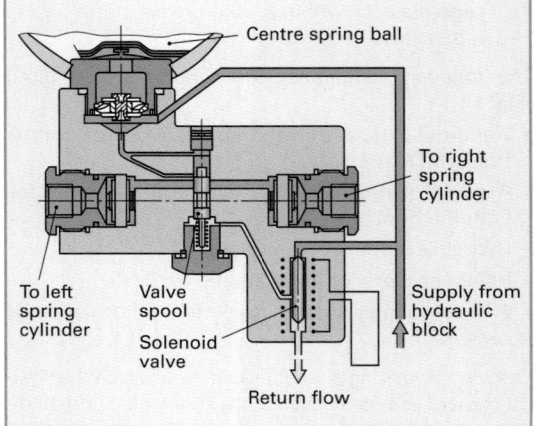

Fig. 2: Hardness controller in sports setting

18.8.4 Vibration dampers

Vibration dampers (shock absorbers) allow vibrations from body and wheels to subside more quickly and therefore increase safety and driving comfort.

These are fitted between the wheel suspension and the body. Vibrations of the wheels and body have different frequencies. A good damper must be set up so that it is effective for the two different vibrations.

Nowadays, hydraulic vibration dampers are almost exclusively used. These consist of a plunger which moves in a cylinder, pushing fluid through small holes or valves (throttle points).

Rebound. The wheel moves downwards and pulls the vibration damper apart telescopically (telescopic shock absorber).

Compression. The wheel moves upwards. As it does so, the vibration damper is pushed back together.

By changing the flow resistance for the fluid as the plunger moves backwards and forwards, it is possible to adjust the vehicle characteristics.

Kinetic energy is converted into thermal energy by vibration dampers.

18.8.4.1 Twin-tube vibration damper

Hydraulic vibration dampers basically consist of a cylinder in which a plunger with plunger rod can move up and down.

In a twin-tube vibration damper **(Fig. 1)**, the plunger rod and protective tube are fixed to the body and the cylinder is fixed to the wheel suspension.

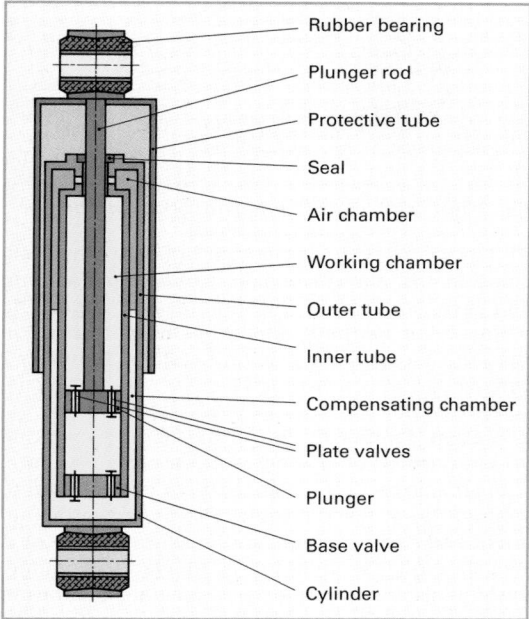

- Rubber bearing
- Plunger rod
- Protective tube
- Seal
- Air chamber
- Working chamber
- Outer tube
- Inner tube
- Compensating chamber
- Plate valves
- Plunger
- Base valve
- Cylinder

Fig. 1: Twin-tube vibration damper

The cylinder consists of an inner and an outer tube. The inner tube contains the working chamber in which the plunger moves. This is completely filled with oil.

Between the inner and the outer tube, there is the compensating chamber. This is only partially filled with oil and is designed to take the oil which is pushed out of the working chamber when the plunger rod moves in.

Valves are fitted in the plunger and the working chamber which throttle the oil flow at differing rates.

During the rebound stage, damping is stronger. As the plunger moves upwards, the oil has to be pressed through fine openings in the plate valves in the plunger. At the same time, oil is sucked back out of the compensating chamber through the base valve.

Installation only with the plunger rod at the top, as otherwise air would be drawn out of the compensating chamber, which would cause the oil to foam and the damping to fail.

18.8.4.2 Single-tube gas-pressure shock absorber

The single-tube gas-pressure shock absorber **(Fig. 2)** behaves exactly the same as the twin-tube vibration damper on the upwards and downwards strokes. However, a special compensating chamber is not required to compensate for the plunger rod volume and so there is no outer tube.

Compensation is achieved with a gas cushion of nitrogen which is usually separated from the oil chamber by a moveable plunger. The gas cushion, which is at a pressure of 20 bar to 30 bar, is squeezed and further compressed by the oil forced out by the plunger rod when the working plunger moves down. The gas cushion and oil are always pressurised, which prevents the oil from foaming and causing a reduction in the damping effect.

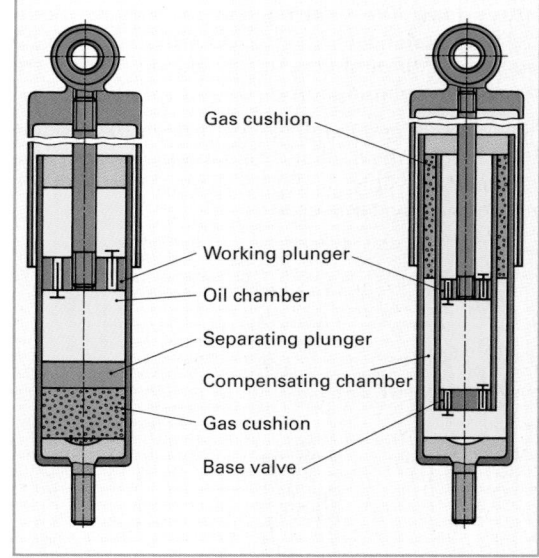

- Gas cushion
- Working plunger
- Oil chamber
- Separating plunger
- Compensating chamber
- Gas cushion
- Base valve

Fig. 2: Single-tube gas-pres-   **Fig. 3: Twin-tube gas-pres-**
sure shock absorber   **sure shock absorber**

Single-tube gas-pressure shock absorbers with a separating plunger can be fitted in any location. For versions with an impact plate, the plunger rod must always be at the bottom.

18.8.4.3 Twin-tube gas-pressure shock absorber

The twin-tube gas-pressure shock absorber **(Fig. 3)** has a similar structure to the twin-tube vibration damper. A ring-shaped compensating chamber contains nitrogen at an initial pressure of 3 bar to 8 bar. This prevents the formation of gas bubbles when the vibration damper moves quickly. The damping forces are improved in nearly all vibration ranges.

Twin-tube gas-pressure shock absorber with variable damping

In the past, it was practically impossible to adjust a shock absorber to the different load conditions of a vehicle. Vehicles with heavy loads (e.g. heavy goods vehicles with trailers) require strong damping, but this results in unpleasant shaking and bouncing when the vehicle is unladen and driven on rough roads.

One or more grooves in the cylinder wall of a twin-tube gas-pressure shock absorber **(Fig. 1)** can provide the desired variable damping characteristics.

Light load. The working plunger moves in the area between the two grooves. The oil can flow through both the plunger valves and the grooves. This additional bypass reduces the damping force, thereby increasing comfort.

Heavy load. The working plunger moves underneath the area with the grooves where there is no additional throughflow cross-section. There are maximum damping forces.

The number and length of the grooves, as well as their height offset, allow the damping forces to be adjusted not only to the load, but also to all the suspension systems used.

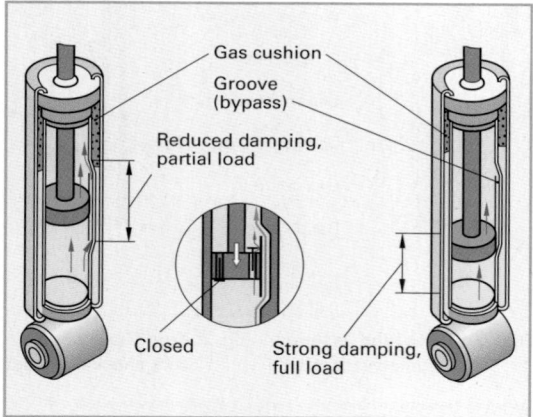

Fig. 1: Twin-tube gas-pressure shock absorber with variable damping

18.8.4.4 Test graphs

Vibration damper removed. In order to obtain the characteristic curves of a vibration damper, the damper must be tensioned in a testing device. The damper is moved by a cranked drive. The damping forces as shown by the plunger travel are measured and plotted on a graph. A constant rebound and compression results in closed curves **(Fig. 2)**. An increase in the radius of the crankshaft on the testing device also increases the rebound and compression of the damper, resulting in further closed curves. The damping force increases because plunger speed in the damper increases if the crankshaft drive is rotating at a constant speed.

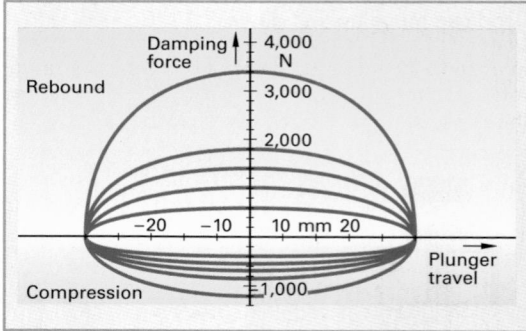

Fig. 2: Test graph of a gas-pressure shock absorber

Fitting valves with various throughflow cross-sections in the plunger results in varying damping forces in the rebound and compression stages. The ratio of the damping forces in the compression stage to those in the rebound stage is between 2 and 5.

Vibration damper fitted. All the dampers on one axle are tested on a shock tester at the same time. The wheels rest on a plate and are each caused to vibrate by an electric motor via an eccentric element and a compression spring. Once the motor is switched off, the vibration is allowed to continue through its entire frequency range until it comes to a standstill and a measuring instrument records this on a disc **(Fig. 3)**.

The greatest amplitude is displayed at the resonance point. This indicates the damping capabilities of the damper concerned. If the measured resonance amplitude is greater than or equal to the limit value given, the damper is faulty. A disc graph can be used to show the damper vibrations on one side of the vehicle.

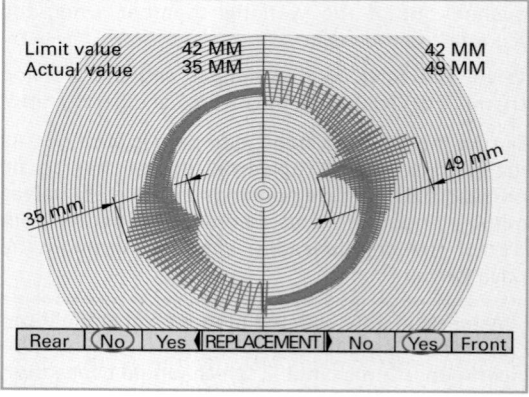

Fig. 3: Vibration patterns of 2 dampers

18.8.4.5 Vibration dampers in the compound suspension system

Suspension strut

> The combination of a vibration damper in a re-inforced construction with a spring, usually a helical spring, is known as a suspension strut.

Suspension struts can also be used as wheel suspension if they have an additional steering knuckle **(Fig. 1)**. Vibration-damper cartridges are used so that the entire suspension strut does not have to be replaced if the vibration damper is faulty. If there is a reduction in damping forces, the cartridge can be changed by opening a threaded connection at the top of the container tube.

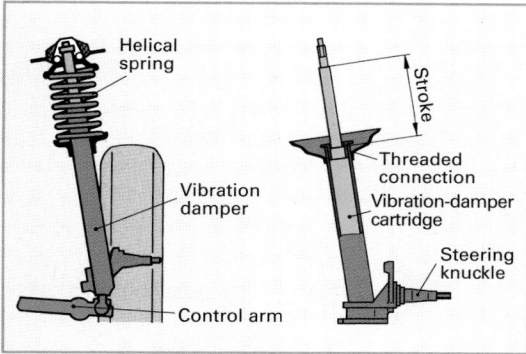

Fig. 1: Suspension strut

Vibration dampers with a level control system

The suspension in a passenger car is usually designed so that the optimum roadholding is achieved when the vehicle has an average load. If the vehicle is fully laden, the rear of the vehicle squats significantly, the ground clearance and spring range are reduced and roadholding is impaired. This often also results in uncontrolled steering characteristics, cross wind sensitivity and glare for oncoming traffic when driving at night. Driving comfort deteriorates because the increased load changes the vibration frequency of steel springs. A constant natural frequency of 1 Hertz (corresponds to a vibration frequency of 60) in all load conditions is only possible with a level-controlled gas spring. This allows the height of the vehicle to be automatically maintained, even when towing a trailer. A distinction is made between pneumatic and hydro-pneumatic systems.

Pneumatic level control system. The system consists of a compressor, an ECU and two air-spring dampers, each with an induction sensor. The air-spring dampers consist of a combination of single-tube gas-pressure shock absorber with air springs **(Fig. 2)**. These bear the entire axle load.

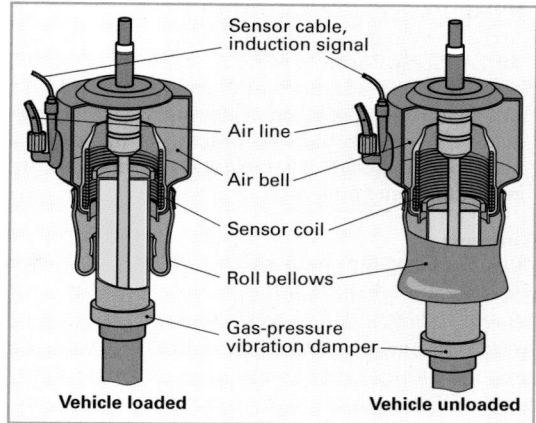

Fig. 2: Air-spring damper

The air spring, which is attached via the gas-pressure shock absorber, consists of an air bell and roll bellows. If the load increases, the shock-absorber tube sinks further into the sensor coil integrated in the air bell and generates an induction voltage which is forwarded to the ECU as a signal. The ECU allows air to flow in via the compressor until the specified vehicle height is reached. The pressure in the suspension air bag is between 5 and 11 bar, depending on the load.

Hydro-pneumatic level control system. The system consists of:

- Suspension struts and spring-type actuators **(Fig. 3)**
- Compressed-oil system with radial-piston pump and oil reservoir
- Control device with level controller and actuation linkage

The spring-type actuators work like a hydro-pneumatic auxiliary spring. If the rear of the vehicle has squatted, the spring element is supplied with pressure oil via the level-control valve until the normal level is reached. The oil is then returned to the container by the pump at nearly zero pressure.

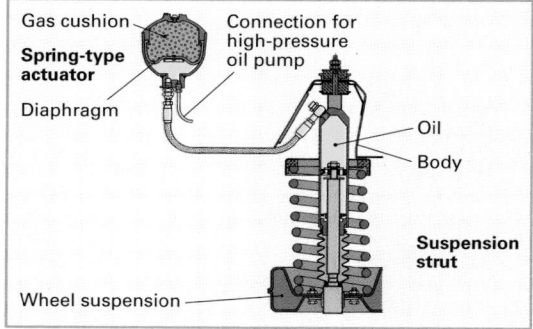

Fig. 3: Suspension strut with spring-type actuator

18.8.5 Active Body Control (ABC)

Active Body Control (ABC) is an electro-hydraulic active chassis system which, in addition to its suspension and damping functions, enables automatic level control while the vehicle is in motion. This maintains the vehicle body at practically the same level at the front and rear axles when the vehicle brakes, accelerates, drives over uneven road surfaces and bends.

Structure

Each wheel is mounted to a suspension strut consisting of a vibration damper and a helical spring.

The plunger is a dynamically adjustable hydraulic cylinder which is able to generate forces which counteract wheel or body movements. To do this, the plunger moves the base of the helical spring and changes the tension. This reduces body movements in the direction of the 3 vehicle axles.

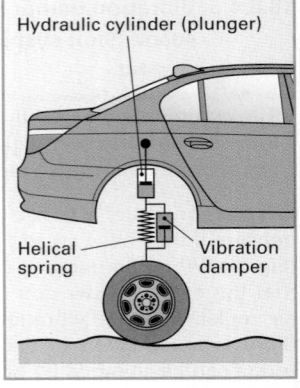

Fig. 1: Suspension strut with plunger

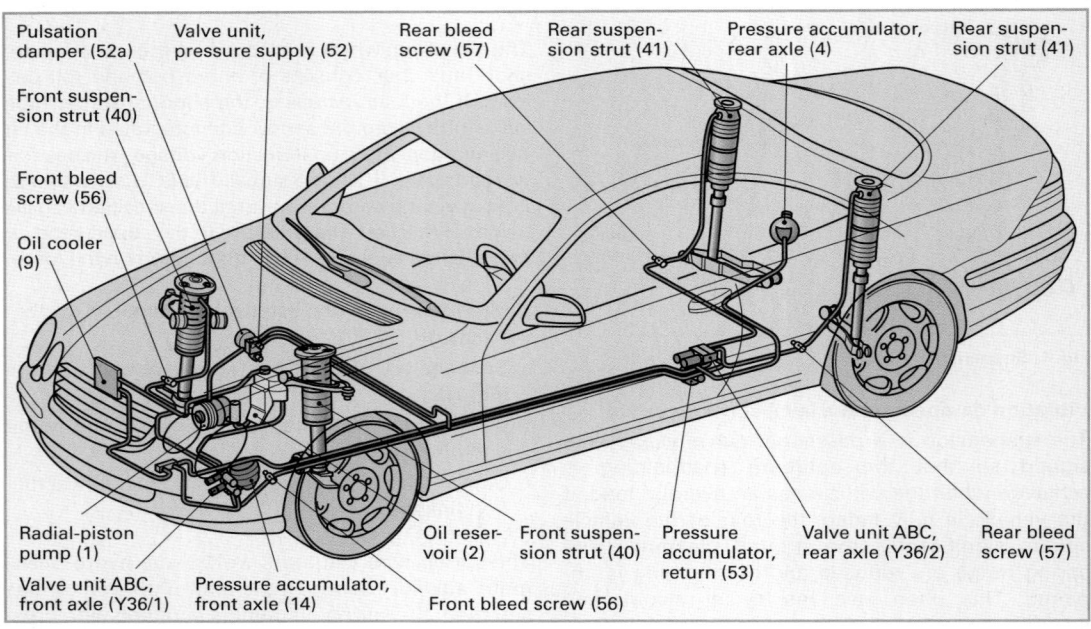

Fig. 2: Active Body Control (layout)

Legend for the ABC diagrams

a	Suction line	53	Pressure accumulator, return	B24/12	Lateral-acceleration sensor
b	Operating pressure	56	Front bleed screw	B24/14	Longitudinal-acceleration sensor
c	Control pressure	57	Rear bleed screw	B24/3	Body acceleration sensor, front left
d	Return line	F1	Fuse 1	B24/4	Body acceleration sensor, front right
		F2	Fuse 2	B24/6	Body acceleration sensor, rear
1	Radial-piston pump	N51/2	ABC ECU	Y36/1	Valve unit ABC, front axle
2	Oil reservoir	N10/6	SAM ECU	y1	Suspension strut control valve, front left
2a	Oil filter	B4/5	ABC pressure sensor	y2	Suspension strut check valve, front left
9	Oil cooler	B22/1	Plunger travel sensor, rear left	y3	Suspension strut control valve, front right
4	Pressure accumulator, rear axle	B22/4	Plunger travel sensor, front left	y4	Suspension strut check valve, front right
14	Pressure accumulator, front axle	B22/5	Plunger travel sensor, front right	y36/2	Valve unit ABC, rear axle
40	Front suspension strut	B22/6	Plunger travel sensor, rear right	y1	Suspension strut control valve, rear left
41	Rear suspension strut	B22/7	Level sensor, rear left	y2	Suspension strut check valve, rear left
52	Valve unit, pressure supply	B22/8	Level sensor, front left	y3	Suspension strut control valve, rear right
52a	Pulsation damper	B22/9	Level sensor, front right	y4	Suspension strut check valve, rear right
52b	Pressure-limiting valve	B22/10	Level sensor, rear right	y86/1	ABC vacuum valve
		B40/1	ABC oil temperature sensor		

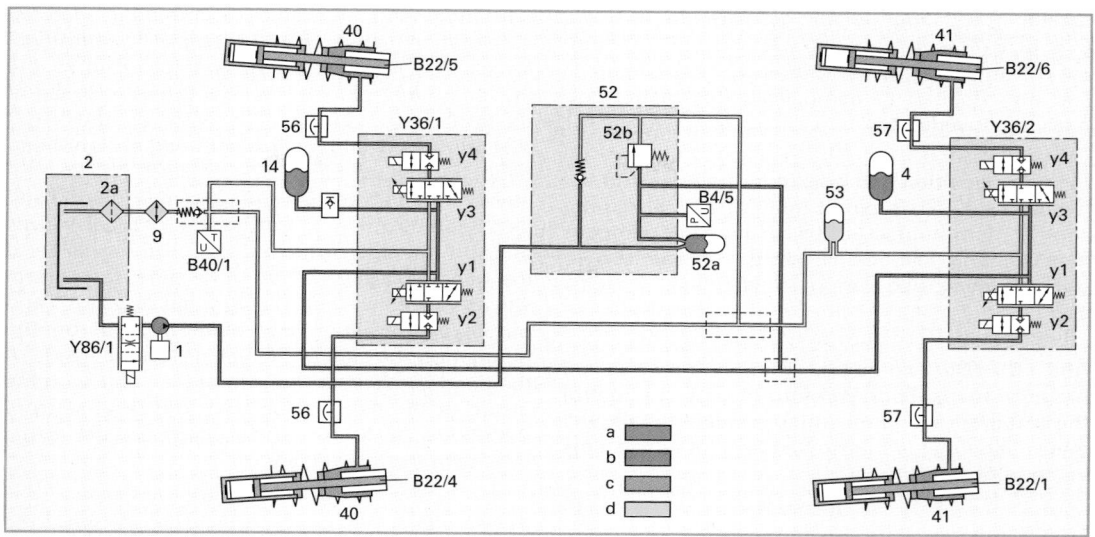

Fig. 1: ABC hydraulic-circuit diagram

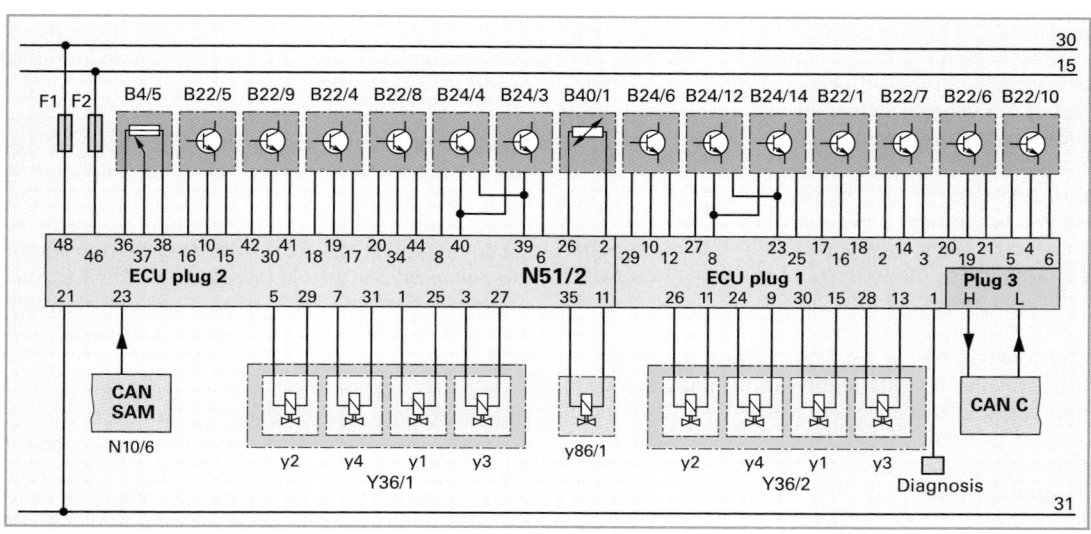

Fig. 2: ABC schematic diagram

Task and function of the sensors

Pressure sensor B4/5 reports the hydraulic pressure to the ECU via pin 36, pin 37 plug 2. This is regulated to 180 to 200 bar by the vacuum valve **y86/1**.

Oil temperature sensor B40/1 measures the hydraulic-oil temperature in the return flow pin 26, pin 2 (plug 2).

Travel sensors in the hydraulic cylinder (plunger) B22/6; B22/1; B22/4; B22/5 transmit the actual position of the positioning cylinder in the suspension strut to the ECU pin 20; pin 17 (plug 1), pin 18, pin 16 (plug 2).

Level sensors B22/7, B22/10, B22/8, B22/9 detect the level of the vehicle body using the relevant control arm pin 2; pin 5 (plug 1), pin 20; pin 42 (plug 2).

Body acceleration sensors B24/3, B24/4, B24/6 measure the vertical acceleration of the vehicle body. They consist of electronic vibration modules which send their signals to the ECU via pin 6, pin 8 (plug 2), pin 29 (plug 1). They are required to be able to record the lifting movements of the body.

18

Lateral and longitudinal acceleration sensors B24/12, B24/14 determine the lateral and longitudinal dynamics of the vehicle pin 27, pin 25 (plug 1) and are required to compensate for rolling and pitching movements.

Signal acquisition and actuation module SAM activates the ECU via pin 23 (plug 2) via the remote control, door contact switch or luggage compartment lighting. The ECU checks the vehicle level in order to lower it to the preselected level if necessary.

ABC ECU N51/2 compares stored and preselected program maps (sport/comfort) in order to control the actuators using incoming sensor signals and information that is transmitted from other systems via the CAN bus.

Task and function of the actuators

The **vacuum valve y86/1** regulates the quantity of oil sucked in by the oil pump so that an oil pressure of 180 to 200 bar can be established and maintained in the ABC system. When it is not energised, the valve is closed in order to maintain the pressure in the system.

Control valves y1, y3. The positioning cylinders are moved when the control valves are actuated. This causes the body to sink or rise at the corresponding wheel. The downforce of the wheels may be briefly increased by this.

Check valves y2, y4 are closed when the engine is off, the vehicle is stationary and if faults occur to prevent pressure loss. This also prevents the positioning cylinders from being pulled apart if the wheel is changed or the vehicle is placed on a lifting platform, for example.

Control procedures

Starting the engine. When the vehicle door is opened, the ABC ECU is activated by the signal acquisition and actuation module pin 23 (plug 2). The level sensors **B22/7 ... 22/10** are used to compare the actual level with the target level. If the actual level is higher than the target level, the control valves **y1, y3** are actuated and the vehicle is lowered to the target level. The ECU is powered with battery + via pin 48 and with battery – via pin 21 in order to carry out this control procedure. Once the ignition has been switched on, there is an additional power supply via pin 46 plug 2.

Cornering. When the vehicle is corning, the lateral-acceleration sensor **B24/12** registers centrifugal forces. The relevant signal is transferred to the ECU via pin 27 plug 1. The ECU uses the speed of the front right and front left wheels from the CAN C to determine whether it is a left-hand or right-hand bend. If it is a left-hand bend, the ECU **N51/2** actuates control valves **y3** via pin 3, pin 27 (plug 2) and pin 28, pin 13 (plug 1), so that the plunger moves out and the side of the vehicle on the outside of the bend is raised. At the same time, the control valves **y1** are switched via pin 1, pin 25 (plug 2) and pin 30, pin 15 (plug 1) so that the load on the plunger on the side of the vehicle on the inside of the bend is relieved. The side of the vehicle on the inside of the bend is lowered. The level sensors **B22/22/7 ... 22/10** are used to compare the actual level with the target level.

Acceleration. When the vehicle accelerates, the longitudinal-acceleration sensor **B24/14** registers acceleration forces on the longitudinal axis of the vehicle. The signal is transferred to the ECU at pin 25 plug 1 which actuates the control valves so that the vehicle body sinks at the front axle and is raised at the rear axle.

Braking. When the vehicle brakes, the ECU receives information that a braking procedure has been commenced from the closed brake-light switch via the CAN C. The longitudinal-acceleration sensor supplies the ECU with information about the deceleration rate via pin 25 plug 1. The ECU actuates the control valves so that the vehicle body is raised at the front axle and lowered at the rear axle.

Driving straight ahead. When the vehicle is driving straight ahead, the ECU receives information about the vehicle speed via the CAN C. The ECU actuates the control valves to automatically lower the vehicle according to the preselected program map. If the driver wishes, the vehicle can be raised by 25 or 50 mm (by pressing the level switch (CAN C)).

Vertical vibrations. If the vehicle vibrates in the direction of the vertical axis due to an uneven road surface, these movements are transferred to the ECU from the body acceleration sensors **B24/3, B24/4, B24/6** via pin 6, pin 8 (plug 2) and pin 29 (plug 1). The level sensors **B22/7, B22/8, B22/9** pin 42, **B22/10** report the amplitude via pin 20 plug 2 and pin 2, pin 5 (plug 1). The ECU actuates the control valves according to the preselected program map (sport/comfort) so that the body vibrations are damped and evened out.

18

18.9 Wheels and tyres

18.9.1 Wheels

Requirements on the wheels

- Low weight
- Large diameter for large brake discs
- High dimensional stability and elasticity
- Good heat dissipation properties (frictional heat)
- Easy replacement of tyres and wheels in the event of damage

Structure of the wheel

The wheel consists of the rim and the wheel disc with a centre hole and bolt holes. Instead of a wheel disc, there may be a wheel spider, or the rim may be connected to the hub by steel spokes. The wheel is secured to the flange of the wheel hub **(Fig. 1),** which pivots about the kingpin, with wheel nuts or wheel bolts. The brake drum or brake disc is also bolted to the wheel-hub flange. If the bearing is open, a hub cover protects the bearing and is also the mounting location for the grease reservoir.

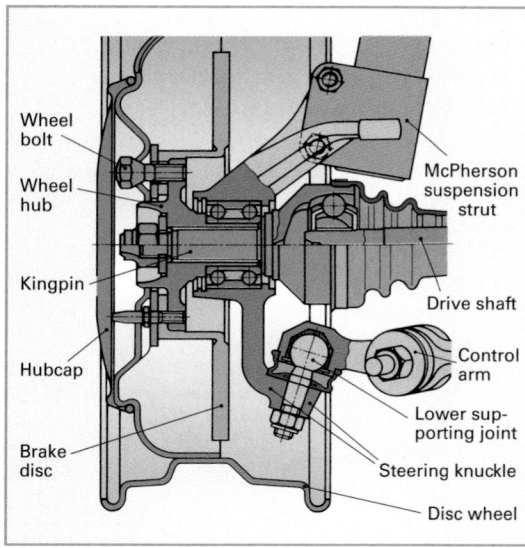

Fig. 1: Passenger-car wheel with bolted wheel hub

Rims

There are rims which are fixed to the wheel disc and those which can be removed. We also make a distinction between single-piece rims (drop-centre rims) and multi-piece rims, which are used on commercial vehicles (see Chapter Commercial vehicle technology).

Drop-centre rims. Single-piece drop-centre rims are used almost exclusively on passenger vehicles. They are a single-piece cast or forged out of light alloy and are riveted, welded or bolted to the

wheel disc or the wheel spider **(Fig. 2).** The cross section of the rim may be symmetrical or asymmetrical.

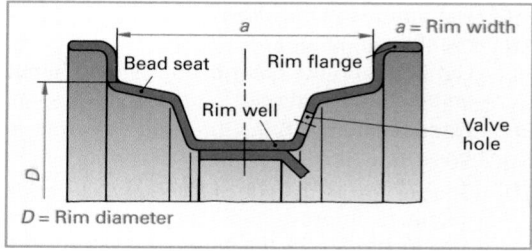

Fig. 2: Single-piece symmetrical drop-centre rim

Hump rim. If tubeless radial-ply tyres are used, drop-centre rims which have a continual raised section = hump (H) on the bead seat near to the rim well **(Fig. 3).**

If this raised section is not rounded, it is known as a flat hump (FH). Both types are designed to prevent the tyre bead from being pushed from the bead seat into the rim well by the large lateral forces which occur when the vehicle is cornering at speed. The air escapes suddenly on tubeless tyres, which could result in a serious accident.

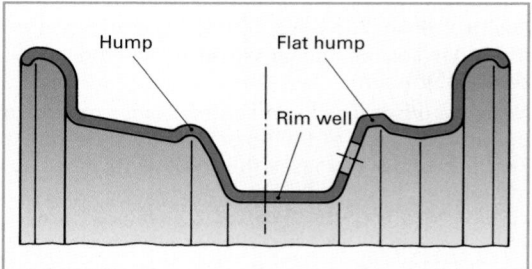

Fig. 3: Asymmetrical hump rim

Dimensions and designations on rims

This data is standardised. The rim designation is stamped on each wheel by the manufacturer. It basically consists of two dimensions: the rim width a in inches and the rim diameter D in inches. The two dimensions are separated by an "x" on drop-centre rims. Code letters after the rim width indicate the shape of the rim flange, code letters after the rim diameter indicate the type of rim.

Example: **6 $^1/_2$ J x 15 H RO 35**

$6^1/_2$	Rim width in inches
J	Code letter for the dimensions of the rim flange
x	Single-piece rim (drop-centre rim)
15	Rim diameter in inches
H	A hump on the outer bead seat
RO 35	Offset 35 mm

18

Other rim designations:

H2   Hump on both sides

FH   Flat hump on the outer bead seat

FH2   Flat hump on both sides

CH   Combination hump:
      Flat hump on the outer bead seat and hump
      on the inner bead seat

EH   Extended hump

SDC   Semi-drop-centre rim

TD   Special rim with reduced a reduced flange
      height to improve the ride comfort of the
      tyre. A groove in the bead seat accommo-
      dates the tyre bead so that the bead cannot
      jump out if the tyre is depressurised. The rim
      width and diameter are given in mm.

Offset

> This is the measurement from the centre of the
> rim to the inner contact face (wheel-mounting
> plane) of the disc wheel **(Fig. 1)**.

Selecting a wheel with a different rim offset may
change the track width.

Note about changing the rim: If the track width
changes, other geometrical dimensions, such as
the kingpin offset and camber, will also change.

Positive offset. The inner contact face is moved to
the outer section of the wheel in relation to the
centre of the rim.

Negative offset. The inner contact face is moved to
the inner section of the wheel. Using rims with a
negative offset increases the track width of a motor vehicle.

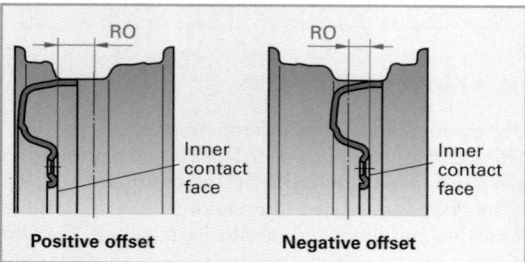

Fig. 1: Rim offset

Types of wheels

Disc wheels are pressed out of steel sheet or cast
or forged out of light alloy, e.g. GK-AlSi 10 Mg.
Benefits of wheels made from light alloy:

● Low weight (small unsprung mass)
● More effective brake ventilation and heat dissipation

Lightweight wheels made from newly developed
steels, e.g. DP 600 or HR 60, can have thinner walls
and have become up to 40 % lighter compared to
previous steel wheels made from RSt 37.

18.9.2 Tyres

Requirements on the tyres

● To support the weight of the vehicle
● To absorb and damping jolts from the road
● To transfer drive, braking and lateral forces
● Low rolling resistance (low friction and heat de-
  velopment)
● Adequate service life
● Quiet and low-vibration rolling

Structure

The tyres include the inner tube and valve, the tyre
itself and the rim band. The latter is now only used
on mopeds and motorcycles with wire-spoked
wheels to protect the inner tube from being damaged by the nipple heads of the wire spokes. The
inner tube must correspond to the tyre size. In this
type of tyre, the inner tube must always be replaced at the same time as the tyre.

The **tyre (Fig. 2)** consists of:

● **Carcass**
● **Protector** with tread
● **Bracing layer** (on radial-ply tyres)
● **Side wall**
● **Beads** with inserted wire-spoked cores
● **Airtight rubber layer**

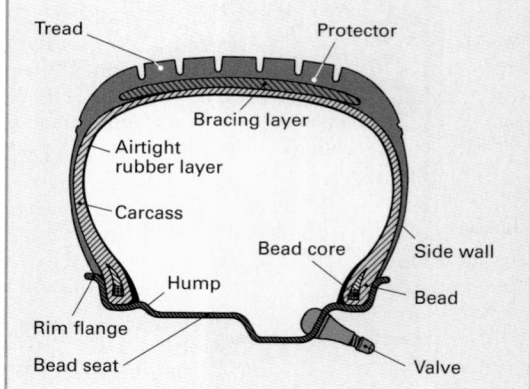

Fig. 2: Structure of a tyre

Carcass. This is constructed of rubberised cord fibres made from nylon, rayon, steel, polyester or
aramid. The fibres are laid on top of each other in
layers, either radially – at right angles to the direction of travel – or diagonally – in a point towards
the running direction. The fibres are wound around
two steel rings (bead cores) and are fixed in place
by vulcanisation.

Protector. This consists of several layers of fabric
and rubber cushions. It damps impacts and protects the carcass.

18

Bracing layer. This consists of several layers of steel wires, textile fibres or aramid fibres embedded in rubber. The bracing layer lies over the carcass and is made in such a way that the wires or fibres cross. In high-speed tyres, the bracing layer may be folded **(Fig. 1)**, thereby increasing the stability.

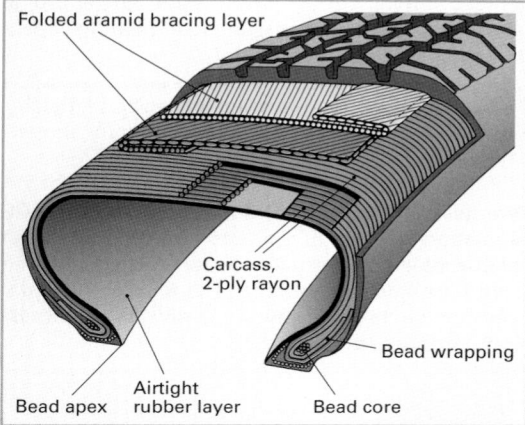

Fig. 1: **Arrangement of the carcass and bracing layer in a tyre**

Tread. This has grooves. The longitudinally grooved tread provides the tyre with cornering stability and the cross-groove tread transfers motive forces. The arrangement of the tread has a considerable impact on aquaplaning, rolling resistance and the noise characteristics of the tyres.

If the road is wet, a wedge of water can form between the tyres and the road surface at high speeds. This eliminates the road-surface adhesion and renders the vehicle unsteerable. The grooves in the tread must be of a certain shape and depth to dissipate the water quickly.

The minimum tread depth of 1.6 mm prescribed by law is not sufficient to prevent aquaplaning in many cases.

Side wall. Lower side walls increase the rigidity of the tyre, which improves steering precision but reduces ride comfort.

Bead. This has the task of keeping the tyre firmly in place on the rim so that braking, motive and lateral forces can be transferred to the road. It is therefore made to be particularly rigid using cables made from steel wire (bead core). On tubeless tyres, it has the additional task of sealing the tyre onto the rim.

Dimensions and designations on the tyre

Tyre size. This is given as 2 measurements: tyre width in inches or mm and rim diameter in inches or mm.

However, these numerical values do not correspond to the actual dimensions of the tyre. Exact values must therefore be taken from the standard table. All measurements apply to tyres that are inflated to the standard pressure and unladen **(Fig. 2)**.

Aspect ratio. In order to distinguish between different tyre types, e.g. balloon tyres and low-cross-section tyres, the ratio of the tyre height H to tyre width W is established. This is given as a percentage in tyre designations.

On modern tyres, the width is greater than the height. If the height of the tyre is 80 % of the width, for example, the ratio height to width = 0.8 : 1. As the percentage is used in the tyre designation, these would be known as 80s tyres.

Effective radius. A vertical tyre under load has a smaller radius (distance from the centre of the wheel to the road surface) than an unladen tyre. This is known as static radius r_{stat} **(Fig. 3)**.

When the vehicle is in motion, the compression of the tyres is eliminated by the centrifugal force and the radius increases again. This is known as dynamic radius r_{dyn}.

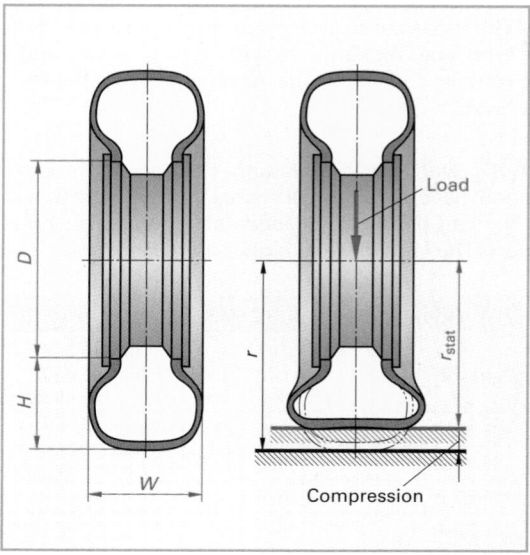

Fig. 2: **Tyre measurements**   Fig. 3: **Tyre under load**

Dynamic rolling circumference U_{dyn}. This describes the distance that the tyre covers with one revolution at a speed of 60 km/h when it is bearing the load specified in the standard and inflated to the specified air pressure. The accuracy of the speedometer reading depends on the rolling circumference. The static radius and the dynamic rolling circumference are given in tyre tables.

Tyre speed category. This classifies tyres for passenger vehicles and motorcycles according to their maximum permissible speed. Each maximum permissible speed is given a code letter, a selection of which is shown in **Table 1**.

Table 1: Speed categories			
Maximum tyre speed in km/h	Speed symbol	Maximum tyre speed in km/h	Speed symbol
160	Q	240	V
180	S	270	W
190	T	300	Y
210	H	over 240	ZR

Tyre load-bearing capacity (Table 2). This is shown by the load index (LI). This is a code number and indicates the maximum load-bearing capacity of the tyre at the standard pressure.

For some tyres for commercial vehicles, the PR designation (ply rating) is also given. 8 PR means that, due to the rigidity of its carcass, a tyre can bear the same load as a tyre with 8 layers of cotton cord even though it has fewer layers.

> The tyre load-bearing capacity depends on the tyre type, maximum speed, tyre pressure and camber. These must be determined from the vehicle.

Tyres with the designation Reinforced or Extra Load have a reinforced carcass. This means that they can bear greater loads at a greater air pressure. The load index is higher on these tyres.

Table 2: Tyre load-bearing capacity LI (selection)						
Tyre size				Reinforced (Extra Load)		
	LI	kg	bar	LI	kg	bar
135/80 R 13	70	335	2.4	74	375	2.8
185/70 R 14	88	560	2.5	92	630	2.9
195/65 R 15	91	615	2.5	95	690	2.9
205/50 R 16	87	545	2.5	91	615	2.9

Examples of tyre designations

195 / 60 R 15 88 H

R = radial-ply tyre
195 = nominal tyre width 195 mm;
60 = aspect ratio 60 %;
15 = rim diameter 15″
88 = load capacity 560 kg
H   = maximum speed 210 km/h.

335 / 30 ZR 18 (102 W)

This tyre has a dual designation for the speed. The part in brackets means that the tyre has a maximum speed of 270 km/h (W) at a load index of 102. If the vehicle is approved for greater speeds, the vehicle manufacturer must issue an approval which sets out the permissible load-bearing capacity and speed.

> **Note:** the tyre load-bearing capacity of these tyres is reduced by 5 % every 10 km/h above 240 km/h.

Tyre designations (Fig. 1). In accordance with ECE regulation no. 20 (ECE = Economic Commission for Europe), the information listed in **Fig. 1** must be used for a tyre designation. (The full tyre designation can be taken from the book "Tabellenbuch Kraftfahrzeugtechnik".)

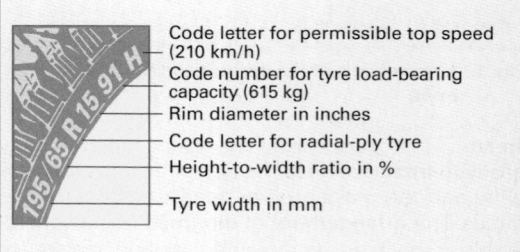

Code letter for permissible top speed (210 km/h)
Code number for tyre load-bearing capacity (615 kg)
Rim diameter in inches
Code letter for radial-ply tyre
Height-to-width ratio in %
Tyre width in mm

Fig. 1: ECE tyre designation

Tyre type
We distinguish between balloon tyres, super low-pressure tyres, low-cross-section tyres, super low-cross-section tyres, 70, 60, 50, 40, 35 tyres, etc., according to the aspect ratio of the tyres **(Fig. 2)**. The ratio of tyre height to tyre width varies between the individual shapes, which results in different handling characteristics. They vary from an almost round profile (balloon) to an ever flatter and wider cross-section. Wider treads and lower side walls result in better driving safety, which is very important as the speed increases.

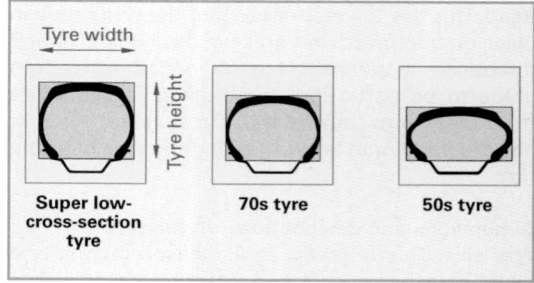

Tyre width
Tyre height
Super low-cross-section tyre
70s tyre
50s tyre

Fig. 2: Tyre cross-sections (selection)

Balloon tyres (height to width = 0.98 : 1) e.g. 4.50-16, have good suspension characteristics but poor cornering stability due to the large tyre height.

Super low-pressure tyres (height to width = 0.95 : 1) e.g. 5.60-15, are distinguished from balloon tyres by their wider shape and smaller inner diameter (up to 15").

Low-cross-section tyres (height to width = 0.88 : 1) e.g. 6.00-14, have a width in 1/2" gradations. They may also be marked with the letter L (low).

Super low-cross-section tyres (height to width ~0.82) e.g. 165 R 13, were manufactured as crossply tyres and, from 1964, as radial-ply tyres (80s tyres).

70s tyre (height to width = 0.70 : 1) have a height that is 70 % of the width. This is what gives the tyres their designation. They have the advantage of increased road grip and vehicle stability. The high lateral forces allow greater cornering speeds.

50s tyres (height to width = 0.5 : 1) e.g. 225/50 R 16, have a height that is only 50 % of the width. The rim diameter is increased since the rolling circumference of the tyre remains constant, compared to 195/65 R 15 tyres.

Advantages:

- Larger and higher performance brake discs with better ventilation can be fitted.
- Not sensitive to lateral deformation as the cross-section is low and flat.
- High lateral stability when steering into bends; occurrence of large lateral forces even at small wheel-slip angles, allowing high cornering speeds.
- Increased resistance to lateral twisting.
- More precise response to steering movements.

Disadvantages:

- Poorer aquaplaning characteristics
- Lower internal suspension, loss of comfort
- Greater steering effort required

Tyre contact patch (tyre contact area, positive tread)

As the tyre width increases, so does the contact patch of the tyre on the ground **(Fig. 1)**. A larger contact area increases the friction force so that the tyre grip is increased under fast cornering and under braking. Coulomb's law, which states that the friction force depends only on the normal force (vertical load) and the friction coefficient, only applies to tyres to a limited extent. When rubber-elastic materials rub against coarse surfaces (roads), the size of the surfaces rubbing against each other due to the indentations is significant.

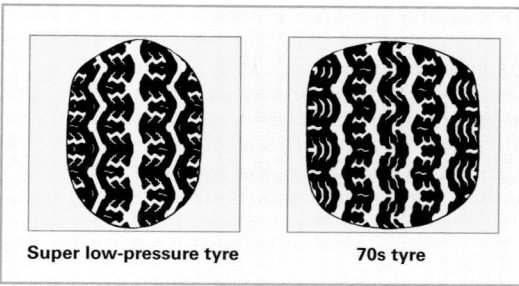

| Super low-pressure tyre | 70s tyre |

Fig. 1: Tyre contact patches

Negative tread. This is made up of lateral, longitudinal and diagonal grooves between the individual tread bars. If the tyre contact patch is large, the proportion of negative volume must be increased in relation to the tyre contact patch to prevent the tyre aquaplaning due to increased water absorption. The effectiveness of the tyre for winter driving is also increased by the higher ground pressure.

Air-pumping effect. Deformation of the tyre contact patch while the vehicle is in motion can create enclosed cavities, depending on the layout of the negative tread, which can abruptly fill with air and then empty again. This results in considerable driving noise.

Tyre construction

A distinction is made between crossply tyres and radial-ply tyres according to the carcass structure.

Crossply tyres. The fabric plies are laid diagonally on top of each other so that the cord fibres form a point (cord angle) of 26° to 40° along the direction of travel **(Fig. 2)**. The smaller the cord angle, the harder the tyres, the better the lateral stability and the greater the possible cornering speeds. Crossply tyres are primarily used on motorcycles (see Chapter Motorcycle technology).

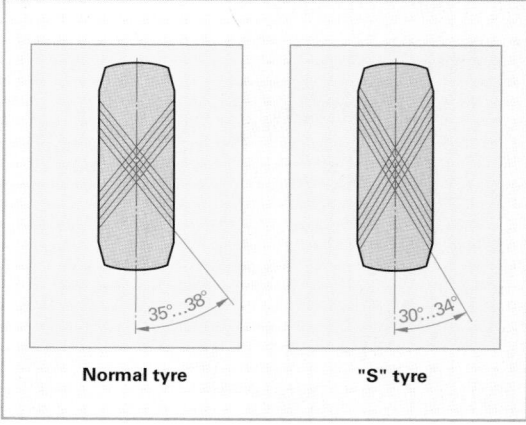

| Normal tyre | "S" tyre |

Fig. 2: Cord angle on crossply tyres

18

Radial-ply tyres (Figs. 1 and 2). All the cord threads on the carcass are situated next to each other and are arranged radially, i.e. at 90° to the direction of travel. A belt made of several layers of fabric or steel cord or aramid at an angle of approximately 20° to the direction of travel is fitted between the carcass and the tread of the tyre, so that the tread becomes only very slightly deformed when the vehicle moves away. **Fig. 1** shows 2 criss-cross steel cord and 2 circumferential nylon belts at 0°. The nylon bracing layers at 0° enable the tyre to withstand higher speeds.

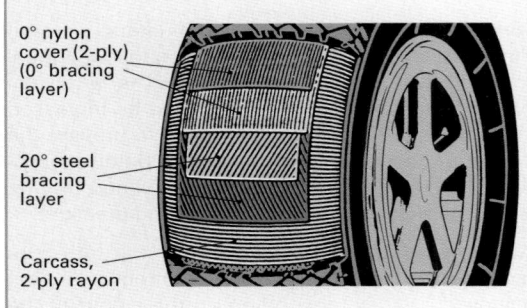

Fig. 1: Structure of a radial-ply tyre

The side walls of radial-ply tyres compress, the deformation is mainly limited to the flexing zone.

At lower speeds, radial-ply tyres run firmer than diagonal tyres, thanks to the reinforcement belt. At greater and higher speeds, the springiness of the soft carcass comes into play, meaning that the radial-ply tyre operates more quietly than the diagonal tyre. In addition, the bracing layer produces good lateral stability and thus high lateral forces.

Tubeless tyres (Fig. 2). An airtight rubber layer made of butyl prevents the air from escaping. Nevertheless, pressure is lost through diffusion of air molecules over time. The tyre pressure must therefore be checked regularly.

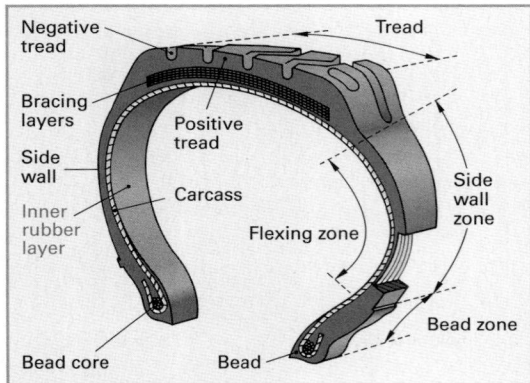

Fig. 2: Tubeless radial-ply tyre

If the tyre is filled with nitrogen instead of air, it will last longer, because nitrogen molecules are larger than air molecules. The rubber valve fitted in the rim must also be perfectly sealed. Tubeless tyres bear the inscription "Tubeless" or "sl".

The benefits of tubeless tyres:

- Less heat build-up because there is no friction between tyre and tube.
- Lighter weight and easier to assemble.

Wheel-slip angle

If a moving vehicle is affected by disruptive forces (wind force, centrifugal force), a wheel-slip angle appears and the lateral forces at work in the tyre contact areas counterbalance these disruptive forces.

> The angle between the actual direction of motion and the rim plane (following the line of the rim) is called the wheel-slip angle α **(Fig. 3)**. A tyre can only transmit lateral force if it runs at an angle to the direction of travel.

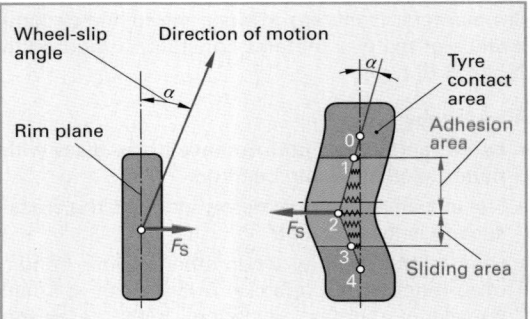

Fig. 3: Wheel-slip angle α

The lateral force in the tyre builds up through the deformation of the tyre contact area, when cornering, for example. As soon as a wheel-slip angle forms, the tyre tread moves further and further away from the line along which it normally lies when in contact with the road surface, in the centre of the tyre **(Fig. 3)**.

This creates a deformation in the tyre. The further away the tread moves from the centre line, the greater the deformation. The sum of these application forces is the lateral force, which takes effect at the centre of gravity of the deformed tyre contact area.

If the wheel-slip angle continues to increase, a sliding friction occurs in the rear section of the tyre and the application force is alleviated. However, the lateral force continues to increase because the adhesion area is still larger than the sliding area. If the wheel-slip angle still continues to increase, the sliding area becomes larger than the adhesion area and the lateral force is alleviated.

When cornering, the wheel load on the outer cornering wheels of an axle is increased, whereas the wheel load on the inner cornering wheels is decreased. The higher the wheel load becomes, the greater the build up of lateral force in the tyre. Strong lateral forces must also be built up in wide tyres in the event of high wheel loads and lateral acceleration, thus increasing safety when cornering, whereas with super low-cross-section tyres, such as 165/80 R 13 tyres, the lateral force is actually decreased (Fig. 1).

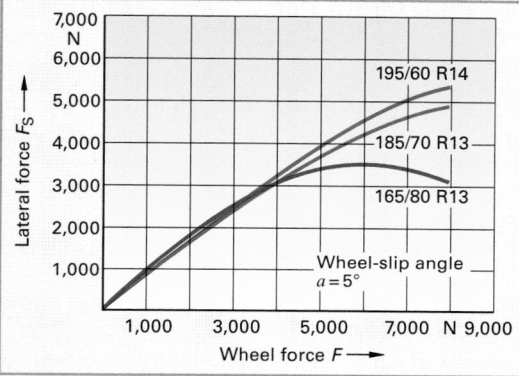

Fig. 1: Lateral force build-up for radial-ply tyres

Winter tyres (M+S tyres). In contrast to the coarse studded tyre treads that were previously used, today's tyre treads have small tyre tread grooves with many fine fins. The fine fins give the tyre a better grip on snowy or slippery road surfaces in winter. To keep the rubber on the tread surface elastic at lower temperatures starting from $\leq 7\,°C$, silicic acid (silica) or natural rubber is added. This has the following benefits:

- Better adhesion between tyre and lining
- Lower rolling resistance
- Good service life of tyre tread (less internal heat formation)

Winter tyres with a tread depth of less than 4 mm are no longer sufficiently winter-proof.

Tread wear indicators (Fig. 2). These are elevations in the tread-groove base. If the tread wears down to the legally prescribed minimum tread depth of 1.6 mm, the height of the tread wear indicators will be level with the tread. The position of these indicators in the tyre tread is marked with the letters TWI (tread wear indicator) or with a triangle, on the tyre wall.

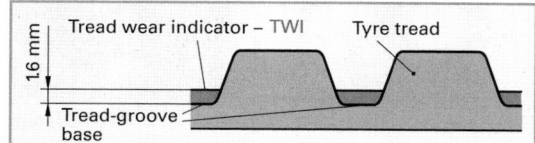

Fig. 2: Tread wear indicator

Due to the high risk of aquaplaning, at high speeds in particular, and due to the increased braking distance on wet roads if tyres have low tread depths (Table 1), it is advisable to change the tyre if the tread wear indicators are in contact with the road surface.

Table 1: Braking distance when braking from 100 km/h to 60 km/h				
Tread depth (mm)	Braking distance in m (wet road surface)			
	20	40	60	80
7				
5				
3				
2				
1.6				

Wheel balancing
The mass of a wheel when turning is never evenly distributed. In the areas where the mass is greater, an imbalance appears, in other words, centrifugal forces develop which increase more, the greater the mass and the higher the engine speed (Fig. 3).

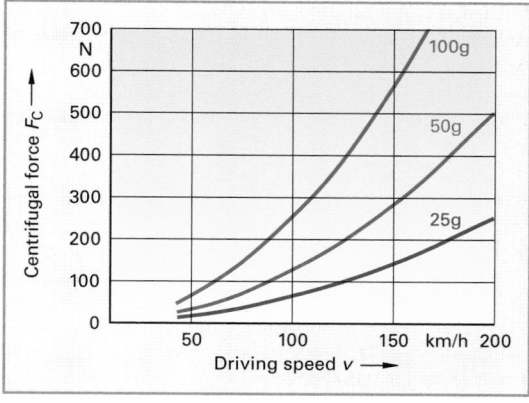

Fig. 3: Centrifugal forces on a tyre with designation 195/65 R 15

Static imbalance. If, for example, rubber is worn on a section of the tread as a result of locking brakes, this produces a centrifugal force in the section opposite, which can cause the wheel to bounce off the road surface at higher engine speeds. This fault can be viewed by spinning the wheel.

To ensure that the wheel remains stationary in each position when spinning, the sum of all the moments of inertia around the wheel rotational axis must equal 0.

$$M_1 = M_2 \qquad G_1 \cdot r_1 = G_2 \cdot r_2$$

A balancing mass m_2 with weight force G_2 must be fixed on the rim opposite the heaviest section of the wheel. This mass must be large enough to make the existing torque M_2 correspond with the torque M_1. The wheel is then statically balanced (Fig. 1).

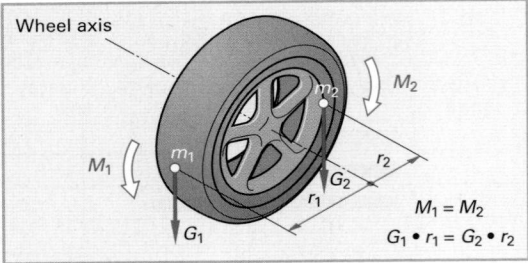

Fig. 1: Balancing (static)

Dynamic imbalance. A wheel's imbalance weight m_1 is seldom at the same level as its balance weight m_2 affixed to the rim. The wheel is statically balanced, but at higher engine speeds, the centrifugal forces m_1 and m_2 produce a torque in line with the axle and cause the wheel to wobble. In this case, the wheel has a dynamic imbalance. If the imbalance weight m_1 is level with the wheel-mounting plane, then only the torque M_{C2} (Fig. 2) will take effect.

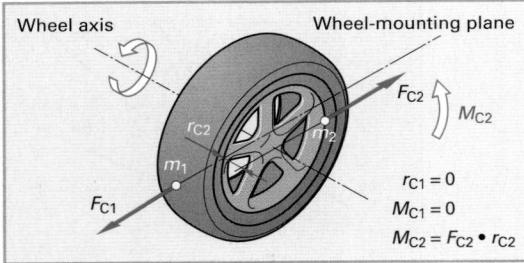

Fig. 2: Dynamic imbalance

Attaching a second balance weight m_3 to the inside of the rim can cause the existing torque M_{C3} to balance the torque M_{C2}, the wheel is then dynamically balanced. (Fig. 3). The size and position of the balance weights m_2 and m_3 are determined on balancers.

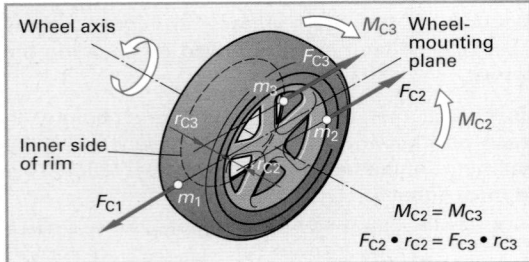

Fig. 3: Balancing (dynamic)

If a wheel is out-of-round despite being balanced, a radial tyre runout could be the problem. If the radial tyre runout protrudes by more than 1 mm from the tread surface, attempts must be made to reduce the radial tyre runout by turning the tyre on its rim (matching).

18.9.3 Run-flat systems

> Run-flat systems can either aggravate or prevent critical driving situations caused by a sudden air loss in the tyre, particularly at higher speeds. It is normally possible to reach the closest workshop without having to change the tyre.

Run-flat systems are wheel/tyre systems with limp-home characteristics.

A distinction can be made between 2 possible applications:

- Systems which can be used with conventional rims.
- Systems composed of special rims and corresponding tyres.

The use of compressed-air monitoring systems is a requirement for both systems. The driver must know about the pressure loss in the tyre in order to adjust the speed and continue driving.

Systems with conventional rims

Conti Support Ring (CSR). A light metal ring with flexible mounting is fitted on the rim (Fig. 4). When air is lost, the tyre is supported against the ring without touching the inside of the tyre walls and causing irreparable damage to the tyre through friction-induced heat. It is possible to drive on for approximately 200 km at reduced speed. The additional weight per wheel is about 5 kg. Tyres with a height/width ratio of >60 are suitable, otherwise they cannot be fitted.

Self-Supporting Run-Flat Tyres (SSR, DSST*). The side walls of these tyres are reinforced with rubber (Fig. 5). At zero pressure, the tyre can still be supported on its bead, so that the bead does not slide down into the rim well. It is possible to drive on for approximately 200 km at a speed of 80 km/h. This side wall reinforcement can also reduce comfort due to the increased transmission of bumps in the road surface.

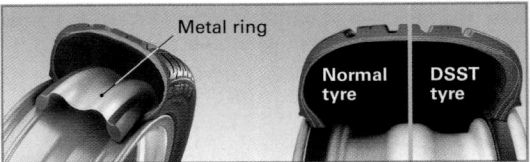

Fig. 4: CSR system   Fig. 5: DSST system

---

* DSST Dunlop Self Supporting Technology

Systems with special rims and tyres

PAX system. The PAX system is composed of a special rim with a flexible insert and the corresponding tyre with vertical anchorage on the rim (**Fig. 1**).

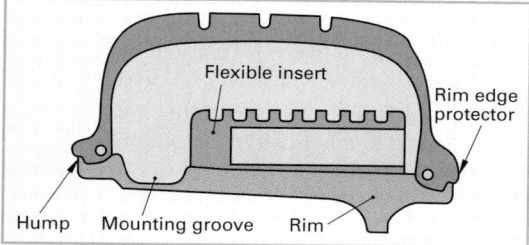

Fig. 1: PAX system

Rim. The rim is very flat and has one small mounting groove in place of the rim well. Rim flanges are not supplied, both humps are on the outside of the rim. Rim diameters for large brake discs can be achieved due to the flat shape of the rim.

Tyre. The tyre has shorter side walls, which increase its rigidity. Lateral forces can cause less deformation to the tyre contact patch, thus improving the road adhesion and reducing the rolling resistance.

The tyre bead is in a groove outside on the humps. All the forces working on the tyres produce a tensile strength in the carcass, meaning that the bead is always pressed into the groove (**Fig. 2**). This vertical anchorage ensures that the bead cannot slip out of the rim, even when the tyre is at zero pressure.

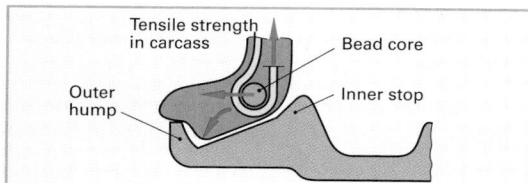

Fig. 2: Vertical anchorage of the tyre bead

Flexible insert. The insert is an elastomer ring which is pushed onto the rim. Thanks to its high load-bearing capacity it supports the tyre when pressure is lost, meaning that at a speed of 80 km/h, approximately 200 km more can still be covered.

Size designations for PAX systems

205/650 R 440 A

205	Tyre width in mm
650	Outside diameter of the tyre in mm
R	Radial structure
440	Average rim seat diameter in mm
A	Asymmetrical seat

18.9.4 Compressed-air monitoring systems

Compressed-air monitoring systems are designed both to recognise the air loss in the tyre and also to warn the driver.

The following types of compressed-air monitoring systems are used in motor vehicles:
- Indirect measuring systems
- Direct measuring systems

Indirect measuring systems

When pressure is lost, the tyre's rolling circumference, which increases the engine speed in relation to the other tyres, is reduced. The engine speeds are determined via the ABS or ESP sensors. However, the driver is not warned until there is a difference in air pressure of more than 30 % between the tyres.

Direct measuring systems

The pressure is measured directly by sensors in the tyre. The following functions are fulfilled:
- Continued monitoring of tyre pressure whilst driving and when the vehicle is stationary.
- The driver is given early warning in the event of a pressure loss, reduced pressure and flat tyre.
- Automatic individual wheel recognition and wheel positioning.
- Diagnostic procedure for systems and components in the workshop.

The system is composed of:
- 1 tyre-pressure sensor per wheel
- Antennae for tyre pressure monitoring
- Instrument panel with display
- ECU for tyre-pressure monitoring
- Function-selector switches

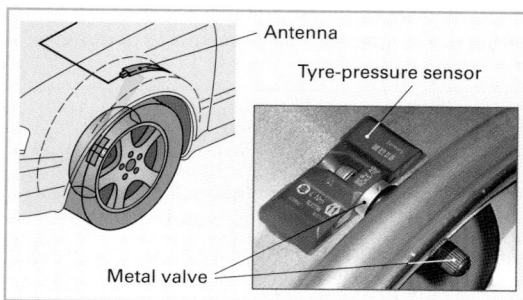

Fig. 3: Tyre-pressure sensor and antenna

Tyre-pressure sensor. This sensor is bolted to the metal valve (**Fig. 3**) and can be reused when changing the tyres or wheel rims. In addition, a temperature sensor, transmitting antenna, measuring and control electronics as well as a battery with a service life of approximately 7 years are integrated. Since the filling pressures are altered by temperature influences, the pressures and temperatures recorded in the ECU are set to a standard temperature of 20°C.

> To avoid damaging the sensor irreparably when changing a tyre, the tyre must be pressed down on the side opposite the valve.

ECU. The ECU obtains the following information from the transmitting antenna:

- Individual identification number (ID code), used for individual wheel recognition.
- Current inflation pressure and current temperature.
- Condition of the lithium battery.

The ECU evaluates the signals transmitted by the antenna for the tyre pressure monitoring and imparts the information for the driver on the display screen, according to the importance of this information. If wheels are changed on the vehicle, for example, from the front axle to the rear axle and vice versa, the ECU must be recoded with the new pressures.

Individual wheel recognition. The sensors belonging to the vehicle are recognised by the ECU and stored. The sensors are recognised when the vehicle is being driven, to avoid interference from sensors on cars parked nearby.

System messages, top priority (Fig. 1). These messages are intended for when driving safety is no longer guaranteed. They are displayed to the driver if, for example:

- … signal threshold 2 is undershot (0.4 bar below the stored setpoint tyre pressure of 2.3 bar).
- … signal threshold 3 is undershot (minimum pressure limit value, 1.7 bar in the diagram).
- … a pressure loss is greater than 0.2 bar/minute.

System messages, second priority (Fig. 1). They are displayed to the driver if, for example:

- … signal threshold 1 is undershot (0.2 bar below the stored setpoint tyre pressure of 2.3 bar).
- … the difference in pressure on the wheels of one axle is 0.4 bar.
- … the system is switched off or has a fault.

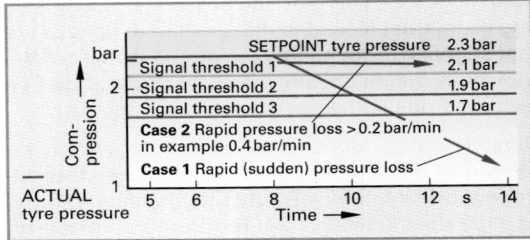

Fig. 1: Diagram of system messages

WORKSHOP NOTES

- Secure the motor vehicle against rolling away before removing the wheels.
- Only use wheel rims which are relevant for the tyre you are using (vehicle documents).
- Check the rims for cracks and remove traces of rust.
- Tighten the wheel nuts to the specified torque in a diagonal pattern.
- Note the specified air pressure to avoid loss of operating life.

- Using a mixture of diagonal- and radial-ply tyres is not permitted for passenger cars.
- Only use tyres of the same type and with the same tread on one and the same axle.
- Label the wheels after removing them and store them in a room that is cool, dry and dark.
- Do not stand wheels up and do not pile more than 4 tyres on top of each other.

18

REVIEW QUESTIONS

1 From which components is a wheel constructed?
2 Which types of rim are there?
3 Why are rims with humps used?
4 What are the benefits of using wheels made of light-metal alloys?
5 From which components are tyres made?
6 What does "dynamic rolling circumference of a tyre" mean?
7 What are the advantages and disadvantages of a 50 series tyre?
8 How are radial-ply tyres designed?
9 What is a "tyre-contact area"?
10 Explain the tyre designation 195/65 R 15 86 T M + S.

11 What is a tread wear indicator and how is its position on the tyre indicated?
12 What is the "wheel-slip angle"?
13 Why do wheels need to be balanced?
14 What does "dynamic imbalance" mean?
15 How can a radial-tyre runout be removed?
16 What are run-flat systems?
17 How is the PAX system designed?
18 What are the tasks of compressed-air monitoring systems and where should they be fitted?
19 What are the benefits of direct-measuring compressed-air monitoring systems in relation to indirect-measuring systems?

18.10 Brakes

Brakes are used in a vehicle for deceleration, for bringing the motor vehicle to a halt and for securing it against rolling away. When a vehicle is braked, the kinetic energy is converted into heat.

Brake systems

Service brake system. This system enables the speed to be reduced, if necessary, until the vehicle is stationary. The vehicle must stay firmly in lane during this process. The service brake is operated continuously with the foot (foot brake) and impacts on all the wheels.

Auxiliary brake system. This system must fulfil the functions of the service brake system when it is malfunctioning, possibly to less effect. It does not have to be an independent third brake, for the intact circuit of a dual-circuit service brake system or a graduated parking brake system is sufficient.

Parking brake system. Its function is to secure a stopping or parked vehicle from rolling away, including on a sloping road surface. Its components must be able to work mechanically for reasons of safety. In passenger cars, it is normally operated in stages by a coupling lever (handbrake) or a pedal via linkage and control cables. It works on the wheels of one axle only.

Continuous brake system. Its function is to keep the speed of the vehicle to a prescribed value when driving downhill (third brake). It is a requirement for motor busses whose weight is $m_{gvwr} > 5.5\,t$ and other vehicles with a weight of $m_{gvwr} > 9\,t$.

Antilock-braking system (ABS), also anti-skip system. The ABS measures the wheel slip automatically during braking, regulates the braking pressure and thus prevents the brakes from locking. The ABS is a legal requirement on vehicles weighing $m_{gvwr} > 3.5\,t$.

Structure of a brake system (Fig. 1)

A brake system consists of:
- Energy supply equipment
- Control equipment
- Transmission equipment
- Possible supplementary equipment for trailer vehicles, for example, trailer control equipment
- Parking brake
- Service brake
- Possibly also brake pressure control, such as an ABS, for example
- Wheel brake on the front axle and rear axle

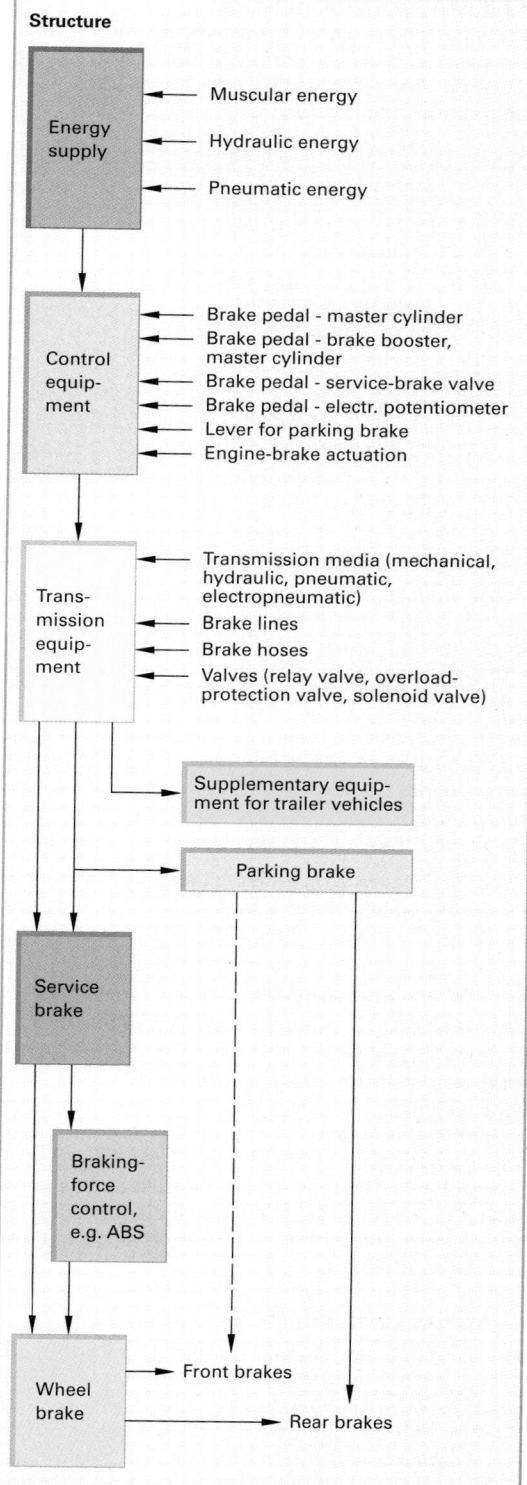

Fig. 1: Structure of a brake system

18

Legal requirements (extracts)

The legal requirements for brakes on motor vehicles are set out in the German National Road Traffic Licensing Regulations (StVZO), in EC directives and the ECE regulations.

Table 1: Motor-vehicle classifications (extracts)		
L		Motorcycles and three-wheelers
M	M_1	Passenger car with up to 9 seats including driver's seat
	M_2	Motor bus with > 9 seats and up to 5 t gross weight
	M_2	Motor bus with > 9 seats and up to 5 t gross weight
N	N_1	Heavy goods vehicle up to 3.5 t gross weight
	N_2	Heavy goods vehicle > 3.5 t and up to 12 t gross weight
	N_3	Heavy goods vehicle > 12 t gross weight
O		Trailer and semi-trailer

Specified brake systems (§ 41 StVZO)

Motor vehicles from classes M and N must have two separate brake systems (service brake systems, parking brake systems) or a brake system with two separate control units. Each control unit must be able to function if the other fails.

One of the brake systems must function mechanically and be able to secure the vehicle against rolling away (parking brake system). If more than two wheels can be braked, the same brake areas and mechanical transmission equipment can be used.

Motor vehicles from classes $M_{2/3}$ and $N_{2/3}$ and with a maximum speed of more than 60 km/h determined by the model must be fitted with an ABS system.

Continuous braking action (Directive on Approximation of European Community Laws RREG 71/320 EC)

Motor vehicles from class M_3 with a permissible total weight from 5.5 t (except for city buses) and vehicles from class $N_{2.3}$ with a permissible total weight of more than 9 t must have a continuous braking action (continuous brake) for long downhill gradients. The braking action must be designed such that it limits a fully-laden vehicle that is being driven on a gradient of 7 % for a distance of 6 km to a speed of 30 km/h.

Brake lights (§ 53 StVZO)

The service-brake operation must be made visible by two red brake lights to the rear on class L (v_{max} > 50 km/h), M, N and O motor vehicles. Since 18.3.93 class M_1 vehicles have been allowed to have a third brake light in the centre at the rear. This third brake light is a legal requirement on all vehicles whose first registration was after 1.1.2000.

Inspection of motor vehicles and trailers

(§ 29 StVZO)

Owners of vehicles and trailers must establish within specified intervals and at their own cost whether the motor vehicles comply with the regulations. A distinction can be made here between:

General inspections GI: to check the vehicle's roadworthiness in accordance with § 29 StVZO (appendix VIII).

Safety inspections SI: the chassis and suspension components are subjected to a comprehensive visual, operation and function check (e.g. brakes, steering, tyres).

Table 2: Type and time interval of inspections (extracts)		
Class of vehicle	Time interval GI	Month SI
L	24	–
M_1 M_1 Passenger transportation (e.g. taxi, hire car)	24 (36) 12	– –
M_2, M_3 in the 1st year in the 2nd and 3rd year from the 4th year	12 12 12	– 6 3
N_1 N_2, N_3	24 12	– 6
O to 750 kg O > 750 kg to 3.5 t O > 3.5 t to 10 t O > 10 t	24 (36) 24 12 12	– – – 6
Values in ( ) are those for the initial inspection after the vehicle was first registered.		

Minimum braking § 29 StVZO (GI guide line) **(Table 3).** The minimum braking rate can be calculated from the measured values determined on the brake dynamometers. Formula:

$$z = \frac{\text{Sum of the wheel brake forces}}{\text{Vehicle weight force}} \times 100\,\%$$

Table 3: Minimum braking z in %		
Class of vehicle	Service brake system	Parking brake system
M_1 Passenger car	50	16
M_2, M_3 Motor bus	50	16
N_1 Heavy goods vehicle, with gross weight up to 3.5 t	50	16
N_2, N_3 Heavy goods vehicle > gross weight 3.5 t	45	16

Types of brake system according to energy supply

Muscular energy braking. The braking force is applied by the driver and enhanced by the mechanical and hydraulic transmission ratio.

Assisted braking (power-assisted brake). In addition to muscular-energy braking, the braking force is also enhanced by other energy sources (vacuum pressure, hydraulic-accumulator pressure, compressed air)

Externally-powered brake (compressed-air brake). The driver controls the braking force. The braking energy (compressed air) is not generated by the driver.

Overrun brake. When the tractor vehicle is braked, the trailer comes closer due to its inactive state (overrun). The braking energy is generated on the trailer wheel brake via the towbars.

Types of energy transmission

Mechanical transmission through pedal, lever, towbar and control cables, for example, when applying the parking brake in a car or the overrun brake when towing a trailer.

Hydraulic transmission caused by fluid pressure in the brake line, e.g. in a passenger-car service brake.

Pneumatic transmission caused by compressed air in the brake line, e.g. in commercial vehicles.

Electrical transmission caused by electrical leads, for example, or the magnetic field in an electric retarder, for commercial-vehicle continuous brake systems.

18.10.1 Braking

Braking duration

During braking, the braking action does not take effect until an obstacle has been recognised. The total duration (stopping time t_S) of a braking procedure is determined by the reaction time t_R and the braking time t **(Fig. 1)**.

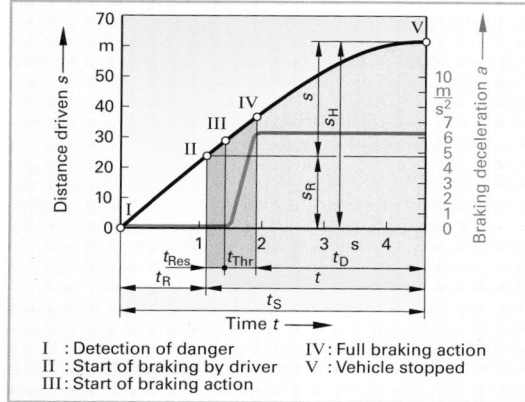

I : Detection of danger   IV : Full braking action
II : Start of braking by driver   V : Vehicle stopped
III : Start of braking action

Fig. 1: Braking

Reaction time t_R. This is the time needed by the driver between recognising a hazard and operating the brake pedal (reaction). The reaction time depends very much on the physical and mental condition of the driver. It can be lengthened signifi-

cantly as a result of alcohol consumption, drug consumption and also tiredness.

Braking time t. The sum of the response, threshold and delay times is known as the braking time t.

Response time t_{Res}. This is generated by the clearance in the brake system, e.g. pedal idle travel, clearances.

Threshold time t_{Thr}. The pressure in the brake system is increased during the threshold time and the desired braking deceleration is achieved.

Delay period t_D. The braking deceleration remains constant until the vehicle comes to a halt.

Braking distance

The braking distance is dependent on the driving speed. Under normal conditions, doubling the speed will increase the braking distance by four times.

Further factors which influence the braking distance are:

- Road conditions, e.g. dry, wet, icy
- Tyre condition, e.g. tread depth, tyre pressure
- The condition of the brakes, e.g. worn, stiff, damaged, corroded
- The condition of the brake pads, e.g. wet, glazed over, oily
- The brake type, e.g. drum or disc brake, compressed-air brake, SBC
- The weight of the vehicle, weight distribution, e.g. when towing a trailer
- The condition of the shock absorbers

18.10.2 Hydraulic brake

Structure

The hydraulic-brake system **(Fig. 2)** consists of the brake pedal, tandem master cylinder with brake booster, wiring system (possibly with brake-pressure reducer), brake cylinder with wheel brakes.

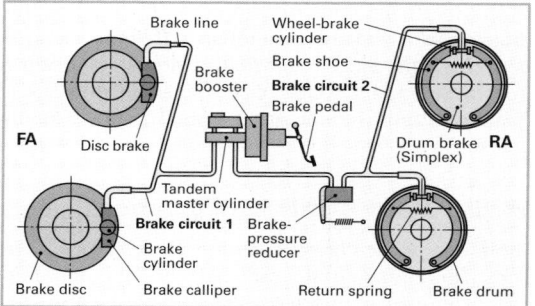

Fig. 2: Hydraulic-brake system

Wheel brakes. Normally, all wheels have disc brakes, older and smaller vehicles have drum brakes on the rear wheels. For reasons of safety, a dual-circuit brake system with tandem master cylinder is a requirement. If a brake circuit fails, it is still possible to brake the vehicle using the other brake circuit.

Operating principle

The operating principle of the hydraulic brake is based on **Pascal's law:**

> The pressure on fluid which is enclosed on all sides acts evenly on all sides.

The force with which the brake pedal presses on the plunger in the master cylinder generates the fluid pressure. The fluid pressure takes effect through the brake lines and generates the application forces (contact pressures).

The hydraulic power transmission normally involves a transmission of force **(Fig. 1)**.

The forces interact like the plunger surfaces, in other words, the strongest force is created on the largest surface. The plunger travels, on the other hand, behave in the opposite way to the forces. So, an actuating force of 1,000 N with a plunger travel of 8 mm on the master cylinder on the four wheel-brake cylinders, for example, produces a total force of 4,000 N and a corresponding plunger travel of 2 mm.

The work performed $(W = F \cdot s)$ is therefore the same on the master cylinder and the wheel-brake cylinders.

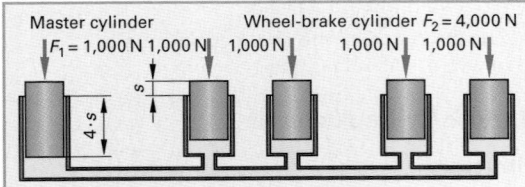

Master cylinder   Wheel-brake cylinder $F_2 = 4,000$ N
$F_1 = 1,000$ N   1,000 N   1,000 N   1,000 N   1,000 N

Fig. 1: Diagram of a hydraulic brake

The hydraulic brake can work at high pressures of up to about 180 bar. This explains the small dimensions of the hydraulic construction components. The hydraulic brake remains maintenance-free for a longer period of time. Since brake fluid is almost impossible to compress and the clearances are small, only small quantities of brake fluid are moved. The pressure increase is very fast and the brakes respond quickly.

18.10.3 Brake-circuit configuration

Hydraulic service brake systems are split into 2 circuits. This means that a sufficient braking action is still produced if one circuit fails. There are 5 designs **(Table 1)**.

Table 1: Brake-circuit configurations

Abbreviation Design		Remark Use
II (TT)		Front/rear axle configuration. Each circuit drives one axle. For rear-wheel drive with ABS. (black-white)
X		Diagonal configuration. Each circuit drives one front wheel and the rear wheel diagonally opposite. All-wheel drive and front drive with ABS and negative kingpin offset.
HI (HT)		One brake circuit drives the front and rear axle, the other drives the front axle only. Seldom used. (4-2)
LL		Each brake circuit drives the front axle and one rear wheel (triangle). Seldom used.
HH		Each brake circuit drives the front and rear axle. Seldom used.

Vehicles with ABS control systems normally use the II (black-white) and X (criss-cross) brake-circuit configurations.

18.10.4. Master cylinder

> Only tandem master cylinders are used, because the law requires that two separate brake circuits are used. This cylinder is operated by the brake pedal via the brake booster.

Its tasks are:

- To achieve a rapid pressure build-up in each brake circuit.
- To achieve a rapid pressure reduction so that the brakes are rapidly released.
- To balance the volume of the brake fluid during a temperature change and when the clearance is increased because the brake pad is worn.

Structure

The tandem master cylinder **(Fig. 1, Page 455)** contains two plungers arranged one behind the other - the push-rod plunger and the intermediate plunger, which is stored in fluid. The plungers form two separate pressure chambers in one housing. Both plungers are designed as double plungers, meaning that there is a ring-shaped castor chamber between the front and rear sealing section of each plunger. This chamber is always filled with brake fluid via the snifter bore. The primary cup seal is located at the front of each plunger and seals the pressure chamber.

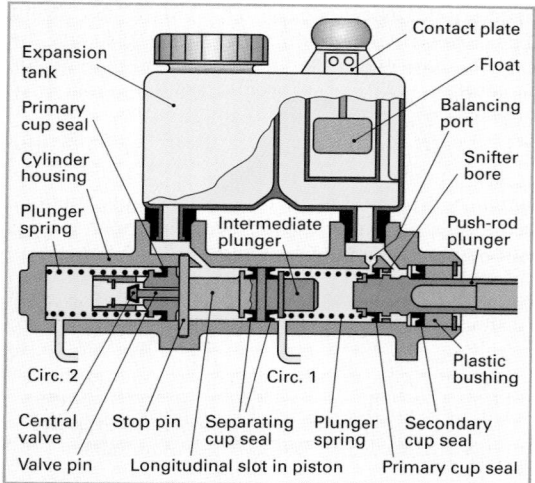

Fig. 1: Tandem master cylinder

The push-rod plunger is sealed at the rear by the secondary cup seal. The separating cup seals the intermediate plunger against the push rod circuit. The intermediate plunger has a slot into which a central bore hole runs at the front. The central valve rests in this bore hole. A stop pin which leads goes through the slot on the intermediate plunger keeps the plunger in the cylinder and forms the front and rear stop.

Central valve. This is used on vehicles with ABS systems and assumes the function of the balancing port. There are also tandem master cylinders which have a central valve on both plungers.

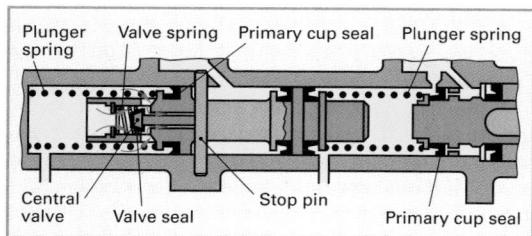

Fig. 2: Rest position

Operating principle

Rest position. The plunger springs press the plungers against their stop. The primary cup seal on the push-rod plunger releases the balancing port and the intermediate plunger is placed at the front of the stop pin. This means that the central valve **(Fig. 2)** is opened by the valve pin which fits into it and assumes the function of the balancing port. Both pressure chambers are now linked to the expansion tank. The volume of the brake fluid can be balanced during a temperature change, for example.

If the balancing port is closed because the push-rod plunger is in the wrong rest position or due to contamination, it will not be possible to balance the brake fluid. The fluid expands due to heat, which then increases or automatically triggers the braking action.

Brake actuation. When the vehicle is braked, the primary cup seal **(Fig. 3)** on the push-rod plunger travels over the balancing port and seals the pressure chamber. The filler shim thus prevents the balancing port from pressing in to the filler bores and from becoming damaged. The intermediate plunger is now somewhat displaced by the brake fluid. The stop pin releases the valve pin and the central valve is closed. Pressure builds up in both brake circuits.

Advantages of the central valve

- The primary cup seal has a longer service life because the sealing lip cannot be damaged by the balancing port.
- In ABS systems, the primary cup seal would be pressed into the balancing port by pressure-peak reverse movements in an ABS control system and would thus be damaged.

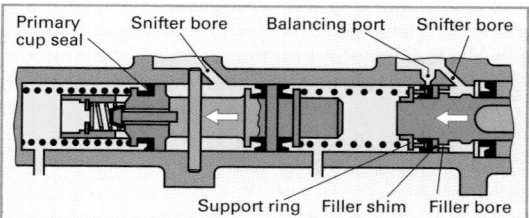

Fig. 3: Brake position

Releasing the brake. The plunger is pushed back up by the fluid pressure and the plunger springs. The primary cup seal on the push-rod plunger folds down, the filler shim lifts up and the brake fluid flows from the castor chamber through the filler bores into the expanding pressure chamber **(Fig. 4)**. The intermediate plunger returns to its original position. The pressure chambers are linked to the expansion tank by the central valve and the balancing port. The pressure falls and the brakes are released.

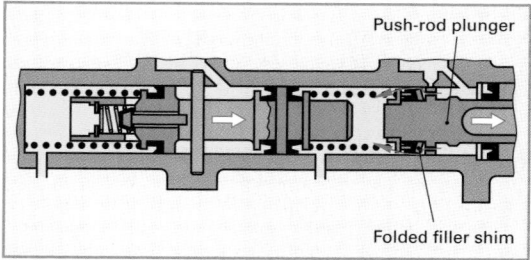

Fig. 4: Release position

Failure of circuit 1 (Fig. 1)

The push-rod plunger is pushed up to the stop on the intermediate plunger. The actuating force now acts directly on the plunger for intact circuit 2, where it generates braking pressure.

Failure of circuit 2 (Fig. 1)

The intermediate plunger is pushed forwards by the fluid pressure in circuit 1 until it meets the stop. It seals the intact circuit 1 to the non-tight circuit 2. The pressure now builds up in circuit 1.

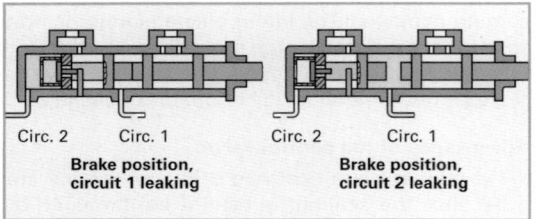

Fig. 1: Failure of a brake circuit

Tiered tandem master cylinder (Fig. 2)

This master cylinder was developed for II systems (TT, black-white) with front axle/rear axle brake-circuit split. The cylinder diameters are graduated, meaning that the diameter of the intermediate plunger, which works on the rear axle brake circuit, is smaller than the diameter of the push-rod plunger. In intact brake circuits, the same pressure is created in both circuits when the vehicle is braked. The larger push-rod plunger diameter in the front axle brake circuit pushes back a greater volume of fluid when the vehicle is braked, causing the brakes to respond faster. If the front axle brake circuit fails, the push-rod plunger is pushed onto the intermediate plunger when the vehicle is braked and the plunger's plunging force now acts directly on the intermediate plunger. The pedal travel is lengthened and a higher pressure is created in the rear axle brake circuit due to the smaller diameter of the intermediate plunger without the pedal force being increased. If the front axle circuit fails, a sufficient braking action is still achieved with the rear axle brakes.

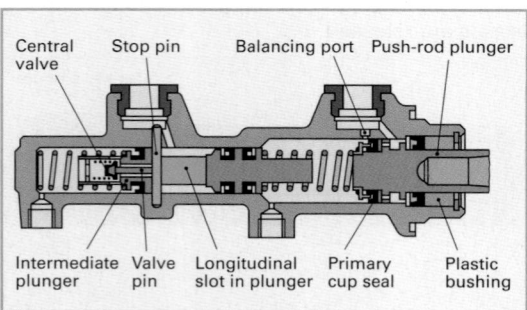

Fig. 2: Tiered tandem master cylinder with central valve

Tandem master cylinder with riveted plunger spring

The screw-riveted compression spring keeps the intermediate plunger and the push-rod plunger the same distance apart when they are in the rest position (Fig. 3).

This causes the pressure to build up evenly in both brake circuits when the brake is actuated. If the braking pressure is increased, the intermediate plunger is no longer moved by the plunger springs but by the brake-fluid pressure.

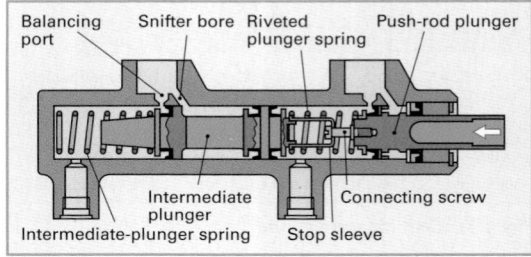

Fig. 3: Tandem-brake master cylinder with riveted plunger spring

18.10.5 Drum brake

Nowadays, drum brakes (Fig. 4) are predominantly used as brakes for rear wheels of passenger cars or in commercial vehicles.

Structure and operating principle

The brake drum fits snugly on the wheel hub. The brake shoes and the components which generate the application force are found on the brake anchor plate. The brake anchor plate is fixed to the wheel suspension. When braking occurs, the brake shoes and their pads are pressed against the brake drum by the clamping fixture, thus generating the friction required. The application force can be created hydraulically by the wheel-brake cylinder (service brake) or mechanically by the control cable and the tension lever, expander lever or brake shoe expander (parking brake).

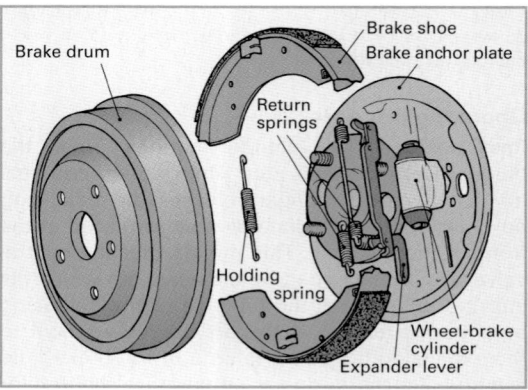

Fig. 4: Parts of the drum brake

Features:
- Self-reinforcement
- Dirt-proof design
- Parking brake easier to use
- Long idle time of brake pads
- Pad replacement and maintenance is costly and time-consuming
- Poor heat dissipation
- Tendency towards fading

Designs

According to the actuation methods and brake-shoe supports, it is possible to distinguish between:
- Simplex brakes
- Duo-servo brakes

Simplex brake (Fig. 2). This brake has **one overrun** and **one trailing brake shoe.** To tension the brake shoes, a **double-acting wheel-brake cylinder,** brake shoe expander, S cam, expanding wedge or expander lever can be used. Each brake shoe has a fixed pivot or fulcrum point, such as a support bearing.

Simplex brakes have the same effect when driving forwards as they do when reversing but have only reduced self-reinforcement **(Fig. 1).** The pad wear on the overrun brake shoe is greater. A parking brake is easy to use.

Duo-servo brake (Fig. 3). The self-reinforcement of the overrun brake shoe is used to press down the second overrun brake shoe. The **support bearing** is **floating.** The support is provided by the **double-acting wheel-brake cylinder.** The braking action is the same when driving forwards or reversing. It is often used as a parking brake in cup washers **(Fig. 6).** A control-cable-actuated brake-shoe expander is then used in place of a wheel-brake cylinder.

> **Self-reinforcement (Fig. 4).** The friction creates torque which pulls the overrun brake shoe into the drum and strengthens the braking effect. This reinforcement is expressed by the brake coefficient C **(Fig. 1).** The pressing force on the trailing brake shoe is then reduced.

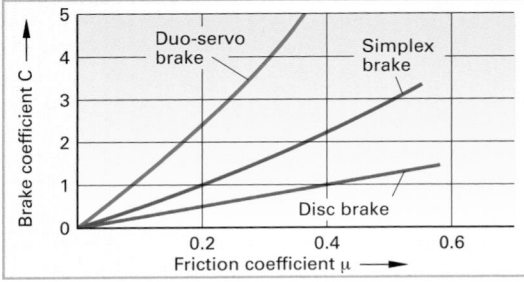

Fig. 1: Brake coefficient C

> **Fading.** This is an abatement of the braking effect caused by overheating, e.g. during long braking. The friction coefficient in the pad decreases at high temperatures or high sliding speeds. The brake drum can also become deformed to a conical shape, because the heat supply to the wheel hub is more efficiently carried off. The brake area then becomes smaller.

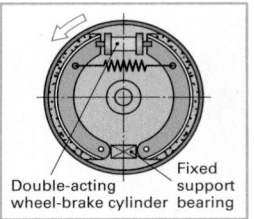

Double-acting
wheel-brake cylinder
Fixed support bearing

Fig. 2: Simplex brake

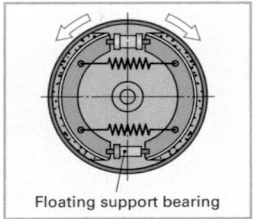

Floating support bearing

Fig. 3: Duo-servo brake

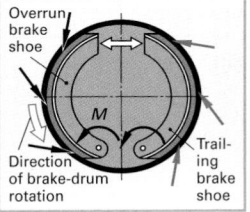

Overrun brake shoe
Direction of brake-drum rotation
Trailing brake shoe

Fig. 4: Self-reinforcement of the drum brake

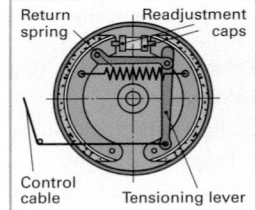

Return spring
Readjustment caps
Control cable
Tensioning lever

Fig. 5: Clamping fixture for parking brake

Clamping fixtures

These are intended to tension or expand the brake shoes and press them onto the brake drum. Wheel-brake cylinders are normally used with hydraulic brakes **(Fig. 1, Page 458).** With mechanically operated parking brakes, a tensioning lever **(Fig. 5)** or a brake-shoe expander **(Fig. 6)** is used.

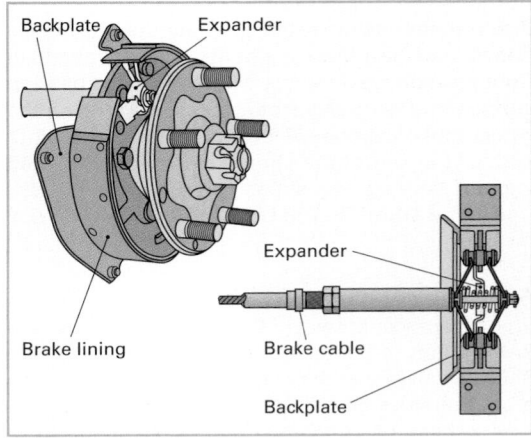

Backplate
Expander
Expander
Brake lining
Brake cable
Backplate

Fig. 6: Parking brake integrated into the cup washer

Wheel-brake cylinder

In the double-acting wheel-brake cylinder (**Fig. 1**), the pressure generated in the master cylinder acts on the plungers and generates an application force. The plungers are sealed by rubber sleeves. Dust caps prevent dirt from entering. On the back of the wheel-brake cylinder are threaded bore holes which fasten it to the brake anchor plate and the brake line connection. A bleeder valve is screwed in at the highest point.

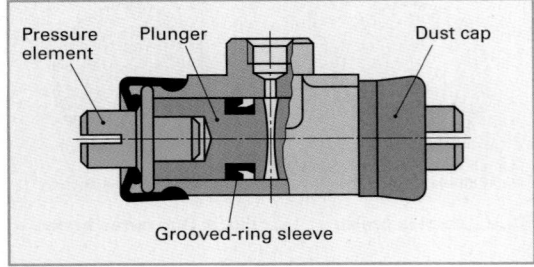

Fig. 1: Double-acting wheel-brake cylinder

Brake drum (Fig. 4, Page 456)

Features:

- High wear resistance
- Inherent stability
- Good heat conductivity

Substances:

- Cast iron with flake graphite
- Malleable cast iron
- Cast iron with nodular graphite
- Cast steel
- Combined casting of light alloy and cast iron

The brake drum must run centrally and free from runout. The brake area is finely spun or ground.

Brake shoes (Fig. 4, Page 456)

Brake shoes maintain their rigidity due to a T-section and are cast from a light metal alloy or welded from pressed steel. At one end they have a bearing surface for the mostly slotted pressure pins on the wheel-brake cylinder. A bolt is fitted at the other end, or the end of the shoe is flush with the fixed support bearing. The shoes can therefore be centred in the drum. They fit better and the pad wear is more even.

Adjusting components

The clearance between the brake pad and the brake drum is increased by the brake pad wear. This also increases the pedal idle travel. The brakes must therefore be adjusted on a regular basis, either by hand or using an automatic adjusting component.

REVIEW QUESTIONS

1 What are brakes for?
2 Which types of brake systems can be distinguished according to their method of use?
3 Explain the structure of a hydraulic brake system.
4 Which brake systems are specified in vehicle classes M and N?
5 How can brake systems be distinguished according to their mode of operation?
6 What are the functions of the master cylinder?
7 How does the primary cup seal work?
8 What are the functions of the central valve?
9 How does the tandem master cylinder work when one of the brake circuits fails?
10 What is the advantage of tiered master cylinders?
11 Which brake-circuit configurations are there?
12 What are the features of drum brakes?
13 List the distinguishing features of different types of drum brakes.
14 What does "fading" mean?

18.10.6 Disc brake

Disc brakes are designed as a **fixed-calliper** or **floating-calliper brake** (**Fig. 1**). The brake plungers are located in the brake calliper. They press the pads against the brake disc when the vehicle is braked.

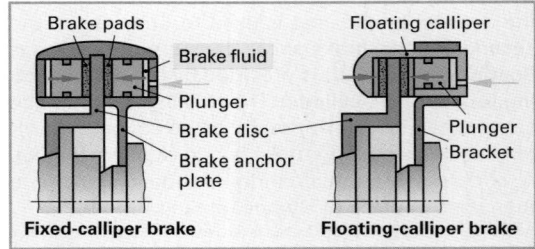

Fig. 2: Disc brakes

Features:

- No self-reinforcement due to the even brake areas. This requires greater downforces and therefore brake cylinders whose diameters (40 mm to 50 mm) are larger than the diameters of the wheel-brake cylinders in the drum brake and additional brake boosters are required.
- Good metering of the braking force, because the absence of self-reinforcement and the minor changes in friction coefficient ensure that hardly any fluctuations occur in the braking.
- Efficient cooling.
- Low tendency towards fading.
- Higher brake-pad wear due to the high downforces.

- Easy maintenance and pad replacement.
- Automatic adjustment of clearance.
- More heat generated by the brake fluid, because the pads fit tightly on the brake plungers. Danger of vapour bubbles.
- Good automatic cleaning due to centrifugal force.
- Tendency of vapour-bubble formation because the brake plungers fit tightly against the brake pad.
- The parking brake requires great effort.

Designs

Fixed-calliper disc brake. Two- and four-cylinder fixed-calliper disc brakes are normally used **(Fig. 1)**.

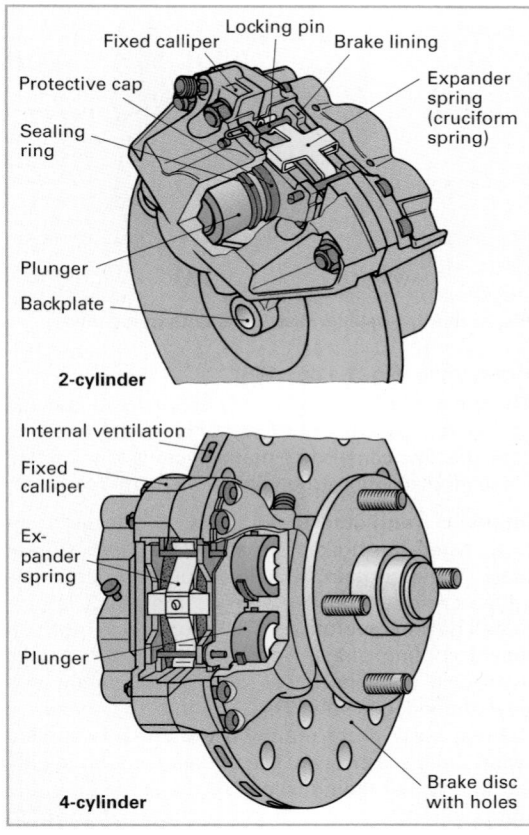

Fig. 1: Fixed-calliper disc brake

The fixed brake-cylinder backplate (fixed calliper) is bolted onto the wheel suspension. This backplate grips the brake disc like pliers. It consists of one two-piece housing. Each housing section contains brake cylinders which are situated opposite each other in pairs. They contain the brake plungers with sealing ring, protective cap and clamping ring. The brake cylinders are linked by channels. The bleeder valve sits on top of the housing.

When the vehicle is braked, the brake-cylinder plungers press against the brake pads. The brake pads are then pushed against the brake disc on both sides.

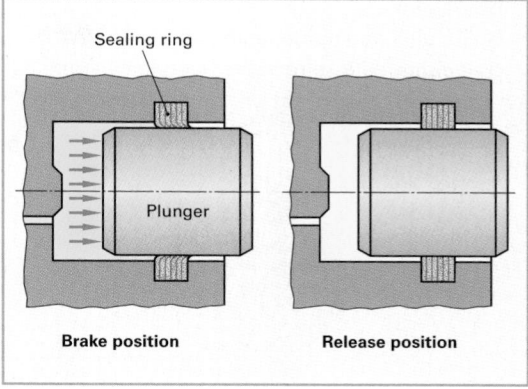

Fig. 2: Plunger reset

Plunger reset (Fig. 2)

A rectangular rubber sealing ring used to seal the plunger is located in a groove in the brake cylinder. The inner diameter of the sealing ring is somewhat smaller than the plunger diameter. It therefore encompasses the plunger with its pretension.

The braking movement of the plunger deforms the sealing ring elastically due to its static friction and the plunger stroke. When the pressure drops in the brake fluid, the sealing ring returns to its starting shape or position. This also removes the plunger from the clearance of about 0.15 mm and releases the brake disc. This is only possible with complete pressure reduction in the wire system and ease of movement of the plunger and pads.

Expander spring. It fits the brake pads onto the plungers and thus prevents the pads from knocking and chattering.

Floating-calliper disc brake (Fig. 1, Page 460)

This consists of two main components, the bracket and the housing or floating calliper and has the following features:

- Low weight
- Small size
- Good heat dissipation
- Large pad surfaces
- Takes up less space.
- Reduced tendency towards vapour-bubble formation, as only one or two of the brake cylinders are on the bracket side.
- Maintenance-free housing versions, therefore not sensitive to dirt and corrosion.

Bracket. The bracket is fixed to the wheel suspension. The housing is fitted within the bracket. Floating-calliper disc brakes with various guides are used, such as:

- Guide teeth
- Guide pins
- Guide pins and guide teeth combined
- Guide pins with retractable floating calliper

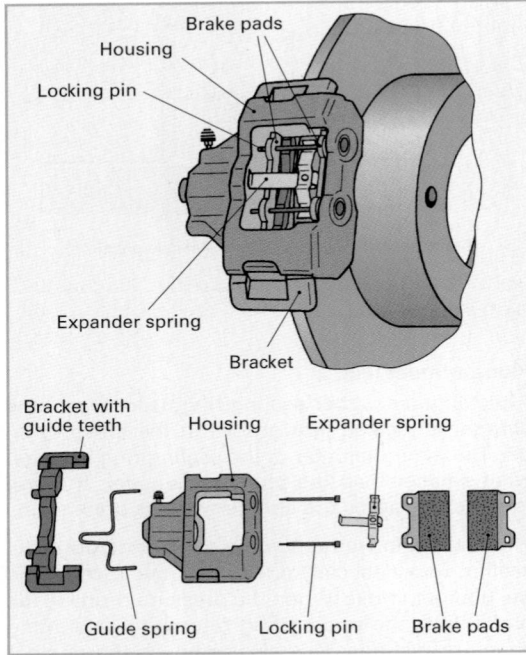

Fig. 1: Floating-calliper disc brake with guide teeth

Floating-calliper disc brake with guide teeth (Fig. 1)

Bracket. The bracket has two teeth on each side.

Housing. The housing is kept in the bracket by the guide teeth which fit into its semicircular grooves, thus enabling it to slide back and forth.

Guide spring. The guide spring presses the housing onto the bracket teeth to prevent clattering noises from occurring.

Floating-calliper disc brake with guide pins (Fig. 2)

This brake has two guide pins bolted onto the housing on the cylinder side of the bracket. The bracket has two bore holes which contain sliding inserts made of Teflon, for example. The housing is kept in these bore holes by the guide pins and can slide back and forth.

Braking. The plunger in the housing presses the inner brake pad against the brake disc once the clearance has been overcome. The reaction force pushes the housing in the opposite direction. The plunger in the housing now also presses the outer brake pad against the brake disc once the additional clearance has been overcome. Both brake pads are pushed against the brake disc with the same amount of force.

The guide teeth support the inner pad directly, the outer pad is supported against the housing by the peripheral force.

If guide pins are used, both brake pads are supported on the housing. When the brake is released, the return forces of the sealing ring restore the clearance, with the support of the expander spring.

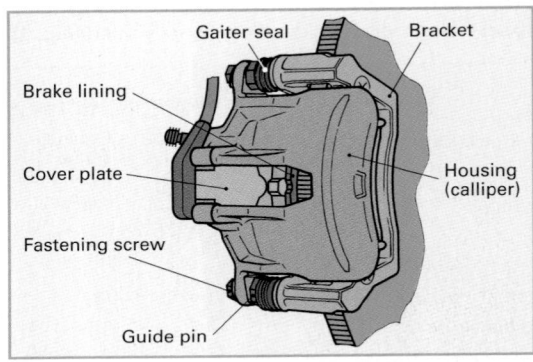

Fig. 2: Floating-calliper disc brake with guide pins

Brake disc (Fig. 1, Page 459)

The brake disc is normally disc-shaped and is made of cast iron, malleable cast iron or cast steel. In racing cars, this disc can also be made of composite materials reinforced with carbon fibres or ceramic carbon.

Internally ventilated brake discs. These discs are used when the brakes are subject to a very great load. They contain radially mounted air ducts which are designed such that a fan effect is produced during revolutions. This produces a more efficient cooling effect. Sometimes, the brake area even also contains bore holes and possibly also oval-shaped grooves. This ensures that water is drained away more rapidly if the brake is applied when the discs are wet. The brakes respond evenly and the risk of fading is low. At the same time, the bore holes also bring about a reduction in weight.

18.10.7 Brake pads

The friction material uses the braking force to generate considerable friction with the brake disc or brake drum. This then converts the kinetic energy generated by the vehicle into heat. With drum brakes, the brake pad is riveted or adhered to the brake shoe. On disc brakes, the pad is adhered to the steel brake-pad backplate.

Electric contacts can be incorporated into disc brake pads for a wear indicator.

Requirements of the friction lining:

- Very stable at high temperatures, considerable mechanical strength and long operating life.
- Constantly high coefficient of friction even at high temperatures and sliding speeds.
- Not sensitive to water and dirt.
- No glazing at high thermal load, good heat conduction.

Brake lining materials. Brake linings contain, for example:

- Metals such as steel wool and copper powder.
- Filler materials such as iron oxide, barite, mica powder and aluminium oxide.
- Anti-friction agents such as coke dust, antimony sulphide and graphite.
- Organic substances such as resin filler material, aramid fibres and binding resin.

Brake pads have a friction coefficient of approximately $\mu = 0.4$. They are heat-resistant to approximately 800 °C.

18.10.8 Diagnosis and maintenance of the hydraulic brake system

Visual check. Monitoring of the brake-fluid level in the expansion tank; looking for damp, dark patches on the brake cylinders and connection points and for corrosion on the brake lines/state of the brake hoses (chafe marks, bubbles, animal bites).

Functional test. This includes checking the pedal travel by activating the service brake system. If the pedal travel slowly increases, this may be a result of a leaking primary cup seal or a leaking central valve. If the pedal travel is too great or it is only possible to build up the pressure by pumping, the cause could be air bubbles or the clearance being too great.

Leak tests (Fig.1). A pressure-tester tool and a pedal holder are required. Before the tests, the brake system and the pressure tester tool filled with brake fluid must be bled.

Low-pressure test. The low/high-pressure manometer combination of the pressure-tester tool is connected to the bleeder valve on a wheel brake and the pedal holder used to apply a pressure of between 2 bar and 5 bar. This pressure should be maintained for 5 minutes. The entire system should be left untouched for this time. If the pressure falls, there is a leak.

High-pressure test. Using the pedal holder, the braking pressure is set to a value of between 50 bar and 100 bar. Within 10 minutes, this set pressure may drop by a maximum of 10 %. If there is a larger pressure drop, this means there is a leak.

Filling and bleeding the brake system (Fig. 2). This work can be carried out by one person with filler and bleeder apparatus. A bleeder pipe and a transparent bleeder hose with collector are the tools required. For vehicles with ABS, observe the brake bleeding instructions.

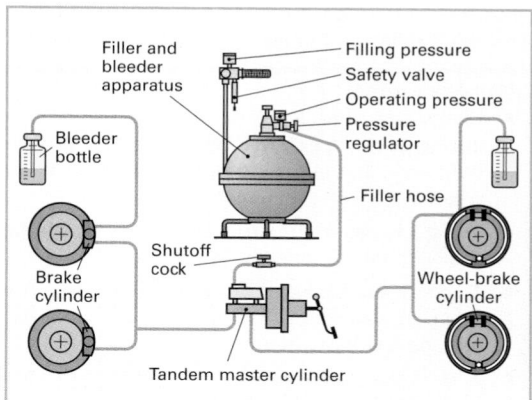

Fig. 2: Bleeding with apparatus

Connect the filler and bleeder apparatus to the bleeder pipe on the expansion tank and attach the bleeder hose with collector to a bleeder valve. Now open the shutoff cock on the filler hose of the equipment and then open the bleeder valve, until new, clear brake fluid flows out without bubbles. Then close the bleeder valve. Repeat the process for all bleeder valves. Finally, close the shutoff cock. Before removing the bleeder pipe, open a bleeder valve briefly and release the pressure.

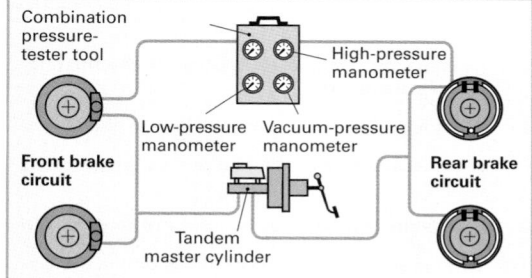

Fig. 1: Leak test

18

Work on the wheel brakes

Brake drums and brake discs. During a brake check, you must check these for ridges, out-of-roundness and knock. Brake discs with lateral runout that is too great must be replaced. The disc-brake pads, sliding calliper and floating calliper must move smoothly. Brake drums and brake discs that are out-of-round or have ridges must be skimmed or turned down. Observe the maximum skimming measurement or the minimum disc thickness, the brake discs and/or brake drums may need replacing. Brake drums or brake discs with cracks and/or damaged brake callipers must be replaced.

Brake pads. The thickness and oiling must be checked and the pads must be replaced if necessary.

Brake test

The brake tests are mainly carried out on brake dynamometers.

The following are measured for each wheel:
- Braking force
- Rolling resistance
- Fluctuation of the braking force, e.g. in the case of an out-of-round drum
- Occurrence of incipient lock

Dynamic brake analyser (Fig. 1). This has two identical sets of rollers so that the brakes for both wheels on an axle can be tested at the same time. These each drive one braked wheel during the test. The drive rollers on one side are driven together. The third roller is a sensor roller. It automatically

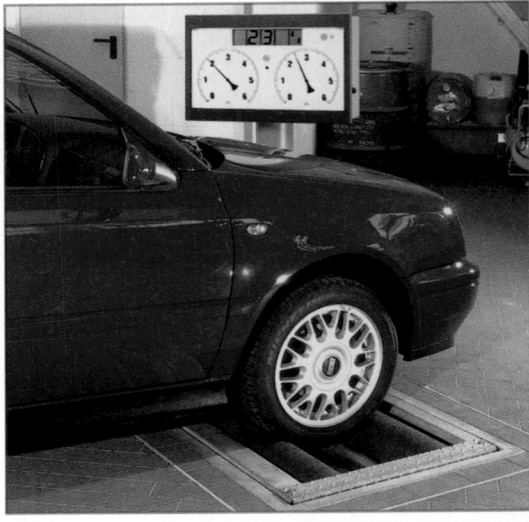

Fig. 1: Dynamic brake analyser

activates the dynamometer and the locking protection. The braking force (peripheral force) of every wheel is measured.

The **braking factor z** is mainly determined **as a percentage (see Page 452).** The brake-force differential (for the service brake system) of an axle must not be greater than 25%. Special test instructions must be observed for motor vehicles with permanent all-wheel drive and variable engine torque distribution.

WORKSHOP NOTES
- Check the fluid level in the expansion tank during each check. On disc brakes, the sunken fluid level can be a sign of considerable lining wear.
- The thickness of the drum brake linings can be checked using inspection holes.
- To check the brake drum, remove it; clean the brake of wear debris. Vacuum off the wear debris, do not blow it out.
- The brake pads must be replaced on the brakes of one axle at the same time.
- Renew the brake fluid in accordance with company regulations, e.g. yearly.
- Do not use drained brake fluid; store it in marked containers and have it disposed of by a disposal or reprocessing company.
- Keep greases and oils away from the brake components.
- Only use the specified brake fluid to refill.
- Only use brake cleaner, possibly alcohol (methylated spirits) for cleaning.

REVIEW QUESTIONS

1 What are the features of disc brakes?

2 How is a fixed-calliper disc brake constructed?

3 How is the clearance set for disc brakes?

4 What are the different types of floating-calliper disc brakes, as distinguished by the calliper guide?

5 How is the braking factor determined?

6 Describe the braking procedure for the floating-calliper disc brake.

7 What requirements are made of brake pads?

8 What inspections are carried out on the hydraulic brake?

9 How is the hydraulic brake filled and bled?

10 What measurements can be taken on the dynamic brake analyser?

18.10.9 Power-assisted brake

> To create the assistance (power-assistance), a vacuum pressure or hydraulic brake booster is connected to the master cylinder of the hydraulic brake.

Vacuum brake booster

For motor vehicles with a spark-ignition engine, the vacuum pressure can generally be taken from the induction pipe. The small pressure difference between the air pressure and the intake manifold pressure of approximately 0.8 bar requires the working plunger to have large surfaces, so that the plunger rod force can be increased fourfold, for example.

For diesel engines, the pressure difference is generated by a vacuum pump driven by the engine.

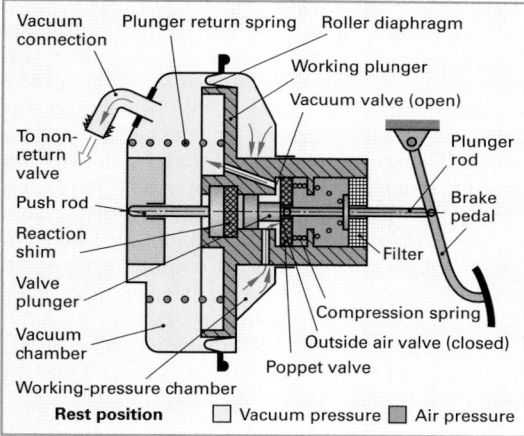

Fig. 1: Vacuum brake booster

Structure (Fig. 1). The master cylinder is usually

flange-mounted to the reinforcement housing. The working plunger divides the housing into a vacuum chamber and working chamber. The working chamber is connected alternately via a vacuum pressure and outside air valve with the outside air or with the vacuum chamber. The double valve is actuated by the brake pedal via the plunger rod. This plunger rod presses on the master cylinder push rod via valve plungers and the rubber reaction shim. The working plunger and its boosting force also presses on the push rod.

Operating principle

Release position (Fig. 1). The outside air valve is closed, the working chamber is connected via the open vacuum valve to the vacuum chamber. Both sides of the working plunger have the same pressure of approximately $p_{abs} = 0.2$ bar.

Partially braked position (Fig. 2). During braking, the push rod is moved forwards and the vacuum valve is closed. The reaction shim is squeezed by the valve plunger and the outside air valve is opened. The pressure difference which arises in the working chamber compared to the vacuum chamber has the effect of a boosting force on the working plunger. This is pushed forwards with the timing case and push rod until the reaction force from the master cylinder is equal. When the push rod is still, the reaction shim expands again and presses on the valve plunger. This closes the outside air valve. The booster force on the working plunger and push rod remains constant.

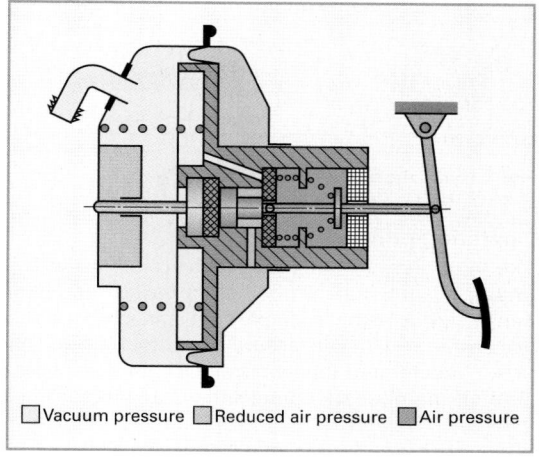

Fig. 2: Partially braked position

Full braking (Fig. 3). At full pedal force, the reaction shim is constantly being squeezed by the plunger rod and the counterforce from the push rod, whereby the outside air valve is constantly open. The pressure difference ($\Delta p = 0.8$ bar) between the two chambers is the largest possible and the largest booster force is therefore exerted on the working plunger and push rod.

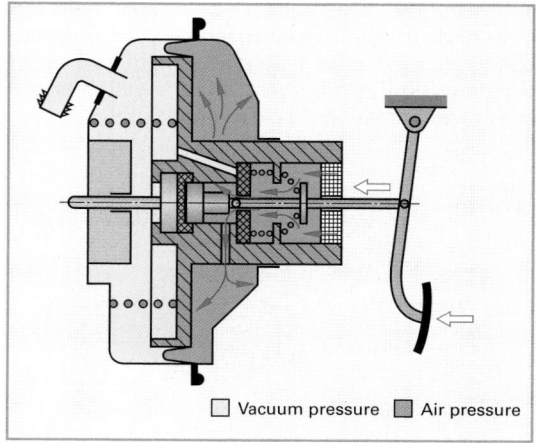

Fig. 3: Fully braked position

18

Hydraulic brake booster (Fig. 2)

The system (**Fig. 1**) consists of the high-pressure oil pump for the power steering, the hydraulic accumulator, the pressure-regulated oil-flow controller and the hydraulic brake booster with tandem master cylinder and the oil supply reservoir.

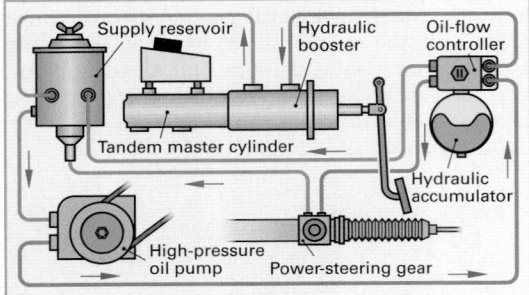

Fig. 1: Hydraulic brake-booster system

Operating principle

The high-pressure oil pump delivers oil to the hydraulic accumulator. The oil compresses the nitrogen inside it using a diaphragm and charges the accumulator with a pressure of up to 150 bar. The brake booster and the hydraulic-accumulator pressure-oil chamber are connected via an electric line.

Brake position. By applying the brake, the control plunger (**Fig. 2**) of the brake booster is moved. It closes the return passage and opens the inlet passage. The working chamber is supplied with pressure oil and assists the working plunger. The moving working plunger closes the inlet passage thus enabling a variable boost depending on the pedal force.

Release position. When the pedal force is released, the control plunger closes the inlet passage and opens the return passage. The hydraulic fluid can flow back to the supply reservoir. The resetting spring pushes the working plunger into its original position. If the engine fails, there is still pressure oil for approximately 10 brake applications.

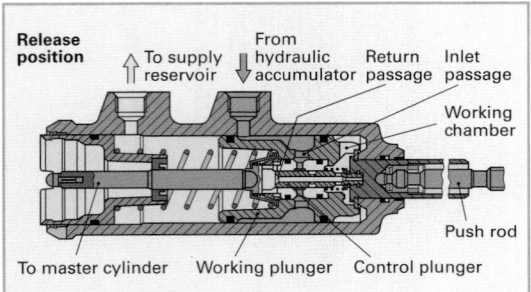

Fig. 2: Hydraulic brake booster

Pneumatic brake booster

Pneumatic brake boosters (**Fig. 3**) can be fitted to vehicles with a combined compressed-air/hydraulic brake system. With an operating pressure of approximately 7 bar, great booster forces can be achieved on small vehicles.

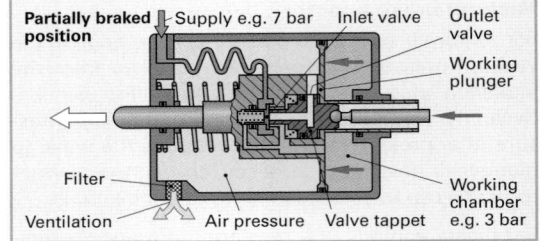

Fig. 3: Pneumatic brake booster

Function. When the brakes are applied, the valve tappet is moved by the plunger rod. The valve tappet comes into contact with the valve plate and therefore closes the outlet valve. Simultaneously, the valve plate is raised and the inlet valve therefore opens. The supply pressure surges into the working chamber and has the effect of a booster force on the working plunger. The moving working plunger closes the inlet valve again. This results in a variable booster force which is directly related to the pedal force. When the pedal force is released, the valve tappet closes the inlet valve and opens the outlet valve. The pressure in the working chamber is discharged and the working plunger is moved back to its original position by the resetting spring.

18.10.10 Braking-force distribution

The axle load displacement that occurs during braking depends on the level of braking deceleration, the load, the load distribution on the vehicle and the height of its centre of gravity. If the brakes are applied when the vehicle is being driven in a straight line, the front wheels are under load and the rear wheels are relieved. If the brakes are applied when the vehicle is cornering, the wheels on the outside of the bend are subjected to an additional load. The brakes are usually designed in such a way to provide optimum effect at medium deceleration and medium load. When braking sharply, however, the rear wheels may lock and the vehicle could skid. Brake-pressure reducers reduce this danger and are used on vehicles without ABS.

Brake-pressure reducer (Fig. 1, Page 465). This controls the braking pressure of the rear wheels in the brake line. They are braked with only slightly increased pressure as of a certain changeover pressure.

Pressure characteristic in the brake system without braking-pressure control. The blue line shows the routing of the braking pressures during actual braking. The same braking pressure is exerted on the front and rear axles until the changeover point (e.g. 40 bar). After the changeover point, further increase of braking pressure on the rear axle is reduced. The rear axle is prevented from locking.

Optimum braking is when the braking pressure on the rear axle increases further at the start of braking than it does on the front axle. This is shown in **Fig. 1** for a laden and an unladen vehicle. When the vehicle is laden, the wheel contact forces are greater and therefore enable stronger braking forces, generated by the higher braking pressures in the wheel brake cylinders.

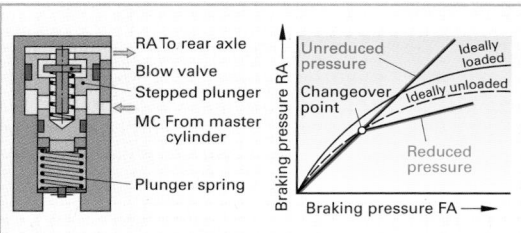

Fig. 1: Brake-pressure reducer

Load-sensitive brake-pressure reducer (Fig. 2). This has the same effect as the normal brake-pressure reducer, but after the changeover point, the braking pressure during braking is controlled according to the load and axle-load shift.

The braking pressure within the control range is always adjusted to the ideal pressure with the load-sensitive shift of the changeover pressure.

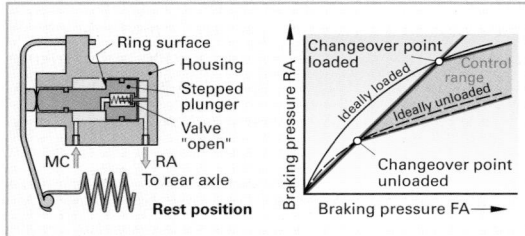

Fig. 2: Load-sensitive brake-pressure reducer

18.10.11 Mechanically operated brake

Mechanically operated brakes are often only still used as the parking brake in vehicles with a hydraulic service-brake system and as the service brake in light motorcycles and single-axle trailers.

The efficiency of the mechanical load transmission is low (according to the maintenance status, only ~50%). In winter, load transmission components may freeze together in wet weather or frost.

Brake cables. These are steel cables which are routed over rollers in pipes or flexible metal hoses (Bowden cables). To reduce friction and protect against icing and corrosion, these are coated with plastic. Tensioning bolts are attached to adjust the brake cables.

Brake compensator (Fig. 3). This is required so that the same forces are exerted on the wheels of one axle.

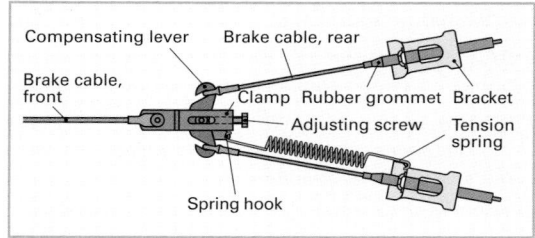

Fig. 3: Brake compensator

Overrun brake (Fig. 4). This is used for trailers. When the tractor vehicle is braked, the trailer runs onto the tractor vehicle. The shear force of the trailer causes the pull rod to be pushed against a compression spring. The movement caused is created by a reversing lever and a control cable on an expander which creates an application force on the brake.

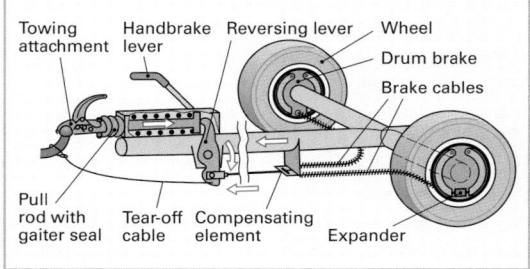

Fig. 4: Overrun brake

REVIEW QUESTIONS

1 What types of brake booster are used in hydraulic brakes?

2 What pressure difference is used in the vacuum brake booster?

3 How does the vacuum brake booster work in emergency braking?

4 What are the components that make up a hydraulic brake-booster system?

5 What is understood by dynamic axle-load displacement when braking?

6 What are the basic types of brake-pressure reducer used?

7 How does the brake-pressure reducer work?

18

18.10.12 Basics of the electronic chassis control systems

> Electronic control systems should guarantee safe control of a motor vehicle during braking, accelerating and steering.

The following control systems are used:

- **ABS** (Antilock-Braking System), prevents wheel locking during braking.
- **BAS** (Brake Assistant), detects emergency situations and brings about shorter braking distances.
- **SBC** (Sensotronic Brake control), reduces braking distances and increases the directional stability when braking in bends.
- **TCS** or **ASC** (Acceleration Skid Control), **ELSD** (Electronic Limited-Slip Differential), prevents wheel spinning when pulling away and accelerating.
- **VDC** (Vehicle Dynamics Controller such as **ESP** or **DSC),** prevents the vehicle from skidding.

Every vehicle movement or change in movement can only be achieved by forces on the wheels. These are:

- Peripheral force as motive or braking force. This acts on the longitudinal direction of the tyre.
- Lateral force, e.g. caused by steering or external interferences such as crosswind.
- Normal force caused by vehicle weight. This acts at right angles to the road surface.

The strength of these forces depends on the road surface, tyre condition/type and weather influences.

The possible load transmission between the tyres and road surface is determined by the friction force. Optimum transmission of the loads can only occur as a result of static friction between the tyres and road surface. The electronic control systems utilise the static friction optimally.

The peripheral force is transferred via static friction as a motive (F_M) or braking force (F_B) to the road surface.
Its size is equal to the normal force F_N multiplied by the coefficient of friction μ_F ($\mu_{Ice} = 0.1$ to $\mu_{Dry} = 0.9$).

$F_{M, B} = \mu_F \cdot F_N$	$F_{M, B}$	Motive force, braking force
	F_N	Normal force
	μ_F	Coefficient of friction

The coefficient of friction μ_F (grip value) is determined by:

- Material pairing, tyres and road surface
- Occurring weather influences

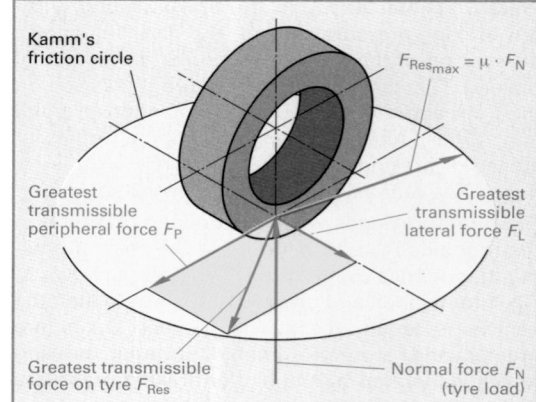

Fig. 1: Forces on the wheel, Kamm's friction circle

Kamm's friction circle (Fig. 1). The largest force transferable onto the road ($F_{max} = F_N \cdot \mu_F$) is shown as a circle. For a stable driving condition, the resulting F_{Res} of peripheral force F_P and lateral force F_L must lie within the circle and therefore be smaller than F_{max}.

If the **peripheral force F_P** reaches its maximum as a result of spinning or locked wheels, no lateral force F_L can be transferred. The vehicle can then no longer be steered.

If when cornering at maximum cornering speed the **lateral force F_L** is at its maximum, the vehicle cannot be braked or accelerated, as it would otherwise break away at the rear.

Slip (Fig. 2). While a tyre rolls, elastic deformations and sliding occur. If, for example, a braked wheel with a rolling circumference of 2 m covers a distance of only 1.8 m during a turn, the travel difference between the tyre circumference and braking distance is 0.2 m. This corresponds to a slip of 10 %.

If a wheel locks or spins when it is being driven, there is a slip of 100 %.

> A slip-free transmission of force between the tyres and the road surface is not possible, because the tyre is not interlocked with the road surface and always slides a little when driving or braking.

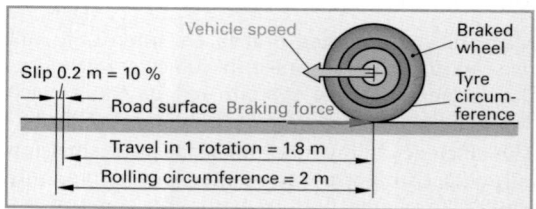

Fig. 2: Slip on the braked wheel

Relationship of forces on the wheel and slip

The relationship between motive force, braking force, lateral force and slip when driving straight ahead is shown in simplified form in **Fig. 1**. Even at low-slip values, the braking force increases steeply to its highest level. It then falls again a little when the slip values increase further. The routing and highest value of the motive/braking force curve depend on the friction coefficient of the tyre on the road surface. The highest value lies between 8 % and 35 %. The first area of the curve is called the stable area because the wheel remains stable for driving and steerable. This is where the wheel has the best force transmission. Electronic control systems therefore work in this control range. For large slip values, the lateral force decreases significantly, the vehicle can no longer be steered and the driving characteristics become unstable. The control systems in the vehicle make sure that the stable area is kept to.

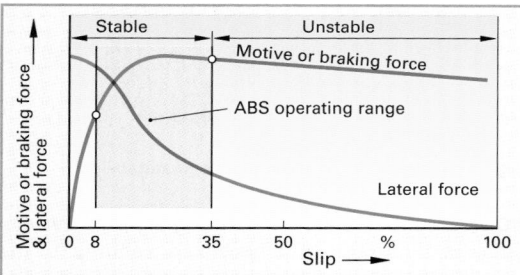

Fig. 1: Forces on the wheel subject to slip

18.10.13 Antilock-braking system (ABS)

Antilock-braking systems (ABS), also known as anti-skip system, are used in hydraulic brake systems and air-brake systems for brake-pressure control.

> During braking, ABS systems control the braking pressure of a wheel according to its grip on the road surface in order to prevent wheel locking.
>
> Only moving wheels can be steered and can transfer lateral forces.

Structure

An ABS consists of the following components:
- Wheel sensors with pulse rings
- ECU
- Hydraulic modulator with solenoid valves

The solenoid valves are selected by the ECU in three control phases; **pressure build-up, pressure holding** and **pressure reduction**. They prevent the wheels from locking.

ABS systems have the following features:
- Lateral forces and directional stability remain the same, whereas the risk of skidding is reduced.
- Vehicle is still steerable and obstacles can thus be avoided.
- An optimum braking distance can be achieved on normal road conditions (no gravel, snow).
- "Flat spots" on tyres are prevented as the wheels do not lock.

Antilock-braking systems. They can be differentiated between according to the **number of control channels or sensors,** and according to the **type of control** in

4-channel system with 4 sensors and X (diagonal) or II (black-white, TT) brake-circuit configuration. Each wheel is controlled individually.

3-channel system with 3 or 4 sensors and X (diagonal) brake-circuit configuration. The front wheels are controlled individually and the rear wheels always together.

Individual Control (IC). The greatest possible braking pressure for each wheel is adjusted here. This means that the braking force is at its maximum. Because the wheels of an axle can be braked with varying forces, e.g. due to a road surface that is icy on one side, there is vehicle torque on the vertical axis (yaw moment).

Select-Low Control (SLC). With SLC, the wheel determines the common braking pressure of an axle with the low road-surface adhesion. The yaw moment when braking on road surfaces with varying road-surface adhesion is lower, because the braking forces on the rear wheels are approximately the same.

> The front wheels are generally controlled individually and the rear wheels are often controlled according to the Select-Low principle.

Operating principle

The brakes are mainly applied when there is low slip. The ABS does not therefore take effect. The ABS closed-loop control circuit **(Table 1, Page 468)** is only activated and wheel locking prevented during sharp braking and when there is significant slip. The ABS control range lies between 8 % ... 35 % slip. Below approximately 6 km/h, ABS is generally deactivated so that the vehicle comes to a stop.

There is a toothed pulse ring around each wheel which creates alternating voltage by induction in a speed sensor. The frequency of the alternating voltage is a measurement for the wheel speed. The ECU can therefore determine the acceleration or deceleration for each wheel.

18

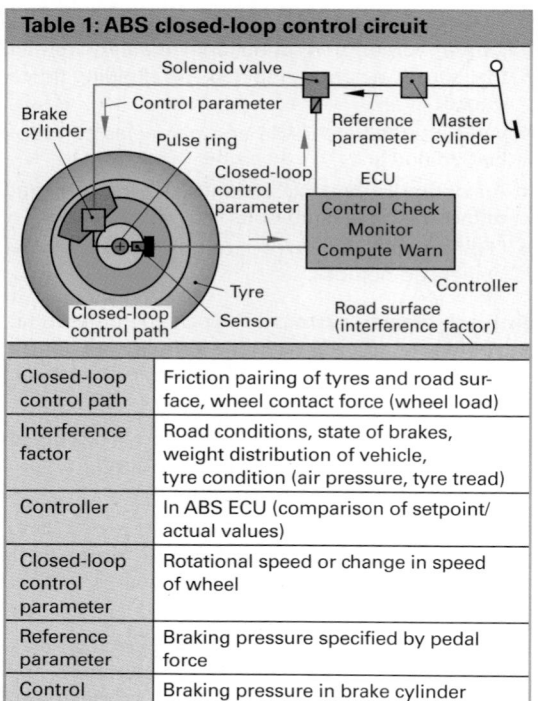

Table 1: ABS closed-loop control circuit

Closed-loop control path	Friction pairing of tyres and road surface, wheel contact force (wheel load)
Interference factor	Road conditions, state of brakes, weight distribution of vehicle, tyre condition (air pressure, tyre tread)
Controller	In ABS ECU (comparison of setpoint/actual values)
Closed-loop control parameter	Rotational speed or change in speed of wheel
Reference parameter	Braking pressure specified by pedal force
Control parameter	Braking pressure in brake cylinder

Pressure build-up. The pressure created in the master cylinder is transferred to the wheel-brake cylinder.

Pressure holding. If a wheel tends to lock during braking and exceeds a predefined slip, this is detected by the ECU. It switches the solenoid valve of the wheel to pressure holding. The connection master cylinder – wheel-brake cylinder is interrupted. The braking pressure remains the same.

Pressure reduction. If the slip and therefore the incipient lock continue to increase, the switch to pressure reduction is made. A connection from the wheel-brake cylinder via the return pump to the master cylinder is therefore made. The slip is reduced. If the slip falls below a particular threshold, then the ECU switches the solenoid valve back to the pressure build-up. The control cycle is repeated (4 … 10 times per second) as long as the brake is applied.

ABS with return in a closed circuit.

During pressure reduction, brake fluid is taken in by a pressure accumulator. At the same time, the return pump pumps it back to the respective master-cylinder brake circuit.

Structure (Fig. 1). This ABS has the following components in addition to the usual brake system:

- Wheel sensors
- Hydraulic modulator
- ECU
- Warning lamp

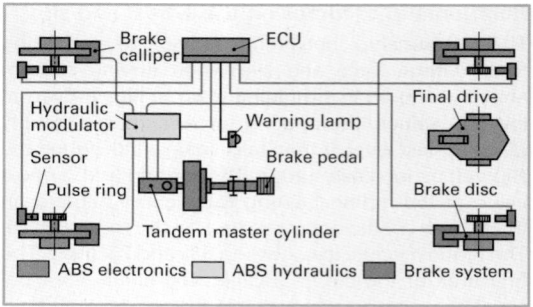

Fig 1: ABS with return in a closed circuit (illustration)

Wheel sensors (Fig. 2). These are on every wheel. For each sensor there is one pulse ring around the wheel. Inductive speed sensors or Hall-effect sensor are used.

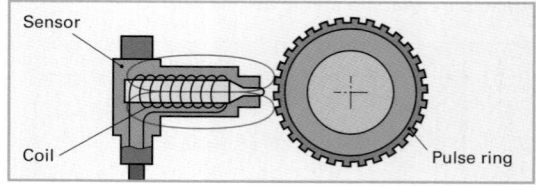

Fig. 2: Inductive wheel sensor (speed sensor)

ECU. This processes the incoming signals from the sensors, determines the necessary settings for the solenoid valves and adjusts these accordingly. The function of the ABS system is constantly monitored by self-diagnosis.

Warning lamp. Upon starting, this signals the operational readiness of the ABS. It lights up should the ABS control system fail. The vehicle can still be braked fully.

Hydraulic modulator with return pump. This contains solenoid valves for control, an accumulator for brake fluid for each brake circuit and an electrically-driven return pump. The pump is activated via a relay and always runs during the ABS control system.

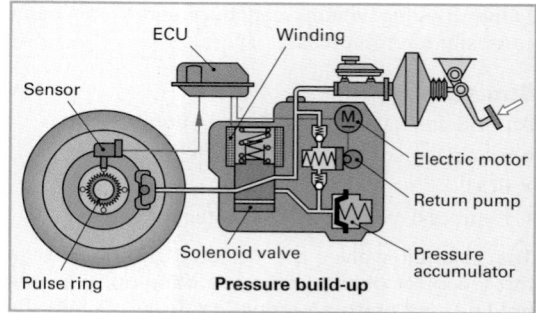

Fig. 3: 3/3 solenoid valve - operating principle

Operating principle with 3/3 solenoid valves (Fig. 3, Page 468).

For braking-pressure modulation in the ABS control system, the ECU triggers a 3/3 solenoid valve in the hydraulic modulator for each channel. In accordance with the three control phases, the master cylinder is connected as follows:

- To the wheel brake cylinder for pressure build-up
- No connection for pressure holding
- To the return pump for pressure reduction

Operating principle with 2/2 solenoid valves (Fig. 1)

In this system, the hydraulic modulator is equipped with smaller, lighter and faster switching 2/2 solenoid valves. Each control channel now requires an inlet valve and an outlet valve.

The ECU switches the solenoid valves in the control phases as follows:

- **Pressure build-up.** Inlet valve (IV) open, outlet valve (OV) closed.
- **Pressure holding.** Both valves closed.
- **Pressure reduction.** Inlet valve closed and outlet valve open. The running return pump pumps the excess brake fluid from the accumulator back into the relevant brake circuit.

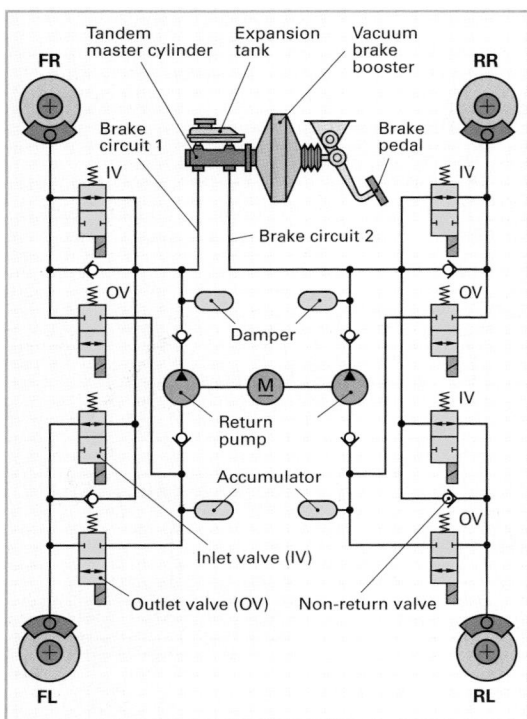

Fig. 1: ABS with closed circuit and 2/2 solenoid valves (hydraulic circuit)

ABS with return in an open circuit and 2/2 solenoid valves (Fig. 2)

During a control action, the excess brake fluid flows back into the expansion tank at zero pressure. The hydraulic pump is selected by the ECU using the position of the pedal-travel sensor. It pumps the missing volume of brake fluid out of the expansion tank at high pressure back into the respective brake circuit and therefore brings the brake pedal to its basic position. The pump is then deactivated.

Structure
The system is composed of:

- ECU
- Wheel sensors
- Actuating unit
- Hydraulic unit
- Warning lamp

ECU. This processes the sensor signals and passes them on as control signals to the solenoid valves. The signals from the travel sensor control the hydraulic pump in the ABS control system. Faults and malfunctions are detected by the ECU, ABS is switched off and the ABS warning lamp is switched on.

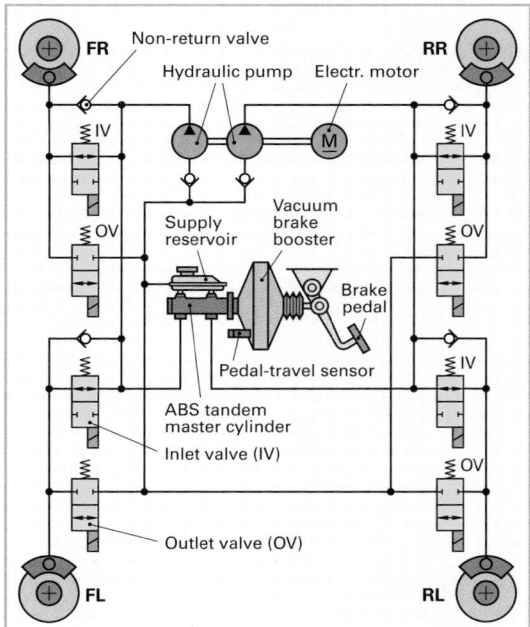

Fig. 2: ABS with open circuit (hydraulic circuit)

Wheel sensors. These are on every wheel and transmit the wheel speed.

Actuating unit. This consists of a vacuum brake booster, which has an integrated pedal-travel sensor, and the ABS tandem master cylinder with expansion tank. The pedal-travel sensor reports the position of the brake pedal to the ECU.

Hydraulic unit. As the **engine-pump unit,** it includes a dual circuit electrically-driven hydraulic pump and the **valve block.** This has two 2/2 solenoid valves for each closed-loop control circuit. An inlet valve (IV) and an outlet valve (OV) with a parallel selected non-return valve.

Operating principle of the ECU

If the ECU detects an incipient lock, e.g. on the front left wheel, then the inlet valve closes and the outlet valve opens. The brake fluid now flows at zero pressure back into the expansion tank. When switching to pressure build-up, the outlet valve closes and the inlet valve opens. The brake fluid missing from the brake cylinder will be added by the master cylinder plunger. The master cylinder plunger and the brake pedal move slightly as a result. The travel sensor informs the ECU. This switches the hydraulic pump on. It pumps fluid back until the original pedal position has been reached again.

Electrical circuit of an ABS

The schematic diagram **(Fig. 1)** shows a 4-channel ABS with return in a closed circuit with eight 2/2 solenoid valves and 4 sensors.

When the ignition switch is switched on, the control winding in the electronic protection relay is supplied with the voltage from terminal 15, the ECU switches and connects to terminal 30 (positive) via pin 1 (plug-in connection on the ECU). At the same time, the warning lamp lights up because it is connected to terminal 15 to the positive and via terminal L1 to the valve relay and via the diode to earth. The ECU now checks the ABS for faults. If everything is OK, it connects across pin 27 and returns the control winding in the valve relay to earth. The valve relay switches. Pin 32 on the ECU receives positive from terminal 30, as does

the cathode of the diode. The warning lamp goes out. The solenoid valves are now at the positive.

Should the ECU detect a risk of locking, pin 28 is returned to earth. The motor relay switches on the return pump. FR can now be switched to the control phases by connecting to earth at pin 35 or 37 in the control phases.

Check of the electrical system

This can be carried out using a voltage or resistance measuring device, a test diode or special test equipment.

> Before disconnecting the ECU, the ignition must be switched off.

1. **Inspection. Power supply ECU:**
   Ignition "on"; between pin 1 and earth, $U > 10\,V$.

2. **Valve relay function:**
   Pin 27 at earth, ignition "on"; senses the switching of the relay, or between pin 32 and earth, $U > 10\,V$.
   Electric circuit control winding: ignition "off", resistance measuring device between pin 1 and 27, $R \approx 80\,\Omega$.

3. **Speed sensor FR resistance:**
   Ignition "off" between pin 11 and 21, $R = 750\,\Omega \ldots 1.6\,k\Omega$.
   Function: turn wheel, between pin 11 and 21 e.g. at 1 rotation of wheel/second $U > 30\,mV$ alternating voltage.

4. **Motor relay function:**
   Ignition "on", pin 28 at earth, senses the tripping function, or between pin 14 and earth, $U > 10\,V$, return pump runs (noise).

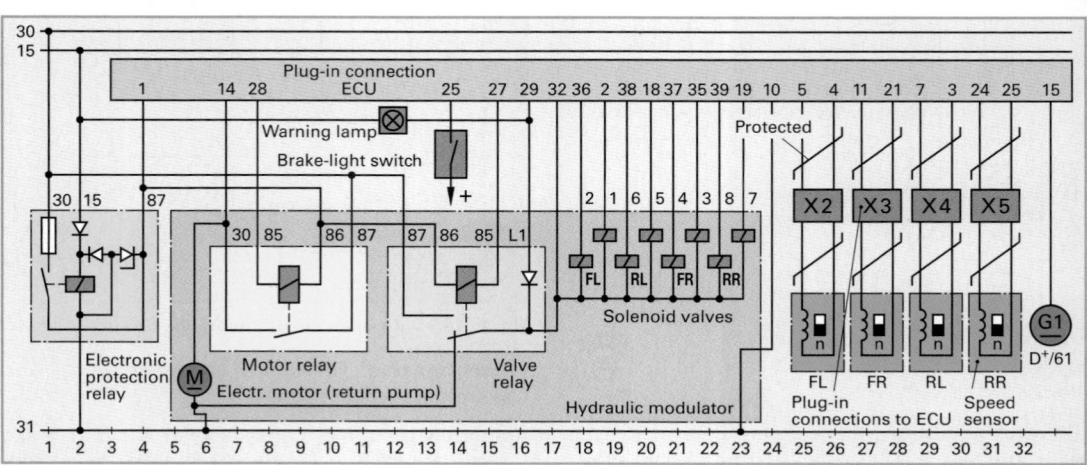

Fig. 1: Schematic diagram for a 4-channel ABS

18.10.14 Brake assistant (BAS)

> The brake assistant immediately makes sure in the case of panic braking that there is maximum brake boosting effect, which means that the braking distance is considerably reduced.

Many drivers brake quickly in critical situations but do not depress the brake pedal enough. The braking distance is therefore longer which can lead to collisions.

Structure
The brake assistant **(Fig. 1)** consists of the following components:

- BAS ECU
- Solenoid
- Travel/pedal sensor
- Release switch

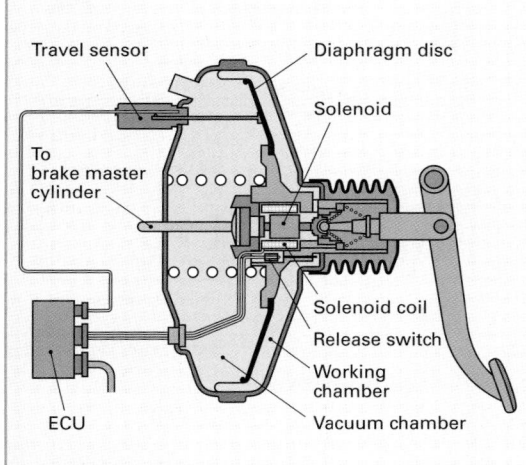

Fig. 1: Brake assistant

Operating principle
The movement of the pedal causes a change in resistance in the pedal sensor. This is reported to the BAS ECU. If the ECU detects that the pedal is suddenly applied, for example during panic braking, then the solenoid is activated. This vents the working chamber of the brake booster to create the full force of the booster. The result is emergency braking. The ABS prevents the wheels from locking. The solenoid is only switched off via the release switch once the brake is released and the brake pedal has returned to its initial position.

For data exchange, the BAS ECU is connected via the CAN bus to the ECUs for other electronic chassis control systems, e.g. ABS, TCS, ESP.

If the ECU detects faults, the brake assistant is switched off. The failure is displayed with a yellow warning lamp.

REVIEW QUESTIONS
1 Which electronic chassis control systems are used?
2 Which forces take effect on the vehicle wheel?
3 What is understood by slip?
4 What is the slip range in which the vehicle remains steerable and stable?
5 What are the tasks of an ABS?
6 Name the components of an ABS.
7 Name and explain the terms of the ABS closed-loop control circuit.
8 Name the control phases for ABS.
9 What are the essential differences of the hydraulic-ABS concepts?
10 How does the brake assistant work?

18.10.15 Traction-Control System (TCS)

> The TCS system prevents the drive wheels spinning when pulling away and accelerating.

This stabilises the vehicle in the longitudinal direction, the cornering stability is maintained and the vehicle is prevented from breaking away at the powered axle.

The TCS is an enhancement of ABS. Both systems use common sensors and actuators and often have a common ECU where the data exchange is usually carried out via a CAN bus. When the vehicle is being driven with snow chains, the TCS can be deactivated. A distinction can be made between:

- TCS systems with engine intervention.
- TCS systems with brake intervention, otherwise known as ELSD Electronic Limited-Slip Differential.
- TCS systems with engine and brake intervention.

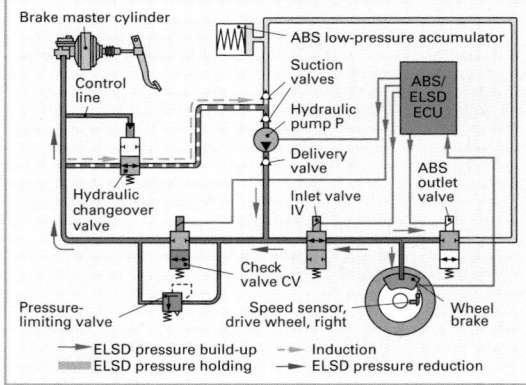

Fig. 2: TCS/ELSD brake circuit of a wheel

Advantages
- Improvement of traction when pulling away or accelerating.

18

- Increase of driving safety at high motive forces.
- Automatic adjustment of engine torque to the grip ratios.
- Driver information about reaching dynamic limits.

TCS with brake intervention/ electronic limited-slip differential ELSD

An electro-hydraulic system is used as a starting-off aid. The lock effect is created as a result of brake intervention on the spinning wheel in order to achieve better traction.

Structure (Fig. 2, Page 471)

Hydraulic system. This is composed of a hydraulic pump with suction and delivery valves, inlet and outlet valves, a hydraulic changeover valve and a check valve with pressure limiter.

Electrical system. This is composed of ABS/TCS (ELSD) ECU and wheel-speed sensors.

Operating principle

Pressure build-up. If a driven wheel spins, this is detected by the ECU with a speed sensor. It activates the hydraulic pump and the check valve. The check valve (CV) closes and the pressure generated by hydraulic pump P brakes the spinning wheel.

Pressure holding. The inlet valve (IV) is closed.

Pressure reduction. If the wheel has stopped spinning, then the inlet and check valves are opened and the pressure is relieved to the expansion tank via the master cylinder.

TCS with engine and brake intervention

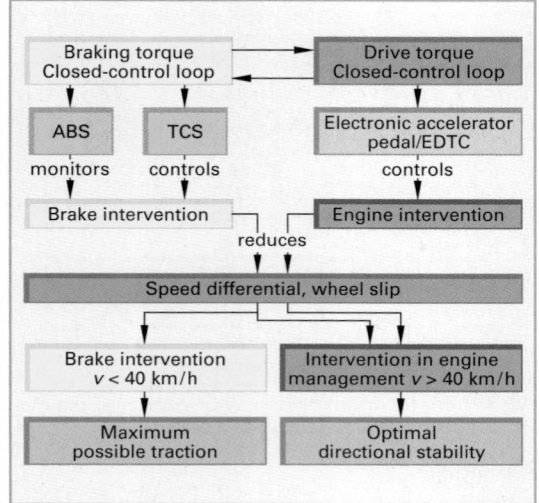

Fig. 1: TCS block diagram

The system works with engine or brake intervention, according to the driving situation. The block diagram **Fig. 1** shows the collaboration of engine and brake intervention for preventing unreliable wheel slip when pulling away (TCS operation/ELSD operation) or in overrun mode (EDTC operation).

Structure (Fig. 2).

- ABS/TCS EDTC ECU
- ABS/TCS hydraulic unit
- Electronic accelerator pedal with ECU
- Setpoint generator, servo-motor and throttle valve

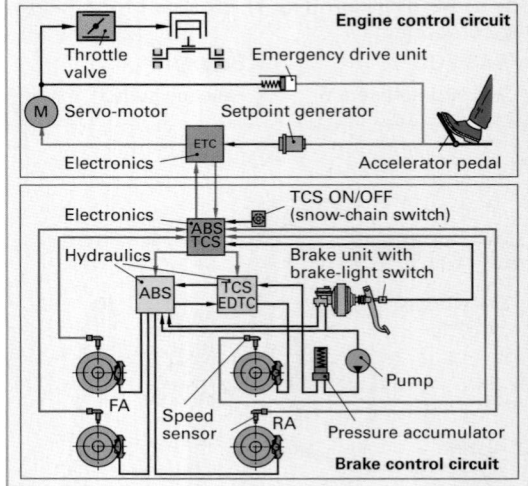

Fig. 2: TCS-system overview

Operating principle (Fig. 1, Page 473)

All wheel speeds are entered and processed in the ABS/TCS ECU. If one or both wheels tend to spin, then TCS control is activated.

Control when pulling away

If a wheel is threatening to spin, then the braking torque control overrides because it is important to have as much traction as possible. If, for example, the rear right wheel (RR) starts to spin, then pump P1 is activated via the ECU. The intake solenoid valve Y15 is opened, the changeover valve Y5 and solenoid valve Y10 on the rear left wheel (RL) are closed. The pump-interior pressure therefore brakes the wheel (RR). With the solenoid valves Y12 and Y13 in the hydraulic unit, the braking torque can be controlled through pressure build-up, pressure reduction and pressure holding.

Control when driving

If, for example, both wheels are threatening to spin, then the drive-torque control overrides to gain optimum traction. The throttle-valve position is returned with a servo-motor and the moment of ignition delayed, whereby the drive torque is reduced.

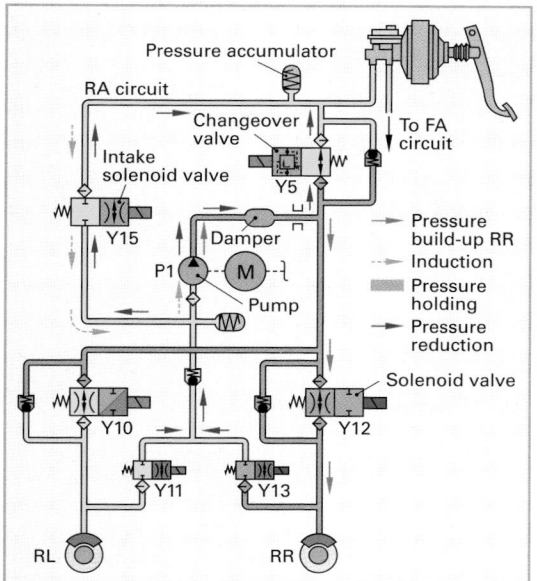

Fig. 1: Hydraulic circuit diagram of a brake circuit

If the wheels spin despite this, the braking torque control is activated by feeding braking pressure from pump P1 via solenoid valves Y10 and Y12 to the rear wheels until the wheels stop spinning. The directional stability is increased.

Overrun

If slip occurs during sudden deceleration caused by the braking effect of the engine on the drive wheels, the ECU detects this and activates the engine-drag torque control (**EDTC**). By activating the servo-motor, the throttle valve is moved to such an extent and the engine speed therefore increased that there is no longer slip on the drive wheels.

TCS warning lamp. This informs the driver in the case of TCS closed-loop control and if the system fails.

18.10.16 Vehicle Dynamics Control ESP, DSC

Through the specific braking of individual wheels, the vehicle can be stabilised laterally and longitudinally. This therefore prevents the vehicle turning on a vertical axis.

In the Electronic Stability Program (**ESP**) (Fig. 2), the following systems work together:

- Antilock-braking system (**ABS**)
- Automatic braking-force distribution (**ABV**)
- Traction-control system (**TCS**) with engine-drag torque control (**EDTC**)
- Automatic regulation of yaw moment (**GMR**).

With a networked data bus, the systems control the brake intervention depending on the wheel speed, braking pressure, yaw rate, steering angle, lateral acceleration and defined program maps.

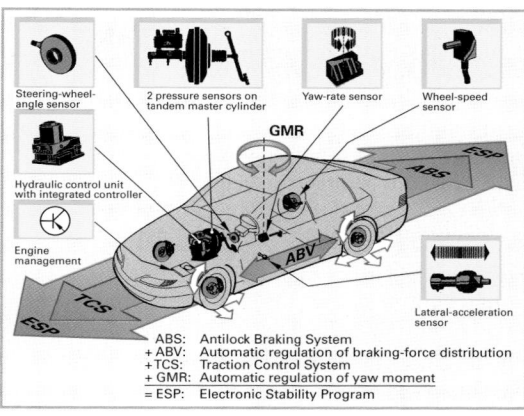

Fig. 2: Components of the ESP system

Operating principle

The signals from the sensors, e.g. wheel speed, steering movement and lateral acceleration are recorded by the ECU as actual values and compared with stored setpoint values. If the actual values deviate from the desired and actual course (setpoint value), then one wheel is braked specifically so that the vehicle remains stable.

The ESP system decides …

- … which wheel is braked and how sharply.
- … whether the engine torque is downrated.

Understeer. If the vehicle tends to understeer when cornering or during a swerve to avoid an obstacle (**Fig. 3**), then it would be pushed straight ahead by the front axle. The ESP system controls uses a presupply pump (**Fig. 1, Page 474**) to control the braking pressure of the rear wheel in the inside of the bend. The yaw moment created as a result twists the vehicle on the vertical axis and counteracts the understeering.

Oversteer. If the vehicle tends to oversteer (**Fig. 3**), then the front wheel on the outside of the bend, for example, is braked by the system, therefore stabilising the vehicle.

Fig. 3: Understeering and oversteering vehicle

Hydraulic circuit diagram (Fig. 1)

The brake circuit of a wheel is shown here.

Pressure build-up

If ESP intervenes in the control, P1 draws in the brake fluid from the supply reservoir and supplies it to pump P2. This guarantees that the system quickly builds up braking pressure in the brake circuit even at low temperatures. The return pump P2 works in the same way, increasing the braking pressure further until the wheel is braked. The high-pressure switching valve Y1 and the inlet valve Y2 are therefore opened. The outlet valve Y3 is closed and the switching valve Y4 is blocked.

Pressure holding

In this control phase, the high-pressure switching valve Y1 and the inlet valve Y2 are closed. The braking pressure remains constant.

Pressure reduction

In this phase, the outlet valve Y3 and the switching valve Y4 are opened. The brake fluid is returned through the return pump back to the master cylinder.

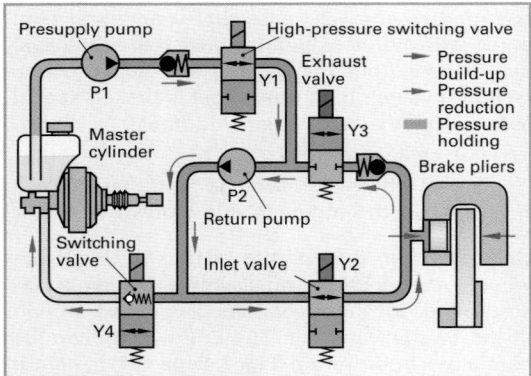

Fig. 1: ESP-system hydraulic circuit diagram

18.10.17 Sensotronic Brake Control (SBC)

The Sensotronic Brake Control SBC (or EHB) is a "Brake by Wire" system. This means that the driver's wish to brake is transmitted via an electrical wire. The system incorporates the functions of ABS, TCS, BAS and ESP.

Structure

The SBC system **(Fig. 2)** essentially consists of the hydraulic unit with pressure accumulator, actuating unit, ECU and speed and yaw-angle sensors.

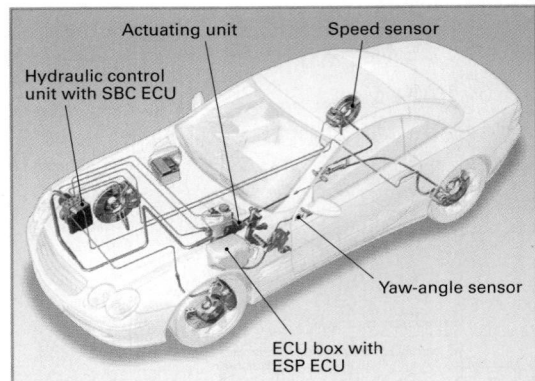

Fig. 2: Components of the SBC system

Unlike a conventional brake system, where all wheels are first subjected to a high braking pressure as quickly as possible and then a pressure control is carried out, with SBC the braking pressure of the individual wheels is controlled individually. Sensors determine the current driving situation and the ECU calculates the optimum braking pressure from this for each wheel. In this way, it is possible to brake the left-hand wheels more sharply which are subjected to heavier loads during a right-hand bend, for example. When braking in bends, this results in an optimum braking factor and stable driving characteristics.

In addition to the functions of a conventional hydraulic brake system, SBC can, for example, adopt the following functions:

- Holding the vehicle on an incline (hill-starting).
- Applying the footbrake and handbrake until the discs and drums are dry in wet conditions.
- Softstop to prevent diving under braking.
- Filling the lines when rapid deceleration is detected, therefore faster pressure build-up during an emergency braking manoeuvre.
- Automatic adaptive speed and distance control (ACC).

The system does not need a brake booster. In the case of an electronics failure, restricted braking can take place via an emergency hydraulic connection.

Function

Fig. 1, Page 475, shows the structure of the SBC hydraulics. The driver operates the brake pedal, therefore generating a braking pressure in both brake circuits in the master cylinder. The pressure is recorded by pressure sensor b1.

SBC normal braking

The ECU closes the hydraulic connection to the front axle by supplying both isolating valves y1, y2.

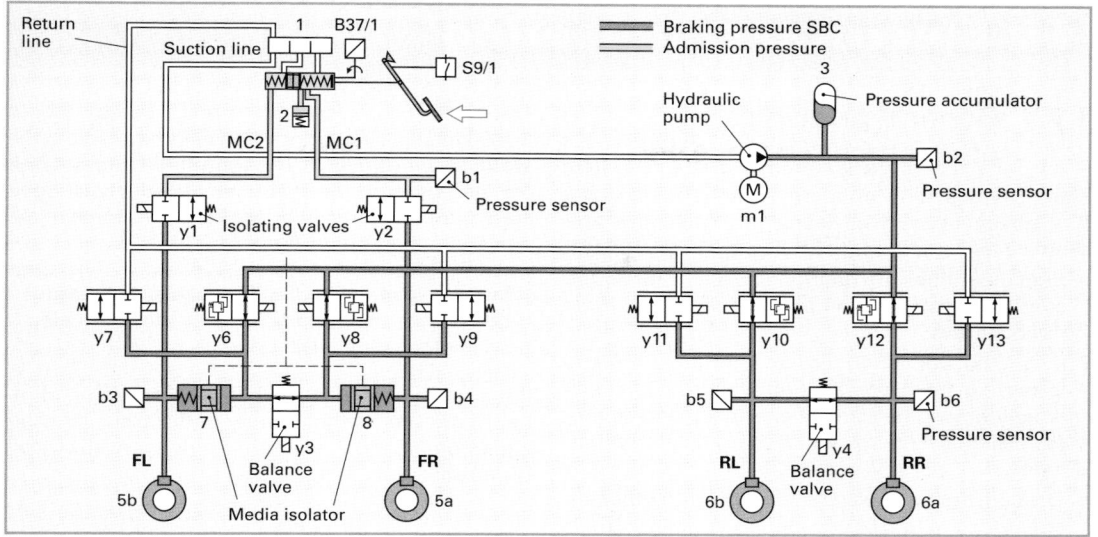

Fig. 1: SBC hydraulic circuit diagram, normal braking

The brake system pressure supply is now provided by the pressure accumulator 3. The storage pressure is generated by the electrically driven hydraulic pump m1 and measured by the pressure sensor b2. This can be up to 150 bar. Should the storage pressure sink below a particular value, then the hydraulic pump is reactivated.

The ECU calculates the optimum braking pressure for each wheel and adjusts it accordingly using inlet valves y6, y8, y10, y12 and outlet valves y7, y9, y11, y13. Pressure sensors b3, b4, b5, b6 report the actual values of the individual wheel brake cylinders to the ECU.

Balance valves y3, y4 balance the pressure for the wheels of one axle during a brake application. They are activated and closed during braking when cornering and in the Electronic Stability Program. It is now possible to regulate the brake pressure individually for each wheel.

The two media isolators 7, 8, prevent nitrogen from entering master cylinder 1 from a leaking pressure accumulator 3.

Emergency-braking manoeuvre if SBC fails
The two isolating valves y1, y2 are not energised and therefore remain open **(Fig. 2)**.

The braking pressure generated by the driver in the master cylinder is directed to the brake cylinder at the front axle. The rear axle is unbraked. Because there is no brake booster, the braking effect is low. The vehicle speed is therefore restricted by the engine ECU to a maximum of 90 km/h.

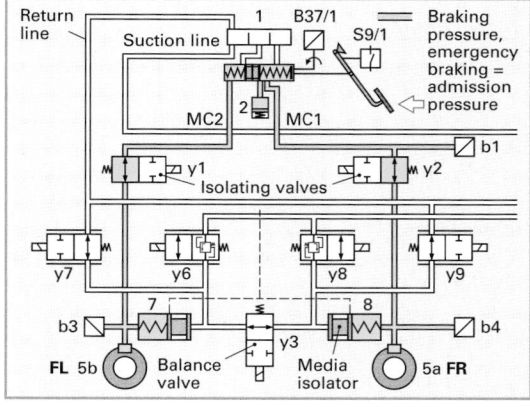

Fig 2: SBC hydraulic circuit diagram, emergency braking manoeuvre

REVIEW QUESTIONS

1 **What is understood by a traction-control system?**
2 **What are the advantages of traction-control systems?**
3 **What are the components in the TCS system required for the braking-torque control loop?**
4 **Describe the function of the TCS system with engine and brake intervention.**
5 **What are the advantages of an Electronic Stability Program?**
6 **How does the ESP/DSC system work if the vehicle is oversteered?**
7 **What are the extra functions of SBC in addition to a hydraulic brake system?**
8 **Explain the function of SBC.**

18

19 Electrical engineering

19.1 General principles of electrical engineering

Electricity is a form of energy: compared to other forms of energy such as heat, light, mechanical and chemical energy, it has the following advantages:

- High quantities of energy can be transported over long distances via power lines to the most remote areas.
- It can be converted easily into other energy forms, e.g. heat in preheating systems, light in filament lamps, mechanical energy in electric motors, chemical energy when charging starter batteries.
- The conversion of electrical energy into other energy forms is to a large extent non-polluting.

The general principle for understanding electrical processes is the Bohr model of the atom (**Fig. 1**). An atom is the smallest, chemically indivisible particle of an element.

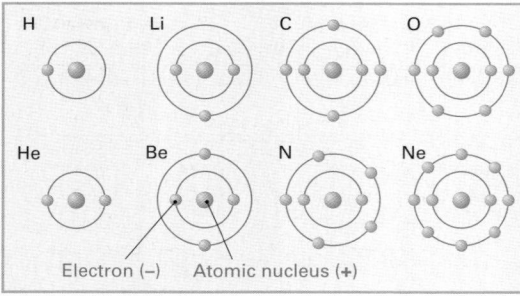

Fig. 1: Structure of atoms

The principal constituents of an atom are the atomic nucleus and the electrons. The atomic nucleus is composed of protons and neutrons.

Protons are positively charged particles of matter. A hydrogen nucleus consists of only one proton, for example. It has the lowest positive quantity of charge, the elementary charge.

Neutrons are particles of matter that have no charge.

Electrons are particles of matter that have a negative charge. A single electron has the lowest negative quantity of charge, the elementary charge.

Electrons are the carriers of the negative elementary charges, protons carriers of the positive elementary charges. The respective elementary charges are the same size.

The electrons move at great speed (about 2,200 km/s) in circular or elliptical orbits around the atomic nucleus (**Fig. 2**). The centrifugal forces which are thereby experienced by the negatively-charged electrons are balanced by the attractive force of the positively charged protons.

Particles of matter with different electric charges attract each other, particles with the same electric charges repel each other.

If the nucleus of an atom contains the same number of protons as the number of electrons circulating around it, then the atom is outwardly electrically neutral.

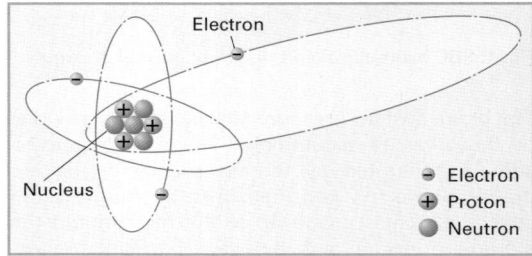

Fig. 2: Structure of a lithium atom

In addition to the electrons which are attached to the atomic nucleus, all matter also has other electrons which temporarily dissociate and are able to move freely between the atoms. These electrons are also known as "free" electrons. Provided no outside energy is applied to the matter, the free electrons move randomly, i.e. it is not possible to identify a preferred direction of motion (**Fig. 3**).

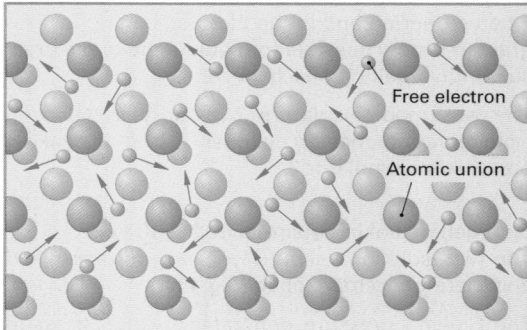

Fig. 3: Random movement of free electrons

Electrical processes depend on the existence and the mobility of free electrons. Electricity is not created, it is present in all matter.

19.1.1 Voltage

Voltage is present when there is a difference in the quantity of electrons between two points, e.g. the terminals of a battery. The amount of voltage depends on the size of the difference in the quantity of electrons. The voltage is produced by charge separation in the voltage source (**Fig. 1**).

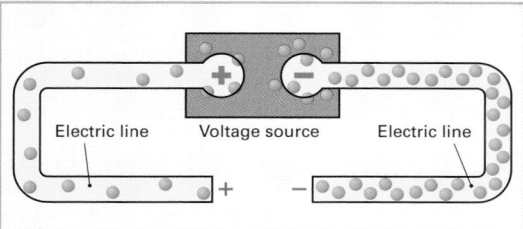

Fig. 1: Voltage generation by charge separation

> At the negative terminal there is an electron surplus, at the positive terminal a deficit of electrons.

Between the negative terminal and the positive terminal, there is a balancing of the electrons, i.e. when the two terminals are connected, electrons flow from the negative terminal via the consumer to the positive terminal and thereby perform electrical work (**Fig. 2**).

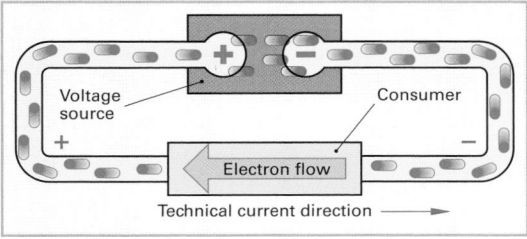

Fig. 2: Electron flow in the electric circuit

> Voltage is the balancing effort of different quantities of charge. It is the cause for the flow of electrical current.

The terminal connectors in the alternator have no voltage when the alternator is non-operative, i.e. the free electrons are evenly distributed in the windings, so that the windings are electrically neutral. If the alternator is actuated, the free electrons are moved to the negative terminal; at the negative terminal, the result is an excess of neutrons compared to the positive terminal and, with it, voltage is produced.

> The unit of voltage U is the volt (V).

19.1.2 Electrical current

The cause of the electrical current is the voltage.

> The electrical current is the directional movement of free electrons.

Electric circuit (Fig. 3). Electrical current can only flow in a closed electric circuit. An electric circuit consists of at least the voltage generator, the consumer and the electric lines. The electric circuit can be closed or interrupted with a switch. Switches are usually shown in an inactive state in circuit diagrams.

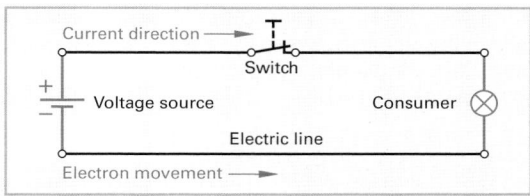

Fig. 3: Electric circuit

Fuses (Fig. 4). These are switched in the electric circuit. **Electric line fuses** protect electric lines against overloading and short circuits. **Equipment fuses** protect individual items of equipment, e.g. ECUs, radios, in the event of faults.

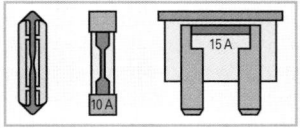

Fig. 4: Fusible links in motor vehicles

Electron conduction (Fig. 5). This occurs in all electrical conductors that are made of a metallic material. The metal atoms release electrons. These "free" electrons can be moved easily between the fixed metal atoms of the metal wire. If the electric circuit is closed, all the conductor's and the consumer's free electrons are forced into a directional movement as a consequence of the applied voltage. An electrical current flows.

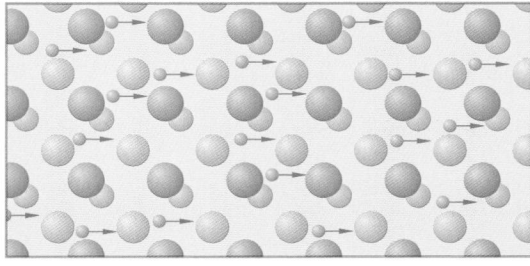

Fig. 5: Directional movement of free electrons

19

Ion conduction (Fig. 1). This permits the current to be carried by the directional movement of charged particles of matter (ions). The positive ions are referred to as cations, because they move to the negative electrode, the cathode. The negative ions are referred to as anions, because they move to the positive electrode, the anode.

> Ion conductors are chemical connections which divide into positive and negative constituents.

The splitting of gases into negative and positive particles of matter is referred to as **ionisation**. It can be caused by irradiation, heating or electric fields.

If the A/F mixture is ionised by the strong electric field in the air gap between the electrodes of a spark plug, it becomes electroconductive and the spark crosses over **(Fig. 1)**.

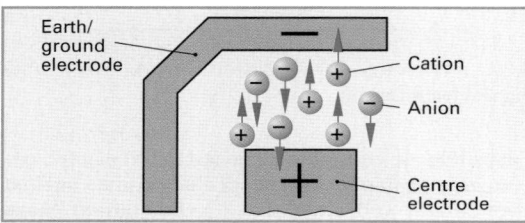

Fig. 1: Ionisation at the spark plug

Current direction
Direction of the electron flow. At the voltage source there is an electron surplus at the negative terminal and a lack of electrons at the positive terminal. If the negative terminal is connected to the positive terminal of the voltage source via a consumer, electrons from the negative terminal flow in the outer electric circuit via the consumer to the positive terminal of the voltage source **(Fig. 2)**.

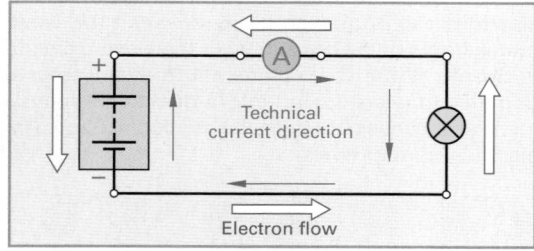

Fig. 2: Battery as electron pump

Technical current direction. Due to the absence of knowledge of the direction of electron flow, it was decided in electrical engineering that the direction of current was from the positive to negative **(Fig. 2)**.

Current intensity *I*. This is the number of electrons which flow through the conductor cross-sectional area per second.

> The unit of current intensity I is amps (A).

Current density *J*. This is the current I which flows per square millimetre of conductor cross-sectional area A.

> $$J = \frac{I}{A}$$
>
> The unit of current density J is amps per square millimetre (A/mm^2).

The permissible current density in electric lines depends on the cooling property of the surface of the conductor in particular **(Table 1)**. Thin wires have a larger surface area in relation to the conductor cross-sectional area than thick wires and can thus carry more current per mm^2 of conductor cross-sectional area.

Table 1: Capacitance of copper electric lines		
A in mm^2	I_{max} in A	J in A/mm^2
1.0	20	20.0
2.5	34	13.6
6.0	57	9.5
16.0	104	6.5

Types of current
Direct current (DC, symbol –). In an electric circuit, in which the voltage and resistance are constant, a direct current flows when the same number of electrons move in the same direction per second **(Fig. 3)**.

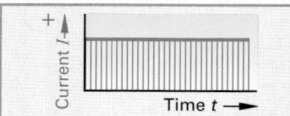

Fig. 3: Direct current

Alternating current (AC, symbol ~). In an electric circuit in which the voltage and resistance are constant, an alternating current flows when the free electrons constantly move the same distance to and fro in both directions **(Fig. 4)**.

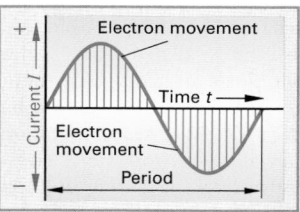

Fig. 4: Alternating current

19.1.3 Electrical resistance

It is useful to distinguish between two related terms in this area of electrical engineering:

- "Electrical resistance", which refers to the physical properties for conducting electrical current in matter
- "Electrical resistor", which refers to a component in electrical engineering and electronics

19.1.3.1 Electrical resistance of matter

If a voltage is applied to an electrical conductor, the electrons can no longer flow unresisted. The restriction of the flow of electrons is known as the electrical resistance R.

> The electrical resistance R is the inhibition of the electrical current in a conductor. It is expressed in ohms (Ω).

Specific electrical resistance ϱ. Each conductive material has its own characteristic specific electrical resistance ϱ[1]. For example, copper has a resistance of 0.01789 Ω for a conductor length of 1 m and a conductor cross-sectional area of 1 mm².

> The specific electrical resistance ϱ is the resistance of a conductor of 1 mm² conductor cross-sectional area and 1 m conductor length.

In electrical engineering, the electric conductivity $\varkappa$[2] is often stated instead of the specific electrical resistance ϱ. It is the reciprocal value of the specific electrical resistance.

$$\varkappa = \frac{1}{\varrho} \qquad \text{Unit:} \quad \frac{m}{\Omega \cdot mm^2}$$

Thus, the numerical value of the electric conductivity of copper is 56, that of aluminium is 36. This means that for conductors with the same dimensions, copper conducts the electrical current approximately 1.5 times better than aluminium (56 : 36 ≈ 1.5).

Conductor resistance R. The resistance R of a conductor is higher, the higher the specific electrical resistance and the length of the conductor, and the smaller the conductor cross-sectional area A.

$$R = \frac{\varrho \cdot l}{A}$$

The unit of resistance R is the ohm (Ω).

---

[1] ϱ (rho, Greek alphabetic character)
[2] $\varkappa$ (kappa, Greek alphabetic character)

Resistance and temperature

The resistance value of a conductor material depends on temperature. The resistance value as the temperature rises can both increase **(PTC resistor)** as well as decrease **(PTC conductor)**, depending on its application.

PTC resistors. These conduct the electrical current better in a cold state than in a warm state, i.e. their resistance increases when the temperature increases. These materials are known as PTC resistors[3] since they have a Positive Temperature Coefficient **(Fig. 1)**. Most metals are PTC resistors.

> The resistance of PTC resistors increases when the temperature increases.

The increased resistance in PTC resistors is caused by the increase in the heat oscillation of the atoms and molecules in the conductor material. This reduces the conductivity of the materials, i.e. the flow of the electrons is impeded.

NTC resistors. These carry the electrical current better as the temperature increases than as the temperature drops. These materials are known as NTC resistors[4] since they have a Negative Temperature Coefficient **(Fig. 1)**. Carbon, some metal alloys and most semiconductor materials are NTC resistors.

> The resistance of NTC resistors decreases when the temperature increases.

The lowering of the resistance on NTC resistors is caused by the electrons detaching from their bonds in atoms and molecules. There are more free electrons available to conduct electrical current. This increases the conductivity of the materials, i.e. the flow of the electrons is impeded to a lesser extent.

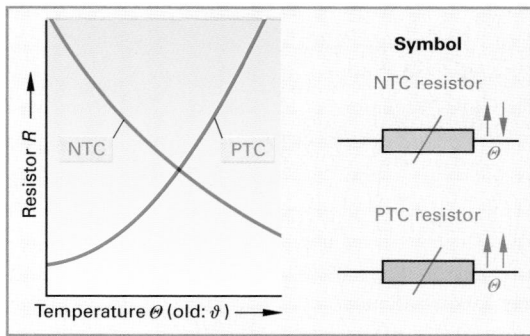

Fig. 1: Temperature dependence of resistors

---

[3] PTC = **P**ositive **T**emperature **C**oefficient
[4] NTC = **N**egative **T**emperature **C**oefficient

19.1.3.2 Resistors

A distinction is made between **fixed resistors** and **variable resistors**. The symbols for the most important resistor structural shapes are shown in **Fig. 1**.

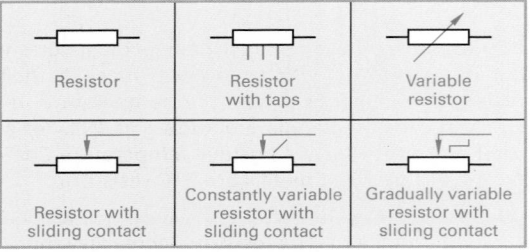

Resistor	Resistor with taps	Variable resistor
Resistor with sliding contact	Constantly variable resistor with sliding contact	Gradually variable resistor with sliding contact

Fig. 1: Switching symbols of resistors

Fixed resistors. The resistance value of fixed resistors is determined during manufacture. Alternative resistance values can be obtained by combining different resistors in a series connection, parallel connection or mixed circuit.

Variable resistors. The respective resistance value of variable resistors can be adjusted by sliders or taps. They are often connected in series to the consumer to adapt the operating voltage.

Potentiometer (Fig. 2). The total resistance of the slideway is present between connections A (start) and end (E). The resistance value between connection points S and E can be infinitely adjusted between zero and the total resistance using the sliding contact (terminal S).

If a voltage U is applied at connection points A and E, the voltage U_2 between connection points S and E can be tapped off. The voltage U_2 can be infinitely adjusted between zero and the applied voltage U using the sliding contact.

In automotive engineering, potentiometers are frequently used to measure angles of rotation on mechanical components, e.g. on an electronic throttle control, potentiometer on the throttle valve (throttle-value potentiometer). Thereby, the angle of rotation of the slider is converted into a voltage value and supplied to the ECU.

19.1.3.3 Electrical behaviour of materials

Materials can be classified according to their electrical behaviour as follows:

- Conducting materials, e.g. copper, aluminium
- Insulating materials, e.g. plastics, porcelain/china
- Semiconducting materials, e.g. silicon, selenium

Metallic conductor materials. These conduct the electrical current very well, since they contain a high number of free electrons. They only have a low degree of resistance to the flow of current.

Insulating materials. These are materials which hardly conduct the electrical current at all. They have high resistance to the electrical current, i.e. they have an electric conductivity of almost zero. The insulating properties of a material are:

- Contact resistance (leakage resistance)
- Dielectric strength

Semiconductor materials. These have a much lower conductivity than that of electrical conductors, but higher than that of insulators. At low temperatures, they have the properties of insulating materials. At temperatures above room temperature, their electrical resistance decreases considerably.

Fig. 2: Potentiometer

$$\frac{U}{U_2} = \frac{R_1 + R_2}{R_2}$$

The total voltage U is proportional to U_2 in the same way as the total resistance $(R_1 + R_2)$ is proportional to the branch resistance R_2.

REVIEW QUESTIONS

1 Give the symbol and unit for voltage, current and current density.
2 What do you understand by voltage?
3 What are the differences between direct current and alternating current?
4 What functions do fuses perform?
5 What do you understand by current density?
6 What are the consequences of a current density that is too high in an electric line?
7 How is specific electrical resistance defined?
8 How does the resistance of PTC resistors change when the temperature increases?

19.1.4 Ohm's Law

In a closed electric circuit, applied voltage U produces a current I through the resistor R **(Fig. 1)**. The ratio of voltage U in volts and current I in amps gives the resistance R in ohms. This rule is known as **Ohm's Law**.

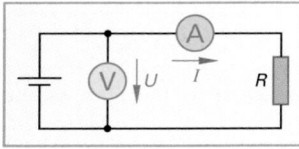

$$I = \frac{U}{R} \qquad \text{Unit: } A = \frac{V}{\Omega}$$

Fig. 1:
Measured variables in the electrical circuit

Table 1. If a variable direct voltage U is applied to resistors $R_1 = 2\,\Omega$ and $R_2 = 1\,\Omega$, different measured currents I_1 and I_2 are obtained for each resistor at the same voltage U.

Table 1: Current depending on voltage

Resistance	U in V	0	2	4	6	8	10
$R_1 = 2\,\Omega$	I_1 in A	0	1	2	3	4	5
$R_2 = 1\,\Omega$	I_2 in A	0	2	4	6	8	10

If currents I_1 and I_2 are plotted against voltage U, two straight lines with different gradients are obtained **(Fig. 2)**. The diagram shows that:

- Current I is proportional to voltage U ($I \sim U$).
- A lower resistance at the same voltage U leads to a higher current I; i.e. the current increases more sharply.

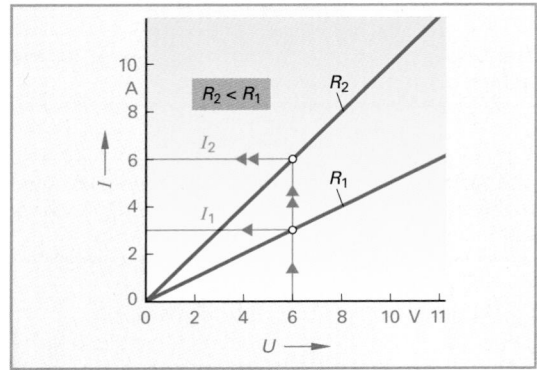

Fig. 2: I as a function of U

Table 2. If a constant direct voltage of $U_1 = 5$ V or $U_2 = 10$ V is applied respectively to resistors R_1 and R_2, different measured currents I_1 and I_2 are obtained for resistance values of the same size of R_1 or R_2.

Table 2: Current depending on resistance

Voltage	R in Ω	0	2	4	6	8	10
$U_1 = 5$ V	I_1 in A	Short circuit	2.5	1.25	0.83	0.675	0.5
$U_2 = 10$ V	I_2 in A	Short circuit	5	2.5	1.66	1.35	1.0

If currents I_1 and I_2 are plotted against resistance R, two hyperbolae **(Fig. 3)** are obtained. The diagram shows that:

- The higher the resistance R becomes at constant voltage U, the lower the current I becomes.
- Current I is inversely proportional to the resistance R ($I \sim 1/R$).

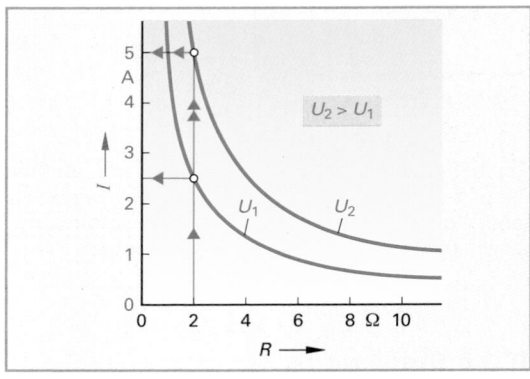

Fig. 3: I as a function of R

19.1.5 Power, work, efficiency

Electrical power with direct current

> The electrical power P is the product of voltage U and current I.

$$P = U \cdot I$$

The unit of electrical power is the watt (W).

1 watt is the power of a current of 1 A at a voltage of 1 V.

$$1\,W = 1\,V \cdot 1\,A = 1\,J/s = 1\,Nm / s$$

19

Electrical work with direct current

> The electrical work W is the product of the electrical power P and the time t in which the power output P is generated.

> $W = P \cdot t$
> $W = U \cdot I \cdot t$
>
> The unit of electrical energy W is the watt second (Ws).

1 watt second is the work exerted at a power of 1 W for 1 s. 3,600,000 Ws is equal to 1 kWh.

> $1\,\text{Ws} = 1\,\text{V} \cdot 1\,\text{A} \cdot 1\,\text{s} = 1\,\text{J} = 1\,\text{Nm}$

Efficiency

> The efficiency η is the ratio of power output P_{out} to the power input P_{in}.

> $\eta = \dfrac{P_{out}}{P_{in}}$

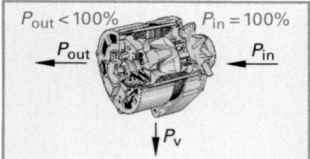

> $P_v = P_{in} - P_{out}$

Since the power supplied P_{in} is always greater than the power output P_{out}, the efficiency η is always less than 1 or less than 100 %. This is due to losses P_v which occur each time energy is converted.

19.1.6 Resistor circuits

Series-wired resistor circuits (Fig. 1)
Series connections aid the distribution of voltage. Electronic devices are connected in series, for example, if the permissible operating voltage of an individual electronic device is lower than the total voltage. In this way, LEDs, for instance, obtain a protective resistance, so that they are able to operate at a nominal voltage of e.g. 2.4 V, even in the 12 V onboard electrical system of a vehicle.

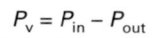
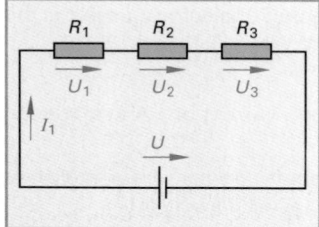

Fig. 1:
Series connection

The following rules apply in a series connection:
The same current flows through all resistors.

> $I = I_1 = I_2 = I_3 = \ldots$

The total voltage is equal to the sum of the branch voltages.

> $U = U_1 + U_2 + U_3 + \ldots$

The branch voltages behave like the branch resistances (distribution of voltage).

> $U_1 : U_2 : U_3 = R_1 : R_2 : R_3$

The total resistance is equal to the sum of the branch resistances.

> $R = R_1 + R_2 + R_3 + \ldots$

Parallel circuits with resistors (Fig. 2)
Parallel circuits aid the division of current. All consumers have the same voltage. The total current I from the voltage source branches out then according to the size of the resistors, i.e. a high current flows in a small resistor, a small current in a large resistor.

In a motor vehicle, all consumers are usually parallel to the voltage source, e.g. the lamps of the lighting system.

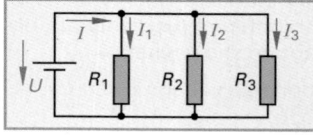

Fig. 2:
Parallel connection

The following rules apply in a parallel connection:
The same voltage source is applied to all resistors.

> $U = U_1 = U_2 = U_3 = \ldots$

The total current is equal to the sum of all the branch currents (current branching).

> $I = I_1 + I_2 + I_3 + \ldots$

The reciprocal value of the total resistance is equal to the sum of the reciprocal values of the branch resistances.

> $\dfrac{1}{R} = \dfrac{1}{R_1} + \dfrac{1}{R_2} + \dfrac{1}{R_3} + \ldots$

The total resistance is always lower than the lowest branch resistance.

Mixed circuits with resistors (Fig. 1)

A circuit in which the resistors are connected both in series and also in parallel is referred to as a mixed circuit or combination circuit with resistors.

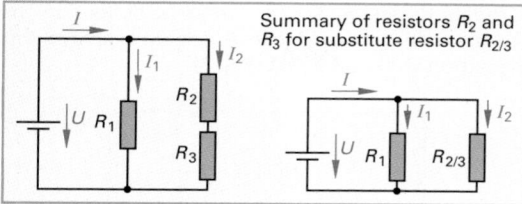

Fig. 1: Mixed circuits with resistors

The principles of series and parallel circuits apply in mixed circuits.

When calculating networks, the circuit must be simplified in steps. To this end, the resistors connected in parallel or in series must be combined as what are called substitute resistors.

19.1.7 Measurements in electrical circuits

Measuring the voltage (Fig. 2)

The voltage is measured using a voltmeter. The voltmeter is connected in parallel to the voltage source taking into account the polarity.

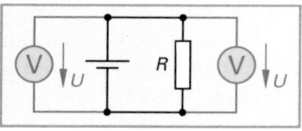

Fig. 2: Voltage measurement

Measuring the intensity of the electrical current (Fig. 3)

The electrical current is measured using an ammeter. The ammeter is connected in the electric circuit, i.e. it is connected in series to the consumer either in the feed line or return line. The positive terminal of the ammeter should be attached to the positive terminal of the voltage source.

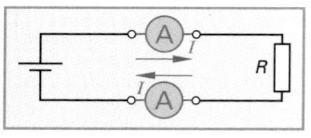

Fig. 3: Current measurement

If an ammeter is inadvertently connected as a voltmeter, the low internal resistance of the measuring element will cause a short circuit. This can irreparably damage the measuring instrument and the electrical and electronic components which are being measured.

Measuring the electrical resistance

The resistance value can be determined by direct or indirect measurement.

Direct measurement using an ohmmeter. For a direct measurement of a resistance value, the component must be disconnected from the voltage source and the ohmmeter must be connected in parallel to the resistor (**Fig. 4**). Failure to observe this can cause damage to the measuring instrument, making it unusable.

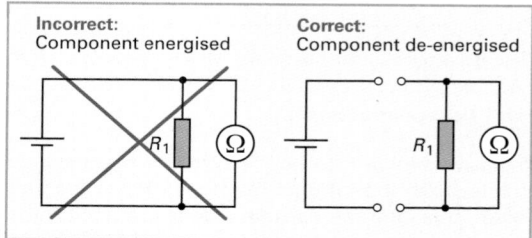

Fig. 4: Direct measurement of a resistance (note the zero potential)

If the resistor to be measured is connected in parallel to a second resistor, the former must be disconnected from the electric circuit (**Fig. 5**). Failure to do so will result in an incorrect measured value.

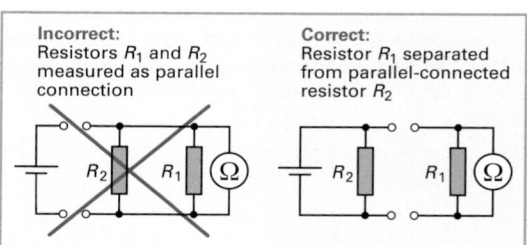

Fig. 5: Direct measurement of a resistance (note the parallel connection)

At low resistances, direct measurement is very inaccurate since the internal resistor of the measuring instrument connected in parallel distorts the measurement.

Indirect measurement. This is performed by measuring the voltage and the current at the resistor. The resistance is calculated from the measured values in accordance with Ohm's Law. For indirect determination of resistance using a current and voltage measurement, two circuits are possible: voltage-error circuit (**Fig. 1, Page 484**) and current-error circuit (**Fig. 2, Page 484**).

With **voltage-error circuit (Fig. 1, Page 484)**, the ammeter measures the current that is actually flowing through the resistor R. The voltmeter, on the other hand, displays a voltage U that is too high by the

19

voltage drop U_{iA} at the ammeter. When calculating resistance using Ohm's Law, the value obtained is therefore too high.

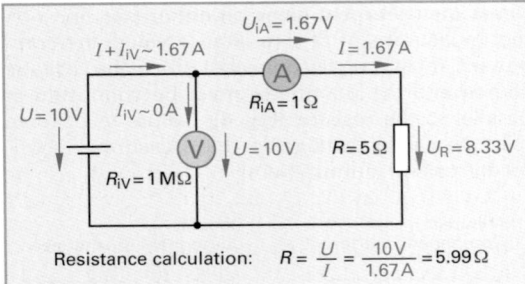

Resistance calculation: $R = \dfrac{U}{I} = \dfrac{10\,V}{1.67\,A} = 5.99\,\Omega$

Fig. 1: Voltage-error circuit

If the resistance R to be measured is significantly greater than the internal resistance R_{iA} of the ammeter, this internal resistance does not need to be taken into account.

> With voltage-error circuit high resistances can be measured precisely.

In the case of **current-error circuit (Fig. 2)**, the voltmeter measures the actual voltage present at the resistor. However, the ammeter displays a current I which is too high by the current I_{iV} through the voltmeter. Hence, when calculating resistance using Ohm's Law, a value which is too low is thus obtained.

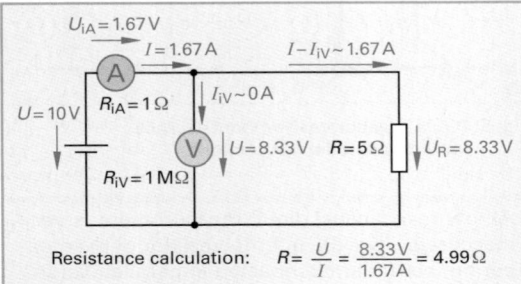

Resistance calculation: $R = \dfrac{U}{I} = \dfrac{8.33\,V}{1.67\,A} = 4.99\,\Omega$

Fig. 2: Current-error circuit

If the current through the voltmeter is considerably lower than the current through the resistor to be measured, e.g. with digital voltmeters, the current through the voltmeter does not need to be taken into account. Only a small part of the current flows through the voltmeter if resistance R is much lower than the internal resistance R_{iV} of the voltmeter. The current leakage can be ignored in this instance.

> With current-error circuit low resistances can be measured precisely.

19.1.7.1 Measuring instruments with analogue display

The measured values to be determined for a measured quantity, e.g. voltage, are converted into a compatible, i.e. analogue[1], needle deflection on the measuring instrument. A needle displays the measured value on a dial **(Fig. 3)**.

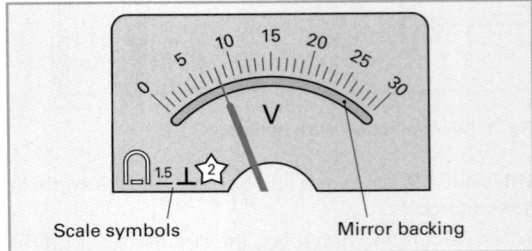

Scale symbols　　　　Mirror backing

Fig. 3: Analogue measured value display

The person observing the needle deflection converts the analogue needle deflection into a numerical, i.e. digital version, thereby carrying out a mental analogue-to-digital conversion.

> Needle gauges and oscilloscopes are analogue display measuring instruments.

Identification of measuring instruments

Measuring instruments with analogue display also carry further information about the measuring element on the dial next to the scale, e.g. operating principle, accuracy class, type of current, status and test voltage. This additional information is represented by symbols and figures **(Table 1)**.

Table 1: Symbols on the dial

∿	For direct current and alternating current	⊥	Vertical nominal position
⌓	Moving-coil measuring element with permanent magnet, universal	⌐	Horizontal nominal position
⌓	Moving-coil measuring element with rectifier	**1.5**	Class parameter, based on the measurement range end value
⌾	Measuring instrument with amplifier	☆2	Test voltage symbol: the number in the star signifies the test voltage in kV (star without number 500 V test voltage)
⌇	Moving iron measuring element		

In automotive engineering, when a measuring instrument with analogue display is being used, generally only moving-coil measuring elements are

---
[1] analogue = equivalent, homogenous

used. These are only suitable for measuring direct voltage or direct current. If alternating current is to be measured, the moving-coil measuring element must have a rectifier circuit.

Accuracy class. This is identified by the class parameter and displays the error margin by which the displayed measured value is permitted to deviate from the actual measured value. The classification categorises into seven accuracy classes:

0.1	0.2	0.5	1	1.5	2.5	5

The figures indicate the permissible margins of error. The error margin of 1.5 signifies that the error is permitted to be ± 1.5 % of the measurement range end value (dial end value), e.g. for a measurement range end value of 100 V, the permissible display error is ± 1.5 V i.e. the display can be between 8.5 V and 11.5 V for a measurement voltage of 10 V.

For a measurement range end value of 100 V and a measurement voltage of 100 V, the display should lie between 98.5 and 101.5 V.

The absolute percentage error for a measurement voltage of 10 V would be ± 15%, for a measurement voltage of 100 V it would be ± 1.5 %.

To keep display error to a minimum, when using multipurpose meters the measurement range should be selected so that the measured value display is in the upper third of the dial.

If a measuring element with an accuracy class of 0.2 were used to carry out the same measurement, the corresponding display values would be 9.8 V/ 10.2 V and 99.8 V/100.2 V. The absolute percentage error would then be ± 2 % or ± 0.2 %.

19.1.7.2 Measuring instruments with digital display

The measured values to be determined for a measured quantity (e.g. voltage) are displayed directly as a number sequence **(Fig. 1).** The measured quantity is thereby converted into a numerical, i.e. digital, display by an analogue-digital converter (AD converter) integrated in the measuring instrument.

The digital display makes it easier to read off the measured value. Furthermore, the resolution is higher compared to an analogue display, i.e. the measured value between two points on the scale no longer has to be estimated.

With measuring instruments with digital display, usually two measurements are performed per second. The measured value is only measured for a fraction of a second and stored temporarily. The average value of the two measurements is displayed. Since the measured value is not usually constant, measuring and displaying continuously would cause the last figure to flicker.

On the other hand, it is not possible to measure sudden changes in the measured quantity nor the extent of deviation. For some measurements, however, the measurement of the extent of deviation for the measured quantity is required. In this case, an analogue measuring instrument or a digital measuring instrument with an additional analogue display must be used.

The additional analogue display **(Fig. 1)** appears in the display panel, e.g. as a black bar, the length of which changes as the measured value changes. In this case, 25 or more measurements per second are performed and displayed. The person observing has the impression that the measurement is continuous.

Measurement error. Measuring instruments with digital display often appear to be accurate due to their numerical display when in fact they are not. Therefore, the permissible measurement error specified by the manufacturer of the measuring instrument must be taken into consideration. This is given as a percentage value, relative to the measurement range end value, e.g. ± 0.5 % of 19.99 V. In addition, the last figure displayed may differ by one digit.

Resolution and number of digits. The simple types of digital measuring instruments have 3 1/2 digits, high-quality instruments have 6 1/2 digits. A 3 1/2-digit display shows 4 figures, however the numeric string for the first digit displayed does not go up to 9.

The first digit only has a limited numeric range, e.g. 0 to 1 or 0 to 3; thus the following numeric strings can appear as the maximum display values respectively: 1999 or 3999.

If these values are exceeded, the measurement range is generally changed automatically.

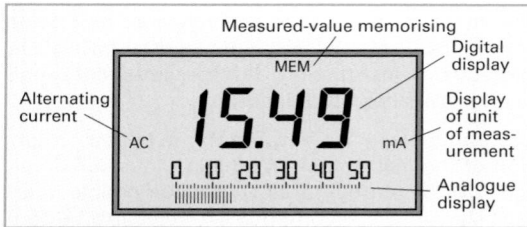

Fig. 1: Digital display with additional analogue linear display

19.1.7.3 Multipurpose meters

Analogue multimeter (Fig. 1)

These are suitable for measuring both voltage and current for direct and alternating voltage. It is only possible to measure resistance values indirectly via a voltage and current measurement; a battery is therefore required to supply power.

The current I that flows though the resistor R is measured. On the basis of the principles in Ohm's Law, $I \sim 1/R$. The scale for the resistance value is designed to be compatible with this rule, so the scale is not linear as a consequence. At infinitely high resistance there is no reading. If the resistance value is zero the reading is off the scale.

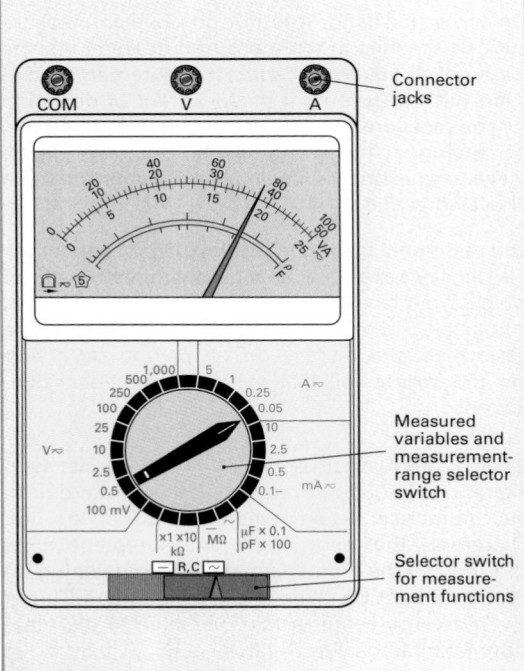

Fig. 1: Analogue multimeter

The measurement range can often be expanded further by a factor of 1,000, so that resistance values in the range Ω-, kΩ- and MΩ can be measured.

When measuring an unknown measured quantity, the largest measurement range should always be selected first, in order to set the measurement range selector switch to the measurement range where the display is in the upper third of the scale.

The measured value is obtained by dividing the number read off the scale divisions by the scale end value and multiplying by the factor and unit given on the range switch, e.g. scale divisions displayed 33 **(Fig. 1)**, measurement range 0.5 V, scale end value 50. This gives a displayed measured value of $(33 : 50) \cdot 0.5\,V = \mathbf{0.33\,V}$.

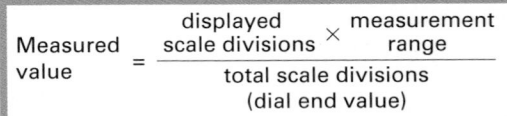

$$\text{Measured value} = \frac{\text{displayed scale divisions} \times \text{measurement range}}{\text{total scale divisions (dial end value)}}$$

Digital multimeter (Fig. 2)

The applications for digital multimeters are the same as those for analogue multimeters. They have the advantage of being relatively robust and can be manufactured at low cost even despite being very accurate.

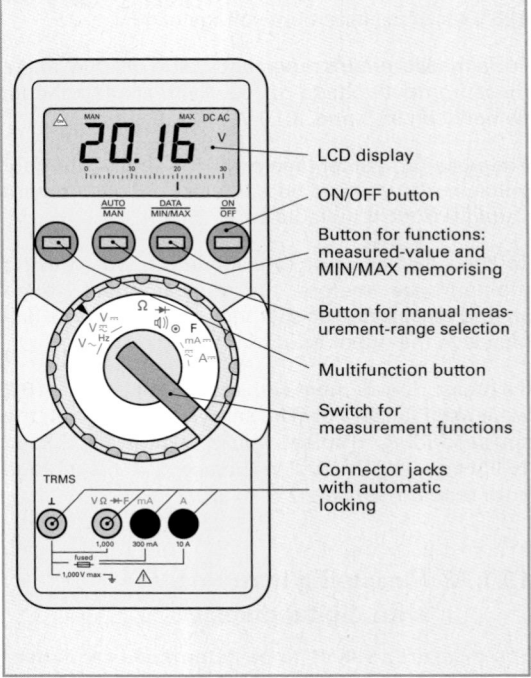

Fig. 2: Digital multimeter

The measurement range and function, e.g. diode testing, can be activated with a central switch. On high-quality instruments, the measurement range is often switched automatically.

Electronic fuses can assume the overload protection of the instrument. In addition, measured values may be stored possibly. If a compatible interface is available, a computer (PC) can also be connected, for example, in order to process the measuring data.

19.1.7.4 Oscilloscope

The oscilloscope is a measuring instrument with analogue display which can measure and graphically display rapid, intermittently recurring electrical processes, such as sensor signals or the flow of the ignition voltage. The display is shown on the display screen of a cathode ray tube **(Fig. 1)**. An oscilloscope that can display two processes at the same time is a two-channel or a dual-trace oscilloscope, depending on the design.

If non-recurring processes need to be displayed, a storage oscilloscope should be used. The measuring process is stored and can be called up again later as a fixed-image.

Design and operating principle

A cathode-ray oscilloscope essentially comprises four components **(Fig. 1)**:

- Oscilloscope picture tube (cathode ray tube)
- Vertical amplifier (Y-amplifier; $Y_1 Y_2$)
- Time-base sweep generator with synchroniser (X-amplifier; $X_1 X_2$)
- Network component for power supply

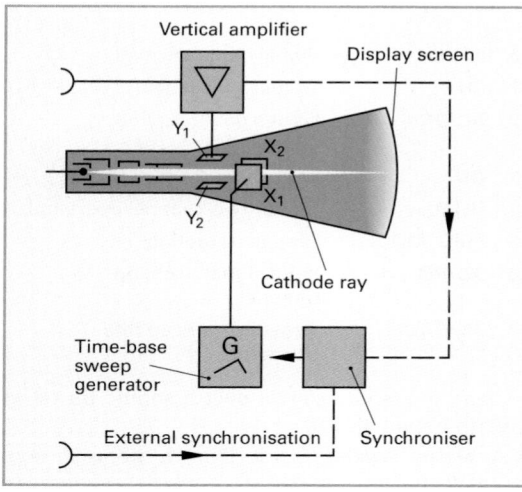

Fig. 1: Block diagram of a cathode-ray oscilloscope

In the cathode ray tube, similar to a television picture tube, a tightly bundled cathode ray flowing in a vacuum is created, which produces a dot on a screen with fluorescent material in the centre of the picture tube.

The cathode ray can be moved in a vertical direction **(Fig. 2)** using the couple $Y_1 Y_2$ and in a horizontal direction **(Fig. 3)** using the couple $X_1 X_2$.

If a direct voltage is applied to the vertical sweep $Y_1 Y_2$, the small luminescent area moves away from the centre upwards if the voltage is positive, downwards if the voltage is negative, and remains there displayed as a dot. If an alternating voltage is applied instead of direct voltage, a vertical bar is generated **(Fig. 2)**.

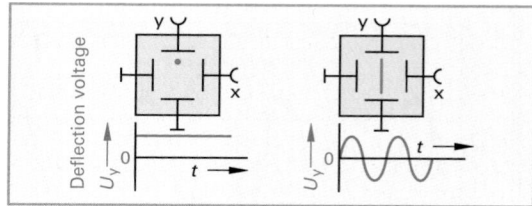

Fig. 2: Vertical sweep $Y_1 Y_2$

If a direct voltage is applied to the horizontal sweep $X_1 X_2$, the small luminescent area moves around on the display screen to the left or to the right depending on the polarity of the voltage and remains there displayed as a dot. If a uniformly changing voltage, e.g. sawtooth voltage, is applied instead of direct voltage, a horizontal bar is generated **(Fig. 3)**.

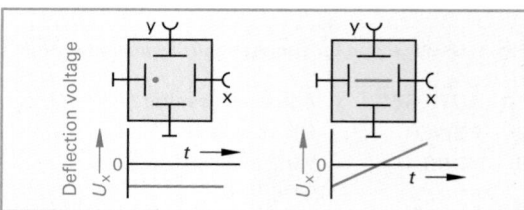

Fig. 3: Horizontal sweep $X_1 X_2$

If the measurement voltage is now applied to the vertical couple $Y_1 Y_2$ and a sawtooth voltage as line sweep voltage is applied to the horizontal couple $X_1 X_2$, the actual measurement voltage process in real time is visible on the screen **(Fig. 1)**.

Triggering. The time-base sweep must always start at the same measurement voltage level (signal voltage) so that a stationary picture is produced on the screen of the oscilloscope. The time-base sweep is triggered by a trigger pulse. The time-base sweep generates a sawtooth, i.e. the cathode ray runs once over the screen and back again over a period. The trigger pulse can either be produced in the time-base sweep generator itself (internal) or it is supplied from the outside as a voltage pulse (external).

Triggering refers to activation of the time-base sweep by a triggering pulse.

19

Control elements of an oscilloscope

The terms on the control panel of an oscilloscope are usually in English, regardless of the country of application, and are standardised to a large extent **(Fig. 1)**.

By way of example, the control panel shown with the individual control elements includes all the important connections and adjusting devices that are required to operate two-trace or two-channel oscilloscopes.

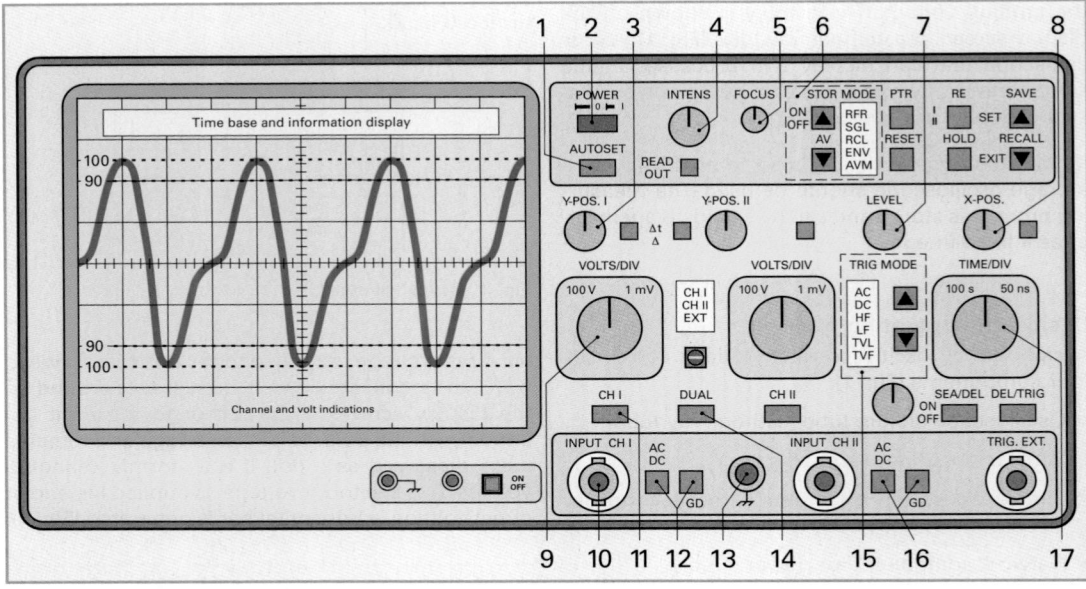

Fig. 1: Oscilloscope for simultaneous representation of two processes (two-channel oscilloscope)

1	**AUTO SET**	Automatic setting
2	**POWER**	On/off switch
3	**Y-POS: I**	Vertical displacement channel I
4	**INTENS**	Brightness setting
5	**FOCUS**	Sharpness setting
6	**STORE MODE**	Storage mode
7	**LEVEL**	Setting the trigger point
8	**X-POS.**	Horizontal beam displacement
9	**VOLTS/DIV.**	Amplitude adjustment for channel I

10	**INP. CH I**	Signal input, channel I
11	**CH I**	Trigger switching
12	**AC/DC**	Input signal coupling Channel I
13	**GD**	Earth jack
14	**DUAL**	One or two-channel operation
15	**TRIG. MODE**	Type of triggering
16	**AC/DC**	Input signal coupling Channel II
17	**TIME/DIV**	Time-base sweep rate Horizontal

Measuring with the oscilloscope

> The oscilloscope essentially measures only voltage.

When taking measurements with the oscilloscope, the following points should be borne in mind:

- To take measurements, a connection must be established from the measurement object to the earth jack and to the channel I or II signal input of the oscilloscope.
- The neutral axis of the display picture can be adjusted on the screen using the rotary knob **(Fig. 1, Item 3)** in accordance with the signals, so that the whole signal is visible. To carry out this adjust-

ment, the input selector switch should be set to earth potential (GD).
- A higher sweep factor **(Fig. 1, Item 9)**, e.g. 100 V/cm, can then be set.
- Then the horizontal time-base sweep speed is changed as far as necessary until the signal picture is stationary **(Fig. 1, Item 17)**.

On many types of oscilloscope, the housing is connected to the protective earth conductor of the mains power supply. In the event that the measured object needs to be operated with an alternating voltage of more than 50 V, a mains adaptor unit or an isolating transformer must be used for safety reasons.

Table 1: Measuring with the oscilloscope (examples)

Assembling a circuit	Displays and settings	Analysis
Measuring a direct voltage *U*		
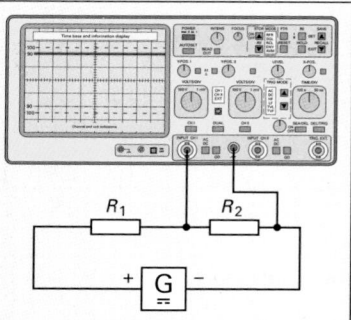	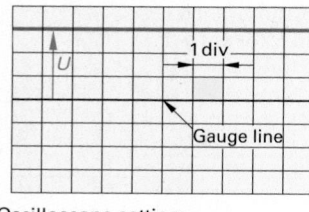 Oscilloscope setting: 5 V / div[1] [1] div: abbreviation for divit (part), raster unit of the display screen	Direct voltages are measured with the DC setting. Example: Direct voltage *U*: $U = \dfrac{5\,\text{V}}{\text{div}} \cdot 3\,\text{div} = 15\,\text{V}$
Measuring alternating current and period duration		
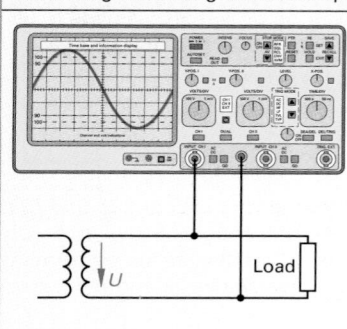	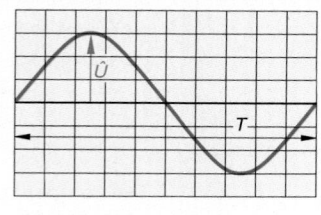 Oscilloscope setting: – Amplitude I: 2 V / div – Time Base: 2 ms / div	AC voltages are measured in the AC setting. Example: $\hat{U} = \dfrac{2\,\text{V}}{\text{div}} \cdot 3\,\text{div} = 6\,\text{V}$ $U = \dfrac{\hat{U}}{\sqrt{2}} = \dfrac{6\,\text{V}}{\sqrt{2}} = 4.2\,\text{V}$ $T = \dfrac{2\,\text{ms}}{\text{div}} \cdot 10\,\text{div} = 20\,\text{ms}$ $f = \dfrac{1}{T} = \dfrac{1}{20\,\text{ms}} = 50\,\text{Hz}$
Measuring currents (indirect measurement)		
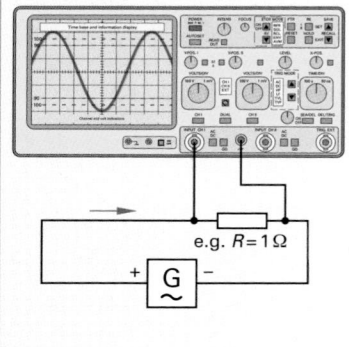	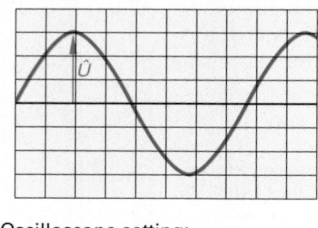 Oscilloscope setting: 50 V/div	The current intensity is determined by measuring the voltage *U* at a known resistance, e.g. 1 Ω, and calculating the current *I* using Ohm's Law. Example: $\hat{U} = \dfrac{50\,\text{mV}}{\text{div}} \cdot 3\,\text{div} = 0.15\,\text{V}$ $U = \dfrac{\hat{U}}{\sqrt{2}} = \dfrac{0.15\,\text{V}}{\sqrt{2}} = 0.1\,\text{V}$ $I = \dfrac{U}{T} = \dfrac{0.1\,\text{V}}{1\,\Omega} = 0.1\,\text{A}$

REVIEW QUESTIONS

1 What do you understand by indirect resistance measurement?

2 How is an ammeter connected?

3 Which electric measuring instruments are analogue display instruments?

4 What should be noted when measuring the AC voltage with the moving-coil measuring element?

5 What does accuracy class 1.5 mean?

6 What is the disadvantage of digital display instruments when measured variables are fluctuating?

7 What rule applies when analogue multimeters are used?

8 What voltage is applied at the vertical and/or horizontal sweep?

9 What do you understand by triggering?

10 Describe the movement of the cathode ray on the display screen, if only the time-base sweep is switched on.

19.1.7.5 Bridge circuits with resistors

Bridge circuits **(Fig. 1)** consist of two series connections with two resistors each that are connected in parallel to a common voltage source. The total current I_G branches at point A into the branch currents I_1 (over resistors R_1 and R_2) and I_2 (over resistors (R_3 and R_4). Resistors R_1 ... R_4 function as voltage dividers.

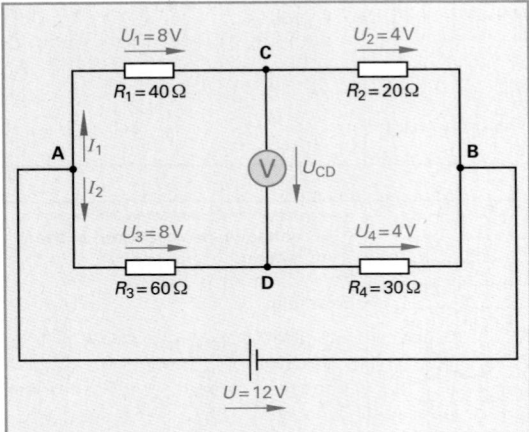

Fig. 1: Bridge circuits with resistors

If the voltage divider R_1-R_2 divides the voltage of the voltage generator in the same ratio as voltage divider R_3-R_4, there is no voltage between points C and D (zero point method). Resistors R_1 and R_2 are therefore in the same ratio to each other as resistors R_3 and R_4.

> A bridge circuit is balanced if there is no current flowing in the bridge diagonals C-D, i.e. when the resistive ratio is the same in both voltage dividers.
>
> $$\frac{R_1}{R_2} = \frac{R_3}{R_4} \quad \Rightarrow \quad \frac{U_1}{U_2} = \frac{U_3}{U_4}$$

Resistance measurements are performed with the aid of a bridge circuit. Resistors R_1 are thereby replaced by the resistor to be measured R_x and R_2 by the variable resistor R_n.

The bridge circuit for measuring resistance is referred to as a Wheatstone measuring bridge **(Fig. 2)**.

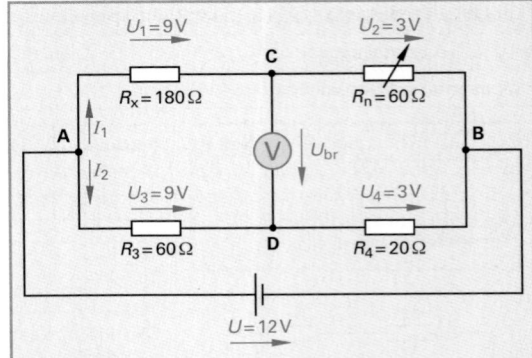

Fig. 2: Wheatstone measuring bridge

In a balanced measuring bridge, to calculate R_x it is sufficient to know R_n and the ratio of R_3 to R_4.

> $$\frac{R_x}{R_n} = \frac{R_3}{R_4} \quad \Rightarrow \quad R_x = R_n \cdot \frac{R_3}{R_4}$$

The comparison resistance of resistors R_2 and R_n is generally adjustable. This enables a level to be obtained for it which does not deviate too far from the level of the unknown resistance R_1 and R_x. This reduces measurement errors.

Resistors R_3 and R_4 can be replaced by an infinitely variable resistor (variable resistor or slide wire with slider) **(Fig. 3)**.

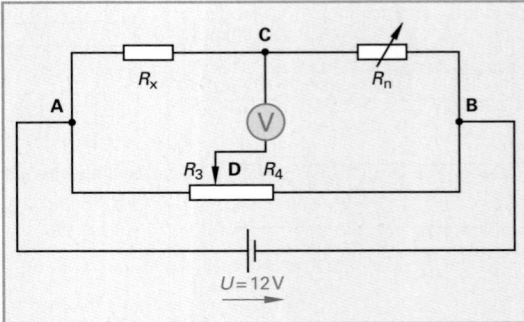

Fig. 3: Slide-wire bridge

> Resistors can be measured very accurately using measuring bridges. The result of the measurement depends on the level of the supply voltage.

The measuring bridge is used on motor vehicles, e.g. in the air-mass meter.

19.1.8 Properties of electrical current

Thermal properties

When current flows in a metallic conductor, the electrons move between the individual atoms. The kinetic energy of the electrons is transferred to the atoms. These begin to oscillate and thereby generate heat (**Fig. 1**). A measurement of the "heat oscillations" is the temperature of the conductor.

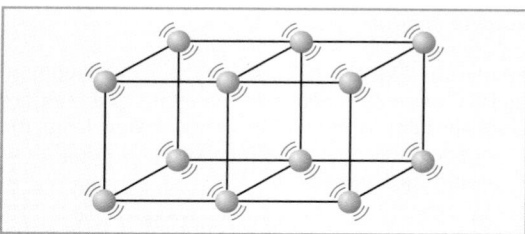

Fig. 1: Heat oscillation of the molecules

> Heat is generated in all current-carrying conductors.

Light properties

Thin wires made from metal are heated by the electrical current to such an extent that they glow. The light yield is greater, the higher the temperature of the spiral-wound filament. For this reason, metals with a high melting point are used, e.g. tungsten. So that the spiral-wound filament does not oxidise, it is allowed to glow in a vacuum or in an inert gas (e.g. nitrogen, krypton). Filament lamps are temperature emitters.

> When current flows through gases, light is generated by the impact of charged gas particles.

Gas-discharge lamps, e.g. fluorescent lamps (**Fig. 2**), are more efficient than filament lamps, since they lose less heat. Gas-discharge lamps are cold emitters.

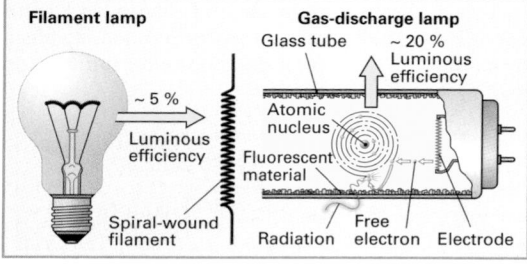

Fig. 2: Light properties

Chemical properties

> Liquids that conduct current are known as electrolytes. Their principal constituents break down when current flows through them. This process is called electrolysis.

Electrolytes are acids, bases, salts and metal oxides in aqueous solution or molten state. The principal constituents produced by degradation as the current flows through migrate to the current inputs (electrodes) and are deposited there. This action of an electrical current is used in copperplating (**Fig. 3**), for example. This process is called galvanising.

Magnetic properties

> A magnetic field is generated in every current-carrying conductor.

A current-carrying conductor can cause a magnetic needle to move from pointing north-south, i.e. a magnetic force comes from a current-carrying conductor (**Fig. 4**). The direction of this force depends on the current direction in the conductor. This electromagnetic phenomenon is used in electrical machines, for instance.

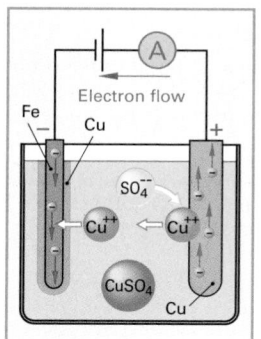
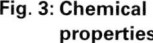
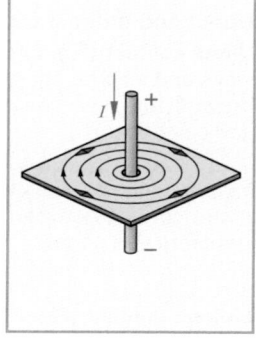

Fig. 3: Chemical properties

Fig. 4: Magnetic properties

Physiological properties

> The physiological effects of electrical current are its impact on living organisms.

A current is able to flow through the human body when voltage sources are touched. The electrical current "electrifies", the person receives an "electric shock". This property of current is used in rodent repellers.

19.1.9 Protection against the hazards of electrical current

Electrical current is life-threatening to people and animals.

When a current with an intensity of more than 50 mA flows through the human body, it can be fatal. An AC voltage above 50 V can produce life-threatening currents in the human body.

Failure modes (Fig. 1). In electric systems, faults to frame, short circuits, conductor faults and earth faults can occur.

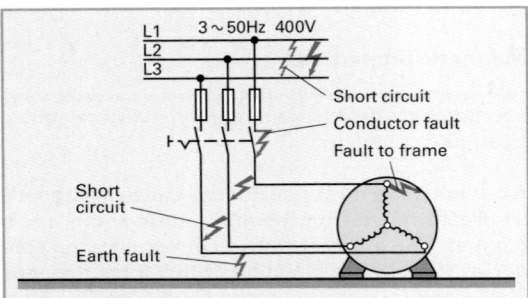

Fig. 1: Failure modes

Direct and indirect contact

Direct contact (Fig. 2) occurs when energised devices and electric lines can be touched directly. To prevent this, live components must be insulated and covered.

Protective measures against direct contact prevent contact between energised electrical conductors and parts.

Indirect contact (Fig. 3) occurs when parts of devices which should not be able to conduct voltage are energised due to a fault, and are touched. This may be the case if the housing of a device is energised due to an insulation fault.

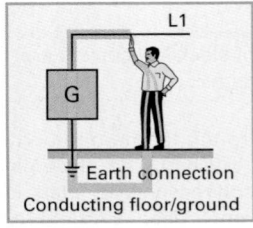

Fig. 2: Direct contact

Fig. 3: Indirect contact

Protective measures against indirect contact prevent impermissibly high error voltages from occurring in equipment parts.

Mains-independent protection measures

In the event of a fault, the device is not switched off; the protection measures take effect without a protective earth conductor. Mains-independent protection measures include insulation, protective low voltage and fuse disconnectors.

Insulation (Fig. 4). All parts that can receive voltage against earth connection in the event of a fault are also encased in insulation or separated from the live part of the device by insulators, in addition to their earthing insulation.

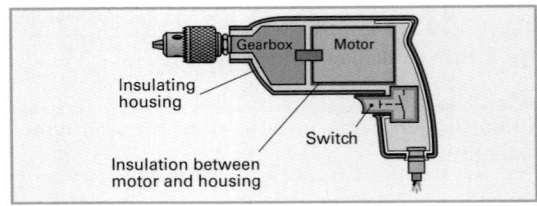

Fig. 4: Insulation

Protective low voltage (Fig. 5). Protective low voltages are AC voltages up to 50 V. These need to be produced in transformers or circumferential transducers, whereby the low voltage side must not have an electrical link with the mains power supply.

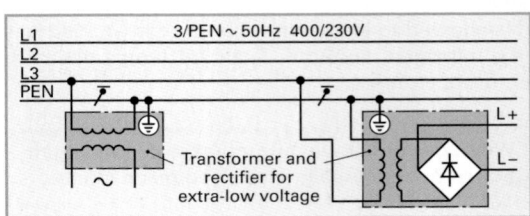

Fig. 5: Protective low voltage

Fuse disconnectors (Fig. 6). A transformer is connected between the power supply and consumer. The output side of the transformer must not have an earth connection, i.e. in the event of a fault, there is also no voltage between the device and earth connection. The isolating transformer usually has a turns ratio of 1 : 1, i.e. it does not alter the voltage level.

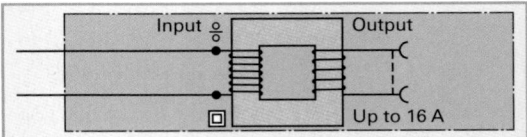

Fig. 6: Fuse disconnectors

Mains-dependent protection measures

These protection measures are only effective with a protective earth conductor, PE (Protection Earth). As protection, overflow protection equipment (fuses, automatic cut-outs) and earth leakage circuit breakers (ELCB) are used, which disconnect the equipment from the mains in the event of a fault.

Protection provided by overflow protection equipment

The degree of protection was previously referred to as "neutral". The voltage generator is directly earthed. The conductive parts of the housing and the device are connected via the PE protective earth conductor (green/yellow-coloured insulation) to the earth connection of the voltage generator. In the event of a fault, the fault to frame causes a short circuit which allows overflow protection equipment (e.g. fuses, automatic cut-out/line safety switch) to respond and disconnect the equipment from the mains within the given time **(Fig. 1)**.

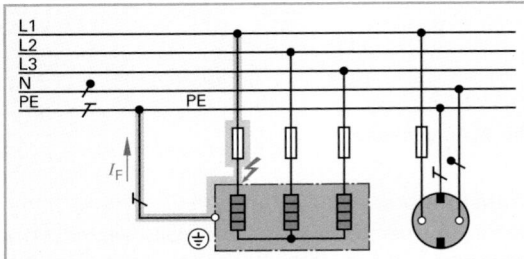

Fig. 1: Overflow protection equipment

On portable equipment, on which the electrical energy is supplied via a plug and socket connection, the PE protective earth conductor must be connected to the respective earthing contacts on the connector and the socket **(Fig. 2)**.

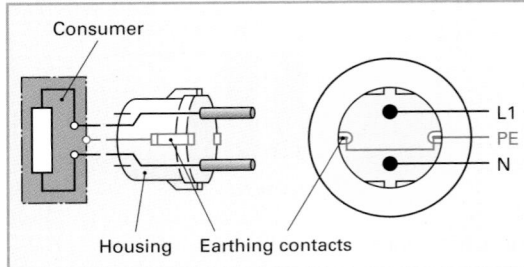

Fig. 2: Portable consumers with protective earth conductor

Residual current protective equipment (Fig. 3). The function of this equipment is to disconnect all poles from the consumers within 0.2 s in the event of a fault. All conductors (e.g. L1, N) that lead from the mains to the device to be protected are routed

through a summation current transformer. The PE protective earth conductor is not routed through the summation current transformer.

As long as there is no fault present, the current flowing in (I_L) is the same as the current flowing out (I_N), i.e. the magnetic fields in the summation current transformer produced in the current-carrying coils cancel out the effect of each other.

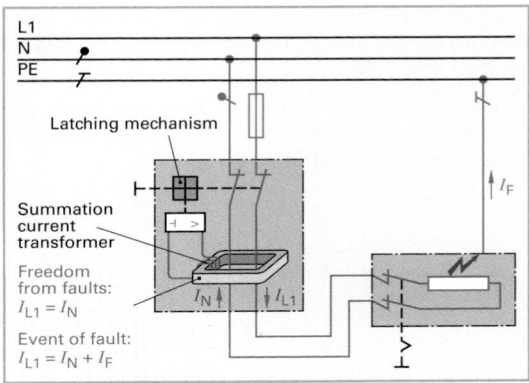

Fig. 3: Earth leakage circuit breaker

In the event of a fault, a partial current (residual current I_F) flows via the PE protective earth conductor, i.e. the currents in the summation current transformer (I_L, I_N) are not the same size.

The magnetic fields in the summation current transformer produced in the current-carrying coils no longer cancel out the effect of each other. A voltage is induced in the secondary winding of the summation current transformer which activates the initiator trigger in the latching mechanism and disconnects the supply leads to the consumers at all poles. The functional efficiency of the earth leakage circuit breaker can be checked using test button P.

REVIEW QUESTIONS

1 **What relationship in the electrical circuit is described by Ohm's Law?**

2 **On what variables does electrical power depend?**

3 **On what variables does electrical work depend?**

4 **What units are used for electrical work?**

5 **What is efficiency?**

6 **Why is efficiency always < 1?**

7 **How do voltages and currents behave in a series connection?**

8 **What is the relationship between total resistance and branch resistances in a parallel connection?**

9 **What properties of electrical current can occur?**

10 **What failure modes can occur in electrical systems?**

19

19.1.10 Voltage generation

Voltage generation by induction

> Induction means the generation of voltage by a change in the magnetic flux that carries through a conductor loop or coil.

If a permanent magnet is moved back and forth inside a coil, an alternating voltage is generated in the coil **(Fig. 1, left)**.

If a conductor loop is moved back and forth in a magnetic field, an alternating voltage is generated in the conductor loop **(Fig. 1, right)**.

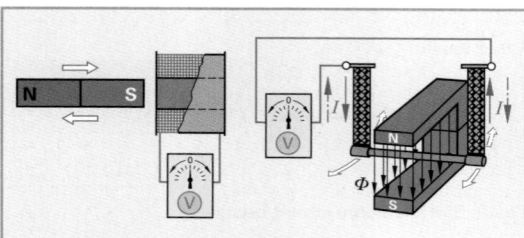

Fig. 1: Induction of the movement

This voltage generation process is referred to as induction. Thereby, a voltage is induced for only as long as the magnetic flux changes in the coil or conductor loop. The term magnetic flux means the total number of magnetic field lines that are surrounded by the coil or conductor loop.

The level of the induced voltage is proportional to the rate of change of the magnetic flux surrounded by the coil (increase or decrease per time unit) and the number of coils.

The manner in which the magnetic flux in the coil changes does not affect the generation of the induction voltage.

The magnetic flux can be changed by:
- Moving or turning a coil in the magnetic field.
- Switching the current on and off in a winding, e.g. excitation current in the excitation winding of an alternator.
- Periodically changing the current intensity, e.g. in the primary winding of a transformer.

The direction of the voltage induced depends on the direction of the movement and on the direction of the magnetic field **(Fig. 1, left)**.

The direction of the current can be determined using the alternator rule **(Fig. 2)**.

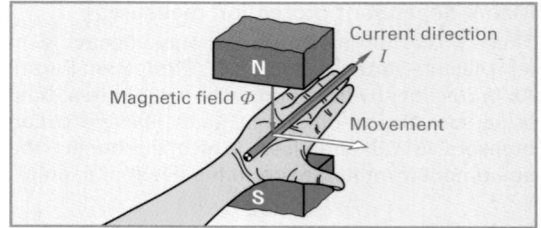

Fig. 2: Alternator rule

Voltage generation by induction has become the general principle for the generation of electrical energy. A rotor with excitation field produces a voltage in the stationary induction coil (stator winding) **(Fig. 3)**.

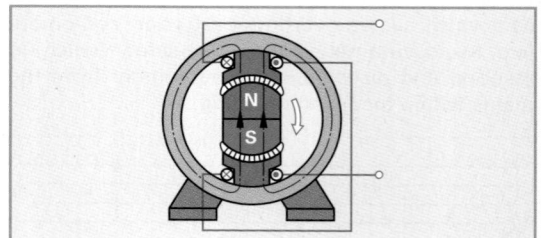

Fig. 3: AC generator

Transformers

A transformer consists of two coils seated on a common iron core **(Fig. 4)**. The input winding (primary winding) takes electrical energy from the alternating-current power line. The alternating current taken by the primary winding creates an alternating magnetic field which is passed to the iron core and from there transmitted to the secondary winding. The varying magnetic field induces a voltage in the secondary winding.

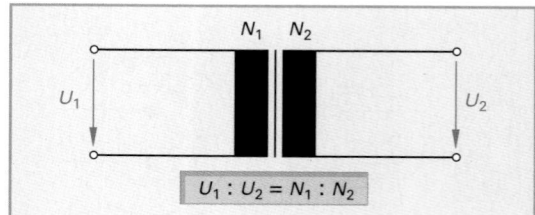

$$U_1 : U_2 = N_1 : N_2$$

Fig. 4: Transformer

The following applies to the transformer:

> The primary voltage U_1 is proportional to the secondary voltage U_2 in the same way as the number of coils of the primary winding N_1 is proportional to the number of coils of the secondary winding N_2.

Voltage generation by electrochemical processes

Galvanic element (Fig. 1). If two different metals are immersed in an electrolyte (acid, base or salt solution), this creates a galvanic element; a direct voltage is present between the two metallic electrodes (poles) **(Fig. 1).** Carbon can also be used instead of a metal.

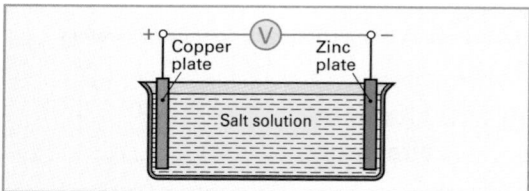

Fig. 1: Galvanic element

> The voltage level of a galvanic element depends on the position of the electrode material in the electrochemical series.

When several galvanic elements are series connected to multiply the voltage, this creates a battery.

When the current is drawn, the current also flows inside the galvanic elements, and from the negative terminal to the positive terminal. The electrolyte decomposes, the negative terminal metal dissolves or is chemically converted. The hydrogen occurring at the positive terminal must be chemically bonded so that the voltage does not drop as the current is drawn. This happens with the zinc-carbon element **(Fig. 2)** by sheathing the positive terminal with materials which allow bonding with the hydrogen, e.g. brownstone (MnO_2). The supply of current ceases when the electrolyte is used up or the negative terminal metal has been chemically converted.

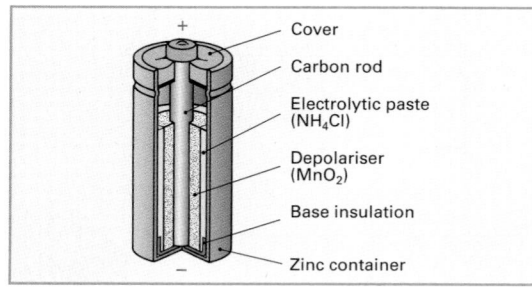

Fig. 2: Zinc-sal ammoniac-brownstone cell

Galvanic elements for which the electrochemical conversion by current input can be reversed again are called accumulators.

Voltage generation by heat

Thermocouple (Fig. 3). When two wires made from different metallic materials are connected and the joint heats up, a direct voltage arises between the free ends of the wires. The size of the voltage depends on the combination of the metallic wires and the temperature. A connected voltmeter can be calibrated to display the temperature in °C. Thermocouples are used to control electrical fans, for example.

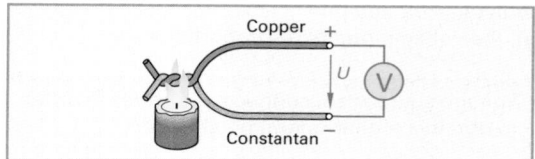

Fig. 3: Thermocouple

Voltage generation by light

Photoelement (Fig. 4). This usually consists of a metallic base plate to which a semiconductor layer, e.g. selenium, is applied. The semiconductor layer is connected to a contact ring. When there is an incidence of light, a direct current is generated between the contact ring and the base plate. Photoelements are used as photoelectric switches, for example.

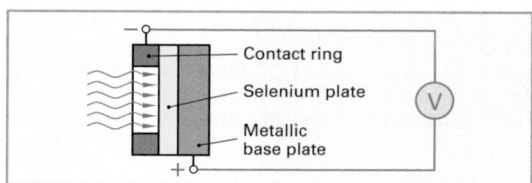

Fig. 4: Photoelement

Voltage generation by crystal deformation

Piezo element (Fig. 5). This consists of a crystal (e.g. silicon dioxide). Pressure variation generates an alternating voltage which is dissipated via conductive coverings. Piezoelectric[1] voltage generators are used as sensors in fast-changing pressure processes, e.g. as knock-sensors on combustion engines.

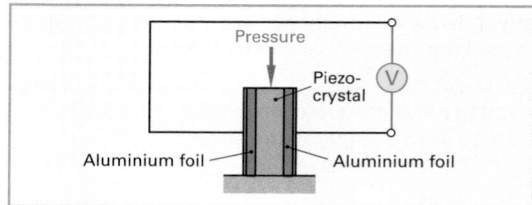

Fig. 5: Piezo element

---
[1] from piedein (Greek) = to press

19

Wiring voltage generators

Voltage generators can be operated in series connections and in parallel connections.

Series connection (Fig. 1). The principles of series-wired resistor circuits apply. Both the internal resistances and the open-circuit voltages are also added. The total current intensity is as high as the current intensity of the individual voltage generator. The capacity of identical starter batteries connected in series is the same as the capacity of the individual battery (no increase in capacity). However, the voltage multiplies proportionately.

> Voltage sources are connected in series in order to obtain a higher operating voltage.

When voltage generators are connected in series, the positive terminal of one battery is connected to the negative terminal of the next.

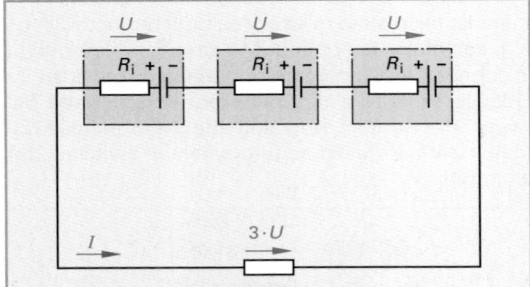

Fig. 1: Series connection with voltage generators

Parallel connection (Fig. 2). The principles of series-wired resistor circuits apply. Both the currents and the conductances are also added. Only voltage sources with the same nominal voltage should be connected in parallel. If batteries with different voltages are connected in parallel, a higher equalising current flows from the battery with the higher voltage to the battery with the lower voltage. Both batteries could be damaged, making them unusable.

The voltage of identical starter batteries connected in parallel is the same as that of an individual battery; however, it multiplies according to the capacity and the current that can be drawn.

> Voltage sources are connected in parallel in order to draw a higher current.

In a parallel connection of voltage generators, all positive terminals must be connected to each other and all negative terminals must be connected to each other.

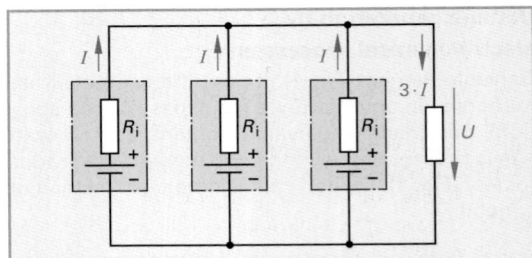

Fig. 2: Parallel connection with voltage generators

19.1.11 Alternating voltage and alternating current

In power engineering, sinusoidal AC voltages are used predominantly. These can be generated easily in alternators, have their level changed with the aid of transformers and be carried over long distances.

When a conductor loop rotates steadily in the magnetic field, a sinusoidal voltage is generated by induction. It is constantly changing its size and periodically changes its direction **(Fig. 3)**.

One complete fluctuation is called a period. The time it takes to occur is the period duration T. The number of periods in a second is the frequency f. The unit of frequency is the Hertz (Hz).

The frequency f of the alternating voltage depends on the rotational speed n and the number of pairs of poles p of the alternator ($f = p \cdot n$). If a double-pole rotor (a pair of poles) of an alternator turns at a rotational speed of 3,000 rpm = 50 rps, the induced alternating voltage has a frequency of 50 Hz.

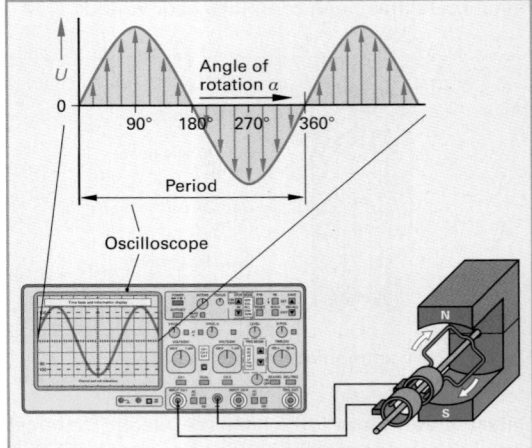

Fig. 3: AC generator

19.1.12 Three-phase AC voltage and three-phase current

The alternator has three windings in the stator (U_1 – U_2, V_1 – V_2, W_1 – W_2) which are spatially offset by 120° **(Fig. 3)**. When the pole wheel or the rotor is rotated (coil with direct current excitation) 360°, three AC voltages or alternating currents which are phase-delayed to each other by 120° are generated in the windings **(Fig. 1)**.

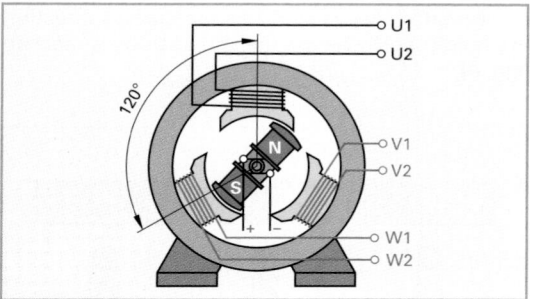

Fig. 1: Alternator

In the line drawing of the three alternating currents **(Fig. 2)**, current I_1 that flows in coil U_1 – U_2 has its highest level in Time 1 (pole wheel position 90°). Current I_2 in coil V_1 – V_2 and current I_3 in coil W_1 – W_2 is half as high as I_1 respectively. Currents I_2 and I_3 also act in the opposite direction to current I_1.

> The sum of the currents I_1, I_2, I_3 is always zero.

This applies to any position of the pole wheel **(Fig. 2)**.

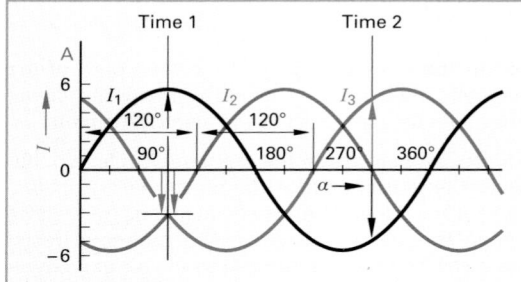

Fig. 2: Graph of the alternator

To conduct these three alternating currents away, six conductors (one supply and one return each) should be available. However, it is possible to manage with only three conductors by connecting (interlinking) the three coils respectively, because these conductors are alternately "supply" and "return" due to the time displacement of the three alternating currents.

Star connection (Fig. 3). This is obtained by connecting the ends of the 3 windings U_2, V_2, W_2 together in the neutral point. The beginnings of the windings U_1, V_1, W_1 are connected with the external conductors L_1, L_2, L_3 of the mains.

Delta connection (Fig. 4). This is obtained when one end of each winding is connected to the start of the next winding, e.g. U_1 to W_2, W_1 to V_2, V_1 to U_2. The junctions are connected to the external conductors of the mains L_1, L_2, L_3.

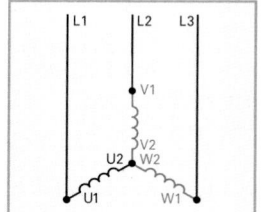 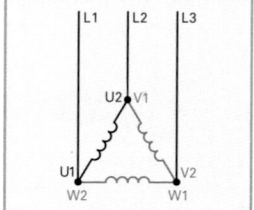

Fig. 3: Star connection   **Fig. 4: Delta connection**

The interconnection to make the star or delta connection is called interlinking.

19.1.13 Magnetism

19.1.13.1 Permanent magnetism

Magnets attract iron, nickel and cobalt. The points of the strongest attraction are the poles of the magnet. Each magnet has a north and south pole.

> Different poles of two magnets attract, the same poles repel.

If a bar magnet is free to move, it adjusts itself to the north-south direction. The pole that points towards the north is the north pole of the magnet, the opposite pole is the south pole. A magnetic field exists in the area surrounding the magnet. The field lines are imaginary lines, which each give the direction of the magnetic force. They are always closed and run outside the magnet from the north to the south pole and inside the magnet from the south to the north pole **(Fig. 5)**.

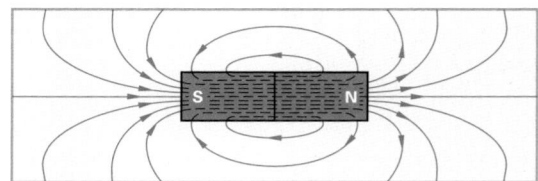

Fig. 5: Field of a bar magnet

19

19.1.13.2 Electromagnetism

A magnetic field is generated around a conductor carrying an electrical current. The magnetic field lines form concentric circles.

The direction of the field lines around a current-carrying conductor can be determined using the screw rule. If one thinks of a screw with right-hand thread screwed in in the direction of the current in a conductor, then the direction of rotation gives the direction of the field lines (Fig. 1).

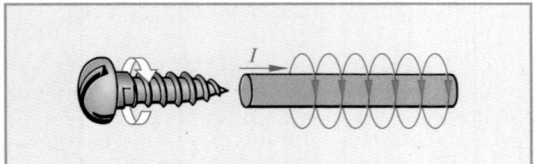

Fig. 1: Magnetic field of the current-carrying conductor

The symbol $\otimes$ is used for current entering the conductor, the symbol $\odot$ for a current leaving the conductor.

If the conductor is wound up into a coil, the magnetic field lines inside the coil are bundled. They run parallel and at the same density here; they are known as a homogenous magnetic field. At the emergence point of the field lines, a north pole is created, at the entry point, a south pole (Fig. 2).

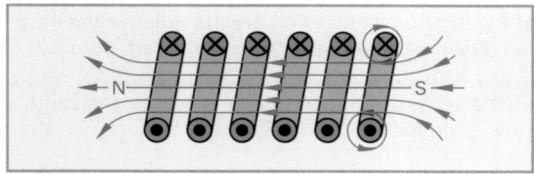

Fig. 2: Magnetic field of a coil

Action of forces between current-carrying conductors. Two current-carrying conductors exert forces on each other due to their magnetic fields (Fig. 3).

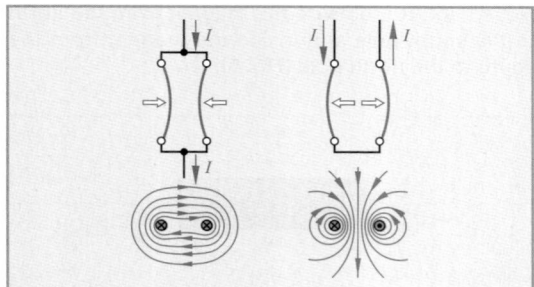

Fig. 3: Current-carrying conductors

Conductors carrying current in the same direction attract, conductors carrying current in the opposite direction repel.

Current-carrying conductors in the magnetic field. A pivoted current-carrying coil in the magnetic field is turned to a specific position until the field it creates is in the same direction as the stationary field. Continuous rotation can be achieved if a commutator (collector) is attached to the moving coil, and the commutator always switches the current direction in the coil shortly before the end position is reached (Fig. 4).

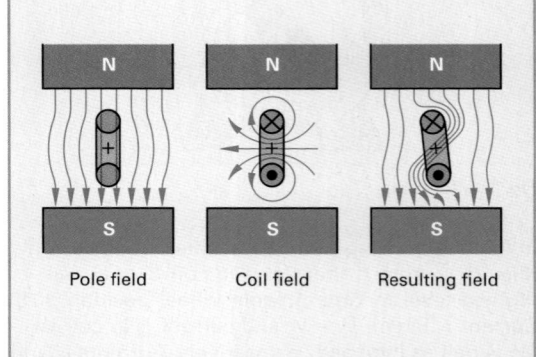

Fig. 4: Conductor and coil in the magnetic field

In the magnetic field, a force is exerted on the current-carrying conductor, which aims to move it out of its idle state.

Iron in the magnetic field. The closed path of the magnetic field lines is called a magnetic circle; which can be compared to the electrical circuit.

If there is an air gap in the magnetic circle, e.g. between the stator and the rotor of alternators or electric motors, the magnetic field lines must overcome a high magnetic resistance. The magnetic resistance can be reduced by making the air gap smaller or by introducing a nucleus made of soft magnetic material into the coil cavity if it is a solenoid coil.

Iron amplifies the magnetic flux Φ of a coil.

The reason for this is the alignment of the elementary magnets in the iron, which produce additional magnetic field lines.

19

19.1.14 Self-induction

This occurs at current-carrying coils, if the coil current changes. This current change causes a change in the magnetic field in the coil, i.e. the size of the magnetic flux in the coil changes. This leads to the induction of a voltage, the **self-induction voltage**.

Test 1 (Fig. 1). A coil with an iron core (N = 1,200 windings) and an adjustable resistance are each connected with a filament lamp (1.5 V/3 W) in series and subjected to a voltage of 6 V. The variable resistor is adjusted, so that the two filament lamps light up equally brightly.

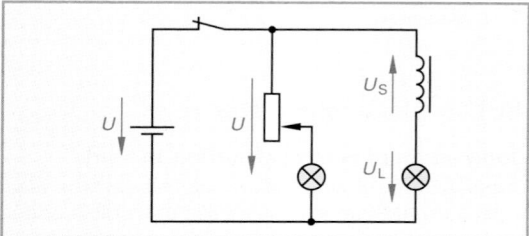

Fig. 1: Switching on a coil

Observation. When closing the electric circuit, the filament lamp connected in series with the coil lights up with a delay.

The current flowing in the coil builds up a magnetic field. The magnetic field building up produces a magnetic flux change in the coil, which in turn induces a voltage U_S in the coil, which acts in the opposite direction to the applied voltage. This means that the applied voltage becomes fully effective gradually **(Fig. 3)**

When the current in a coil is switched on, the self-induction causes a slowing down of current and magnetic field generation.

Test 2 (Fig. 2). A coil with an iron core (N = 1,200 windings) and a glow lamp with an ignition voltage of approximately 150 V are connected in parallel and subjected to a voltage of 6 V.

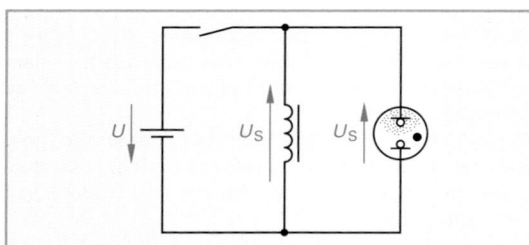

Fig. 2: Switching off a coil

Observation. When opening the electric circuit, the glow lamp connected in parallel to the coil immediately lights up briefly.

Once the voltage source is switched off, current no longer flows through the coil. The magnetic field built up before falls very quickly, i.e. it changes direction compared to the build up phase; a very high voltage is induced in the coil (self-induction voltage, **Fig. 3**).

When the current in a coil is switched off, the self-induction causes a slowing down of current and magnetic field dissipation.

This induced voltage (self-induction voltage) is rectified to the previous applied voltage. It continues to support a flow of current in the coil for a short time, which prevents the sudden collapse of the magnetic field **(Fig. 3)**.

The self-induction voltage is always rectified such that it acts in the opposite direction to the change of current.

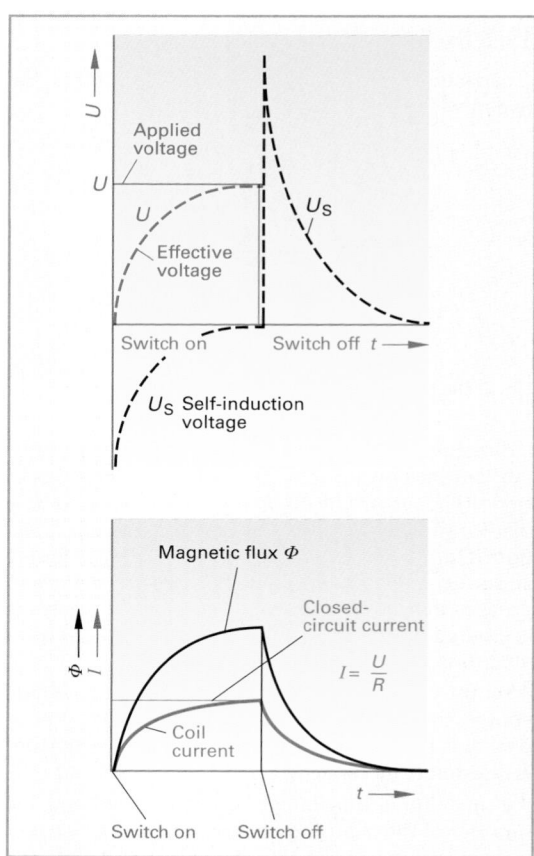

Fig. 3: Voltage and current curves

Since the high self-induction voltage that results when a coil is switched off has the same direction as the applied voltage beforehand, an arc of light is produced as soon as the contact is opened slightly.

If alternating voltage is applied to a coil, the self-induction voltage increases with increasing frequency, so that the mean value of the current per time unit decreases. This means that the current in the coil thereby decreases, i.e. the resistance of the coil seems to increase (**inductive reactance**) **(Fig. 1)**.

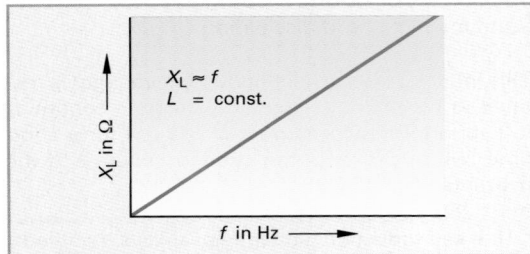

Fig. 1: Reactance of a coil

19.1.15 Capacitor

A capacitor consists of two metallic conductors, between which there is an insulator **(Fig. 2)**.

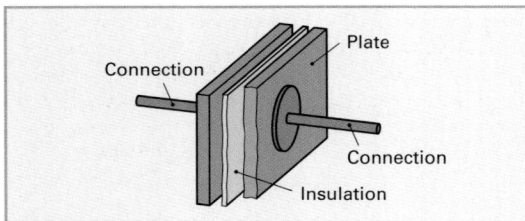

Fig. 2: Design of the capacitor

If direct voltage is applied to the capacitor, a charging current flows briefly. The capacitor then blocks the direct current. If the capacitor short-circuits, a discharge current then flows in the opposite direction **(Fig. 3)**. During charging, the voltage source sucks up electrons from one capacitor plate and presses them onto the other, i.e. a lack of electrons is created on one side and on the other, a surplus of electrons.

After the capacitor is disconnected from the voltage source, this electron differential remains, i.e. the capacitor is charged. The storage ability of the capacitor is referred to as capacity C. The unit is the farad (F).

If a capacitor is subjected to alternating voltage, the number of the charging and discharging processes increases with increasing frequency, so that the mean value of the current per time unit also in-

creases. This means that the current in the capacitor thereby increases, i.e. the resistance of the capacitor seems to decrease (capacitive reactance).

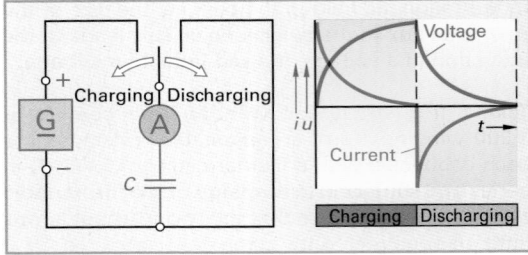

Fig. 3: **Charging and discharging behaviour of a capacitor**

19.1.16 Electrochemistry

Conduction of electrical current in fluids

Chemically pure water is an insulator for the electrical current. If an acid or a base or a salt is added to chemically pure water, it becomes conductive.

Liquids that conduct electricity are electrolytes.

In an electrolyte, e.g. H_2SO_4, a specific part of the molecule is split into its principal constituents 2 H^+ and SO_4^-. This process is called dissociation. These principal constituents, atoms and molecules, have different electrical charges; they are referred to as ions[1].

When a voltage is applied to the electrolyte, the ions are moved by the force of the electric field **(Fig. 4)**.

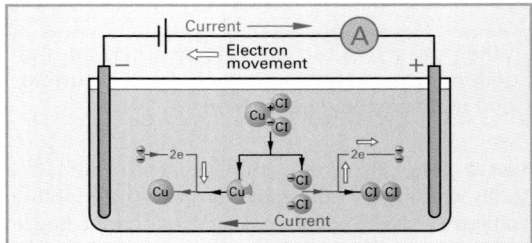

Fig. 4: **Electrolysis of copper chloride**

Positively charged ions migrate to the cathode (negative terminal). There, they take up the electrons, become electrically neutral and break away from the cathode.

Negatively charged ions migrate to the anode (positive terminal). There they release their excess electrons, become electrically neutral and break away from the anode.

---

[1] Ion (Greek) = migrating

Electrolysis

> During electrolysis, the principal constituents of the electrolyte break down when a direct current passes through. This process is called electrolysis.

The principal constituents liberate themselves from the current inputs (electrodes) and can form compounds with the latter.

Galvanising. By electrolytic means, workpieces can be given a thin metal plating, e.g. to protect against corrosion or to produce electrically conductive surfaces on plastics (printed-circuit boards).

If a direct voltage is applied in the test arrangement **(Fig. 1)**, the positively charged copper ions (Cu^{++}) migrate to the negative electrode and release their charge there; the copper breaks away from the negative electrode (cathode) and forms a deposit.

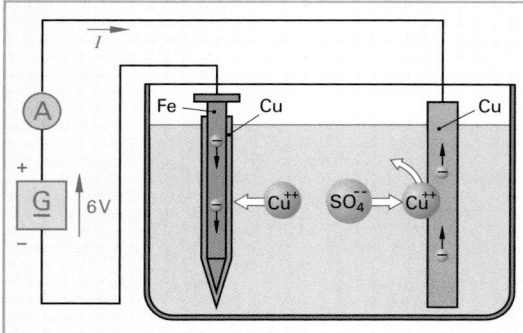

Fig. 1: Galvanising

The surplus of negatively charged acid ions (SO_4^{--}) migrate to the positive copper electrode (anode) and release their charge (electrons) there. A copper sulphate molecule ($CuSO_4$) is created at the same time. This can dissociate in turn. This process continues until the copper anode is used up. As this happens, pure copper is deposited on the cathode (–pole). This process is used to produce high-purity non-ferrous metals, e.g. electrolyte copper 99.98 %. Electrolysis can be used to apply a layer of zinc of a precisely defined thickness to a body sheet.

Galvanic elements

> They consist of two different metal electrodes or a metal electrode and a carbon electrode and an electrolyte.

The voltage is generated between the electrodes by electrochemical processes.

The voltage generated depends on the position of the electrodes in the electrochemical series **(Fig. 2)** and on the type and concentration of the electrolyte.

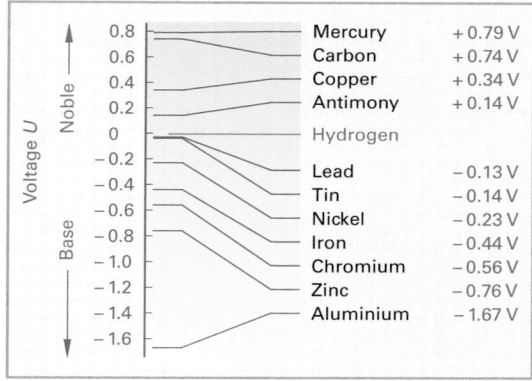

Fig. 2: Electrochemical series

Galvanic elements can be classified into **primary elements** and **secondary elements**.

Primary elements. The electrochemical processes occurring during the conversion of energy are irreversible. The negative terminal, which is always composed of the base metal, is irreparably damaged; the electrolyte can dry up or run out.

Secondary elements. When these are used, it is possible to reverse the electrochemical processes by charging with direct current, e.g. as with starter batteries. During charging, the electrical energy is stored in the form of chemical energy. During discharging, this is converted into electrical energy again.

> All electrolytic elements contain matter harmful to the environment, such as acids, bases, lead and other heavy metals. These must be disposed of specially and must not be added to the household rubbish.

19

REVIEW QUESTIONS

1 How do the poles of two magnets affect each other?

2 What is the effect of an iron core in a current-carrying coil?

3 How do two current-carrying conductors behave when they carry current in the same direction or carry current in the opposite direction?

4 What does self-induction mean?

5 How does a capacitor behave at an alternating voltage with increasing frequency?

6 What occurs in a galvanising process?

19.1.17 Electronic components

Semiconductor materials are used in the manufacture of electronic components, e.g. diodes, transistors. Close to absolute zero (– 273 °C ≙ 0 K), these materials behave like electrical insulators, i.e. they have a high specific electrical resistance.

At room temperature, the specific electrical resistance of semiconductor materials is between that of insulating materials and metallic conductors (**Fig. 1**).

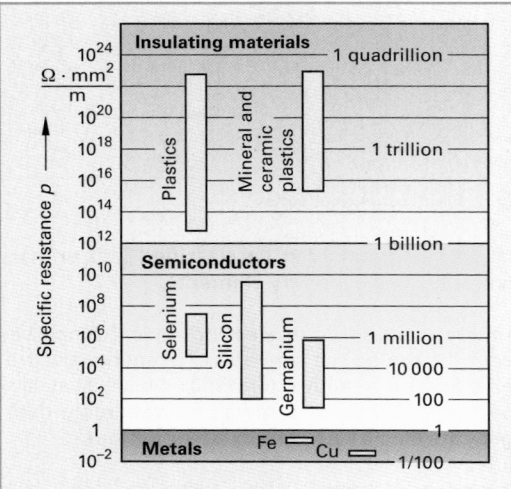

Fig. 1: Specific electrical resistance of materials at room temperature

The resistance of semiconductor materials decreases as temperature rises, their conductivity increases.

Semiconductor materials are highly temperature-sensitive. This property is utilised on thermistors, for example. As the temperature increases, the resistance of semiconductor materials drops and they exhibit NTC behaviour. At increased temperatures, there is therefore a larger flow of current given the same voltage, which can irreparably damage the semiconductor components. For this reason, semiconductor components are often assembled on a cooling plate. In a motor vehicle, the ECUs must be installed so that they are not exposed to intensive heat radiation, for example.

The resistance value of semiconductor materials can also depend on the applied voltage, on the incident light, on the mechanical pressure acting on them or on the strength of the magnetic field. The resistance characteristics are affected by material additives (contamination).

Table 1 lists commonly used semiconductor materials and their application.

Table 1: Semiconductor materials	
Description	Use
Silicon Si Germanium Ge	Rectifier diodes Transistors Photodiodes Phototransistors
Selenium Se	Rectifier diodes Photoelements
Gallium arsenide GaAs	Photodiodes

N-conductors and P-conductors

A slight "contamination" with impurities greatly increases the conductivity of the purest silicon. Depending on the material which is incorporated into (contaminates) the crystal lattice of the silicon base material, for example, either N-conductive semiconductor materials or P-conductive semiconductor materials are obtained (**Fig. 2**).

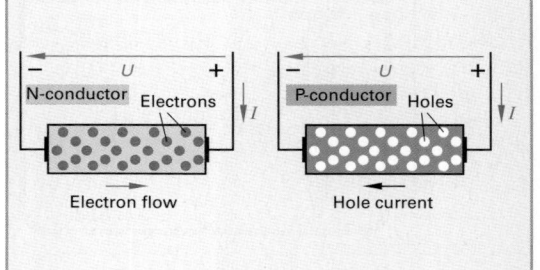

Fig. 2: N-conductor and P-conductor (system diagram)

N-conductors (N for negative). These are semiconductor materials which have a surplus of electrons. If a voltage is applied to an N-conductor, the free electrons move as in a metallic conductor.

N-conductors have electrons as charge carriers.

P-conductors (P for positive). These are semiconductor materials which have a lack of electrons. Where electrons are missing there is an electron deficit, i.e. the semiconductor material has a positive charge. The point where electrons are missing is also referred to as a hole. If a voltage is applied to the P-conductor, an adjacent free electron can jump into the hole. The hole, however, has migrated to the atom which has released an electron.

P-conductors have holes as charge carriers.

PN junction. If a P-conductor and an N-conductor are adjoining, a PN junction is formed. The free electrons in the boundary layer migrate from the N-conductor into the holes of the P-conductor. Thus, there are hardly any free charge carriers (electrons and holes) left in the boundary layer **(Fig. 1)**.

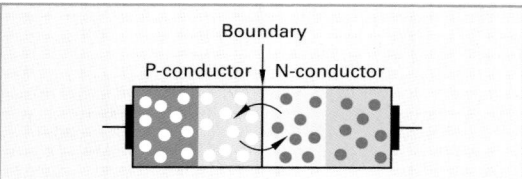

Fig. 1: PN junction

A depletion layer is generated at the PN junction of semiconductors.

19.1.17.1 Diodes

These are semiconductor devices which consist of a P-conductor and an N-conductor; these form a PN junction. They have two connections.

If the diode is installed in an electric circuit, a distinction is made between the conducting state and blocking state, depending on polarity **(Fig. 2)**.

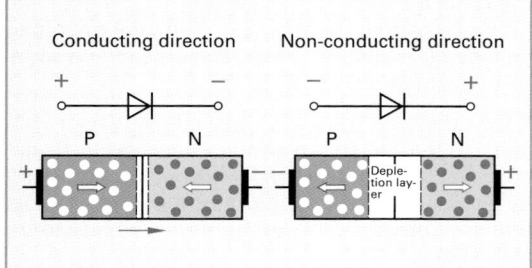

Fig. 2: Circuit with diodes

Diodes only let the current through in one direction and block it in the opposite direction. They act like valves.

Conducting-state region (Fig. 3 and 4). For diodes operated in conducting direction (forward biassed $\xrightarrow{+} \triangleright\!\!\vdash \xleftarrow{-}$), as voltage U_F increases, conducting-state current I_F increases sharply above the threshold voltage. The threshold voltage for germanium diodes is approximately 0.3 V, for silicon diodes approximately 0.7 V.

A diode operated in the conducting-state region is high-ohmic below the threshold voltage and low-ohmic above the threshold voltage.

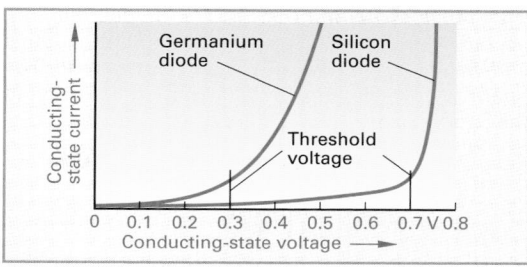

Fig. 3: Conducting-state region of diodes

Blocking-state region (Fig. 4). For diodes operated in non-conducting direction (reverse direction $\xrightarrow{-} \triangleright\!\!\vdash \xleftarrow{+}$), even when reverse voltage increases U_R, only a low locking current I_R flows.

Breakdown region (Fig. 4). If the reverse voltage is increased even more, the diode becomes conductive; the blocking current that now increases suddenly very rapidly becomes a breakdown current that can irreparably damage diodes.

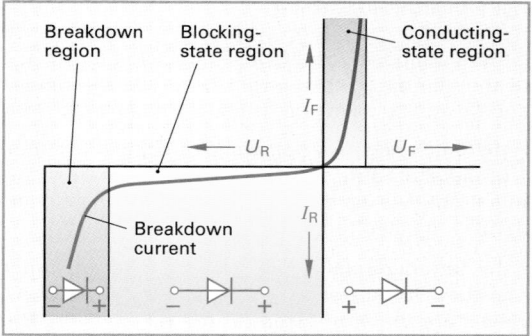

Fig. 4: Characteristic of a diode

Rectifier circuits

Diodes are used for AC voltage rectification.

Single-pulse control (Fig. 5). If there is positive half-wave directly at terminal 1 of the alternator, the diode is switched in the conducting direction; the positive half-wave is conducted by the diode. If there is a negative half-wave, the diode is switched in the non-conducting direction, the negative half-wave is inhibited and the voltage is zero during this time.

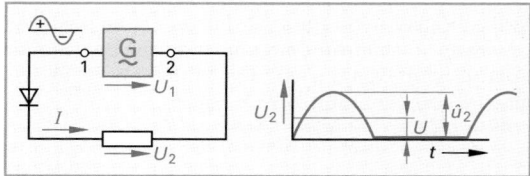

Fig. 5: Single-pulse control

Dual-pulse switching (Fig. 1). The diodes are connected so that both the positive and the negative half-waves can be drawn on for rectification. The operating principle for rectification can be seen in the circuit diagram **(Fig. 1)**.

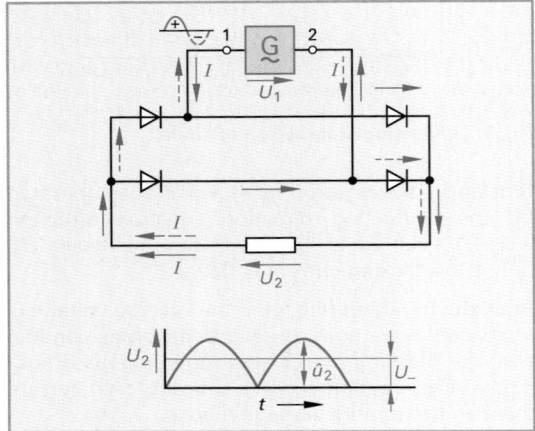

Fig. 1: Dual-pulse switching

If there is a positive half-wave directly at terminal 1 of the alternator, the current flows via the diodes and the consumers to terminal 2 (red arrow).

If there is a negative half-wave directly at terminal 1 of the alternator, the current now flows from terminal 2 out via the diodes and the consumers to terminal 1 (dashed red arrow).

If the current direction is observed in the consumer resistor R, it is the same in both cases. The two half-waves are therefore used for the rectification. The resulting direct voltage is more steady than with single-pulse control **(Fig. 1)**.

Z-diode

Z-diodes (Zener[1] diodes) are usually reverse operated, i.e. they are connected in the non-conducting direction. Their characteristic shows a sharp kink in the transition from the blocking-state region into the breakdown region. At the same time, the breakdown current (Zener current I_Z) increases sharply **(Fig. 2)**.

> The operating range of the Z-diodes is the breakdown region.

In the breakdown region, Z-diodes act as switches or valves. In electronic circuits, they can be used for voltage stabilisation, voltage limitation or as desired-value generators, for instance.

---
[1] G.M. Zener, American Physicist

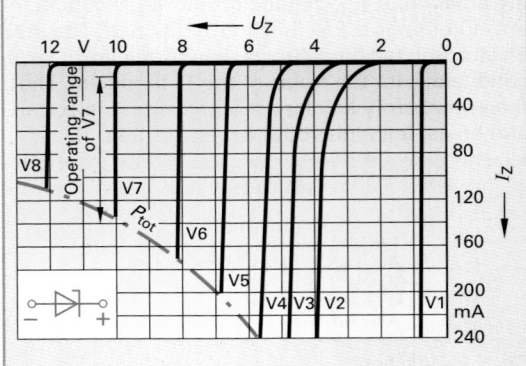

Fig. 2: Working characteristics of Z-diodes

The Z-diode model V6, for example, **(Fig. 2)** is conductive with a Zener voltage U_Z of between 8.0 V and 8.1 V. The maximum permissible current I_Z through this Zener diode is approximately 170 mA. A higher current would overload it thermally and damage it irreparably.

> Each Zener diode requires a protective resistor for current limitation.

Voltage stabilisation (Fig. 3). Until the Zener voltage is reached at the Zener diode, the resistance of the Zener diode R_Z is considerably greater than that of the protective resistance R_1. The total operating voltage U_1 is applied to the Zener diode and so also to the load resistor R_L.

If the operating voltage U_1 exceeds the Zener voltage U_Z, the resistance of the Zener diode R_Z decreases very significantly. The Zener current thereby also flows through the protective resistor R_1, so that the voltage drop U_a at the protective resistor R_1 increases.

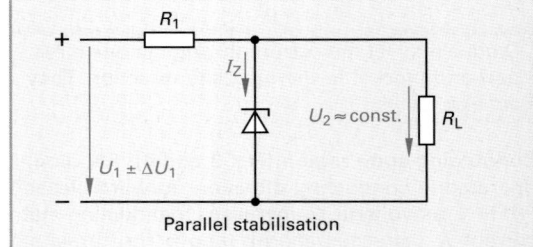

Fig. 3: Voltage stabilisation

> During voltage stabilisation by Zener diode, an almost constant output voltage U_2 is achieved as a result of the voltage drop U_a at the protective resistor R_1.

19.1.17.2 Transistors

These consist of three superimposed semiconductor layers that each have an electrical connection. In the sequence of semiconductor layers, the transistor can be compared to two diodes interconnected against each other. **PNP** transistors and **NPN** transistors differ according to the arrangement of the semiconductor layers. The semiconductor layers with their connections are referred to as emitter **E**, collector **C** and basis **B (Table 1)**.

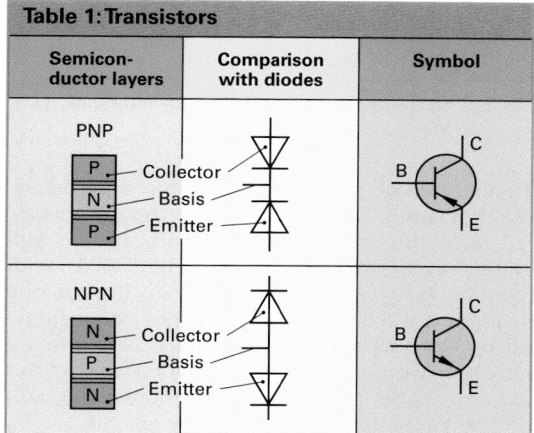

Table 1: Transistors		
Semicon-ductor layers	Comparison with diodes	Symbol
PNP		
NPN		

Transistors can be used as switches with relay function, amplifiers and as controllable resistors.

Transistor as a switch (Fig. 1)

It enables contactless switching of a high working current at a low control current; since there are no mechanically moved parts, it is wear-and-tear-resistant, noiseless and functions with no spark gap. The switching processes occur in a matter of microseconds without any time lag. The transistor has a relay function here.

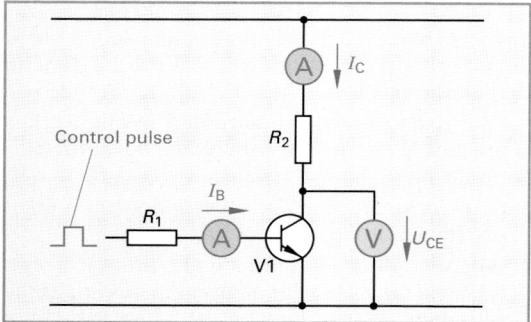

Fig. 1: Transistor as a switch (principle)

PNP transistor as a switch (Fig. 2)

Switch status "On". When a PNP transistor is used, the basis and collector always have negative polarity compared to the emitter **(Fig. 2)**. If a direct voltage is applied between emitter E and basis B, a low base current I_B (control current) flows which interconnects the transistor: a high emitter-collector current I_C (working current) can now flow through the consumer to be operated (filament lamp). The base current I_B is restricted by a resistor at the same time.

Switch status "Off". If the base current I_B is interrupted, the collector current I_C is also interrupted, i.e. the transistor blocks the working current. The collector current is also interrupted if the basis has a positive polarity **(Fig. 2)**.

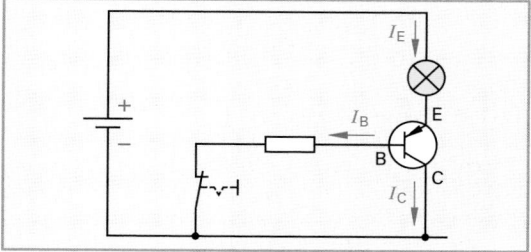

Fig. 2: PNP transistor as a switch

NPN transistor as a switch (Fig. 3)

Switch status "On". When a NPN transistor is used, the basis and collector always have a positive polarity compared to the emitter **(Fig. 3)**.

Switch status "Off". The collector current is interrupted by interrupting the base current or by negative polarity of the basis. All other processes run as with a PNP transistor.

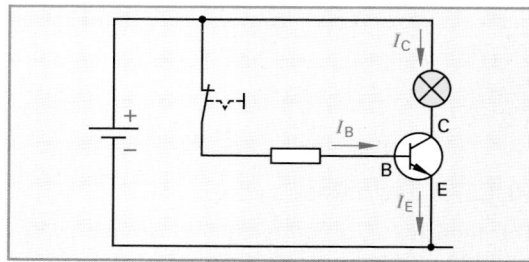

Fig. 3: NPN transistor as a switch

A low control current between emitter and basis (base current) causes a high working current between emitter E and collector C (emitter-collector current).

19

Transistor as an amplifier (Fig. 1)

The load resistor R_L and the collector-emitter resistor of the transistor R_{CE} form a voltage divider. If the resistance of the transistor is changed, the ratio of the voltage division changes $U_L : U_{CE}$.

If the voltage U_{BE} is increased, the resistance of the transistor decreases. A higher current flows in the voltage divider. The division of voltage in the voltage divider changes; at the load resistor R_L a higher voltage drop U_L results.

A small change in the basis-emitter voltage U_{BE} causes a large increase in the voltage U_L at the load resistor R_L. This process is called voltage gain.

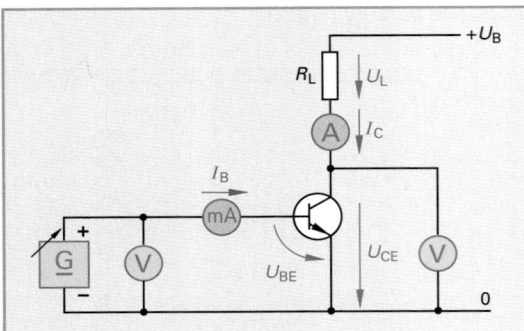

Fig. 1: Transistor as an amplifier (test circuit)

If the voltage U_{BE} is increased slightly, the base current I_B also increases. The considerable decrease in the transistor resistance R_{CE} which thereby occurs leads to a significant increase in the collector current I_C. This process is called current gain.

Transistor as a variable resistor. The operating principle is the same as when using the transistor as an amplifier (**Fig. 2**). However, it is important to make sure that the heat loss occurring in the "transistor resistor" does not irreparably damage the transistor.

19.1.17.3 Thyristors

The thyristor is a controllable electronic switch with the property of a rectifier. It consists of four semiconductor layers connected in series. Three of these semiconductor layers have connections (**Fig. 2**):

- Anode (**A**)
- Cathode (**C**)
- Gate (**G**)

The gate, also referred to as a gateway, is the control electrode. P-gate thyristors and N-gate thyristors differ according to the arrangement of the semiconductor layers. The most common type of thyristor is a PNPN-semiconductor component with P-gate.

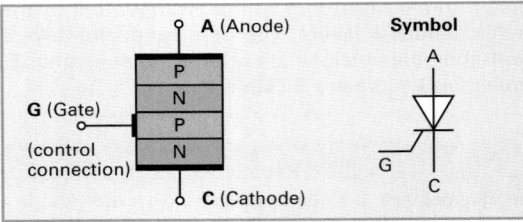

Fig. 2: Basic set-up of a P-controlled thyristor and symbol

Thyristor conducts. For a P-gate thyristor, activation (ignition), i.e. making the four semiconductor layers conductive, occurs by a short positive voltage pulse, which is given at the gate (**Fig. 3**). After ignition, the thyristor remains conductive as long as there is a slight voltage difference between anode (A) and cathode (C). This is the case when a minimum working current (holding current) is flowing. In contrast to the transistor, the working current at the thyristor itself cannot be adjusted.

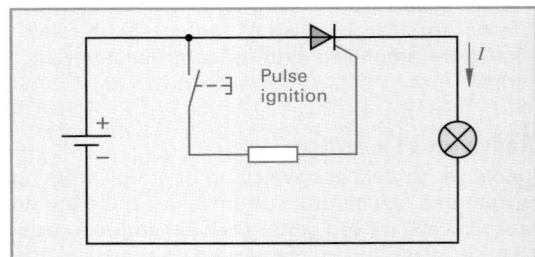

Fig. 3: Thyristor as a switch

After ignition, a thyristor acts like a diode.

Thyristor blocks. If a thyristor is supposed to block, i.e. interrupt the current flow, this can be done in the following ways:

- The load current is interrupted briefly. This is practically impossible with high load currents.
- When the holding current is flowing, it is suppressed for fractions of a second when a short negative pulse is applied to anode (A).
- When the direction of the load current is inverted, as is the case with alternating current at the zero point. It must then be re-ignited again after each zero point.

19

Thyristors can be used in the following areas:
- Rectification (alternating voltage into direct voltage), e.g. in large alternators in buses.
- Conversion (direct voltage into alternating voltage, e.g. in transducers).
- Adjustable rectification. The direct voltage level can be controlled and regulated.
- AC power controllers, e.g. dimmers. The voltage level can be adjusted.
- Frequency converters. The frequency of the alternating voltage generated from the direct voltage can be altered. This enables the revolution speed of alternating current motors to be controlled.
- In power electronics at nominal inverse voltages of between 50 V and 8,000 V and rated currents of between 0.4 A and 4,500 A.

19.1.17.4 Semiconductor resistors

These are components with two terminals, which always require a power supply in the electrical circuit.

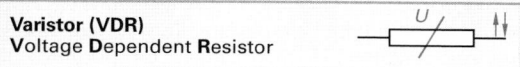

Varistor (VDR)
Voltage **D**ependent **R**esistor

They are voltage-dependent resistors. As the voltage increases, their resistance suddenly decreases, i.e. the current in the varistor then rises sharply. Their characteristic curve is similar to that of a Z-diode, however the curve of the varistor characteristic is dependent on the current direction (polarisation).

Varistors (VDR) have a high resistance at low voltage, at high voltages they have a low resistance.

Use. They are used to prevent overvoltages, e.g. in electronic components. These types of overvoltage occur when the current in coils changes suddenly. High self-induction voltages can occur when this happens.

To protect the electronic component, the varistor must be connected in parallel to the voltage source which generates the high voltage peaks (coil) **(Fig. 1)**. When the voltage peak occurs, it short-circuits the coil.

In addition, VDR resistors are used for voltage stabilisation. They take over the function of a Z-diode in the circuit.

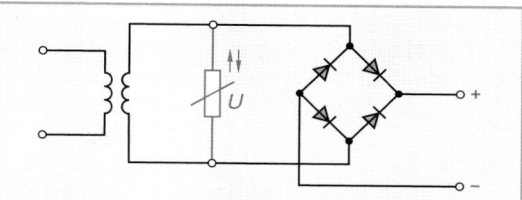

Fig. 1: Protective circuit with varistor

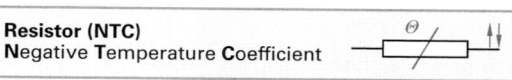

Resistor (NTC)
Negative **T**emperature **C**oefficient

They are also referred to as NTC resistors or NTC thermistors.

NTC resistors have a high resistance at low temperatures, at high temperatures they have low resistance.

They have a negative temperature coefficient. When the temperature increases, their resistance decreases **(Fig. 2)**. The resistance change does not run linearly in relation to the change in temperature.

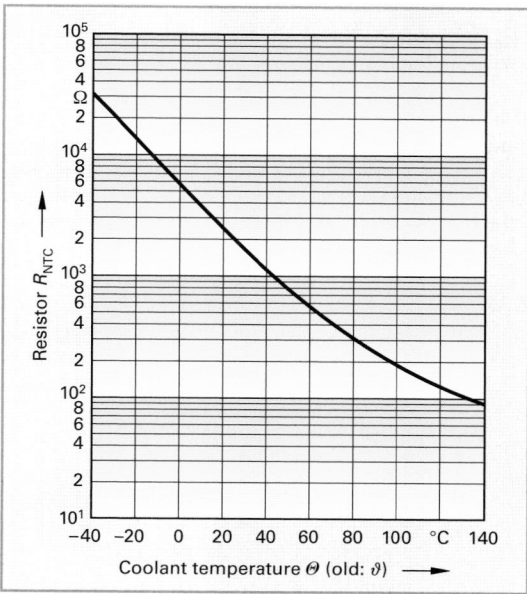

Fig. 2: Resistance/temperature behaviour of an NTC resistor

Use. They are used as pickups on systems in which the temperature needs to be measured.

The temperature information is converted into a voltage value which can then be used to display the temperature or to control and regulate.

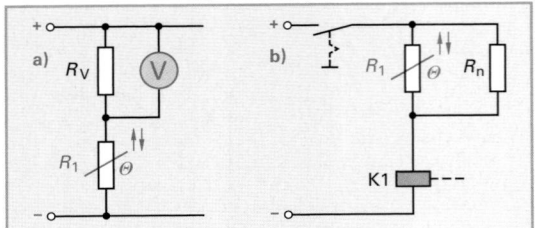

Fig. 1: Examples of uses for NTC resistors

They have a positive temperature coefficient. When the temperature rises, their resistance increases (**Fig. 2**). The resistance change does not run linearly to the change in temperature.

Use. The application areas for PTC resistors are the same as for NTC resistors. However, it is important to make sure when assembling a circuit that the resistance/temperature behaviour is reversed.

Temperature detection (Fig. 1a). As the temperature increases, the resistance value of the NTC (R_1) decreases. The voltage drop in the voltage divider at the resistor R_v increases. The voltage displayed U_v can be calibrated in °C.

On-delay (Fig. 1b). When switched on, the resistance value of the NTC (R_1) is high, the resistance connected in parallel R_n is also so high that the pick-up current in the relay K_1 cannot be achieved. The NTC is heated by the current flow, its resistance decreases. The current increases until the pickup current is achieved and the relay K_1 switches. A blower motor, for example, can be switched via the relay contacts.

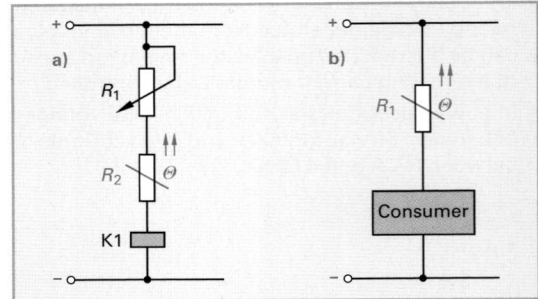

Fig. 3: Examples of use for PTC resistors

Temperature-sensitive control (Fig. 3a). In a control circuit, the holding current of the relays is adjusted to a specific temperature by using the variable resistor, e.g. as with icing protection for air conditioning systems. If this preset temperature is exceeded, the resistance of the PTC resistor R_2 increases significantly, the relay becomes currentless. The desired switching procedure is triggered via the relay contacts.

Resistor (PTC)
Positive **T**emperature **C**oefficient

They are also known as PTC resistors or PTC thermistors.

> PTC resistors have a low resistance at low temperatures, at high temperatures they have a high resistance.

Overload protection (Fig. 3b). A PTC resistor is installed in the electric circuit of the consumer. If the current exceeds a permissible level, the PTC resistor heats up. It increases its resistance value and so limits the current to a permissible level, e.g. as with exterior mirror heating.

19.1.17.5 Optoelectronics

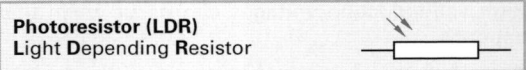

Photoresistor (LDR)
Light **D**epending **R**esistor

These are light-depending resistors. They reduce their resistance value as brightness increases.

Photoresistors are used as flame monitors in heating systems and alarm systems, in dimmer switches and in photoelectric light barriers (e.g. car washes, ignition pickups).

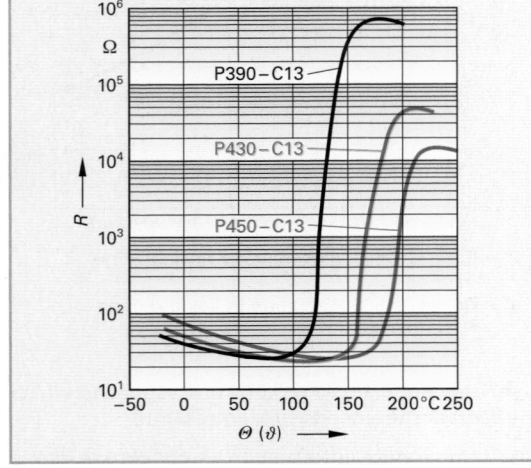

Fig. 2: Resistance/temperature behaviour of a PTC resistor

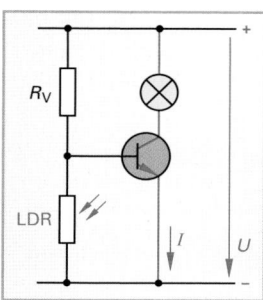

Fig. 1:
Light-sensitive control

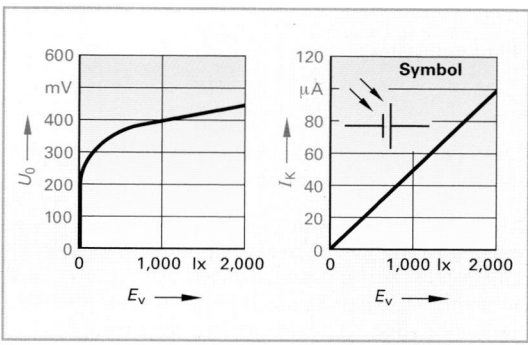

Fig. 3: Characteristics of a photoelement

Light-sensitive control (Fig. 1). If the resistance value R_1 of the LDR increases when it is dark, the basis B of the transistor becomes positive due to the distribution of voltage between the resistors R_1 and R_v. The transistor connects through, the filament lamp lights up.

Photodiodes. These are semiconductor devices which …

- … by means of a voltage source, function as light-depending resistors **(Fig. 2).**
- … without a voltage source, function like a photoelement **(Fig. 3).**

Photodiodes can be very small and can be used as photoelectric converters in open control loops and closed-loop control circuits.

Photodiode as light-depending resistor	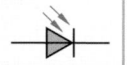

If light hits the photodiode, its resistance falls as the light intensity increases. It is possible for current to pass through the photodiode and the relay K_1 responds. The desired switching procedure is triggered. Photodiodes are operated in non-conducting direction (reverse bias) **(Fig. 2).**

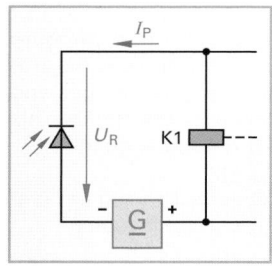

Fig. 2:
Principle circuit
of a photodiode

Photodiode as photoelement	

When photovoltaic cells are hit by light, they give out a voltage, depending on the semiconductor material and the level of brightness. They are used as the voltage source in watches, pocket calculators and light sensors.

The characteristics of a photoelement are the open-circuit voltage U_0 and the short-circuit current I_k **(Fig. 3).** On Si-photoelements the open-circuit voltage at 1,000 lx is approximately 0.4 V, with selenium approximately 0.3 V. The intensity of illumination E_v is given in lux (lx).

> When photoelements are hit by light, they give out a voltage, depending on the semiconductor material and the intensity of the illumination.

Use. A large surface-area of interconnected silicon photoelements can be used to exploit solar energy (solar cells). They have an efficiency of approximately 20 %, i.e. they convert 20 % of the light energy into electrical energy. They function as voltage generators in photovoltaic systems, in particular as the power supply for parking machines, mountain huts, transmitters and satellites.

Light-emitting diode (LED)	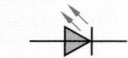

When a voltage is applied, these diodes light up in green, yellow, orange, red or blue depending on the diode material. The operating voltage is between 1.5 V and 3 V. If operated with other voltages, a protective resistor must be fitted to limit the current **(Fig. 4).**

In motor vehicles, they are only used as displays and indicator lamps because of their low own consumption of only a few mW.

Light-emitting diodes are used in conducting direction (forward-biased).

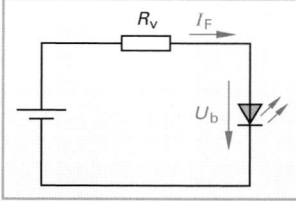

Fig. 4:
LED with
protective resistor

19

Phototransistor
(luminous and infrared radiation)

A transistor is usually activated via the basis with a positive or negative voltage.

On the phototransistor, light or infrared radiation reaches the collector-basis depletion layer via an optical lens and generates there a photoelectrical current I_p which increases in proportion to the light intensity E_v **(Fig. 1)**. It acts as a base current.

The collector current of a phototransistor increases with the light intensity.

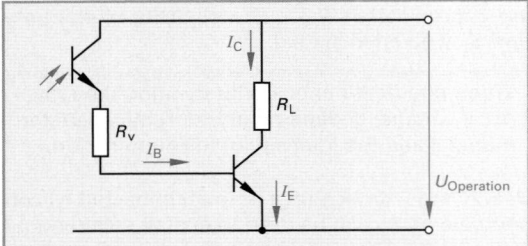

Fig. 1: Phototransistor with amplifier transistor

Use. They are used in automotive engineering on light-dependent control systems, e.g. anti-dazzle control on rear-view mirrors, optoelectronic couplings.

Optoelectronic couplings. These consist of a radiation transmitter and a radiation receiver which are both integrated in a common, light-proof housing, so that the receiver only receives radiation from the transmitter **(Fig. 2)**. It is preferable to use infrared light-emitting diodes as transmitters.

Photodiodes, phototransistors or photothyristors can be used as receivers (detectors), depending on field of application.

Optocouplers couple two electric circuits usually by infrared radiation and isolate them galvanically from each other.

The output side voltage is about 5 V, i.e. all signals which enter the optocoupler are converted and are between 0 V and 5 V at the output. They are supplied to the appropriate ECUs in the motor vehicle as the input signal for processing.

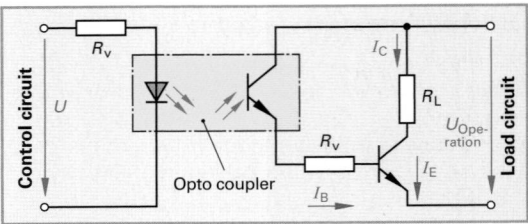

Fig. 2: Optoelectronic couplings with photodiode and phototransistor

19.1.17.6 Semiconductor elements influenced by magnetic field

Hall generator

The Hall effect occurs at a semiconductor layer carrying a supply current I_v **(Fig. 3)**. If a magnetic field is present at a right angle to the semiconductor layer, the Hall voltage U_H is generated between the contact areas A. Its level depends on the strength of the magnetic field.

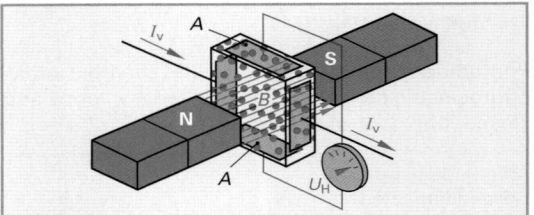

Fig. 3: Hall effect

19.1.17.7 Semiconductor elements influenced by pressure

Piezo element

When subjected to load by tractive forces, pressure forces or shear forces, piezoelectric sensors produce a displacement of the electrical charges and with it a voltage at the connector electrodes. This piezoelectric effect can be observed in quartz crystals (SiO_2), **for example (Fig. 4).**
In motor vehicles, Piezo elements are used as pressure sensors and knock-sensors, for instance.

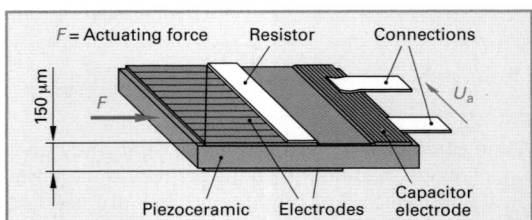

Fig. 4: Piezoelectric sensor

19.1.17.8 Integrated circuits

Planar technology enables all the components of a circuit (resistors, capacitors, diodes, transistors, thyristors), including the conductive connections, to be manufactured together in a common production process on a single (monolithic)[1] silicon chip[2].

In doing so, individual integrated control circuits known as ICs[3] are combined into monolithically integrated circuits (**Fig. 1**).

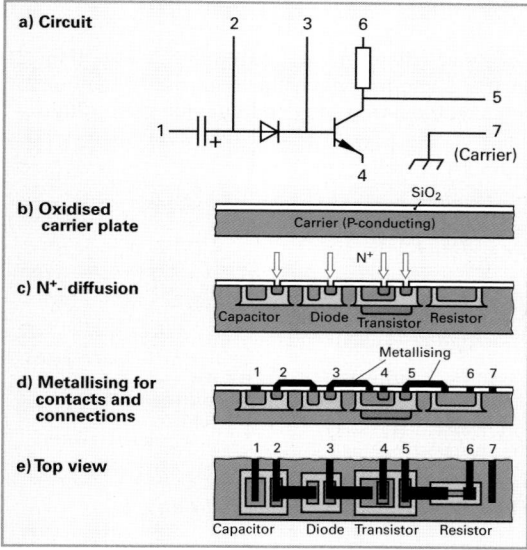

Fig. 1: Example of an integrated circuit in monolithic technology (selection of production steps)

As there are no "independent" components (components having external connections) in an **IC**, one refers to switching elements or operational elements.

Planar technology. This is a procedure in semiconductor engineering for producing semiconductor devices and chips. Layers insulated from each other, contained in the components in addition to connecting lines and terminals, are superimposed in consecutive work steps. This can be carried out by screen process in the thick-film technique or by vapour deposit in the thin-film technique. This means that a chip can therefore contain more than 100,000 active functions (e.g. transistors, diodes) and passive functions (e.g. resistors, capacitors).

[1] monolithic (Greek) = of one stone
[2] Chip = disc/record, block
[3] Integrated Circuits

Hybrid circuits. These are a combination of integrated control circuits and individual components (**Fig. 2**). They are connected together on a carrier plate by means of plugs, soldering or other methods. Thereby, specific circuits can be manufactured with special properties, e.g. an ignition control unit.

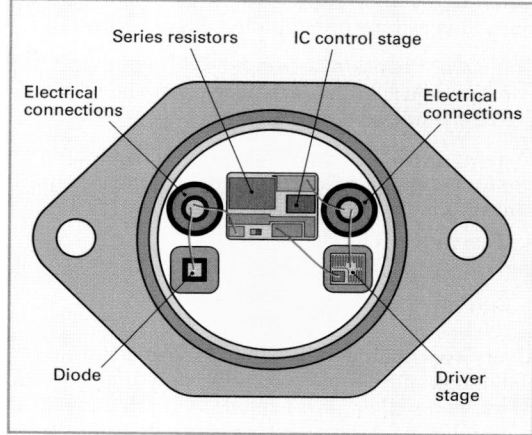

Fig. 2: Voltage regulator in hybrid technology

Rᴇᴠɪᴇᴡ ǫᴜᴇꜱᴛɪᴏɴꜱ

1 **Which charge carriers do N-conductors and P-conductors have?**

2 **How should a PN junction be poled so that a conducting-state current flows?**

3 **What do you understand by the threshold voltage?**

4 **Which part of the characteristic of a Z-diode is used for voltage stabilisation?**

5 **What is a single-pulse control?**

6 **How is an NPN transistor designed?**

7 **What are the electrode connections of a transistor called?**

8 **How should an NPN junction be poled so that it is conductive?**

9 **How does a NTC resistor behave when it is heated?**

10 **How does the resistance of a varistor change when the voltage increases?**

11 **How does an LDR resistor behave when light falls on it?**

12 **What is an LED?**

13 **What functions do optoelectronic couplings have?**

14 **How are hybrid circuits designed?**

19

19.2 Applications of electrical engineering

19.2.1 Circuit diagrams

Classification of circuit diagrams

A circuit diagram is the diagrammatical representation of electrical components by means of symbols, illustrations or simplified design drawings.

The circuit diagram shows how different electrical components are related to each other and how they are connected to each other.

In automotive electrical engineering, depending on the function, the following types of circuit diagram are used:

- Overview circuit diagrams
- Terminal diagrams
- Current flow diagrams

Overview circuit diagram (Fig. 1). This is the simplified illustration of a circuit, with only the essential parts being taken into consideration. It shows the method of operation and structure of an electrical system.

The devices are represented by squares or rectangles with drawn in labels or by symbols or designations.

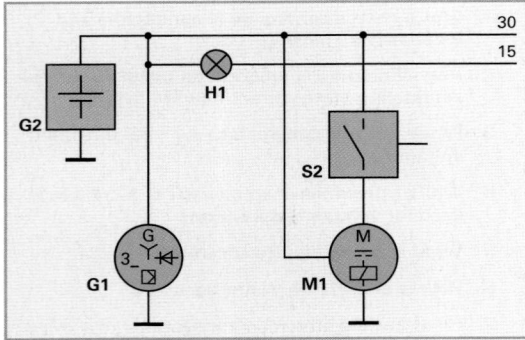

Fig. 1: Overview circuit diagram

Terminal diagrams

A distinction can be made between:

- Terminal diagrams in connected view
- Terminal diagrams in exploded view

Terminal diagrams in connected view (Fig. 2). This shows the terminal locations of an electrical system and the inbound and outbound connections connected to it. For this purpose, the individual components are usually illustrated true to their actual positions with the cable routing, all terminal locations and terminal designations.

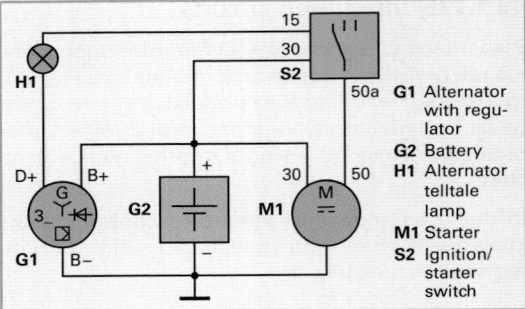

Fig. 2: Terminal diagram (connected view)

G1 Alternator with regulator
G2 Battery
H1 Alternator telltale lamp
M1 Starter
S2 Ignition/starter switch

Terminal diagram in exploded view **(Fig. 3).** In the exploded view, the end-to-end connection lines (cables) from device to device are omitted.

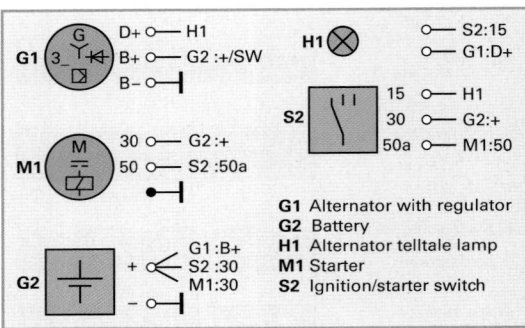

G1 Alternator with regulator
G2 Battery
H1 Alternator telltale lamp
M1 Starter
S2 Ignition/starter switch

Fig. 3: Terminal diagram (exploded view)

Device labelling. For easier recognition, the symbols are still provided with a device label. This consists of a string of defined signs, letters and numbers, e.g. G1 for the alternator.

Destination labels. All electric lines from the device are provided with a destination label **(Fig. 3)**, consisting of:

- The designation for the terminal from which the electric line comes, e.g. on alternator B+.
- The line symbol ○—.
- The target device to which the line leads, e.g. G2 for the starter battery.
- The terminal designation at the target device to which the line leads. This is always separated from the device designation of the target device by a colon (:), e.g. G2:+ means that the line leads to the positive terminal of the starter battery.
- The line colour, if prescribed. The colour labelling is always separated from the terminal designation on the target device by a slash (/), e.g. +/sw means that the connecting line is black.

In the example in **Fig. 3, Page 512**, the labelling at alternator G1 means that the following connections are made to the alternator:

D+ o— **H1**   Terminal D+ is connected to the alternator telltale lamp H1.

B+ o— **G2:+/sw**   Terminal B+ is connected to the positive terminal of the starter battery G2. Black line.

B– o—| Terminal B– is connected to earth.

The course of the above-mentioned connecting lines can also be taken from the terminal diagram in connected view. **(Fig. 2, Page 512).**

Current flow diagrams

These are detailed illustrations of the individual components of circuits. They show the operating principle of the circuit by means of an overview of the individual electric circuits. A current flow diagram contains the electric circuit, the device labelling and the terminal designation.

Based on the symbol allocation, the following can be distinguished:

● Current flow diagrams in connected view

● Current flow diagrams in exploded view

Current flow diagram in connected view (Fig. 1). All components contained in a circuit diagram are displayed connected directly next to each other. No consideration of the spatial position of the individual components and their terminal locations is necessary. Mechanical connections are indicated by discontinuous connecting lines.

Current flow diagram in exploded view (Fig. 2). The symbols of the electrical components are arranged so that the individual current paths are as easy as possible to follow, irrespective of the spatial togetherness and mechanical interrelationship of the individual components and assemblies.

A clear, straight-line, junction-free arrangement of the individual circuits has priority. Usually, the positive and negative lines are shown as horizontal parallel lines. The individual current paths then run from positive to negative, i.e. from top to bottom. If unavoidable, parts of a current path can also be shown horizontally.

The section labelling at the top edge of the circuit diagram helps make it easier to find circuit positions **(Fig. 2).** There are three possible ways to illustrate this:

● Consecutive numbers (1, 2, 3, …) at equal intervals from left to right.

● Designation of the circuit sections, e.g. power supply.

● Combination of consecutive numbers and designated circuit sections.

The current flow diagram can be viewed in a simplified form or in a detailed form with internal circuits.

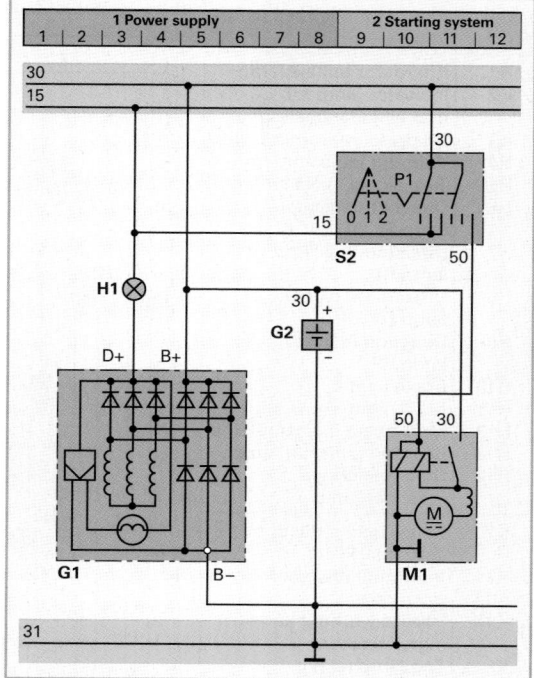

Fig. 2: Current flow diagram (exploded view)

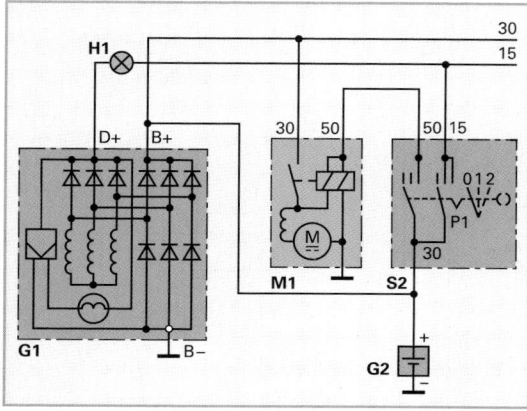

Fig. 1: Current flow diagram (connected view)

19.2.1.1 Current flow diagram for a passenger car in exploded view

Label	Device
A1	Glow control unit
A2	Car alarm control unit
A3	Car radio
B1	Loudspeaker
B2	Temperature regulator
B3	Signal horn
B4	Supertone horn, fanfare
B13	Speed sensor
B16	Thermostatic switch
E3	Interior light with switch
E4	Rear window heating
E5	Reversing lamp L and R
E7	Instrument panel lighting
E9	Licence-plate lamp L
E10	Licence-plate lamp R
E11	Clearance lamp L
E12	Tail lamp L
E13	Clearance lamp R
E14	Tail lamp R
E15	Main/dipped-beam headlights L
E16	Main/dipped-beam headlights R
E17	Fog lamp L
E18	Fog lamp R
E19	Rear fog lamp L
E20	Rear fog lamp R
F2 ... 26	Fuses
G1	Alternator (with regulator)
G2	Battery
H1	Alternator telltale lamp
H2	Indicator lamp for rear window heating
H3	Oil pressure warning lamp
H4	Warning indicator lamp
H5	Turn signal indicator lamp
H6	Direction-indicator lamp FL
H7	Direction-indicator lamp RL
H8	Direction-indicator lamp FR
H9	Direction-indicator lamp RR
H10	Brake light L
H11	Brake light R
H12	Main beam indicator lamp
H13	Rear fog indicator lamp
H14	Standby mode indicator lamp
K1	Relay, load reduction Terminal 15
K2	Intermittent wiper relay
K3	Horn relay
K4	Hazard warning and turn signal flasher
K5	Fog lamp relay with diode

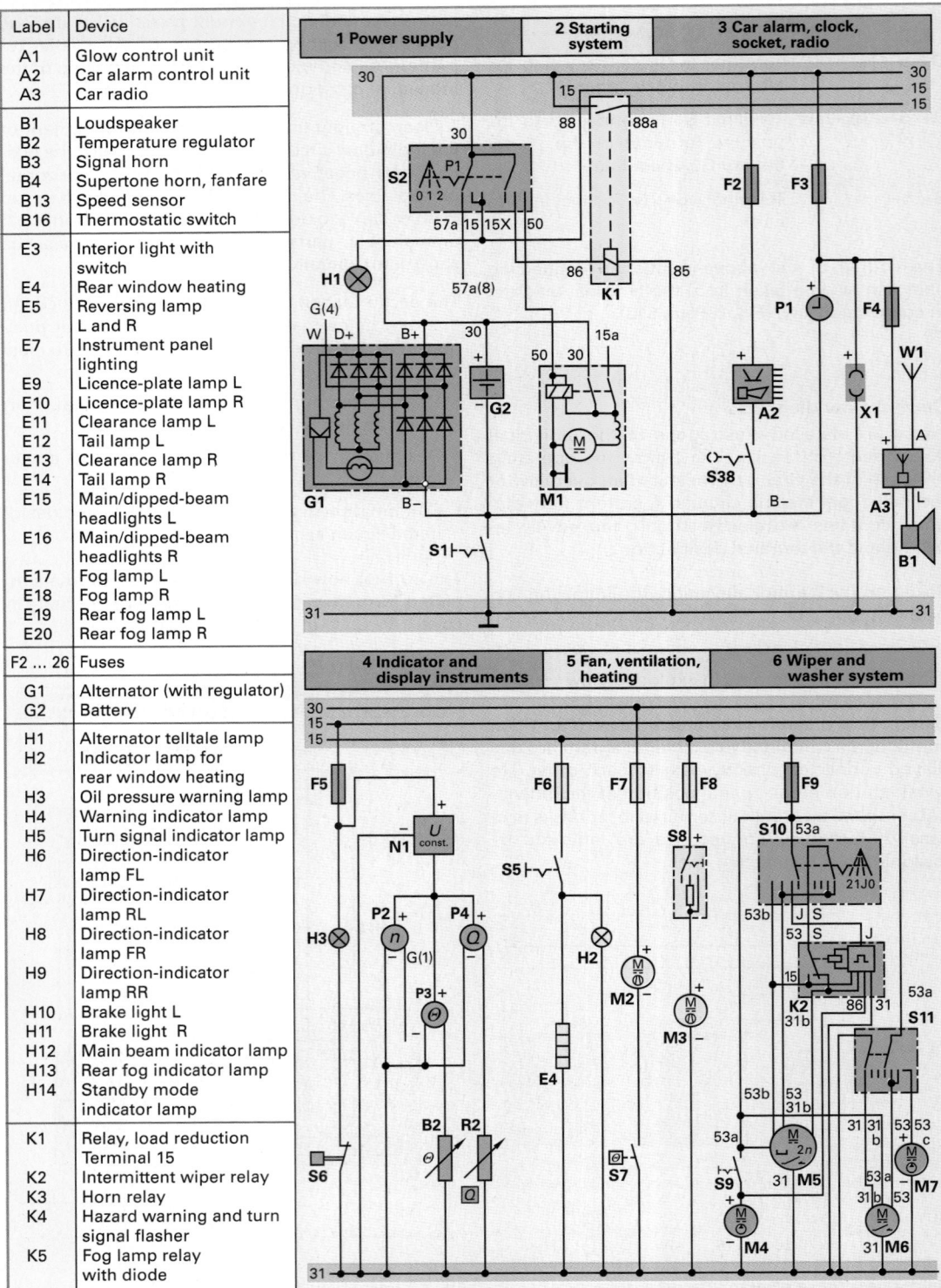

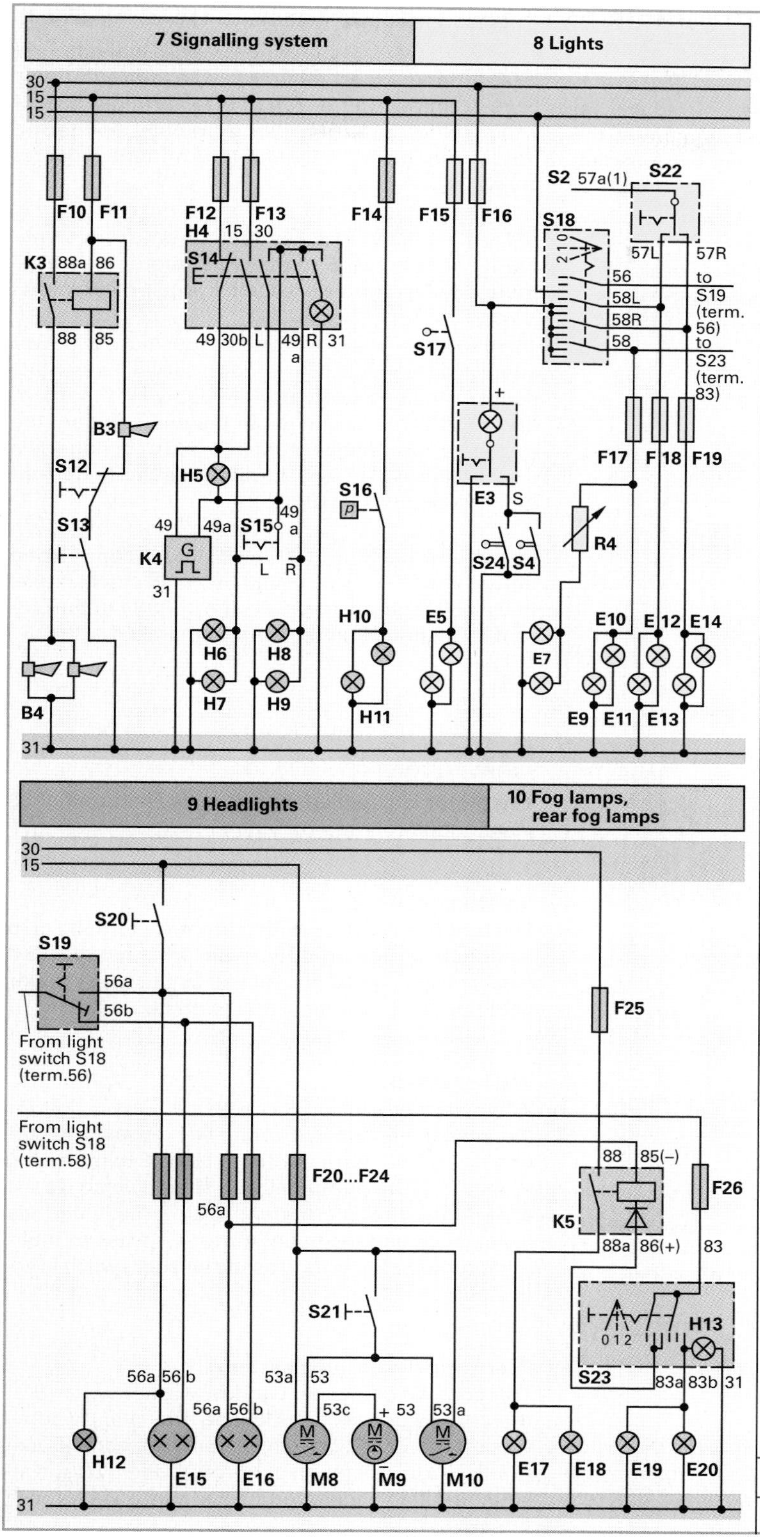

7 Signalling system | 8 Lights

9 Headlights | 10 Fog lamps, rear fog lamps

Label	Device
M1	Starter
M2	Radiator fan motor
M3	Fresh-air blower motor
M4	Windscreen washer motor
M5	Wiper motor
M6	Rear wiper motor
M7	Rear window washer motor
M8	Light wiper motor with pump control
M9	Washer motor for M8 and M10
M10	Light wiper motor
N1	Voltage stabiliser
P1	Clock
P2	Rev counter
P3	Coolant temperature display
P4	Fuel gauge display
R2	Fuel level sensor
R4	Regulating resistor for E7
S1	Battery master switch (mech.)
S2	Ignition/starter switch (driving starting switch)
S4	Door contact switch for E3, R
S5	Rear window heating switch
S6	Oil pressure warning switch
S7	Thermostatic switch (cooling)
S8	Fan switch
S9	Washer switch
S10	Wiper switch
S11	Rear window washer switch
S12	Horn changeover switch
S13	Horn switch
S14	Warning light switch
S15	Turn signal switch
S16	Brake-light switch
S17	Reversing lamp switch
S18	Light switch
S19	Dimmer switch
S20	Headlamp flasher switch
S21	Light wiper switch
S22	Parking light switch
S23	Fog lamp switch
S24	Door contact switch for E3, L
S38	Car alarm switch
W1	Car aerial
X1	Socket (internal)

Sectional current flow diagrams

19.2.1.2 Wiper and washer system

This consists of wipers and the associated washer systems for the windscreen and rear window. In the overall circuit diagram, it is assigned section identification 6.

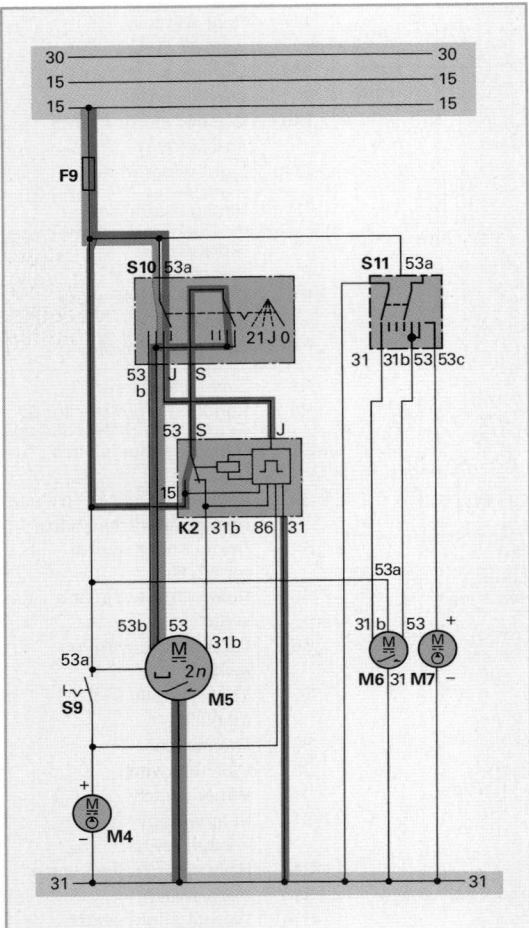

Fig. 1: Wiper and washer system

The wiper motors M5, M6 are connected so that they return to the initial position after they are deactivated.

The entire wiper and washer system is fused via fuse F9.

Wiper system. In the diagram shown, the windscreen wiper circuit is connected via wiper switch S10. Depending on the switch position, different wiper speeds can be set.

Switch position J (intermittent switching) **(Fig. 1).**
Wiper motor M5 of the windscreen is activated via intermittent wiper relay K2. The current flows from terminal 15 via fuse F9 to terminal 53a of wiper switch S10. Between terminal J and terminal 31, voltage is present for intermittent wiper relay K2.

Relay K2 briefly switches the current from terminal 15 via terminal S to the wiper motor input (terminal 53). The motor executes a wiper movement.

To return the wiper to the end position, a power supply via contact 53a is necessary. This contact supplies the necessary power supply for the return line in all wiper positions. In the end position, the motor is no longer supplied with voltage via 53a, as the wiper would otherwise continue running beyond this position.

In the end position, the motor must come to an abrupt stop. This is achieved by an electrical brake. In the still rotating armature, the polarity of the current is reversed and the armature braked.

Switch position 1 (Fig. 1).
The motor is supplied with current via terminal 53 (continuous-running-duty type).

Switch position 2.
The motor is supplied with current via terminal 53b (continuous-running-duty type fast mode). The quicker wiper speed is achieved by the wiper motor being operated via the shunt winding.

Washer system.
When washer switch S9 is activated, the circuit for the windscreen washer motor M4 closes. The current flows from terminal 15 via fuse F9 to motor M4 to earth 31. The pump delivers screen wash. At the same time, the intermittent relay is activated via terminal 86 and the wiper motor switches to intermittent mode for as long as the washer switch is activated.

Rear window cleaning system.
This consists of the rear window washer switch S11, the rear wiper motor M6 and the rear window washer motor M7. Switch S11 enables continuous-running-duty operation of the rear wiper (W) and also intermittent operation of the pump via a touch function (S).

19.2.1.3 Signalling system

In the overall current flow diagram, this is designated with section label 7.

> The signalling system consists of various devices which emit acoustic and visual signals.

In road traffic, these devices perform important warning and information functions for other road users and make a significant contribution to road safety.

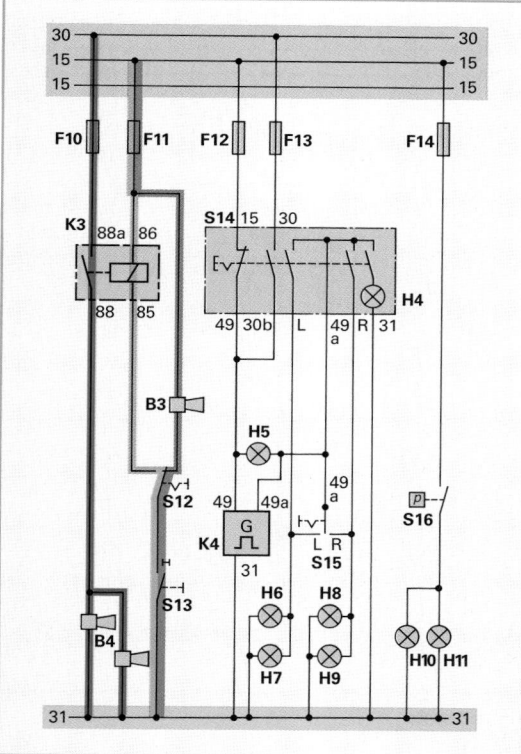

Fig. 1: Signalling system

Acoustic signal transmitters. These include horns and fanfare horns. In the circuit shown, the driver can operate either signal horn B3 or fanfare horn B4.

As simultaneous operation of both signal transmitters is not permissible according to the German Road Traffic Licensing Regulations (see Paragraph 55), the horn changeover switch S12 is fitted.

Signal horn B3. When horn switch S13 is operated, the current flows (blue) from terminal 15 via fuse F11 to signal horn B3 to earth 31.

Fanfare horn B4. The horn changeover switch S12 is reversed. When S13 is operated, the control current flows from terminal 15 to relay K3 via F11. The relay connects the working current from terminal 30 via F10 to the fanfare horn B4 to earth 31.

This switching is necessary, as the currents for a fanfare horn are higher than for a signal horn.

In the circuit diagram, the control circuit for the fanfare horn is light green and the working current is dark green.

Visual signal transmitters. These include brake lights, flashing lights for the direction indicator and the hazard warning system.

Brake lights H10, H11. When the brake pedal is applied, the brake-light switch S16 is closed. The current flows from terminal 15 via F14 to the brake lights to earth 31. New vehicles must have a third brake light. Brake lights show the traffic behind that the brakes have been applied.

Direction-indicator lamps for the direction indicator function. Turn signalling is activated by the driver via turn signal switch S15 and is operated on a timed cycle by turn signal flasher K4. The turn signal indicator lamp H5 shows the driver that the turn signal is working.

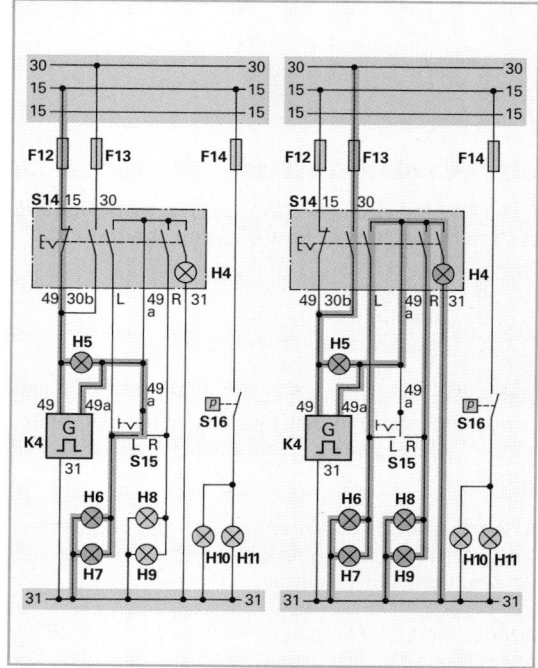

Fig. 2: Current flow during direction indication and during use of hazard warning lamps

If turn signal switch S15L is pressed, the current flows from terminal 15 via fuse F12 through the closed contact in warning light switch S14 through terminal 49 to the turn signal flasher K4 to earth. Output 49a of turn signal flasher K4 leads to turn signal switch S15. The circuit of the direction-indicator lamps H6, H7 is closed via terminal 31 **(Fig. 2, Page 517)**. The indicator lamp connected between terminals 49 and 49a lights alternately with the turn signals.

The failure of a direction-indicator lamp is indicated either by the indicator lamp flashing at a higher rate or going out.

Flashing lights for the hazard warning function. Multilane vehicles require a hazard warning system. It must be possible to activate and deactivate this system independently of the turn signal system. To ensure that the hazard warning system can be operated even with the ignition switched off, it is connected directly to terminal 30 via fuse F13. A red warning indicator lamp H4 must show the driver that all direction-indicator lamps are operating.

If warning light switch S14 is activated, current flows from terminal 30 via F13 to terminal 49 of the turn signal flasher K4. Via terminal 49a, all four direction-indicator lamps and the indicator lamp H4 are supplied with current **(Fig. 2, Page 517)**.

19.2.1.4 Lighting system

The following circuit sections belong to the lighting system of the current flow diagram shown:

    8   Lights
    9   Headlights
    10 Fog lamps and rear fog lamps

Lights (Fig. 1). Circuit section 8 shows the reversing lamps and interior lights, instrument panel lighting, licence plate lamps, clearance lamps and tail lamps.

Reversing lamps E5. When reverse gear is selected, current is routed from terminal 15 via F15 to the reversing lamps E5 to earth.

Interior light E3. This can be switched on and off via an integrated switch or via the door contacts S24 or S4. When the door is opened, a connection between S and earth is established via S24 or S4. So that the

interior light can be activated independently of the ignition/starter switch, it is supplied with voltage via terminal 30. It is fused via F16.

Instrument panel lighting E7. If the light switch S18 is in position 1 or 2, current flows from terminal 30 via 58, F17 and the regulating resistor R4 via E7 to earth. The regulating resistor allows the voltage at E7 to change, so that the brightness of the instrument panel lighting can be changed.

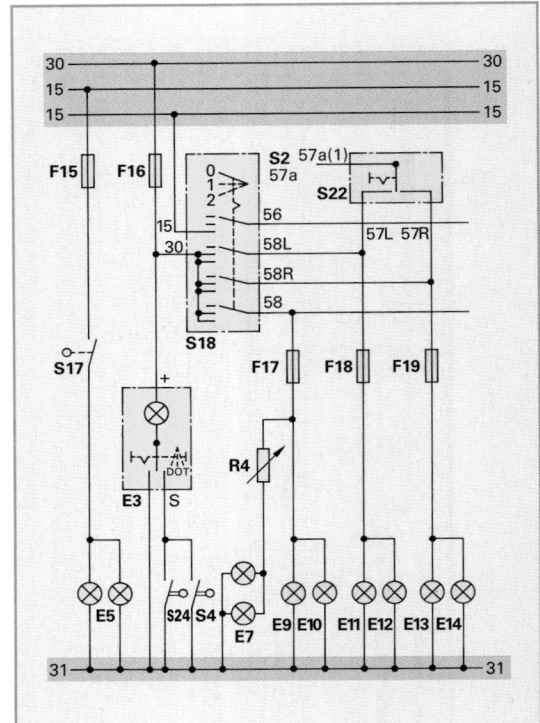

Fig. 1: Lights

Licence plate lamps E9, E10, clearance lamps E11, E13, tail lamps E12, E14. These are fused via fuses F17, F18 or F19 and are supplied with voltage in switch positions 1 and 2 of the light switch via terminals 58L, 58R and 58.

Parking light circuit. The tail lamps and clearance lamps E11/E13 or E12/E14 can be switched on using parking light switch S22 either via position 57L or 57R. The power is supplied to terminal 57a via the ignition/starter switch S2 in switch position 0.

Headlights. Section 9 of the overall current flow diagram shows main-beam and dipped-beam headlights.

If light switch S18 is in position 2, power is supplied to terminal 56 of the dimmer switch S19 when the ignition is on.

Dipped-beam headlight (Fig. 1, red background). If S19 is set to 56b, the dipped-beam headlight lights up, fused via F21/F23.

Main-beam headlight (Fig. 1, blue background). If S19 is set to 56a, the main-beam headlight lights up, fused via F22/F24. At the same time, the blue indicator lamp H12 lights up.

Headlamp flasher. If headlamp flasher switch S20 is activated, power is supplied to the main-beam headlights, and they light up.

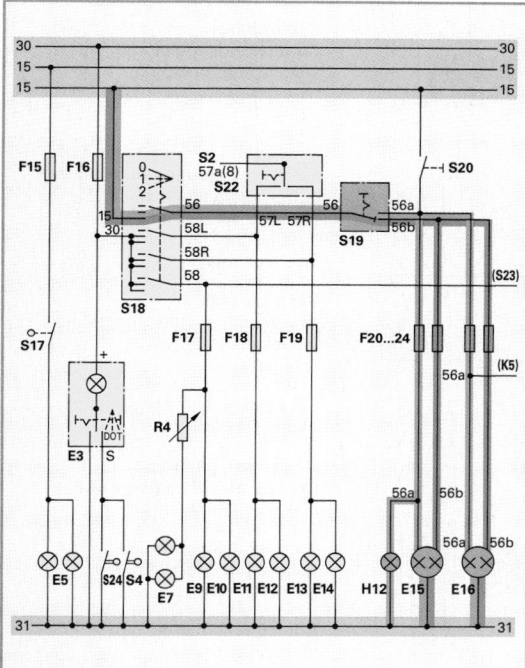

Fig. 1: Current flow at switch position 2 of the light switch

Fog lamps, rear fog lamps. These are illustrated in section 10 of the overall current flow diagram.

Fog lamps E17/E18. There is voltage at terminal 83 of fog lamp switch S23 if light switch S18 is in position 1 or 2.

With the fog lamp switch in position 1, the current flows from 83a via the control winding of fog lamp relay K5 to terminals 56a and E16 to earth 31, if the main beam is switched off. The relay operates and the fog lamps are supplied with power from terminal 30.

In the circuit diagram **(Fig. 2)** the control circuit of fog lamp relay K5 is light green and the working current is highlighted in dark green.

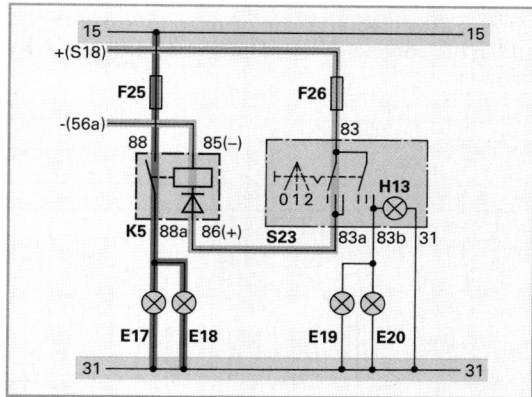

Fig. 2: Fog lamps, rear fog lamps

If the main beam is switched on, terminal 56a is positive. Therefore, no more power is supplied between terminals 85 and 86 in the fog lamp relay and the magnetic field of the coil collapses. The switch in the fog light relay opens and switches the fog light off.

Rear fog lamps S23. With fog lamp switch S23 in position 2, the rear fog lamps E19, E20 and the indicator lamp H13 are supplied with power via terminal 58 of light switch S18.

Headlamp wash-wipe system (Fig. 3). Via push-button S21, the motors of the headlight wash-wipe system M8, M9, M10 are supplied with current from terminal 15 when the ignition is on. The return of the wipers to the initial position takes place via terminal 53a. See wiper and washer system.

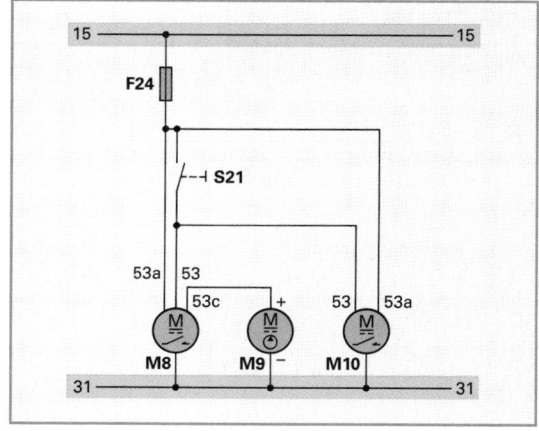

Fig. 3: Headlight wash-wipe system

Additional information and labelling options

Depending on the vehicle manufacturer, the current flow diagrams may differ. They can be provided with various additional information (**Fig. 1**).

Current flow diagrams can be used for malfunction diagnosis on the electrical systems of motor vehicles and in the context of retrofitting auxiliary equipment, e.g. auxiliary heaters, navigation systems or mobile phones.

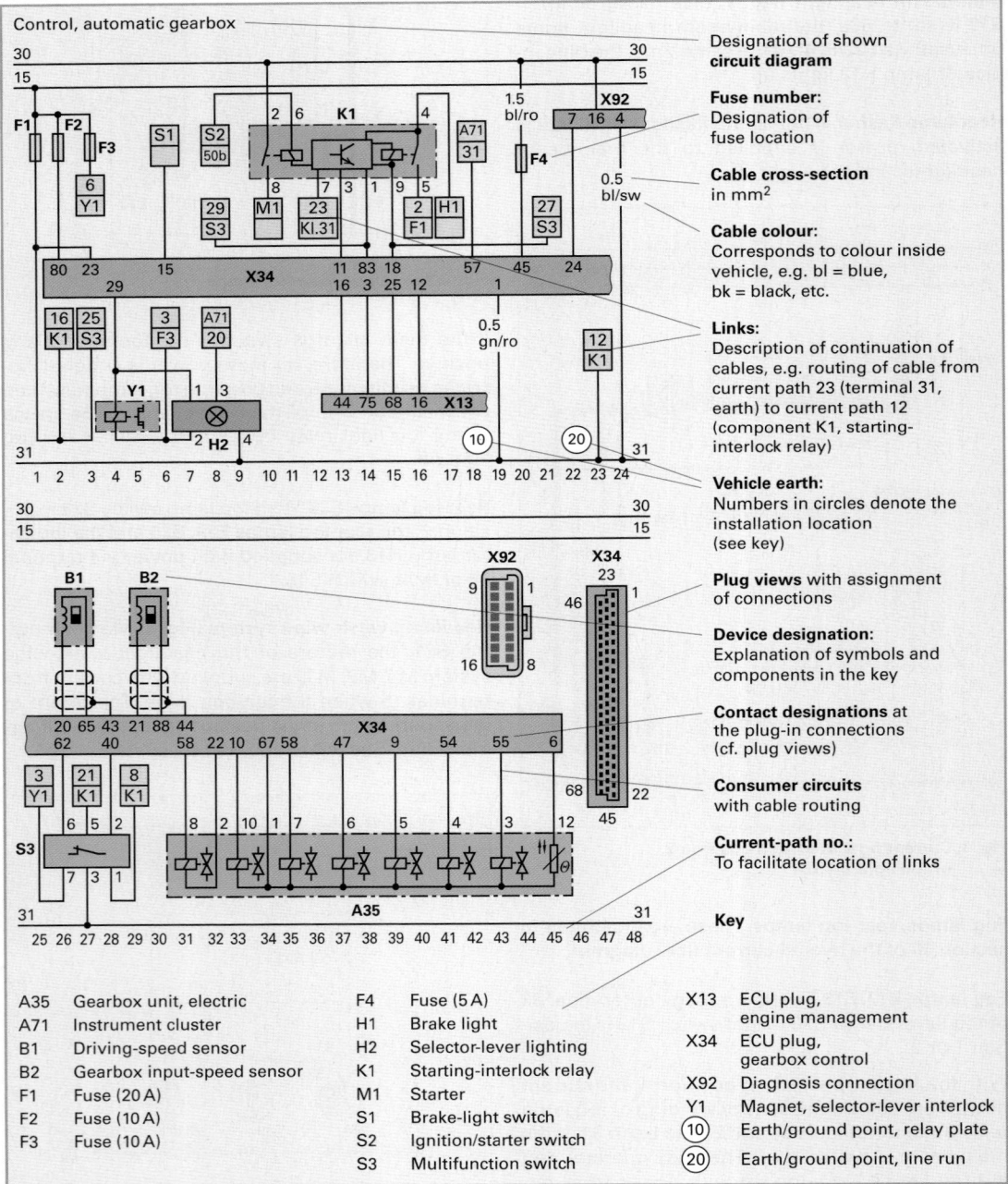

A35	Gearbox unit, electric	F4	Fuse (5 A)	X13	ECU plug, engine management
A71	Instrument cluster	H1	Brake light		
B1	Driving-speed sensor	H2	Selector-lever lighting	X34	ECU plug, gearbox control
B2	Gearbox input-speed sensor	K1	Starting-interlock relay	X92	Diagnosis connection
F1	Fuse (20 A)	M1	Starter	Y1	Magnet, selector-lever interlock
F2	Fuse (10 A)	S1	Brake-light switch	(10)	Earth/ground point, relay plate
F3	Fuse (10 A)	S2	Ignition/starter switch	(20)	Earth/ground point, line run
		S3	Multifunction switch		

Fig. 1: Additional information and labelling options in current flow diagrams

Using current flow diagrams

Current flow diagrams are used for malfunction diagnosis in the vehicle electrical system or on electrical components. The example (Fig. 1) shows the malfunction diagnosis process on the electrical gearbox control using a current flow diagram.

The manufacturer's instructions must be observed during malfunction diagnosis.

1. Read out fault memory:

The fault code indicates a fault on the component or on the cables between the component and the ECU.

2. Disconnect ECU plug from ECU:

Checking of components and cables between the component and the ECU is often performed in the form of a resistance measurement. The plug must be detached from the ECU for this purposes.

Warning: Turn off ignition beforehand!

In the circuit diagram the detached plug is indicated by the interrupted cables to the ECU.

3. Determine test points:

The test points for the resistance measurement in the ECU plug must then be determined using the circuit diagram.

Here: Measurement of the conductor and coil resistance between contact numbers 22 and 58 with an ohmmeter.

When using a test box, bear in mind that the contact numbers of the plug generally correspond to the contact numbers of the test box.

Important:
Follow manufacturer's instructions!

4. Reference/actual-value comparison:

If the manufacturer's specified reference value is not achieved, a cable fault must be ruled out by way of a resistance measurement of the cable between the component and the ECU. If the resistance value of the cable and the cable contacts are OK, then the component is faulty.

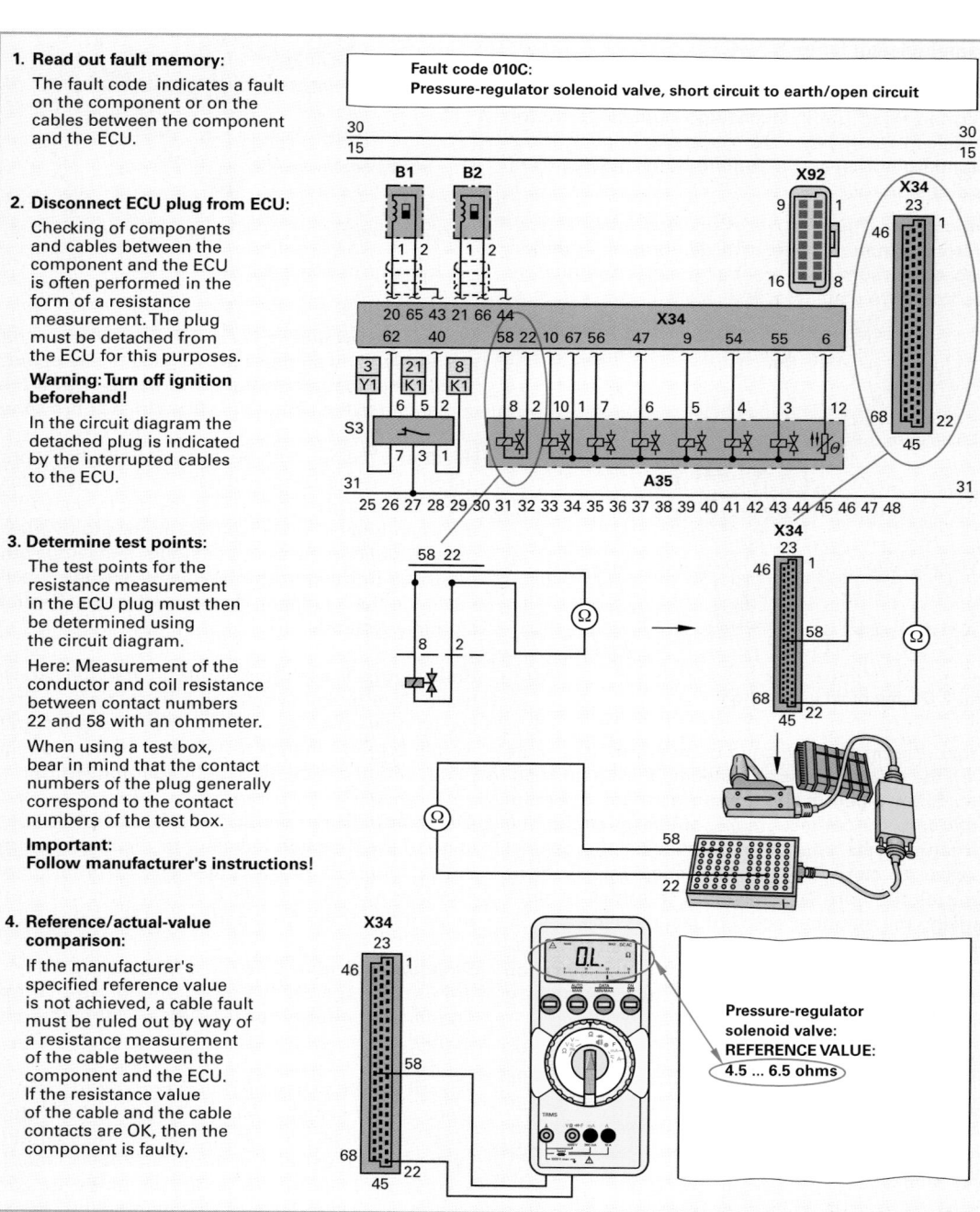

Fig. 1: Malfunction diagnosis with current flow diagram on the electric gearbox control

19.2.2 Signal transmitters

The function of signal transmitters is to provide warnings to other road users (signal horn, headlamp flasher), to indicate that the vehicle is slowing down (brake lights), or to indicate changes in the driving direction and make the vehicle visible in dangerous situations (direction-indicator lamps).

Signal horn. In accordance with paragraph 55 of the German Road Traffic Licensing Regulations motor vehicles must have a horn system. Impact horns (standard horns) and/or fanfare horns may be used as signal horns.

Impact horn. This consists of an electromagnet, a vibratory armature plate with diaphragm, vibration disc, diaphragm and a contact breaker operated by the armature plate **(Fig. 1)**.

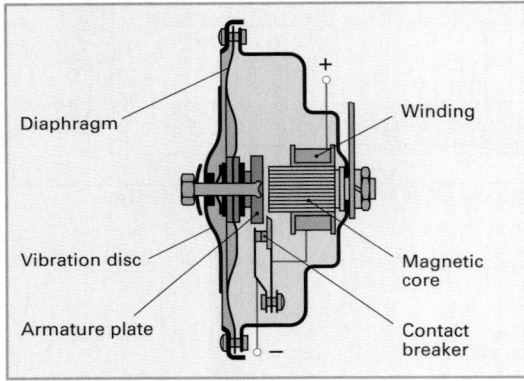

Fig. 1: Design of an impact horn

When the horn is switched on, the armature plate with the diaphragm is attracted by the electromagnet. Shortly before the armature strikes the magnetic core, the contact breaker opens the circuit. The armature plate springs back, causing the contact breaker to close. The process repeats itself for as long as the horn is activated. The armature plate striking the magnetic core (impact horn) causes oscillations in the vibration disc connected to the diaphragm. The air column in front of the vibration disc also begins to oscillate and creates a constant signal tone.

Supertone horn. This horn is more powerful than a standard horn. As it may only be used outside built-up areas, the horn system must also have a standard horn. The driver can choose between the standard and supertone horn via a switch (see Section Signalling system).

Fanfare horn. This can be used instead of the supertone horn. Here, as with the impact horn, an elec-

tromagnet causes a diaphragm to oscillate. The oscillations of the air column in the spiral horn cone create the characteristic fanfare sound **(Fig. 2)**.

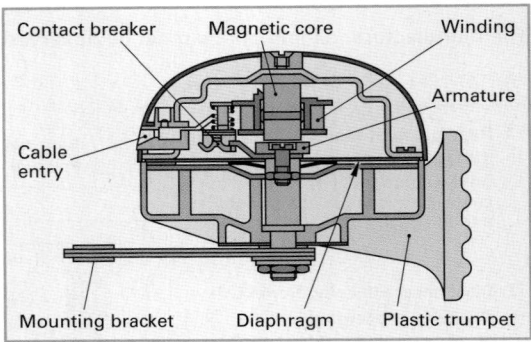

Fig. 2: Design of a fanfare horn

Compressed-air horns. These are fanfare horns that are operated by air pressure. An electrically operated compressor creates the compressed air which is connected to the air outlet in front of the horn cone.

Headlamp flasher. With the headlamp flasher (paragraph 16 of the German Road Traffic Regulations) the driver can give light signals by briefly activating the main-beam headlights if he wishes to overtake outside built-up areas or considers himself or others to be in danger. The headlamp flasher is activated via the headlamp flasher switch (see Section Lighting system).

Brake lights. These must light up when the service brake (foot brake) is activated. Brake lights with a red light beam are prescribed. They must be significantly brighter than the other rear lighting equipment, with the exception of the rear fog lamps.

Direction-indicator lamps. These are used for operation of the direction indicator or the hazard warning system (see Section Signalling system). Direction-indicator lamps with a yellow light beam are prescribed for the front and rear of the vehicle.

Direction-indicator. Electronic turn signal flashers are used to operate the direction-indicator lamps. The flash frequency must be 90 +/– 30 pulses per minute.

Hazard warning system. This is prescribed for multilane motor vehicles. The hazard warning system must work independently of the vehicle's lighting system and must always be in standby mode. To achieve this, all direction-indicator lamps are connected in parallel. Their activation must be indicated by a red indicator lamp.

19

19.2.3 Relays

> This is an electromagnetically operated switch in which the relay contacts are operated by a solenoid coil.

Design (Fig. 1). A relay consists of a relay coil, a relay armature with resetting spring and the relay contacts.

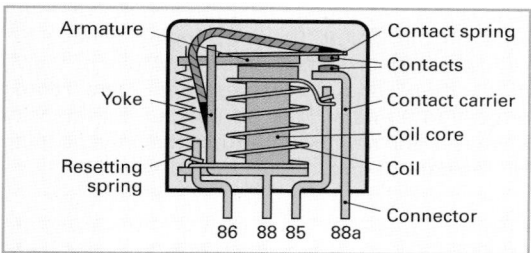

Fig. 1: Design of a relay (make contact)

Types of relay. Depending on the type and layout of the switch contacts, a distinction is made between make relays, break relays and changeover relays.

Make relay (Fig. 2). This closes the control circuit between the voltage source and consumer, i.e. the consumer is switched on. The control current through the electromagnets (terminal connectors 85 and 86) causes the switch armature to be activated. This closes the switch contact. Thus, the working current circuit (terminal connectors 88 and 88a) is connected. A comparatively small control current is sufficient to connect the working current circuit.

Example of use: main beam headlamps and auxiliary lamps, signal horn, fan motor, window regulators.

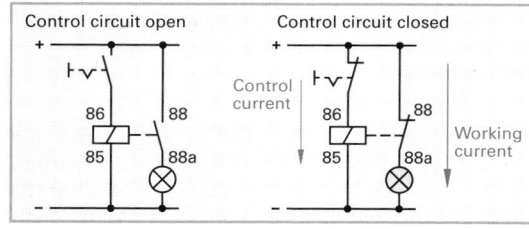

Fig. 2: Make relay with open and closed control circuit

Break relay (Fig. 3). This opens the control circuit between the voltage source and consumer, i.e. the consumer is switched off.

Example of use: breaking electric circuits of consumers during the start process, e.g. main beam headlamps, rear window heating, radio, etc.

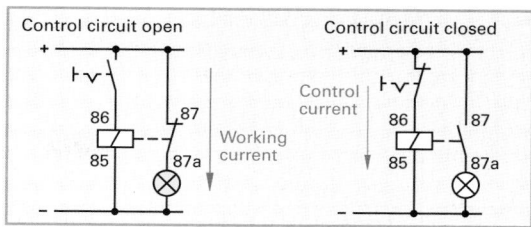

Fig. 3: Break relay with open and closed control circuit

Changeover relay (Fig. 4). This is a combination of a make relay and a break relay, i.e. it activates two control circuits at the same time. The relay switches the current path from one consumer to the other by changing the break contact of one circuit to the make contact of the other circuit.

Example of use: switching between two consumers (bulbs).

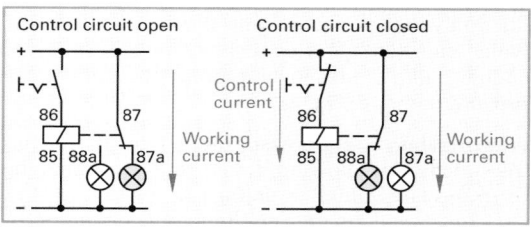

Fig. 4: Changeover relay with open and closed control circuit

The relay fulfils the following tasks:

- Large load currents (e.g. up to 2,000 A in starters) are switched by means of a small control current (approximately 0.15 A to 1 A).

- The electric line for the working current circuit between the voltage source and consumer can be kept short. This reduces the voltage drop. The low-load control line across the switch to the relay coil can be the same length. There are savings in terms of cost and weight as the control line requires a significantly smaller conductor cross-section compared with the working current line.

- The contacts of the switch for the control current are only placed under low load. This means consumers with a higher initial current load, e.g. bulbs and starters, can be connected without problems.

Relay terminal designations. These are standardised according to DIN 72552 (**Table 1, Page 524**). In practice, the old terminal designations are frequently still used. Some manufacturers also use their own system.

19

Table 1: Relay terminal designations in accordance with DIN 72552

Terminal designation	Meaning	Terminal designation (old)
85	Control circuit (–) Coil winding end	85
86	Control circuit (+) Coil winding start	86
87	Input terminal Working circuit (break and changeover contacts)	30 / 51
87a	Output terminal Working circuit (break contact side)	87a
88	Input terminal Working circuit (make contact)	30 / 51
88a	Output terminal Working circuit	87

Safety equipment in the relay. In relay coils, high induction voltages occur when the circuit is broken, which can damage electronic components in the circuit. Using diodes in an appropriate manner can prevent damage **(Fig. 2).**

Freewheeling or suppressor diode (Fig. 1). This has the function of allowing the current flow resulting from the self-induction voltage of the coil to flow back to the coil. As the self-induction voltage of the applied voltage acts in the opposite direction, the diode is connected in the conducting direction at the moment of deactivation.

Reverse polarity protection diode (Fig. 1). This has the function of protecting the freewheeling diode against polarity reversal and thus against damage. Without a reverse polarity protection diode and with incorrect polarity (terminal 85 (+) and 86 (–)), there is full voltage at the freewheeling diode and the diode can be destroyed by the short-circuit current.

Suppressor resistor (Fig. 1). This can be used instead of a freewheeling or suppressor diode. In this case, no reverse polarity protection diode is necessary. However, the stray power that occurs during operation is disadvantageous.

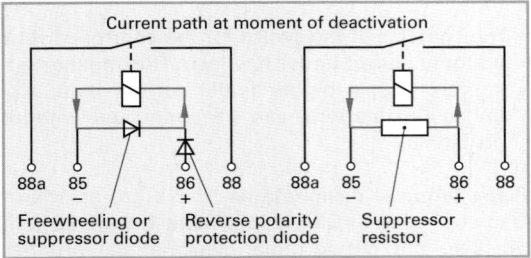

Fig. 1: Make relay with freewheeling and reverse polarity protection diode

Overvoltage protection relay. This has the task of supplying electronic control units with voltage and protecting them against overvoltage.

Operating principle (Fig. 2). When the ignition is switched on (terminal 15), the control circuit for the relay is closed and its contact is activated. Via the closed contact, the control unit is supplied with voltage (positive potential from terminal 30 to terminal 87). In this status, the Z-diode is operated in the blocking-state region and has no influence on the circuit. If the operating voltage exceeds the Zener voltage, the resistance of the Z-diode drops and the voltage peaks are routed directly to terminal 31/earth. The current intensity increases and the fuse interrupts the power supply for the control unit.

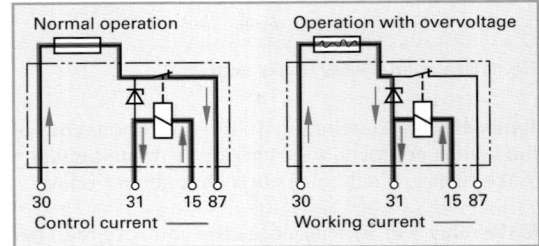

Fig. 2: Overvoltage protection relay

Reed relay (Fig. 3). The reed relay consists of a glass tube filled with inert gas, with two installed contact tongues (contact pair). A coil of a few windings of thick wire is wound around the glass tube.

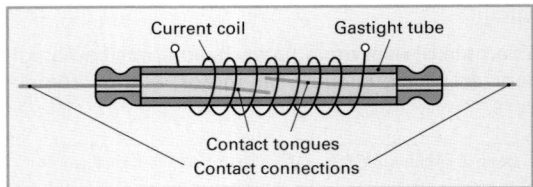

Fig. 3: Reed relay

Operating principle. If a current is passed through the coil, a magnetic field forms. This is concentrated in the contact tongues, which are similar to the coil core. The field lines try to shorten themselves and in doing so close the contact tongues. When the current flow is interrupted, the magnetic field breaks and the spring action of the contacts opens the contact tongues, e.g. as with lamp current monitoring.

In addition to operation by the magnetic field of a coil charged with current, the Reed relay can also be operated by the field lines of a permanent magnet, e.g. as with fill level monitoring.

19

19.2.4 Lighting in the motor vehicle

The functions of the lighting systems in the motor vehicle are:

- To illuminate the road, e.g. main-beam headlights, dipped-beam headlights.

- To make the contours of the vehicle visible in darkness, e.g. clearance lamps, and parking lights, reflectors.

- To indicate intended manoeuvres to other road users, e.g. direction-indicator lamps, brake lights.

- To warn other road users, e.g. hazard warning system.

- To make the driver aware of certain switch statuses of the lighting system, e.g. main-beam indicator lamp.

The legal requirements for lighting systems distinguish between headlights, lights and reflective agents, e.g. reflectors (Fig. 1).

Headlights. These are used to illuminate the road.

Lights. These should make it possible for other road users to see the vehicle and to signal the intentions of the driver in terms of driving behaviour.

All motor vehicles must have the prescribed lighting systems, and additional lighting systems that have been declared permissible may also be fitted.

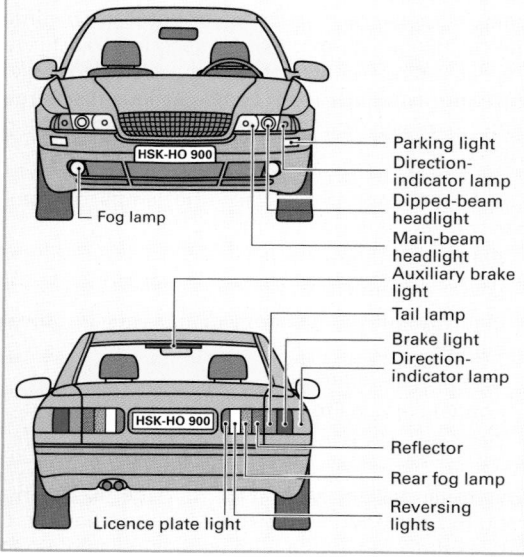

Parking light
Direction-indicator lamp
Dipped-beam headlight
Main-beam headlight
Auxiliary brake light
Fog lamp
Tail lamp
Brake light
Direction-indicator lamp
Reflector
Rear fog lamp
Reversing lights
Licence plate light

Fig. 1: Lighting systems

Layout possibilities for main-beam and dipped-beam headlight (Fig. 2).

Dual-headlight system. The dipped beam and main beam are beamed into a common reflector. The light is created by bulbs with two filaments, which use a common reflector (Bilux lamp, H4 halogen lamp).

Quad-headlight system. One headlight pair is designed either for dipped beam and main beam or for dipped beam only. The second headlight pair is only designed for main beam.

Six headlight system. In addition to the quad-headlight system, another headlight pair may also be provided for front fog lamps or main-beam headlights, depending on the headlight layout.

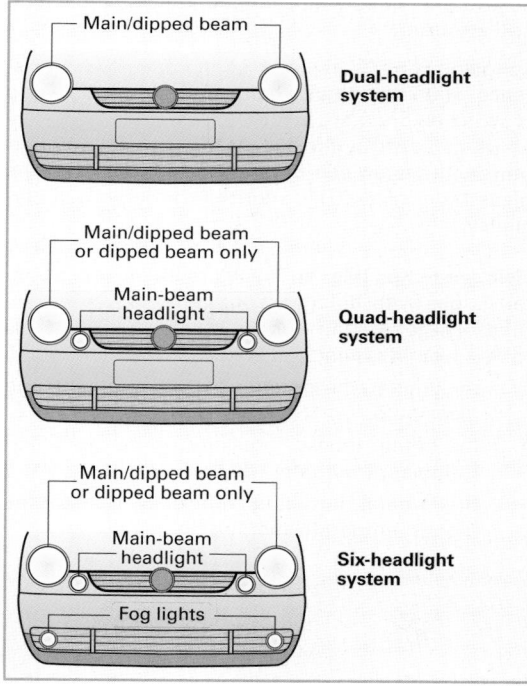

Main/dipped beam
Dual-headlight system

Main/dipped beam or dipped beam only
Main-beam headlight
Quad-headlight system

Main/dipped beam or dipped beam only
Main-beam headlight
Fog lights
Six-headlight system

Fig. 2: Headlight systems

Lighting systems must ...

- ... be fixed (except concealed headlights).
- ... be fitted so that they cannot mutually influence each other.
- ... always be ready for operation.

Lighting systems available in pairs must ...

- ... be symmetrical.
- ... be ready for operation together and at the same time.
- ... have the same colour and brightness.

19

19.2.4.1 Illuminators

In motor vehicles, the following lamp types can be used in headlights and lights:

- Metal filament lamps
- Neon gas-discharge lamps
- Halogen lamps
- Gas-discharge lamps
- Light emitting diodes

Metal filament lamps. The light source (filament, coil) is made of tungsten, which has a melting point of approximately 3,400 °C. The coil itself can reach temperatures of up to 3,000 °C. To prevent oxidation (combustion) at these high temperatures and to be able to conduct away the heat produced more easily, the glass bulb is first evacuated and then filled with small quantities of nitrogen and krypton.

Tungsten is a PTC resistor, i.e. it has a lower resistance when cold than when warm. When the lamp is switched on, there is therefore a brief high surge which can destroy the coil. At the high coil temperatures, tungsten can break away and blacken the glass bulb inside. This reduces the luminous efficiency.

Halogen lamps (Fig. 1). These are bulbs which contain a gas with halogen admixtures (bromine, iodine). The operating behaviour of halogen lamps differs from filament lamps in ...

- ... the higher temperature of the filament and glass bulb.
- ... the higher interior pressure of the gas filling (up to approximately 40 bar).
- ... the greater luminous efficiency due to the higher temperature of the filament.

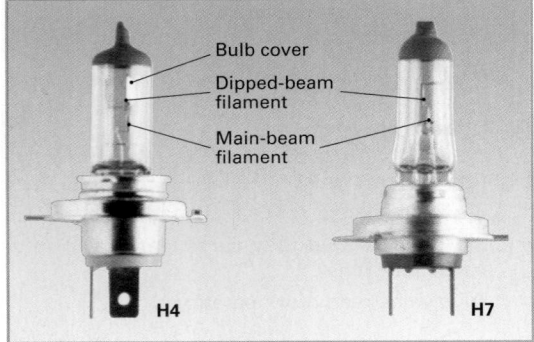

Fig. 1: Halogen lamps type H4 and H7

The bulb of the halogen lamp is made of quartz glass. It has very small dimensions so that it can heat up to approximately 300° C during operation.

The evaporated tungsten particles undergo a chemical reaction and are deposited on the hottest part of the filament again (cyclic process).

> With halogen lamps, the glass bulb remains clear as no evaporated tungsten can break away here due to the cyclical process.

Gas-discharge lamps. Between two electrodes located in a small ball-shaped glass bulb, an arc is ignited in a xenon gas atmosphere by a high voltage pulse. The metal salts within the glass bulb evaporate and ionise the spark gap. In doing so, they emit light and prevent the electrodes from wearing.

In contrast to the gas-discharge lamps for reflection systems, the lamps for projection systems **(Fig. 2)** do not have any shadowing effects on the glass bulb. The gas-discharge lamps for reflection systems require the shadowing effect to create the light-dark cut-off.

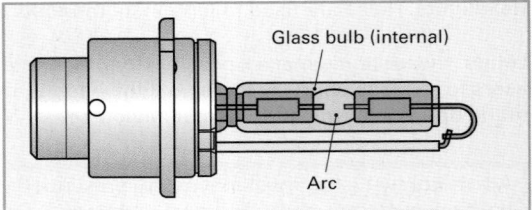

Fig. 2: Gas-discharge lamp for projection systems

The disadvantage of gas-discharge lamps by comparison with halogen lamps is that it takes approximately 5 seconds to reach full brightness. The time for the halogen lamp is just 0.2 seconds. To achieve the desired operating status as quickly as possible, the control unit upstream of the gaseous-discharge lamp increases the lamp current in the warm-up phase.

The advantages of gas-discharge lamps by comparison with halogen lamps are:

- Better road illumination
- Lower current consumption
- Light output independent of the vehicle power supply
- Lower heat generation
- Longer service life
- The light colour is almost that of daylight

Electronic ballast unit (Fig. 1, Page 527). This required for operation of gas-discharge lamps. It ignites the lamp with a high voltage pulse of up to 24 kV, which causes a spark to jump between the electrodes of the lamp. After the lamp is switched on, the lamp output is constantly regulated at 35 W with a spark voltage of approximately 85 V (300 Hz alternating voltage).

> Due to the high voltages that occur during ignition and the high operating voltage, there is a danger to life if the headlights are not maintained properly or if they are damaged.
> **Safety regulations must be observed.**

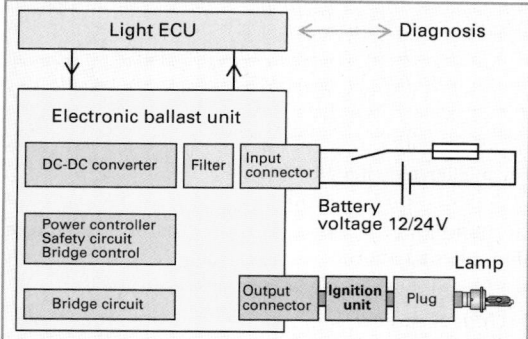

Fig. 1: Electronic ballast unit

Control and safety circuits. The ballast unit is able to detect a rupture in the arc of the gaseous-discharge lamp during ignition and in operation. In this event, it first makes several attempts at reignition. If this does not succeed due to a lamp or line defect, the voltage is switched off. In diagnostics-enabled systems, an entry is stored in the control unit's fault memory. Defects in the headlight system can cause leakage currents. If these are more than 20 mA, the ballast unit interrupts the supply voltage to the lamps.

Neon gas-discharge lamps. These are gas-discharge lamps that achieve full brightness in approximately 0.2 ms. They are therefore primarily used in auxiliary brake lights.

Light emitting diodes (LED). A certain number of diodes are interconnected with a physical unit according to the brightness required and the desired light colour. The multiple allocation reduces the probability of failure of the overall function. Light emitting diodes have a service life of approximately 10,000 hours. They are used in particular for brake lights, as they achieve their maximum brightness in a significantly shorter time than filament lamps or halogen lamps (approximately 2 ms) **(Fig. 2)**.

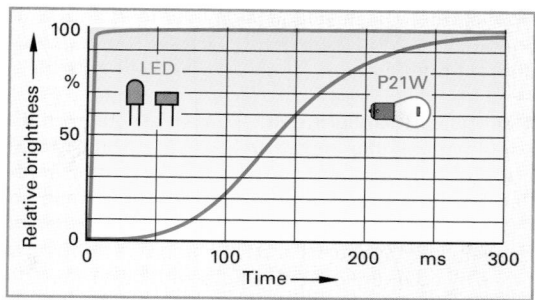

Fig. 2: Switch-on behaviour of light emitting diodes

19.2.4.2 Headlight systems with halogen lamps (Fig. 3)

These consist primarily of:
- **Housing.** This holds the reflector with the lens, the light source and the headlight adjustment mechanism.
- **Reflector.** This reflects and focuses the light in the bulb. Paraboloidal, ellipsoid-shaped or freeform reflectors are used.

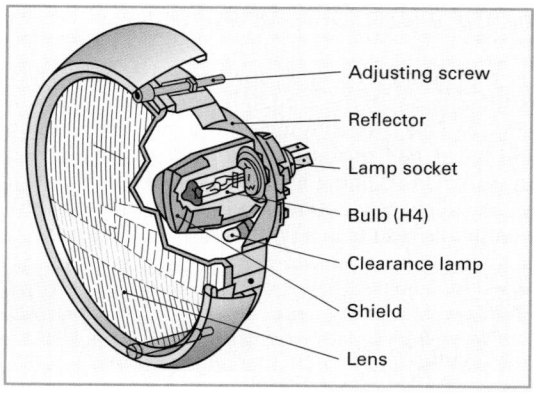

Fig. 3: Design of the H4 headlight

Headlight systems with paraboloidal reflectors (Fig. 4)

The shape of a paraboloid results from a parabola rotating around its axis. The rotational axis is also the optical axis. There is a focal point. These reflectors are suitable for single-wire and dual-wire lamps.

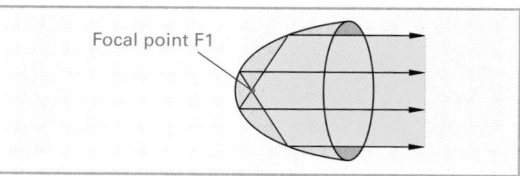

Fig. 4: Paraboloidal reflector

19

Application. These headlight systems are used in connection with H4 dual-wire lamps for the generation of main-beam and dipped-beam headlight.

Main-beam headlight (Fig. 1). The main-beam radiant filament lights up, which lies exactly in the focal point of the paraboloidal reflector. The light is reflected and focused so that it is emitted parallel to the headlight axis. Focusing increases the luminous intensity in the beam area by about a thousand times by comparison with a bulb with no a reflector.

Dipped-beam headlight (Fig. 1). The dipped-beam radiant filament lights up, which is positioned in front of the focal point of the paraboloidal reflector. The light beams are emitted at a downward angle in relation to the mirror axis.

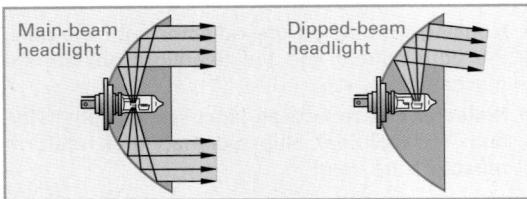

Fig. 1: Main-beam and dipped-beam headlight

So that no light escapes downwards, a bulb cover is fitted beneath the dipped-beam radiant filament **(Fig. 2)**. This prevents the light beam from hitting the lower half of the reflector and being beamed upwards. In addition, in dipped-beam headlights, it acts as a sharply defined limit to the course of the beam; this results in a light-dark cut-off **(Fig. 3)**.

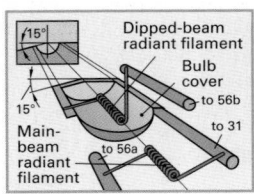

Fig. 2: Bulb cover for dipped-beam headlight   **Fig. 3: Light-dark cut-off**

Stepped reflector (Fig. 4). The reflector surface consists of several paraboloidal sub-reflectors with different focal lengths (multifocus reflector).

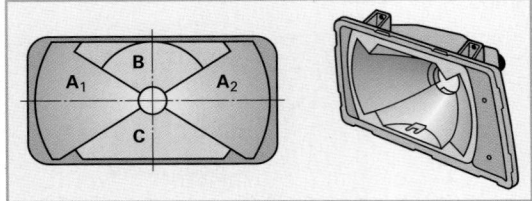

Fig. 4: Stepped reflector

The stepped reflector achieves higher luminous efficiency and better illumination of the road ahead.

Headlight systems with ellipsoid-shaped reflectors (Fig. 5).

The shape of the reflector results from an ellipsis rotating around its axis. This is also the optical axis. There are two focal points.

Application. These reflectors are suitable for dipped-beam headlights or fog lights with single-wire lamps.

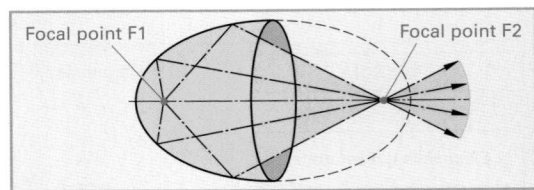

Fig. 5: Ellipsoid-shaped reflector

A headlight system with ellipsoid reflectors **(Fig. 6)** consists of:

- Ellipsoid-shaped reflector
- Optical screen
- Shutter
- Convergent lens

In focal point F1, there is a halogen single-wire lamp. The light beams emitted from F1 are reflected by the reflector to focal point F2 and are beamed to the convergent lens from there. The convergent lens focuses the light into an almost parallel light band.

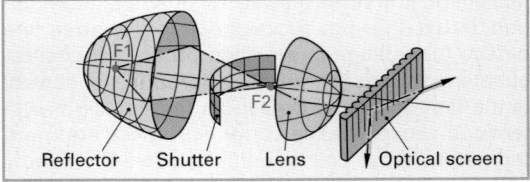

Fig. 6: Ellipsoid reflector with lens

The shutter in front of focal point F2 creates a sharply defined light-dark cut-off. The optical screen causes equal light distribution. Compared with paraboloidal reflectors, the luminous efficiency is higher.

Multiple axis ellipsoid reflector (Fig. 1, Page 529).

This is a reflector whose basic shape is formed by two ellipses with a joint apex, joint main axis and separate minor axes (manufacturer's designations: DE reflector = triple axis ellipsoid reflector; PES reflector = poly ellipsoid reflector). They consist of the reflector, the shutter and the convergent lens. The reflector is made from plastic due to its complicated shape.

Due to their geometric construction, these reflectors have a very high luminous efficiency with little stray light. The shutter in front of the focal point creates a sharply defined light-dark cut-off. The optical screen ensures equal light distribution.

Application. They are suitable for dipped-beam headlights or fog lights with single-wire lamps or gas-discharge lamps.

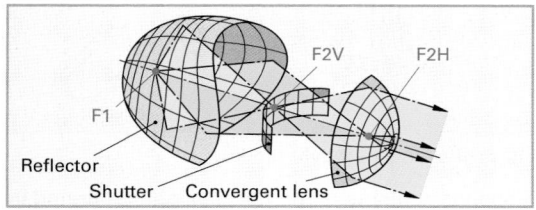

Fig. 1: **Multiple ellipsoid reflector with lens**

Headlight systems with freeform reflectors.

These are reflectors with an infinitely variable focal point (focus). The reflector is freely formed in shape. Each point of the freely calculated reflector is assigned a specific part of the road to be illuminated **(Fig. 2)**.

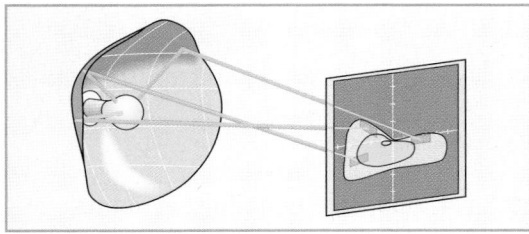

Fig. 2: **Freeform reflector**

With this design, nearly all reflector surfaces can be used for dipped-beam headlights. They are designed so that light from all reflector segments is reflected downwards to the road.

Manufacturer's designations are:
- Freeform reflector (FF reflector)
- Variable Focus reflector (VF reflector)
- Homogeneous Numerically calculated Surface (HNS)

The reflector surface is designed in accordance with the manufacturer's requirements for light distribution and illumination of the road **(Fig. 3)**. The individual zones are assigned the following functions:

- **Zone I:** Asymmetrical sector; illumination of the distant zone of the right-hand side of the road.
- **Zone II:** Symmetrical sector; illumination of the zone directly under the light-dark cut-off.

- **Zone III:** Near field sector; primarily for road illumination.
- **Zone IV:** Near field sector; primarily for illuminating the edge of the carriageway.

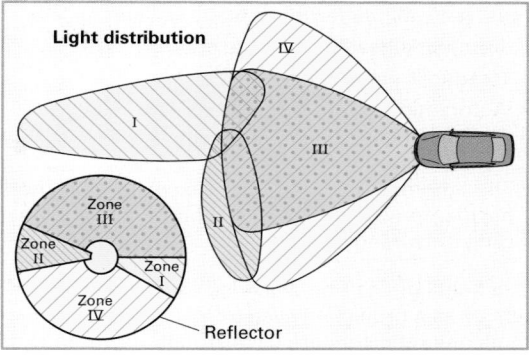

Fig. 3: **Freeform reflector – Light distribution**

Application. Freeform reflectors can be used for all types of headlights with single-wire lamps or gas-discharge lamps. For dipped-beam headlights, the bulb cover can be omitted from the lamp. The overall light generated is available for illuminating the road.

In addition, the refraction elements in the lens can be omitted. The reflector can be covered with an unprofiled glass or plastic lens.

Headlight systems with freeform reflectors and projection lens (Fig. 4).

The reflector surfaces are designed with the help of freeform technology. The light that strikes the reflector is aimed in such a way that as much as possible of it passes through the lens via a shutter.

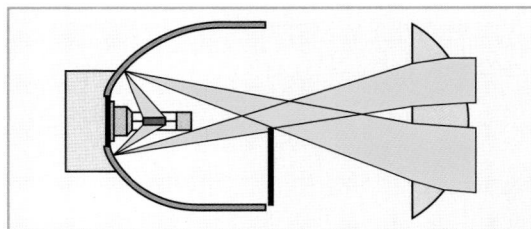

Fig. 4: **Freeform reflector with projection lens**

The reflector aims the light so that the light is distributed at the height of the shutter and is projected onto the road by the lens (manufacturer's designation: Super DE). This technology allows for wide ranges and better illumination of the sides of the road. The light is largely concentrated at the light-dark cut-off.

Application. The system can be used for headlights for dipped-beam light with single-wire lamps and for gas-discharge lamps.

19

19.2.4.3 Headlight systems
with gas-discharge lamps

Motor vehicles with headlights for dipped-beam light, which are fitted with gas-discharge lamps, must have all the technical features listed below:

- Automatic headlamp range control
- Headlight wash-wipe system
- Automatic operation of the dipped-beam headlight when the main beam is switched on

> The automatic headlamp range control and the headlamp wash-wipe system help to avoid dazzling oncoming traffic.

The headlight can be designed as a reflection system or as a projection system. With projection systems, the reflectors are usually manufactured with freeform technology.

Automatic headlamp range control. Automatic headlight range control ensures that the headlights are always automatically correctly adjusted irrespective of the laden state of the vehicle. Axle sensors on the rear axle measure differences in suspension compression, e.g. as a consequence of load. A servo-motor adjusts the headlights to the correct downward angle.

Dynamic headlamp range control (Fig. 1). This also takes into consideration the vehicle speed and processes the signals of the axle sensors on the front and rear axles. A control unit adjusts the downward angle of the headlights using stepping motors. This makes it possible to compensate for rapid changes in the vehicle's inclination angle caused by braking or acceleration.

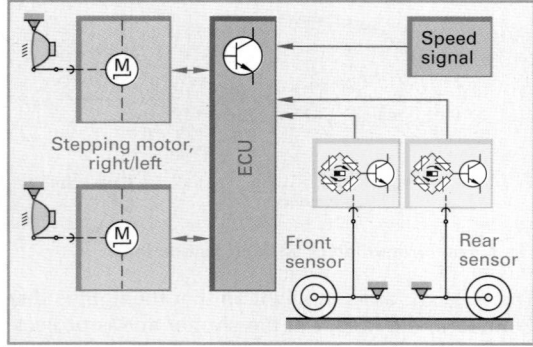

Fig. 1: Dynamic headlamp range control

Headlight systems with gas-discharge lamps that create both main-beam and dipped-beam light (manufacturer's designation: Bi-Xenon, Bi-Litronic), usually have an additional main-beam headlight in addition to the headlight unit with gas-discharge lamp, e.g. with an H7 lamp.

The changeover between main-beam and dipped-beam headlight takes place in the bi-xenon headlight module with the help of a mechanical shutter. This is adjusted with an electromagnet **(Fig. 2)**.

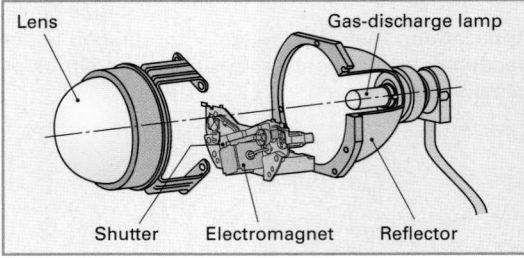

Fig. 2: Bi-xenon headlight module

The shutter blocks off part of the light generated in the lamp for the dipped-beam headlight and thus the required light-dark cut-off is created. With the mechanism adjusted to the main-beam position, all the light created in the lamp is allowed through **(Fig. 3)**.

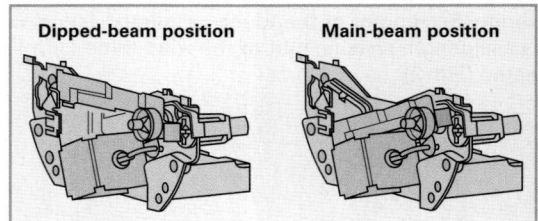

Fig. 3: Mechanical shutter

19.2.4.4 Adaptive headlight systems

These are able to adapt to different traffic situations.

The system for the dynamic cornering light enables this adaptation when cornering. The headlights are swivelled about the vertical axis to such an extent as determined by the radius of the bend. For very tight corners, e.g. intersections, an auxiliary lamp is switched on for the main light function (static cornering light or turning light) **(Fig. 4)**.

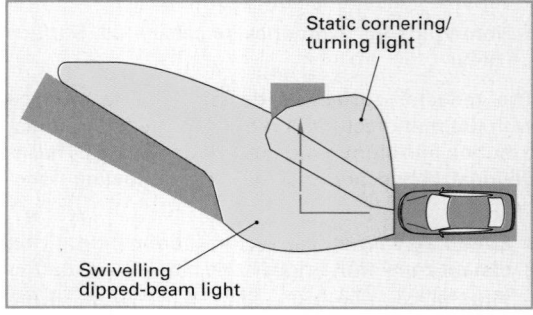

Fig. 4: Cornering light

Design. The headlight system consists of an auxiliary lamp with a halogen lamp and a projector module with a swivel device **(Fig. 1)**.

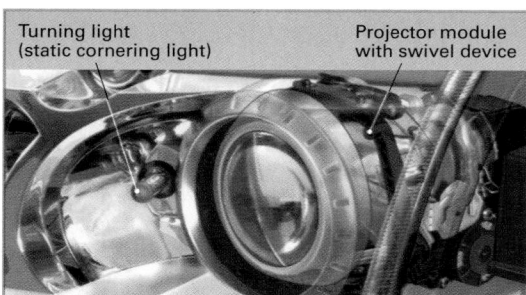

Fig. 1: Headlight system with dynamic cornering light

Operating principle. The projector module, as a bi-xenon system, has a moving shutter for switching between the main-beam and dipped-beam headlight. For the cornering light, the projector is pivoted about the vertical axis by a worm-gear pair driven by a stepping motor to such an extent as determined by the radius of the bend **(Fig. 2)**. The radius of the bend is detected by sensors, which either measure the steering wheel input or the turn-

ing of the vehicle about the vertical axis. A control unit processes the sensor signals and controls the stepping motor for the swivel device.

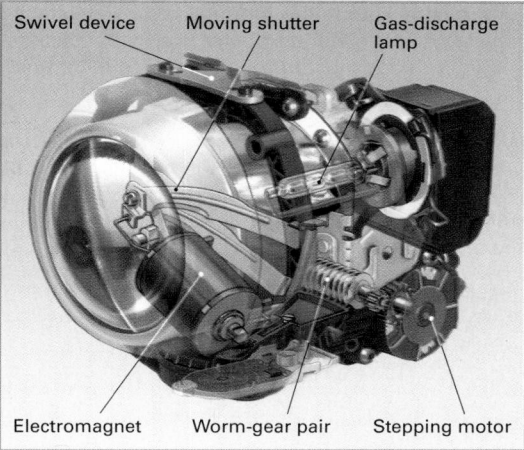

Fig. 2: Projector module with swivel device

The turning light is switched on as soon as the control unit detects a turning motion, i.e. when the vehicle moves through a correspondingly small bend radius.

WORKSHOP NOTES

Fault clearance in lighting system (bulbs)

Problem: Bulb does not light up		
Possible cause	**Test**	**Fault correction**
Filament burnt through	Visual check	Replace bulb
Defective fuse	Visual check	Replace fuse
Line break	Measure resistance or voltage	Repair line

Problem: Bulb light is weak		
Possible cause	**Test**	**Fault correction**
Contact resistance in lines or at contacts	Resistance or voltage measurement	Eliminate contact resistance
Battery not fully charged	Check battery voltage	Charge or replace battery
Incorrect bulb used (24 V bulb in 12 V system)	Visual check	Replace bulb

Setting headlights with automatic range control

In vehicles with automatic headlamp range control, the basic setting of the headlights is carried out with the help of diagnostics devices in setting mode. To do this, the vehicle must be prepared as when setting standard headlamp systems. The control position of the headlights is taught in using the diagnostic tool. This task must be carried out after the headlamp system has been replaced, for example. An incorrect basic setting, or no basic setting can be the cause for a faulty headlamp range control system.

Other possible causes are:
- Defective servo-motor in headlight
- Damaged wiring / corroded connector
- Defective vehicle level sensors
- Bent or damaged steering rods between sensors and chassis
- Defective control unit for headlight range control

19

19.2.5 Power supply and vehicle electrical system

> To reliably supply the motor vehicle's electrical systems with electrical power, a system consisting of a voltage generator, voltage regulator and a charge accumulator is used.

With the engine running, the alternator supplies the vehicle electrical system with electrical energy and charges the battery. The alternator generates a three-phase current which is rectified and supplies the electrical consumers **(Fig. 1)**.

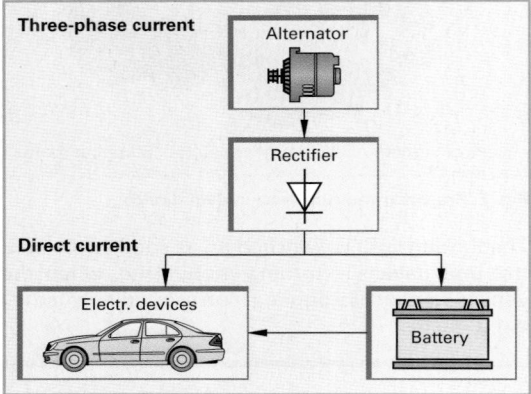

Fig. 1: Vehicle electrical system

The vehicle electrical system is supplied with power by the battery when the engine is idling or, where necessary, switched off. The battery also supplies all of the power required for the starting procedure.

The demand for electrical energy has increased greatly in motor vehicles. The alternator currently satisfies a demand for electrical power of up to approximately 2,000 W.

This demand results from increasing electronic regulations as well as safety and comfort and convenience electronics, such as

- electronic ignition and fuel-injection systems
- self-monitoring
- seat and auxiliary heating
- window and mirror heating
- electrically driven fans
- air conditioning systems with up to 10 electric motors
- ABS, ESP

The average efficiency of an alternator in a motor vehicle is approximately 60%. If it outputs power of 2,000 W for an hour under these conditions, it consumes approximately 1 litre of fuel to generate this power.

19.2.5.1 Starter battery

> The starter battery supplies and stores the energy for the electrical systems in a motor vehicle. As it can be recharged, it is also referred to as an accumulator.

Design (Fig. 2)

Battery cell. The starter battery consists of several cells. A cell consists of the positive and negative lead plates. To avoid contact and short circuit, they are separated by separators. To enable a high current to flow, there must be a large plate surface. Therefore, the cells are equipped with as many thin electrode plates as possible. The cells are grouped into blocks. One cell provides a nominal voltage of 2 volts. Series connection of the cells results in 6 or 12 volts, depending on the number of cells.

Electrolyte. This fills the space between the lead plates and the separator. The electrolyte reservoir is located above the plate groups. Beneath the plates, there is a sludge compartment that collects the lead that breaks away.

Housing. The cells are enclosed in a block box. The battery is closed with the block cover, which contains the drain and filler plugs. To enable the battery to be fixed to the vehicle body, a guide rail is cast on the side at the bottom.

Terminals. The terminals are used to collect the total voltage (terminal voltage) and to connect the battery to the vehicle electrical system. The terminals are labelled with + and –. To avoid confusing them for each other, the positive terminal has a wider diameter **(Fig. 2)**.

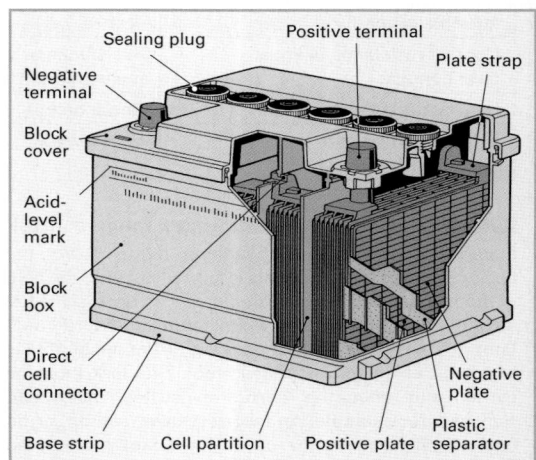

Fig. 2: Starter battery

Electrochemical processes

Charged status (Fig. 1a)
The active materials of the positive plates consist of brown lead dioxide (PbO_2), and the active materials of the negative plate consist of grey leads (Pb). The electrolyte is diluted sulphuric acid (H_2SO_4) with a concentration of 37% and a density of $\varrho = 1.28$ g/cm^3.

Discharging (Fig. 1b)
The drawing of current causes an electrochemical reaction in the cell. The brown lead dioxide of the positive plates and the grey lead (Pb) of the negative plates are converted into white lead sulphate ($PbSO_4$). During the process, sulphuric acid (H_2SO_4) is converted, resulting in the formation of water (H_2O) and a lower electrolyte density **(Fig. 1c)**.

$$PbO_2 + 2\,H_2SO_4 + Pb \;\Rightarrow\; PbSO_4 + 2\,H_2O + PbSO_4$$

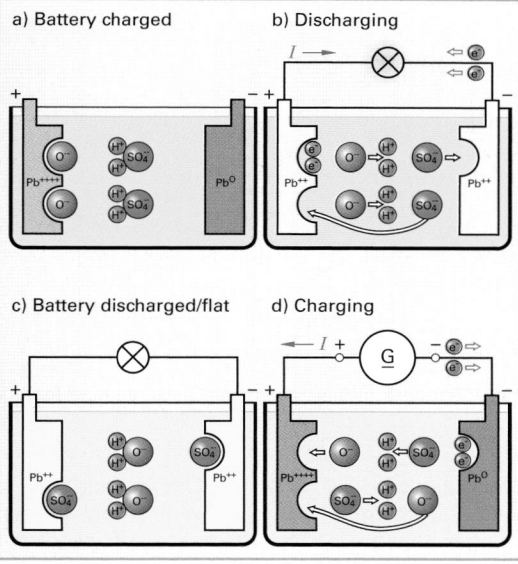

Fig. 1: Electrochemical processes during charging and discharging

Discharged status (Fig. 1c)
Both plates have converted into lead sulphate ($PbSO_4$). The electrolyte density has dropped to approximately 1.12 g/cm^3, the specific gravity of the electrolyte to 12 %.

Charging (Fig. 1d)
When current is supplied, the electrochemical reaction reverses itself. The white lead sulphate ($PbSO_4$) of the positive plate turns into brown lead dioxide (PbO_2), and the white lead sulphate of the negative plate turns into grey lead (Pb). During the process, water (H_2O) is converted, resulting in the formation of sulphuric acid (H_2SO_4). The electrolyte density increases **(Fig. 1a)**.

$$PbSO_4 + 2\,H_2O + PbSO_4 \;\Rightarrow\; PbO_2 + 2\,H_2SO_4 + Pb$$

Battery labelling
To make it possible to compare and exchange starter batteries of different manufacturers, a type designation on the battery housing is prescribed according to European standard EN 60095-1.

The label **(Fig. 2)** consists of the nine-digit European type number ETN (e.g. **544 105 045**). This includes

- the nominal voltage,
  e.g. reference number 5: 12 V,
- the nominal capacity,
  e.g. reference number 44: 44 Ah,
- information on the type of battery and how it is fixed into the vehicle, reference number 105, and
- the cold test current,
  e.g. reference number 045: 450 A.

In Germany, the type designation is also specified in accordance with DIN 72310.

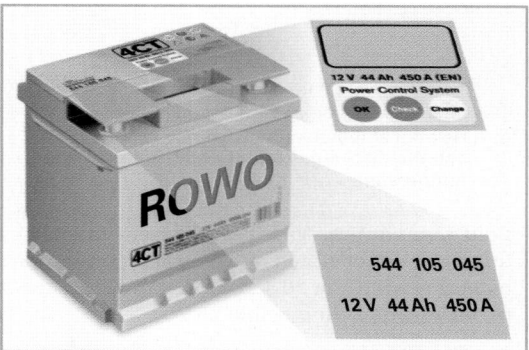

Fig. 2: Labelling of starter batteries

> **Nominal voltage.** This is defined as 2.0 V per cell (DIN 40 729). The nominal voltage of the entire starter battery results from the number of cells connected in series. It is 12 V for 6 cells.

Capacity. Capacity $K = I \cdot t$ means the stored electrical energy in ampere hours (Ah) which can be supplied to or drained from a starter battery **(Fig. 2)**.

This depends on …
- … the size of the discharge current.
- … the density and temperature of the electrolyte.
- … the state of charge of the battery.
- … the age of the battery.

> **Nominal capacity K20.** This is the capacity that a starter battery can hold with 20-hour discharge with a discharge current of 1/20 of the numerical value of the nominal capacity. The voltage must not fall below the cut-off voltage of 10.5 V.

The temperature of the electrolyte must be approximately +25°C. At higher temperatures, the discharge capacity rises compared with the nominal capacity **(Fig. 1)**. This direct relationship between capacity and temperature is due to the fact that electrochemical processes take longer at lower temperatures.

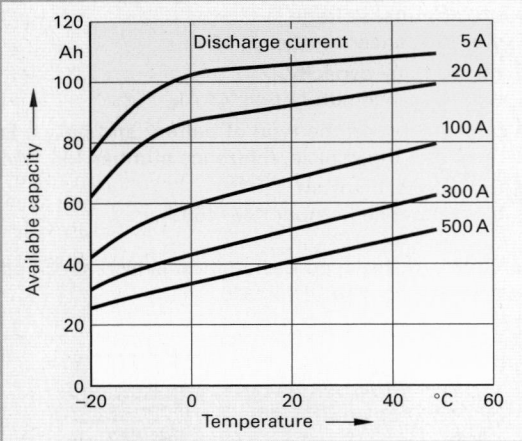

Fig. 1: Relationship between discharge capacity and the discharge current and the electrolyte temperature

Cold test current. This is the current intensity that must be generated by a fully charged starter battery at –18°C for 10 seconds, without the terminal voltage falling below 7.5 V.

With the cold test current, the starting response is assessed at low temperatures. If the voltage falls below the specified voltage values, the starter battery can no longer provide full output.

Additional battery characteristics

Steady-state voltage U_0 (open-circuit voltage). This is measured between the battery terminals of a starter battery under no load. It depends on the state of charge and the electrolyte temperature. It can be used as a first point of reference for a capacity check. A fully charged battery at approximately 25°C has a steady-state voltage of approximately 12.8 V, while the steady-state voltage in a discharged battery is approximately 12.0 V **(Table 1)**. This can only be discovered via a test under load with consumers.

Table 1: Steady-state voltage – State of charge

Steady-state voltage	State of charge
Less than 12.2 V	Discharged
12.2 V – 12.5 V	Half charged
12.5 V – 12.8 V	Charged

Terminal voltage (U_K). This is the voltage between the battery terminals under load. The terminal voltage is less than the steady-state voltage (U_0), because a voltage drop takes place inside the battery (U_i). If a current (I_E) flows through the battery, this creates internal resistance inside the battery (R_i) **(Fig. 2)**.

This resistance is connected in series with the resistors of the consumers (R_v). The falling voltage at this resistor (U_i) is then no longer available for the power supply. This internal resistance is highly dependent on the temperature of the electrolyte. At low temperatures, the electrochemical reaction has greater losses than at high temperatures. The voltage drop is also highly dependent on the state of charge. The lower the state of charge, the higher the internal resistance.

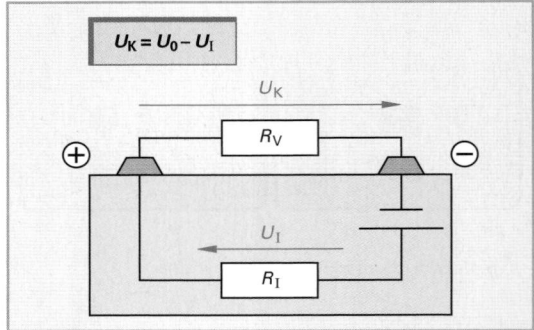

Fig. 2: Terminal voltage of a starter battery under load

Charging voltage. This is the voltage with which a battery is charged by the alternator or an external charger. When a battery reaches a voltage of approximately 14.4 V during charging, it begins to gas strongly if charging continues **(gassing voltage)**. This process leads to water losses in the battery. Some of the water breaks down into hydrogen and oxygen. This results in highly explosive detonating gas. Therefore, the charging voltage during charging by the alternator or suitable chargers (I_U curve) is regulated and kept below the gassing voltage. The gassing voltage depends on the temperature of the electrolyte. As the electrolyte temperature rises, the gassing voltage falls and thus the maximum charging voltage falls. Full charging is achieved when the electrolyte density no longer increases at the end of the charging process.

Discharge voltage. A starter battery is discharged when the cell voltage falls to the **cut-off voltage** of 1.75 V. The electrolyte density falls to around 1.12 g/cm³.

Starter current. When starting a passenger car engine, the battery briefly supplies a very high current of up to 400 A. Approximately 1% of the nominal capacity is required for this.

Short-circuit current. This is the maximum current flow that the battery can supply. This depends on the plate surface. A battery must only be laden with the short-circuit current for a maximum of 2 seconds.

Ageing processes

Self-discharge. This takes place inside the battery without the external circuit being closed. After a certain time, even without external consumers, the battery is discharged. Heat and contamination in the electrolyte speed up this process. Self-discharge also leads to gassing which consumes water and reduces the electrolyte level. Self-discharge can take up to 1% of the nominal capacity each day depending on the design. It depends on …

- … the specific gravity of electrolyte.
- … the electrolyte level.
- … the electrolyte temperature.
- … the age and external condition of the battery.

Fully charged, new, maintenance-free starter batteries have a state of charge of less than 40 % after approximately 2-3 months out of use at room temperature. Used batteries reach this value correspondingly sooner.

Capacity loss and cell short circuit. This can result from certain loads on the battery, such as

- **sulphation**
- **cyclic discharging**
- **exhaustive discharging**

There is significant wear on the thin positive plates through loosening of the active materials. The capacity falls and the starter battery may become suddenly unusable if the accumulated lead sludge leads to a short circuit.

Sulphation. Sulphation can occur if the starter battery is in a discharged state for a long time. Here, the fine crystalline lead sulphate is converted into a rough crystalline lead sulphate. If the reaction process has not yet progressed far, it can be reversed by charging with a small current intensity of approximately 0.2 A.

Cyclic discharging is frequently repeated discharging of 60% to 80% of the nominal capacity. Starter batteries must therefore not be used for applications in which they would be subjected to this type of load.

Exhaustive discharging. This is discharging of over 80% of the nominal capacity.

Construction designs
Maintenance-free starter battery (conforming to European standards)

> Under normal conditions, the electrolyte level should only drop to the extent that it does not have to be topped up within 2 years.

These have filler plugs for filling with electrolyte and for topping up the electrolyte level with distilled water. The lead lattice plates of these lead storage batteries contain antimony which gives them the required solidity. However, the antimony causes the battery to self-discharge and thus to gas, with a correspondingly high water consumption. When new, this battery is therefore stored dry. When filled for the first time, it must be filled with 37% sulphuric acid at least 20 minutes before first use.

Absolutely maintenance-free starter battery

> It is tightly sealed and does not require the acid level to be topped up. A battery tilt angle of 70 degrees is permissible.

These batteries do not have a filler plug and are stored full. The lead lattice plates contain calcium instead of antimony, which significantly reduces the self-discharge and water consumption. This means that the electrolyte supply via the plates lasts for the entire service life of the battery and the batteries are suitable for vehicles with long periods out of use. The batteries only have ventilation openings which allow a tilt angle through a safety labyrinth system. The state of charge can no longer be determined via an electrolyte density test. Therefore, many of these batteries have integrated control displays that give information on the charge status of the starter battery, e.g.

- green: OK
- grey: recharge (check)
- white: change

Heavy duty batteries

> These are batteries with a high service life, which have an extremely high vibration and cycle resistance.

Separators with a glass fibre mat support the positive plate and reduce erosion. Securing with cast resin or plastic prevents loosening of the plate blocks. These batteries are particularly suitable for construction machines.

19

Batteries with bonded electrolyte.

> This type of battery is completely fail-safe and can be fitted in any position. It has an increased cycle resistance and allows complete discharging.

The electrolyte is no longer liquid. It can be bonded into gel or in fibres. The distance between the positive and negative plates is smaller, which reduces the internal resistance. A chemical reaction prevents gassing through an internal oxygen recombination. That is, the gases are converted back into water within the cell. Features are:

- Very low self-discharge.
- Compact construction concept as no electrolyte reservoir.
- Higher short-circuit current.
- A safety valve opens if battery is overcharged.

These batteries are suitable for vehicles with long periods out of use and can also be used as general-purpose batteries. A distinction is made between gel and glass-mat battery technology.

Gel batteries. With these, the electrolyte is bonded into a fixed multi-component gel. Adding silica to the sulphuric acid results in a gel-type material in which the electrochemical reaction takes place.

Glass-mat technology. With this design, intermediate layers of glass-fibre fleece surround and bind the acid. The capillary attraction and the wetting cause the electrolyte liquid to be absorbed into the networked microfibres (AGM, Absorbing Glass Mat). At the same time, the layers act as the separator and exert an equal and high contact pressure on the plate surface. The active material is thus bound solidly between the fleece. This prevents the active materials from breaking away and at the same time improves the stability through the support measures. In a special design construction, the lead plates and the glass-fibre fleece are wound up. This leads to a further increase in the packing density.

Battery sensor

> This enables a permanently high state of battery charge thanks to regulation of the charging voltage depending on temperature and charge state.

A small control unit integrated in the battery sensor determines the optimum charge voltage reference value from the currently measured battery temperature and the state of charge. To calculate the battery state of charge (SoC), the charging and discharging current, terminal voltage and electrolyte temperature are recorded and saved in every operating status of the vehicle.

The sensor is integrated in the terminal slot of the battery negative line **(Fig. 1)**.

Fig. 1: Battery sensor

Battery safety terminals

> In case of accidents, this separates the starter/alternator cable from the battery within milliseconds.

Separation prevents the starter cable being jammed in electrically conductive body parts and causing a short circuit. The safety terminal is screwed together with the positive terminal of the battery. There is a pyrotechnic ignition element in the contact bushing combined with the terminal. When triggered, this forces the starter cable out of the bushing **(Fig. 2)**. The rest of the vehicle electrical system supply is excluded from the open circuit.

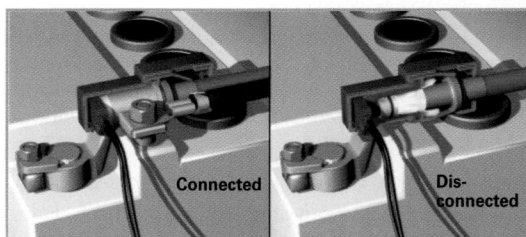

Fig. 2: Breaking of the starter line

Disposal

> Used or defective starter batteries are waste for recycling requiring special supervision.

Starter batteries must be recycled and are therefore labelled with the ISO 7000 sign. Batteries are collected and stored in special recycling containers in workshops and points of sale.

Workshop notes

Maintenance-free batteries (conforming to EN).
In starter batteries with filler plugs, you can check the electrolyte level and determine the state of charge via the acid.

Electrolyte level. This should be around 10 mm to 15 mm above the top edge of the plate. If water is lost, only distilled or demineralised water (no acid) may be used to top up.

State of charge. You can check the electrolyte density using an acid density meter (areometer) **(Fig. 1)**. In a fully charged battery and at a temperature of around + 25 °C, it should be about 1.28 g/cm³, and in a discharged battery it should be about 1.12 g/cm³. There is an approximate relationship between the electrolyte density and the steady-state voltage U_0 **(Fig. 1)**.

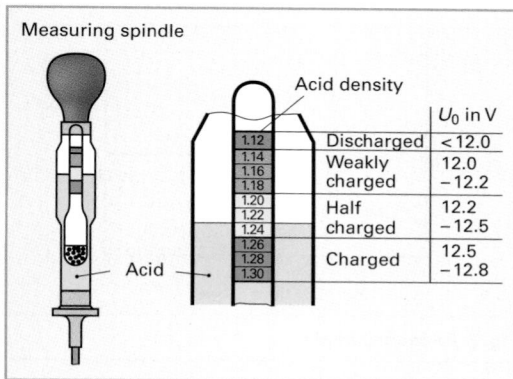

Measuring spindle

Acid density

Acid

	Acid density	U_0 in V
1.12 – 1.18	Discharged	< 12.0
1.14 – 1.18	Weakly charged	12.0 – 12.2
1.20 – 1.24	Half charged	12.2 – 12.5
1.26 – 1.30	Charged	12.5 – 12.8

Fig. 1: Acid density meter (areometer)

Battery types without filler plugs
State of charge. This can be determined approximately by measuring the steady-state voltage with the multimeter.

Performance check. The battery is charged with a current for approximately 5 seconds, which is approximately the cold start current of the starter. The average cell voltage must not fall below 1.1 V. This performance check can be carried out using electronic battery testers. These are connected to the battery to be tested with the corresponding cables and insulated pliers. The battery can remain fitted in the vehicle. The entire test process runs automatically and the results of the respective individual tests appear on a readable display. The following displays are usual:

- The battery voltage level
- The state of charge as a percentage
- The start capability as a percentage
- The qualitative battery status

Jump starting with booster cables
The battery can also be jump started from an external vehicle. This procedure must only be used with both batteries installed, and in compliance with the manufacturer's instructions. To jump start effectively, only standardised booster cables (DIN 72 553) with a conductor cross-section of at least 16 mm² for petrol engines and 25 mm² for diesel engines should be used. Both batteries must have the same nominal voltage.

The following steps are necessary:
- Connect the positive terminal of the discharged battery to the positive terminal of the external current source.
- Connect the negative terminal of the external current source to a suitable and permissible earth point on the vehicle to be jump started, at a sufficient distance from the battery. Near the battery, sparks may form which could cause detonating gas.
- Check the contact points of the booster cable to ensure they are firmly connected and have good contact.
- Start the vehicle with the battery assisting the vehicle to be jump started. After a short period of time, start the vehicle with the discharged battery.
- After successful jump starting, disconnect the cables in reverse order.

Charging starter batteries
The following types of charging can be distinguished: **Normal charging, boost charging, trickle charging.**

- **Normal charging.** The battery charge current is about 10% of the numerical value of the nominal capacity.
- **Boost charging.** The battery charge current is a maximum of 80% of the numerical value of the nominal capacity. Boost charging, however, may only be carried out until the gassing voltage is reached. The electrolyte temperature must not exceed 55°C.
- **Trickle charging.** Idle starter batteries discharge themselves automatically. The trickle charging current intensity is about 0.1% of the numerical value of the nominal capacity. If trickle charging is not possible, normal charging is required at intervals of approximately every two months.

Using chargers
Chargers are distinguished by their characteristics, which can also be combined in one device.
- *W*-characteristic   Resistance characteristic
- *U*-characteristic   Constant voltage characteristic
- *I*-characteristic   Constant current characteristic

Unregulated charger with W-characteristic

During operation, this outputs an unregulated voltage. As the charging duration increases, the internal resistance of the battery increases. This leads to a fall in the charging current. The charging voltage U_L of the battery, on the other hand, increases into the gassing voltage range **(Fig. 1)**. Therefore, the battery can become overcharged and detonating gas may form. This is to be avoided through careful monitoring of the charging process. The formation of gas or an electrolyte temperature of more than 55°C (battery housing hotter than lukewarm) are signs of overcharging.

Absolutely maintenance-free batteries and batteries with bonded electrolyte must not be connected to these chargers. Devices with this design are usually simple workshop chargers or even small chargers.

Regulated charger with IU-characteristic

Until the limit voltage is reached, which is less than the gassing voltage, the charging current I_L is kept constant through regulation of the device voltage.

When the limit voltage is reached, the charging voltage U_L is kept constant by changing the resistance in the charger. The charging current I_L then drops sharply in accordance with the W-characteristic. Devices with this design are suitable for charging absolutely maintenance-free starter batteries, as it is ensured that they cannot be charged to within the gassing range **(Fig. 2)**.

For batteries with bonded electrolyte, only special devices may be used in which the limit voltage is less than 13.2 V or less than 2.2 V per cell.

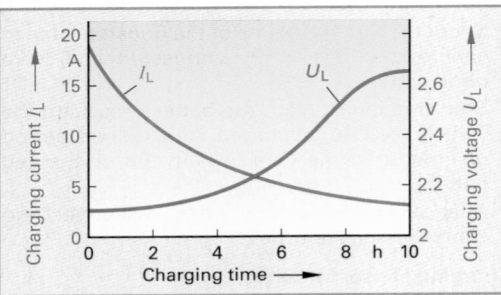

Fig. 1: W-characteristic

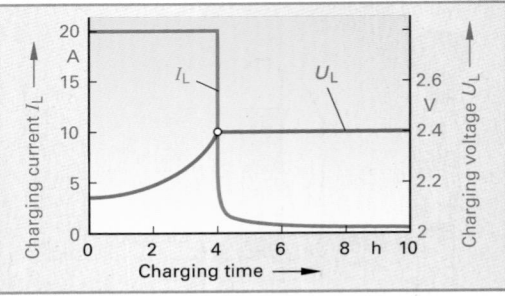

Fig. 2: IU-characteristic

Safety instructions

Before removing the battery, all consumers must be deactivated. Then the earth cable must be released first as short circuits (e.g. with the tool) can generate sparks and cause burns.

Particular care must be taken when connecting and disconnecting a charging or jump starting cable as there is a risk of short circuit.

The following safety principles must be complied with when working with batteries:

- When handling sulphuric acid or when topping up with water, protective goggles and rubber gloves must be worn.

- Do not fill acid over the max. mark.

- Do not tip the battery excessively or for a long time.

- Due to the risk of detonating gas forming during charging, avoid naked flames and the creation of sparks and refrain from smoking (connection and disconnection in specified sequence with the charger deactivated).

- The room in which the battery is being charged should be well ventilated.

REVIEW QUESTIONS

1 What are the processes involved when a starter battery is charged or discharged?

2 What is the electrolyte density in a fully charged or fully discharged starter battery?

3 What are the most important specifications of a starter battery?

4 What is the cold test current and how is it defined?

5 Why must the charging voltage always be below the gassing voltage?

6 What are the advantages of batteries with bonded electrolyte?

7 To which ageing processes are starter batteries subjected?

19.2.6 Alternator

During operation of the motor vehicle, this supplies the electrical consumers with energy and charges the starter battery.

In motor vehicles, the alternators used are almost exclusively claw-pole rotor alternators (**Fig. 1**) in compact design.

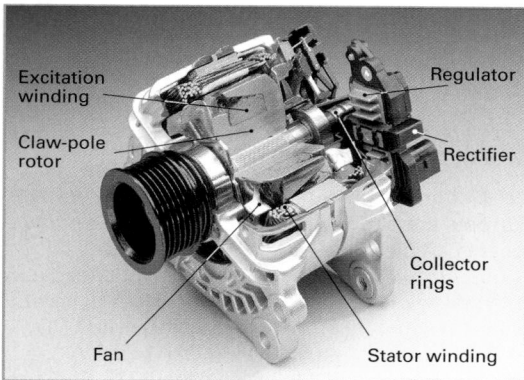

Fig. 1: Alternator

Design (Fig. 2)
An alternator consists of:

- Stator (housing) with three-phase stator winding as the induction coils
- Diode plate with 6 power diodes and 3 excitation diodes to rectify the voltage
- 12-pole, rotating claw-pole rotor with excitation winding for generating the magnetic field and collector rings and carbon brushes for current supply
- Fan for cooling
- Voltage regulator for constant operating voltage
- Connections B+/B– for voltage discharge
- A pulley for driving the alternator at a speed of 2-3 times the engine speed

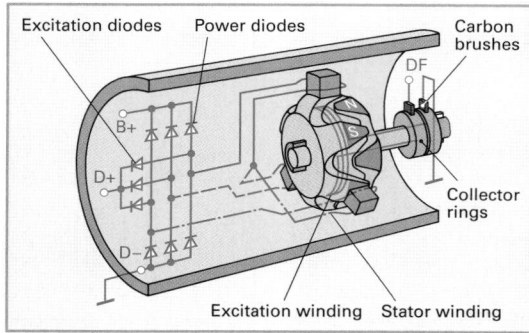

Fig. 2: Alternator (basic diagram)

Operating principle.
Voltage generation in the alternator is based on the principle of induction. When a magnetic field in a conductive loop (coil or winding) changes, voltage is generated in the conductive loop. If a rotating magnetic field with a north and south pole passes through a coil, this results in a sinusoidal alternating voltage (**Fig. 3**).

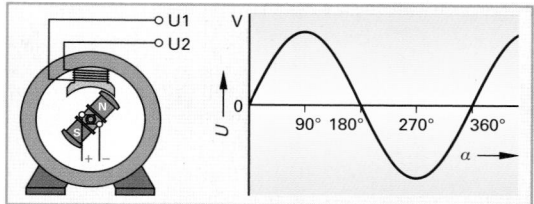

Fig. 3: Sinusoidal alternating voltage

If this rotating magnetic field passes through three coils, three sinusoidal alternating voltages occur. If these are arranged at 120° from each other, the 3 alternating voltages are phase-delayed by 120° (**Fig. 4**). Three-phase voltages (U_P) are generated at all times. In **Fig. 4**, this is shown with an angle of rotation of 90° and 300°.

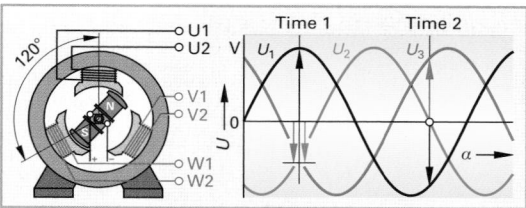

Fig. 4: Three-phase alternating voltage

Rectification.
An alternating voltage is unsuitable for charging the battery and the motor vehicle's electrical system also requires direct voltage to supply the consumers.

The current is rectified with diodes after the dual-pulse switching (see **Page 504** for a description). The negative half-waves are reformed into positive half-waves and their energy is used for the power supply. A single rectification of the three AC voltages requires a significant construction effort with 12 diodes and 6 lines (phases). The three windings are therefore connected to each other in a bridge circuit.

The three-phase current is rectified through 6 power diodes and 3 phases. In each phase, one diode is assigned to the positive side (positive diode) and one diode is assigned to the negative side (negative diode) (**Fig. 1, Page 540**). Depending on the position of the magnetic field, each phase is one feed and return of the current.

19

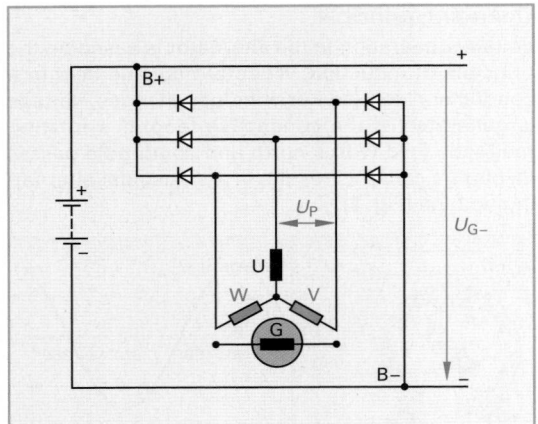

Fig. 1: Rectification with star connection

The positive half-waves produced in the three winding branches are allowed to pass through by the positive diodes, the negative half-waves are allowed to pass through by the negative diodes. The resultant is derived from an addition of the positive and negative envelopes. It still has a slight residual ripple.

Table 1 shows the formation of the envelopes at the 90° and 300° angles of rotation of the magnetic field. To simplify the example, a phase voltage U_P = 1 V, a resistance R = 1 Ω and a current of 1 A are assumed. At a 90° angle of rotation, the individual coils generate a voltage of U_U = 1 V, U_V = 0.5 V and U_W = 0.5 V. Accordingly, the phase currents are I_U = 1 A, I_V = 0.5 A and I_W = 0.5 A.

Star connection. With this type of connection, the phase voltages are added in accordance with the equivalent circuit diagram. $U_G = U_U + U_V$ = 1 V + 0.5 V = 1.5 V. U_W is not added as it is parallel to U_V and the current flow is blocked by the diode connection. This amplifies the alternator voltage (U_G) compared with the phase voltage (U_P). At 300° $U_G = U_U + U_W$ = 0.86 V + 0.86 V = 1.72 V.

Delta connection. With this type of connection, the currents are added in accordance with the equivalent circuit diagram. $I_G = I_U + I_V$ = 1 A + 0.5 A = 1.5 A. I_W is not added as it is in series with I_U. The alternator current (I_G) is amplified in comparison with the phase current (I_P). At 300° $I_G = I_U + I_W$ = 0.86 A + 0.86 A = 1.72 A. This circuit is used for alternators with high currents.

Table 1: Generation of the alternator voltage and the alternator current		
	Star connection	**Delta connection**
$U_{p\,max}$ = 1 V R = 1 Ω $I_{p\,max}$ = 1 A	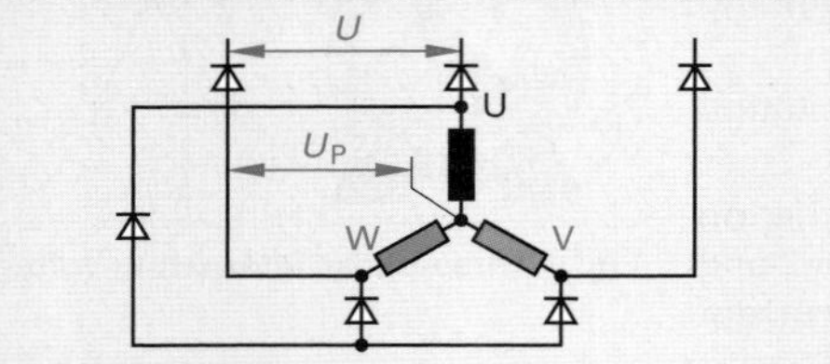	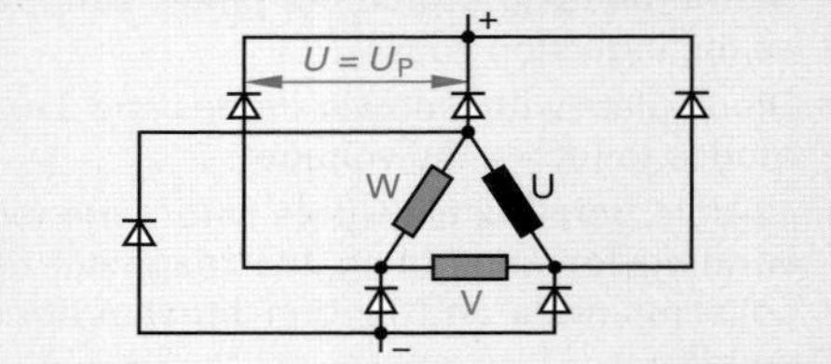
Resultant phase voltages and currents	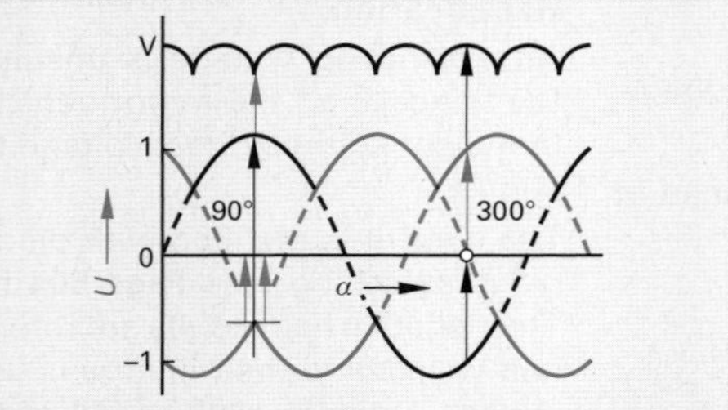	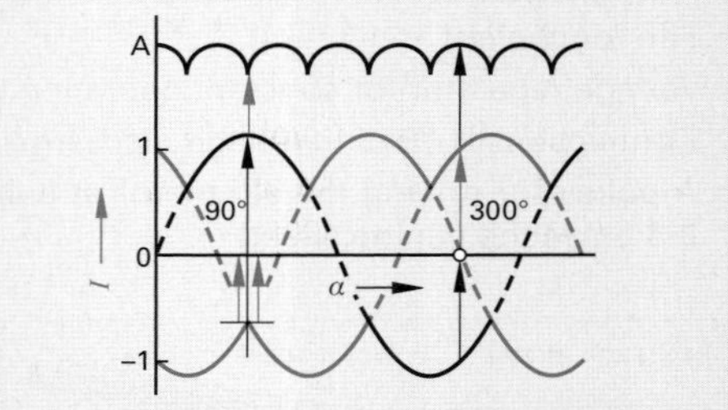
Equivalent circuit diagram: Induced phase voltages and currents in the individual windings U, V, W 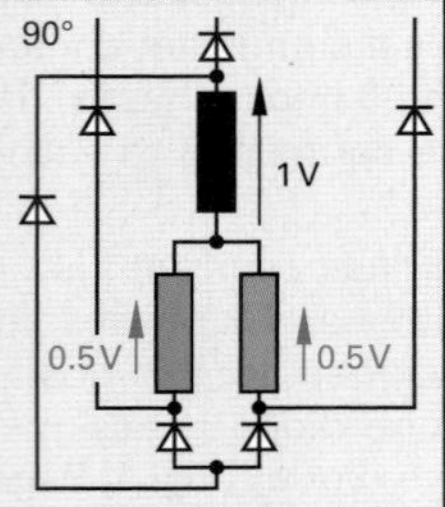 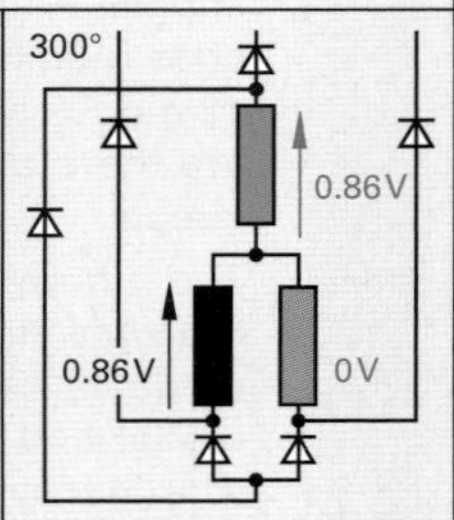 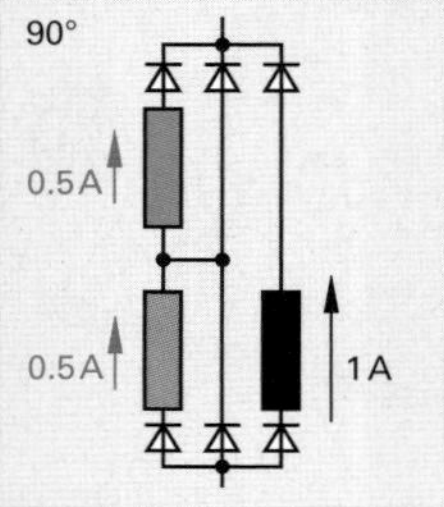 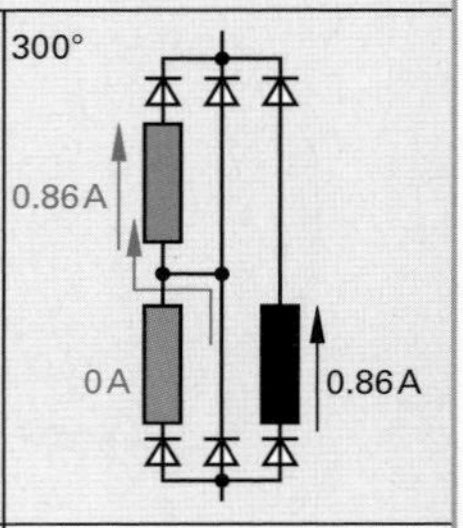	90°: 1 V, 0.5 V, 0.5 V 300°: 0.86 V, 0.86 V, 0 V	90°: 0.5 A, 0.5 A, 1 A 300°: 0.86 A, 0 A, 0.86 A
U_G / I_G	1.5 V / 1 A 1.72 V / 0.86 A	1 V / 1.5 A 0.86 V / 1.72 A

Voltage generation with claw-pole rotor

Instead of a magnet with a north and south pole, a claw-pole rotor with 6 north and 6 south poles **(Fig. 2, Page 539)** is used. The spatial layout of the stator coils is adapted to this. Thus, instead of 6 half-waves (2 poles $\times$ 3 lines) for every rotation of the rotor, there are 36 half-waves (12 poles $\times$ 3 lines) **(Fig. 1)**. After rectification, the larger number of poles leads to an output voltage with less residual ripple. It can be further smoothed by a smoothing condenser which is parallel to the rectifier circuit.

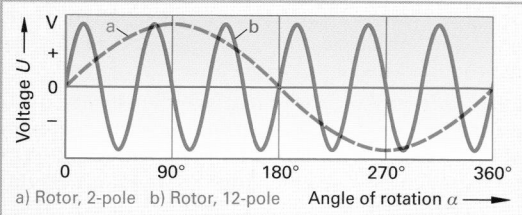

Fig. 1: Induced voltage of a winding

The phase voltage (U_P) and thus the output voltage (U_G) of the alternator, are determined by 3 factors:

- Rotational speed of the magnetic field
- Intensity of the magnetic field
- Number of windings on the coils

Diode design. The diodes are fixed on a diode plate. Silicon diodes are used. They let the current flow in one direction above the threshold voltage. If the diodes are connected in the conducting direction, there is a voltage drop in each diode of approximately 0.7 V. The resulting heat losses are dissipated by cooling plates.

Reverse-current block. The current may only flow from the alternator to the battery. The positive diodes prevent a current flow from the battery to the alternator and thus also prevent discharging of the battery. This would be the case if the alternator voltage were less than the battery voltage.

Overvoltage protection. The power diodes used in current alternators are Z-diodes. These limit the energy-rich voltage peaks that arise from, e.g. failure of the regulator, disconnection of induction currents or line breaks. They offer central overvoltage protection for the rectifier itself and for the overall vehicle electrical system. In contrast to the conventional circuit, the Z-diodes are installed in the conducting direction **(Fig. 4, Page 543)**.

Voltage regulation

> The alternator voltage is kept at the required level through a regulator at all speeds and under all load states.

The electrical consumers must not be exposed to high voltage fluctuations. The regulator is configured so that it adjusts the alternator voltage in 12-V systems to a nearly constant 14 V, and in 24-V systems to nearly 28 V. The alternator voltage remains just under the gassing voltage of the starter battery. This ensures sufficient charging and prevents damage through overcharging.

Closed-loop control process. The level of voltage induced in the alternator depends on the alternator speed and the intensity of the magnetic field or the excitation current I_E. As the alternator speed changes continuously due to the different driving conditions, the voltage regulation can only be carried out by changing the excitation field or the excitation current.

The size of the required excitation current depends on the current load and the alternator speed **(Fig. 2)**.

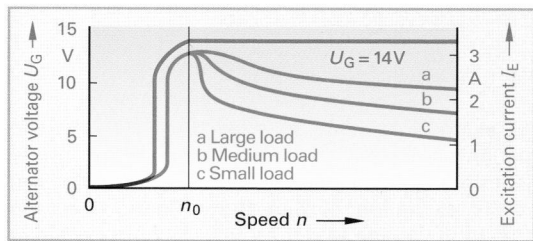

Fig. 2: Excitation current I_E at different loads

The regulator changes the level of the average excitation current I_E through continuous activation and deactivation (operating time t_E, downtime t_A). This amplifies or weakens the excitation field and thus changes the induced voltage **(Fig. 3)**.

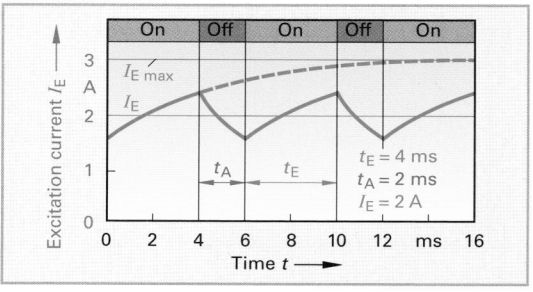

Fig. 3: Excitation current I_E depending on t_E/t_A

Electrical internal circuit and electric circuits (Fig. 1)

> In the alternator, a distinction is made between three circuits: **pre-excitation circuit, excitation circuit, charging circuit**.

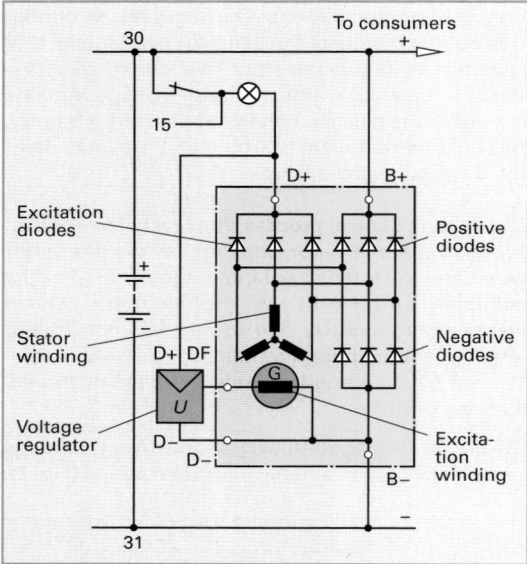

Fig. 1: Electrical internal circuit

Pre-excitation circuit (Fig. 2a). After activation of the ignition switch, the pre-excitation circuit establishes a magnetic field in the excitation winding until the excitation current flows. To do this, the threshold voltage ($2 \times 0.7\,V = 1.4\,V$) of the positive and negative diodes must be exceeded.

After starting the engine, the alternator excites itself. The excitation current flows and the alternator lamp goes out because the potential at both ends is the same.

The pre-excitation circuit runs via the starter battery +/30 → ignition switch/telltale lamp D+ → regulator D+ → regulator DF → excitation winding DF → earth D–/B– to the starter battery –/31.

If the alternator charge lamp is defective, there is no pre-excitation because there is a break in the pre-excitation circuit.

Excitation circuit (Fig. 2b). The excitation current establishes the magnetic field in the excitation winding of the rotor. The regulator supplies the respective required excitation current.

If the three-phase current bridge circuit is also used to rectify the excitation current, three special excitation diodes are available on the positive side. On the negative side, rectification is carried out via the negative diodes.

The excitation current runs from alternator D+ → regulator D+ → regulator DF → excitation winding DF → earth D–/B– → negative diodes → stator winding → excitation diodes to terminal D+.

Charging circuit (main circuit Fig. 2c). The charging circuit supplies the vehicle electrical system with electrical energy.
It runs from the stator winding → positive diodes → terminal B+ → battery/consumer → earth B– → negative diodes to the stator winding.

Part of the current generated by the alternator flows via the excitation diodes and the voltage regulator to the excitation winding, in order to generate the required magnetic field for induction.

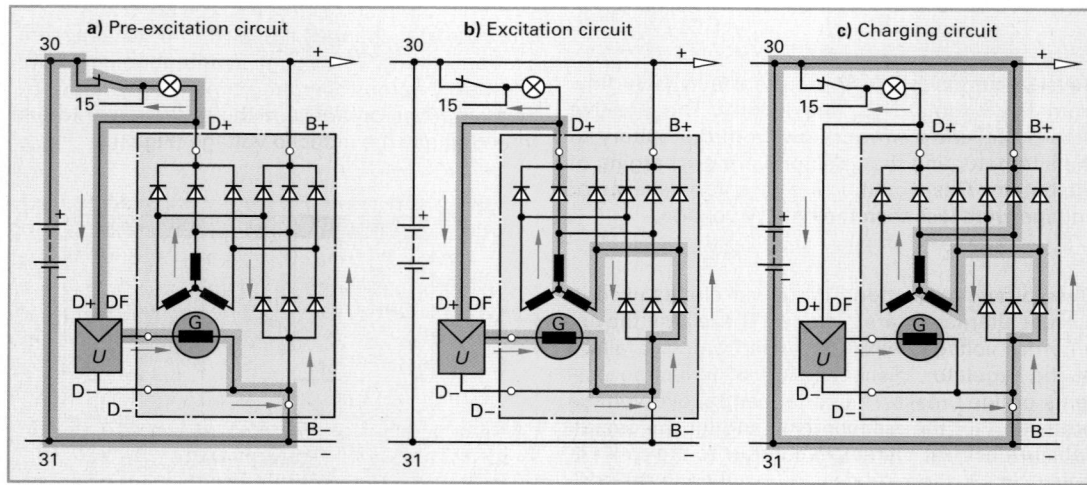

Fig. 2: Pre-excitation circuit, excitation circuit, charging circuit

Types of voltage regulators. Initially, these functions were undertaken by mechanical regulators which were positioned outside the alternator. Today, the function is undertaken by electronic regulators.

Hybrid regulator. This contains all control circuits (IC) in an encapsulated housing and is assembled directly onto the alternator without wiring.

Operating principle. The excitation current is switched "on" and "off" via the control circuit, independently of the alternator reference voltage.

Connection status "On" (Fig. 1). If the actual value of the alternator voltage is below the reference value of e.g. 14.2 V, the Z-diode blocks transistor T_2. Thus, basis B of T_1 is above resistance R_3 at D-, and T_1 becomes conductive. The excitation current is fed from T_1 to the excitation winding via excitor E and collector C.

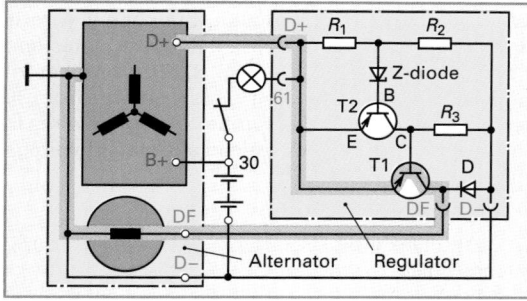

Fig. 1: Connection status "On"

Connection status "Off". If the alternator voltage exceeds the prescribed reference value, the Z-diode becomes conductive, whereas the basis B of transistor T_2 becomes negative. T_2 becomes conductive, while basis B of T_1 receives a positive voltage. T_1 is not conductive and blocks the excitation current **(Fig. 2)**.

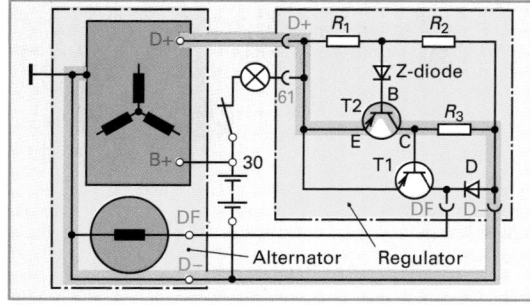

Fig. 2: Connection status "Off"

Regulator curve. As the gassing voltage falls as the temperature rises, the generator reference voltage is adapted to this behaviour **(Fig. 3)**. Thus, at a temperature of 0 °C, there is a reference value of between 14.6 V and 14.9 V. At a temperature of 50 °C it is between 14 V and 14.4 V.

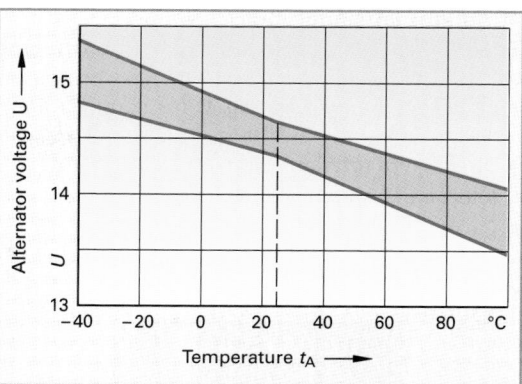

Fig. 3: Alternator voltage depending on the temperature

Monolithic controller. All the functions required for voltage regulation are accommodated in a chip.

Multifunction regulator. This is also a one-chip regulator. The alternator no longer has any excitation diodes, as the excitation current is not taken from the rectifier circuit. It is drawn directly from terminal B+ **(Fig. 4)**.

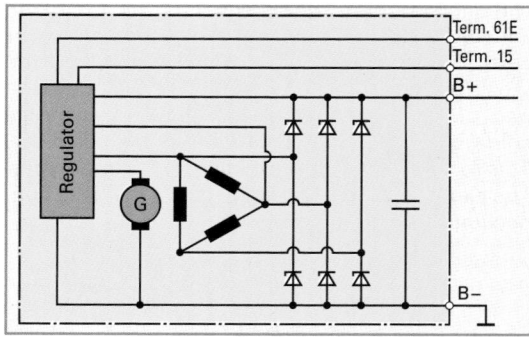

Fig. 4: Multifunction regulator

Multifunction regulators offer the following advantages:

- Fault diagnosis and display by LED
- Load response start. The load on the alternator is connected with a delay during the starting process.
- Load response drive. With a high change in load through electrical consumers, the excitation current is only increased slowly. To stabilise engine idling, the excitation current can be reduced.
- Temperature-controlled charging voltage through battery management and battery sensor
- Selection circuit. After querying the system load, the engine idling speed can be increased or individual consumers can be switched off.

19

Alternator labelling

The following technical data is indicated on the type plate **(Fig. 1)**:

- Construction concept, e.g. C for compact alternator
- Direction of rotation, e.g. clockwise
- Alternator nominal voltage, e.g. 14 V
- Current at engine idle speed, e.g. 70 A
- Nominal current at alternator nominal speed (6,000 rpm) e.g. 140 A

Fig. 1: Alternator nameplate

Alternators with windingless rotor (Fig. 2)

> This alternator contains a fixed excitation winding, no brushes and no collector rings. It is wear-and-tear-resistant.

These alternators have an integrated non-rotating windingless rotor, which contains the excitation winding. In addition to the housing with the stator core, the cooling plates with the power diodes and the fitted transistor regulator, it belongs to the fixed part of the alternator. The rotating part consists solely of the rotor with pole wheel and guide piece. This means there are no collector rings or brushes. The excitation current is supplied via the fixed connections.

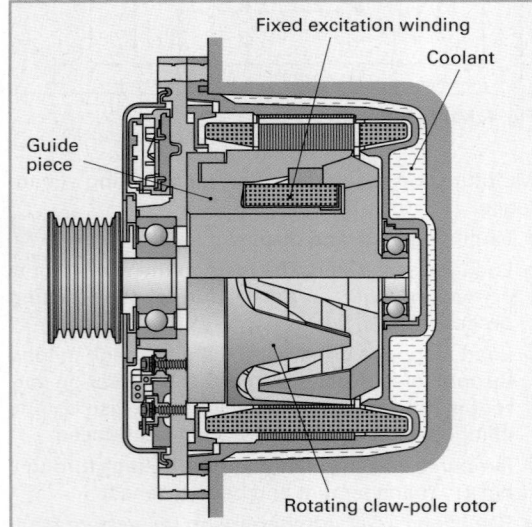

Fig. 2: Liquid-cooled alternator with windingless rotor

Guide piece alternator with liquid cooling

> With this alternator, the heat is dissipated via the engine's coolant.

The alternator is surrounded by a coolant jacket. The occurring heat, which results at high current outputs of up to 200 amps, can be dissipated by the coolant. In addition, the coolant jacket reduces the running noise of the alternator.

Alternator with flat-pack winding

> This alternator generates a voltage of 14 V, 28 V or optionally 42 V, with a maximum power output of 4 kW.

With this type of construction, the copper wire is wrapped around the stator in a flat pack status and then closed into a ring **(Fig. 3)**. The number of grooves fitted with winding wires increases from 36 to 48. This increases the number of copper wires in the stator winding. The induced alternating voltage increases. This allows for:

- Higher power output
- Lower alternator speed
- Less noise
- Smaller size
- Greater efficiency

The alternator has a multifunction regulator and a rectifier with a delta circuit.

Fig. 3: Alternator with flat-pack winding

Circuit variants

There are different types of circuit for the rectifier depending on the requirements:

- Parallel connected power diodes with high power output.
- Additional diodes between the neutral point and the positive and negative terminals to minimise the output losses at a high alternator speed.

Vehicle electrical systems

Single battery vehicle electrical systems. This vehicle electrical system with one alternator and one battery is found in most passenger cars.

Dual battery vehicle electrical systems. Vehicle electrical systems with 2 batteries can be used in series on vehicles with a high energy requirement. The high energy requirement results from driving operation management, safety and security systems, comfort and convenience electronics and infotainment systems. In order to guarantee a reliable cold start, the vehicle starting function is undertaken by a separate starter battery. The electrical system battery supplies all electrical consumers with energy. Battery management monitors and regulates the optimum state of charge of the batteries.

When cold starting or with a low state of charge of the starter battery, a parallel circuit can also be realised for the batteries via a high load relay, in order to increase the starting power.

A second battery for supplying the consumers can even be retrofitted. It must be connected in parallel to the first battery. In order to prevent compensating currents between the two batteries, an isolating relay must be connected between them **(Fig. 1)**.

The batteries should have the same nominal capacity. If the general purpose battery is installed in the passenger compartment, a non-gassing battery (gel or glass-fibre fleece technology) should be used.

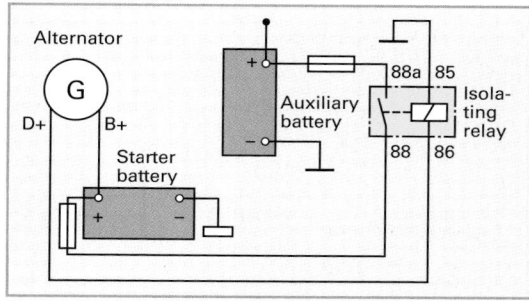

Fig. 1: Batteries with isolating relay

Dual voltage vehicle electrical system. This consists of two separate voltage networks with voltages of 42 V and 14 V. The alternator supplies the 42 V power consumers directly and the 14 V network via a DC-DC converter. The electronic energy management regulates the entire energy balance and controls the corresponding functions.

The 42 V supply also enables operation of an integrated starter alternator (ISG, ISAD).

WORKSHOP NOTES

Checking the alternator by observing the alternator telltale lamp

Error	Failure cause	Remedy
Alternator telltale lamp does not light up when the engine is stationary and the ignition switch is on	Telltale lamp burnt out	Replace telltale lamp
	Battery discharged	Charge battery
	Battery damaged	Replace battery
	Lines released or damaged	Replace lines
	Regulator damaged	Replace regulator
	Short circuit of a positive diode	Disconnect charging cable, repair alternator
	Carbon brushes worn	Replace carbon brushes
	Oxide layer on collector rings, Open circuit in rotor winding	Repair alternator
At high alternator speeds, the alternator telltale lamp burns consistently brightly	Line D+ /61 has short to earth	Replace line
	Regulator damaged	Replace regulator
	Diodes damaged, collector ring dirty, short to earth in the DF line or rotor winding	Repair alternator, or replace DF line
With the engine stationary and the ignition switched on, the alternator telltale lamp burns brightly but glows when the engine is running	Contact resistances in the charging circuit or in the line to the lamp	Replace lines, clean and tighten connections
	Regulator damaged	Replace regulator
	Alternator damaged	Repair alternator

19

WORKSHOP NOTES

Malfunction diagnosis with the oscilloscope

Conclusions about the status of the alternator and the installed diodes can be drawn from the shape of the voltage curve **(Fig. 1)**.

For the assessment, the voltage occurring at terminal D+/61 is illustrated with the oscilloscope. The alternator is subjected to a load of approximately 15 A and the engine speed kept at approximately 2,500 rpm.

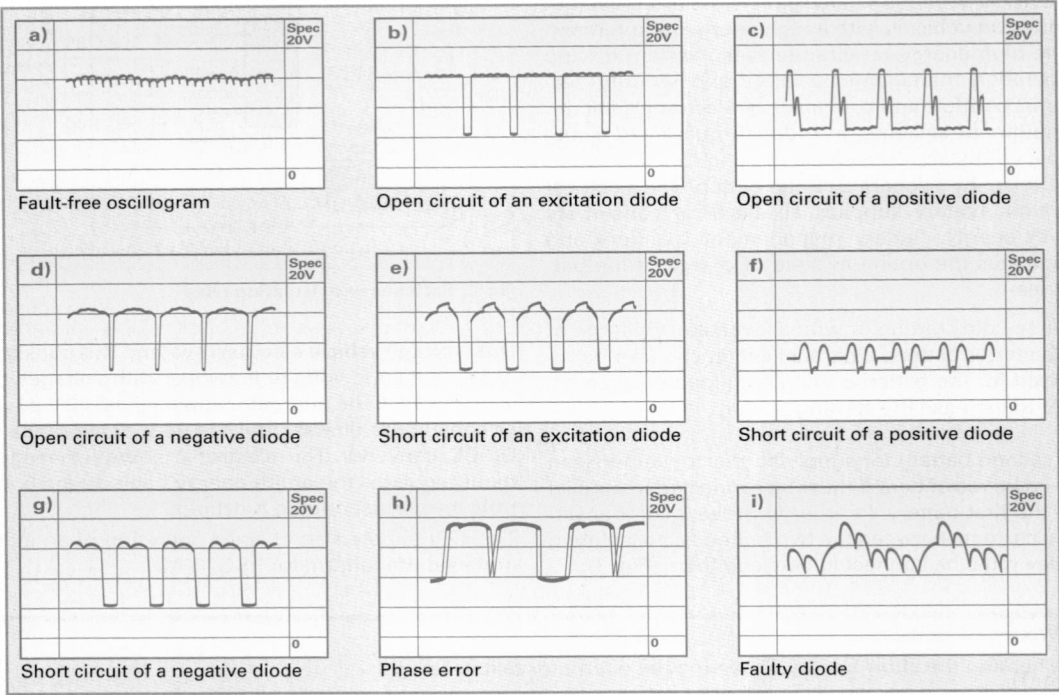

Fig. 1: Oscillogram of an alternator

a) Fault-free oscillogram
b) Open circuit of an excitation diode
c) Open circuit of a positive diode
d) Open circuit of a negative diode
e) Short circuit of an excitation diode
f) Short circuit of a positive diode
g) Short circuit of a negative diode
h) Phase error
i) Faulty diode

Measurements with the multimeter. The alternator voltage is measured directly on the alternator terminals B+ and D–.

Regulating voltage:
- Start engine.
- Engine speed between 3,500 and 4,000 rpm.
- Read test specifications.
- Depending on the alternator type and temperature, values must be between 13.7 and 14.7 V.

Alternator voltage under load:
- Start engine.
- Engine speed between 1,800 and 2,200 rpm.

- Switch on all consumers, if possible.
- Depending on the alternator type and temperature, the voltage values must not fall below 13.0 to 13.5 V.

Closed-circuit current:
- Switch off engine.
- Connect the ammeter between the battery negative terminal and the earth strap.
- Depending on the vehicle type, there must only be a slight current up to approximately 60 mA from the battery to the alternator.
- If a larger current flows, there is a defect.

REVIEW QUESTIONS

1 What is the design of the alternator?
2 What is the operating principle of the alternator?
3 What additional task does the positive diode have?
4 How is the current rectified?
5 What are the differences between star and delta connections?

6 What does the alternator's output voltage depend on?
7 How is the alternator voltage regulated?
8 What functions are possible with a multifunction regulator?
9 What must be taken into account when installing an auxiliary battery for supply purposes?

19.2.7 Electric motors

In motor vehicles, DC motors are used as starters and as auxiliary drives, e.g. for fans, wipers, seat adjustments.

If devices in the motor vehicle are to be adjusted by precisely defined distances or angles, e.g. for the idle charge adjustment, stepping motors can be used.

In the field of electrical vehicle drives, three-phase asynchronous motors and synchronous motors are used in addition to DC motors.

19.2.7.1 DC motors

Operating principle. The principle of the DC motor is based on the fact that a force is exerted on a current-carrying conductor in the magnetic field.

This force depends on ...

- ... the intensity of the electrical current in the conductor.
- ... the intensity of the magnetic field (magnetic flux density).
- ... the effective conductor length (number of coils).

With a DC motor, there is a swivelling coil in a magnetic pole field with distinct north and south poles **(Fig. 1)**.

If voltage is applied to the coil, the current flowing in the coil causes a magnetic field (coil field), which runs vertically to the winding surface **(Fig. 2)**.

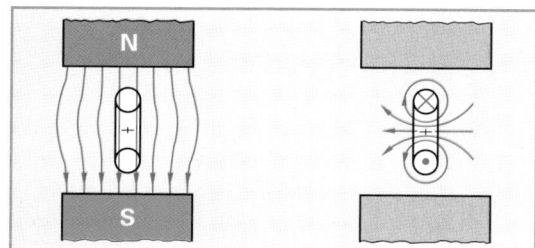

Fig. 1: Pole field   Fig. 2: Coil field

The pole field (main field) and the coil field (armature field) give a resultant magnetic field. Depending on the current direction in the conductor loop, the resulting torque is either clockwise or anticlockwise **(Fig. 3)**. The coil turns until the coil field is in the same direction as the pole field, and then it remains in the so-called neutral zone of the pole field.

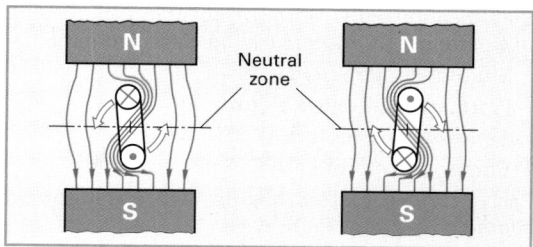

Fig. 3: Resulting field and rotational movement

Commutation. If there is to be a continuous rotation, the current direction in the armature coil must be changed if it is in the neutral zone. The current direction is changed by a commutator. The starts or ends of the coils are connected to this. This means that the current in the coil sides always has the same current direction under a specific pole **(Fig. 4)**. The current is supplied via two fixed carbon brushes that form a friction contact with the commutator.

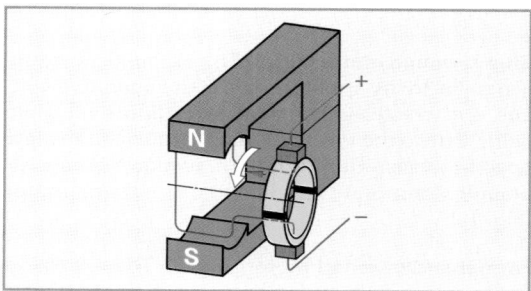

Fig. 4: Commutator

The armature usually has several coils. This results in a torsional force around the entire circumference of the armature and the resulting torque is more even with a rotation around 360°.

If a multiple coil armature winding is used instead of an armature coil, the commutation is also such that the current in the coil sides is always in the same direction under a certain pole **(Fig. 5)**.

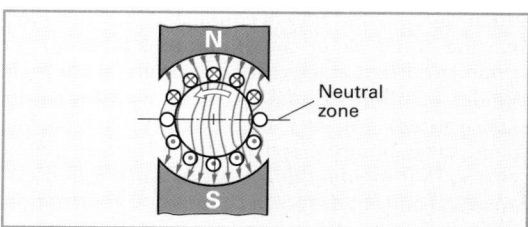

Fig. 5: Several coils in one armature

Classification of DC motors. Depending on the type of excitation, DC motors can be classified into:

- Shunt-wound motors
- Permanently excited motors
- Series-wound motors
- Compound motors

Each of these motors has its own speed-torque characteristic **(Fig. 1)**.

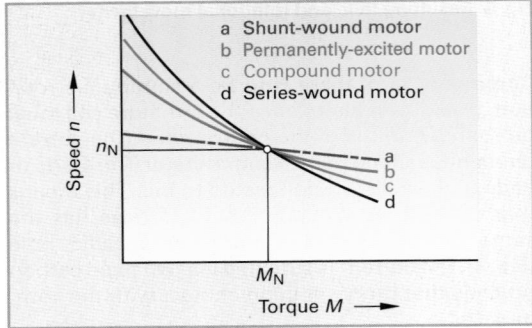

	a Shunt-wound motor
	b Permanently-excited motor
	c Compound motor
	d Series-wound motor

Fig. 1: Speed-torque characteristic

Shunt-wound motor (Fig. 2). The excitation winding is parallel to the armature. It is at the operating voltage and creates a constant excitation field (pole field). Due to the low tightening torque and the low speed increase during load reduction, the shunt-wound motor is not very suitable as a starter motor **(Fig. 1a)**.

Permanently-excited motor (Fig. 3). The excitation field is created by strong permanent magnets. This results in a speed-torque characteristic that is between that of the shunt-wound motor and the series-wound motor **(Fig. 1b)**.

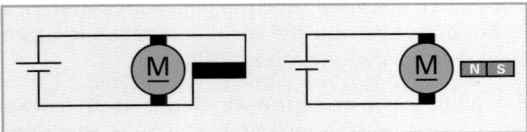

Fig. 2: Shunt-wound   Fig. 3: Permanently-excited
          motor   motor

Series-wound motor (Fig. 4). The excitation winding and armature winding are connected in series (one after the other). When the armature is blocked, the armature current and excitation current are at their highest, i.e. the torsional force and the torque (release torque) are very high. With an increasing speed, the countervoltage in the armature becomes greater. This means the armature current and the excitation current drop, which makes the excitation field smaller. The reduction of the excitation field results in a significant increase in the speed during

load reduction **(Fig. 1d).** This speed-torque characteristic is advantageous for starter motors because the starting speed is reached quickly due to the rapid increase in revolutions.

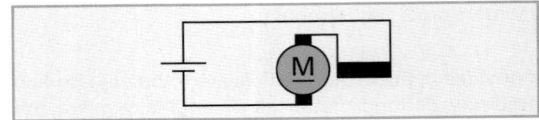

Fig. 4: Series-wound motor

Compound motor (Fig. 5). This has both a series winding and a shunt winding. Due to the high winding effort, it is only used in large starters. The shunt winding amplifies the torque in the series winding but reduces the impermissibly high revolution of the armature during load reduction **(Fig. 1c)**.

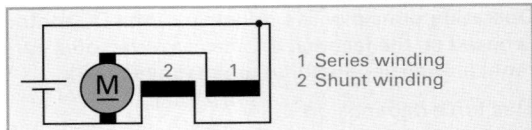

	1 Series winding
	2 Shunt winding

Fig. 5: Compound motor

19.2.7.2 Stepping motors

With stepping motors, the rotor, and thus the drive shaft are each turned through a specific angle or step. Depending on the design of the stepping motor, small angle steps of up to 1.5° can be achieved.

Design. The rotor of a stepping motor is a toothed rotor made from a permanent magnet, with the teeth being magnetised in an axial direction. The teeth continuously change between the north and south pole **(Fig. 6)**. Between two teeth, there is a gap of a half tooth width.

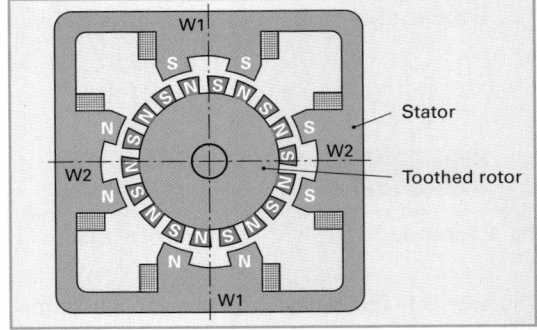

Fig. 6: Stepping motor – Basic structure

In the example in **Fig. 6**, the laminated stator has two excitation windings W1 and W2 (phases). They form the two stator pole pairs, with the north and south poles of a pole pair lying opposite each other in each case. The tooth distribution of the stator corresponds to that of the pole wheel.

Operating principle. The pole wheel is always adjusted so that, e.g. one north pole of the toothed rotor is opposite a south pole of the stator **(Fig. 1a)**. When the polarity of the current in winding W1 is reversed, the polarity in the vertical pole pair **(Fig. 1b)** changes, but remains the same in the horizontal pole pair. The rotor rotates by a half tooth.

The next reversal of polarity takes place in winding W2, with the polarity in the horizontal pole pair changing. The rotor rotates by a further tooth **(Fig. 1c)**. Each further polarity reversal in the sequence W1, W2, W1, etc. causes a further rotation.

The direction of rotation can be changed through corresponding polarity of the stator windings W1 and W2 **(Fig. 1d)**.

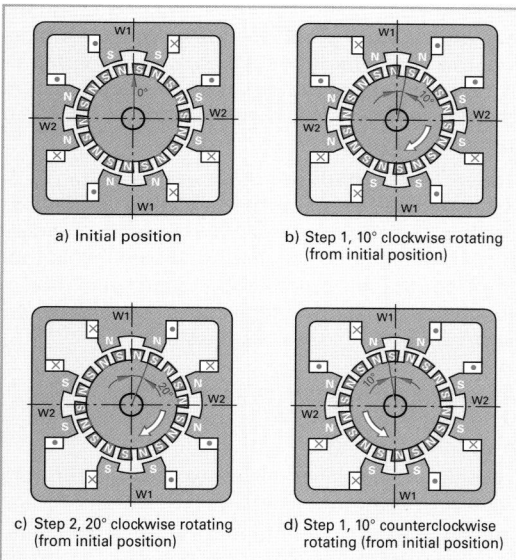

a) Initial position

b) Step 1, 10° clockwise rotating (from initial position)

c) Step 2, 20° clockwise rotating (from initial position)

d) Step 1, 10° counterclockwise rotating (from initial position)

Fig. 1: Stepping motor – Operating principle

Based on sensor information on the required adjustment, the following values are defined in the control unit:

- Number of steps
  (corresponds to angle of rotation)
- Required direction of rotation
- Rotation or adjustment speed

If the stator winding is not charged with current, the rotor remains in its last position between the magnetic pole wheel and the laminated stator due to the magnetic effect (lock-in effect).

A stepping motor can carry out any number of steps in both directions.

Stepping motors are used for e.g.:
- Automatic throttle valve adjustment
- Fan valve adjustment in air conditioning systems
- Electrical door mirror adjustment
- Seat adjustment with memory function

Stepping motors can also be provided with a worm-gear pair whose transmission ratio switches to "low" ($i > 1$). Thus, the rotor can execute a large number of steps with each control or regulation process, with the actuator, e.g. throttle valve, only moving by a small, accurately defined angle.

With quicker pulse trains, the stepping motor works as a synchronous motor. The armature turns at the same time as the stator magnetic field (synchronous).

19.2.7.3 Brushless motors

This type of construction reverses the construction principle of the mechanically commutated motor. With brushless motors, the windings are applied to the stator and the rotor has the rotating permanent magnets **(Fig. 2)**.

An electronic commutation device supplies the windings with current depending on the angle of rotation and causes the rotor to turn.

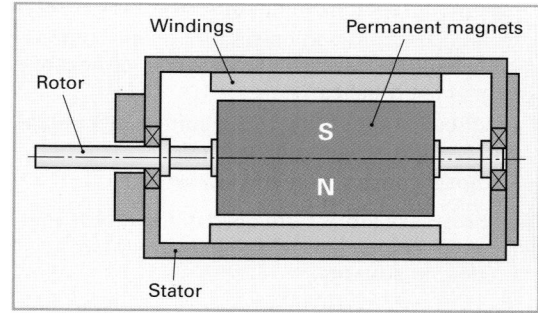

Fig. 2: Brushless motor (principle)

External rotor motor. With this type of motor, the fixed stator windings are inside and the rotating rotor jacket surrounds the windings.

Internal rotor motor. With this type of motor, the magnetic rotor is inside the motor and the stator surrounds the rotor.

Electronic commutation. In brushless motors, the activation and deactivation of the individual windings is carried out with the help of control electronics. The control electronics receive information on the position of the rotor from the rotor position sensor, e.g. a Hall-effect sensor, and activate or deactivate the corresponding windings in the stator **(Fig. 1)**. The speed of the motor depends on the commutation frequency of the commutation device. Electronically commutated DC motors usually have three or more windings.

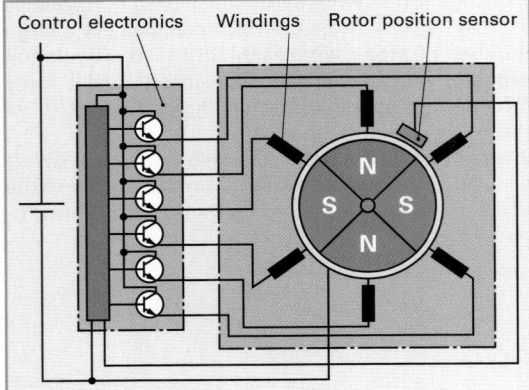

Fig. 1: Electronic commutation (principle)

Advantages of brushless motors compared with mechanically commutated motors:

- Possibility of higher rotational speeds, as the maximum speed of the motor only depends on the way in which it is mounted and on the centrifugal forces acting on the magnet mounting.
- Regulation of the motor speed by measuring the rotational speed with the rotor position sensor.
- Quiet operation, good electromagnetic compatibility and reduced maintenance due to the omission of the brushes.
- Diagnostics capability of the control electronics and smooth mechanical running.
- Compact construction and low weight.

Use. Brushless motors are used in the motor vehicle, e.g. for engine fans and as blower motors in air conditioning systems.

19.2.7.4 Starter

Combustion engines must be started with external energy. During the start process, the mass inertia and the frictional and compression resistances of the engine must be overcome.

Design of the starter

A starter **(Fig. 2)** usually consists of:

- Starter motor (electric motor)
- Engagement relay (relay, starting-motor solenoid)
- Meshing drive (pinion, roller-type one-way clutch)

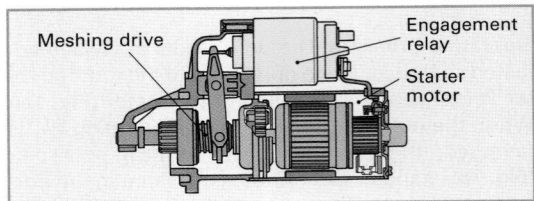

Fig. 2: Assembly of a starter

Starter motor. The pole housing consists of a tube in which the pole shoe is fitted with the excitation windings or permanent magnets. It is also used as the magnetic yoke of the magnetic field lines and is therefore produced from a magnetically conductive steel.

The armature carries the armature windings. The continuous changing of the current direction in the armature coils results in an alternating magnetic field which would produce eddy currents in a massive iron core, which could lead to an impermissibly high heating of the armature. It therefore consists only of individual sheets which are insulated from each other.

Grooves are etched into the armature core discs, which are used to mount the armature winding. In addition, the armature is used to improve conduction of the magnetic field lines of the pole field running from the north pole to the south pole. Therefore, the air gap between the pole shoes and the armature must be as small as possible.

Engagement relay (solenoid switch) (Fig. 3). This is a combination of a relay and starting-motor solenoid. It has the task of

- pushing forward the pinion to mesh in the ring gear of the motor
- closing the contact bridge to switch on the main starter current.

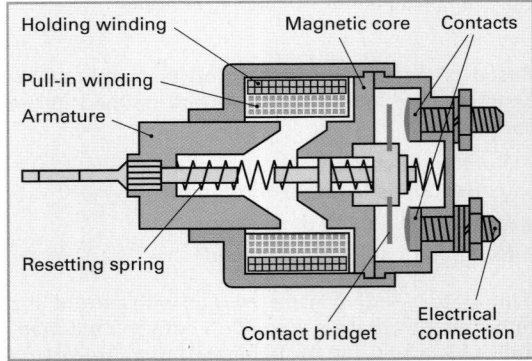

Fig. 3: Engagement relay (solenoid switch)

Meshing drive (Fig. 1). This essentially consists of:

- Pinion for force transfer and for speed and torque conversion.
- One-way clutch as overrunning clutch after starting.
- Engagement lever for the meshing stroke.
- Meshing spring, enables meshing in the event of tooth-on-tooth contact.

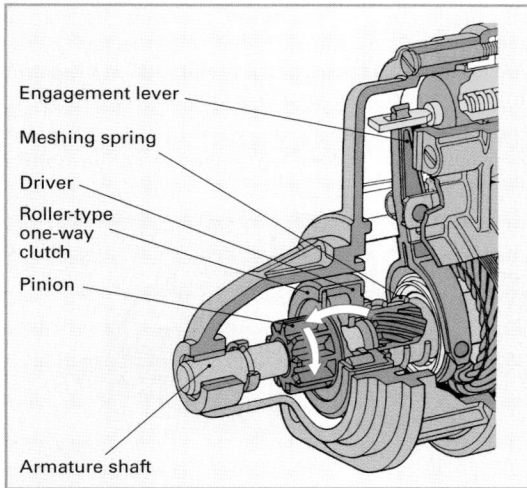

Fig. 1: Meshing drive of a pre-engaged drive starter

During the starting process, the pinion meshes into the ring gear which is located on the flywheel. The turns ratio is in the range between 10 and 15. This increases the torque acting on the flywheel. The pinion is highly mechanically stressed during the meshing process and operation.

During the starting process, the one-way clutch has the function of transferring the drive torque of the starter to the pinion, actuated by positive engagement. When the engine starts, i.e. if the armature is driven by the engine, it must separate the still meshed pinion from the starter. This prevents an impermissibly high revolution and thus destruction of the armature.

A distinction can be made between:

- Roller-type one-way clutch. This is used with smaller starters for passenger cars and vans.
- Multi-plate overrunning clutch. This is used with larger starters for trucks.

The roller-type one-way clutch consists of the clutch shell with the roller races, the rollers and the coil springs **(Fig. 2)**. The rollers slide on the pinion shaft. The roller races narrow in one direction.

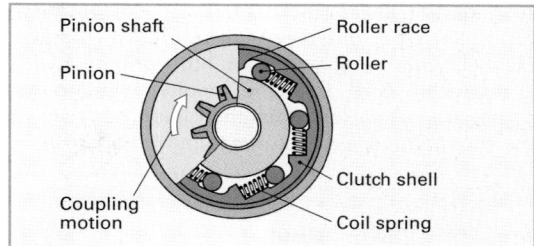

Fig. 2: Roller-type one-way clutch

If the clutch shell is driven by the starter, the rollers are pushed into the narrowing part of the roller races. The pinion shaft is coupled with the starter. After the engine starts, the rollers are pushed into the further part of the roller races by the overrunning pinion, which is now driven by the engine; the friction connection is released.

Pre-engaged drive starter

During meshing, the driver coupled with the pinion via a roller-type one-way clutch moves on the helical spline of the armature shaft **(Fig. 3)**.

The driver is pushed forward under spring force by the engagement lever moved by the solenoid switch and caused to rotate by the helical spline. If one tooth of the pinion moves in front of the tooth space, it meshes immediately. If a tooth hits a tooth, the meshing spring is compressed until the solenoid switch switches on the primary current. The armature rotates and the pinion pushes against the end face of the gear until it can mesh.

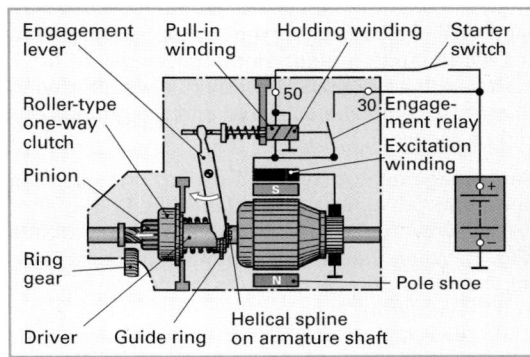

Fig. 3: Pre-engaged drive starter

The solenoid switch has two windings, one pull-in winding and a holding winding. Two windings work together to pull in. When the starter current is switched on, the pull-in winding is short-circuited. The winding receives a positive from both sides. The solenoid switch is only held by the holding winding **(Fig. 1)**.

After starting the engine, the pinion overruns due to the roller-type one-way clutch, but it remains meshed with the ring gear until the ignition switch is activated. Only after the holding winding becomes free of current does the pinion demesh, as the engagement lever returns to the initial position.

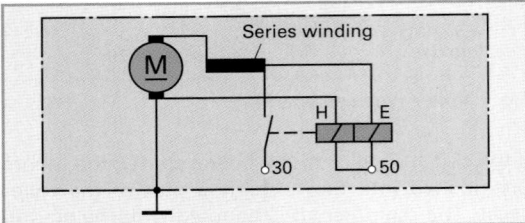

Fig. 1: Internal circuit of a pre-engaged drive starter

Pre-engaged drive starter with permanent-magnet field and reduction gear

With this type of starter, the excitation winding is replaced with permanent magnets (Fig. 2).

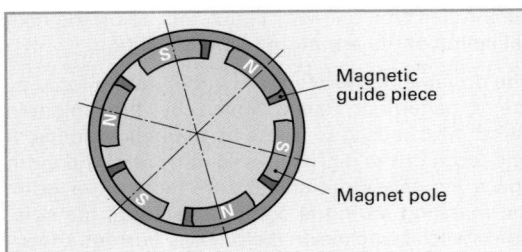

Fig. 2: Pole housing with permanent magnets

The permanent magnets are fixed in a thin-walled pipe. This is also used as the starter housing. With the same power output as the pre-engaged drive starter with an excitation winding, a weight saving of up to 20% can be achieved and the dimensions can be reduced.

The engagement relay and meshing drive and the operating principle are the same for both starter types. Only the electrical internal circuit differs (Fig. 3). When the starter current circuit is connected, the current flows directly to the carbon brushes and the armature.

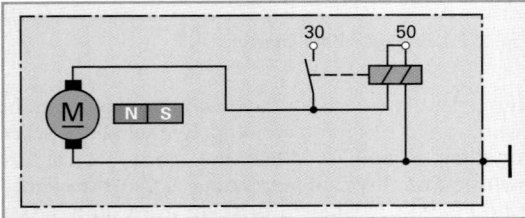

Fig. 3: Internal circuit of a starter with permanent excitation

Pre-engaged drive starters with permanent magnetic field have a shunt-wound characteristic.

As the release torque in shunt-wound type motors is comparatively low, they are provided with a planetary gearbox which is used as a reduction gear.

The meshing, demeshing and overrunning processes are the same as for the pre-engaged drive starter.

The planetary gearbox fitted between the starter and pinion as a reduction gear has the task of reducing the high speed of the starter and at the same time increasing the torque on the pinion (Fig. 4).

The meshing, demeshing and overrunning processes are the same as for the pre-engaged drive starter.

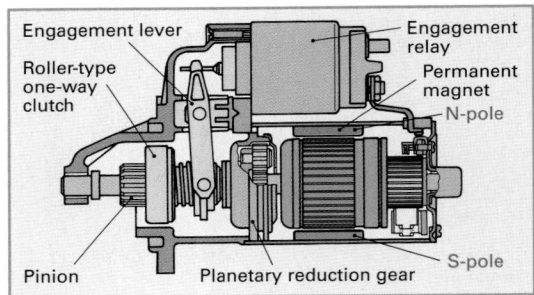

Fig. 4: Reduction gear starter

The planetary reduction gear consists of the ring gear, the planetary gears with planetary carriers and the sun gear (Fig. 5).

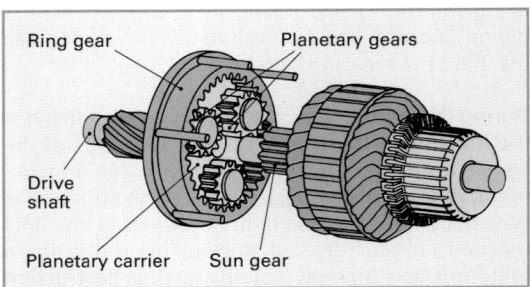

Fig. 5: Armature with planetary reduction gear

The sun gear sits on the armature shaft and forms the drive gear of the planetary reduction gear.

The planetary gears are mounted on the planet carrier by bearings. The planetary carrier is connected to the drive shaft which is provided with a helical spline on which the pinion is mounted by means of a slide bearing.

The ring gear is mounted in the starter housing and is made of plastic.

In motor vehicles with manual transmission, a short circuit test can be carried out on the starter. To do this, the armature of the starter is blocked and the short circuit current draw is measured using a current clamp.

The level of the short circuit current depends on the capacity and the state of charge of the battery and on the level of the power input of the starter.

- The difference between the voltage at the battery and the voltage at the starter is the voltage drop in the starter main line.
  Permissible voltage drop: 0.25 V in 6 V systems, 0.5 V in 12 V systems, 1 V in 24 V systems.
- When charged with the permissible short circuit current, the terminal voltage of the battery must not fall below 3.5 V in 6 V systems, 7 V in 12 V systems and 14 V in 24 V systems.

Short circuit test

- Connect one current meter and two voltmeters according to **Fig. 1**, apply the highest gear, tighten the hand brake and apply the foot brake.

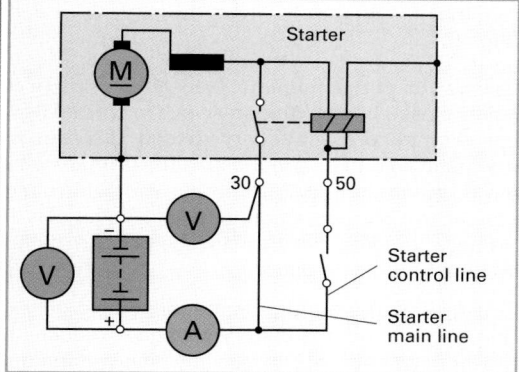

Fig. 1: Test circuit for starter with solenoid switch

- Activate the starter briefly (maximum of 5 seconds).
- During starting, switch off all other electrical consumers, if possible.
- Read the short circuit current, terminal voltage at the battery and voltage at the starter.

Diagnosis

- If the short circuit current is less than prescribed, with correct terminal voltage at the battery, then there are additional resistances in the circuit, e.g. increased contact resistance at the positive pole head, cross-section reduction in the starter main line or earth cable.
- If the starter does not reach the prescribed short circuit current even though the battery voltage and the voltage drops are within the permissible limits, the starter is defective.
- If the starter does not reach the prescribed short circuit current and the battery voltage falls below the permissible minimum limit, even though the voltage drops are within the permissible limits, the fault could be either in the starter or in the battery.

Maintenance notes

- Oxidised pole terminals of the battery, loose terminal connectors, dirty switch contacts and damaged lines increase the line resistance. Remove oxide layers, tighten connections, replace components with dirty contacts. Protect terminal connectors from corrosion using suitable grease.

REVIEW QUESTIONS

1 What are the main parts of a DC motor?

2 How can DC motors be classified according to the type of excitation?

3 What is the task of the commutator in a DC motor?

4 Describe the tasks and operating principle of a stepping motor.

5 How is a brushless DC motor constructed?

6 Explain the electronic commutation of the brushless DC motor.

7 What are the main parts of a starter?

8 Describe the task and operating principle of a roller-type one-way clutch.

9 How is the meshing process for a pre-engaged drive starter carried out?

10 What are the advantages of starters with a planetary reduction gear compared with those without reduction gear?

19

19.2.8 Ignition systems

> The ignition system has the task of igniting the mixture
> - at the right time (ignition point)
> - with the required ignition energy
>
> so that complete combustion of the A/F mixture takes place.

These can achieve the following goals:
- maximum torque,
- maximum power with
- minimum fuel consumption, and
- low exhaust gas values.

19.2.8.1 Generation of the ignition spark

An electric ignition spark can be generated by a simple construction **(Fig. 1)**.

A battery (voltage source) provides the energy required to generate the ignition spark. If both the driving switch and the ignition switch, e.g. contact breaker, are closed, electrical energy is stored in the coils of the ignition coil (transformer). When the ignition switch is opened, there is a brief ignition spark at both electrodes of the spark plugs.

In vehicles, this design has been realised in a simple battery coil ignition system **(Fig. 1)**. For multi-cylinder engines, a distributor has been added. This has the task of conducting the ignition energy to the cylinders whose pistons are each just before ignition TDC.

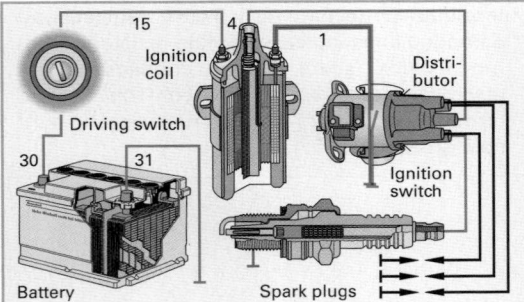

Fig. 1: Design of a battery coil ignition system

Physical processes in the generation of the ignition spark

> In all coil ignition systems, the electrical energy is stored in the form of a magnetic field. This is generated by current flow in the primary winding of the ignition coil and amplified by the iron core of the ignition coil.

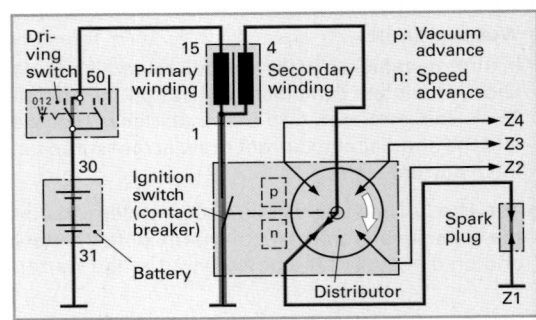

Fig. 2: Circuit diagram of a battery coil ignition system

Processes in the primary circuit

Current course in the primary circuit: earth – terminal 31 – battery – terminal 30 – driving switch – terminal 15 – primary winding of the ignition coil – terminal 1 – ignition switch – terminal 31 – earth **(Fig. 2)**.

Generation of the magnetic field. If the primary circuit is closed by the ignition switch, a current flows based on the present battery voltage. This causes a magnetic field to be produced in the primary winding of the ignition coil. The magnetic field change created during generation induces a voltage in the primary winding which acts against the battery voltage (mutual induction).

This means that during the magnetic field generation, the effect of the battery voltage is reduced by the induced voltage. The less effective voltage only allows for a reduced current flow. The magnetic field is thus generated with a delay.

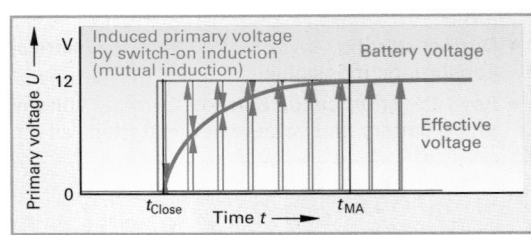

Fig. 3: Effective voltage in the primary circuit when the circuit is closed

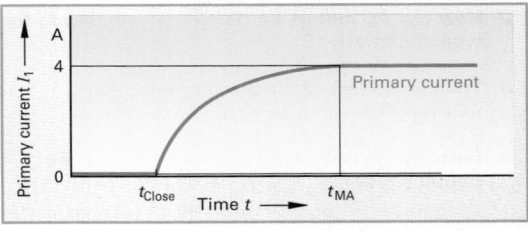

Fig. 4: Primary voltage curve when the circuit is closed

If the magnetic field generation ends at time t_{MA} and the magnetic field change is zero, there is no longer any mutual induction. The current flow is now only determined by the ohmic resistance of the coil and by the present battery voltage, e.g. U = 12 V; R = 2 Ω; I = 6 A.

Magnetic field collapse. By opening the ignition switch at time t_{Open}, the current flow in the primary circuit is interrupted. The magnetic field stored in the winding breaks down abruptly. The large magnetic field change in a very short time induces a high voltage. The quicker the magnetic field collapses, the higher the induced voltage. Depending on the type and design of the ignition coil, it can be up to 400 V.

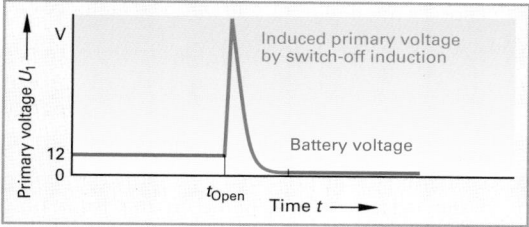

Fig. 1: Voltage in the primary circuit when the circuit is opened

Processes in the secondary circuit (ignition switch closed)

Earth – terminal 31 – ignition switch – terminal 1 – secondary winding of the ignition coil – terminal 4 – ignition distributor – positive electrode – negative electrode – terminal 31 – earth **(Fig. 2, Page 554)**.

The processes described in the primary circuit are transformed to the secondary circuit. The ignition coil is designed so that it amplifies the voltage in the secondary circuit to the same extent as it reduces the current intensity, barring internal losses. The transformer equation applies:

$$n = \frac{U_1}{U_2} = \frac{I_1}{I_2} = \frac{N_1}{N_2}$$

n Turns ratio
U_1 Primary voltage
U_2 Secondary voltage
I_1 Primary current
I_2 Secondary current
N_1 Number of primary windings
N_2 Number of secondary windings

Magnetic field generation with closure of the primary electric circuit (Fig. 2).

The primary side mutual induction voltage generates a voltage in the secondary winding which is amplified in accordance with the turns ratio. With a turns ratio of e.g.

n = 150 and U_{Bat} = 13.5 V, the secondary voltage can reach values of up to 2,000 V. With magnetic field generation completed at time t_{MA}, this voltage is zero. As the generated voltage does not release any sparks, the secondary circuit is not electrically closed due to the air gap of the spark plug. No current flows. The energy in the coil dissipates in the form of damped oscillations.

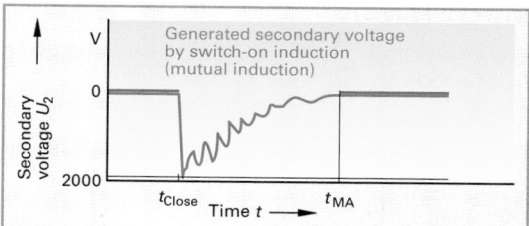

Fig. 2: Voltage in the secondary circuit when the circuit is closed

Magnetic field collapse (contact breaking on the primary circuit by opening the ignition switch).

The rapidly collapsing magnetic field results in a high induction voltage. This is increased by the transformer acting as an ignition coil as a function of the turns ratio n. Values of up to 40,000 V can be achieved.

The generated high voltage ionises the gas cloud between the electrodes and the spark plug. The gas molecules which previously acted as an insulator are at least partially electrically conductive. A spark jumps over. The electrical current is thus closed by the distributor, the windings of the ignition coil and the battery. A current flows.

The energy released by the sparkover ignites the A/F mixture in the air gap. The previous ionisation reduces the voltage during the sparkover (spark voltage) compared with the ignition voltage. After approximately 1 ms - 2.5 ms, the energy stored in the ignition coil is dissipated to such an extent that the ignition spark breaks down at time t_{BE}. The energy still remaining in the ignition coil dissipates in the form of damped oscillations (oscillation decay process).

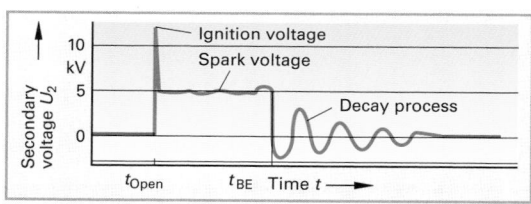

Fig. 3: Secondary voltage curve when the circuit is opened

19.2.8.2 Standard oscillogram

Ignition oscillograms show the voltage curves for closed and open ignition switches. They are required in workshop practice to diagnose any faults that occur. To be able to assess and evaluate the oscillograms clearly, the standard oscillograms of a faultless ignition system must be known. These all have roughly the same structure, but illustrate specific differences depending on the ignition system (see also **Page 570**).

The following terms and key values can be taken from standard oscillograms (**Figs. 1, 2**).

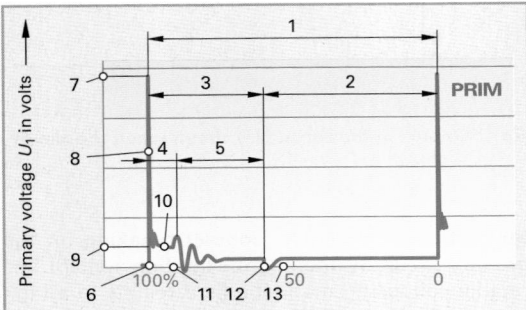

Fig. 1: Standard oscillogram of the primary circuit

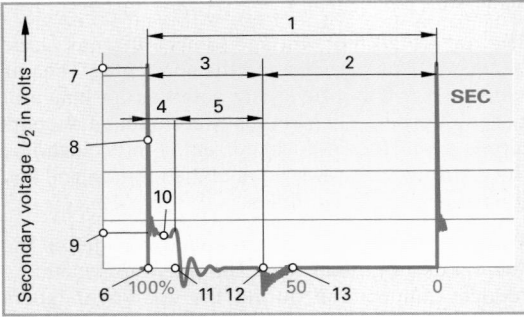

Fig. 2: Standard oscillogram of the secondary circuit

1 **Ignition interval** γ. This is the angle through which the crankshaft turns between ignition sparks. It is calculated using the following formula:

$$\gamma = \frac{720 \, °CA}{\text{Number of cylinders}}$$

The ignition interval consists of the dwell angle α and opening angle β.

$$\gamma = \alpha + \beta$$

2 **Dwell angle α (dwell section).** This is the angle of rotation of the crankshaft in which the primary electric circuit is closed and the magnetic field can be generated. It is specified as °CA or as a % of the ignition interval γ, where the value of γ corresponds to 100%.

e.g.: 4-cylinder four-stroke petrol engine,
$$\alpha = 55 \, \%$$

Ignition interval:　　$\gamma = 720 \, °CA / 4 = 180 \, °CA$

Dwell angle α in %:　$100 \, \% = 180° \, CA$

$$55 \, \% = \frac{180° \cdot 55}{100} = 99 \, °CA$$

3 **Opening angle β (opening section).** This is the angle of rotation of the crankshaft in which the primary circuit is open. It consists of the combustion time and the oscillation decay process.

4 **Combustion time (spark duration).** This specifies how long a standing ignition spark is available at the electrodes of the spark plug (approximately 1 ms).

5 **Oscillation decay process.** The energy remaining in the system is dissipated in the form of a damped oscillation.

6 **Ignition point (t_{Open}).**

> The ignition is triggered by breaking the primary circuit. The ignition point is related to TDC and specified in °CA.

7 **Ignition voltage (8 voltage pin).** This represents the voltage that is necessary to trigger a spark transfer. Under all circumstances, the ignition system must generate enough voltage for the ignition spark to jump. If a sparkover has taken place, the voltage drops to the spark voltage.

9 **Spark voltage (10 spark voltage line).** This shows the voltage required to maintain the ignition spark. Due to the ionised gas cloud between the electrodes of the spark plug, this voltage is less than the ignition voltage.

11 **End of burning (t_{BE}).** At this time, the energy stored in the magnetic field of the ignition coil dissipates to such an extent that the ignition spark breaks down. The oscillation decay process begins.

12 **Closing time (t_{Close}).** The primary circuit is closed. The magnetic field is generated.

13 **End of magnetic field generation (t_{MA}).** The secondary voltage is zero due to the missing magnetic field change in the primary circuit. The primary voltage is 12 V.

19.2.8.3 Ignition coils

> These have the task of storing energy in the form of a magnetic field and transforming the induction voltage into the ignition voltage when the magnetic field collapses.

Ignition coils are transformers with a turns ratio n of 60 ... 150 (**Fig. 2**). Their essential components are the primary winding, the secondary winding, the electrical connections and the iron core. This is formed of several layers of thin iron sheet metal and has the task of amplifying the generated magnetic field.

Primary winding. This consists of a thick, insulated copper wire with very few windings (N_1 = 100 - 500). The low number of coils reduces the inductance of the coil. The short wire length and the large cross-section cause a small ohmic resistance (R = 0.3 Ω ... 2.5 Ω), so that a higher current can flow through the winding. Low inductance and low ohmic resistance result in a quick magnetic field generation and a high level of energy in the primary winding (**Fig. 1**).

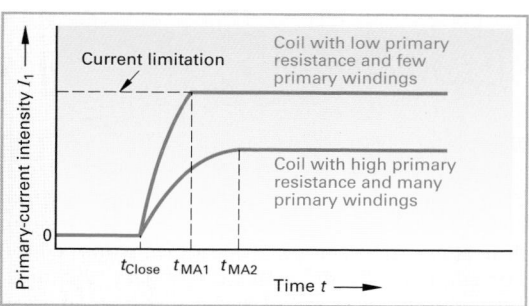

Fig. 1: Current flow and magnetic field generation of different ignition coils

Secondary winding. This consists of a very thin, insulated copper wire with N_2 = 15,000 - 30,000 windings. Depending on the type of ignition coil, the ohmic resistance is 5 kΩ ... 20 kΩ.

To ensure reliable ignition of the A/F mixture, an ignition energy of at least 6 mWs is required. In actual fact, however, the ignition coils are designed for a total energy of up to 120 mWs. This is required because only a fraction of the energy stored can be used for ignition of the A/F mixture. In addition, ignition of the A/F mixture must be guaranteed under all circumstances, i.e. even with a poorly maintained ignition system. Depending on the system, the high voltage offer is 25,000 V - 40,000 V.

Cylinder ignition coil (Fig. 2)

This is used for ignition systems with rotating high voltage distribution (distributor).

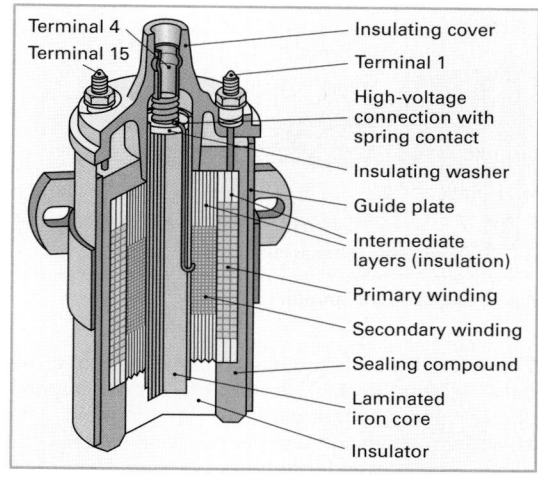

Fig. 2: Cylinder ignition coil

Saver circuit. These ignition coils are arranged to form a saver circuit, i.e. the primary and secondary winding have a joint winding connection. This is connected to the ignition switch (terminal 1). The primary positive supply is via terminal 15; the high-voltage connection is at terminal 4 (**Fig. 3**).

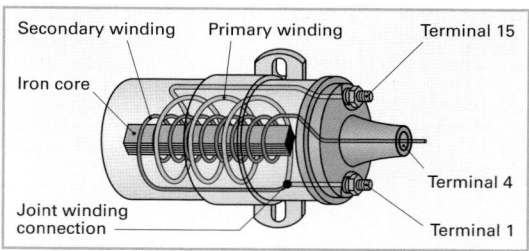

Fig. 3: Internal circuit of an cylinder ignition coil

For engines with a high speed and many cylinders, there may be ignition problems if cylinder ignition coils are used. With a high sparking rate, the dwell period becomes too short; the magnetic field generation is not complete, which means that the stored energy is too low. If there is a sufficient dwell period, a magnetic field can be generated in full. A sufficient ignition power with adequately high ignition voltage is guaranteed.

A measure against the formation of ignition sparks when the primary circuit is closed is not necessary, as the distributor rotor is between the contacts in the distributor cap at the closing time. Thus the spark gap is too wide to allow a sparkover.

19

Single-spark ignition coils (Fig. 1)

Each individual cylinder has its own ignition coil with primary and secondary winding. The coils are usually positioned directly on the spark plugs.

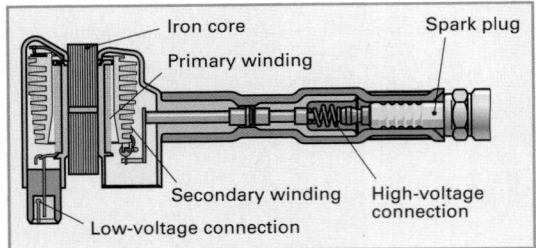

Fig. 1: Single-spark ignition coil

These ignition coils too have connections to terminal 1 (ignition switch), terminal 15 (power supply) and terminal 4 (spark plug). For ignition coils with an additional 4th connection, this is labelled as terminal 4b. It is used to monitor misfiring. Terminal 4 is then renamed terminal 4a (Fig. 2).

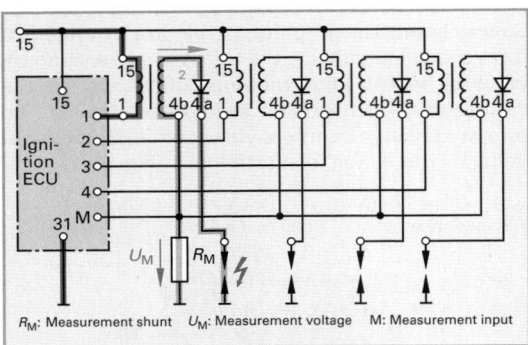

R_M: Measurement shunt U_M: Measurement voltage   M: Measurement input

Fig. 2: Single-spark ignition coil circuit

An ignition spark is triggered on the low voltage side by a power module with distribution logic. This connects the primary current of the ignition coil in question, based on the reference mark signal from the crankshaft and camshaft, which clearly specifies which cylinder is at ignition TDC.

Due to their electrical design, these ignition coils very quickly generate a magnetic field. This can lead to an undesired sparkover already at the end of the intake stroke or at the start of the compression stroke during magnetic field generation. With the help of the diode cascade (a group of several diodes) connected in the secondary circuit, this sparkover can be suppressed. The diodes block it, as the current flow generated by the switch-on induction flows in the opposite direction to the current flow generated by the switch-off induction I_2.

Dual-spark ignition coils (Fig. 3)

These can only be used for engines with an even number of cylinders. Two spark plugs in each case are supplied with high voltage by them.

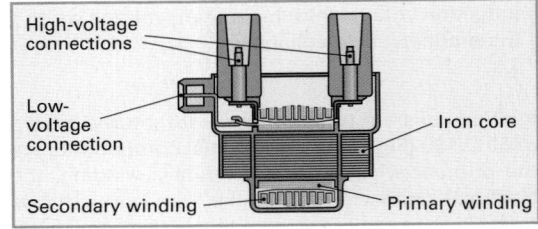

Fig. 3: Dual-spark ignition coil

Dual-spark ignition coils have a secondary and a primary winding, each with two connections. For the primary winding, this is terminal 15 (power supply +) and terminal 1 (connection to the ignition unit). On the secondary circuit, one spark plug is connected to each of the two outputs (Figs. 3, 4).

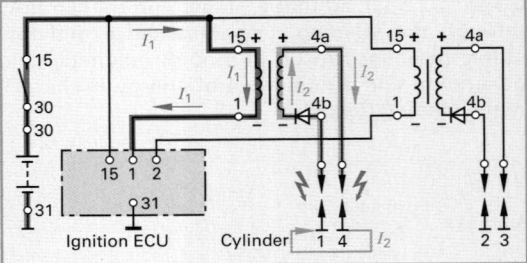

Fig. 4: Dual-spark ignition coil circuit

With each crankshaft revolution, these ignition coils generate one ignition spark each at the same time at two spark plugs. For engines with the ignition sequence 1-3-4-2, one spark, e.g. in cylinder 1, starts combustion at the end of the compression stroke (main spark). The other spark (support spark) deflagrates at the end of the exhaust stroke in cylinder 4, which has moved through 360 °CA. After one revolution of the crankshaft, cylinder 4 is ignited while the support spark in cylinder 1 deflagrates. This can be seen in the secondary oscillogram (Fig. 1, Page 559).

Here, you can see that the ignition voltage at the support sparks is significantly lower than at the main sparks. The reason for this is that at the end of the compression stroke (main spark), significantly more insulated gas molecules are between the electrodes of the spark plug than in the exhaust stroke (support spark). Therefore, a higher voltage is required to generate the ignition spark.

You can also see that the voltages from cylinder 1 and cylinder 4 are in the opposite direction. The prescribed current direction in the secondary winding

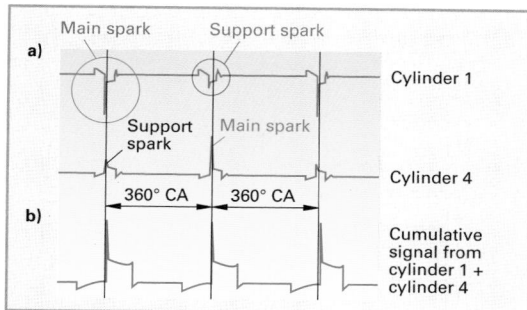

Fig. 1: Secondary oscillogram for a dual-spark ignition coil

means that the ignition spark in one of the spark plugs jumps from the centre electrode to the earth electrode; at the other spark plug, it jumps from the earth electrode to the centre electrode.

Depending on the design of the coils, measures to suppress spark formation through switch-on induction may be required for these ignition systems.

Dual-spark ignition. With these ignition systems, 2 spark plugs are used for each cylinder. When using dual-spark ignition coils, the spark plugs connected to an ignition coil ignite in two different cylinders whose ignition points are displaced by 360 °CA. With the ignition sequence 1 – 3 – 4 – 2, for example, ignition coil 1 and ignition coil 4 each ignite a main spark in cylinder 1 and the support sparks in cylinder 4. 360 °CA later, both ignition coils generate the main sparks in cylinder 4 and support sparks in cylinder 1. In the process, the two ignition coils can be activated displaced by 3 °CA - 15 °CA depending on the load and rotational speed.

Dual ignition leads to a cleaner and faster combustion and thus to fewer harmful exhaust gases.

Four-spark ignition coils (Fig. 2)
These were developed for use in 4-cylinder engines and replace 2 dual-spark ignition coils.

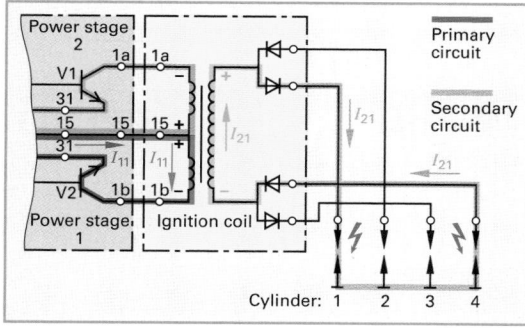

Fig. 2: Four-spark ignition coil circuit

As with dual-spark ignition coils, here too an ignition spark is generated simultaneously in two cylinders whose ignition points are displaced by 360 °CA.

The four-spark ignition coil consists of two primary windings which are activated by one power stage each. In the secondary circuit, there is only one winding. At the two outputs of the winding, there are in turn two connections which are provided with diode cascades and are connected in the opposite direction. These form the connections for the four spark plugs.

19.2.8.4 Primary current circuit

To generate the ignition spark, the primary circuit must be broken to induce a high voltage.

In the first **contact-controlled battery coil ignition systems** the primary current was connected through mechanically activated **contact breakers** (**Fig. 3**). These were replaced because they ...
- ... can only connect primary currents up to a maximum of 5 A.
- ... are subject to high levels of wear (melting loss).
- ... are too slow to activate the required number of ignition trigger processes, due to their mechanics.
- ... cannot accurately maintain the ignition trigger times.

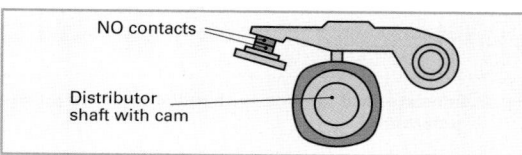

Fig. 3: Contact breakers

In **transistor coil ignition systems,** contact breakers were replaced by electronic ignition trigger boxes. These originally consisted essentially of a relatively simple transistor circuit, which was activated by an induction-type pulse generator or Hall-effect sensor.

Transistorised ignition (TSZ-i, TZ-i) with induction-type pulse generators (Fig. 4) use an induction-type pulse generator housed in a distributor to activate

19

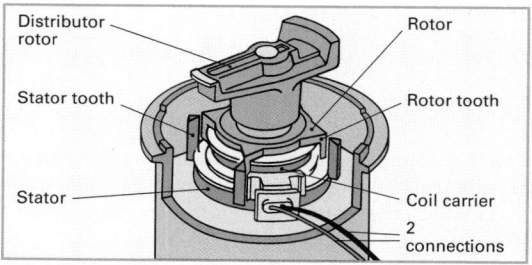

Fig. 4: Induction-type pulse generators in the distributor

the transistor circuit and thus to trigger ignition. The induction-type pulse generator generates an alternating voltage signal U_i according to the alternator principle (**Fig. 1**).

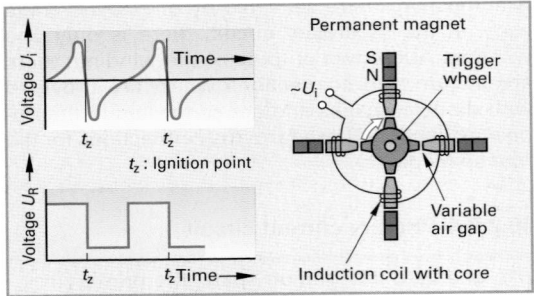

Fig. 1: Sensor signal and ignition point

The signal generated by the sensor is transferred via terminals **0** and **–** to terminals **7** and **31d** on the trigger box and used to activate the transistor (**Fig. 2**).

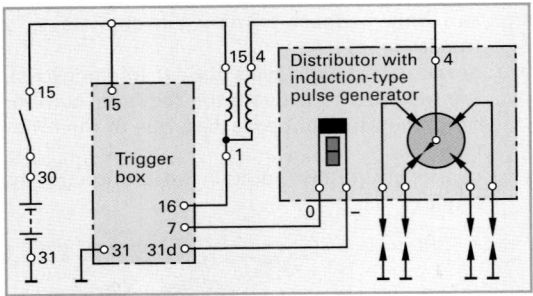

Fig. 2: Transistorised ignition with induction-type pulse generator

Transistorised ignition (TSZ-h, TZ-h) with Hall-effect sensor (**Fig. 3**) use a Hall-effect sensor housed in a distributor to activate the transistor circuit and thus to trigger ignition.

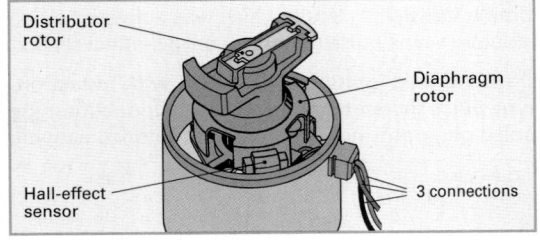

Fig. 3: Hall-effect sensor in the distributor

The Hall voltage generated by the Hall-effect sensor U_H is converted into a generator signal U_G and used to connect the transistor in the primary circuit (**Fig. 4**).

The sensor is connected via terminals **0**, **+** and **–** to terminals **7**, **8h** and **31d** of the trigger box (**Fig. 5**).

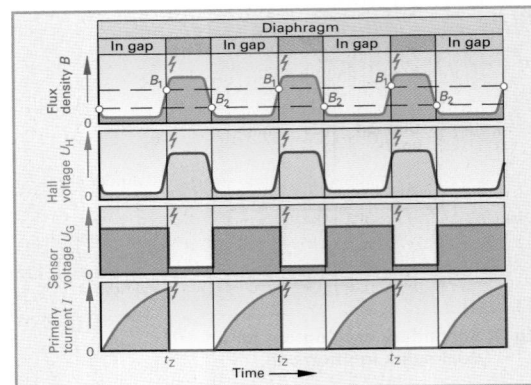

Fig. 4: Sensor signal and ignition point

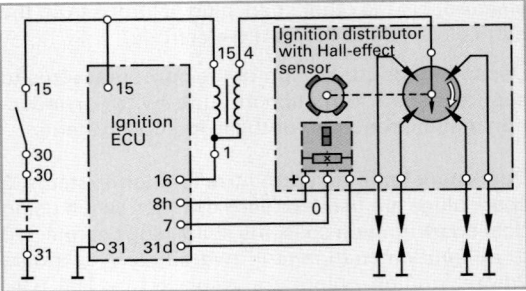

Fig. 5: Transistorised ignition with Hall-effect sensor

Through ongoing development, these trigger boxes have been given more and more additional functions, so that complex control units have been derived from the originally simple trigger boxes.

With the introduction of **Motronic**, the mixture formation and ignition are now controlled by **a joint engine control unit**. This connects the primary circuit via power stages. This is understood as transistor trigger boxes that are activated by a control unit. In ignition systems, they are either in the control unit itself, but usually due to the high heat development they are located outside the control unit at the ignition coils (**Fig. 6**).

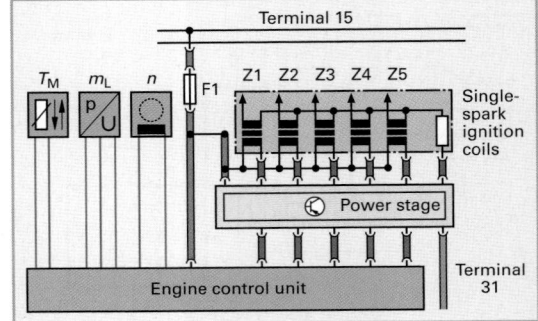

Fig. 6: Ignition coils connected by Motronic

19.2.8.5 Adaptation of the ignition point

> The A/F mixture must be ignited so that the maximum combustion pressure occurs shortly (10 °CA … 20 °CA) after TDC.

Between triggering the ignition and complete combustion of a stoichiometric A/F mixture and thus reaching the maximum combustion pressure, there is a time interval of 1 ms - 2 ms. With constant charge, this time interval is always the same. As the piston moves towards BDC during this time, ignition must be triggered before TDC so that the maximum combustion pressure reaches the piston shortly after TDC (**Fig. 1**). The earlier ignition is triggered, the higher the power output by the engine.

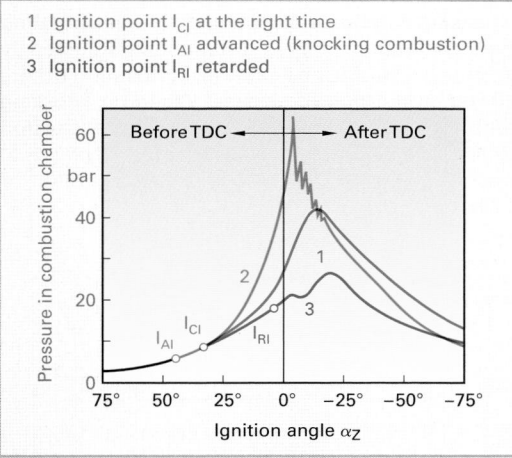

1 Ignition point I_{CI} at the right time
2 Ignition point I_{AI} advanced (knocking combustion)
3 Ignition point I_{RI} retarded

Fig. 1: Pressure characteristic depending on ignition point

If the **ignition point is too early**, however, there are uncontrolled combustion processes with high pressure and temperature peaks. Knocking combustion through auto-ignition of the A/F mixture can occur in this case. This damages the engine, or at least leads to a significant performance drop and deteriorations in exhaust gas values.

If the **ignition point is too late**, the piston has already moved far towards BDC before the A/F mixture is combusted. The size of the combustion chamber increases, leading to a reduction in the pressure acting on the piston and the resulting piston force. Therefore, the piston is only weakly and briefly accelerated on its remaining course to the BDC. This results in a performance drop, increased fuel consumption, higher exhaust gas values and an increased thermal load on the engine.
It is therefore important to strictly adhere to the optimum ignition point.

However, the optimum ignition point changes in different load statuses. Likewise, with an increasing rotational speed, ignition must be triggered increasingly early, as the ignition angle is passed through more quickly, but the combustion time remains almost the same. For these reasons, the ignition point must be adapted to the load and rotational speed.

Centrifugal force and vacuum control. With transistorised ignition systems (**TSZ and TZ systems**), this is controlled mechanically by the two mapped systems, which work independently of each other. The effect of the centrifugal force control is based on the fact that as the rotational speed increases, the flyweights move against the spring force to the outside and turn the driver against the carrier plate (**Fig. 2**).

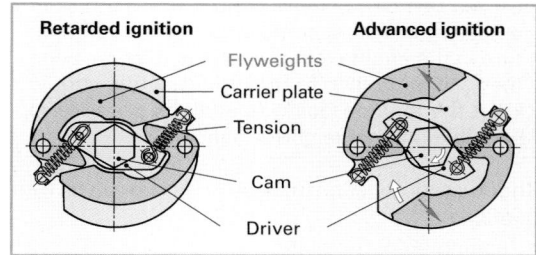

Fig. 2: Centrifugal force control

With vacuum control, the breaker plate is turned by a diaphragm and pull rod through vacuum pressure (**Fig. 3**).

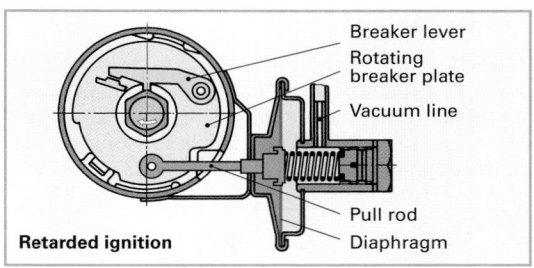

Fig. 3: Vacuum control

Program map. In contrast to this, with electronic ignition and fully electronic ignition systems (**EZ and VZ systems**), the optimum ignition points based on load and rotational speed are determined on a test bench and saved in a program map (**Fig. 4**) in the control unit.

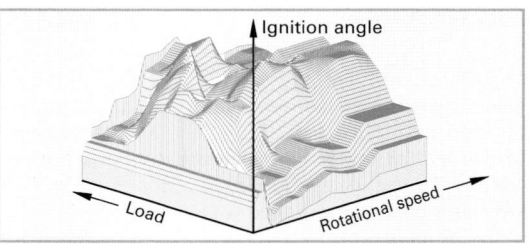

Fig. 4: Ignition map

19

If the control unit receives the reference mark signal from the crankshaft and/or camshaft, the value for the optimum ignition point is taken from the program map, in addition to the rotational speed values (from engine speed sensor) and load (from air-mass sensor or throttle-valve potentiometer). When the corresponding crankshaft position is reached, the control unit activates the ignition power stage. This interrupts the primary circuit, which triggers the ignition.

Knock control in electronic or fully electronic systems

On the one hand, knock control has the task of detecting knocking combustion of the A/F mixture in the engine and preventing it by delaying the ignition point. On the other hand, it should keep the ignition point as early as possible to enable high engine performance.

Knocking combustion of the A/F mixture can be triggered by:
- Ignition point too early
- Fuel with too low an octane number
- Engine overheating
- Compression ratio too high
- Incorrect mixture composition
- Engine overloading

If knocking combustion occurs in the cylinder, high pressure variations occur in the combustion chamber which cause the engine block to vibrate. These vibrations are recorded by a knock sensor mounted on the engine block **(Fig. 1)**.

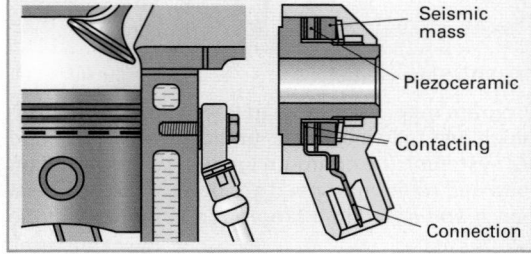

Fig. 1: Knock sensor

The piezoceramic of the sensor is pressurised by a seismic mass (excited by vibrations). The piezocrystal then generates voltages which are transmitted to the evaluating circuit. If certain values are exceeded, knocking combustion is signalled **(Fig. 2)**. If knocking combustion occurs several times, the ignition point is retarded by e.g. 3 °CA compared with the program map value.

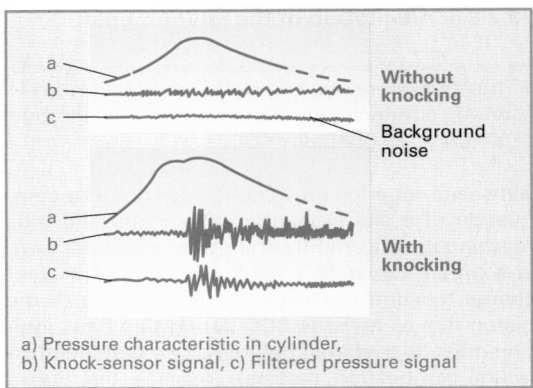

a) Pressure characteristic in cylinder,
b) Knock-sensor signal, c) Filtered pressure signal

Fig. 2: Knock-sensor signals

If knocking combustion continues for this operating point, the ignition point is continuously adjusted by a further 3 °CA until no knocking combustion occurs or the control limit is reached. When no knocking combustion is registered any more, the control unit attempts to move the ignition point forward in small steps until the saved program value is reached **(Fig. 3)**.

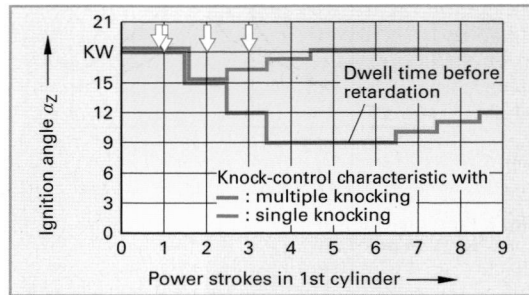

Fig. 3: Adjustment of the ignition point by the knock control

If the control unit continues to register knocking combustion through the knock sensor, a second saved program map is used. This was calculated for operation with low octane fuel. However, the later ignition point of the second program map leads to performance drops and higher fuel consumption. Knock control means that it is possible for engines that were designed for operation with e.g. Super Plus 98 RON to be used with premium grade 95 RON.

When a system fault is detected, e.g. failure of the background noise, the system switches to limp-home-mode function. The ignition point is set to an extremely late value to avoid knocking combustion under all circumstances.

19

Originally, the ignition point was adjusted for all cylinders, even if knocking combustion only occurred in one cylinder.

Cylinder-selective knock control. With this, only the ignition point of the cylinder in which knocking combustion is actually registered is adjusted. To this end, it is necessary for the control unit to be able to specifically assign the occurrence of knocking combustion to a specific cylinder. This assignment is carried out by comparing the time of the ignition points of the individual cylinders via the reference mark signal with the occurrence of the knock signals.

19.2.8.6 Adaptation of the primary current

> The level of the primary current essentially determines the time for the generation of the magnetic field in the ignition coil and the energy that can be transferred by the ignition sparks.

For these reasons, the highest possible current intensity is desired on the one hand. On the other hand, the current flow must not become too high as the ignition coil could possibly heat up so much that it would be thermally destroyed. This risk exists in particular with a low rotational speed and high dwell angle. Therefore, various mechanisms have been developed to adapt the primary current.

Limiting the ignition current by means of a protective resistor. In battery coils and transistorised ignition systems, the current flow in the primary winding is limited by a protective resistor. This can be bridged when the engine is started in order to compensate for the voltage drop caused by the excessively high loading of the vehicle electrical system through the starter (start lifting).

Electronic primary-current limitation. In order to achieve as fast as possible an increase in the primary current, and at the same time as quick as possible generation of the magnetic field, the primary winding is designed so that its closed-circuit current would end at approximately 30 A. However, this current intensity must not be reached, as both the ignition stage and the power transistor would be thermally destroyed.

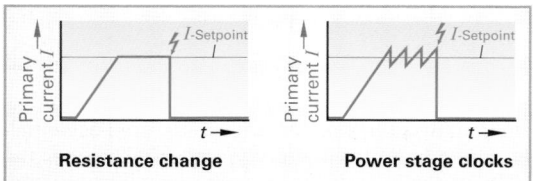

Fig. 1: Types of primary-current limitation

If the reference primary current of 10 A - 15 A is reached due to the specified dwell angle, regulation of the current limitation system begins **(Fig. 1)**. Limitation can result from the fact that the power stage (power transistor) in the control unit ...

- … increases its resistance.
- … clocks the primary current.

A clocked primary-current limitation is visible in the secondary diagram at very low rotational speeds **(Fig. 2)**. Through the clocking of the primary coil, there is a small magnetic field change in the primary winding which is transformed to the secondary side.

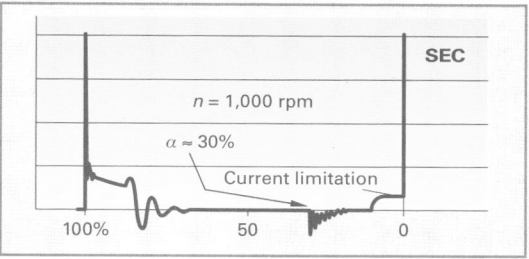

Fig. 2: Secondary diagram at 1,000 rpm and clocked primary-current limitation

Closed-circuit current deactivation. In order not to thermally overload the ignition coil in a stationary engine with the ignition switched on, the primary current is deactivated after approximately one second, if the control unit does not receive a speed pulse.

Dwell angle control. In the control unit, a dwell angle map is saved to determine the closing time. This depends on the respective battery voltage and the current engine speed. In order to reach a sufficient primary-current intensity with a low voltage and high speed, the dwell angle must be increased. With a high voltage and low speed, the dwell angle is reduced to avoid thermal overloading of the ignition coil **(Fig. 3)**.

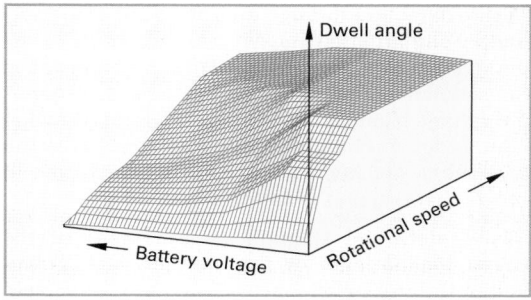

Fig. 3: Dwell angle map

Dwell angle regulation. This works in the same way as the dwell angle control, by calculating the switch-on time of the primary current. This must be designed so that the dwell time (time between making and breaking the primary circuit) is sufficient to achieve the required primary-current intensity **(Fig. 1a)**.

If the required primary-current intensity is not achieved the dwell angle must be increased, which occurs through a pre-assignment of the closing time **(Fig. 1b)**. If the primary current has to be limited for too long, the dwell angle is reduced. The closing time is defined later **(Fig. 1c)**.

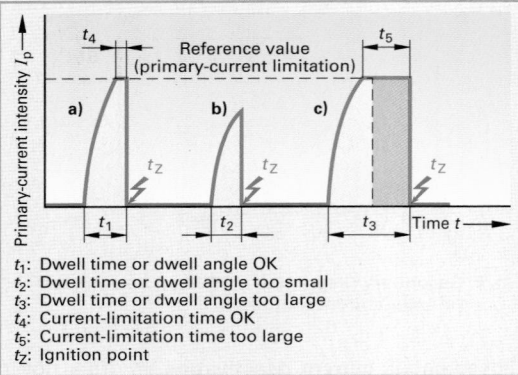

t_1: Dwell time or dwell angle OK
t_2: Dwell time or dwell angle too small
t_3: Dwell time or dwell angle too large
t_4: Current-limitation time OK
t_5: Current-limitation time too large
t_Z: Ignition point

Fig. 1: Primary current characteristic depending on dwell angle

19.2.8.7 Detection of misfiring

> Detection of misfiring is required as part of On-Board-Diagnosis (OBD) because the catalytic converter can be destroyed by misfiring.

If the system detects a certain number of misfires, the fuel injector of the corresponding cylinder is deactivated so that uncombusted fuel does not reach the catalytic converter and destroy it.

Misfire detection through measurement of the secondary current intensity. With this system, a measuring shunt of approximately 240 Ω is connected in the secondary circuit. The control unit determines the voltage falling at the resistor. If a certain voltage limit is not reached, the primary current was too low. The associated fuel injector is deactivated to prevent damaging the catalytic converter.

Misfire detection through measurement of the speed fluctuations. At the start of each power stroke, the crankshaft is accelerated by the incipient combustion pressure. The current higher engine

speed caused by this generates a higher voltage with higher frequency in the engine speed signal of the engine speed sensor. If a cylinder is not ignited, the current speed of the crankshaft falls, causing the amplitude and frequency of the engine speed signal to fall **(Fig. 2)**. These differences in voltage are evaluated by the control unit. The cylinder, which the crankshaft did not accelerate due to the misfiring ignition, is deactivated.

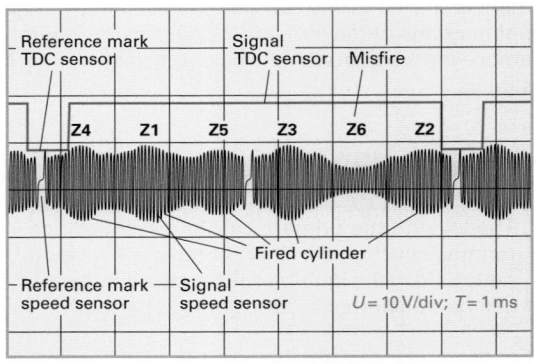

Fig. 2: Engine speed signal during misfiring

19.2.8.8 Multiple-stage ignition

Multiple-stage ignition means that a spark is generated several times one after the other at a spark plug at the end of the compression stroke. This ignition variant is used especially during starting and at very low engine speeds. This should make it possible to reliably ignite A/F mixtures that are difficult to ignite, such as during cold starting, and should burn them with as little residue as possible.

Multiple-stage ignition with up to seven consecutive ignition sparks is possible because ...

- ... the lower rotational speeds, e.g. up to 1,300 rpm, provide sufficient time for the repeated generation of the magnetic field.

- ... the ignition coils used have a low level of mutual induction and thus allow for rapid magnetic field generation.

- ... the primary windings allow for a very high primary current.

- ... the combustion time per ignition spark is in the range from 0.1 ms ... 0.2 ms.

- ... the ignition sparks are generated up to 20 °CA after TDC.

How many sparks are actually generated essentially depends on the battery voltage. The higher the voltage, the quicker the magnetic field is generated, and the more sparks can be generated.

19.2.8.9 Ignition distribution

> The ignition distribution should transfer the high voltage generated by the ignition coil to the spark plug of the cylinder to be ignited.

Components of the ignition distribution are **(Fig. 1)**:
- Spark-plug connector
- High-voltage cables
- Distributor with distributor rotor and contacts
- Protective caps to prevent the penetration of water and dirt

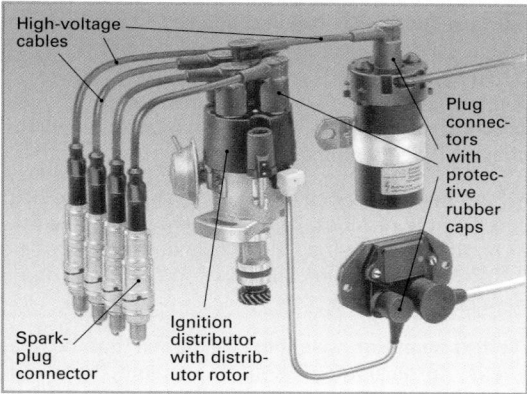

Fig. 1: Ignition distribution

ROV. The type of high voltage distribution shown in **Fig. 1** is also called **rotating high voltage distribution (ROV).**

The high voltage generated in the ignition coil is forwarded to the distributor via a high voltage cable. The distributor rotor fitted on the distributor shaft turns at half the crankshaft rotational speed = camshaft rotational speed. The positioning of the rotor is determined so that at the ignition point, the contact of the rotor is exactly opposite the cable contact which leads to the spark plug of the cylinder to be ignited.

RUV. For systems with multiple spark ignition coils, only high-voltage cables from the ignition coil to the corresponding spark plug and the associated connector are necessary. This is called **static high voltage distribution (RUV).** In single-spark ignition coils, the ignition coils usually sit directly on the spark plugs, so that no other connection components are necessary.

As no moving, mechanical parts are required with static high voltage distribution, such systems are wear-and-tear resistant and fail-safe. However, the control unit must be able to detect which cylinder is to be ignited, in order to be able to activate the corresponding ignition coil.

19.2.8.10 Spark plugs

> Spark plugs have the task of igniting the A/F mixture through a high voltage pulse. To this end, a sparkover occurs between the electrodes of the spark plug after the ignition voltage is reached.

In the process, a spark plug is subject to several stresses:
- Pressure variations between the induction and power stroke of approximately 0.9 bar and 60 bar.
- Temperature fluctuations between the induction and power stroke of approximately 100 °C to 2,500 °C.
- Up to 4,000 sparkovers per minute or up to 66 sparkovers per second at an engine speed of 8,000 rpm.
- Ignition voltages of up to 40 kV with brief current peaks of up to 300 A in the spark head, which lead to erosion of the electrodes.
- Chemical processes that change the properties of the spark plug materials and promote corrosion.

Design of spark plugs
The materials used to manufacture spark plugs are metal, ceramic and glass. The design of a spark plug is shown in **Fig. 2**.

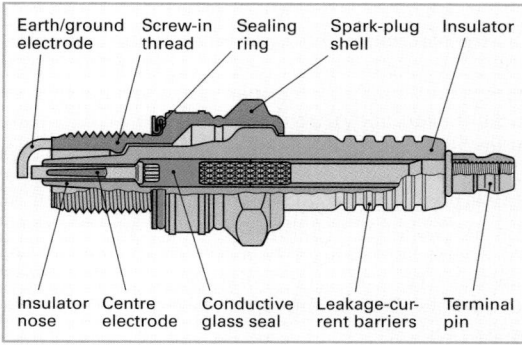

Fig. 2: Design of a spark plug

Insulator (Fig. 2). For the insulator body, a special ceramic from aluminium oxide is used. To increase the leakage current resistance, the insulator is glassed outside the housing and provided with leakage-current barriers.

Seal seat (Fig. 1, Page 566). Depending on the engine construction, the seal between the cylinder head and spark plug is provided by a
- flat seat seal with sealing ring **(Fig. 1a, Page 566)**
- conical seat with conical surface as the sealing element, without sealing ring **(Fig. 1b, Page 566)**.

19

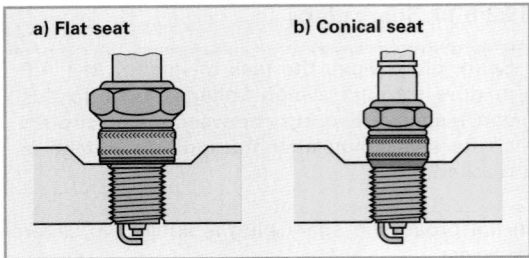

Fig. 1: Seal seat

Electrodes (Fig. 2). Spark plugs have a centre electrode (positive) and one or more earth electrodes.

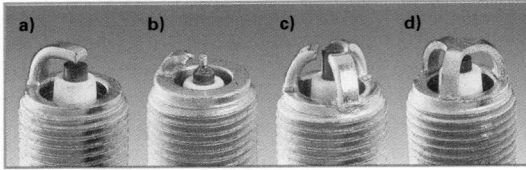

Fig. 2: Electrode shapes

Earth electrodes. These are fixed to the housing. Depending on the structural shape of the spark plugs, they can be designed differently, e.g.

- front electrode **(Fig. 2a)**
- side electrode of a platinum spark plug **(Fig. 2b)**
- multiple pole side electrode **(Fig. 2c)**
- triangular earth electrode **(Fig. 2d)**

Centre electrode. The cylindrical centre electrode juts out of the insulator nose. It consists either of a compound material (copper core, wrapped in a nickel-based alloy), or a precious metal, e.g. platinum.

Spark position (Fig. 3). This is the arrangement of the spark gap in the combustion chamber. The electric sparks should jump to wherever the airflow behaviour of the A/F mixture is particularly favourable. Depending on the engine, there are spark plugs with

- normal spark position **(Fig. 3a)**,
- advanced spark position **(Fig. 3b)** and
- withdrawn spark position **(Fig. 3c)**.

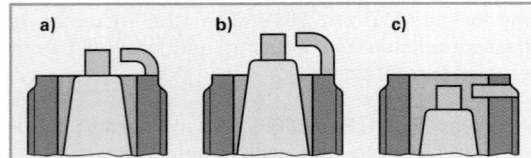

Fig. 3: Spark positions

Spark gap (Fig. 4). The spark usually jumps directly from electrode to electrode (air spark, **Fig. 4a**).

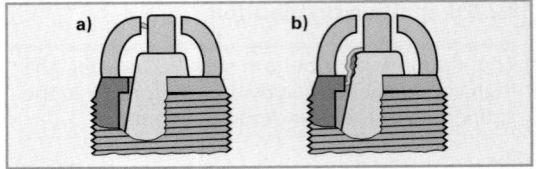

Fig. 4: Spark gap

However, if a conductive layer of soot has formed at the insulator nose through combustion residues, the resistance between the narrow air gap and the insulator nose is less than that of the sparking gap. The spark discharges over the insulator nose peak **(surface spark, Fig. 4b)**. In doing so, it burns the deposits away. The clean spark plug again acts as an air spark plug.

Calorific value. This is essentially determined by the shape of the insulator nose.

A **long insulator nose (Fig. 5a)** means that the heat is difficult to dissipate, on the one hand. On the other hand, the large insulator nose surface absorbs a lot of heat. The spark plug becomes hot. It has a high calorific value.

A **short insulator nose (Fig. 5b)** means that the heat is easy to dissipate. The small insulator nose surface absorbs little heat. The spark plug remains cold. It has a low calorific value.

The **correct calorific value** is selected if the spark plug quickly reaches its self-cleaning temperature of more than 450°C during operation, and does not exceed a temperature of 850°C under full load. When the spark plug reaches its self-cleaning temperature, the residues such as oil soot burn.

The **calorific value is too high** if the insulator nose reaches temperatures of more than 850°C. Auto-ignitions can then trigger uncontrolled combustion which destroys the engine.

If the **calorific value is too low** the spark plug remains too cold to clean itself. The insulator nose becomes dirty and there is no ignition spark, or only a very weak spark.

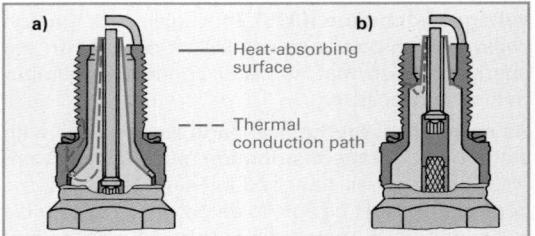

Fig. 5: Heat conduction in insulator nose

19

19.2.8.10 Overview of conventional ignition systems

System / Features	Transistorised ignition system Transistorised ignition (TZ)	Electronic ignition system (map-controlled ignition) Electronic ignition (EZ)	Fully electronic ignition system Fully electronic ignition (VZ)
Introduction of the system from approximately	1976	1987	1988
Visible characteristic for detection of the system	Distributor with vacuum unit, Ignition sensor in distributor	Distributor without vacuum unit, Pure distribution function	• Single-spark ignition coil • Dual-spark ignition coil
Type of ignition coils used	Ignition coil	Ignition coil	• Single-spark ignition coil • Dual-spark ignition coil
Secondary voltage diagram			
Resistance values of the ignition coils (primary, secondary)	$0.5\ \Omega \ldots 2.0\ \Omega$ $8\ k\Omega \ldots 19\ k\Omega$	$0.5\ \Omega \ldots 2.0\ \Omega$ $8\ k\Omega \ldots 19\ k\Omega$	$0.3\ \Omega \ldots 1\ \Omega$ $8\ k\Omega \ldots 15\ k\Omega$
Connection of the primary current by	Trigger box/Ignition-control unit with actuation by induction or Hall-effect sensor	Motronic-control unit	Motronic-control unit
Adaptation of the ignition point through	Centrifugal adjustment mechanism (speed) and vacuum control (load)	Program map depending on speed and load	Program map depending on speed and load
Type of knock control	None	Simple knock control	Cylinder-selective knock control
Adaptation to poor fuel quality	Manual conversion by coding plug	Automatic conversion to second saved program map	Automatic conversion to second saved program map
Adaptation of the primary current	Primary-current limitation Standby current deactivation Dwell angle control / Dwell angle regulation	Primary-current limitation Standby current deactivation dwell angle regulation depending on voltage and load	Primary-current limitation Standby current deactivation Dwell angle regulation depending on voltage and load
Detection of ignition or engine misfires through	None	None	Measurement of the secondary current or measurement of the speed fluctuations
Spark distribution	Rotating high voltage distribution through distributor and high voltage lines	Rotating high voltage distribution through distributor and high voltage lines	Static high voltage distribution through use of single or multiple spark ignition coils

19

19

REVIEW QUESTIONS

1   What is the function of the ignition system?

2   What are the main components of each coil igni-tion system?

3   What happens in the primary circuit of a coil igni-tion system when the circuit is opened and closed?

4   What is the dwell angle? How big is the dwell an-gle in a six-cylinder engine in %, if it has an angle of 60° CA?

5   What is the function of the ignition coil?

6   What types of ignition coils are used and what are their respective characteristics?

7   Sketch the primary and secondary circuit of a fully electronic ignition system when using single-spark and dual-spark ignition coils.

8   Which components connect the primary circuit in current standard ignition systems?

9   What variables determine the ignition point?

10 Describe the operating principle of a cylinder-selective knock control.

11 What are the features of a fully electronic ignition system?

WORKSHOP NOTES

 **Accident prevention**
Electronic ignition systems work in a performance range which is haz-ardous for people if they come into contact with live parts.

This applies both to the primary cir-cuit and the secondary circuit. Dan-gerous voltages can occur not only in the ignition system components, but also at wiring harnesses and plug-in connections, for example.

Safety measures

● Do not touch or disconnect high-tension igni-tion cables when the engine is running or at cranking speed.

● Only connect and disconnect the ignition sys-tem cables when the ignition is switched off.

● If the engine is to be operated at cranking speed without the engine starting, e.g. during com-pression pressure measurement, disconnect the connectors from the ignition coils and fuel injectors. After carrying out the work, call up and clear the fault memory.

● Only wash the engine with the ignition off.

● The battery must only be connected and discon-nected with the ignition off, as otherwise the en-gine control unit could become damaged.

Visual check of the ignition system

● All components of the ignition system must be checked for cleanliness. Dirt and wetness can cause leakage currents, which can lead to mis-firing.

● Connections, plugs and cables must be checked for faultless condition. Damage could lead to sparkovers or could cause higher levels of re-sistance.

● In ignition systems with distributors, the dis-tributor caps are to be cleaned inside and out. They must be checked for cracks and damage.

Spark plugs

● Spark plugs are to be replaced at regular inter-vals according to the manufacturer's specifica-tions (usually after 60,000 km ... 100,000 km). **The spark plugs recommended by the vehicle manufacturer must be used.**
The different spark plug types can be compared using a comparison table.

● The spark plugs must only be replaced when the engine is cold. When the engine is hot, there is a risk that the spark plugs will shrink.

● After releasing the spark plug, any dirt must be blown away before removing the spark plug, in order to avoid contamination in the combustion chamber.

● Spark plug conditions **(Table 2, Page 569)** are to be checked for each cylinder in order to draw conclusions about any faults that are present.

● When reusing spark plugs, the electrode gap must be checked and adjusted if necessary with a spark plug gap gauge.

● The use of grease or oil is not permitted when fitting spark plugs as there is a risk that it could burn onto the cylinder head.

● The spark plug should be easy to screw in. If it is difficult to screw in, there is a risk that it has been inserted at an angle and the thread could become damaged.

● The spark plug must only be fitted tightened to the prescribed tightening torque or the pre-scribed tightening angle **(Table 1, Page 569)**.

Table 1: Tightening torque of spark plugs			
	Thread	Light-alloy	Grey cast iron
Flat seat	M 10 · 1	10 ... 15 Nm	10 ... 15 Nm
	M 12 · 1.25	15 ... 25 Nm	15 ... 25 Nm
	M 14 · 1.25	20 ... 40 Nm	20 ... 30 Nm
	M 18 · 1.5	30 ... 45 Nm	20 ... 35 Nm
	or: For new spark plugs, turn through a further 90° after rotation stop and for used spark plugs, turn through a further 30°		
Conical seat	M 14 · 1.25	20 ... 25 Nm	15 ... 25 Nm
	M 18 · 1.5	20 ... 30 Nm	15 ... 23 Nm
	or: Turn through a further 15° after stop		

Note: See manufacturer's instructions.

Malfunction diagnosis with the engine analyser

When performing malfunction diagnosis on ignition systems using an engine analyser, the fault memory must always be read out first. Entries are stored in the fault memory if, for example, sensors such as the air-mass meter or the knock sensor do not transmit any signals to the control unit, or only transmit implausible signals. The faults entered must be checked and eliminated in the system. The fault memory must then be cleared. The fault memory must be read out again after operation of the engine. There must no longer be any new entries.

No fault display. If the fault in the ignition system is due to a plausible but incorrect sensor value, no fault will be entered in the fault memory. Example: Due to increased contact resistance at the plug-in contact of the air-mass meter, an incorrect value for the air-mass is signalled to the control unit. In such a case, the sensor signals must be read out with the tester or displayed on the oscilloscope and checked. The results are to be compared with the corresponding manufacturer's specifications.

Faults on the secondary side. If there are faults on the secondary side of the ignition system, the display of the secondary voltage curve can often give an indication of the fault at hand. However, this requires exact knowledge of the standard oscillogram for the respective ignition system. The fault can be deduced from the deviation between the actual image and the reference image. In practice, it is the case that some faults can only be detected by means of very accurate observation, a high level of detailed knowledge and extensive experience **(see Table 2, Page 570).**

Malfunction diagnosis on the ignition system by checking the electrical resistances

Often, incorrect electrical resistances are the cause of faults in the ignition system. When measuring the resistance, the following guide values apply (note manufacturer's specifications):

High-voltage cable	Approximately 5 kΩ/m - 6 kΩ/m
Spark-plug connector	Approximately 5 kΩ
Distributor rotor	Approximately 3 kΩ - 5 kΩ
Ignition coils	< 2.0 Ω (primary) < 19 kΩ (secondary)
EF and DF ignition coils	< 1.0 Ω (primary) < 15 kΩ (secondary)

> **It is not possible to measure the resistance of the secondary winding if there are integrated diodes that suppress sparks by means of switch-on induction.**

Engine diagnostics and malfunction diagnosis by spark plug conditions (Table 2)

Conclusions about the condition of the engine, and thus about any faults that may be present, can be drawn from the condition of the spark plugs.

Table 2: Spark plug conditions				
Normal	Sooted	Oiled	Centre electrode melted	Severe wear of earth electrode
Insulator nose in grey-white grey-yellow colour	**Causes:** Mixture too rich, predominantly used in short-distance traffic	**Causes:** Oil level too high, highly worn piston rings or valve guides	**Causes:** Thermal overloading caused by auto-ignition	**Causes:** E.g. engine knocking

Table 1: Malfunction diagnosis

Fault \ Test steps	Ignition point	Firing order	Spark plug	High voltage cable	Distributor cap	Distributor rotor	Distributor	Ignition coil	Ignition sensor	Cables and connectors	Trigger box	Engine speed sensor	TDC sensor	Knock sensor	Engine-temperature sensor	Air-temperature sensor	Load sensor
Engine does not start	•	•	•	•	•	•	•	•	•	•	•	•	•	•			
Engine stalls after starting	•							•		•	•	•			•		•
Engine does not start when warm			•	•				•		•	•						
Engine difficult to start when warm			•	•				•		•	•				•	•	•
Engine does not start when cold			•	•				•		•	•						
Engine difficult to start when cold			•	•				•		•	•				•	•	
Engine stalls			•	•				•		•	•				•	•	
Engine after-fires	•		•														
Engine knocks	•		•											•			
Engine becomes too hot	•			•													
Irregular idle	•		•	•	•	•	•	•		•	•	•			•	•	•
Fuel consumption too high	•		•											•	•	•	•

Table 2: Malfunction oscillograms (secondary voltage)

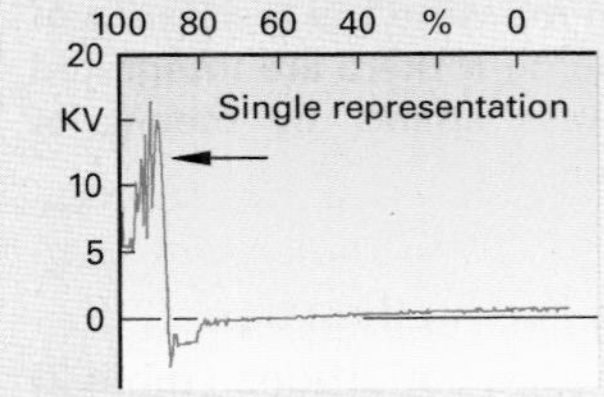

System: Transistorised ignition (TZ)
Fault: Severe deformation of the combustion voltage cable, e.g. broken connection in the distributor cap

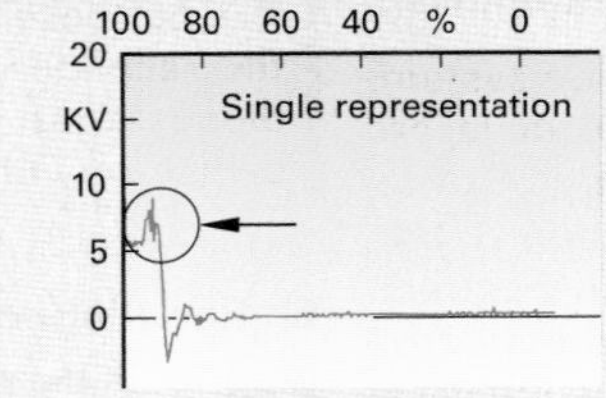

System: Transistorised ignition (TZ)
Fault: Deformation of the combustion voltage cable, fault in the secondary circuit, e.g. corrosion

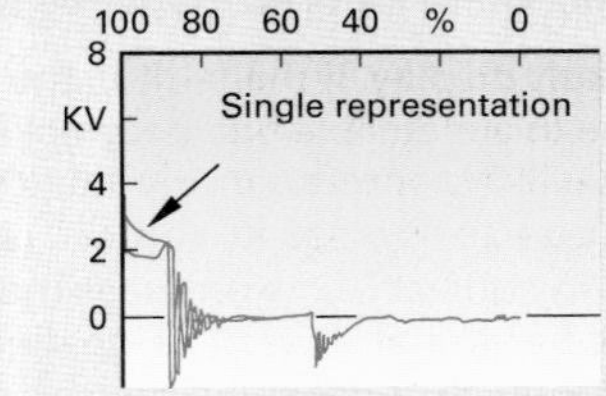

System: Transistorised ignition (TZ)
Fault: Combustion voltage cable jumps due to dirty spark plug, for example

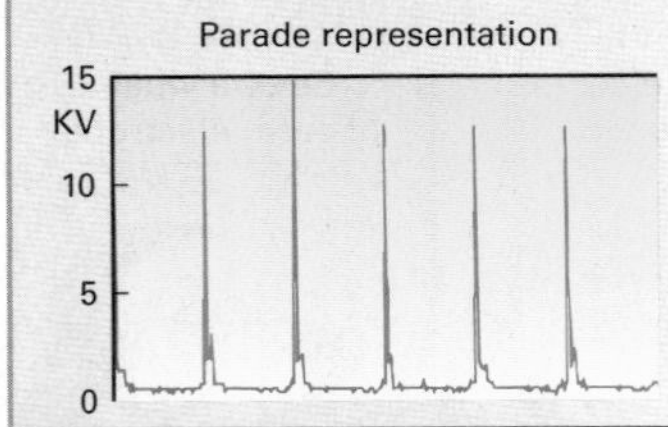

System: Fully electronic ignition system (VZ) with single-spark coil
Fault: Increased ignition voltage at cylinder 3 due to electrode gap being too wide, for example

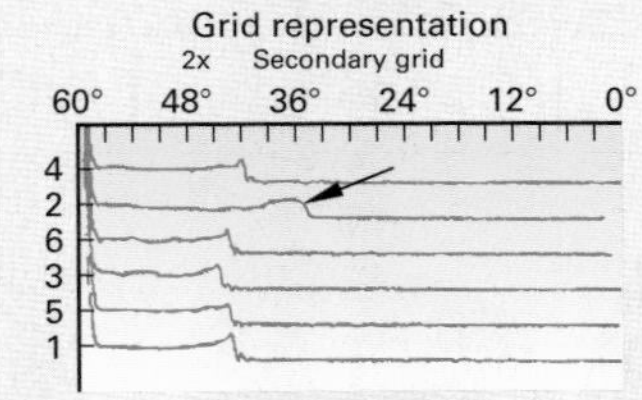

System: Fully electronic ignition system (VZ) with single-spark coil
Fault: Extended combustion voltage cable due to electrode fault on the spark plug, for example

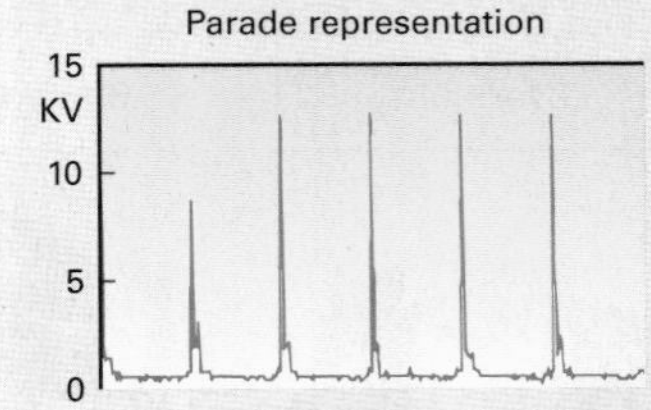

System: Fully electronic ignition system (VZ) with single-spark coil
Fault: Reduced ignition voltage at cylinder 5 due to compression being too low, for example

19

19.2.9 Sensors

> Sensors have the task of recording the operating statuses in electronically controlled systems and converting them into electrical signals.

19.2.9.1 Classification of sensors

Sensors are distinguished according to:
- Function (e.g. calculation of speeds, temperatures, pressures),
- Type of output signal (e.g. analogue, binary, digital),
- Type of characteristic curve (e.g. constantly linear, constantly non-linear, non-constant),
- Physical operating principle (e.g. inductive, capacitive, optical, thermal),
- Number of integration levels **(Fig. 1),**
- Active or passive.

Integration levels. This means that several steps, which are required before the signal is utilised in the control unit, have already been executed within the sensor housing. For sensors in the 3rd integration level, for example, the information is recorded by the sensor and converted into an electrical voltage. This is prepared, e.g. amplified and then digitalised. The signal is then processed in evaluation electronics until it can be used directly in the control unit. A high level of integration leads to the following **advantages:**
- The sensor can serve several control units by transferring the signal via a bus system.
- The signal only has to be prepared once for use in several control units.

- The signal is relatively interference-proof due to the digitalisation.
- Control units can be more easily adapted to different sensors, as the signal is prepared in the sensor.
- The sensor signal information can be called up by the control unit as required.

Disadvantage: Sensors in the 2nd and 3rd integration levels can no longer be checked using standard workshop means, such as multimeter or oscilloscope. They can only be checked using an engine analyser.

Active sensors. These are sensors that require their own power supply to record the physical values. Examples or active sensors: hot-film air-mass meter, vacuum pressure sensor, Hall-effect sensor.

Passive sensors. In contrast to active sensors, these do not require a separate power supply. Examples of passive sensors: NTCs, potentiometers, knock sensors.

19.2.9.2 Examples of conventional sensors

Switches
The most simple form of sensors are switches. These can be activated mechanically, e.g. throttle-valve switches, **(Fig. 1, Page 572),** pneumatically, e.g. warning pressure switches in compressed-air brakes, hydraulically, e.g. oil pressure switches, thermally, e.g. thermoswitches or electrically, e.g. relays. Switches can provide the control unit with two statuses, i.e. switch closed or switch open. The information is provided by the voltage drop at the switch **(Fig. 2, Page 572).**

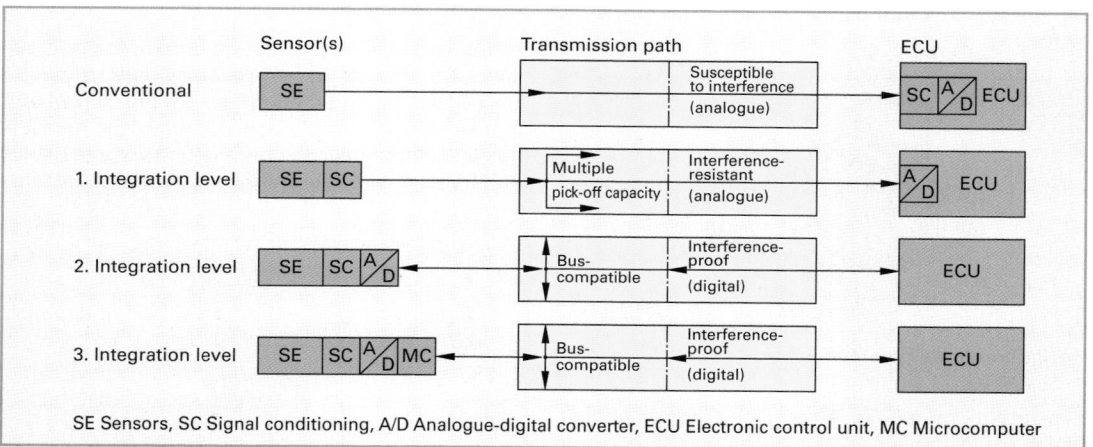

SE Sensors, SC Signal conditioning, A/D Analogue-digital converter, ECU Electronic control unit, MC Microcomputer

Fig. 1: Integration levels of sensors

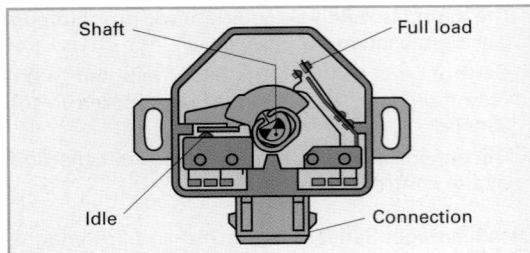

Fig. 1: Throttle-valve switch

When the switch is open, the measuring device (voltmeter or oscilloscope) shows a voltage of 5 V, for example, if it is connected to pin 4 (earth) and pin 5 (positive). If the switch is then closed, the voltage drops to 0 V. The logical unit (LU) in the control unit receives the same values and "recognises" whether or not the engine is idling.

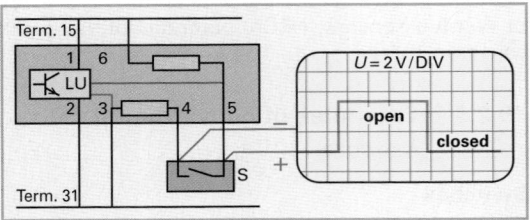

Fig. 2: Signal illustration on switch

Potentiometers

These are used to signal the angle position or the position of shafts or valves to the control unit. In the vehicle, they are used, for example, as throttle-value potentiometers **(Fig. 3)**, potentiometers on accelerator pedal or tank fill level sensors. Potentiometers work like voltage distributors. To do this, a slider activated by a shaft scans resistor paths. The changing length of the resistor paths changes the resistance and thus the voltage drop at the resistor.

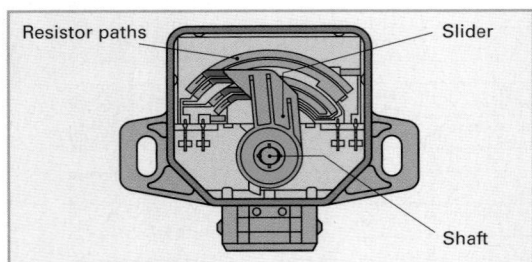

Fig. 3: Throttle-valve potentiometer

The potentiometer **(Fig. 4)** receives a supply voltage of e.g. 5 V from pin 4 of the control unit. If the shaft or sliding contact is in the initial position, there is a

voltage of e.g. 4.2 V at pin 3. If the shaft is moved to the stop position, the output voltage falls continuously to e.g. 0.7 V. The voltage drop at pin 3 to earth is evaluated by the logical unit (LU) in the control unit. Each occurring voltage can be assigned to a very specific position of the shaft or the associated valve. When checking the signal, it is important that the voltage course is constant and without interruptions (noise checking).

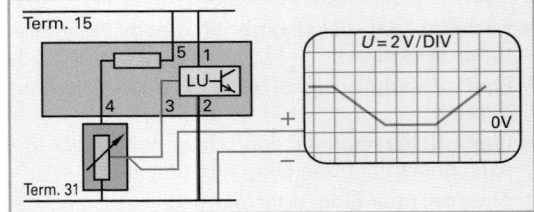

Fig. 4: Signal illustration on potentiometer

Temperature sensors

These are used to record temperatures electronically. They are found in the vehicle as engine **(Fig. 5)**, air or fuel-temperature sensors, for example. The sensor housing contains a measuring shunt made from semiconductor material. This is an NTC (Negative Temperature Coefficient), i.e. the resistance falls as the temperature rises.

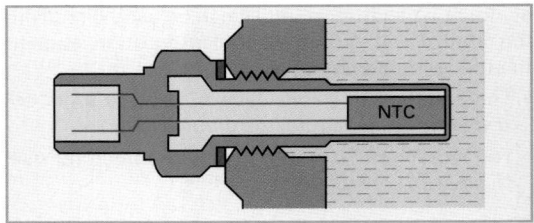

Fig. 5: Engine-temperature sensor

As with resistance, the voltage drop in the NTC becomes smaller as the temperature rises. It can be checked with a multimeter **(Fig. 6)**. If the control unit applies a power supply of 5 V at pin 5, the measuring instrument must display a falling voltage of less than 5 V as the temperature rises. At very high tem-

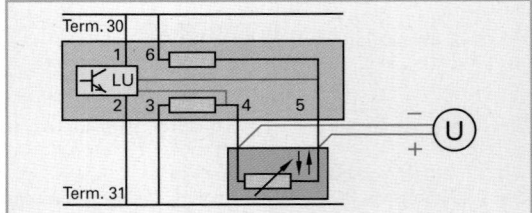

Fig. 6: Voltage drop with resistance

19

peratures, the resistance of the NTC drops to near zero. In this case, the measuring instrument records a value of just over 0 V.

If the resistance is to be measured instead of the voltage drop, the plug-in connection to the control unit must be detached. The resistance values measured must be compared with the values prescribed by the manufacturer. Usually the temperature-resistance behaviour of the NTC is shown as a characteristic curve **(Fig. 1)**.

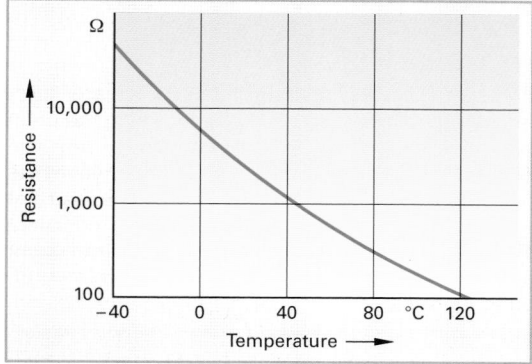

Fig. 1: Characteristic curve of an NTC

19.2.9.3 Examples of sensors in the 2nd or 3rd integration level

Angle sensors

These are used to determine the rotational angle of shafts. The Hall principle is usually used to do this. One or more Hall ICs are fitted so that they are penetrated by corresponding magnetic fields when the shaft rotates. From the generated Hall voltages, the microprocessor integrated in the sensor calculates the angle of rotation and prepares the signal for transfer to the CAN bus. This principle is applied in accelerator pedal sensors in Motronic systems **(Fig. 2)**, in steering-angle sensors for the ESP and in axle sensors for dynamic headlamp range control.

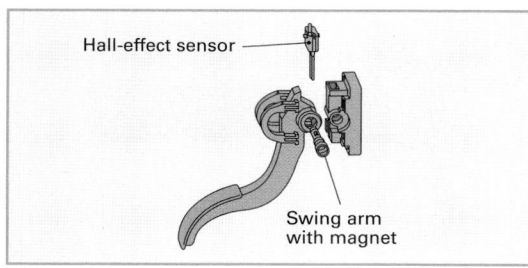

Fig. 2: Pedal travel sensor

Ultrasonic sensors

These are used to monitor spaces and distances to obstacles. A sensor consists of evaluation electronics and a transmit-receive unit which emits ultrasound waves and receives the reflected waves **(Fig. 3)**. With the parking assistance, the space in front of or behind a passenger vehicle can be monitored at intervals of approximately 0.25 m ... 1.5 m by using 4 ... 6 sensors, for example, in the bumper of the vehicle. For intrusion detection, the sensors are fitted in the passenger compartment in order to detect unauthorised entry into the vehicle.

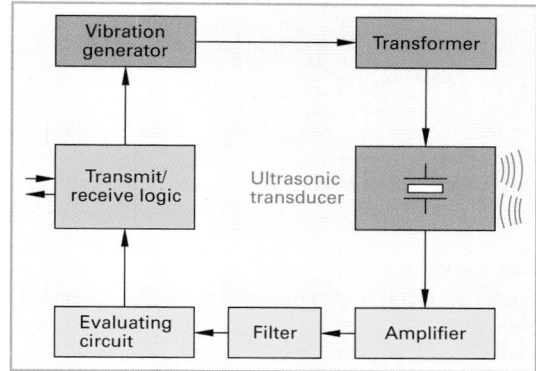

Fig. 3: Assemblies of an ultrasonic sensor

Rotation rate sensors

These sensors are either piezoelectric or capacitive **(Fig. 4)**. They are used to determine the rotational movement of the vehicle about its vertical axis. The sensors are able to record the yaw moment occurring when cornering or skidding. They are used as yaw sensors for ESP and in navigation systems.

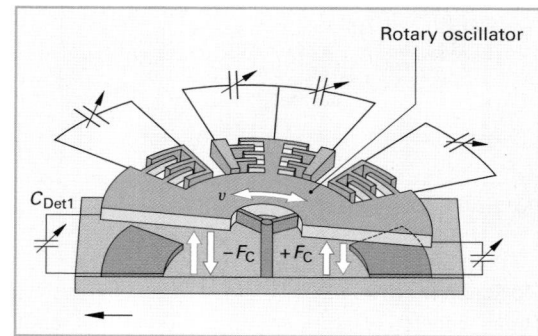

Fig. 4: Rotation rate sensor

Acceleration sensors

These are for recording the acceleration of a vehicle during a collision and trigger the passenger restraint systems via the control unit. To do this, a seismic (freely oscillating) mass is shifted during the collision, which results in a capacitive change. This is amplified by the evaluation electronics, filtered and digitalised for processing in the control unit. Other types of design use a piezoelectrical seismic body clamped on one side. These sensors are required to trigger belt tensioning systems, airbags or rollover bars, for example.

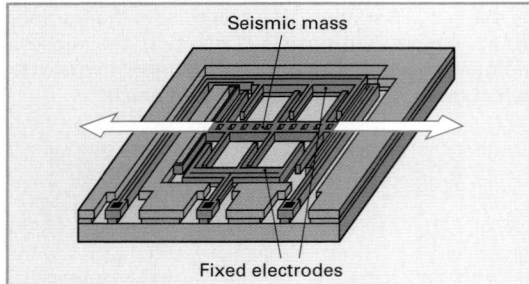

Fig. 1: Acceleration sensor

Gas sensors

These are used to monitor NO_x and CO concentrations and the air humidity. They consist of thick-film resistors, which contain stannous oxide. If the materials to be measured settle reversibly, the resistances change. These sensors are used to monitor the humidity and the air quality in air conditioning systems. They are also used as NO_x sensors in vehicles with direct petrol injection.

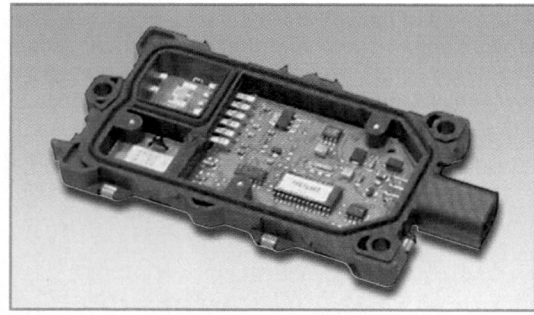

Fig. 2: Air quality sensor

Optical sensors

These consist of light emitting diodes that emit light, and photodiodes that receive light. Based on the changed reflection, the control unit detects dirt on the headlights, broken glass or rain on the windscreen as a consequence of reduced incoming light. These sensors are used as rain sensors **(Fig. 3),** for example, for automatic activation of the wipers or as dirt sensors for automatic cleaning of the xenon headlight lenses.

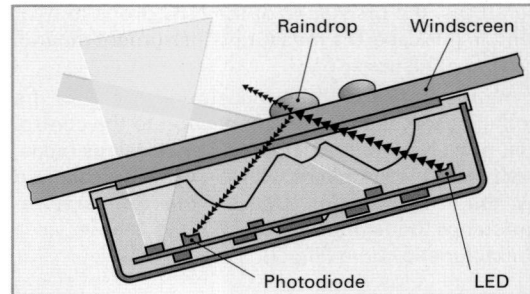

Fig. 3: Rain sensor

Force sensors

Pressure-sensitive resistance elements are combined to make a sensor mat **(Fig. 4)**. From the pressure distribution on the mat, the control unit can calculate the weight, position and movement of the passengers, for example, and trigger the restraint system in the vehicle in the event of a collision. These sensors are used as the basis for intelligent airbag deployment. Child seat detection is integrated in the system.

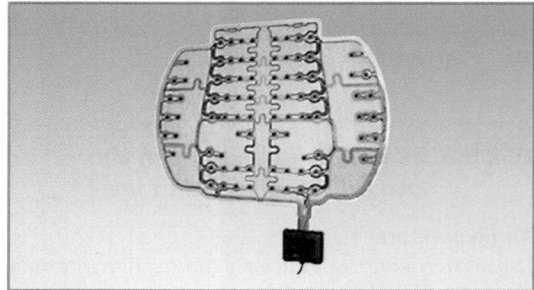

Fig. 4: Sensor mat for passenger classification

Oil sensor (Fig. 5)

This sensor is able to record both the quality (age) and temperature of the available engine oil, as well as the volume of engine oil. In addition to the conventional temperature measurement by NTCs, the conductivity of the engine oil is also evaluated. This means that the technical condition of the vehicle can be accurately monitored, which enables more flexible inspection intervals to be introduced.

Fig. 5: Oil sensor

19.2.10 High-frequency technology

Tasks

> High-frequency technology enables the exchange of information in the form of sounds, pictures and data without a line connection (wireless).

Applications (Fig. 1)

Sound transmission:

- Radio, television, telephone

Picture transmission:

- Television

Data transmission:

- Internet, telematics (e.g. emergency call, traffic information)
- Vehicle-internal systems such as tyre-pressure check, radio remote control, wireless telephones
- Global Positioning System (GPS) for navigation systems

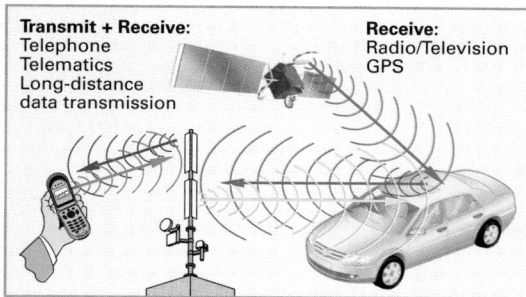

Transmit + Receive:
Telephone
Telematics
Long-distance
data transmission

Receive:
Radio/Television
GPS

Fig. 1: HF technology in modern motor vehicles

The information can only be transmitted in its original form (e.g. audible sounds with a frequency of 16 – 20,000 Hz) as sound waves or along electric lines. This frequency range is called low frequency (LF).

High-frequency technology transmits the information using the radiation and the receipt of electromagnetic waves with frequencies higher than 30 kHz. This frequency range is called high frequency (HF). **Fig. 2** shows the frequency ranges used in the motor vehicle and their applications.

Design of transmitter or receiver

> This consists of the transmitter or receiver, the antenna line and the antenna.

In order to transmit or receive electromagnetic waves, HF technology requires an antenna. In small

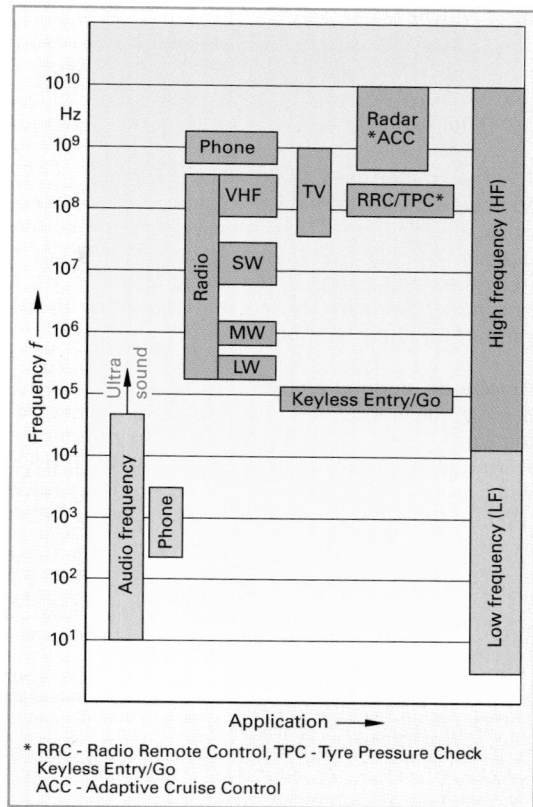

Fig. 2: Frequency ranges in motor vehicles

* RRC - Radio Remote Control, TPC - Tyre Pressure Check
Keyless Entry/Go
ACC - Adaptive Cruise Control

devices, this is built into the housing, e.g mobile phone, radio remote control. For telephone systems, radio and navigation systems, the antenna is also designed as an additional component which is connected to the transmitter or receiver by the antenna cable. **Fig. 3** shows the structure of a car phone as an example of a transmitter/receiver.

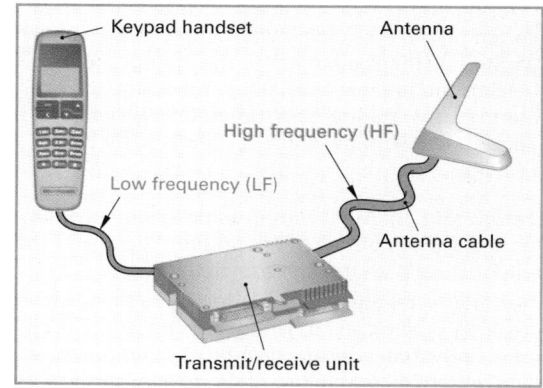

Fig. 3: Design of a car phone

Transmitter

> This has the function of creating a sinusoidal alternating voltage (carrier frequency) and transmitting the useful signal on this.

The transmitter uses a frequency generator to create a sinusoidal alternating voltage with a frequency in the HF range. This is called the carrier frequency. The useful signal (e.g. the sound signal from the microphone in the keypad handset on a telephone) is transferred by the transmitter to the carrier frequency by modulation.

Amplitude modulation (AM). With this type of modulation, the transmitter changes the amplitude of the carrier frequency within the course of the useful frequency **(Fig. 1).** Amplitude modulation is used when transmitting radio signals in the long wave (LW), short wave (SW) and medium wave (MW) range.

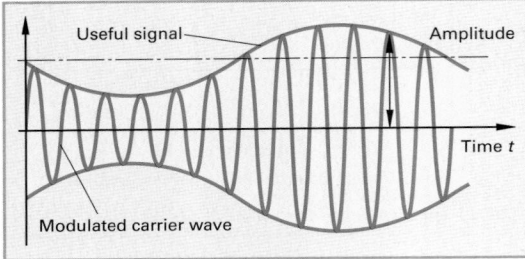

Fig. 1: Amplitude modulation (AM)

Other types of modulation are **frequency modulation (FM),** which is used in the very high frequency (VHF) range, and **phase modulation,** which is used for transmitting digital signals (e.g. telephone, navigation).

Antenna

> During transmission, this has the function of emitting the modulated alternating voltage generated by the transmitter to the environment as electromagnetic waves. When receiving, the antenna converts the electromagnetic waves into alternating voltage which is evaluated by the receiver.

Design (Fig. 2). Most motor vehicles have rod antennas. These consist of the antenna base and the antenna rod with the antenna tip.

In modern day vehicles, window antennas are increasingly being used. These use the heating element of the rear window heating or additionally applied antenna conductors in the rear window, side windows or windscreen. The main component is the antenna module with antenna amplifier which establishes the connection between the antenna conductors on the window and the antenna cable. The power supply for the antenna amplifier is provided by the antenna cable.

The basic functional method of the window antenna is the same as that of a rod antenna.

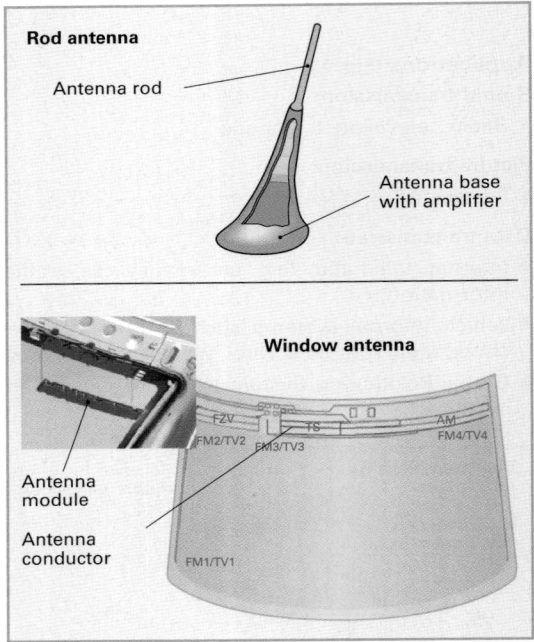

Fig. 2: Design of rod and window antenna

Functional method of a transmitting antenna (Fig. 1, Page 577)

> The voltage creates an electrical field in the antenna and the current creates a magnetic field.

The antenna tip together with the antenna base acts as a capacitor with plates located far apart from each other. If voltage is applied between the antenna tip and the antenna base, an **electrical field** is generated. The electrical field lines are emitted parallel to the antenna rod.

If the voltage rises or falls during the alternating voltage period, a current flows through the antenna rod. The antenna rod acts as a coil and generates a **magnetic field.** The magnetic field lines are arranged as rings around the antenna rod.

The electrical and magnetic fields alternate and are emitted vertically to each other by the transmitting antenna. Together, they are called **electromagnetic waves.**

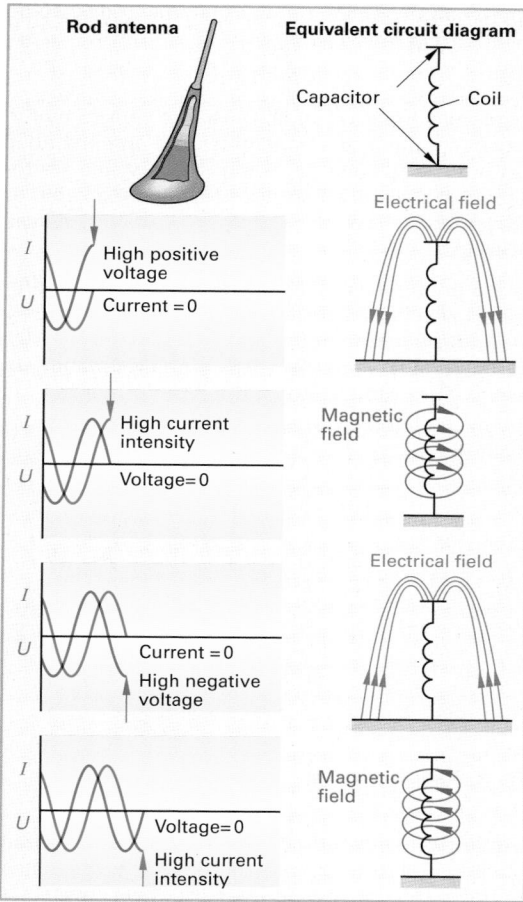

Fig. 1: Operating principle of transmitting antenna

Receiving antenna

High frequency alternating voltage is generated:
- By the electrical field in a rod antenna,
- By the magnetic field in a coil antenna.

Rod antenna (Fig. 2). If the electromagnetic wave of a vertical transmitting antenna reaches an also vertical receiving antenna, the electrical field creates a high frequency alternating voltage in this.

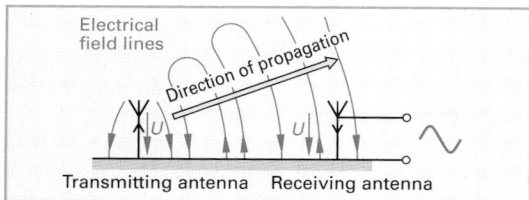

Fig. 2: Receive process with a rod antenna

Coil antenna. This receives the magnetic waves of a radio signal and creates a high frequency alternating voltage. As it has a directional effect, it is not suitable for use as a receiving antenna in a motor vehicle. It is used as a transmitting antenna in special cases (e.g. keyless central locking system).

Tuned antenna (Fig. 3)

The length of the antenna must be tuned to the wavelength of the carrier wave. If the length of the antenna is a quarter of the wavelength (λ), this is a tuned antenna.

Optimal antenna (Fig. 3a). This antenna has the highest voltage at the antenna tip. The highest current flows at the antenna base. It is in resonance with the receive frequency. Thus, it has the highest reception performance.

Antenna too short (Fig. 3b). There is a lower voltage in the antenna tip. This leads to a lower current in the antenna base. The reception performance is reduced. In today's vehicles, shorter antennas are fitted for design reasons. This makes antenna amplifiers in the antenna base or antenna module necessary.

Antenna too long (Fig. 3c). With a long antenna, the voltage falls again at the antenna tip. The reception performance is also reduced.

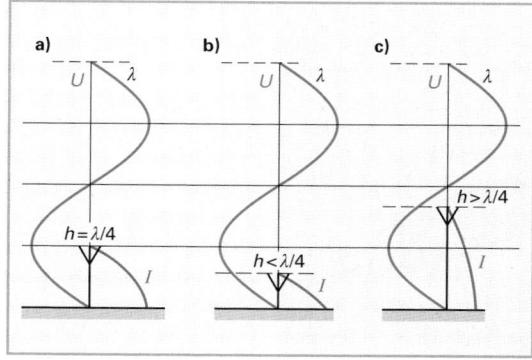

Fig. 3: Current and voltage distribution on antennas of different lengths

Wavelength.

The wavelength λ specifies the path travelled by an alternating voltage signal during a period. The greater the frequency f, the smaller the wavelength.

This is calculated from the velocity of propagation of the waves (c) and the frequency (f) and is specified in metres. Electromagnetic waves travel through the air at almost the speed of light.

$$\lambda = \frac{c}{f} \qquad \text{Unit: m}$$

Example for calculating a tuned antenna for VHF radio reception

Velocity of propagation $c = 300 \cdot 10^6$ m/s
Frequency $f = 100$ MHz (VHF)
Antenna length $= h$

$$\lambda = \frac{c}{f} = \frac{300 \cdot 10^6 \text{ m/s}}{100 \cdot 10^6 \text{ 1/s}} = 3 \text{ m}$$

$h = \lambda/4 = 3$ m $/ 4 =$ **0.75 m**

Antenna cable (Fig. 1)

This has the task of transmitting the alternating voltage to the antenna in a transmitter.

Design (Fig. 1). The antenna cable is designed as a coaxial cable and is protected against electromagnetic interference by a shield connected to the vehicle earth. The coaxial cable opposes the alternating current with a **wave resistance** and thus influences the transmission and reception performance.

Wave resistance. If a high frequency alternating current flows through a coaxial cable, it acts as a control circuit with series connected coils and parallel connected capacitors.

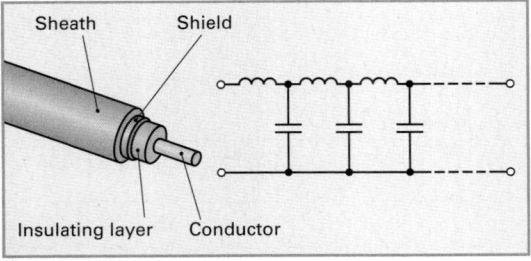

Fig. 1: Design and equivalent circuit diagram of antenna cable

The wave resistance consists of the ohmic resistance of the conductor, the capacitive reactance and the inductive reactance. The wave resistance is crucial for the transmission or reception quality.

Tuned antenna system. With this type of system, the wave resistance of the antenna cable is the same as the base resistance of the antenna. When transmitters and receivers are developed, the components are tuned to each other. Therefore, only the

spare parts approved by the manufacturer may be used during repair work.

Stationary waves (Fig. 2). These are caused by the incorrect tuning of a transmitter. The antenna does not emit its full energy. Some of the waves are reflected at the transition from the antenna cable to the antenna. The incoming and outgoing waves overlap.

In the antenna cable, the overlapping waves form waves with higher amplitude and wave crests ($U_h + U_r$) and lower amplitude and wave troughs ($U_h - U_r$) at regular intervals ($\lambda/2$). These are called stationary waves. They reduce transmitting performance.

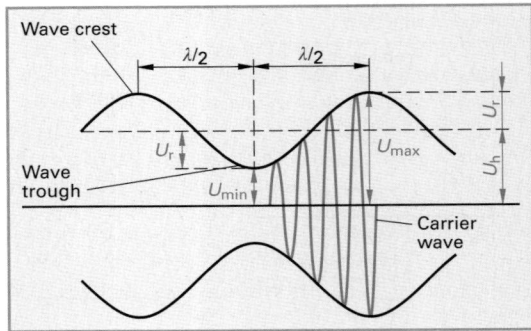

Fig. 2: Stationary waves

Stationary wave ratio (SWR). This is the measured quantity for tuning transmitters. It is calculated from the relationship between the wave crest ($U_h + U_r$) and the wave trough ($U_h - U_r$).

$$\text{SWR} = \frac{U_h + U_r}{U_h - U_r}$$

In a perfectly tuned transmitter, there are no wave crests or troughs. The voltage ur is thus 0.

Hence:

$$\text{SWR} = \frac{U_h + 0\,\text{V}}{U_h - 0\,\text{V}} = 1$$

This means that the optimum stationary wave ratio is 1.

In workshop practice, SWR measuring instruments are used for malfunction diagnosis on transmitters (e.g. radio, telephone).

Receivers

> The receiver has the task of evaluating the high frequency alternating voltages created by the antenna, acquiring the useful signal from this (sounds, pictures, data) and controlling actuators with it (loudspeakers, display screens, etc.).

The receiver accepts the alternating voltage of the antenna and separates the useful signal from the carrier frequency using demodulation. The useful signal acquired is then used to activate actuators such as loudspeakers, display screens, engines, etc.

Wave propagation (Fig. 1)

> Depending on their frequency, the radio waves propagate differently. A distinction is made between ground waves and sky waves.

Ground waves. These propagate along the earth's surface. In the long wave (LW) range, the ground waves can propagate up to 1,000 km. As the frequency increases, however, the losses along the earth's surface increase. Thus, the ranges of the ground waves in the short wave (SW) range fall to approximately 100 km.

Sky waves. These propagate in straight lines and would therefore inevitably leave the earth. In certain frequency ranges, however, conductive layers of the earth's atmosphere at 50 to 300 km altitude reflect the sky waves back to the earth's surface. The reflective capacity depends on the wave length and the time of day. The reflection of the sky waves extends the ranges of the medium (MW) and short waves (SW).

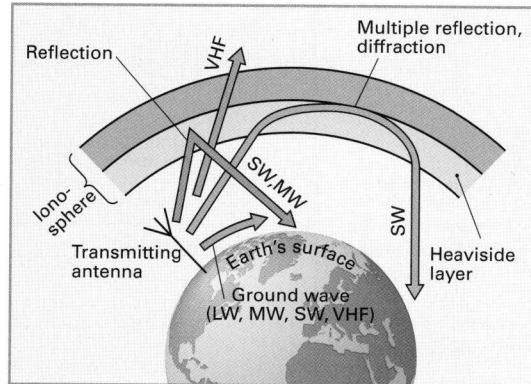

Fig. 1: Propagation of radio waves

In the very high frequency (VHF) and television range, the waves propagate almost linearly. The best reception conditions occur when the line of sight between the transmitter and the receiver is unobstructed. Very high frequency waves make radio traffic to space ships and satellites possible by penetrating the ionosphere.

Polarisation direction

> This describes the direction of the electrical field of a transmitting antenna in relation to the propagation direction and thus determines the installation position of an optimal receiving antenna.

A vertically installed transmitting antenna emits an electromagnetic field whose electrical field direction runs vertical to the propagation direction and whose magnetic field direction runs horizontal to it. This is called **vertical polarisation**. Accordingly, with a horizontally installed transmitting antenna, there is **horizontal polarisation**. Both are called **linear polarisations**. **Fig. 2** shows vertical polarisation.

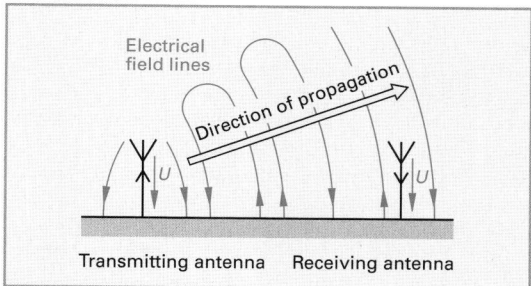

Fig. 2: Vertical polarisation

Circular polarised wave. This consists of two waves, each with linear polarisation in levels vertical to each other, with a phase displacement of 90° from each other. Depending on the direction of this phase displacement, the resulting wave is either clockwise or anticlockwise. This technique is used predominantly in satellite radio (e.g. GPS).

Diffraction of the radio waves by the vehicle body (Page 580, Fig. 1)

VHF signals in Europe are emitted with horizontal polarisation. It is only the influence exerted on the field by the metal skin of the vehicle body that enables VHF reception using a rod antenna, which is usually fitted vertically.

In addition, places with different concentrations of electromagnetic fields form on the vehicle body. On the vehicle shown, the highest field concentrations occur on the front and rear edges of the roof. These are the optimum installation locations for a receiving antenna.

19

Fig. 1: Diffraction by vehicle body

Causes of reception interference
Radio shadow (Fig. 2)

> This occurs if the line of sight between the transmitter and receiver is obstructed by obstacles.

Obstacles like mountains and buildings cannot be penetrated by the radio waves. The reception performance drops.

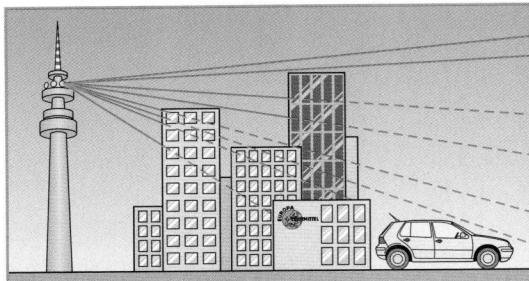

Fig. 2: Radio shadow

Multipath reception, multipath (Fig. 3)

> This occurs if a radio signal hits the receiving antenna directly and reflected once or several times.

The reflected signal hits the antenna at a later time, i.e. with a phase displacement. The signals partially cancel each other out and reduce the reception performance.

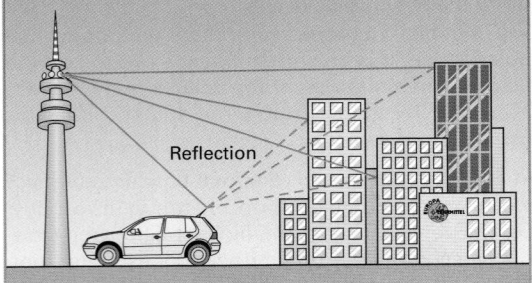

Fig. 3: Multipath reception, multipath

Interference sources in the motor vehicle

> Switching and sliding contacts in the motor vehicle create electromagnetic waves that cause interference in radio systems.

Interference sources in the motor vehicle are, in particular:
- Ignition system
- Loose contacts in current carrying lines or their connections
- Alternator
- Starter
- Electric motors
- Electrostatic charges (e.g. tyres)
- Poor or alternating metallic contact of large metal parts of the vehicle (e.g. bonnet)

19.2.11 Electromagnetic compatibility (EMC)

> This indicates the capacity of an electric or electronic device to operate properly within its electromagnetic environment (**immunity to interference**) and at the same time not to exert an influence on other devices (**emitted interference, Fig. 4**).

The systems fitted in the vehicle, such as the ignition system, fuel-injection system, antilock-braking system, mobile telephone, etc. must not influence each other. The systems must behave neutrally in terms of the environment.

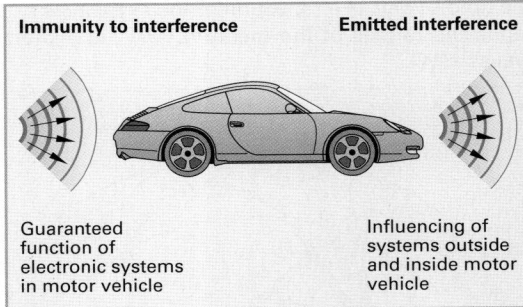

Immunity to interference   Emitted interference

Guaranteed function of electronic systems in motor vehicle

Influencing of systems outside and inside motor vehicle

Fig. 4: Immunity to interference and emitted interference

Suitable measures for achieving a high level of EMC are the installation of:
- Noise suppression capacitors
- Inductance coils
- Shielded lines
- Shields made of sheet steel
- Twisted lines

Diagnosis on receivers

Field strength measurement (level measurement)

This is carried out for malfunction diagnosis on receivers. The sinusoidal alternating voltage created in the antenna is measured using field strength or level measuring instruments.

The unit of measurement is the decibel microvolt (db μV).

Field strength measuring instruments are seldom used in the dealership.

The field strength can be checked using the diagnosis tester in the context of self-diagnosis.

The field strength is highly dependent on external influences. This means it is not possible to define a reference value. The procedure whereby comparisons are made with vehicles with no fault is used for malfunction diagnosis. Here, the field strength is measured on a working vehicle and compared with the measured value on the problem vehicle.

The following conditions must be noted:

- Field strength measurements are in principle only to be carried out in the open.
- Both vehicles must be of the same type and equipment.
- The same frequency must be set on both vehicles.

- Both vehicles must be positioned at the same place and in the same direction one after the other.

If the measured value differs from the value of the comparison vehicle, the cause of the fault can be determined in a further test step.

- Check the antenna cable duct.
- Check the antenna earth connection.
- Measure the supply voltage and current draw of the antenna amplifier.
- Try replacing components of the receiver (e.g. antenna, antenna amplifier, antenna cable, receiver) and repeat the field strength measurement.

In the new generation of receivers, the power is supplied to the antenna amplifier via the antenna cable. The self-diagnosis of the receiver monitors the line connection and current draw of the amplifier using the current intensity drawn in by the antenna amplifier. If there are deviations compared with the reference values, entries are stored in the fault memory. These can be read out and cleared using the diagnosis tester.

In addition, the current draw of the antenna amplifier can be checked with the diagnosis tester.

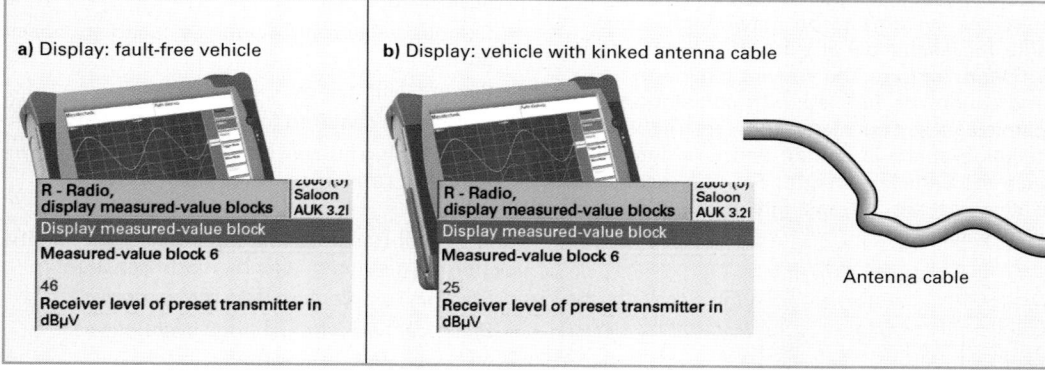

a) Display: fault-free vehicle

b) Display: vehicle with kinked antenna cable

R - Radio, display measured-value blocks | 2003 (3) Saloon AUK 3.2I

Display measured-value block

Measured-value block 6

46

Receiver level of preset transmitter in dBμV

R - Radio, display measured-value blocks | 2003 (3) Saloon AUK 3.2I

Display measured-value block

Measured-value block 6

25

Receiver level of preset transmitter in dBμV

Antenna cable

Fig. 1: Field strength measurement with diagnosis tester   **Fig. 2: Display of kinked antenna cable**

REVIEW QUESTIONS

1 What does a transmitter consist of?

2 What field is emitted by a current carrying transmitting antenna?

3 What is the task of the receiving antenna?

4 When does radio shadow occur?

5 What is the effect of multipath reception?

6 How long is a tuned receiving antenna?

7 What are the interference sources in the motor vehicle?

8 What is the stationary wave ratio?

9 What is electromagnetic compatibility?

10 How is the field strength measured in receivers?

11 What is polarisation?

19.2.12 Data transmission in motor vehicles

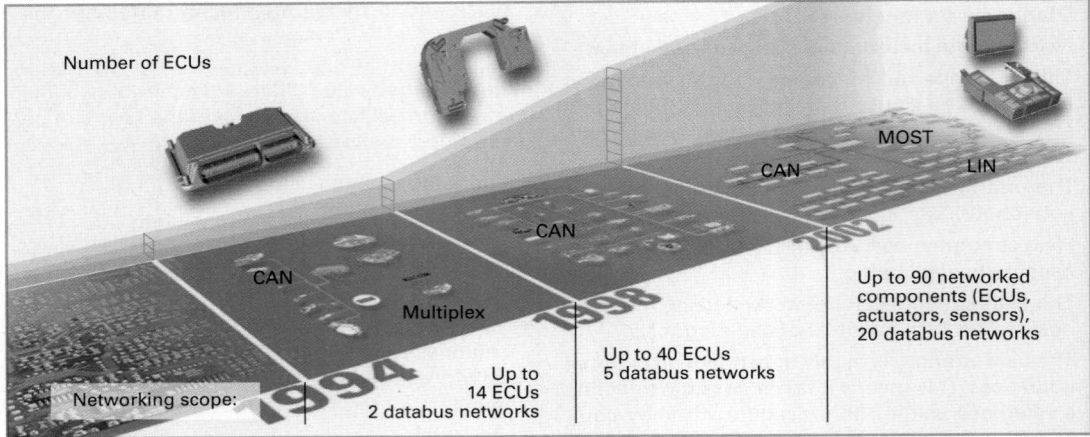

Fig. 1: Development of the networking scope in the motor vehicle in the last 10 years

> This enables the transport and exchange of information between the electrical components in the form of data and signals.

Conventional data transmission (Fig. 2). With conventional data transmission, at least two lines are required for each piece of information, e.g. air-mass meter, Hall-effect sensor. The increasing requirements of vehicle electronics in the fields of vehicle safety, comfort and convenience, communication, fuel consumption, exhaust reduction and diagnosis require the systems to be networked in order to enable the exchange of a lot of information in a short period of time. The above-mentioned requirements cannot be fulfilled with conventional data transmission as too many lines are required, the installation space is limited and it is desirable to reduce the vehicle's weight. For this reason, data bus systems are increasingly being installed in motor vehicles.

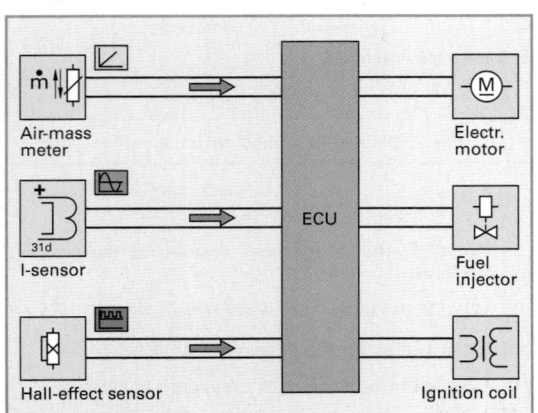

Fig. 2: Conventional data transmission

Data bus systems (Table 1, Page 583)

> These transmit information using information units (Bits). The information units are grouped into data packets.

Advantages of data bus systems compared with conventional data transmission technology:

- Joint use of sensors and calculated physical values (e.g. outside temperature, engine speed) by several control units.
- Extended electronic diagnosis capability of actuators.
- Fewer electric lines.
- Less space required for wiring harnesses.
- Smaller plug housings and multiway connectors, meaning smaller control units.

The type of transmission depends on the data bus system. A distinction can be made between:

- Electrical single-wire data bus systems
- Electrical two-wire data bus systems
- Optical data bus systems
- Wireless data bus systems

The use of the different data bus systems is based on properties and requirements, such as

- Data transmission speed, which is specified in bauds (bd)*, bits per second
- Electromagnetic compatibility (EMC)

---

* Baud (bd): Named after Jean Baudot (Communications engineer, France 1845 – 1903)

Table 1: Overview of the most important data bus systems					
Transmission type	Bus system	Maximum transmission speed	Application	Data transmission process	System
Electrical single-wire	Multiplex	100 kbd	Simplest control	Asynchronous	Bus structure
	Local Interconnect Network (LIN)	19.2 kbd		Asynchronous	Bus structure
Electrical two-wire	Controller Area network	1 Mbd	Drive and comfort	Asynchronous	Bus structure or passive star structure
Light pulses	Domestic Digital Bus (D²B)	5.6 MBd	Information, communication, entertainment	Synchronous	Ring structure
	Media Oriented System Transport (MOST)	21.2 Mbd	Information, communication, entertainment	Synchronous	Ring or active star structure
Light pulses or radio waves	Byteflight	100 Mbd	Security, information, entertainment	Synchronous	Ring, tree, star structure
Wireless	Bluetooth™	1 Mbd	Communication	Asynchronous	

- Real-time compatibility
- Synchronous or asynchronous data transmission
- Expense and costs

Real-time compatibility.

This is the ability of an electronic system to calculate processes in real time or to transmit them as they occur in the real world. This means, for example, that a very quick acting system is required to control combustion in the engine, as the processes there take place within a very short time. On the other hand, a slower computer system is sufficient for controlling the electric power windows.

Synchronous, time-controlled data transmission (Fig. 1a)

This is used to transmit the data in fixed time intervals. It is also called time-controlled data transmission.

Examples A, B and C are messages that are sent in a defined order. One message each is transmitted in certain time intervals, e.g. oil temperature, engine speed. As the engine speed is subject to more frequent time fluctuations than the oil temperature, it is transmitted to the data bus in shorter time intervals.

Asynchronous, event-driven data transmission (Fig. 1b)

This is used to transmit the message to the data bus after events occur, if the data bus is free. If several control units try to transmit to the data bus at the same time, the most important message is transmitted first.

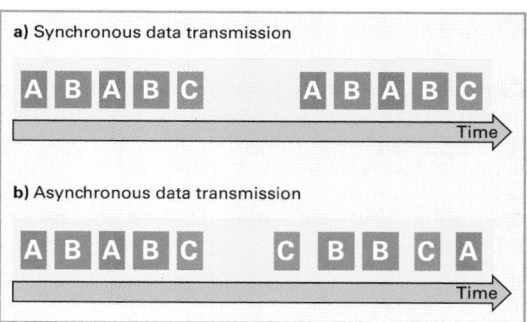

a) Synchronous data transmission

b) Asynchronous data transmission

Fig. 1: Synchronous and asynchronous data transmission

Serial data transmission

In the bus systems currently in use, data transmission is serial. This means that the bits are transmitted one after the other on the transmission agent (electric line, fibre-optic cable, radio waves).

The information can be transmitted as follows:
- In electrical data bus systems by a voltage change on the transmission line.
- In optical data bus systems by changing the light waves, light wave modulation.
- In wireless data bus systems by changing the radio waves, pulse/frequency modulation.

So that the control units can communicate with each other, each bit is assigned a fixed time value (pulsing). This time is the bit time t_{bit}.

19.2.12.1 Electrical data bus systems

Design (Fig. 1). Electrical data bus systems consist of at least two nodes, bus stations (e.g. ABS control unit, engine control unit) with separate power supply and one or two bus lines on which the data is transmitted.

Nodes, bus stations
These consist of
- the evaluation electronics of the sensor signals,
- the control electronics for the actuators,
- the microprocessor for calculating the functions,
- the controller for controlling the data bus communication, and
- the transceiver for transmitting the data to the bus line.

Transceiver. This is an electronic component in the node and performs two functions. It transmits and receives the data on the bus line. It receives the data to be sent from the controller. It forwards the received data to the controller.

Controller. This prepares the data calculated by the microprocessor so that the transceiver can transmit it on the data line. If another node transmits data on the data bus line, the transceiver forwards all data to the controller. It filters the data required for the control unit and forwards it to the microprocessor.

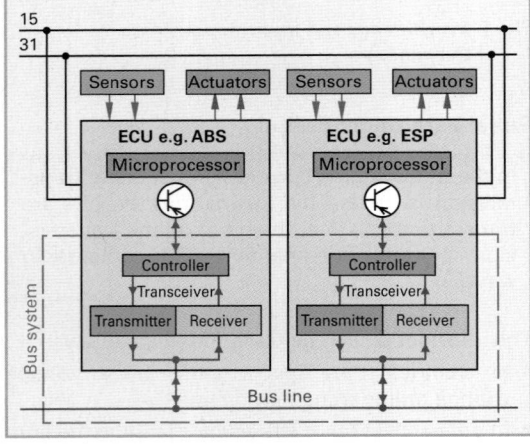

Fig. 1: Design of an electrical data bus system

All data transmitted on the bus line is always received by every node. In the controller, each node decides on the use of the individual data.

Multiplex process (Fig. 2)

Multiplexers are electronic circuits that record different input signals and transmit them via one cable either through time phasing or at different frequencies.

After transmission, the signals are separated by a further electronic circuit, the demultiplexer, and delivered to their receivers. Several input signals can be transmitted to the control unit using one line. The form (protocol) of the digital information is not uniformly defined by the system manufacturers. Different signal levels and protocols are used.

Applications. Multiplex systems are used in the hydroactive suspension system, for example.

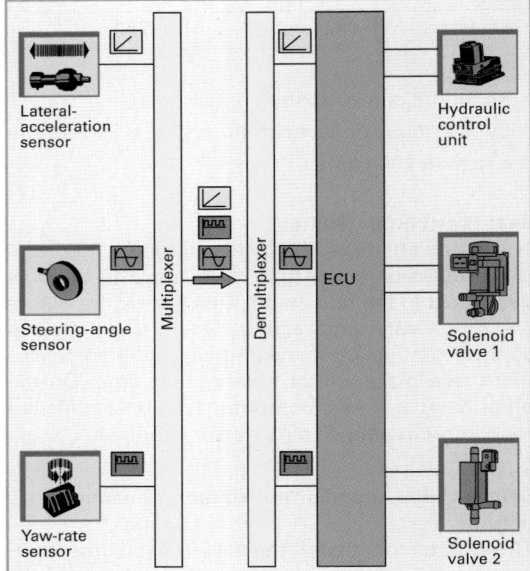

Fig. 2: Principle of the multiplex method

Local Interconnect Network (LIN)

This is predominantly used for data transport between the control unit and active sensors and actuators. It works according to the Master-Slave principle. The signal shape and protocol are standardised.

Active sensors, actuators. These are components with a separate power supply and evaluation or control electronics.

Features of the LIN bus system:

- Maximum data transmission speed 19.2 kbd
- Data transmission via one line
- Address-based data transmission

Address-based data transmission. With this, the recipient's ID is included in the message.

Data definition. Unlike in the multiplex system, the uniform definition of the data transmission enables the use of LIN component by many system and vehicle manufacturers, as well as the use of standardised diagnostics tools, software and simplification of malfunction diagnosis.

Design (Fig. 1)

LIN data bus systems connect a master control unit with up to 16 slave control units (Master-Slave principle).

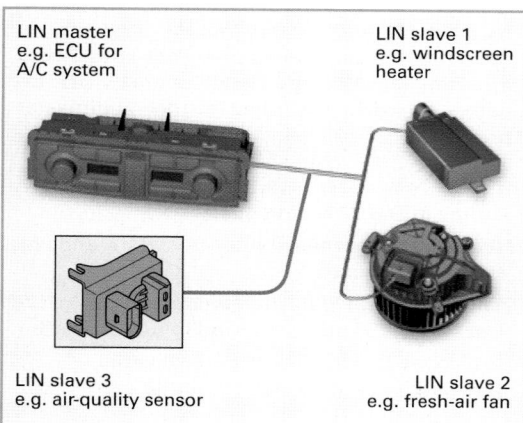

LIN master
e.g. ECU for
A/C system

LIN slave 1
e.g. windscreen
heater

LIN slave 3
e.g. air-quality sensor

LIN slave 2
e.g. fresh-air fan

Fig. 1: Design of an LIN data bus system

Master control unit. This transmits a header to the data bus line and is the interface to other data bus systems. It synchronises the remaining bus stations (slaves) via the bit time, and is the diagnosis interface between the diagnosis tester and the slave control units.

Header (Fig. 2). This contains data such as

- **Start bits.** These signal the start of a new message to all slave control units.
- **Synchronisation.** This is a bit sequence for configuring the bit time so that the message can be read.
- **Message identifier.** This contains the recipient address and an initial indication of the master to the slave, e.g. "Transmit actual speed" or "Configure target speed".

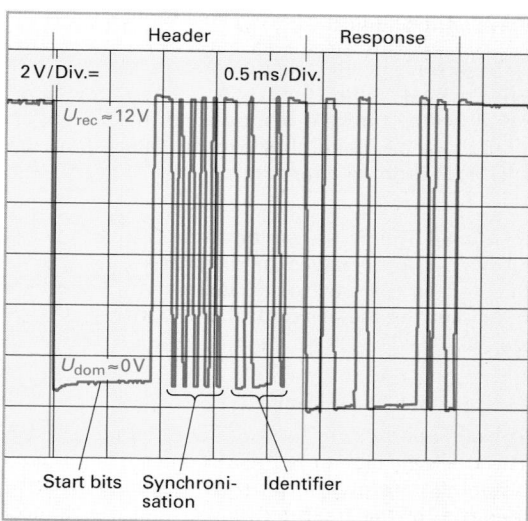

Header   Response

2 V/Div.=   0.5 ms/Div.

$U_{rec} \approx 12V$

$U_{dom} \approx 0V$

Start bits   Synchroni-   Identifier
sation

Fig. 2: Oscillogram of the voltage level on the LIN bus line

Voltage level in the LIN bus (Fig. 2, Fig. 3)

To transmit the data, the control units connect the recessive, logical bit value = 0, or the dominant level, logical bit value = 1 to the LIN bus line.

Recessive level, U_{rec} (Fig. 3a). This occurs on the data bus lines to which no bit is transmitted. The transistor T1 in the transceiver is not conductive. As there is no current, there is no voltage drop at resistor R. There is battery voltage on the bus line (approximately 12 V).

Dominant level. U_{dom} (Fig. 3b). This occurs when a node switches transistor T to conductive and thus transmits to the bus. The transistor connects the bus line to earth. The voltage on the bus line is 0 V. The voltage levels can be displayed and checked using an oscilloscope.

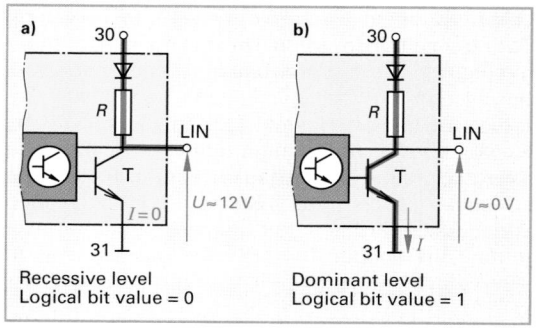

a)   30

R   LIN

T

$I = 0$ $U \approx 12V$

31

Recessive level
Logical bit value = 0

b)   30

R   LIN

T

$U \approx 0V$

31   I

Dominant level
Logical bit value = 1

Fig. 3: Operating principle of the LIN transceiver

Slave control units

These execute the instructions transmitted by the master control unit. If the master sends a query, they return a response. In the case of a function command, they execute the command without sending a response.

Response. This is transmitted from the addressed slave if current values or diagnosis data are requested. In the case of an instruction to execute functions, the master control unit transmits the response.

Example of LIN communication

The air conditioning control unit is the LIN master. The fresh-air blower is the LIN slave. The air conditioning control unit transmits a header with the identifier of the fresh-air blower and the request "Send current speed". The fresh-air blower transmits the response immediately after the end of the header with the information "Current speed = 200 rpm". If the fresh-air blower speed changes, the air conditioning control unit transmits a header with the identifier of the fresh-air blower and the instruction "Configure set speed". Now, however, the air conditioning control unit sends the response, with the information "Set speed = 500 rpm". The fresh-air blower increases the speed accordingly to 500 rpm. In order to close the control loop, the air conditioning unit makes a further request for the fresh-air blower speed, as previously described.

Sleep and wake up mode

If no communication is required on the LIN data bus, the LIN master stops sending the header. The LIN data bus goes to sleep. If communication is required again, a node (Master or Slave) wakes the LIN data bus up.

Sleep mode. This is necessary because the current consumption of electronic systems is very high during data transmission. This means that, in order to ensure that the starter battery can start, the data bus system must be deactivated when the alternator is switched off. However, the functions, such as electric boot opening, must remain available long after the ignition is switched off. For this reason, the LIN master continuously checks if communication is necessary. If this is not the case, it stops transmitting the header. The LIN data bus switches to sleep mode.

If a node, master or slave requires communication, it briefly connects its transistor in the transceiver.

The voltage on the LIN bus line is 0 V for a short time. The master then begins transmitting the header. The LIN data bus is woken up. The process is called wake up mode.

WORKSHOP NOTES

If data bus systems are active, they increase the closed-circuit current. In the event of increased discharge of the starter battery, the sleep and wake up mode function must be checked first.

Controller Area Network (CAN)

The CAN data bus is predominantly used for data transmission between the control units of systems for vehicle safety, comfort and convenience, and for controlling information, communication and entertainment systems. It transmits the data using two data lines which are either shielded by a sheath or are twisted together. It works according to the multimaster principle.

Features of the CAN data bus

- A distinction is made between class B and class C CAN data buses.
- The maximum data transmission rate in CAN class B is 125 kbd, and 1 Mbd in CAN class C.
- CAN class C is not single-wire suitable.
- CAN class B is single-wire suitable.

Single-wire suitability. This occurs if the data bus system can still correctly receive the communication when one bus line fails, e.g. open or short circuit. If the bus system is in single-wire mode, however, interference resistance is no longer available. There may be occasional malfunctions.

Design (Page 587, Fig. 1). CAN data bus systems consist of at least two nodes, the CAN-low line, the CAN-high line and at least two terminal resistors.

Nodes. The inner layout of the nodes is the same as the layout of the LIN bus nodes. In order to be able to transmit the higher data transmission rates and the voltage level, which is different to that of the LIN data bus, on the bus lines, higher quality controllers and transceivers are installed.

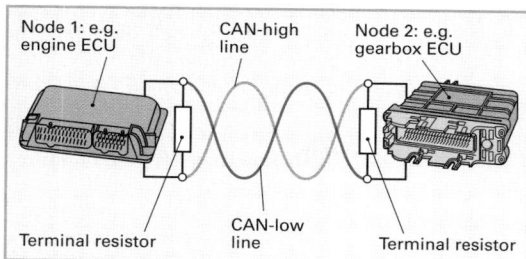

Fig. 1: Design of the CAN data bus

Bus lines, CAN high, CAN low (Table 1).

If a dominant level (U_{dom}) is created by the transceiver of a node on the CAN-high line, the voltage on this line increases. At the same time, the voltage on the CAN-low line is reduced. The two lines are either twisted together or shielded by a wire mesh.

As a consequence of the conflicting voltage change, the magnetic fields of both CAN bus lines occurring with each switchover cancel each other out. The lines are electromagnetically neutral to the outside and do not cause any interference; interference resistance is guaranteed.

Table 1: Voltage level on the bus lines

Logical values	LIN	CAN class B		CAN class C	
0	approx. 12 V	low	1 V	low	1.5 V
		high	4 V	high	3.5 V
1	0 V	low	5 V	low	2.5 V
		high	0 V	high	2.5 V
red = dominant level / blue = recessive level					

Terminal resistors. These close the circuit between the CAN-high line and the CAN-low line. This prevents reflections in the CAN bus lines. They are predominantly fitted in the nodes.

WORKSHOP NOTES

CAN bus lines without terminal resistors can cause function failures in particular in CAN class C systems. Therefore, the terminal resistors must be checked if faults occur.

Operating principle
Multimaster principle. With this, each node can transmit a message on the data bus line if it is free. If several control units want to transmit a message

at the same time, the most important message is sent first by means of arbitration.

Arbitration. This controls access to the data bus line when several nodes attempt to transmit a message at the same time. The importance (priority) of a message is defined by the identifier. The lower the identifier, the higher the priority.

Design of the data protocol (Fig. 2)

The data protocol determines the structure of the data message and is standardised.

In the CAN data bus, the length of a message can be up to 128 bits. These are split into consecutive fields.

Start field. This identifies the start of a message and informs all nodes of the start of message transmission.

Status field. This consists of the identifier. Based on the identifier, the nodes recognise the content of the message. In addition, arbitration is carried out using the identifier.

Control, safety and acknowledgement field. These are used to guarantee data transmission. The transmitter of the message uses the acknowledgement field to detect if the message has been correctly read by the receivers. If this is not the case, it repeats the message. If receipt is not acknowledged after several attempts, the transmitter stops sending the message. This prevents failure of the entire bus system if there is a faulty node.

Data field. This contains the useful data of the message, e.g. engine speed, coolant temperature.

End field. This marks the end of the message and releases the bus for the next message.

19

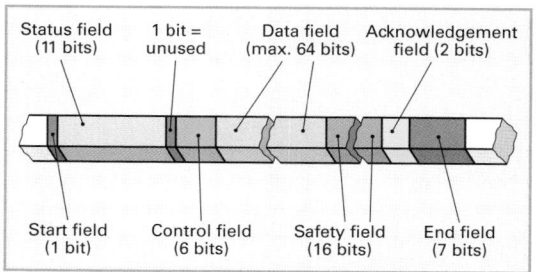

Fig. 2: Design of the CAN message

WORKSHOP NOTES

Oscillograms of CAN data bus signals

These can be used to identify failure causes on CAN data lines.

Oscillogram CAN class B

Fault-free signal (Fig. 1)

The following points must be noted during the check:

- The voltage levels on both lines must be compared. CAN high goes from 0.2 - 3.8 V; CAN low goes from 5.0 - 1.0 V.
- The chronological courses of the voltage changes must be synchronous.

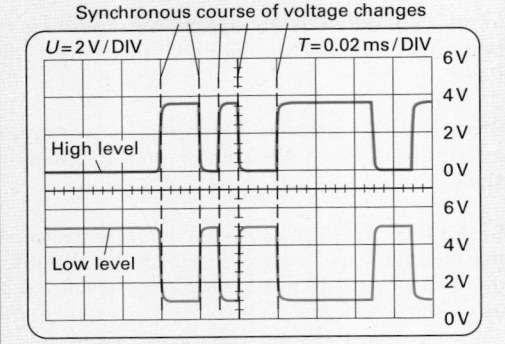

Fig. 1: Oscillogram of fault-free signal CAN class B

CAN class B systems transmit the information even if one of the two data bus lines fails. This status is called single-wire mode. Causes for single-wire mode can be:

- Open circuit in a CAN line to one or more nodes (control units). This cannot be unambiguously shown with the oscilloscope. In the measurement data blocks, the node(s) working in single-

wire mode are shown through self-diagnosis. In order to determine the cause, the CAN lines must be checked using a resistance measuring instrument.

- Short circuit of a CAN line to earth **(Fig. 2)**. With this measurement, instead of the signal curve on the CAN-low line, for example, a continuous line at 0 volts is visible in the oscillogram. The high signal is retained.

- Short circuit to battery positive terminal of a CAN-high line **(Fig. 3)**. In this oscillogram, the incorrect voltage level, which is the same as the battery voltage, approximately 12 to 14 volts, can be identified on the CAN-high line. The low signal is retained.

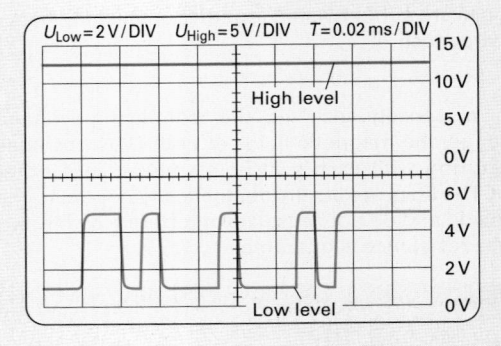

Fig. 3: Example short circuit CAN-high line to battery positive terminal

- Short circuit between the CAN lines **(Fig. 4)**. The CAN-low and CAN-high voltage curves are the same. The high signal remains, the low signal is inverted. This high signal is transmitted by both lines.

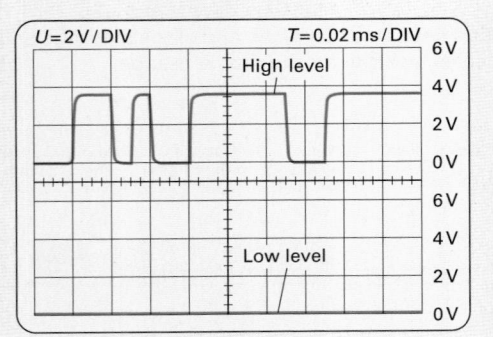

Fig. 2: Example short circuit CAN-low line to earth

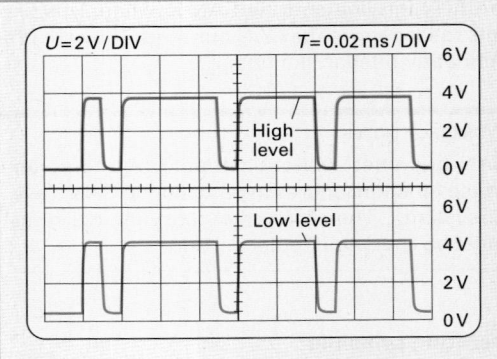

Fig. 4: Short circuit CAN-high line to CAN-low line

19.2.12.2 Optical data bus systems (D²B, MOST)

They are used in information, communication and entertainment systems (e.g. radio, navigation). Using light waves, they enable the transmission of large volumes of data, which is required for the transport and exchange of moving images (video) and sounds.

Features of optical data bus systems

- High data transmission rates, Digital Domestic Bus (D²B) 5.6 Mbd, Media Oriented Systems Transport (MOST) 21.2 Mbd.
- Ring structure or active star structure.
- Transmission of light waves using plastic optical fibres (POF).
- High interference resistance.

For digital transmission of a stereo sound signal, a data transmission rate of 1.54 Mbd is required, while the transmission of an MPEG video requires 4.4 Mbd. As the transmission rates of electrical data bus systems are a maximum of 1 Mbaud, optical data bus systems are increasingly being used in motor vehicles.

Optical data bus systems also offer the option of synchronous data transmission, which is necessary for music and video data.

Interference resistance. By transmitting data via light waves, the optical data bus systems do not transmit any electromagnetic interference waves. Optical data bus systems are also insensitive to electromagnetic interference waves.

Design of optical data bus systems (Fig. 1)

The nodes of optical data bus systems are predominantly arranged in a ring structure.

The light waves are sent from one node to the next. In the process, the information content of the light is checked and evaluated by every node. It is possible for the message to be filled with additional information by nodes and then forwarded to the next control unit as a new light signal. The interference resistance is increased, because the processing reduces the attenuation of the light waves in the entire ring.

It is possible to extend the system by inserting a further node in the ring. The node must be tuned to the

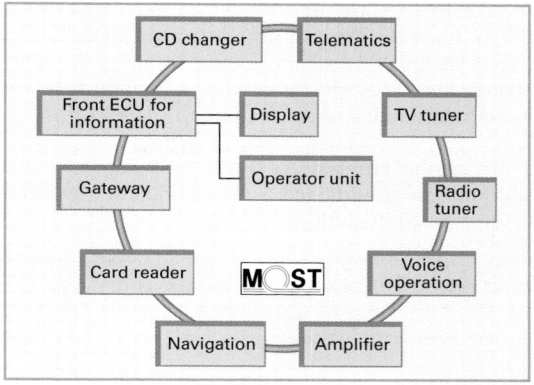

Fig. 1: Ring structure MOST bus

system configuration. If the data transmission between two nodes fails, the entire bus communication is disrupted.

Design of a node in the optical data bus (Fig. 2)

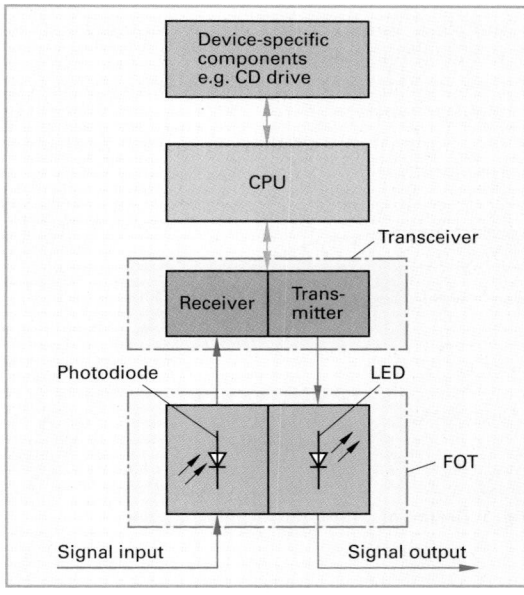

Fig. 2: Design of the control unit in the optical data bus

Device-specific components (e.g. CD player, radio module). These are activated by the microcontroller and execute the functions assigned to the node.

Microcontroller (CPU). The central processing unit (CPU) is the central computing unit. It controls all significant functions of the control unit.

19

Transceiver component. This receives the messages from the fibre optical transceiver (FOT) and forwards the required message content to the CPU. It fills the messages with the information to be transmitted and transmits them to the FOT.

Fibre optical transceiver (FOT). This consists of a light emitting diode and a photodiode. It transmits and receives the light wave signals.

The light emitting diode in the FOT transmits the light waves. The photodiode in the FOT converts the received light waves into electrical signals. They are processed in the node.

Fibre-optic cable, plastic optical fibre (FOC, POF)

> This has the function of transferring the light waves from the transmitter to the receiver with the least possible losses.

Design (Fig. 1)

The fibre-optic cable consists of
- the outer sheath for colour identification and to protect against damage,
- the black, inner sheath for protection against external light irradiation,
- the core of transparent plastic for transmission of the light waves
- the transparent coating to assist light-wave transport.

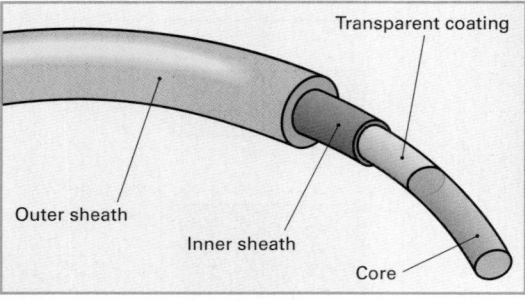

Fig. 1: Design of a fibre-optic cable

Operating principle of light-wave transmission (Fig. 2)

Total reflection. The operating principle of the fibre-optic cable is based on the physical principle of total reflection.

If a light beam hits a boundary layer between an optically dense and an optically thinner material at a flat angle, the beam is reflected with nearly no losses (total reflection).

The core in the fibre-optic cable is the optically denser material and the transparent coating is the optically thinner material. Thus, total reflection takes place at the boundary between the core and the coating inside the core. The light waves are transmitted inside the core.

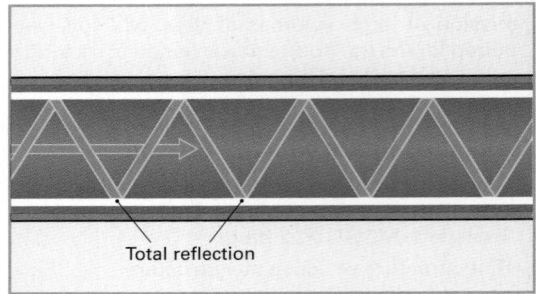

Fig. 2: Transport of the light waves by means of total reflection

Causes of increased attenuation of the fibre-optic cable

Total reflection depends on the angle of the light waves hitting the boundary from the inside. If this angle is too acute, the light waves exit the core. This results in higher losses and thus increased attenuation of the light waves. This happens when the fibre-optic cable is bent or folded too sharply **(Fig. 3)**.

The fibre-optic cable is laid in a corrugated pipe to avoid sharp bending radii.

Fig. 3: Escaping of light waves if bending radius too narrow

Other causes of increased attenuation are **(Fig. 1, Page 591)**:
- Kinked fibre-optic cable
- Damage to the sheath
- Scratched end face of the fibre-optic cable
- Dirt on the end face
- Dislocated coupling of the fibre-optic cable in the plug connection
- Skewed coupling
- Air gap between the fibre-optic cable and the plastic optical transceiver (POT)
- Incorrectly fitted crimp sleeve

19

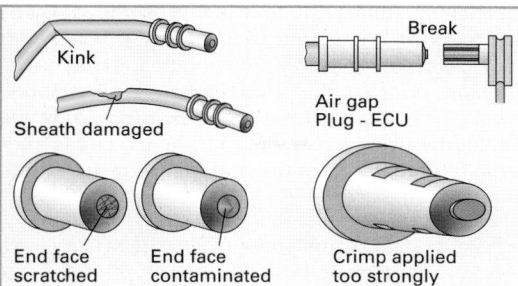

Fig. 1: Causes of increased attenuation of the fibre-optic cable

Self-diagnosis of the data bus nodes

Transmission of diagnosis data
In networked vehicles, this is not carried out using a diagnostic cable which is connected to every control unit, but via the corresponding data bus. The diagnosis tester is connected to a node in the vehicle network by the diagnosis connector, the gateway.

Gateway
This has the task of establishing the connection between the bus systems in order to enable the exchange of data between the different bus systems. It also establishes the connection between the diagnosis tester and the nodes.

Fault memory entries
Entries are saved in the fault memory if the node does not receive the messages from one or more nodes within a fixed period of time.

Measurement data blocks
These enable the technician to check the current status of the data bus communication. The following operating statuses can be displayed, for example:

- The entire data bus is active or not active.
- Communication to a node is active or not active.
- The entire bus or individual nodes are ready to switch to sleep mode.
- The entire bus is in single-wire mode.
- An individual node is in single-wire mode.

The measurement data blocks can also be used to display the transmitted useful data.

Open ring diagnosis in optical data bus systems
This has the task of supporting malfunction diagnosis in the search for the cause of an open ring.

Self-diagnosis of the nodes in the optical bus system cannot be supported, as the transmission of the diagnosis data from the nodes to the diagnosis tester also fails if there is an open ring.

Open ring diagnostics cable
This establishes an electrical connection between the gateway and the nodes.

The open ring diagnosis is activated by the technician with the help of the diagnosis tester.

The gateway sends a start signal to the nodes via the open ring diagnostics cable. Following this, all nodes switch on their light emitting diodes in their fibre optical transceiver.

In addition, they check if the light signal of the predecessor arrives at the photodiode via the fibre-optic cable. If this is the case, each node sends the information "optically ok" to the open ring diagnosis cable.

The diagnosis tester then shows the result of the test in the form of a list of nodes.

Using this list, the technician can tell between which nodes the light wave connection is broken.

1 What are the advantages of data bus transmission compared with conventional data transmission?

2 How big are the voltages of the recessive level on the lines of a CAN class B data bus system?

3 What are the causes for increased attenuation of the fibre-optic cable?

4 What structure is predominantly used in the design of optical data bus systems?

5 Where are LIN data bus systems predominantly used?

6 What does "sleep mode" mean?

7 What are the functions of the gateway?

8 What are the features of optical data bus systems?

19

19.2.13 Measuring, testing, diagnosis

Motor vehicles consist of subsystems, such as the engine, gearbox and suspension. The different fields of engineering, such as mechanics, hydraulics, electrics and electronics, work together in these. If faults occur, suitable measurement and test procedures must be able to detect their causes. There are two types of fault:

- **Permanent faults.** These are faults that are present until they are rectified.
- **Sporadic faults.** These are faults that only occur under certain operating conditions, are not shown on the tester or are incorrectly shown, or cannot be identified from a unique symptom. They are difficult to identify.

Systematic malfunction diagnosis. In order to find the faults quickly and cost-effectively, systematic malfunction diagnosis must be carried out:

- **Vehicle data.** It must be possible to collect the vehicle data, such as the chassis number or first registration, from the vehicle certificate.
- **Customer information.** The fault must be narrowed down based on a fault description provided by the customer and by means of targeted questioning. For example, under what conditions and when does the fault occur.
- **Visual checks and noise-level testing.** Visual impression of fault-related features and components, such as leakage of liquids or grease, loose linkages, broken cables or connectors. Rattling noises must be localised.
- **Test drive.** This may have to be carried out with the customer.
- **Fault memory readout.** For systems with self-diagnosis, the fault memory must be read out.
- **Other malfunction diagnosis.** By dismantling parts or by means of electrical measurements, further malfunction diagnosis must be carried out if the self-diagnosis system does not show the faults or if the system is not capable of self-diagnosis.

Test procedure

In order to be able to test the function of the subsystems or their components, setting values must be tested and adjusted, if necessary, or components must be replaced. Conclusions for diagnosis and repair can be drawn from the measured values. To this end, overview plans of the location of components, function plans for understanding, manufacturer's specifications and instructions, circuit diagrams and suitable measuring and test equipment are required. Here, a distinction is made between mechanical or electrical measuring and test procedures.

Mechanical measuring and test procedures. These are used for faults that occur in the vehicle's mechanical systems, such as the engine, gearbox, suspension, body **(Table 1)**. The causes can be cracks, breaks or bending of parts, for example. They are caused by wear, excessive mechanical stress, material fatigue, faults in the material or overheating, for example.

Table 1: Checks and tests with mechanical test equipment

Measuring and test equipment	Use
Internal measuring instrument with dial gauge	Cylinder wear check
Gauge	Check valve clearance
Dial gauge	Check true running
Compression pressure tester	Compression pressure test
Pressure gauge	Check oil pressure

Electrical measuring and test procedures. The faults occur in the vehicle's electric and electronic systems, such as lighting system, comfort and convenience systems, engine or gearbox management. They are caused by cable breaks, failure of electrical components or corrosion, for example. With these faults, electric and electronic measuring and test instruments are used, such as

- diode lamp
- multimeter (multipurpose meter)
- oscilloscope
- diagnosis tester (vehicle system tester)

Diode lamp.

It enables a quick check of the voltage (from 4 volts) and polarity of lines or plug-in connections.

Due to the low current draw, damage to electronic components is avoided.

Multimeter (Fig. 1 and Fig. 2, Page 486).

This is usually used to measure voltages U, currents I and resistances R on the motor vehicle.

Voltage measurement. One of the most frequent causes (up to 60%) for the failure of electric and electronic systems is faulty plug-in connections. By measuring the voltage, it is possible to ascertain if a plug-in connection is corroded, for example. The voltage drop at the plug-in connection is measured. If it is 0 volts, for example, the plug-in connection is ok. If the voltage drop is greater than 0 volts, the plug-in connection is corroded and must be replaced.

Example: Corroded plug-in connections in the circuit of a windscreen wiper motor. The performance drop at the plug-in connection of 11.5 W, for example, is converted into heat and can cause cable fires. In addition, the operation of the windscreen wiper motor is significantly impaired by the performance drop (**Fig. 1**). Due to the increased contact resistance, the current is low.

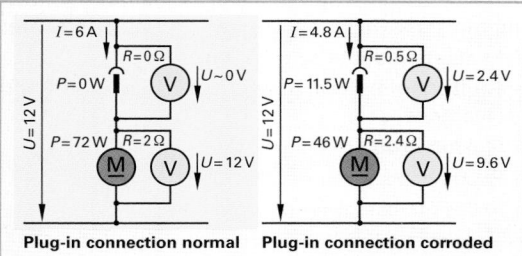

Fig. 1: Voltage measurement at plug-in connections

Resistance measurement. Often, incorrect electrical resistances are the cause of faults. The resistance of electrical components, such as ignition coil, induction-type pulse-generator, fuel injector, relay, is measured. If the measured resistance value is significantly higher than the value specified by the manufacturer, there is an open circuit. If it is lower, there is a coil-winding short circuit.

Example: The resistance values of an ignition coil are measured between terminal 1 and terminal 15 for the primary winding, and between terminal 1 and terminal 4 for the secondary winding (**Fig. 2**).

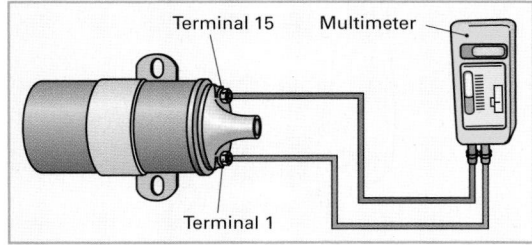

Fig. 2: Resistance measurement at an ignition coil

Current measurement. This is carried out using a current measuring instrument connected in series or usually by using a current clamp on the cable which does not have a line break.

Example. To check the alternator, the battery charging current can be measured with a current clamp on line B + (**Fig. 3**). When applying the current clamp, the arrow on the current clamp must point in the current direction.

Fig. 3: Measuring the charging current with a current clamp

Oscilloscope

This is used for graphical illustration of voltages, signals and frequencies on a screen.

Measuring signals. When testing the ignition system of a four-stroke petrol engine, the signal of a Hall-effect sensor can be displayed on the screen of the oscilloscope, for example.

Measuring process. Using a suitable measuring circuit (**Fig. 4**) the signal is displayed on the screen of the oscilloscope.

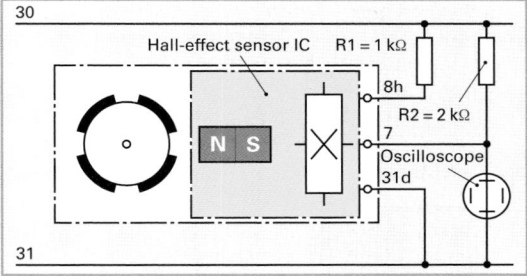

Fig. 4: Circuit for recording a Hall signal

Measuring result (Fig. 5). Information on the opening time, dwell period, ignition interval and voltage level can be read from the screen of the oscilloscope. The opening time is 0.8 ms, the dwell period is 1.8 ms, the ignition interval is 2.6 ms and the voltage on the Hall-IC is 12 V.

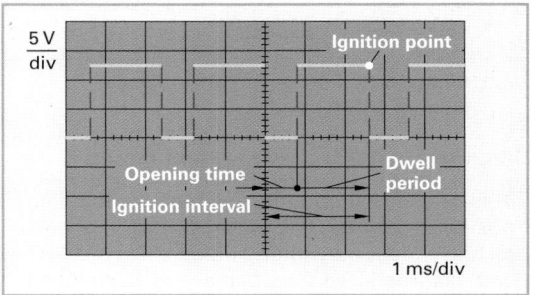

Fig. 5: Hall signal on the oscilloscope

Pinbox (test box). This enables measuring instruments to access signal and voltage cables without damaging them.

Adapter leads. Depending on the type, these connect the pinbox with:
a) the control unit and the ECU plug (wiring harness of sensors and actuators) (Y cable),
b) only the ECU plug (I cable),
c) only the control unit (Y cable with the ECU plug connection not plugged in).

Version a (Fig. 1a): This enables voltage measurements and signal recordings of the sensors and actuators during operation.

Version b (Fig. 1b): With this, the power supply of the control unit and the resistances of the sensors and actuators can be measured.

Version c (Fig. 1c): This represents a special application case. With this, for example, the terminal resistance of the CAN bus can be measured.

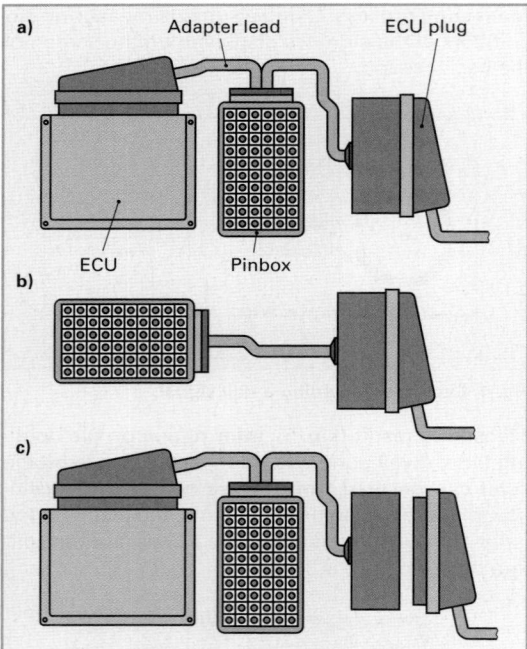

Fig. 1: Connection of the pinbox with different types of adapter lead

Diagnosis tester (vehicle system tester)

This provides support during malfunction diagnosis and provides information on components, technical data and setting values, electrical circuit diagrams, service plans and system descriptions.

This consists of (**Fig. 2**):
- Diagnosis computer with monitor
- Printer
- Measuring instruments
- Adapter

The vehicle system tester is a computer with a monitor, DVD-ROM drive, floppy disk drive, possibly an infrared interface for data transmission to peripheral devices such as a printer and connection to the workshop IT network. It contains all the measuring technologies, such as multimeter, digital memory oscilloscope and can execute the following functions in connection with suitable software:
- Evaluation of the self-diagnosis
- Fault memory readout (fault code readout) and clearance
- Guided problem diagnosis
- Reset inspection intervals
- Access to customer data
- Target/actual comparison
- Actuator diagnosis
- Coding of control units
- Vehicle identification
- Learning programs
- Access to manufacturer's database
- Measured-value block readout

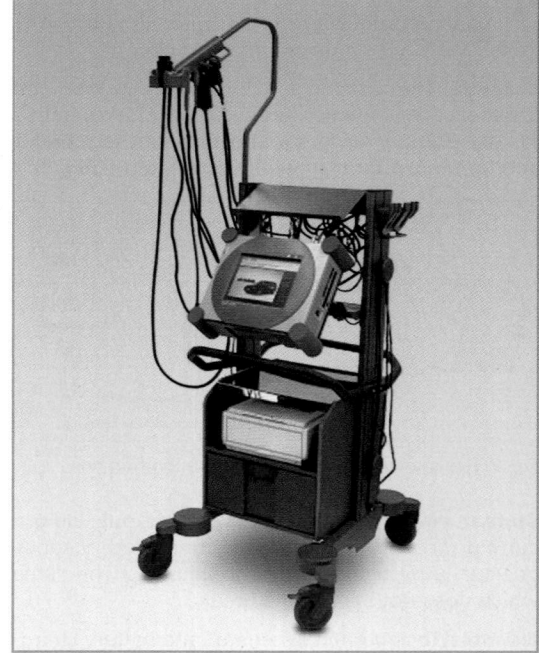

Fig. 2: Vehicle system tester

Diagnosis. For malfunction diagnosis, the vehicle must be connected to the tester and identified by a systematic query for the vehicle type, model year and engine code letter, for example **(Fig. 1)**. This ensures unique assignment of all documents and test specifications.

Fault memory readout. The fault memory readout provides information on faults that have occurred in the relevant system, such as engine, body, suspension, convenience and comfort. The fault saved in the fault memory can be read out in the form of a code or in plain text. During querying of the individual control units, all faults found are displayed one after the other: e.g. the engine-temperature sensor or the speed sensor do not transmit any signals, or transmit implausible signals. The faults entered must be checked and eliminated. The user can be assisted by a guided problem diagnosis by the tester. Then the fault memory must be cleared. After operation of the engine, the fault memory must be read out again. There must be no new entries.

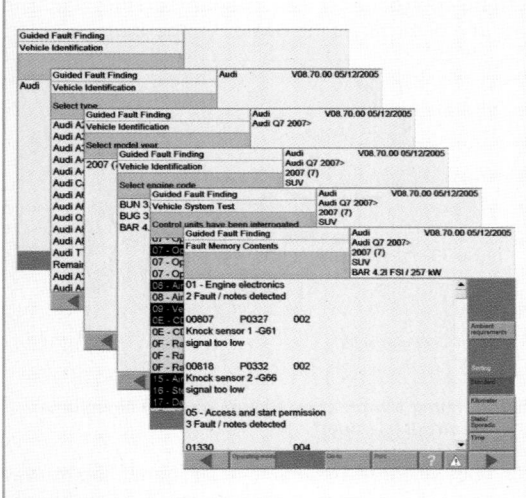

Fig. 1: Fault memory readout

Guided problem diagnosis. This makes it possible to find a fault following a procedure specified by the tester. Starting with a fault memory entry, e.g. intake air-temperature sensor, the tester selects a test program. The user is guided in the dialogue until the fault is found. All test requirements and testing steps required until the fault is eliminated and all measuring and test equipment and activities required for the individual tests are specified by the tester and displayed on the monitor. E.g.:

- Testing the line for an open circuit
- Measuring the voltage
- Measuring line + to test box bushing 35
- Measuring line - to test box bushing 19
- Reference value must be between 10 and 15 V.

If the fault can be traced back to a plausible but incorrect sensor value (e.g. as a consequence of increased contact resistance), no fault will be displayed in the fault memory. The user can now select a formulated problem using a menu, e.g. engine juddering between 1,500 and 3,000 rpm. The tester executes the appropriate test program according to the selection.

Reset inspection intervals. If the workshop carries out an inspection on the vehicle, it resets the inspection interval display in the instrument cluster.

Actuator test. The tester can check the operation of the actuators by means of defined activation, e.g. fuel injectors, idle speed actuator, ABS solenoid valves, tank ventilation valve, door locks, servo-motors in seats.

Target/actual comparison. This is used to measure the actual value and compare it with the prescribed reference values. Examples are the power supply, air flow rate/air-mass signal, lambda sensor voltage. If the actual value deviates from the prescribed reference value, there is a fault.

Measured-value blocks readout. In connection with diagnosis-compatible control units, current data is transmitted to the tester. The signals of several components can be monitored and recorded at the same time **(Fig. 2)**.

Example: Checking the overrun fuel cut-off during vehicle operation

- Throttle-valve angle goes to lowest value
- Duration of injection must drop to zero
- Speed drops
- Sensor voltage < 100 mV
- Overrun fuel cut-off "active"

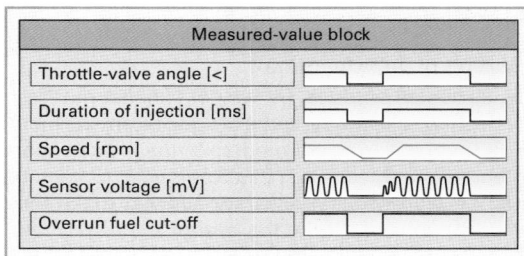

Fig. 2: Measured-value block

REVIEW QUESTIONS

1 **What measuring and testing options are available?**

2 **What measurements can be carried out with a multimeter?**

3 **Describe the procedure for a guided problem diagnosis.**

19

20 Comfort and convenience technology

20.1 Ventilation, heating, surrounding air, air conditioning

The temperature and composition of the surrounding air has a substantial influence on the performance and attention levels of vehicle occupants. It is therefore necessary to supply the passenger compartment with filtered fresh air, which has to be heated or cooled according to the outside temperature.

Ventilation equipment

It should be designed so that ...
- ... there is sufficient fresh (and heated) air available for all occupants.
- ... the spent air is purged through outlets.
- ... no dust and water can enter the vehicle's interior.
- ... the air is guided to prevent the windows from misting up.
- ... no cold air can gather anywhere.
- ... the exchange of air takes place with as little draft as possible.

The unaided flow of fresh air into the vehicle is only possible once road speed reaches about 60 km/h. At lower road speeds the fresh air supply is controlled by a blower. The inlets for incoming air should be positioned as high as possible on the vehicle, in the zone with the lowest possible level of exhaust gases and contaminants. Slightly higher air pressure within the vehicle is advantageous. Open windows create a vacuum under normal circumstances. The result is that a higher level of exhaust gases, dust and insects can make their way into the passenger compartment. Furthermore, vehicle and road noise become far more audible.

Heating of vehicle interior

On air-cooled engines. By way of exhaust gas/fresh air heating. In this way, some of the air from the blower is channelled off, warmed up by heat exchangers integrated in the exhaust pipes and used to heat the interior. No exhaust gases are allowed make their way with the hot air into the interior.

On water-cooled engines. The heat from the coolant is used for heating. A difference is made between the following three types of heater temperature change:
- Coolant volume control (coolant side)
- Fresh air volume control (air side)
- Electronically controlled heater

Heater temperature change by coolant volume control (Fig. 1). The volume of coolant that flows to the heat exchanger (heater matrix) can be altered using the heater valve. The flow volume determines the temperature of the heating air.

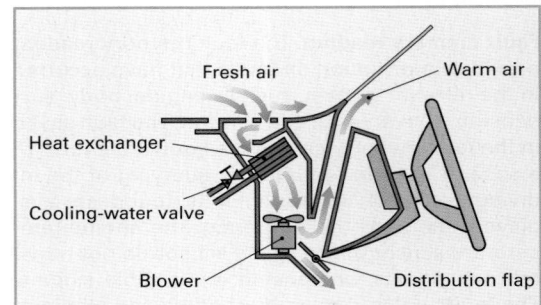

Fig. 1: Heater temperature change by coolant volume control (coolant side)

Heater temperature change by fresh air volume control (Fig. 2). The volume of fresh air that is heated up by the coolant flowing through the heat exchanger can be controlled by way of a temperature flap. The volume determines the heater temperature.

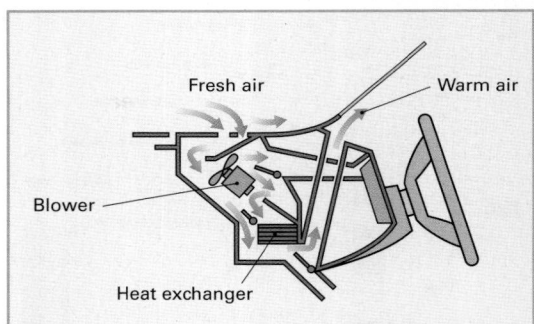

Fig. 2: Heater temperature change by fresh air volume control (air side)

On both systems, fresh air can be channelled by way of the fresh air flap positions via the heat exchanger and directed to the windscreen, the front side windows or the footwell as heating air. If there is not enough air flow to supply the air, a blower can be switched on. If the heat exchanger stays switched off, e.g. in summer, the fresh air is channelled directly to the interior or the windscreen.

Electronically controlled heater. The temperature in the interior of a motor vehicle can be adjusted by way of a rotary switch. With the aid of temperature sensors, the set temperature (actual value) is detected and compared with the stored temperature (specified value) in a control unit. If the two do not match, the system regulates the heating temperature appropriately. With coolant volume control, a coolant valve (solenoid valve) is electro-mechanically actuated. With fresh air volume control a fresh-air flap is electro-mechanically actuated.

Auxiliary heating systems

Auxiliary heater (Fig. 1). This is an additional heater that provides heat for the interior of the vehicle when the engine is switched off. Petrol, diesel, heating oil or gas is burnt in a fan burner. The heat generated is transferred to a stream of air flowing through a heat exchanger to furnish heat for warming the interior.

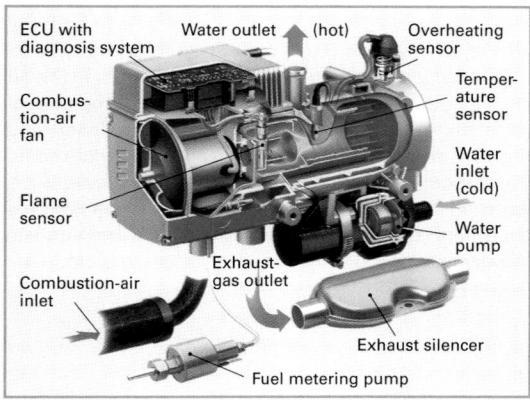

ECU with diagnosis system — Water outlet (hot) — Overheating sensor — Temperature sensor — Combustion-air fan — Water inlet (cold) — Flame sensor — Water pump — Exhaust-gas outlet — Combustion-air inlet — Exhaust silencer — Fuel metering pump

Fig. 1: Auxiliary heater

On engines with low fuel consumption, e.g. direct-injection diesel engines, the combustion heat transferred to the coolant is low. Sufficient heat to warm the vehicle's interior is not available under all operating conditions. To improve heater output, the following auxiliary heating systems can be used in the motor vehicle:

- Fuel-fired auxiliary heater
- Exhaust-gas heat exchanger
- PTC heater
- Electric auxiliary heater

Fuel-fired auxiliary heater. In this type of heater, fuel is burnt in a combustion chamber surrounded by flowing coolant from the engine. The heated coolant flows through the heat exchanger of the heater. The input air for the vehicle interior is heated up on its cooling fins. In addition the coolant can be preheated. The fuel-fired auxiliary heater can be housed in the coolant radiator of the vehicle.

Electric auxiliary heater. This type of heater can comprise, for example, six heater elements, similar to glow plugs, which are installed in the coolant circuit. During the warm-up phase, these heater elements heat up the coolant. Not only is operating temperature reached more quickly, but the interior of the vehicle can also be heated immediately.

PTC heater (Fig. 2). This type of heater is normally downstream from the heat exchanger in an air conditioning system. This unit converts DC electrical energy from the vehicle's on-board electrical system into heat.

Design and operating principle. The PTC heater consists of single ceramic semiconductor resistors, or PTC chips (PTC conductors). They are supplied with electricity via contact rails made from aluminium. The contact rails simultaneously transmit heat from the PTC chip to the corrugated fins of the PTC heater. When electric current flows through the PTC chips they heat up to approx. 120 °C. The contact rails and corrugated fins transfer the heat thus generated to the air flowing into the interior. Overheating of the PTC chips is prevented by an increase in their electrical resistance as temperature rises. This reduces the current flow through the chips.

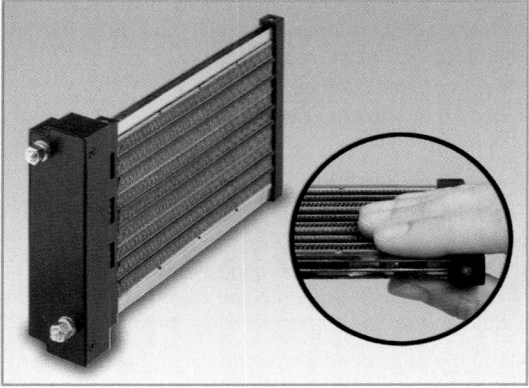

Fig. 2: PTC auxiliary heater

The engine control unit activates the PTC heater under the following conditions:

- Air conditioning system switched off
- Ambient temperature below 5 °C
- Coolant temperature less than 80 °C
- Engine running

Exhaust-gas heat exchanger (Fig. 3). This transfers exhaust heat to the coolant. In this way, part of the exhaust energy is recovered and can be utilised to heat the interior of the vehicle.

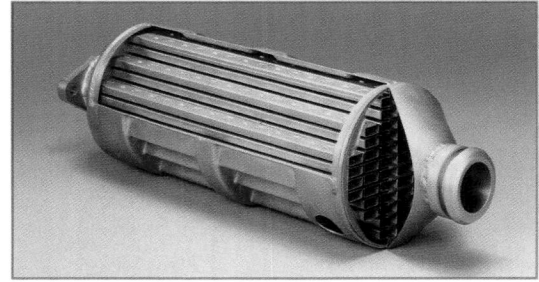

Fig. 3: Exhaust-gas heat exchanger

20

Air conditioning in motor vehicles

Certain demands are placed on the air conditioning system for the interior of a motor vehicle, e.g.:

- The passenger compartment must be heated or cooled to a comfortable temperature.
- A comfortable temperature must be maintained under all outside weather conditions.
- For every occupant, a comfortable flow of air and air temperature has to be generated.
- The air quality has to be improved.
- Simple operation must be possible.
- There must be no ill effects from air flow out of the vehicle.

In order for the above-mentioned demands to be met, the air conditioning system must fulfil the following roles: The air must be

- supplied and cleaned,
- heated or cooled and also
- dehumidified.

Types of air conditioning systems

The following types of air conditioning systems can be found:

- Manually controlled air conditioning systems
- Thermostatically regulated air conditioning systems
- Fully automatic air conditioning systems

Manually controlled air conditioning systems. Temperature, air distribution and blower speed are set by hand.

Thermostatically regulated air conditioning systems. Once the temperature is set, it is maintained at that level in the vehicle while air distribution and blower speed can be adjusted manually.

Fully automatic air conditioning systems (automatic climate control). The preselected temperature in the vehicle interior is maintained at a constant level automatically. It is continuously monitored by several temperature sensors, while the air distribution and blower speed are regulated fully automatically so there is an optimal temperature balance, e.g. head area 23 °C, chest area 24 °C, footwell 28 °C. On multizone climate control systems, the temperature can be adjusted separately for each seat **(Fig. 1)**.

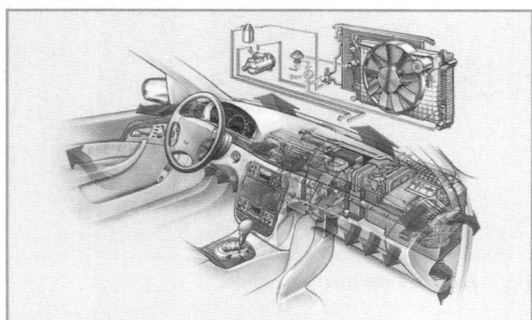

Fig. 1: Multizone climate control

Components of an air conditioning system

An air conditioning system comprises three areas:
- Air flow in motor vehicle with means of heating
- Refrigerant circuit   • Temperature control

Air flow in the motor vehicle. A difference can be made between two operating modes:
- Fresh air mode   • Recirculated-air mode

Fresh-air mode (Fig. 2). The blower draws in ambient air via the fresh air flap. From there the air makes its way to the dust filter, in which contaminants in the air such as dust, pollen, etc., are removed. At the evaporator the air is cooled, the water is condensed and thereby removed. The condensation is released to the atmosphere via drain hoses. The dry, cool air is warmed to the selected temperature in the heat exchanger. From there it is guided via flaps and vents to the desired points in the interior of the vehicle.

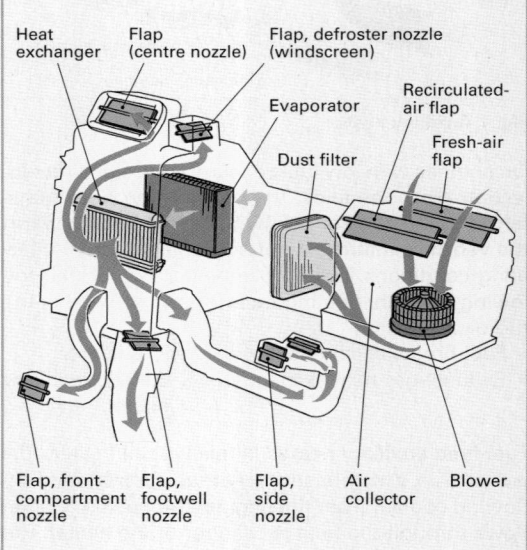

Fig. 2: Air flow path in fresh-air mode

Recirculated-air mode. Here the air is drawn out of the vehicle interior, cleaned in a dust filter, heated up at the evaporator and then fed back to the inside of the vehicle. The recirculated-air mode can be selected via a switch by the driver, e.g. in heavy traffic.

Air-quality sensor. This measures the concentration of harmful substances, e.g. unburnt hydrocarbons from the ambient air in the heater box. The sensor's resistance falls as the level of harmful substances rises. The rise in current in the sensor serves as an index of the concentration of harmful substances. A medium level of air quality is assumed in the interior. If the concentration of harmful substances in the fresh air is markedly higher than the assumed interior air quality, the automatic air conditioning system switches over to 100 % recirculated air.

From this point on, a continual decrease in the quality of interior air is assumed by the electronics of the automatic air conditioning control unit. If the interior air quality is worse than the exterior air quality measured, the automatic air conditioning system switches over to 100 % fresh air mode.

Refrigerant circuit

> Gaseous refrigerant is compressed, cooled down and liquified, restricted at the expansion element, and vaporises during heat absorption before being recompressed **(Fig. 1)**.

The refrigerant circuit comprises the following components:

- Compressor
- Condenser
- Expansion element (valve or restrictor)
- Evaporator
- Receiver/drier with desiccant insert
- Safety devices (pressure switch or high-pressure sensor and temperature sensor)
- Open and closed-loop control systems
- Hose and pipe assemblies
- Refrigerant

Compressor. Responsible for refrigerant circulation. To do this, the compressor draws in gaseous refrigerant and compresses it. The gaseous refrigerant, now heated from pressurisation, is fed to the condenser at a pressure of about 16 bar.

The piston stroke is generated by a swash plate that, depending on its design, actuates 3 to 10 pistons. If the angle of the swash plate changes, the length of stroke is altered and with it the delivery rate of the compressor (volume control).

The cooling output of the compressor is influenced by the delivery rate. It is determined by the following variables:

- Temperature selected by driver
- Exterior and interior temperature
- Evaporator temperature
- Refrigerant pressure, refrigerant temperature

The compressor operates only when the engine is running and the air conditioning system is switched on. Only gaseous refrigerant may be drawn in. If fluid refrigerant were to be drawn in, the compressor would be damaged beyond repair because fluid cannot be compressed. For lubrication of the compressors, oil is added to the refrigerant.

The piston-type or swash plate compressors used for climate control in motor vehicles can be categorised, depending on the type of control system, into internally controlled compressors and externally controlled compressors.

Internally controlled a/c compressor (Fig. 1, Page 600). Driven by a poly V-belt drive. Integrated in the belt pulley is a solenoid clutch, with which the compressor can be switched on or off as required. The position of the swash plate and thereby the delivery rate of the refrigerant is determined by the pressure conditions within the compressor.

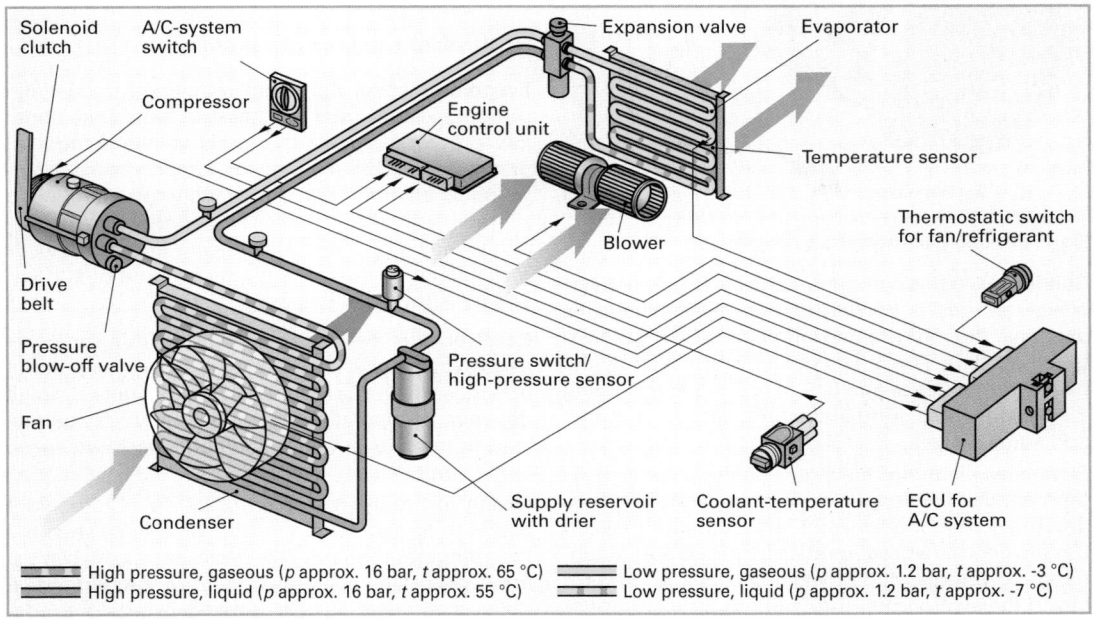

| High pressure, gaseous (p approx. 16 bar, t approx. 65 °C) | Low pressure, gaseous (p approx. 1.2 bar, t approx. -3 °C) |
| High pressure, liquid (p approx. 16 bar, t approx. 55 °C) | Low pressure, liquid (p approx. 1.2 bar, t approx. -7 °C) |

Fig. 1: Refrigerant circuit with expansion valve

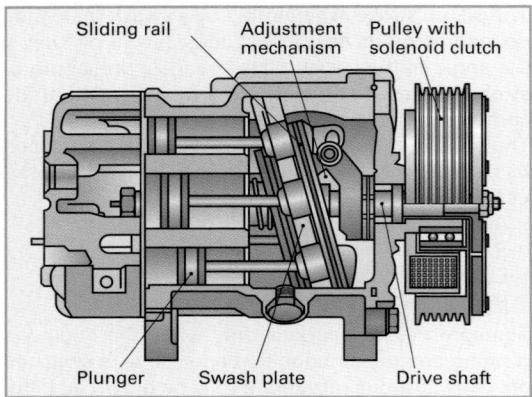

Fig. 1: Internally controlled a/c compressor

Externally controlled a/c compressor (Fig. 2). Integrated in the belt pulley of the compressor is an overload protection device. It does not feature a solenoid clutch and continues running even when the cooling function is switched off. The delivery volume of the refrigerant is restricted to under 2%. The pressure conditions inside the compressor are determined by a control valve, which is actuated by the air conditioning control unit. In this way, the position of the swash plate and thereby the delivery rate of the refrigerant can be influenced.

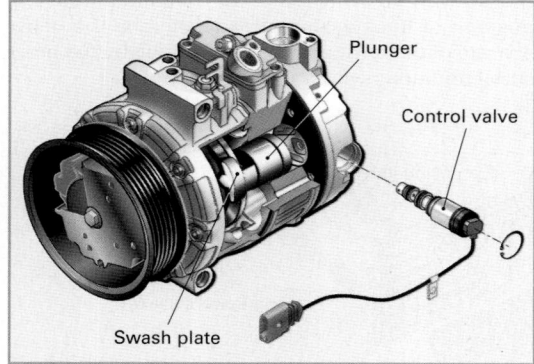

Fig. 2: Externally controlled a/c compressor

Condenser. In this component the 60 °C to 100 °C hot refrigerant gas is cooled down rapidly. When this happens, the refrigerant condenses from a gaseous to a fluid state. The rapid cooling is achieved by the heat from the pipes and fins of the condenser being dissipated by the wind and by the air flow from the additional blower.

Expansion element. This regulates the volume of refrigerant that is injected into the evaporator. The optimum volume of refrigerant is the amount that can be converted to a gaseous state in the evaporator under the given operating conditions. It varies according to the suction pressure and the temperature of the refrigerant downstream from the evaporator.

The following different types of expansion element can be found:
- Expansion valve
- Restrictor

Expansion valve. With this component the volume of refrigerant converted to a gaseous state within the evaporator can be precisely adapted to current operating conditions. In this way, the cooling output of the refrigerant circuit is optimal in every stage of operation. The accumulator is installed on the high-pressure side of the air-conditioning circuit **(Fig. 1, Page 599).**

Restrictor. This sprays the fluid refrigerant into the evaporator. The calibrated restrictor determines the flow rate of the refrigerant for an operating status predetermined by the manufacturer. Under these conditions the refrigerant circuit works optimally. The accumulator can be found in the low pressure part of the refrigerant circuit **(Fig. 3).**

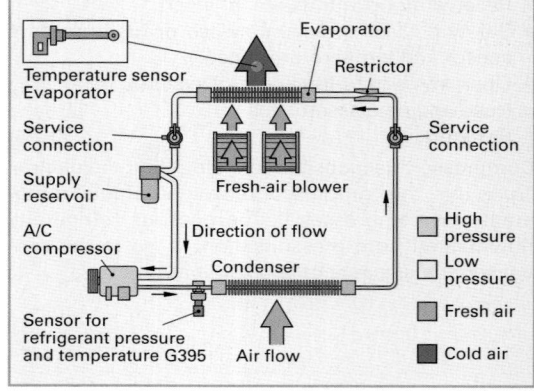

Fig. 3: Refrigerant circuit with restrictor

Evaporator. Here, the liquid refrigerant under high pressure is converted to a gaseous substance under low pressure. During this process, the refrigerant draws the heat required for evaporation from the surrounding environment. The quantity of heat required is drawn from the air, which is channelled by a blower depending on the mode of operation – fresh-air or recirculated-air mode – over the evaporator surface. The air is cooled in this way.

Receiver/drier with desiccant insert. This serves as an expansion tank and supply reservoir. The amount of refrigerant that is required by the refrigerant circuit depends on operating conditions, such as thermal load from the evaporator and condenser and speed of the compressor.

Owing to the hygroscopic properties of refrigerant and leaks in the system, water can find its way into the refrigerant circuit. The desiccant insert can remove water and contaminants from the refrigerant. Between 6 g and 12 g of water can be stored, depending on the design.

20

Safety devices. These consist of high-pressure senders and temperature sensors. The microprocessor from the high-pressure sender sends a pulsed signal to the air conditioning control unit. The pulse width is dependent on the pressure: Low pressure, low pulse width; high pressure, high pulse width. The control logic of the air conditioning system evaluates this information and switches the compressor on or off depending on the pressure in the refrigerant. For example, the compressor is switched off if the pressure in the refrigerant rises to approx. 30 bar in order to prevent permanent damage to the air conditioning system, or at pressures below 2 bar as the system assumes there is a leak in the lines. At excessive refrigerant temperatures (above 60 °C), the temperature sensor activates a supplementary fan at the condenser.

On systems with two pressure switches, the pressures are monitored by a high-pressure and a low-pressure switch. When the pressure reaches 40 bar a pressure dump valve on the compressor **(Fig. 1, Page 599)** discharges pressure for reasons of safety.

Hose and pipe assemblies. High-pressure lines have a small cross-section and heat up during operation of the air conditioning system.

Low-pressure lines have a large cross-section and become cold during operation of the air conditioning system.

Refrigerant. This circulates in an enclosed circuit in the lines of the air conditioning system and transports heat from the interior of the vehicle to the outside. In this process it changes continuously between fluid and gaseous state. In today's air conditioning systems, only the refrigerant R134a is used. Charging air conditioning systems with R12 refrigerant is forbidden (see also **Page 33**).

Open and closed-loop control systems. These are control elements in the vehicle interior with which the desired climatic conditions in the motor vehicle can be set.

Temperature control (Fig. 1, Fig. 2). This controls the temperature control loop for the interior of a motor vehicle and also influences the refrigerant circuit. The air conditioning control unit (X4) uses different temperature sensors, such as the evaporator temperature sensor (B2), air temperature sensor (B3) and interior sensor (B4), to monitor all important temperatures and interference factors. From the setpoint control device the control unit receives the temperature chosen by the occupants. The specified temperature is compared with the actual temperature. The difference determined generates reference variables for heater control (heat exchanger, solenoid valve), cooling regulation (evaporator, compressor), air volume control (blower, M3) and air distribution control (flap position for fresh air, recirculated air, defrosting, bypass, footwell). All control circuits can be influenced by manual entry.

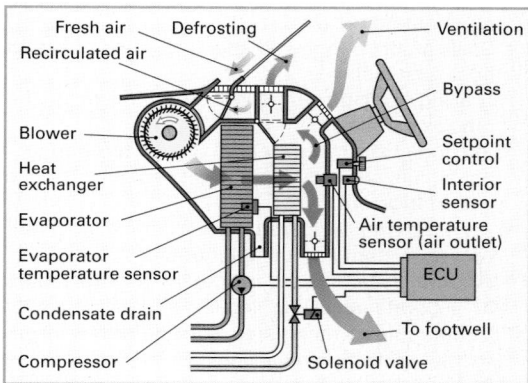

Fig. 1: Electronically controlled air conditioning system

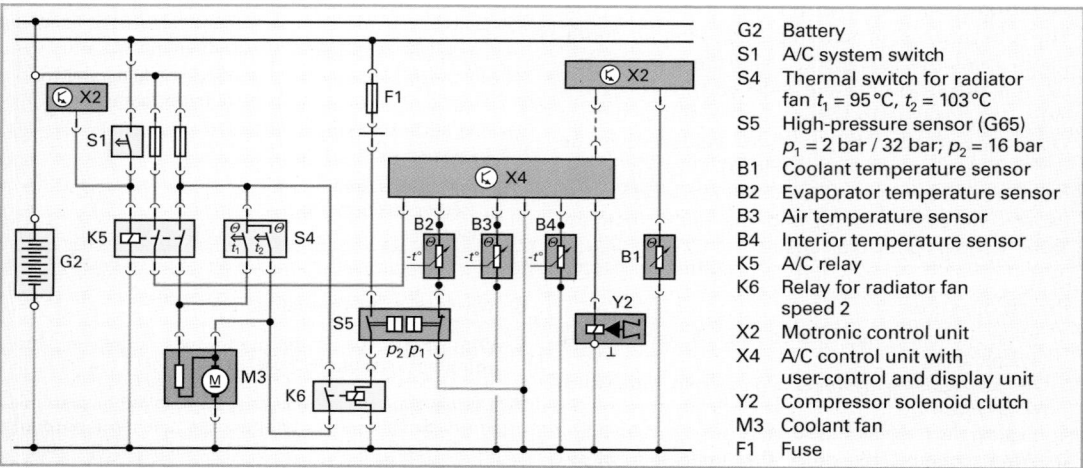

G2	Battery
S1	A/C system switch
S4	Thermal switch for radiator fan t_1 = 95 °C, t_2 = 103 °C
S5	High-pressure sensor (G65) p_1 = 2 bar / 32 bar; p_2 = 16 bar
B1	Coolant temperature sensor
B2	Evaporator temperature sensor
B3	Air temperature sensor
B4	Interior temperature sensor
K5	A/C relay
K6	Relay for radiator fan speed 2
X2	Motronic control unit
X4	A/C control unit with user-control and display unit
Y2	Compressor solenoid clutch
M3	Coolant fan
F1	Fuse

Fig. 2: Circuit diagram of air conditioning system

The air's flow volume can be set via blower speeds or throughout an infinitely variable range without direct reference to monitored conditions. At high road speeds, however, pressure builds up and this causes an increase in the delivery volume. With a special closed-loop control system, the blower speed can be reduced as road speed increases in order to keep the air flow constant.

The defrost mode (DEF setting) makes it possible to remove condensation and ice from window surfaces quickly.

To do this, the thermostatic controller has to be set to maximum heat, the fan has to be set to maximum speed and air distribution has to be directed upwards. On automatic climate control systems, this is done at the press of a button.

In winter or inclement weather conditions, the blower is stopped by the control unit at cold start to prevent drafts from the unheated air until a moderate coolant temperature is reached. If the defrost function is activated, this setting does not apply.

WORKSHOP NOTES

- Work of a general nature on the vehicle should be prepared and carried out so that the refrigerant circuit of the vehicle is not opened, for example, for radiator removal, engine removal.
- Replacement parts for the air conditioning system should be stored in dry and sealed packaging.
- If the refrigerant circuit of the air conditioning system has to be opened, seal off all openings immediately (refrigerant is hygroscopic).
- The expansion valve cannot be adjusted and must not be repaired.
- Seals must be renewed after loosening pipe and hose assemblies.
- Brazing or welding must not be carried out on the pipe assemblies.
- To check, extract, purge and charge the refrigerant circuit, use only equipment designed for use with refrigerant.
- Do not pump refrigerant back from the air conditioning system into the filler bottle.
- Heavily contaminated refrigerant is extracted into separate recycling bottles and disposed of through the correct channels in line with environmental protection measures.
- Empty refrigerant bottles must always be kept closed.
- Observe safety regulations for handling refrigerants.

Error diagnosis by way of pressure test

The pressure test is carried out with the air conditioning system switched on. Using the pressure values on the high-pressure side and low-pressure side, it is possible to check whether the system is working correctly. The pressure values must be within the tolerance ranges **(Fig. 1)**.

If the pressures are outside the tolerance range, the causes should be found and the faults rectified. **Table 1** shows possible causes.

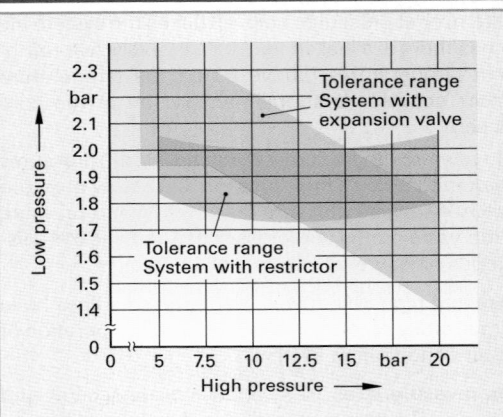

Fig. 1: Tolerance ranges of pressure test

Table 1: Possible malfunction sources	
Low pressure and/or high pressure	
too high	too low
• Plate valve of compressor defective	• Humidity or dirt in system causing blockage of expansion valve
• Piston clearance too great or piston ring defective	• Not enough refrigerant in system
• Expansion valve defective	• Evaporator iced up or contaminated or filter contaminated
• Expansion valve sensor loose or poorly insulated	• Blockage in condenser or in refrigerant line
• Too much refrigerant in system	
• Condenser contaminated or auxiliary filter defective	

REVIEW QUESTIONS

1 What demands are placed on air conditioning systems?

2 Air conditioning systems can be split into which three areas?

3 Explain the fresh-air mode.

4 What components does the refrigerant circuit include?

5 What tasks does the refrigerant have?

6 Why must the compressor only draw in and compress gaseous refrigerant?

7 Explain the operating concept employed in systems in which changes in temperature selection, entered by the user for relay to the thermostatic controller, form the basis for electronic closed-loop control of interior temperature.

20.2 Antitheft systems

Included here are all assemblies that serve to protect the vehicle and vehicle parts from theft and unauthorised use.

Examples of these are
- Central locking system
- Vehicle immobiliser
- Alarm system

20.2.1 Central locking system

This allows locking, unlocking and securing of all doors, boot lid/tailgate and fuel filler door.

This can be initiated from a central locking point, e.g. the driver's door, front passenger's door or the boot lid/tailgate.

Depending on the convenience and safety equipment in the motor vehicle, the central locking system may allow the sunroof or power windows to continue operating for a certain time, such as 60 seconds, even after the key has been withdrawn.

In order that the locks in the doors, the boot lid/tailgate and fuel filler door can be locked and unlocked, actuators, or servo-motors, are necessary.

Depending on actuation, a distinction is made between the:
- Electric central locking system
- Electro-pneumatic central locking system

Electric central locking system

With this system, the basic lock engagement and release functions, for instance, at vehicle doors through activation of the servo-motors, are carried out by the electrically controlled actuators. Actuation is often by way of two changeover contacts, whereby one is located in the door lock and the other in the actuator.

The simplified circuit diagram in **Fig. 1** shows how they work together. When the key is turned, the lock and changeover contact S1 are actuated mechanically. It can be found at the respective locking points, e.g. driver and front passenger door. In this way, all servo-motors that are part of the central locking system can be actuated via the control unit. The changeover contact S1 has two switch positions: Lock (V) and unlock (E). The changeover contact S2 is often integrated in the actuator and activated by the servo-motor via a linkage or a gear mechanism. The changeover contact switches the servo-motor on or off with two switching positions as an end point switch. The control signals are transmitted via wiring or a bus system (CAN bus, multiplexer) to a control unit.

Principle of locking (Fig. 1). With a turn of the key, Term. 30 and Term. V in the changeover contact are joined together. This control pulse causes the control unit to supply Term. 83a with voltage. The servo-motor M1 runs. In changeover contact S2, Term. 83a and 83 remain connected until the locking process has reached its end position and the connection between 83a and 83 is interrupted by servo-motor M1. The servo-motor stops.

Principle of unlocking. With a turn of the key in the opposite direction, Term. 30 and Term. E in the changeover contact are joined together. This control pulse causes the control unit to supply Term. 83b with voltage. Servo-motor M1 now runs in the opposite direction. In changeover contact S2, Terminals 83b and 83 remain connected until the unlocking process reaches its end position and the connection between 83b and 83 is interrupted by servo-motor M1. The servo-motor stops.

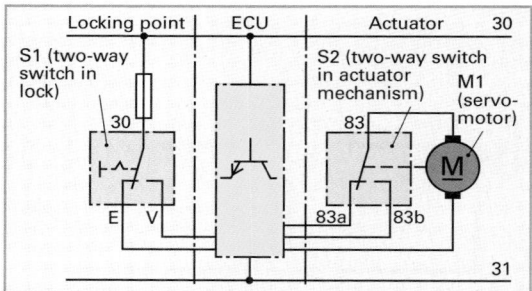

Fig. 1: Simplified circuit diagram of servo-motor with two two-way switches

Electric actuator (Fig. 2). This controls locking and unlocking. The pinion of the servo-motor is joined mechanically via a gear mechanism to the drive pinion of the toothed rack. If the lock is actuated mechanically by means of the key in the locking point of the central locking system, e.g. unlocked, the pull/push rod transfers the movement via a toothed rack and several gear wheels in the actuator. When this happens, the two-way switch (S2) is set mechanically in end position to unlock. The servo-motor remains unenergised.

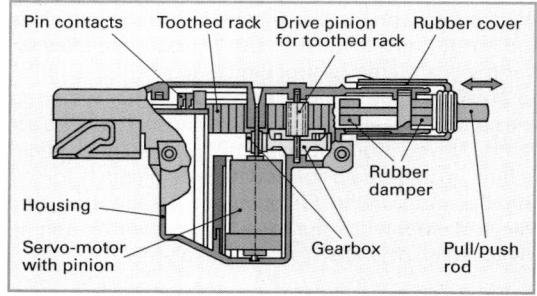

Fig. 2: Electric actuator

The unlock control pulse is sent to the control unit via the pin contacts. The servo-motors of all other actuators are supplied with current and carry out the unlocking procedure.

Electro-pneumatic central locking system

This comprises an electric control circuit and a pneumatic single-loop control circuit (Fig. 1).

Electric control unit. This controls the pneumatic single-loop control circuit. When the key is turned in the door lock, a microswitch is actuated. The control unit receives this signal and activates the pneumatic control unit of all other locks (electro-pneumatic control circuit).

Pneumatic single-loop control circuit. This actuates the actuators by way of vacuum or pressure in a line. When the vehicle is unlocked, for example, there is pressure in the line. When the vehicle is locked, there is vacuum.

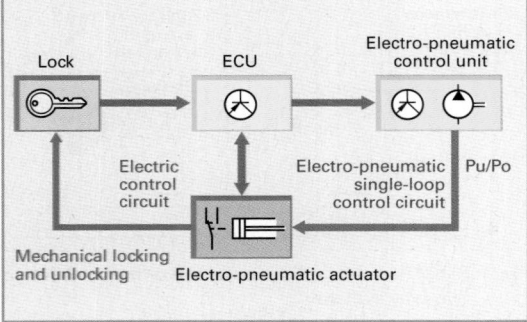

Fig. 1: Schematic diagram of electro-pneumatic central locking system

Electro-pneumatic actuator (Fig. 2). This is designed to carry out the locking procedure and is present at every door that needs to be closed. If, depending on the lock/unlock procedure, pressure or vacuum is generated by the pneumatic control unit, this affects the diaphragm in the vacuum or pressure chamber of the actuator. The diaphragm and the lock are joined to the pull/push rod. In this way, the locking procedure can be carried out by using a key to activate the linkage or by pneumatic means. When the vehicle is locked the microswitch in the actuator supplies an earth signal to the control unit. If an attempt is made to break into the vehicle, a positive signal is sent to the control unit via the relevant microswitch in the actuator. The control unit reacts to this. A magnetic field is built up in the safe coil and the locking pin is forced into the recess in the pull/push rod. At the same time, vacuum supply in the pneumatic control unit is switched on. The lock remains engaged.

Operating principle (Fig. 2)

Pneumatic unlocking. The vacuum displaces the diaphragm to press the pull/push rod upwards. In this way, the lock is disengaged mechanically via a linkage.

Pneumatic locking. The vacuum affects the diaphragm and pulls the pull/push rod downwards. The lock is engaged mechanically via the linkage.

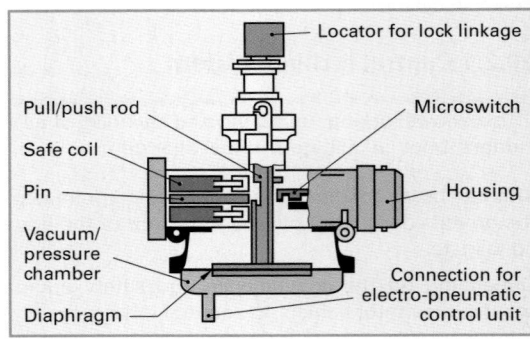

Fig. 2: Electro-pneumatic actuator

Pneumatic control unit. This comprises an electronic circuit (interface) and a dual-pressure pump. The electronic circuit receives the commands from the control unit and passes them on to the dual-pressure pump.

Dual-pressure pump. This is a vane pump that generates either pressure or vacuum by reversing the impeller's direction of rotation. Anticlockwise rotation causes pressure, for example, and clockwise rotation generates vacuum.

Operation of the central locking systems

For operation of the central locking system there are four different systems:

- Mechanical key system
- Infrared remote control system
- Radio-frequency remote control system
- Radio-frequency remote control system with self-actuation (keyless-go)

Mechanical key system. With this system the key is inserted at one location (usually a front door or the boot lid) and then rotated within the lock's tumbler mechanism to manually engage or release the lock at this location. At the same time, an electric switch supplies the control signal for the servo-motors or pneumatic actuators. These lock or unlock the other vehicle opening points.

Infrared remote control system. Here, the locking or opening procedure can be initiated via an infrared signal up to a distance of approx. 6 metres.

20

The system **(Fig. 1)** can consist of the following components:

- Transmitter key
- Infrared control unit
- Control unit with combined functions
- Relay for lock status
- Receiver unit, e.g. in interior rearview mirror
- Pneumatic control unit
- Actuators

Operating principle. The infrared sender, e.g. to the key, sends light signals within the infrared frequency range to the receiver. The receiver is connected to the infrared control unit. It detects via the lock response relay whether the vehicle door locks are engaged or disengaged. If the vehicle is unlocked, the driver is informed of this fact, for example, by flash code via the direction-indicator lamps.

This information is also sent to the actual control unit with combined functions. It is joined via a CAN bus to the pneumatic control unit. On the electro-pneumatic central locking system, this generates the pressure or vacuum necessary to make the locking/unlocking procedures possible.

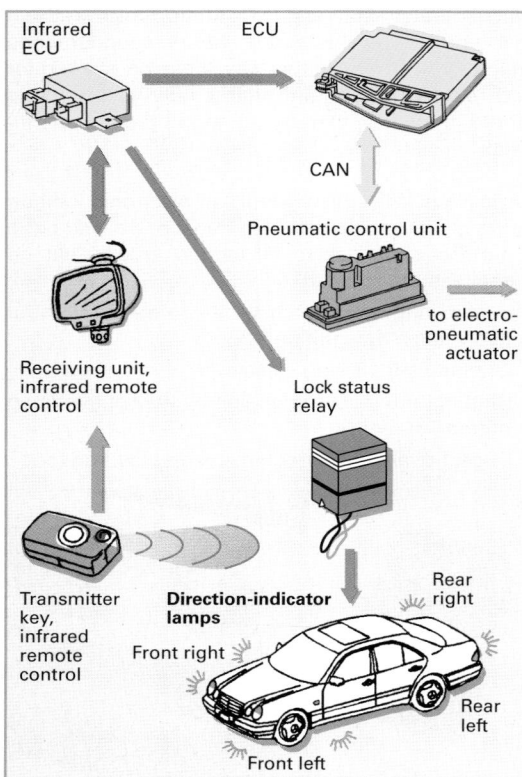

Fig. 1: Structure of infrared remote control system

Radio-frequency remote control system. For actuation of the positioning elements, a radio-wave signal can also be used. With radio waves, the sender need not be directed at the receiver. Initiation of the locking procedure and activation of the alarm system can be carried out covertly. Radio waves offer higher protection against decoding of the signal by unauthorised persons. The code itself can also be given a more complex structure.

Radio-frequency remote control with self-actuation (keyless-go). The driver does not have to touch the key to gain access. All that is required is for the electronic key to be on his/her person and one of the door handles to be touched. The capacitive sensor in the door handle detects an attempt to gain access to the vehicle and sends a signal to the access and start authorisation control unit **(Fig. 2)**. This starts an inductive interrogation of the transponder in the electronic key. If access authorisation is granted, the vehicle is opened. The locking procedure is triggered by pressing the locking button.

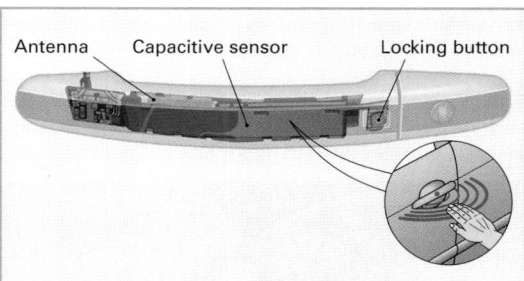

Fig. 2: Unlocking by touching door handle

20.2.2 Vehicle immobiliser

> This is an electronic system that prevents unauthorised persons from bringing the vehicle into operation.

Vehicle immobilisers to supplement the locking system have been legally mandated for vehicles initially registered on or after 1.1.1995. This is stipulated by Paragraph 38a of the German Road Traffic Regulations and EU Directive ECE R18, et al.

Activation

Locking with the vehicle key. The control unit receives the information for activation of the antitheft system from the door contact switch.

Locking with radio-frequency remote control. The receiver converts the infrared or radio signal from the transmitter to an electrical signal and relays it to the control unit, which activates the antitheft system.

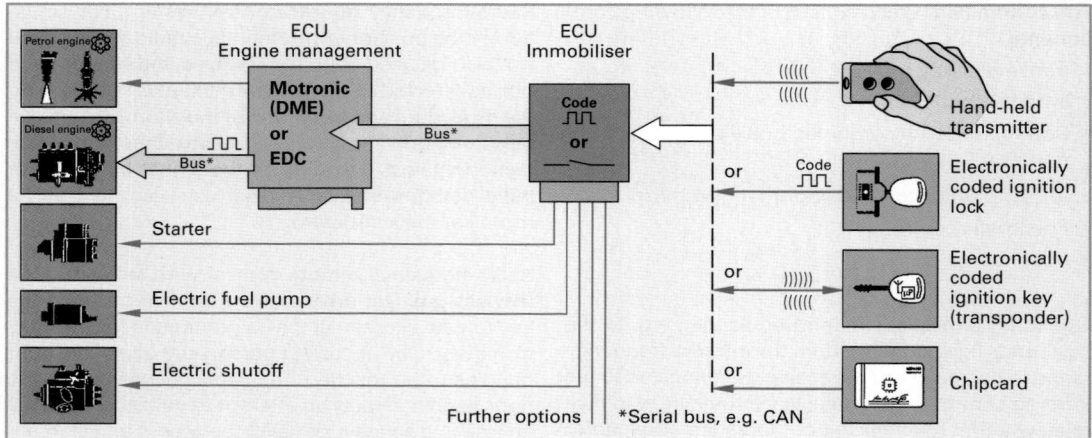

Fig. 1: System overview of immobiliser

Design (Fig. 1)

The immobiliser consists of a control unit and, depending on the manufacturer, either a hand-held transmitter or a transponder. This can be integrated in an electronic key or chipcard.

Transponder (made up from the words transmitter and responder, Fig. 2). It comprises a microchip, which is encapsulated in glass, and an induction coil.

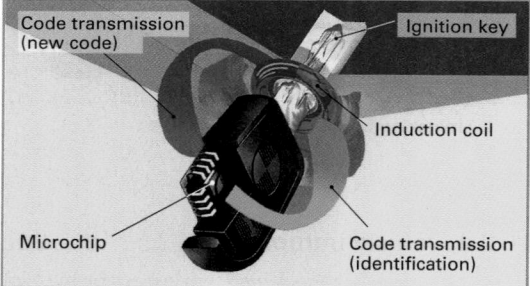

Fig. 2: Transponder

This is supplied with energy from an induction coil in the ignition lock. During production, the microchip is allocated a unique, undeletable code number (identification code, ID code). At the same time, a programmable memory (EEPROM) is reserved for the changeable code.

Operating principle. When the key is turned in the ignition lock, energy is transferred and the interrogation procedure of the vehicle immobiliser control unit (**Fig. 2**) is initiated. The transponder detects the interrogation signal and releases its identification code. This code is compared with that stored in the memory.

If the code is valid, the vehicle immobiliser control unit passes on a coded digital signal, e.g. via a CAN bus, to the engine management control unit. If this signal is accepted by the DME control unit, the engine can be started.

If the code is invalid, the engine will not start.

At the same time, the vehicle immobiliser control unit generates a new code at random, which is stored in the programmable part of the transponder (code change procedure). In this way, it can be assured for each start of the engine that a new valid code is stored in the key, thereby rendering the old one invalid.

Keyless-go. With this system, an electronic ignition key, e.g. key or chipcard with transponder and radio remote control, has to be carried around but no longer actuated (**Fig. 3**).

The following functions can be carried out without actively using the ignition key:

- Opening and locking vehicle
- Starting and stopping engine by way of start/stop button
- Engaging and disengaging steering column lock

Fig. 3: Chip card and start/stop button

Operating principle

Detection. This means detection of the ignition key with access authorisation. To do this, there are antennae on the inside and outside of the vehicle **(Fig. 1).**

They send a radio signal with identification number and coded identification request to the transponder. Following verification of identity, the doors, for example, are unlocked.

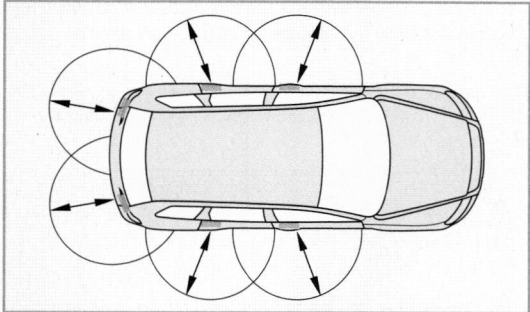

Fig. 1: Interior and exterior area of detection antennae

Exterior area. Responsible for detection during locking and unlocking procedure. Can only be carried out if the key is within this area.

Interior area. Responsible for starting and vehicle operation. For this function, the ignition key with valid transponder can be inserted in the slot for access and start authorisation. However, all that is needed is for the key to be in the vehicle so that when the start button is pressed, an inductive interrogation via the interior antennae can be initiated. Once positive identification has been verified, the engine will start. As soon as the engine starts, the electric steering column lock releases the steering. The driver has to depress and maintain pressure on the brake or clutch pedal during starting.

20.2.3 Alarm system

> An alarm system triggers optical and acoustic warning signals in the event of unauthorised intervention or impact.

This comprises the following components **(Fig. 2):**

- Remote control
- Control unit with power supply
- Contact switch, for example, for doors, bonnet, boot lid/tailgate, luggage or glove compartment
- Infrared sensor or ultrasonic transponder for interior monitoring
- Position sensor for wheel theft and tow-away protection
- Status display
- Signal horn
- Starting system

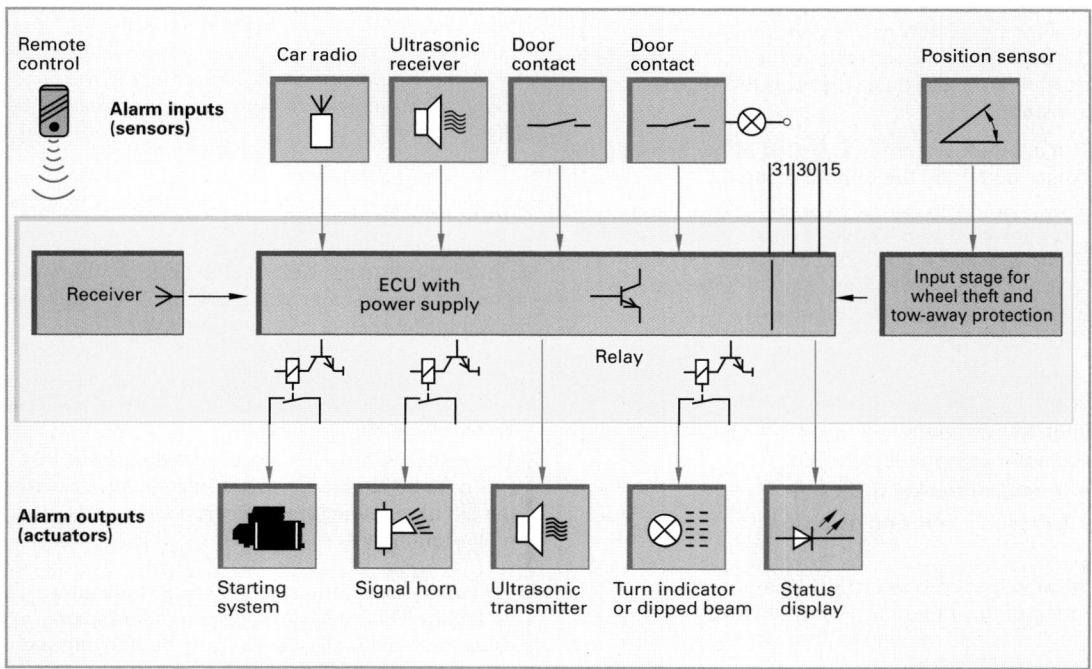

Fig. 2: Systematic view of alarm system

Operation principle. When the alarm system is activated, the control unit checks by way of the contact switches that the doors, windows, sliding sunroof, bonnet and boot lid/tailgate are closed. If all requirements are met for the locked status, the alarm can be activated after a delay of 10 to 20 seconds. A flashing LED, for example, indicates that the alarm system is armed.

The alarm will be triggered by:
- Unauthorised opening of doors, boot lid, tailgate or bonnet
- Intrusion into interior
- Unauthorised activation of ignition
- Key with invalid transponder code inserted in ignition lock
- Intrusion sensor removed
- Radio removed
- Centre console stowage compartment opened
- Alarm horn removed
- Temporary interruption of control unit voltage supply
- Change to vehicle position

When the system is active, the alarm signal can be supplemented by way of integrated signal horn, flashing signals from the hazard warning system and the interior lighting. The period during which the alarm stays on is predetermined by each country. For example, the signal horn can emit an acoustic signal for 30 seconds and the turn-indicator system with flashing headlights can remain active for longer than 30 seconds. At the same time, the immobiliser (fitted as standard these days) prevents the engine from being started.

The antitheft system is switched off by pressing the unlock button on the remote control.

> In the event of emergency opening by unlocking mechanically, the key has to be inserted in the ignition after a set period, e.g. 15 seconds, otherwise the alarm will be triggered.

Intrusion detection. On activation of the alarm system, the interior monitoring sensors are also activated.
A difference is made between:
- Infrared intrusion detection
- Ultrasonic intrusion detection

Infrared intrusion detection. With this system, monitoring of the interior is performed by an infrared sensor. It reacts to moving heat sources, e.g. people, and triggers the alarm.

Ultrasonic intrusion detection (Fig. 1). An ultrasonic transmitter in the interior of the vehicle generates an ultrasonic field with a frequency of approx. 20 kHz. Pressure fluctuations in this field, for instance, in response to intrusion or a window being broken, will be detected by the ultrasonic sensor. The electronic processing circuit then triggers the alarm.

> If an auxiliary heater is installed, the sensitivity of the system has to be adapted again because the warm air flow is enough to trigger an alarm.

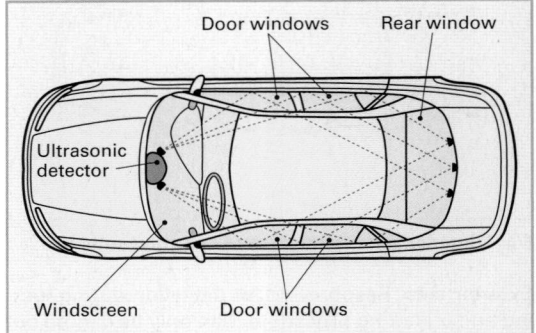

Fig. 1: Ultrasonic field in vehicle interior

Wheel theft and tow-away protection. This comprises a tilt sensor and a processing unit. The position of the vehicle when parked is recorded as the zero position when the alarm system is activated. Any change leads to triggering of the alarm system. Normal changes to the position (e.g. air loss in the tyres, rocking of the vehicle, soft ground) will not cause the alarm to go off.

Component protection

Following unauthorised removal of a component, such as the engine control unit, the component's complete functionality will no longer be available. For reinstallation it is rendered useless.

> **WORKSHOP NOTES**
>
> The systems are capable of self-diagnosis and can only be checked, repaired or initialised with the aid of manufacturer-specific testing and diagnosis equipment.
>
> For service work, matching or repair measures to an antitheft alarm system, certain manufacturer's rules and safety standards must be applied and also greater care exercised:

CONTINUED FROM PAGE **608**

- If the key, lock cylinders or engine control unit, for example, have to be renewed on a vehicle, the order form has to be accompanied by copies of documentation of owner identification and vehicle documentation. The procedure is carefully documented and the paperwork has to be stored with the date and certificate of proof.
- Following installation of the new replacement component it will be necessary to restore the system to operational status (with "system enable" command and/or initialisation) using the diagnosis tester (approval or initialised). On newer generation vehicles, it is also necessary to assign other control units with a new electronic identity. In addition, all car keys have to be renewed and matched.
- Depending on the type of vehicle, the mechanic requires a PIN from the manufacturer or an online connection to the manufacturer during repair.
- **PIN.** This is a secret number, which contains the dealer code and also the date. It has to be requested by the workshop from the manufacturer, e.g. per fax or Internet. It is valid only for this workshop and only for a set period of time, e.g. 24 hours.

- Online approval. This is given only if the vehicle is connected via the diagnosis tester to the manufacturer online.
- The mechanic that carries out the online request must have a personal user ID and password. The vehicle and the replacement part installed identify themselves to the manufacturer via the diagnosis tester. In addition, the customer's name and passport number, including nationality, have to be entered.
- Approval is given and confirmation sent to the diagnosis tester on completion.
- When this happens, all keys have to be initialised by inserting them in the ignition lock and switching on the ignition.
- Additional keys can also be initialised through this approval procedure.
- If new car keys are needed, the reordered keys are cut to the right profile in the factory and coded to the respective vehicle. The new keys can only be approved for this vehicle.

REVIEW QUESTIONS

1 What two central locking systems can commonly be found?

2 What are the tasks of the actuators on central locking systems?

3 What advantage does radio-frequency remote control have over infrared remote control?

4 Which assemblies can antitheft alarm systems include?

5 What is a transponder?

6 Which components of an antitheft system can trigger the alarm?

7 What types of intrusion detection (interior monitoring) are there?

8 Explain the function of an immobiliser.

9 What functions can be performed on a keyless-go system without using the ignition key?

20.3 Comfort and convenience systems

20.3.1 Electric power windows

This allows electric opening and closing of the windows and, where fitted, sliding sunroof via a rocker switch (pushbutton switch).

To power the windows up and down, a cable drive system is normally used (**Fig. 1**). The drive motor actuates a cable via a worm gear mechanism, which will open or close the window depending on the direction of motor rotation. The self-locking effect of the worm gear mechanism prevents the window from opening in response to application of force.

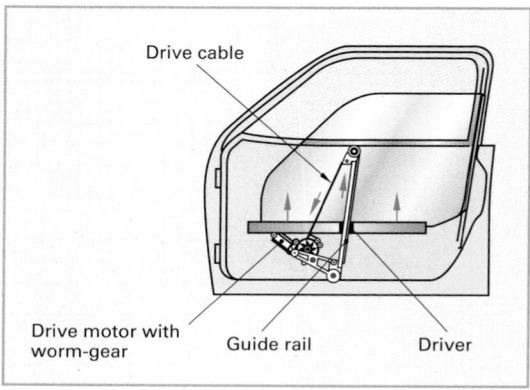

Fig. 1: **Cable drive**

Electric actuation of the windows. This can be performed by

- Rocker switch (manual actuation)
- Control electronics combined with rocker switch

Actuation via rocker switch. Using the rocker switch, which is allocated to the relevant power window motor, the window can be closed or opened. On a central locking system, all windows can also be closed at the same time **(Fig. 1)**.

Operating principle (Fig. 1). The main relay receives power from Terminal 15 whenever the ignition switch is on. It pulls in and connects Term. 30 with Term. 87 and applies voltage to Term. d from switch S1 and S2. With the aid of switch S5, which is integrated in switch S1 of the driver's door, switches S3 and S4 for the rear doors can be supplied with energy or the power supply can be switched off.

Switch position: Open window (Fig. 2). On actuation of the rocker switch (S1, S2, S3, S4) Term. d (+) is joined to Term. b and Term. c (–) to Term. a. The power window drive unit lowers the respective window.

Switch position: Close window (Fig. 2). If the window is to be closed, Term. d (+) in the rocker switch is joined to Term. a and Term. e (–) to Term. b. The drive motor is driven in the opposite direction because the terminal polarity of a and b is reversed. The window closes.

Closing windows on central locking system. The central locking control unit connects the control coil of the control relay's Term. 85 to earth. In this way, Term. 87 and Term. 30 are joined in the relay. Switch terminal a is connected to positive via Term. c. This earths the circuit through Terminals b and e. The windows are closed.

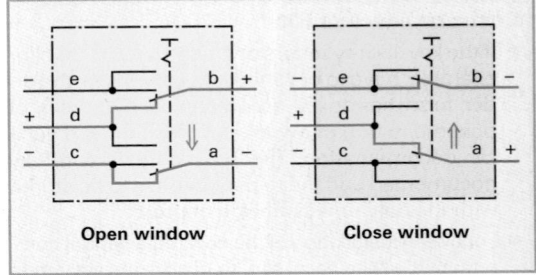

Fig. 2: Switch positions for rocker switch window actuation

Combination of window control by rocker switch and control electronics. The control electronics can be housed centrally in one control unit for this purpose. In order to keep the wiring as simple as possible, it can be integrated in the power window motor. If the control button of the power window motor is touched briefly, the control electronics cause the window to close. If the control button is pressed for a longer period, the window can be moved to the desired position. If the vehicle is locked via central locking, all windows close at the same time or are closed up to a ventilation gap.

Finger-protection feature. To prevent body parts such as hands and arms from becoming trapped, the force with which the window closes is limited to a specified level. The finger-protection feature functions electrically, by deactivating the electric motor once current draw reaches a specific level, or mechanically, using load-sensitive couplings in the drive system.

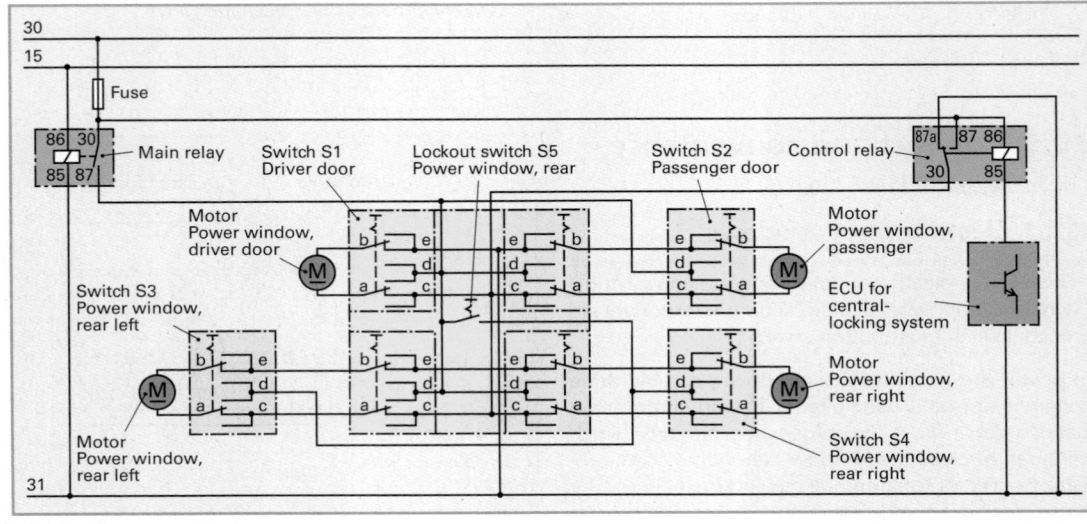

Fig. 1: Circuit diagram for rocker switch actuation

20.3.2 Convertible roof actuation

> This allows the convertible roof to be opened and closed electrically via a pushbutton switch.

The convertible roof actuation shown in **Fig. 1** is of the electro-hydraulic type. By way of a switch, an electric motor is activated that sets a rotary pump in motion. This generates pressure. Double-action hydraulic cylinders cause the roof to be opened or closed.

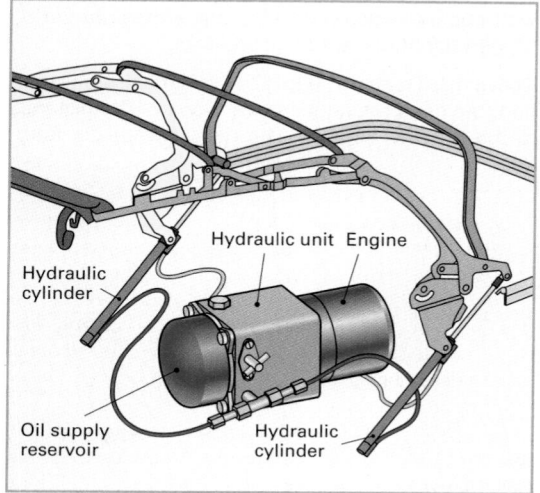

Fig. 1: Convertible roof actuation

Hydraulic device

This consists of:
- Fluid reservoir
- Electrically driven hydraulic rotary pump
- Hydraulic control unit
- 2 double-action hydraulic cylinders

Electrical device (Fig. 2)

This consists of the following main components:
- Convertible roof button (E137)
- Control unit (J256)
- Hydraulic pump motor (V82)
- Ignition/starter switch (D)
- Thermal cut-out (S68)
- Contacts for fitted convertible roof covering (F155, F156)

Principle of convertible roof operation

First, the convertible hood must be unlocked manually. Then, in the interests of safety, activation of the electro-hydraulic convertible hood is possible only with the ignition key inserted and the ignition switched off.

To do this, PIN S of ignition/starter switch D applies voltage to the convertible roof button E137. Once this is pressed, the control unit J256 receives voltage at

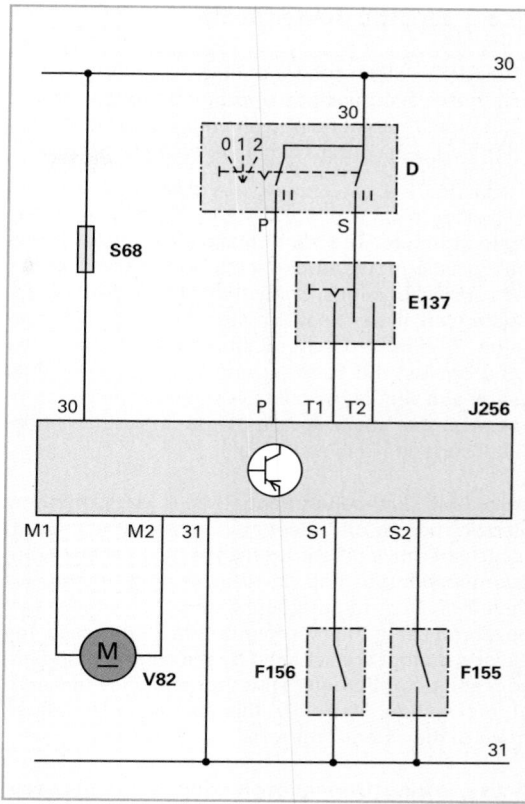

Fig. 2: Current flow diagram for convertible roof actuation

PIN T1. It receives voltage in the same way from ignition/starter switch D via PIN P.

The working current supply for the system is via a thermal cut-out S68 from Terminal 30 to PIN 30 of the control unit J256.

Since there is voltage at PIN T1 of control unit J256, the control unit controls direction of rotation of the electric motor V82 for the hydraulic pump so the piston rod of the hydraulic cylinder retracts. The convertible roof is opened.

Closing convertible roof

If the convertible roof is to be closed, the control unit J256 receives voltage from convertible roof button E137 at PIN T2. It triggers the electric motor for hydraulic pump V82 to extend the hydraulic cylinder's piston rod. The convertible roof is closed.

To prevent the roof from being activated when the top is up, safety switches are installed. Switch F155 and F156 are reed contact switches that are switched without physical contact by magnets on the roof cover studs. Switches F155 and F156 apply earth to PIN S2 and PIN S1 of control unit J256 when a roof cover stud is engaged. Control unit J256 now no longer actuates the electric motor V82.

20.3.3 Electric power seats

> In order to allow people of all types of stature to sit in a relaxed position appropriate to the traffic situation, the seat height, tilt angle, backrest position and seat contour can be adapted to personal requirements.

Basic structure. The seat consists of a steel frame with plastic moulded elements and assemblies for the various functions. Beneath the seat's padded upholstery an air-permeable material ensures air circulation. Rubberised natural fibre mats support the bond against the backrest and the seat. Individual heater and ventilation controls provide optimal comfort. A lumbar support, with massaging system if desired, supports active seating.

Individual seat adjustment. This is performed by electric motors and electronic control units. Occupants can move different sections of the seat in various directions to find an ideal seating posture for their individual requirements. Personal settings can be stored using the memory button. In this case, the electric motors are actuated by the control unit. Sensors and potentiometers monitor and relay positional information. Normally, this also includes adjustment of the exterior mirrors.

Active seating. During long periods when the occupant sits in a rigid position, the vertebrae are protected against muscular tension and stiffness by continual, minimal alteration of the seat position. To support this procedure, there is a variable lumbar support with pulse-controlled air chamber (**Fig. 1**).

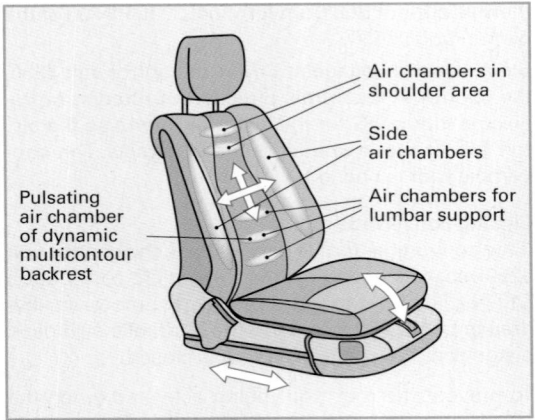

Air chambers in shoulder area

Side air chambers

Air chambers for lumbar support

Pulsating air chamber of dynamic multicontour backrest

Fig. 1: Seat for individual and active seating

Seat heating. Central heating elements provide quick warmth for the seat's centre section in winter. The edges are adjusted slowly to the right temperature afterwards.

Ventilation. The seat surfaces heated by the sun in the summer are cooled down and also ventilated by means of small fans installed below the surface.

Dynamic seat adjustment. Inflatable air chambers in the seat cushion's side bolsters increase lateral support during cornering (**Fig. 1**).

Seat occupant detection sensor. Integrated in the seat is a sensor mat that allows occupant classification, e.g. child seat detection, based on the weight being applied to the seat's surface. The system decides with this information whether the airbag should be triggered in the event of an accident.

Active head restraint (Fig. 2). In a rear collision situation, the head restraints move forward. The distance to the head is reduced. The risk of whiplash is reduced.

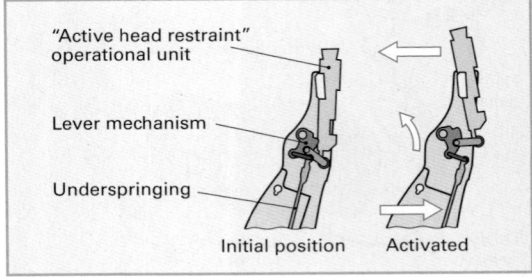

"Active head restraint" operational unit

Lever mechanism

Underspringing

Initial position   Activated

Fig. 2: Active head restraint

20.3.4 Electronic windscreen wiper

> These allow electronic control of the wiper motion and parallel running of the wiper arms.

Design. The electronic windscreen wiper system comprises one or two reversing DC motors with a small crank drive and no wiper reverse linkage (**Fig. 3**). The control electronics are integrated in the gear mechanism cover of the wiper motor.

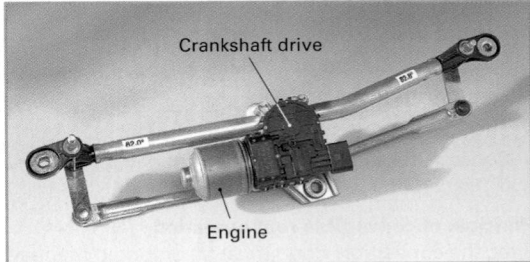

Crankshaft drive

Engine

Fig. 3: Electronic window wipers

Operating principle. The up and down motion of the wiper arms is carried out by changing the rotational direction of the twin brush wiper motors (reversing). The voltage applied to the brushes is reversed in po-

20

larity electronically at the reverse positions of the wiper arms **(Fig. 1).** To detect the position and speed there are Hall-effect sensors attached to the wiper motor and gear mechanism.

Clockwise rotation. The driver stage triggers and controls the transistors T1 and T4 through Pins 5 and 8. The working current flows from + via T1 to the motor and via T4 to earth.

Anticlockwise rotation. The transistors T2 and T3 are actuated and switched by the driver stage circuit via Pin 6 and Pin 7. The working current flows from + via T2 to the motor and via T3 to earth.

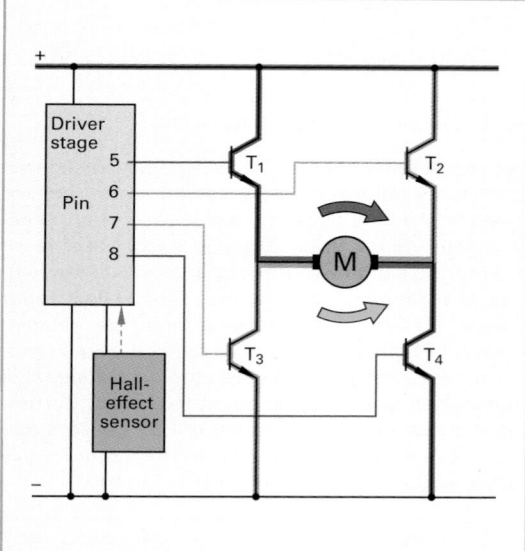

Fig. 1: Principle of rotational-direction change

The wiper motor receives the wipe commands via a CAN interface. If a second motor (slave) is installed, this receives the wipe command via a serial, single wire interface from the first motor (master).

The following advantages are given:

- Space-saving design
- Improved field of vision (wiping angle approx. 150°)
- Lowered park position of wiper arms outside field of vision in heated area depending on vehicle speed
- Low noise from change in direction of wiper blade rubber by means of speed reduction in wiper motor
- Alternating rest position in up and down direction
- Seizure protection with snow guard feature

To change to the wiper blades, the wiper arms have to be brought to the vertical position using the central control unit!
Never employ force to pull the wiper arms up from their parked position!

20.3.5 Electric adjustable exterior mirrors

These are brought into the optimum position via a pushbutton switch on the inside of the vehicle.

Design. If the driver presses the setting switch to adjust the mirror, the information is sent to the door control unit. This actuates two DC motors with clockwise and anticlockwise rotation. They adjust the mirror via worm gears and adjusting screws in one of the four directions of motion. Using a selector switch, the exterior mirror on the driver's or front passenger's side can be selected. Normally a heating element is integrated to defrost the mirror's lens **(Fig. 2).**

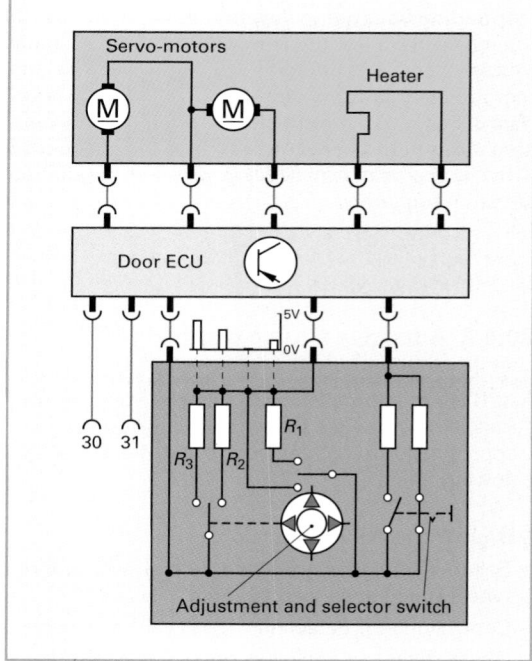

Fig. 2: Circuit diagram for control of exterior mirror

Operating principle. Actuation of the door control unit is often by way of voltage coding. Using the control switch, one of the four parallel-wired closing mechanisms is actuated. This closes the circuit to the door control unit. Each circuit has a resistor of different size (wire only, R1, R2, R3). In each circuit, they produce a different voltage drop from e.g. 0 V if the wire has no resistance, 1.3 V for R1, 2.7 V for R2 and 4 V for R3. The door control unit reads the desired direction of motion from the voltage that is applied and switches the working current to the correct servo-motor. It runs as long as the control switch is pressed. Only one positive wire is required from the control switch to the door control unit.

20

20.4 Driver assistance systems

20.4.1 Cruise control system

> It automatically keeps the vehicle at a speed set by the driver.

Design. The cruise control system consists of the:

- Speed sensor
- Throttle valve with servo-motor
- Controller
- Means of entering commands

Operating principle. The driver sets the desired speed by pressing the control lever. The throttle valve is adjusted to furnish the engine with the corresponding volumetric flow of induction mixture. If the speed changes, the controller receives a signal to this effect. It alters the throttle valve angle and thereby the mixture volume using the servo-motor **(Fig. 3, Page 71)**. The vehicle accelerates or decelerates. Automatic brake intervention does not occur. If the brake or clutch pedal is pressed, regulation stops immediately.

20.4.2 Adaptive Cruise Control (ACC)

> ACC is an automatic speed and distance control system. It works in a speed range from 30 km/h to approx. 200 km/h. It serves to aid the driver in flowing traffic.

Design. The system consists of:

- Sensors for radar, yaw rate, lateral acceleration, wheel speed and steering angle
- Control unit for detection of own vehicle motion
- Object detection and allocation
- Adaptive cruise control
- Control units for engine, gearbox and ESP with actuators

Operating principle. With the aid of radar sensors, vehicles travelling ahead and their speed are detected up to a distance of approx. 100 m. ACC differentiates between the two operating modes 'clear driving' and 'following driving' **(Fig. 1)**.

Clear driving. If the road ahead is clear, it works as a cruise control system.

Following driving. If ACC detects a vehicle in the same lane, it adapts the speed to that of the vehicle travelling directly ahead. It maintains the distance specified by the driver by automatic actuation of the brakes and automatic acceleration. ACC lowers the speed of the vehicle by reducing engine torque and, if necessary, brake intervention. If the brake pedal is pressed, the system switches off automatically.

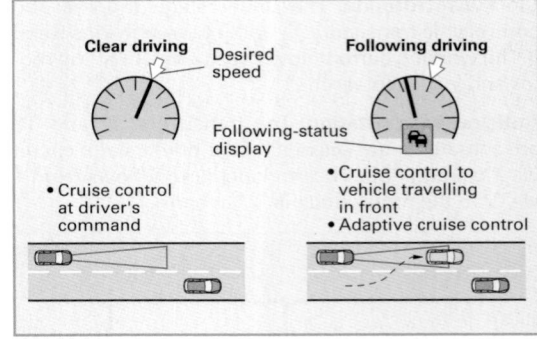

Fig. 1: Clear and following driving on ACC

Object detection (Fig. 2). To pick up the vehicle travelling ahead there is a radar sensor in the radiator grille. It contains three transceivers with an effective aperture of three degrees each, capable of monitoring a 3-lane motorway and preceding vehicles at a range of 100 metres. They reflect radar pulses (77 Ghz). The distance and the relative speeds of both vehicles are calculated based upon the time that elapses between the time a signal is transmitted and its subsequent reception. Cornering is determined with the aid of ESP sensors and the relevant vehicle travelling in the same lane is picked up.

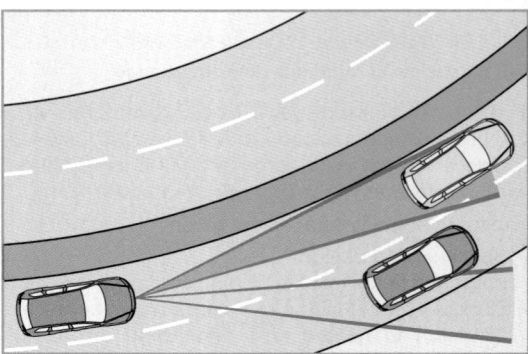

Fig. 2: Object detection of vehicle travelling ahead

WORKSHOP NOTES

If on-vehicle repairs or adjustments (e.g. running gear, cross members) are carried out that affect the position of the radar sensor, the radar sensor has to be readjusted. Readjustment is absolutely essential whenever major service operations are performed on the vehicle.

20.4.3 Parking assistance system

> They indicate to the driver the distance to an obstacle during parking or reversing and give an optical and acoustic warning.

Design. At the front and rear of the vehicle are ultrasonic sensors. They are actuated by the control unit and indicate the distance via indicator lamps and warning buzzers.

Operating principle. System operation is based on the echo-sounding concept. The system periodically activates its entire complement of peripheral transmitters, which respond by transmitting 30 kHz ultrasonic signals. Thereafter, all sensors change to receiver mode and record the sound waves reflecting off the obstacles. The system calculates both the distances to obstacles and their locations relative to the vehicle based on the bounce times that elapse between transmission and reception of the echo-sounding signals. If the distance is too small, the system warns the driver **(Fig. 1)**.

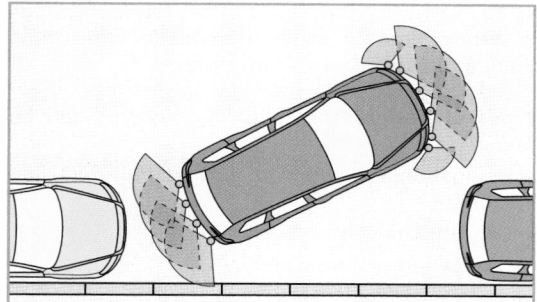

Fig. 1: Parking assistance with ultrasonic signals

20.5 Infotainment system

> The details in question cover **info**rmation (data presentation), communication and enter**tainment**. They can be operated using a central unit or be installed independently of each other in the vehicle.

Infotainment systems offer access to:
- Operating and travel data display
- Navigation with telemetry
- Suspension settings and service functions
- Mobile phone and Internet
- Audio/TV

It comprises, for example:
- Instrument cluster
- Display and user-control units
- Multifunction steering wheel
- Navigation system
- Mobile phone with antennae and microphone
- TV-radio tuner and CD player

20.5.1 Operating and travel data display

> In the instrument cluster, the most important information is displayed for the driver.

Examples of these displays are road speed, engine speed, engine temperature, oil pressure, alternator warning lamp or lighting and self-diagnosis error message. The instrument cluster is in the central field of vision of the driver. Using an onboard computer, which evaluates data and sensor signals and communicates them to other control units (e.g. engine, gearbox, ABS), further information can be displayed. Examples are:
- Trip data, such as average fuel consumption and cruising range
- Inspection and service intervals
- Wear limits, e.g. brake pads
- Fluid levels, e.g. fuel and oil
- Bulb function check

20.5.2 Navigation systems

> These offer help in finding the right route to the target destination and orientation in unfamiliar areas.

Navigation systems can take on the following tasks:
- Vehicle positioning
- Position-data transfer
- Calculation of optimum route based on current traffic conditions
- Guidance to destination with route recommendations

Shown in **Fig. 2** are all components and subsystems involved. The input signals are processed by the navigation computer. Output is visual on the display and vocal.

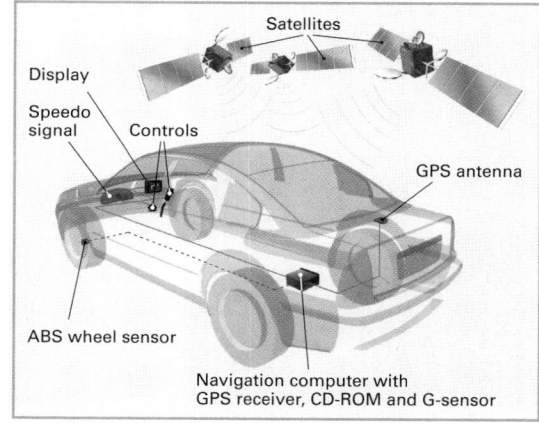

Fig. 2: Components of a navigation system

Vehicle positioning. This forms the basis for calculation of a route. With the aid of the Global Positioning System (GPS), the actual position of the vehicle can be determined. GPS comprises 24 geostationary satellites, which follow different orbits around the earth. These emit identification, time and position signals at uniform intervals. For determination of the vehicle position using the vehicle's navigation computer, the signals from at least three satellites are required. The position can be calculated with the data from GPS to an accuracy of about 10 metres. In order to increase the accuracy, the vehicle movements are supplemented by the speedometer signal and the signals from a yaw rate sensor. This makes it possible to monitor distances while also distinguishing between linear progress and cornering. Any corrections to the localisation result that might be necessary owing to external factors, such as tunnels, bridges, etc., are carried out by the navigation computer.

Position data transfer. This serves as a means of sending the location of the vehicle to rescue services in the event of an accident, or breakdown recovery services if the vehicle has broken down. Furthermore, the vehicle can be located more quickly in the event of theft.

Optimal route calculation. When the driver enters his or her target destination using the control elements or voice input, the navigation equipment determines the vehicle's current location. From this point the navigation computer can calculate the optimum route to the target destination using data from the map memory. The navigation computer can calculate distances and angles on curving stretches of road based on data supplied by the speedometer signal and the yaw-rate sensor. The

details recorded from the sensors over the distance travelled are compared with data from the CD-ROM, DVD or software in the map memory and adjusted as necessary (map matching). In this way, the actual position of the vehicle over a set route can be determined with great accuracy. If the GPS signal is available, the position can also be verified.

Dynamic route guidance. The latest traffic situations, e.g. congestion, road maintenance sites, road blocks, from communication sources such as TIM (Traffic Information System), RDS (Radio Data System) or the Internet can be considered in the route calculation.

Guidance to destination by recommendations on where to turn. The navigation system guides the vehicle with recommendations on where to turn on the calculated route to the target destination. Normally, these recommendations are given vocally by a computer so the driver is not distracted. A route map **(Fig. 1)** or direction arrows on a display can also be shown as a supplement. The system responds to wrong turns by issuing immediate instructions guiding the user to an alternate route.

Types of design. The following systems can be found:
- Navigation system with permanently installed monitor
- Car radio navigation systems
- Navigation system with PDA

Navigation system with permanently-installed monitor (Fig. 1a). This offers full functionality as previously described. It is often offered as optional equipment in new vehicles.

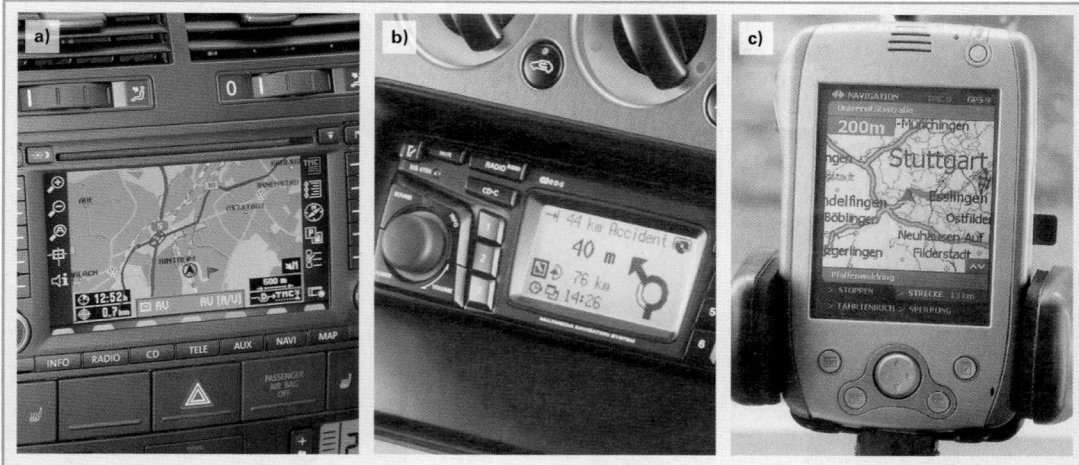

Fig. 1: Types of navigation systems: a) with permanently installed monitor, b) with car radio, c) with PDA system

Car radio navigation systems (Fig. 1b, Page 616).
A map cannot be shown due to the small dimensions of the displays. The instructions for the calculated route on these units are given acoustically and visually by way of arrows. The systems use the GPS and speedometer and yaw rate sensor signals for calculation of the route.

PDA navigation system (Fig. 1c, Page 616).
On this system a small pocket computer (Personal Digital Assistant) is mounted in the vehicle. Its display serves as the monitor. The computer receives the signal from a GPS antenna. The map details are transferred from a PC. PDA systems are not very accurate as they only refer to the GPS signal.

Additional functions. With the navigation system an electronic road atlas can be carried onboard. Features such as voice guidance, voice information, folding monitor, touchscreen or information about speed limits for the road the vehicle is on are available as options.

20.5.3 Mobile phones

> These allow calls to be made and taken in the vehicle.

Permanently-installed mobile phones. These consist of a transmitter and receiver (up to 8 W) and a handset, including handsfree system with microphone and loudspeaker. On handsfree systems, there is a risk of feedback if the echoes from the electronic voice are not suppressed. Therefore, complicated echo compensating solutions have to be realised that enable simultaneous talking and listening of the people at each end of the line. This occurs on the basis of digital signal processors. In addition, measures are necessary to suppress the driving noise in the microphone signal.

Mobile phones. These are mobile units not primarily intended for automotive use. The main components of a GSM cell phone are a 2 watt transceiver, a digital signal processor for channel and voice coding and a control circuit to co-ordinate communications with the rest of the network. In addition, there are a microphone, speaker, antenna, keypad, display, rechargeable battery and SIM card reader.

Installation kits for the vehicle consist of a cradle with power supply, microphone and antenna connection **(Fig. 1)**. Since use of the phone while driving is banned, most of the systems feature a handsfree device or headphones. Because the inside of a vehicle works like a Faraday cage, a shielded cable leads to an externally mounted antenna.

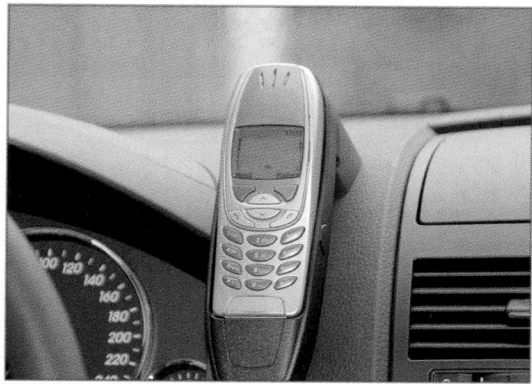

Fig. 1: In-vehicle mobile phone

Internet. With the aid of a cell phone (UMTS) or PDA with integrated cell phone (GPRS), a wireless connection to the internet can be created. With Bluetooth technology, networking of these mobile units is possible with communication systems, such as Personal Digital Assistant (PDA), car radio, CD changer or navigation system in the car. Each Bluetooth-compatible unit features a unique equipment code, which means it can be identified clearly among others.

Additional mobile equipment, such as a computer, notebook, PDA, can thereby be connected to each other without cables and without complex configuration. This opens a wide range of opportunities, for example transfer of addresses and maps to the navigation system or downloading of music or videos from the internet.

REVIEW QUESTIONS

1 Why is it necessary to install a finger-protection feature with electric power windows?

2 How are the windows closed with a central locking system?

3 What is the function of the seat occupant sensor?

4 How is the direction of rotation changed on windscreen wipers?

5 What are the two operating modes for adaptive cruise control?

6 Describe the park distance monitor's operating concept.

7 What information does the driver receive from the operating and travel data display?

8 Explain the principle of operation of a navigation system.

20

21 Motorcycle technology

21.1 Types of motorcycle

Motorcycles are single-track vehicles with two wheels. It is permissible to attach a trailer to them. They can be fitted with a sidecar without the motorcycle classification being affected.

> Motorcycles must be ridden with a crash helmet.

A difference is made between:
- Bicycles with engine assistance (motor bicycles)
- Low-powered motorcycles (e.g. monkey bike)
- Lightweight motorcycles
- Scooters
- Motorcycles, motorcycles with sidecar

21.1.1 Bicycles with engine assistance

These are single-track, one-seat vehicles, the cubic capacity of which must not exceed 50 cm^3. The vehicle-specific top speed is set at 25 km/h. The low powered motor bicycles (Fig. 1) can be propelled with engine assistance and also with pedals.

> Motor bicycles, mopeds, mopeds w/kickstarter may ...
> - ... not be ridden on the German motorways.
> - ... only be operated in Germany if an annual (1.3. to 29.2.) insurance permit is obtained.

Trade names for these types of vehicle are city bike, fun bike, naked bike or enduro.

Fig. 1: Motor bicycle (city bike)

The motor bicycle can only be ridden with a licence. However, it does not have to be registered, thereby making it tax-free. A full driving licence is not necessary. These vehicles can be ridden by persons from the age of 16 years. The only stipulation is that persons who reached their 16th birthday after 1.4.1980 must carry with them a traffic competency certificate. Scooters with a cubic capacity below 50 cm^3 can be converted to a licence-free vehicle by reducing the bench seat to one seat and installing a modified ignition control unit that governs the speed. However, these changes must

be carried out by an authorised workshop and the alterations entered in the vehicle document.

Engines. The types used are generally two-stroke, single reciprocating-piston engines with choke. The common outputs are from 0.5 kW to 3.7 kW at speeds of up to 4,000 rpm. Power transmission is either by one or two-speed automatic gearboxes or manual gearboxes with 2 or 3-speed hand or foot gear selection. **Fig. 2** shows a single-cylinder, low-powered engine for motor bicycles with integrated manual gearbox.

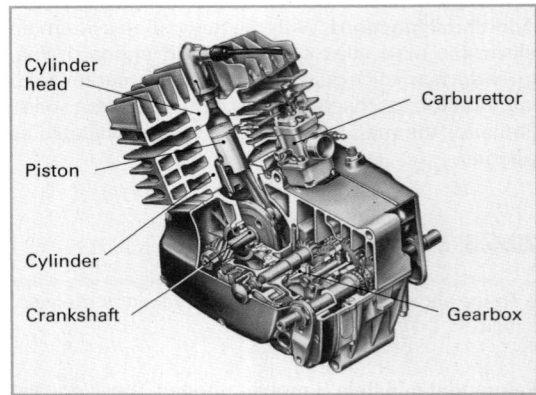

Cylinder head

Carburettor

Piston

Cylinder

Crankshaft

Gearbox

Fig. 2: Motor bicycle engine with gearbox

High-powered moped. This is a bicycle with engine assistance with a cubic capacity restricted to 45 cm^3. Its vehicle-specific top speed must not exceed to 50 km/h. For additional means of propulsion it has pedals.

21.1.2 Low-powered motorcycles

These have foot rests at the side or in front of the seat, a kickstarter and an electric starter. The cubic capacity is 50 cm^3 and their top speed is restricted to 50 km/h. These motorcycles can only be ridden with a licence or compulsory bike test, but they are tax-free and do not have to be registered. The rider must have a driving licence of category A. This can be issued at age 16.

Engines. The types used in this vehicle are generally two or four-stroke, single reciprocating-piston engines. The common outputs are up to 3.2 kW at speeds of up to 7,500 rpm. Power transmission is by chain drive from 2 to 3-speed hand or foot selected gearboxes mounted on the rear wheel. Automatic gearboxes (two-speed or infinitely variable) can also be found.

21

21.1.3 Lightweight motorcycles

These are built as motorcycles or higher powered scooters with a cubic capacity greater than 50 cm³ but no more than 125 cm³. Their rated output must not exceed 11 kW. These vehicles can only be ridden with a licence. However, they do not have to be registered in some countries, thereby making them tax-free. They must have a valid registration plate and are therefore subject to a regular general inspection (GI). The rider must have a driving licence of category A. This can be issued at age 16. For riders under the age of 18 years, the vehicle-specific top speed is restricted to 80 km/h.

21.1.4 Scooters (Fig. 1)

These are a special type of low powered motorcycle that can be ridden without sitting astride the bike. They have smaller wheels, no pedals and a smaller wheelbase. The power unit is covered and can be found at the rear of the vehicle in a housing designed to incorporate power unit and suspension as one assembly. Scooters have different trade names depending on their intended use: city, fun, sport, classic, all-round or comfort.

Fig. 1: Scooter (sports scooter, 49 cm³, 3.2 kW)

Engines. In general, two and four-stroke, single reciprocating-piston engines with choke are installed. Engine data:

Capacity	Output	Speed
49 cm³	to 3.9 kW	7,500 rpm
to 125 cm³	to 14 kW	8,500 rpm
to 250 cm³	to 15.5 kW	7,500 rpm
to 500 cm³	to 29.4 kW	7,250 rpm

As alternate propulsion concepts there are also direct current electric motors with up to 4.8 kW output. The energy is drawn from four 12-volt lead-fleece rechargeable batteries. With this electrically stored energy, a top speed of 50 km/h can be reached and a distance of approx. 40 to 60 km.

Drive train. Power transmission on modern scooters is normally by a compact power unit and suspension assembly, comprising of:

- Engine
- Clutch
- Variator
- Rear wheel transmission

Power unit and suspension assembly (Fig. 2). The design is normally made up of engine and transmission housing and is swivel mounted in a frame. It also serves as a swing arm to locate the rear wheel. Drive is normally from a single cylinder engine. The crankshaft is joined to the driving pulley pair, also referred to as a variator.

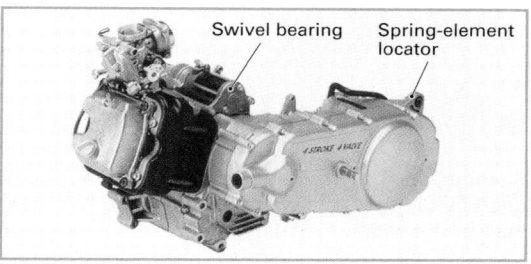

Fig. 2: Power unit and suspension assembly

Variator (Fig. 3). This comprises a driving and driven pulley pair. The different infinitely variable gear reduction ratios are given by the change in effective input and output diameter of the pulleys.

Operating principle. When starting, the drive plate pair has a small effective plate diameter due to lower centrifugal forces. In this way, a high to low ratio is given.

If the input speed is increased, the rise in centrifugal force pushes the rollers outwards and the effective diameter of the drive plate pair becomes greater. At the same time, there is a smaller diameter at the output plate pair. In this way, the gear ratio is lower and output speed increases.

Starting clutch. This is designed as a centrifugal clutch and sits on the output shaft.

Rear wheel transmission. This can be found on the output shaft and converts the motive force via a gear mechanism with spur gear teeth to a low ratio.

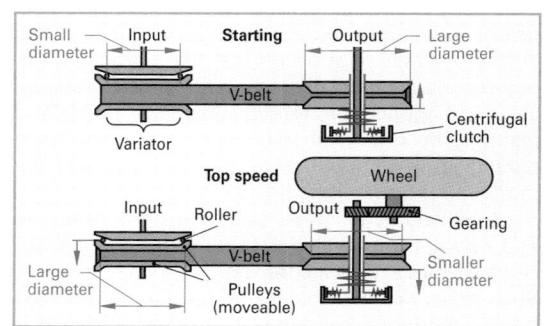

Fig. 3: Gear ratios when pulling away

21

Frame (Fig. 1). This is normally a bent pipe frame with attachment elements to which the power unit and suspension assembly with mono shock absorber, telescopic fork and trim parts can be fixed.

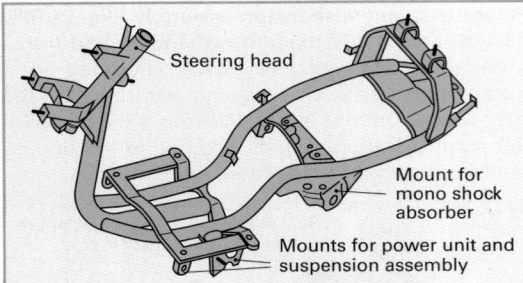

Fig. 1: Tubular frame for scooters

Suspension. The front wheel is normally dampened by an upside down telescopic fork. The main load of the scooter is suspended at the rear wheel by a mono shock absorber with external coil spring. This is attached between the frame and the power unit and suspension assembly.

Brakes. Normally, scooters are decelerated at the front wheel with single or dual disc brakes that are hydraulically actuated. The rear wheel can be braked on small scooters with a foot lever by drum brakes and disc brakes where performance is higher.

21.1.5 Motorcycles

These are motor-powered bikes that the rider sits astride. The engine capacity is over 50 cm^3 and the vehicle-specific top speed is above 45 km/h. A motorbike licence is required to ride these vehicles, they have to be registered, road tax has to be paid on them and they are subject to periodical road worthiness inspections. The rider must be over 18 years of age and have passed a compulsory bike test. This allows the person to ride motorcycles with an engine output of up to 25 kW. The power to weight ratio must not exceed 0.16 kW/kg in this instance. Not until two years´ practical riding experience have elapsed can this restriction be lifted, on application, without further training and testing. The rider will then receive the full European class A driving licence that allows him or her to ride all motorcycles.

A classification is made between light, medium and heavy motorcycles, whereby the type of use can also be used as a criterion.

Such machines are sold as enduros or motocross bikes, choppers or cruisers, touring bikes and sports bikes.

Enduro-, motorcross machine (Fig. 2). These have a high ground clearance, large suspension travel, a high mounted exhaust system and the tyres have a coarse, stud-like tread. They are normally powered by single cylinder, two or four-stroke engines that have a cubic capacity of up to 650 cm^3. The engines have an output of up to 47 kW at speeds up to 9,000 rpm.

Fig. 2: Enduro-, motorcross machine

Choppers, cruisers (Fig. 3). These have high handle bars that stretch far back and the front forks are tilted at a sharp angle and stretched to the front. The seat is designed as a stepped bench seat. The power units and components are on show and chrome-plated. The rear wheel is normally wider. The engines have a cubic capacity of up to 1,450 cm^3 and return up to 54 kW of power at engine speeds of up to 5,500 rpm.

Fig. 3: Motorcycle (chopper)

Touring bike (Fig. 1, Page 621). These have high handle bars and a comfortable seat for rider and pillion passenger. The motorcycle is normally fitted with part or full fairing that serves to protect against wind and weather. For luggage there are panniers and luggage racks available. The engine capacity can range up to 1,800 cm^3 with engine performances of up to 112 kW and speeds up to 10,000 rpm.

Fig. 1: Touring bike

Sports bike (Fig. 2). These feature flat handle bars, in many cases just one seat and full aerodynamic fairing. This is intended to offer the rider wind protection at high speeds and provide, above all, a very low drag coefficient (c_d factor) of the motorcycle. The engines have cubic capacities up to 1,300 cm^3 and return up to 130 kW of power with speeds up to 12,500 rpm.

Fig. 2: Sports bike

21.2 Motorcycle engines

For smaller capacities of up to about 650 cm^3 normally one or two-cylinder, two or four-stroke engines are used. On units with greater capacity, multiple-cylinder engines with 2, 3 or 4 cylinders are common. These are installed as inline, boxer or V-engines. The crankcase of the motorcycle engine shown in **Fig. 3** is made from a light aluminium pressure cast alloy. The cylinder linings have a highly resistant, low friction, nickel silicon carbide dispersion coating. The crankshaft is cast from a high quality steel alloy and has five mounting points in the crankcase with trimetal bearings. From the crank drive, the torque is transferred via a bevel tooth gear primary drive to the clutch. Inlet and exhaust phases are controlled by two overhead camshafts that are driven by a timing chain. These have five bearing points in a one-piece light metal cylinder head and are made us-

ing a shell cast procedure. They actuate the valves in a V-layout via barrel tappets. To improve the charge, four valves per cylinder are used. The angled position provides a compact, apex shaped combustion chamber, in which the spark plug is located centrally. Valve guides and valve seat rings are made from sintered metal and shrunk to fit.

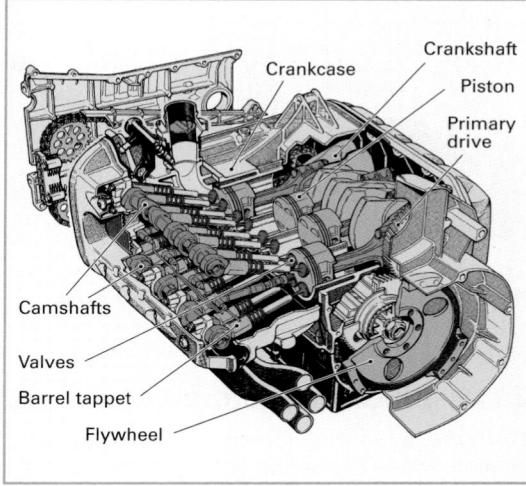

Fig. 3: Four cylinder inline engine 1,200 cm^3

21.3 Exhaust system

On two-stroke engines in particular, exhaust pipe with silencer and intake line with air filter are carefully adapted to each other. Modifications can lead to deregistration and cause performance drops in certain speed ranges.

The exhaust systems are made from painted or chrome-plated sheet steel, in some cases also stainless steel. If the exhaust is located in the vicinity of foot rests, it is fitted with a heat shield. The two-stage scooter sports exhaust system shown in **Fig. 4** comprises a front silencer and screw-fitted carbon-fibre rear silencer.

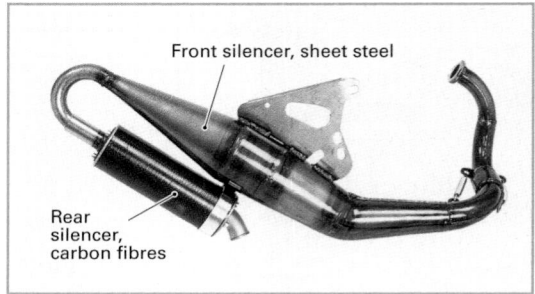

Fig. 4: Racing exhaust for scooters

21

To reduce the harmful emissions, unregulated or regulated metal carrier 3-way catalysts are fitted in the exhaust system. On engines with regulated catalyst, the heated λ sensor picks up the unburnt oxygen level and the control unit regulates the mixture composition based on this value, so a stoichiometric mixture of $\lambda = 1$ is given. In this way, an optimal conversion rate can be assured.

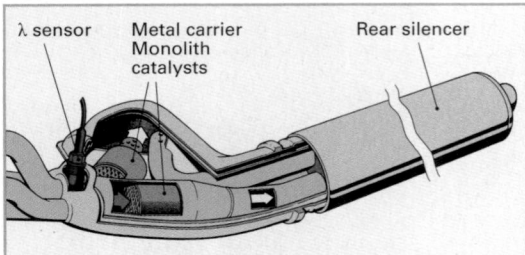

Fig. 1: **Exhaust system with 3-way catalyst**

21.4 Mixture formation

The following mixture-formation systems are used on motorcycles depending on the engine capacity:
- **Single or multiple carburettor systems** with secondary air system or unregulated catalyst
- **Injection systems** (engine management systems)

The following motorcycle carburettors are used:
- **Flat-flow** and **angled-flow carburettors**
- **Slide-valve carburettor** with
  - mechanical actuation of the throttle slide valve
  - pneumatic actuation of the plunger

Flat-flow slide-valve carburettor (Fig. 2)

Operating principle.
At idle, fuel is drawn in via the idling jet with the plunger downwards and mixed with air in the in-

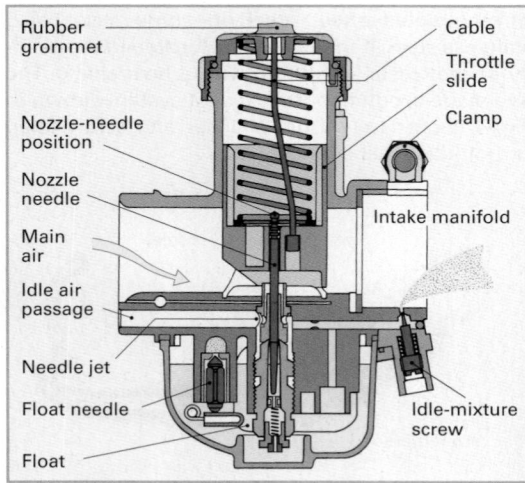

Fig. 2: **Flat-flow slide-valve carburettor**

take manifold. When the accelerator is actuated, the plunger is lifted up by the throttle cable and fuel is drawn in by the prevailing vacuum via the needle jet. The tapered nozzle needle opens up the annular orifice in relation to the throttle slide valve position, whereby the fuel supply is adapted to the volume of air drawn in. At part throttle, the contour of the needle jet of the fuel inlet is slightly restricted because less fuel is required than at full throttle.

Acceleration enrichment system. During acceleration, the nozzle needle is lifted together with the throttle slide valve. The small plunger located in the nozzle block is pushed upwards by spring pressure. The fuel above the plunger is thereby forced through the annular orifice, resulting in enrichment.

Cold start. Mixture enrichment is carried out as follows:
- Mechanically via cable and slide valve. Here, the air slide of the intake cross-section of the carburettor is reduced and a higher vacuum is generated at the nozzle block.
- Electro-hydraulically by a cold start valve. On a cold engine it opens a bypass channel that feeds additional air/fuel mixture behind the slide valve.

WORKSHOP NOTES

Vehicle does not start.

If the engine does not start after a long period out of use, the fault could be in a non-combustible air/fuel mixture due to components in the fuel with a low boiling point having evaporated.

The following measures must then be applied:
- Remove carburettor and dismantle.
- Visually check nozzles, nozzle seat and float chamber for deposits.
- Visually check moving parts for mechanical wear.
- Clean idler/main jet and nozzle seat, if necessary, in an ultrasonic bath.
- Check float condition and adjust as necessary.
- Assemble carburettor with new gasket/seal set and install in the correct position.
- Basic setting of idler jet: screw in idler jet onto stop and unscrew 2 to 2.5 turns.
- Start engine and adjust idling speed with lights switched on (800 to 1,200 rpm). This is carried out by lifting the throttle slide using the idling speed screw or throttle cable.
- Adjust CO content using mixture screw to between 1.5 and 4 % depending on manufacturer. Readjust idling speed as necessary.

21

Motronic system (Fig. 1)

On motorcycles with large cubic capacity, control of the timing and injection system is by way of Digital Motor Electronics **(DME)**. The control unit receives the following **input signals**:

- Engine speed
- Air and coolant temperature
- λ signal

The control unit calculates the quantity to be injected and regulates the mixture in the λ-1 range. The following **actuators** are activated:

- Fuel pump
- Automatic cold-starting device
- Fuel injectors
- Ignition system
- Electric fan

Fuel pump. This provides the constant fuel system pressure of 3.5 bar.

Fuel injectors. These are actuated on the earth side and inject once per crankshaft revolution half the volume of fuel. For cold starting, injection occurs twice.

Automatic cold-starting device. This works with electronic regulation by adjusting the throttle valve strip using a servo-motor with worm gear assembly. In this way, a stable idling speed is set for all operating conditions.

Ignition system. The control unit interrupts the primary current, which triggers ignition sparks.

Emergency running. In the event of input signal failure, this is detected by the control unit, stored in the fault memory and a substitute value is made available. In the event of engine speed signal failure, it can no longer be replaced and the engine will not run.

> **WORKSHOP NOTES**
>
> **Self-diagnosis.** If faults occur, e.g. in the lambda closed-loop control, in sensors or actuators, these are stored in the fault memory of the control unit. By way of the diagnosis interface, a fault reader can be used to interrogate the fault memory. Once the faults have been rectified, the fault memory has to be erased.
>
> **Actuator diagnosis.** Components can be actuated with this test to check their function.

21.5 Engine cooling

Motorcycle engines of up to about 800 cm³ are normally air-cooled. Cooling is from the wind against the cylinders. These are made from an aluminium alloy and have large cooling fins for greater heat dissipation. For motorcycles and scooters on which the engine is covered by panels or fairing, forced-air cooling is generally employed. Liquid-cooled engines can be found in all engine size classes. These engines are a lot quieter and less sensitive to thermal load. On large-sized motorcycles with higher output, pump circulation cooling systems with radiator, additional electric fan, thermostat control and expansion tank are employed.

21.6 Engine lubrication

Mixture lubrication. This is normally used on smaller, two-stroke engines. A self-mixing, two-stroke oil is added to the fuel during refilling. Mixture ratio 1:20 to 1:100.

Pressurised lubrication (Fig. 2). This is employed on large, four-stroke engines and has a similar structure to that of passenger vehicles. An oil pump draws the oil out of the oil pan and pumps it to the bearing points via the oil filter. An oil-pressure switch informs the driver about faults in the system by way of a warning lamp in the cockpit.

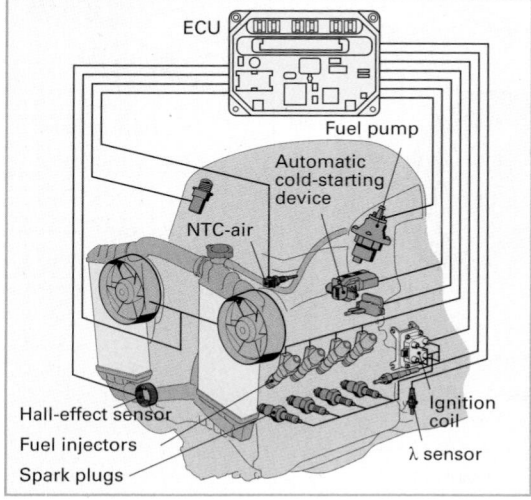

Fig. 1: Digital engine electronics (motorcycle engine)

ECU
Fuel pump
Automatic cold-starting device
NTC-air
Hall-effect sensor
Fuel injectors
Spark plugs
Ignition coil
λ sensor

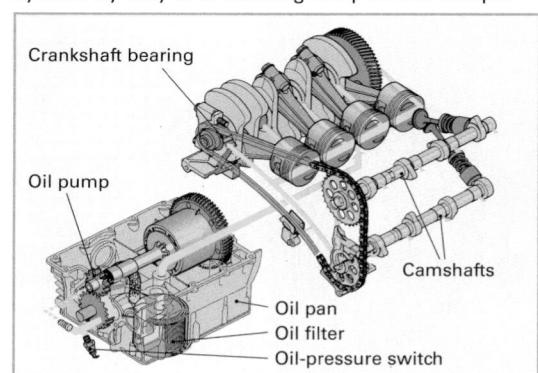

Fig. 2: Pressurised lubrication, oil circuit

Crankshaft bearing
Oil pump
Camshafts
Oil pan
Oil filter
Oil-pressure switch

21

21.7 Clutch

This serves to transfer power and assist in pulling away. Low- and high-powered mopeds normally have a self-actuating clutch (one-speed automatic) **(Fig. 1)**. As speed increases, the flyweights move outwards and the drive carrier bonds with the clutch drum, which is joined to the output shaft.

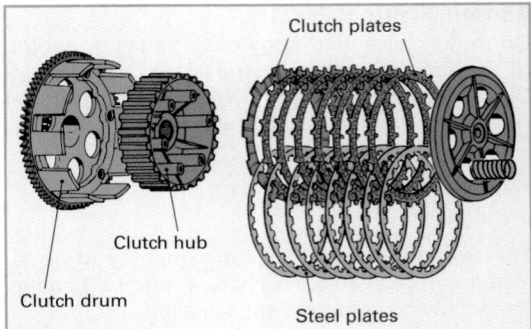

Fig. 2: Multi-plate clutch

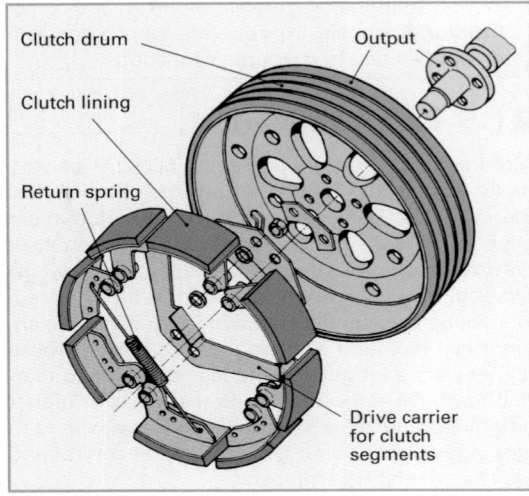

Fig. 1: Centrifugal clutch

Multi-plate clutch (Fig. 2). This is the type of clutch construction that can normally be found on motorcycles. It comprises several successively and alter- nately located clutch friction plates with outer splines and steel plates with inner splines. When engaged, these plates join the drive clutch drum to the clutch hub. The friction plates in the clutch housing are submersed in oil; in rare cases they are of the dry type.

Single-plate clutch with hydraulic actuation (Fig. 3). This is installed on motorcycles with high output and large capacity. It is designed similar to that of a passenger vehicle. Actuation is by way of a self-adjusting hydraulic system. The master cylinder is actuated using the hand lever and pres- sure is imparted on the slave cylinder. This presses the push rod against the diaphragm spring. The pressure plate is relieved of pressure and the clutch released.

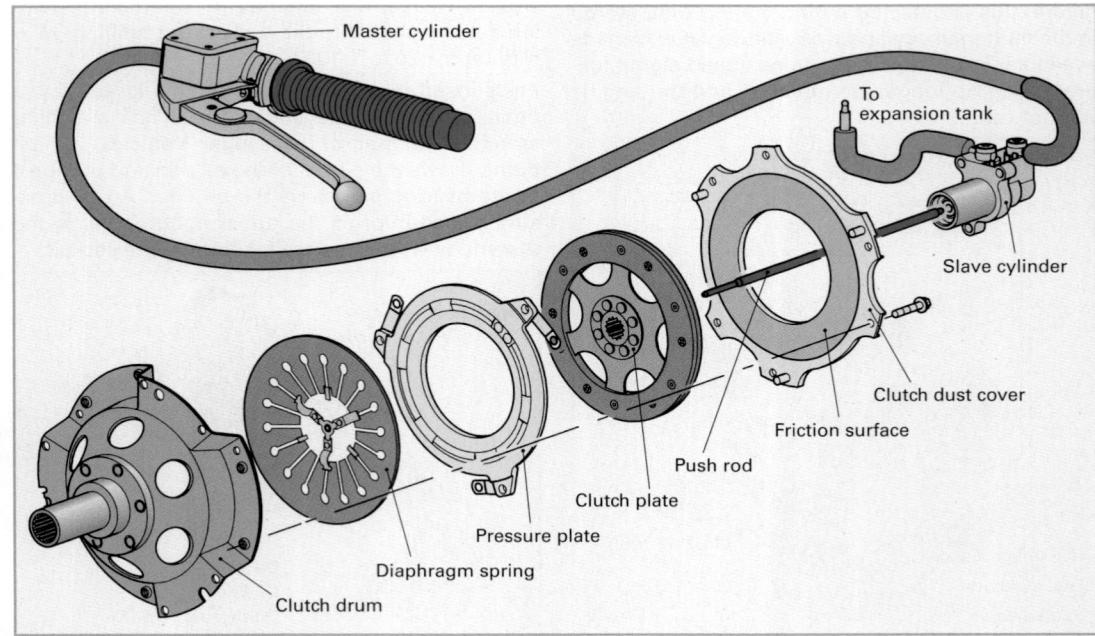

Fig. 3: Single-plate clutch with hydraulic actuation

21.8 Drive train

The torque generated by the engine is transferred from the **primary drive (Fig. 1)** to the clutch. It comprises a toothed gear pair or a toothed chain, which converts the torque and the speed. In the 4-speed gearbox with foot selection there is a further power conversion.

Secondary drive (Fig. 1). This transfers the motive force from the gearbox to the rear wheel. Chain drives, propellor shaft drives **(Fig. 3)** or toothed-belt drives are used.

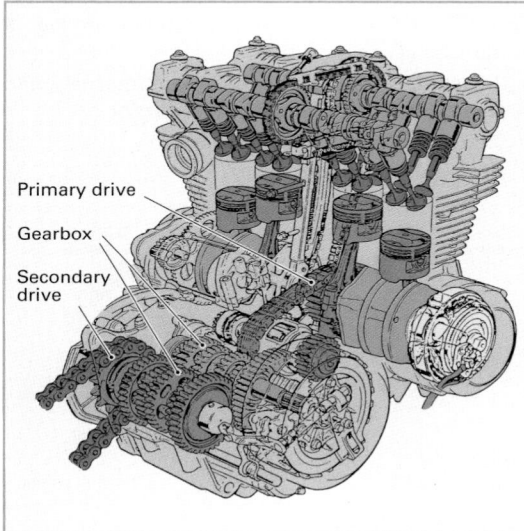

Fig. 1: Motorcycle engine with primary drive

Chain drive (Fig. 2). With this type of drive system, normally roller chains with O-rings or sleeves are used. Endless chains are installed or split chains with chain lock.
The rollers are permanently filled with a lubricant. The O-rings between roller and outer link plate are designed to prevent the lubricant from escaping. These are used on road motorcycles. For offroad or motocross bikes, sleeve chains are fitted.

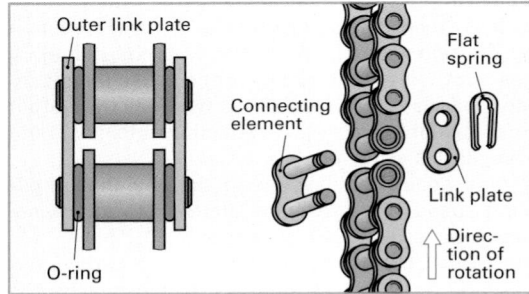

Fig. 2: O-ring roller chain and chain lock

Propellor shaft drive. This type of drive system is used mainly on motorcycles with high engine output. The motive force is transferred from the gearbox via a propeller shaft from the drive pinion to the crown wheel. The gear reduction to the rear wheel is normally $i \approx 3.0$. Despite being more complicated in design, this type of drive has the following advantages:

- Maintenance-free
- Less running noise
- High functional security
- Not sensitive to dirt

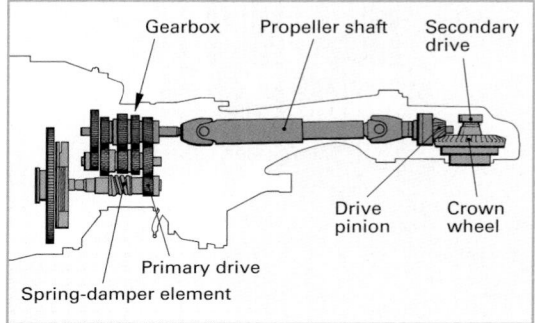

Fig. 3: Drive with propeller shaft

Gearbox (Fig. 4). Motorcycles normally have a straight-toothed, foot-actuated, shift dog gearbox. The input shaft, which features a drive off damper (spring-damper element) **(Fig. 3)**, drives the auxiliary shaft. The selector gears sit on the auxiliary shaft and the output shaft and are moved by gearshift forks. The driver actuates the pawl during gear selection by way of the foot lever. This in turn turns the gear selector drum and moves the gearshift forks and the selector gears via the gearshift gates. The gears are changed up by lifting the foot lever and changed down by pushing the foot lever (sequential selection).

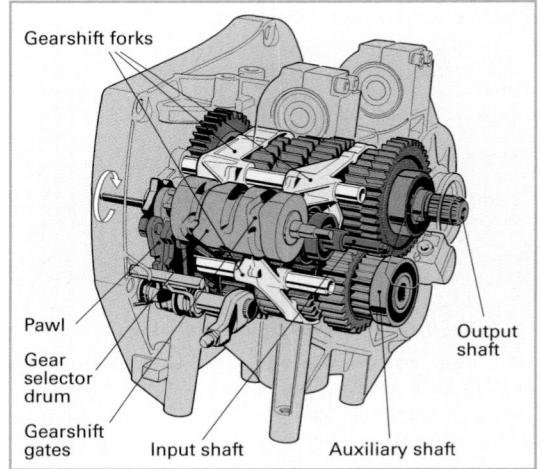

Fig. 4: 6-speed motorcycle gearbox

21

5-speed shift dog gearbox

The 5-speed gearbox shown **(Fig. 1)** is an identical-shaft gearbox and comprises in essence of the following components:

- Input shaft
- Output shaft (main shaft)
- Countershaft
- Gear wheels

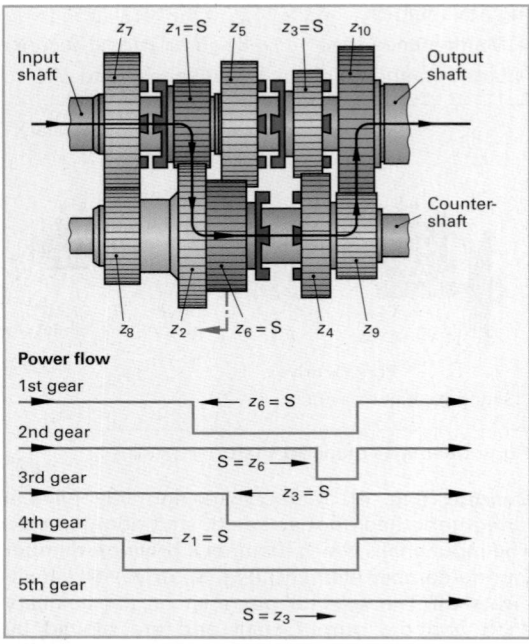

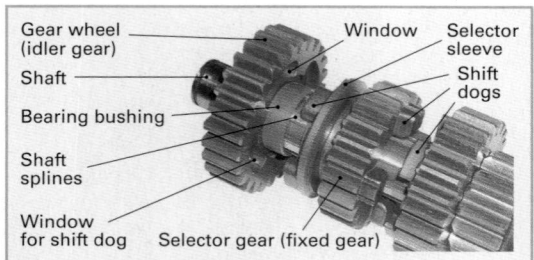

Fig. 1: 5-speed shift dog gearbox (1st gear selected)

Operating principle. The torque and speed conversion is brought about by paired gear wheels, of which one gear wheel is mounted on the output shaft (main shaft) and one on the countershaft.

Since all gear wheels are always engaged, one of the gear wheels from each of the gear wheel pairs must be free to rotate on its shaft.

Fig. 2: Shaft, fixed and idler gears, shift dogs

The torque transfer, from the main shaft to the countershaft for example, does not occur until a freewheeling gear wheel (idler gear) is joined to a selector gear S (e.g. z_6) and thereby its shaft. To do this, a selector gear S is moved to the left or right and engages with its front shift dogs in the windows of the idler gear.

Power flow, e.g. in 1st gear:

Input shaft → Fixed gear z_1 → Idler gear z_2 → Selector gear $z_6 = S$ → Countershaft → Fixed gear z_9 → Fixed gear z_{10} → Output shaft.

Fixed gears are: z_1, z_3, z_6, z_8, z_9, z_{10}.

Idler gears are: z_2, z_4, z_5, z_7.

Gear wheels z_9 and z_{10} form a constant part gear reduction ($i_1 = z_{10}/z_9$). This reduction is always effective except in 5th gear.

Automatic gearbox

These are installed mostly on low-powered motorcycles. The 2-speed automatic shown in **Fig. 3** comprises in essence of:

- Planetary gearbox with sun gear, internal gear, planetary gears, planetary gear carrier
- Centrifugal clutch with flyweights
- One-way clutch and pulley

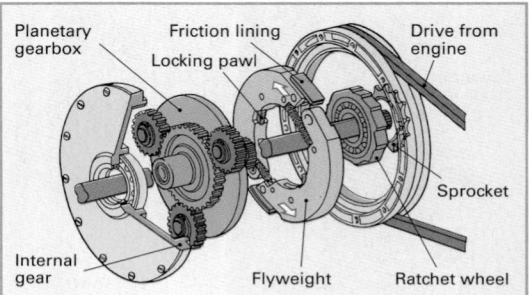

Fig. 3: 2-speed automatic

Planetary gearbox: This is where the speed and torque conversion is carried out. 2 gears are automatically selected.

Internal gear. This is joined firmly to the pulley and turns at engine speed.

Sun gear. This is joined via the one-way clutch to the pedal crank bearing shaft. The one-way clutch locks one direction of the sun gear.

Planetary gear carrier. This forms an output to the chain sprocket.

Operating principle. When starting, the internal gear is driven, the sun gear is held in this direction by the one-way clutch. In this way, the planetary gears turn about the sun gear and cause the planetary gear carrier to rotate. This provides the output to the chain sprocket. A reduction into low ratio (1st gear) is achieved.

If road speed increases, and thereby the output shaft speed, the flyweights attached to the planetary gear carrier move outwards.

In this way, the planetary gear carrier and the internal gear are blocked together. A direct transmission ratio of 1 : 1 is achieved, 2nd gear is selected.

21

21.9 Electrical system

This normally comprises the following main components on motorcycles:

- Cockpit
- Central electrics
- Start system
- Alternator
- Ignition system
- Headlight system

Cockpit (Fig. 1). This informs the driver about almost all the important functions of the vehicle, such as road speed, engine speed, oil pressure, tank volume, charging system, ABS system, turn-signal indicator system, lighting.

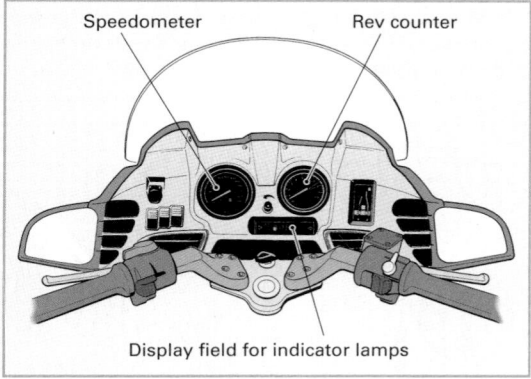

Fig. 1: Motorcycle cockpit

Central electrics. These are located in a housing. Fuses and relays can be found here for equipment such as: starter motor, petrol pump, horn, turn-signal indicator, motronic and ABS system.

Starting system (Fig. 2). In addition to the mechanical kickstart system, which is still installed on most vehicles, an electric starter system is generally used these days. It comprises an electric motor fitted with a reduction gear and a pinion with one-way clutch. On larger motorcycles, reduction-gear starters are installed similar to those used on passenger vehicles.

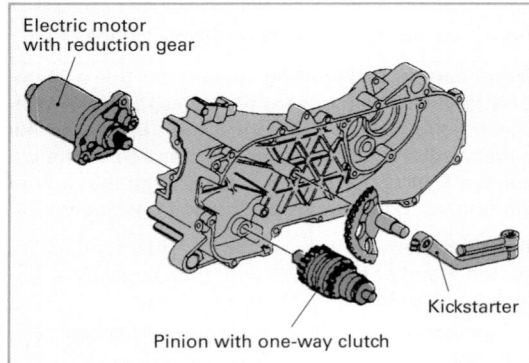

Fig. 2: Starter motor of a scooter

Voltage generation (Fig. 3). On small engines, alternator systems in combination with solenoid ignition systems are used. The pole wheel with permanent magnet sits firmly on the crankshaft and turns with it without slipping thanks to a Woodruff key. In the coils installed in the engine frame, an AC voltage is generated which is converted to direct current. The ignition and lighting system is powered with this energy and the battery (if fitted) is charged.

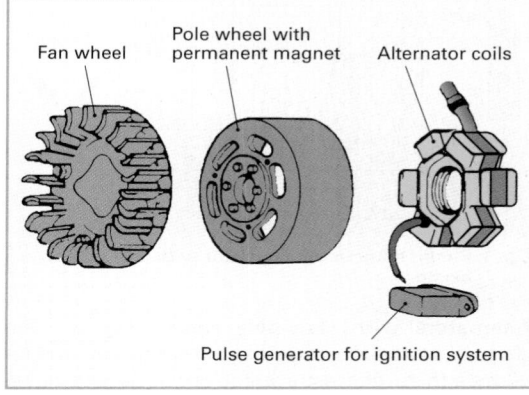

Fig. 3: Alternator system of a scooter

The following alternators can be found almost exclusively on large motorcycles these days: Alternators with

- permanent excitor
- electromagnetic claw-pole rotors.

Alternators with permanent exitor (Fig. 4). On this type, the permanent-magnet rotor is driven by the crankshaft. Three-phase current is generated in the fixed three-phase stator winding. This is converted to DC outside in an electronic assembly, whereby the charge voltage is restricted to 14 V. Outputs of up to about 300 W are reached.

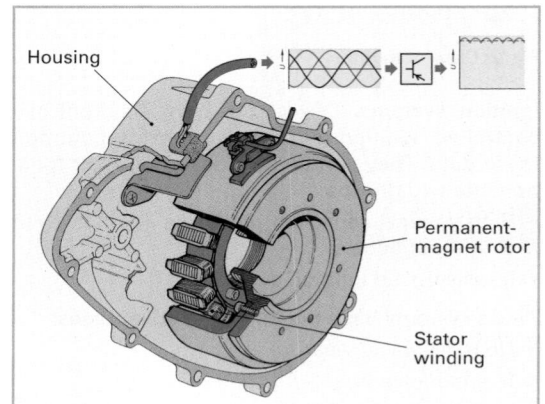

Fig. 4: Alternator with permanent exitor

21

Regulation (Fig. 1). Beneath the regulator response voltage, the generated and converted voltage charges the battery and supplies the consumers. Once the specified maximum charge is reached, the electronic regulator actuates the thyristors, which causes a short circuit of the stator windings to earth, thereby interrupting voltage output.

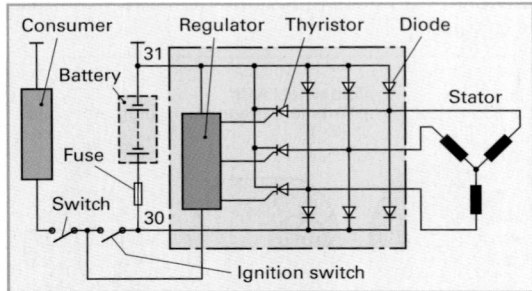

Fig. 1: Circuit diagram of alternator with permanent excitor

Alternators with claw-pole rotors (Fig. 2). The claw-pole rotor is screwed to the crankshaft and induces a three-phase current in the three-phase stator winding. This is converted to DC via a transistor bridge circuit, as on passenger vehicles. A regulator controls the excitor current. A charge voltage of 14 volts at outputs of up to 850 watts is made available.

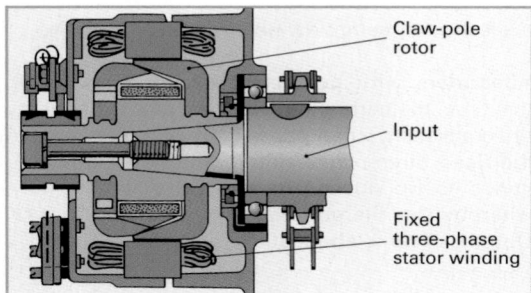

Fig. 2: Claw-pole rotor

Ignition systems. On motorcycles, electronically controlled ignition systems are most common these days. They can be split based on their function into two groups:

- High-voltage capacitor discharge ignition with or without battery
- Transistorised ignition systems

These systems have the following **advantages:**

- No mechanical wear
- Maintenance-free
- High secondary voltage at high speed
- Not sensitive to spark plug soiling

High-voltage capacitor discharge ignition (Fig. 3). These are also referred to as capacitive discharge ignition **(CDI)** systems.

Operating principle. This system features a capacitor discharge coil and an ignition pulse generator. If the pole wheel turns with the permanent magnet, a voltage of 100 V to 400 V is induced in the capacitor charging coil. This voltage is converted to DC and charges the capacitor. The thyristor is actuated at the gate at ignition point by the ignition pulse generator coil. It switches to positive and becomes conductive. The capacitor, which is connected in series to the primary winding, discharges itself rapidly and results in a high ignition voltage on the secondary side. The ignition timing is based on speed and is controlled by the CDI control unit.

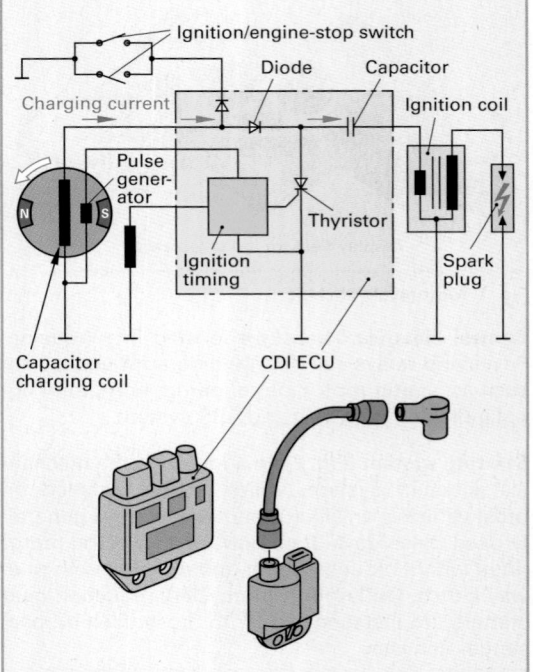

Fig. 3: High-voltage capacitor discharge ignition

Direct current CDI ignition system. On this system, a battery is used for charging the capacitor. A voltage transformer in the control unit increases the battery voltage to 220 V, which is stored in the capacitor. This system has the advantage that a high ignition voltage is available even at low speeds.

Transistorised ignition systems. This type is installed mainly on engines with high capacity. A difference is made between two systems:

- Transistorised ignition systems with pulse generator
- Digitally controlled transistorised ignition systems

Transistorised ignition systems with pulse generator (Fig. 1)

Operating principle. On this system, the basis of the transistor is actuated by the pulse generator and the primary current can flow. If the basic current of the transistor is interrupted, the primary current is switched off and high voltage is induced in the ignition coil on the secondary side. An ignition timing device in the control unit controls the ignition point. The period of the primary current is determined by way of dwell angle control.

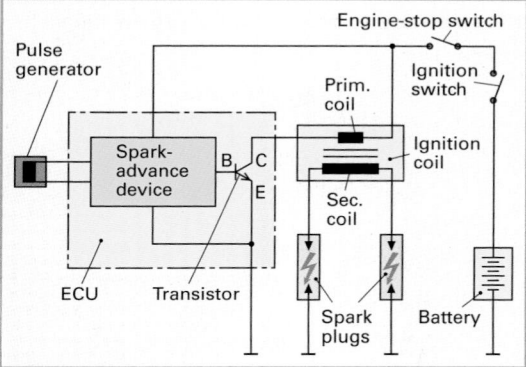

Fig. 1: Transistorised ignition

Digitally controlled transistorised ignition systems.

Design. These comprise one or two pulse generators, ignition control unit, ignition coil(s) and spark plug(s).

Function. A pulse generator informs the control unit about the engine speed and the crankshaft position. Using this information, the control unit calculates the optimum ignition point from the stored ignition maps.

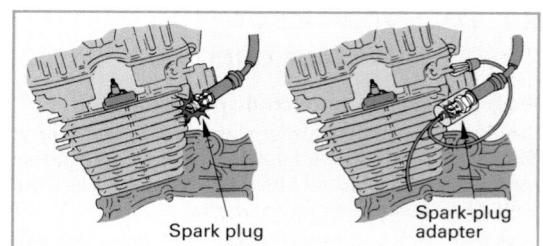

Fig. 2: Spark probe

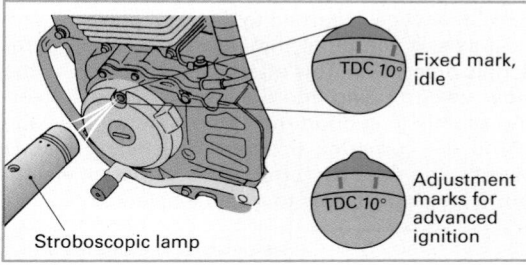

Fig. 3: Checking ignition point

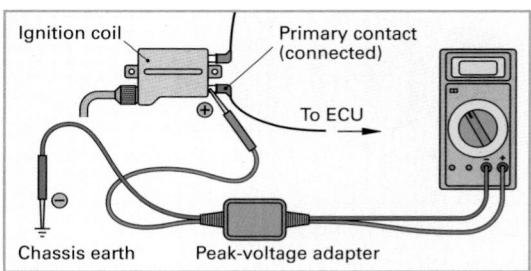

Fig. 4: Measurement of primary current of ignition coil with peak-voltage adapter

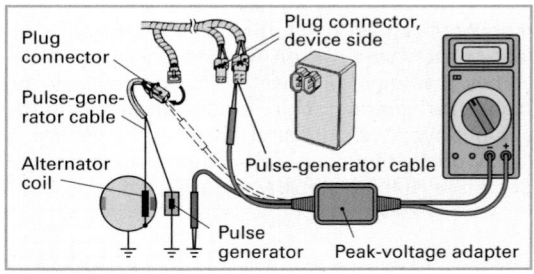

Fig. 5: Measurement of pulse generator and alternator coil

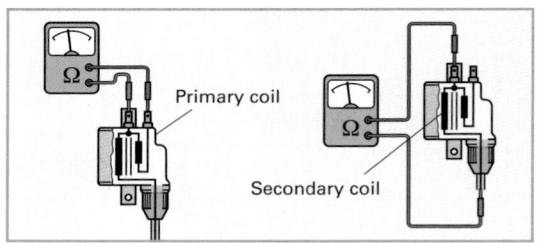

Fig. 6: Resistance measurement from primary and secondary coil

21

21.10 Dynamics of vehicular operation

Stabilisation by gyroscopic precision

Single-track vehicles are unstable when stationary; they tip over. As speed increases, gyroscopic forces are imparted on the wheel and stabilise the vehicle as it is travelling forwards.

Test (Fig. 1): If a rotating wheel is tilted to the left, for example, gyroscopic forces tip it to the right about its vertical axis until it has regained a balance. If a wheel is turned to the right about its vertical axis, it will tilt to the left as a result of gyroscopic precession. This effect can be utilised by the rider when driving into a corner by countersteering slightly to support the tilting effect of the vehicle in the cornering position. By shifting his/her weight, the rider can bring the centrifugal, weight and gyroscopic forces to a stable balance.

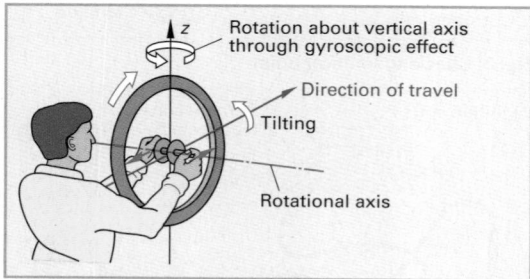

Fig. 1: Gyroscopic precession

Castor effect (Fig. 2). The castor is the distance from the intersection point of the steering axis to the wheel contact point on the road surface. Due to friction forces or braking power that are effective at the wheel contact point, there is a self-centring effect M_R, for example, on the steered wheel. It becomes greater the higher the castor angle or the higher the steering angle. It pulls the wheel in the straight-ahead position and stabilises the vehicle. A low wobble tendency, self-centring steering and good true running stability are achieved.

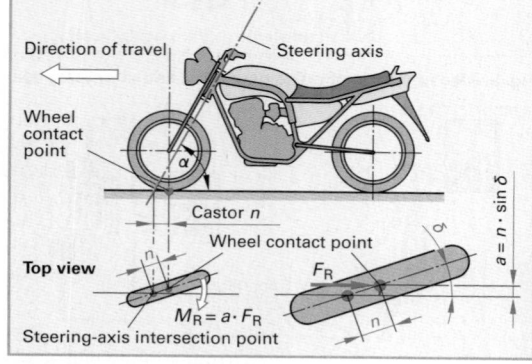

Fig. 2: Castor on motorcycle

Cornering. When riding a motorcycle, cornering is made possible at high speed by steering slightly. If a left-hand corner is taken, for example, the rider steers briefly to the right and the vehicle tilts to the left by way of gyroscopic precession. Stabilisation in an ideal cornering situation on a motorcycle is from the following moment balance **(Fig. 3)**.

$$G \times l_1 = F_z \times h_S$$

The moment, formed from weight and centre of gravity, is the same as the moment from the centrifugal force and centre point of gravity height.

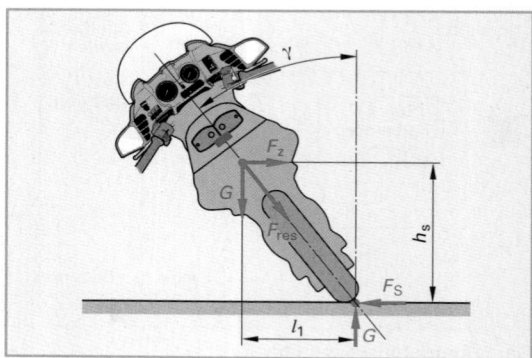

Fig. 3: Moment balancing during cornering

However, since the wheel contact point on wide tyres is not always in the centre, as shown in **Fig. 3**, but more towards the inside, there is a smaller moment $G \times l_2$ **(Fig. 4)** because l_2 is less than l_1. In order that the balance can be restored, the motorcycle must be ridden at more of an angle.

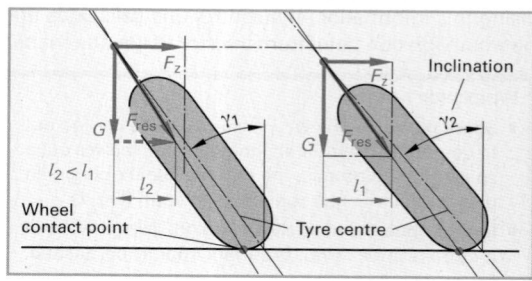

Fig. 4: Real cornering

REVIEW QUESTIONS

1 What types of motorcycle are there?
2 What does a multiple clutch consist of?
3 How does total-loss lubrication work?
4 Explain the design and function of a HV capacitor ignition system.
5 What is understood by primary and secondary drive?
6 What advantages do O-ring roller chains have?

21.11 Motorcycle frames

This is the supporting element of a motorcycle and is designed to provide a torsionally rigid connection between the front wheel and the rear wheel suspension.

Demands on frames:
- Low weight
- Torsionally rigid
- Low-vibration mounting of engine
- Appealing design
- High load capacity
- High elongation at fracture

A difference is made between pipe, pressed steel, light-metal pressure case frames and light-metal frames with profile-section structure.

Types of frame. There are many very different types of frames on motorcycles depending on the requirements. The single-loop tubular frame shown in **(Fig. 1)** is made from square-steel tube. The engine is integrated as a supporting element within the structure. An additional frame truss strengthens the construction.

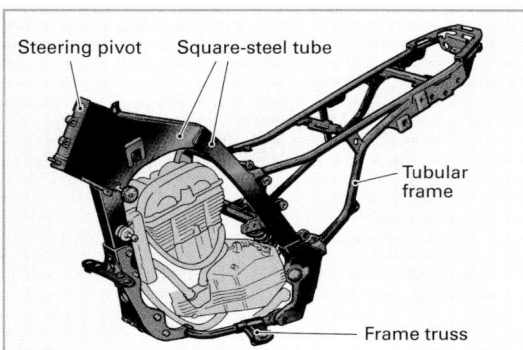

Fig. 1: Single-loop tubular frame

Double-loop frame (Fig. 2). This is a welded structure with steel tubes and steel cast elements and offers higher stability than single-loop frames.

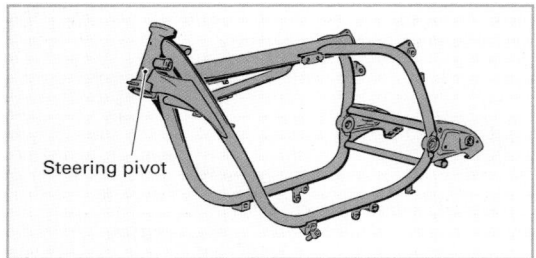

Fig. 2: Double-loop frame

Bridge tubular frame (Fig. 3). This is welded from steel tubes and is often referred to as a naked frame. Isolating the engine vibrations can be very difficult.

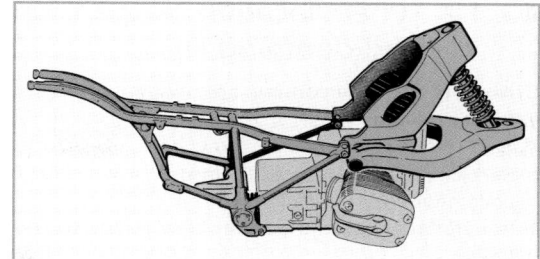

Fig. 3: Bridge tubular frame

Bridge frame with box sections (Fig. 4). This very flexible and torsionally rigid structure consists of an aluminium welded cast construction in a honeycomb design with webs and cavities. In this way, the highest rigidity is achieved in a compact structure.

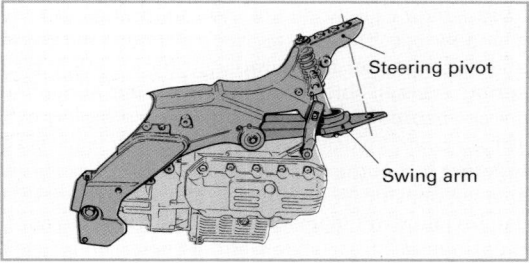

Fig. 4: Bridge frame with box sections

Aluminium frame with box-section design (delta box frame) (Fig. 5). This design is optimised for weight and rigidity. The box-section design means that the frame can be adapted optimally.

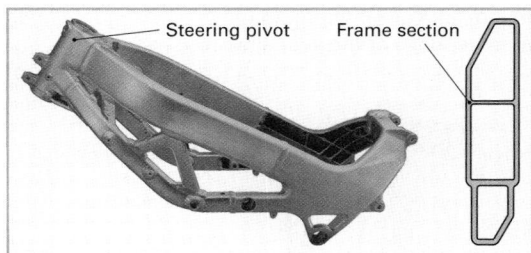

Fig. 5: Aluminium frame

Space frame (Fig. 6). This is a welded space-frame structure made from steel tubes and is thereby extremely torsionally rigid.

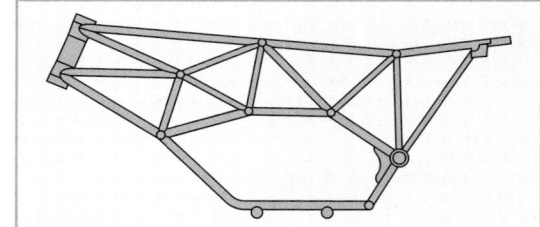

Fig. 6: Space frame

21

632 Motorcycle technology

21.12 Wheel location, suspension and damping

Drivability and ride comfort are dependent on the vehicle's design.

Tasks
- Reduce and dampen road bumps
- Locate wheels
- Transfer braking and acceleration forces to the frame

Front wheel location. The following designs are used:
- Telescopic forks
- Telelever system
- Upside down forks
- Steering knuckle

Telescopic fork (Fig. 1). The steering tube is mounted in the steering pivot of the frame. These forks have a high rigidity due to the fork bridge and the clamped axle. The two telescopic tubes (stanchion and sliding tube) are dampened by an integrated spring. A small spring or a rubber buffer above the shock-absorber rod prevents it from springing out. The air cushion located above the piston is pressed together on compression and results in progressive suspension. The hydraulic shock-absorber unit can be found in the lower part of the forks. On compression, the shock-absorber fluid is forced into the lower area and flows through holes in the valve unit. This has a lower shock absorbing effect in the compression stage, so the wheel suspension works softly and without jolts. On rebound, the oil has to flow back again. This is restricted by damper valves, which cause the pull stage to be harder. This results in a good damping effect and a more sensitive response of the suspension.

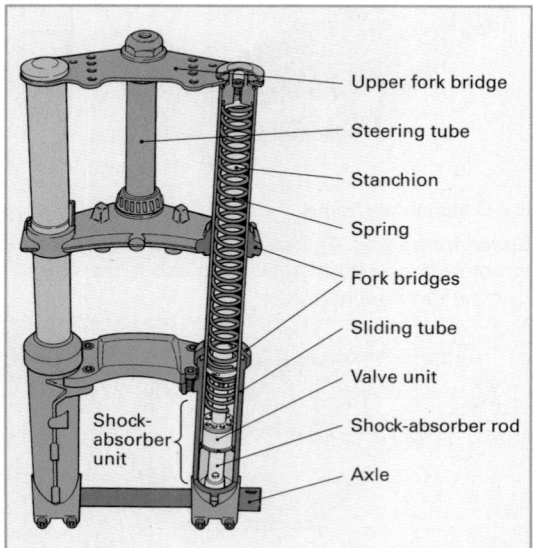

Fig. 1: Telescopic fork

- Upper fork bridge
- Steering tube
- Stanchion
- Spring
- Fork bridges
- Sliding tube
- Valve unit
- Shock-absorber rod
- Shock-absorber unit
- Axle

Upside down forks. This type of design is based on the same principle but the other way around. The more stable outer tube takes on the role of stanchion. The sliding tube, to which the axle is attached, compresses on this version. It is used on scooters, motocross and sports bikes. It has a high bending strength and rigidity but a markedly lower unsprung mass.

Telelever system (Fig. 2). On this system, the fork bridge is mounted at the top in a ball joint in the frame. The swivel mounted trailing arm is responsible for the location of the front wheel. It is suspended by a damped suspension strut. The system has the following advantages:
- More sensitive response due to less friction
- Greater directional stability on compression by increase in castor
- Anti-dive effect when braking

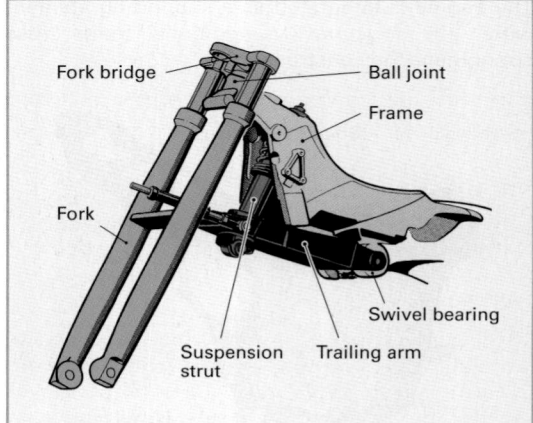

- Fork bridge
- Ball joint
- Frame
- Fork
- Swivel bearing
- Suspension strut
- Trailing arm

Fig. 2: Telelever system

Steering knuckle. On this version, the wheel is located by two swing arms. The suspension and damping work in much the same way as a telelever system via a central suspension strut. On compression, the castor is increased and steering stability is improved.

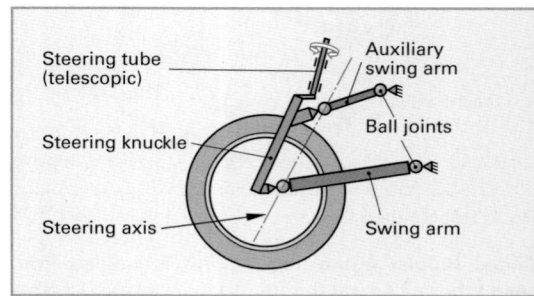

- Steering tube (telescopic)
- Auxiliary swing arm
- Ball joints
- Steering knuckle
- Steering axis
- Swing arm

Fig. 3: Steering knuckle

21

Rear wheel location. The wheels can be located by the following systems:

- Double swing arm
- Single swing arm
- Cantilever suspension
- Swing arm with lever

The swing arms are supported against the frame by suspension struts. The relevant interplay between suspension strut and lever influences the suspension response, ride comfort and road handling.

Double swing arm (Fig. 1). This type of design can be either a welded tubular steel structure or, as more commonly found these days, made from aluminium box sections. It is mounted via a swivel point in the frame and supports the suspension strut centrally on a transverse strut and holds the wheel in the rear part. This structure has a high level of rigidity. Removal of the wheel, however, is more complicated than on a single swing arm design.

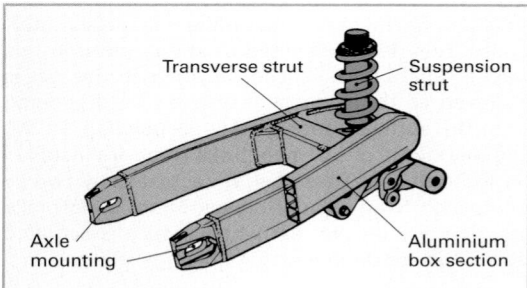

Fig. 1: Double swing arm

Single swing arm. The asymmetrical swing arm in aluminium box section design is mounted in the frame or on the engine with a swivel point and suspended via a central suspension strut. The wheel is attached by a central bolt connection. Removal of the wheel is therefore easier.

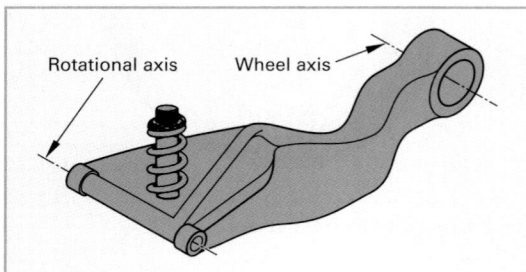

Fig. 2: Single swing arm

Paralever system (Fig. 3). This comprises a single swing arm and a compression rod. The swing arm locates the wheel and the compression rod positively influences suspension response under load change. Large down forces are prevented (elevator effect). A central suspension strut can be infinitely adjusted in its characteristic and damping.

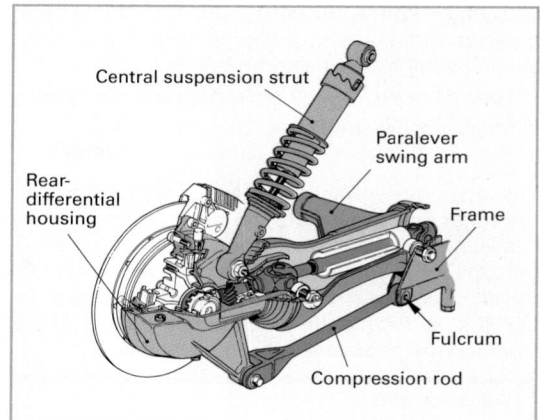

Fig. 3: Paralever system

Pro-Link system (Fig. 4). The swing arm of this suspension system is mounted in the frame. The suspension strut supports it via a lever system. On compression, there is a small amount of spring travel at the suspension strut bearing. If compression increases, the lever system stretches and the excursion and suspension characteristics thereby increase progressively.

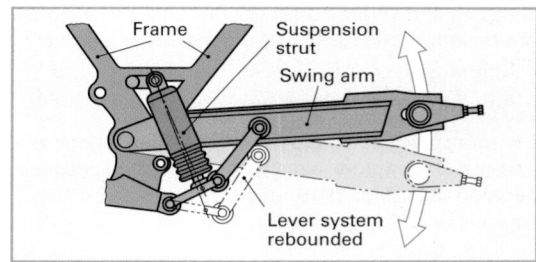

Fig. 4: Pro-Link system

Cantilever suspension (Fig. 5). This is a rear-wheel suspension system with angled swing arm and a centrally located suspension strut in the tank tunnel. High spring travel is possible on this system and good damping of the road imperfections can be achieved. The rigid rear wheel swing arm results in stable directional guidance of the motorcycle.

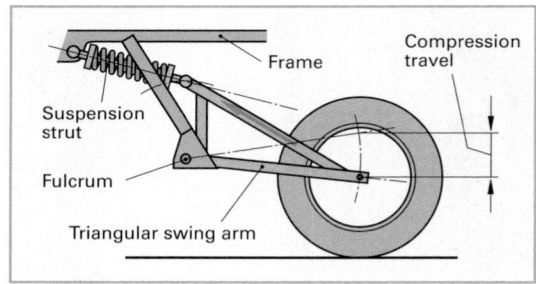

Fig. 5: Cantilever suspension

21

Damping response of front forks. The spring compression/rebound and damping response is influenced by the following components:

- Spring length
- Spring characteristic
- Shock-absorber fluid
- Unsprung mass of wheel and tyre

The longer the spring is, on front forks for example, the softer the suspension. The suspension response is adjusted by the manufacturer by altering the spiral gaps for example, so that a progressive characteristic is achieved. It is also possible by increasing the fluid to reduce the air chamber volume in the front forks to attain a harder characteristic.

> **WORKSHOP NOTES**
> - Change fork suspension fluid in accordance with manufacturer's specifications in order that wear resistance and damping properties remain intact.
> - Fill in correct amount of fluid as too much fluid will result in a harder response of the suspension.
> - Check fork seals for leaks.

Damping response on rear suspension struts. The damping and spring compression/rebound response can be influenced further by the following factors:

- Adjustment of the initial spring tension
- Adjustment of compression and pull-stage level

On motorcycles, single-tube, gas-filled shock absorbers or single-tube shock absorbers with expansion tank **(Fig. 1)** are used mostly.

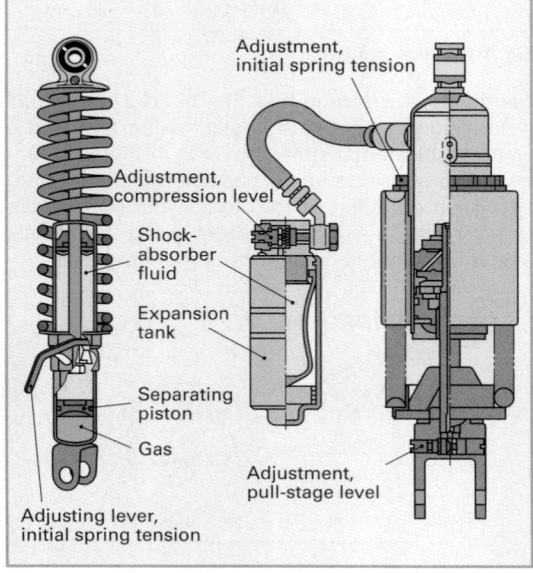

Fig. 1: Single-tube gas-filled shock absorber

Single-tube shock absorber. On this type, just the compression level of the damper can be adjusted. On single-tube shock absorbers with expansion tank, both the compression and pull-stage level can be adjusted. Adjusting the compression level means the downward response (shortening) of the suspension is affected, adjusting the pull-stage level means the upward response (elongation) of the suspension is affected.

21.13 Brakes

Disc brakes (Fig. 1). These are installed in most cases both on the front wheels and also rear wheels of motorcycles and scooters. Normally, the handbrake works the front wheel and the foot brake works on the rear wheel. Actuation is by hydraulic means.

Brake discs. These are manufactured from stainless steel and have floating mountings on large motorcycles. They feature slotted or spiral holes. In this way, a rapid and balanced response in the wet is achieved as the water and dirt can be dissipated from the surface more quickly. Depending on the engine output, one or two brake discs are installed on front wheels. These can be actuated by two or four-piston fixed calliper brakes. There is normally just one disc on the rear wheel, which is actuated by a single- or double-piston floating calliper brake.

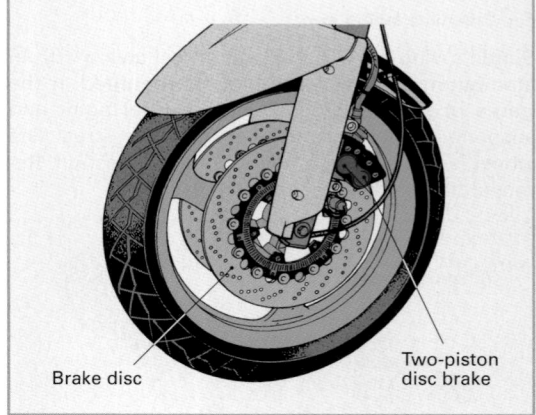

Fig. 2: Disc brake on front wheel

Brake pads. These are manufactured from sintered metal or semimetal. The materials have an equally high friction coefficient under all operating conditions.

Drum brake. These can still be found as internal-shoe brakes on the front and rear wheels of motor bicycles. The mechanically actuated brake lever turns the brake cam, which in turn presses the brake shoes against the inside of the brake drum.

Brake-pressure-controlled servo-mechanic braking system CBS (Combined Brake System) with **TCS** (Traction Control System).

> This **CBS TCS combined system (Fig. 1)** is installed on motorcycles with high output and provides optimal driving safety and directional stability during braking and acceleration.

- **CBS system** adapts braking force distribution optimally to the front and rear wheel based on the driving and load status.
- **ABS system** prevents blocking of the wheels during braking.
- **TCS system** is designed to prevent the driving wheel from slipping during acceleration.

CBS system. This is a servo-mechanical setup that features no electrical components whatsoever. The short wheelbase and the high centre point of gravity position of the motorcycle cause a pronounced dynamic wheel load distribution when braking from high speeds. Therefore, a clever servo-mechanical system is used to distribute brake pressure to the front and rear wheel in relation to the wheel load. Brake pressure is apportioned to achieve a balanced response. The rider can brake the motorcycle using the hand and/or the foot lever. Brake pressure distribution apportioned by the system depends on:

- Speed
- Road condition
- Vehicle weight
- Centre point of gravity

The system has the following **advantages**:

- Two independent brake circuits
- Balanced, simple operation of the brakes
- No counteracting disturbance from hand and foot brake
- Brake feeling from hand and foot brake remains intact

Operating principle

Braking with hand lever only. The brake pressure works on both the outer brake pistons of the front brake. As this happens, sophisticated mechanics transfer part of the braking force to the secondary master cylinder. This generates pressure that is imparted to both outer pistons of the rear brake via the proportional-control valve in between. The valve can perform a 3-staged braking force distribution.

Braking with foot lever only. The brake pressure works on the middle piston of the rear wheel, and on the left middle piston of the front wheel via the deceleration valve in the first phase only. In this way, there is a softer response of the brakes as pressure at the front wheel is reduced by about 50 %. As brake pressure rises, the deceleration valve regulates pressure to the right brake calliper of the front wheel. The typical dive effect that can occur when braking with the front brake is reduced to a great degree by this measure.

ABS TCS system. This comprises the following assemblies:

- Wheel-speed sensors
- Pressure modulators
- Driver unit
- Control unit

Operating principle. When the ignition is switched on, self-diagnosis is initiated by the control unit. In the event of defects, the system is switched off and the rider is informed by a warning lamp. If drive slip or a blocking tendency is detected by the control unit, the pressure modulator is actuated by the driver unit. An integrated electric motor actuates a control piston if a tendency to block is detected, which regulates brake pressure so that blocking is prevented. If the rear wheel develops a tendency to slip during acceleration, the TCS system intervenes and the control unit retards ignition until slip is no longer present. A control lamp informs the rider of this occurrence.

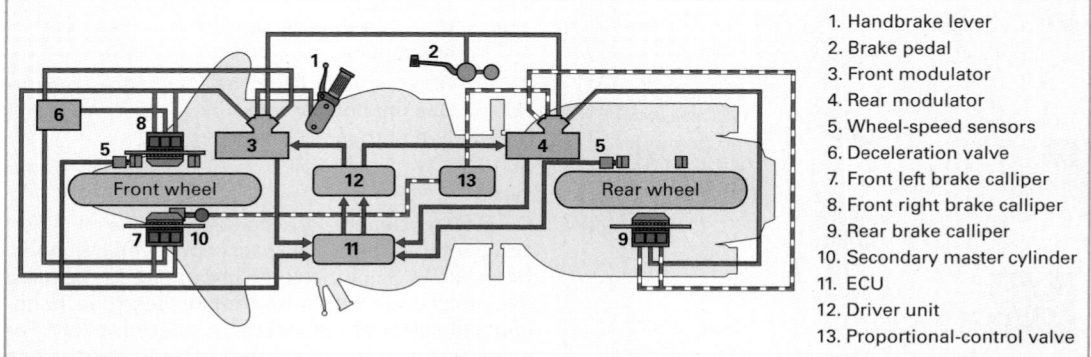

1. Handbrake lever
2. Brake pedal
3. Front modulator
4. Rear modulator
5. Wheel-speed sensors
6. Deceleration valve
7. Front left brake calliper
8. Front right brake calliper
9. Rear brake calliper
10. Secondary master cylinder
11. ECU
12. Driver unit
13. Proportional-control valve

Fig. 1: CBS ABS TCS system

Antilock braking system (Fig. 1)

It is installed in addition to the hydraulic braking system on motorcycles with higher output to increase directional stability.

Design: The system comprises the electrical and hydraulic components shown in **Fig. 1**.

Operating principle. When the ignition switch is turned, a self-test is carried out. If all components are intact, the system is made ready for operation. The wheel-speed sensors detect the wheel speed and the control unit calculates the slip. For braking without a tendency to block, the ABS does not influence either brake circuit (front and rear circuit). The relevant brake pressure generated by the rider works on both brake callipers. For braking with a tendency to block, the electromagnet in the respective pressure modulator is actuated by the electronic control unit. The control plunger is pulled downwards and the ball valve interrupts flow to the brake calliper. As a result of the further descending control plunger, a volume increase is given and thereby a rapid pressure drop in the brake circuit. The wheel is then accelerated. This closed-loop control process is repeated until the tendency to block has subsided. The pressure modulator is then changed back to continuity by the control unit.

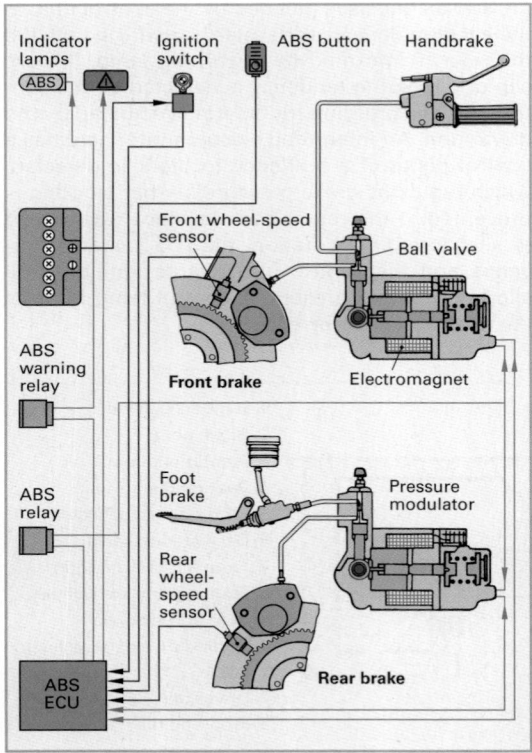

Fig. 1: Antilock braking system

21.14 Wheels, tyres

Wheels

These have to support the tyres and transfer braking and acceleration forces. To do this, they have to meet the following requirements:

- Low mass
- High structural strength and elasticity
- Good true running

The following types of wheel are fitted to motorcycles:

Spoked wheels (Fig. 2). The rim is made from steel or aluminium, the spokes are made from steel. Depending on the type of design, tyres with inner tubes or tubeless tyres can be fitted. These wheels are used mainly on cross-country machines these days as they have high elasticity with low weight.

Light alloy wheels (Fig. 3). These are manufactured as one-piece wheels from pressure-cast aluminium for tubeless tyres and are used on motor bicycles, scooters and motorcycles.

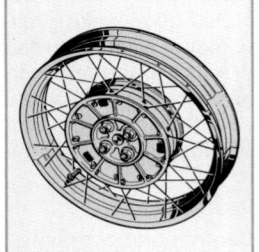 

Fig. 2: Spoked wheel   **Fig. 3: Cast wheel**

Composite wheels. These are manufactured from pressure cast aluminium, feature good true running properties and maintain their shape very well. They can be composed of 2 or 3 parts: the rim, the spokes and the hub.

Rim designation. This is structured much the same way as on passenger vehicle rims. It means e.g.
3.50 × 17 MT H2
3.50 = Rim width in inches
× = Non-split rim
17 = Rim diameter in inches
MT = Designation for motorcycle rim
H 2 = Two humps

Tyres

On motorcycles, the tyre contact patch is considerably smaller than on passenger vehicle tyres. However, it is particularly important for guiding the motorcycle wheel and influences road handling and safety of the vehicle in a decisive way. For these reasons, manufacturers specify dimensions for tyres that are allowed to be fitted, and sometimes even specify makes of tyre. More often than not, different tyre sizes and treads at the front and rear wheels are used. The front wheel must transfer mainly steering and lateral guiding forces. Therefore, the unsprung masses must be kept as low as possible. The rear wheel is considerably wider due to the high motive and lateral guiding forces. The tread of motorcycle tyres is shown in **Fig. 1**.

The tread of the front wheel normally takes the shape of longitudinal grooves or is arrow-shaped in the direction of rotation. This tread shape counteracts shark fin formation during wear. The tendency for rear wheels of high-powered road machines is towards wide tyres. The contour **(Fig. 1)** of these tyres is designed so that the tyre contact patch and thereby adhesion to the road surface is increased as inclination rises. The tread profile of the rear wheel has a parabolic arrow design, which prevents stepped formation even after high mileage. Rubber composites specially from motor racing increase adhesion of the tyre to the road in the threshold area.

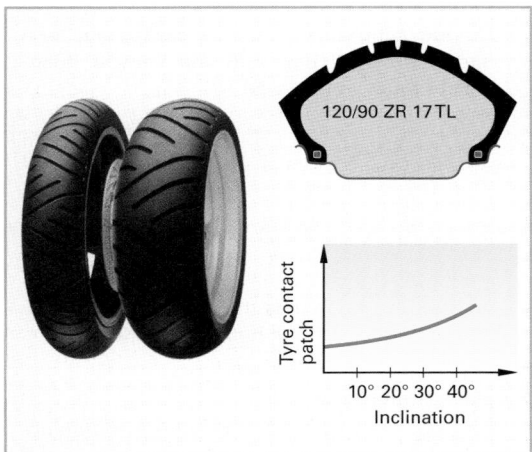

Fig. 1: Motorcycle tyre and tyre contour

Motorcycle tyres should have the following properties:

- Good adhesion regardless of the tread depth.
- High lateral stability and lateral guidance.
- Good straight running stability.
- Good road compatibility or offroad compatibility depending on intended use.

Types of tyre. There are four types of motorcycle tyre on offer:

- Bias ply tyre
- Bias belted tyre
- Radial ply tyre with bias belt
- Radial ply tyre with 0° steel belt

Bias ply tyre (Fig. 2). On this type, the 4 nylon or polyamide carcass plies are fitted at an angle of about 45° diagonally to each other and folded over the steel bead wires. Lateral guiding forces can be supported more or less depending on the fold over height of the side carcasses.

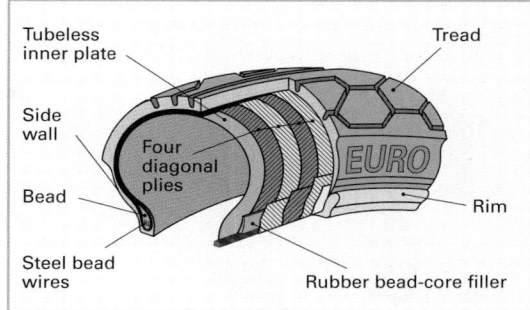

Fig. 2: Bias ply tyre

Bias belted tyre (Fig. 3). The fabric underlay comprises two diagonally layered carcass plies and 2 bracing layers, e.g. made from Kevlar fibres. As a result, the tyre features good true running properties and good lateral guidance.

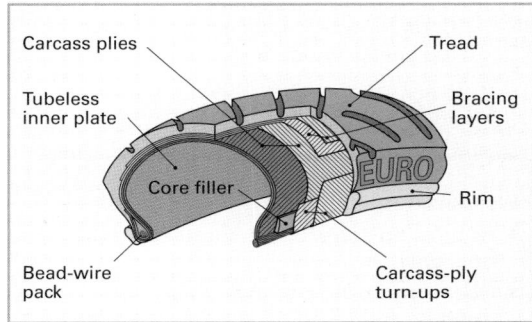

Fig. 3: Bias belted tyre

Radial ply with bias belt (Fig. 4). This type of tyre has a single layered 90° radial carcass and a two-layer diagonally located bracing layer made from aramide fibres.

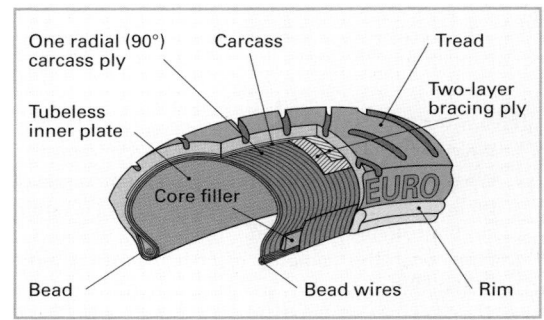

Fig. 4: Radial ply tyre with bias belt

21

Radial ply with 0° steel belt (Fig. 1). On this version, there is a single layered 0° steel belt above the radial carcass ply. This type is particularly suitable for high speeds as the contour remains very stable.

Rubber compositions. Tyres are available from the trade in up to 3 different rubber compositions from hard to soft. In this way, the optimal tyre can be fitted for any use.

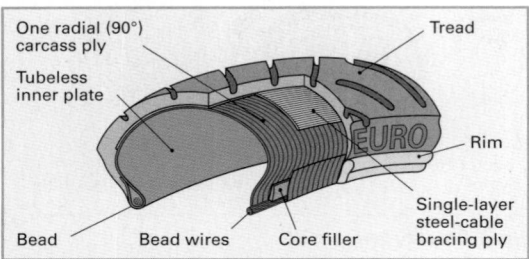

Fig. 1: Radial ply with 0° steel belt

Tyre designations.

Bias ply tyre: 4.10 – 18 60 P

4.10	= Tyre width in inches
18	= Rim diameter in inches
60	= Code for tyre load-bearing capacity
P	= Top speed 150 km/h

Low-profile tyre: 120/50 ZR 17TL

120	= Tyre width in mm
50	= Height/width ratio 50 %
Z	= Top speed is over 240 km/h
R	= Radial ply tyre
17	= Rim diameter in inches
TL	= Tubeless

Tyre registration data

The tyre sizes entered in the vehicle documents must be adhered to. On many motorcycles there is an option of fitting alternative tyres. However, a check should be made as to whether an approval certificate has been issued.

Details on this could read:

- **Custom alterations do not necessitate a vehicle document update.** The rider must carry the approval certificate with him/her.
- **Custom alterations must be carried out in a specialist workshop.** Approval certificates must be kept with the vehicle.
- **Custom alterations must be carried out by a specialist workshop and the vehicle documents must be updated.**

WORKSHOP NOTES

- Fit tyres on correct, non-corroded, undamaged rims only.
- Observe direction arrows when fitting, if featured.
- When changing tyres with inner tubes, always use new tubes to avoid folds.
- On spoked wheels, always fit new rim bands.
- On tubeless tyres, always fit new rubber valves.

- Inflate tyres to 1.5 times the operating pressure until the tyre sits in the bead properly.
- Adjust the inflation pressure.
- Balance wheel and tyre.
- From 2.5" rim width, balance wheel on balancing machine dynamically.
- Due to roughening, tyres must be run in for about 200 km in a moderate manner, so the optimal adhesion properties can be attained.

21

REVIEW QUESTIONS

1. What demands are placed on motorcycle frames?
2. What is the role of wheel location on motorcycles?
3. What types of front wheel and rear wheel location are there on motorcycles?
4. What does the damping response on front forks depend on?
5. Explain the CBS, ABS and TCS system on motorcycles.
6. Explain the rim designation 3.25-17 MT-H2.
7. What properties should motorcycle tyres have?
8. What types of motorcycle tyres are there?
9. Explain the tyre designation 160/60 ZR 18TL.

22 Commercial vehicle technology

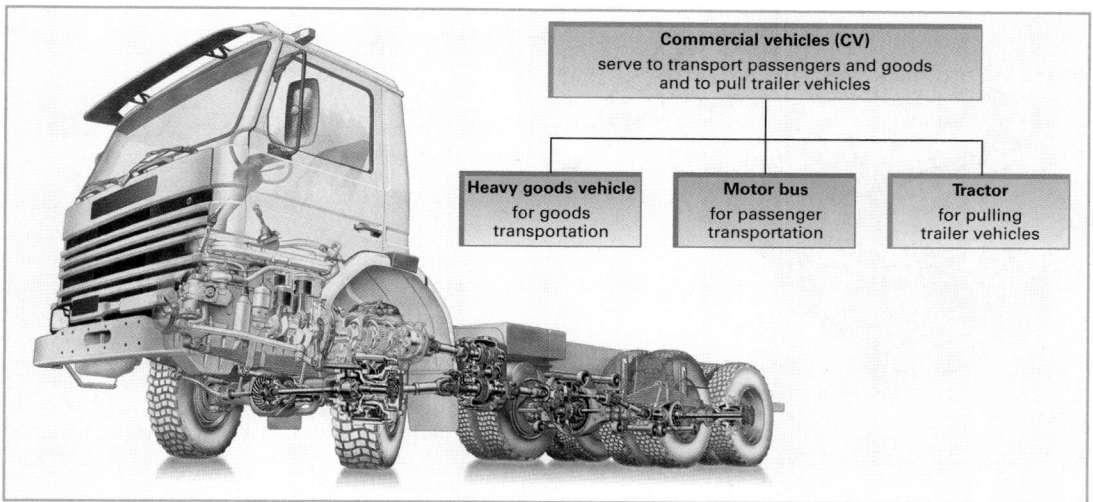

Commercial vehicles (CV)
serve to transport passengers and goods
and to pull trailer vehicles

| **Heavy goods vehicle** for goods transportation | **Motor bus** for passenger transportation | **Tractor** for pulling trailer vehicles |

Fig. 1: Categorisation of commercial vehicles

22.1 Categorisation

Commercial vehicles have the following main assembly groups:

- **Engine**, with fuel system and injection equipment
- **Power transmission,** with clutch, gearbox and final-drive unit
- **Chassis,** with frame, ancillaries, suspension, wheels, tyres, steering and brake system
- **Vehicle electrics,** with batteries, alternator, starting system, additional equipment

Differentiation of commercial vehicles according to their intended use:

Multipurpose heavy goods vehicle (Fig. 2). These can be used to transport goods on a flat bed, e.g. drop-side, or in a closed structure, e.g. box body.

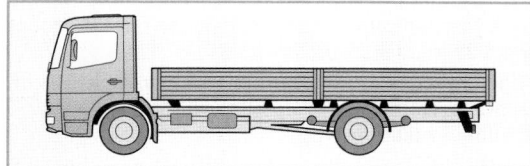

Fig. 2: Multipurpose heavy goods vehicle

Special heavy goods vehicle (Fig. 3). These vehicles have a custom body. There may also be additional special equipment or devices attached according to the intended use, e.g. tanker or silo truck, waste disposal truck, etc.

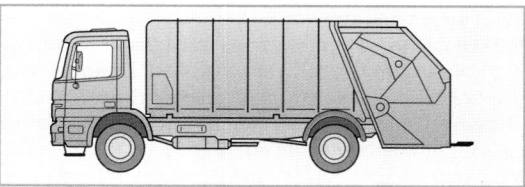

Fig. 3: Special heavy goods vehicle

Motor bus (Fig. 4). This can be used as a touring bus, regular bus or special-purpose bus depending on the equipment.

Fig. 4: Touring bus

Tractors (Fig. 5). Tractor units are equipped with a fifth wheel or coupling unit to attach a semi-trailer. Together, these form an articulated truck. Tractors are used to draw trailer vehicles.

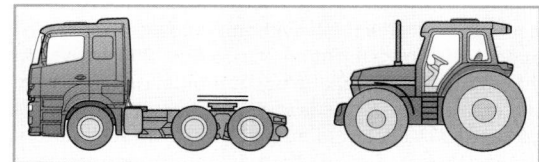

Fig. 5: Tractors

22

22.2 Engines

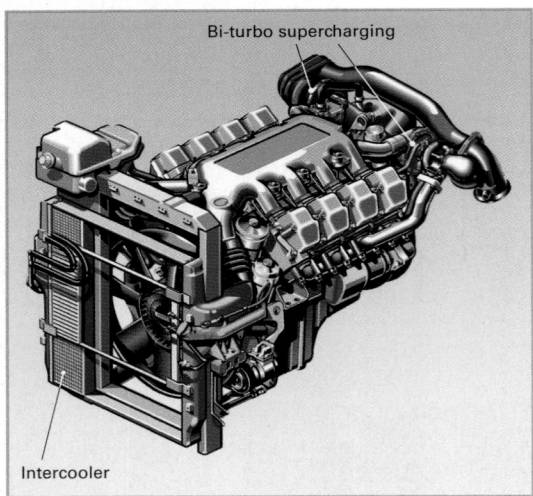

Fig. 1: Engine for heavy commercial vehicle

Commercial vehicles are generally equipped with direct injection diesel engines. They are mostly charged by exhaust gas turbochargers (**Fig. 4, Page 243**). Depending on the maximum permissible weight and the use of the vehicle, engines from about 3 l capacity to about 16 l capacity are installed. Their cylinder displacements are normally greater than 600 cm^3. The performance per cylinder is above 25 kW in most cases. Engines are installed with up to 16 cylinders. Most of the time engines have 6 to 8 cylinders (**Fig. 1**). Light commercial vehicles have about 70 kW. Heavy commercial vehicles, tractor units and buses/coaches have about 450 kW.

> Due to legal requirements, the minimum engine output of heavy goods vehicles, motor buses and articulated trucks must be at least 4.4 kW per tonne of the maximum permissible gross weight of the motor vehicle and the relevant trailer load.

The maximum torque from commercial vehicle diesel engines is in a range between 1,500 Nm and 3,000 Nm. At this level, the engines normally work at a speed between roughly 1,200 rpm and 2,400 rpm. The engine torque stays practically constant over a wide speed range (**Fig. 2**).

Modern commercial vehicle diesel engines run particularly cost-effectively with full-load best figures for specific fuel consumption at less than 200 g/kWh. At 40 t total weight, trailer/truck combinations and articulated trucks have medium fuel consumption figures of about 32 l/100 km to about 40 l/100 km with potential mileages of more than 1,000,000 km without large scale service measures being needed.

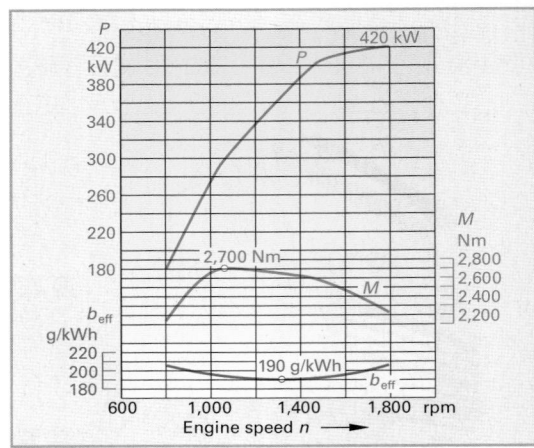

Fig. 2: Engine data and engine performance curves

Fig. 2 shows the engine performance curve of an 8-cylinder engine with unit pump system injection (**UPS**) and exhaust gas turbocharging with 2 inlet valves and 2 exhaust valves per cylinder.

Engine data: $V_H = 15,928$ cm^3; $d = 130$ mm; $s = 150$ mm; $\varepsilon = 17.25$. $P_{eff} = 420$ kW at $n = 1,800$ rpm; $M_{max} = 2,700$ Nm at $n = 1,080$ rpm; $b_{eff} = 190$ g/kWh at 1,300 rpm.

22.3 Injection systems for CV diesel engines

> CV injection systems perform the following tasks:
> - Provide the necessary injection pressure available.
> - Inject the necessary quantity of fuel (fuel delivery control).
> - Adjust the required start of injection (start of injection control).

As in the passenger vehicle sector, mechanically controlled inline and distributor-type injection pumps have been replaced almost completely by map-controlled fuel injection systems in the heavy goods vehicle sector as well. They inject at increasingly higher pressures and with increasingly greater accuracy and thereby comply with the stricter emissions thresholds. Mechanically controlled injection pumps cannot meet the demands set.

A difference is made between the following fuel injection systems:
- Inline injection pumps
- Distributor-type injection pumps as reciprocating plunger or radial plunger injection pumps with solenoid valve control (**see Page 299**)
- Unit injector systems with solenoid valve control (**UIS, see Page 302**)
- Unit pump systems with solenoid valve controlled injectors (**UPS, see Page 647**)
- Common rail injection with solenoid valve controlled injectors (**CR, see Page 304**)

22.3.1 Injection system with inline injection pump

Items that belong to a fuel system with inline injection pump are: fuel primer pump, fuel filter with fuel heater if necessary and high pressure equipment. This comprises injection pump, high pressure injector lines, nozzle holders with injection nozzles and return line **(Fig. 1)**.

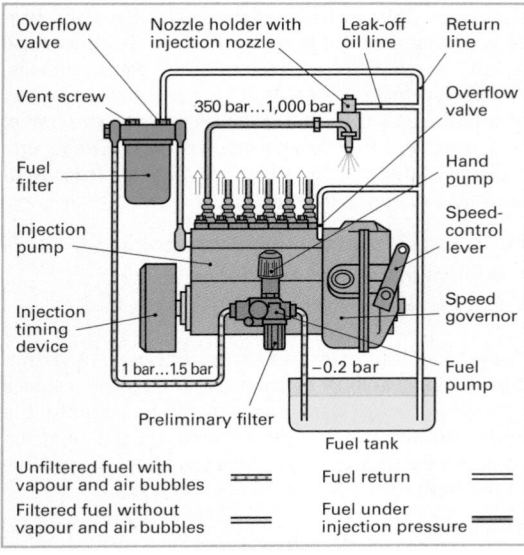

Fig. 1: Fuel circulation of an inline injection pump

Fuel primer pump.

This is normally flanged directly to the injection pump and is driven by the camshaft of the injection pump via an eccentric plate **(Fig. 2)**.

> The fuel primer pump delivers fuel at approx. 1 ... 1.5 bar to the inline injection pump.

The fuel is taken in from the fuel tank by the fuel primer pump and delivered via the fuel filter to the high pressure pump. Approx. 1/4 of the fuel volume serves as a means of cooling and lubrication of the inline injection pump. The fuel flows through the pump and back via the overflow valve and the return line to the fuel tank. To circulate the necessary injection quantity and the necessary volume of fuel for cooling of the inline injection pump, the primer pumps have outputs of 150 ... 200 l/h. Often they also have to be able to cope with long distances along which fuel has to be pumped. If the tank is too far away, additional electric fuel pumps are installed.

For bleeding of the injection system, e.g. after the filter has been renewed, the fuel pump is normally equipped with a hand pump. This can be actuated once the handle has been released. After the hand pump has been used, it is important to screw the handle back in place.

Primary strainer filter. This captures coarse particles of dirt and water.

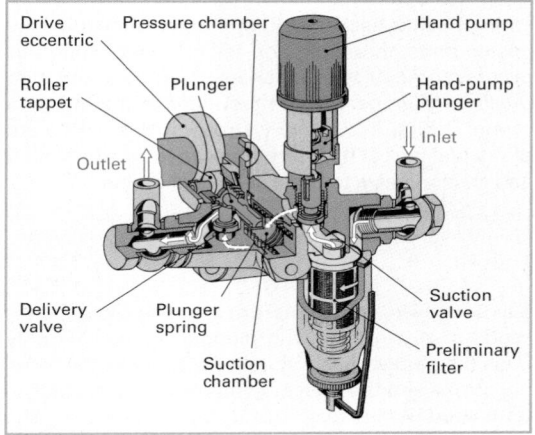

Fig. 2: Fuel primer pump

Function. As can be seen in **Fig. 3**, the delivery stroke and suction chamber filling (intake stroke) occur at the same time. The pressure chamber is filled during the intermediate stroke.

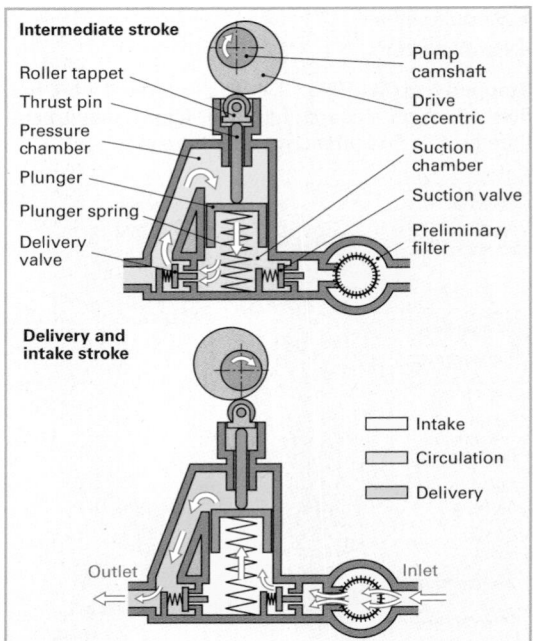

Fig. 3: Function of the fuel primer pump

Intermediate stroke. During this stroke, the eccentric element moves the plunger forwards via the roller tappet and thrust pin. With the suction valve closed, fuel is thereby delivered via the delivery valve to the high pressure chamber. The plunger spring is pressed together as this happens and the spring loaded delivery valve closes at the end of the stroke.

Delivery and intake stroke. Once the eccentric element has completed its largest stroke, the plunger spring pushes back the plunger and the loosely connected parts, thrust pin and roller tappet. When this occurs, some of the fuel is delivered from the high pressure chamber via the fuel filter to the injection pump. During this delivery stroke, fuel is drawn out of the fuel tank at the same time via the primary filter and suction valve into the suction chamber.

> Only every 2nd stroke of the plunger is a delivery stroke.

Elastic delivery. If the pressure in the delivery line exceeds a certain level, the plunger spring can only push the plunger back partially. This means the delivery stroke and the delivery volume are reduced. The term used to described this is "elastic delivery". The lines and filter are protected in this way against excessive pressures.

Commercial vehicle fuel filter

These have the task of separating dirt, e.g. fine particles of dust, and water out of the fuel.

A difference is made between two types of filter:
- Stepped filter
- Parallel filter

Stepped fuel filter (Fig. 1). With this type of filter, fuel flows through a coarse filter (1st filter housing) and then on to a fine filter (2nd filter housing).

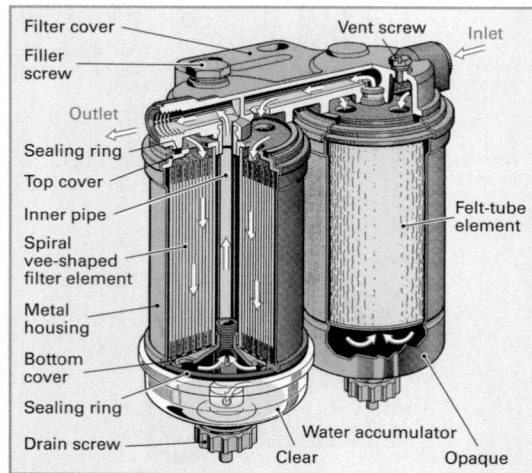

Fig. 1: Stepped box filter with water separator

Parallel fuel filter. These are used on larger sized diesel engines. They are no different in appearance to the stepped fuel filters. However, the fuel supply is distributed in the filter cover so that each of the two filter boxes equipped with fine filter elements have fuel flowing through them at the same time. In this way, the effective filter surface is doubled and the potential flow rate of the fuel is increased.

Fuel heaters. These are normally installed in the fuel system right in front of the fuel filter. They prevent blocking of the filter element through precipitation of paraffin at low outside temperatures. Where "summer diesel fuel" is used (no additive), wax-like scales (paraffin) precipitate from the fuel at fuel temperatures below approx. 4°C. These can block the filter cartridges. To prevent this from happening, the fuel is heated.

Two types of heating systems are common.
- Coolant in the heat exchanger
- Electric heater elements

Preheating fuel using a heat exchanger. A thermostat with expansion element regulates the volume of fuel to the heat exchanger. If the fuel is cold, the entire quantity of fuel flows through the heat exchanger. As temperature increases, the supply line to the heat exchanger is closed (Fig. 2).

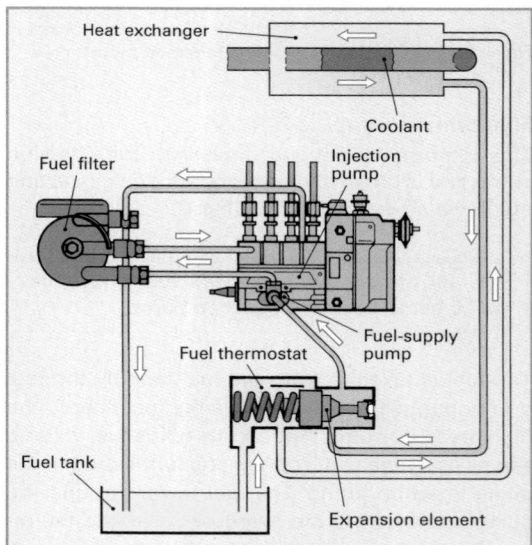

Fig. 2: Preheating fuel using heat exchanger

Electric heater elements. There is a preference for installation of self-governing PTC resistors as heater elements. These can be found in the intermediate flange between filter cover and filter box, for example **(Fig. 1, Page 22).**

Standard inline injection pump (Fig. 1)

Functions. Inline injection pumps are designed to …
● … generate the injection pressure required.
● … meter the injection quantity exactly according to the accelerator pedal position.
● … adapt the injection point to the engine speed.
● … regulate idling speed and maximum speed.

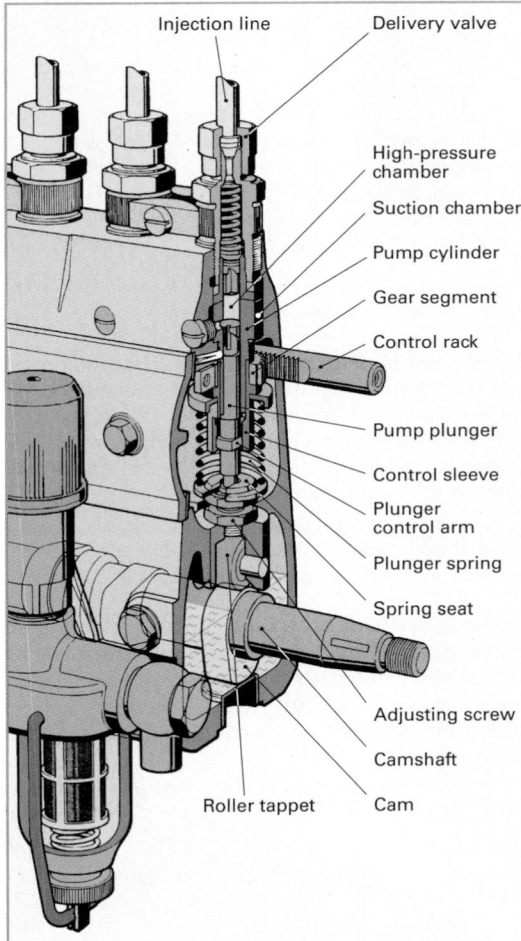

Fig. 1: Inline injection pump (cross section)

Design and operating principle.

> The inline injection pump **(Fig. 1)** is a piston pump with one pump element for each engine cylinder. Each pump element **(Fig. 2)** comprises a pump cylinder and a pump plunger.

The individual pump elements are driven via roller tappets by a camshaft integrated in the pump housing. The plunger spring presses the roller tappets against the cams. The pump plunger is fitted so close-

ly in the pump cylinder that its sealing properties are effective even at very high pressures and low speeds. This very low play (2 μm … 3 μm), which is required due to the high pressures that occur, makes it necessary to exchange both the pump cylinder and pump plunger in the event of renewal. Diesel fuel lubricates and cools the pump element and ensures fine sealing. In the lower part of the pump, engine oil is responsible for lubrication of the camshaft and the roller tappets.

Fuel metering (fuel delivery control, Fig. 2). Apart from a longitudinal groove and a ring groove, the jacket of the pump plunger has a screw-like milling. It forms the helix. With its help, it is possible to regulate the delivery volume. Fuel flows through the inlet passage at a pressure of approx. 1 bar … 1.5 bar into the high-pressure chamber.

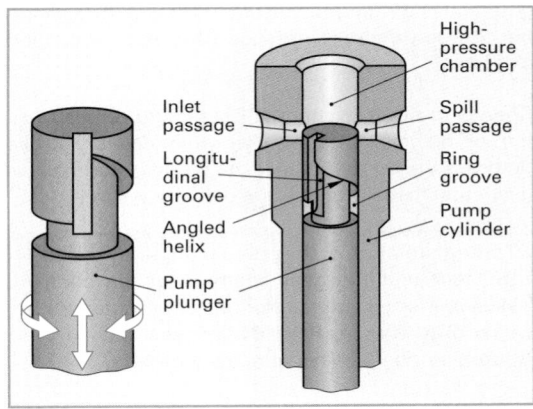

Fig. 2: Pump element

Fuel inlet (pre-stroke). From the suction chamber, fuel flows via the inlet passage to the high-pressure chamber of the pump cylinder. As soon as the upper edge of the plunger frees the inlet passage **(Fig. 3)**, the fuel subjected to pre-delivery pressure flows in the high-pressure chamber above the pump plunger.

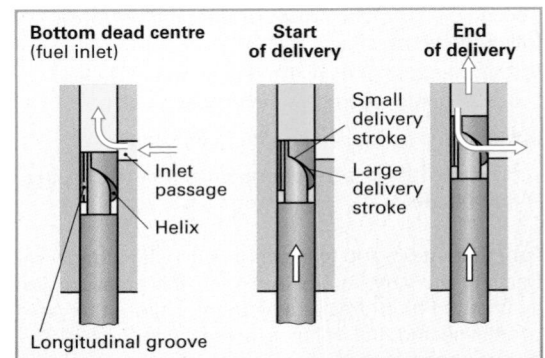

Fig. 3: Fuel supply and delivery in pump element

Start of delivery. With the upward stroke of the pump plunger, the inlet passage is sealed by the upper edge of the plunger and the delivery stroke begins. The pressure is built up in the pump element via the delivery valve in the high-pressure injection line and in the nozzle holder to the injection valve in front of the injection nozzle. If a certain pressure is exceeded, the injection valve opens and fuel is injected at high pressure (up to about 1,200 bar) through the nozzle into the combustion chamber.

End of delivery. End of delivery is reached as soon as the helix frees the inlet passage. From this point on, the high-pressure chamber of the pump cylinder is joined to the suction chamber via the longitudinal groove and ring groove. Pressure falls, the injection valve and the delivery valve close. The plunger continues the upward stroke and pushes fuel via the longitudinal and ring grooves out of the pressure chamber through the inlet passage back into the suction chamber.

Change in delivery volume. Depending on the position of the helix, a different length in delivery stroke between the upper edge of the plunger and helix is given and thereby a different delivery volume.

> Through rotation of the pump plunger by the control rack and the gear segment of the control sleeve, the delivery stroke (injection quantity) is controlled **(Fig. 1, Page 643)**. The stroke of the pump plunger always remains the same.

In the end positions of the plunger rotation there is either **maximum delivery** or **zero delivery**.

Zero delivery. This is necessary to switch off the engine. The control rack is pulled to the "stop" position and the longitudinal groove aligns with the spill passage. The high-pressure chamber can no longer be closed.

Delivery valve (Fig. 1). This can be found in the connecting pieces of the pressure lines and performs the following tasks:

● Relieves load in injection line at end of delivery to assure rapid closure of the injection valve.

● Maintains residual pressure in injection line.

● Prevents fuel dribble or post-injection of the injection nozzle.

Function. At the end of injection, the relief plunger retracts in the valve guide and seals the pressure line against the high-pressure element. Unless the valve cone is seated, the volume increases in the fuel line by the retraction volume. This causes a rapid pressure drop in the line and thereby rapid closure of the

injection nozzle. Fuel dribble that causes harmful emissions is thus prevented.

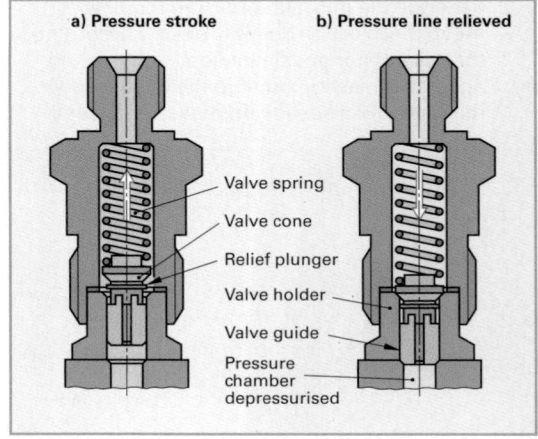

a) Pressure stroke   b) Pressure line relieved

Valve spring
Valve cone
Relief plunger
Valve holder
Valve guide
Pressure chamber depressurised

Fig. 1: Delivery valve with relief plunger

Speed governor (Fig. 2). Mechanical speed restriction with centrifugal weights. This works in relation to the engine speed and alters the quantity of fuel to be injected by the injection pump. Normally, a two-point device is used as a minimum/maximum speed governor on CV inline injection pumps.

> The speed governor keeps the idling speed constant and governs the maximum speed.

The idling speed must be kept constant so the engine does not run "lumpy" or even cut-out completely as a result of the minimum injection volumes. The maximum speed must be governed to prevent the engine from "running away", which could lead to irreparable engine damage.

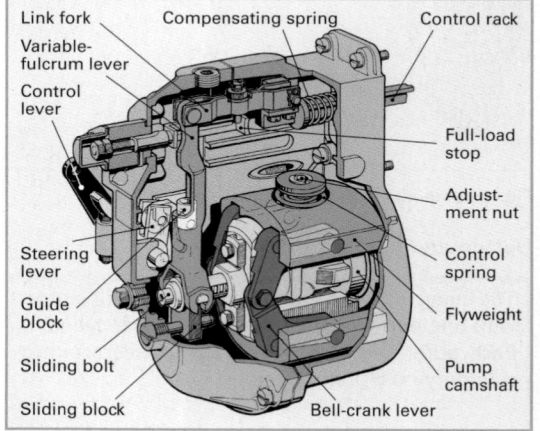

Link fork
Variable-fulcrum lever
Control lever
Compensating spring
Control rack
Full-load stop
Adjustment nut
Steering lever
Control spring
Guide block
Flyweight
Sliding bolt
Pump camshaft
Sliding block
Bell-crank lever

Fig. 2: Speed governor

In the range between idling speed and end speed (maximum permissible speed), speed restriction is not necessary because in this range the driver actuates the control rack via the accelerator pedal and thereby adjusts the throttle condition. In this way, the injection volume is determined and with it the necessary engine torque.

Function. The governor is equipped with two flyweights and is driven by the camshaft of the injection pump. Housed in each flyweight are one idling spring and two maximum speed springs. The radial paths of the flyweights are converted by two pairs of bell-crank levers into axial motions of the sliding bolt. This transfers them to a sliding block. The sliding block, which is mounted in the bottom end of the variable-fulcrum lever, is guided in a straight line by the sliding bolt and makes the connection to the control rack via the variable-fulcrum lever and the link fork. Since the variable-fulcrum lever **(Fig. 1)** has a variable pivot point, the turns ratio of the lever can be altered.

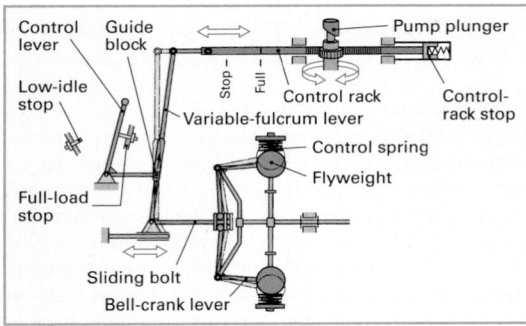

Fig. 1: Governor – end speed

Control-rack stop. There is one end stop each for the minimum and maximum control lever angle. One stop for the idling speed, one for the full throttle volume or rated output. These end stops are sealed and may only be adjusted by trained specialists.

Spring-type control-rack stop. These are used on CV engines that require a greater amount of fuel to start than for full throttle operation.

Mechanical timing adjuster (start of injection control) (Fig. 2). The automatic injection timing device comprises a housing with flyweights and an adjuster mounted in the housing to allow rotational movement. The adjusting plate joined to the hub uses centrifugal force to adjust the injection point and therefore works in relation to the speed.

> As speed increases, two flyweights overcome spring pressure and move outwards, thereby causing start of injection to be "advanced".

The further the flyweights move outwards, the more the adjusting plate, and with it the pump camshaft, is turned with the direction of rotation. The pump plunger starts its upward stroke earlier and seals the inlet passage earlier. Start of delivery is advanced.

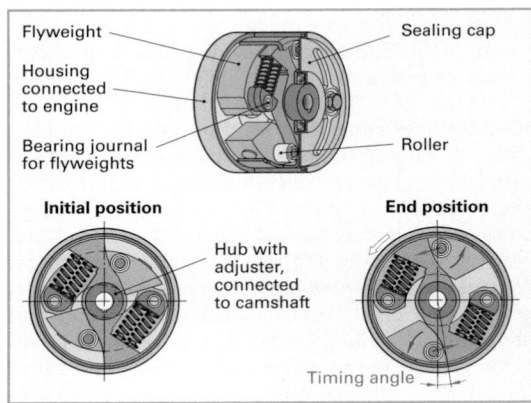

Fig. 2: Mechanical timing adjuster

22.3.2 Control sleeve inline fuel-injection pump

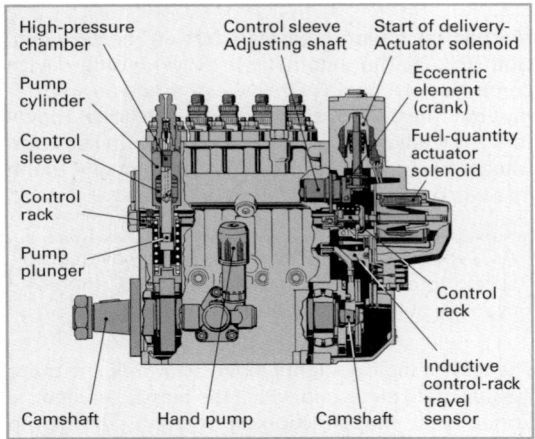

Fig. 1: Control sleeve inline fuel-injection pump

The control sleeve inline fuel-injection pump **(Fig. 1)** differs from the mechanically controlled inline injection pump mainly by a map-controlled start of injection control system (EDC). It reaches injection pressures of up to 1,350 bar.

Injection flow control. From the main control variables (driver command and engine speed) and the correction control variables, e.g. engine temperature, intake air temperature, fuel temperature, charge pressure, the control unit calculates the control rack travel necessary for the specified injection volume. The **fuel-quantity actuator solenoid (Fig. 1)**, energised by the control unit, moves the control rack and brings about a change in the injection volume by rotation of the pump element. The inductive travel sender informs the control unit of the respective control rack position. This readjusts the position as necessary. In the event of signal failure, the resetting spring pushes the control rack to zero delivery.

Start of injection control. Start of injection is changed by the control sleeves **(Fig. 2)**. They are moved at the same time by an actuator solenoid for start of delivery via an eccentric element and an adjusting shaft. The earlier the start of injection, the earlier the bottom edge of the control sleeve seals the spill passage.

Adjustment of the control sleeve towards BDC means a small pre-stroke and thereby advanced start of injection and vice versa.

The actual start of injection is transmitted from a needle motion sensor to the control unit. This continually alters the adjustment until the specified values and actual values tally.

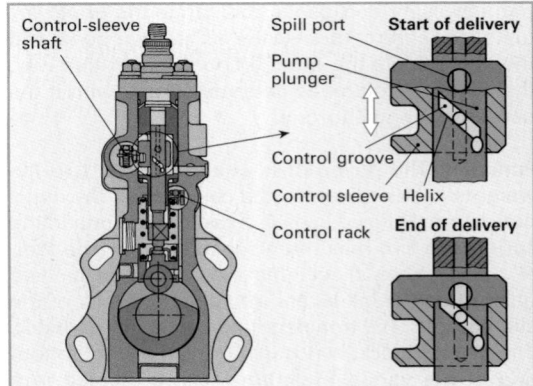

Fig. 2: Start of injection control by way of control sleeve

22.3.3 Unit pump systems

A difference is made between two systems:

- Mechanically controlled individual injection pumps
- Solenoid valve controlled unit pump systems

Individual injection pumps (Fig. 3).
These can be found in commercial vehicles, locomotives, agricultural and construction machines and also in stationary and shipping machinery. They reach injection pressures of up to 1,500 bar and are distinguished by their sturdy and service-friendly properties and also their especially high max. injection volume of up to 18,000 mm³ per cylinder. Up to 1,000 kW output per cylinder is reached in this way.

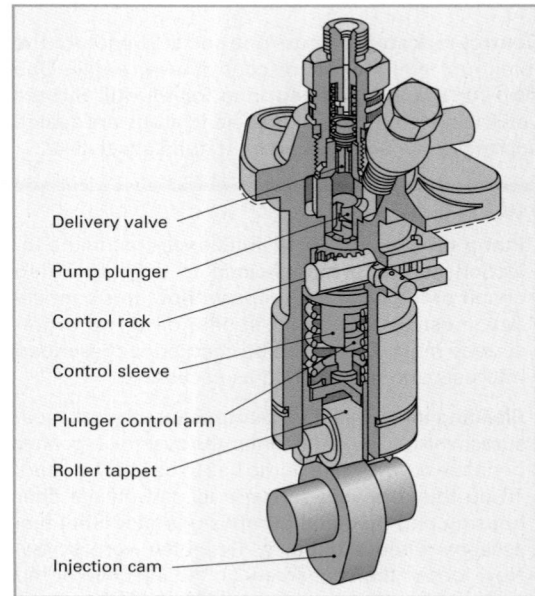

Fig. 3: Individual injection pumps

For each engine cylinder, a pump is inserted in the engine block on the lower mounted engine camshaft. They are therefore sometimes referred to as insert pumps. The pump is joined to the injection nozzle via a high pressure line.

Function. They have the same operating principle as the standard inline injection pump. The only difference is that they are driven not by a camshaft integrated in the pump housing but by the camshaft of the diesel engine. They therefore do not require an additional injection cam per engine cylinder.

Start of injection control. Since the drive cam of the individual injection pump sits on the engine camshaft and this also takes on valve control of the engine, it cannot simply be rotated to adjust start of injection. By adjustment of an intermediate member, e.g. an eccentrically mounted rocker arm **(Fig. 1)**, an adjustment angle of several degrees can be realised.

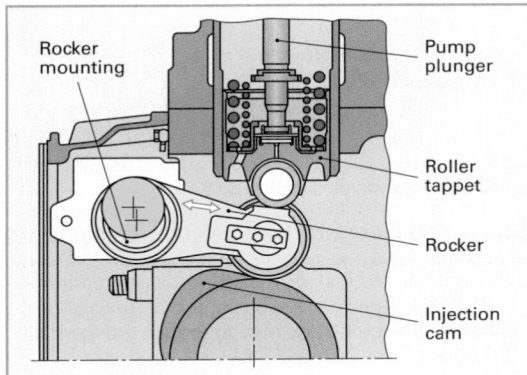

Fig. 1: Individual injection pump – injection setting

Injection flow control. This is made possible by rotation of the pump plunger via the control sleeve **(see Inline injection pump on Page 643 and 644).**

Unit pump systems (Fig. 2)

This is an electronically controlled fuel injection system. Each engine cylinder features an insert pump with solenoid valve control. It supplies the injection nozzle with fuel via a line. Max. injection pressures of up to 1800 bar are reached.

Design. This is a further development of the individual injection pump. High pressure generated in the pump unit is supplemented by a map-controlled high pressure solenoid valve. A short high pressure line joins it to the injection nozzle in the cylinder head.

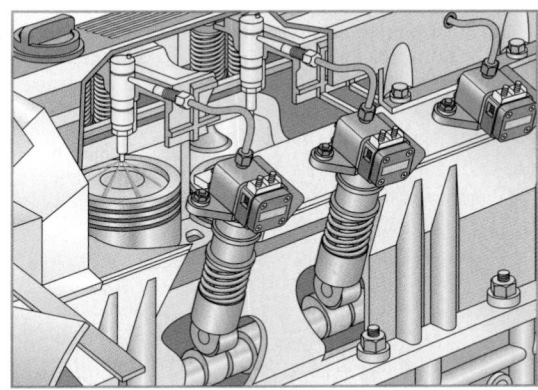

Fig. 2: Unit pump system (UPS)

Operating principle. It is controlled by 4 successive delivery phases:

1. Intake stroke phase
2. Pre-stroke phase
3. Delivery stroke phase
4. Rest stroke phase

Intake stroke phase (Fig. 3/1). The pump plunger moves downwards in direction of BDC as the cam lobe becomes smaller. As this occurs, the high-pressure chamber fills with fuel via the open solenoid valve (de-energised).

Pre-stroke phase (Fig. 3/2). When the roller tappet runs over the rising cam lobe, the pump plunger is pushed in direction of TDC. Fuel is forced out of the high-pressure chamber and flows via the open valve to the fuel return line.

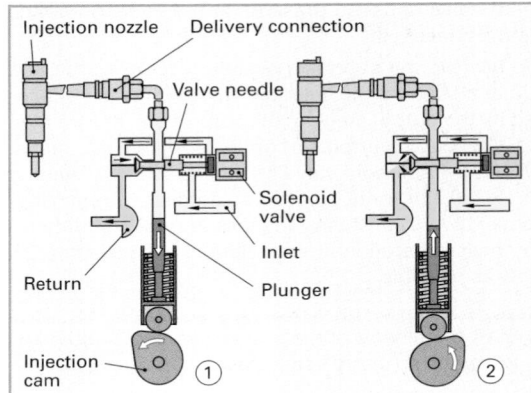

Fig. 3: Suction and pre-stroke phase of the UPS

Delivery phase (Fig. 1/3, Page 648). The solenoid valve is energised, therefore the valve needle is closed. The pump plunger moves by the rising roller tappet further in direction of TDC. Fuel is subjected to pressure in the high-pressure chamber. Injection commences once it exceeds the opening pressure of the nozzle holder.

Start of injection and **period of injection** (injection quantity) are determined for each cylinder individually by the EDC. A voltage pulse at the solenoid valve causes start of injection. A break in voltage causes end of injection.

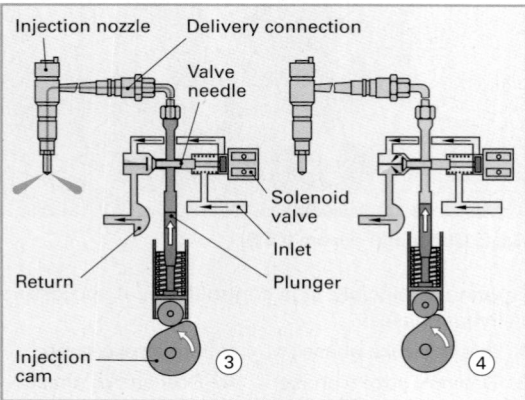

Fig. 1: Delivery and rest stroke phase of UPS

Rest stroke phase (Fig. 1/4). Until the top dead centre position is reached, fuel under pressure is forced through the open solenoid valve in the return line. The pump plunger moves further downwards and initiates a new intake stroke phase.

Pilot injection. This can only be realised on this system with the aid of a two-spring nozzle holder (see **Page 292**).

Fig. 2 shows the voltage pattern, current at the solenoid valve and also pressure at the injection nozzle and its needle lift.

To achieve the shortest possible response time and quick closure of the solenoid valves, they are actuated by the output stage of the control unit at a voltage of 70 V. Cut-in currents of up to 18 A flow as a result. When the solenoid valve strikes its seat, i.e. when it is closed, the control unit sets a holding current of approx. 12 A. In this way, any loss in power and thereby heat generation in the solenoid valves is kept as low as possible.

Start of delivery of the pump element is initiated when the solenoid valve closes.

The holding time of the solenoid valve determines the injection volume. The effective stroke ends when the solenoid valve opens. The injection nozzle closes. During the residual stroke, fuel delivered by the pump plunger flows in the return line. From the speed signal of the crankshaft sender, the control unit calculates the start of injection. With the position sender of the camshaft it generates the cylinder allocation (injection sequence).

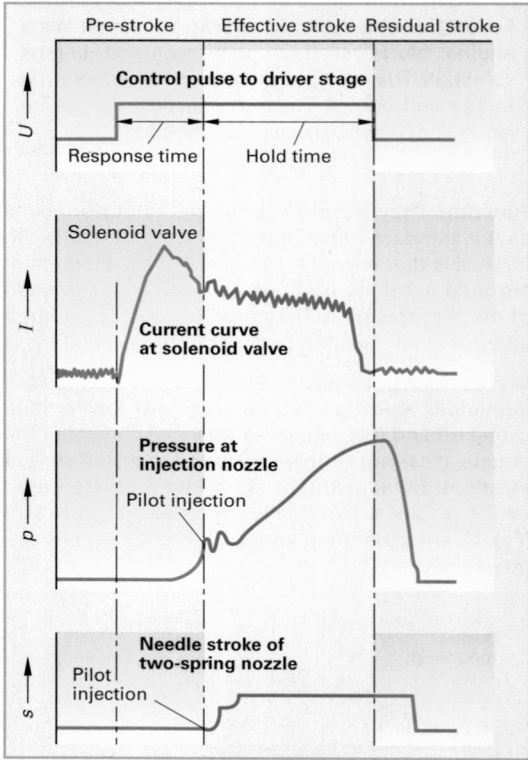

Fig. 2: Solenoid valve control of UPS

22.3.4 Auxiliary starting assistance systems

These have the task of making it easier to start a cold diesel engine, to keep running smooth and to stabilise the engine and also to reduce the harmful emissions.

In commercial vehicles, as in passenger vehicles, sheathed-element glow plugs (see **Page 290**) and also flame-start systems are used.

Flame-start system. This is installed in the charge air housing of the engine. It consists of a combustion chamber, a nozzle, a solenoid valve and sheathed-element glow plug (**Fig. 3**).

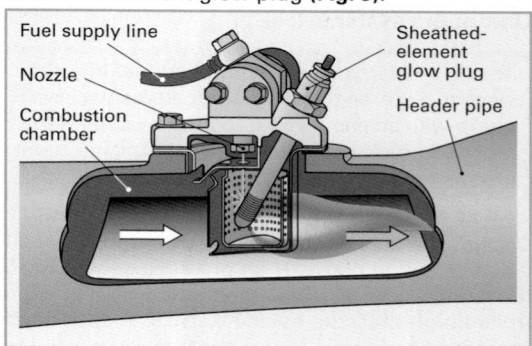

Fig. 3: Flame start system

The use of a flame-start system is only necessary for cold starting at extreme temperatures below –15 °C.

Function. If coolant temperature drops to below – 4 °C, the flame-start system is activated automatically when the ignition key is turned to the "driving position". On completion of the voltage-dependent preglow period of 20 ... 25 seconds, it is ready for operation and the engine can be started. The flame-start system is supplied with filtered fuel via the presupply pump. The fuel injected in the combustion chamber ignites itself at the hot sheathed-element glow plug. The air flowing past the combustion chamber is heated up in this way briefly to 800 °C. The conditions for ignition in all cylinders are thereby improved. To control the fuel inlet, a solenoid valve energised by the glow control unit is installed in the fuel supply line in front of the flame plug.

Fuel supply is interrupted when ...

● ... the coolant temperature in the running engine reaches approx. 0 °C.

● ... the engine is not started within 30 seconds of the control lamp going out.

Diagnosis. The glow control unit monitors the sheathed-element glow plug, the solenoid valve and the wiring connections. If a fault occurs, the driver is given a warning via the display with a fault code, e.g. **"FLA"**.

22.3.5 Reduction of harmful emissions on CV diesel engines

Commercial vehicles have to comply with increasingly strict exhaust emission thresholds. In Europe, the thresholds for commercial vehicles have been gradually reduced in accordance with 91/542/EEC **(Table 1)**.

Table 1: Limit valves for commercial vehicles
(m_{gvwr} > 3.5 t)

Standard	Year	CO	HC	NO_x	Particles	Smoke emission
Euro 0	1988	12.3	2.6	15.8	–	–
Euro 1	1992 < 85 kW	4.5	1.1	8.0	0.612	–
	> 86 kW	4.5	1.1	8.0	0.36	–
Euro 2	1996	4.0	1.1	7.0	0.25	–
	1998	4.0	1.1	7.0	0.15	–
Euro 3	2000	2.1	0.66	5.0	0.20/0.13	0.8
Euro 4	2005	1.5	0.46	3.5	0.02	0.5
Euro 5	2008	1.5	0.46	2.0	0.02	0.5

To comply with the stricter exhaust emission thresholds, measures inside and outside of the engine have to be adapted to each other optimally. To do this, the use of electronic fuel injection systems is essential.

External engine measures
To reduce harmful emissions on commercial vehicles, the following systems are used:

● Oxidation catalyst
● Exhaust-gas recirculation
● Particulate filter
● SCR catalyst

Oxidation catalyst (see Page 324).
Due to the excess air in the diesel exhaust gas, this can only reduce the HC and CO emissions through oxidation. A reduction of the NO_x emissions cannot take place.

Exhaust-gas recirculation (see Page 324).
Combustion with low NO_x figures is possible at low combustion temperatures and low flame velocities. On cooled EGR systems, the fresh A/F mixture can be mixed with up to 40 vol.%. The increased fuel consumption of 1 ... 2 % associated with this due to the reduced output means the system is only really effective on commercial vehicles with low annual mileage.

SCR process (Fig. 1, Page 650).

With the SCR procedure (**S**elective **C**atalytic **Re**duction), ammonia is used to convert nitrogen oxide to nitrogen and water on the catalyst surface. This reduces the emission of nitrogen oxide by up to 80%. The emission of particles is reduced slightly in this way.

Design. Fig. 1, Page 650 shows the design of an SCR system. It comprises a combination of oxidation catalyst and SCR catalyst and also a metering device. The SCR catalyst is installed behind the oxidation catalyst. The metering device injects the reducing agent in relation to the engine load in front of the SCR catalyst by way of compressed air. The function of the system is monitored by an exhaust-gas sensor (broadband lambda probe, **see Page 318**).

SCR catalyst. The wash coat has a titanium, tungsten and vanadium coating. Together with ammonia (NH_3), these precious metal compounds are suitable to (selectively) cause reduction of the nitrogen oxide (NO_x) to nitrogen (N_2) and water (H_2O).

Reducing agent. This comprises a watered down urea solution with a concentration of 32.5 vol.%. Poisonous ammonia is formed on the catalyst surface from the harmless urea solution. Calculation of the injected urea solution must therefore be very precise so that discharge of the poisonous ammonia to the surrounding air is prevented.

22

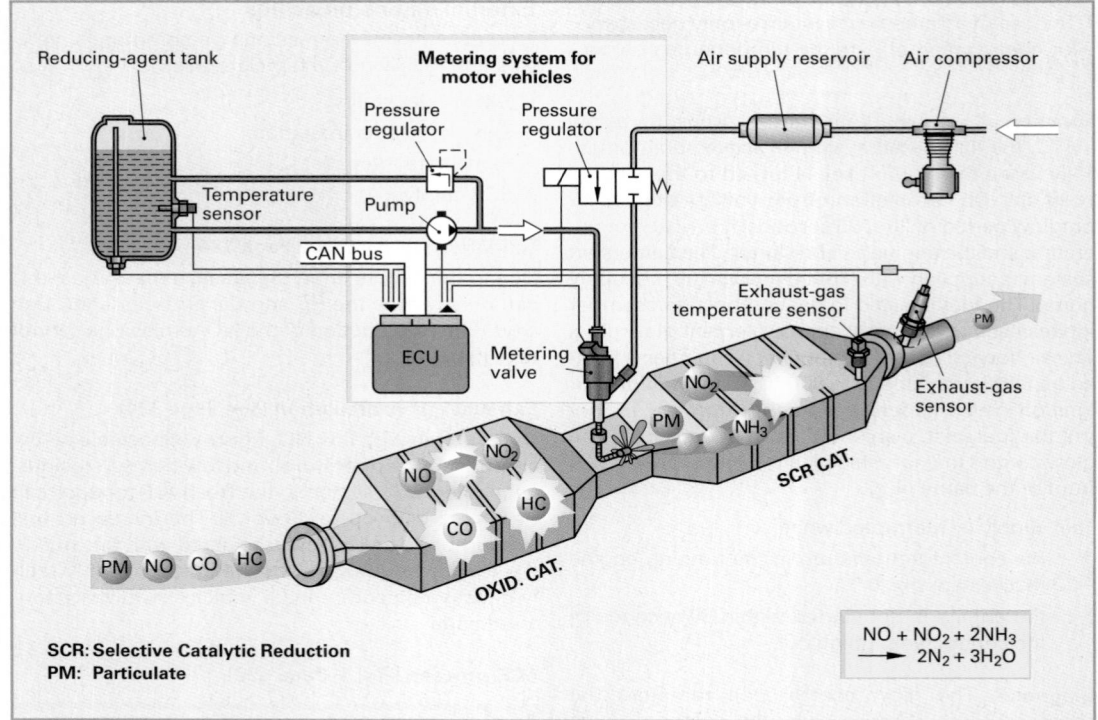

SCR: Selective Catalytic Reduction
PM: Particulate

$$NO + NO_2 + 2NH_3 \longrightarrow 2N_2 + 3H_2O$$

Fig. 1: Design of an SCR system with urea metering device

The ammonia reacts in the SCR catalyst with NO_x to N_2 and H_2O.

Fuel saving. The amount of NO_x is heavily reduced by the SCR system.

For this reason, start of injection can be advanced, resulting in a reduction in fuel consumption by approx. 6%. The higher level of NO_x caused as a result is then reduced in the SCR system.

REVIEW QUESTIONS

1 What diesel injection systems are used mostly in commercial vehicles?

2 Describe the path the fuel takes from the fuel tank to the combustion chamber.

3 Describe the operating principle of the fuel-supply pump from an inline injection pump.

4 After renewing the filter, the fuel system of an inline injection pump has to be bled. Describe the procedure.

5 The overall stroke of a pump element from an inline injection pump is split into what phases?

6 What are the functions of the delivery valve?

7 For what engines are inline injection pumps with spring-loaded control-rack stop used?

8 Explain the timing adjustment for a standard inline injection pump.

9 Describe the procedure for changing the delivery volume of a pump element from an inline injection pump.

10 What is the difference between control sleeve inline fuel-injection pumps and standard inline injection pumps?

11 Describe the operating principle of the speed control device of a control sleeve inline fuel-injection pump.

12 What are the effects of adjusting the control sleeve upwards?

13 What is the difference between the individual injection pump and the pump/line/nozzle system?

14 Explain the start of injection adjustment for an individual injection pump.

15 How is pilot injection realised on a pump/line/nozzle system?

16 Describe the start of injection control and the injection volume control of a pump/line/nozzle system.

17 Explain the operating principle of a flame start system.

18 By which procedures can the nitrogen oxide emissions be reduced in the exhaust gas?

19 Describe the operating principle of the SCR procedure.

22.4 Drive train

22.4.1 Drive concepts

These are determined according to the drive and non-driven axles based on the following pattern:

- First digit: overall number of wheels
- Second digit: number of driven wheels

A vehicle with the designation 4 × 2 has four wheels (possibly twin tyres) and two wheels are driven.

22.4.2 Types of drive

The following types of drive can be found:

- Rear wheel drive with one drive axle
- Rear wheel drive with trailing axle **(Fig. 1)** or leading axle
- Rear wheel drive with two driven axles
- All-wheel drive

The vehicle shown in **(Fig. 1)** has the designation 6 × 2. It has 6 wheels, 2 of which are driven.

On an unladen or partly laden vehicle, the trailing or leading axle with lifting axle design can be raised to reduce rolling resistance and tyre wear.

When laden, it can take up to 10 t of load in order to relieve the drive axle. To pull away, it can be lowered to give the drive axle more load and thereby more traction.

Fig. 1: Three-axle vehicle 6 x 2

All-wheel drive systems (Fig. 2). A transfer case is required on this 6 × 6 configuration. The first drive axle has a drive-through to the second driven axle.

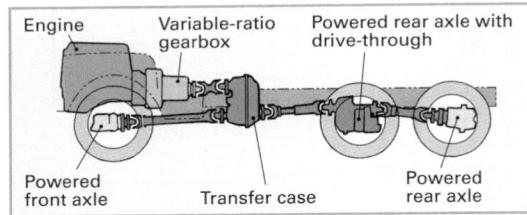

Fig. 2: All-wheel drive

22.4.3 Steering axles

Non-driven steering axles. These are installed in commercial vehicles as

- stub axle **(Fig. 3)**
- forked (Elliot) axle **(Fig. 4)**

The axle body has a stub or fork design. The steering knuckle joins the axle body to the swivel mounted hub.

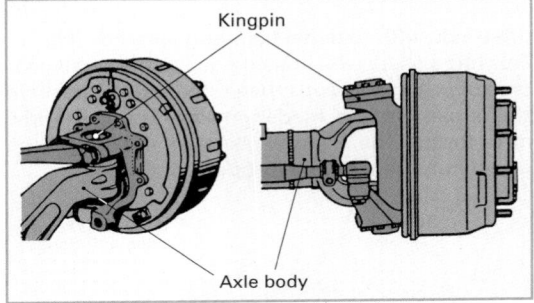

Fig. 3: Stub axle   **Fig. 4: Forked (Elliot) axle**

Driven steering axles. (Fig. 5). The steerable drive axle illustrated has a drive-through, e.g. for an 8 × 8 drive. The motive force is transferred through the drive shaft to the front axle. A bevel reduction gear mounted on it transfers the motive force to the bevel gear ring with differential. The axle shaft drives the wheel hub via double universal joints. Within this hub is a planetary gear set that converts the speed to low ratio and thereby increases the drive torque.

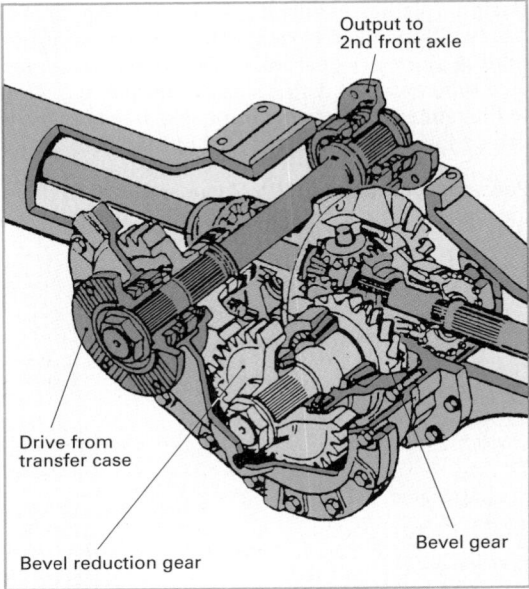

Fig. 5: Powered front axle with drive-through

22

22.4.4 Drive axles

These have a hypoid-gear unit design with bevel gear and crown wheel and external planetary axles with bevel gear drive and planetary gearbox. (Used in construction site vehicles).

External planetary axle with hypoid-gear unit. This runs very quietly, has a large axle ratio of $i = 6.0$ to 8.0 and therefore features a large axle housing with low ground clearance.

Drive axle with external planetary gearbox (Fig. 1). This has a small axle housing due to the small gear ratio of $i = 1.1$ to 1.3 in the bevel gear drive. The drive shafts can also be made smaller due to the low drive torque. The external planetary gearbox in the wheel hubs convert the torque 5- or 6-fold.

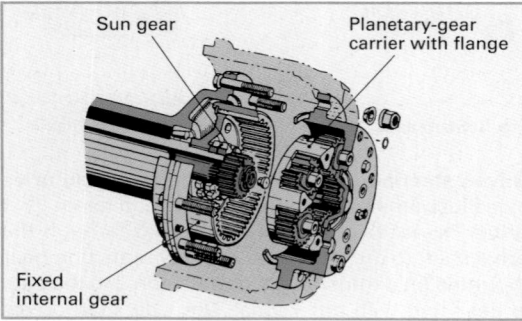

Fig. 1: External planetary axle

22.4.5 Transfer case

Design. This has a 3-shaft gear design **(Fig. 2)** and has two transmission ranges for driving on or off-road. A planetary gearbox in the drive train serves as a differential. It distributes the torque, e.g. 65% to the rear and 35% to the front axle. If necessary, it can be locked longitudinally for offroad use.

Power takeoff units (PTO). These are used to drive ancillaries, such as hydraulic pumps, winches and so on.

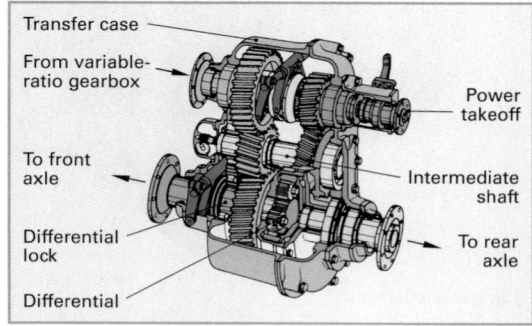

Fig. 2: Transfer case with compensator

Operating principle (Fig. 3).

Road gear. The motive force is transferred via the constant input gear pair to the planetary gear carrier. The sun gear drives the front axle, the ring gear and rear axle.

Off-road gear. With the upper locking collar, the motive force is converted to low ratio via the right gear wheel pair to $i = 1.44$, for example. In addition, the lower locking collar can be used to block the planetary gear set and thereby lock it longitudinally.

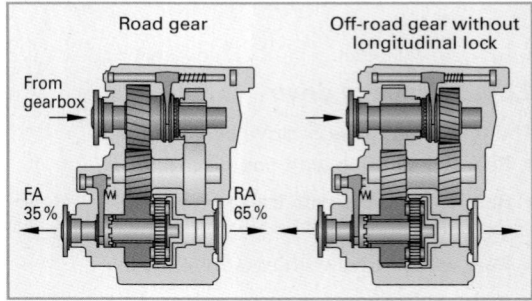

Fig. 3: Transfer case power flow

22.4.6 Auxiliary gearbox (Fig. 3)

These are installed in the drive train after the engine. With this the engine can be operated in the low fuel consumption range and also in the upper performance range.

On the **advanced shift auxiliary gearbox** (split groups) there is a finer distribution of the gears and smaller transmission ratio gaps.

On the **post shift auxiliary gearbox** (range groups), the transmission ratio range and, with it, the number of the gears is extended.

22.4.7 Variable-ratio gearbox

These are normally designed as a three shaft gearbox with common axle and can have a splitter box and a range-change box **(Fig. 4)** installed in them.

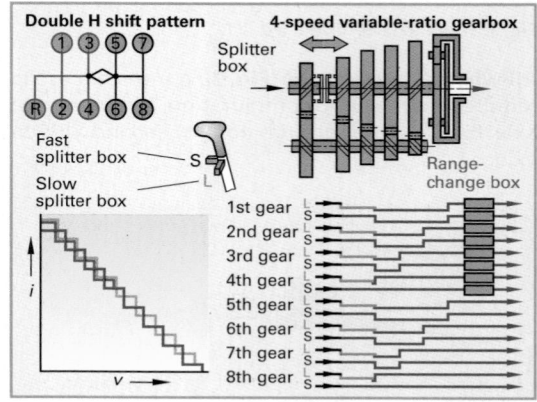

Fig. 4: Variable-ratio gearbox with splitter box and range-change box

22.4.8 Electro-pneumatic transmission control (Fig. 2)

On this system, the correct gear is engaged according to the driving speed, load and laden state automatically or manually by electro-pneumatic selector cylinders.

Design

An EPS transmission control comprises the following components:

- Mechanical clutch, operated pneumatically via the clutch actuator.
- Mechanically synchronised 4-shaft gearbox with attached pneumatic selector cylinders and solenoid valves for engagement of the gears.
- Sender unit for detection of the driver command.
- Gearbox control unit with program map for calculation of the correct gear and actuation of the solenoid valves and selector cylinders.
- Sensors for data pickup.
- Multifunction display for gear indication.

Operating principle

Control unit (Fig. 1). This features an additional changeover switch for manual or automatic operating mode.

Manual mode. Gears are changed by the driver. With the half-gear lever **(HG)**, the driver can change up half a gear by lifting up and change down half a gear by pushing down. If the function button **(FB)** is pressed and the selector lever is moved forwards, the gearbox changes up a gear. If the selector lever is pulled back, the gearbox changes down a gear. If the neutral button **(N)** is pressed, the gearbox switches to the idle position. When the vehicle is brought to a halt, the system disengages the clutch just before idling speed is reached; the selected gear remains engaged.

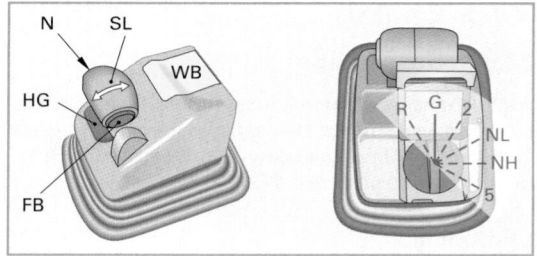

Fig. 1: Sender unit for gear change

Automatic mode. All pulling-away, manoeuvring and selection processes are carried out automatically. The optimal gear is calculated and selected specifically according to the driving condition, accelerator pedal position and engine operating condition. Clutch actuation is also controlled fully automatically.

Multifunction display. This informs the driver of the selected or preselected gear and split group.

Warning buzzer. This tells the driver that the gearbox cannot be changed up a gear as the permissible engine speed would be exceeded.

A difference is made between 2 emergency running functions:

- **Emergency running mode**
- **Gear selection using emergency switch**

Emergency running mode. If there is a fault in the system, the driver is informed by a warning buzzer and visual display that he/she must fold out the clutch pedal. It is now possible to actuate the clutch mechanically and select the gears manually.

Gear selection using emergency switch. If the control unit detects a fault that only permits continued operation via emergency selection, all automatic selection processes are blocked. Emergency selection can only be carried out with the vehicle stationary. With the emergency switch, 2nd and 5th gear, reverse and the range-change box can be actuated. With this operating condition, the driver must actuate the clutch mechanically.

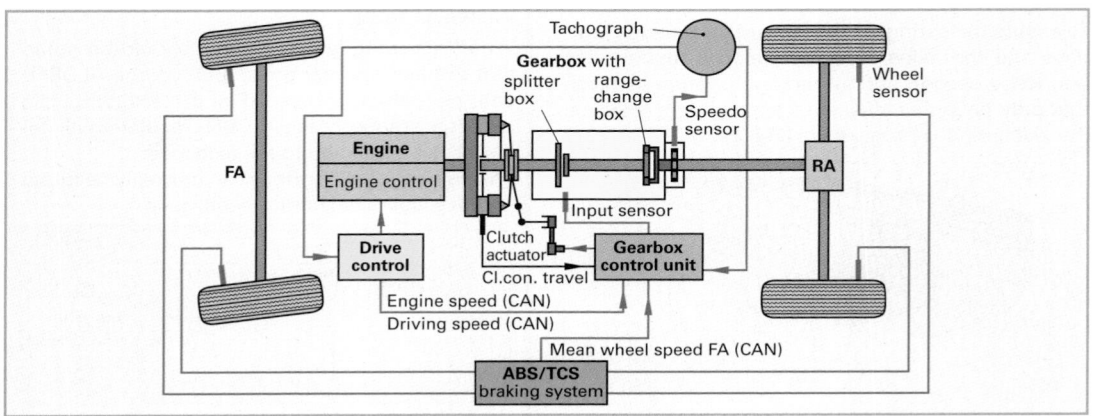

Fig. 2: Electro-pneumatic transmission control

22.5 Chassis

22.5.1 Suspension

For suspension elements in commercial vehicle construction, **leaf springs** are used much more often than coil springs. **Air suspension** is very common, particularly in buses/coaches.

Leaf spring

This is a bending spring and is used in layers or leaves to form a leaf spring. Types of design:

- Semi-elliptical spring or trapezoid spring
- Parabolic spring

Semi-elliptical spring. This is made of flat steel which has a semi-elliptical shape. Several spring leaves are joined to form a package that then takes the shape of a trapezoid **(Fig. 1)**.

The spring leaves are drilled through in the middle and are held together by the spring bolt (heart bolt), which also serves to prevent the individual leaves moving in the longitudinal plane. To prevent them moving in the lateral plane there are spring retainers.

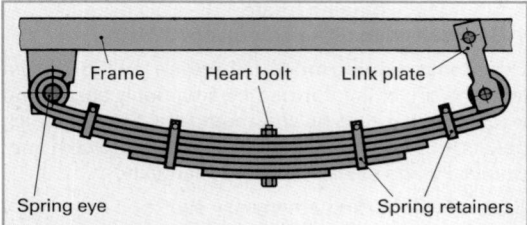

Fig. 1: Semi-elliptical spring, trapezoid spring

Trapezoid springs are hard springs. The thicker the spring leaves, the harder the spring pressure. Spring pressure also increases the greater the number of spring leaves placed on top of each other.

To ensure the spring is not too hard for unladen journeys and the axles do not develop a tendency to bounce, a second spring package is often installed that only becomes effective from a certain load, i.e. the leaf spring is progressive **(Fig. 2)**.

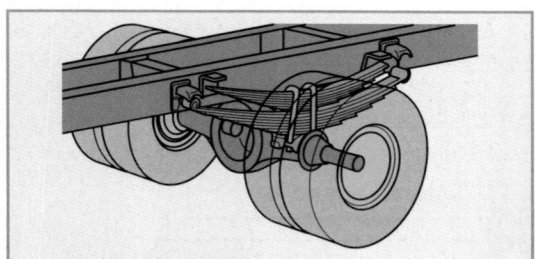

Fig. 2: Double trapezoid spring

Through friction of the individual spring leaves against each other during compression and rebound, there is a high level of self damping which in turn helps to absorb oscillations. No rust must form between the spring leaves. There must always be a lubricating layer in between.

Leaf springs can transmit braking and acceleration forces and also side forces. The components for attaching the spring to the frame or chassis, such as spring clips, pins, bushing and spring eyes, are thereby placed under a very high load. To prevent the axle becoming loose from the frame in the event of upper spring leaf breakage, the front end of the second spring leaf is also bent over the spring pin. The rear spring eye has a lashing mounting that can compensate for the spring extension during compression.

Parabolic spring. The individual spring leaves become narrower, starting from the centre to both ends, in a parabolic fashion.

The parabolic spring comprises just a few strong spring leaves with intermediate layers made from plastic, so the spring leaves cannot rub against each other **(Fig. 3)**. Due to the long spring travel and the low inner friction, the parabolic spring works more softly and offers more comfort.

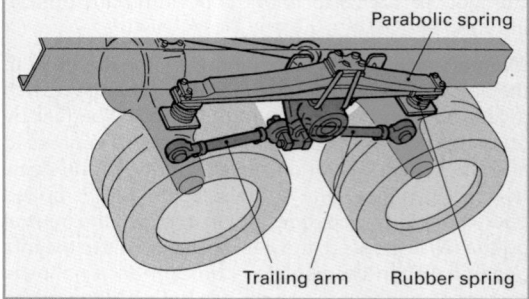

Fig. 3: Parabolic spring

WORKSHOP NOTES

When renewing leaf springs, it should be noted that the leaf springs prescribed on the ALDBFR data plate must be used **(Fig. 4)**. Otherwise, the output pressures of the ALDBFR regulator will not be correct in relation to the axle load.

The number of leaf springs can normally be found on the second leaf spring bearing.

Rear spring no.	Input pressure		8	bar			
81.43402.6504 .6505	Output pressure at the ALDBFR				$l =$	100	mm
Rear axle load kg	To the front axle bar		To the rear axle bar		Stroke s at lever mm		
2,000	4.5/5.5*		1.8		117		
2,500	4.6/5.6*		2.0		112		
3,000	4.8/5.7*		2.4		106		
3,500	4.9/5.8*		2.7		101		

Fig. 4: Extract from ALDBFR data plate

22

Air suspension

> Used mainly in buses/coaches, commercial vehicles and trailers. Here, the compressibility of gases is utilised for purposes of suspension.

Compared with leaf springs, air suspension has the following features:

- Improved ride comfort and greater safety of the load thanks to lower spring rate and lower natural frequency.
- No self-damping.
- Progressive spring characteristic.
- Constant vehicle height regardless of the load.
- Suspension height adjustment possible, e.g. in conjunction with load platform.
- Lifting axles can be controlled simply, e.g. to aid pulling away or protect against overload.
- Cannot support wheel guiding forces.

Design. As suspension elements, roll bellows or air bags are used that are made from rubber with a fabric inlay **(Fig. 1)**. A hollow rubber spring limits the spring buffer and allows manoeuvrability of the vehicle at walking pace in the event of total air loss.

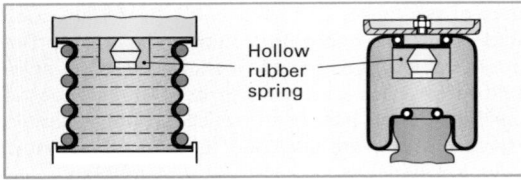

Fig. 1: a) Gaiter seal   b) Roll bellows

Hollow rubber spring

Wheel guidance (Fig. 2). The wheels are guided by trailing and transverse control arms or stabiliser arms. The vibrations are dampened by shock absorbers.

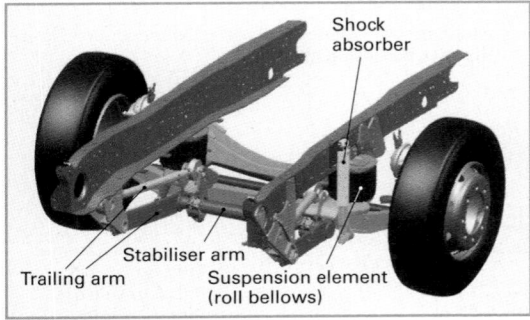

Shock absorber

Trailing arm

Stabiliser arm

Suspension element (roll bellows)

Fig. 2: Wheel suspension on front axle with air suspension

Electronically controlled air-suspension systems.
These can be used to fulfil a whole range of functions, such as suspension height control, suspension height adjustment, height restriction, lifting axle control, pressure control, fault detection and storage.

Specified suspension height control (Fig. 3). The travel sensors fixed to the vehicle frame record the suspension height of the vehicle on a continual basis and send the data to the control electronics. If deviations from the specified level are detected, solenoid valves on the front axle and the rear axle are actuated. By inflating or deflating the air-suspension bellows, the actual level of the vehicle is brought to the specified height within the specific tolerance limits. Depending on the number of travel sensors, a difference is made between two-, three- and four-point regulation. The most common is three-point regulation. With this system, for example, two travel sensors are used for the steering axle and one travel sensor for the drive axle.

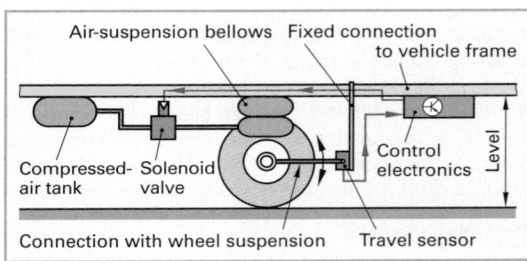

Air-suspension bellows Fixed connection to vehicle frame

Compressed-air tank Solenoid valve

Control electronics Level

Connection with wheel suspension Travel sensor

Fig. 3: Diagram showing principle of electronic suspension height control

Switch selected suspension height adjustment. Here, pre-programmed suspension heights can be selected using a switch to lower or raise the vehicle, for example, where interchangeable bodies such as containers are used.

Height restriction. If the upper or lower height threshold (rubber buffer stop) is reached, height adjustment is cancelled.

Lifting axle control/regulation. The lifting axles can be raised using a switch to improve traction of the drive axle, for example. From a certain axle load, e.g. 11 t, the lifting axle lowers itself automatically.

Pressure control. To measure the actual air-suspension pressure and to keep the pressure within certain limits, pressure sensors are joined to the air-suspension elements. These convert the measured pressure figure into a voltage signal. The voltage signal is sent to a control unit, which actuates solenoid valves respectively in the event of thresholds being reached (maximum pressure; minimum pressure).

Fault detection and fault storage. In the event of fault detection, a warning lamp is activated on the inside of the vehicle. At the same time, the fault is stored in the control unit. The vehicle can continue to be driven providing the prescribed suspension height is adhered to.

Electronically controlled air-suspension system (Fig. 1)

Design. The air-suspension system shown is from a HGV with full air suspension and lifting axle. With the ignition switched on, three travel sensors **(3)** send information about the suspension height of the vehicle to the control unit on an ongoing basis. Inflation and deflation of the air-suspension bellows is by way of 3/2 and 2/2 solenoid valves. Pressure sensors monitor the pressures in the air bags. With the drive-off button **(7)**, the lifting axle air bags are deflated briefly to improve traction of the drive axle. With the "raise" **(8)** and "lower" **(9)** pushbuttons, the driver can control the lifting axle manually. If the maximum permissible single axle load is exceeded, the electronics prevent the lifting axle from being raised. A pressure switch **(5)** monitors the system pressure. If system pressure drops below the required level, the **warning lamp 4a** is activated. **Fault lamp 4b** indicates faults in the electrics/electronics. These faults are stored in the electronic system. With a remote control **(2)**, the specified suspension height of the vehicle can be altered in conjunction with a loading platform, for example. However, this is only possible below a certain speed set by the manufacturer.

Operating principle

Air supply. The air-suspension system is supplied with up to 12.5 bar of compressed air via a compressor installed on commercial vehicles. The compressed air is extracted either at the secondary consumer connection of the four-circuit protection valve 23 or 24 or from behind the pressure regulator.

Specified suspension height control. The actual values supplied by the travel sensors are compared with the specified values stored in the system. If deviations from the specified suspension height exceed the predetermined tolerance range, the solenoid valves are actuated by the control unit respectively so that the necessary alteration in suspension height is achieved at the front and drive axle by deflation or inflation of the air bags. Regulation takes a matter of seconds when the vehicle is stationary. Brief oscillations in the axles from uneven road surfaces, for example, are not com-

pensated for as the control system does not become active until after a specific time delay.

Inflation of air-suspension bellows on front and drive axle. To do this, the solenoid valves **a, b, c** and **d** are actuated electrically. Solenoid valve **c** switches to pressure build-up. Solenoid valves **a, b, d** switch to through-flow. Once the specified suspension height has been reached, the solenoid valves are switched back to the initial position.

Deflation of air-suspension bellows on front and drive axle. Solenoid valves **a, b** and **d** are actuated electrically and switched to through-flow. The air bags are deflated to the open air by way of a muffler via solenoid valve **c** until the specified suspension height is reached again.

Lifting axle control. Using pressure sensors **(6)** on the drive axle, automatic lifting axle lowering can be carried out. If air bag pressure on the drive axle rises above a certain limit, e.g. 5.3 bar, due to a heavy load, the lifting axle is lowered automatically. There is no automatic lowering of the lifting axle from peaks in pressure while the vehicle is in motion.

Reducing compressed-air supply of lifting axle. To do this, solenoid valve **f** is actuated electrically, the lifting bellows is deflated via solenoid valve **c**, the axle lowers itself. To supply the air-suspension bellows of the lifting axle with compressed air, solenoid valves **c, e** and **g** are actuated electrically. Solenoid valve **c** switches to pressure build-up, solenoid valves **e** and **g** switch to through-flow. Once the specified level has been reached, the solenoid valves **e** and **g** are switched to blocked. Solenoid valve **c** is switched back to the initial position.

REVIEW QUESTIONS

1 **What advantages and disadvantages are there for air suspension on commercial vehicles compared with leaf spring suspension?**

2 **How is the specified suspension height reached on a vehicle with air suspension?**

3 **How are the solenoid valves activated as per Fig. 1 to raise the lifting axle?**

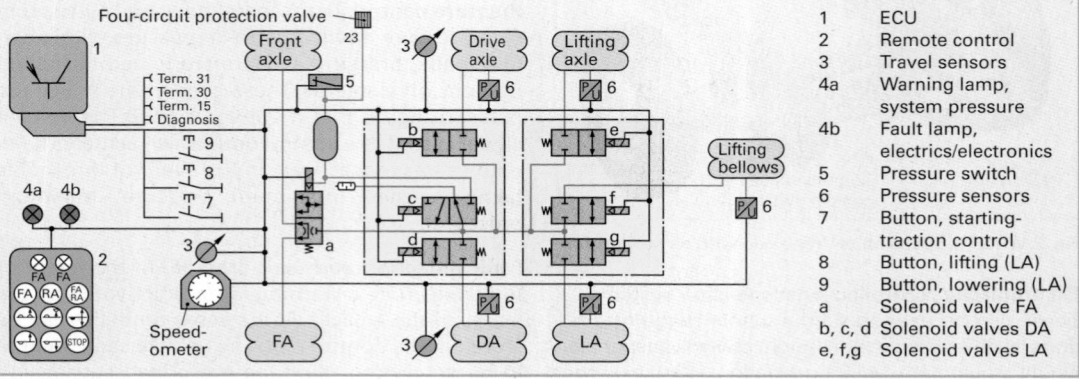

1	ECU
2	Remote control
3	Travel sensors
4a	Warning lamp, system pressure
4b	Fault lamp, electrics/electronics
5	Pressure switch
6	Pressure sensors
7	Button, starting-traction control
8	Button, lifting (LA)
9	Button, lowering (LA)
a	Solenoid valve FA
b, c, d	Solenoid valves DA
e, f, g	Solenoid valves LA

Fig. 1: Functional diagram of an air-suspension system from a fully air sprung 6×2 HGV

22.5.2 Wheels and tyres

Tyres

In order to achieve low wear, good traction, high load ratings and large inner wheel diameters on commercial vehicles tyres, low profile tyres with radial ply design are used almost exclusively **(Fig. 1)**. Here, the carcass and bracing layers are normally manufactured from steel cord due to the high load rating of heavy goods vehicles.

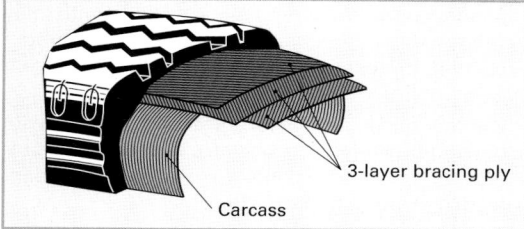

Fig. 1: Structure of a HGV radial ply tyre

Designations on the tyre **(Table 1, Fig. 2)**.

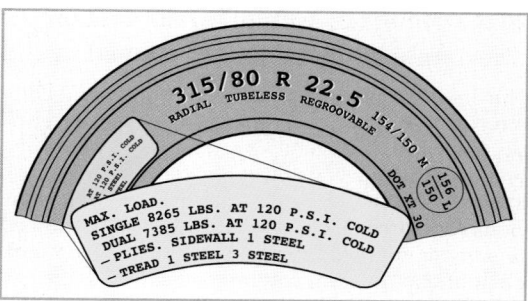

Fig. 2: Tyre dimensions and additional details

Wheels

The wheels comprise a rim to support the tyre and a wheel rim to attach the wheel to the hub. On commercial vehicles, the following types of wheel design are common:

Unsplit 15° tapered bead rim (Fig. 3). These can be identified by ".5" in the rim diameter designation, e.g. 22.5.

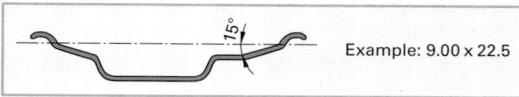

Fig. 3: 15° tapered bead rim

Longitudinally split 5° tapered bead rim (Fig. 4) and **semi-drop centre rim** (SDC; H – code for rim flange dimension) **(Fig. 5)**. Split rims can be identified by a hyphen "–" between the rim flange width and rim diameter dimensions.

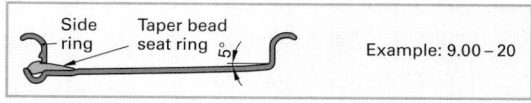

Fig. 4: 5° tapered bead rim

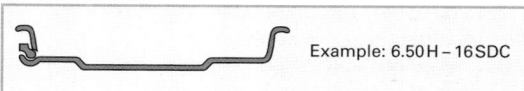

Fig. 5: Semi-drop centre rim

Table 1: Tyre details

Example: 315/80 R 22.5 154/150M $\left(\frac{156}{150}\right)$ L ...	
315	Tyre width in mm
80	Tyre height ≠ 80 % of rim width
R	Radial design
22.5	Tyre inner diameter in inches
154/150 M	Load rating code for single and twin tyres at speed according to speed index M (v_{max} = 130 km/h) and manufacturer prescribed inflation pressure, e.g. 8.5 bar.
156/150 L	Additional service code for single and twin tyres at low speeds (L: v_{max} = 120 km/h).
Regrooveable	Tyres can be recut by specialists based on manufacturers specifications.
Tread: 3 steel 1 steel	Beneath the contact surface are four steel cord plies. Three steel cord radial plies and one steel cord carcass ply.
Sidewall 1 steel	Located in the sidewall is a one-layer steel cord (tyre casing).
Single 8265 LBS. AT 120 P.S.I	Single tyre, load and pressure code designation for USA/Canada 1 LBS (pound) = 0.4536 kg; 1 P.S.I. (pounds per square inch) = 0.06897 bar).

Tyre safety and service life are dependent on load, inflation pressure and speed. Therefore, the inflation pressure specified by the manufacturer should be adhered to at all times. Precise inflation pressure figures can be gleaned from the table.

WORKSHOP NOTES

- Side and sealing rings on split wheel rims must fit in dimension to the rim.
- Wheel rims must always be clean and free of rust.
- On new tyres, new valves i.e. new tubes and bead belts have to be used.
- When inflating HGV tyres, 150% of the standard air pressure must not be exceeded. **However, 10 bar maximum!**
- On twin tyre systems, the tyre with the largest diameter must be fitted on the inside.

22

22.5.3 Air-brake system (brake system with external power source)

The air-brake system is used on medium and commercial vehicles. It is powered by an external source, which is actuated by the driver by way of a brake valve thus causing the external source, e.g. 8 bar to 10 bar of compressed air, to apply the clamping forces at the wheel brakes. Combined pneumatic-hydraulic brake systems are common on light and medium commercial vehicles.

Illustration of air-brake systems

For standardised illustration of the equipment, graphic symbols can be used. Numbers/codes can be used for the equipment connections.

Equipment connections

These are designated by one or two digit numbers. The first digit means

0 intake connection	5 not assigned
1 energy input	6 not assigned
2 energy outlet (not to atmosphere)	7 antifreeze connection
3 ventilation, atmosphere	8 lubricant connection
4 control connection	9 coolant connection

If several connections of the same type are present in the system, e.g. more than one circuit, an additional number is added. This should be started with 1 without gaps, e.g. 21, 22, 23. Several connections of the same type from one chamber are given the same code.

Example of use

The numbers in **Fig. 1** mean

1	energy inlet from compressor
1-2	charge connection e.g. using auxiliary compressor used for tyre-inflation connection
3	ventilation to atmosphere
21	energy outlet (first connection)
22	energy outlet (second connection, switch connection)

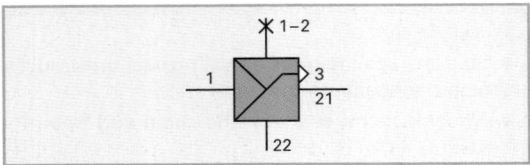

Fig. 1: Pressure regulator

22.5.3.1 Dual-circuit, dual-line air-brake system

Fig. 1, Page 660, shows a dual-circuit, dual-line air-brake system configured according to EU guidelines for "brake systems". The equipment from the same component groups are colour coded.

Component groups

- **Compressed-air supply system** (energy supply) with compressor, pressure regulator, air drier, regeneration reservoir or antifreeze pump, four-circuit protection valve, 3 air reservoirs with drain valve, pressure display equipment and pressure warning system.
- **Dual-circuit service brake system for pulling vehicle/tractor unit** with service-brake valve and proportioning pressure regulator, automatic load-dependent brake force regulator (ALDBFR) with relay valve, combination brake cylinder with diaphragm element for rear axle, diaphragm cylinder for front axle.
- **Parking brake and auxiliary brake system** with parking-brake valve, relay valve with overload protection, combination brake cylinder with spring-loaded element for rear axle.
- **Trailer control system** with trailer control valve, coupling heads "supply" and "brake".
- **Dual line trailer brake system** with supply line and brake line, trailer brake valve, automatic load-dependent brake force regulator (ALDBFR), brake cylinder.
- Continuous brake system with selector valve, working cylinder with actuation for exhaust flaps and control racks.
- Trailer parking brake system (mechanical) with handbrake lever, linkage, brake levers on wheel brakes.

General operating principle of the air-brake system (Fig. 1, Page 660)

Compressed-air supply system

The compressor draws in surrounding air via an air filter, compresses it and forces it via the pressure regulator to the air drying equipment. The pressure regulator automatically controls the pressure in a range between 7 bar and 8.1 bar, for example. The air drier uses a filter to clean the compressed air and extracts any water from it. To do this, it is passed through an air drying agent (desiccant), which traps any moisture. The dried air then flows to the regeneration reservoir and four-circuit protection valve. This distributes the compressed air to four supply circuits and makes these individual circuits safe. These are:

- Circuit I   (21) (service brake – rear axle)
- Circuit II (22) (service brake – front axle)
- Circuit III (23) (parking-brake system, trailer)
- Circuit IV (24) (retarder, secondary consumers)

Once the shutoff pressure has been reached, the pressure regulator dumps air to the open, the non-return valve closes and maintains pressure in the system. The dry air stored in the regeneration reservoir flows via the drying agent back to the open. In this way, any moisture is drawn out with it and the drying agent is dried and regenerated for the next charging process. A heater element in the area of the cut-out plunger prevents freezing of the damp air being dumped and thereby protects against malfunctions. A dual pressure gauge shows the driver the supply pressure in both service brake circuits. If pressure drops below a warning pressure of about 5.5 bar, a warning lamp will light up.

Compressed-air supply system without air drier.

On this system an antifreeze pump is installed after the pressure regulator. This injects antifreeze into the system during the charging procedure.

Service brake system of pulling vehicle/tractor unit (Fig. 1, Page 660)

This includes a service-brake valve with integrated proportioning pressure regulator for load-dependent front axle control. The system is regulated by control connection 4, which is actuated by the ALDBFR regulator of the rear axle. The ALDBFR regulator adapts brake pressure of the rear axle to the load. The brake pressure from the front axle (connection 22) is regulated depending on the pressure of the ALDBFR regulator by the service-brake valve. Regulation is also load-dependent in this case. If the vehicle is unladen, the pressure output is less than the pressure that otherwise equates to the brakes applied pressure of the service-brake valve. Not until the vehicle is fully laden is reduction of the brake pressure output stopped.

Driving position (released). In both circuits of the service-brake valve (connections 21, 22), the inlet is closed and the outlet is open. The brake cylinders of the front axle and the control line to the relay valve with overload protection (connections 41, 42) and control line to the ALDBFR regulator (control connection 4) are vented to the open by their open outlets. Furthermore, the spring-loaded actuators of the combination brake cylinder (connections 12) are vented via the relay valve with overload protection. The springs are tensioned and all brakes of the engine carriage are released.

Brake position. In the service-brake valve, the outlets are closed and the inlets (connections 11 and 12) are opened. As the pedal is pressed, compressed air is now applied proportionally from the service-brake valve in the control line to the ALDBFR regulator (connection 21 after 4) for the rear axle. The ALDBFR regulator actuates its relay valve and this charges the diaphragm cylinder of the rear axle (connection 2 after 11) with supply pressure in relation to the braking force and the load status. The front axle receives its brake pressure from the service-brake valve (connection 22). This adapts the brake pressure proportionally to the vehicle load with the integrated proportion-

ing pressure regulator. In addition, two control lines from the service-brake valve (connection 21 after 41 and 22 after 42) actuate the trailer control valve. With a trailer attached, the trailer brake line is now charged with pressure proportionally and the trailer brakes are actuated via the trailer control valve.

On brake systems with a service-brake valve without integrated proportioning pressure regulator, a special proportioning pressure regulator is installed for load-dependent regulation of the front axle brake pressure **(Fig. 1)**.

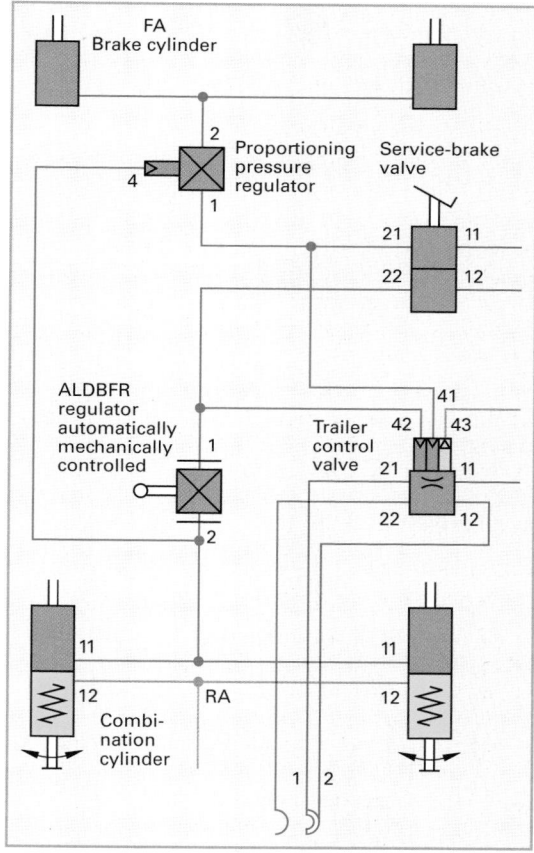

Fig. 1: Service brake system with proportioning pressure regulator

Parking and auxiliary brake system

A control line runs from the parking-brake valve to the relay valve with overload protection (connection 21 after 42) and a second to the trailer control valve (connection 22 after 43). With the aid of these, the spring-loaded actuators of the rear axle can be applied in proportion as auxiliary or parking brake for the engine carriage and the service brake for the trailer. The parking-brake circuit is safeguarded against pressure loss in supply circuit III by a non-return valve.

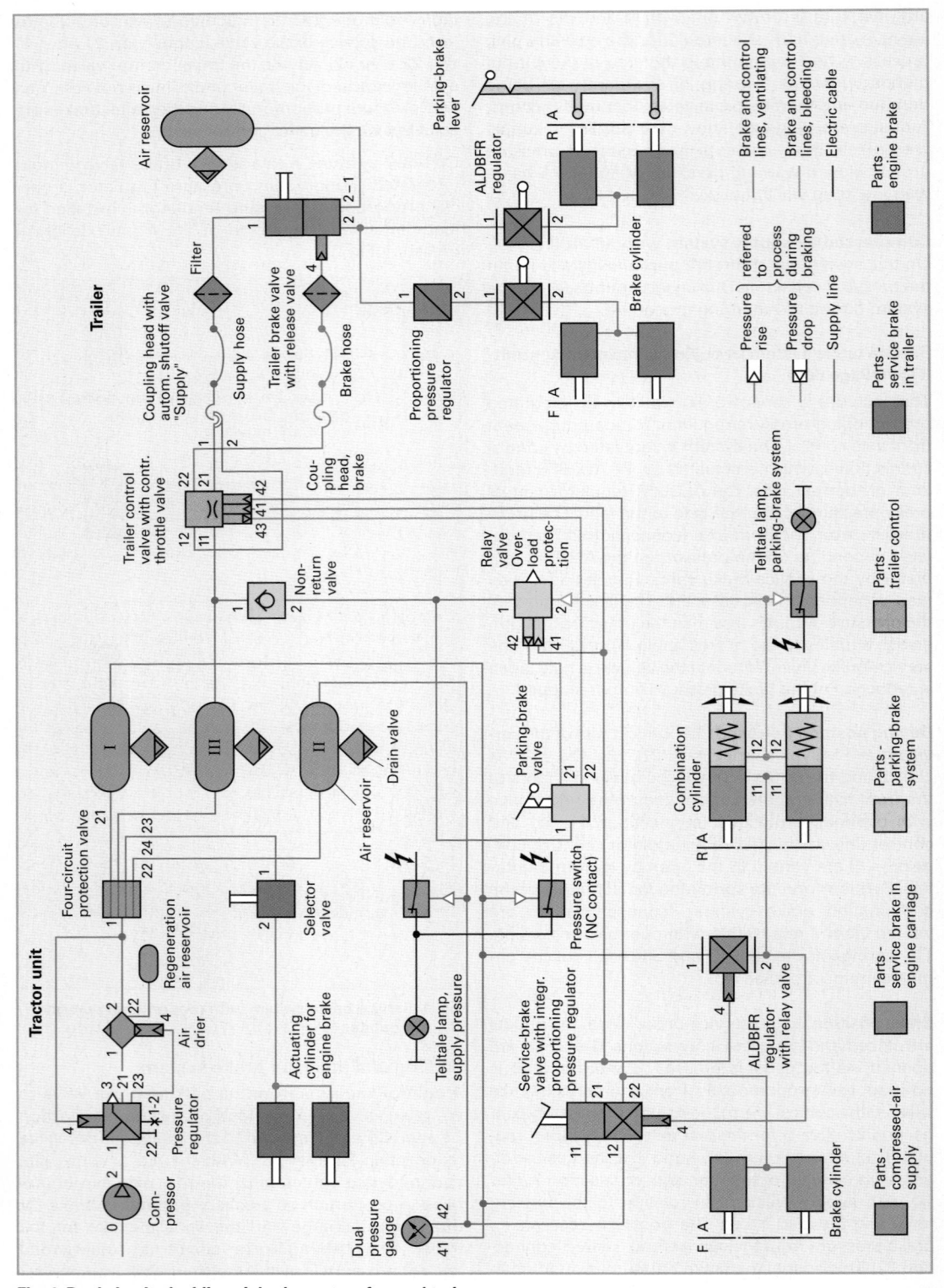

Fig. 1: Dual-circuit, dual-line air-brake system for road train

Control position. In accordance with regulations, the parking brake of the engine carriage must be in a position to hold the truck and trailer combination on a hill even with the trailer brake released. As a functional test, the parking-brake valve has a control position with which the spring-loaded brake is actuated and the trailer brake released.

Driving position. The parking-brake valve charges the control line to the relay valve (connection 21 after 42). This switches over and the spring-loaded actuators (connection 2 after 12) are applied with supply pressure. The springs are tensioned and the brakes released. At the same time, the control line is charged to the trailer control valve (connection 22 after 43). The brake line (connection 22) is now depressurised. The trailer brakes are thus released.

Brake position. By actuation of the parking-brake valve, the control lines to the relay valve (connection 21 after 42) and to the trailer control valve (connection 22 after 43) can be relieved of pressure proportionally. The relay valve switches over, the spring-loaded actuators in the combination brake cylinders are relieved of pressure and the brakes are applied by the springs. The trailer control valve passes reservoir air proportionally to the trailer brake valve at connection 22 via the brake line. This now brakes the trailer appropriately.

Overload protection. Protection against overload is provided if, for example, the service brake is pressed when the parking brake is also applied. The parking brake is then relieved of pressure and released in relation to the excess pressure rising in the service brake system. Therefore, where parts of the brake system could be overloaded, the full forces in the diaphragm and spring-loaded cylinder are not effective at the same time.

Continuous brake system. If the driver presses the selector valve, compressed air flows from connection 43 of the four-circuit protection valve to the working cylinder that actuates the engine brake (exhaust flap).

Trailer brake system
Supply line. A red coupling head on the engine carriage with automatic shutoff element supplies the trailer via a connecting hose and line filter from circuit 3. The valve in the trailer brake valve does not open and charge the air reservoir until the coupling process is complete.

Brake line. The yellow coupling head supplies the trailer during braking with brake pressure applied by the service brake circuit.

Released position. The brake line is relieved of pressure by the trailer control valve via connection 22. As a result, the trailer brake valve relieves the trailer brakes of pressure and releases them.

Brake position. The trailer control valve is actuated proportionally with compressed air by the service-brake valve via connections 41 and 42. It charges the brake line via connection 22. As a result of the pressure rise, the trailer brake valve is actuated proportionally and releases compressed air from the reservoir tank in the trailer to both ALDBFR regulators of the trailer axles. These now regulate the brake pressure for the brake cylinders in relation to the axle load. A proportioning pressure regulator reduces the braking power of the front axle if the vehicle is unladen or partly laden to prevent overbraking. The trailer is therefore braked in relation to the braking power and the load.

Supply line shear. Pressure in the supply line drops. The trailer brake valve thereby triggers full braking power in the trailer. This is also the case during decoupling. To move the decoupled trailer, a release valve on the trailer brake valve must be pressed.

Defect in brake line. The brakes stay released to start with. Not until one of the brakes is applied in the engine carriage does supply pressure escape from the defective brake line and connection 22 on the trailer control valve. This is joined via connection 12 to connection 2 on the "supply" coupling head. Pressure in the supply line drops and the trailer brake valve triggers full braking power in the trailer. When the brakes in the engine carriage are released, the trailer brakes are also released again.

Parking-brake system in trailer
This works purely mechanically. When the parking-brake lever is actuated, the rear axle brakes of the trailer are applied via linkage and levers.

22.5.3.2 Components of air-brake system
Compressor (Fig. 1, Page 662)

Function
● Supplies brake system with compressed air

Operating principle. The compressor is a one- or two-cylinder piston compressor that is driven by and runs permanently with the engine. On the intake stroke, it draws in fresh air via an air filter and compresses it. The valves seated on the cylinder head control the air inlet and outlet. Lubrication is normally force-fed from the engine.

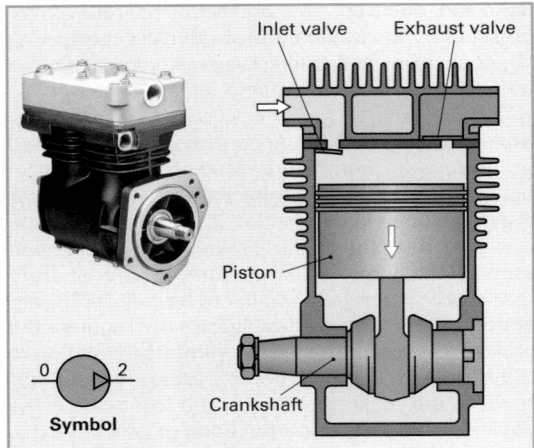

Fig. 1: Compressor

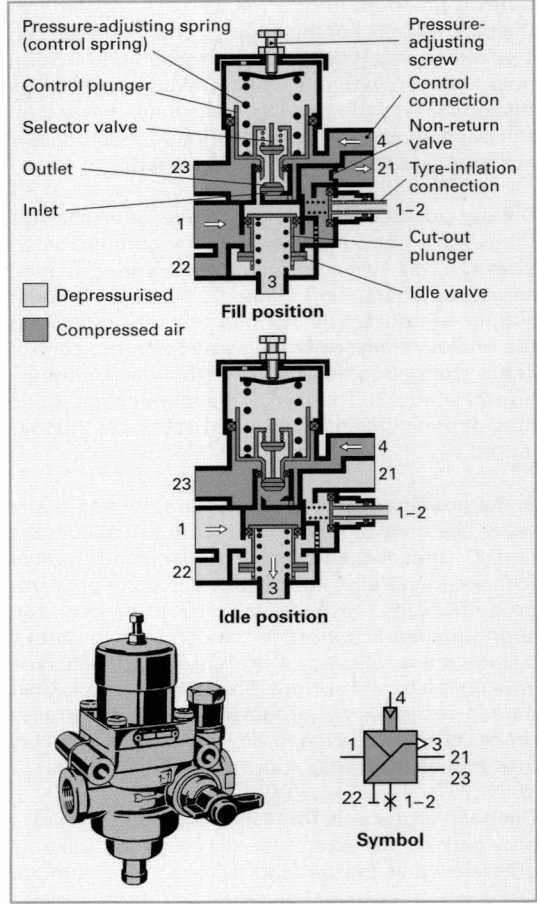

Fig. 2: Pressure regulator

Pressure regulator (Fig. 2)

Functions

- Automatically regulates operating pressure between cut-out and cut-in pressure.
- Safeguards the system against dirt (filter).
- Allows compressed air to be accessed from tyre-inflation connection, e.g. for inflation of tyres or to charge system, e.g. from external source.
- Safeguards system against overpressure (idle valve works as safety valve).
- Controls air drier or frost protection.

Operating principle

Fill position. Compressed air from the compressor flows from connection 1 to connection 21 in the system. This pressure is also applied at connection 4 and works upwards against the control plunger. When cut-out pressure is reached (e.g. 8.1 bar), this works upwards against spring pressure from the control spring. The outlet valve closes and the inlet valve opens. Compressed air now pushes the cut-out plunger downwards and opens the idle valve. The compressor now delivers air from 1 → 3 to the open. The non-return valve maintains pressure in the system.

Idle position. If supply pressure drops to cut-in pressure due to air being drawn out of the system, the control spring pushes the control plunger downwards. The inlet closes and the outlet opens. The cut-out plunger is now relieved of pressure and is pushed upwards by its spring. The idle valve closes and the air reservoirs are filled again.

Tyre-inflation connection. The tyre inflation hose moves the valve body in the tyre-inflation connection during attachment and blocks connection 2. Compressed air can now be withdrawn or added to the system.

Four-circuit protection valve (Fig. 1, Page 663)

Functions

- Distributes compressed air to the four brake circuits.
- Retains pressure in intact circuits in event of pressure drop in one or more brake circuits.
- Charges service brake circuit where necessary.

Operating principle

The compressed air flows in from the compressor via connection 1. Once the opening pressure is reached, e.g. 7 bar, the two overflow valves open to the connections of the service brake circuits 21 and 22. Air can now flow into the connected tanks. Compressed air is also applied at the overflow valves to the connections of circuits 23 and 24 at the same time via the non-return valves. When the pressure reaches 7.5 bar, for example, the overflow valves open. The air reservoirs of the service brakes are already practically charged. These two circuits are now also charged.

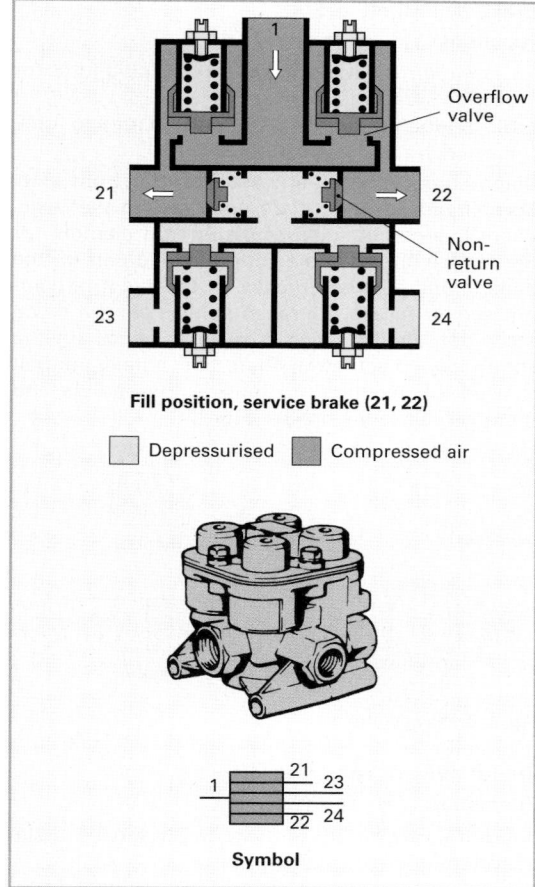

Fig. 1: Four-circuit protection valve

Protection measure

If circuit 21 has a leak, for example, air will escape. Pressure and also that in circuit 22 drops to the closing pressure of approx. 5.5 bar. The overflow valve from circuit 21 closes. The compressor then charges circuit 22 again until the opening pressure of the overflow valve in circuit 21 is reached (e.g. 7 bar). Pressure in circuits 23 and 24 remains constant as the non-return valves prevent air escaping through the leak. The parking brake (circuit 23) therefore stays released.

Service-brake valve with proportioning pressure regulator

Functions

- Finely proportions increase and decrease of pressure in dual-circuit service brake system in pulling vehicle/tractor unit.
- Controls trailer control valve.
- Load-dependent control of front axle brake pressure with proportioning pressure regulator as necessary.

Operating principle

The actuation device, e.g. foot plate or brake pedal, works on two successively located valves.

Driving position (Fig. 2). The inlet at connections 11 and 12 is closed. The service brake circuits are not supplied with reservoir pressure. The outlet of connections 21 and 22 is open. They are vented to the open air via connection 3.

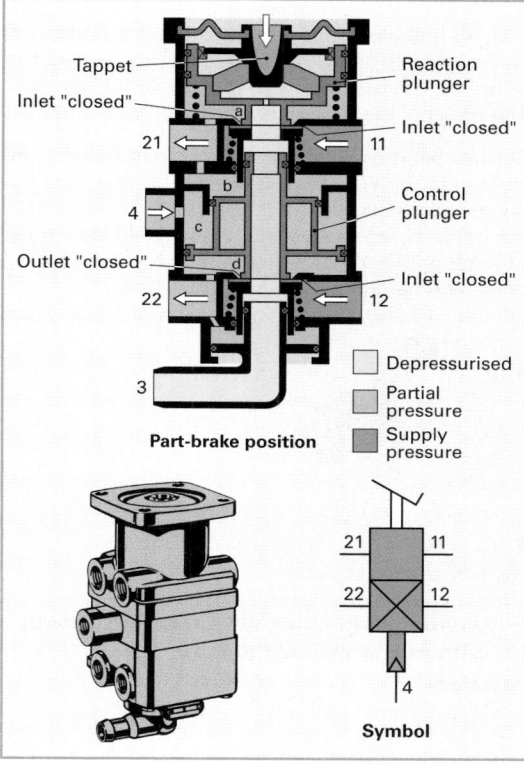

Fig. 2: Service-brake valve with proportioning pressure regulator – partially braked position

Partially braked position. On actuation of the brakes, the inlet is opened via the reaction plunger. Compressed air flows in and affects the control plunger. The lower inlet is opened in this way. The end brake position is reached when the pressure in chamber 'a' pushes the reaction plunger upwards until both forces are balanced. The outlet valves are now closed. Partial pressure is applied in service brake circuits 21 and 22.

Fully braked position. To achieve this, the brake pedal is fully depressed and the reaction plunger is thereby moved to the end position. In this way both inlet valves are fully opened. The maximum operating pressure is applied in the brake circuits.

Parking brake and auxiliary brake valve (Fig. 1)
Functions
- Actuates parking and auxiliary brake proportionally using spring-loaded cylinders.
- Provides control setting to check effect of parking brake in pulling vehicle/tractor unit.

Driving position. Spring-loaded actuators of combination cylinders and control line to trailer control valve are charged. The springs are tensioned.

Parking-brake position. Spring-loaded actuator and control line to trailer control valve are relieved of pressure. The brakes of the spring-loaded actuators and those of the trailer are applied.

Control position. The rear axle of the pulling vehicle/tractor unit is braked by the vented spring-loaded actuators; the trailer brakes are released via the trailer control valve. The whole road train must now be held on a 12% gradient using the parking brakes of the pulling vehicle/tractor unit.

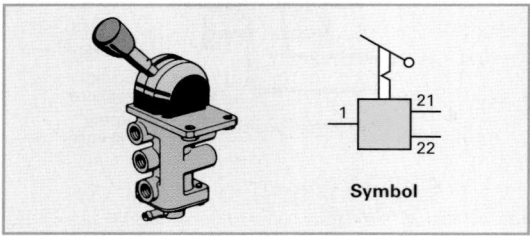

Fig. 1: Parking-brake valve

Automatic load-dependent brake force regulator with relay valve (ALDBFR, Fig. 2).
Functions
- Automatically regulates brake force based on load.
- Provides control on air suspension vehicles through pressure in air bag, on mechanical suspension vehicles through range of spring.
- Relay valve to rapidly charge or relieve pressure.

In unladen state, the brake pressure is reduced e.g. by about 5:1, i.e. at 6 bar brake pressure 1.2 bar works on the wheel cylinders, at full load there is 6 bar.

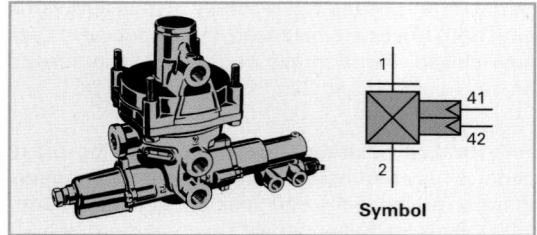

Fig. 2: ALDBFR with relay valve with air suspension

Brake cylinder (Fig. 3)
Functions
- Diaphragm cylinders generate clamping force for service brakes.
- Spring-loaded cylinders generate clamping force for parking and auxiliary brakes.

Diaphragm cylinders are used on the front axle. Combination cylinders are used on the rear axle. There is a combination of diaphragm element for the service brakes and a spring-loaded part for the parking and auxiliary brakes. In the event of compressed-air failure, a braked vehicle can be made ready for towing using a release device on the spring-loaded actuators. To do this, a hexagon bolt is used, for example, to tension the springs and thereby release the brakes (**Fig. 3**).

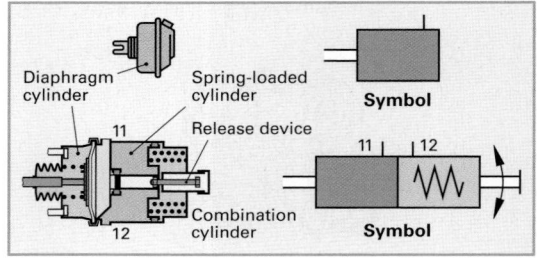

Fig. 3: Brake cylinder

Trailer control valve (Fig. 4)
Functions
- Controls trailer brake system via service brake system of engine carriage.
- Controls trailer brake system via parking and auxiliary brake valve.
- Supplies trailer brake system with compressed air.

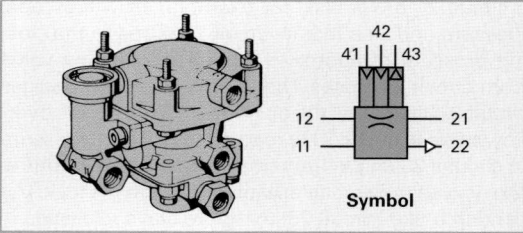

Fig. 4: Trailer control valve with throttle valve

Driving position. Connection 41 and connection 42 are depressurised. Connection 43 is under supply pressure. Outlet 22 is also depressurised.

Brake position. If brake pressure is applied by service-brake valve 41, the brake line to the trailer is charged with pressure respectively.
If the handbrake is pressed, connection 43 is depressurised and operating pressure is applied to connection 21.

Wheel brakes

These are friction brakes that convert brake force into heat as a result of friction.

On buses/coaches and also quite commonly on heavy goods vehicles, disc brakes are fitted to all wheels. Drum brakes are also used occasionally, e.g. on construction site vehicles or on rear axles.

Drum brakes (Fig. 1). In most cases, simplex brakes are used that are actuated using S-cams or wedges. The wedge is actuated directly from the diaphragm cylinder. There is no brake lever or brake shaft. Automatic adjustment of the friction lining to compensate for wear is normally integrated. The wedge brake has practically replaced the S-cam brake.

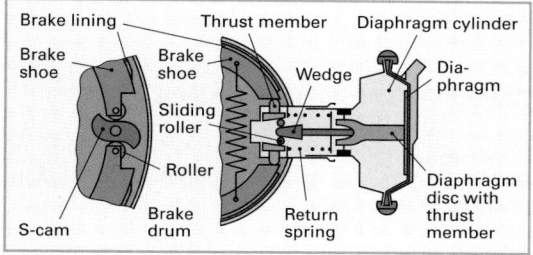

Fig. 1: S-cam brake and wedge brake

Disc brakes (Fig. 2). These are very widespread due to

- good proportioning capability,
- good heat dissipation,
- good dirt removal,
- low fading and
- equal braking effect.

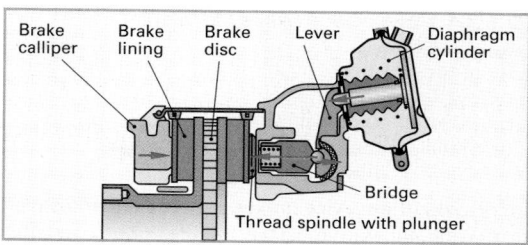

Fig. 2: Pneumatically actuated disc brake

22.5.3.3 Combined pneumatic-hydraulic brake system (Fig. 3)

These types of brake systems are used in medium weight goods vehicles and buses/coaches (6 t to 13 t maximum permissible weight) without trailer operation.

Advantages as a result of hydraulic transfer of braking forces:

- High brake pressures with small components.
- Short threshold peaks and direct response of brakes.

Design

- Compressed-air supply system as on air-brake system with four supply circuits, but only two air reservoirs.
- A service-brake valve controls the booster cylinder of the tandem master cylinder pneumatically with two circuits. This actuates the wheel cylinder hydraulically via an ALDBFR.
- A parking brake and auxiliary brake valve controls the spring-loaded actuators pneumatically on the rear axle without linkage.

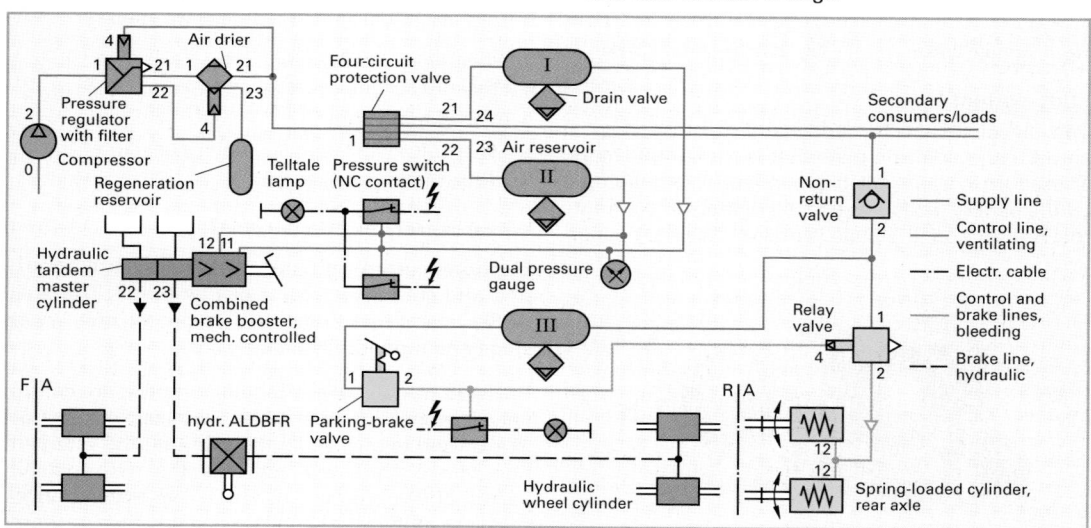

Fig. 3: Combined pneumatic-hydraulic brake system (schematic representation)

22.5.3.4 Continuous brake systems

> Retarders are brake systems that do not wear and operate as long as the vehicle is rolling. Their main purpose is to decelerate the vehicle on long downhill stretches.

In this way, the service brakes are relieved and saved from wear and damage. They are also often used during vehicle operation as a means of normal deceleration. When actuated, the brake lights can also light up.

Engine brake
This is actuated by a hand lever and 2 engine brake stages can be selected.

Stage 1: On actuation, the constant throttle located in the compression chamber is opened pneumatically. Only a small amount of air escapes during the compression stroke via an additional valve in the outlet channel. In this way, compression takes place with engine braking output. The remaining residual pressure escapes at TDC and no accelerating force is imparted due to compressed air at the piston.

Stage 2: Constant throttle and exhaust flap brake.
Here, a flap in the exhaust channel is closed in addition, which brings about an additional braking force due to back-pressure of the exhaust gases. At the same time, fuel supply is interrupted when the engine brake is applied.

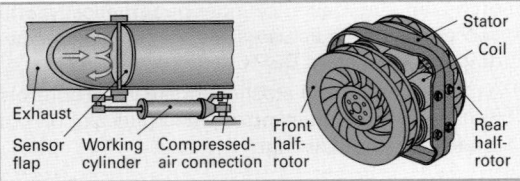

Fig. 1: Engine brake   Fig. 2: Eddy-current brake

Eddy-current brake
The air-cooled electric eddy-current brake **(Fig. 2)** comprises a soft iron plate that is formed of rotors. These turn in a controllable magnetic field generated by the battery (coils). As a result of the eddy currents that are formed, the plate is braked. The heat gener-

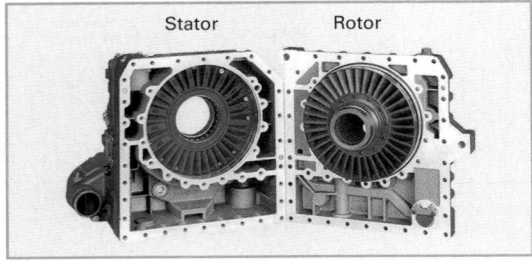

Fig. 3: Retarder

ated by the eddy currents in the rotors is dissipated by the wind. Brake control is by way of altering the excitation current taken from the battery.

The eddy-current brake is installed between the gearbox and differential.

Hydrodynamic brake (retarder)
This converts the brake energy into heat by fluid friction. The brake comprises a fixed stator and a rotor driven by the differential. Both have paddles in the same way as a hydrodynamic coupling. In between these paddles, hydraulic fluid is accelerated by the rotor and decelerated by the stator. Control is by way of a pump. With this pump the oil volume can be altered. The heat generated is transferred by the hydraulic fluid to the coolant of the engine via a heat exchanger.

22.5.3.5 ABS for air-brake systems
(Fig. 1, Page 667)
Heavy goods vehicles, tractor units $N_{2/3}$, trailers and buses/coaches $M_{2/3}$ must be equipped with a compressed-air ABS system.

Advantages of compressed-air ABS:
- Vehicle remains directionally stable by yaw rate deceleration.
- Vehicle remains steerable.
- Trailers from a truck/trailer combination do not break away.
- Optimum decelerations can be attained.
- Driver does not have to countersteer on one-sided slippery road surface.

Components of compressed-air ABS:
- Wheel sensors with pulse rings on wheels
- Electronic control unit
- Pressure control valves
- ABS warning light
- Electronic switch unit for trailer detection
- ABS connection to trailer

Wheel sensors pick up the wheel speeds.

Electronic control unit. This works as a central control unit and comprises four areas as on the hydraulic ABS system: input amplifier, computer unit, driver stage and monitoring circuitry.

Pressure control valves. Single port pressure control valves are used in most cases. Each controlled wheel is allocated to a valve. In the event of ABS intervention, the control unit activates the two solenoid valves of the pressure control valve and regulates pressure in the wheel brake so that blocking cannot occur.

ABS warning light. This is activated by the control unit and goes out when the vehicle is travelling at a speed greater than approx. 7 km/h. The ABS lamp lights up if there is a malfunction.

Electronic switch unit for trailer detection.
There is a red warning lamp in the pulling vehicle/tractor unit and a yellow information lamp.

They inform the driver as follows:
● Red and yellow lamps light up:
  Malfunctions in trailer ABS

● Yellow information lamp lights up:
  Trailer attached does not have ABS

ABS connection to trailer.
The pulling vehicle/tractor unit has a 5-pin ABS socket for connection of the ABS lead for the trailer. On an articulated truck, the ABS socket is located on the semi-trailer.

22.5.3.6 TCS (Traction Control System) for air-brake systems (Fig. 1)

When pulling away on a road surface that is slippery on just one side or both sides, or when accelerating in corners, no or little lateral force (depending on the degree of slip) can be transferred by the skidding wheels. Road handling therefore becomes unstable. Furthermore, skidding wheels can cause greater wear of the tyres and differential. TCS increases the traction and brings the vehicle safely on track.

TCS closed-loop control circuit. A difference is made between:
● TCS brake closed-loop control circuit
● TCS engine closed-loop control circuit

TCS brake closed-loop control circuit. This consists of:
● ABS component on rear axle
● ABS/TCS control unit
● Two-way changeover valves
● TCS solenoid valve

Operating principle
If a wheel develops a tendency to slip when the vehicle is pulled away, it is braked in modulation by the control unit. The TCS works in this way like an automatic differential lock. At the same time, the control unit reduces the engine torque to an optimum level for the overall drive torque by way of the engine closed-loop control circuit. The brake closed-loop control circuit works up to a speed of about 30 km/h. With this, just the engine torque, for example, is reduced by the engine closed-loop control circuit.

TCS engine closed-loop control circuit. This can be made up of the following systems:
● Electronic engine management system (EMS)
● Electronic diesel control (EDC)
● Proportioning valve with positioning cylinder (P)
● Servo-motor and linear positioner (M)

The engine management system EMS or EDC influences the engine torque. The control unit reduces the injection quantity directly and thereby the engine torque via a valve with positioning cylinder on positioning system **P** and via a servo-motor on the **M** system.

The TCS information lamp reports that the TCS is working and indicates that there is slip.

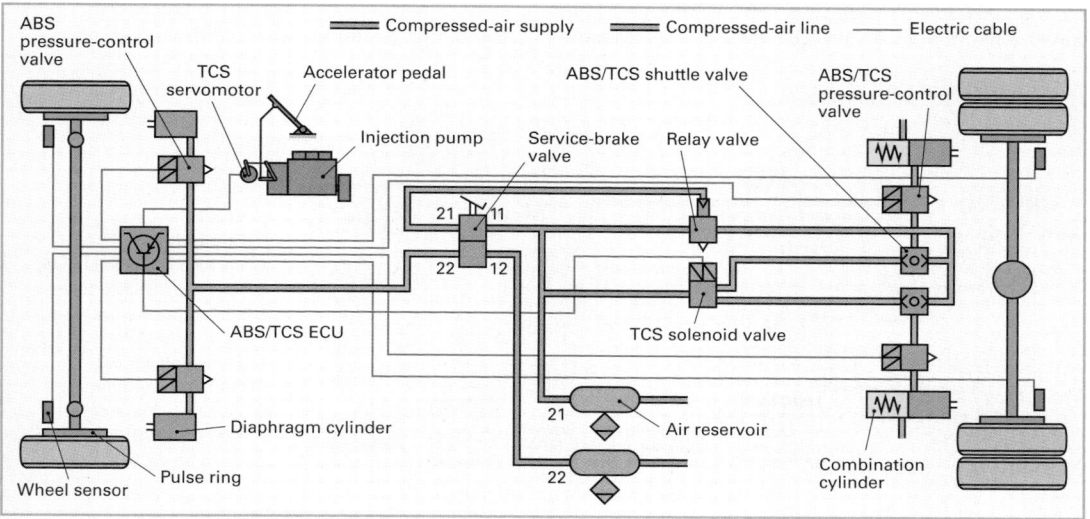

Fig. 1: Compressed-air ABS/TCS system

22.5.3.7 EBS with ESP (electronic braking system with electronic stabilisation program)

The electronic braking system **EBS (Fig. 1)** is a further development of the electro-pneumatic air-brake system. Through the individual electronic and electro-pneumatic components, the system can fulfil the following subroutines:

- **Electronically controlled braking.** Shorter stopping distances are made possible and brake lining service life is extended.
- **Integrated coupling force control.** This changes the braking effect of the trailer so that coupling forces during braking between trailer and engine carriage are practically zero.
- **Braking with ABS control.** The pulling vehicle remains steerable and stable.
- **Pulling away with TCS control.** The traction on one side or both sides of a slippery road surface is increased.
- **Track stabilisation with ESP.** The system detects unstable driving conditions and allows the vehicle to be brought back under control in critical situations.
- **Adaptive cruise control (ACC).** This maintains a set distance from the vehicle travelling ahead, depending on the driving speed.

Electronic braking system (Fig. 1 and 2)

Design. This is an air-brake system, which is controlled electropneumatically. It comprises two service brake circuits for the front axle **(FA)** and rear axle **(RA)** and a parking-brake circuit **(PBC)**.

Operating principle

EBS braking. When the driver presses the brake pedal, the central control unit is informed of his/her input by the foot-brake module, which is equipped with travel sensors. Via CAN the EBS/ABS modulators and

the trailer are actuated. Brake pressure in the brake cylinders is regulated from the central control unit by the solenoid valves in accordance with driver input. The wheel-speed sensors calculate the deceleration. Friction lining wear is detected by friction lining sensors. The brake system of the trailer is actuated by the trailer control module.

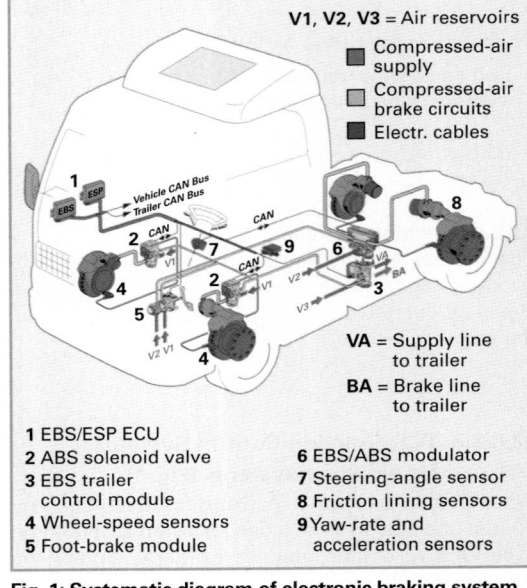

V1, V2, V3 = Air reservoirs

- Compressed-air supply
- Compressed-air brake circuits
- Electr. cables

VA = Supply line to trailer

BA = Brake line to trailer

1 EBS/ESP ECU
2 ABS solenoid valve
3 EBS trailer control module
4 Wheel-speed sensors
5 Foot-brake module
6 EBS/ABS modulator
7 Steering-angle sensor
8 Friction lining sensors
9 Yaw-rate and acceleration sensors

Fig. 1: Systematic diagram of electronic braking system

Advantages

- Faster and simultaneous pressure build-up in all brake cylinders, thereby shorter stopping distance.
- Higher braking comfort through good proportioning properties.
- Precise determination of brake forces from pulling vehicle and trailer in relation to each other.
- Equal friction lining wear for whole road train.

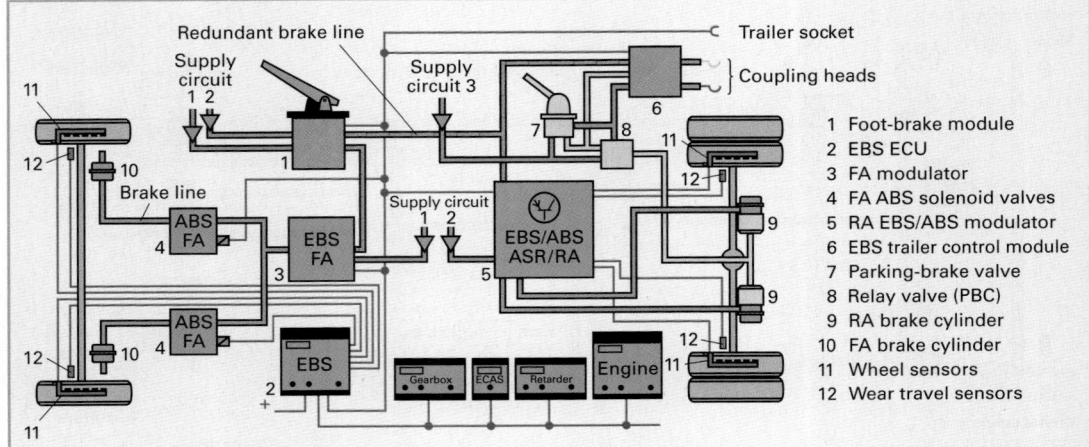

1 Foot-brake module
2 EBS ECU
3 FA modulator
4 FA ABS solenoid valves
5 RA EBS/ABS modulator
6 EBS trailer control module
7 Parking-brake valve
8 Relay valve (PBC)
9 RA brake cylinder
10 FA brake cylinder
11 Wheel sensors
12 Wear travel sensors

Fig. 2: Electronic braking system

- Greater service life of brake linings thanks to fully automatic use of retarder and engine brake.
- If EBS fails, brakes can be applied pneumatically as a substitute using the two service brake circuits (auxiliary brake function).

Integrated coupling force control. When the truck/trailer combination is braked, the brake pressure of the pulling vehicle and trailer is optimised so that coupling forces between the trailer and pulling vehicle are practically eliminated. To do this, it is necessary for the overall mass and the load distribution of the road train to be calculated. To calculate the overall mass, the acceleration response is evaluated. The load distribution is determined from the braking response of the pulling vehicle and trailer. These figures are stored and used as a basis for successive braking. If the calculated deceleration is not reached, brake pressure is increased in increments. This procedure is stored and used again for successive braking.

ABS braking. If there is a tendency during braking to block, this is detected by the speed sensors and the ABS solenoid valves of the front axle and the ABS modulators are actuated. As on passenger vehicle antilock braking, the control phases pressure increase, pressure decrease and pressure retention are carried out. When pressure is built up (increase), the compressed air is modulated to the open until there is no further risk of blocking.

TCS drive-off control. TCS control can be applied by intervention in the brake system or engine management.

TCS braking control. If a wheel slips when the vehicle is pulled away or during acceleration at speeds below 40 km/h, the slipping wheel is braked in pulses until it turns at the same speed as the opposing wheel. The pressure sensor and the wheel-speed sensor give the control unit the necessary information to do this.

TCS engine control. If both wheels slip during acceleration, the engine torque is decreased until the circumferential speed calculated from the speed of the wheels is slightly above the driving speed. TCS engine control is effective at all speeds.

Offroad TCS. With the aid of a switch, for example for driving with snow chains or driving offroad, the ACS can be switched off so the vehicle can be rocked free.

ESP control with engine and brake intervention. The system works in two ways:
- At low and medium friction coefficient, it counteracts under- and oversteer and jack-knifing of the articulated truck.
- At medium and high friction coefficient, it counteracts tilting over.

Control during oversteer. If, for example, an articulated truck develops a tendency during cornering or steering **(Fig. 1)** to oversteer or to jack-knife, the ESP brakes the outer front wheel in the corner (as an example). The yaw rate from this action stabilises the vehicle. To counteract jack-knifing of the vehicle, the trailer is also braked in certain circumstances. In this way, sliding out or skidding of the trailer is prevented. For this ESP function, a trailer with ABS is necessary.

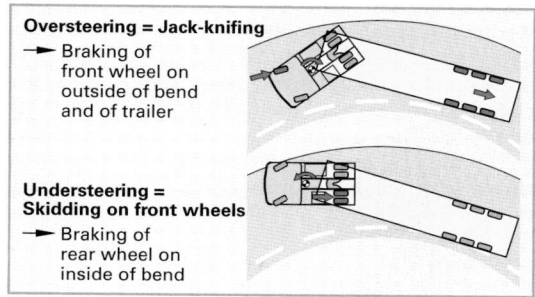

Oversteering = Jack-knifing
→ Braking of front wheel on outside of bend and of trailer

Understeering = Skidding on front wheels
→ Braking of rear wheel on inside of bend

Fig. 1: EPS control during over and understeer

Control during understeer. If a vehicle develops a tendency to understeer, ESP control attempts to counteract skidding or sliding out by braking several wheels of the pulling vehicle/tractor unit and by pulsed periodic brake intervention on the trailer. When a left-hand corner is taken, for example, the rear inner wheel of the corner is braked to generate a counteracting yaw rate, which should stabilise the vehicle.

In order that the control unit can calculate the individual brake pressures for ESP intervention, various input variables such as friction coefficient, load status, steering angle and yaw tendency are required. In certain situations, the engine performance is also reduced by engine intervention in order to control slip at the drive axle.

ROP (rollover protection). With this function, tipping over of the vehicle at high and medium friction coefficient is avoided. To do this, the vehicle speed is first reduced and the brakes are applied if necessary until the critical tipping situation is brought under control.

REVIEW QUESTIONS

1   Of which components does a dual-circuit air-brake system consist?
2   Explain the function of a compressed-air ABS system for commercial vehicles.
3   Which subroutines can an EBS system with ESP control fulfil?
4   What are the advantages of an EBS system?

22

22.6 Starting systems for commercial vehicles

Starting systems on engines up to about 12 l capacity can be configured for 12 V or 24 V. Large starter motors that are designed for capacities up to 24 l are always set up for a voltage of 24 V. The reason for this is that only half the current flows with the same output of two starters of 24 V systems than 12 V systems ($P = U \cdot I$). The main starter lead can be made smaller with a higher voltage. As leads with small cross sections are able to dissipate heat better than those with large cross sections due to their relatively large surface area, the change of resistance from heat generation is lower as a result. The voltage drop in the main starter lead can therefore be brought better under control. However, the risk of contact corrosion on 24 V systems is greater.

22.6.1 Starter types

In light commercial vehicles, **pre-engaged drive starters** are generally used as in passenger vehicles. On heavy commercial vehicles, two-stage sliding-gear starters **(Fig. 1 and 2)** are more common in different designs, e.g. with inline or compound motor.

The cross sectional diagram **(Fig. 1)** shows the design of a two-stage sliding-gear starter with compound motor.

The armature shaft has a hollow design and takes the form of a drive flange on the pinion side. This supports the multi-plate overrunning clutch.

Attached to the collector side are the starting-motor solenoid and control relay.

Starting-motor solenoid and control relay. Due to the location of the starting-motor solenoid, the pinion has to be moved via an engagement rod, which is guided through the hollow armature shaft, in axial direction towards the ring gear.

Furthermore, the starting-motor solenoid actuates the contact bridge of the control relay via a release lever, locking pawl and stop plate. The control relay switches the starter motor on in two stages.

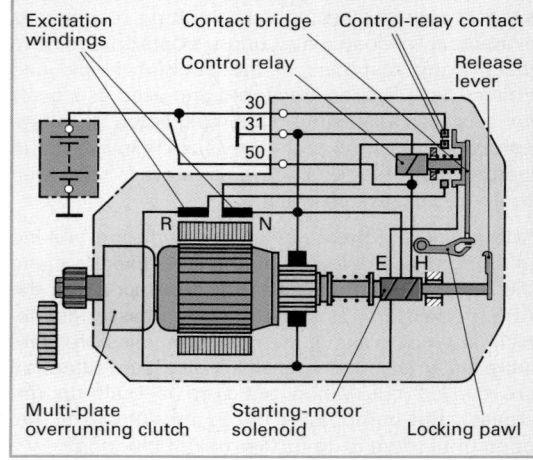

Fig. 2: **Sliding gear starter (rest position)**

Meshing drive. The multi-plate overrunning clutch sits on the helical spline of the gearbox spindle. The multi-plate overrunning clutch can be pressed together on the helical spline and allows the flow of power between starter armature and pinion.

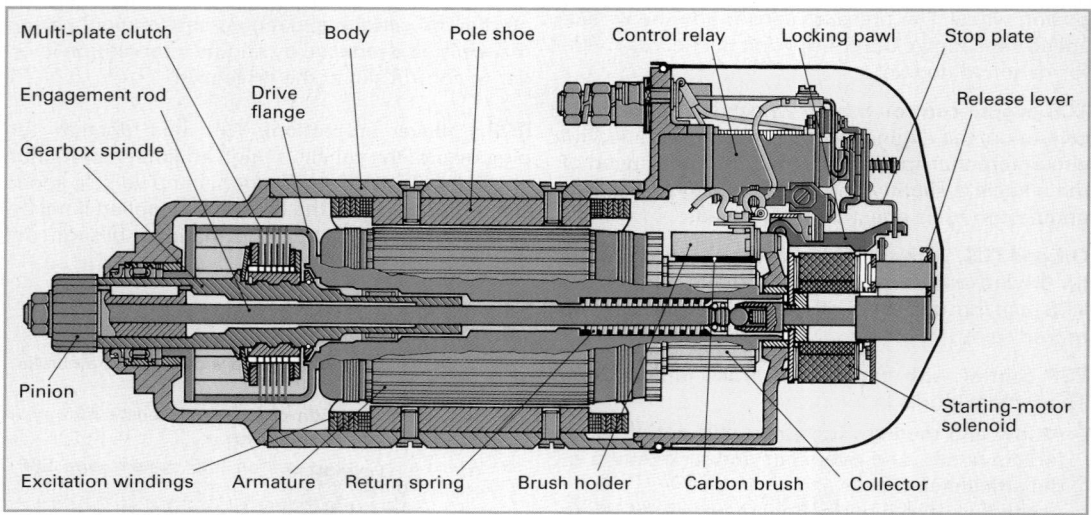

Fig. 1: Two stage sliding gear starter

1st selection stage – initial stage (Fig. 1). When the starting switch is pressed, the solenoid valve of the control relay and the holding winding (H) of the starting-motor solenoid are actuated. The pull-in winding (E) of the starting-motor solenoid also receives voltage via the control-relay contact (K). As a result of the magnetic force from the starting-motor solenoid, the engagement rod is now moved axially, which causes the pinion to be pushed towards the ring gear.

The shunt winding (N) of the starter motor is connected initially in series with the starter armature. A weak magnetic field is built up within, which generates a small amount of torque. The starter motor turns slowly.

> In the 1st selector stage, the starter pinion is moved axially by slow rotation to allow soft engagement.

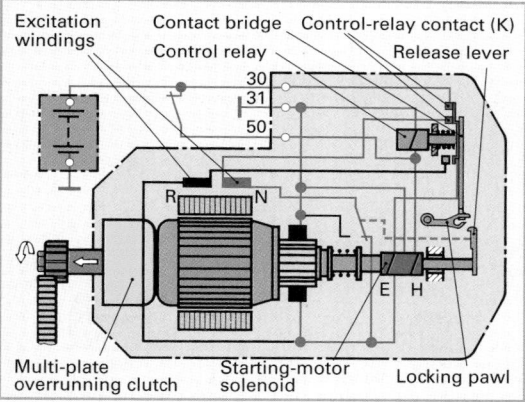

Excitation windings   Contact bridge   Control-relay contact (K)
Control relay   Release lever
30
31
50
R   N
E   H
Multi-plate overrunning clutch   Starting-motor solenoid   Locking pawl

Fig. 1: Sliding gear starter (selector stage 1)

2nd selection stage – main stage (Fig. 2). Just before the end of the pinion engagement process, the locking pawl on the control relay is lifted up, which closes the contact bridge in the control relay. The series winding is activated, the starter motor now draws the full amount of current and passes on its drive torque via the multi-plate overrunning clutch (which is now engaged) to the pinion.

During the changeover process in the control relay, the shunt winding (N) is switched to earth. The torque is increased. The pull-in winding (E) of the starting-motor solenoid, previously connected to it in series, is de-energised as both connections are connected to positive (+). This selection process means that ...

- ... the series-wound motor becomes a compound motor
- ... a strong magnetic field is generated in the shunt winding and the torque is thereby increased

- ... when the starter motor load is reduced, the speed of the armature is restricted and it thereby prevents unacceptably high speeds.

> In 2nd selector stage, changeover from series-wound motor to compound motor with increased torque and engine speed limitation with load reduction.

As soon as the release torque of the engine is overcome, the starter motor begins to turn; this occurs with the speed/torque characteristic of a series-wound motor, i.e. as torque load is reduced, its speed increases rapidly.

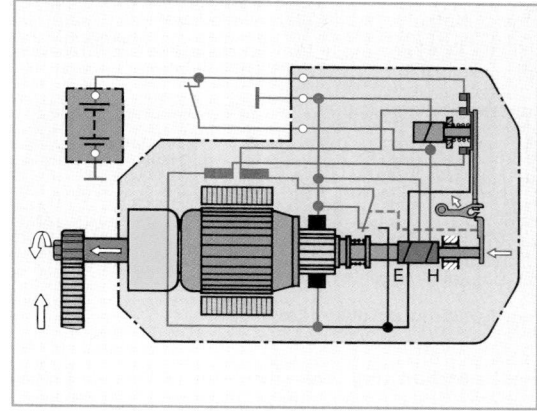

E   H

Fig. 2: Sliding gear starter (selector stage 2)

Overtake procedure. As a result of the reverse in force direction, the power transmission of the multi-plate overrunning clutch is released; starter motor and engine are no longer connected to each other.

Disengagement procedure (Fig. 2, Page 670). When the control relay is de-energised, the starter motor solenoid is also de-energised. The return spring **(Fig. 1, Page 670)** in the hollow shaft pushes the pinion out of the ring gear via the engagement rod; when the starter is de-energised, it prevents the pinion from engaging in the rotating ring gear from mechanical jolts.

Multi-plate overrunning clutch

Functions. During the starting procedure ...

- ... it joins the pinion to the starter motor.
- ... it interrupts the flow of power when the pinion is engaged once the engine has started.
- ... it restricts the transfer of torque from the pinion to the ring gear if the engine is blocked (overload protection).

Design (Fig. 1). This is a type of coupling that is comprised essentially of metallic outer and inner plates. To transfer the drive torque, these are pressed firmly together. The outer plates are connected to the drive flange, the inner plates are connected to the coupler; they can therefore move slightly in axial direction when they are pressed together. The external coupler sits on a helical spline, which is attached to the gearbox spindle.

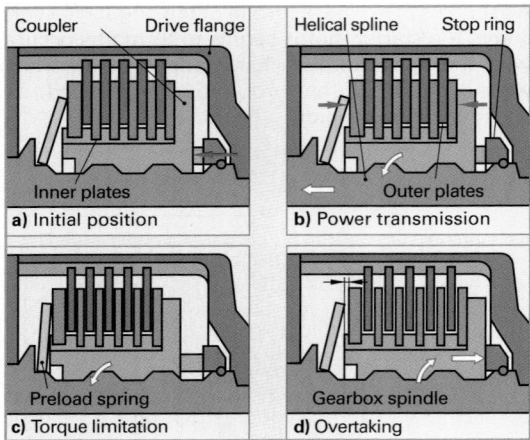

Fig. 1: Multi-plate overrunning clutch

Rest position (Fig. 1a). The outer and inner plates are placed under a small amount of initial tension. This assures that the coupler is picked up during engagement.

Power transmission (Fig. 1b). If the pinion is engaged and thereby held by the ring gear, the inner plates are pressed with greater force against the outer plates via the coupler, which is moved on the helical spline. Pressing together is continued until the starting torque necessary to turn over the engine can be transferred (release torque).

Torque limitation (Fig. 1c). To prevent the starter, pinion and ring gear from being subjected to unacceptably high loads, the coupling is designed to slip when the maximum permissible torque is reached (overload coupling).

Overtaking (Fig. 1d). If during the starting procedure the ring gear turns faster than the pinion, the inner plates are relieved of pressure and the multiplate clutch is released. It acts in this way like a freewheel. No dangerous acceleration forces can therefore be transferred to the armature after the engine has been started.

22.6.2 Additional relay in starter systems

These are used mainly in starter systems of commercial vehicles. There are additional relays for the following important range of tasks:

- Battery changeover relay
- Starter repeat relay
- Starter lock relay
- Battery relay

Battery changeover relay

This is used in starting systems if the vehicle power supply is 12 V and the starter voltage is 24 V for reasons of current reduction during starting. The system generally features two starter batteries of the same size that are connected either in parallel or in series (Fig. 2).

Basic setting (Fig. 2). Both starter batteries are connected in parallel. The vehicle electrical system and the alternator voltage work with 12 V.

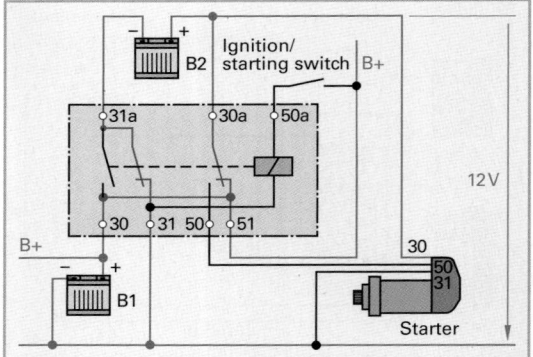

Fig. 2: Battery changeover relay - basic setting

Start setting (Fig. 3). During the starting procedure, both starter batteries are connected in series. The starter now works with 24 V, the vehicle electrical system still works with 12 V.

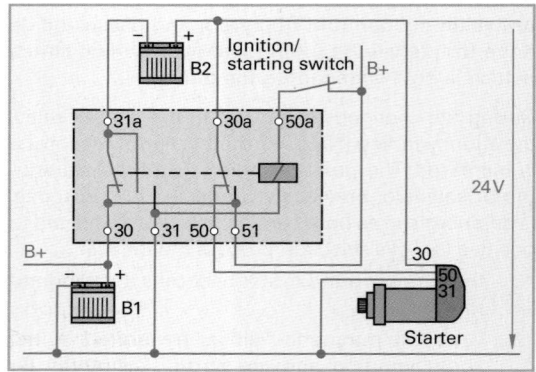

Fig. 3: Battery changeover relay - start setting

Starter lock relay

This is used if the starting procedure is not immediately perceptible.

This could be the case, for example, on:
- Vehicles with underfloor or rear engine
- Starting systems with remote control
- Fully automatic starting systems

The following functions must be in place with regards to the starter motor:
- Shutoff once start is successful
- Starter lock when engine is running
- Starter lock when engine is running down
- Starter lock after faulty start

The starter lock relay **(Fig. 1)** then only joins term. 30 to term. 50f if no voltage is present at terminal D+. A time block is still integrated in the electronic component, which prevents a repeat of the starting sequence until after a few seconds have elapsed.

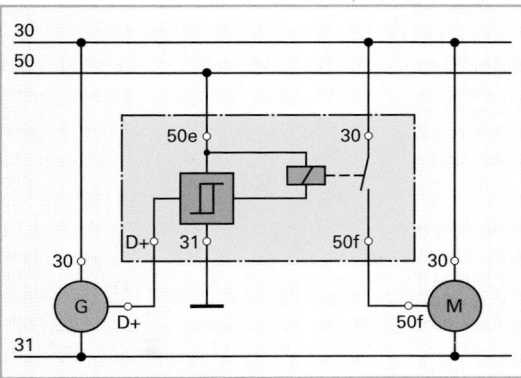

Fig. 1: Starter lock relay

Starter repeat relay

This is used exclusively on heavy commercial vehicles that are equipped with two-stage sliding gear starters, on which the starting procedure is not immediately perceptible. The use of a starter repeat relay is only possible if the starter has a term. 48 **(Fig. 2)**.

During normal starting the relay does not respond. If, however, a tooth hits a tooth, i.e. the ring gear cannot engage in a gap, there is no contact release for the main current circuit despite the engagement relay being activated because, with the starter blocked, the engagement relay could be subjected to excessive thermal load if activated for too long.

With the aid of a delayed opening relay, the engagement relay is switched off and then back on again. This procedure is repeated until the pinion engages and the main current circuit is closed.

During an attempt at starting, the relay separates the connection between term. 50g and 50h if at least 20 V is not applied to starter terminal 48 after a few seconds. This is the case if the pinion cannot engage and the control relay cannot close the main current circuit (term. 30) as a result.

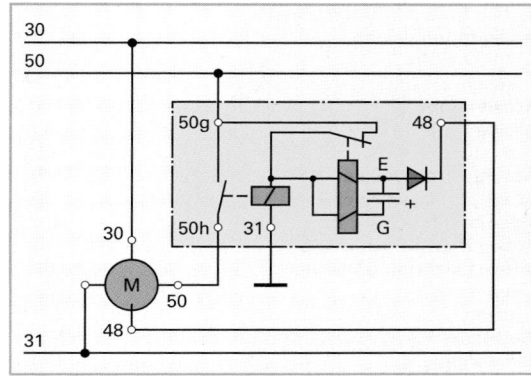

Fig. 2: Starter repeat relay

Combined systems. There are starter systems that have both starter lock relays and also starter repeat relays. Here, a starter repeat relay is located after the starter lock relay in the circuit.

Term. 50f of the starter lock relay actuates term. 50g of the starter repeat relay (see Fig. 1 and 2). The individual functions of both relays remain fully intact.

Battery relay (battery master switch)

For electrical systems in buses/coaches and tankers, a battery master switch is prescribed by law. It is used to isolate the vehicle electrical system from the batteries. In this way, the risk of short circuits and fires are reduced when working on the vehicle and in the event of accidents.

When parked, isolation of the vehicle electrical system, in particular on 24 V systems, reduces the electrochemical corrosion of conductive parts that are subjected to salty water during winter operation.

REVIEW QUESTIONS

1 What procedures occur in the 1st and 2nd selector stage?

2 What are the functions of the multi-plate overrunning clutch?

3 What are the advantages of a battery changeover relay?

4 Which functions are taken over by the starter lock relay?

22

23 Keyword index

23

23

N

O

23

23

23